T0186864

EUROPEAN SYMPOSIUM ON ENVIRONMENTAL BIOTECHNOLOGY, ESEB 2004

PROCEEDINGS OF THE EUROPEAN SYMPOSIUM ON ENVIRONMENTAL BIOTECHNOLOGY, ESEB 2004, 25–28 APRIL 2004, OOSTENDE, BELGIUM

European Symposium on Environmental Biotechnology, ESEB 2004

Edited by

W. Verstraete
University of Ghent, Belgium

A.A. BALKEMA PUBLISHERS LEIDEN / LONDON / NEW YORK / PHILADELPHIA / SINGAPORE

Published by: A.A. Balkema Publishers, a member of Taylor & Francis Group plc
www.balkema.nl and www.tandf.co.uk

ISBN 90 5809 653 X

European Symposium on Environmental Biotechnology, ESEB 2004 - Verstraete (ed)
© 2004 Taylor & Francis Group, London, ISBN 90 5809 653 X

Table of Contents

Water purification

Biosafety

Solid waste, waste recycling and biofuels

Composition and function of biofilms and consortia

Innovative in situ remediation

Biodegradation and bioremediation

Soil clean-up

Gas treatment

POSTERS

Microbial diversity

Water purification

Biosafety

Solid waste, waste recycling and biofuels

Composition and function of biofilms and consortia

Innovative in situ remediation

Environmental genomics

Mobile genetic elements

Biosensing

Novel nitrogen removal

Managerial aspects and policies

Solid waste and composting

Biodegradation and bioremediation

Soils clean-up

Gas treatment

TBT (Tributyltin)

Late posters

European Symposium on Environmental Biotechnology, ESEB 2004 - Verstraete (ed)
© 2004 Taylor & Francis Group, London, ISBN 90 5809 653 X

Introduction

In the past years, we all have observed staggering progress in molecular methodologies to analyse and characterise microbial communities. These changes certainly have a major impact on how Environmental Biotechnology is evolving.

Moreover, there are a series of startling new developments in the field of microbiology: anaerobic ammonium oxidation, dehalorespiration, quorum sensing, to name a few. These phenomena become implemented into the technologies of water, soil, air and waste treatment.

The European Federation of Biotechnology (EFB) has seen major changes over the past years. There is now a Section dedicated to Environmental Biotechnology within the Federation. In accordance with the past years, the EFB will continue to organise its top-level symposium series focussed on environmental processes. The first symposium in the series was organised in 1991 in Oostende. The 5th event is again scheduled in Oostende and is an activity of EFB and the Section on Environment (Genootschap Milieutechnologie) of the Technological Institute of the Royal Flemish Society of Engineers (K VIV-TI). The previous events were known as 'International Symposium Environmental Biotechnology'. To avoid confusion with a parallel series of activities; the events organised by the EFB are from now on labelled as 'European Symposium Environmental Biotechnology' (ESEB).

This book represents the papers and posters submitted to the ESEB 2004. The topics range from fundamental to applied, from microbial diversity to water purification, from biosafety to solid waste, waste recycling and biofuels, from composition and function of biofilms and consortia to innovative in situ remediation, from gas treatment to soil cleanup.

We trust that these papers will be an impetus for all dealing with new frontiers in environmental issues.

W. Verstraete, Chairman ESEB 5

Keynote lectures

European Symposium on Environmental Biotechnology, ESEB 2004 - Verstraete (ed.)

The horizontal catabolic gene pool, or how microbial communities can rapidly adapt to environmental pollutants

E.M. Top
Department of Biological Sciences, University of Idaho, Moscow, USA

D. Springael
Laboratory of Soil and Water Management, Department of Land Management, Faculty of Agricultural and Applied Biological Sciences, Catholic University of Leuven, Heverlee, Belgium

ABSTRACT: Retrospective studies such as comparative DNA sequence analyses and characterization of mobile genetic elements (MGEs) in bacteria clearly indicate that horizontal gene transfer (HGT) mechanisms play an important role in the exchange and even *de novo* construction of degradation pathways in bacterial communities. Such HGT processes thus allow these communities to rapidly adapt to xenobiotic compounds by providing the members with the genetic information required to cope with these novel, often toxic, compounds. Direct evidence for the importance of MGEs in bacterial adaptation to xenobiotics stems from observed correlations between catabolic gene transfer and accelerated biodegradation in several habitats, or from studies that monitor catabolic MGEs in polluted sites. The combination of these different studies demonstrate that HGT and pathway assembly are key to the adaptation of bacterial communities to xenobiotics, but many questions remain about the rates at which these processes occur and the selection on the MGEs that drive this genetic exchange.

1 INTRODUCTION

Recently published analyses of prokaryotic gene and whole genome DNA sequences have led investigators to speculate that horizontal gene transfer (HGT) may have played a far more important role in the evolution of microorganisms than had been previously recognized. Indeed, analyses of whole genome sequences suggest that the fraction of foreign DNA obtained by HGT varies between 0% and 17% of prokaryotic genomes (Daubin et al. 2003; Kurland et al. 2003). Consistent with this, studies done in the laboratory and in the field have shown that the horizontal transfer of genes among prokaryotic lineages readily occurs in various habitats (Hill & Top 1998), and results in the wholesale acquisition of genes encoding traits that are sometimes essential (e.g., antibiotic or heavy metal resistance) or otherwise confer a selective advantage (e.g., the ability to metabolize a novel carbon source). While there is no doubt about the occurrence of HGT and its role as a driving force in prokaryotic evolution and adaptation (Kurland et al. 2003), the diversity of the agents of this genetic exchange, and the mechanisms that govern their dispersal in the microbial world are poorly understood.

Three main mechanisms for horizontal gene transfer between prokaryotes have been identified: (a) transformation, a process whereby cell-free DNA is taken up and integrated into the genome; (b) transduction, wherein gene transfer is mediated by bacteriophages; and (c) conjugation, a process whereby plasmids or conjugative transposons mediate cell–cell contact and transfer DNA from donor to recipient cells (Dröge et al. 1999). The genetic information that is thus shared by a broad spectrum of prokaryotic species representing distinct, phylogenetic lineages is referred to as the horizontal gene pool. The MGEs involved are transferable plasmids, (conjugative) transposons, integrons, genomic islands, or phages, which are able to move within and/or between genomes, thus allowing ‘evolution in quantum leaps’. Conjugative gene transfer mediated by plasmids with broad host-ranges (BHR) is generally believed to be the most common and widespread mechanism for the transfer of genes among prokaryotic species, in particular across wide phylogenetic ranges (Thomas 2000a). Plasmids are extrachromosomal replicons that replicate independently from host chromosomes, seem to be very common in bacteria, and can confer a variety of phenotypes to the host organism (Table 1).

Table 1. Examples of phenotypes encoded by plasmids [summarized from Top et al. (2000)].

Phenotype	Plasmid (Inc group)*
Resistance to detrimental conditions:	
Antibiotic resistance	RP4 (IncP-1)
Bacteriophage resistance	pNP40
Heavy metal resistance	R831b(IncM)
UV protection	R46 (IncN)
Production of toxic compounds:	
Antibiotic production	SCP1
Enterotoxin production	pTP224
Uptake, metabolism and growth:	
Biodegradation of 1,3-dichloropropene	pPC170
Biodegradation of 2,4-D	pJP4 (IncP-1)
H_2S production	pNH223
Iron uptake	pJM1
Thiamine synthesis (vitamin B_1)	PSym
Interactions with eukaryotes:	
Nodulation of plant root	pPN1
Virulence	pX01
Genomic modifications:	
Restriction/modification	pRleVF39b
Other properties:	
Sex pheromones	pAD1

* Inc group: incompatibility group.

One of the many plasmid-encoded traits is the ability to degrade various organic compounds, called xenobiotics. Xenobiotics have been strictly defined as man-made molecules, foreign to life and as such never encountered by bacterial populations before their introduction by man. For the purpose of this review, we will used the broader definition, i.e., 'all compounds that are released in any compartment of the environment by the action of man and thereby occur in a concentration that is higher than natural' (Leisinger 1983). The first degradative plasmids were described ca. 30 years ago (Worsey & Williams 1975), and many more have been detected and characterized to date (Top et al. 2002). During the last decade, the types of MGEs that carry genes for organic pollutant degradation have been extended from plasmids to transposons and even to the increasing array of mosaic MGEs such as genomic islands, conjugative transposons and integrative plasmids (Springael & Top 2004).

The ability of microbial communities to adapt to anthropogenic chemicals by HGT first became evident from the observation that rapid dissemination of antibiotic resistance determinants by means of MGEs accompanied the widespread use of antibiotics (Cohen 1992). During the last 15 years, several groups have provided strong indications that MGEs also play an important role in the horizontal spread of existing catabolic pathways, as well as in the natural construction of novel ones, and that this allows bacteria to rapidly adapt to new xenobiotic compounds entering their habitats. Recent studies pointed out two major breakthroughs regarding the molecular basis of bacterial adaptation to pollutants, i.e., the definition of new types of MGEs that encode xenobiotic degradation, and findings that more convincingly suggest that such adaptation events actually occur in a polluted environment in the field. This paper will summarize some of this evidence.

2 ADAPTATION TO XENOBIOTICS: RETROSPECTIVE EVIDENCE

Several key findings provide evidence for the importance of HGT in the wholesale acquisition and de novo construction of novel catabolic pathways by bacterial populations. (i) First of all, findings of new associations between several degradative genes and the wide range of MGEs listed above, are continuously being reported; (ii) Secondly, the same catabolic genes or entire operons are sometimes found on different replicons (Top et al. 1995); (iii) Furthermore, evolutionary related catabolic genes and gene clusters are found in bacteria that were isolated from very distant locations; and (iv) in some cases lateral genetic exchange is revealed by showing clear incongruence between the phylogeny of catabolic genes and that of the 16S rRNA genes of the corresponding hosts, since the 16S rRNA sequence is considered to be representative of the phylogeny of the organism (McGowan et al. 1998);

There are three main groups of MGEs that carry catabolic genes. The first one consists of the catabolic plasmids. Interestingly, genes that encode degradation of mostly man-made compounds, such as several chloroaromatics, are often encoded by the well-known broad-host-range IncP-1 plasmids (Top et al. 2000). These plasmids are the most promiscuous self-transmissible plasmids characterized to date, with a host range that is much wider than that of other catabolic plasmids. Their genetic structure shows clear evidence of their role as shuttle vectors: e.g., the atrazine degradative plasmid pADP-1 and the antibiotic resistance plasmid pB10 have very similar backbone regions but encode very different traits, suggesting that the replicons act as shuttles for genes from diverse sources (Schlüter et al. 2003, Figure 1). The correlation between the high promiscuity of IncP-1 plasmids and their ability to acquire and transfer genes encoding various (temporarily) essential traits, suggests that these MGEs are very well suited to facilitate rapid local adaptation of distinct populations within a microbial community.

The second group of catabolic MGEs are the type I and type II transposons, which are often found on

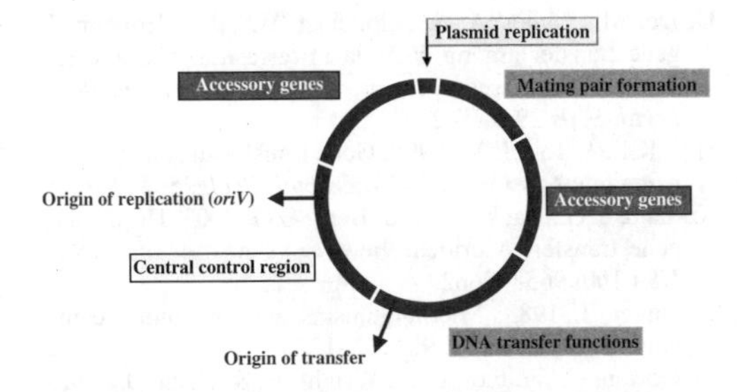

Figure 1. Common genetic structure of IncP-1 plasmids.

plasmids (Tsuda et al. 1999). The third, most recently described group includes the genomic islands. The generic term 'genomic island' has been proposed for a wide range of large mobile or potentially mobile defined DNA segments that are found integrated into the chromosome or other replicons. They typically carry an integrase gene and gene modules related to conjugation modules of plasmids. Well-known examples of genomic islands are pathogenicity islands and symbiotic islands. Some of them such as the symbiosis gene island of *Mesorhizobium loti* is able to transfer by conjugation. The best described examples of catabolic gene islands are the *clc*-element and Tn*4371* (Springael & Top 2004). Both elements display a gene organization typical for gene islands. The *clc*-element was shown to be self-transmissible by conjugation while transfer of Tn*4371* has only been observed after integration into a conjugative plasmid.

These genomic analyses, although very informative, only provide retrospective evidence of the mechanisms and extent of gene exchange. Questions about the environmental conditions that favor HGT, the rates at which these HGT events occur, and the selection required to maintain them in the gene pool, can only be answered by conducting ecological studies using lab microcosms and field plots.

3 ADAPTATON TO XENOBIOTICS: EVIDENCE FROM ECOLOGICAL STUDIES

Very few studies have provided clear direct evidence that a microbial community adapts to a xenobiotic by means of HGT, i.e. that the overall degradative capacity has improved due to gene exchange. We restrict our summary here to the most recently published studies in this area, and refer to a recent mini-review for a more complete and extensive description of case studies (Top et al. 2002).

In studies on catabolic gene transfer in soils, Newby & Pepper (2002), Newby et al (2000) and Dejonghe et al. (2000) have used the herbicide 2,4-dichlorophenoxyacetic acid (2,4-D) as a model compound, and the IncP-1β plasmid pJP4 or a non-IncP-1β plasmid pEMT1, which both encode 2,4-D degradation, as model plasmids. In some but not all cases, a clear correlation was shown between transfer of the plasmids from inoculants to indigenous bacteria and accelerated degradation of 2,4-D. Another study on gene transfer in soil, showed adaptation of a rhizosphere bacterial community from pine seedlings mycorrhized with *Suillus bovinus* to a pollutant by acquisition of the self-transferable TOL plasmid from an inoculated *P. fluorescens*, thereby protecting the plant and fungus from metatoluate (Sarand et al. 2000). *In situ* spread of the *tfdA* gene, encoding the first step in the degradation of 2,4-D and phenoxyacetic acid (PAA), from an inoculum into indigenous phenol degraders in soil was shown to result in vertical expansion of the phenol degradation pathway to now include PAA. Similarly, Springael and co-workers also observed transfer of the *clc*-element in an inoculated model biofilm membrane reactor treating chlorobenzoate-containing influent. The original inoculum carrying the *clc*-element was often replaced by other chlorobenzoate degrading bacteria. Since control reactors without inoculum never developed a chlorobenzoate degrading microbial community nor a detectable degradation activity, it was suggested that contaminant bacteria had acquired the *clc*-element from the inoculum strain and became more competitive under the implied bioreactor conditions (Springael et al. 2002).

Several of these reports show that the spread of catabolic genes on MGEs enabled specific indigenous bacterial populations to acquire new catabolic characteristics. The disadvantages of these studies are the unnatural aspect of inoculation of high numbers of donor cells, and the relatively high concentrations of contaminants added. These manipulations also often entail mixing and thus disturbing the samples. There is no evidence yet that HGT would be as clearly responsible for community adaptation without this disturbance and at lower, more environmentally relevant xenobiotic concentrations.

Recent field studies provide more evidence that gene transfer and even pathway assembly occurs in natural environments. For example, after isolating a collection of naphthalene degraders, natural horizontal transfer of the *phnAc* gene, encoding an enzyme involved in naphthalene degradation, was suggested to have occurred among different *Burkholderia* strains because of the observed incongruence between the phylogeny of the catabolic genes and that of the 16Sr RNA genes of the corresponding hosts (Wilson et al 2003). Similar observations were made for the *pcpB* gene, which encodes initiation of the degradation of pentachlorophenol, in a phylogenetically diverse group of *Sphingomonas* strains isolated from contaminated groundwater (Tiirola et al. 2002). *In situ* pathway assembly was recently demonstrated in a chlorobenzene degrading aquifer by Muller et al. (2003). Bacterial chlorobenzene mineralization occurs in two steps. The first step is encoded by a gene cluster that is evolutionary related to gene clusters involved in the initial attack of toluene. The second step involves a chlorocatechol modified *ortho*-cleavage pathway specified by *clcRABD*-related genes. Genetic analyses of the chlorobenzene degradation pathways have clearly indicated that this pathway originated by combining both gene modules in the same strain. In the aquifer, chlorobenzene-degrading bacteria, were shown to have assembled a similar pathway. Moreover, the authors also isolated the toluene degraders from which the gene module encoding the first step presumably originated. Interestingly, the pathway in this aquifer strain was assembled on a genomic island strongly related to the *clc*-element, showing that genomic islands are indeed able to recruit genetic information. This study is the first to report strong evidence of the *de novo* construction of a catabolic pathway by assembly of different catabolic genes in a natural setting.

4 CONCLUSIONS

While there is no doubt that HGT and pathway assembly are key to the adaptation of bacterial communities to xenobiotic compounds that enter their environment, many questions remain about the exact mechanisms, the rates at which these processes occur, the environmental conditions that influence them, and the diversity and existence conditions of the MGEs that drive this genetic exchange in natural environments.

ACKNOWLEDGEMENTS

E.M. Top is in part supported by NIH Grant P20 RR 16448 from the COBRE Program of the National Center for Research Resources.

REFERENCES

Cohen, M.L. 1992. Epidemiology of drug resistance: implications for a post-antimicrobial era. *Science* 257: 1050–1055.

Daubin, V., Moran, N.A. & Ochman, H. 2003. Phylogenetics and the Cohesion of Bacterial Genomes. *Science* 301: 829–832.

Dejonghe, W., Goris, J., El Fantroussi, S., Hofte, M., De Vos, P., Verstraete, W. & Top, E.M. 2000. Effect of dissemination of 2,4-dichlorophenoxyacetic acid (2,4-D) degradation plasmids on 2,4-D degradation and on bacterial community structure in two different soil horizons. *Appl. Environ. Microbiol.* 66: 3297–3304.

Dröge, M., Pühler, A. & Selbitchka, W. 1999. Horizontal gene transfer among bacteria in terrestrial and aquatic habitats as assessed by microcosm and field studies. *Biol. Fertil. Soils* 29: 221–245.

Hill, K.E. & Top, E.M. 1998. Gene transfer in soil systems using microcosms. *FEMS Microbiol. Ecol.* 25: 319–329.

Kurland, C.G., Canback, B. & Berg, O.G. 2003. Horizontal gene transfer: A critical view. *Proc. Nat. Academ. Sci. USA* 100: 9658–9662.

Leisinger, T. 1983. Microorganisms and xenobiotic compounds. *Experientia* 39: 1183–1220.

McGowan, C., Fulthorpe, R., Wright, A. & Tiedje, J. 1998. Evidence for interspecies gene transfer in the evolution of 2,4-dichlorophenoxyacetic acid degraders. *Appl. Environ. Microbiol.* 64: 4089–4092.

Muller, T.A., Werlen, C., Spain, J. & van der Meer, J.R. 2003. Evolution of a chlorobenzene degradative pathway among bacteria in a contaminated groundwater mediated by a genomic island in *Ralstonia*. *Environ. Microbiol.* 5: 163–173.

Newby, D.T. & Pepper, I.L. 2002. Dispersal of plasmid pJP4 in unsaturated and saturated 2,4-dichlorophenoxyacetic acid contaminated soil. *FEMS Microb. Ecol.* 39: 157–164.

Newby, D.T., Gentry, T.J. & Pepper, I.L. 2000. Comparison of 2,4-dichlorophenoxyacetic acid degradation and plasmid transfer in soil resulting from bioaugmentation with two different pJP4 donors. *Appl. Environ. Microbiol.* 66: 3399–3407.

Sarand, I., Haario, H., Jørgensen, K.S. & Romantschuk, M. 2000. Effect of inoculation of a TOL plasmid containing mycorrhizosphere bacterium on development of Scots pine seedlings, their mycorrhizosphere and the microbial flora in m-toluate-amended soil. *FEMS Microbiol. Ecol.* 31: 127–141.

Schlüter, A., Heuer, H., Szczepanowski, R., Forney, L.J., Thomas, C.M., Pühler, A. & Top, E.M. 2003. The 64,508 bp IncP-1beta antibiotic multiresistance plasmid pB10 isolated from a waste water treatment plant provides evidence for recombination between members of different branches of the IncP-1β group. *Microbiology* 149: 3139–3153.

Springael, D. & Top, E.M. 2004. Horizontal gene transfer and microbial adaptation to xenobiotics : new types of mobile genetic elements and lessons from ecological studies. *Trends Microbiol.*: in press.

Springael, D., Peys, K., Ryngaert, A., Van Roy, S., Hooyberghs, L., Ravatn, R., Heyndrickx, M., van der Meer, J.-R., Vandecasteele, C., Mergeay, M. & Diels, L. 2002. Community shifts in a seeded 3-chlorobenzoate degrading membrane biofilm reactor: indications for involvement of in situ horizontal transfer of the *clc*-element from inoculum to contaminant bacteria. *Environ. Microbiol.* 4: 70–80.

Thomas, C.M. 2000. *The horizontal gene pool. Bacterial plasmids and gene spread.* Amsterdam: Harwood Academic Publishers.

Tiirola, M.A., Wang, H., Paulin, L. & Kulooma, M.S. 2002. Evidence for natural horizontal transfer of the *pcpB* gene in the evolution of polychlorophenol-degrading Sphingomonads. *Appl. Environ. Microbiol.* 68: 4495–4501.

Top, E.M., Holben, W.E. & Forney, L.J. 1995. Characterization of diverse 2,4-dichlorophenoxyacetic

acid-degradative plasmids isolated from soil by complementation. *Appl. Environ. Microbiol.* 61: 1691–1698.
Top, E.M., Springael, D. & Boon, N. 2002. Catabolic mobile genetic elements and their possible use in bioaugmentation of polluted soils and waters. *FEMS Microbiol. Ecol.* 42: 199–208.
Top, E.M., Moënne-Loccoz, Y., Pembroke, T. & Thomas, C.M. (2000). Phenotypic traits conferred by plasmids. In C.M. Thomas (ed): *The horizontal gene pool: bacterial plasmids and gene spread*: 249–285. Amsterdam: Harwood Academic Publishers.
Tsuda, M., Tan, H.M., Nishi, A. & Furukawa, K. 1999. Mobile catabolic genes in bacteria. *J Biosci Bioeng* 87: 401–410.
van der Meer, J.R., de Vos, W.M., Harayama, S. & Zehnder, A.J.B. 1992. Molecular mechanisms of genetic adaptation to xenobiotic compounds. *Microbiol. Rev.* 56: 677–694.
Wilson, M.S., Herrick, J.B., Jeon, C.O., Hinman, D.E. & Madsen, E.L. (2003). Horizontal transfer of *phnAc* dioxygenase genes within one of two phenotypically and genotypically distinctive naphthalene-degrading guilds from adjacent soil environments. *Appl. Environ. Microbiol.* 69: 2172–2181.
Worsey, M.J. & Williams, P.A. (1975). Metabolism of toluene and xylenes by *Pseudomonas putida (arvilla)* mt-2: evidence for a new function of the TOL plasmid. *J. Bacteriol.* 124: 7–13.

European Symposium on Environmental Biotechnology, ESEB 2004 - Verstraete (ed)
© 2004 Taylor & Francis Group, London, ISBN 90 5809 653 X

Biological nitrogen and phosphorus removal – processes of amazing diversity and complexity

J. Keller & R.J. Zeng
Advanced Wastewater Management Centre, The University of Queensland, QLD, Australia

ABSTRACT: Nitrogen and phosphorus removal from wastewater is now considered an essential element of the protection of our waterways. Biological nutrient removal processes are generally the most efficient and cost-effective solution to achieve this. While the principles of these processes are well known, intriguing and valuable details are being discovered with the recent advances in bio-process engineering and microbial sciences. Phosphorus accumulating organisms have only been identified in recent years, and there are now competing glycogen accumulating organisms being found in biological phosphorus removal systems. These can possibly explain the reasons for the variable phosphorus removal performance of certain systems, and can help in the development of more stable and higher performing processes. Detailed investigations of the traditional nitrification-denitrification systems, but also of novel developments for nitrogen removal, reveal a more complex and diverse range of processes involved in these transformations. Increasingly, linked phosphorus and nitrogen removal processes are being developed, creating further opportunities to optimise the technologies. However, this might also bring certain risks such as the potential to produce the greenhouse-gas nitrous oxide (N_2O) rather than nitrogen gas as the final denitrification product. A range of recent developments in these areas is covered in this paper.

1 CONVENTIONAL BIOLOGICAL NITROGEN AND PHOSPHORUS REMOVAL

Nitrogen (N) and phosphorus (P) are the key nutrients causing eutrophication in waterways. In conventional biological nutrient removal (BNR) systems, nitrogen removal is accomplished by a two-stage treatment, nitrification and denitrification. During nitrification, autotrophic nitrifiers aerobically oxidise ammonium, the major form of nitrogen in wastewater, to nitrite, then to nitrate. During denitrification, heterotrophic denitrifiers reduce nitrate to nitrite and then finally to nitrogen gas under absence of oxygen (so called anoxic conditions), using organic carbon (referred to as COD) as electron donor.

Phosphorus removal is achieved through enhanced biological phosphorus removal (EBPR) under alternating anaerobic–aerobic conditions using polyphosphate-accumulating organisms (PAOs) (Comeau et al., 1986). In EBPR, PAOs store volatile fatty acids (VFA) as poly-hydroxyalkanoates (PHA) and release phosphate under anaerobic conditions, followed by accumulation of excess amounts of phosphate in the form of polyphosphate (polyP) under aerobic conditions. The accumulated phosphorus is then removed from the system together with the excess sludge that is produced in the process.

While these process principles have been known for a long time, recent research has generated a large range of new knowledge, which has helped in improving the BNR process operation and has revealed a number of new and improved process options. This paper will focus on some of these recent findings and demonstrate the microbial diversity and process complexity of these systems, which were thought to be well understood for some time already.

2 SIMULTANEOUS NITRIFICATION AND DENITRIFICATION (SND)

Despite the different process requirements, a number of studies have shown that the two processes of nitrification and denitrification can occur concurrently in one well-mixed reactor under aerobic conditions, which has been termed simultaneous nitrification and denitrification (SND) (e.g. von Münch et al., 1996; Helmer and Kunst, 1998). The SND process simplifies treatment plant operation and reduces capital costs. Additionally, it has been reported that nitrogen removal can be achieved by partial oxidation of ammonium to nitrite with direct reduction to nitrogen gas (Surmacz-Gorska et al., 1997; Yoo et al., 1999). This mechanism is called SND via nitrite (Figure 1). This process can save 25% oxygen and 40% of the COD required compared to conventional nitrification and denitrification, respectively (Turk and Mavinic, 1986).

As nitrification requires oxygen, but denitrification is inhibited by the presence of oxygen, suitable conditions for SND need to be created for both processes within the one process tank. This is achieved at a microscopic level through mass-transfer limitations in activated sludge flocs (Pochana and Keller, 1999; Wilen et al., accepted). When oxygen is transferred from the liquid phase into the flocs for nitrification, it leads to a diffusion-reaction competition (Figure 2). If oxygen consumption in the outer zone of the flocs is greater than diffusion through the floc oxygen penetration is limited. This can create an aerobic zone on the edge of the floc where nitrification can take place, and an anoxic zone in the centre of the floc where heterotrophic bacteria can denitrify. In addition to the floc-internal mass transfer limitations, it is now increasingly recognized that flocs larger than approx. 100–150 μm also have a significant external concentration boundary layer which can reduce the oxygen concentration at the surface of the floc considerably compared to the bulk liquid (Wilen et al., accepted).

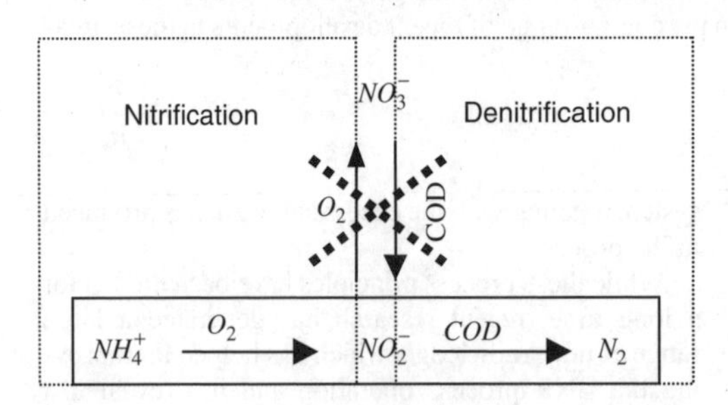

Figure 1. Schematic of mechanism of simultaneous nitrification and denitrification via nitrite pathway.

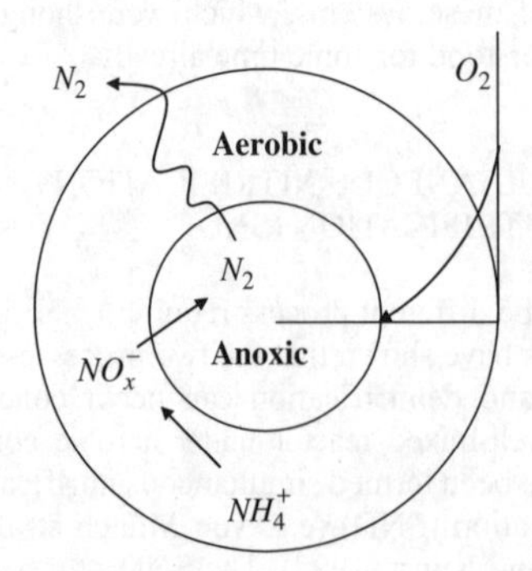

Figure 2. Schematic of activated sludge floc having internal aerobic/anoxic zones.

3 GLYCOGEN ACCUMULATING ORGANISMS COMPETING IN BIOLOGICAL P REMOVAL

While the enhanced biological phosphorus removal (EBPR) process has been utilised for many years in full-scale treatment plants, it was not until recently that the relevant polyphosphate accumulating organism (PAO) has been identified and named as *Candidatus Accumulibacter phosphatis* henceforth called *Accumulibacter* (Crocetti et al., 2000). However, another group of microorganisms, called glycogen accumulating organisms (GAOs), have been found to also have the capacity for anaerobic VFA uptake that was once thought to be unique to PAOs (Liu et al., 1996). Unlike PAOs, GAOs do not store poly-P aerobically and hence do not contribute to P removal in EBPR systems. They instead use glycogen as their sole energy source for anaerobic VFA uptake (Figure 3). In a full-scale EBPR plant survey, Saunders et al. (2003) found *Accumulibacter* (only identified PAO) and *Competibacter* (only identified GAO, Crocetti et al., 2002) coexisted in 5 of 6 plants (Table 1), showing

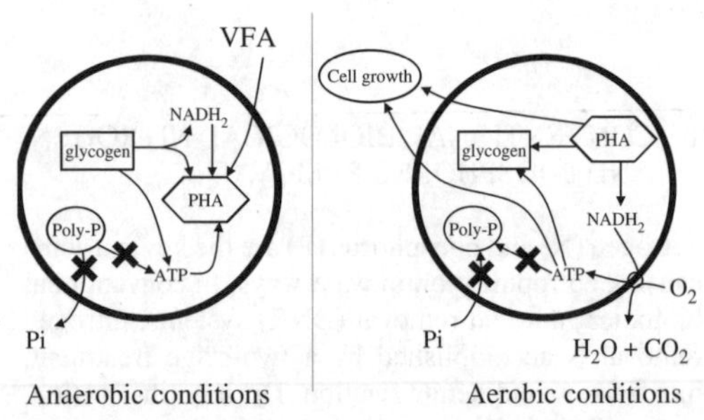

Figure 3. Schematic diagram of the metabolisms of PAO (without crosses) and GAO (with crosses) under anaerobic and aerobic conditions.

Table 1. P release to acetate uptake ratio and FISH quantification (from Saunders et al., 2003).

Plant	P_{rel}/HAc_{up} Pmol/Cmol	Accumulibacter (% of bacteria) mean	Competibacter (% of bacteria) mean
A	0.49	7	<1
B	0.33	12	12
C	0.29	8	4
D	0.40	11	3
E	0.51	Present	Present
F	0.41	Present	Present

GAO was not an artifact of lab reactor systems. The survey further revealed that GAOs competed with PAOs for the limiting VFA substrates present in EBPR systems.

4 SIMULTANEOUS DENITRIFICATION AND PHOSPHORUS REMOVAL

It has been shown that denitrification can be accomplished by the so-called denitrifying PAOs (DPAOs) in anaerobic-anoxic EBPR systems, allowing simultaneous nitrate/nitrite reduction and phosphorus uptake using the same COD (Kerrn-Jespersen and Henze, 1993; Kuba et al., 1993). DPAOs perform exactly the same processes as PAOs except they are using nitrate or nitrite as electron acceptor instead of oxygen. Under anoxic conditions, DPAOs generate 40% less energy and thus have a 20–30% lower cell yield (Kuba et al., 1996). Therefore, the use of DPAOs in BNR systems is highly beneficial in terms of a lower COD demand, reduced aeration cost and less sludge production. Recently, Zeng et al. (2003a) found the PAOs and DPAOs were indeed the same organisms. Potentially, most currently existing BNR plants could be modified to promote simultaneous denitrification and phosphorus removal. However, recent lab-scale studies by Zeng et al. (2003b) showed that GAOs, the competitor of PAOs, can denitrify under anaerobic-anoxic conditions, which is termed as denitrifying GAOs (DGAOs). The existence of DGAOs makes the BNR system more complex again as competition for both VFAs and nitrate exists between PAOs and GAOs.

5 SIMULTANEOUS NITRIFICATION DENITRIFICATION AND PHOSPHORUS REMOVAL (SNDPR)

Both SND and anaerobic-anoxic EBPR processes can save significant capital and operating costs. It would be ideal to combine these processes to achieve simultaneous nitrification, denitrification and phosphorus removal, which can be called SNDPR process.

Zeng et al. (2003c) demonstrated in a lab-scale study that it is feasible to achieve SNDPR with alternating anaerobic and low-DO aerobic (0.5 mg/L) stages. Figure 4 shows that during the anaerobic stage, acetate was completely consumed, which was accompanied by the consumption of glycogen, production of PHA and release of phosphorus. In the subsequent aerobic stage, ammonium was nitrified but without nitrite or nitrate accumulation. Meanwhile, PHA was oxidized, glycogen was replenished and phosphorus was taken up. In another batch study with only aerobic conditions and without COD addition, nitrite instead of nitrate was found to be the major nitrification product, which suggested that the nitrogen removal in this SNDPR process was via the nitrite pathway. However, this study also showed that DGAOs rather than DPAOs were responsible for the denitrification activity, limiting the benefits in terms of COD utilization for both phosphorus and nitrogen removal.

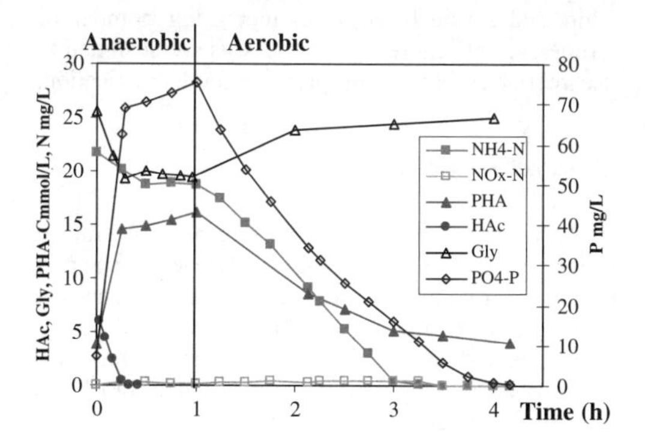

Figure 4. Nutrient and carbon profiles measured during a typical cycle of SNDPR reactor (from Zeng et al., 2003c).

6 PAO-GAO COMPETITION

It is clear that GAOs can compete with PAOs for the limiting VFA substrates present in anaerobic–aerobic or anaerobic-anoxic biological phosphorus removal systems. However, the factors affecting the competition between PAOs and GAOs are largely unknown. Since anaerobic VFA uptake has been shown to be a critical parameter that enables the growth and proliferation of both PAOs and GAOs, it is desirable to know which particular type of VFAs is more readily metabolised by PAOs or GAOs. The primary VFA present in most full-scale BNR systems is acetate, and thus, most experimental studies in this field have focused largely on the use of acetate as the sole carbon substrate. There are, however, many other types of VFAs present in the influent wastewater to BNR plants, and their effects have been less widely studied. The second most common VFA is propionate, which has been shown to be present in domestic and industrial wastewater and is also generated in high quantities specific prefermenters employed for VFA generation (von Münch, 1998; Thomas et al., 2003). The impact of propionate on biological P removal systems has attracted recent interest because of suggestions that it can be a favorable substrate for successful EBPR operation (Thomas et al., 2003; Chen et al., 2002). Oehmen et al. (2004) has revealed that a culture of GAOs enriched with acetate was unable to anaerobically metabolise propionate in short-term experiments, while a similarly enriched culture of PAOs was able to immediately take up propionate in identical batch tests.

In the study of SNDPR with propionate as sole carbon source, Zeng et al. (submitted) found that GAOs were able to out-compete PAOs in this process. A microbial study on the population shift showed that *Accumulibacter* was the organism responsible for P removal in propionate feed. However, a previously unknown GAO from the *Alpha-proteobacteria* phylum was dominant in the biomass instead of *Competibacter* which is usually found in acetate-fed reactors. This suggests that *Accumulibacter* was the common PAO in acetate or propionate system but GAOs were more diverse. Additionally, the low-DO conditions used in the SNDPR process (0.5 mg/L) may be a critical factor in the PAO/GAO competition.

7 N_2O EMISSION

The recognised denitrification steps from nitrate to nitrogen gas are:

$$NO_3^- \rightarrow NO_2^- \rightarrow NO \rightarrow N_2O \rightarrow N_2$$

Most denitrification studies simply assume that the final denitrification product is nitrogen gas without actual gas measurements. This assumption is not always correct. Figure 5 shows a detailed study of the processes under anoxic conditions including gas measurements in the DGAO culture previously described (Zeng et al., 2003b). Data from the prior anaerobic stage (2.25 hr) is not shown, but it confirmed that all acetate was fully taken up and converted to PHA. After nitrate addition, it was found that the final product of denitrification in these studies was mainly nitrous oxide (N_2O) rather than nitrogen gas (N_2), which was only produced as the beginning and the end of the experiment. N_2O production is certainly a significant environmental concern as it is a major greenhouse gas with a global warming potential 296 times that of CO_2 (Levine, 1984).

The reason why denitrification stops at N_2O is still unclear. It has been reported in literature that an elevated level of nitrite (over 2 mg/L) could lead to N_2O production (von Schulthess and Gujer, 1996). This report seems to be confirmed by the results shown in Figure 5. N_2 was produced at both the beginning and the end of the anoxic period, when the nitrite concentration was low. N_2O was produced during the rest of the anoxic period, when an elevated nitrite concentration was measured.

Figure 6 shows both N_2 and N_2O were produced during the low-DO aerobic period in the SNDPR process (Zeng et al., 2003c). No nitrite accumulation was detected in this study, indicating nitrite may not be the only trigger for N_2O production. The common elements between the DGAO and SNDPR studies are that DGAOs were the main organisms responsible for denitrification and that internally stored PHA was the

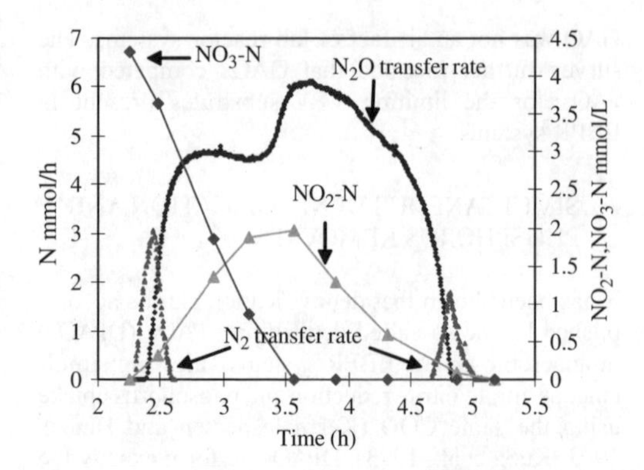

Figure 5. Complete nitrogen profiles during the anoxic phase in DGAO reactor (from Zeng et al., 2003b).

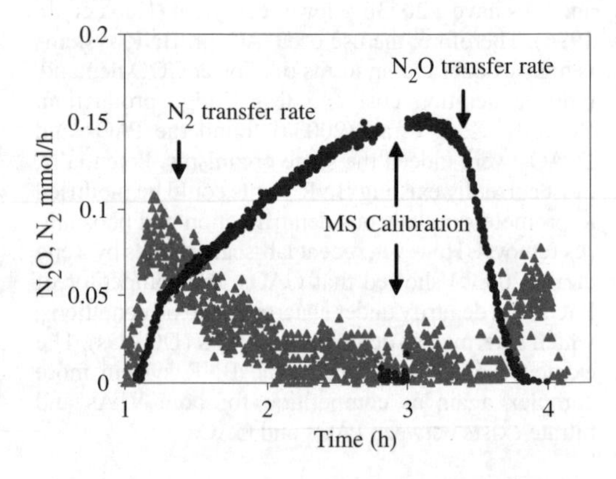

Figure 6. Off gas analysis of a cyclic study during aerobic period in the SNDPR reactor (from Zeng et al., 2003c).

carbon source for denitrification. It is unknown if N_2O can be produced in DPAO cultures. Nevertheless, the question whether PHA (or other cell-internal storage products) can trigger N_2O production needs to be addressed seriously since an increasing number of biological nutrient removal processes are designed to use internal carbon storage products for denitrification.

8 CONCLUSIONS

The processes for nitrogen and phosphorus removal in wastewater treatment systems have been known in principle for many years. However, recent advances in process engineering and microbial science allow far more detailed investigations of these systems now. What we are finding is an intricate system of biochemical processes and a diverse microbial community, generally much more complex than many researchers or process designers have been aware of. This greatly enhanced knowledge can and should be used for improved process design, operation and optimisation. This has the potential to replace the generic "black-box" approach used in the past and will likely enable the development of better performing and more efficient treatment plants in future.

However, there is still a major need for further research. In particular, the co-existence of GAOs and PAOs in biological phosphorus removal processes can cause major performance deterioration, but the deciding factors in this competition are still unknown. On the nitrogen removal processes, the production of N_2O during certain denitrification conditions (and possibly even during nitrification) will need to be investigated further to avoid transferring an environmental problem from the waterways to the atmosphere.

REFERENCES

Chen, Y., Trujillo, M., Biggerstaff, J., Ahmed, G., Lamb, B., Eremektar, F.G., McCue, T., Randall, A.A. 2002. In *Water Environment Federation 75th Annual Conference and Exposition (WEFTEC)* Chicago, Illinois, U.S.A.

Comeau, Y., Hall, K., Hancock, R., Oldham, W. 1986. Biochemical model for enhanced biological phosphorus removal. *Water Research* 20: 1511–1521.

Crocetti, G.R., Hugenholtz, P., Bond, P.L., Schuler, A., Keller, J., Jenkins, D., Blackall, L.L. 2000. Identification of polyphosphate accumulating organisms and the design of 16S rRNA-directed probes for their detection and quantitation. *Applied and Environmental Microbiology* 66(3): 1175–1182.

Crocetti, G.R., Banfield, J.F., Keller, J., Bond, P.L., Blackall, L.L. 2002. Glycogen accumulating organisms in laboratory-scale and full-scale activated sludge processes. *Microbiology* 148: 3353–3364.

Helmer, C., Kunst, S. 1998. Simultaneous nitrification/denitrification in an aerobic biofilm system. *Water Science and Technology* 37(4–5), 183–187.

Levine, J.S. 1984. Water and the photochemistry of the troposphere. In: Satellite sensing of a cloudy atmosphere: observing the third planet, A. Handerson-Sellers (Ed.). Tayler & Francis, Ltd. London.

Keller, J., Subramaniam, K., Gosswein, J., Greenfield, P.F. 1997. Nutrient removal from industrial wastewater using single tank sequencing batch reactors. *Water Science and Technology* 35(6): 137–144.

Kerrn-Jespersen, J.P., Henze, M., Strube, R. 1994. Biological Phosphorus Release and Uptake under Alternating Anaerobic and Anoxic Conditions in a Fixed-Film Reactor. *Water Research* 28(5): 1253–1255.

Kuba, T., Smolders, G.J.F., van Loosdrecht, M.C.M., Heijnen, J.J. 1993. Biological phosphorus removal from wastewater by anaerobic and anoxic sequencing batch reactor. *Water Science and Technology* 27: 241–252.

Kuba, T., Murnleitner, E., van Loosdrecht, M.C M., Heijnen, J.J. 1996. A metabolic model for biological phosphorus removal by denitrifying organisms. *Biotechnology and Bioengineering* 52(6): 685–695.

Oehmen, A., Yuan, Z., Blackall, L.L., Keller, J. 2004. Short-term effects of carbon source on the competition of PAO and GAO. *3rd IWA conference on SBR3*, Noosa, Australia.

Pochana, K., Keller, J. 1999. Study of factors affecting simultaneous nitrification and denitrification (SND). *Water Science and Technology* 39(6): 61–68.

Saunders, A.M., Oehmen, A., Blackall, L.L., Yuan, Z., Keller, J. 2003. The effect of GAOs (glycogen accumulating organisms) on anaerobic carbon requirements in full-scale Australian EBPR (enhanced biological phosphorus removal) plants: *Water Science and Technology* 47(11): 37–43.

Surmacz-Gorska, J., Cichon, A., Miksch, K. 1997. Nitrogen removal from wastewater with high ammonia nitrogen concentration via shorter nitrification and denitrification. *Water Science Technology* 36(10): 73–82.

Thomas, M., Wright, P., Blackall, L., Urbain, V., Keller, J. 2003. Optimisation of Noosa BNR plant to improve performance and reduce operating costs. *Water Science Technology* 47(11): 141–148.

Turk, O., Mavinic, D. 1986. Preliminary Assessment of a shortcut in Nitrogen Removal from Wastewater. *Can. J. Civil Eng.* 13: 600–605.

von Münch, E. 1998. DSP Prefermenter Technology Book, Science Traveller International, CRC WMPC Ltd., Brisbane, Qld, Australia.

von Münch, E., Lant, P., Keller, J. 1996. Simultaneous nitrification and denitrification in bench-scale sequencing batch reactors. *Water Research* 30(2): 277–285.

von Schulthess, R., Gujer, W. 1996. Release of nitrous oxide (N_2O) from denitrifying activated sludge: verification and application of a mathematical model. *Water Research* 30(3): 521–530.

Wilén, B.M., Gapes, D., Blackall, L.L., Keller, J. (accepted). Determination of external and internal mass transfer effects in nitrifying microbial aggregates. *Biotechnology and Bioengineering*.

Yoo, K., Ahn, K.H., Lee, H.J., Lee, K.H., Kwak, Y.J., Song, K.G. 1999. Nitrogen removal from synthetic wastewater by simultaneous nitrification and denitrification (SND) via nitrite in an intermittently-aerated reactor. *Water Research* 33(1): 145–152.

Zeng, R.J., Saunders, A.M., Yuan, Z., Blackall, L.L., Keller, J. 2003a. Identification and comparison of aerobic and denitrifying polyphosphate-accumulating organisms. *Biotechnology and Bioengineerin* 83(2): 140–148.

Zeng, R.J., Yuan, Z., Keller, J. 2003b. Enrichment of denitrifying glycogen-accumulating organisms in anaerobic/anoxic activated sludge systems. *Biotechnology and Bioengineering* 81(4): 397–404.

Zeng, R.J., Lemaire, R., Yuan, Z., Keller, J. 2003c. Simultaneous Nitrification and Denitrification and Phosphorus Removal in a Lab-Scale Sequencing Batch Reactor. *Biotechnology and Bioengineering* 84(2): 170–178.

Zeng, R.J., Lemaire, R., Crocetti, G.R., Keller, J. (submitted). Effect of Carbon Source on the Enrichment Process of Simultaneous Nitrification, Denitrification and Phosphorus Removal in a Lab-Scale Sequencing Batch Reactor. *Environmental Science & Technology*.

European Symposium on Environmental Biotechnology, ESEB 2004 - Verstraete (ed)
 ISBN 90 5809 653 X

Bioremediation, microbial pathogenicity and regulation of mammalian cell growth and death

T. Yamada, Y. Hiraoka & A.M. Chakrabarty
Dept. of Microbiology and Immunology, University of Illinois, College of Medicine, Chicago, USA

ABSTRACT: Microorganisms of the genus *Pseudomonas* and *Burkholderia* are efficient in degrading various toxic chemicals and are potential candidates for bioremediation of toxic chemical dump sites. Both are, however, opportunistic pathogens and their pathogenic potentials must be fully evaluated before they can be released in an open environment. The secretion of a redox protein azurin as a potential virulence factor by members of *Pseudomonas* and *Burkholderia* species and how azurin modulates target mammalian host cells through interaction with the tumor suppressor protein p53 is described here as an example of complex nature of virulence factors.

1. Biodegradation of toxic chemicals by Pseudomonas and Burkholderia species

The release of synthetic chemicals, mostly chlorinated compounds, in the form of herbicides/pesticides, solvents, fire retardants, refrigerants, etc., has created major environmental pollution problems that require urgent remediative measures. Bioremediation where microorganisms are used for enhanced biodegradation and removal of many synthetic, toxic chemicals, is one such remedy. Among microorganisms, the *Pseudomonas* and *Burkholderia* species are well known for their nutritional versatility and for their ability to consume both natural and synthetic chemicals (Ogawa et al., 2003). While the pseudomonads are very efficient in metabolizing many different compounds (Ramos, 2004), *Burkholderia* species such as *B. cepacia* are becoming adept in utilizing many of such toxic chemicals as well (Lessie & Gaffney, 1986; Nelson et al., 1987; Daubaras et al., 1996; Kumamaru et al., 1998). Thus many *B. cepacia* strains are ideal candidates for bioremediation. In addition, *B. cepacia* strains are often effective in the control of plant diseases by killing plant pathogenic soil fungi and nematodes, and can thus substitute for chemical pesticides. For example, *B. cepacia* is quite effective in reducing corn damping off caused by *Pythium* and *Fusarium* species and is known to produce high amounts of anti-fungal pyrrolnitrin for control of plant pathogenic fungi (Mao et al., 1997; Upadhyay et al., 1991). Use of *B. cepacia* strains both in green house and field tests has shown comparable results with standard fungicide treatment such as captan, thiram and benlate and has significantly increased crop growth compared to no treatment (Bowers and Parke, 1993).

A problem with the environmental application of *B. cepacia* either as biopesticides or in toxic chemical pollution control is the concern about their potential pathogenicity. Human infections caused by *B. cepacia* are prominent in patients with cystic fibrosis (CF) and chronic granulomatous disease (Thomassen et al., 1985; O'Neill et al., 1986). Infection with *B. cepacia* may also occur in children with sickle cell anaemia, in oncology patients, in those undergoing cardiac treatment, as rare instances of pediatric meningitis and as nosocomial outbreaks or outbreaks associated with contaminated therapeutics/devices. Due to the intrinsic antibiotic resistance of *B. cepacia*, all of these types of infections are very difficult to treat and often fatal (LiPuma, 1998a; LiPuma, 1998b).

2. Defining microbial pathogenicity: the role of a bacterial redox protein in the modulation of mammalian cell growth and death

Essentially all infectious diseases have environmental etiology. The infecting agents, whether bacteria, fungi or viruses, usually have an environmental reservoir and may enter the human body through air, water or contaminated food. Yet the outcome of a successful infection is very much dependent not only on the initial load, that is the number of infecting agents, but also the nature of virulence factors that ultimately enable the infectious agents to evade the host immune system and proliferate or colonize the human tissues.

The role played by the host immune system and the genes that encode the system are thus of paramount importance. Nevertheless, pathogens have evolved various mechanisms over a long evolutionary period that help them evade host defenses and establish themselves as successful infectious agents both in humans and animals (Finley & Falkow, 1997; Falkow, 1998). An understanding of the mechanisms by which the pathogens gain entry into the human body and establish a foothold may therefore help in the development of drugs that inactivate or render harmless these pathogenic traits.

We have recently demonstrated the secretion of two types of enzymes, ATP-utilizing enzymes and redox proteins, by clinical isolates of *P. aeruginosa* (Zaborina et al., 2000). We have shown that these enzymes demonstrated high cytotoxicity towards macrophages, which are the body's first line of defense. The redox proteins consisted of azurin and cytochrome c_{551}.

Azurin is a member of a family of blue copper proteins known as cupredoxins which are small, soluble proteins (10–14 kDa) whose active site contains a type I copper. Members of the cupredoxin family in bacteria or plants may show low (<20%) to moderate sequence identity but have highly conserved active site structures (Rienzo et al., 2000; Murphy et al., 2002) which allow them to accept or donate electrons from diverse sources. Using purified azurin and cytochrome c_{551} from *P. aeruginosa*, Zaborina et al. (2000) demonstrated induction of apoptosis in macrophages by the purified proteins. Similar secretion of azurin and its associated cytotoxicity has been reported for *B. cepacia* (Punj et al., 2003a). Azurin was found to be more cytotoxic to macrophages than cytochrome c_{551}. More recently, Yamada et al. (2002a) reported that azurin, purified after hyperexpression of the *P. aeruginosa* azurin gene in *E. coli*, can form a complex with and enhance the intracellular level of the tumor suppressor protein p53 in macrophages. p53 is a known inducer of apoptosis (Vogelstein et al., 2000; Schuler & Green, 2001) and hence elevated levels of p53 allowed high levels of apoptosis in macrophages. A higher level of reactive oxygen species (ROS), generated during treatment of macrophages with wild type (wt) azurin, correlated with its cytotoxicity. Treatment with some ROS-removing antioxidants greatly reduced azurin-mediated cytotoxicity, thus demonstrating a novel apoptogenic property of this bacterial redox protein (Yamada et al., 2002a). The cytotoxicity was also examined in some human cancer cells, using both wt and a double methionine mutant Met44LysMet64Glu (M44KM64E) azurin where the Met residues in the hydrophobic patch were replaced by polar amino acids. Azurin and M44KM64E mutant azurin were shown to enter the cytosol of a p53-positive human melanoma cell line UISO-Mel-2 and p53-null cell line UISO-Mel-6 cells; however, while wt azurin could traffic to the nucleus in p53-positive UISO-Mel-2 cells, it could not traffic to the p53-null UISO-Mel-6 cells, thus demonstrating a role of p53, which can traffic between the cytosol and the nucleus because of the presence of nuclear import/export signals, in the trafficking of azurin to the nucleus (Yamada et al., 2002b). p53 is a transcriptional activator of the gene for the proapoptotic protein Bax (Vogelstein et al., 2000; Schuler & Green, 2001). Accordingly, treatment of UISO-Mel-2 cells with azurin led to a higher level of Bax in the mitochondria, allowing significant release of mitochondrial cytochrome c into the cytosol, thus initiating the onset of apoptosis (Yamada et al., 2002b). Interestingly, the M44KM64E double mutant with charged Lys and Glu residues in the hydrophobic patch, was shown to be deficient in cytotoxicity towards cancer cells, deficient in forming a complex with oligomeric p53, and was less efficient in stabilizing p53 than wt azurin (Yamada et al., 2002b). More interestingly and of some potential clinical significance, azurin was shown to allow regression of human UISO-Mel-2 tumors xenotransplanted in immunodeficient nude mice (Yamada et al., 2002b). Thus azurin, a member of the cupredoxin family that has been studied extensively for electron transfer property (Rienzo et al., 2000) but was not known for cytotoxicity, seems to be a potential virulence factor elaborated by pathogenic bacteria for the killing of phagocytic cells such as macrophages (and also cancer cells). The redox activity of azurin is not necessary for its cytotoxic action (Goto et al., 2003) and the physical presence of azurin in the mammalian cell cytosol is believed to trigger apoptosis in mammalian cells (Punj & Chakrabarty, 2003b).

3. Azurin interacts with mammalian tumor suppressor protein p53

p53 is normally a labile protein and is proteolytically cleaved via ubiquitination in presence of MDM2 (Haupt et al., 1997), which binds in the N-terminal region of p53. Certain mammalian redox proteins are known to stabilize p53 through binding in the N-terminal region and competing with MDM2 binding, thus reducing ubiquitination and proteolytic cleavage of p53 (Asher et al., 2001). p53 has two major functions in the cell. It is an inducer of apoptosis and it also causes growth arrest through transcriptional activation of a gene encoding an inhibitor (p21/CIP1/Waf1) of the cell cycle progression. p53 is also capable of oligomerization and this multimer formation occurs at the C-terminal end of p53 (Delphin et al., 1999). As mentioned previously, wt azurin forms a complex with p53 that stabilizes p53 and thereby enhances apoptosis through ROS generation, while the M44KM64E mutant azurin does not appear to form a stable complex with p53, does not stabilize p53, generates less ROS and has low apoptosis inducing activity, presumably due to its complex formation at the C-terminal.

In order to determine which domain of p53 might be involved in interacting with wt azurin, we have used a GST pull-down assay (Punj et al., 2004), using full length and truncated derivatives of p53. When GST alone or GST-full length p53 fusion proteins were immobilized on agarose beads, and various concentrations of wt azurin were mixed with the beads, the beads washed and the proteins analyzed by SDS-PAGE, azurin was found to be retained only on the beads with immobilized GST-p53 but not with GST alone, suggesting that azurin was retained as part of p53-azurin complex. When such GST pull-down assays were conducted with truncated p53 derivatives as well, azurin was found to form complexes and retained in the beads with p53 containing the N-terminal and the middle part, but not the C-terminal part (Punj et al., 2004).

The above observation raises another important question: if the mutant M44KM64E azurin forms a different type of complex with p53 than the wt azurin, will that change the nature of p53 as a transcriptional activator? p53 is a transcriptional activator of proapoptotic genes such as *bax* and *NOXA*, which promote apoptosis (Schuler & Green, 2001). It is also a transcriptional activator of *p21* which is an inhibitor of cell cycle progression by interfering in DNA replication from Go/G1 to S and M phases. p21, also called Cip1 or Waf1, is an inhibitor of CDK/cyclin complexes that phosphorylate retinoblastoma (Rb) at the G1 to S phase during cell cycle progression (Sherr, 2000). Excess p21, therefore, inhibits CDK2/cyclinA or other CDK/cyclin complexes to inhibit G1 to S phase transition and is an essential component in inducing p53-mediated growth arrest as a response to the DNA damage (Sherr, 2000). p53 is known to modulate either of p21 or Bax level, depending upon the cell type or the presence of activated oncogenes. Indeed, analysis of p53-regulated gene expression patterns using oligonucleotide arrays has shown such expression to be highly heterogeneous (Zhao et al., 2000). To examine if complex formation by M44KM64E mutant azurin at the putative C-terminal of p53 might have altered p53's specificity for induction of apoptosis or inhibition of cell cycle progression, we measured the effect of wt azurin and M44KM64E mutant azurin on cell cycle progression as measured by the DNA content at Go/G1, S and M phases with or without treatment with wt azurin and M44KM64E mutant azurin. While wt azurin is an effective inducer of apoptosis, enhancing the levels of intracellular p53 and Bax (Yamada et al., 2002a), it has very little effect on the inhibition of cell cycle progression. In contrast, the M44KM64E mutant azurin, which had low cytotoxicity and apoptotic activity, significantly inhibited cell cycle progression in J774 cell line-derived macrophages, demonstrating how small changes in the protein have profound effect on its activity.

Since azurin is believed to form a complex with p53 at its N-terminal end, thereby stabilizing it, while M44KM64E mutant azurin is thought to form the complex at its C-terminal end, thereby preventing its oligomerization, it was deemed possible that the two p53 complexes might activate their target promoters differently. We, therefore, determined the levels of p53, p21 and Bax by Western blotting using monoclonal antibodies against these proteins in the crude extracts of macrophages treated with wt azurin and M44KM64E mutant azurin for 0, 6, 18 and 24 hours. The p53 level was elevated at 18 to 24 h during treatment of the macrophages with wt azurin, while such effect was mimimal during treatment with M44KM64E mutant azurin. Similarly, the Bax level was higher in the crude extracts of macrophages treated with wt azurin but not in the extracts of M44KM64E mutant azurin-treated macrophages. In contrast, the level of p21 protein appeared to be elevated in the crude extracts of macrophages treated with M44KM64E mutant azurin for 18 to 24 h, compared to macrophages treated with wt azurin. The levels of actin, used as an internal control whose expression is independent of p53 transcriptional activity, did not change during treatment with either wt azurin or M44KM64E mutant azurin. Thus wt azurin activates the *bax* promoter leading to enhanced level of Bax and apoptosis, while the M44KM64E mutant azurin allows activation of the *p21* promoter leading to enhancement of p21 accumulation and consequent inhibition of cell cycle progression.

Based on the above observations, we have constructed a model on the mode of action of wt azurin and M44KM64E mutant azurin on mammalian cells, particularly J774 cells (Fig. 1). Highly purified wt azurin has been shown to enter such cells and bind to the N-terminal to mid part, but not the C-terminal part, of p53, thereby likely competing with MDM2 binding. MDM2 binding to p53 allows ubiquitination of p53, accelerating its proteolytic cleavage; wt azurin thus stabilizes p53 and enhances its intracellular level. p53 is a transcriptional activator and two of its major target promoters are the *bax* and *p21* genes. In different cellular backgrounds, p53 activates either the *bax* or *p21* genes but seldom both (Zhao et al., 2000). The reason for this differential activation is not known. Apparently wt azurin-p53 complexes activate the *bax* promoter, enhancing Bax levels which then target the mitochondria and allow release of mitochondrial cytochrome c to the cytosol. Release of cytochrome c then leads to the formation of a multimeric complex with other proteins such as Apaf-1 and procaspase 9, called an apoptosome, which activates

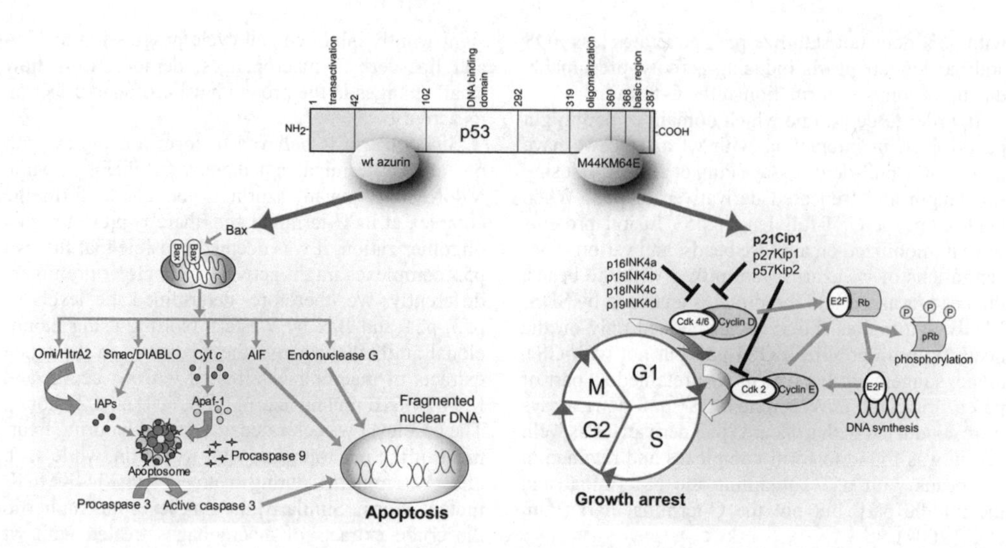

Figure 1. Modulation of apoptosis and growth arrest pathways by the tumor suppressor protein p53, when complexed with wt azurin. or M44KM64E mutant azurin complex formation of p53 with wt azurin leads to activation of bax and consequent apoptosis while complex formation with M44KM64E mutant azurin leads to activation of *p21*, triggering inhibition of cell cycle progression and growth arrest.

procaspase 3 to active caspase, leading to apoptosis (Fig. 1, left panel). When J774 cells are incubated in presence of purified M44KM64E mutant azurin, the mutant azurin enters the cells and appears to form an unstable complex with p53 that prevents p53 oligomerization (Yamada et al., 2002a). The oligomerization domain of p53 is in the mid to C-terminal (Fig. 1, right panel). This complex is capable of activating the *p21*, but not the *bax*, promoter. The enhanced transcription of the *p21* gene leads to enhanced p21 levels. p21 is a member of a family of proteins called cip (another name for p21), kip family (Sherr, 2000). They ($p21^{cip1}$, $p27^{Kip1}$, $p57^{Kip2}$) are potent inhibitors of the formation of complexes of cyclin D (D1, D2 etc) with their kinases called cyclin dependent kinases (CDKs) such as CDK4, CDK6, etc. These complexes phosphorylate the tumor suppressor protein Rb. Rb normally forms a complex with the transcriptional activator E2F, keeping it inactive. Phosphorylation of Rb allows dissociation of the complex, releasing active E2F. E2F activates the genes for cyclins such as cyclin E and other proteins involved in DNA synthesis. Thus CDKs are very important in the phosphorylation of Rb which causes release of active E2F from the inactive Rb.E2F complex and initiates DNA synthesis (start of S phase from the gap phase called G1). Once DNA replication is completed, the cells enter a second gap phase called G2 before mitosis (M phase). Mitosis completes the cell division. M44KM64E mutant azurin inhibits cell division by hyperexpressing p21, which inhibits cyclin/CDK formation, and thereby prevents phosphorylation of Rb, required for release of active E2F and DNA synthesis (Fig. 1, right panel). This is an interesting example where a single bacterial protein, with or without a couple of mutations in the hydrophobic patch, modulates the transcriptional specificity of a tumor suppressor protein p53.

ACKNOWLEDGEMENT

This work was supported by PHS grant ES04050-18 from the National Institute of Environmental Healthn Sciences.

REFERENCES

Asher, G., Lotem, J., Cohen, B., Sachs, L. & Shaul, Y. 2001. Regulation of p53 stability and p53-dependent apoptosis by NADH quinone oxidoreductase 1 *Proc. Natl. Acad. Sci. USA* 98: 1188–1193.

Bowers, J.H. & Parke, J.L. 1993. Epidemiology of pythium damping-off and aphanomyces root rot of peas after seed treatment with bacterial agents for biological control. *Phytopathol.* 83: 1466–1473.

Daubaras, D.L., Danganan, C.E., Hubner, A., Ye, R.W., Henderickson, W. & Chakrabarty, A.M. 1996. Biodegradation of 2,4,5-trichlorophenoxyacetic acid by *Burkholderia cepacia* strain AC1100: evolutionary insight. *Gene* 179: 1–8.

Delphin, C., Ronjat, M., Deloulme, J.C., Garin, G. et al. 1999. Calcium-dependent interaction of S100B with the C-terminal domain of the tumor suppressor p53. *J. Biol. Chem.* 274: 10539–10544.

Falkow, S. 1998. What is a pathogen? *ASM News* 63: 359–365.

Finley, B.B. & Falkow, S. 1997. Common themes in microbial pathogenicity revisited. *Microbiol. Mol. Biol. Rev.* 61: 136–169.

Goto, M., Yamada, T., Kimbara, K., Horner, J. et al. 2003. Induction of apoptosis in macrophages by *Pseudomonas aeruginosa* azurin: tumour suppressor protein p53 and reactive oxygen species, but not redox activity, as critical elements in cytotoxicity. *Mol. Microbiol.* 47: 549–559.

Haupt, Y., Maya, R., Kazaz, A. & Oren, M. 1997. MDM2 promotes the rapid degradation of p53. *Nature* 387: 296–299.

Kumamaru, T., Suenaga, H., Mitsuoka, M. Watanabe, T. & Furukawa, K. 1998. Enhanced degradation of polychiorinated biphenyls by directed evolution of biphenyl dioxygenase. *Nature Biotechnol.* 16: 663–666.

Lessie, T.G. & Gaffney, T. 1986. Catabolic potential of *Pseudomonas cepacia*. In the Bacteria, Vol. 10, The Biology of *Pseudomonas*. Sokatch, J.R. and Ornston, L.N., Eds., Academic Press, Orlando, Fl., pp. 439–481.

LiPuma, J.J. 1998a. *Burkholderia cepacia* – management issues and new insights. *Clin. Chest Med.* 19: 473–486.

LiPuma, J.J. 1998b. *Burkholderia cepacia* epidemiology and pathogenesis: implications for infection control. *Curr. Opin. Pulm. Med.* 4: 337–341.

Mao, W., Lewis, J.A., Hebbar, P.K. & Lumsden, R.D. 1997. Seed treatment with a fungal or a bacterial antagonist for reducing corn damping-oft caused by species of *Pvthium and Fusarium. Plant Dis.* 81: 450–454.

Murphy, L.M., Dodd, F.E., Yousafzai, F.K., Eady, R.R. & Hasnain, S.S. 2002. Electron donation between copper containing nitrite reductases and cupredoxins: The nature of protein-protein interaction in complex formation. *J. Mol. Biol.* 315: 859–871.

Nelson, M.J., Montgomery, S.O., Mahaffey, W.R. & Pritchard, P.H. 1987. Biodegradation of treichloroethylene and involvement of an aromatic biodegradative pathway. *Appl. Environ. Microbiol.* 53: 949–954.

Ogawa, N., Miyashita, K. & Chakrabarty, A.M. 2003. Microbial genes and enzymes in the degradation of chlorinated compounds. *Chem. Rec.* 3: 158–171.

O'Neill, K.M., Herman, J.H., Modlin, J.F., Moxon, E.R. & Winkelstein, J.A. 1986. *Pseudomonas cepacia*: an emerging pathogen in chronic granulomatous disease. *J. Pediatr.* 108: 940–942.

Punj, V., Sharma, R., Zaborina, O. & Chakrabarty, A.M. 2003a. Energy-generating enzymes of *Burkholderia cepacia* and their interactions with macrophages. *J. Bacteriol.* 185: 3167–3178.

Punj, V. & Chakrabarty, A.M. 2003b. Redox proteins in mammalian cell death: an evolutionarily conserved function in mitochondria and prokaryotes. *Cellular Microbiol.*, 5: 225–231.

Punj, V., Das Gupta, T.K. & Chakrabarty, A.M. 2004. Bacterial cupredoxin azurin and its interactions with the tumor suppressor protein p53. *Biochem. Biophys. Res. Communs.*, in press.

Ramos, J.L. 2004. The Pseudomonads, Vol. III, Kluwer Academic/Plenum Publishers, Dordrecht, The Netherlands, in press.

Rienzo, F.D., Gabdoulline, R.R., Menziani, M.C. & Wade, R.C. 2000. Blue copper proteins: a comparative analysis of their molecular interaction properties. *Protein Sci.* 9: 1439–1454.

Schuler, M. & Green, D.R. 2001. Mechanisms of p53-dependent apoptosis. *Biochem. Soc. Trans.* 29: 684–688.

Sherr, C.J. 2000. The Pezcoller lecture: Cancer cycles revisited. *Cancer Res.* 60: 3689–3695.

Thomassen, M.J., Demko, C.A., Klinger, J.D. & Stern, R.C. 1985. *Pseudomonas cepacia* colonization among patients with cystic fibrosis: a new opportunist. *Am. Rev. Respir.* Dis 131: 791–796.

Upadhyay, R.S., Visintin, L. & Jayaswal, R.K. 1991. Environmental factors affecting the antagonism of *Pseudomonas cepacia* against *Trichoderma viridie. Can. J. Microbiol.* 37: 880–884.

Vogelstein, B., Lane, D. & Levine, A.J. 2000. Surfing the p53 network. *Nature*. 408: 307–310.

Yamada, T., Goto, M., Punj, V., Zaborina, O. et al. 2002a. The bacterial redox protein azurin induces apoptosis in J774 macrophages through complex formation and stabilization of the tumor suppressor protein p53. *Infect. Immun.* 70: 7054–7062.

Yamada, T., Goto, M., Punj, V., Zaborina, O., et al. 2002b. Bacterial redox protein azurin, tumor suppressor protein p53, and regression of cancer. *Proc. Natl. Acad. Sci. USA* 99: 14098–14103.

Zaborina, O., Dhiman, N., Chen, M.L. et al. 2000. Secreted products of a nonmucoid *Pseudomonas aeruginosa* strain induce two modes of macrophage killing: external ATP-dependent, P2Z-receptor-mediated necrosis and ATP-independent, caspase-mediated apoptosis. *Microbiology* 146: 2521–2530.

Zhao, R., Gish, K., Murphy, M., Yin, Y. et al. 2000. Analysis of p53-regulated gene expression patterns using oligonucleotide arrays. *Genes Dev.* 14: 981–993.

European Symposium on Environmental Biotechnology, ESEB 2004 - Verstraete (ed)

Biodiversity and application of anaerobic ammonium-oxidizing bacteria

Mike S.M. Jetten[1,2], Markus C. Schmid[2], Ingo Schmidt[1,5], Mark van Loosdrecht[2], Wiebe Abma[3], J. Gijs Kuenen[2], Jan-Willem Mulder[4] & Marc Strous[1]

[1]*Dept. of Microbiology, University of Nijmegen, Toernooiveld, Nijmegen, The Netherlands*
[2]*Dept. of Biotechnology, Delft, The Netherlands*
[3]*Paques BV Balk, The Netherlands*
[4]*ZHEW Dordrecht, The Netherlands*
[5]*Dept. of Microbiology, University of Bayreuth, Germany*

ABSTRACT: In a denitrifying pilot plant reactor, a new obligately anaerobic ammonium oxidation (anammox) process with great potential for nitrogen removal for high strength wastewater was discovered. After transfer of the complex microbial community to a laboratory SBR system, a highly enriched population, dominated by a single anaerobic chemolithoautotrophic bacterium related to the Planctomycetes was obtained. The bacterium was purified via percoll centrifugation and characterized as '*Candidatus* Brocadia anammoxidans'. Survey of different wastewater treatment plants using anammox specific 16S rRNA gene primers and anammox specific oligonucleotide probes revealed the presence of at least three other anammox bacteria, tentatively named '*Candidatus* Kuenenia stuttgartiensis', '*Candidatus* Scalindua wagneri' and '*Candidatus* Scalindua brodae'. A close relative of the two Scalindua species, '*Candidatus* Scalindua sorokinii' was found to be responsible for about 50% of the nitrogen conversion in the anoxic zone of the Black Sea. Electron microscopic studies of all 5 anammox bacteria showed that several membrane-bounded compartments are present inside the cytoplasm, which are surrounded by unique ladderane lipids. Hydroxylamine oxidoreductase, a key anammox enzyme, was present exclusively inside one of these compartments, named the 'anammoxosome'. The implementation of the anammox process in the treatment of wastewater with high ammonium concentrations was started at the treatment plant in Rotterdam, the Netherlands, where it is combined with the partial nitrification process SHARON. Gas lift reactors could sustain the highest anammox capacity at 8.9 kg N removed per m^3 reactor per day. Alternative configurations of anammox are the oxygen-limited CANON or OLAND processes in which aerobic ammonium-oxidizing bacteria protect anammox bacteria from oxygen and produce the necessary nitrite. Maximum nitrogen removal with CANON in gas lift reactors was 1.5 kg N per m^3 reactor per day. The estimated price for nitrogen removal with partial nitrification and anammox is about 0.75 euro per kg N.

1 INTRODUCTION

Anaerobic ammonium oxidation (anammox) is a pathway in the nitrogen cycle that has long been overlooked. Over the last decade, feasibility studies (Jetten et al 1997; van Dongen et al 2001) showed that anammox is an attractive alternative for nitrogen removal in waste water treatment plants (Jetten et al 2001; 2002) and recently field experiments in marine ecosystems revealed that anammox can contribute as much as 70% to dinitrogen gas production (Kuypers et al 2003; Thamdrup and Dalsgaard 2002).

In the anammox process, one mole of ammonium and one mole of nitrite are directly converted to dinitrogen gas, and hydrazine is an important intermediate (Jetten et al 1998). Due to the extremely low growth rate, the cultivation of the anammox bacteria required very efficient biomass retention (Strous et al 2002). The bacteria responsible for the process have been physically purified from the highly enriched retentostat cultures by a percoll density gradient centrifugation (Strous et al 1999; 2002). The chemolithoautotrophic life style of the anammox bacteria was established by $^{14}CO_2$ incorporation into the cells and confirmed by reactor balances (Strous et al 1998). The first purified anammox planctomycete-like bacterium was named '*Candidatus* Brocadia anammoxidans', (Kuenen and Jetten 2001).

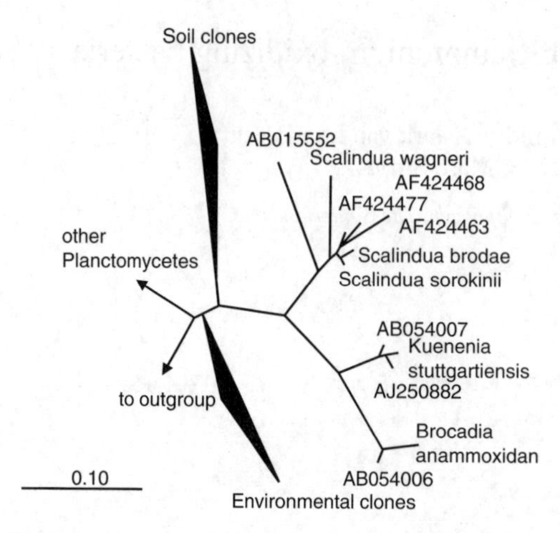

Figure 1. Phylogenetic (on 16S rRNA gene basis) relation of 3 groups of anammox bacteria (Scalindua, Kuenenia and Brocadia) to other planctomycete bacteria and environmental clones (Schmid et al 2003).

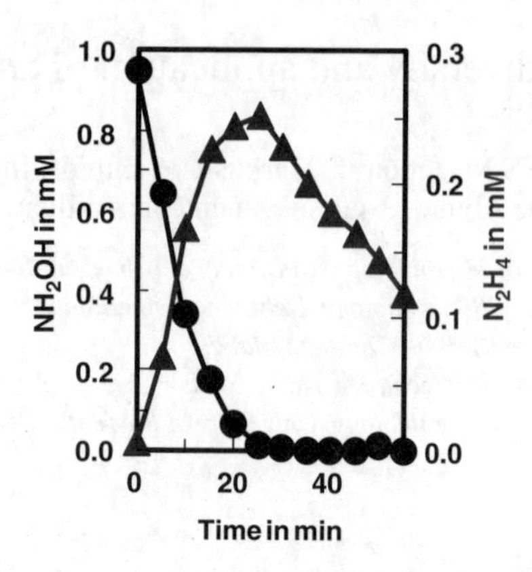

Figure 2. Hydrazine accumulation by anammox bacteria.

2 BIODIVERSITY OF ANAMMOX BACTERIA

Many water treatment systems and fresh water ecosystems also appeared to contain significant populations of anammox bacteria, which were only 2 distantly related (less than 90% similarity on 16S rRNA gene; (Fig. 1)) to the Brocadia branch (Fuji et al 2002; Helmer et al 2002; Jetten et al 2003). These anammox bacteria were named '*Candidatus* Kuenenia stuttgartiensis' (Egli et al 2001; Schmid et al 2000). Furthermore a new group of 'Scalindua' anammox bacteria was discovered in the Black Sea (Kuypers et al 2003) and in a UK treatment plant (Schmid et al 2003). In this plant about 20% of the popula-tion consisted of two new anammox species, named '*Candidatus* Scalindua wagneri' and '*Candidatus* Scalindua brodae'. Further more '*Candidatus* Scalindua sorokinii' in the Black Sea was the first anammox bacterium directly linked to the removal of fixed inorganic nitrogen from a natural ecosystem. Also other studies indicate that marine anammox bacteria play a very important role in the oceanic nitrogen cycle (Dalsgaard et al 2002; 2003; Trimmer et al 2003).

3 PROPERTIES OF THE ANAMMOX BACTERIA

The anammox bacteria are very well suited to convert their substrates. They have K_s values (Strous et al 1999) for ammonium and nitrite well below the chemical detection level (<5 micromol/l). However, they are reversibly inhibited by very low levels (<1 micromol/l) of oxygen and irreversibly inhibited by high nitrite (>10 mM) concentrations (Strous et al 1997; 2002). The anammox pathway was elucidated using ^{15}N-labelling experiments, which showed that hydrazine (Fig. 2) was an important intermediate. As far as we know, the occurrence of free hydrazine in microbial nitrogen metabolism is rare, if not unique (Jetten et al 1998). The HAO enzyme hydroxylamine/hydrazine oxidoreductase was purified from Brocadia anammoxidans (Schalk et al 2000). Unique peptide sequences of tryptic fragments were used to locate the *hao* gene in the Kuenenia stuttgartiensis genome. The purified enzyme was also used to raise polyclonal antibodies for localization studies (Lindsay et al 2001). Using immunogold electron microscopic analysis, the enzyme was found to be present exclusively inside a membrane bounded organelle (the anammoxosome), that made up more than 30% of the cell volume. In a follow up study, this dedicated intracytoplasmic compartment was found to be surrounded by a membrane nearly exclusively composed of unique ladderane lipids (Sinninghe Damste et al 2002; Schmid et al 2003), composed of 5 linearly concatenated cyclobutane rings. The ladderane lipids occurred both as ether and ester lipids in all three groups of anammox bacteria (Jetten et al 2003). The identification of mixed ether–ester lipids in the anammox planctomycetes is puzzling. These bacteria are neither thermophiles nor related to Archaea, but the presence of mixed ester– ether lipids might reflect their early divergence in the bacterial lineage (Brochier and Philippe 2002).

4 APPLICATION OF ANAMMOX BACTERIA

Both in natural and man-made ecosystems anammox bacteria have to be provided with their substrates ammo-nium and nitrite. Ammonium is abundantly

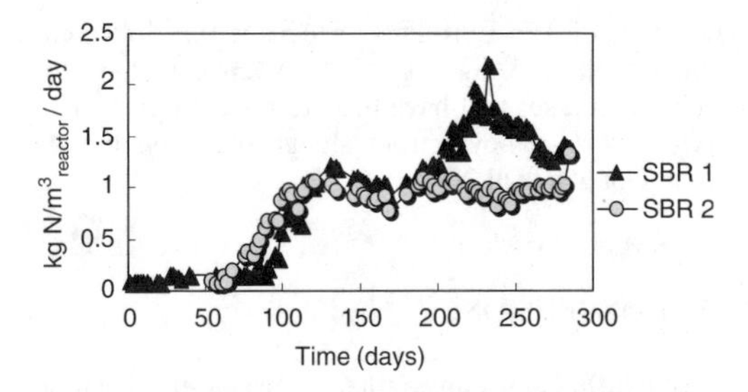

Figure 3. Enrichment of anammox biomass from activated sludge on Sharon effluent (Van Dongen et al 2001).

available in anoxic ecosystems via degradation of organic matter. Nitrite can be provided by nitrate reducing bacteria as is most likely the case for marine sediments and water columns (Dalsgaard et al 2003; Kuypers et al 2003). Nitrite can also be produced by aerobic ammonium-oxidizing bacteria operating at the oxic–anoxic interface of many ecosystems (Schmidt et al 2002). Such cooperation between aerobic and anaerobic ammonium-oxidizing bacteria is the microbial basis of one integrated reactor system (CANON) or the OLAND process in which oxygen-limited aerobic ammonium-oxidizing bacteria (AOB) and anammox planctomycetes perform two sequential reactions simultaneously (Kuai & Verstraete 1998; Philips 2002 et al; Pynaert et al 2002abc; 2003; Schmidt et al 2003; Sliekers et al 2002). Under oxygen limitation, the supplied ammonium is partly oxidized to nitrite. The produced nitrite is utilized with the remainder of the ammonium by anammox and converted into dinitrogen gas. The feasibility of the CANON concept was established by carefully introducing limited amounts of oxygen into anammox SBR systems. Within 2 weeks a new stable consortium of AOB and anammox was operative (Sliekers et al 2002). FISH analysis of the CANON biomass showed that about 40% of the community consisted of AOB, also the anammox cells made up about 40% of the community. No aerobic nitrite oxidizing bacteria (NOB, *Nitrospira* or *Nitrobacter*) were ever detected. Activity tests confirmed the absence of NOB.

The NOB were only able to develop in the CANON reactor after prolonged exposure (>1 month) to ammonium limitation (Third et al 2001). Biofilm modeling studies could confirmed the observed phenomena (Hao et al 2002). The upper limits of nitrogen loading of both anammox and CANON processes were explored in gaslift reactors (Sliekers et al 2003; Fig. 4). In this type of reactor anammox planctomycetes were able to remove 8.9 kg N $m^{-3}{}_{reactor}$ day^{-1}. In the same set up the combined action of AOB and anammox planctomycetes

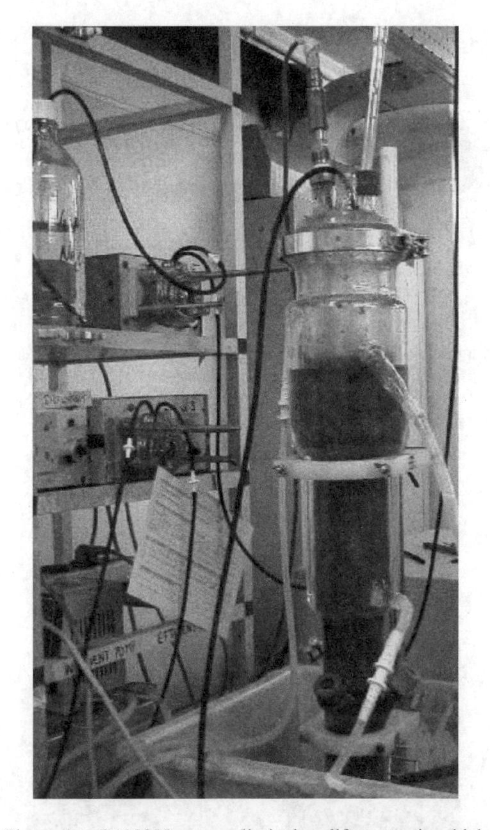

Figure 4. CANON oxygen-limited gaslift reactor in which aerobic ammonium-oxidizing bacteria and anammox bacteri cooperate to remove ammonium directly into dinitrogen gas (Sliekers et al 2003).

achieved 1.5 kg N removal $m^{-3}{}_{reactor}$ per day, which is more than sufficient to start application trials.

The introduction of anammox to N-removal would lead to a reduction of operational costs of up to 90%. Anammox would replace the conventional denitrification step completely and would also save half of the nitrification aeration costs. In feasibility studies with sludge digestor effluents on laboratory scale we showed that the effluents did not negatively affect the anammox activity (van Dongen et al 2001), and that anammox biomass could be enriched from activated sludge within 100 days (Fig. 3). In a separate study we investigated the possibility to use the Sharon process in combination with anammox. Sharon was developed for the removal of ammonium via the so called nitrite route. It was tested for 2 years in the laboratory and successfully scaled-up from two liters to 1800 m^3 full scale plant (Mulder et al 2000). In the Sharon process, the ammonium was oxidized for 53% to nitrite at 1.2 kg N m^{-3} day^{-1}, without pH control. The ammonium/nitrite ratio in the effluent of the

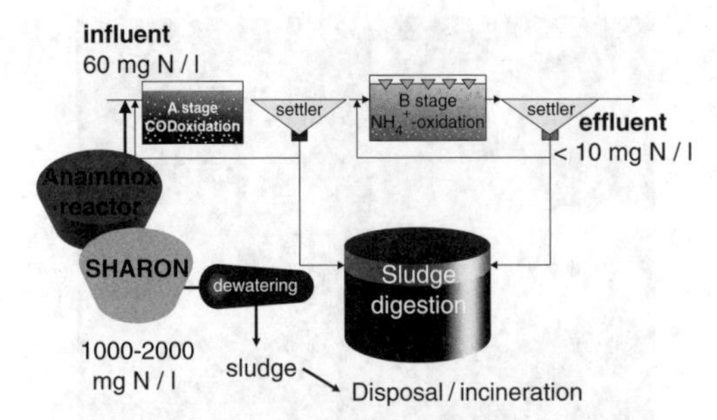

Figure 5. Scheme of the Sharon – anammox plant Dokhaven-Sluisjesdijk, Rotterdam, NL (van Dongen et al 2001).

Figure 6. Anammox reactor at Rotterdam WWTP (Courtesy Paques and ZHEW) in EU-icoN project.

Sharon process could be fine-tuned by adjusting the pH between 6.5 and 7.5. The effluent of this Sharon reactor was fed to an anammox SBR and all nitrite was removed. The specific activity of the anammox biomass was very high: 0.8 kg N (kg dry weight)$^{-1}$ day^{-1} and the load could be increase to more than 2 kg N m^{-3} day^{-1}. The Sharon-anammox process was patented and implemented in the WWTP Rotterdam (Fig. 5 and Fig. 6). Based on the design of the combined Sharon anammox process we made a cost estimate of 0.75 Euro kg^{-1} N. This is very low compared to the 2–5 Euro kg^{-1} N that were calculated for other processes that have been tested on a pilot plant scale for N-removal from sludge digestion liquors (van Dongen et al 2001).

5 CONCLUSIONS

Three different groups of planctomycete-like bacteria (Brocadia, Kuenenia, and Scalindua) are responsible for the anaerobic oxidation of ammonium. They contain a unique set of enzymes located in a specialized compartment. This anammoxosome is surrounded by an exceptionally dense membrane composed of ladderane lipids. The planctomycete anammox bacteria are natural partners of aerobic ammonium oxidizers in oxygenlimited systems, and together they convert ammonium directly into dinitrogen gas. Both in natural and man-made ecosystems anammox bacteria can contribute significantly to the loss of fixed nitrogen. The combination of partial nitrification and anammox is ready for full scale implementation in nitrogen removal which will lead to substantial savings in energy and resources.

ACKNOWLEDGEMENTS

The research on anaerobic ammonium oxidation and partial nitrification was financially supported by the Foundation for Applied Sciences (STW), the Foundation of Applied Water Research (STOWA), the EU research project IcoN EVK1-2000-00054, the Netherlands Foundation for Life Sciences (NWO-ALW), the Royal Netherlands Academy of Arts and Sciences (KNAW), Gist-brocades, DSM, Paques BV, Grontmij, and ZHEW. We gratefully acknowledge the contributions of various coworkers and students over the years.

REFERENCES

Brochier C, Philippe H (2002) Phylogeny – A non hyperthermophilic ancestor for bacteria. Nature 417: 244

Dalsgaard T, Thamdrup B (2002) Factors controlling anaerobic ammonium oxidation with nitrite in marine sediments. Appl Environ Microbiol 68: 3802–3808

Dalsgaard T, Canfield DE, Petersen J, Thamdrup B, Acuña-González J (2003) Anammox is a significant pathway of N_2 production in the anoxic water column of Golfo Dulce, Costa Rica. Nature 422: 606–608

Egli K, Franger U, Alvarez PJJ, Siegrist H, Vandermeer JR, Zehnder AJB (2001) Enrichment and characterization of an anmmox bacterium from a rotating biological contractor treating ammonium-rich leachate. Arch Microbiol 175: 198–207

Fujii T, Sugino H, Rouse JD, Furukawa K (2002) Characterization of the microbial community in an anaerobic ammonium-oxidizing biofilm cultured on a non-woven biomass carrier. J Biosci Bioeng 94 (5): 412–418

Hao X, Heijnen JJ, van Loosdrecht MCM (2002) Sensitivity analysis of a biofilm model describing a one stage completely autotrophic nitrogen removal (CANON) process. Biotechnol. Bioeng. 77: 266–277

Helmer-Madhok C, Schmid M, Filipov E, Gaul T, Hippen A, Rosenwinkel KH, Seyfried CF, Wagner M, Kunst S (2002) Deammonification in biofilm systems: population structure and function. Water Sci Technol 46: 223–231

Jetten MSM, Horn SJ, van Loosdrecht MCM (1997) Towards a more sustainable municipal wastewater treatment system. Water Sci Techol 35: 171–180

Jetten MSM, Schmid MA, Schmidt I, Wubben M, Van Dongen U, Abma W, Sliekers AO, Revsbech NP, Beaumont HJE, Ottosen L, Volcke E, Laanbroek HJ, Campos-Gómez JL, Cole JA, Van Loosdrecht MCM, Mulder JW, Fuerst J, Richardson D, Van de Pas-Schoonen KT, Mendez-Pampín, R, Third K, Cirpus IY, Van Spanning R, Bollmann A, Nielsen LP, Op den Camp HJM, Schultz C, Gundersen J, Vanrolleghem P, Strous M, Wagner M, Kuenen JG (2002) Improved nitrogen removal by application of new nitrogen-cycle bacteria. Reviews in Environ Sci Biotechnol 1: 51–63

Jetten MSM, Sliekers AO, Kuypers MMM, Dalsgaard T, Van Niftrik L, Cirpus I, Van de Pas-Schoonen KT, Lavik G, Thamdrup B, Le Paslier D, Op den Camp HJM, Hulth S, Nielsen LP, Abma W, Third K, Engström P, Kuenen JG, Jorgensen BB, Canfield DE, Sinninghe Damste JS, Revsbech, NP, Fuerst J, Weissenbach J, Wagner M, Schmidt I, Schmid M, Strous M (2003) Anaerobic ammo-nium oxidation by marine and freshwater planctomycete-like bacteria. Appl Microbiol Biotechn. doi: 10.1007/ S00253-003-1422-4

Jetten MSM, Strous M, Van de Pas-Schoonen KT, Schalk J, Van Dongen L, Van de Graaf AA, Logemann S, Muyzer G, Van Loosdrecht MCM, Kuenen JG (1998) The anaerobic oxidation of ammonium. FEMS Microbiol Reviews 22: 421–437

Jetten MSM, Wagner M, Fuerst J, van Loosdrecht MCM, Kuenen JG Strous M (2001) Microbiology and application of the anaerobic ammonium oxidation (anammox) process. Curr Opinion Biotechnol 12: 283–288

Kuai L, Verstraete W (1998) Ammonium removal by the oxygen-limited autotrophic nitrification–denitrifcation system. Appl Environ Microbiol 64: 4500–4506.

Kuenen JG, Jetten MSM (2001) Extraordinary anaerobic ammonium oxidizing bacteria. ASM News 67: 456–463

Kuypers MMM, Sliekers AO, Lavik G, Schmid M, Jørgensen BB, Kuenen JG, Sinninghe Damsté JS, Strous M, Jetten MSM (2003) Anaerobic ammonium oxidation by Anammox bacteria in the Black Sea. Nature 422: 608–611

Lindsay MR, Web RI, Strous M, Jetten M, Butler MK Fuerst JA (2001) Cell compartmentalization in planctomycetes: novel types of structural organization for the bacterial cell. Arch Microbiol 175: 413–429

Mulder JW, Van Loosdrecht MCM, Hellinga C, Van Kempen R (2000) Full scale application of the Sharon process for treatment of rejection water of digested sludge dewatering. *Proceedings First IWA conference* pp 267–274, IWAQ London UK

Philips S, Wyffels S, Sprengers R, Verstraete W (2002a) Oxygen-limited autotrophic nitrification/denitrification by ammonia oxidisers enables upward motion towards more favourable conditions. Appl Microbiol Biotechnol 59(4–5): 557–566

Pynaert K, Wyffels S, Sprengers R, Boeckx P, Van Cleemput O, Verstraete W (2002b) Oxygen-limited nitrogen removal in a lab-scale rotating biological contactor treating an ammonium-rich wastewater.Water Sci Technol 45(10): 357–363

Pynaert K, Sprengers R, Laenen J, Verstraete W (2002c) Oxygen-limited nitrification and denitrification in a lab-scale rotating biological contactor. Environ Technol 23(3): 353–362

Pynaert K, Smets BF , Wyffels S, Beheydt D, Siciliano SD and Verstraete W (2003) Characterization of an autotrophic Nitrogen-removing biofilm from a highly loaded lab-scale rotating biological contactor. Appl Environ Microbiol 69: 3626–3635

Schalk J, Devries S, Kuenen JG, Jetten MSM (2000) A novel hydroxylamine oxidoreductase involved in the anammox process. Biochemistry 39: 5405–5412

Schmid M, Schmitz-Esser S, Jetten MSM, Wagner M (2001) 16S–23S rDNA intergenic spacer and 23S rDNA of anaerobic ammonium oxidizing bacteria: implications for phylogeny and *in situ* detection. Environmental Microbiology 7: 450–459

Schmid M, Walsh K, Webb R, Rijpstra WIC, van de Pas-Schoonen KT, Verbruggen MJ, Hill T, Moffett B, Fuerst J, Schouten S, Sinninge Damsté JS, Harris J, Shaw P, Jetten MSM, Strous M (2003) Candidatus "Scalindua brodae", sp. nov., Candidatus "Scalindua wagneri", sp. nov., two new species of anaerobic ammonium oxidizing bacteria. Syst Appl Microbiol 26: 529–538

Schmid M, Twachtmann U, Klein M, Strous M, Juretschko S, Jetten M, Metzger J, Schleifer KH Wagner M (2000) Molecular evidence for genus level diversity of bacteria capable of catalyzing anaerobic ammonium oxidation. Sys Appl Microbiol 23: 93–106

Schmidt I, Sliekers AO, Schmid MC, Cirpus IY, Strous M, Bock E, Kuenen JG, Jetten MSM (2002) Aerobic and anaerobic ammonia oxidizing bacteria – competitors or natural partners? FEMS Microbiol Ecology 39: 175–181

Schmidt I, Hermelink C, Van de Pas-Schoonen KT, Strous M, Op den Camp HJM, Kuenen JG, Jetten MSM (2002) Anaerobic ammonia oxidation in the presence of nitrogen oxides (NOx) by two different lithotrophs. Appl Environ Microbiol 68: 5351–5357

Schmidt I, Sliekers O, Schmid MC, Bock E, Fuerst J, Kuenen JG, Jetten MSM, Strous M (2003) New concepts of microbial treatment processes for the nitrogen removal in wastewater. FEMS Microbiol. Rev. 27: 481–492

Sinninghe Damsté JS, Strous M, Rijpstra WIC, Hopmans EC, Geenevasen JAJ, Van Duin ACT, Van Niftrik LA, Jetten MSM (2002) Linearly concatenated cyclobutane

lipids form a dense bacterial membrane. Nature 419: 708–712
Sliekers AO, Derwort N, Campos L, Kuenen JG, Strous M, Jetten MSM (2002) Completely autrophic ammonia removal over nitrite in a single reactor system. Water Res 36: 2475–2482
Sliekers AO, Third K, Abma W, Kuenen JG, Jetten MSM (2003) CANON and Anammox in a gas-lift reactor. FEMS Microbiol. Lett. 218: 339–344
Strous M, Gerven E, Kuenen GJ and Jetten MSM (1997) Effects of aerobic and microaerobic conditions on anaerobic ammonium-oxidizing (anammox) sludge. Applied and Environmental Microbiology 63: 2446–2448
Strous M, Heijnen JJ, Kuenen JG, Jetten MSM (1998) The sequencing batch reactor as a powerful tool to study very slowly growing micro-organisms. Appl Microbiol Biotechnol 50: 589–596
Strous M, Kuenen JG, Jetten MSM (1999) Key physiology of anaerobic ammonium oxidation. Appl Environ Microbiol 65: 3248–3250
Strous M, Fuerst J, Kramer E, Logemann S, Muyzer G, van de Pas K, Webb R, Kuenen JG, Jetten MSM (1999) Missing lithotroph identified as new planctomycete. Nature 400: 446–449
Strous M, Kuenen JG, Fuerst JA, Wagner M, Jetten MSM (2002) The anammox case – A new experimental manifesto for microbiological eco-physiology. Antonie van Leeuwenhoek 81: 693–702
Thamdrup B, Dalsgaard T (2002) production of N_2 through anaerobic ammonium oxidation coupled to nitrate reduction in marine sediments. Appl Environ Microbiol 68: 1312–1318
Third K, Sliekers AO, Kuenen JG, Jetten MSM (2001) The CANON System (Completely Autotrophic Nitrogen-removal Over Nitrite) under Ammonium Limitation: Interaction and Competition between Three Groups of Bacteria. Sys Appl Microbiol 24(4): 588–596
Trimmer M, Nicholls JC, Deflandre B (2003). Anaerobic ammonium oxidation measured in sediments along the Thames estuary, United Kingdom. Appl Environ Microbiol 69: 6447–6454
Van Dongen U, Jetten MSM, van Loosdrecht MCM (2001) The Sharon-anammox process for the treatment of ammonium rich wastewater. Wat Sci Technol 44: 153–160

European Symposium on Environmental Biotechnology, ESEB 2004 - Verstraete (ed)

Biofilm development by pseudomonas

T. Tolker-Nielsen & S. Molin
BioCentrum-DTU, The Technical University of Denmark

ABSTRACT: Bacteria in natural, industrial and clinical settings live in surface-associated communities called biofilms. The study of biofilms has become a highly significant topic in microbiology with relevance for many important areas in industry, nature, and disease. Detailed knowledge about the developmental process from single bacteria scattered on a surface to the formation of thick structured biofilms is essential in order to create strategies to control biofilm development. *Pseudomonas putida* and *Pseudomonas aeruginosa* are used as model organism for the study of structural biofilm development. *P. putida* represent an example where biofilm development occur in distinct phases and involves programmed release of planktonic cells. The control of *P. putida* biofilm development involves regulated expression of a cellulase-degradable cell-to-cell interconnecting substance. *P. aeruginosa* represent an example where the biofilms are very dynamic, and extensive type IV pili-driven motility occur during development. The control of *P. aeruginosa* biofilm development appears to be complex and includes differentiated twitching motility in distinct subpopulations.

1 INTRODUCTION

Although the study of bacteria in planktonic culture has been the norm in microbiology from the time of Pasteur to the present, and has provided an increasingly accurate understanding of prokaryotic physiology and genetics, it is now clear that bacteria in natural, industrial and clinical settings live in surface-associated communities, and that knowledge about this lifestyle is essential in order to understand bacterial biology in nature and disease (Costerton et al. 1995). These structured communities of cells which are enclosed in a self-produced polymeric matrix and adherent to an inert or living surface are termed biofilms.

Biofilms growing in clinical settings can withstand host immune responses, and they are much less susceptible to antibiotics than their nonattached individual planktonic counterparts (Costerton et al. 1999). Biofilms growing in natural and industrial environments are resistant to bacteriophage, to predation, and to chemically diverse biocides used to combat biofouling in industrial processes (Costerton et al. 1987). Biofilm formation therefore leads to various persistent and sometimes lethal infections in humans and animals, and to a variety of problems in industry where solid–water interfaces occur.

Although biofilms often are detrimental in clinical and industrial settings there are many cases where biofilms are useful. For examples, many of the processes involved in the environmental cycling of the elements are governed by bacteria in biofilms, and processes occurring in bioremediation and wastewater treatment are to a large extent carried out by bacteria in biofilms. It is thus evident that knowledge about the sessile lifestyle of bacteria is needed in order to understand bacterial biology in these environments.

Accordingly, during the last ten years there has been a rapidly increasing recognition of biofilms as a highly significant topic in microbiology with relevance for many important areas in modern society such as drinking water supply systems, industrial settings, waste water treatment, bioremediation, chronic bacterial infections, nosocomial infections, and dental plaque. It has become apparent that detailed knowledge about the developmental process from single bacteria scattered on a surface to the formation of thick structured biofilms is essential in order to create strategies to control biofilm development.

Here we present our current understanding of structural biofilm development by *Pseudomonas putida* and *Pseudomonas aeruginosa*. Although the focus is limited to just two pseudomonas species it is evident that these organisms use very different mechanisms during biofilm development, and thus illustrate that biofilm formation can occur through multiple pathways. Initially we will briefly describe some experimental methods which are used for the study of microbial biofilms.

2 METHODS USED TO STUDY STRUCTURAL BIOFILM DEVELOPMENT

The flow-chamber, where the bacteria grow under controlled hydrodynamic conditions and form biofilm on a glass surface, is a very useful in vitro system for microscopic studies of biofilm formation (e.g. Woolfardt et al. 1994) (see the flow-chamber setup in figure 1). The confocal laser scanning microscope is an excellent tool for capturing three-dimensional images and time-lapse image series of biofilms grown in flow-chambers (e.g. Lawrence et al. 1991). The fluorescence necessary for confocal laser scanning microscopy (CLSM) is conferred to the bacterial cells either via genetically tagging with fluorescent proteins (such as Gfp, Cfp, Yfp, and DsRed), or by staining with certain fluochromes (such as acridine orange, Syto9, or LIVE/DEAD *Bac*Light) or fluorescently labeled ribosomal RNA-targeting oligo-nucleotide probes (FISH). In addition, the exocellular polymeric substances surrounding the cells in a biofilm can be stained with fluorescently labeled lectins. The creation of unstable variants of Gfp (Andersen et al. 1998) has enabled the use of Gfp as an in situ gene expression reporter (e.g. Møller et al. 1998). The use of flow-chamber technology, fluorescent reporter genes, and confocal laser scanning microscopy (CLSM) has enabled very detailed studies of specific and general interactions between the members in biofilm communities (e.g. Nielsen et al. 2000), and has permitted non-destructive studies of the dynamics and developmental steps occurring during biofilm formation (e.g. Tolker-Nielsen et al. 2000). CLSM images can be analyzed objectively by the use of digital image analysis programs such as COMSTAT (Heydorn et al. 2000). Knock out of specific genes by the use of allelic displacement techniques and comparative analysis of biofilm formation in flow-chambers, which may include digital image analysis, is useful for studying roles of specific genetic elements in biofilm formation. A microtitre assay, where the bacteria grow under static conditions in the wells of microtitre plates and form a biofilm on the well surface if they possess the necessary genetic elements, has greatly facilitated genetic screens of factors involved in biofilm formation (e.g. Mack et al. 1994, O'Toole and Kolter, 1998).

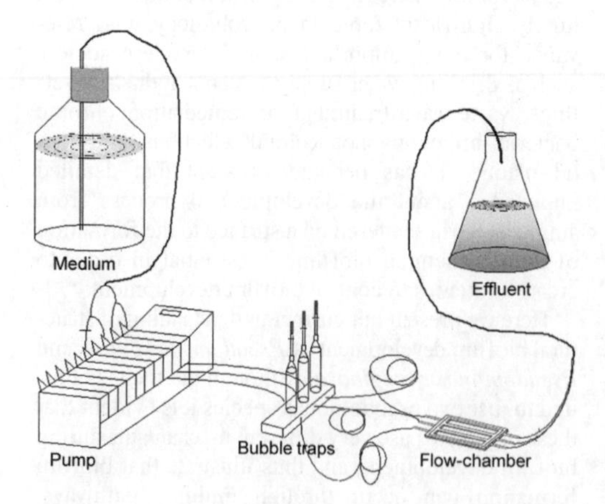

Figure 1. Schematic drawing of the flow-chamber setup. Reproduced with permission from the American Society for Microbiology (Microbial Biofilms, G.A. O'Toole (Ed.)).

3 STRUCTURAL BIOFILM DEVELOPMENT

Biofilm formation is believed to occur in a sequential process of (i) transport of microbes to a surface, (ii) initial attachment, (iii) formation of microcolonies, and (iv) further biofilm proliferation and maturation (e.g. O'Toole et al. 2000), and the involvement of cellular motility, adhesins, exopolymers, and cell-to-cell signalling has been demonstrated (e.g. Mack et al. 1994; Heilmann et al. 1996; Pratt and Kolter, 1998; O'Toole and Kolter, 1998; Davies et al. 1998; Prigent-Combaret et al. 1999; Yildiz and Schoolnik, 1999; Danese et al. 2000; Whitchurch et al. 2002). Although this general description of biofilm formation is widely accepted it is clear that different bacteria form biofilms by the use of different mechanisms. Below we will describe biofilm formation by *P. putida* as an example where regulation of the adhesiveness of the bacteria and flagellum driven motility play important roles during biofilm development, and biofilm formation by *P. aeruginosa* as an example where, besides cellular adhesiveness, regulated type IV pili-driven motility play an important role during biofilm development.

3.1 *Structural biofilm development by* P. putida

Using flow-chambers for controlled cultivation of biofilms, and CLSM for investigations of the three-dimensional biofilm-structures formed by bacteria genetically tagged with different fluorescent proteins, we have found that *P. putida* biofilm development occurs through at least three phases (Tolker-Nielsen et al. 2000). Initially compact microcolonies form at the substratum via clonal growth of sessile bacteria. When the microcolonies have reached a certain size the bacteria inside the microcolonies degrade or modulate their interconnecting substances and begin to move rapidly in circles driven by flagella. Further dissolution of the microcolonies then result in a shift in spatial organization from compact microcolonies to loose protruding structures which are characteristic of the mature *P. putida* biofilm (see figure 2).

Knowledge about factors involved in *P. putida* biofilm formation and dissolution have been obtained in a recent study (Gjermansen et al. 2004). Enzymatic analysis indicated that a cellulase-degradable exopolymer functions as cell-to-cell interconnecting substance in *P. putida* biofilms. Genetic analysis indicated that production of the cellulase-degradable exopolymer is controlled by a GGDEF-domain containing regulatory protein. The local biofilm dissolution process occurring during *P. putida* biofilm development was found to be induced by carbon starvation. When a flow-chamber-grown *P. putida* biofilm was subjected to carbon starvation the entire biofilm dissolved within ten minutes. A biofilm formed by a non-flagellated *P. putida* mutant also dissolved in response to carbon starvation supporting the suggestion that *P. putida* biofilm dissolution occurs through degradation or modulation of the cell-to-cell interconnecting substances rather than by flagellum-driven movements that potentially could overcome weak interconnecting forces. The dissolution of *P. putida* biofilm occurring in response to carbon starvation was used for the isolation of a mutant which is deficient in carbon starvation-induced biofilm dissolution. The biofilm formed by this mutant consisted of compact microcolonies and had higher substratum coverage and a higher density than the wildtype biofilm, and it did not dissolve in response to carbon starvation. The step with rapid cellular movements inside microcolonies, and the transition from compact microcolonies to loose protruding structures, did not occur in the mutant biofilm, supporting the assumption that the local and over-all biofilm dissolution processes involve the same mechanisms, and suggesting that local biofilm dissolution is an integrated part of the *P. putida* wildtype biofilm development which results in the formation of the characteristic loose protruding structures.

3.2 *Structural biofilm development by* P. aeruginosa

Structural biofilm development by *P. aeruginosa* appears to be conditional. For example, in flow-chambers irrigated with citrate minimal medium *P. aeruginosa* forms a flat biofilm, while in flow-chambers irrigated with glucose minimal medium it forms a heterogeneous biofilm with mushroom-shaped multicellular structures (Klausen et al. 2003b).

The formation of the flat *P. aeruginosa* biofilm in flow-chambers irrigated with citrate minimal medium was shown to occur via initial formation of microcolonies by clonal growth of sessile bacteria at the substratum, followed by expansive migration of the bacteria out on the substratum, resulting in the formation of a very dynamic flat biofilm (Klausen et al. 2003b). Since biofilm formation by a *P. aeruginosa* *pilA* mutant (which is deficient in biogenesis of type IV pili) occurred without the expansive phase, and resulted in discrete protruding microcolonies, it was suggested that the expansive migration of the bacteria out on the substratum was type IV pili-driven. CLSM time-lapse microscopy of the biofilm development indicated that the shift from sessile to migrating cells became pronounced when the initial microcolonies reached a certain size, suggesting that the shift may be induced by some sort of limitation arising in the initial microcolonies.

The formation of the mushroom-shaped structures in the heterogeneous glucose-grown *P. aeruginosa* biofilm was shown to occur in a sequential process involving a non-motile bacterial subpopulation which formed the initial microcolonies by growth in certain foci of the biofilm, and a migrating bacterial subpopulation which initially formed a monolayer on the substratum, and subsequently formed the mushroom caps by climbing the microcolonies (which then

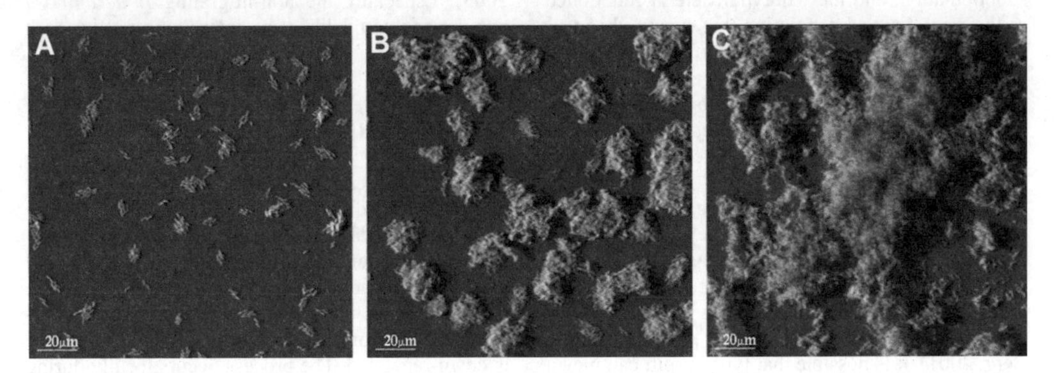

Figure 2. Structural development of a *P. putida* biofilm grown in a flow-chamber irrigated with citrate minimal medium. CLSM shadow projection micrographs recorded in a one (A), three (B), and five (C) day old biofilm. Reproduced with permission from Journal of Bacteriology (182: 6482–6489).

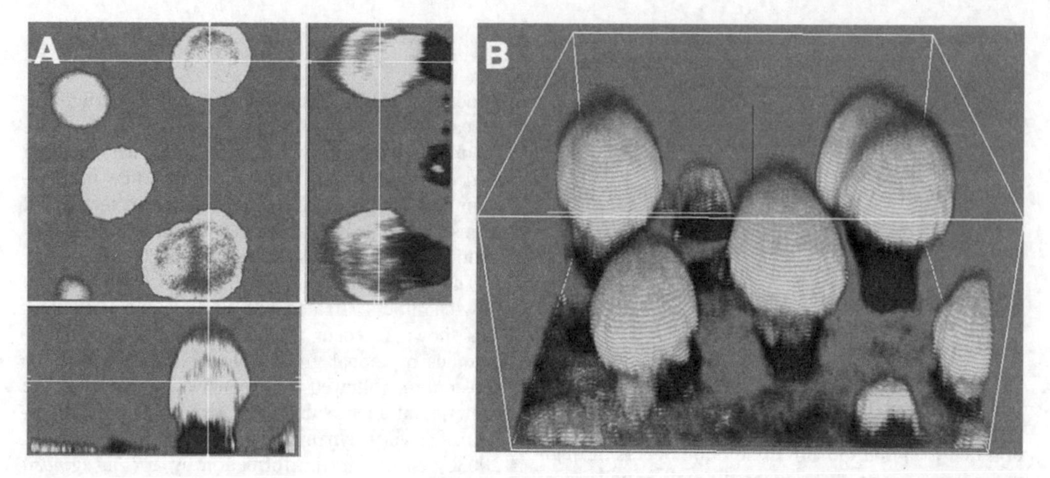

Figure 3. Mushroom-shaped multicellular biofilm structures with caps composed of yellow fluorescent protein-tagged *P. aeruginosa* PAO1 wildtype (appear white on the image), and stalks composed of cyan fluorescent protein-tagged *P. aeruginosa pilA* mutants (appear black on the image). CLSM images were acquired in a 4 days old biofilm which was initiated with a 1:1 mixture of yellow fluorescent protein-tagged *P. aeruginosa* PAO1 wildtype and cyan fluorescent protein-tagged *P. aeruginosa pilA* mutants, and grown on citrate minimal medium. The central picture in panel A shows a horizontal CLSM section, while the two flanking pictures in panel A show vertical CLSM sections. The thin white lines in the central picture indicate the positions of the vertical sections, while the thin white lines in the flanking pictures indicate the position of the horizontal section. Panel B shows the corresponding 3-D CLSM image. Reproduced with permission from Molecular Microbiology (50: 61–68).

become mushroom stalks) and aggregating on the tops in a type IV pili-driven process (Klausen et al. 2003a). Growth of the initial microcolonies in the glucose-grown biofilms continued past the point where spreading by twitching motility prevented further microcolony formation in the citrate-grown biofilms. In biofilms containing a 1:1 mixture of the wildtype and *pilA* mutants, formation of mushroom-shaped structures could occur also with citrate as the carbon source. In these mixed biofilms the *pilA* bacteria functioned as the non-motile mushroom stalk-forming subpopulation, and the wildtype bacteria functioned as the migrating subpopulation that eventually accumulated on top of the *pilA* stalks and formed the mushroom caps (see figure 3).

The nature of the cell agglutinating factor(s) in the very dynamic *P. aeruginosa* biofilms is not known at present. A role of alginate as cell-to-cell interconnecting substance has been proposed previously (e.g. Boyd and Chakrabarty, 1994), but has recently been refuted (Wozniak et al. 2003). Some cell populations are apparently kept in the biofilm by substances that allow type IV pili-driven migration. Since twitching motility is powered by a mechanism involving extension, grip, and retraction of type IV pili (Skerker and Berg, 2001), it is possible that type IV pili can play a role as cell-to-cell and cell-to-substratum interconnecting compound. Evidence has been provided that extracellular DNA may play a role as cell-to-cell interconnecting substance in *P. aeruginosa* biofilms (Whitchurch et al. 2002, Nemoto et al. 2003), and interestingly there is evidence that type IV pili binds to DNA (Schaik and Irvin, 2003). Yet, other cell-to-substratum and cell-to-cell connections keep the *pilA* mutant bacteria substratum-associated and agglutinated in the biofilms. Evidence is emerging that a novel type of fimbriae may function as adhesin in *P. aeruginosa* biofilms (Vallet et al. 2001), and that certain exopolysaccharides may function as cell-to-cell interconnecting substance (Friedmann and Kolter, 2003). Especially the non-migrating *P. aeruginosa* populations could likely be interconnected by such compounds.

4 CONCLUDING REMARKS

P. putida represent an example where biofilm development is controlled by the regulated expression of a cellulase-degradable cell-to-cell interconnecting substance. The interconnecting substance is expressed when the cells are supplied with nutrients. When the cells experience starvation the production or agglutinating activity of the cellulose-degradable substance is down-regulated. The process occurs locally during *P. putida* biofilm development, and the release of planktonic cells leads to characteristic structural changes, which presumably ensure that all the cells in the biofilm are supplied with nutrients. The process may also act globally and enable the biofilm cells to leave a nutrient-depleted environment as planktonic cells for colonization of new surfaces.

P. aeruginosa represent an example where the biofilms are very dynamic, and extensive type IV pili-driven motility occur during development. The migrating cells may to some extent be interconnected by the type IV pili and perhaps extracellular DNA, and the non-motile cells may be interconnected by exopolysaccharides, fimbrial appendages, and extracellular DNA. The control of *P. aeruginosa* biofilm development appears to be complex, and includes regulation of twitching motility and cellular adhesiveness.

REFERENCES

Andersen, J.B., Sternberg, C., Poulsen, L.K., Bjørn, S.P., Givskov, M., & Molin, S. 1998. New unstable variants of green fluorescent protein for studies of transient gene expression in bacteria. Appl. Environ. Microbiol. 64: 2240–2246.

Boyd, A., & Chakrabarty, A.M. 1994. Role of alginate lyase in cell detachment of *Pseudomonas aeruginosa*. Appl. Environ. Microbiol. 60: 2355–2359.

Costerton, J.W., Cheng, K.-J., Geesey, G.G., Ladd, T.I., Nickel, J.C., Dasgupta, M., & Marrie, T.J. 1987. Bacterial biofilms in nature and disease. Annu. Rev. Microbiol. 41: 435–464.

Costerton, J.W., Lewandowski, Z., Caldwell, D.E., Korber, D.R., & Lappin-Scott, H.M. 1995. Microbial biofilms. Annu. Rev. Microbiol. 49: 711–745.

Costerton, J.W., Stewart, P.S., & Greenberg, E.P. 1999. Bacterial biofilms: a common cause of persistent infections Science 284: 1318–1322.

Danese, P.N., Pratt, L.A., & Kolter, R. 2000. Exopolysaccharide production is required for development of *Escherichia coli* K-12 biofilm architecture. J Bacteriol 182: 3593–3596.

Davies, D.G., Parsek, M.R., Pearson, J.P., Iglewski, B.H., Costerton, J.W., & Greenberg, E.P. 1998. The Involvement of Cell-to-Cell Signals in the Development of a Bacterial Biofilm. Science 280: 295–298.

Friedmann, L. & R. Kolter. 2003. The exopolysaccharide content of the *Pseudomonas aeruginosa* biofilm matrix. ASM conference: Biofilms 2003. Poster 2(B).

Gjermansen, M., Ragas, P., Sternberg, C., Molin, S., & Tolker-Nielsen, T. 2003. Programmed transition of *Pseudomonas putida* from the biofilm state to the planktonic state. Submitted for publication.

Heilmann, C., Gerke, C., Perdreau-Remington, F., & Götz, F. 1996. Characterization of Tn917 insertion mutants of *Staphylococcus epidermidis* affected in biofilm formation. Infect Immun 64: 277–282.

Heydorn, A., Nielsen, A.T., Hentzer, M., Sternberg, C., Givskov, M., Ersbøll, B.K., & Molin, S. 2000. Quantification of biofilm structures by the novel computer program COMSTAT. Microbiology 146: 2395–2407.

Klausen, M., Aaes-Jørgensen, A., Molin, S., & Tolker-Nielsen, T. 2003a. Involvement of bacterial migration in the development of complex multicellular structures in *Pseudomonas aeruginosa* biofilms. Mol. Microbiol. 50: 61–68.

Klausen, M., Heydorn, A., Ragas, P., Lambertsen, L., Aaes-Jørgensen, A., Molin, S., & Tolker-Nielsen, T. 2003b. Biofilm formation by *Pseudomonas aeruginosa* wild-type, flagella, and type IV pili mutants. Mol. Microbiol. 48: 1511–1524.

Lawrence, J.R., Korber, D.R., Hoyle, B.D., Costerton, J.W., & Caldwell, D.E. 1991. Optical sectioning of microbial biofilms. J. Bacteriol. 173: 6558–6567.

Mack, D., Nedelmann, M., Krokotsch, A., Schwarzkopf, Heesemann, J., & Laufs, R. 1994. Characterization of transposon mutants of biofilm-producing *Staphylococcus epidermidis* impaired in the accumulative phase of biofilm production: Genetic identification of a hexosamine-containing polysaccharide intercellular adhesin. Infect Immun 62: 3244–3253.

Møller, S., Sternberg, C., Andersen J.B., Christensen, B.B., & Molin, S. 1998. In situ gene expression in mixed-culture biofilms: evidence of metabolic interactions between community members. Appl. Environ. Microbiol. 64: 721–732.

Nemoto, K., Hirota, K., Murakami, K., Taniguti, K., Murata, H., Viducic, D., & Miyake Y. 2003. Effect of Varidase (streptodornase) on biofilm formed by *Pseudomonas aeruginosa*. Chemotherapy 49: 121–125.

Nielsen, A.T., Tolker-Nielsen, T., Barken, K.B., & Molin, S. 2000. Role of commensal relationships on the spatial structure of a surface-attached microbial consortium. Environ. Microbiol. 2: 59–68.

O'Toole, G., Kaplan, H.B., & Kolter, R. 2000. Biofilm formation as microbial development. Annu. Rev. Microbiol. 54: 49–79.

O'Toole, G.A., & Kolter, R. 1998. Flagellar and twitching motility are necessary for *Pseudomonas aeruginosa* biofilm development. Mol. Microbiol. 30: 295–304.

Pratt, L.A., & Kolter, R. 1998. Genetic analysis of *Escherichia coli* biofilm formation: roles of flagella, motility, chemotaxis and type I pili. Mol. Microbiol. 30: 285–293.

Prigent-Combaret, C., Vidal, O., Dorel, C., Hooreman, M., & Lejeune, P. 1999. Abiotic surface sensing and biofilm-dependent regulation of gene expression in *Escherichia coli*. J Bacteriol 181: 5993–6002.

Schaik, E.J. & Irvin, R.T. 2003. DNA uptake in *Pseudomonas aeruginosa*: an additional type IV pilus function. ASM conference: *Pseudomonas* 2003. Poster.

Skerker, J.M., & Berg, H.C. 2001. Direct observation of extension and retraction of type IV pili. Proc. Natl. Acad. Sci. USA 98: 6901–6904.

Tolker-Nielsen, T., Brinch, U.C., Ragas, P.C., Andersen, J.B., Jacobsen, C.S., & Molin, S. 2000. Development and dynamics of *Pseudomonas* sp. biofilms. J Bacteriol 182: 6482–6489.

Vallet, I., Olson, J.W., Lory, S., Lazdunski, A., Filloux, A. 2001. The chaperone/usher pathways of *Pseudomonas aeruginosa*: identification of fimbrial gene clusters (cup) and their involvement in biofilm formation. Proc. Natl. Acad. Sci. USA. 98: 6911–6916.

Whitchurch, C.B., Tolker-Nielsen, T., Ragas, P.C., & Mattick, J.S. 2002. Extracellular DNA is required for bacterial biofilm formation. Science. 295: 1487.

Wolfaardt, G.M., Lawrence, J.R., Robarts, R.D., Caldwell, S.J., & Caldwell, D.E. 1994. Multicellular Organization in a Degradative Biofilm Community. Appl. Environ. Microbiol. 60: 434–446.

Wozniac, D.J., Wyckoff, T.J.O., Starkey, M., Keyser, R., Azadi, P., O'Toole, G.A. & Rarsek, M.R. 2003. Alginate is not a significant component of the extracellular polysaccharide matrix of PA14 and PAO1 Pseudomonas aeruginosa biofilms. Proc. Natl. Acad. Sci. USA 100: 7907–7912.

Yildiz, F., & Schoolnik, G.K. 1999. Vibrio cholerae O1 El Tor: Identification of a gene cluster required for the rugose colony type, exopolysaccharide production, chlorine resistance, and biofilm formation. Proc. Natl. Acad. Sci. USA 96: 4028–4033

European Symposium on Environmental Biotechnology, ESEB 2004 - Verstraete (ed)
© 2004 Taylor & Francis Group, London, ISBN 90 5809 653 X

In situ soil bioremediation and permeable reactive barriers: risk, control and sustainability

L. Diels, K. Vanbroekhoven & L. Bastiaens
Vito, Mol, Belgium

ABSTRACT: Permeable reactive barriers have been under development since the early ninety's. The first PRBs were designed for the reductive dehalogenation of chlorinated solvents by zero valent iron granules. Since the beginning questions arose about the possible clogging due to inorganic or organic precipitates. More than ten years later more knowledge exists concerning the long term performance of these barriers. It has been observed that the clogging is not that problem as foreseen and that iron barriers are not just based on a chemical process but can be a combination of chemical, physico-chemical and biological processes. Also much more information does exist on the different organic processes that can occur and the transformation of certain precipitates. E.g., iron reducing bacteria can introduce an *in situ* biological renewal of the iron granules. Based on these observations PRB technology is now extended to the development of an *in situ* treatment technology applicable for remediation of all kind of pollutants. PRBs are developed for the sorption (on zeolites, silicates, compost etc.) and *in situ* bioprecipitation of heavy metals (as in soluble metal sulphides). Sorption on granular activated carbon can be used for the removal of all kinds of hydrophobic contaminants as PAHs.

Permeable barriers are restricted in their use by the depth (10–15 m) as construction costs and technical problems will be to high in case of larger depths. Therefore an increasing interest is going into the direction of the development of virtual barriers or *in situ* reactive zones. These barriers are induced by creating specific conditions that induce pollutant removal processes. These can be obtained by the induction of biological processes in so called bioscreens (sulphate reduction, denitrification, biodegradation or biodehalogenation) or by the injection of nano-materials (e.g., nano-iron, or ultramicrobacteria) that migrate with the same velocity as the groundwater flow rate.

Further a combination of these activities with zerovalent iron systems (or adsorption systems) can lead to the treatment of mixed pollution in a so called 'Multibarrier' system. This allows to treat mixtures of pollutants as is e.g., the case when landfill leachates are entering the groundwater.

1 INTRODUCTION

Permeable reactive barriers (PRBs) emerged during the 1990s as a potentially effective remedial technology for treating contaminated groundwater. Over 100 pilot scale and field scale PRBs have been constructed in North America and Europe. A permeable reactive barrier prevents or reduces contaminant flux whilst allowing groundwater to flow through the barrier. This is achieved by the placement of reactive materials in the flowpath of contaminated groundwater. These reactive materials either immobilize or transform (biologically or abiotically) the pollutants such that the treated groundwater down hydraulic gradient of the PRB should not represent an unacceptable risk to water resources or other receptors.

These mechanically simple barriers may contain metal-based catalysts for degrading volatile organics, chelators for immobilizing metals, nutrients and oxygen for micro-organisms to enhance bioremediation, or other agents. Degradation reactions break down the contaminants in the plume into harmless byproducts. Precipitation barriers react with contaminants to form insoluble products that are left in the barrier as water continues to flow through. Sorption barriers adsorb or chelate contaminants to the barriers surface. The reactions that take place in barriers are dependent on parameters such as pH, oxidation/reduction potential, concentrations, and kinetics. Thus, successful application of the technology requires characterization of the contaminant, ground-water flux, and subsurface geology. Although most barriers are designed to operate *in situ* for years with minimal maintenance and without an external source, the stability of aging barriers is still under study.

The development of treatment technology for the clean up of contaminated groundwater resources has expanded in the past few years. The main perceived advantage of this technology over *ex situ* and other *in situ* groundwater remediation approaches is reduced operation and maintenance costs.

PRBs are a relatively new technology and future developments are likely to increase the range of contaminants that can be treated, and the range and effectiveness of reactive media. PRBs will typically need to be operated over extended periods, being decades due to the low rate of groundwater/contaminant flow in many aquifers.

Although, considerable design details have already been developed through field- and pilot-scale applications of this technology, some critical issues still remain to be solved (e.g., establishing tested and proven design procedures, documenting long-term performance, and evaluating synergy with other groundwater remediation technologies). Currently planned field-scale tests and any ongoing laboratory studies are designed to address these issues and facilitate wider implementation of this technology.

2 RESULTS AND DISCUSSION

2.1 *Permeable reactive barriers*

The most common design of PRBs used to date are 'funnel and gate' and 'continuous' reactive barriers. 'Funnel and gate' PRBs comprise impermeable walls, such as sheet piles and slurry walls, which direct contaminated groundwater to 'gate(s)' containing the reactive material. 'Continuous' PRBs transect the pollutant plume flow-path with an unbroken wall of permeable materials, which are combined with the reactive materials (e.g., a pea-gravel and reagent filled trench that is constructed across the groundwater flow direction). The majority of PRBs have been placed to relatively shallow depths, around 5 to 20 m deep. For deeper applications reactive zones are more applicable.

To date, PRBs have been used to treat a range of contaminants in groundwater such as organohalogen compounds (e.g. PCE, TVE, DCE), metals (e.g., chromium, arsenic, zinc etc.), nitrate and radionuclides (e.g., uranium). In treating these contaminants a range of processes have been used such as manipulation of redox potential to enhance biological reductive dechlorination and to change the chemical speciation of metals), chemical degradation, precipitation, sorption (such as organic matter partitioning and ion exchange) and biodegradation/biotransformation. Treatment processes can be described as:

Destructive, as is the case for biodegradation, abiotic oxidation or reduction and hydrolysis whereby the contaminants are permanently transformed into less harmful substances;

Non-destructive, as is the case of sorption, precipitation or change in chemical state, where the contaminants are immobilised or fixed to reactive media.

A range of reactive materials have been used to treat contaminants such as metals (e.g., zero valent iron), granular activated carbon (GAC), organic materials (e.g. wood chips, compost), chelators, zeolites, chemical oxidants and reducing agents.

2.2 *Long term performance of PRBs*

Since the inception of permeable reactive barriers, there has been considerable speculation as to the long-term performance of each PRB that has been installed (Diels et al., 2003a). The key to an effective PRB application is to understand both the groundwater chemistry and flow characteristics of the system, by completing adequate site characterization along the line of the PRB installation. The PRB design is dependent on having adequate dimensions to both provide sufficient residence time within the reactive zone and to ensure complete plume capture. The primary issue associated with PRBs has been with hydraulic performance. Factors such as variation in flow velocity, variable seasonal flow direction, and inadequate plume delineation have all caused hydraulic issues at some sites (i.e., by pass of VOCs). Issues such as compaction or densification of the aquifer material during vibration of equipment, flowing sands and the smearing of clay materials along the PRB walls are some examples. These factors have been addressed during PRB construction on a site specific basis, but have added time and cost to system installations. However these lessons learned have helped improve the design and cost effectiveness of subsequent PRB installations, making the technology an even more attractive remedial alternative.

Most PRB installations occur within a plume and as the clean water emerges from the PRB formation, it flushes the contaminated downgradient aquifer. Consequently, downgradient concentrations persist due to the compounds desorbing from the aquifer solids. The rate of desorption is dependent upon the aquifer properties such as fraction organic carbon content and the presence of the low hydraulic conductivity horizons, but the most important fact is the flow velocity.

Based on laboratory predictions, the greatest factor with respect to precipitate formation now appears to be the loss of reactivity on the iron surface (Gillham et al., 2001; Battelle, 2002). This reactivity loss will be very dependent upon the groundwater flow velocity and the inorganic constituents of the site water, especially the calcium, magnesium and alkalinity concentrations. To date no iron PRB has required rejuvenation with respect to precipitate formation. The use of technologies such as ultrasound, to rejuvenate granular iron surfaces appears promising. These rejuvenation techniques represent less costly alternative to partial PRB replacement, and will therefore improve the technologies cost effectiveness even further.

It seems that the hydrogen, produced by the anaerobic corrosion of the iron, can be used as electron donor by a wide variety of microbial phenotypes, including denitrifying bacteria transforming nitrates in the groundwater into nitrogen gas.

2.3 *PRBs and reactive zones for remediation of heavy metal containing groundwaters*

PRBs and reactive zones are being developed (Diels et al., 2003b, 2003c) and in many cases demonstrated to be efficient for the treatment of dissolved metals and dissolved non-metallic species typically associated to metal pollution problems as could be sulfur containing species (mainly SO_4^{-2}), arsenic (III, V) containing species ($HAsO_4^{-2}$, $HAsO_3^{-2}$), selenium (IV, VI) containing species (SeO_3^{-2} and SeO_4^{-2}) (Blowes, 2000; Sherer, 2000).

Selection of the treatment technology for those type of pollution scenarios depends upon the type, water flux and pollutant concentration reduction or removal requirements. Reactive materials can be classified into the following types:

- inducing the precipitation of a relatively insoluble mineral phase
- promoting adsorption onto mineral and adsorbent phases
- supplying carbon sources and nutrients for bioprecipitation
- modifying the redox potential and acidity to promote any of the previous processes.

Several remediation projects have used calcareous surface barriers to neutralize acidity and remove metallic species, generated by the oxidation of sulfidic minerals in mines however limited success was achieved when applied to sub-surface barrier. The reduction in hydraulic conductivity resulting from metallic oxy-hydroxide precipitation has been viewed as an impediment in sub-surface streams because flow is diverted around, rather than continuing through the calcareous barrier. The chemical evolution within a ferric-calcareous barrier, proceeds with the hydrolysis of Fe^{3+} generating H^+, which is neutralized by reaction with calcite, causing poorly crystalline ferric oxy-hydroxide to precipitate. The deposition of ferric oxy-hydroxide peaked the precipitate front, where the neutralization of acidity generated by hydrolysis was completed and the hydraulic conductivity decrease achieve up to a factor of 100 to 1000 times (Fryar, 1994) .

Typically, waters, from landfills or mining operations often contain high concentrations of iron and other metals at low pH. As the pH is neutralized, multivalent metal ions tend to hydrolyze and polymerize, and precipitate poorly crystalline or amorphous hydrous oxides, such as oxy-hydroxides. Because of their relatively disordered structure and high surface area, iron oxy-hydroxides can incorporate other metals through co-precipitation (Mallants et al., 2002) or adsorption, a feature exploited in the use of ferric chloride as an additive in water treatment systems. Almost all the materials that have been suggested to be used in chemical barriers are particles (e.g., zero valent Fe(Gillham, 1994), anoxic limestone (Hedin, 1994), and peat (Longmire, 1991). These materials require emplacement in an excavation such as trench because injection would clog the aquifer causing flow to be diverted around the chemical barrier.

In the last decade an extensive work have been focused on above ground-surface treatment of oxidized mine drainage water characterized by low pH and elevated concentrations of dissolved SO_4^{2-}, Fe(III) and other metal ions. Under favorable conditions sulfate-reducing bacteria (SRB) convert sulfate to sulfide by catalyzing the oxidation of organic carbon coupled with the reduction of sulfate. In the same time they increase alkalinity and pH. Increases in dissolved H_2S concentration enhance the precipitation of metal (M^{n+}) as metal sulfides $M_2S_n(s)$ (Diels et al., 2003b). Additionally the increase of total alkalinity will produce the precipitation of the metal carbonates and hydroxides, and also the dissolution of gypsum.

2.4 *PRBs based on sorption processes with granular activated carbon*

At many contaminated sites groundwater has been found to be contaminated with hydrophobic organic compounds. The sorption capacity of GAC for organic contaminants mainly depends on the chemo-physical properties of the various compounds and also on the type and concentration of dissolved organic carbon in general (usually determined as DOC). K_d-values, which are determined from sorption isotherms, can be used to quantify the sorption behavior of organic contaminants. Especially PAHs but also PCBs and compounds can be eliminated from the water phase very effectively over many years without changing the sorbent.

Most important, above-ground treatment using GAC in many cases is performed under aerobic conditions, i.e. dissolved iron or manganese either are removed before the sorption filter or they are not present at significant concentrations. In contrast, contaminated ground-water to be treated in-situ typically contains considerable amounts of dissolved iron which may precipitate within the GAC barrier. This does not only reduce the sorption capacity but more important such precipitates reduce the permeability of the barrier.

Other processes which have to be considered for *in-situ* applications can be summarised under the term 'biofouling'. This may occur in particular when groundwater is treated coming from different parts of a site containing different amounts of oxygen or e.g. nitrate mixes within the barrier. Microbiological activity then may develop within the barrier leading to sliming and clogging.

2.5 *Long-term monitoring of PRB activity*

The overall purpose of monitoring is to demonstrate whether the PRB is effective as a remedial measure for the polluted groundwater and that its application does not cause additional pollution or harm (M.A. Carey et al., 2002). A typical monitoring network includes:

A. Monitoring points up-gradient of the contaminant source and PRB to determine background water quality and any changes in this reference during operational life of the PRB;
B. Monitoring points up-gradient of the PRB to provide information on groundwater quality/contaminant flux. These boreholes can be located within a few metres of the PRB to provide accurate data on the contaminant flux;
C. Monitoring point within the PRB to provide information on processes within the reactive material. As the majority of changes in a PRB tend to occur within the first few centimetres of the reactive material, the installation of monitoring points in this zone will be extremely important;
D. Monitoring points down-gradient of the PRB to provide confirmation of the effectiveness of the PRB in treating contaminated groundwater, i.e., there is no unacceptable breakthrough of the contaminant (or metabolites) and that the PRB does not result in a detrimental change in groundwater quality;
E. F. Side-gradient monitoring points to provide assurance that contaminated groundwater does not flow around the PRB and monitoring points below the PRB to provide assurance that contaminated groundwater does not flow under the PRB;
G. Boreholes located between the plume and the identified receptors, for which exceedence of an operational control level will require implementation of additional mitigation measures.

A high frequency of sampling would be required during the first year of monitoring to confirm the treatment process and to establish trends. Thereafter, the frequency could be reduced, subject to agreement with the authorities. Annual monitoring or even less frequent for particularly slow-moving groundwater) may be acceptable for some sites, once baseline conditions and trends in the PRB behaviour have been established with confidence. The design should also consider whether the use of field measurements coupled with less frequent laboratory measurements is an effective approach.

3 CONCLUSIONS

A Permeable Reactive Barrier (PRB) is an engineered zone of reactive material(s) that is placed in the subsurface in order to remediate contaminated fluids as they flow through it. A PRB has a negligible overall effect on bulk fluid flow rates in the subsurface strata, which is typically achieved by construction of a permeable reactive zone, or by construction of a permeable reactive 'cell' bounded by low permeability barriers that direct the contaminant towards the zone of reactive media.

In most environments these PRBs will last a long time. However, depending on the inorganic geochemistry, issues arise related to the long term impact of the inorganic precipitates on the iron surface. Initially, it was thought the these precipitates would cause significant losses in permeability, however recent research indicates that only slight permeability losses will be observed over time. The knowledge of processes occurring in the reactive material and interactions with microbiology and inorganic elements allowed to understand and design the barriers that they could work for a longer period than expected until now. The combination between bacterial activity and the chemical processes on the iron surface or the physical processes of adsorption on the GAC surface seem to play an important role in the longevity of the barrier. This knowledge also lead to the understanding that the combination of different chemical, physical and biological processes must make that also mixed pollutants (inorganics and organics) could be treated in one single measure in a MULTIBARRIER (see other presentations at ESEB).

Almost all the materials that have been suggested to be used in chemical barriers are particles (e.g. zero valent Fe (Gillham, 1994), anoxic limestone (Hedin, 1994), and peat (Longmire, 1991). These materials require emplacement in an excavation such as a trench because injection would clog the aquifer causing flow to be diverted around the chemical barrier. Therefore several attempts are going on to introduce compounds that are either soluble or can be transported at the water flow rate. In this was the possibility of injecting $FeCl_3$ into the subsurface through wells was investigated. A chemical barrier emplaced through wells could be used to treat deep contaminated aquifers and at sites where surface disturbances (such as trenching) are unacceptable. AFO could be emplaced beneath structures or landfills by horizontal drilling wells.

For the chemical barrier to be effective, AFO must spread throughout the aquifer during injection and must remain immobile after emplacement to prevent the transport of contaminants. Excessive clogging of the subsurface will hinder injection by decreasing hydraulic conductivity and may cause groundwater to be diverted around the chemical barrier.

An other way is the injection of nano-iron particles. These can be injected in the wells and will proceed with the water flow rate. At the moment no information exists about the behavior of such iron particles in the aquifer (possible aggregation, co-precipitation and clogging).

REFERENCES

Battelle Press, 2002. Evaluating the longevity and hydraulic performance of permeable reactive barriers at Department of Defense Sites. Columbus, Ohio.

Gillham, R.W., Ritter, K., Zhang, Y., Odziemkowski, M.S., 2001. Factors in the Long-Term Performance of Granular Iron PRBs. Proceeding of the 2001 International Containment and Remediation Technology Conference, Orlando, Florida.

Mallants, D., Diels, L., Bastiaens, L., Vos, J., Moors, H. Wang, L., Maes, N. Vandenhove, H., 2002. Interantional conference ion Uranium mining and hydrogeology UMHIII, Freiberg 15–21 Septemùber 2002, 565–571.

Reynolds, T.J., Gillham, R.W., 2002. Use of the *In Situ* Microcosm Technique for evaluating the long-term Performance of a Granular Iron PRB. Presented at 2nd Annual Oxidation-Reduction Technology Conference, Toronto, Ontario, Canada, Nov. 18–21.

Sorel, D., Warner, S.D., Longino, B.L., Hamilton, L.A., Vogan, J.L., 2000. Performance monitoring of the first commercial permeable zero-valent iron reactive barrier – Is it still working? Presented at the 2000 Theis Conference 'Iron in Ground Water', Jackson Hole, Wyoming, September 15–18.

Blowes, D.W., Reardon, E.J., Jambor, J.L., Cherry, J.A. The formation and potential importance of cemented layers in active sulfide mine tailings. Geochim. Cosmochim. Acta., 1991, 55, 965–978.

Blowes, D.W., Ptacek, C. In Short Course Handbook on Environmental Geochemistry of Sulfide Mine-Waste. Jambor, J.L., Blowes, D.W., eds.; Mineralogical Association of Canada: Nepean, ON. 1994a, 22, 271–292.

Blowes, D.W., Ptacek, C.J. Proceedings of Subsurface Restoration Conference, 3rd International Conference on Ground Water Quality Research, June 21–24, Dallas, TX. 1992, pp. 214–216.

Blowes, D.W., Ptacek, C.J., Bain, J.G., Waybrant, K.R., Robertson, W.D. Proc. Sudbury 1995: Mining and the Environment; CANMET: Ottawa, 1995a, Vol. 3, 979–987.

Blowes, D.W., Ptacek, C.J., Jambor, J.L. In-Situ Remediation of Cr(VI)-Contaminated Groundwater Using Permeable Reactive Walls: Laboratory Studies. Environmental Science & Technology, 1997, 31(12), 3348–3357.

Blowes, D.W., Ptacek, C.J. Benner, S.G., Mc Rae, C., Bennett, T., Puls R.W., 2000. Treatment of inorganic contaminants using reactive permeable barriers, J. of Cont. Hydr., 45, 123–137.

Bostick, W.D., Jarabek, R.J., Bostick, D.A., Conca, J. 1999. Phosphate-Induced Metal Stabilization: Use of Apatite and Bone Char for the Removal of Soluble Radionuclides in Authentic and Simulated DOE Groundwaters. Advances in Environmental Research, 3, 488–498.

Bowman, R.S., Haggerty, G.M., Huddleston, R.G., Neel, D., Flynn, M.M., 1997. Sorption of nonpolar organic compounds, inorganic cations, and inorganic oxyanions by surfactant-modified zeolite. In D. A.

Bowman, R.S., Sullivan, E.J., 1995. Surfactant-modified zeolites as permeable barriers to organic and inorganic groundwater contaminants, in 'Proceedings, Environmental Technology Development through Industry Partnership Conference,' 2, 392–397.

Carey, M.A., Fretwell, B.A., Mosley, N.G., Smith, J.W.N., 2002. Guidance on the Design, Construction, Operation and Monitoring of Permeable Reactive Barriers, Environment Agency, Bristol, UK, NGWCLC Report NC/01/51.

Conca, J.L., Lu, N., Parker, G., Moore, B., Adams, A., Wright, J.V., Heller, P., 2000. PIMS – Remediation of Metal Contaminated Waters and Soils, in 'Remediation of Chlorinated and Recalcitrant Compounds' (G.B. Wickramanayake, A.R. Gavaskar and A. Chen, eds.) Battelle Memorial Institute, Columbus, Ohio, 7, 319–326.

Diels, L., Bastiaens, L., O'Hannessin, S., Cortina, J.L., Alvarez, P.L., Ebert, M., Schad, H., 2003a. Permeable Reactive Barriers: a multidisciplinary approach of a new emerging sustainable groundwater treatment technology, Consoil 2003.

Diels, L., Geets, J., Vos, J., Vanbroekhoven, K., Bastiaens, L., 2003b. Remediation of sites contaminated by heavy metals: sustainable approach for unsaturated and saturated zones. International Biohydrometallurgy meeting, Athens, 14–19 September 2003.

Diels, L., Geets, J., Vos, J., Van Roy, S., Bossus, A., Bastiaens, L., 2003c. Remediation of sites contaminated by heavy metals: sustainable approach for unsaturated and saturated zones. In situ and on site bioremediation Orlando, 2–5 June 2003.

Fryar, A.E., Schwartz, F.W. Modeling the removal of metals from ground water by a reactive barrier: experimental results. Water Resour. Res., 1994, 30 (12), 3455–3469.

Fryar, A.E., Schwartz, F.W. Hydraulic-conductivity reduction, reaction-front propagation, and preferential flow within a model reactive barrier. J. of Contaminant Hydrology, 1998, 32, 333–351.

Naftz, D.L., Fuller, C.C., Davis, J.A., Piana, M.J., Morrison, S.J., Freethey, G.W., Rowland, R.C., 2000. Field Demonstration of PRBs to Control Uranium Contamination in Groundwater, in Remediation of Chlorinated and Recalcitrant Compounds, (Wickramanayake, G.B., Gavaskar, A.R., Chen, A., eds.) Battelle Memorial Institute, Columbus, Ohio, 6, 281–290.

Powell, R.M., Puls, R.W., Hightower, S.K., Sabantini, D.A., Environ Sci. Technol., 1995, 29, 1913–1922.

Ritcey, G.M. Tailings Management. Elsevier Ed. Amsterdam, 1989.

Robertson, W.D., Cherry, J.A., Ground Water, 1995, 33, 99–111.

Sabatini et al. (ed.) Surfactant-enhanced subsurface remediation. ACS Symp. Ser. 594. Am. Chem. Soc., Washington, DC. 1995, 54–64.

Scherer, M., Richter, S., Richard, L.V., Alvarez, P.J., 2000. Chemistry and Microbiology of Permeable Reactive Barriers for In situ Groundwater clean-up, Critical Reviews in Microbiology, 26(4), 221–264.

Tratnyek, P.G., Johnson, T.L., Scherer, M.M., Eykholt, G.R., 1997. Remediating groundwater with zero-valent metals: Kinetic considerations in barrier design. Groundwater Monitor. Remed., 17, 108–114.

Vilesky, M., Berkovitz, B., Warshawsky, A., 2002. In situ remediation of groundwater contaminated by heavy and transition metal ions by selective Ion exchange methods, Env. Sci. and Tech., 36, 1851–1855.

Hedin, R.S., Watzlaf, G.R., Nairn, R.W. Passive treatment of acid mine drainage with limestone. J. Environ. Qual. 1994, 23, 1338–1345.

Longmire, P.A., Brookings, D.G., Thomson, B.M., Eller, P.G. Application of sphagnum peat, calcium carbonate, and hydrated lime for immobilizing uranium tailings leachate. In : Scientific Basis for Nuclear Waste Management XIV, ed. by T. Abrajano, Jr. and L.H. Johnson, Materials Research Society, Pittsburgh, PA. 1991, 212. 623–631.

Wildman, M. J., Alvarez, P.J.J., 2001, RDX degradation using an integrated Fe(0)-microbial treatment approach. Water Science and Technology, 43(2): 25–33.

European Symposium on Environmental Biotechnology, ESEB 2004 - Verstraete (ed)

Biosensing systems for monitoring aerobic wastewater treatment processes

R.M. Stuetz
Centre for Water & Waste Technology, School of Civil & Environmental Engineering, The University of New South Wales, Sydney, Australia

J.E. Burgess
Dept of Biotechnology, Rhodes University, Grahamstown, South Africa

W. Bourgeois & T. Stephenson
School of Water Sciences, Cranfield University, Cranfield, UK

ABSTRACT: Biological wastewater treatment processes are dynamic systems which are often exposed to environmental changes resulting in process failure or partial inhibition. Monitoring strategies traditionally used to determine process failure are often inadequate. The analyses of wastewater off-gases using specific and non-specific gas sensors can provide a non-contact measurement technique for continuous wastewater monitoring. The measurement of the production of dinitrogen oxide (N_2O) during wastewater nitrification can be directly related to the degree of nitrification inhibition, whereas the correlation of non-specific gas sensors so called "electronic noses" can be used to discriminant the presence or absence of organic pollutants, such as diesel within a wastewater matrix. These studies have demonstrated that biosensing systems could be further developed to serve a number of potential applications within wastewater treatment, such as early warning of treatment failure and inlet protection for toxicity monitoring.

1 INTRODUCTION

Biological wastewater treatment processes are dynamic systems which are often exposed to changing environmental conditions, such as organic loads, wet weather inflows and toxic discharges. These environmental changes can impose a range of process responses, with the impact of toxic loads on aerobic treatability resulting in process failure or partial inhibition.

Monitoring strategies traditionally used to determine process failure typically involves the infrequent collection of samples followed by their retrospective analyses for specific organic pollutants using conventional analytical methods (Bourgeois *et al.*, 2001). These measurements provide no early warning of wastewater treatment failure, but are major methods used to control the intermittent or accidental discharging of pollutants into the environment. More recently, the application of toxicity monitors using UV absorbance, biosensors and respiration kinetics have permitted early warning of process failure (Vanrolleghem *et al.*, 1994; Hayes *et al.*, 1998; Ingildsen *et al.*, 2001), however these systems are often unreliable due to fouling of the sidestream systems and the variability of results inherent in direct biological toxicity assessment (Burgess *et al.*, 2001). Therefore, there exists a need to develop non-contact measurement technologies that are capable of continuous wastewater treatment monitoring and control (Fuerhacker *et al.*, 2000; Bourgeois and Stuetz, 2002).

The analysis of the exhaust gas above a wastewater treatment process could offer a non-invasive solution for early warning of wastewater treatment failure. However, off-gas analysis for online monitoring and control of wastewater treatment plants has until recently gained rather limited attention (Hellinga *et al.*, 1996) compared to other wastewater contact based measurements such as dissolved oxygen and more recently ammonia monitoring (Vanrolleghem and Lee, 2003).

The direct measurement of the O_2 and CO_2 concentrations using a variety of infra-red, paramagnetic and gas sensors has been widely reported in fermentation processes (Hellinga *et al.*, 1996). However, their application to wastewater measurements have had limited. More recently, the application of headspace analysis using non-specific gas sensors has shown that the variability of wastewater can be correlated to either the concentration of specific gas or the response patterns between different types or groups of sensor measurements (Stuetz *et al.*, 1999a). These measurements could then be used for the assessment of real-time changes in the wastewater matrix as it arrives at a sewage treatment plant or during the treatment process itself (Stuetz *et al.*, 1999b).

The objectives of this paper are to review two types of biosensing systems for wastewater monitoring and discuss how these systems can be developed to serve a number of potential applications in wastewater treatment.

2 SPECIFIC GAS SENSING

Dinitrogen oxide (N_2O) an intermediate compound in the biological transformation of NH_3 to NO_3^- was investigated as an indicator of wastewater nitrification failure. Experimental studies (Burgess *et al.*, 2001) were used to establish if by directly measuring the concentration of N_2O emitted from an activated sludge process the stability of the process could be accessed. The dosing of varying concentrations of NH_3 into the wastewater influent was used to simulate NH_3 shock loads. These studies demonstrated that a strong correlation exists between NH_3 shock loads and the concentration of N_2O in the off-gas from the aeration tank (Fig. 1).

The results show that the appearance of N_2O in the off-gas was produced several hours prior to ammonia emerging in the wastewater effluent and was equivalent to between 0.75 and 0.90 of the hydraulic retention time (HRT) of the activated sludge process. Similar dosing studies have also been undertaken using allylthiourea (a nitrification inhibitor) and when the aeration was depleted of oxygen to stimulate operational failure (Burgess *et al.*, 2001). Both resulted in similar increases in the concentration of N_2O in the

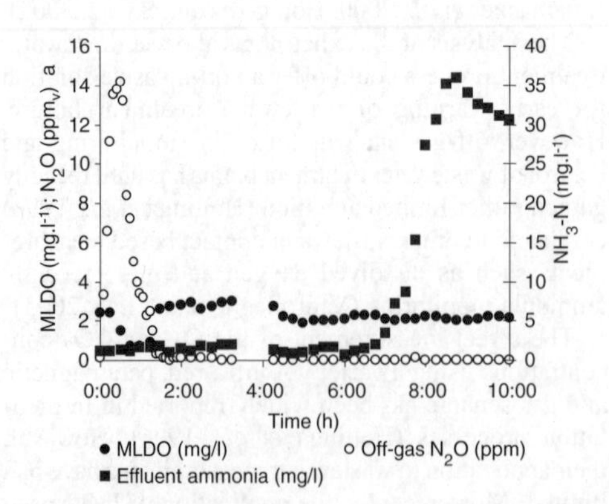

Figure 1. Nitrogen profile showing increase in N_2O upon the addition of NH_3 into the influent and the subsequent increase in effluent NH_3 (after Burgess *et al.*, 2001).

process off-gas prior to the appearance of ammonia in the wastewater effluent.

These observations show that the concentration and changes in the concentration of N_2O in the off-gas from an activated sludge process can be used as indicator of nitrification failure when an aeration process has been subjected to high ammonia loads, inhibition by toxic chemicals and the plant aeration problems.

3 NON-SPECIFIC GAS SENSING

Non-specific sensor array, so called "electronic noses" were investigated to detect the presence or absence of organic pollutants (such as diesel) in wastewater effluents. Electronic noses use an array of similar or different sensors to produce a response pattern or fingerprint between the sensors. These response patterns or individual sensor responses can then be correlated to changes within a wastewater matrix.

Bourgeois & Stuetz (2000) had previous demonstrated that a headspace gas generated from an external flowcell could be used to produce a reproducible headspace gas from a wastewater sample for subsequent sensor array analyses. This concept was used to continuously monitor for changes within wastewater for 6 months (Bourgeois & Stuetz, 2002). Figure 2 shows a photograph of the operational system.

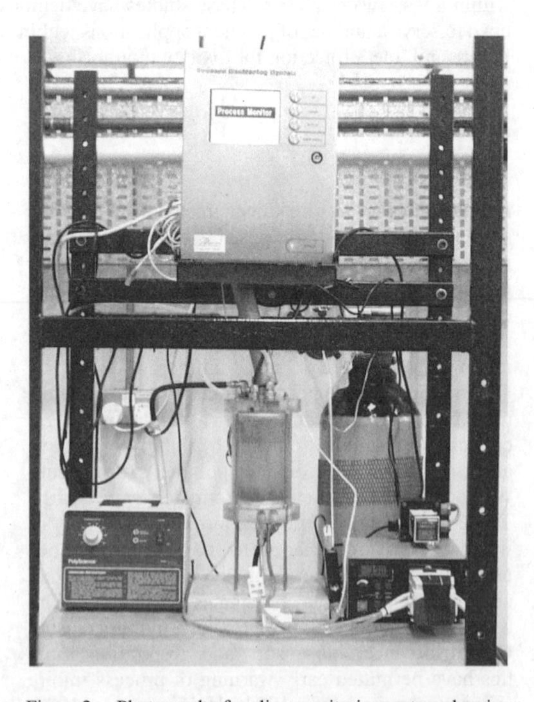

Figure 2. Photograph of on-line monitoring system showing sampling system and sensor array module (Bourgeois & Stuetz, 2002).

Experimental results (Bourgeois & Stuetz, 2002) from the continuous monitoring of a wastewater stream has shown good repeatability for the sensor responses and suggested that by superimposing the responses profiles over a specific time period, for example 24 hrs (Fig. 3), the data points that lay outside some predetermined limits could be used to indicate abnormalities in the wastewater matrix. Figure 4 shows an example of such a pollutants event, were an unknown pollutants is detected by some of sensors within a sampling window.

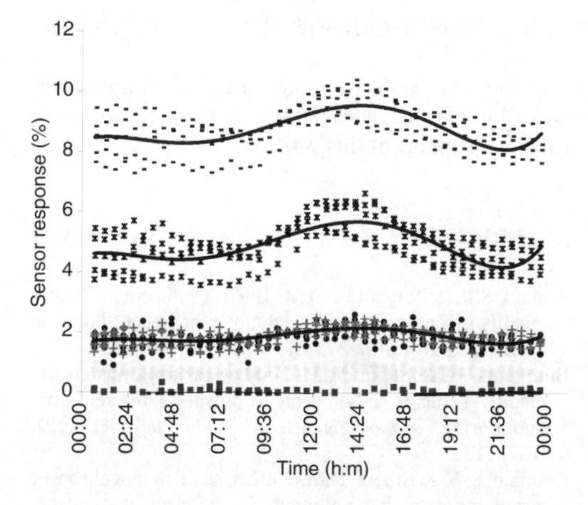

Figure 3. Plot of sensor responses showing repeatability of sensor responses for detecting abnormality (after Bourgeois & Stuetz, 2002).

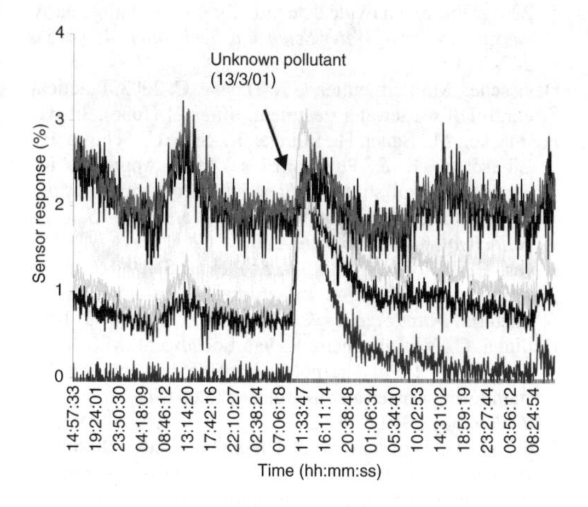

Figure 4. Plot of sensor responses showing the detection of an unknown pollutant in wastewater (after Bourgeois & Stuetz, 2002).

4 BIOSENSING SYSTEMS

In attempting to manufacture reliable wastewater monitoring systems using different chemical, biochemical and spectroscopic techniques, many analytical instruments have failed in there development, due to the poor integration and grouping of the devices as an integrated analytical system (IAS). Some of principle reasons for the high failure rate in applying wastewater monitoring systems is either because the analytical process needs the incorporation of sampling handling sequence to introduce the sample into the analyses module or that the measurement devices required constant contact with the liquid environment, resulting in instrument and sample collection fouling, causing a loss of measurement sensitivity and reproducibility, requiring the need for frequent cleaning and re-calibration of the analytical instrument (Ahmad & Reynolds, 1997).

Preliminary studies have shown that off-gas analysis using specific and non-specific sensors has the potential to be developed into a number of non-contact monitoring and control applications within wastewater treatment. Odorous emissions have long been used by process operators to identify types and stages of process as well as to detect abnormalities within a process (Namdev *et al.*, 1998). In wastewater treatment, specific odours such as earthy smells have traditionally been used by wastewater operators to indicate that an activated sludge process is operating effectively, whereas other sulphurous emissions have suggested that some form of process malfunction is occurring. To date these qualitative observations have not developed into a monitoring approach, but have been previously cited as a potential valuable measurement system that could be able to identify sudden changes in the wastewater as it arrives at the sewage treatment works (Ives *et al.*, 1994).

The current scientific challenge in the application of these two biosensing approaches in wastewater monitoring is the transformation of these laboratory based monitoring systems into integrated non-contact monitoring systems. The ideal monitoring configuration would be to have a continuously analyses of the off-gas above a wastewater stream. This could be achieved by measuring above the aeration tank or at the inlet of treatment plant. The different sampling configurations will be dependent on the objective of the monitoring system, which could be either to detect the presence or absence of specific pollutants in the wastewater or to use the system as a process control tool to adjust the treatment process in order to prevent some operational malfunction. An additional application could be to use the monitoring system as an alarm tool in order to increase the probability of obtaining a sample of an offending discharge by signaling a sample collection or online measurement system.

4.1 *Off-gas monitoring*

Continuous off-gas analysis has many obvious sampling advantages over liquid based measurement systems such as no contact between the analytical instrument and the liquid phase. Other advantages of a gas phase based sampling system are that the samples to be analysed are taken from a large and well mixed gas flow as it leave the wastewater surface, this allows the concentration of the sample to reflect the overall performance of the wastewater, whereas a contact probe in the liquid phase or a sidestream based monitoring system only reflects a point measurement of the process (Hellinga *et al.,* 1996). In addition, unlike point measurements such as grab samples that provide only a snapshot of a wastewater process, continuous measurements offer a complete picture of the variations in measured parameters over a specified time period such as a day or week (Devisscher *et al.*, 2003). These measurements help to provide a better insight into short- and long-term dynamics of the wastewater treatment process, and enhance our understanding of the underlying biological mechanisms in wastewater treatment. This is particularly true in wastewater plants that are running at the limit of their capacity, where even small variations in influent flow and composition can translate into large effluent fluctuations.

4.2 *Process monitoring and control*

Process monitoring and control enables plant operation to be continuously and dynamically adapted to the actual needs of the treatment process (Devisscher *et al.*, 2003) by detecting influent fluctuations so that appropriate control actions can be implemented and environmental damage avoided (Love & Bott, 2000). The benefits of process control are increased process performance and reduced operational costs (Ahmad & Reynolds, 1997; Ingildsen *et al.*, 2001). However, to date the implementation of automated control systems in wastewater treatment has contrasted poorly with the chemical process industry (Devisscher *et al.*, 2003). Sensor reliability has been previously identified as the main resistance point to preventing the implementation of online process control of wastewater treatment plants, but more likely today the barrier to widespread acceptance of new monitoring system is that existing treatment plants were not designed for the application of real-time process control (Vanrolleghem & Lee, 2003).

Future implementation of new monitoring systems such as off-gas analysis will require better integration of the technology as an alternative or an addition to current nutrient and oxygen based sensor process control.

5 CONCLUSIONS

Off-gas measurements are capable of detecting changes in wastewater treatment performance and should be considered as an alternative or as an addition to liquid phase monitoring. Potential application of biosensing systems are to the monitoring of wastewater as it arrives at the sewage work or during aerobic wastewater treatment. Further work is needed to directly couple changes in off-gas measurements with operational parameters, in order to enhance our understanding of the underlying biological process.

ACKNOWLEDGEMENTS

The authors wish to thank the UK Engineering and Physical Sciences Research Council for their financial support of this work.

REFERENCES

Ahmad, S.R. & Reynolds, D.M. 1997. Monitoring of water quality using fluorescence technique: prospect of on-line process control. *Water Research* 33: 2069–2074.

Bourgeois, W. & Stuetz, R.M. 2000. Measuring wastewater quality using a sensor array – prospects for real-time monitoring. *Water Science & Technology* 41 (12): 107–112.

Bourgeois, W. & Stuetz, R.M. 2002. Use of chemical sensor array for detecting pollutants in domestic wastewater. *Water Research* 36: 4505–4512.

Bourgeois, W., Burgess, J.E. & Stuetz, R.M. 2001. On-line monitoring of wastewater quality: a review. *J Chemical Technology & Biotechnology* 76: 1–12.

Burgess, J.E., Stuetz, R.M., Morton, S. & Stephenson, T. 2001. Dinitrogen oxide detection for process failure early warning systems. *Water Science & Technology* 45 (4–5): 247–254.

Devisscher, M., Parmentier, G & Thoeye, C. 2003. Practical control of wastewater treatment. *Water* 21 (June): 39–41.

Fuerhacker, M., Bauer, H., Ellinger, R., Sree, U., Schmid, H., Zibuschka, F. & Puxbaum, H. 2000. Approach for a novel control strategy for simultaneous nitrification/denitrification inactivated sludge reactors. *Water Science & Technology* 34 (9): 2499–2506

Hayes, E., Upton, J., Batts, R. & Pickin, S. 1998. Online nitrification inhibition monitoring using immobilised bacteria. *Water Science & Technology* 37 (12): 193–196.

Hellinga, C., Vanrolleghem, P., Van Loosdrecht, M.C.M. & Heijen, J.J. 1996. The potential of off-gas analyses for monitoring wastewater treatment plants. *Water Science & Technology* 33 (1): 13–23.

Ingildsen, P., Jeppsson, U. & Olsson, G. 2001. Dissolved oxygen controller based on on-line measurements of ammonia combining feed-forward and feedback. *Water Science & Technology* 45 (4–5): 453–460.

Ives, K.J., Hammerton, D. & Packham, R.F. 1994. The river Severn pollution incident of April 1994 and it

impact on public water suppliers. Report for Severn Trent Water Ltd.
Love, N.G. & Bott, C.B. 2000. *A Review and Needs Survey of Upset Early Warning Devices*. Water Environment Research Foundation, Report No.99-WWF-2. Alexandria.
Namdev, P.K., Avoy, Y. & Singh, V. 1998. Sniffering out trouble: use of an electronic nose in bioprocesses. *Biotechnology Progress* 14: 75–78.
Stuetz, R.M., Fenner, R.A. & Engin, G. 1999a. Characterisation of wastewater using an electronic nose. *Water Research* 33 (2): 442–452.
Stuetz, R.M., George, S., Fenner, R.A. & Hall, S.J. 1999b. Monitoring wastewater BOD using a sensor array, *J Chemical Technology & Biotechnology* 74(11): 1069–1074.
Vanrolleghem, P.A., Kong, Z., Rombouts, G. & Verstraete, W. 1994. An on-line respirographic biosensor for the char-acterisation of loads and toxicity of waste-water. *J Chemical Technology & Biotechnology* 59: 321–333.
Vanrolleghem, P.A. & Lee, D.S. 2003. On-line monitoring equipment for wastewater treatment processes: state of the art. *Water Science & Technology* 47 (2): 1–34.

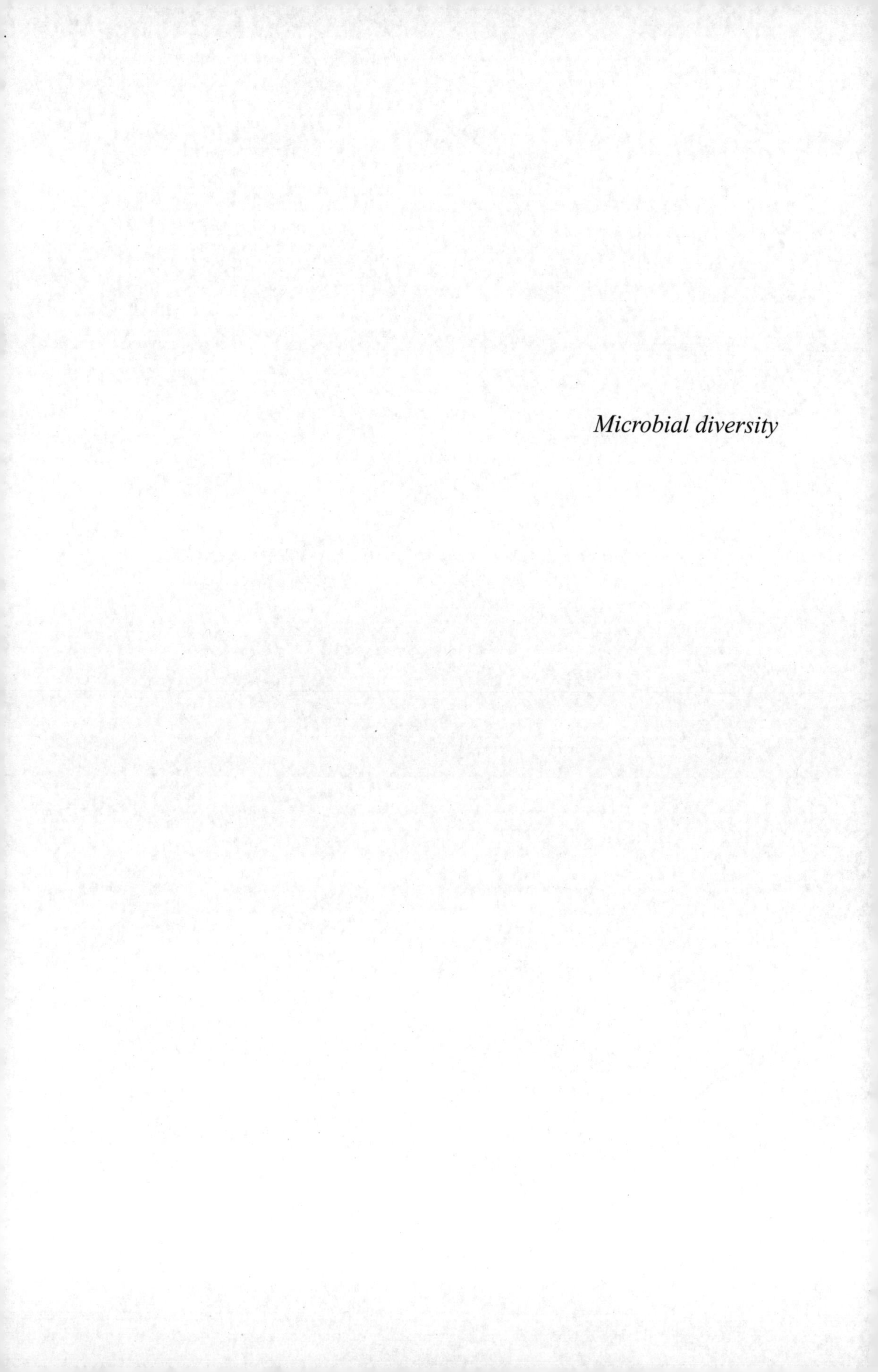

Microbial diversity

European Symposium on Environmental Biotechnology, ESEB 2004 - Verstraete (ed)
© 2004 Taylor & Francis Group, London, ISBN 90 5809 653 X

Analysis of sulfate reducing bacterial diversity in anaerobic wastewater treatment reactors

S.A. Dar, J.G. Kuenen & G. Muyzer
Department of Biotechnology, Delft University of Technology, The Netherlands

ABSTRACT: Here we describe the development and application of a nested PCR-DGGE approach to study the diversity of sulfate reducing bacteria (SRB) in different industrial bioreactors. The approach combines the use of general and specific primer sets to generate PCR products suitable for DGGE analysis. Subsequently the individual DGGE bands were excised and sequenced to determine the phylogenetic affiliation of the SRB. Preliminary results indicate that SRB belonging to the *Desulfovibrio-Desulfomicrobium* group were predominant in most of the bioreactors analysed, while populations of the *Desulfotomaculum* and *Desulfosarcina-Desulfonema-Desulfomonas* group could only be observed when a nested PCR approach was applied. *Desulfobulbus* species were only identified in some of the reactors.

1 INTRODUCTION

Sulfate reducing bacteria form a phylogenetically diverse and heterogeneous group of anaerobic bacteria that have the ability to use sulfate as a terminal electron acceptor in the degradation of organic matter, with the subsequent production of H_2S. They are known to be ubiquitous and play an important role in the biogeochemical sulfur cycle. Sulfate reduction dominates the organic matter degradation in high sulfate environments. It has been estimated that sulfate reduction accounts for up to 50% of the total organic matter degradation in marine sediments (Jørgensen, 1982). Besides marine sediments, SRB have been found in other environments, such as fresh water lake sediments (Sass et al., 1997), anaerobic biofilms (Raskin et al., 1996), and wastewater treatment facilities (Okabe et al., 1999). Although SRB are generally considered to be obligatory anaerobic bacteria, their occurrence has also been reported from aerobic environments, such as oxic zones of cyanobacterial mats (Krekeler et al., 1997) and wastewater biofilms grown under oxic conditions (Ramsing et al., 1993).

Because of the importance of SRB to critical processes in ecosystem functioning and environmental remediation increasing interest in SRB has been shown over the last decade. Combinations of different culture-independent methods have been used to study SRB populations in various ecosystems, resulting in an increased knowledge about the diversity of sulfate reducing bacteria. 16 S rRNA targeted oligonucleotide probes specific for SRB have been used in fluorescence *in situ* hybridization (FISH) in the detection of these microorganisms in a variety of environments (Amann et al., 1992). Probes targeting the functional genes have also been used to detect SRB (Voordouw et al., 1990). 16 S rDNA targeted PCR primer sequences specific for SRB subgroups have been designed and used to detect phylogenetic sub-groups of SRB (Daly et al., 2000). Recently DNA microarray suitable for SRB diversity analysis has been used to detect SRB in complex environmental samples (Loy et al., 2002).

Denaturing gradient gel electrophoresis (DGGE) is another molecular tool that has been used to determine the presence and distribution of SRB in natural and engineered environments (Santegoeds et al., 1999). Separation of PCR amplified, DNA fragment in DGGE is based on the decreased electrophoretic mobility of partially melted double stranded DNA molecules in polyacrylamide gels containing a linear gradient of DNA denaturants (a mixture of urea and formamide). Molecules with different sequences may have different melting behaviors, and will therefore, stop migrating at different positions on the gel. The banding pattern on DGGE gel thus represents the major constituents of the analysed community.

This study describes the diversity of sulfate reducing bacteria in different anaerobic wastewater treatment reactors using a nested PCR-DGGE.

2 MATERIALS AND METHODS

2.1 *Reference strains and sludge samples*

The six strains of SRB used as control in this study were *Desulfotomaculum nigrificans, Desulfobulbus propionicus, Desulfobacterium autotrophicum, Desulfobacter postgatei, Desulfosarcina variabilis* and *Desulfovibrio desulfuricans.* These strains represent the main phylogenetic groups of SRB.

Samples were obtained from four different anaerobic bioreactors (B1, B2, B3 and B4).

2.2 *Oligonucleotides used in this study*

Details concerning the different oligonucleotides used in this study are shown in Table 1. Oligonucleotides DFM140 to DSV838 are specific for the 16 S rDNA of different phylogenetic groups of SRB (Daly et al., 2000). The primer pair GM3-GM4 amplifies nearly the complete sequence of 16 S rDNA of the domain Bacteria. The primer 341F(G + C) can be used in combination with 518 R for amplification of bacterial 16 S rDNA (Muyzer et al., 1993).

Table 1. Oligonucleotides used in this study.

Primer	Target group	Reference
DFM140	*Desulfotomaculum*	(Daly et al., 2000)
DFM842	*Desulfotomaculum*	(Daly et al., 2000)
DBB121	*Desulfobulbus*	(Daly et al., 2000)
DBB1237	*Desulfobulbus*	(Daly et al., 2000)
DBM169	*Desulfobacterium*	(Daly et al., 2000)
DBM1006	*Desulfobacterium*	(Daly et al., 2000)
DSB127	*Desulfobacter*	(Daly et al., 2000)
DSB1273	*Desulfobacter*	(Daly et al., 2000)
DCC305	*Desulfonema-Desulfosarcina-Desulfococcus*	(Daly et al., 2000)
DCC1165	*Desulfonema-Desulfosarcina-Desulfococcus*	(Daly et al., 2000)
DSV23	*Desulfovibrio-Desulfomicrobium*	(Daly et al., 2000)
DSV838	*Desulfovibrio-Desulfomicrobium*	(Daly et al., 2000)
GM3	Bacteria	(Muyzer et al., 1995)
GM4	Bacteria	(Muyzer et al., 1995)
341F(G + C)	Bacteria	(Muyzer et al., 1993)
518R	Bacteria	(Muyzer et al., 1993)

2.3 *Nucleic acid extraction*

Genomic DNA was isolated from the reference strains and the reactor samples using the Ultra Clean Soil DNA isolation kit (MOBIO Laboratories) according to the manufacturers protocol. The yield and quality of the DNA was analysed electrophoretically on 1% (w/v) agarose gel.

2.4 *PCR amplification*

Two approaches were used to amplify 16 S rDNA from the reactor samples. First a direct amplification of 16 S rDNA was attempted on the DNA extracted from the samples using SRB group specific primers (Table 1). Reactions were carried out as follows: 95°C for 1 minute, annealing for 1 minute, and 72°C for 1 minute for 30 cycles and a final extension at 72°C for 7 minute. Secondly a nested amplification of 16 S rDNA was performed, whereby the whole sequence of 16 S rDNA was amplified using the universal primer pair GM3 and GM4 at low stringency (annealing temperature 45°C) in the first step. The product generated thereby was used as template for the nested amplification with the group specific SRB primers. A third round of amplification was applied using the DGGE primers (341F G + C and 518 R) and the product of the nested amplification as template to generate SRB group specific 16 S rDNA fragment suitable for gradient gel analysis. This reaction was carried out using a so called Touchdown protocol: 5 minutes at 94°C, followed by 11 cycles of 1 minute at 94°C, 1 minute 65°C (lowering the annealing temperature by 1°C in every cycle), and 1 minute at 72°C, followed by 19 cycles of 1 minute at 94°C, 1 minute at 55°C, and 1 minute at 72°C. The reaction was completed with a final extension step for 7 minutes at 72°C.

2.5 *Denaturing gradient gel electrophoresis (DGGE)*

A denaturing gradient of 35 to 75% of denaturant concentration (urea and formamide) was used for 6% (wt/vol) polyacrylamide gel. DGGE was performed in 1% TAE buffer at 60°C and with a constant voltage of 70 V for 16 hours.

3 RESULTS AND DISCUSSION

3.1 *Direct amplification of 16 S rDNA using group specific primers*

Direct amplification of DNA extracted from the reactor samples was attempted using the primers specific for different groups of SRB (Table 1). *Desulfovibrio-Desulfomicrobium* like products were amplified from three of the four reactor samples (Figure 1). No amplification was observed when other group specific primers were used for the amplification (data not shown). The results are summarized in (Table 2).

3.2 *Nested amplification of 16 S rDNA*

The nested amplification of the SRB specific 16 S rDNA yielded additional products that corresponded

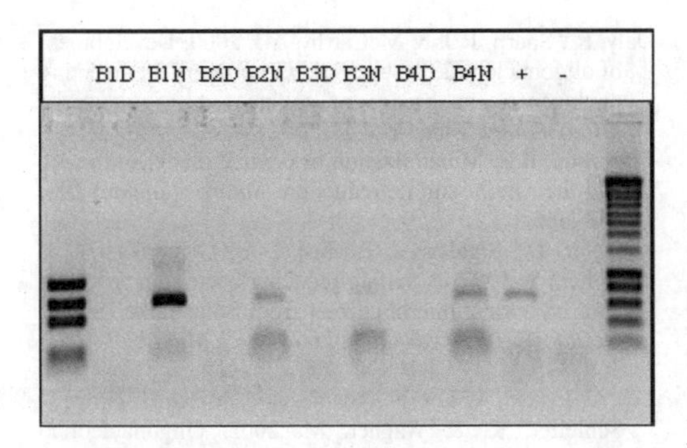

Figure 1. Direct and indirect amplification of 16 S rDNA from the genomic DNA extracted from reactor samples B1 to B4 using *Desulfotomaculum* group specific primers. 'D' stands for direct amplification 'N' for nested amplification, + for positive control and – for negative control.

Table 2. Results of a direct and nested PCR on the genomic DNA extracted from wastewater treatment reactors.

SRB group	Detection in reactor samples			
	B1	B2	B3	B4
(a) Direct amplification				
1. DFM	–	–	–	–
2. DBB	–	–	–	–
3. DBM	–	–	–	–
4. DSB	–	–	–	–
5. DCC-DNM-DSS	–	–	–	–
6. DSV-DMB	–	+	+	+
(b) Nested amplification				
1. DFM	+	+	+	+
2. DBB	+	–	–	–
3. DBM	–	–	–	–
4. DSB	–	–	–	–
5. DCC-DNM-DSS	+	+	+	+
6. DSV-DMB	+	+	+	+

DFM: *Desulfotomaculum*, DBB: *Desulfobulbus*, DBM: *Desulfobacterium*, DSB: *Desulfobacter*, DCC-DNM-DSS: *Desulfococcus-Desulfonema-Desulfosarcina*, DSV-DMB: *Desulfovibrio-Desulfomicrobium*.

to *Desulfotomaculum* (Figure 2), *Desulfobulbus* and *Desulfococcus-Desulfonema-Desulfosarcina* (data not shown) like amplification products. Amplification of 16 S rDNA products corresponding to *Desulfobacterium* and *Desulfobacter* was never achieved with either of the approaches (data not shown).

The results obtained using the two approaches of amplification suggest that there is only one dominant group of SRB i.e., the *Desulfovibrio-Desulfomicrobium* group in each of the reactors. The nested amplification revealed the presence of other groups that were not detected by the direct PCR. It may be inferred that the SRB groups that were detected only when a second

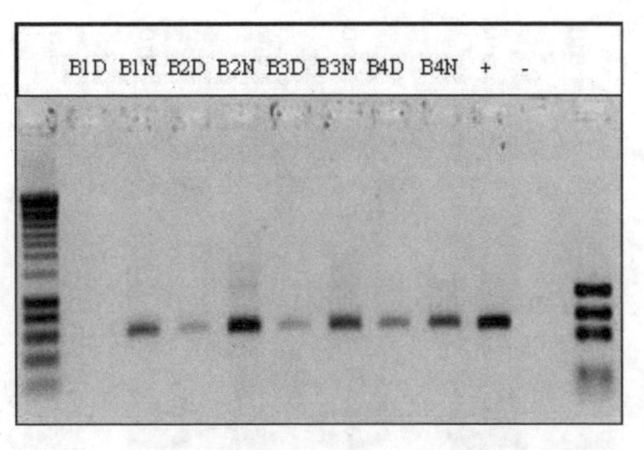

Figure 2. Direct and indirect amplification of 16 S rDNA from the genomic DNA extracted from reactor samples B1 to B4 using *Desulfovibrio-Desulfomicrobium* group specific primers. 'D' stands for direct amplification, 'N' for nested amplification, + for positive control and – for negative control.

round of nested PCR was employed are present in low numbers than the *Desulfovibrio-Desulfomicrobium* group detected by direct PCR.

3.3 *SRB diversity analysis using nested PCR DGGE*

The banding patterns generated by DGGE of the PCR products obtained by direct and indirect amplification were more or less identical, with only a slight degree of difference in the intensity of bands. The indirect amplification gave a more intense pattern. The indirect nested amplification used to generate the SRB 16 S rDNA fragments for DGGE analysis enabled the comparison of the profiles of different groups of SRB with that of the total bacterial community.

The DGGE pattern obtained gives an insight into the overall bacterial diversity within the reactors (Figure 3). It was found that SRB were not among the dominant microbial species. Only a few low intense bands could be identified as those belonging to SRB when direct amplification of 16 S rDNA approach was used. Since the reactors are methanogenic, the sulfate reducing bacteria were indeed expected to be low in number.

Figure 3A gives a comparison between the profiles of 16 S rDNA products obtained by the direct and indirect approach. The bands numbered were excised and sequenced. It was found that the two patterns are more or less identical. The indirect approach helped in resolving more and intense bands than the direct one. Figure 3B and 3C show a comparison in the profiles of whole microbial community and *Desulfotomaculum* and *Desulfovibrio-Desulfomicrobium* groups in reactor B1.

By using the nested PCR-DGGE approach it was possible to obtain an enhanced view of the genetic

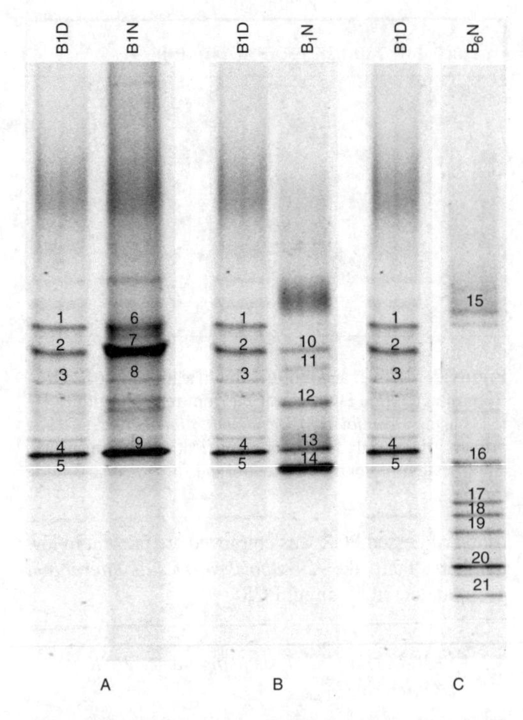

Figure 3. (A) Comparison of DGGE patterns of PCR-amplified 16 S rDNA fragments obtained from DNA extract of reactor B1 by direct (B1D) and nested (B1N) approach. (B) Comparison of the DGGE profile of whole microbial community (B1D) and *Desulfotomaculum* group of SRB (B_1N). (C) Comparison of the DGGE profile of whole microbial community (B1D) and *Desulfovibrio-Desulfomicrobium* group of SRB (B_6N).

diversity of SRB. By using the nested PCR DGGE approach we could detect the SRB, which was not possible through direct method. Similar findings were made with other groups of SRB and for other reactor samples, (data not shown). The presented approach is a welcome addition to the genomic toolbox for studying the diversity of sulfate reducing bacteria in natural and engineered environments.

REFERENCES

Amann, R.I., Stomley, J., Devereux, R.K. & Stahl, D.A. 1992. Molecular and microscopic identification of sulfate reducing bacteria in multispecies bioflims. *Appl. Environ. Microbiol.* 58: 614–623.

Daly, K., Sharp, R.J. & McCarthy, A.J. 2000. Development of oligonucleotide probes and PCR primers for detecting phylogenetic subgroups of sulfate reducing bacteria. *Microbiology.* 146: 1693–1705.

Jørgensen, B.B. Mineralization of organic matter in the sea bed-the role of sulfate reduction. *Nature* (London) 296: 643–645.

Krekeler, D., Sigalevich, P., Teske, A., Cypionka, H. & Cohen, Y. 1997. A sulfate-reducing bacterium from the oxic layer of a microbial mat from Solar Lake (Sinai), *Desulfovibrio oxyclinae* sp.nov. *Arch.Microbiol.* 167: 369–375.

Loy, A., Lehner, A., Lee, N., Adamczyk, J., Meier, H., Ernst, J., Schleifer, K. & Wagner, M. 2002. Oligonucleotide microarray for 16 S rRNA gene-based detection of all recognised lineages of sulfate-reducing prokaryotes in the environment. *Appl. Environ. Microbiol.* 68: 5064–5081.

Muyzer, G., De Waal, E.D. & Uitterlinden, A.G. 1993. Profiling of complex microbial populations by denaturing gradient gel electrophoresis analysis of polymerase chain reaction-amplified genes coding for 16 S rRNA. *Appl. Environ. Microbiol.* 59, 695–700.

Muyzer, G., Teske, A., Wirsen, C.O. & Jannasch, H.W. 1995. Phylogenetic relationships of *Thiomicrospira* species and their identification in deep-sea hydrothermal vent samples by denaturing gradient gel electrophoresis of 16 S rDNA fragments. *Arch. Micrbiol.* 164: 165–172.

Okabe, S., Itoh, T., Satoh, H. & Watanabe, Y. 1999. Analysis of spatial distribution of sulfate reducing bacteria and their activity in aerobic wastewater biofilms. *Appl. Environ. Microbiol.* 65: 5107–5116.

Ramsing, N., Kühl, M. & Jørgensen, B.B. 1993. Distribution of sulfate reducing bacteria, O_2 and H_2S in photosynthetic biofilms determined by oligonucleotide probe and microelectrodes. *Appl. Environ. Microbiol.* 59: 3840–3849.

Raskin, L., Amann, R.I., Poulsen, L.K., Rittmann, B.E. & Stahl, D.A. 1995. Use of ribosomal RNA based molecular probes for characterization of complex microbial communities in anaerobic biofilms. *Water Science Technology.* 31: 261–272.

Santegoeds, C.M., Damgaard, L.R., Hesselink, G., Zopfi, J., Lens, P., Muyzer, G. & de Beer, D. 1999. Distribution of sulfate reducing and methanogenic bacteria in anaerobic aggregates determined by microsensor and molecular analysis. *Appl. Environ. Microbiol.* 65: 4618–4629.

Sass, H., Cypionka, H. & Babenzien, H.D. 1997. Vertical Distribution of sulfate reducing bacteria at the oxic-anoxic interface in sediments of the oligotrophic Lake Stechlin. *FEMS Microbiology Ecology.* 22: 245–255.

Voordouw, G., Niviere, V., Ferris, F.G., fedorak, P.M. & Westlake, D.W.S. 1990. Distribution of hydrogenase genes in *Desulfovibrio* spp. and their use in identification of species from the oil field environment. *Appl. Environ. Microbiol.* 56: 3748–3754.

European Symposium on Environmental Biotechnology, ESEB 2004 - Verstraete (ed)
 ISBN 90 5809 653 X

Microbial characterization of TNT-contaminated soils and anaerobic TNT degradation: high and unusual denitration activity

L. Eyers, B. Stenuit, S. El Fantroussi & S.N. Agathos
Unit of Bioengineering, Catholic University of Louvain, Louvain-la-Neuve, Belgium

ABSTRACT: 2,4,6-trinitrotoluene (TNT) is one of the most common explosives. In spite of its known toxicity and mutagenicity for many organisms, soils and groundwater are still frequently contaminated at manufacturing, disposal and TNT-destruction sites. Thirteen soil samples collected from a TNT-destruction field were investigated for their concentrations of TNT and its derivatives as well as for their microbial diversity. They contained TNT at various concentrations, ranging from 4 to 36 g TNT/kg soil. Various metabolites were also detected, including 2,4,6-trinitrobenzaldehyde, 4-amino-2,6-dinitrotoluene, 2-amino-4,6-dinitrotoluene and 2,4-dinitrotoluene. Microbial communities of these soil samples were characterized by 16S rRNA PCR-DGGE using bacterial universal primers. The DGGE patterns of these soil communities were compared with non-contaminated soils found at the same TNT-destruction site. Non-polluted soils revealed complex fingerprints of microorganisms while the contaminated samples showed the presence of dominant bands, indicating that a strong selection had occurred. The biodegradation capacities of a non-polluted and a TNT-contaminated soil were evaluated by enrichment cultures with TNT as sole N-source. High and unusual denitration activities were detected with the contaminated soil under defined anaerobic conditions.

LIST OF ABBREVIATIONS

TNT – 2,4,6-trinitrotoluene
TNBA – 2,4,6-trinitrobenzaldehyde
4-A-2,6-DNT – 4-amino-2,6-dinitrotoluene
2-A-4,6-DNT – 2-amino-4,6-dinitrotoluene
2,6-DA-4-NT – 2,6-diamino-4-nitrotoluene
2,4-DA-6-NT – 2,4-diamino-6-nitrotoluene
2,4-DNT – 2,4-dinitrotoluene
DGGE – denaturing gradient gel electrophoresis
HPLC – high performance liquid chromatography

1 INTRODUCTION

2,4,6-trinitrotoluene (TNT) is a common explosive that has been extensively used both for military and civilian applications. Soils and groundwater are polluted by this compound and its reduced derivatives at manufacturing, disposal and TNT-destruction sites. Moreover, it has been described as toxic for many living organisms, from microorganisms to humans. Mutagenic effects have also been demonstrated by the Ames test. Therefore, efficient remediation technologies are needed for the decontamination of TNT-polluted sites.

In this respect, we first evaluated the impact of the toxicity of TNT and its derivatives on soil microbial communities by comparing fingerprints of non-contaminated and polluted soils using DGGE (Denaturing Gradient Gel Electrophoresis). Clear discrimination between microbial populations of polluted and non-polluted soils was observed.

Given these previous results, we investigated the possible selection of TNT-degrading strains in polluted soils. This was done by enrichment cultures under various electron acceptor conditions, in the presence of a mixture of C-sources, and with a non-polluted soil (control) or a contaminated soil as inoculum. Under defined anaerobic conditions, high denitration activities were observed. This process offers the advantage of preferential denitration of the TNT molecule to easily mineralizable metabolites, instead of reduction of the nitro groups to toxic and bottleneck compounds.

2 MATERIALS AND METHODS

2.1 *Chemicals*

TNT was obtained from Nobel Explosives (Chatelet, Belgium). 2-A-4,6-DNT and 4-A-2,6-DNT were synthesized as previously described (Van Aken and Agathos 2002). 2,4-DA-6-NT was graciously donated by Dr. D. Bruns-Nagel (University of Marburg, Germany). TNBA was kindly supplied by M. Régis (GIAT Industries, France). Other nitroaromatic compounds were purchased from Sigma (St Louis, MO).

2.2 *Soil samples*

TNT-contaminated soils of various compositions were collected in a field located in France that had been used for TNT destruction over the past two decades. A non-contaminated soil of the same composition was sampled in the same location, a few meters from a contaminated one. All these samples were collected at the surface of the soil.

2.3 *Analytical methods*

TNT and related aromatic compounds were extracted from soil samples using a protocol adapted from the US-EPA 8330 method. Briefly, soil samples were dried overnight on silica gel at room temperature. Next, 10 ml of acetonitrile was added to 2 g of soil and the mixture was vigorously shaken for 18 hr. Five ml of the supernatant was withdrawn and mixed with 5 ml of a 5 g/l $CaCl_2$ solution. After centrifugation, the samples were analyzed by HPLC as described below. For liquid cultures, TNT and other derivatives were directly extracted after sampling by mixing 1 ml of the liquid culture with 1 ml of acetonitrile. After centrifugation, samples were analyzed by HPLC.

Quantitative determination of TNT and related aromatic compounds was done on a HPLC system (Waters 600E Multisolvent Delivery System, Waters 660 Controller and Waters 717 Autosampler (Waters, Milford, MA)). The column was a 4 μm C18 Nova-Pak, 3.9 ×300 mm (Waters). The mobile phase consisted of 30% acetonitrile and 70% phosphate buffer (10 mM, pH 3.2) and the flow rate was maintained at 1 ml/min. After 2 min, a linear gradient was applied to reach a composition of 90% acetonitrile and 10% phosphate buffer in 18 min. Then, a linear gradient was set for 5 min to reach the initial mobile phase composition and this composition was maintained for 15 min. Compounds were monitored by a Waters 770 photodiode array detector. TNT metabolites were identified by comparing their UV spectra with authentic standards. These standards were also used for quantification studies.

For nitrite, samples were centrifuged and the concentration of nitrite was colorimetrically determined by reaction with sulphanilic acid and N-1-naphthylethylenediamine dihydrochloride using a nitrite Spectroquant® kit from Merck (Darmstadt, Germany). A nitrite standard (Merck) was used for quantification.

2.4 *Denaturing Gradient Gel Electrophoresis*

Nucleic acids were extracted from soil samples with a UltraClean Soil DNA kit from MoBio (MoBio, Solana Beach, CA). The 16S rRNA genes were amplified by PCR using primers P338f (5′-ACTCCTACGGG-AGGCAGCAG-3′, forward) and P518r (5′-ATTAC-CGCGGCTGCTGG-3′). The forward primer contained a 40 bp GC-clamp attached to its 5′ end. The PCR mixture contained 50 μl of 2 X Red'y'Star mix (Eurogentec, Seraing, Belgium), 0.25 μM of each primer, approximately 50 ng of DNA and sterile water to a final volume of 100 μl. A Gene Amp 2400 PCR thermal cycler was used (Perkin-Elmer, Norwalk, CT). The program of temperature was as follows: 10 min at 95°C, followed by 30 cycles of 94°C for 1 min, 55°C for 1 min, and 72°C for 1 min, with a final extension at 72°C for 10 min.

DGGE (Denaturing Gradient Gel Electrophoresis) was realised on a D-Code System (Bio-Rad, Hercules, CA). PCR samples were loaded on a 6% polyacrylamide gel with a denaturing gradient from 35% to 55%. The electrophoresis was run for 7 hr at 150 V in a 0.5 X TAE buffer. During the run, the temperature of the buffer was maintained at 60°C. After the electrophoresis, the gels were soaked for 15 min in 0.5 X TAE buffer and ethidium bromide (0.5 mg/l), followed by 15 min in 0.5 X TAE buffer. The stained gels were immediately photographed on a transillumination table.

2.5 *Media*

M8 minimal medium (Haïdour and Ramos 1996) was prepared by using standard anaerobic techniques. Argon was passed through heated reduced copper filings to remove traces of oxygen. This oxygen-free argon was used for flushing boiling M8 medium to remove oxygen. The medium contained 1 g/l TNT as sole N-source (i.e. in excess of its solubility in water). The C-sources were glucose, acetate, succinate and pyruvate (6 mM each). The inoculum consisted of 2 g of soil for 150 ml of medium. The biotic degradation of TNT was investigated in the presence of different electron acceptors. For abiotic controls, $HgCl_2$ (200 ppm) was provided together with TNT, soil inoculum and electron acceptors, except for C-sources that were omitted. After addition of all components mentioned above, the pH was adjusted to 6.8. Each bottle of enrichment medium was sealed with a rubber stopper and an aluminium crimp. The enrichments were incubated in the dark at 35°C and shaken at 170 rpm. Experiments were done in triplicate.

3 RESULTS

3.1 *Concentration of TNT and derivatives in soil samples*

TNT and its relative metabolites (2,4-dinitrotoluene, 4-amino-2,6-dinitrotoluene, 2-amino-4,6-dinitrotoluene and 2,4,6-trinitrobenzaldehyde) were extracted from various soil samples collected at a TNT-destruction site (Table 1).

All samples were highly contaminated (TNT concentration ranging from 4 to 36 g/kg of soil), except soil samples 'KF4' and 'Ref'. No TNT metabolites were detected for these two samples, whereas they were measured at various concentrations in other soil samples. These TNT metabolites were at concentrations three orders of magnitude lower than the concentration of TNT itself.

3.2 *Denaturing Gradient Gel Electrophoresis*

DNA from soil samples was extracted using a bead-beating method (Materials and Methods). Universal and bacterial specific primers were used for the PCR and PCR products were loaded on a DGGE (Figure 1).

Non-polluted ('Ref') and faintly polluted ('KF4') soil samples featured a complex fingerprint (Figure 1). However, a strong bacterial selection was apparent in the TNT-polluted soil samples, as indicated by the presence of fewer but brighter bands. Interestingly, some polluted soil samples showed the presence of dominant bands at the same position.

Table 1. Concentration of TNT and related metabolites.

	Concentration (mg/kg soil)				
Sample	TNT	2,4-DNT	4-A-2,6-DNT	2-A-4,6-DNT	TNBA
KF1	13 333	21.2	3.7	5.0	11.8
KF2	4 116	3.7	3.6	4.9	4.4
KF3	5 207	7.4	1.6	2.4	9.7
KF4	2	ND	ND	ND	ND
KF5	6 386	4.7	4.7	7.3	10.0
KF6	19 534	11.5	23.4	34.1	23.9
KF6b	12 982	8.5	11.3	15.7	19.9
KF6c	19 151	12.1	15.5	21.4	26.3
Ref*	ND	ND	ND	ND	ND
KX1*	26 562	19.1	13.9	21.0	33.8
KL1	36 175	13.5	15.7	20.8	28.8
KL2	23 023	21.9	39.7	64.6	17.9
KFc1	28 030	38.1	10.0	4.9	59.9

ND: Not detected.
* Samples of same composition.

3.3 *Enrichment cultures under defined anaerobic conditions*

Enrichment cultures were set up under different electron acceptor conditions with the 'Ref' (non-polluted soil, Table 1) and the 'KX1' (TNT-contaminated soil, Table 1) samples as inocula. TNT was the sole N-source. Under defined anaerobic conditions, nitrite release was significant. Indeed, up to 4869 μM of nitrite was detected with the polluted soil as inoculum (Figure 2).

With the non-polluted soil inoculum, nitrite release was 475 μM after 112 days. Abiotically and with the non-polluted and the contaminated inocula, nitrite release was respectively 98 and 293 μM after 111 days. This suggests that nitrite release also occurred abiotically. However, these abiotic denitration activities were much lower than what was observed biotically.

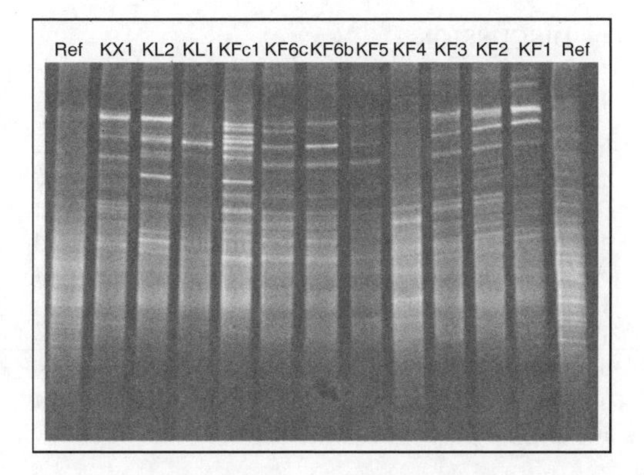

Figure 1. 16S PCR-DGGE of extracted DNA from polluted soil samples.

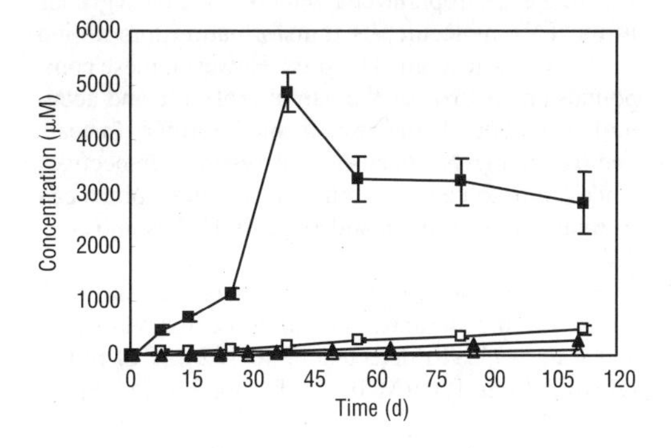

Figure 2. Nitrite release under defined anaerobic conditions with TNT (4405 μM) as sole N-source and non-polluted (□) or polluted soil samples as inoculum (■). Abiotic control (200 ppm $HgCl_2$ and omission of C-sources), non-polluted (Δ) and polluted soil sample (▲). Error bars represent the standard deviation of triplicate cultures (not visible when the value is low).

After 56 days, TNT was not detected anymore in the enrichment cultures with the contaminated soil inoculum whereas TNT was still present with the non-polluted inoculum. Aminodinitrotoluene (ADNT) and diaminonitrotoluene (DANT) compounds were also produced in both enrichments and accumulated in the medium. Mass balances after complete disappearance of TNT indicated that 30% of it was transformed to these aminodinitro- and diaminonitrotoluene compounds.

Among other identified TNT metabolites, 3,5-dinitroaniline and trinitrobenzaldehyde were detected. 2,4-dinitrotoluene was also observed abiotically in low concentrations with the non-polluted and the contaminated soil inocula (i.e., respectively 1.1 μM and 0.9 μM after 50 days). Other polar and apolar peaks were produced but could not be identified. 2,6-dinitrotoluene and 2-nitrotoluene were not detected.

4 DISCUSSION

4.1 *Microbial ecology of TNT-contaminated soils*

In the present study, clear discrimination between contaminated and non-polluted soils could be achieved by DGGE. These evident modifications of the bacterial diversity in the presence of TNT might be explained by the toxic effects of TNT and the selection of TNT-resistant strains and/or TNT-degrading strains. Therefore, DGGE might be utilized during remediation processes to evaluate the extent of the decontamination and restoration of the bacterial diversity to a state such as that in uncontaminated areas.

4.2 *Denitration of TNT under defined anaerobic conditions*

Due to the electrophilic character of three nitro groups in the TNT molecule, its transformation into amino derivatives is generally favoured. However, these compounds are as toxic as the parent molecule and accumulate without being transformed further. Instead, denitration of TNT opens up interesting perspectives since dinitrotoluene and nitrotoluene metabolites can be mineralized (Haigler and Spain 1993; Nishino et al. 2000).

Few reports describe the release of nitrite from TNT under aerobic or anoxic conditions. Esteve-Nunez and Ramos (1998) detected 19 to 102 μM of nitrite released from 100 μM of TNT, Fiorella and Spain (1997) 40 μM and Kalafut et al. (1998) 13 μM (data adapted from the literature). To the best of our knowledge, denitration activities have never been reported under strict anaerobic conditions. In the present study, a maximum of 4869 μM of nitrite was produced with the polluted inoculum under defined anaerobic conditions. If the quantity of TNT exogenously added and the TNT initially present in the soil used as inoculum are considered, this is 82 μM of nitrite released from 100 μM of TNT. Since part of the TNT was also reduced to non productive ADNT and DANT compounds that accumulate, not all of this 100 μM of TNT undergoes denitration. If the amount of TNT leading to TNT-reduced compounds is taken into account, the denitration observed amounts to 117 μM of nitrite released per 100 μM of TNT.

The denitration mechanism taking place under our defined anaerobic conditions is far from clear. However, the process of the present study involved both high initial TNT concentrations and high nitrite release and therefore offers an exciting opportunity for the treatment of sites highly contaminated with TNT.

REFERENCES

Esteve-Nunez, A. and Ramos, J.L. 1998. Metabolism of 2,4,6-trinitrotoluene by *Pseudomonas* sp. JLR11. *Environ Sci Technol* 32: 3802–3808.

Fiorella, P.D. and Spain, J.C. 1997. Transformation of 2,4,6-trinitrotoluene by *Pseudomonas pseudoalcaligenes* JS52. *Appl Environ Microbiol* 63: 2007–2015.

Haïdour, A. and Ramos, J.L. 1996. Identification of products resulting from the biological reduction of 2,4,6-trinitrotoluene, 2,4-dinitrotoluene, and 2,6-dinitrotoluene by *Pseudomonas* sp. *Environ Sci Technol* 30: 2365–2370.

Haigler, B.E. and Spain, J.C. 1993. Biodegradation of 4-nitrotoluene by *Pseudomonas* sp. strain 4NT. *Appl Environ Microbiol* 59: 2239–2243.

Kalafut, T., Wales, M.E., Rastogi, V.K., Naumova, R.P., Zaripova, S.K. and Wild, J.R. 1998. Biotransformation patterns of 2,4,6-trinitrotoluene by aerobic bacteria. *Curr Microbiol* 36: 45–54.

Nishino, S.F., Paoli, G.C. and Spain, J.C. 2000. Aerobic degradation of dinitrotoluenes and pathway for bacterial degradation of 2,6-dinitrotoluene. *Appl Environ Microbiol* 66(5): 2139–2147.

Van Aken, B. and Agathos, S.N. 2002. Implication of manganese (III), oxalate, and oxygen in the degradation of nitroaromatic compounds by manganese peroxidase (MnP). *Appl Microbiol Biotechnol* 58: 345–351.

European Symposium on Environmental Biotechnology, ESEB 2004 - Verstraete (ed)

Interactions between microorganisms and heavy metals

A.C.M. Toes, B.A. Maas, J.S. Geelhoed, J.G. Kuenen & G. Muyzer
Dept. of Biotechnology, Delft University of Technology, Delft, The Netherlands

ABSTRACT: Many sediments in coastal waters are contaminated with various metals. Microorganisms in these sediments can exert great influence on the bioavailability of metal compounds. The European Union project TREAD (Transport, REActions and Dynamics of heavy metals in marine sediments) aims to identify the physical, chemical and biological parameters influencing heavy metal mobility. This paper is focusing on the microorganisms that play a role in the transformations of heavy metals in marine sediments. The typical characteristics of heavy metal compounds are introduced, followed by a discussion on their toxic effects and the strategies microbes employ in order to gain a certain level of resistance. Possible interactions between microbe and metal within marine sediments are considered, including the decomposition of organic matter, reduction and reoxidation of metal oxides, as well as the formation of sulfides. Subsequently, the TREAD project is described in more detail followed by a discussion of comparable studies. This paper is concluded with an overview of the experimental tools that will be applied to study the microorganisms involved in heavy metal transformations.

1 HEAVY METALS AND MICROBES

1.1 *Characteristics of heavy metals*

Heavy metals are metals with a density above 5 g/cm^3. This definition leads to the conclusion that of the 90 naturally occurring elements, 53 (with As included) belong to the notorious family of heavy metals. Of practical interest are especially chromium, manganese, iron, cobalt, nickel, copper, zinc, cadmium, mercury and lead and the metalloid arsenic, due to their abundance in the environment as a result of industrial activities. These metals and metalloids exist mostly as cations and/or oxyanions, in aqueous solution, and as crystalline salts, oxides or as amorphous precipitates in insoluble form (Ehrlich 1997).

1.2 *Microbial toxicity*

Though all microbes employ metal species to some extent for structural and/or catalytic functions, high intracellular concentrations of these metals cause formation of complex compounds, which in turn leads to toxic effects. This toxicity lies foremost in the incapacity of passive transport systems within the cell membrane to distinguish between structurally very similar compounds. This is the case for divalent cations such as Mn^{2+}, Fe^{2+}, Co^{2+}, Cu^{2+} and Zn^{2+} for example, which carry the same charge and possess roughly the same ionic diameter. But the observation is also valid for oxyanions like chromate, which resembles sulfate with the four tetrahedrally arranged oxygen atoms, and for arsenate and phosphate (Nies 1999). Especially the heavy metals with a high atomic number, e.g. Hg^{2+} and Cd^{2+}, exhibit another toxic effect on microbial cells, namely the specific binding to SH groups, causing inactivation of enzymes. Additionally, heavy metal ions may interact with physiologically important ions, for example, Cd^{2+} with Ca^{2+}, and Ni^{2+} or Co^{2+} with Fe^{2+}, thereby inhibiting their function in the cell. Occasionally, radical formation occurs by oxygenation of metal complexes, or by reduction of the oxyanion, as is the case for chromate.

1.3 *Resistance mechanisms*

In general, there are three possible strategies for a microorganism to decrease the toxicity of heavy metals. Firstly, specific efflux pumps can decrease the internal metal concentration at the cost of ATP hydrolysis. Secondly, heavy metal cations with high atomic numbers can be complexed with thiol-containing molecules. Thirdly, some metal ions may be reduced to less toxic species (Nies 1999).

Lately, a lot of research has been focused on the isolation of bacteria with high tolerance or resistance to metal species (Ahmed et al. 2001, Ivanova et al. 2002, Rasmussen & Sörensen 1998, Rosen 2002, Shi et al. 2002 and Silver 1996), therewith increasing our knowledge on specific transport systems and detoxification mechanisms. Moreover, the possibility to detect enzymes involved in detoxification on the DNA level, as demonstrated for a cadmium transporter by Oger et al. (1997), and the availability of genomic sequence information of a multiple metal-resistant bacteria, i.e. *Ralstonia metallidurans* (Mergeay et al. 2003), will facilitate a more wide spread and efficient screening.

1.4 *Microbial interactions with heavy metals*

Despite the briefly discussed toxic effects of heavy metal ions on the microbial cell, microorganisms need small amounts of certain metal species for basic metabolism. Besides that, an increasing number of microbes are being described that are actually dependent on metal ions for energy source, albeit on a facultative basis. Additionally, there are some nonspecific and/or indirect biochemical reactions that can take place which also influence the speciation of metal ions and therewith their bioavailability (Unz & Shuttleworth 1996).

The main mechanisms by which microorganisms affect changes in the speciation and mobility of metals are listed below, including some examples (Gadd et al. 2001). These processes can be summarized in a scheme as depicted in Figure 1.

- Enzymatic detoxification mechanisms:
  – Transformation reactions, including oxidation (AsO_2^- to AsO_4^3), reduction (CrO_4^{2-} to Cr^{3+} or Hg^{2+} to Hg^0) and methylation
  – Modification/elimination of membrane transport systems into or out of the cell

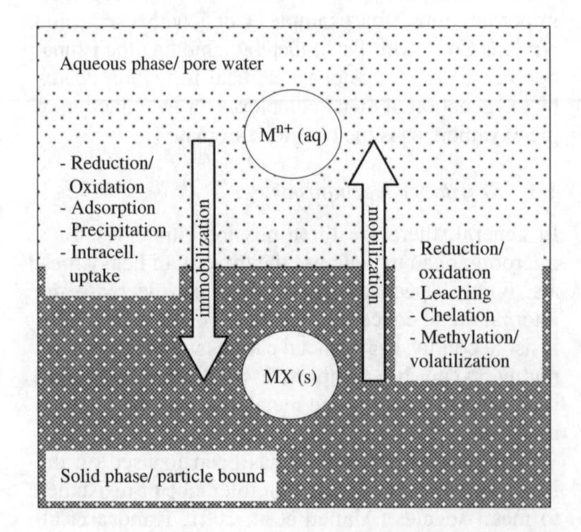

Figure 1. Schematic presentation of microbial influences on processes leading to either mobilization or immobilization of heavy metals.

- Incorporation into metalloenzymes:
  – Uptake may require specific or non-specific transport systems e.g. corA Mg^{2+} or siderophores
- Utilization as electron donor or acceptor:
  – Oxidation by chemolithotrophs e.g.: Fe^{2+} to Fe^{3+} (*T. ferrooxidans, L. ferrooxidans, A. brierleyi, S. acidocaldarius*), Sb_2O_3 to Sb_2O_5 (*S. senarmontii*), AsO_2^- to AsO_4^{3-} (*P. arsenitoxidans*)
  – Reduction by heteretrophs or autotrophs e.g.: Fe^{3+} to Fe^{2+}, Fe_3O_4, $FeCO_3$ (*S. putrefaciens*) MnO_2 to Mn^{2+}, $MnCO_3$ (*G. metallireducens*), SeO_4^{2-}, SeO_3^{2-} to S^0 (*T. selenati*), MoO_4^{2-} to a lower oxidation state (*Sulfolobus* sp.), CrO_4^{2-} to Cr^{3+} (*P. fluorescens*)
- Non-enzymatic processes:
  – Binding or accumulation of metal cations to the cell surface
  – Promotion of leaching by metabolite products (acids or ligands)
  – Precipitation of metals by metabolic products as sulfide, carbonate or phosphate ions

A number of recent publications describe the characterization of bacteria that show interactions with heavy metals belonging to the two latter categories listed above. The utilization of arsenite as electron donor by a chemoautotroph, for instance, was described by Oremland et al. (2002). Characterization of the typical parameters that induce the process of chromate reduction in *Shewanella oneidensis* was performed by Viamajala et al. (2002). More general information on metal reduction by microorganisms can be found in the publications of Lloyd et al. (2001) and Nealson & Cox (2002).

Among the non-enzymatic processes involving changes in metal mobility, the immobilization of cobalt by sulfate reducing bacteria (Krumholz et al. 2003), the mobilization of lead cations by microbial leaching processes (Fein et al. 1999) and the precipitation of cadmium sulfides by a *Klebsiella planticola* strain (Sharma et al. 2000) have been described. Application of these types of microbes in bioremediation strategies has received ample attention in a number of reviews (Lovley & Coates 1997) and (Macaskie et al. 2001).

2 TREAD: TRANSPORT, REACTIONS AND DYNAMICS OF HEAVY METALS IN CONTAMINATED SEDIMENTS

2.1 *Problem description*

Many of the sediments in coastal waters are contaminated with various metals. Microorganisms in these sediments can exert great influence on the bioavailability of these metal compounds, as discussed previously. The transformations of heavy metals in marine sediments depend on the biological activity affecting decomposition of organic matter, reduction and re-oxidation of metal oxides, and the formation of sulfides. While marine sediments generally serve as an important sink for dissolved heavy metals, sediment-bound heavy metals may be resupplied to interstitial and overlying waters if physical or chemical conditions change. These changes are important in coastal regions and even more in harbors, because these are areas of high sedimentation and erosion caused by industrial activities. Concomitantly, dredging of sediments may have a large impact on the mobility of heavy metals.

2.2 *TREAD consortium*

This EU project involves several partners from different institutes and universities in Europe. The Max-Planck-Institute for Marine Microbiology in Bremen, Germany, provides knowledge and expertise in the construction and maintenance of mesocosms, used to simulate the natural habitat as much as possible. Research at the Environmental Science department of Lancaster University, UK, has specialized in the detection of metals and metalsulfides at the microscale range. Scientists at the Marine Biological Laboratory, University of Copenhagen, Helsingör, Denmark, have developed a number of devices such as specialised microsensor equipment and planar optrodes to enable detailed in situ measurements. Finally, research at the department of Environmental Biotechnology, University of Technology, Delft, the Netherlands, is dedicated to the integration of information from the fields of molecular ecology and microbial physiology.

2.3 *Objectives*

The overall objective of this research project is to understand and numerically model heavy metal interactions and transport in three dimensions at the microscale that controls the processes. This will enable regulations and practice to be founded on process understanding by establishing a network between decision makers, end-users and scientists.

The objectives of the microbiological part of this research are (i) to characterize bacterial communities from both pristine and contaminated marine sediments, (ii) to isolate species involved in heavy metal transformations, and (iii) to assess the changes in microbial community upon heavy metal stress.

3 SCIENTIFIC APPROACH

3.1 *Comparable research*

One of the earlier publications on the subject of heavy metals and the effects on a microbial population describes a rather general investigation of the consequences of zinc and copper pollution in soils on enzyme activity (Hemida et al. 1995). Roane & Kellog (1995) performed a thorough investigation on the effects of the presence of cadmium and lead on bacterial enumeration studies, metabolic activity and community response in soils with different physical and chemical characteristics. Restriction fragment length polymorphism analysis was applied to detect shifts in a microbial community after copper contamination (Smit et al. 1997). An array of different molecular tools was applied by Sandaa et al. (1999a) in order to assess the differences in bacterial soil communities upon heavy metal contamination. In another publication they describe the abundance and diversity of *Archaea* in heavy-metal contaminated soils (1999b). Kandeler et al. (2000) analyzed enzyme activities, amounts of biomass and diversity of molecular community of soils that had been exposed to a combination of four heavy metals over a period of 10 years.

Despite these efforts to describe possible interactions between metals and microbial populations, no previous research has focused on these processes in a marine environment, where the high sulfate and salt concentrations may affect the importance of the various interactions listed in paragraph 1.4 significantly. Additionally, the correlation between the culturable isolates and the molecularly detected species is often lacking. This is of course partially a consequence of "the Great Plate Count Anomaly" (Staley & Konopka, 1985), which will especially apply to marine bacteria due to the oligotrophic conditions and the fact that marine (pelagic) bacteria continuously seem to exist in a low-activity state (Eguchi 1999, Zengler 2002). Therefore, the work of Ellis et al. (2003), showing that a combination of cultivation-dependent and independent approaches provided the most reliable results regarding metal-induced diversity analysis, will serve as an example for our study.

3.2 *Experimental outline*

Our approach will be to study bacterial communities at different levels of resolution. At the MPI in Bremen four mesocosms were constructed (large scale, i.e. $V \approx 160\,dm^3$) and filled with sediments from the metal-contaminated harbor of Bremerhaven and the pristine coast of the Isle of Sylt (both in Germany). In Delft, microcosms (small scale, i.e. $V \approx 8\,dm^3$) were manufactured, that were also filled with the above-mentioned sediments. For comparative reasons, marine sediments from two additional metal-contaminated sites were sampled, i.e. the marina in Helsingör (Denmark) and the industrial harbor of Rotterdam (the Netherlands). Anaerobic batch flasks and agar gradient systems were inoculated with samples from these four sediments ($V \approx 0.02\,dm^3$).

Molecular tools, including PCR-DGGE (Polymerase Chain Reaction Denaturing Gradient Gel Electrophoresis (Schäfer & Muyzer 2001)) and two modified FISH protocols (i.e. CARD-FISH, CAtalyzed Reporter Deposition Fluorescent In Situ Hybridization (Pernthaler et al. 2002) and FISH-MAR, FISH-Micro Auto Radiography (Lee et al. 1999)), will be used to analyze samples from the environment directly and from the mesocosms, microcosms and cultures.

This experimental set-up facilitates a multidisciplinary approach, in which both reproducibility and flexibility are warranted, due to the continuous nature of the larger scale systems on the one hand, and the possibility to vary experimental conditions in replicates on the other hand. The challenge lies in linking the structure of microbial communities to the actual role microbes play in marine heavy metal transformations.

REFERENCES

Ahmed, N. et al. (2001). Resistance and accumulation of heavy metals by indigenous bacteria: bioremediation. In N. Ahmed et al. (eds) *Industrial and environmental biotechnology*: 81–102. Wymondham, Horizonpress.

Eguchi, M. 1999. The nonculturable state of marine bacteria. In *Microbial biosystems: New frontiers, proceedings of the 8th international symposium on microbial ecology*, Halifax, Canada.

Ehrlich, H.L. 1997. Microbes and metals. *Applied Microbiology Biotechnology* 48: 687–692.

Ellis, R.J. et al. 2003. Cultivation-dependent and -independent approaches for determining bacterial diversity in heavy-metal-contaminated soil. *Applied Environmental Microbiology* 69(6): 3223–3230

Fein, J.B. et al. 1999. Bacterial effects on the mobilisation of cations from a weathered Pb-contaminated andesite. *Chemical Geology* 158: 189–202.

Gadd, G.M. et al. (2001). Microbial processes for solubilisation or immobilisation of metals and metalloids and their potential for environmental bioremediation. In N. Ahmed et al. (eds) *Industrial and environmental biotechnology*: 55–80. Wymondham, Horizonpress.

Hemida, S.K. et al. 1997. Microbial populations and enzyme activity in soil treated with heavy metals. *Water, Air and Soil Pollution* 95: 13–22.

Ivanova, E.P. et al. 2002. Tolerance to cadmium of free-living and associated with marine animals and eelgrass marine marine gamma-proteobacteria. *Current Microbiology* 44: 357–362.

Kandeler, E. et al. 2000. Structure and function of the soil microbial community in microhabitats of a heavy metal polluted soil. *Biology and Fertility of Soils* 32: 390–400.

Krumholz, L.R. et al. 2003. Immobilisation of cobalt by sulphate-reducing bacteria in subsurface sediments. *Geomicrobiology Journal* 20: 61–72.

Lee, N. et al. 1999. Combination of fluorescent in situ hybridisation and microautoradiography – a new tool for structure-function analysis in microbial ecology. *Applied Environmental Microbiology* 65(3): 1289–1297.

Lloyd, J.R. et al. 2001. Metal reduction by sulphate-reducing bacteria: physiological diversity and metal specificity. *Hydrometallurgy* 59: 327–337.

Lovley, D.R. et al. 1997. Bioremediation of metal contamination. *Current Opinion in Biotechnology* 8: 285–289.

Macaskie, L.E. et al. (2001). Application of microorganisms to the decontamination of heavy metal bearing wastes. In N. Ahmed et al. (eds) *Industrial and environmental biotechnology*: 103–112. Wymondham, Horizonpress.

Mergeay, M. et al. 2003. *Ralstonia metallidurans*, a bacterium specifically adapted to toxic metals: towards a catalogue of metal-responsive genes. *FEMS Microbiology Reviews* 27: 385–410.

Nealson, K.H. & Cox, B.L. 2002. Microbial metal-reduction and Mars: extraterrestrial expectations? *Current Opinion in Microbiology* 5: 296–300.

Nies, D.H. 1999. Microbial heavy-metal resistance. *Applied Microbiology Biotechnology* 51: 730–750.

Oger, C. et al. 2003. Distribution and diversity of a cadmium resistance (cadA) determinant and occurrence of IS257 insertion sequences in Staphylococcal bacteria isolated from a contaminated estuary (Seine, France). *FEMS Microbiology Ecology* 43: 173–183.

Oremland, R.S. et al. 2002. Anaerobic oxidation of arsenite in Mono Lake water and by a facultative, arsenite-oxidising chemoautotroph, strain MLHE-1. *Applied Environmental Microbiology* 68(10): 4795–4802.

Pernthaler, A. et al. 2002. Fluorescence In Situ Hybridisation and Catalyzed Reporter Deposition for the Identification of Marine Bacteria. *Applied Environmental Microbiology* 68: 3094–3101

Rasmussen, L.D. & Sörensen, S.J. 1998. The effect of long-term exposure to mercury on the bacterial community in marine sediment. *Current Microbiology* 36: 291–297.

Roane, T.M. et al. 1996. Characterisation of bacterial communities in heavy metal contaminated soils. *Canadian Journal of Microbiology* 42: 593–603.

Rosen, B.P. 2002. Transport and detoxification systems for transition metals, heavy metals and metalloids in eukaryotic and prokaryotic microbes. *Comparative Biochemistry and Physiology part A* 133: 689–693.

Sandaa, R.-A. et al. 1999a. Analysis of bacterial communities in heavy metal-contaminated soils at different levels of resolution. *FEMS Microbiology Ecology* 30: 237–251.

Sandaa, R.-A. et al. 1999b. Abundance and diversity of *Archaea* in heavy metal-contaminated soils. *Applied Environmental Microbiology* 65(8): 3293–3297.

Sharma, P.K. et al. 2000. A new *Klebsiella planticola* strain (Cd-1) grows anaerobically at high cadmium concentrations and precipitates cadmium sulphide. *Applied Environmental Microbiology* 66(7): 3083–3087.

Shi, W. et al. 2002. Association of microbial community composition and activity with lead, chromium and hydrocarbon contamination. *Applied Environmental Microbiology* 68(8): 3859–3866.

Schäfer, H. & Muyzer, G. 2001. Denaturing gradient gel electrophoresis in marine microbial ecology. *Methods in microbiology*, 30: 426–468.

Silver, S. 1996. Bacterial resistances to toxic metal ions – a review. *Gene* 179: 9–19.

Smit, E. et al. 1997. Detection and shifts in microbial community structure and diversity in soil caused by copper

contamination using amplified ribosomal DNA restriction analysis. *FEMS Microbiology Ecology* 23: 249–261.
Staley, J.T. & Konopka, A. 1985. Measurements of in situ activities of nonphotosynthetic microorganisms in aquatic and terrestrial habitats. *Annu. Rev. Microbiol.* 39:321–346.
Unz, R.F. & Shuttleworth, K.L. 1996. Microbial mobilization and immobilization of heavy metals. *Current opinion in Biotechnology* 7:307–310.
Viamajala, S. et al. 2002. Chromate reduction in *Shewanella oneidensis* MR-1 is an inducible process associated with anaerobic growth. *Biotechnology Progress* 18: 290–295.
Zengler, K. et al. 2002. Cultivating the uncultured. *Proceedings of the National Academy of Sciences* 99(24): 15681–15686.

European Symposium on Environmental Biotechnology, ESEB 2004 - Verstraete (ed)
© 2004 Taylor & Francis Group, London, ISBN 90 5809 653 X

Dynamics and composition of filamentous micro-organism communities in industrial water systems

Janneke Krooneman
Bioclear, The Netherlands

Per Nielsen
Aalborg University, Denmark

Caterina Levantesi
CNR, Italy

Hermie Harmsen
RUG, The Netherlands

Claudia Beimfohr
Vermicon, Germany

Dick Eikelboom
TNO, The Netherlands

Bert Geurkink & Jaap van der Waarde
Bioclear, The Netherlands

ABSTRACT: Filamentous micro-organisms are an important group of bacteria in activated sludge. Excessive growth of these organisms results in poor settling of activated sludge in a process called bulking sludge, leading to poor effluent quality and high costs for the industry. Little is known about filamentous micro-organisms in industrial water systems and this severely limits the use of prevention strategies for sludge bulking. Specific identification, monitoring techniques and better understanding of the behaviour of these organisms is a first step towards improving waste water treatment process performance and the prevention of sludge bulking. A survey of the presence of filamentous bacteria in a large number of European waste water treatment plants (WWTPs) was carried out, within the scope of the EU-Dynafilm project. The survey revealed many new hitherto non-described filamentous micro-organisms revealing a large unknown biodiversity.

1 BIODIVERSITY

There exist many different types of filamentous bacteria that can cause sludge bulking in WWTPs. Each type of filament will have its own specific needs to develop growth and activity. Knowledge about these physiological characteristics is necessary to allow accurate process performance of WWTPs directed towards prevention of sludge bulking. Although many bacteria can be identified and characterized by microscopic analyses showing their main morphological and some physiological characteristics, still many species cannot be identified and remain unknown. The availability of a variety of molecular techniques added new information about the identity and biodiversity of this group of filamentous micro-organisms. The use of genotyping techniques for instance, revealed that their biodiversity is much more complex than initially thought. In some cases morphological identical filaments differ genetically and thus represent a different genus or species. The other way around can also be true: some filaments that obviously differ based on their morphology can be the same species, because they are genetically identical. Figure 1 shows an

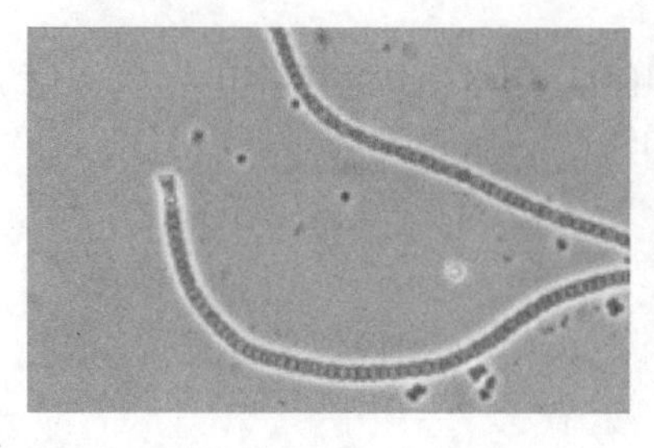

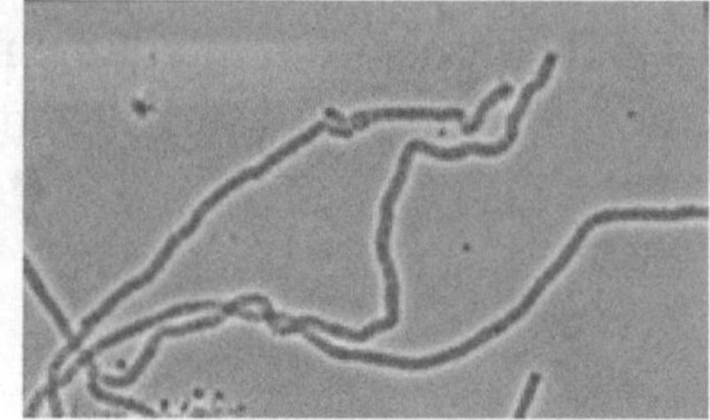

Figure 1. Two morphotypes of *Alisphaera europea,* a filamentous species which responds to probe Noli 644.

example of two completely different filaments that were identified as the same species, based on 16S rRNA-analyses.

2 DYNAFILM PROJECT

In the Dynafilm project a multidisciplinary approach was chosen to investigate microbial population structure and dynamics with respect to the composition and abundance of the microbial filament population in industrial water systems. Techniques for morphological characterization, isolation and physiological studies have been combined with innovative molecular techniques to develop reliable molecular markers for detection, identification, quantification and further physiological characterization of these microbes in active sludge.

3 IDENTIFICATION OF FILAMENTS, MORPHOLOGY AND FISH ANALYSES

More than 125 sludge samples from 107 European WWTPs were collected during a period of two years. All samples were initially characterized focussing on morphology and Fluorescence In Situ Hybridisation (FISH) using 16S rRNA fluorescently labelled oligonucleotide probes (Figure 2). Some filaments that could not be identified based on their morphology, did hybridise with an available group-specific or species-specific FISH-probe, revealing their species or genus identity.

4 DENTIFICATION OF FILAMENTS, EX-SITU PHYSIOLOGY AND MICROMANIPULATION

For a better understanding of the ecology and thus the function of the filamentous bacteria in mixed communities, the physiological characteristics of important species of filaments need to be investigated. The

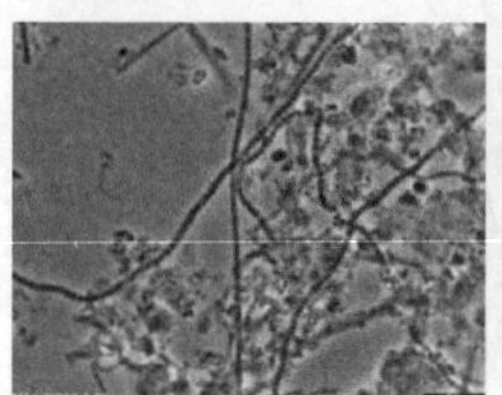

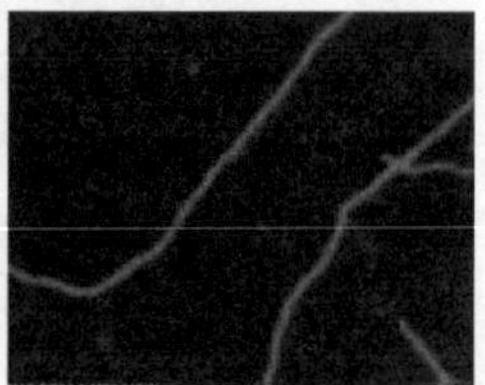

Figure 2. Example of species specific detection using FISH analysis. Left panel: phase contrast photo filaments in activated sludge sample. Right panel: fluorescence microscopic view of the same sample hybridized with species specific probe.

filaments that remained completely unknown, but were present in high amounts were selected and subsequently isolated through dilution culturing procedures or with micro-manipulation. Once in culture, these organisms were further characterized physiologically. Knowledge was obtained about growth characteristics and substrate use. In addition, these isolates were genetically identified as well. As a result the number of unknown filaments decreased and the knowledge about their main characteristics and needs for growth and activity increased.

5 IDENTIFICATION OF FILAMENTS, REVERSED TRANSCRIPTASE PCR, PHYLOGENY AND PROBE DESIGN

Through micro-manipulation many filaments could be selectively isolated. These filaments appeared to be non-culturable under the applied laboratory conditions, but could be subjected to genetic identification studies. By applying reversed transcriptase-PCR (RT-PCR) on the micro-manipulated isolates followed by cloning and sequencing procedures more filaments were genetically identified. Then FISH-probes could be easily designed targeting the 16S rRNA target sites. By applying these newly designed probes on both the original sample material from which the filament was isolated as well as on all other samples with unknown

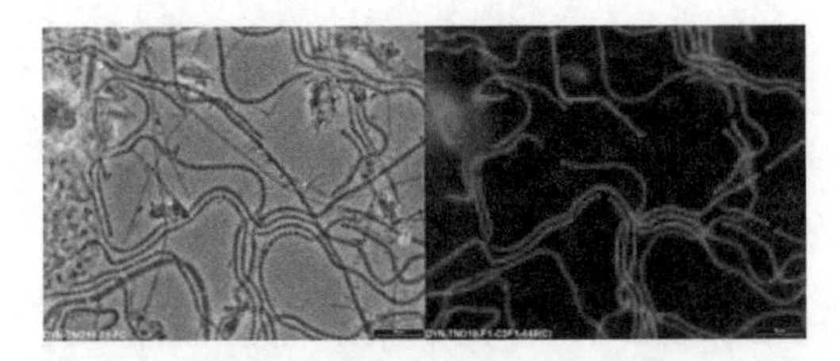

Figure 3. Left panel: phase contrast of sludge sample containing unknown filament belonging to the alfa-subclass. Right panel: FISH-analysis with a species specific probe that was designed upon RT-PCR and sequence analyses of the unknown filament that was micromanipulated. Phylogenetic analysis showed resemblance with filaments from the *Alisphaera*-group.

filaments, specific detection and identification of these formerly unknowns became possible. As a result the percentage unknown filaments could be reduced from 70% towards 50%. Figure 3 shows an example of an unknown filament that was identified through RT-PCR, cloning and sequencing, and finally detected using a newly designed FISH-probe.

In order to get a better understanding of filament biodiversity it is essential to complement the morphological and physiological description of filaments with a phylogenetic study, to properly describe species identity and nomenclature. Therefore, the newly retrieved 16S rRNA sequences formed the basis for phylogenetic analyses, showing the relatedness of the newly identified filaments. During the project 8 different sequences were phylogenetically analysed, whereupon 8 new probes were designed. The identified filaments seemed to belong to different phylogenetic groups.

6 IDENTIFICATION OF FILAMENTS, IN SITU PHYSIOLOGY USING FISH-MAR, ELF AND MAC STUDIES

For a better understanding of the ecology and thus the function of the filamentous bacteria in the mixed communities, the physiological characteristics of important species of filamentous micro-organisms need to be investigated in situ as well. Various techniques were used for studies of the physiology (substrate uptake, exoenzymatic activity and surface properties) directly in the complex microbial systems. Uptake of radiolabelled substrates was detected by microautoradiography (MAR), exoenzymatic activity by enzyme labelled fluorescence (ELF) and surface properties by microspheres adhesion to cells (MAC). The key physiological properties of the genetically identified filaments were described by applying these methods directly in situ and combined with FISH and analyzed using a confocal laser scanning microscopy (CLSM). Figure 4 shows an example of an in situ physiological study using MAR-FISH, performed with two FISH-probes; Pp $\times$ 3-1428 (appearing yellow) specifically targeting the *Alisphaera* Pp $\times$ 3 and EUB targeting most Eubacteria (in green). The radioactive labeled substrate used here was ^{3}H-Acetate, which is seen as silver grains on top of the bacteria indicating substrate utilization under aerobic conditions.

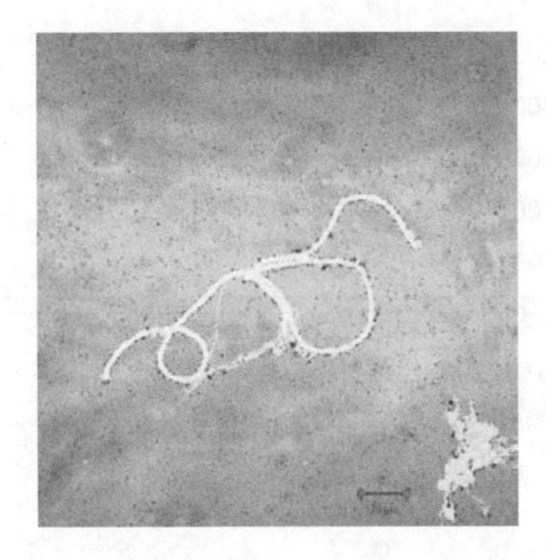

Figure 4. MAR-FISH image of *Alisphaera* Pp $\times$ 3 taking up radiolabelled ^{3}H-Acetate under aerobic conditions. The probe for $P_P \times$ 3-1428 appears yellow and the probe for Eubacteria (EUB) is green.

7 IDENTIFICATION OF FILAMENTS, QUANTITATIVE PCR

In order to properly describe microbial population structure and dynamics, detection and identification methods need to be quantitative. Real-time PCR was set up, using species specific PCR primers and newly developed Taqman probes. At first, quantitative PCR analysis (Q-PCR) of different micro-organisms in complex microbial communities was set up. The developed Q-PCR method was subsequently applied to sludge samples for accurate and quantitative detection of the filaments *Alisphaera europea* and *Microthrix parvicella*. The treshold was 1×10^4 bacteria per 1 ml sample of a complex environment with high numbers of biomass, and calibration was done with standard curves. Figure 5 shows an example of Q-PCR detection in which the fluorescence is measured on-line during the PCR-reaction. Figure 6 shows Q-PCR results of a dilution series of a *B. essensis* culture in Feaces. The specificity of Q-PCR is more specific than FISH, since it relies on 3 oligonucleotides, instead of 1 oligonucleotide as holds for FISH analyses.

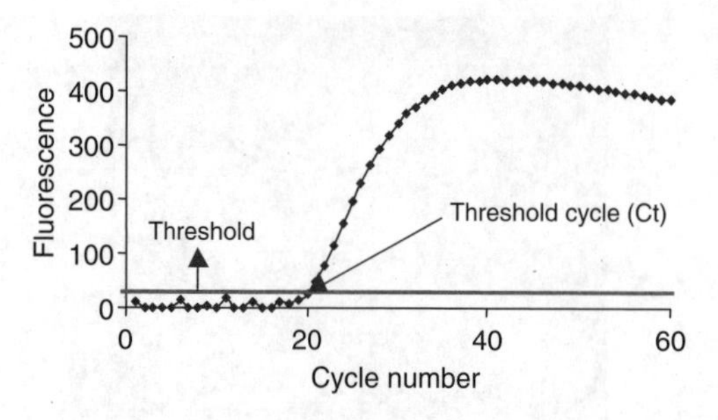

Figure 5. An example of Q-PCR using real time PCR detection.

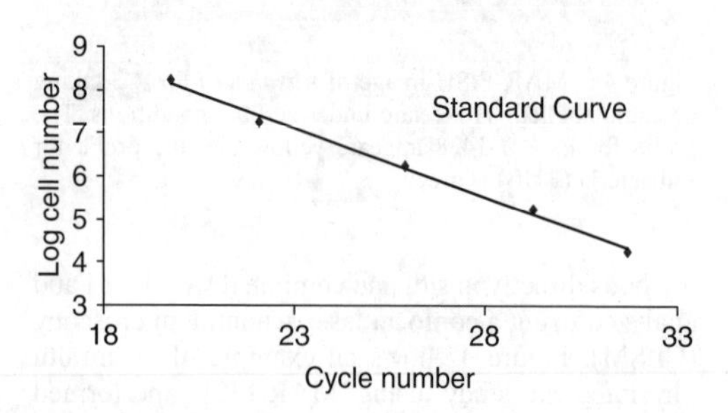

Figure 6. Dilution series (10-fold) in faeces of a *B.essensis* culture containing 1.74×10^9 cells/ml.

For application in environmental samples it can therefore also be too specific, meaning that very closely related organisms (one or two mismatches in one of the oligonucleotide-primers or probes) may not be detected with Q-PCR but will be detected using FISH.

6 DYNAMICS OF FILAMENTOUS POPULATIONS IN WWTPS

The ultimate aim of the project was to monitor the dynamics of filamentous populations in industrial waste water treatment plants using the knowledge and methods and methodologies that have been developed in this project. This includes on one hand molecular identification, quantification using molecular methods and confirmation of filament identity using conventional microscopy to get a complete description of biodiversity, population structure and dynamics of the filamentous population. On the other hand analysis of waste water characteristics, process conditions and configuration of waste water treatment plant were analysed and potential correlations between filament dynamics and the industrial processes were retrieved. As a result industrial waste water treatment processes can be improved by better understanding and management of the deteriorating process of sludge bulking. In addition, new insights in the relevance of filamentous micro-organisms in water reuse systems, cooling water systems and other water systems prone to biofouling may lead in the future to more efficient and sustainable production methods.

European Symposium on Environmental Biotechnology, ESEB 2004 - Verstraete (ed)
© 2004 Taylor & Francis Group, London, ISBN 90 5809 653 X

PHA synthases from a mixed culture performing enhanced biological phosphorus removal

S. Yilmaz & D. Jenkins
Department of Civil and Environmental Engineering, University of California at Berkeley, Berkeley, USA

J.D. Keasling
Department of Chemical Engineering, University of California at Berkeley, Berkeley, USA

ABSTRACT: Culture-independent molecular techniques were used to identify genes coding for polyhydroxyalkanoate synthase (PhaC) from a mixed culture carrying out enhanced biological phosphorus removal (EBPR). An acetate-fed lab-scale sequencing batch reactor was operated to enrich for microorganisms responsible for EBPR. It had previously been shown that the dominant organisms in the mixed culture performing EBPR were novel, yet uncultured β-*Proteobacteria* related to *Rhodocyclus*; therefore, our work concentrated on finding the key enzyme of PHA synthesis (PHA synthase, *pha*C) belonging to the dominant organism in this mixed culture. A library of *pha*C fragments of genomic DNA extracted from EBPR sludge was generated by PCR amplification and cloning into *E. coli.* Nine distinct sequence types were obtained. An internal fragment of *R. tenuis pha*C was also amplified, cloned, and sequenced. These PhaC sequences were mapped with known synthase sequences and with the *R. tenuis* PhaC sequence on a phylogenetic tree.

1 INTRODUCTION

Enhanced biological phosphorus removal (EBPR) is an activated sludge process modification widely used to remove inorganic phosphorus (P_i) from wastewaters. EBPR is achieved by cycling influent wastewater and microbial biomass through alternating anaerobic then aerobic conditions (Mino 2000, Blackall et al. 2002). Under these transient conditions polyphosphate accumulating organisms (PAOs) are selectively enriched. According to current metabolic models of EBPR, under carbon-rich anaerobic conditions, PAOs rapidly take up organic substrates such as volatile fatty acids (VFAs) and accumulate them as polyhydroxyalkanoates (PHAs). Energy for this conversion is largely supplied by intracellular polyphosphate (polyP) hydrolysis, which is accompanied by P_i release into the surrounding liquid and by glycogen degradation. The reducing power required for PHA synthesis is provided by glycogen degradation and TCA cycle (Pereira et al. 1996, Mino et al. 1998, Hesselmann et al. 2000). Under the subsequent carbon-poor aerobic conditions, stored PHAs are used as a carbon and energy source for cell growth and for replenishment of polyP and glycogen reserves. Because cell growth occurs in the aerobic zone, more P_i is taken up in the aerobic cycle than is released in the preceding anaerobic cycle, resulting in a net P removal from solution.

Despite extensive efforts it has not been possible to isolate PAOs in pure culture. Nevertheless, recent culture independent studies showed that the microbial communities of lab-scale, acetate-fed EBPR systems are dominated by novel, yet uncultured β-Proteobacteria related to *Rhodocyclus* (Bond et al. 1995, Hesselmann et al. 1999, Crocetti et al. 2000). It has been shown that these *Rhodocyclus*-like bacteria (proposed name *Candidatus Accumulibacter phosphatis*) possess typical characteristics of EBPR (Bond et al. 1999, Crocetti et al. 2000). Recent findings indicate that *Rhodocyclus*-like bacteria also contribute to P_i removal in pilot- and full-scale EBPR plants (Lee et al. 2002, Zilles et al. 2002).

While several studies have been carried out on the enzymes involved in the anaerobic metabolism of EBPR (Mino et al. 1998, Bond & Rees 1999, Seviour et al. 2003), and one study has been conducted on the enzymes of polyP synthesis during aerobic phase (McMahon et al. 2002), no studies have been done on the enzymes involved in the PHA metabolism.

The key enzymes of PHA metabolism (PHA synthases, PhaC), have been studied extensively in many microorganisms (Madison and Huisman 1999,

Rehm & Steinbuchel 2001, Steinbuchel & Hein 2001). In total, 54 different PHA synthases from 44 microorganisms have been cloned, and the primary structures of 44 of these are available (Rehm & Steinbuchel 2001). PCR-based molecular techniques have been used to retrieve fragments of conserved metabolic genes from uncultivated microorganisms. Examples include functional genes involved in ammonia oxidation (*amoA*) (Rotthauwe et al. 1997), denitrification (*nirK*, *nirS*) (Braker et al. 1998), sulfate respiration (*dsr*) (Minz et al. 1999), and polyphosphate synthesis (*ppk*) (McMahon et al. 2002). Since PHA synthases share enough homology to form a distinctive group, PCR-based molecular techniques could be used to survey PHA synthases from uncultivated organisms carrying out EBPR.

The goal of this study was to retrieve PHA synthase genes from the dominant organism in EBPR and to isolate novel PHA synthases from a mixed culture performing EBPR. We used a PCR-based clone library technique to retrieve fragments of PhaC genes, and restriction fragment length polymorphism (RFLP) analysis to identify unique PHA synthases in an acetate-fed lab-scale sequencing batch reactor enriched to perform EBPR.

2 METHODS

2.1 *Sequencing batch reactor operation*

A sequencing batch reactor (SBR) with a working volume of 4 L was operated under alternating anaerobic/aerobic conditions to achieve EBPR. The reactor operated on a 6-h cycle with 30 min draw-and-fill, 120 min anaerobic phase, 180 min phase, and 30 min settling. The hydraulic residence time was maintained at 12 h and mean cell residence time was 4 days. The reactor was fed a mineral salts medium and a carbon feed. The composition of the mineral salts medium was similar to that reported by Schuler et al. (2001), phosphate was added as $NaH_2PO_4 \cdot H_2O$ to give a P concentration of 12.6 mg/L. The carbon feed contained 92 mg acetate/L of total feed and 15 mg casamino acids/L of total feed to maintain a fixed COD loading of 115 mg/L of mixed liquor/cycle. Mineral salts feed was added immediately after the draw phase, and carbon feed was added after 40 min of N_2 (g) stripping. The pH was controlled at 7.0–7.3 by addition of HCl or NaOH, and temperature was 23 ± 3°C.

Rhodocyclus tenuis (ATCC 25093) was grown anaerobically under the light at 30°C with succinate and acetate as carbon sources as recommended by Malik (1983).

2.2 *Genomic DNA extraction*

The community genomic DNA for use in PCR was extracted from EBPR sludge using a modified bead beating protocol consisting of phenol-chloroform extraction with bead-beating, precipitation of DNA with sodium acetate and isopropanol, and purification by passage of DNA through a size selection column (Dojka et al. 1998). Genomic DNA from pure culture of *R. tenuis* was extracted by method of Marmur (1961).

2.3 *PCR amplification of putative phaC gene fragments*

Oligonuclotide primers targeting conserved regions of PHA synthase (*phaC*) were designed by using an alignment of known β-Proteobacteria PHA synthase amino acid sequences. The web-based consensus-degenerate hybrid oligonucleotide primer algorithm (CODEHOP) was used to generate hybrid primers with a short 3′ degenerate core and a longer 5′ non-degenerate consensus clamp region (Rose et al. 1998). These primers, phbCf (5′-GAT GGT GCC GTG YAT HAA YAA-3′) and phbCr (5′-CCT TCA GGT AGT TGT CCA CCA CRT ART TCC-3′), should amplify an internal ~500 bp fragment of *phaC*.

PCR was carried out on 40 ng of bulk DNA in a 30-μL reaction volume using either an Advantage-GC cDNA PCR kit (Clontech, CA) or an Amplitaq Gold PCR kit (Perkin Elmer, CA) on an MJ Research PTC-200 model thermal cycler. The Advantage GC PCR reaction mixture contained 1× Advantage GC cDNA PCR buffer, 0 mM GC-melt, 300 μM of each dNTP, 400 nM of each forward and reverse primer and 1× Advantage GC cDNA Polymerase mix. The Amplitaq Gold PCR reaction mixture contained 1× PCR buffer II, 2.5 mM $MgCl_2$, 300 μM of each dNTP, 400 nM of each forward and reverse primer and 0.025 U μL^{-1} AmpliTaq Gold. The PCR was conducted with an initial denaturation step at 94°C for 12 min, followed by 30 cycles of denaturation at 94°C for 1 min, annealing for 45 s, and extension at 72°C for 2 min. The initial annealing temperature was 58.5°C, and stepped down to 53.5°C in 1°C increments every two cycles followed by 22 cycles with an annealing temperature of 58.5°C, and a final extension at 72°C for 12 min. An internal ~500-bp *phaC* fragment was amplified from *R. tenuis* genomic DNA by using the Amplitaq Gold PCR kit with touchdown PCR as described above.

2.4 *Clone library construction, clone screening by restriction fragment length polymorphism analysis and sequencing*

500-bp PCR products were cloned into vector pCR®4-TOPO by using the TOPO TA cloning kit for sequencing (Invitrogen) in accordance with the manufacturer's instructions. Plasmids were prepared with a 96-well alkaline lysis procedure using a MultiScreen Resist Vacuum Manifold (Millipore) according to manufacturer's instructions. Approximately 270 clones were screened by using restriction fragment length polymorphism (RFLP) analysis with MspI (New England Biolabs). Representatives of unique RFLP types were sequenced with vector primers by the DNA Sequencing Facility of UC, Berkeley.

2.5 *Phylogenetic analyses*

Sequences were edited and assembled using the ABI Autoassembler software package and compared to available databases using the BLAST (Basic Local Alignment Search Tool) network service on the NCBI website (Altschul et al. 1990). PhaC fragments were translated and the amino acid sequences were aligned against 10 PhaC sequences from Genbank using the Seqlab program in the GCG software package, version 10.0 (Genetics Computer Group). The alignment was masked manually to exclude positions with gaps in more than 15% of sequences. Phylogenetic analyses were carried out with amino acid alignment by using the software package PHYLIP, version 3.6a (J. Felsenstein, Department of Genetics, University of Washington, Seattle). The maximum likelihood program was used to reconstruct the evolutionary relationships between PhaC sequences.

3 RESULTS AND DISCUSSIONS

Fragments of PHA synthase genes have been retrieved from enriched EBPR biomass using a culture-independent, PCR-based clone library technique. Since recent studies have shown that the dominant organism in lab-scale EBPR sludges is *Candidatus* Accumulibacter phosphatis (a β-Proteobacterium closely related to Rhodocyclus) (Hesselmann et al. 1999, Crocetti et al. 2000, Zilles et al. 2002), all known β-Proteobacteria PHA synthase sequences were included in the protein alignment used for primer design. Primers were designed to target conserved regions of PHA synthase (*phaC*). Using these primers, a 500 bp fragment of *phaC* from a pure culture of *Rhodocyclus tenuis* was amplified, cloned, and sequenced. A library of approximately 500-bp *phaC* fragments of genomic DNA extracted from EBPR sludge was generated by PCR amplification and cloning into *E. coli*. Approximately 270 clones were screened using restriction fragment length polymorphism (RFLP) analysis to identify unique clones for sequencing. Representative types in the library were sequenced. Nine distinct types were obtained. Phylogenetic analyses were carried out to compare the sequences obtained to all available amino acid sequences in the databases and to *R. tenuis* PhaC. Comparative sequence analyses using the online tool BLAST showed that 8 of the sequences were homologous to PHA synthases from β-proteobacteria. One of the sequences clustered

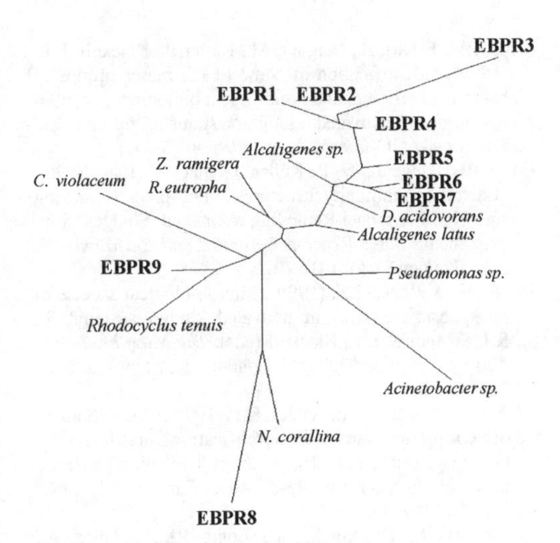

Figure 1. Phylogram indicating inferred relatedness of *phaC* genes based on amino acid sequences. This tree was generated using the PHYLIP package, version 3.6a, and maximum likelihood analysis for tree construction was carried out using the program PROML. Confirmed *phaC* sequences were obtained from GenBank, *R. tenuis phaC* and novel *phaC*s retrieved from EBPR sludge (EBPR 1–9) all from this study. Full organism names and the accession numbers of confirmed *phaC*s are as follows: *Acinetobacter* spp. (ASU04848), *Alcaligenes latus* (ALU47026), *Alcaligenes* spp. (ASU78047), *Chromabacterium violaceum* (AF061446), *Comamonas acidovorans* (ABOO9273), *Nocardia corallina* (AF019964), *Pseudomonas* spp. (AB014757), *Ralstonia eutropha* (AFAPHBAA), *Zooglea ramigera* (ZRU66242).

closely to PhaC from *R. tenuis* on a PhaC phylogenetic tree suggesting that this sequence might belong to *A. phosphatis* (Fig. 1).

4 CONCLUSIONS

Nine novel PHA synthase gene fragments were obtained from an acetate-fed lab-scale sequencing batch reactor enriched to perform EBPR. One sequence type was related to a PhaC homolog from *R. tenuis*.

REFERENCES

Altschul, S.F., Gish, W., Miller, W., Myers, E.W., Lipman, D.J. (1990). "Basic local alignment search tool." *J Mol Biol* 215(3): 403–10.

Blackall, L.L., Crocetti, G.R., Saunders, A.M., Bond, P.L. (2002). "A review and update of the microbiology of enhanced biological phosphorus removal in wastewater treatment plants." *Antonie Van Leeuwenhoek* 81(1–4): 681–91.

Bond, P.L., Erhart, R., Wagner, M., Keller, J., Blackall, L.L. (1999). "Identification of some of the major groups of bacteria in efficient and nonefficient biological phosphorus removal activated sludge systems." *Applied and Environmental Microbiology* 65(9): 4077–84.

Bond, P.L., Hugenholtz, P., Keller, J., Blackall, L.L. (1995). "Bacterial Community Structures of Phosphate-Removing and Non-Phosphate-Removing Activated Sludges from Sequencing Batch Reactors." *Applied and Environmental Microbiology* 61(5): 1910–16.

Bond, P.L. & Rees, G.N. (1999). Microbiological aspects of phosphorus removal in activated sludge systems. In R.J. Seviour & L.L. Blackall (eds), *The Microbiology of Activated Sludge*: 227–56. London: Kluwer Academic Publishers.

Braker, G., Fesefeldt, A., Witzel, K.P. (1998). "Development of PCR primer systems for amplification of nitrite reductase genes (nirK and nirS) to detect denitrifying bacteria in environmental samples." *Appl Environ Microbiol* 64(10): 3769–75.

Crocetti, G.R., Hugenholtz, P., Bond, P.L., Schuler, A., Keller, J., Jenkins, D., Blackall, L.L. (2000). "Identification of polyphosphate-accumulating organisms and design of 16S rRNA-directed probes for their detection and quantitation." *Applied and Environmental Microbiology* 66(3): 1175–82.

Dojka, M.A., Hugenholtz, P., Haack, S.K., Pace, N.R. (1998). "Microbial diversity in a hydrocarbon- and chlorinated-solvent-contaminated aquifer undergoing intrinsic bioremediation." *Appl Environ Microbiol* 64(10): 3869–77.

Hesselmann, R.P.X., von Rummell, R., Resnick, S.M., Hany, R., Zehnder, A.J.B. (2000). "Anaerobic metabolism of bacteria performing enhanced biological phosphate removal." *Water Research* 34(14): 3487–94.

Hesselmann, R.P.X., Werlen, C., Hahn, D., van der Meer, J.R., Zehnder, A.J.B. (1999). "Enrichment, phylogenetic analysis and detection of a bacterium that performs enhanced biological phosphate removal in activated sludge." *Systematic and Applied Microbiology* 22(3): 454–65.

Lee, N., Jansen, J.C., Aspegren, H., Henze, M., Nielsen, P.H., Wagner, M. (2002). "Population dynamics in wastewater treatment plants with enhanced biological phosphorus removal operated with and without nitrogen removal." *Water Sci Technol* 46(1–2): 163–70.

Madison, L.L. & Huisman, G.W. (1999). "Metabolic engineering of poly(3-hydroxyalkanoates): From DNA to plastic." *Microbiology and Molecular Biology Reviews* 63(1): 21–53.

Malik, K.A. (1983). "A Modified Method for The Cultivation of Phototrophic Bacteria." *Journal of Microbiological Methods* 1: 343–52.

Marmur, J. (1961). "A Procedure for Isolation of Deoxyribonucleic Acid from Micro-Organisms." *Journal of Molecular Biology* 3(2): 208-&.

McMahon, K.D., Dojka, M.A., Pace, N.R., Jenkins, D., Keasling, J.D. (2002). "Polyphosphate kinase from activated sludge performing enhanced biological phosphorus removal." *Applied and Environmental Microbiology* 68(10): 4971–78.

Mino, T. (2000). "Microbial selection of polyphosphate-accumulating bacteria in activated sludge wastewater treatment processes for enhanced biological phosphate removal." *Biochemistry-Moscow* 65(3): 341–48.

Mino, T., Van Loosdrecht, M.C.M., Heijnen, J.J. (1998). "Microbiology and biochemistry of the enhanced biological phosphate removal process." *Water Research* 32(11): 3193–207.

Minz, D., Flax, J.L., Green, S.J., Muyzer, G., Cohen, Y., Wagner, M., Rittmann, B.E., Stahl, D.A. (1999). "Diversity of sulfate-reducing bacteria in oxic and anoxic regions of a microbial mat characterized by comparative analysis of dissimilatory sulfite reductase genes." *Appl Environ Microbiol* 65(10): 4666–71.

Pereira, H., Lemos, P.C., Reis, M.A.M., Crespo, J.P.S.G., Carrondo, M.J.T., Santos, H. (1996). "Model for carbon metabolism in biological phosphorus removal process based on in vivo 13C-NMR labelling experiments." *Water Research* 30(9): 2128–38.

Rehm, B. & Steinbuchel, A. (2001). PHA synthases:The Key Enzymes of PHA Synthesis. In A.Steinbuchel (ed) *Biopolymers*: 3a: 173–215. Weinheim, Wiley-VCH.

Rose, T.M., Schultz, E.R., Henikoff, J.G., Pietrokovski, S., McCallum, C.M., Henikoff, S. (1998). "Consensus-degenerate hybrid oligonucleotide primers for amplification of distantly related sequences." *Nucleic Acids Research* 26(7): 1628–35.

Rotthauwe, J.H., Witzel, K.P. & Liesack, W. (1997). "The ammonia monooxygenase structural gene amoA as a functional marker: molecular fine-scale analysis of natural ammonia-oxidizing populations." *Appl Environ Microbiol* 63(12): 4704–12.

Schuler, A.J., Jenkins, D., Ronen, P. (2001). "Microbial storage products, biomass density, and settling properties of enhanced biological phosphorus removal activated sludge." *Water Sci Technol* 43(1): 173–80.

Seviour, R.J., Mino, T., Onuki, M. (2003). "The microbiology of biological phosphorus removal in activated sludge systems." *FEMS Microbiol Rev* 27(1): 99–127.

Steinbuchel, A. & Hein, S. (2001). Biochemical and Molecular Basis of Microbial Synthesis of Polyhydroxyalkanoates in Microorganisms. In T. Scheper (ed) *Advances in Biochemical Engineering/Biotechnology*: 71: 81–123. Berlin: Springer-Verlag.

Zilles, J.L., Peccia, J., Kim, M.W., Hung, C.H., Noguera, D.R. (2002). "Involvement of Rhodocyclus-related organisms in phosphorus removal in full-scale wastewater treatment plants." *Applied and Environmental Microbiology* 68(6): 2763–69.

European Symposium on Environmental Biotechnology, ESEB 2004 - Verstraete (ed)
© 2004 Taylor & Francis Group, London, ISBN 90 5809 653 X

Comparative evaluation of methanogenic communities in psychrophilic and mesophilic UASB reactors treating domestic wastewater

C. Yangin Gomec, V. Eroglu & I. Ozturk
Dept. of Environmental Engineering, Istanbul Technical University, Maslak, Istanbul, Turkey

B. Calli & B. Mertoglu
Dept. of Environmental Engineering, Marmara University, Goztepe, Istanbul, Turkey

ABSTRACT: Methanogenic communities in two identical Upflow Anaerobic Sludge Bed Reactors (UASBR) seeded with the same sludge and operated at psychrophilic and mesophilic temperatures were comparatively investigated by using Fluorescent In-Situ Hybridization (FISH) technique. Furthermore, effects of temperature on the reactor performances were correlated with population dynamics. Initially, until steady state conditions, synthetic wastewater and then raw domestic sewage were fed to the reactors. Archaeal and bacterial communities in the reactors were investigated by using Cy-3 and Fluorescein labelled rRNA targeted oligonucleotide probes for Bacteria (Eub338) and Archaea (Arch915) domains and for *Methanosaeta* (MX825) and *Methanosarcina* (MS821) species as acetoclastic methanogens. According to in-situ hybridization results with Arch915 and Eub338 probes, archaeal cells representing the methanogens were found intensively dominant in the bottom sampling ports of the reactors fed with synthetic wastewaters at all sampling days. Besides, *Methanosaeta* sp. (MX825) abundant in seed sludge was identified as the major methanogenic archaea in samples taken from both psychrophilic and mesophilic reactors as granules remained undisturbed and kept their rigidity even under psychrophilic conditions. On the other hand, this is an indication of temperature adaptation for originally mesophilic *Methanosaeta* species in psychrophilic reactor. Comparing the results of FISH experiments for the anaerobic treatability studies with synthetic wastewater and raw domestic sewage the microbial diversity did not significantly change in both of the reactors, while COD removal efficiencies dropped dramatically.

1 INTRODUCTION

Anaerobic treatment over aerobic treatment for low-strength wastewaters has advantages that include decreased sludge production, lower energy requirements and decreased operating costs (Angenent et al. 2001). The most appropriate anaerobic system to treat domestic wastewater has been considered as the UASBR because of its simplicity, low investment and operation costs and the long favorable experience in the treatment of a wide range of industrial wastewaters Elmitwalli (2000). It is reported that if the organisms are acclimatized to desired temperatures, the reactors can be operated even under psychrophilic conditions Patel & Madamwar (2002). Psychrophilic anaerobic treatment is an attractive option for wastewaters, which are discharged at moderate to low temperature (Lettinga et al. 2001). Generally, domestic sewage has a temperature lower than 30°C that necessitates heating during the mesophilic anaerobic treatment. Langenhoff & Stuckey (2000).

Four different groups of microorganisms, primary fermenting bacteria, secondary fermenting bacteria, and two types of methanogens degrade organic matter to methane and carbon dioxide in discrete steps in anaerobic treatment. Among them, methanogens produce methane as an internal part of their energy metabolism. Methanogens are found at large numbers and belong to Archaea. Methanogens are slower growing and tend to be rate limiting in anaerobic processes treating sewage, sewerage sludges and most industrial wastewaters. Methanogens may represent all basic morphological types such as; cocci and packets of cocci, rods of different shape and size, spirillum, and filamentous forms. They are dependent on other organisms for the supply of substrates in most anaerobic environments. Acetate using methanogens consist of members of the genera *Methanosarcina* and

Methanosaeta. Generally, the granules are composed of *Methanosaeta* sp. rather than *Methanosarcina* sp. in high rate anaerobic sludge bed reactors (Raskin et al. 1994).

Instability in anaerobic reactors at low temperatures is generally considered to be the result of inadequate adaptation of the microbial population. Besides, monitoring of bacterial communities in the start-up is considered to be very useful in evaluating and interpreting the adaptation. A desirable microbial monitoring analysis in anaerobic reactors should be directly associated with the active population of microorganisms especially of the populations responsible for the critical steps including methanogens. Microbial identification tools for such analyses are not available for routine use due to the limitations of traditional microbiological techniques. However, through the recent 16 S rRNA/rDNA based molecular methods; it is now possible to identify methanogens present in anaerobic reactors regardless of their cultivation capabilities (Raskin et al. 1994). Therefore, such monitoring analyses are becoming progressively more essential in the control of anaerobic treatment systems. Although methanogens in anaerobic reactors have been studied widely in more recent studies by using molecular methods, our understanding in their adaptation to psychrophilic temperatures is still limited.

In this study, performances and methanogenic population dynamics of two identical UASBR's seeded with the same sludge and operated at psychrophilic and mesophilic temperatures were comparatively investigated by using conventional monitoring parameters and Fluorescent In-Situ Hybridization (FISH) technique, respectively.

2 MATERIALS AND METHODS

2.1 *Mesophilic and sub-mesophilic UASB reactors*

Two identical lab-scale anaerobic UASBR with an effective volume of 6.45 lt were set up for the anaerobic treatability tests which operated as continuous systems. The reactors consisted of plexiglass column of 1.0 m in height and 90 mm in diameter, on the top of which special gas-solids-liquid separators were installed. The mesophilic UASBR operated at 35°C ± 2.0°C (Figure 1a) during the study. On the other hand, the psychrophilic reactor operated at room temperature at sub-mesophilic conditions during the start-up and reduced to around 10°C at the end of the study for the estimation of process performance of a UASBR when treating sewage under relatively cold temperature conditions. The temperature in the sub-mesophilic reactor varied between 21.5–26°C during summer and reduced to around 17–19°C during

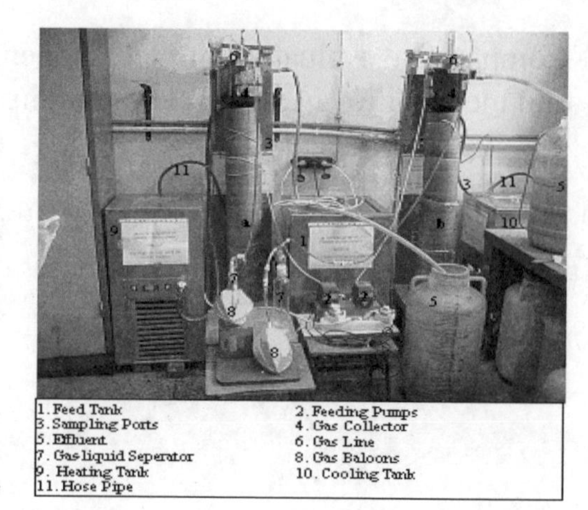

Figure 1. (a) Mesophilic UASBR (b) Psychrophilic UASBR.

autumn. The continuous experiments were done at different temperatures in later studies, reducing from sub-mesophilic temperatures to psychrophilic temperature (10°C) in small decrements by the hose-pipe in which cooled tap water traveled and the psychrophilic conditions were maintained (Figure 1b). The reactors were inoculated (20%) by the granular sludge taken from Pasabahce Tekel Raki (an alcoholic drink) Industry in Istanbul having a VSS/SS ratio of 91%. In both reactors, around 21.2 gr VSS were added per liter of reactors as seed.

2.2 *Wastewater source*

2.2.1 *Synthetic wastewater*

Since the composition of raw domestic sewage varies considerably in time and from one place to another and the great fluctuations in flow, concentration and composition of domestic sewage may make the comparison between the performances of the mesophilic and the psychrophilic reactors difficult, synthetic wastewater characterizing domestic sewage was fed to the reactors in the first part of the study. Synthetic wastewater was employed by preparing daily of a solution containing in each liter of tap water the substances according to International Organization for Standardization (ISO). Total COD (COD_{tot}) concentration of the synthetic wastewater was calculated as 300 mg/l, which represented weak sewage. Since small portions of leachate are added into sewerage, the COD concentration mostly reflects medium domestic wastewater in Istanbul. Thus, in this study the amounts of the substances given in Table 1 doubled in the synthetic wastewater to represent medium domestic effluent.

Table 1. The composition of the synthetic wastewater (pH = 7.5 ± 0.5).

Components	Amount (mg/l)
Peptone	160
Meat extract	110
Urea	30
NaCl	7
$CaCl_2.7H_2O$	4
$MgSO_4.7H_2O$	2
K_2HPO_4	28

Table 2. Raw domestic sewage characterization.

Parameter	Unit	
COD_{tot}	mg/l	222
COD_{sol}	mg/l	117
TKN	mg/l	42
NH_3-N	mg/l	33
TP	mg/l	9.5
PO_4^{2-}	mg/l	6.8
pH	–	7.45
Alkalinity	mg $CaCO_3$/l	345
Conductivity	mmho/cm	6030

2.2.2 *Domestic sewage*
Raw domestic sewage was taken from Baltalimani Preliminary Treatment Plant located in Istanbul. The plant is planned to treat the domestic wastewater in flow rate capacities of 400,000 m^3/day (equivalent to one million residents) in the first stage, and 760,000 m^3/day (equivalent to two million residents) in the second stage. There are coarse screen building, pumping station, and a pretreatment plant available in the plant. It was reported that one of the most convenient methods for leachate control is to treat landfill leachates with domestic wastewaters and mixing 2% of landfill leachates by volume with domestic wastewaters does not lead to important operational problems in Municipal Treatment Plants (Yangin et al. 2002). Very small percentages of leachate are being added into sewerage and pretreated in Baltalimani pretreatment plant. The wastewater used in this study was taken from the outlet of the grit chamber. The characterization of the wastewater is given in Table 2.

2.3 *Fish analyses*

FISH was carried out in three steps; fixation, hybridization and identification by the microscope. For fixation, the sludge samples were fixed overnight in 4% paraformaldehyde-phosphate-buffered saline at 4°C. Gelatin-coated multiwell glass slides were used for the fixed cells to be spotted. On each well of each slide 4 μl of sample was put and allowed drying at room temperature. For hybridization, each labeled probe (30 ng/well) were hybridized by a hybridization buffer (0.9 M NaCl, 20 mM Tris/HCl, pH 8.0, 0.01% SDS) at 46°C for 2 h. For ensuring the optimal hybridization stringency, formamide was added to the final concentrations of about 20% for Arch 915 and 35% for other probes. Then, a washing buffer was used to rinse and remove the unbound oligonucleotides after hybridization. 4,6-diamidino-2-phenylindole (DAPI) was added to the wash buffer (100 μl of 0.1% DAPI) to detect all DNA in the dark. 10 min incubation at 48°C was applied to the slides with washing buffer and than rinsed briefly with double distilled water and immediately air-dried. The slides were investigated with an Olympus BX50 microscope and digital images of the slides were taken with a digital camera (Calli et al. 2003).

3 EXPERIMENTAL STUDY

3.1 *Anaerobic treatability studies*

Anaerobic treatability studies were conducted for about 12 months with synthetic wastewater in both reactors whereas they operated for about 3 months with raw sewage. At start-up, COD_{tot} removals could be obtained as 86% and 82% for the mesophilic and sub-mesophilic reactors respectively when both reactors operated at a HRT of 0.89 d with synthetic sewage. Results indicated at a HRT of 0.45 d that COD_{tot} removals were obtained as 87% and 76% for the mesophilic and sub-mesophilic reactors (19 ± 2°C), respectively. COD_{tot} were obtained as 88% and 80% when HRT reduced to 0.21 d for the mesophilic and sub-mesophilic reactors, respectively. The temperature of the reactor decreased to around 16 ± 2°C in the psychrophilic reactor whereas the temperature of the mesophilic reactor was at 35 ± 2°C. After finalizing the study on the performance of UASBR at sub-mesophilic condition, the temperature of the sub-mesophilic reactor decreased and kept at 10 ± 1°C (psychrophilic) starting with a HRT of 0.52 d. In the first week, the COD_{tot} removal decreased significantly to less than 68% in the psychrophilic reactor whereas around 82% COD removal was observed in the mesophilic reactor. However, COD_{tot} removal improved in the following days indicating that the microorganisms could well adapt to such low temperatures. COD removal increased up to 83% at a HRT of 0.52 d at 10 ± 1°C. At the same HRT, it reached up to 90% in the mesophilic reactor treating synthetic sewage.

On the other hand, results of the anaerobic treatability studies with pre-settled and raw sewage did not indicate high COD_{tot} removals in both reactors owing to the fact that the influent COD_{tot} values were very low. The influent sewage was diluted by seawater during the sampling period, which could be understood from high conductivity value. COD_{tot} removals with pre-settled sewage were obtained as around 54% and 32% at a HRT of 0.35 d for the mesophilic and psychrophilic (13 ± 1°C) reactors, respectively. On the other hand, COD_{tot} removals with raw sewage did not show any improvement in the mesophilic reactor whereas around 37% removal was obtained in psychrophilic (13 ± 1°C) reactor at a HRT of 0.26 d.

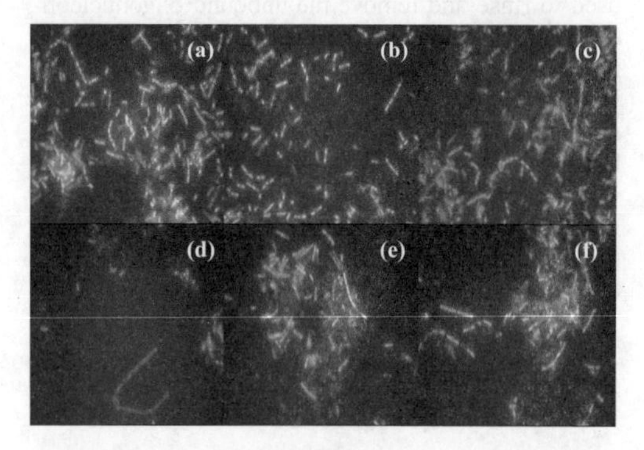

Figure 2. FISH results, *Methanosaeta* in UASBR's fed with synthetic wastewater (a) at 35°C (HRT = 0.21) (b) at 35°C (HRT = 0.52) (c) at 35°C (HRT = 0.26) (d) at 16°C (HRT = 0.21) (e) at 11°C (HRT = 0.52) (f) at 10°C (HRT = 0.26 d).

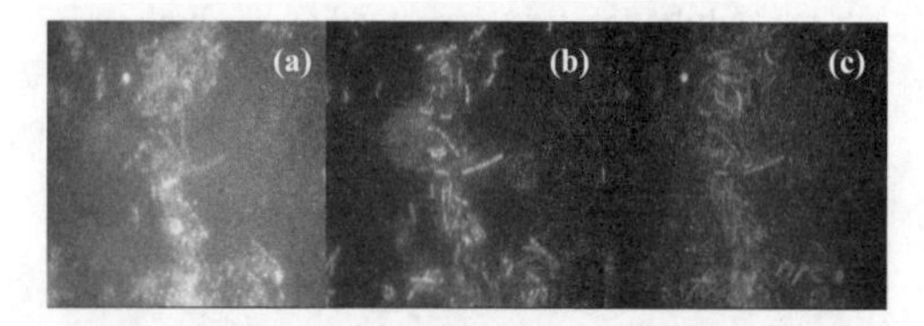

Figure 3. DAPI staining and FISH results (Arch 915-Eub 338) of UASBR's fed with synthetic wastewater at 10°C (a) All microorganisms (b) Methanogens (c) Bacteria.

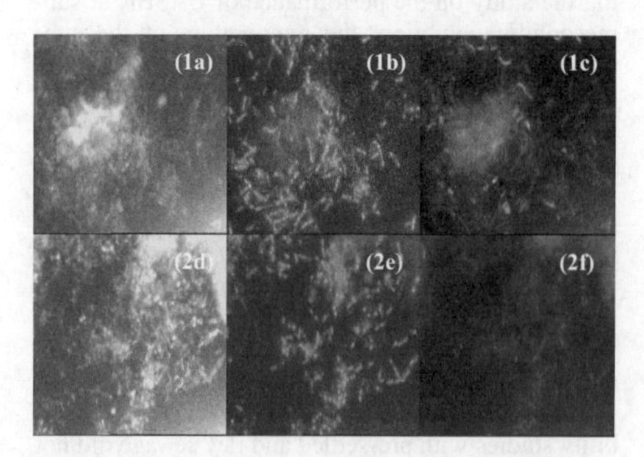

Figure 4. DAPI staining and FISH results of UASBR's fed with raw domestic sewage at 13°C (1) and 35°C (2), (a, d) All microorganisms (b, e) Methanogens (c, f) Bacteria.

3.2 *Results of FISH*

In-situ hybridization results have shown that archaeal cells representing the methanogens (ARC 915) were intensively dominant in the bottom sampling ports of both mesophilic and psychrophilic UASBR's treating synthetic sewage. Comparing the results of FISH experiments carried out with oligonucleotide probes specific for the different groups of methanogens, it is observed that *Methanosaeta* (MX825) species were the major methanogenic archaea present in both reactors fed with synthetic wastewater at all sampling times (Figure 2a–f) and methanogens (ARC 915) were intensively dominant in psychrophilic reactor at Day 357 (Figure 3a–c).

The results obtained investigation of hybridization in psychrophilic and mesophilic UASBR's treating raw domestic sewage; methanogenic achaea were intensively dominant population even the durations of low COD removal efficiencies. However it was observed that psychrophilic reactor has more bacterial communities than mesophilic reactor (Figure 4a–f).

The results of FISH experiments have showed that in both reactors treating synthetic wastewater and raw domestic sewage, *Methanosaeta* species were the major methanogenic archaea present in both reactors at all sampling times.

4 CONCLUSIONS

Effects of low temperatures on UASBR performances treating low strength wastewaters were investigated for correlating to the variations in methanogenic diversity by using 16 S rRNA/DNA based FISH technique. According to in-situ hybridization results with Arch915 and Eub338 probes, archaeal cells representing the methanogens were found intensively dominant in the bottom sampling ports of the reactors fed by synthetic wastewater at all sampling days. Besides, *Methanosaeta* sp. (MX825) abundant in seed sludge was identified as the major methanogenic archaea in samples taken from both psychrophilic and mesophilic reactors as granules remained undisturbed and kept their rigidity even under psychrophilic conditions. Calli (2003) reported that *Methanosaeta* sp. mainly prevailing in sludges were replaced by *Methanosarcina* sp. under unfavorable conditions. On the contrary, results indicated well temperature adaptation for originally mesophilic *Methanosaeta* species in psychrophilic reactor fed by synthetic sewage.

Comparing the results of FISH experiments for UASBR's fed with synthetic wastewater and raw domestic sewage show that archaeal cells representing the methanogens were intensively dominant in both reactors. It is observed that in both reactors the microbial diversity did not significantly change, while COD removal efficiencies dropped dramatically.

REFERENCES

Angenent, L.T., Banik, C.G. & Sung, S. 2001. Anaerobic migrating blanket reactor treatment of low strength wastewater at low temperatures, *Water Environment Research*, 73(5): 567–574.

Calli, B. 2003. Elucidation of ammonia inhibition in anaerobic treatment process by using 16 S rRNA/DNA based microbial identification techniques, *PhD Thesis,* Boğiçi University, Istanbul, Turkey.

Calli, B., Mertoglu, B., Tas, N., Inanc, B., Yenigun, O. & Ozturk, I. 2003. Investigation of variations in microbial diversity in anaerobic reactors treating landfill leachate, *Water Science and Technology*, 48(4): 105–112.

Elmitwalli, T.A. 2000. Anaerobic treatment of domestic sewage at low temperature, *PhD Thesis,* Wageningen University, Wageningen, The Netherlands.

Langenhoff, A.A.M. & Stuckey, D. 2000. Treatment of dilute, wastewater using an anaerobic baffled reactor: effect of low temperature, *Wat. Res*, 34(15): 3867–3875.

Lettinga, G., Rebac, S. & Zeeman, G. 2001. Challenge of psychrophilic anaerobic wastewater treatment, *TRENDS in Biotechnology*, 19(9): 363–370.

Patel, H. & Madamwar, D. 2002. Effects of temperatures and organic loading rates on biomethanation of acidic petrochemical wastewater using an anaerobic upflow fixed-film reactor, *Bioresource Technology*, 82: 65–71.

Raskin, L., Poulsen, L.R., Noguera, D.R., Rittman, B.E. & Stahl, D.A., Quantification of methanogenic groups in anaerobic biological reactors by oligonucleotide probe hybridization. Appl Environ Microbiol 60: 1241–1248 (1994).

Yangin, C., Yilmaz, S., Altinbas, M. & Ozturk, I., 2002. A new process for the combined treatment of municipal wastewaters and landfill leachates in coastal areas, *Water Science and Technology*, 46(8): 111–118.

European Symposium on Environmental Biotechnology, ESEB 2004 - Verstraete (ed)
© 2004 Taylor & Francis Group, London, ISBN 90 5809 653 X

Diversity of the microflora of a compost-packed biofilter treating benzene-contaminated air

S. Borin, M. Marzorati, L. Cavalca, C. Sorlini & D. Daffonchio
Università degli Studi di Milano, Dipartimento di Scienze e Tecnologie Alimentari e Microbiologiche, Milano, Italy

M. Zilli & A. Converti
Università degli Studi di Genova, Dipartimento di Ingegneria Chimica e di Processo, Genova, Italy

H. Cherif & A. Hassen
Institut National Recherche Scientifique & Technologique, Laboratoire Eau & Environnement, Tunis, Tunisia

ABSTRACT: A laboratory scale compost-packed upflow biofilter was fluxed with benzene contaminated air over a period of 240 days with a progressive increase of concentration, resulting in an efficiency of benzene removal of 95–100%. Samples of compost from different depths of the biofiltering column were analysed for total bacterial counts and eubacterial diversity with ARISA fingerprinting, showing that the treatment established a rich bacterial community adapted to increasing benzene concentrations. Identification of the strains cultured from the compost showed that besides low G + C Gram positives and actinomycetes, typical of compost, during the treatment new strains appeared, belonging to β- and γ-proteobacteria, and high G + C Gram+. Among benzene-degraders strains affiliated to *Neisseria* and *Bordetella* genera known to include pathogenic species were identified. 37.5% of the strains were able to use benzene as a substrate, and several resulted to have a toluene dioxygenase-like gene. Several peaks of cultivatable benzene users, mainly represented by *Rhodococcus* sp. were present in the ARISA fingerprint, which described the total bacterial diversity of the biofilter, showing that they constituted a significant part of the microflora.

1 INTRODUCTION

Biofilters are used for the removal of volatile, toxic or odorous compounds which contaminate air emissions from industrial plants treating wastes, food or animal-rendering, or for the removal of toxic compounds stripped from contaminated soils during *in situ* bioremediation [Leson and Winer 1991, Devinny et al. 1999]. Microorganisms grow as a biofilm on packed organic material flushed by the contaminated air, and oxidize the pollutants or convert them into biomass [Møller et al. 1996]. The packing medium guarantees conditions suitable for biomass development, immobilizes the microbial cells preventing washout, constitutes a nutritious and humid reservoir as well as a mechanical odorless support, minimizes the overall reactor volume, energy consumption, and ensures high removal yields.

Although biofiltration technologies have been well established and optimised [Sene et al. 2002], microbiology of biofiltration has been studied only in the past two decades. Several authors focused their research on the cultivatable fraction of the microflora, by the identification and chemo-taxonomical description of strains isolated from the biofiltering material [Ahrens et al. 1997, Lipski and Altendorf 1997, Juteau et al. 1999], and the study of their degrading potential [Hanson et al. 1999]. Recent studies applied molecular culture-independent methods, and could describe the composition of the whole microbial community developed during the biofiltration. [Møller et al. 1996, Sakano & Kerkhof 1998, Friedrich et al. 2002].

We studied a laboratory-scale compost biofilter describing the shifts in the total diversity of the microbial community during the biofiltration process and along the filtering column using a cultivation-independent approach, followed by the characterisation of abundance, species diversity and degrading potential of isolated strains. The biofilter analysed was optimised for the removal of air contaminated with benzene, a typical petroleum component highly toxic and volatile, deliberately or accidentally released into the environment in large amounts, and on the pollutants priority list of most countries' environmental protection agencies. Benzene degraded in aerobic conditions can be completely mineralised to carbon dioxide, by extensively studied dioxygenases [Gibson et al. 1968].

2 MATERIAL AND METHODS

2.1 *Biofiltration process*

A laboratory-scale biofilter was designed, made of a 1 m height column, packed with powdered compost from the organic fraction of municipal solid waste, which was fluxed with air artificially contaminated with benzene. The process was monitored over a period of 240 days; every 60 days the concentration of benzene in the upflowing air was progressively increased from 0.01 to a maximum of 0.3 gm^{-3} (Table 1). The removal efficiencies were measured as described elsewhere [Sene et al. 2002].

2.2 *Analysis of microflora*

The packed compost was sampled at different depths of the biofiltering column (0.25, 0.5, 0.75, 1 m from the bottom). Total DNA was extracted with a commercial kit (QBiogene) from 0.5 g of compost. ARISA fingerprint was performed on the extracted DNA as described elsewhere [Brusetti et al. 2003].

Serial dilutions of compost were plated on Plate Count Agar (Merk) for the total microbial counts. Isolated colonies were subcultured in the same medium. Bacterial DNA was extracted with the method of Ausubel et al. [1994], and RISA analysis was performed as described by Daffonchio et al. [1998]. Selected strains were identified by partial sequencing of the 16S rDNA as described in Urzì et al. [2001].

2.3 *Test for growth on benzene and screening for* todC1 *gene*

Pure cultures were streaked on agarised M9 medium [Sambrook et al. 1999]. Inoculated plates were incubated for 10 days at 30°C in closed boxes in the presence and absence of benzene vapours. Strains exhibiting growth only in presence of benzene were scored as positive for benzene utilisation.

The *TodC1* gene was amplified with specific primers from the genomic DNA of the strains, and homology with *todC1* amplicon was verified by Southern Blot hybridisation as described by Cavalca et al. [2000].

Table 1. Biofilter efficiency, number of ITS groups found in the compost, and ITS groups which comprise strains with the capability of growing on benzene.

Sampling days	Benzene load (g/m^3)	Removal efficiency (%)	n° ITS groups isolated	n° ITS groups with growth on benzene
0	0	–	3	1
60	0.01	100	8	2
120	0.02	100	6	0
180	0.05	100	10	2
240	0.3	95	20	12

3 RESULTS

3.1 *Biofiltration process*

Biofiltration of benzene contaminated air was monitored for a period of 240 days, at a gas velocity in the biofilter column of 31 ms^{-1}. During this period the removal efficiency of the pollutant was 100% for the first 180 days, after which it slightly decreased to 95% (Table 1). The process was very efficient, and the maximum removal capacity has been calculated as g benzene/$m_{\text{packing material}}^{-3}$ h^{-1} 20.1 at a gas velocity of 31 mh^{-1} and an influent benzene concentration of 0.4 gm^{-3} (data not shown).

3.2 *Bacterial diversity during the biofiltration*

Two grams of compost were sampled before the biofilter packing (0 days), and during the process after 60, 120, 180 and 240 days, at 4 different depths of the biofiltering column. The overall diversity of the total eubacterial microflora was analysed with a cultivation-independent approach. The total DNA was extracted from the samples, and the microbial diversity described with the ARISA method, which targets the spacer regions between the 16S and the 23S rRNA producing highly informative fingerprints (figure 1).

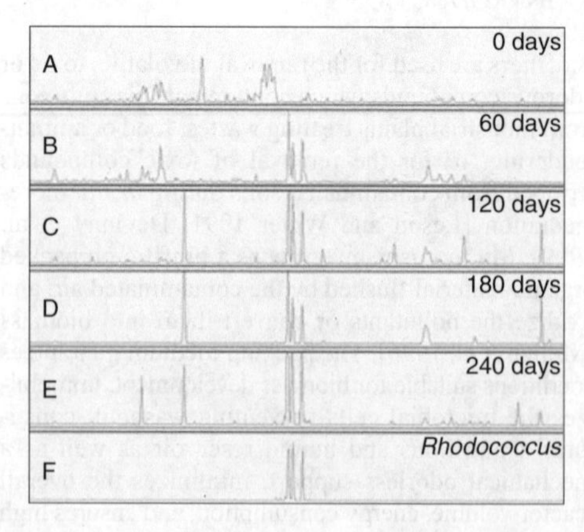

Figure 1A, B, C, D, E. ARISA fingerprinting of eubacterial communities of compost at the different sampling days in the central position of the biofilter column (0.5 m). 1F: ARISA profile obtained from a pure culture identified as *Rhodococcus* isolated from the compost.

ARISA profiles of community DNA showed that during the treatment the microbial community in the biofilter shifted and selected specific new populations which predominated in the last sampling times.

Besides the analysis of the total DNA, we also analysed the cultivatable fraction of microflora. Microbial counts (figure 2) showed that the cultivatable microorganisms increased about one order of magnitude during the last 60 days of biofiltration, in parallel with the increase of the concentration of benzene in the upflowing gas. In the first 120 days of the treatment, the benzene concentration was kept low, to permit to the microflora to adapt at the presence of the pollutant, and then the concentration was increased to $0.3\,\mathrm{gm}^{-3}$ to obtain an efficient remediating process. After the increase, the efficiency of the biofiltration remained high, and this could lead to the hypothesis that there was a selection of microbial strains with good degrading capability using benzene as carbon source. Species diversity of the cultivatable microflora was analysed on a collection of about 250 strains isolated from the different samples of compost.

The strains have been analysed by RISA fingerprint which produced from each strain a profile of bands species/subspecies specific [Daffonchio et al. 1998]. This analysis could divide the collection into groups of the same species/subspecies, which were identified by partial sequencing of the 16S rDNA of one or two strains for each group. In figure 3 is reported the species/subspecies variability observed along the treatment, where it was apparent that in the last 60 days of biofiltration new populations appeared in the compost, in parallel with the increase in benzene concentrations, cultivatable bacterial counts and of the total populations observed with the analysis of total DNA by ARISA.

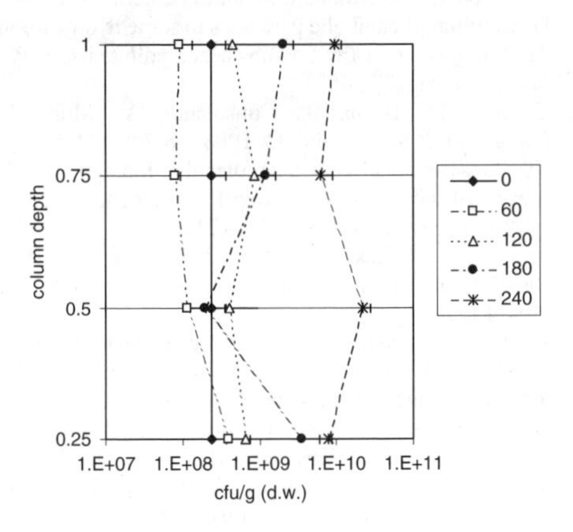

Figure 2. Microbial counts measured along the column depth (0.25, 0.5, 0.75, 1 m from the column bottom) at the different days of treatment (0, 60, 120, 180, 240 days of biofiltration).

In the original unpacked compost, characterised by a low level of diversity in cultivatable microflora, we identified species typical of this material, such as actimomycetes and aerobic sporeformers [Peters et al. 2000]. During the process new species appeared belonging to other taxonomic groups including α, β and γ proteobacteria, and high G + C gram positives (Table 2). These bacteria were either present in the original compost under the detection limit, or have been vehiculated with the air flowed into the biofilter. The most abundant isolates were analysed by ARISA, and their profile compared with the profile of the total microbial community (figure 1).

The results confirmed that some of the most abundant strains isolated at the end of the treatment, belonging to the genus *Rhodococcus* and *Neisseria*, constituted a significative part of the total microflora selected in the packing compost during the biofiltration. Moreover, PCR amplification of the gene *asd* specific for the genus *Neisseria* [Lansac et al. 2000] was performed on the total DNA of compost, obtaining positive signals only in the last days of biofiltration (data not shown), confirming that culturable *Neisseria* have been positively selected by the treatment, nevertheless they are not typically found in environmental habitats.

3.3 *Study of benzene degradation*

We studied in more detail the degrading potential of strains isolated at 240 days of biofiltration, when the benzene load was of $0.3\,\mathrm{gm}^{-3}$. 37.5% of the strains were able to grow on benzene as unique carbon source, and 32.5% harboured the *tod*C1 gene codifying for the toluene-monoxygenase, as confirmed by specific amplification and Southern Blot hybridisation with the gene of *Pseudomonas putida* F1 [Cavalca et al. 2000]. Not all the strains which grew on benzene have the *tod*C1 gene, and not all which

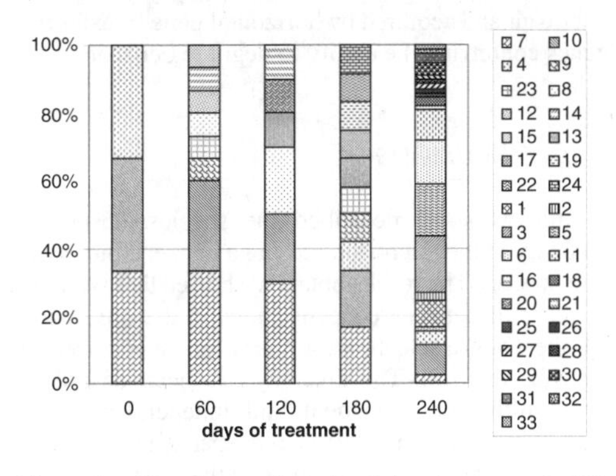

Figure 3. Percentage of strains belonging to each ITS group at the different sampling time.

Table 2. Main characteristics of strains belonging to selected ITS groups: phylogenetic identification, days of isolation, number of strains of the group which grow on benzene.

ITS group	n° of strains	rDNA sequence affiliation (% sequence similarity)	Samples of isolation	Growth on benzene
1	6	*Rhodococcus* (96%)	240	3
2	2	*Bordetella* (98%)	240	1
3	13	*Rhodococcus* (96%)	240	11
4	5	*Bacillus thuringiensis* (99%)	0, 180, 240	0
5	9	*Neisseria* (97%)	240	0
6	10	*Variovorax* (98%)	240	1
7	11	*Staphylococcus* (97%)	0, 60, 120, 180, 240	0
8	9	*Bacillus subtilis* (96%)	60, 120, 240	0
11	7	*Rhodococcus* (93%)	240	7
12	1	*Pseudoxanthomonas* (97%)		
16	1	*Neisseria* (98%)	240	1
17	3	*Alcaligenes* (98%)	240	0
18	2	*Agromyces* (91%)	0, 180	1

have the gene grew on benzene as unique carbon source. Moreover, not all the strains belonging to the same species/subspecies, *i.e.* to the same ITS group (Tables 1 and 2), grow on benzene and/or had the *tod*C1 gene. The capacity to grow on benzene was mainly associated with members of the genus *Rhodococcus*, known to have degrading activity on aromatic compounds [Cavalca et al. 2000]. These strains were prevalently isolated in the upper part of the biofiltering column, and have been revealed also among the total community analysed by ARISA (Figure 1). Interestingly, some strains affiliated to the genera *Neisseria* and *Bordetella* known to include pathogenic species not correlated with benzene degradation, did grow on this compound. To our knowledge, there is only one report of *Bordetella* associated with degradation of polycyclic aromatic hydrocarbons [Eriksson et al. 2003]. This finding led us to hypothesise that several of the benzene-degrading bacteria were carried into the compost by the atmospheric air and acquired by horizontal gene transfer the genes conferring the ability to degrade benzene.

4 CONCLUSIONS

In this work we described the species/subspecies diversity of the microflora during a benzene biofiltration process. The results obtained showed that with the increase of benzene concentrations in the upflowing air, there was a selection for microflora specialised in benzene removal. The diversity has been described with cultivation–dependent and independent techniques, indicating that several of the benzene degrading bacteria were cultivatable, and that the process conditions selected a degrading microflora with relatively high phenotypic and genetic diversity.

REFERENCES

Ahrens, A., Lipski, A., Klatte, S., Busse, H.J., Auling, G. and Altendorf, K. 1997. Polyphasic classification of proteobacteria isolated from biofilters. *System. Appl. Microbiol*. 20: 255–267.

Ausubel, F.M., Brent, R., Kingston, R.E., Moore, D.D., Seidman, J.G., Smith, J.A. and Struhl, K. 1994. *Current protocols of molecular biology*. John Wiley and Sons, Inc. U.S.A.

Brusetti, L., Francia, P., Bertolini, C., Pagliuca, A., Borin, S., Sorlini, C., Abruzzese, A., Sacchi, G., Viti, C., Giovannetti, L., Giuntini, E., Bazzicalupo, M. and Daffonchio, D. 2003. Bacterial rhizosphere community of transgenic Bt176 maize and its non transgenic counterpart. *Plant and soil*. Accepted for publication.

Cavalca, L., DiGennaro, P., Colombo, M., Androni, V., Bernasconi, S., Ronco, I. and Bestetti, G. 2000. Distribution of catabolic pathways in some hydrocarbon-degrading bacteria from a subsurface polluted soil. *Res. Microbiol.* 51: 877–887.

Daffonchio, D., Borin, S., Consolandi, A., Mora, D., Manachini, P. and Sorlini, C. 1998. 16-23S rRNA internal transcribed spacers as molecular markers for the species of the 16S rRNA group I of the genus *Bacillus. FEMS Microbiol. Lett.* 163: 229–236.

Devinny, J.S., Deshusses, M.A. and Webster, T.S. 1999. Biofiltration for air pollution control. Boca Raton: CRC Press LLC.

Eriksson, M., Sodersten, E., Zhongtang, Y., Dalhammar, G. and Mohn, W.W. 2003. Degradation of polycyclic aromatic hydrocarbons at low temperaure under aerobic and nitrate-reducing conditions in enrichment cultures from Northern soils. *Appl. Environm. Microbiol*. 69: 275–284.

Friedrich, U., Prior, K., Altendorf, K. and Lipski, A. 2002. High bacterial diversity of a waste gas-degrading community in an industrial biofilter as shown by a 16S rDNA clone library. *Environmental Microbiology.* 4(11): 721–734.

Gibson, D.T., Koch, J.R. and Kallio, R.E. 1968. Oxidative mineralization of hydrocarbons by microorganisms. I.

Enzymatic formation of catechol from benzene. *Biochemistry*. 7: 2653–2662.

Hanson, J.R., Ackerman, C.E., and Scow, K.M. 1999. Biodegradation of methyl tert-butyl ether by a bacterial pure culture. *Appl. Environm. Microbiol*. 65: 4788–4792.

Juteau, P., Larocque, R., Rho, D. and LeDuy, A. 1999. Analysis of the relative abundance of different types of bacteria capable of toluene degradation in a compost biofilter. *Appl. Microbiol. Biotechnol.* 52: 863–868.

Lansac, N., Picard, F.J., Ménard, C., Boissinot, M., Ouellette, M., Roy, P.H. and Bergeron, M.G. 2000. Novel genus-specific PCR-based assays for rapid identification of *Neisseria* species and *Neisseria meningitidis*. *Eur. J. Clin. Microbiol. Infect. Dis*. 19: 443–451.

Leson, G. and Winer, A.M. 1991. Biofiltration: an innovative air pollution control technology for VOC emissions. *J. Air Waste manag. Assoc*. 41: 1045–1054.

Lipski, A. and Altendorf, K. 1997. Identification of heterotrophic isolated from ammonia-supplied experimental biofilters. *System Appl. Microbiol.* 20: 448–457.

Møller, S., Pedersen, A.R., Poulsen, L.K., Arvin, E. and Molin, S. 1996. Activity and three-dimensional distribution of toluene-degrading. *Pseudomonas putida* in a multispecies biofilm assessed by quantitative *In situ* hybridisation and scanning confocal laser microscopy. *Appl. Environm. Microbiol*. 62(12): 4632–4640.

Peters, S., Koshkinski, S., Schwiegher, F. and Tebbe, C.C. 2000. Succession of microbial communities during hot composting as detected by PCR- Single-Strand-Conformation-Polymorphism-based genetic profiles of small-subunit rRNA gene. *Appl. Environm. Microbiol.* 66(3): 930–936.

Sakano, Y. and Kerkhof, L. 1998. Assessment of changes in microbial community structure during operation of an ammonia biofilter with molecular tools. *Appl. Environm. Microbiol.* 64(12): 4877–4882.

Sambrook, J., Fritsch, E.F. and Maniatis, T. 1989. *Molecular cloning: a laboratory manual. 1,2,3,II ed.* Cold Spring Harbor, NY: Cold Spring Harbor Laboratory Press.

Sene, L., Converti, A., Felipe, M.G.A., Zilli, M. 2002. Sugarcane Bagasse as Alternative Packing Material for Biofiltration of Benzene Polluted Gaseous Streams: A Preliminary Study. *Biores. Technol*. 83: 153–157.

Urzì, C., Brusetti, L., Salamone, S., Sorlini, C., Stackebrandt, E. and Daffonchio, D. 2001. Frequency and biodiversity of *Geodermatophilaceae* isolated from altered stones in the Mediterranean basin. *Environmental Microbiology*. 3, 471–479.

European Symposium on Environmental Biotechnology, ESEB 2004 - Verstraete (ed)
© 2004 Taylor & Francis Group, London, ISBN 90 5809 653 X

Microbial population dynamics in a CANOXIS and a UASB TCE-degrading bioreactor

O. Tresse
LGPTA/INRA, Villeneuve d'Ascq, France

F. Mounien, M.-J. Lévesque & S.R. Guiot
Environmental Bioengineering Group, BRI/NRC, Montreal (QC), Canada

ABSTRACT: Trichloroethylene (TCE), a common soil and groundwater pollutant, can be biodegraded either in aerobic and anaerobic conditions. Tartakovsky *et al.* (2003) have shown recently that the TCE degradation performances were higher in the coupled aerobic/anaerobic bioreactor (CANOXIS) oxygenated with hydrogen peroxide (H_2O_2) at a sub-bactericidal concentration than in an upflow anaerobic sludge bioreactor (UASB). To understand the microbial ecology underlying the TCE degradation in those reactors, a molecular study of the microbial population dynamics and a phylogenetic characterization were conducted using polymerase chain reaction-denaturing gradient gel electrophoresis (PCR-DGGE). The two bacteria domains and the presence of functional genes for TCE degradation were investigated using appropriate primers.

1 INTRODUCTION

One of the solutions to decrease the TCE contamination in the environment is to exploit the microbial potential for TCE degradation. TCE can be biodegraded in either aerobic or anaerobic conditions. The anaerobic dehalogenation of TCE can be performed by reductive dechlorination or by halorespiration (van Eekert & Schraa, 2001). The aerobic bacterial biodegradation of TCE was found to occur only by co-metabolism (fortuitous transformation) in the presence of growth substrates such as ammonia, methane, butane, toluene, chloroethane, propylene, isoprene, and phenol. The most studied enzymes responsible for the TCE co-metabolism are the soluble (sMMO) and particulate (pMMO) methane monooxygenase found in methane consuming bacteria (e.g. *Methylosinus trichosporium* and *Methylomonas methanica*) (Arp *et al.*, 2001).

The aerobic and anaerobic catabolisms of chlorinated compounds have limitations in their biodegrading abilities. Most chlorinated chemicals are incompletely transformed under anaerobic conditions (Freedman & Gosset, 1989; Wild *et al.*, 1995). In contrast, aerobic microorganisms are efficient degraders of less chlorinated or dechlorinated compounds and can cause complete mineralization (Arp, Yeager & Hyman, 2001). For instance, reductive dechlorination of TCE leads readily to dichloroethylene (DCE) and vinyl chloride (VC) that are more amenable to mineralization via aerobic pathways. Therefore, a combination of aerobic and anaerobic conditions is a good alternative to complete degradation of TCE.

Recently, combined anaerobic and aerobic conditions were achieved in a single-stage bioreactor system by oxygenating microorganisms from anaerobic sludge immobilized on peat granules. This coupled anaerobic/aerobic reactor performed a near-to-complete degradation of TCE at high rates as compared with an incomplete TCE degradation obtained without oxygenation (Tartakovsky, Manuel & Guiot, 2003). In the present study, we propose a microbial ecology of these reactors by analyzing the bacterial population dynamics by polymerase chain reaction-denaturing gradient gel electrophoresis. This study was completed by a phylogenetic characterization to determine the possible TCE degraders and their growth conditions.

2 MATERIALS AND METHODS

2.1 *Bioreactor setup, operating conditions and performances*

The experimental setup and operating conditions were described previously by Tartakovsky *et al.* (2003). Briefly, both reactors were loaded with 300 mL of

Table 1. Bioreactor operating conditions and performances. UASB: Upflow anaerobic sludge bed reactor, CANOXIS: Coupled anaerobic/aerobic single-stage reactor.

	1		2		3	
Phase reactor	UASB	CANOXIS	UASB	CANOXIS	UASB	CANOXIS
COD load (g COD L_R^{-1} d^{-1})	1.0	1.0	1.0	1.0	1.0	1.0
H_2O_2 load (mL L_R^{-1} d^{-1})	0	0.75	0	1.5	0	1.5
TCE load (mg L_R^{-1} d^{-1})	0	0	0	0	18	18
Duration (d)	30	30	34	34	31	31
TCE mineralization* (mg g_{peat}^{-1} d^{-1})	0.77 ± 0.05	0.89 ± 0.31	0.49 ± 0.01	0.94 ± 0.30	0.62 ± 0.04	1.95 ± 0.25

* Results are expressed as mean ± SD for triplicates.

pre-soaked granulated peat and inoculated with 200 mL of anaerobic sludge obtained from an upflow anaerobic sludge bed (UASB) reactor treating wastewater from a food plant (Sensient Flavors Canada Inc., Cornwall, Ontario, Canada) and operated with a hydraulic retention time of 17–20 h, a liquid upflow velocity of 4–6 m h^{-1} and at 25°C. The operating conditions and the performances of the two bioreactors during the 3 experimental phases are shown in Table 1. At the end of each experimental phase, samples of biomass were collected through a sampling port at the bottom of the reactors and conserved at −80°C.

2.2 *DNA extraction*

Genomic DNA was extracted as described before by Tresse *et al.* (2002). After extraction, the DNA was purified using polyvinylpolypyrrolidone spin columns (Berthelet *et al.*, 1996) and subjected to PCR amplification.

2.3 *PCR amplification of 16S rRNA gene*

Fragments of 16S rDNA were amplified for *Bacteria* with primers 341f and 758r (Muyzer *et al.,* 1993), for *Archaea* with primers 344f and 915r (Dumestre *et al.*, 2002) and *pmoA* gene with primers A189 and Mb661 (Costello & Lisdstrom, 1999). The GC-clamp described by Muyzer *et al.* (1993) was included on the 5′ end of the forward primers for DGGE analysis. The PCR mixture preparation, the DNA amplification and the DNA quantification were performed as described by Tresse *et al.* (2002). DNA from pure cultures of *Methanosaeta concilii* GP6 (Patel, 1984) and *Methylosinus sporium* (ATCC 35070) was used as positive controls for *Archaea* and *pmoA* gene fragments, respectively.

2.4 *DGGE and Phylogenetic analyses*

The protocols for DNA-fragment isolation and sequencing were those described by Tresse *et al.* (2002) except that the urea-formamide denaturant gradient gels were from 40% to 60%. Each DGGE gel was done in duplicate. The sequencing was done in both directions for each DNA fragments and the sequences were compared to Genbank databases. Numerical analysis of DGGE banding pattern was conducted using GelcomparII software package (Applied Maths Inc., Austin, TX). The 16S rDNA sequences obtained from the DGGE bands were aligned with complete 16S rDNA sequences of the most related organism and some well-known organisms obtained from NCBI Genbank. Phylogenetic trees were constructed from the evolutionary distances using the Neighbor-Joining algorithm. Bootstrap analyses (1000 data sets) were performed using the program SEQBOOT to provide confidence estimates of the tree topologies.

3 RESULTS AND DISCUSSION

The TCE-degradation performances showed a near-to-complete degradation of the TCE in the coupled reactor as compared to a low and incomplete TCE degradation found in the anaerobic reactor (Tab. 1). The DNA was extracted from samples collected in the anaerobic sludge (UASB) reactor and the coupled anaerobic/aerobic single-stage (CANOXIS) reactor at the end of each operating phase.

The banding profiles obtained with the archaeal primer set showed a major band corresponding to *Methanosaeta concilii* (bands #B, #D and #E: 100% identity) in all the samples and a minor band corresponding to *Methanobacterium formicicum* (band #A: 89% identity) only in the inoculum (SeS) and the UASB at the end of phase 3 (Fig 1a). As shown previously, *Methanosaeta*-like organisms are commonly found in high density. The above two methanogens were also shown to be present in brewery-treating granular sludge (Chan *et al.*, 2001) and in the core of the anaerobic granules of the agrofood industry wastewater-treating UASB reactor (Rocheleau *et al.*, 1999; MacLeod *et al.*, 1990). The presence of *M. concilii* in the coupled reactor confirms the methanogenic activity (CH_4 production) described by Tartakovsky *et al.* (2003) in the coupled reactor.

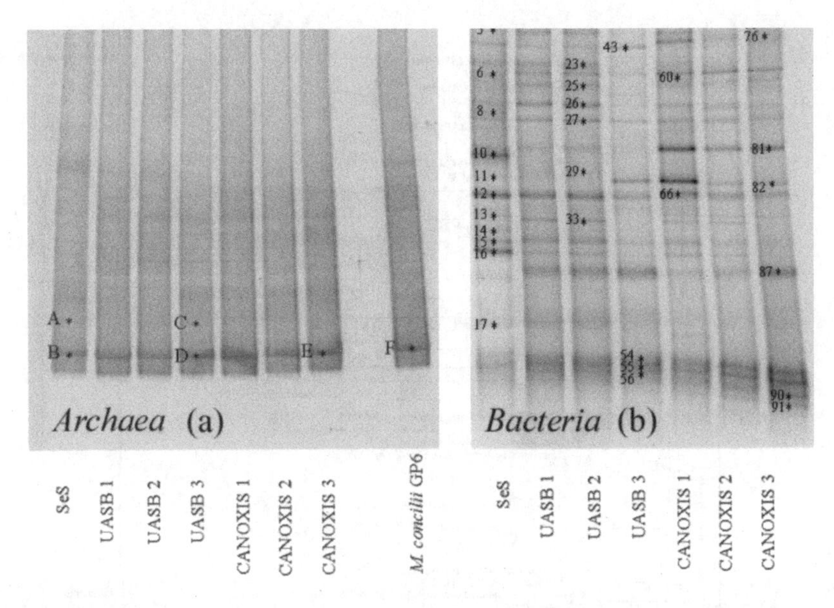

Figure 1. DGGE banding profiles of *Archae* (a) and *Bacteria* (b). 1–3: Operating phases, SeS: Sensient flavors sludge, UASB: Upflow anaerobic sludge bed reactor, CANOXIS: Coupled anaerobic/aerobic single-stage reactor. *Methanosaeta concilii* GP6: Positive control. Each star associated to a letter or a number represents a DNA band that was subjected to sequencing.

The analysis of the eubacterial populations showed that the DNA extracted from bands present only in the inoculum (SeS) belonged essentially to the γ-proteobacteria (Fig. 1b). Both reactors were characterized by bacteria belonging to the CFB group and the δ-proteobacteria. However, Firmicutes and γ-proteobacteria were only detected in the anaerobic reactor whereas α- and β-proteobacteria were only detected in the coupled reactor. Consequently, the bacterial populations evolved differently in the anaerobic and the coupled reactor.

The introduction of the TCE in phase 3 was followed in both reactors by the disappearance of two bands (#16 and #33), related to an Uncultured bacterium Kg-Gitt2 (93% identity, Firmicutes) and an Uncultured eubacterium AA26 (97% identity, CFB group) indicating a toxic effect of the TCE (Figs 1b, 2). On the other, the introduction of H_2O_2 in the coupled reactor, as compared to the anaerobic reactor, favored the appearance of microorganisms related to the *Dechloromonas* genus (bands #75: 98%, #76: 96%, #90 and #91: 100% identity), Rhizobiaceae family (*Rhizobium* sp. (band #81: 99% identity), *Sinorhizobium* sp. (band #66: 99% identity), *Agrobacterium tumefaciens* (band #82: 98% similarity)) (Fig. 1b, 2) and *Methylomonas* sp. LW21 with 100% identity (accession number: AF150801) detected in phase 2 and 3 when the reactor was fed with H_2O_2 at a rate of $1.5\,mL\,L_R^{-1}d^{-1}$, with specific primers targeting the *pmo*A gene (Fig. 3). No *smmo* gene was detected in both reactors (data not shown). Among these oxygen-stimulating bacterial species, only *Methylomonas* sp. LW21, a methanotrophic strain, is known to be able to co-metabolized the TCE. The presence of this strain confirms the methanotrophic activity detected by Tartakovsky *et al.* (2003) in the coupled reactor during phase 2 and 3. However, the fact that *Methylomonas* sp. LW21 was detected by gene-targeting primer and not with eubacteria-targeting primers indicates that this stains was not among the dominant bacterial populations into the coupled reactor. It is likely that other aerobic heterotrophic species in the coupled reactor out-competed the methanotrophic populations for the oxygen. Nevertheless, the study clearly indicates the possibility to have both methanogens and methanotrophs tightly associated in a single microsystem such as granular biofilms. This combination reduced the influence of toxic degradation intermediates, such as DCEs and VC on the overall degradation rate and provided a near complete mineralization of TCE.

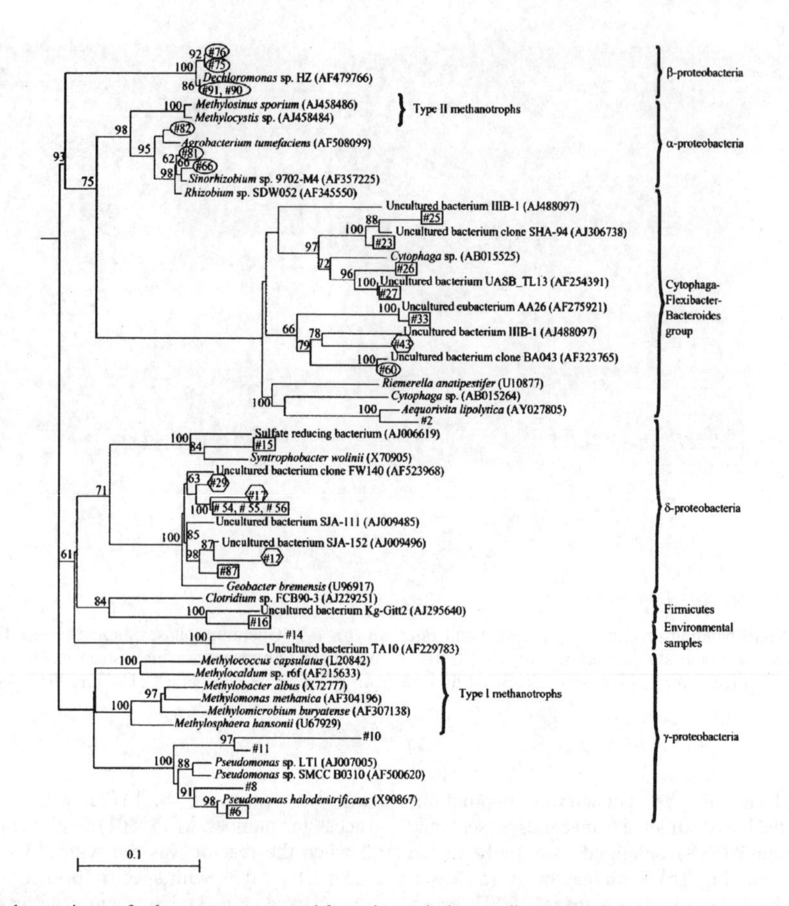

Figure 2. Phylogenetic tree for *bacteria* constructed from the evolutionary distances using the Neighbor-Joining algorithm. Bootstrap analyses (1000 data sets) were performed using the program SEQBOOT to provide confidence estimates of the tree topologies. Numbers in parenthesis correspond to the GeneBank accession numbers. (⬭): Bands detected in the CANOXIS reactor, (⬡) : Bands detected in UASB reactor, (▭): Bands detected in both bioreactors.

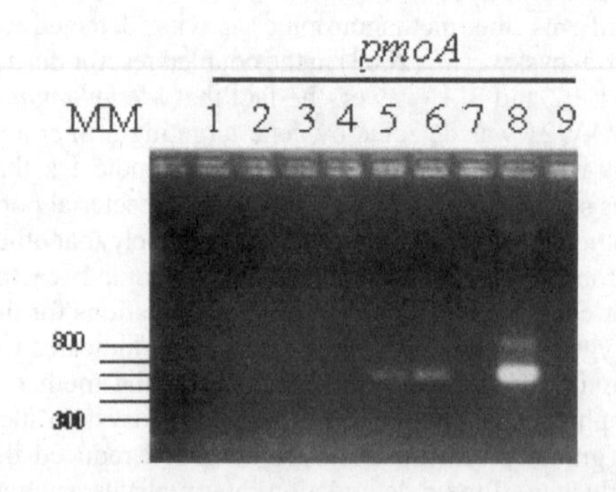

Figure 3. PCR amplification of *pmoA* genes in the reactor biomass. MM: 100 pb Molecular weight marker; 1, 2, and 3: UASB in phase 1, 2, and 3, respectively; 4, 5, and 6: CANOXIS in phase 1, 2, and 3, respectively, 7: Sensient flavors sludge, 8: PCR positive control (*Methylosinus trichosporium*) 9: Negative control.

ACKNOWLEDGMENTS

The authors thank Boris Tartakovsky and Michelle-France Manuel who provided the biological material and performance data from the reactors, Manon Lalumière for the DNA sequencing and Roland Brousseau for assisting with the realization of the phylogenetic tree.

REFERENCES

van Eekert, M. & Schraa, G. 2001. The potential of anaerobic bacteria to degrade chlorinated compounds, *Wat. Sci. Technol.* 44: 49–56.

Arp, D., Yeager, C. & Hyman, M. 2001. Molecular and cellular fundamentals of aerobic cometabolism of trichloroethylene, *Biodegradation* 12: 81–103.

Freedman, D. & Gosset, J. 1989. Biological reductive dechlorination of tetrachloroethylene and trichloroethylene to

ethylene under methanogenic conditions, *Appl. Environ. Microbiol.* 55: 2144–2151.
Wild, A., Winkelbauer, W. & Leisinger, T. 1995. Anaerobic dechlorination of trichloroethene, tetrachloroethene, and 1,2-dichloroethene by an acetogenic mixed culture in a fixed-bed reactor, *Biodegradation* 6: 309–318.
Tartakovsky, B., Manuel, M. & Guiot, S. 2003. Trichloroethylene degradation in a coupled anaerobic/aerobic reactor oxygenated using hydrogen peroxide, *Env. Sci. Technol.* (in press).
Tresse, O., Lorrain, M. & Rho, D. 2002. Population dynamics of free-floating and attached bacteria in a styrene-degrading biotrickling filter analyzed by denaturing gradient gel electrophoresis, *Appl. Microbiol. Biotechnol.* 59: 585–590.
Berthelet, M., Whyte, L. & Greer, C. 1996. Rapid, direct extraction of DNA from soils for PCR analysis using polyvinylpolypyrrolidone spin columns, *FEMS Microbiol. Lett.* 138: 17–22.
Muyzer, G., de Waal, E. C. & Uitterlinden, A. G. 1993. Profiling of complex microbial populations by denaturing gradient gel electrophoresis analysis of polymerase chain reaction-amplified genes coding for 16S rRNA, *Appl. Environ. Microbiol.* 59: 675–700.
Dumestre, J., Casamayor, E., Massana, R. & Pedrós-Alió, C. 2002. Changes in bacterial and archaeal assemblages in an equatorial river induced by the water eutrophication of Petit Saut dam reservoir (French Guiana), *Aquatic Microb. Ecol.* 26: 209–221.
Costello, A. & Lisdstrom, M. 1999. Molecular characterization of functional and phylogenetic genes from natural populations of methanotrophs in lake sediments, *Appl. Environ. Microbiol.* 65: 5066–5074.
Patel, G. 1984. Characterization and nutritional properties of *Methanothrix concilii* sp. nov., a mesophilic acetoclastic methanogen, *Can. J. Microbiol.* 30: 1383–1396.
Chan, O., Liu, W. & Fang, H. 2001. Study of microbial community of brewery-treating granular sludge by denaturing gradient gel electrophoresis, *Wat. Sci. and Technol.* 43: 77–82.
Rocheleau, S., Greer, C., Lawrence, J., Cantin, C., Laramee, L. & Guiot, S. 1999. Differentiation of *Methanosaeta concilii* and *Methanosarcina barkeri* in anaerobic mesophilic granular sludge by fluorescent in situ hybridization and confocal scanning laser microscopy, *Appl. Environ. Microbio.* 65: 2222–2229.
MacLeod, F., Guiot, S. & Costerton, J. 1990. Layered structure of bacterial aggregates produced in an upflow anaerobic sludge bed and filter reactor, *Appl. Environ. Microbiol.* 56: 1598–1607.

European Symposium on Environmental Biotechnology, ESEB 2004 - Verstraete (ed)
© 2004 Taylor & Francis Group, London, ISBN 90 5809 653 X

Diversity of pJP4-transconjugants from activated sludge

S. Bathe, M. Lebuhn & M. Hausner
Institute of Water Quality Control and Waste Management, Technical University of Munich, Am Coulombwall, Garching, Germany

S. Wuertz
Department of Civil and Environmental Engineering, University of California, Davis, USA

ABSTRACT: To extend the knowledge about the possible host range of IncP1-β plasmids in the environment, an agar surface mating was carried out with a donor strain carrying the 2,4-D degradative plasmid pJP4 and an undefined recipient community derived from activated sludge. Culturable transconjugants were isolated from this conjugation experiment, and a subset of isolates was identified by partial sequencing of their 16S rRNA genes. All were members of the proteobacteria, particularly of *Rhizobiaceae, Comamonadaceae,* and the γβ-branch of proteobacteria. The isolates were screened for their ability to grow on 2,4-D as the sole source of carbon and energy, but only members of the genera *Pseudomonas* and *Delftia* were able to grow under these conditions.

1 INTRODUCTION

To date, there is only limited information available about the host range of conjugative plasmids in a natural microbial community. Many potential hosts of a self-transferable plasmid have been determined by successful transfer of the plasmid from a pure donor to a pure recipient culture, as was done with pJP4 (Don & Pemberton 1981) and pWW0 (Ramon-Gonzalez et al. 1991). To gain a more detailed knowledge about the behaviour of a conjugative plasmid in the environment, isolation and identification of transconjugants from an undefined recipient community is necessary, and a number of recent studies have addressed this issue using different methodological approaches (Geisenberger et al. 1999, Newby et al. 2000, Goris et al. 2002, Goris et al. 2003). However, not all plasmid recipients will be able to profit from plasmid-encoded genetic traits, e.g. from plasmids carrying genes for the catabolism of xenobiotic compounds. Reasons may be lack of expression of these genes or – if the plasmid only carries parts of a catabolic operon – incomplete assembly of a catabolic pathway.

The experiment described here tries to extend the knowledge about the possible host range of the 2,4-dichlorophenoxyacetic acid degradative plasmid pJP4. This plasmid belongs to the incompatibility group IncP1-β, has a size of approx. 80 kb and carries a nearly complete pathway for the degradation of the herbicide 2,4-D. Additionally, genes for the partial catabolism of 3-chlorobenzoate and a mercury resistance operon are located on the plasmid.

In our study, we conducted an agar surface mating of a *Pseudomonas putida* donor strain carrying a pJP4-derivative with an undefined recipient community derived from activated sludge. We used the wildtype pJP4-plasmid tagged with a transposon carrying the gene for the red fluorescent protein *dsRed* under the control of a *lac*-promotor together with a gentamycin resistance gene. Cells carrying this plasmid displayed red fluorescence, as well as gentamycin and mercury resistance. To be able to differentiate between the donor strain and transconjugants, the donor strain was tagged with two copies of a wildtype *gfp* gene. Additionally, it carried a chromosomally encoded *lacI*-gene which repressed expression of the *dsRed* gene in the donor strain. Thus, donor cells were only green fluorescent, whereas transconjugant cells displayed only a red fluorescence.

2 MATERIALS AND METHODS

2.1 *Donor strain*

The donor strain used was *Pseudomonas putida* SM1443 carrying a *lacI* gene on the chromosome (Christensen et al. 1998). This strain was labelled with two copies of a wildtype *gfp* gene using the transposon plasmid pUT*gfp*2x, additionally carrying a kanamycin resistance (Unge et al. 1997).

In order to label the plasmid pJP4 with a *dsRed* gene, a transposon labelling of the original plasmid host *Ralstonia eutropha* JMP134 using pUTGm-*dsRed* (Tolker-Nielsen et al. 2000) was carried out. The resulting colonies consisted of cells with the transposon either integrated into the chromosome or into the plasmid. To select for labelled plasmids, a conjugation with these colonies as donor cells and the above constructed strain SM1443::*gfp*2x as a recipient was carried out and gentamycin and kanamycin-resistant colonies were selected. One representative colony was chosen as the donor strain. This strain showed a constitutive green fluorescence, an IPTG-inducible red fluorescence, and was able to grow on 2,4-D as the sole source of carbon and energy.

2.2 *Mating and transconjugant isolation*

To obtain transconjugants, 1 ml of activated sludge was cultured overnight in a synthetic wastewater (Nancharaiah et al. 2003). Approx. 10^9 cells of this culture were mixed with an equal amount of cells of the donor strain, grown in LB medium. The mixture was suspended in 100 μl PBS and spotted on the surface of an R_2A agar plate. The plate was incubated overnight at 30°C and stored at 4°C for several days to overcome the slow maturation of DsRed in transconjugants.

Then the mating patch was scraped from the surface of the plate, suspended in PBS, and a preparative flow cytometry targeting red fluorescent cells was carried out. Samples of the target fraction were plated on R_2A agar containing $HgCl_2$. The developing colonies were restreaked on R_2A containing gentamycin, and screened for red fluorescence by using an epifluorescence microscope. This procedure yielded 56 colonies which were assumed to be transconjugants.

2.3 *Phylogenetic analysis of the transconjugants*

Extracts of heat-denatured cells of the isolates were used as template-DNA in a genetic fingerprint PCR with the primer $(GTG)_5$ (Couto et al. 1996). Twenty four isolates showing different profiles were then selected, and approx. 500 bp of their 16S rRNA genes, starting around *E. coli* position 360, were sequenced. Identification of their phylogenetic identity was then carried out by subjecting the sequences to a database search at GenBank. Finally, strains showing identical or nearly identical 16S sequences were again subjected to a genetic screening using the primer BoxA1R (Rademaker et al. 1998) to exclude the possibility that the same strain had been isolated more than once.

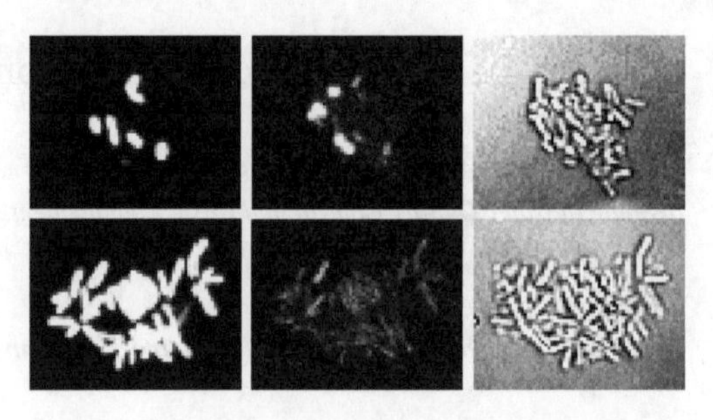

Figure 1. Two cell clusters in the flow cytometer target fraction. Transconjugants (left), donor cells (middle), and a transmission picture showing all cells (right) are shown.

2.4 *Utilization of 2,4-D*

The ability of the isolates to grow on 2,4-D was checked by using an indicator broth (DiGiovanni et al. 1996) containing 2,4-D and bromothymol blue in a weakly buffered solution. Growth in this medium was indicated by a colour change from green to yellow due to acidification during degradation of 2,4-D.

3 RESULTS

3.1 *Transconjugant isolation*

Microscopic investigation of flow cytometry fractions revealed that not single cells, but cell clusters consisting of transconjugants, donor cells, and recipients had been sorted (Figure 1). In order to obtain pure transconjugant cultures, flow cytometry fractions were plated on agar selective for cells carrying the pJP4 plasmid. The number of colonies developing on these plates agreed well with the number of "fluorescent events" counted by the flow cytometer.

3.2 *Phylogeny of the transconjugants*

All isolated transconjugants were members of the proteobacteria and displayed a broad taxonomic distribution. Table 1 shows a list of the isolates and their phylogenetic affiliations.

Almost all isolates showed sequence identity of more than 99% to the most closely related GenBank sequence. Most transconjugants affiliated with *Rhizobiaceae* (8 isolates), *Comamonadaceae* (4 isolates), and the γβ-subclass (Anzai et al. 2000) of the proteobacteria (7 isolates). The taxonomy within this branch containing a number of strains designated to *Stenotrophomonas maltophilia* is not clear and requires further analysis.

A number of strains with identical or nearly identical 16S rDNA sequences were obtained: within the *Rhizobiaceae* transconjugant strains 4, 7, 22, 23, and

Table 1. The isolated transconjugants and their phylogenetic affiliations.

Isolate no.	Best match GenBank	Group of proteobacteria	Growth on 2,4-D
4,7,22,23,29	100% *Rhizobium radiobacter* CFBP2243 (AJ389893)	α	−
14	100% '*Ochrobactrum anthropi*' HAMBI2402 (AF501340)	α	−
24	100% '*Agrobacterium tumefaciens*' C58 (AE009348)	α	−
47	100% *Ensifer adhaerens* LMG20582	α	−
2	100% *Alcaligenes faecalis* M3A (AF155147)	β	−
33	99,6% *Alcaligenes faecalis* sp. *parafaecalis* G (AJ242986)	β	−
17	99,8% *Comamonas terrigena* (AJ430346)	β	−
41	98,5% *Comamonas* sp. 12022 (AF078773)	β	−
31,36	100% *Delftia* sp. AN3 (AY052781)	β	+
40	99,8% *Stenotrophomonas maltophilia* HK40 (AJ011332)	γβ	−
45	99,8% *Stenotrophomonas maltophilia* C6 (AJ293468)	γβ	−
49,50	100% *Stenotrophomonas maltophilia* e-a21 (AJ293470)	γβ	−
64	100% *Stenotrophomonas maltophilia* LMG11087 (X95924)	γβ	−
57	99,8% '*Ultramicrobacterium*' str. 12–3 (AB008507)	γβ	−
59	99,8% *Xanthomonas* group bacterium LA37 (AF513452)	γβ	−
20	100% *Pseudomonas putida* PB4 (D37925)	γ	+
38	100% *Pseudomonas putida* #HR9 (AJ132993)	γ	+
44	98,7% *Klebsiella oxytoca* JCM1665 (AB004754)	γ	−

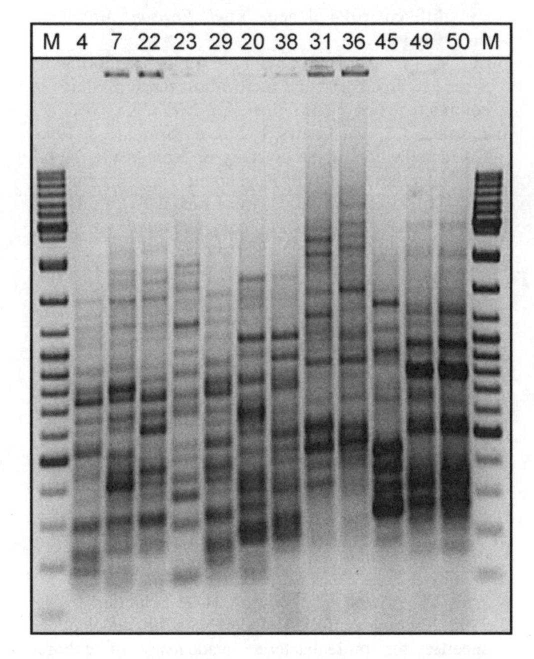

Figure 2. Box-PCR-profiles of selected transconjugant strains. M: 100 bp ladder, numbers indicate transconjugant isolates.

29, two strains of *Pseudomonas putida* (20 and 38), two *Delftia* strains (31 and 36), and three strains of the γβ branch (45, 49 and 50). To ensure that these isolates were genetically different strains, an additional screening by Box-PCR was carried out. The results are shown in Figure 2. All of the tested isolates differed in at least one band and thus represent different strains.

3.3 *Growth on 2,4-D*

Only 4 of the 24 sequenced isolates showed a positive in the reaction in the 2,4-D indicator broth, suggesting utilization of the compound: *Pseudomonas putida* strains 20 and 38, and *Delftia* strains 31 and 36.

Pseudomonas spp. have already been described earlier as being able to degrade 2,4-D after they had received pJP4 (e.g. Newby et al. 2000). There is no report on 2,4-D degradation by strains of the genus *Delftia*, but among 2,4-D degrading pJP4-transconjugants, related genera such as *Burkholderia* and *Ralstonia* have frequently been found.

4 DISCUSSION

The results of our study show a high phylogenetic diversity of pJP4-transconjugants obtained in this single experiment. The diversity of genera is considerably higher than that obtained in other studies (e.g. Goris et al. 2002, Newby et al. 2000). However, most of these studies were carried out with 2,4-D present during the mating, whereas this selective pressure was not applied in our experiment. When 2,4-D was present, the isolated transconjugants were nearly exclusively strains of the genera *Ralstonia, Burkholderia*, and *Pseudomonas*. Particularly strains of these genera seem to be able to express the catabolic genes residing on the plasmid and thus can degrade 2,4-D. Most of these investigations were carried out in a soil environment, whereas our experiment dealt with a recipient community derived from activated sludge. One of these differences could explain why we did not isolate transconjugants belonging to the genera *Ralstonia* and *Burkholderia*.

The results of the study may be further influenced by the conditions under which the conjugation experiment was carried out: Goris et al. (2003) compared the transconjugant diversity obtained by transfer of the IncP1-β plasmid pC1gfp in liquid activated sludge with the diversity obtained from matings on LB agar. In sludge, 15 of 20 transconjugant isolates were strains of *Delftia acidovorans*, whereas 53 out of 67 isolates obtained from the agar mating were *Aeromonas* spp. This difference can be explained by the growth conditions of the two environments: Wagner et al. (1993) showed in an earlier study that β-proteobacteria had a higher abundance in liquid activated sludge than γ-proteobacteria, whereas LB agar strongly selected for γ-proteobacteria.

In general, our data are in good agreement with the results obtained in an earlier study by Pukall et al. (1996), who investigated transfer of an IncP1-β plasmid under nonselective conditions and obtained a broad taxonomic distribution of transconjugants within the proteobacteria.

REFERENCES

Anzai, Y., Hongik, K., Park, J. -Y., Wakabayashi, H. & Oyaizu, H. 2000. Phylogenetic affiliation of the pseudomonas based on 16S rRNA sequence. *Int. J. Syst. Evolution. Microbiol.* 50:1563–1589.

Christensen, B. B., Sternberg, C., Andersen, J. B., Eberl, L., Moller, S., Givskov, M. & Molin, S. 1998. Establishment of new genetic traits in a microbial biofilm community. *Appl. Environ. Microbiol.* 64:2247–2255.

Couto, M. M. B., Eijsma, B., Hofstra, H., Huis in't Veld, J. H. J. & van der Vossen, J. M. B. M. 1996. Evaluation of molecular typing techniques to assign genetic diversity among *Saccharomyces cerevisiae* strains. *Appl. Environ. Microbiol.* 62:41–46.

DiGiovanni, G. D., Neilson, J. W., Pepper, I. L. & Sinclair, N. A. 1996. Gene transfer of *Alcaligenes eutrophus* JMP134 plasmid pJP4 to indigenous soil recipients. *Appl. Environ. Microbiol.* 62:2521–2526.

Don, R. H. & Pemberton, J. M. 1981. Properties of six pesticide degradation plasmids isolated from *Alcaligenes paradoxus* and *Alcaligenes eutrophus*. *J. Bacteriol.* 145:681–686.

Geisenberger, O., Ammendola, A., Christensen, B. B., Molin, S., Schleifer, K. -H. & Eberl, L. 1999. Monitoring the conjugal transfer of plasmid RP4 in activated sludge and in situ identification of the transconjugants. *FEMS Microbiol. Lett.* 174:9–17.

Goris, J., Boon, N., Lebbe, L., Verstraete, W. & de Vos, P. 2003. Diversity of activated sludge bacteria receiving the 3-chloroaniline-degradative plasmid pC1gfp. *FEMS Microbiol. Ecol.* 46:221–230.

Goris, J., Dejonghe, W., Falsen, E., de Clerck, E., Geeraerts, B., Willems, A., Top, E. M., Vandamme, P. & de Vos, P. 2002. Diversity of transconjugants that acquired plasmid pJP4 of pEMT1 after inoculation of a donor strain in the A- and B-horizon of an agricultural soil and description of *Burkholderia hospita* sp. nov. and *Burkholderia terricola* sp. nov. *Syst. Appl. Microbiol.* 25:340–352.

Nancharaiah, Y. V., Wattiau, P., Wuertz, S., Bathe, S., Wilderer, P. A. & Hausner, M. (2003). Dual labelling of *Pseudomonas putida* with fluorescent proteins for in situ monitoring of conjugal transfer of the TOL plasmid. *Appl. Environ. Microbiol.* 69:4846–4852.

Newby, D. T., Gentry, T. J. & Pepper, I. L. 2000. Comparison of 2,4-dichlorophenoxyacetic acid degradation and plasmid transfer in soil resulting from bioaugmentation with two different pJP4 donors. *Appl. Environ. Microbiol.* 66:3399–3407.

Pukall, R., Tschäpe, H. & Smalla, K. 1996. Monitoring the spread of broad host and narrow host range plasmids in soil microcosms. *FEMS Microbiol. Ecol.* 20:53–66.

Rademaker, J. L. W., Louws, F. J. & de Bruijn, F. J. 1998. Characterization of the diversity of ecologically important microbes by rep-PCR genomic fingerprinting. In A. D. L. Akkermans, J. D. van Elsas & F. J. de Bruijn (eds), *Molecular microbial ecology manual*: vol. 3.4.4. Kluwer, Dordrecht, The Netherlands.

Ramon-Gonzalez, M. -I., Duque, E. & Ramos, J. L. 1991. Conjugational transfer of recombinant DNA in cultures and in soils: host range of *Pseudomonas putida* TOL plasmids. *Appl. Environ. Microbiol.* 57:3020–3027.

Tolker-Nielsen, T., Brinch, U. C., Ragas, P. C., Andersen J. B., Jacobsen, C. S. & Molin, S. 2000. Development and dynamics of *Pseudomonas* sp. biofilms. *J. Bacteriol.* 182:6482–6489.

Unge, A., Tombolini, R., Müller, A. & Jansson, J. K. 1997. Optimization of GFP as a marker for detection of bacteria in environmental samples. In J. W. Hastings, L. J. Kricka & P. E. Stanley (eds), *Bioluminescence and chemiluminescence, molecular reporting with photons*: 391–394. Whiley & Sons, Sussey, United Kingdom.

Wagner, M., Amann, R., Lemmer, H. & Schleifer, K. -H. 1993. Probing activated sludge with oligonucleotides specific for proteobacteria: inadequacy of culture-dependent methods for describing microbial community structure. *Appl. Environ. Microbiol.* 59:1520–1525.

Water purification

European Symposium on Environmental Biotechnology, ESEB 2004 - Verstraete (ed)
© 2004 Taylor & Francis Group, London, ISBN 90 5809 653 X

A novel slow release approach to bioaugment more effectively activated sludge systems

N. Boon, L. De Gelder, H. Lievens & W. Verstraete
Laboratory of Microbial Ecology and Technology (LabMET), Faculty of Agricultural and Applied Biological Sciences, Ghent University, Coupure Links; Ghent, Belgium

S.D. Siciliano
Department of Soil Science, University of Saskatchewan, Saskatoon, Saskatchewan, Canada

E.M. Top
Department of Biological Sciences, University of Idaho, Moscow

ABSTRACT: The survival and activity of microbial degradative inoculants in bioreactors is critical to obtain successful biodegradation of non- or slowly degradable pollutants. Achieving this in industrial wastewater reactors is technically challenging. We evaluated a strategy to obtain complete and stable bioaugmentation of activated sludge, used to treat a 3-chloroaniline (3-CA) contaminated wastewater in a lab-scale Semi-Continuous Activated Sludge system. A 3-CA metabolizing bacterium, *Comamonas testosteroni* strain I2, was mixed with molten agar and encapsulated in open ended silicon tubes. The tubes containing the immobilized bacteria represented about 1% of the volume of the mixed liquor. The bioaugmentation activity of a reactor containing the immobilized cells was compared with a reactor with suspended I2*gfp* cells. From day 25–30 after inoculation, the reactor with only suspended cells failed to degrade 3-CA completely, due to a decrease in specific metabolic activity. In the reactors with immobilized cells however, 3-CA continued to be removed. A mass balance indicated that ca. 10% of the degradation activity was due to the immobilized cells. Slow release of the growing embedded cells from the agar into the activated sludge medium, resulting in a higher number of active 3-CA degrading I2 cells, was responsible for ca. 90% of the degradation. Our results demonstrate that this simple immobilization procedure was effective to maintain a 3-CA degrading population within the activated sludge community.

1 INTRODUCTION

Xenobiotics transiently present in sewage persist for extended periods of time and disrupt reactor function. Because of slow microbial adaptation and growth, there is seldom sufficient metabolic capacity to protect reactors from these toxicants that only rarely appear. The removal of these chemicals is often enhanced by inoculating specialized xenobiotic degrading bacteria. These laboratory-cultured inocula, which can transform the xenobiotica very efficiently in pure cultures, are however usually of little effectiveness once they are inoculated into an established microbial community. It is hypothesized that bioaugmentation failures arise because, (i) the contaminant concentration is too low to support microbial metabolism; (ii) microbial inhibitors are present in the environment; (iii) the growth rate of the degrading organism may be slower than the rate of cell removal, for example by predation or wash-out; (iv) the inoculum may use substrates other than the pollutant whose destruction is desired, or finally; (v) the organism may physically fail to come in contact with the pollutant; (vi) the pollutant may not be bioavailable due to sorption (Goldstein et al., 1985).

As a result of these practical problems, few successful cases of lab-scale activated sludge bioaugmentation have been described (Top et al., 2002), and most of these report only partial success. These studies are limited to the inoculation of suspended cells into the mixed liquor. After the introduction, cell densities decline and there is relatively minor influence on reactor functions (McClure et al., 1989).

Our research group reported successful removal of 3-chloroaniline (3-CA) in activated sludge. 3-CA is a chemical used in industry during the production of polyurethanes, rubber, azo-dyes, drugs, photographic chemicals, varnishes and pesticides and are often transiently detected in wastewaters at concentrations ranging from 12 mg/L to 230 mg/L (Jen et al., 2001). These compounds can cause reactor failure when shock loads of 250 mg/L are applied (Boon et al., 2003). A 3-CA degrading strain, *Comamonas testosteroni* I2*gfp* was inoculated into the activated sludge and during 14 days, the 3-CA was completely degraded (Boon et al., 2000). From day 14 on, the degradation rate declined, resulting in an accumulation of 3-CA in the reactor and only 50% of 3-CA removal was achieved after these two weeks.

Faced with the same difficulties encountered by other investigators, we evaluated a novel encapsulation method with silicone tubes as a means to protect the 3-CA degradative activity. These tubes were open at their ends, so that release of cells from inside the tubes could occur. The hypothesis of this work was that a continuous seeding of metabolically active bacteria would increase the 3-CA degradation.

2 MATERIALS AND METHODS

2.1 *Bacterial strains*

In order to monitor the bacterial inocula, the *Comamonas testosteroni* I2 cells were chromosomally marked with the genes encoding for the Green and Red Fluorescent Protein. The marked strains were designated respectively I2*gfp* (Boon et al., 2000) and I2*rfp* (Boon et al., 2003). Strains I2*rfp* and I2*gfp* mineralizes 3-CA completely and are rifampin and kanamycin (50 μg/mL) resistant (100 μg/mL). Strains I2, I2*gfp* and I2*rfp* are deposited in the BCCM™/LMG – Bacteria Collection (Ghent, Belgium) under the numbers LMG 19554, LMG 21409 and LMG 21402, respectively.

2.2 *Bioaugmentation experiment*

Semi-continuous activated sludge (SCAS) reactors have been used previously as a model for activated sludge systems and they allow to examine different reactor set-ups simultaneously in a reliable and reproducible way. The experiments were conducted in duplicate with sludge freshly collected from a domestic wastewater treatment plant (Bourgoyen-Ossemeersen, Ghent, Belgium), according to a SCAS (Semi-Continuous Activated Sludge) procedure as described previously (Boon et al., 2002). All reactors received a loading rate of 37 mg 3-CA/L.d, added as a single dose every other day.

The inocula were grown overnight in LB medium. The I2*gfp* or I2*rfp* containing silicon tubes were prepared as follows: an overnight grown culture of I2*gfp* or I2*rfp* cells was centrifuged and resuspended in saline and mixed with liquid LB agar at 40°C. Subsequently, while the mixture was still fluid, it was injected into the sterile silicone tubes. When the mixture was solidified, the tubes were cut into pieces of 3 cm long and placed in the reactors (100 cm per reactor); the silicone pieces remained open at both ends. The tubes containing the immobilized bacteria represented about 1% of the volume of the mixed liquor.

Bacterial counts, 3-CA and chloride analysis were performed, as described previously (Boon et al., 2002).

3 RESULTS

Cells marked with the green fluorescent protein (I2*gfp*) were freely suspended and cells marked with the red fluorescent protein (I2*rfp*) were encapsulated. Four types of reactors were used: (A) non-inoculated control reactor; (B) only suspended I2*gfp*; (C) suspended I2*gfp* and encapsulated I2*rfp*; (D) only immobilized I2*rfp*.

Complete removal of 3-CA was obtained by day 2 in all inoculated reactors (Figure 1). From day 26 on, 3-CA degradation decreased to only 24 mg/L mixed liquor·d in the reactors inoculated with suspended I2, compared to a rate of at least 37 mg/L mixed liquor·d for the reactors with encapsulated cells for the duration of the experiment. Reactors containing encapsulated and suspended cells degraded 3-CA at a lower rate for a four day period from day 20 until day 24 but reactors containing only encapsulated cells continuously removed all 3-CA throughout the experiment.

Suspended *C. testosteroni* I2*gfp* inoculated in the reactors B remained quite constant at ca. 10^7 CFU/mL mixed liquor throughout the study period despite the loss of 3-CA biodegradation activity (Figure 2). In contrast, I2*gfp* in the presence of encapsulated I2*rfp*

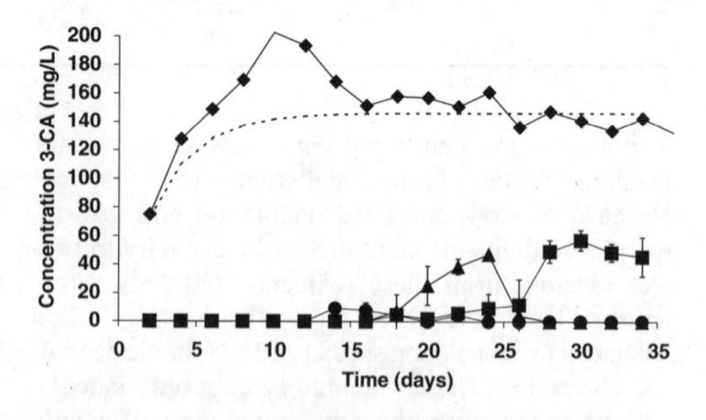

Figure 1. Concentration of 3-CA in the reactor A (non-inoculated control reactor) (◆), reactors B (suspended I2*gfp*) (■), the reactors C (suspended I2*gfp* and immobilized I2*rfp*) (▲) and the reactors D (immobilized I2*rfp*) (●). The striped line represents the theoretical 3-CA concentration at the imposed loading rate of 37 mg 3-CA/L mixed liquor•d. Values represent the mean ± error bars ($n = 2$); in some cases, the error bars were too small to be visible.

in reactors C only remained at 10^7–10^8 CFU/mL mixed liquor until day 15 when the number of cells declined drastically to 4.0×10^3 CFU/mL mixed liquor. The encapsulated I2*rfp* cells of reactors D remained quite stable at 10^7 CFU/mL mixed liquor except for a brief period between day 15 and 25. This corresponds to the observed lapse in 3-CA biodegradation activity. By transferring one tube out of the reactors to saline for one hour and plating the amount of released cells, it was observed that the silicone tubes released ca. 10^6 cells/h·silicone tube via their open ends. These measurements were repeated at regular intervals and throughout the entire period of the SCAS tests this rate of release of cells remained essentially unchanged (average release rate over was log 6.0 ± 0.3 CFU/tube·h).

Based on the data of the bioaugmentation test, bacterial kinetics were calculated and the relative proportion of each mode of action estimated. The calculations of the specific activity of cells present in the sludge for some time (aged cells) are based on reactors B (with only free suspended cells) and on the period where some 3-CA accumulation occurred. During this period (reference day 32) the removal rate was for the suspended cells 2.5 mg 3-CA/mg CDW_{I2gfp}·d. This was not sufficient to remove the daily dosed 37 mg 3-CA/L, so accumulation occurred.

In the reactors D with only the immobilized I2*rfp* cells, four different modes of 3-CA removal are possible: absorption, diffusion with degradation by the immobilized cells inside the tubes, degradation by the freshly released cells in the mixed liquor and by the aged cells already present in the mixed liquor. The absorption effect is minimal since only 3.3 mg 3-CA was able to bind to the tubes the first day and once the tubes are saturated they will no longer absorb 3-CA.

The second effect of the tubes on the degradation rate is the 3-CA degradation by immobilized cells inside the tubes. This estimated biodegradation rate potential, mediated by diffusion through the silicone walls, was minor, since the encapsulated cells in the batch experiment had the potential to degrade 9.8 mg/L mixed liquor·d.

A third fraction of the degradation can be considered to be due to the ongoing released I2*rfp* cells in the mixed liquor. Based on the 3-CA degradation rates of freshly released (within 24 h), active I2*rfp* cells in mineral medium, the 3-CA removal rate of the freshly released cells in reactor D was approximated at maximum 54.1 mg 3-CA/L mixed liquor·d. The aged cells already present in the activated sludge (from earlier release) also counted for 33.0 mg 3-CA/L mixed liquor·d degradation. The 3-CA degradation rate of the aged suspended cells from reactor was assumed to be the same as for the aged suspended cells from reactor B.

In conclusion, four different mechanisms played a role in the 3-CA removal of reactors (Figure 3). The prolonged 3-CA degradative activity in reactor D may have been caused by a combination of minor adsorption on the silicone tubes (0.1%), and degradation activities by three different types of I2*rfp* cells: immobilized cells in the agar (estimated to be responsible for only 10% of the 3-CA removal); freshly released active cells (56% of the degradation capacity), and aged cells present in the sludge (34% of the degradation capacity).

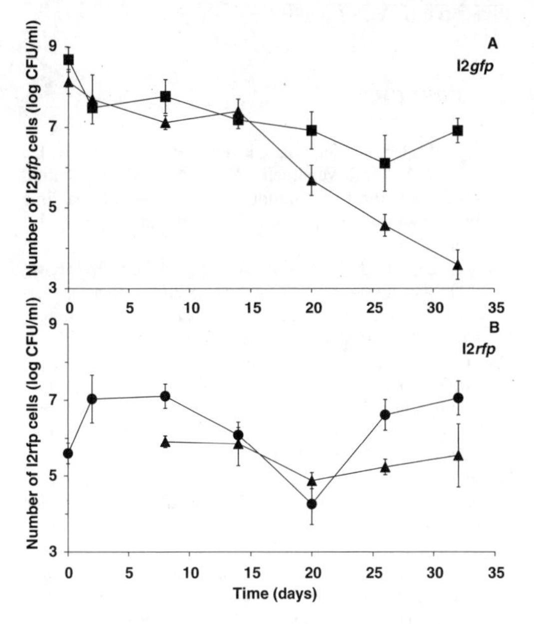

Figure 2. Number of I2*gfp* (A) and I2*rfp* (B) cells in the mixed liquor the reactors B (suspended I2*gfp*) (■), the reactors C (suspended I2*gfp* and immobilized I2*rfp*) (▲) and the reactors D (immobilized I2*rfp*) (●).

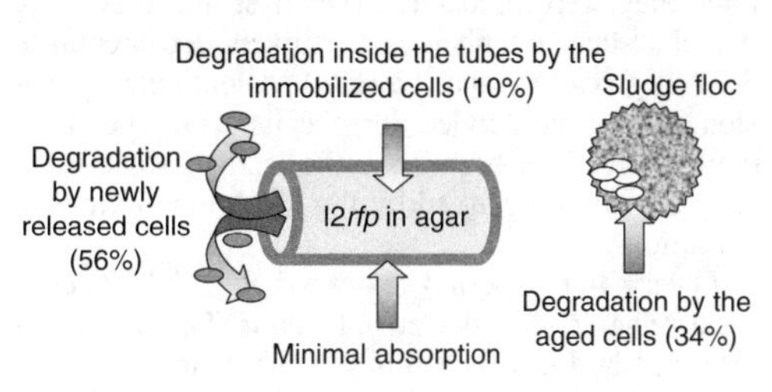

Figure 3. Schematic representation of the second bioaugmentation experiment. The modes of action and respective removal rates of 3-CA by agar immobilized *C. testosteroni* I2 cells in silicone tubes are schematized. The value represented between brackets is the estimated relative contribution of each mode of action to the total 3-CA removal in the reactor experiment.

4 DISCUSSION

The slow release concept is to our knowledge a new bioaugmentation strategy for bioreactors. In several studies where the use of carrier materials was examined, positive effects on survival and activity of inoculated strains have been reported (van Veen et al., 1997). Our work identifies an additional mechanism by which encapsulation improves biodegradation, namely continuous but relatively low level release of actively growing cells. In our set-up in reactors D, cell release could account directly or indirectly for 90% of the degradation occurring in a bioaugmented reactor, due to the aged cells in the sludge flocs (34%), but also due to the freshly released cells (56%). The release of free suspended cells may become of critical importance for trace toxicants that need to be removed from industrial waste water streams because, typically, biodegradation rates at low concentrations are better in suspended compared to immobilized cultures (Juarez-Ramirez et al., 2001). Thus, a slow release bioaugmentation set-up is at least as effective as the systems with complete immobilization or massive inoculation of specialized cells.

There was a significant difference between the capability of newly released or inoculated I2 cells to degrade 3-CA and aged I2*gfp* cells. While newly released cells continuously degraded 3-CA, the biodegradative activity of the original inoculum decreased substantially with time. In our previous work, lower degradation activity was also observed for I2*gfp* cells continuously exposed to 3-CA (Boon et al., 2000), resulting in an accumulation of 3-CA in the effluent. The situation for the inoculated I2 cells in the activated sludge environment can be considered as unfavorable. It has been shown that starvation can induce flocculation by *Ralstonia eutropha*, resulting in a complete loss of phenol-oxygenation activity (Watanabe et al., 2000). The I2 cells also showed a similar formation of flocs in the activated sludge, once they were inoculated (Boon et al., 2000). We postulate that this phenotypic shift is detrimental to 3-CA degradation capacity. Thus, the continuous provision of new suspended I2*rfp* cells from the tubes could result in a higher metabolic rate, since these cells are not yet starved by the harsh environmental conditions.

Over a 44 day period, about 3.5×10^{10} I2*rfp* cells were released into the mixed liquor. The substrates provided by the LB medium in the agar and by the 3-CA diffused into the silicone tubes were sufficient to sustain growth of the immobilized cells, which were subsequently released. These slowly released I2*rfp* cells in reactors C2 survived better (3.5×10^5 CFU/mL mixed liquor at day 32) compared to the suspended I2*gfp* cells (3.9×10^3 CFU/mL mixed liquor at day 32). Thus, I2*rfp* cells embedded in silicone tubes grew and slowly diffused out of the silicone tubes into the mixed liquor where they rapidly degraded 3-CA. The dual labeling strategy allowed to monitor the replacement of the I2 cells that were inoculated as a suspension by the continuously released cells, as is shown in Figure 3. The newly released I2*rfp* cells thus seemed to be more competitive in the reactors than the aged I2*gfp* cells. In these experiments, the positive effect lasted for at least 45 days. Longer experimental set-ups are needed to verify how long the slow release seeding can be effective.

In most activated sludge bioaugmentation experiments, inoculation with a specific strain generally has no or only a transient effect. This study has demonstrated that encapsulation of the inoculum improves biodegradation of 3-CA for an extended period of time in (industrial) wastewaters. The slow release of specialized, metabolic active cells was essential to maintain prolonged bioaugmentation, in contrast to their 'aged' counterparts which lost biodegradation activity after 30 days in the reactor.

ACKNOWLEDGMENTS

This work was supported by the project grant G.O.A. (1997–2002) of the "Ministerie van de Vlaamse Gemeenschap, Bestuur Wetenschappelijk Onderzoek" (Belgium) and by the Flemish Fund for Scientific Research (FWO-Vlaanderen).

REFERENCES

Boon, N., De Gelder, L., Lievens, H., Siciliano, S.D., Top, E.M. and Verstraete, W., 2002. Bioaugmenting bioreactors for the continuous removal of 3-chloroaniline by a slow release approach. Environ. Sci. Technol., 36: 4698–4704.

Boon, N., Goris, J., De Vos, P., Verstraete, W. and Top, E.M., 2000. Bioaugmentation of activated sludge by an indigenous 3-chloroaniline degrading *Comamonas testosteroni* strain, I2*gfp*. Appl. Environ. Microbiol., 66(7): 2906–2913.

Boon, N., Top, E.M., Verstraete, W. and Siciliano, S.D., 2003. Bioaugmentation as a tool to protect the structure and function of an activated sludge microbial community against a 3-chloroaniline shock load. Appl. Environ. Microbiol., 69: 1511–1520.

Goldstein, M.G., Mallory, L.M. and Alexander, M., 1985. Reasons for possible failure of inoculation to enhance biodegradation. Appl. Environ. Microbiol., 50(4): 977–983.

Jen, J.-F., Chang, C.-T. and Yang, T.C., 2001. On-line microdialisis-high-performance liquid chromatographic determination of aniline and 2-chloroaniline on polymer industrial wastewater. J. Chromatogr. A, 930: 119–125.

Juarez-Ramirez, C., Ruiz-Ordaz, N., Cristiani-Urbina, E. and Galindez-Mayer, J., 2001. Degradation kinetics of phenol by immobilized cells of *Candida tropicalis* in a

fluidized bed reactor. World J. Microbiol. Biotechnol., 17(7): 697–705.
McClure, N.C., Weightman, A.J. and Fry, J.C., 1989. Survival of *Pseudomonas putida* UWC1 containing cloned catabolic genes in a model activated-sludge unit. Appl. Environ. Microbiol., 55(10): 2627–2634.
Top, E.M., Springael, D. and Boon, N., 2002. Mobile genetic elements as tools in bioremediation of polluted soils and waters. FEMS Microbiol. Ecol., 42: 199–208.
van Veen, J.A., van Overbeek, L.S. and van Elsas, J.D., 1997. Fate and activity of microorganisms introduced into soil. Microbiol. Mol. Biol. Rev., 61: 121–135.
Watanabe, K., Miyashita, M. and Harayama, S., 2000. Starvation improves survival of bacteria introduced into activated sludge. Appl. Environ. Microbiol., 66(9): 3905–3910.

European Symposium on Environmental Biotechnology, ESEB 2004 - Verstraete (ed)
© 2004 Taylor & Francis Group, London, ISBN 90 5809 653 X

Removal of perchlorate and nitrate in an ion exchange membrane bioreactor

S. Velizarov, C. Matos, J.G. Crespo & M.A.M. Reis
CQFB/REQUIMTE, Department of Chemistry, FCT, Universidade Nova de Lisboa, Caparica, Portugal

ABSTRACT: The ion exchange membrane bioreactor (IEBM) concept is based on the use of a non-porous ion exchange membrane, acting as a barrier between water, containing one or more target inorganic charged pollutants, and a biological compartment, containing a microbial culture able to degrade (transform) these pollutants to harmless products. In this study, the IEMB process potential for treating drinking water streams, in which ClO_4^- may be present as a trace ion was evaluated. The results showed that ClO_4^- and NO_3^- ions could be simultaneously transported and biologically reduced. For a model polluted stream containing 100 ppb of ClO_4^- and 60 ppm of NO_3^-, the concentrations of both ions in the treated stream dropped below the recommended levels of 18 ppb for ClO_4^- and 25 ppm for NO_3^-, which corresponded to a treated water production rate of about 30 L/(m^{2}h).

1 INTRODUCTION

Perchlorate in drinking water supplies represents a serious concern and its possible removal has been the subject of considerable recent research efforts for its removal since otherwise it can persist for many decades under typical groundwater and surface water conditions (Urbansky 2000). Perchlorate can interfere with the production of human thyroid hormones (Christen 2000) and, therefore, a provisional action level of less than 18 μg/L for perchlorate has been already established by the California Department of Health Services and is already informally applied by other governmental agencies in USA.

The primary known sources of perchlorate contamination are industrial operations and military units using ClO_4^- as an oxidising agent in the form of ammonium, sodium, and potassium salts. However, perchlorate has been also identified in fertilizers at levels up to 0.84% (w/w) (Susarla et al. 1999), thus, considerably extending the possible affected areas. Since nitrate-rich fertilizers are the primary sources of nitrate contamination, it may be expected that both, NO_3^- and ClO_4^- ions could be present in irrigation and/or drinking water supplies. This leads to the emerging issue of their simultaneous removal under conditions, in which ClO_4^- is expected to be present at trace levels, thus requiring also analytical efforts.

Biological reduction of perchlorate (Wallace et al. 1998) allows for effective depollution by ensuring its reduction to chloride, while in physical processes this anion is just transferred and concentrated in a receiving phase. However, the main disadvantage of the biological treatment is possible microbial and secondary pollution of the treated water by excess substrates and metabolic by-products.

The ion exchange membrane bioreactor (IEMB) concept, schematically presented in Figure 1 was re-recently patented (Crespo & Reis 2001) and a

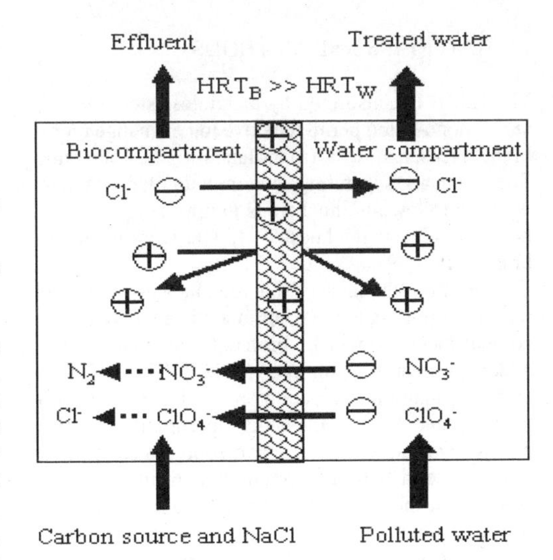

Figure 1. Schematic diagram of the ion transport mechanism.

mathematical model has been proposed for analysis of the trace ion transport rates in this type of reactor, based on the Donnan dialysis principles (Velizarov et al. 2003). Besides complete isolation of the microbial culture in a biocompartment, transport of co-ions (indicated as + containing circles on Figure 1) through the membrane is negligible due to their electrostatic repulsation (Donnan exclusion) from the positively charged membrane polymeric material. A suitable driving counter-ion (chloride in the case studied) is added to the biocompartment, providing the coupled counter-diffusion of the target counter-ions (all shown as – containing circles), where they undergo bioreduction to harmless products in the presence of a suitable electron donor (ethanol).

An important feature of Neosepta ACS membrane utilised in our process is its mono-anion perm-selective property, which may be attributed to the presence on the surface of a cross-linked modified top layer, with a very low concentration of fixed anions. Therefore, bivalent anions like SO_4^{2-} and HPO_4^{2-} are also excluded from the membrane, thus preserving the original water composition with respect to these anions.

Removal of trace charged pollutants from water streams represents a challenge due to both mass transfer (generally diffusion coefficients through ion exchange membranes) and bioreaction rate limitations, due to concentrations close to or lower than the Monod kinetic K_s saturation values.

Therefore, the objective of this study was to evaluate the IEMB potential for simultaneous removal of and nitrate and perchlorate from model water streams if perchlorate is present as a trace pollutant (in the ppb concentration range).

2 MATERIALS AND METHODS

The IEMB is constituted by a stainless steel module and a mono-anion perm-selective ion exchange membrane Neosepta ACS (manufactured by Tokuyama Soda, Japan) separating two rectangular channels with the following dimensions in mm: length – 260, height – 15 and thickness – 3. The working membrane area is 39 cm^2.

One of the channels of this module was connected through a re-circulation loop to a stirred vessel (biocompartment), to which "biofeed" was continuously added (Velizarov et al. 2001, 2002). The Cl^- ions served as major driving counter-ions for the transport of ClO_4^- and NO_3^- to the biocompartment, to which ethanol was added as a carbon source and electron donor for enriched on perchlorate and nitrate mixed culture. The inoculum was prepared in a sealed oxygen free flask under batch conditions (20 h, 24°C).

The other module channel was connected to a second vessel, to which a model solution composed of 100 ppb of ClO_4^- and 60 ppm of NO_3^-, both added as their sodium salts, was continuously fed and withdrawn. The two liquid phases (water and biomedium) were re-circulated through the corresponding module channels co-currently by means of gear pumps, providing Reynolds numbers of 3000 for both flows.

The temperature was maintained at 23 ± 0.5°C. The concentration of ClO_4^- was measured with a fully automated DIONEX ion exchange chromatography system, equipped with an Ionpac AS16 column. The system allows anion determination in the ppb range (detection limit of 4 ppb of ClO_4^-). The concentrations of NO_3^- and NO_2^- ions were followed by segmented flow analysis (Skalar Instruments).

3 RESULTS AND DISCUSSION

In order to evaluate the IEMB system performance, experiments were carried out with the same ClO_4^-/NO_3^- water composition but changing the hydraulic residence time in the water compartment, which had a volume of 0.06 L. It has to be mentioned that the combination of 60 ppm of nitrate and 100 ppb of perchlorate in the model water to be treated was chosen to mimic the ranges that could be encountered in practice as suggested by a drinking water supplying company.

The results of these experiments are reported in Table 1, which presents the pseudo-steady-state ion concentration values achieved at the two outlets, treated water and biocompartment effluent. The fluxes are calculated through the respective mass balances in the water compartment.

As can be seen, the nitrate fluxes to the biocompartment are about three orders of magnitude higher, a result of the correspondingly much higher nitrate loading rates. The specific ion flux values and removal degree depend on the process conditions (membrane to water compartment volume ratio; feed concentration and flow rate). Since the inlet water composition was kept constant, the target ion loading rate is proportional

Table 1. Effect of water hydraulic residence time in the water compartment of IEMB on the target ion fluxes.

	Ion concentration in treated water		Target ion flux to biocompartment	
HRT* (h)	ClO_4^- ($mg\,L^{-1}$)	NO_3^- ($mg\,L^{-1}$)	ClO_4^- ($g\,m^{-2}h^{-1}$)	NO_3^- ($g\,m^{-2}h^{-1}$)
1	0.017	11.5	$1.3 \cdot 10^{-3}$	0.75
0.5	0.018	10.1	$2.5 \cdot 10^{-3}$	1.50
0.25	0.033	19.8	$4.1 \cdot 10^{-3}$	2.50

* Hydraulic residence time (HRT) of the biomedium in the biocompartment was kept at about 120 hours.

to the volumetric flow rate, which was varied from 0.06 to 0.24 L/h.

In the IEMB system, development of a biofilm occurs on the membrane surface in contact with the mixed culture after the process start-up. In this case, the steady-state bulk nitrate concentration measured in the biocompartment was always close to zero (0.14 ± 0.12 mg NO_3^- L^{-1}). Obviously, an efficient nitrate bioreduction reaction was occurring within the biofilm at the biomedium/membrane interface. Thus, complete denitrification was achieved without accumulation of NO_3^- in the biocompartment. Such accumulation is highly undesirable and has greatly limited the use of biotechnological processes in drinking water denitrification since nitrite, a highly toxic metabolic intermediate, is usually temporarily observed. Since in the IEMB process NO_3^- ions were not accumulated in the biocompartment, formation of NO_2^- ions, if any, was not able to induce nitrite transport to the treated water.

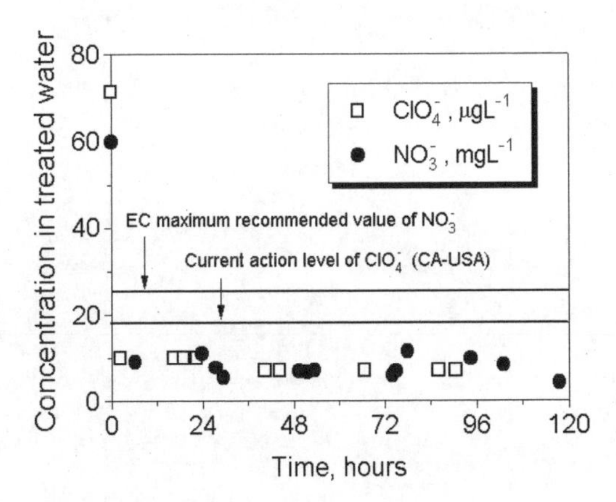

Figure 2. Time course of perchlorate and nitrate concentrations in the treated water. The water HRT was equal to 2 hours.

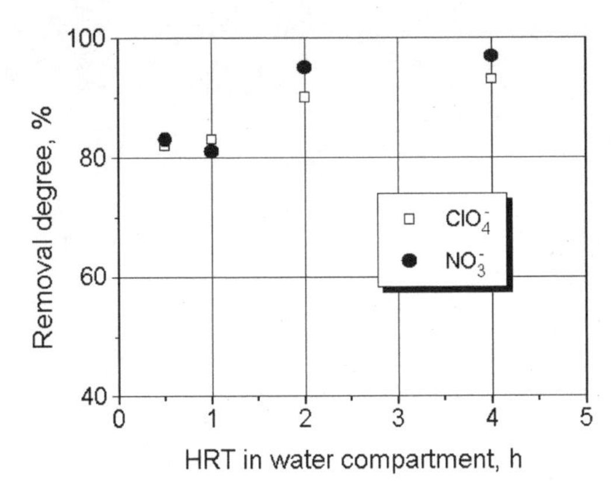

Figure 3. Dependence of target ion removal degree on the water hydraulic residence time.

The process dynamics with respect to the two target ions is presented in Figure 2. The perchlorate data show that a value, much lower than the aimed value of 18 ppb, can be rapidly achieved in the treated water stream and maintained for the time course of the experiment (~120 h).

Nitrate concentration was also reduced to a level well below the maximum of 50 ppm of NO_3^- allowed in the European Community and is even below the value of 25 ppm of NO_3^-, which is informally recommended. Our former experience with nitrate removal from drinking water (Velizarov et al. 2001, 2002) allows us to anticipate that much longer IEMB operation periods can be achieved (up to 2 months) before membrane washing and system re-start become necessary, due to reduction of the nitrate flux and excessive biofilm accumulation.

Figure 3 shows the dependence of the target ion removal degree on the water hydraulic residence time. As it can be observed HRT >30 min was enough to reach reasonably high removal degrees of more than 80% under the operating conditions studied.

4 CONCLUSIONS

The IEMB process, combining the Donnan equilibrium principle to extract a target inorganic anionic pollutant from water, with its simultaneous biodegradation (biotransformation) to one or more harmless products, proved to be an efficient solution to the simultaneous removal of target polluting ions, present in significantly different concentration in water. This behavior is a consequence of both the non-porous mono-anion perm-selective membrane used and future work will be focused on attempting even higher target ions transport rates by intensifying the hydrodynamics in the IEMB water compartment.

ACKNOWLEDGMENTS

The financial support by *Fundação para a Ciência e Tecnologia* (FCT), Lisbon, Portugal to this work through projects No. POCTI EQU/39482/2001 and POCTI/BIO/43625/2001 is gratefully acknowledged. Cristina Matos acknowledges FCT for grant SFRH/BD/9087/2002.

REFERENCES

Christen K. 2000. Surprising human health – perchlorate link. *Environmental Science and Technology News*, September 1: 374A–375A.

Crespo J.G. & Reis A.M., 2001. Treatment of aqueous media containing electrically charged compounds, Patent PCT-WO 01/40118 A1.

Susarla S., Collette T.W., Garrison A.W., Wolfe N.L. & Mccutcheon S.C. 1999. Perchlorate identification in fertilizers. *Environmental Science and Technology* 33: 3469–3472.

Urbansky E.T. 2000. Quantitation of perchlorate ion: practices and advances applied to the analysis of common matrices. *Critical Reviews in Analytical Chemistry* 30: 311–343.

Velizarov S., Rodrigues C.M., Reis A.M. & Crespo J.G. 2001. Mechanism of charged pollutants removal in an ion exchange membrane bioreactor: Drinking water denitrification. *Biotechnology and Bioengineering* 71(4): 245–254.

Velizarov S., Crespo J.G. & Reis A.M. 2002. Ion exchange membrane bioreactor for selective removal of nitrate from drinking water: Control of ion fluxes and process performance. *Biotechnology Progress* 18(2): 296–302.

Velizarov S., Reis A.M. & Crespo J.G. 2003. Removal of trace mono-valent inorganic pollutants in an ion exchange membrane bioreactor: analysis of transport rate in a denitrification process. *Journal of Membrane Science* 217(1-2): 269–284.

Wallace W., Beshear S., Williams D., Hospadar S. & Owens M. 1998. Perchlorate reduction by a mixed culture in an up-flow anaerobic fixed bed reactor. *Journal of Industrial Microbiology and Biotechnology* 20: 126–131.

European Symposium on Environmental Biotechnology, ESEB 2004 - Verstraete (ed)
© 2004 Taylor & Francis Group, London, ISBN 90 5809 653 X

Process intensification of municipal wastewater treatment by aerobic granular sludge systems

M.K. de Kreuk
Department of Biotechnology, TU Delft, Delft, The Netherlands

L.M.M. de Bruin
DHV Water BV, Leusden, The Netherlands

M.C.M. van Loosdrecht
Department of Biotechnology, TU Delft, Delft, The Netherlands

ABSTRACT: Stable aerobic granulation at low oxygen concentrations can take place using a discontinuously system, with anaerobic feeding followed by an aerobic reaction period. High COD-, N- and P-removal efficiencies are obtained in a laboratory scale reactor, fed with synthetic wastewater. In a short experiment, formation of aerobic granules on pre-settled sewage is shown too. A technical and economical feasibility study for the treatment of municipal sewage has been carried out. Depending on the chosen process configuration, a high effluent quality is viable. Based on total annual costs, the aerobic granular sludge system was more attractive (7–17%) than the reference alternative, a conventional activated sludge plant. Because of the high permissible load, the footprint of the aerobic granular sludge variant is only 25% compared to the reference. The aerobic granule based system can be attractive, especially when the expected more stringent effluent demands have to be fulfilled.

1 INTRODUCTION

Present wastewater treatment plants, like activated sludge systems, have the disadvantage of a large area requirement. Moreover these processes have to deal with a large number of conversion processes (COD-oxidation, ammonium oxidation, nitrate reduction, biological phosphate removal etc.). Traditionally, flocculated sludge with low settling velocities is applied, which leads to a low hydraulic load on the settling tanks. Therefore, large settling tanks are needed to separate clean effluent from the organisms. Besides large settling tanks, separate tanks are needed to accommodate the different treatment processes. Conventional processes need many steps for nitrogen, COD and phosphate removal, with large recycle flows and a high total hydraulic retention time. Surplus sludge from a municipal wastewater plant needs different steps to dewater (e.g. thickening and filterpressing) before it can be processed.

To overcome the disadvantages of a conventional wastewater treatment plant, biomass has to be grown in a compact form. Applying a proper feeding and selection regime to the biomass in a Granule Sequencing Batch Reactor (GSBR), organisms will grow in a granular structure, without using carrier material (Beun et al. 2000, Etterer & Wilderer 2001, Tay et al. 2002). The extraordinary settling properties of these granules (>10 m/h) eliminate the use of large settling tanks and allow high biomass concentrations in the reactor. The separation between effluent and biomass can take place in the reactor itself, during a short settling phase (Fig. 1). Because of the rapid settling, the total treatment process can be performed in one reactor.

This contribution presents a summary of the laboratory results that focussed on an optimal nutrient removal with aerobic granules. Furthermore, a design of a process for municipal sewage, in which the GSBR forms the central treatment, is described. This is economically evaluated in a feasibility study in which this new technology is compared to traditional activated sludge systems.

2 GRANULE REACTOR TECHNOLOGY

Granular growth of aerobic biomass has many advantages. Due to diffusion gradients, the various process conditions, usually accommodated in various tanks are now accommodated inside the granular sludge. Therefore effectively only one tank is needed without the need for large recycle flows. In order to obtain the right circumstances for aerobic granular growth and excessive simultaneous COD-, N- and P-removal, temporarily high substrate concentrations under anaerobic circumstances and low oxygen concentrations are needed. This was shown in a laboratory scale 3-l reactor, in which synthetic influent was added via a plug-flow through the settled bed during an anaerobic period of 60 minutes (total cycle 180 minutes). The composition of the wastewater was similar to municipal wastewater according to COD (acetate as C-source), P-total and N_{kj}. Contrary to municipal wastewater, all nutrients were only available as solubles. The exact method and influent composition is described in De Kreuk & Van Loosdrecht (in press). Using an anaerobic feeding, followed by an aeration period (112 minutes) stimulates the selection of slow growing phosphate accumulating organisms (PAOs).

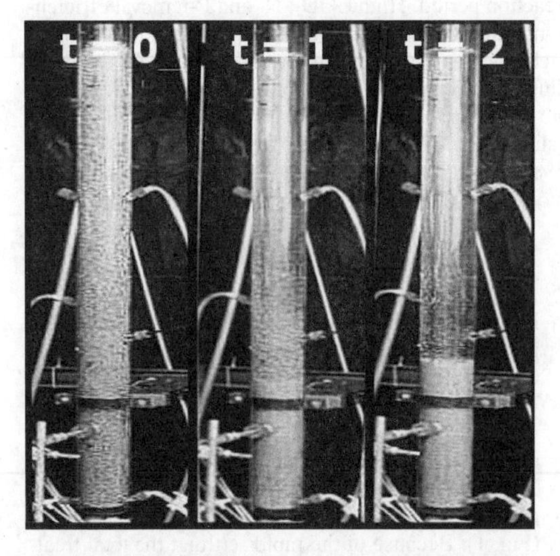

Figure 1. Fast settling of granules (t is settling time in minutes).

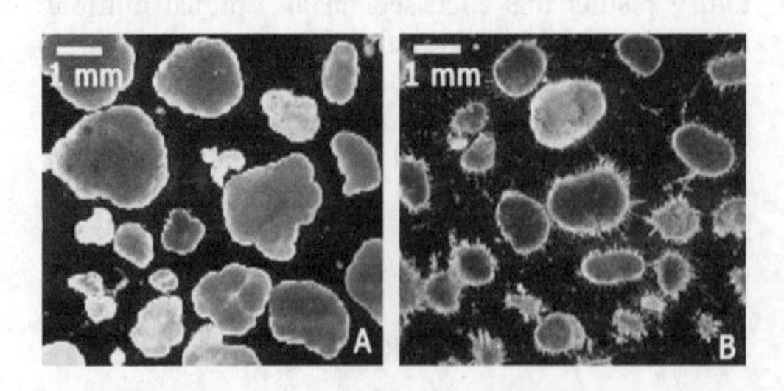

Figure 2. Granules grown on synthetic influent (A) and on pre-settled municipal wastewater (B).

A slow growth rate in combination with high substrate concentration generally promotes the formation of smooth and dense biofilms or granules (Picioreanu et al. 1998). PAOs also have the capability of storing substrate anaerobically as a storage polymer (as polyhydroxybutyrate or PHB), while poly-P is released into the bulk liquid. During the aeration period, nitrification takes place in the outer aerobic layers of the granules. Formed nitrite can diffuse to the inner anaerobic layers, where denitrification takes place using the stored substrate. During this aerobic period, PAO take up phosphate from the bulk aerobically or anoxically. By controlling the oxygen level in the bulk liquid, the anoxic volume of the granules can be maintained as large as possible, in order to obtain a high N-removal efficiency in the system. An optimum has to be found in which the aerobic biomass volume is large enough for nitrification and the anoxic biomass volume is large enough for denitrification. Laboratory scale experiments showed average COD-, N- and P-removal efficiencies for over 300 days of respectively 100%, 94% (of which 100% ammonia removal by nitrification) and 95%.

The compactness of the system, is besides the superfluity of large settling tanks, due to the high biomass concentration in the system. In activated sludge systems, the maximum biomass concentrations are 3–5 g/l, while in aerobic granule reactors at laboratory scale 15–20 g/l is feasible. This leads to an equivalent decreased reaction time, and thereby reactor volume. Because of this high density, also settleability increases. The average settling velocity of flocculated sludge is 1 m/h, while granules settle with a velocity of 12 to 20 m/h (Fig. 1). This means a short settling time can be included in the process cycle of a GSBR, in order to keep the biomass in the system. A small lamella settler or continuous sand filtration would be sufficient to remove small particles that leave the reactor with the effluent.

Granulation was also experimentally verified on pre-settled sewage. Granules could be formed with this complex substrate, although the granulation process itself was slower than on acetate. This was mainly caused by the lower readily available COD load. The granules grown on sewage had more irregular and filamentous outgrowth (Fig. 2).

3 DESIGN2

Several other compact systems investigated nowadays, are difficult to scale-up to a size necessary for municipal wastewater treatment plants (20,000 m^3/day). For example membrane bioreactors need equally dividing the flow over the many membrane modules and scale up is obtained by increasing the number of process units.

The GSBR can consist of 2 or 3 simple and robust airlift or bubble column reactors, in which all processes take place. Well-considered batch scheduling of the total plant could lead to a continuous influent flow. Assuming that only one reactor at the time is fed with influent, the number of parallel treatment lines is determined by the duration of the feeding phase and the total cycle length. In that case, no buffer tanks are needed, which makes the system simple and compact. In case of a post treatment, the effluent will be buffered. Since the effluent is discharged by gravity, the initial flow is high. Thus, in order to reduce the dimensions of the post treatment step, effluent from the GSBR has to be buffered.

The design of the process is based on the standard wastewater composition, as defined by the Dutch foundation for applied research for water management (STOWA); 600 mg COD/l (216 mg/l dissolved 384 mg/l suspended); 55 mg N_{kj}/l (45.7 mg/l dissolved; 9.3 mg/l suspended); 9 mg P_{total}/l and 250 mg/l suspended solids. The average wastewater flow is 160 l/(pe.d) with a peak flow (RWA) of 34 l/(pe.h). The effluent requirements are based on the discharge regulations in The Netherlands (10 mg N_{total}/l and 1 mg P_{total}/l). If a pre-treatment is applied, it consists of a conventional settler, with or without flocculants dosage. Removal efficiencies for suspended solids are 50% without flocculants and 80% with flocculants. No metal salts will be dosed, since PAO have to be selected.

In order to guarantee a plug flow through the settled bed during filling, influent is supplied via diffusers at the bottom of the reactor. In that case, the laboratory scale anaerobic feeding regime is approximated and phosphate will be removed by bio-P, which makes chemical phosphate removal superfluous. The maximum hydraulic load during filling is maximum 7.5 m/h to avoid fluidisation of the granules. This leads to a construction height of 5–6 m. The total cycle time amounts to 60 minutes: 20 minutes anaerobic feeding; 27 minutes aeration; 5 minutes sedimentation

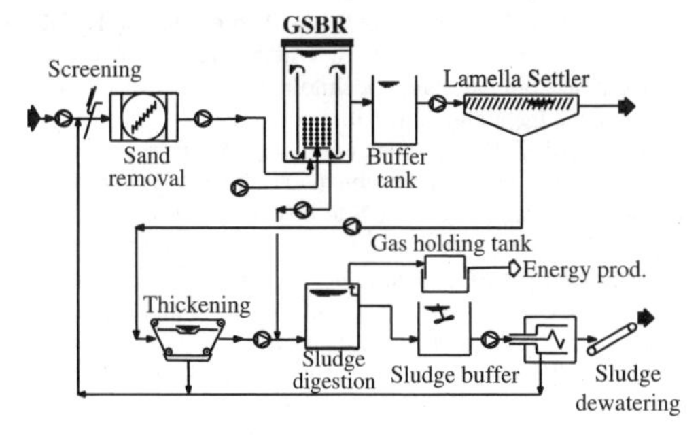

Figure 3. Global process diagram of a GSBR system with post-treatment.

(designed settling velocity of the particles is 15 m/h) and 7.5 min decantation. The design load of the GSBR is chosen at 0.3 kgCOD/(kgDW.d). The net sludge production is almost zero, meaning equal sludge content in influent and effluent.

Two types of GSBR systems are designed and evaluated in the feasibility study: 1) a GSBR with pre-settling with poly-electrolyte dosage, in order to remove as much suspended solids as possible to make a post-treatment redundant; 2) a GSBR with only post-treatment, in order to remove the suspended solids from the effluent flow. Figure 3 gives a process flow diagram of this second GSBR treatment system.

4 FEASIBILITY STUDY

In order to analyse the feasibility of the two different variants of the GSBR systems described above, these systems are economically evaluated and compared to a conventional activated sludge system with Bio-P (design sludge load aeration tanks 0.14 kgCOD/(kgDW.d); SVI 150 ml/g; process temperature 10°C). Sludge treatment of the reference and both GSBR systems consists of gravitational thickening of primary sludge, mechanical thickening of surplus sludge, digestion and dewatering.

For calculation of the capital costs depreciation periods for civil parts and mechanical/electrical parts of respectively 30 and 15 years are used. Capital costs are calculated based on annuities (interest 6%). Operational costs are based on main cost factors as are given in table 1.

Based on the influent characteristics and technological starting points, the feasibility study showed that the GSBR system with post-treatment could meet the present effluent requirements. The GSBR-alternative with primary treatment did not meet this quality with respect to suspended solids, which are insufficient removed during pre-treatment. This also leads to increased N_{kj} and P_{total} concentrations. Effluent requirements can only be met if the suspended solids

Table 1. Starting points costs calculations.

Cost factor	Unit	Value
Electricity*	€/kWh	0.054
Sludge disposal	€/ton dry solids	320
Iron chloride (41%)	€/ton	115
Poly-electrolyte (liquid, 50%)	€/kg active	3.6
Land price	€/m^2	22.7
Maintenance:		
– Civil parts	% of investments	0.5
– Mechanical parts	% of investments	2
– Electro technical parts	% of investments	2

* Power use is corrected for power production from biogas.

concentration in the feed of the GSBR is less than 10–30 mg/l. However, a GSBR system with primary treatment could be an attractive concept when a more stringent effluent quality is required (e.g. N_{total} 2.2 mg/l; P_{total} 0.15 mg/l). In the Netherlands, this is an actual situation for a growing number of treatment plants discharging on sensitive surface waters. In that case, conventional activated sludge systems will be extended with a post-treatment step (e.g. with sand filtrations). A GSBR with primary treatment as well as post-treatment could be an attractive alternative, assuming that the costs for a post-treatment system are similar in both cases.

When surface area requirements of the GSBR alternatives are compared to the reference, it was shown that an installation based on aerobic granulation requires 25% of the surface area needed for a conventional activated sludge system. This is calculated by the sum of all net surfaces of all process units and buildings, multiplied by a factor 1.3. It can be concluded that the GSBR technology is very compact, which is an important advantage in relation to activated sludge technology, especially in densely populated areas.

In Figure 4 the total specific annual costs (sum of capital and operational costs) are given. As can be seen from this figure, the total annual costs for the GSBR-alternatives are lower than for the reference system; on average respectively 7% and 17% for the systems with post- and pre-treatment. This is mainly due to lower investment costs (on average 15% and 30% lower), even despite the higher share of investment costs for mechanical/electrical works of the GSBR-alternatives (40–45%) compared to the reference system (25–30%).

Because of the compactness of the GSBR system, the total specific annual costs are much less sensitive towards land price than the reference system. When land price increases, the difference of the annual costs can increase from 7% to 16% (GSBR with post-treatment, land price 500 €/m^2 and 120.000 p.e.). On the other hand, the GSBR technology is more sensitive to an increasing RWF/DWF ratio. The maximum batch

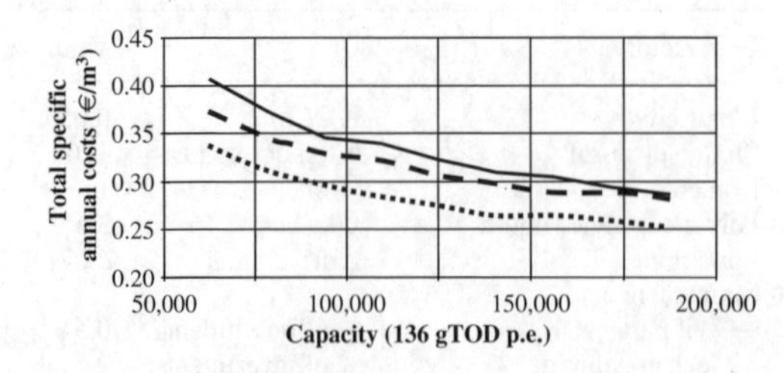

Figure 4. Total specific annual costs of the GSBR with primary treatment (......); the GSBR with post-treatment (– –) and the reference system (——).

to be treated increases at high RWF, which results directly in a larger needed volume for the GSBR.

Because of the promising results of the laboratory study as well as the feasibility study, a pilot plant is started-up at the wastewater treatment plant of Ede, The Netherlands. This pilot plant forms the second step from laboratory experiments to a full-scale installation.

5 CONCLUSIONS

The aerobic granule technology can be seen as a promising technology. Granulation on acetate is stable during long-term experiments, especially when selection for PAO's takes place. At low oxygen concentrations, high simultaneous COD-, N- and P-removal was obtained. Granulation was also shown using more complex pre-settled sewage as influent. Based on total specific annual costs, aerobic granule sludge technology appeared to be an attractive alternative for the conventional activated sludge system. In addition, the footprint of the GSBR system is only 25% of the activated sludge system. This means that the aerobic granule sludge technology could be an economically attractive as well as a compact alternative for present wastewater treatment systems.

ACKNOWLEDGEMENTS

This research was funded by the Dutch foundation for Water Research (STOWA) within the framework of the "Aerobic Granule Reactors" project (TNW99.262) as well as by the Dutch Technology Foundation (STW) under project number DPC.5577.

REFERENCES

Beun, J.J., Van Loosdrecht, M.C.M. & Heijnen, J.J. 2000. Aerobic Granulation. *Wat. Sci. Techn.* 41(4–5): 41–48.

De Bruin, L.M.M., De Kreuk, M.K., Van der Roest, H.F.R., Van Loosdrecht, M.C.M. & Uijterlinde, C. (in press). Aerobic granular sludge technology, alternative for activated sludge technology? *Wat. Sci. Techn.*

De Kreuk, M.K. & Van Loosdrecht, M.C.M. (in press). Selection of Slow Growing Organisms as a Means for Improving Aerobic Granular Sludge Stability. *Wat. Sci. Techn.*

Etterer, T. & Wilderer, P.A. 2001. Generation and properties of aerobic granular sludge. *Wat. Sci. Techn.* 43(3): 19–26.

Picioreanu, C., Van Loosdrecht, M.C.M. & Heijnen, J.J. 1998. Mathematical modeling of biofilm structure with a hybrid differential-discrete cellular automaton approach. *Biotechn. Bioeng.* 58(1): 101–116.

Tay, J.H., Liu, Q.S. & Liu, Y. 2002. Aerobic granulation in sequential sludge blanket reactor. *Wat. Sci. Techn.* 46(4–5): 13–18.

European Symposium on Environmental Biotechnology, ESEB 2004 - Verstraete (ed)
© 2004 Taylor & Francis Group, London, ISBN 90 5809 653 X

The effect of a quinone-based redox mediator on the decolourisation of textile wastewaters: a comparative study between mesophilic (30°C) and thermophilic (55°C) anaerobic treatments by granular sludge

A.B. dos Santos & J.B. van Lier
Sub-department of Environmental Technology, Wageningen University, Wageningen, Netherlands

I.A.E. Bisschops
Lettinga Associates Foundation, Wageningen, Netherlands

F.J. Cervantes
Departamento de Ciencias del Agua y del Medio Ambiente, Instituto Tecnológico de Sonora, Mexico

ABSTRACT: A comparative study between mesophilic (30°C) and thermophilic (55°C) anaerobic treatments by granular sludge was performed on the decolourisation of a reactive textile wastewater, both in presence and absence of the redox mediator anthraquinone-2-sulfonate (AQS). The investigations evidenced the advantage of colour removal at 55°C compared with 30°C. In the presence of anthraquinone-2,6-disulfonate (AQDS), the generation of the hydroquinone form AH_2QDS, i.e. the reduced form of AQDS, was greatly accelerated at 55°C compared in comparison with 30°C. Furthermore, no lag-phase was observed at 55°C. Based on the present results we postulate that the production/transfer of reducing equivalents is the process rate-limiting step, which was accelerated by the temperature increase. It is conclusively stated that 55°C is a much better temperature for colour removal than 30°C, which on the one hand can be attributed to the faster production/transfer of reducing equivalents, but also due to the decrease of activation energy requirements.

1 INTRODUCTION

Azo dyes are the most common synthetic colourants released in the environment through textile, pharmaceutical and chemical industries. Almost 10^9 kg of dyes are produced annually, of which azo dyes represent about 70% by weight. The discharge of azo dyes in water bodies is problematic not only for aesthetic reasons, but also because azo dyes and their cleavage products (aromatic amines) may be carcinogenic (Hao et al. 2000).

The biological treatment of azo dye wastewaters under anaerobic conditions has been extensively researched. The anaerobic microorganisms not only generate the electrons to cleave the azo bond, but also maintain the low redox potential (<-50 mV) required for the electron transfer to the dye molecule. However, during anaerobic azo dye reduction, reducing equivalents transferred from a primary electron donor (co-substrate) to a terminal electron acceptor (azo dye) is mostly the process rate-limiting step. The addition of redox mediators has been shown to accelerate electron transfer and higher decolourisation rates can be achieved in bioreactors. However, the impact of redox mediators on the decolourisation rates has generally been investigated with azo model compounds. Their effectiveness in enhancing the decolourisation of real textile wastewaters is still unclear due to the wide range of redox potentials among azo dyes (-180 to -430 mV). Moreover, a comparative study between mesophilic and thermophilic treatments of textile wastewater in the presence of a redox mediator has never been conducted. In the current investigation, a comparative study between mesophilic (30°C) and thermophilic (55°C) decolourisation by anaerobic granular sludge was performed using a wastewater originating from a textile industry, either in the presence or absence of the redox mediator AQS.

2 MATERIALS AND METHODS

2.1 *Chemicals*

Reactive Red 2 (RR2) was selected as azo dye model compound. Anthraquinone-2-sulfonate (AQS) was used as redox mediator model compound during decolourisation investigations, as well as anthraquinone-2,6-disulfonate (AQDS) was used in the AQDS reducing capacity test. The chemicals were purchased from Aldrich (Gillingham, UK) and used without additional purification.

2.2 *Seed inoculum and basal medium for decolourisation assays*

Anaerobic granular sludge was collected from a full-scale mesophilic upflow anaerobic sludge blanket (UASB) reactor treating paper mill wastewater (Eerbeek, The Netherlands). The mesophilic sludge was acclimated for 3 months at 55°C as described before (Dos Santos et al. 2003a). For batch tests at 30°C the same mesophilic granular sludge was first acclimated to steady state in an EGSB reactor (30°C) with the same co-substrate and hydraulic conditions previously described.

The basal media for the tests with model compounds and AQDS reducing capacity test have been previously described (Dos Santos et al. 2003a; Cervantes et al. 2000). Both media were buffered with 6.21 g L^{-1} of sodium bicarbonate at pH around 7.1. When a wastewater derived from cotton processing textile factory was used, it was used undiluted and without addition of nutrients or trace elements. The pH was adjusted to 7 with NaOH or HCl.

2.3 *Activity test*

Inoculation took place by adding 1.3 ± 0.1 g volatile suspended solids (VSS) L^{-1} of the previously described stabilized sludge to 117-mL serum bottles with 50 mL basal medium and sealed with butyl rubber stoppers. Anaerobic conditions were established by flushing the headspace with N_2/CO_2 (70% : 30%) and 2 g COD L^{-1} (glucose:VFA mixture at a COD ratio of 1:3) co-substrate was added as electron donor and carbon source. After a pre-incubation time of 2 days RR2 (variable) and AQS (variable) were added. Sterile controls were autoclaved once at 122°C for 240 min and again following a 5 days incubation period, after which sterile co-substrate, mediator and dye stock solutions were added. Afterwards, co-substrate, redox mediator and azo dye were added to the bottles from sterile stock solutions. The pH and the amount of VSS were determined after completion of the experiment.

2.4 *Comparative study between mesophilic and thermophilic anaerobic treatments*

2.4.1 *Decolourisation of reactive dye wastewater*

A comparative study between mesophilic (30°C) and thermophilic (55°C) conditions was conducted on anaerobic azo dye reduction with a reactive dyeing wastewater. The wastewater was derived from a Belgian cotton processing textile factory and had a light brown colour. The redox mediator AQS was used at a concentration of 0.5 mM, either in presence or absence of 2 g COD L^{-1} of co-substrate, i.e. glucose : VFA mixture on a COD ratio of 1 : 3. Both previously stabilized mesophilic and thermophilic sludges were incubated with the wastewaters.

2.4.2 *AQDS as an electron acceptor during mesophilic and thermophilic incubations with granular sludge*

To compare the capacity of mesophilic and thermophilic sludge samples to utilize AQDS as a terminal electron acceptor, an AQDS reducing capacity test (production of AH2QDS) was performed. AQDS (1mM) was added to the basal medium incubated at both 30°C and 55°C, in which 2 g COD L^{-1} of a glucose : VFA mixture at a COD ratio of 1 : 3 was either present or absent. Sludge-free and autoclaved sludge bottles were used as controls for the abiotic AQDS reduction. The pH of the medium was 7 in all incubations.

2.5 *Activation energy (Ea) determination during chemical decolourisation*

The activation energy requirement of the chemical reduction of the model compound, Reactive Red 2 (RR2, 0.3 mM), was determined in sludge-free incubations. Sulfide (4.5 mM) was selected as a reducing agent and incubated with RR2 and AQS (variable) at temperatures of 30, 45 and 55°C. AQS (0.012 mM) was supplied to some of the bottles. The concentration of sulfide was measured initially and upon completion of the experiment.

2.6 *Analysis*

Colour removal for RR2 was determined photometrically (Spectronics 60, Milton-Roy Analytical Products Division, Belgium). The absorbance was read at the maximum absorbance wavelength, i.e. at 539 nm. The extinction coefficient used for RR2 was 34.3 AU cm^{-1} M^{-1}.

Wastewater decolourisation was determined in a 1 cm quartz cuvette by scanning the UV/VIS spectra (Perkin-Elmer UV/VIS Lambda 12, Rodgau-Jügesheim, Germany) and comparing the wavelength of two absorbance peaks.

AH_2QDS was determined anaerobically in a Type B Coy anaerobic chamber (Coy Laboratory Products Inc., USA) under N_2/H_2 (96% : 4%) atmosphere, according to Cervantes et al. (2000). An extinction coefficient of 2.08 AU mM^{-1} cm^{-1} at 450 nm was obtained during chemical reduction of AQDS under hydrogen atmosphere in the presence of a hydrogenation catalyst according to Kudlich et al. (1999).

Sulfide was determined photometrically by using the Dr. Lange cuvette method.

Volatile Suspended Solids (VSS) were analyzed according to APHA standard methods (1998).

3 RESULTS

3.1 *Comparative study between mesophilic and thermophilic anaerobic treatments*

3.1.1 *Decolourisation of reactive dyeing wastewater*

Batch assays showed that colour removal under thermophilic conditions was distinctly faster than under mesophilic conditions (Figure 1a and 1b). Moreover, the impact of the external redox mediator AQS (0.5 mM) on colour removal was considerably decreased under thermophilic conditions.

Colour removal by anaerobic granular sludge was accelerated by addition of co-substrate either in the presence or absence of mediator. However, incubations supplied just with mediator in the absence of co-substrate showed no impact on colour removal (Figure 1a and 1b). This was an indication that the wastewater was free of a suitable primary electron donor to promote the microbial generation of reducing equivalents. Thus, in the case of separate treatments of dyeing- and rinsing-step wastewaters the addition of co-substrate should be considered.

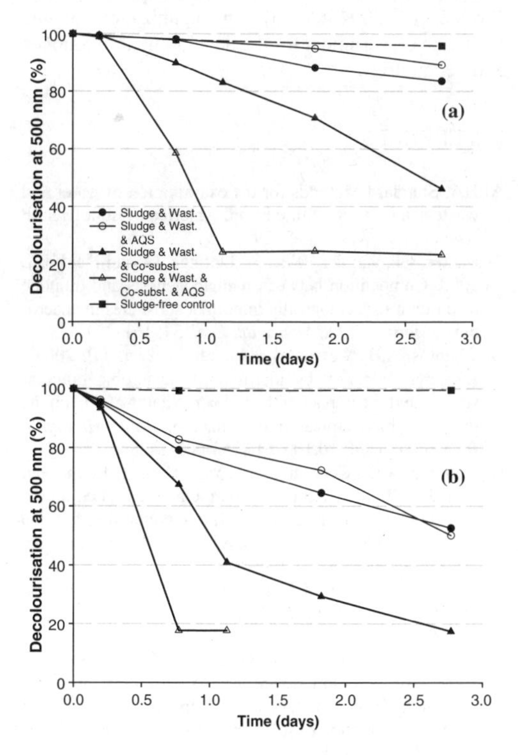

Figure 1. Decolourisation of textile wastewater at 500 nm by mesophilic (30°C) (a) and thermophilic (55°C) (b) granular sludge (1.3 g VSS L^{-1}). AQS (0.5 mM) was added to some bottles and 2 g COD L^{-1} co-substrate. The results are means of duplicate incubations. The standard deviations were lower than 5% in all cases.

Negligible (<4%) colour removal occurred in sludge-free controls in the presence of AQS during the incubation time of 2.7 days.

3.1.2 *AQDS as an electron acceptor during mesophilic and thermophilic microbial incubations with granular sludge*

Anaerobic granular sludge incubated under mesophilic and thermophilic conditions was capable to use AQDS as an electron acceptor for the primary electron donor oxidation (co-substrate). The generation of AH_2QDS, the reduced form of AQDS, was greatly accelerated at 55°C (Figure 2) compared with 30°C. Furthermore, no lag-phase was observed under thermophilic conditions. For instance, about 1 mM AQDS was completely reduced at 55°C after 0.7 days of incubation, whereas the reduction at 30°C after 0.7 days represented just 12.9% of this value (Figure 2).

AH_2QDS formation in endogenous controls, autoclaved sludge and sludge-free incubations was negligible during the 2 days experiment under both mesophilic and thermophilic conditions.

3.2 *Activation energy (Ea) determination during chemical decolourisation of RR2*

The chemical decolourisation of RR2 by sulfide followed a first-order reaction with respect to the dye concentration. In order to calculate the activation energy (Ea) values, ln (k_1) versus 1000/T was plotted, the slope Ea/R being obtained by the linear regression.

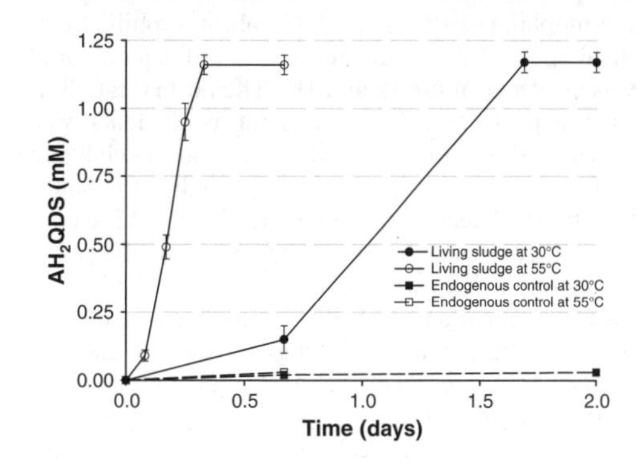

Figure 2. AH_2QDS formation (1 mM) at both 30°C and 55°C temperatures. Measurements were conducted under anaerobic conditions, in which the samples were diluted in a bicarbonate buffer (60 mM, pH 6.8 ± 0.1), and the absorbance was read at 450 nm. The results are means of triplicate incubations.

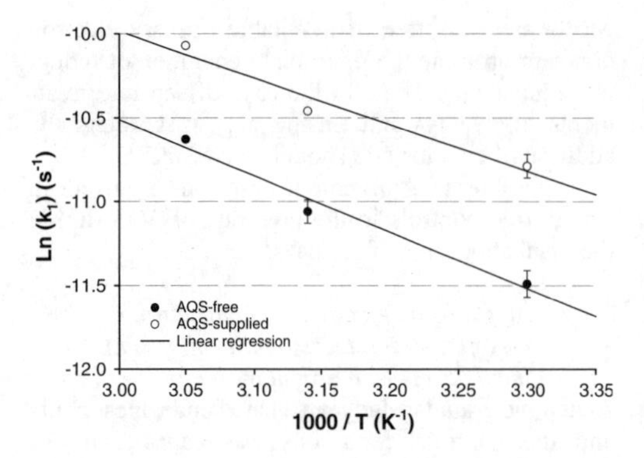

Figure 3. Arrhenius plot and linear regression for the chemical decolourisation of RR2 (0.3mM) by sulfide (4.5 mM) in sludge-free incubations. AQS (0.012 mM) was added to some of the bottles. The value of the ratio Ea/R was obtained by the slope of the linear regression, in which the parameter R was the universal gas constant (8.314 J K^{-1} mol^{-1}).

This ratio was multiplied by the universal gas constant R (8.314 J $K^{-1}mol^{-1}$) to obtain the Ea value. Figure 3 shows that the Arrhenius equation could describe the chemical decolourisation of RR2 by sulfide at different temperatures. The slopes of the AQS-free and AQS-supplied incubations are indeed different (Figure 3). The calculated Ea values were 27.9 kJ mol^{-1} and 22.9 kJ mol^{-1} for the AQS-free and AQS-supplied incubations, respectively. Therefore, the activation energy was decreased 1.2-fold with the addition of 0.012 mM-AQS.

4 DISCUSSION

The present work clearly evidences the advantage of thermophilic treatment at 55°C over mesophilic treatment at 30°C for the decolourisation of azo coloured wastewaters (Figure 1a and 1b). The normal rate limiting step, the transfer of reducing equivalents, was accelerated under thermophilic conditions. Both biotic and abiotic mechanisms may contribute to enhance the observed decolourisation under thermophilic conditions. The faster biological reduction of the redox mediator, AQDS, achieved by sludge incubations at 55°C in comparison with 30°C, evidenced the biological contribution in enhancing the rate of electron transfer. On the other hand, AQDS-supplemented incubations presented a lower activation energy requirement during the chemical reduction of RR2 by sulfide. Therefore, at 55°C, the external mediator dosage can be decreased to achieve the colour removal requirements. This is in agreement with Dos Santos et al. (2003b) who verified in continuous flow bioreactors at 55°C that the redox mediator AQDS slightly enhanced the decolourisation rates of Reactive Red 2.

The impact of temperature on colour removal also corroborated Willetts et al. (2000), who reported faster decolourisation rates at 55°C compared to 35°C while treating the azo dye Reactive Red 235 in UASB bioreactors free of external redox mediator. Furthermore, Laszlo (2000) reported a higher decolourisation rate of the azo dyes Orange II and Remazol Red F3B at 43°C than at 28°C, by anaerobically incubating the facultative organism *Burkholderia cepacia* NRRL B-14803. Based on these reports and our present results we postulate that the transfer of reducing equivalents to the azo dye is the process rate-limiting step, which obviously is accelerated by the temperature increase. Thus, it is conclusively stated that 55°C is a more effective temperature than 30°C for colour removal, which brings good prospects on the application of thermophilic treatment for decolourisation processes.

ACKOWLEDGEMENTS

This work was supported by "Conselho Nacional de Desenvolvimento Científico e Technológico – CNPq" (Project n° 200488/01-5), an organization of the Brazilian Government for the development of Science and Technology.

REFERENCES

APHA. Standard Methods for the examination of water and wastewater, 1998. 20th edition, American Public Health Association, Washington DC.

Cervantes, F.J., van der Velde, S., Lettinga, G. and Field, J.A. 2000. Competition between methanogenesis and quinone respiration for ecologically important substrates in anaerobic consortia. *FEMS Microbiol. Ecol.* 34: 161–171.

Dos Santos, A.B., Cervantes, F.J. and Van Lier, J.B. 2003a. Azo dye reduction by thermophilic anaerobic granular sludge, and the impact of the redox mediator AQDS on the reductive biochemical transformation. *Appl. Microbiol. Biotechnol.* DOI: 10.1007/s00253-003-1428-y.

Dos Santos, A.B., Cervantes, F.J., Yaya-Beas, R.E. and van Lier, J.B. 2003b. Effect of redox mediator, AQDS, on the decolourisation of a reactive azo dye containing triazine group in a thermophilic anaerobic EGSB reactor. *Enz. Microb. Technol.* 33(7): 942–951.

Hao, O.J., Kim, H. and Chiang, P.C. 2000. Decolourisation of wastewater. *Crit. Rev. Environ. Sci. Technol.* 30: 449–505.

Kudlich, M., Hetheridge, M.J., Knackmuss, H.J. and Stolz, A. 1999. Autoxidation reactions of different aromatic *o*-aminohydroxynaphthalenes that are formed during the anaerobic reduction of sulphonated azo dye. *Environ. Sci. Technol.* 33: 896–901.

Laszlo, J. 2000. Regeneration of azo-dye-saturated cellulosic anion exchange resin by *Burkholderia cepacia* anaerobic dye reduction. *Environ. Sci. Technol.* 34: 167–172.

Willets, J.R.M. and Ashbolt, N.J. 2000. Understanding anaerobic decolourisation of textile dye wastewater: mechanism and kinetics. *Water Sci. Technol.* 42: 409–415.

European Symposium on Environmental Biotechnology, ESEB 2004 - Verstraete (ed)
 ISBN 90 5809 653 X

Biological organomercurial removal from vaccine production wastewaters by a *Pseudomonas putida* strain

Raquel Fortunato, João G. Crespo, Maria A. Reis
REQUIMTE/CQFB, Departamento de Química, FCT
Universidade Nova de Lisboa, Caparica, Portugal

ABSTRACT: Vaccine production effluents are frequently polluted with thiomersal, a highly toxic organomercurial compound, for which there is presently no remediation technology available. This work proposes a biotechnological process, for the remediation of vaccine production wastewaters, based on the biological degradation of thiomersal to metallic mercury, under aerobic conditions, by a mercury resistant bacterial strain. The kinetics of thiomersal degradation by a pure culture of *Pseudomonas putida* spi3 strain, in batch culture and using a synthetic wastewater, is presented and discussed. Additionally, a continuous stirred tank reactor (CSTR) fed with the synthetic wastewater was operated, and the bioreactor performance and robustness, when exposed to thiomersal shock loads, was evaluated. Finally, a CSTR for the biological treatment of a real vaccine production effluent was set-up and operated at different dilution rates.

1 INTRODUCTION

Thiomersal is a toxic organomercurial with a strong bactericide effect, used as additive to biologicals and cosmetics since 1930. It is the most widely used preservative in vaccine production, in order to prevent growth of bacteria in cell culture and media and/or in the final container (Keith et al. 1992). Therefore, the wastewaters resulting from vaccine production are frequently strongly polluted with a thiomersal concentration that ranges from 25 mg/l to 50 mg/l, well above the European limit for mercury effluents discharges (0.05 mg/l Hg ⇔ 0.1 mg/l thiomersal), and for which there is, presently, no remediation technology available.

This work proposes a biotechnological process, for the remediation of vaccine production wastewaters, based on the biological degradation of thiomersal to metallic mercury, under aerobic conditions, by a mercury resistant pure culture of *Pseudomonas putida*. The metallic mercury produced can be stripped from the media by aeration, concentrated and recovered.

Bacterial resistance to organomercurials and inorganic mercury compounds is widely observed in nature and the latter has been successfully applied in the remediation of Hg^{2+} containing wastewaters (Wagner-Döbler 2003). The observed bacterial resistance is due to the production, induced by low mercury levels, of the intracellular enzymes organomercurial lyase and mercuric ion reductase (Misra 1992). The former cleaves the C-Hg bond, releasing Hg^{2+} and the respective organic compound, while the latter reduces Hg^{2+} to the less toxic Hg^0. Cleavage of the S-Hg bond in thiomersal is thought to occur spontaneously in the presence of exogeneous thiols (R-SH) (Elferink 1999). Figure 1 depicts the proposed mechanism for thiomersal biodegradation.

In this work, the kinetics of thiomersal biodegradation by a *P. putida* spi3 strain, in batch culture and using a synthetic wastewater, is presented and discussed. Based on the results obtained, a continuous stirred tank reactor (CSTR) was set-up and operated. Thiomersal pulses were applied to the system, at steady state, in order to evaluate the process dynamics and the bioreactor performance and robustness when exposed to thiomersal shock loads. Finally, in order to test the process feasibility, the bioreactor was fed with a real thiomersal contaminated vaccine production effluent and operated at two different dilution rates ($D = 0.03\,h^{-1}$ and $0.05\,h^{-1}$).

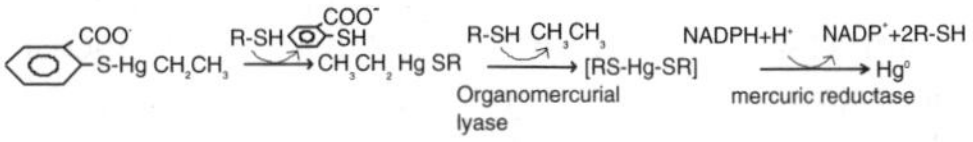

Figure 1. Mechanism for thiomersal biodegradation.

2 MATERIALS AND METHODS

2.1 *Microbial culture and culture media*

The *Pseudomonas putida* spi3 strain used was isolated from sediments of the Spittelwasser river by the Molecular Microbial Ecology Group of GBF-Braunschweig Germany.

In the studies with synthetic wastewater, a mineral medium (3.7 g/l Na_2HPO_4, 3 g/l KH_2PO_4, 5 g/l NaCl, 1 g/l NH_4Cl) supplemented with thiomersal, and with a concentrated micronutrients solution (1:400) (7.23 g/l $MgCl_2 \cdot 6H_2O$, 1 g/l $CaCO_3$, 0.72 g/l $ZnSO_4 \cdot 7H_2O$, 0.42 g/l $MnSO_4 \cdot H_2O$, 0.125 g/l $CuSO_4 \cdot 5H_2O$, 0.14 g/l $CoSO_4 \cdot 7H_2O$, 0.01 g/l H_3BO_3, 0.06 g/l $MgSO_4 \cdot 7H_2O$, 2.5 g/l $FeSO_4 \cdot 7H_2O$) was used. The medium pH was adjusted to pH 7 with sodium hydroxide, and glucose was used as carbon source.

In the studies with the vaccine wastewater a thiomersal contaminated vaccine effluent, kindly supplied by a multinational pharmaceutical company, was used. The wastewater was sterilized by ultrafiltration using a hollow fibber polysulfone membrane module with a molecular weight cut-off of 500 kDalton. The sterile effluent was also supplemented (1:400) with the concentrated micronutrients solution described above. The bioreactor pH was kept constant at pH7 (addition of hydrochloric acid), and glucose was used as carbon source.

Oxygen concentration, temperature and pH in the reactor were continuously measured and controlled at 5 ± 0.5 ppm O_2, which corresponds to 60% saturation (aeration with compressed air), 25°C and pH 7 (addition of NaOH or HCl), respectively. A respirometer, through which the culture medium was continuously recirculated, was coupled to the reactor. The oxygen uptake rates (OUR) were measured, taking in account the decay of dissolved oxygen concentration in the respirometer, when the recirculation pump was stopped.

2.2 *Kinetic studies of thiomersal biodegradation*

In the batch kinetics studies a 1.5 l reactor was used, and the culture was grown aerobically, at 25°C, in the synthetic wastewater described above. Thiomersal was added to the culture medium (25 mg/l) immediately before inoculation. A glucose concentration of 2.5 g/l was used.

2.3 *Biodegradation of thiomersal in a CSTR fed with a synthetic wastewater*

The 2 l bioreactor was continuously fed with the synthetic wastewater described above, supplemented with a thiomersal concentration of 192 mg/l and a glucose concentration of 1 g/l. The system was operated at a dilution rate equal to $0.05\,h^{-1}$. Thiomersal pulses were applied to the system, at steady state. After the first thiomersal pulse (instantaneous addition of 25 mg/l to the bioreactor) the system was allowed to reach again steady state, after which a new thiomersal pulse (addition of 50 mg/l to the bioreactor) was applied.

2.4 *Biodegradation of thiomersal in a CSTR fed with a real vaccine wastewater*

The 1 l bioreactor was continuously fed with the real vaccine wastewater. The thiomersal concentration in the effluent was constant and equal to 48.4 mg/l. A glucose solution with 3.6 g/l was added separately, to the bioreactor, at a constant flow rate of 0.18 ml/min. The bioreactor was operated at two different dilution rates ($0.03\,h^{-1}$ and $0.05\,h^{-1}$). The effluent flow rate was adjusted to the dilution rate chosen.

2.5 *Analytical methods*

Thiomersal concentration was determined by HPLC, using a reverse phase column (Nucleosil 100-5 C18, MN). The eluent was a methanol/10mM phosphate buffer (pH6) solution (55:45). A UV detector (Merck, Hitachi, Japan) set at a wavelength of 250 nm was used. The concentration of glucose was determined by HPLC, using a Aminex HPX-87H column and 0.01N solution of sulphuric acid as eluent. A IR detector (Merck, Hitachi, Japan) was used. The total mercury content of the samples was measured by inductively coupled plasma spectroanalysis (ICP-JYultima 238). Cell concentration was determined by optical density (OD) measurements at 600 nm and compared with a OD vs. cell dry weight calibration curve.

3 RESULTS AND DISCUSSION

3.1 *Kinetic studies of thiomersal biodegradation*

The kinetics of thiomersal biodegradation by a *P. putida* spi3 strain was evaluated in a batch reactor (experiment 1) with an initial glucose and thiomersal concentration of 3 g/l and 25 mg/l, respectively.

The results obtained, showed that thiomersal was totally degraded during the first three hours. While thiomersal was consumed, both growth and glucose consumption rates were much lower than the ones observed afterwards. Therefore, thiomersal degradation

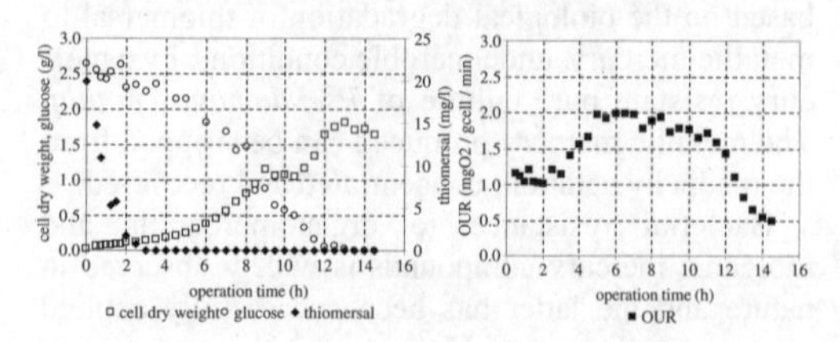

Figure 2. Kinetics of growth, glucose, thiomersal and oxygen consumption in a batch culture of *P. putida* spi3.

is likely to occur as detoxification mechanism: bacteria transform the highly toxic thiomersal in a less toxic mercury form (Hg^0), which is excreted from inside the cell, thus allowing the bacteria to grow using glucose as carbon source. The initial thiomersal degradation rate ($r_{thiomersal}$) was almost constant and equal to $9.37\,mgl^{-1}h^{-1}$ (see Table 1). The specific oxygen uptake rate (OUR) remained steady at $1.0\,mgO_2/g_{cell}/min$ while thiomersal was being consumed, after which increased and reached $2.0\,mgO_2/g_{cell}/min$. The increase on OUR after thiomersal degradation can, probably, be explained by a higher energy requirement to reduce Hg^{2+} to Hg^0. The discontinuity in the growth rate observed between 9 and 10.5 hours was due to a micronutrients limitation. Addition of a pulse of micronutrients (at 10.5 h) to the media caused a growth re-start, that ended only when all carbon source had been consumed.

Kinetic of *P. putida* spi3 in the absence of thiomersal (experiment 2), and using the same glucose concentration of experiment 1, was also evaluated (data not shown). In this case, the specific oxygen uptake rate (OUR) remained constant, and lower ($\sim 1.25\,mgO_2/g_{cell}/min$) than the maximum value obtained in experiment 1 with thiomersal, which may be due to the fact that in this case no extra energy is required to reduce Hg^{2+}. Furthermore, the maximum observed growth/substrate and biomass/oxygen yields ($Y_{x/S}$ and $Y_{x/O2}$) are lower in the experiment with thiomersal (see Table 1).

These results support the idea that more energy is required when thiomersal is present in the media. The maximum specific growth rate (μ_{max}) is similar in both experiments. This result is consistent with the detoxification mechanism proposed, where bacteria first transform the highly toxic thiomersal in a less toxic mercury form (Hg^0), after which growth proceeds, using glucose as carbon source, and reaching the same μ_{max} than in the experiment without thiomersal.

3.2 *Biodegradation of thiomersal in a CSTR fed with a synthetic wastewater*

The ability of the microbial strain to remediate a thiomersal contaminated synthetic wastewater was evaluated in a continuous stirred tank reactor. The thiomersal concentration fed to the bioreactor was deliberately set to a value four times higher than the one observed in the real effluent. In this way, it was possible to test the system at a more unfavorable situation. The results obtained for $D = 0.05h^{-1}$ are presented in Figure 3.

The average biomass concentration reached at steady state was 1 g/l. The glucose concentration in the bioreactor outlet was 0.4 g/l and the thiomersal outlet concentration was below 5 mg/l, except when the thiomersal pulses were applied. The results obtained show that the microbial culture used possess the ability to reduce the high inflow thiomersal concentration of 192 mg/l to an average outlet concentration of 3 mg/l. The thiomersal degradation rate was similar to the one observed in batch studies and equal to $9.45\,mgl^{-1}h^{-1}$.

In order to evaluate the process dynamics, two different thiomersal pulses (Pulse 1 and Pulse 2: addition of 25 and 50 mg/l to the bioreactor, respectively) were applied to the system, at steady state.

The system adapted quickly to the applied transient thiomersal shock loads. The microorganisms were able to degrade the excess of thiomersal and return to the thiomersal steady state residual concentration. Moreover, an increase in the thiomersal degradation rate was observed, for both pulses, immediately after

Table 1. Kinetic and stoichiometric parameters for a *P. putida* spi3 strain grown in the presence and absence of thiomersal.

Parameter	With thiomersal	Without thiomersal
μ_{max} (h^{-1})	0.42 ± 0.01	0.44 ± 0.01
$Y_{x/s}$ (g_x/g_s)	0.96 ± 0.04	1.47 ± 0.03
Y_{s/O_2} (g_s/g_{O2})	3.52 ± 0.18	3.62 ± 0.10
Y_{x/O_2} (g_x/g_{O2})	3.19 ± 0.06	4.61 ± 0.31
$r_{thiomersal}$ ($mgl^{-1}h^{-1}$)	9.37 ± 0.35	—

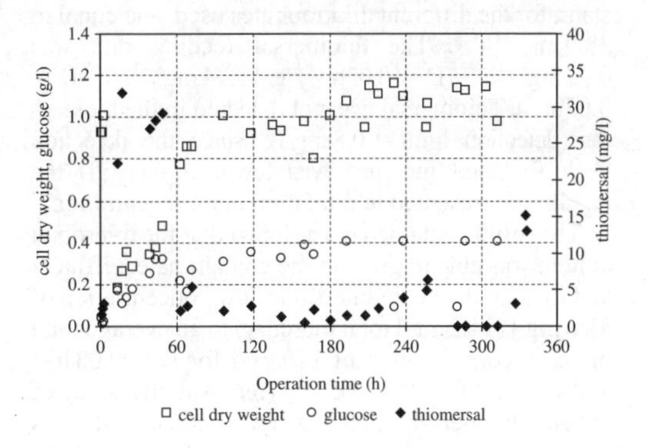

Figure 3. Evolution of biomass, glucose and thiomersal concentration in a CSTR fed with a synthetic wastewater.

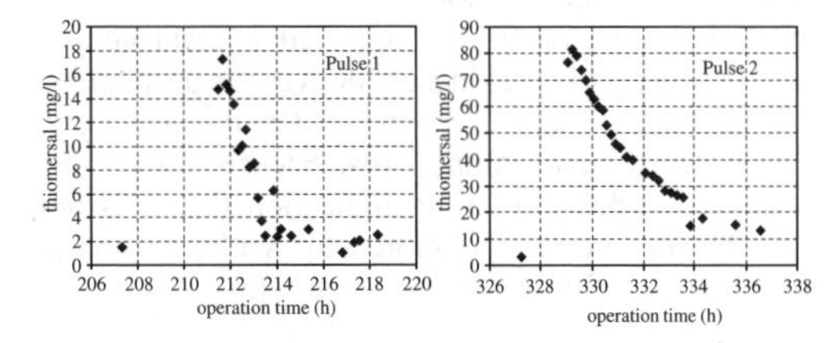

Figure 4. Evolution of thiomersal concentration after the aplication of the thiomersal pulses.

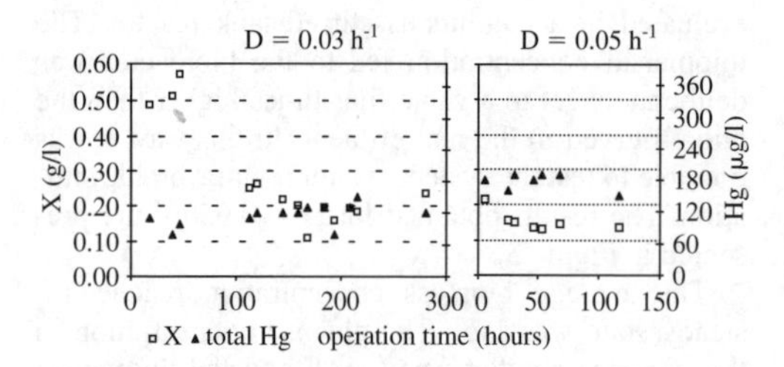

Figure 5. Evolution of biomass and total mercury concentration in a CSTR fed with a real vaccine wastewater.

the thiomersal shock load. During the stabilization period after the pulse, the degradation rate decreased almost linearly with the thiomersal outflow concentration. These results suggest a 1st order degradation kinetics, and that the bioreactor posses a very good ability to deal with sudden fluctuations in the thiomersal inflow concentrations.

3.3 *Biodegradation of thiomersal in a CSTR fed with a real vaccine wastewater*

The ability of the microbial strain to remediate a real thiomersal contaminated vaccine effluent was evaluated in a continuous stirred tank reactor. The thiomersal concentration in the effluent was constant and equal to 48.4 mg/l. The glucose-feeding rate was kept constant, for the different dilution rates used, and equal to 38.9 $mgl^{-1}h^{-1}$. The thiomersal feeding rate was 0.93 $mgl^{-1}h^{-1}$ ($D = 0.03h^{-1}$) and 1.74 $mgl^{-1}h^{-1}$ ($D = 0.05h^{-1}$). Thiomersal was not detected in the bioreactor (detection limit = 0.8 mg/l). Since the detection limit for total mercury was lower (0.05 µg/l) the results are presented in total mercury concentration.

The results obtained have shown that the microbial culture was able to grow in the contaminated effluent and to degrade the thiomersal inflow concentration of 48.4 mg/l (24.2 mg/l total mercury) to an average total mercury concentration of 120 µg/l for $D = 0.03\,h^{-1}$ and 180 µg/l for $D = 0.05\,h^{-1}$ (removal efficiency of 99%). Although the outlet mercury concentration is higher than the admissible mercury discharge limit (50 µg/l), the results obtained suggest a high potential for the remediation of thiomersal containing effluents. In an attempt to improve the process economic feasibility, the system was operated without adding micronutrients to the effluent. However, the decrease on biomass concentration, accompanied by the increase in the mercury concentration, that occurred after this change, constitute a strong indication that it is not possible to operate the bioreactor without supplementing the effluent with micronutrients.

4 CONCLUSIONS

The results obtained have shown that the microbial culture used has the capacity to degrade the thiomersal present in a synthetic wastewater, both in batch and continuous culture. When operated in continuous, the bioreactor was able to sustain the transient thiomersal shock loads applied, by degrading the excess thiomersal. The thiomersal degradation process by a *P. putida* strain proved to have a high degree of self-regularity and adjustability. Moreover, when the bioreactor was fed, in continuous mode, with the real vaccine effluent, contaminated with 50 mg/l of thiomersal, the microbial culture was able to grow in the effluent, and thiomersal was not detected in the outflow. Despite the fact that the outlet mercury concentration is still higher (120 µg/l for $D = 0.03\,h^{-1}$) than the admissible mercury discharge limit (50 µg/l), the results obtained suggest a high potential for the remediation of thiomersal containing wastewaters. Future work will be focused on the development of strategies to decrease the mercury concentration in the outflow to values below the maximum admissible levels. Three main approaches can be followed: increase the mercury stripping, operate the reactor at lower dilution rates or include a carbon filter in the outflow for polishing.

ACKNOWLEDGMENTS

The authors acknowledge funding from the European Commission through project QLK3-1999-01213. Raquel Fortunato acknowledges the research grant PRAXIS XXI/BD/21618/99 from Fundação para a Ciência e a Tecnologia.

REFERENCES

Elferink J.G.R. 1999. Thimerosal A versatile sulphydryl reagent, calcium mobilizer, and cell function-modulating agent. *General Pharmacology, 33:1.*

Keith L.H., Walters D.B. 1992. The National Toxicology Program's Chemical Data Compendium, Vol I-VIII. *Boca Raton, FL: Lewis Publishers, Inc.*

Misra T.K. 1992. Bacterial Resistance to Inorganic Mercury salts and Organomercurials. *Plasmid, 27:4.*

Wagner-Döbler I. 2003. Pilot plant for bioremediation of mercury containing industrial wastewater. *Appl Microbiol Biotechnol, 62:124.*

European Symposium on Environmental Biotechnology, ESEB 2004 - Verstraete (ed)
© 2004 Taylor & Francis Group, London, ISBN 90 5809 653 X

Optimization of metal dosing in anaerobic bioreactors: effect of dosing strategies on the performance and metal retention of methanogenic granular sludge

M.H. Zandvoort, E. van Hullebusch, G. Lettinga & P.N.L. Lens
Sub-department of environmental technology, Wageningen University, Wageningen, The Netherlands

ABSTRACT: The effect of trace cobalt dosage and dosing strategies on the performance and cobalt retention of methanol fed upflow anaerobic sludge bed (UASB) reactors were investigated. The three dosing strategies tested (continuous, pre-loading and pulse) showed that continuous dosing at 0.33 μM was effective in minimizing cobalt losses from the granular sludge but only moderately increased the specific methanogenic activity from 110 to 210 mg CH_4-COD g VSS^{-1} d^{-1}. Pre-loading sludge results in high specific methanogenic activities increasing form 155 to 906 mg CH_4-COD. g VSS^{-1} d^{-1}, within 30 days of operation. However, cobalt losses are much bigger, 54% of the original 1667 μg g TSS^{-1} was lost after 77 days of operation. By pulse dosing, 16% of the amount of cobalt present in the pre-loaded sludge, the methanogenic activity of the sludge increased from 155 to 631 mg CH_4-COD g VSS^{-1} d^{-1}, while the cobalt content increased to only 107 μg g TSS^{-1}, indicating a relatively effective use of the cobalt compared to pre-loading.

1 INTRODUCTION

Dosage of trace metals to anaerobic bioreactor influents is essential to maintain their treatment performance (Zandvoort *et al.*, 2003). Although being essential for good reactor operation, trace metal dosing needs to be optimised in order to reduce the costs, minimise their introduction into the environment and maximise the effect on the biological activity (Zandvoort *et al.*, 2002a). For this optimisation, it is important to understand the chemical and microbial behaviour of metals (e.g. cobalt, iron and nickel) in both the solid (granular sludge) and liquid phase of the bioreactor (Fig. 1). Important to link the metal speciation to the bioavailability of metals and the effect on the biological activity. From this novel trace metal dosing strategies can be developed which lead to optimal reactor performance, with a minimal amount of metal added.

The presented research focuses on the trace metal cobalt and the model substrate methanol. Among the trace elements, cobalt plays a key role as it regulates the methanol degradation pathway (under mesophillic conditions) and the competition between the different trophic groups (acetogens or methanogens) involved in methanogenic methanol conversion (Zandvoort *et al.*, 2002b).

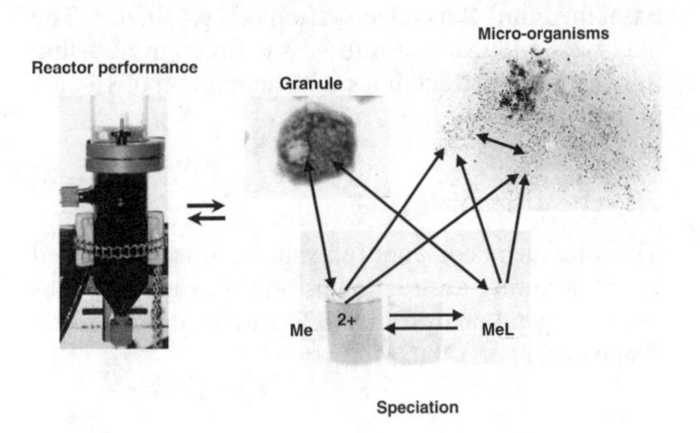

Figure 1. Interactions of metals with the solid and liquid phase of anaerobic bioreactors, which influence the efficiency of the metal dosage.

The response of the methanogenic activity towards cobalt addition was studied using batch tests. The effect of different cobalt dosing strategies (continuous, pulsed and pre-loading) on the reactor performance, methanogenic activity and metal retention was investigated using upflow anaerobic sludge bed (UASB) reactors inoculated with cobalt deprived granular sludge (Zandvoort *et al.*, 2002b; Zandvoort *et al.*, in prep.).

2 MATERIALS & METHODS

2.1 *Source of biomass*

Methanogenic granular sludge was obtained from a full-scale UASB reactor treating alcohol distillery wastewater of Nedalco (Bergen op Zoom, The Netherlands). The sludge has a relatively low initial cobalt content (17 μg g TSS^{-1}).

2.2 *UASB reactor operation*

For the reactor experiments lab-scale UASB reactors with a volume of 7.25 L, except for the loading experiment described in paragraph 4 were reactors with a volume of 0.75 L were used. The reactors were operated in a temperature controlled (30 ± 2°C) room.

A basal medium was fed to the reactors consisting of methanol, macronutrients and a trace element solution as described by Zandvoort *et al.* (2002a). To ensure pH stability 2.52 g (30 mM) of $NaHCO_3$ was added with the influent.

2.3 *Specific maximum methanogenic activity test*

The specitfic methanogenic activity of the granular biomass was determined in duplicate using on-line gas production measurements as described by Zandvoort *et al.* (2002a). The substrate methanol was added to the batches (117 ml serum bottles, 50 ml basal medium) at a concentration of 4 g $CODl^{-1}$. The data were plotted in a rate versus time curve, using moving average trend lines with an interval of 15 data points.

2.4 *Metal analyses*

The total metal content of the sludge was determined after microwave destruction using aqua regia and subsequent metal analyses by ICP-MS as described by Zandvoort *et al.* (2002a).

3 COBALT DEPRIVATION

3.1 *Cobalt limitation in full-scale bioreactors*

Full-scale UASB reactors treating different types of wastewater can already show a limitation for an essential trace metal when they are not supplied with the effluent. This was for instance observed when granular sludge from a UASB reactor treating groundwater contaminated with chlorinated compounds (influent cobalt concentration below the detection limit) was tested for its response to cobalt using methanol as the substrate (Fig. 2). The sludge, initial cobalt content of 22 μg g TSS^{-1}, showed a clear increase in methane formation rate when 5 μM of cobalt was added to the batch. This increased the methanogenic activity of the sludge from 155 to 334 mg CH_4-COD g VSS d^{-1}.

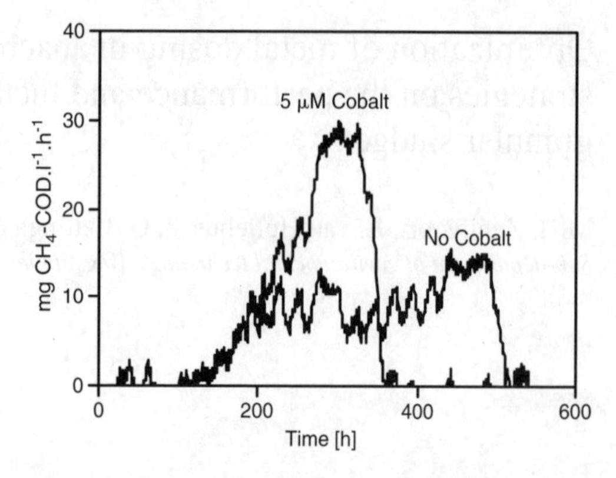

Figure 2. Methane formation rate with methanol as the substrate of granular sludge from a UASB reactor treating contaminated groundwater, with addition 5 μM cobalt and without cobalt.

3.2 *Induced cobalt limitation in lab scale UASB reactor*

A similar response could be induced in a lab-scale bioreactor were a granular sludge, initial cobalt concentration of 17 μg g TSS^{-1}, from a full-scale bioreactor treating distillery wastewater, was subjected to a prolonged operation in the absence of cobalt. A clear response to cobalt was observed in batch tests only after 28 days of operation with methanol as the substrate. The methanogenic activity could be increased from 110 to 491 mg CH_4-COD g VSS d^{-1} by the addition of 0.84 μM of cobalt.

4 DOSING STRATEGIES

As shown in the previous paragraph, limitations for cobalt can already be present in the sludge or can be induced by absence of cobalt in the influent, showing the importance of cobalt dosing to methanol fed bioreactors. Therefore different dosing strategies were applied to study the improvement of the performance and the metal retention.

4.1 *Continuous dosing*

Cobalt was continuously dosed with the influent of the cobalt deprived reactor at a relatively low concentration of 0.33 μM for 31 days resulted in an improvement of the reactor performance and doubling of the methanogenic activity from 110 to 210 mg CH_4-COD g $VSS^{-1} d^{-1}$ and a 2.7 times increased

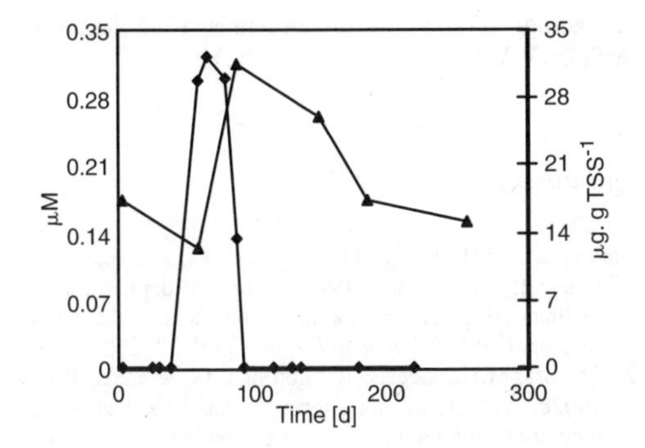

Figure 3. Evolution of the cobalt concentration in the influent (◆) and granular sludge (▲) as a function of time.

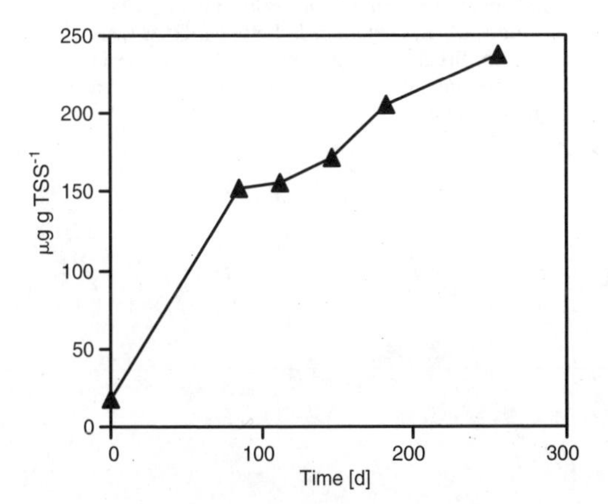

Figure 4. Evolution of the cobalt concentration in granular sludge (▲) of a UASB reactor treating methanol with an influent cobalt concentration 0.84 μM.

cobalt concentration (32 μg g TSS d^{-1}, Fig. 2) of the sludge. Cobalt was found to wash-out from the sludge at a rate of 0.1 μg g TSS d^{-1} but the methanogenic activity remained constant for more than 100 days after cobalt dosing was stopped.

A parallel reactor was operated for 267 days with the same inoculum and an influent cobalt concentration of 0.84 μM (Fig. 4). This reactor showed a continuous accumulation of cobalt at a rate of 0.82 μg g $TSS^{-1} d^{-1}$, resulting in final concentration of 237 μg g TSS^{-1}. This indicates that the cobalt content of the sludge could be well maintained and increased at this influent concentration.

4.2 *Cobalt pre-loading*

Another dosing strategy could be providing the metals all at once by pre-loading the granular sludge, and

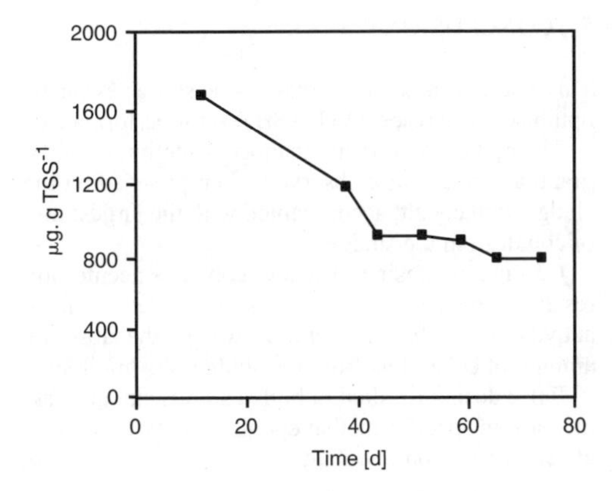

Figure 5. Evolution of the cobalt concentration in granular sludge (▲) of a UASB reactor treating methanol with cobalt pre-loaded sludge (1 mM $CoCl_2$).

thus creating a metal stock (Zandvoort *et al.*, in prep.). To study the effect of this strategy on reactor performance and metal retention. A reactor was inoculated with cobalt pre-loaded sludge (24 h; 30°C; 1 mM $CoCl_2$). This increased the cobalt concentration from 17 to 1667 μg g TSS^{-1}. With time the methanogenic activity increased considerably as well after 30 days of operation from 155 (unloaded) to 906 mg CH_4-COD g $VSS^{-1} d^{-1}$ and increased even further to 1200 mg CH_4-COD g $VSS^{-1} d^{-1}$ at day 77.

However, the losses of cobalt from the sludge were considerable until day 44 the sludge concentration decreased at a rate of 22 μg g TSS d^{-1}. From this day onwards, this rate became 5 μg g TSS d^{-1} (Fig. 5). At the end of the reactor run (day 74) the cobalt concentration in the sludge had decreased to 763 μg g TSS^{-1}, 54% of the initial concentration.

4.3 *Pulse dosing*

Also pulse dosing (in situ loading) was considered as an intermediate dosing strategy between continuous and pre-loading of cobalt (Zandvoort *et al.*, in prep.). This strategy was applied by dosing 31 μM of cobalt (16% of the amount present in the pre-loaded sludge) with the influent for 24 h to a reactor operated 59 days without addition of cobalt. This resulted in a considerable increase in the methanogenic activity from 155 to 631 mg CH_4-COD g $VSS^{-1} d^{-1}$ at day 77. Although this increase is less than with the pre-loaded sludge, the amount of cobalt accumulated in the sludge was only 107 μg g TSS^{-1} (15.6 times lower than pre-loaded) and the cobalt was lost at a rate of 4 μg g TSS d^{-1}. In other words, the relative increase in activity achieved with the retained cobalt is larger and the relative loss of cobalt is lower with this dosing strategy compared to the pre-loading strategy.

5 CONCLUSIONS

This research shows that metal dosing strategies can be optimised with respect to UASB reactor performance.

The highest specific methanogenic activities of the granular sludge were observed when pre-loading the sludge with cobalt, accompanied with the largest loss of cobalt from the sludge.

Continuous dosing at a low cobalt concentration resulted in moderate increases in specific methanogenic activity and cobalt content. However, the absolute amount of cobalt lost from the sludge was minimal.

Pulse dosing resulted in both a significant and fast increase in specific methanogenic activity as well an efficient retention of cobalt.

ACKNOWLEDGEMENTS

This research is supported by the Technology Foundation STW, applied science division of NWO and the technology programme of the Ministry of Economic Affairs (WWL 4928).

REFERENCES

Zandvoort, M.H., Osuna M.B., Geerts, R., Lettinga, G. & Lens P.N.L. 2002a, Effect of nickel deprivation on methanol degradation in a methanogenic granular sludge reactor. *J. of Ind. Micorbiol. Biotechnol.* 29: 268–274.

Zandvoort, M.H., Geerts, R., Lettinga, G. & Lens, P.N.L. 2002b, Effect of long-term cobalt deprivation on methanol degradation in a methanogenic granular sludge bioreactor. *Biotechnol. Prog.* 18: 1233–1239.

Zandvoort, M.H., Geerts, R., Lettinga, G. & Lens, P.N.L. 2003, Methanol degradation in granular sludge reactors at sub-optimal metal concentrations: role of iron nickel and cobalt. *Enz. Microbial. Techn.* 33: 190–198.

Zandvoort, M.H., Gieteling, J., Lettinga, G. & Lens, P.N.L. In prep. Stimulation of methanol degradation in UASB reactors: in situ versus pre-loading cobalt on anaerobic granular sludge.

European Symposium on Environmental Biotechnology, ESEB 2004 - Verstraete (ed)
© 2004 Taylor & Francis Group, London, ISBN 90 5809 653 X

Kinetic characterization in dynamically operated CST-reactors: ferrous iron oxidation by *Leptospirillum ferrooxidans*

R. Kleerebezem & M.C.M. van Loosdrecht
Delft University of Technology, Department of Biotechnology, Julianalaan Delft, The Netherlands

ABSTRACT: The Fe^{II} oxidation kinetics by *L. ferrooxidans* have been investigated in dynamic Continuous Stirred Tank Reactor (CSTR) experiments that allow for rapid and accurate determination of kinetic parameter values. By on-line measurement of oxygen and carbon dioxide in the off-gas of the reactor, and the Fe^{II}:Fe^{III} concentration ratio in the medium, mass balances for electron donor, acceptor and biomass can be established. As opposed to steady state measurements, we conducted experiments in CSTR's that are operated at a variable dilution rate. This dynamic approach allows for kinetic parameter determination within a few day period. It furthermore enables quantitative measurement and subsequent modeling of the response of the system on short-term perturbations.

1 INTRODUCTION

Kinetic characterization of slow growing organisms in chemostat systems is a laborious task. Mesophilic aerobic ferrous iron oxidizing bacteria for example have a maximum growth rate of approximately 0.1 hr^{-1} (Boon *et al*, 1999; Van Scherpenzeel *et al*, 1998). Kinetic characterization in continuous experiments requires at least 5 data points and some measurements at low dilution rate. It can readily be calculated that full kinetic characterization will therefore take at least 2 to 3 months. The investigation of the impact of an operational variable like the temperature or the pH on the kinetics of the process requires a range of kinetic characterizations. Therewith it requires even much more time and effort (Breed and Hansford, 1999; Breed *et al*, 1999).

In this paper we will describe a method that allows for rapid kinetic characterization. The method relies on dynamic operation and data registration of a continuous reactor. The slow growing, ferrous iron (Fe^{II}) oxidizing culture *Leptospirillum ferrooxidans* has been used to demonstrate the concept.

2 MATERIAL AND METHODS

2.1 *Experimental set-up*

The experiments with the acidophilic iron oxidizing *L. ferrooxidans* culture have been conducted in completely stirred tank reactors (CSTR) with a liquid volume of 2 dm^3. Experiments were conducted at a pH of 1.4–1.5 and a temperature of 30°C. A schematic overview of the experimental set-up is shown in Figure 1.

The experimental set-up consists of two reactors. O_2 and CO_2 concentrations in the off-gas of both reactors and the incoming gas are alternately measured for periods of typically 10 to 50 minutes. Switching the gas composition measurement between the two reactors and the incoming gas is accomplished by computer-controlled valves. Generated data are stored in the computer and a continuous data set for both reactors and the influent gas is obtained by linear interpolation.

On-line measurement of the O_2 and CO_2 concentration in the gas entering and leaving the reactor allows for direct estimation of the electron acceptor uptake rate (R_{O2}, mol hr^{-1}) and the biomass formation rate ($R_X = -R_{CO2}$, mol hr^{-1}). Direct measurement of R_X is possible because at the operational pH of 1.4 no bicarbonate is formed and the liquid concentration dissolved CO_2 can be neglected. Measurement of the redox potential (*ORP*) allows for estimation of the actual Fe^{II}:Fe^{III}-concentration ratio in the reactor and the actual electron donor uptake rate (R_{FeII}, mol hr^{-1}). Herewith a full mass balance can be established over the reactor based on on-line measurements.

2.2 *Calculations*

On-line measurements allow for direct calculation of various variables. Based on the actual oxygen uptake rate (R_{O2}, mol hr^{-1}) and carbon dioxide uptake rate (R_{CO2}, mol hr^{-1}), first of all the actual biomass yield

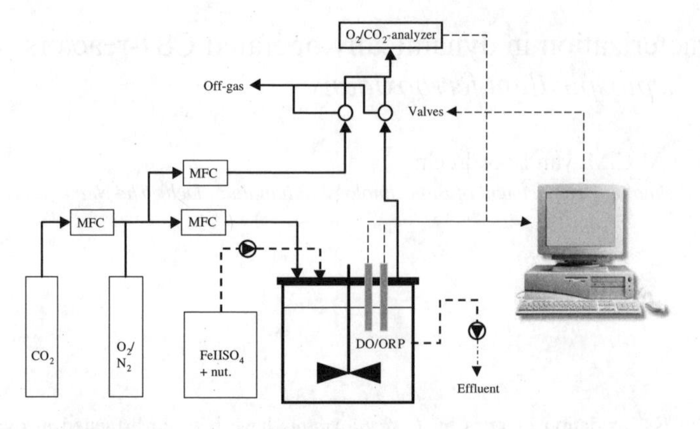

Figure 1. Schematic overview of the experimental set-up.

can be calculated according to:

$$Y_{X/O2} = \frac{R_X}{R_{O2}} = \frac{-R_{CO2}}{R_{O2}}$$

By convention the biomass yield on a substrate has a positive value and consequently the minus sign before R_{CO2} is omitted.

The experiment described here was conducted starting from steady state. In case of steady state the biomass concentration (X^*) can be calculated from R_{CO2} and the influent liquid flow rate (Q_L, dm^3 hr^{-1}):

$$X^* = \frac{-R_{CO2}}{Q_L}$$

During the dynamic (non-steady state) experiment the biomass concentration (X) is calculated by integration in time of the biomass mass balance:

$$X_{t+\Delta t} = X_t - \frac{\left(Q_L \cdot X_t + R_{CO_2}\right) \cdot \Delta t}{V_L}$$

During the experiments biomass concentrations calculated based on R_{CO2} were verified by *TOC*-measurements at regular intervals. Knowing the actual biomass concentration allows for calculation of the actual specific oxygen uptake rate (q_{O2}, mol C-mol^{-1} hr^{-1}) and the actual specific growth rate (μ, hr^{-1}):

$$q_{O_2} = \frac{R_{O_2}}{X \cdot V_L}$$

$$\mu = \frac{R_X}{X \cdot V_L} = \frac{-R_{CO_2}}{X \cdot V_L},$$

Calculation of the actual ferrous iron concentration for comparison with the ORP-based Fe^{II}-concentration as a function of time is based on the mass balance for ferrous iron (*FeII*, mol dm^{-3}):

$$FeII_{t+\Delta t} = FeII_t + \left\{ \begin{matrix} Q_L \cdot (FeII_i - FeII_t) + \\ R_{O_2} \cdot Y^{CAT}_{FeII/O_2} + R_{CO_2} \cdot Y^{AN}_{FeII/CO_2} \end{matrix} \right\} \cdot \frac{\Delta t}{V_L},$$

where $Y^{CAT}_{FeII/O2}$ and $Y^{AN}_{FeII/CO2}$ are the stoichiometric coefficients catabolic and anabolic Fe^{II} oxidation. This equation is primarily used to check the mass balances.

3 RESULTS AND DISCUSSIONS

3.1 *Kinetic parameter determination*

In order to determine kinetic parameter values the dilution rate in the reactor was increased from a (steady state) low value to a value exceeding the maximum growth rate of the culture within a number of days. After initial wash-out was observed, the dilution rate was decreased to a low value. Results are shown in Figure 2.

Evidently, the increase in the dilution rate imposed on the system resulted an increase in the actual O_2 and CO_2 uptake rate. Throughout the experiments no limitations in dissolved oxygen or carbon dioxide were evident. Calculated biomass concentrations could adequately be verified with the *TOC*-measurements. From the results shown in Figure 2 parameter values for q^{max}_{O2} (1.22 mol C-mol^{-1} hr^{-1}) and μ^{max} (0.056 hr^{-1}) can be derived directly.

For estimation of K_{FeII}, q_{O2} needs to be expressed as a function of the concentration Fe^{II} (Figure 4). The results suggest that *L. ferrooxidans* has a high affinity for Fe^{II} ($K_{FeII} = 0.05$ mM), but seems to be moderately inhibited at elevated concentrations *FeII*. The inhibition by *FeII* could adequately be described using a competitive substrate inhibition model for substrate uptake:

$$q_{O2} = q^{max}_{O2} \frac{FeII}{K_{FeII} + FeII + \frac{FeII^2}{Ki_{FeII}}},$$

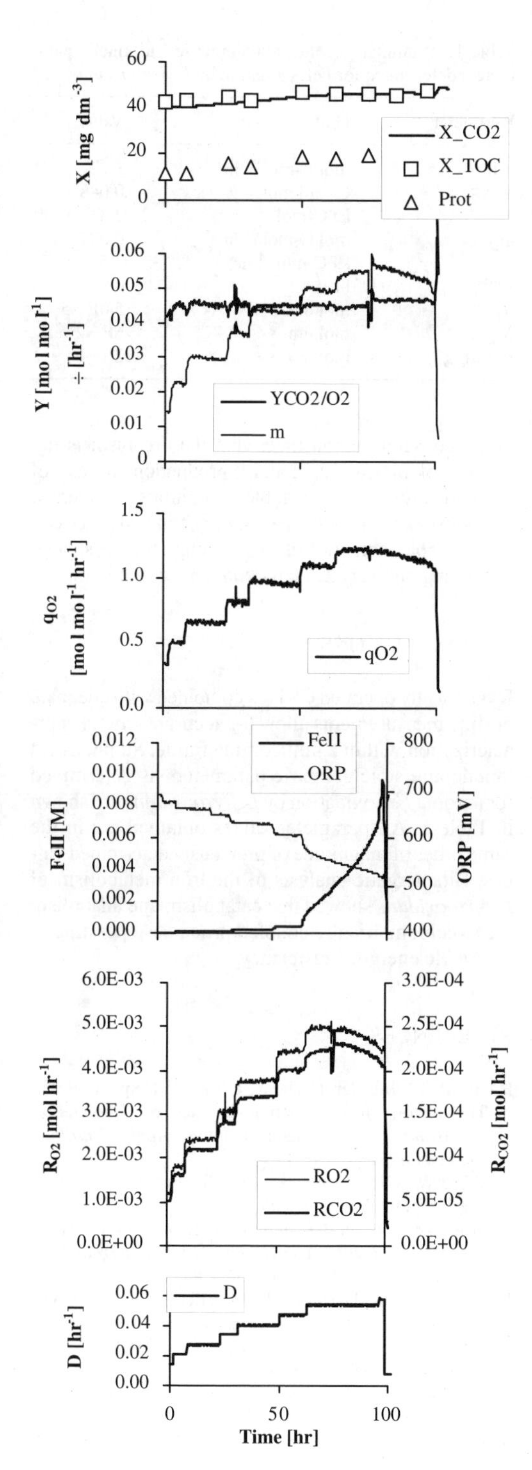

Figure 2. Measured system response to variations in the dilution rate in a Fe^{II} fed CST-reactor with *L. Ferrooxidans.*

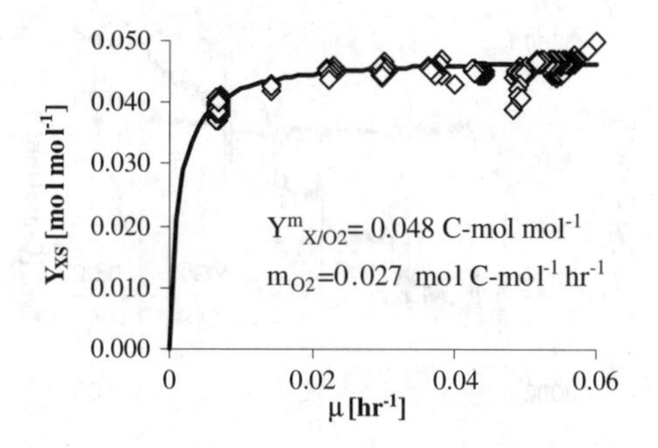

Figure 3. Estimation of $Y^{max}_{X/O2}$ and m_{O2} from the $Y_{X/O2}$-values shown in Figure 2.

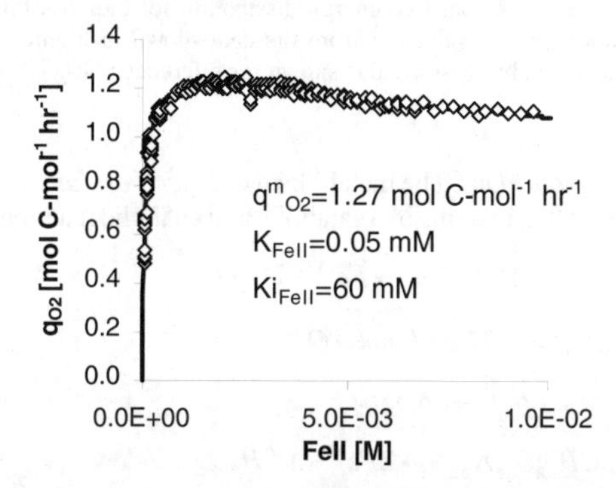

Figure 4. Estimation of the affinity constant for FeII-uptake (K_{FeII}), FeII substrate inhibition constant (Ki_{FeII}), and the maximum FeII-uptake rate (q^m_{O2}) from the data shown in Figure 5.

using an substrate inhibition coefficient (Ki_{FeII}) of 60 mM.

For estimation of the oxygen based maintenance coefficient (m_{O2}) and the maximum biomass yield ($Y^{max}_{X/O2}$), the Pirt-equation was used:

$$\frac{1}{Y_{X/O2}} = \frac{1}{Y^{max}_{X/O2}} + \frac{m_{O2}}{\mu}$$

An optimized description of the measurements was obtained using a $Y^{max}_{X/O2}$-value of 0.048 C-mol mol^{-1} and a m_{O2} value of 0.027 mol C-mol^{-1} hr^{-1} (Figure 3). Herewith the maintenance rate corresponds to only approximately 2% of the maximum specific oxygen uptake rate (q^{max}_{O2}). Assuming furthermore an average actual ΔG^{CAT}-value for iron oxidation of -130 kJ mol-O_2^{-1}, the energy requirement for maintenance ($m_{\Delta G}$) purposes amounts approximately 3.5 kJ C-mol^{-1} hr^{-1}. This value is in the same order of magnitude as the average value predicted by Heijnen (1999); 4.9 kJ C-mol^{-1} hr^{-1} at 30°C.

3.2 *Bioenergetics*

The experimental data shown in Figure 5 furthermore allow for bioenergetic analysis of growth of

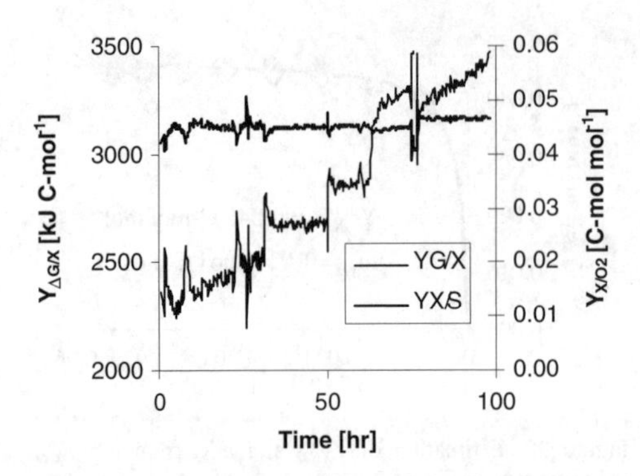

Figure 5. Actual free energy dissipation for biomass formation ($Y_{G/X}$) as derived from the data shown in Figure 5. The actual biomass yield is shown for reference ($Y_{X/O2}$).

L. ferrooxidans. The metabolism of *L. ferrooxidans* can be written in terms of a catabolic and anabolic reaction:

$$4Fe^{+2} + 4H^{+} + O_2 \rightarrow Fe^{+3} + 2H_2O$$

$$\Delta G^0_{CAT} = -177.2\ kJ\ mol - O_2$$

$$CO_2 + 4.2Fe^{+2} + 0.2NH_4^{+} + 4H^{+} \rightarrow$$

$$CH_{1.8}O_{0.5}N_{0.2} + 4.2Fe^{+3} + 1.5H_2O$$

$$\Delta G^0_{AN} = 299.5\ kJ\ mol - X$$

Energy generated in the catabolic reaction is utilized for biomass production in the energetically unfavorable anabolic reaction. However, part of the metabolic energy is dissipated. The amount of free energy that is dissipated as a function of the actual activities of the species involved was calculated. First of all, the actual free energy change of the catabolic (and anabolic reaction) was calculated according to:

$$\Delta G_{CAT} = \Delta G^0_{CAT} + R \cdot T \cdot \ln \frac{a_{Fe+3}^4}{a_{Fe+2}^4 \cdot a_{O2} \cdot a_{H+}^4}$$

The actual activities of Fe^{+2} and Fe^{+3} were calculated from the ORP-based concentrations Fe^{II} and Fe^{III} using activity correction factors. All other activities were estimated from the measured data and the medium composition.

Now the actual free energy dissipation per mole biomass formed ($Y_{G/X}$) can be calculated using:

$$Y_{G/X} = \frac{-R_{O2} \cdot \Delta G_{CAT} - R_{CO2} \cdot \Delta G_{AN}}{R_{CO2}},$$

The results are shown in Figure 5. The $Y_{G/X}$-value increases strongly in time as a result of the changes in Fe^{II}:Fe^{III}-ratio. The observation that $Y_{X/O2}$ remains rather constant demonstrates that the organism is not capable of increasing biomass production in case of energetically more favorable conditions. Evidently, catabolism and anabolism are tightly coupled processes and variable amounts of free energy are dissipated depending on the actual environmental conditions.

Table 1. Summary of the stoichiometric and kinetic parameters determined for FeII oxidation by *L. ferrooxidans*.

Parameter	Unit	Value
q_{O2}^{max}	mol C-mol^{-1} hr^{-1}	1.27
$Y_{X/O2}^{max}$	C-mol mol^{-1}	0.048
$Y_{\Delta G/X}$	kJ C-mol^{-1}	2300–3500
m_{O2}	mol C-mol^{-1} hr^{-1}	0.027
$m_{\Delta G}$	kJ C-mol^{-1} hr^{-1}	3.5
μ^{max}	hr^{-1}	0.057
K_{FeII}	mol dm^{-3}	5.0E−5
Ki_{FeII}	mol dm^{-3}	6E−2
$FeII_{threshold}$	mol dm^{-3}	2E−6

4 CONCLUSIONS

Dynamically operated CSTR's combined with adequate on-line measurements allow for accurate kinetic characterization within a limited time frame. Summarized kinetic and stoichiometric parameters as determined for aerobic Fe^{II} oxidation by *L. ferrooxidans* are shown in Table 1. All parameter values obtained are in the same order of magnitude of previously determined values. Bioenergetic analysis of the iron metabolism of *L. ferrooxidans* showed that catabolism and anabolism are stoichiometrically coupled processes operating at a variable energetic efficiency.

REFERENCES

Boon, M, TA Meeder, C Thone, C Ras, JJ Heijnen (1999). The ferrous iron oxidation kinetics of *Thiobacillus ferrooxidans* in continuous cultures. *Appl. Microbiol. Biotechnol.* **51**: 820–826.

Breed, AW, CJN Dempers, GE Searby, MN Gardner, Gardner, DE Rawlings, GS Hansford (1999). The effect of temperature on the continuous ferrous-iron oxidation kinetics of a predominantly *Leptospirillum ferrooxidans* culture. *Biotechnol. Bioeng.* **65**(1): 44–52.

Breed, AW and GS Hansford (1999). Effect of pH on ferrous-iron oxidation kinetics of *Leptospirillum ferrooxidans* in continuous culture. *Biochem. Eng. J.* **3**: 193–201.

Heijnen, JJ (1999). Bioenergetics of microbial growth. Encyclopedia of Bioprocess Technology. M.C. Flickinger and S.W. Drew. New York, J. Wiley.

Van Scherpenzeel, DA, M Boon, C Ras, GS Hansford, JJ Heijnen (1998). Kinetics of ferrous iron oxidation by *Leptospirillum* bacteria in continuous cultures. *Biotechnol. Prog.* **14**: 425–433.

European Symposium on Environmental Biotechnology, ESEB 2004 - Verstraete (ed)
© 2004 Taylor & Francis Group, London, ISBN 90 5809 653 X

Effect of salts on a partial nitrification reactor

F. González, A. Mosquera-Corral, J.L. Campos & R. Méndez
Department of Chemical Engineering, School of Engineering, University of Santiago de Compostela, Santiago de Compostela, Spain

ABSTRACT: Wastewaters with high salt concentrations are produced in anaerobic digesters as it is the case of some fish canning factories. In the present study a partial nitrification reactor has been evaluated as post-treatment of wastewater with similar features to effluents from anaerobic digester. The effect on biomass of different salt concentrations (NaCl, KCl and Na_2SO_4) was firstly evaluated by mean of batch respirometric assay. The results show that for each salt concentration evaluated the effects were similar. During the continuos operation of the reactor, the salt concentration (NaCl) was increased from 0 to 513 mM. An increase of the ammonia oxidation activity was observed when concentration of 85 mM was added, while higher salt concentrations provoked a small decrease of activity. An important effect of the adaptation of the biomass to the saline medium was observed. For salt concentrations of 342 mM the effect of different pH was evaluated. The pH was found as an accurate control parameter of the process in order to produce an effluent with an equimolar composition of ammonia and nitrite.

1 INTRODUCTION

In the search of improving the sustainability of nitrogen removal from wastewater, partial nitrification techniques have been denoted for quite a while as very promising. During partial nitrification, ammonia is converted to nitrite and further oxidation of nitrite to nitrate is prevented, thus realizing aeration costs savings in comparison with conventional nitrification to nitrate.

In the SHARON (Single reactor High activity Ammonia Removal Over Nitrite) process, partial nitrification is established by working at high temperature (about 35°C), maintaining an appropriate sludge retention time (SRT). The SHARON reactor is operated as a continuously stirred tank reactor (CSTR, chemostat) without biomass retention, so the sludge retention time equals the hydraulic retention time.

In its original configuration, the SHARON process is operated under alternating aerobic and anoxic conditions (Hellinga et al. 1998). In the last few years, coupling of the SHARON process with a so-called Anammox (ANaerobic AMMonia OXidation) process, in which ammonium and nitrite are converted to nitrogen gas under anaerobic conditions by autotrophic bacteria, has gained a lot of interest (van Dongen et al. 2001). With the combined SHARON-Anammox process, low nitrogen effluent concentrations can be obtained, while aeration costs are significantly reduced, no additional carbon source is needed and sludge production is very low.

When the SHARON reactor is used to provide the feed for the Anammox process only 50% of the ammonium needs to be converted to nitrite:

$$NH_4^+ + HCO_3^- + 0.75O_2 \rightarrow 0.5NH_4^+ + 0.5NO_2^- + CO_2 + 1.5H_2O$$

In Spain the anaerobic digestion is widely used to treat effluents generated in fish canneries. These effluents present in many cases high salts concentrations due to the use of sea water in the industrial process. The combination of the SHARON process to the Anammox process can be very useful for the treatment of effluents with low COD to N ratio. The partial nitrification of ammonia in a SHARON reactor is recommended in the cases of wastewater with ammonia nitrogen concentrations from hundreds to thousands milligrams (Hellinga et al. 1998).

The objectives of the present study were:

To study the effects of different salts on the activity of the biomass from a SHARON reactor by mean of batch respirometric assays.

To study the effects of the presence of concentration of salts in the influent to the reactor in order to simulate the effect of a real wastewater coming from an industrial anaerobic digester.

To study the effects of the pH as control parameter in order to produce an effluent with an equimolar composition of ammonia and nitrite.

2 MATERIALS AND METHODS

2.1 *Respirometric assay*

Respirometric batch assays were performed in a BOM5300 device at 35°C and pH 7 in hermetically closed vials of 10 mL. Between 100 and 200 mg VSS/L were suspended in a buffer solution (1.25 mL/L traces; NaCl, 0.2; K_2HPO_4, 1.0; KH_2PO_4, 0.8; $MgSO_4$, 0.2; $MgCl_2 \cdot (H_2O$, 0.3; g/L) of pH 7. The liquid mixture inside the reactor was gasified with air for 15 minutes to reach oxygen saturation (8.2 mg O_2/L at 35°C). To begin the experiment aeration was removed and oxygen depletion was monitored along the time by means of an oxygen electrode connected to a data acquisition program. Firstly the endogeneous respiration was measured and then the substrate was injected to the vial and the oxygen consumption due to complete biomass activity was determined (López-Fiuza et al. 2002).

2.2 *Continuos operation*

The process of partial nitrification was carried out in a continuous stirred tank reactor (Figure 1). The reactor was provided with a thermostatic jacket and temperature was maintained at 35°C using a thermostatic bath. Complete mixture inside the reactor was achieved with a mechanical stirred provided with one blade of 7 × 2 cm (diameter × height) at a rotating speed of 300 rpm.

The pH value was kept at the desired value by means of a control system (pH rocon Mod 18, ORION) connected to the pH electrode. The pH regulation was made by means of addition of acid (H_2SO_4 1 N) or base (NaOH 1 N) solutions, respectively.

The feeding solution, described in Table 1, and the effluent were respectively added and removed from the reactor using peristaltic pumps. Oxygen was supplied by means of air gasification through the liquid phase using a diffuser to obtain small air bubbles. Concentration was kept over 2 mg O_2/L during the whole operation period.

2.3 *Analytical methods*

Ammonia was measured with a DN 1900 (Rosemount, Dohrmann) analyzer. Nitrite and nitrate were measured by using a Water Capillary Ion Analyzer, provided with a capillary column (i.d. 75 μm, length 60 cm, filling of melt Si covered by poliimide) and UV detection system set at 214 nm. The experimental conditions were set at 20 kV(voltage), 25°C temperature and 4 seconds of sampling time. An electroosmotic solution (CIA-Pak™ OFM Anion BT) and Na_2SO_4 (100 mM) were used as recommended by Vilas-Cruz et al. (1994). Total Organic Carbon (TOC) and Inorganic Carbon (IC) were analyzed by using a Total Carbon Analyzer (Shimadzu TOC-5000) equipped with a non-dispersive infra-red (NDIR) detector. Helium was used as carrier gas, at a flow rate of 150 ml/min. pH was measured with an INGOLD U-45 electrode connected to a CRISON 506 pH/mV voltmeter and calibrated with a CRISON buffer of pH 4.00 and 7.02 at 20°C. Dissolved oxygen was measured with a CRISON oxymeter. The concentration of biomass was measured as volatile suspended solids (VSS) according to the standard methods (APHA, 1985).

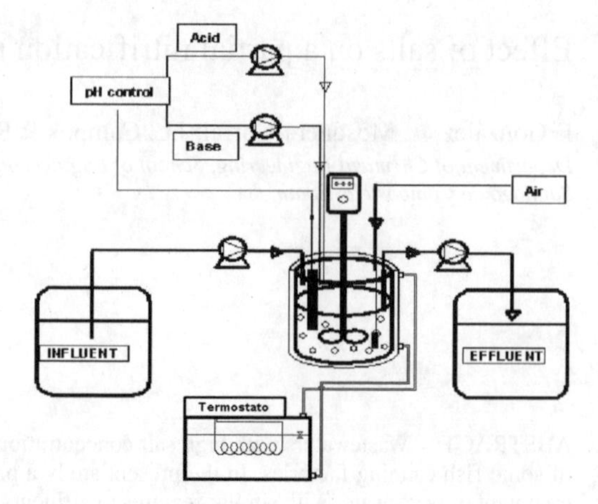

Figure 1. SHARON reactor layout.

Table 1. Feeding composition.

Comopunds	(g/L)
$(NH_4)_2SO_4$	4.71
$NaHCO_3$	6.56
$CaCl_2$	0.30
KH_2PO_4	0.07
$MgSO_4$	0.02
$FeSO_4 \cdot (7\ H_2O$	0.009
EDTA	0.006
H_2SO_4	0.005
Traces solution	1.25 mL/L

3 RESULTS AND DISCUSSION

3.1 *Effects on specific activity of different salts*

The effects of the presence of different salt concentrations (NaCl, KCl and Na_2SO_4) over the maximum specific ammonia oxidising activity, were studied using batch respirometric experiments. Sludge samples were collected from the SHARON reactor treating a synthetic media with low salt content (Table 1). Inhibitory percentages were calculated as the ratio between the ammonia oxidation specific activity for every salt concentration and the maximum specific activity without salt addition (A/A_{max}) (Figure 2).

The results indicated that the three salts provoked similar effects over the ammonia oxidising activity at similar molar concentrations. This effect may be attributed to an increase of osmotic pressure (Hunik et al. 1992), which affect the bacterial activity instead to chemical components that constituted the different salts.

Inhibition values around 60% of ammonia oxidising activity were measured for 100 mM of every salt. The effect of higher concentrations of salts was not studied. Similar results were observed by Hunik et al. (1992) working with *Nitrosomonas europaea* and by Campos et al. (2002) in batch experiments with non-acclimated nitrifying activated sludge. The last authors obtained even the similar percentage of inhibition at the same molar concentration of 100 mM although in the present work only ammonia oxidising bacteria were present in the sludge and no complete nitrification occurred.

3.2 *Effect of different concentrations of NaCl in a SHARON reactor*

Due to the similar effects obtained with every salt only NaCl was tested in the reactor. The reactor

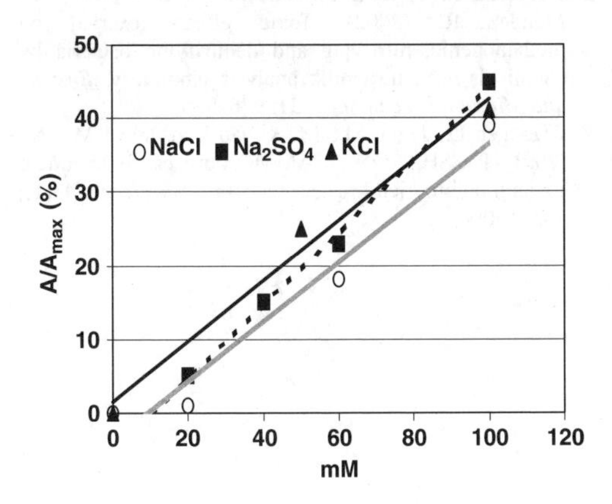

Figure 2. Inhibitory percentages as the ratio between the ammonia oxidation specific activity for every salt concentration and the maximum activity without salt addition.

was continuously fed with the autotrophic solution described in Table 1 and different concentrations of NaCl (Table 2), pH was maintained at 7.00. Concentrations of NaCl of 85, 171, 256 and 342 mM were tested during periods of 7 days (7 SRT) and concentrations of NaCl of 427 and 513 mM were tested during periods of 22 days.

When a concentration of 85 mM of NaCl was added to the feed of the reactor a stimulatory effect was observed and 68% of the ammonia in the medium was oxidised to nitrite. This effect was reduced with the increase of salt concentration. When 342 mM NaCl were added, the ammonia oxidation decreased to 47%, which represents only a decrease of 10% compared to the operation without salt. The salt concentration was then increased to 427 mM, during 22 days and a similar production of nitrite was observed. Total inhibition of ammonia oxidation occurred at 513 mM of NaCl. Therefore it could be said that the process of the SHARON reactor is stable working high salt concentrations up to 427 mM of NaCl.

This adaptation to high salt concentration was already observed by Campos et al. (2002) in a nitrifying system with a synthetic wastewater containing up to 525 mM of salts. Dahl et al. (1997) in an alternating SBR for nitrification/denitrification found adaptation of the biomass up to 563 mM of NaCl. In this work inhibition began at lower values, probably due to the low biomass concentration in the reactor beneath 100 mg VSS/L compared with the other studies from the literature.

The difference of nitrite production between batch experiments and continuos operation with similar salt concentrations can be attributed to biomass adaptation. This difference also was observed by Campos et al. (2002).

3.3 *Effect of pH at high salt concentration*

The reactor was operated (Table 3) at high salt concentration (342 mM of NaCl) to evaluate the parameter pH as a control for setting a desired ammonium/

Table 2. Percentage of partial nitrification during assay with NaCl.

NaCl (mM)	% Partial nitrification	Duration (d) (HRT = 1 day)
0	54 ± 8	7
85	68 ± 9	7
171	56 ± 2	7
256	52 ± 1	7
342	47 ± 2	7
427	51 ± 8	22
513	0	22

Table 3. Percentage of partial nitrification during assay with different pH and 342 mM of NaCl.

pH	% Partial nitrification	Ammonia oxidation activity [gNH_4^+ − N/(gSSV*d)]
7.55	71	7.7
7.25	68	7.5
7.00	60	6.5

nitrite ratio in the effluent. The pH control is based on the principle of the chemostat system: at a constant dilution rate the effluent substrate concentration will be constant (Van Dongen et al. 2001).

Increases of ammonia oxidation activity and nitrite production were observed when pH value was increased. A little variation of the pH have an important effect over the ammonium oxidizing activity, maybe the biochemically active form of the compounds is subject to the chemical acid-base equilibria. It has been shown that NH_3 rather than NH_4^+ is the active substrate (Anthonisen et al. 1976). The pH could be used as control parameter of the process in order to produce an effluent with an equimolar composition of ammonia and nitrite.

4 CONCLUSIONS

- Similar values of the activity were observed to equal concentrations of the different salts (NaCl, KCl and Na_2SO_4) in batch assay.
- Low concentrations of salt (100 mM of NaCl) produced inhibition of up 60% of the ammonia oxidizing activity in the batch experiments with biomass non-acclimated to the salt.
- Concentrations of NaCl of 85 mM cause an increase of 30% in the nitrite production percentage at continuous operation.
- The difference of nitrite production between batch experiments and continuos operation with similar salt concentrations can be attributed to biomass adaptation.
- The SHARON reactor works stable at high salt concentrations up to 427 mM of NaCl.
- pH could be used as control parameter of the process in order to produce an effluent with an equimolar composition of ammonia and nitrite.

ACKNOWLEDGEMENTS

This work was funded by the European Commission through the ICON project (Project EKV-CT-2000-0054), Xunta de Galicia (PGIDT10XJ 120904 PM) and the Spanish CICYT which funded this research through the Oxanamon project (PPQ-2002-00771).

REFERENCES

Anthonisen A.C., Loehr B.C., Prakasam T.B.S., Srinath E.G. (1976). Inhibition of nitrification by ammonia and nitrous acid. *J. Wat. Pollut. Control Fed.*, **48**(5), 835–852.

APHA-AWWA-WPCF. (1985). Standard Methods for examination of water and wastewater. 16th Ed. Washington.

Campos J.L., Mosquera-Corral A., Sánchez M., Méndez R. & Lema J.M. (2002). Nitrification in saline wastewater with high ammonia concentration in an activated sludge unit. *Wat. Res.*, **36**, 2555–2560.

Dahl C., Sund C., Kristensen G.H. & Verdenbregt L. (1997). Combined biological nitrification and denitrification of high-salinity wastewater. *Wat. Sci. Tech.*, **36**(2–3), 345–352.

Hellinga C., Schellen A.A.J.C., Mulder J.W., van Loosdrecht M.C.M. & Heijnen J.J. (1998). The SHARON process: an innovative method for nitrogen removal from ammonium-rich waste water. *Wat. Sci. Tech.*, **37**(9), 135–142.

Hunik J.H., Meijer H.J.G., Tramper J. (1992). Kinetics of *Nitrosomonas europaea* at extreme substrate, product and salt concentrations. *Appl. Microbiol. Biotechnol.*, **37**, 802–807.

López-Fiuza J., Buys B., Mosquera-Corral A., Omil F., Méndez R. (2002). Toxic effects exerted on methanogenic, nitrifying and denitrifying bacteria by chemicals used in a milk analysis laboratory. *Enzyme and Microbial Technology*, **31**, 976–985.

Van Dongen U., Jetten M.S.M. & Van Loosdrecht M.C.M. (2001). The SHARON-ANAMMOX process for treatment of ammonium rich wastewater. *Wat. Sci. Tech.*, **44**(1), 153–160.

Biosafety

European Symposium on Environmental Biotechnology, ESEB 2004 - Verstraete (ed)
© 2004 Taylor & Francis Group, London, ISBN 90 5809 653 X

Gene expression in *Ralstonia metallidurans* CH34 in space flight

N. Leys, S. Baatout, P. De Boever, A. Dams & M. Mergeay
Belgian Nuclear Research Centre (SCK•CEN), Mol, Belgium

R.Wattiez
University of Mons-Hainaut (UMH), Mons, Belgium

P. Cornelis & S. Aendekerke
University of Brussels (UB), Brussel, Belgium

ABSTRACT: The early detection of changes in bacterial communities and single bacterial cells that are present in space ships or space stations is crucial for a variety of biosafety issues (*i.e.* pathogenicity for crew, biodegradation of materials, *etc.*). Bacteria will also be essential for long-term manned missions (e.g. to Mars) for the recycling of waste and the production of food. The ESA-sponsored research program MESSAGE (Microbial Experiments in the Space Station About Gene Expression) studied the effects of space conditions such as microgravity and cosmic radiation on board the ISS on gene expression in a model bacterium, *i.e.*, R. metallidurans CH34 (ATCC 43123). Proteomic analysis clearly showed a differential gene expression between space and ground grown cultures, producing different physiological characteristics of the cells as was observed by flowcytometry analysis. The results of the MESSAGE experiments will be applied to improve further development of "microbial life support systems" and systems to detect micro-organisms in air for application in the International Space Station and future space vehicles.

1 INTRODUCTION

Space is very specific ecological environment. Biological research in space has so far mainly concentrated on animal and human cell behavior during space flights for medical applications. However, with the growing interest in prolonged missions, it becomes more important to study also in detail the different aspects of microbial cell behavior in closed space environments. Micro-organisms that are present in space vehicles, space stations and future interplanetary bases could be of danger for the crew or could cause bio-corrosion of materials. The study of bacterial activity under space conditions is therefore highly important for the early detection of changes in bacteria communities and single bacterial cells with medical or environmental consequences. Nevertheless, bacteria will also be essential for long-term missions for the recycling of waste and the production of food.

The ESA-sponsored research program MESSAGE stands for "Microbial Experiments in the Space Station About Gene Expression" and studied the effects of the specific environmental space conditions on board the International Space Station (ISS) bacterial gene expression. A well-known bacterium, *i.e., R. metallidurans* CH34 (ATCC 43123), was used as a model organism to study the synergetic effects of microgravity and higher cosmic radiation doses on general known bacterial stress response systems and to search for new stress responses. Bacterial cultures of this strain CH34 were allowed to grow during 10 days in the ISS. Post-flight proteins and RNA from cells grown in space flight and ground were extracted and analyzed.

The scientific results obtained with a first MESSAGE experiment (Odissea Mission, Belgian Taxi Flight, October 2002) were re-evaluated with a second space experiment MESSAGE 2 (Cervantes Mission, Spanish Soyuz Mission, 18–28 October 2003).

2 MATERIALS AND METHODS

2.1 *Strain and growth conditions*

R. metallidurans CH34 (ATCC43123) is a flexible, versatile and robust well-studied non-pathogenic bacterium which can grow at room temperature and survive in harsh environmental conditions. A draft version of its otal genome sequence is available on the web (http://www.jgi.doe.gov/programs/genomes.html).

CH34 was cultivated in aerobic rich and poor solid agar medium and rich liquid medium. The cultures were incubated at 4°C until launch, incubated in a closed container in the dark at room temperature (circa 22°C) in the Service Module of the ISS, transported at 4°C after landing, and analyzed within 36 hours after landing. A ground control experiment was inoculated at the same time and maintained at identical conditions parallel in time.

2.2 *Molecular analysis*

The molecular stress response of the bacteria was analyzed using full protein and RNA extracts from space and ground grown cells.

3 RESULTS AND DISCUSSIONS

A limited number of genes were significant over-expressed in space grown cells in comparison to ground grown cells of *Ralstonia metallidurans* CH34.

There was an induction of a few proteins possible involved in "house keeping functions". However, most over produced proteins are possibly representing a specific stress response to space conditions and/or metabolic changes. The transcription of other well-known stress related genes did not seem to be extra (de)activated.

In the second experiment, space grown mutants with the genes coding the proteins which are over expressed in space inactivated, were further analyzed for gene expression and physiology.

To confirm these protonomic results transcriptomic (RNA) analysis is going on using a full genome DNA gene chip of *R. metallidurans* CH34.

4 CONCLUSION

The MESSAGE experiments led to an unique view on the molecular and the physiological response of a whole organism to the specific growth condition of space. The combined application of different molecular and physiologic microbial methods revealed valuable information to understand better the microbial adaptation to space environments.

Bringing a known earth bacterium in space conditions can significantly change its gene expression. Space flight conditions had a limited effect on known stress response *R. metallidurans* CH34 but induced possibly new unknown stress responses.

A international publication about the final results of this experiment is in preparation.

ACKNOWLEDGEMENTS

This project was funded by ESA via the contract for the BTF-MESSAGE project (Prodex agreement No. 90037) and the SSM-MESSAGA project (Prodex agreement No. 90094).

CORRESPONDENCE

N. Leys,
Belgian Nuclear Research Center (SCK•CEN)
Boeretang 200, B-2400 Mol, Belgium
Tel. +32 14 33 27 26, Fax. +32 14 31 47 93
nleys@sckcen.be

European Symposium on Environmental Biotechnology, ESEB 2004 - Verstraete (ed)

Survival, death and fate of microbial biomass in soil

R. Kindler, A. Miltner, H.H. Richnow & M. Kästner
Bioremediation Department, UFZ – Centre for Environmental Research Leipzig-Halle GmbH, Leipzig, Germany

T. Lüders & M.W. Friedrich
Department of Biogeochemistry, Max Planck Institute for Terrestrial Microbiology, Karl-von-Frisch-Strasse, Marburg, Germany

ABSTRACT: Soil microorganisms contribute to the formation of refractory soil organic matter by their metabolism and to a certain extent also by their biomass. However, to determine the fate of soil microorganisms and their cell components is rather difficult since soil microorganisms exhibit (I) an enormous biodiversity, (II) are part of a complex food web and (III) most soil microorganisms are viable but not culturable.

To elucidate the fate of microbial biomass compounds we investigated the contribution of microbial biomass to humus formation during incubation of soil with isotopically (^{13}C) and genetically (*lux* genes) labelled *Escherichia coli* cells. We traced the survival of the active cells and the fate of nucleic acids, amino acids, amino sugars and fatty acids. The incubation experiment proved the decrease of cultivable cells and after 15 weeks no luminescent *E. coli* cells could be determined. However, the *lux* genes were still extractable and could be amplified (by PCR) after 32 weeks. At this time about 50% of the added biomass carbon was mineralized and lost as CO_2, but the genetic information and a significant part of the carbon originating from biomass were therefore retained in the soil or were transformed into other cells. The ^{13}C label of the *E. coli* cell constituents (DNA, fatty acids, amino acids, amino sugars) decreased rapidly, and the carbon was distributed among a net of soil microorganisms and was found in rRNA of other bacteria and fungi by use of RNA-Stable Isotope Probing. The bacteria were identified as *Myxobacteria*, *Bacteroidetes* and *Lysobacter* species that are known as micropredators or organotrophic degraders of biomass constituents in soils.

The distribution of the ^{13}C label on the rRNA level and within the fatty acid patterns of soil microorganisms showed similar results. Thus, we were able to trace a microbial food web in complex environmental systems by use of a combination of isotope and nucleic acid markers.

European Symposium on Environmental Biotechnology, ESEB 2004 - Verstraete (ed)
© 2004 Taylor & Francis Group, London, ISBN 90 5809 653 X

Membrane bioreactor for wastewater treatment and reuse: virus removal

H.M. Wong, C. Shang & G.H. Chen
Department of Civil Engineering, Hong Kong University of Science and Technology, Hong Kong

ABSTRACT: Membrane bioreactor (MBR) is a unit process consisting of a biological reactor with suspended biomass and a microfiltration/ultrafiltration membrane. In addition to its biological degradation and solid-liquid separation functions, MBR may potentially provide a non-hazardous disinfection means. In this paper, an experimental study on virus removal was carried out in a bench-scale MBR unit, using MS-2 bacteriophage as a viral indicator. The membrane itself without the presence of the mixed liquor and the attached biofilm could achieve only 80% virus removal. With the additional presence of suspended biomass in the MBR unit, the total virus removal was enhanced to 93.7%. When sufficient time was given to allow biofilm accumulation on the membrane surface, the removal further increased to 99.9%. Submerging this biofilmed membrane module into a clear water tank gave evidence that the major contribution to the virus removal in an MBR system was the biofilm.

1 INTRODUCTION

The membrane bioreactor (MBR), a unit process consisting of a biological reactor with suspended biomass and microfiltration membrane, is finding its wide application to replace the conventional activated sludge processes in wastewater treatment. MBR has demonstrated outstanding performance with respect to its biological conversion and solid removal capabilities. MBR is now being considered as an effective, non-hazardous alternative disinfection means to achieve effluent pathogen control, since it may also be able to remove bacteria and viruses to a significant extent. If so, the dosage of disinfectants applied afterwards could be much reduced.

Meanwhile, the importance of removing viruses from wastewater is increasingly being recognized because of the epidemiological significance of viral pathogens. Viruses are also much smaller and harder-to-straining than bacteria and considered to be more resistant to disinfectants than bacteria (Leong 1983). Therefore, a suitable viral indicator should be applied to guarantee the MBR effluent microbiological quality. Bacteriophages have been suggested as viral indicators because they closely resemble enteric viruses in terms of the structure, morphology, size and behavior (Maier et al. 2000). Bacteriophage MS-2 was chosen to be the indicator in this study because of its small sizes (0.02–0.025 μm) and hence it is representative to address the ability of viral removal by MBR. Although the average pore size of the microfiltration membrane (0.4 μm) is much larger than that of the bacteriophage MS-2, the removal may be significant and upgraded with the presence of biomass and the visible, sticky biofilm developed on the membrane surface.

This paper therefore describes an experimental study on the contributions of the membrane, the suspended biomass and the biofilm grown on the membrane surface to the virus removal using a bench-scale MBR unit. The significance of internal blockage inside the membrane fibers and the effects of cleaning were also examined.

2 MATERIALS AND METHODS

2.1 *Membrane and bench-scale MBR*

Membrane modules used in this study were provided by the Mitsubishi Rayon Corporation (MRC). Table 1 shows the characteristics of the membrane modules.

Figure 1 shows the configuration of the MBR unit in which the membrane (1) is submerged in the reactor and an air diffuser (2) is equipped at the base of the

Table 1. Characteristics of the MRC membrane.

Membrane type	Microporous hollow fiber
Material	Polyethylene
Average pore size (μm)	0.4
Maximum allowable flux ($m^3 m^{-2} day^{-1}$)	0.4
Total surface area (m^2)	0.2

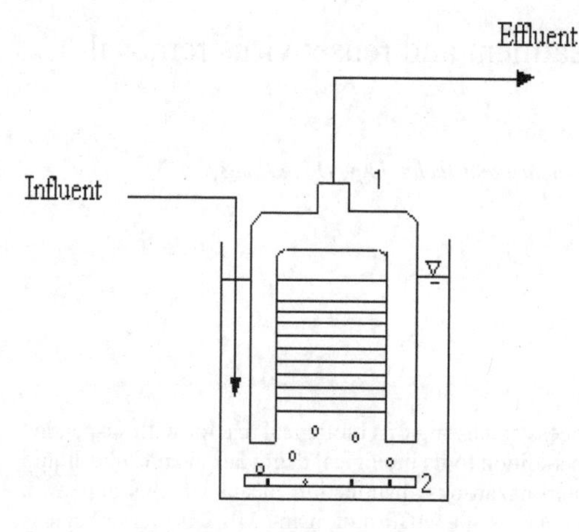

Figure 1. Configurations of the laboratory reactor.

reactor to achieve both aeration and membrane surface cleaning. The aeration tank has an operating volume of 19 L and a footprint area of 26 cm × 25 cm. The effluent pump was operated at a constant suction force throughout the operation and the initial flow rate in a clear water tank with a clean membrane module was 50 L/d.

2.2 *Experimental procedures*

At the beginning of a test cycle, a clean membrane was submerged in a clear water tank and the unit was fed continuously with synthetic water containing only the nutrient broth (peptone and yeast extract), mineral water (phosphate buffer, $CaCl_2$, $MgSO_4$, and $FeCl_3$), and one dose of bacteriophage MS-2 (ATCC 1559-B1) solution at concentrations around 10^7 pfu/ml. Samples were periodically collected from the tank and the outlet of the MBR unit after initiating the test run with manual, rapid mixing for MS-2 enumeration to study the contribution of the sole membrane.

The clean membrane was then submerged in a reactor containing suspended biomass stabilized at a mixed liquor suspended solid (MLSS) level of 6000 mg/L with a sludge age of 200 days. The MBR unit was operated continuously for a period to allow the growth of membrane-attached-biofilm. Same as the previous case, the unit was fed continuously with the synthetic water and shots of the MS-2 solution were dosed to maintain the MS-2 concentration. Again, Samples were periodically collected from the tank and the outlet of the MBR unit for virus enumeration.

The biofilmed membrane was then cleaned by two different methods, water wash using tap water stream and backwashed with a 600 mg/L hypochlorite solution for 15 minutes. Tests of virus rejection were conducted in the same manner as described in the first paragraph of the Section 2.2 except the cleaned membrane modules were used.

2.3 *Sampling and MS-2 assay*

MS-2 concentrations in the samples were numerated using the Double Agar Layer Method (Adams, 1959) with *Escherichia coli* (ATCC 15597) as the host. In this study, the water samples were filtrated through 0.45 μm filters before the viral assay.

3 RESULTS AND DISCUSSION

3.1 *MS-2 removal by MBR*

The main components contributing to virus removal in MBR were postulated to include the sole microfiltration membrane, the suspended biomass, and the biofilm developed on the membrane fiber surface. This study evaluated the significance of each component and the transitional performance in virus removal of the MBR unit before it reached a stable efficiency.

Figure 2 shows the virus rejection in the first few hours when a clean membrane module was submerged in the aeration tank at a MLSS concentration of 6000 mg/L. As shown, more than 1.36 log (or 95%) removal was observed in the first hour and the efficiency dropped to 0.4 log (or 60%) removal in following hour before it stabilized at about 1.25 log removal after 6 hour of operation. Figure 3 shows the same set of operation but the removal efficiency was checked daily for 14 days. As shown, the removal efficiencies increased from 1.29 log (after 24 hours) to 2.91 log (after 48 hours), indicating the first two days of operation is a critical period for virus removal. The removal efficiency fluctuated in the following few days until it reached a relatively constant value of around 3 log after 11 days. The discussion and interpretation of the results are in the Section 3.3.

3.2 *Contribution from physical straining by the sole membrane*

The efficiency of the sole membrane on bacteriophage MS-2 removal was studied by conducting the tests with a clean membrane module submerged in a clear water tank containing the synthetic solution and the MS-2 solution. The results are presented by the dash line in Figure 4 (denoted as "original membrane"). As shown, the reduction of bacteriophage MS-2 by the sole membrane is about 0.54 log (or 84%) removal. Such leakage of bacteriophage MS-2 in the clear water system in the absence of suspended solids is expected since the average pore size of the membrane fibers (0.45 μm) is much larger than the size of bacteriophage MS-2. This 0.54 log reduction can also serve as the baseline to study the contributions of biomass and

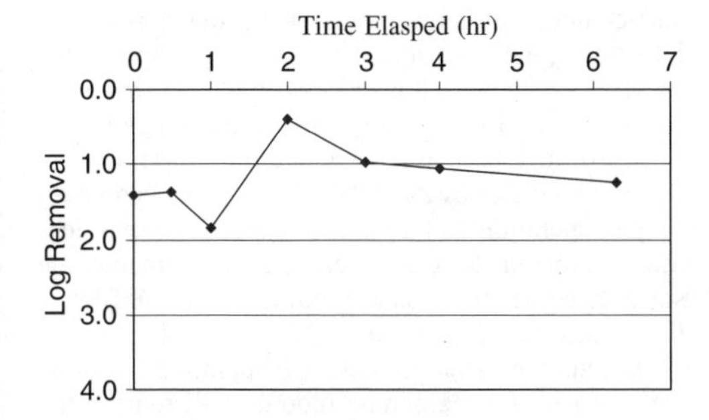

Figure 2. Short-term MS-2 removal by the MBR.

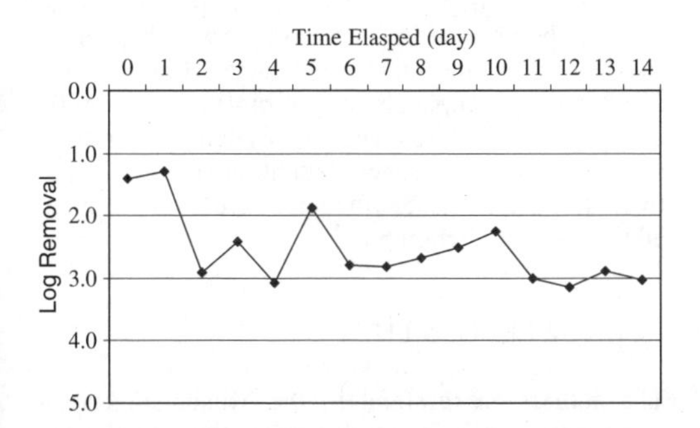

Figure 3. Long-term MS-2 removal by the MBR.

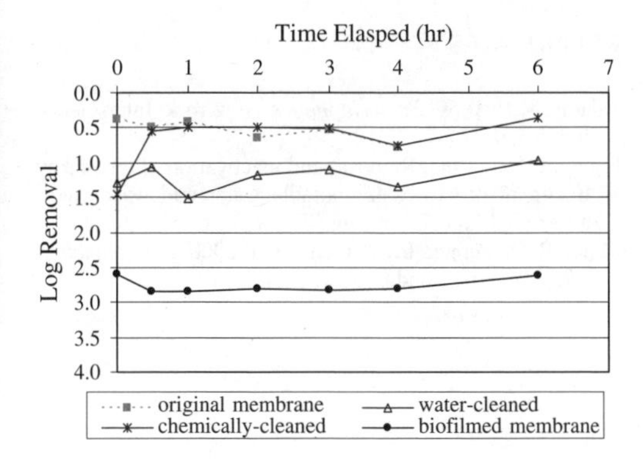

Figure 4. MS-2 rejection in clear water tank and effects of membrane cleaning.

biofilm to virus removal and the effect of cleaning on virus removal.

3.3 *Contributions from biomass and biofilms*

With the presence of suspended biomass (MLSS), virus rejection got improved over the testing period of 14 days (see Figures 2 and 3). The virus rejection achieved on the first 6 hours of the test cycle (Figure 2) can be considered to be attributable to the contributions of both the MLSS and the sole membrane. As shown, the presence of biomass increased the phage reduction to 1.2 log removal on an average, compared to the results using the sole membrane. Therefore, approximately 0.66 log can be assigned to the contribution from the suspended biomass. The main reason to account for the reduction is cell adsorption. The phages are believed to attach on the biomass flocs and then get removed by either flocculation or physical straining provided by the membrane filters. Virus rejection abruptly rose from 1.29 to 2.91 log on the second day, as shown in Figure 3. The rapid increase of the rejection may due to the deposition of biomass flocs on the membrane surface and hence reduce the effective pore size of the membrane fibers. The rejection fluctuated for the following few days. The fluctuation may due to the random flocculation between viruses and sludge flocs. By allowing the development of membrane-attached biofilms on the membrane surface for 14 days, the virus removal efficiency was further improved to a relative constant value of around 3 log. This indicates that virus removal attributed to the attached biofilm is of significance.

After 14 days of operation, the membrane module was taken out from the MBR unit and was immersed in a clear water tank to examine the assistance of the biofilm developed on the membrane surface on virus removal. Since biofilm is porous in nature, hence the development of the biofilm can prevent further deposition of solid inside the membrane pores while virus can adhere on the gluey slime layer on the membrane surface and get rejected. As shown in Figure 4 (denoted as "biofilmed membrane"), the virus removal efficiency can still maintain at 2.8 log without the presence of suspended biomass (only 0.2 log less compared to the results with the presence of suspended biomass). This again suggests that the suspended biomass plays a less significant role in removing the phage than the attached biofilm layer developed on the membrane surface. In fact, although the general term "biofilmed membrane" is used in this paper, the roles of the deposited solids and the attached biomass cannot be distinguished and the enhanced virus removal efficiency is likely attributable to both.

3.4 *The effect of cleaning*

The membrane was then cleaned by two different methods: (1) tap water stream wash to remove the surface attachment and (2) backwash with a 600 mg/L hypochlorite solution for 15 minutes, denoted as "water-cleaned" and "chemically-cleaned", respectively, in Figure 4. The effects of the two cleaning methods on virus removal were studied by submerging the washed membrane modules in a clear water tank (without mixed liquor) and feeding with synthetic water. After water stream wash to remove the external biofilm, the virus removal efficiency dropped down to 1 to 1.5 log, which is still much higher than the virus removal efficiency achieved by the original membrane. The difference in efficiency between using the "water-cleaned" membrane and the "original membrane" implies that the presence of internal blockage is possible. The building block of this internal blockage is believed to be biofilm only since suspended solids cannot pass through the micropores of the membrane fibers. By subtracting the efficiencies between using the "water-cleaned" membrane and the "biofilmed membrane", the contribution of virus removal by the external, visible biofilm is approximately 1.3 to 1.8 log. These results indicate both the external biofilm and the internal blockage are essential to virus removal and the contribution of the latter is relatively smaller than the former.

Figure 4 also shows the virus removal efficiency of the membrane module after chemical cleaning. By comparing the lines of "chemically-cleaned" and "original membrane" in Figure 4, it is evident that, after chemical cleaning with a hypochlorite solution, the membrane resumed its initial condition and achieved the same level of virus removal as did the original membrane. The initial high removal (at time zero) using the chemically cleaned membrane module is attributable to the chlorine residuals remained in and/or on the membrane fibers.

4 CONCLUSIONS

In the absence of biomass and biofilm, the leakage of phages was serious and only about 0.54 log reduction of bacteriophage MS-2 could be achieved. The removal increased to 1.2 log reduction after immersing the membrane into a tank of 6000 mg/L MLSS for 6 hours. It further increased to 3 log reduction when sufficient time was given to allow the development of biofilm on the membrane surface. This implies that an MBR can achieve higher viral removal efficiencies by prolonging its operation between cleanings.

Suspended biomass plays an important role in removing viruses by an MBR system at the beginning of the operation but its significance becomes less after a biofilm layer is developed on the membrane surface. Under the tested conditions, the biofilm is the most essential component in contributing to the virus rejection after 14 days of operation. Routine water wash of a membrane module will remove the visible, surface attached biofilm and reduce the virus removal efficiency to some extent. After a long period of operation, due to the flux decline, the membrane needs to be cleaned by backwashing with hypochlorite to destroy the internal blockage. However, this results in the completely loss of biofilm and significantly reduce the virus removal efficiency. Therefore, after each cleaning, special attention should be paid to control the effluent pathogen concentrations by an additional disinfection process.

ACKNOWLEDGEMENTS

The authors are grateful to the Mitsubishi Rayon Corporation (MRC) for providing the bench-scale MBR modules.

REFERENCES

Adams, M.H. 1959. *Bacteriophages.* New York: Interscience Publishers.

Leong, L.Y.C. 1983. Removal and inactivation of viruses by treatment processes for potable water and wastewater: a review. *Water Science and Technology* 15(5): 91–114.

Maier, R.M., Pepper, I.L. & Gerba, C.P. 2000. *Environmental Microbiology.* London: Academic Press.

European Symposium on Environmental Biotechnology, ESEB 2004 - Verstraete (ed)
© 2004 Taylor & Francis Group, London, ISBN 90 5809 653 X

Putative LuxR homologues influence virulence of *Legionella pneumophila*

I. Lebeau, E. Lammertyn, E. De Buck & J. Anné
Laboratory of Bacteriology, Rega Instituut, KU Leuven, Minderbroedersstraat, Leuven, Belgium

ABSTRACT: *Legionella pneumophila,* the causative agent of Legionnaires' disease, is a ubiquitously found facultative intracellular parasite of several protozoa and human macrophages. *L. pneumophila* pathogenesis is not yet completely understood, but more genes responsible for infection and intracellular replication are becoming known. On the other hand, knowledge how these genes are controlled is still very limited. To get more insight in this, we searched the partially sequenced genome of *L. pneumophila* for the presence of LuxR homologues, a superfamily of transcriptional regulators. We were able to identify four putative *luxR* homologues, which were designated LpnR. Two of these LuxR homologues, LpnR 280/5 and LpnR 355/26 seem to play a role in infection, since deletion of the encoding genes caused an impaired infection capacity and a delayed generation time in *Acanthamoeba castellanii,* respectively. However, none of the LpnR proteins affected uptake or replication in differentiated U937-cells. Therefore it can be concluded that LpnR 280/5 and LpnR 355/26 influence regulation of virulence in *L. pneumophila* towards *A. castellanii.*

1 INTRODUCTION

L. pneumophila is a Gram-negative, rod-shaped bacterium which can be found in freshwater environments like lakes, rivers, streams and mud (Fritsche et al. 1993), were it survives free-living, intracellularly in protozoa or in association with biofilms. The multiplication of *L. pneumophila* within these protozoa is probably the primary means of proliferation.

L. pneumophila is the causative agent of community acquired pneumonia, known as Legionnaires' disease, which can be fatal if it is not promptly and correctly diagnosed. It also causes Pontiac fever, a milder flu-like self-limiting disease. Elderly people, smokers and immunocomprised persons are most susceptible for infection with this pathogen, which is acquired through inhalation of contaminated aerosols. When the bacteria reach the alveolar parts of the lungs, they are taken up by pulmonary macrophages through normal phagocytosis or by a special coiling mechanism (Horwitz 1984). The resulting phagosome does not fuse with the lysosomes and as a consequence *L. pneumophila* is not destroyed. Hence, it can amplify within the macrophages, as it does within amoebae and some other protozoa, hereby overwhelming the host's immune system. New rounds of infection will start after multiplication in and lysis of the host cell (Horwitz 1983, Horwitz & Silverstein 1980).

Already many factors important for infection of and replication in host cells are identified through mutagenesis and complementation experiments. These include for example Mip (macrophage infectivity potentiator) (Fischer et al. 1992), type IV pili (Liles et al. 1998), flagellae (Pruckler et al. 1995) and proteins encoded by *dot* (defect in organelle trafficking) (Vogel et al. 1998), *icm* (intracellular multiplication) (Segal et al. 1998), *mil* (macrophage-specific infectivity) (Gao et al. 1998a) and *pmi* (protozoa and macrophage infectivity) (Gao et al. 1998b). However, little is known about the possible regulation of expression of these genes.

LuxR proteins contain a C-terminal helix-turn-helix motif enabling them to bind DNA and as a result affect transcription. The family of LuxR proteins is divided in two subgroups. One group consists of regulators that play a role in quorum sensing by binding signal molecules. The other group consists of regulators belonging to a two-component transduction system.

We screened the still incomplete genome sequence of *L. pneumophila* for the presence of putative *luxR* homologues. This revealed four open reading frames designated *lpnR 259/6, lpnR 280/5, lpnR 355/26* and *lpnR 47/2*. We focused on the characterization of LpnR 259/6, LpnR 280/5 and LpnR 355/26, since LpnR 47/2 was recently identified as LetA (Hammer et al. 2002, Gal-Mor & Segal 2003, Lynch et al. 2003).

Single deletion mutants of *lpnR 259/6, lpnR 280/5* and *lpnR 355/26* were made and the effect of *lpnR* inactivation on infection of and replication in amoebae and macrophages was examined. LpnR 280/5 and

LpnR 355/26 showed a clearly impaired infection capacity and a delayed generation time in *Acanthamoeba castellanii*, respectively. None of the LpnR proteins affected uptake or replication in differentiated U937-cells.

2 GENERATION OF SINGLE DELETION MUTANTS OF *LPNR*

Characterization of *lpnR 259/6, lpnR 280/5* and *lpnR 355/26* was initiated by generation of single deletion mutants of these three *lpnR* genes (Figure 1).

Using PCR, the upstream and downstream flanking sequences of *lpnR 259/6* (1.5 kb and 0.6 kb, respectively) were amplified and cloned at both sides of the *neo* gene in pBSKan, a pBluescriptII-KS (Stratagene)

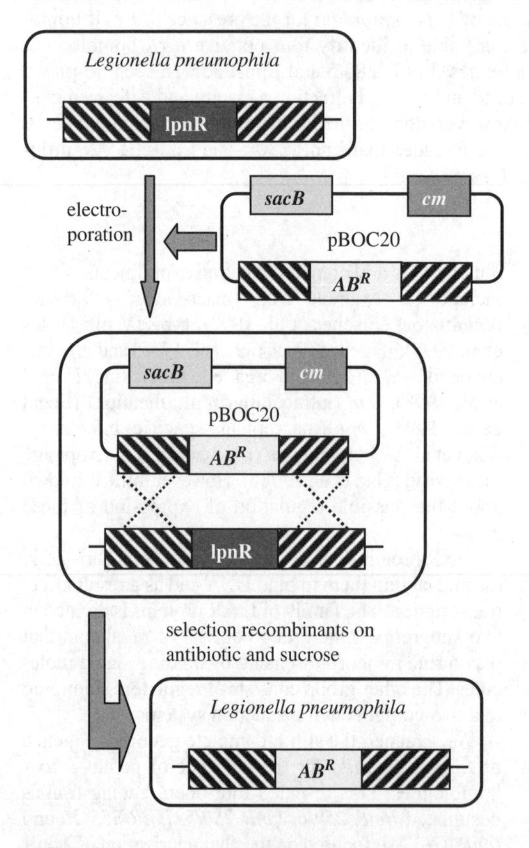

Figure 1. Schematic representation of the construction of a single deletion mutant of *lpnR* in *L. pneumophila. sacB* encodes levansucrase, *cm* is the chloramphenicol resistance gene and AB^R stands for antibiotic resistance gene which is kanamycin for *lpnR 259/6* and *lpnR 355/26* and streptomycin for *lpnR 280/5*.

derivative containing the kanamycin resistance gene. Next, the entire cassette was transferred to pBOC20, an *E. coli-Legionella* shuttle vector (O' Connell et al. 1996) and subsequently introduced in *L. pneumophila* by electroporation. This plasmid carries a chloramphenicol resistance gene and *sacB* of *Bacillus subtilis,* which encodes the secreted enzyme levansucrase. This enzyme catalyzes hydrolysis of sucrose and synthesis of levans. These high-molecular-weight fructose polymers are toxic to the cell. *Legionella* clones carrying the plasmid were first selected on plates with chloramphenicol. Afterwards, double crossover mutants resulting in loss of the plasmid, but retention of the *neo* gene, were selected on plates with sucrose and kanamycin.

A deletion mutant of *lpnR 355/26* was generated in a similar way.

In order to generate a deletion mutant of *lpnR 280/5,* upstream and downstream flanking sequences (both 1 kb) were amplified by means of PCR and cloned into pBluescriptII-KS. The streptomycin resistance gene was isolated from pHP45Ω (Prentki & Krisch 1984) and cloned into pBluescriptII-KS, already carrying the flanking sequences. This cassette was transferred into pBOC20. Electroporation and the selection for mutants were carried out as described above.

PCR and Southern blot analysis were performed in order to confirm the deletion of the *lpnR* genes in the three single deletion mutants which were designated Δ*lpnR 259/6,* Δ*lpnR 355/26* and Δ*lpnR 280/5,* respectively.

3 EFFECT OF *LPNR* INACTIVATION ON INFECTION OF AND REPLICATION IN *A. CASTELLANII* AND DIFFERENTIATED U937-CELLS

In nature, *L. pneumophila* is able to invade and replicate in amoebae, while in humans replication occurs in the alveolar macrophages. *A. castellanii* and differentiated U937-cells were used to follow growth in amoebae and macrophages, respectively. Cells of the human macrophage-like cell line U937 were treated with phorbol 12-myristate 13-acetate (PMA) 72 hours before infection in order to differentiate.

For both cell types infectivity of *L. pneumophila* was tested as follows (Ciancotto & Fields 1992). After addition of the bacteria at a multiplicity of infection (MOI) of 2 to the cells and incubation for 2 hours to allow bacterial uptake, the remaining extracellular bacteria were removed by washing with buffer. The cultures were sampled every 24 hours during 3 consecutive days. Therefore bacteria were released from the host cells by hypotonic lysis and mixed with those present in the cell culture supernatant. Bacterial cells were quantified by plating serial dilutions. Since

L. pneumophila is not able to replicate in the culture medium of the host cell, all colony forming units (CFU) counted represent intracellularly replicated bacteria.

In *A. castellanii, L. pneumophila* Δ*lpnR 259/6* replicated to the same extent as *L. pneumophila* wild type (Figure 2A) indicating that LpnR 259/6 is dispensable for growth in this host. However, replication of Δ*lpnR 280/5* and Δ*lpnR 355/26* was significantly reduced compared to replication of the wild type *L. pneumophila* strain (Figure 2A). Early stages of infection were analyzed to determine whether these effects were due to a decrease in infection or a decrease in replication rate. In order to analyze adhesion of *L. pneumophila* to *A. castellanii*, without entry into this host, the amoebae were first treated with methylamine (King et al. 1991). Early replication stages were analyzed by allowing only 15 min, 30 min, 60 min or 120 min incubation after infection (MOI = 2), after which the cells were washed and sampled as described above. Δ*lpnR 280/5* and Δ*lpnR 355/26* were able to adhere to the host cells to the same extent as *L. pneumophila* wild type, but invasion of Δ*lpnR 280/5* in the host was less efficient than that of the wild type strain (Figure 2B). The amount of CFUs recovered for the Δ*lpnR 280/5* strain remained consistently lower during the later stages of infection. This suggests that LpnR 280/5 plays a role in invasion of *A. castellanii.* There was no difference in entry into *A. castellanii* between Δ*lpnR 355/26* and wild type *L. pneumophila* (Figure 2B). It therefore appeared that LpnR 355/26 only affects replication. Since wild type and mutant *L. pneumophila* strains showed the same growth properties when cultured in liquid medium, the observed effects were not due to differences in intrinsic growth.

Differentiated U937-cells were used to analyze the effect of the three single *lpnR* deletion mutants on infection of and replication in macrophages. Replication in these host cells was not affected by inactivation of *lpnR 259/6, lpnR 280/5* or *lpnR 355/26* (Figure 2C).

It could therefore be concluded that LpnR 259/6, LpnR 280/5 and LpnR 355/26 are dispensable for replication in differentiated U937-cells, whereas LpnR 280/5 is required for efficient invasion of *A. castellanii* and LpnR 355/26 for efficient intracellular replication in this host.

4 DISCUSSION AND CONCLUSIONS

In the framework to further unravel the regulation of virulence in *L. pneumophila,* the database containing the incomplete genome sequence was searched for the presence of putative LuxR homologues. Some LuxR proteins are involved in quorum sensing, a phenomenon in which bacteria are able to sense their cell density through signal molecules and to control gene expression when a certain critical concentration of

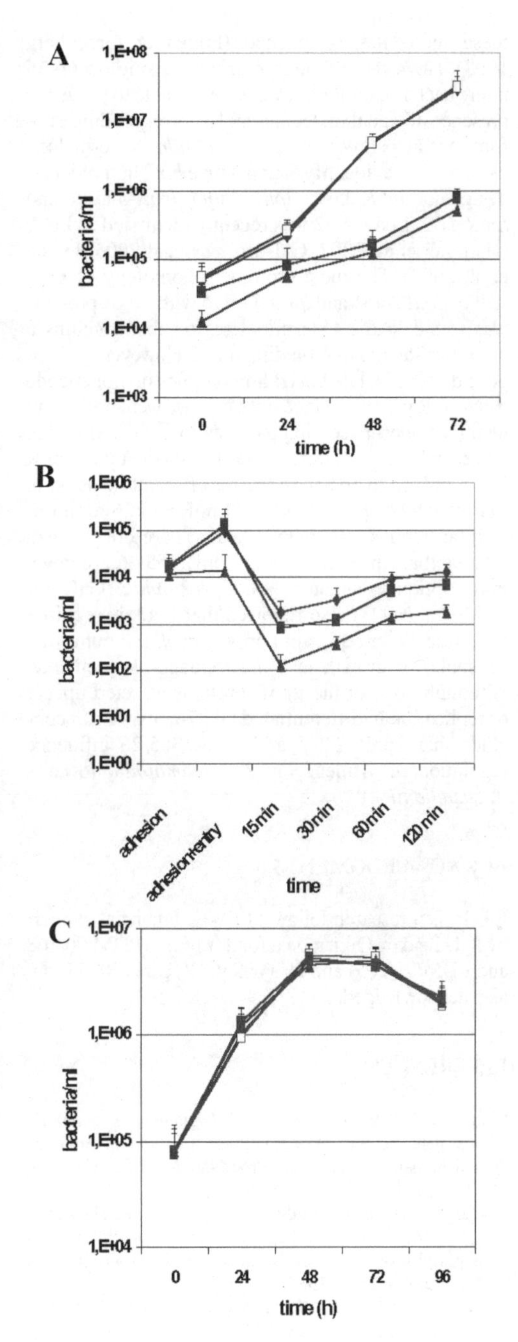

Figure 2. (A) Intracellular growth and (B) invasion efficiency of *L. pneumophila* in *A. castellanii.* (C) Intracellular growth of *L. pneumophila* in differentiated U937-cells. Each time point represents the mean of at least three independent experiments. Error bars indicate the standard deviation. *L. pneumophila* wild type (◆), *ΔlpnR 259/6* (□), *ΔlpnR 280/5* (▲) and *ΔlpnR 355/26* (■).

these molecules is reached (Fuqua & Greenberg 1998). However, although quorum sensing occurs in many bacteria, until now, we were not able to prove the presence of signal molecules and a corresponding quorum sensing pathway in *L. pneumophila.* Nevertheless, we were able to identify four putative *luxR* homologues, designated *lpnR 259/6, lpnR 280/5, lpnR 355/26* and *lpnR 47/2*. LpnR 47/2 was recently identified as LetA (Hammer et al. 2002, Gal-Mor & Segal 2003, Lynch et al. 2003). The most prominent homology of *lpnR 259/6, lpnR 280/5* and *lpnR 355/26* with LuxR proteins was found in the C-terminal region that contains a helix-turn-helix DNA binding motif. However, we suppose that the isolated *luxR* homologues do not encode true LuxR proteins, since until now we were not able to identify a quorum sensing pathway in *L. pneumophila*. Nevertheless, we were able to show that LpnR 280/5 proved to be necessary for efficient invasion in *Acanthamoeba castellanii* and LpnR 355/26 for intracellular replication in this host. Therefore we can assume that LpnR 280/5 and LpnR 355/26 are novel transcriptional regulators in *L. pneumophila*. Gal-Mor and Segal (2003) also reported that LetA is required for intracellular replication in *A. castellanii* but is dispensable for growth in a macrophage-like cell line. Although none of the LpnR proteins affected uptake or replication in differentiated U937-cells we can conclude that LpnR 280/5 and LpnR 355/26 influence regulation of virulence in *L. pneumophila* towards *A. castellanii*.

ACKNOWLEDGMENTS

E.D.B. is a research fellow of FWO. Financial support by K.U.Leuven Onderzoeksfonds to E.L. (PDM/00/168 and PDM/01/163) and by IWT (GBOU nr. 20153) are also acknowledged.

REFERENCES

Cianciotto, N.P. & Fields, B.S. 1992. *Legionella pneumophila mip* gene potentiates intracellular infection of protozoa and human macrophages. Proc Natl Acad Sci USA 89: 5188–5191.

Fischer, G., Bang, H., Ludwig, B, Mann, K. & Hacker, J. 1992. Mip protein of Legionella pneumophila exhibits peptidyl-prolyl-cis/trans isomerase (PPIase) activity. Mol Microbiol 6: 1375–1383.

Fritsche, T.R., Gautom, R.K., Seyedirashti, S., Bergeron, D.L. & Lindquist, T.D. 1993. Occurence of bacterial endosymbionts in Acanthamoeba spp. isolated from corneal and environmental specimens and contact lenses. J Clin Microbiol 31:1122–1126.

Fuqua, C. & Greenberg, E.P. 1998. Self perception in bacteria: quorum sensing with acylated homoserine lactones. Curr Opin Microbiol 1: 183–189.

Gal-Mor, O. & Segal, G. 2003. The *Legionella pneumophila* GacA homolog (LetA) is involved in the regulation of *icm* virulence genes and is required for intracellular multiplication in *Acanthamoeba castellanii*. Microb Pathog 34: 187–194.

Gao, L., Harb, O.S. & Abu Kwaik, Y. 1998a. Identification of macrophage-specific infectivity loci (mil) of *Legionella pneumophila* that are not required for infectivity of protozoa. Infect Immun 66: 883–892.

Gao, L., Stone, B.J., Brieland, J.K. & Abu Kwaik, Y. 1998b. Different fates of *Legionella pneumophila pmi* and *mil* mutants within macrophages and alveolar epithelial cells. Microb Pathog 25: 291–306.

Hammer, B.K., Tateda, E.S. & Swanson, M.S. 2002. A two-component regulator induces the transmission phenotype of stationary-phase *Legionella pneumophila*. Mol Microbiol 44: 107–118.

Horwitz, M.A. 1983. The Legionnaires' disease bacterium (*Legionella pneumophila*) inhibits phagosome-lysosome fusion in human monocytes. J Exp Med 158: 2108–2126.

Horwitz, M.A. 1984. Phagocytosis of the Legionnaires' disease bacterium (*Legionella pneumophila*) occurs by a novel mechanism: engulfment within a pseudopod coil. Cell 36: 27–33.

Horwitz, M.A. & Silverstein, S.C. 1980. Legionnaires' disease bacterium (*L. pneumophila*) multiplies intracellularly in human monocytes. J Clin Investig 60: 441–450.

King, C.H., Fields, B.S., Shotts, E.B. Jr & White, E.H. 1991. Effects of cytochalasin D and methylamine on intracellular growth of *Legionella pneumophila* in amoebae and human monocyte-like cells. Inf Immun 59: 758–763.

Liles, M.R., Viswanathan, V.K. & Ciancotto, N.P. 1998. Identification and temperature regulation of *Legionella pneumophila* genes involved in type IV pilus biogenesis and type II protein secretion. Infect Immun 66: 1776–1782.

Lynch, D., Fieser, N., Glöggler, K., Forsbach-Birk, V. & Marre, R. 2003. The respons regulator LetA regulates the stationary-phase stress response in *Legionella pneumophila* and is required for efficient infection of *Acanthamoeba castellanii*. FEMS Microbiol Lett 219: 241–248.

O'Connell, W.A., Hickey, E.K. & Ciancotto, N.P. 1996. A *Legionella pneumophila* gene that promotes hemin binding. Infect Immun 64: 842–848.

Pruckler, J.M., Benson, R.F., Moyenuddin, M., Martin, W.T. & Fields, B.S. 1995. Association of flagellum expression and intracellular growth of *Legionella pneumophila*. Infect Immun 63: 4928–4932.

Segal, G., Purcell, M. & Shuman, H.A. 1998. Host cell killing and bacterial conjugation require overlapping sets of genes within a 22-kb region of the *Legionella pneumophila* genome. Proc Natl Acad Sci USA 95: 1669–1674.

Vogel, J.P., Andrews, H.L., Wong, S.K. & Isberg, R.R. 1998. Conjugative transfer by the virulence system of *Legionella pneumophila*. Science 279: 873–876.

Solid waste, waste recycling and biofuels

European Symposium on Environmental Biotechnology, ESEB 2004 - Verstraete (ed)
© 2004 Taylor & Francis Group, London, ISBN 90 5809 653 X

Humic acid formation in composts – the role of microbial activity

E. Smidt, E. Binner & P. Lechner
Institute of Waste Management, BOKU – University of Natural Resources and Applied Life Sciences, Vienna, Austria

ABSTRACT: Five composting processes of yard and kitchen waste were compared with regard to humic substance formation. Humic acids were considered to be a reliable parameter to describe the humification process. The composition of the input material and process conditions influence the synthesis of humic substances. The microbial activity that depends on these conditions plays a key role concerning the balance between humification and mineralization. Respiration activity is an indicator for the microbial activity. Metabolic activities are also reflected by IR spectroscopic characteristics of the composted material. Lab scale experiments showed that a slight enhancement of microbial activity by adding easily degradable carbon and nitrogen sources could improve humic acid formation.

1 INTRODUCTION

Composting is a very complex process due to extensive chemical changes such as transformation, mineralization and resynthesis that take place simultaneously. Mineralization causes a release of CO_2, whereas synthesis of stable humic substances results in carbon preservation and leads to decreasing carbon turnover rates – a contribution to greenhouse gas reduction. CO_2 is a main component of the aerobic compost treatment, but composting conditions that prevent too fast mineralization can support humic substance formation.

Due to their favorable properties humic substances provide a lot of benefits for plants and soils: suppression of plant diseases, slow release of nitrogen and nutrients, positive effects on moisture, temperature and structure of soils (Schachtschabel et al. 1998). Therefore the formation of humic substances is a target to reach during composting (Tomati et al. 2000). Humic substances are defined as dark colored, amorphous chemical compounds of high molecular weight, but without stoichiometric formula (Ziechmann 1994). Humic substances comprise a heterogeneous mixture of different fractions. The fraction of extractable humic acids is a reliable parameter to assess the composting process and the quality of the final product. Compost humic acids are very "young" compared to soil humic acids. During the composting process they undergo a change and become similar to soil humic acids. This development was revealed by FT-IR spectroscopy (Sanchez-Monedero 2002, Smidt 2002). Microorganisms have a leading part in the composting process with regard to degradation and transformation of biomolecules. The shift from degradable organic molecules towards stable humic acids is based on microbial activity and chemical mechanisms.

The objective of these investigations was to shed light on the contribution of microbial activity in different industrial composting processes. Microbial activity was determined using respiration activity and IR spectroscopic characteristics of the samples. Lab scale experiments should provide more information about the balance between mineralization and humification. The main interest focused on biowaste materials (mixture of yard and kitchen waste) that were found to build up more humic acids than other materials such as sewage sludge or municipal solid waste.

2 MATERIALS AND METHODS

Yard and kitchen waste (fruits, plants, vegetables) originated from 5 composting processes. Process A (only yard waste): aerated open windrow process; Process B: closed reactor system (mainly kitchen waste), 1 week anaerobic, then aerated; Process C: open windrow (4 weeks) with mechanical rotation, curing phase in a pile; Process D: closed reactor system, curing phase in a pile; Process E: open window with mechanical rotation, curing phase in a pile. The processes C, D and E were carried out with the same mixture of input material. D and E took place at the same time.

Lab-scale experiments were carried out with the material from the outdoor composting plant (processes C, D and E). Five cylindric glass vessels, each containing 150 g of the 14-day-old compost were put on the punched tile in a glass container (volume of 10 l above the punched tile). The bottom (volume of 5 l) was filled with water to avoid the drying up of the samples. The container was covered by a lid where the gap in the center stayed open to guarantee air supply. In two containers biowaste (approximately 3 kg) was put on the punched tile and aerated from the bottom (40 l air/h at the beginning to 20 l/h at the end). The containers were stored in a climate chamber at 45°C. The material was rotated daily. One day prior to sampling the first sample set was fed by 10 ml of deionized water (reference value), the second one by 10 ml of the extract (2 l of hot water/5 kg of fresh biowaste material) and the third one by a glucose solution (20% of glucose/4% of glycine).

2.1 *Chemical investigations*

Chemical investigations were carried out from the air-dried, milled and sieved through 0.63 mm samples. Total organic matter was determined by combustion at 545°C. Optical densities of humic acids were determined photometrically at 400 nm after alkaline extraction (pH 10.5) according to Gerzabek et al. (1993). Investigations were carried out twice.

2.2 *Biological investigations*

For respiration activity fresh materials were used. The oxygen uptake was measured continuously in a Voith Sulzer sapromat (Binner et al. 1998) over a period of 7 days. The respiration activities were calculated as mg O_2/g dry matter (DM) for 4 days (RA 4 d) and 7 days (RA 7 d).

2.3 *IR spectroscopic investigations*

Two milligrams of the prepared material (as described for chemical analyses) were pressed with (FT-IR grade) KBr (1:100) to a pellet. These pellets were immediately measured in the spectrometer after preparation under ambient conditions using the transmission mode. The samples were measured in the mid-infrared range from 4000 cm^{-1} to 400 cm^{-1} with a Bruker Equinox 55 FT-IR spectrometer. The resolution was set to 4 cm^{-1}, 32 scans were recorded, averaged for each spectrum and corrected against ambient air as background.

3 RESULTS AND DISCUSSION

Figure 1 shows the development of humic acid contents. The humic acid content of process A remained at a steady low level, neither was there a considerable mineralization (Fig. 2). A small amount of easily degradable molecules was consumed very fast also causing the respiration activity to decrease (Fig. 3). Due to the composition of the input material and the resulting low level of microbial activity the more resistant molecules were not cracked.

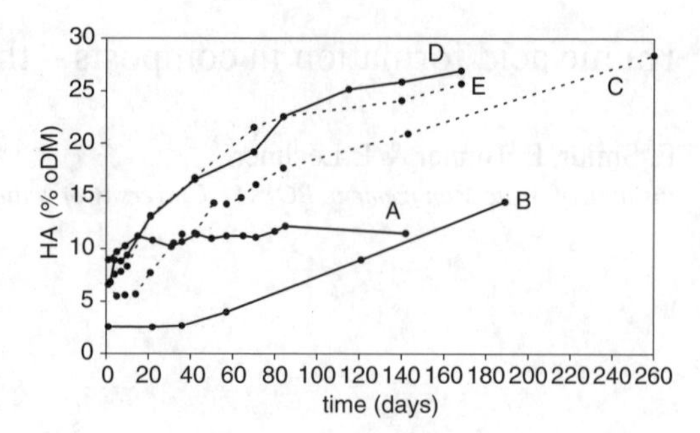

Figure 1. Development of humic acid contents (% of organic dry matter = oDM) in different composting processes.

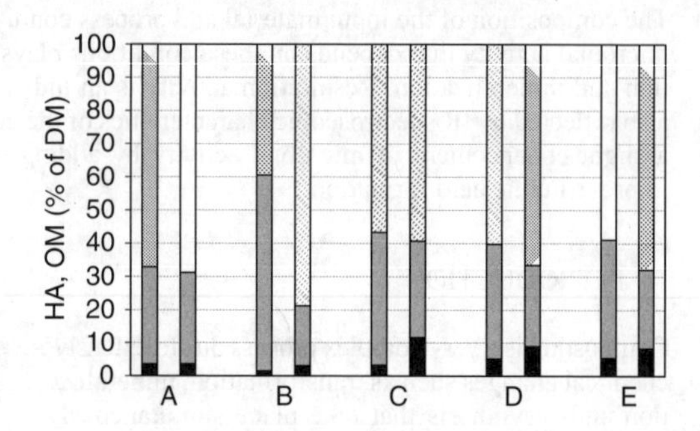

Figure 2. Contribution of humic acids (HA) and the remaining organic matter (OM) fraction at the beginning (21 d) and at the end of the processes (A: 142 d, B: 189 d, C: 260 d, D and E: 168 d).

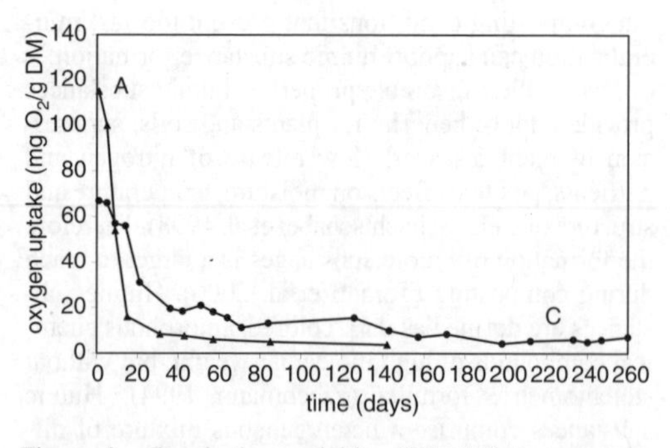

Figure 3. Respiration activities (RA 7 d) in processes A and C.

The humic acid content in process B was very low at the beginning. One week of anaerobic conditions prevented humic acid formation. However, after a lag phase of 5 weeks a continuous increase of humic acid contents was observed, but at a lower level than in the processes C, D and E. In process B mineralization was the highest (Fig. 2). In the processes C, D and E high humic acid contents were achieved.

The mixture of yard and kitchen waste seems to provide the necessary building blocks for humic acids and the required microbial activity that guarantees the transformation of organic matter to the stable fraction of humic acids. The decrease of organic matter was lower than in process B.

Figure 2 shows the contribution (%) of both, the humic acid (black) and the remaining organic matter fractions (gray).

The processes A and C were picked out to demonstrate the different patterns of microbial activities (Fig. 3). As mentioned above in process A the respiration activity (RA) decreased very fast from a high to a low level. In contrast, in process C the course of respiration activity was moderate. Starting at a lower level of oxygen uptake (RA 4 d: 54 mg O_2/g DM, RA 7 d: 70 mg O_2/g DM) a level at around 10 mg/g DM was maintained for months. However, transformation to stable humic acids exceeded mineralization. The pattern of respiration activities of the processes B, D and E were more similar to A. In process B microbial activity was put in mineralization. Controlled aeration supported such development. In processes D and E the favorable composition of the waste material allowed a continuous increase of humic acids, despite a faster decrease and resulting lower respiration activities than in process C. In addition mineralization was low compared to process B (Fig. 2).

Metabolic activities were also reflected by IR spectroscopic characteristics. The band at 1320 cm^{-1} that can be assigned to aromatic amines shows a typical behavior of increase and decrease during the composting process and reveals kinetics of the process (Smidt 2002). In process A only a weak shoulder is visible (days: 1, 22, 43, 142) at this position, indicating the low transformation rate. In process B the band at 1320 cm^{-1} is weak or only present for a short time between two sampling dates. However, microbial activities are indicated by the rising amide II band at 1540 cm^{-1} (Smith 1999, Naumann et al. 1996) in the IR spectrum (days: 1, 22, 36, 189). When conditions became aerobic the increase of the band was observed. After having reached a maximum it decreased and disappeared after 10 weeks. This band can be attributed to microbial biomass that was built up under aerobic conditions and to the degraded components of waste materials that are rich in proteins. The processes C, D and E showed the typical behavior of the 1320 cm^{-1} band. In process C (days: 1, 23, 59, 260) increase and decrease took place over a period of 10 weeks, as well as in the processes D and E. Temporary anaerobic conditions retarded the turnover. Nevertheless, humic acids were built up continuously. Investigations about organic matter in soils verify the relation between a high content of humic substances and the moderate microbial activity. Soils in climatic zones with considerable changes in temperature have higher contents of humic substances due to

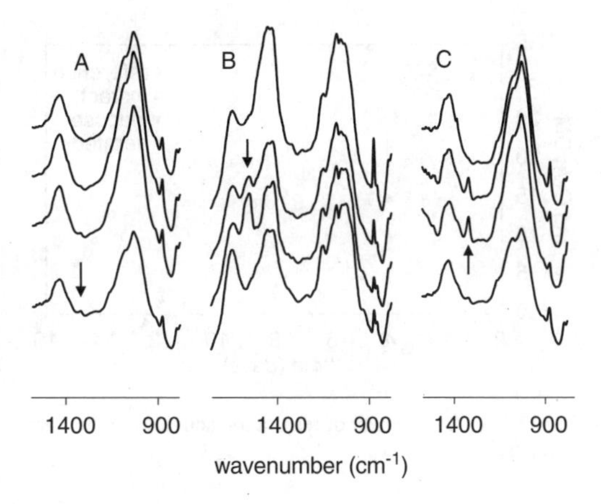

Fgure 4. Characteristic behavior of IR bands (process B: 1540 cm^{-1}, process C: 1320 cm^{-1}), indicating metabolic activities; no changes in process A.

alternate presence and absence of stimulating effects (Kononova 1975).

3.1 *Lab scale experiments*

Previous results from real composting processes had shown that the mixture from one composting plant (as processes C, D, E) achieved very good results with regard to humic acid formation. To avoid the influence of different input materials lab scale experiments were only carried out with the mixture mentioned above. Due to the composition all necessary ingredients for humic acid formation, especially aromatic building blocks were present.

Due to the age of 2 weeks the oxygen uptake has already been at a lower level. As expected, respiration activity increased by addition of glucose/glycine solution. The extract from fresh material also caused an increase compared to the reference. The oxygen uptake of the aerated sample was slightly above the reference, but on day 4 at the lowest level. On day 8 and 12 all samples had reached a similar low level of respiration activity as shown in Figure 5. After two weeks a slight increase took place again, mostly in the samples with the glucose/glycine solution. An occasional increase is frequently observed due to the degradation of more resistant molecules. However, the increase never reaches the high level of the initial phase.

Data from humic acid contents showed that the increased respiration activity had a positive effect on humic acid formation. At this level of microbial activity transformation of organic matter towards humic acids was improved. No significant difference was found regarding the decrease of organic matter.

The addition of extract solution led to higher values at the beginning. After 2 weeks a decrease was observed. Due to extracted dark colored substances

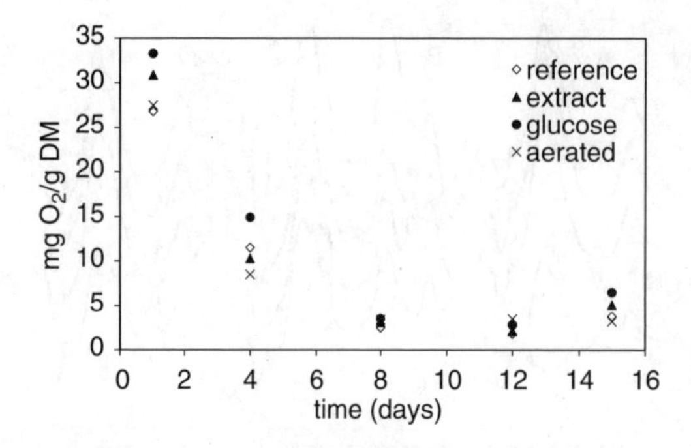

Figure 5. Development of respiration activities in the four variants.

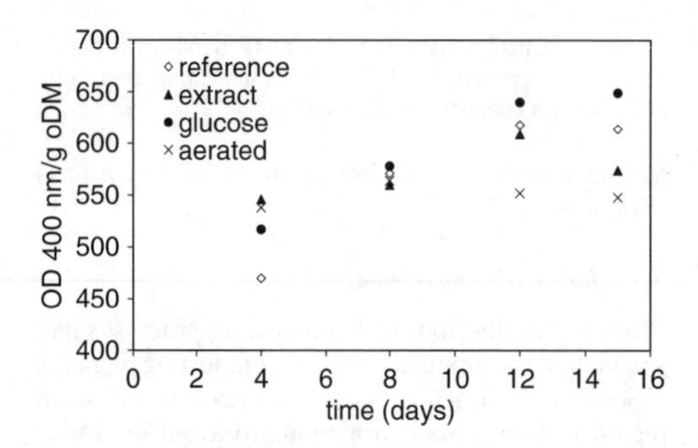

Figure 6. Development of humic acid formation in the four variants.

fresh materials often present incorrect higher values. During the first 2 weeks of composting a decrease can be noticed. In the aerated sample higher humic acid contents were determined on day 4 compared to the reference. On day 8 all samples reached a similar level. The second week the aerated sample remained at a constant level, the reference increased until day 12 and then remained at a constant level. From the second week the samples with glucose/glycine addition showed higher humic acid contents than the other variants. An increasing trend was observed.

4 CONCLUSIONS

Results from investigations in composting plants demonstrated the relation between microbial activity and humic acid formation. With regard of the input material a balanced mixture of easily degradable substances, usable aromatic building blocks and more resistant molecules is necessary. Best results were achieved with a mixture of yard and kitchen waste with an organic matter content at around 50% at the beginning of the composting process. Microbial activity is necessary to carry out the shift to the stable humic acid fraction. A balanced mixture of input materials regulates the velocity of degradation.

Lab scale experiments confirmed the role of the microbial activity. Respiration activity stimulated by addition of easily degradable carbon and nitrogen sources, resulted in higher humic acid contents. It is noteworthy that mineralization was not enhanced compared to the reference sample. Further investigations should elucidate to which extent humic acid formation can be promoted by stimulating the respiration activity, without directing the development of organic matter to mineralization.

The findings of the lab scale experiments have to be verified in real composting processes. Due to the necessity to define quality features of composts in the future the content of humic acids could be a reliable parameter. The knowledge about the balance of suitable substrates and microbial activity will make the influence and control of composting processes possible in order to get higher yields of humic acids.

REFERENCES

Binner, E., Zach, A., Widerin, M. & Lechner, P. 1998. Auswahl und Anwendbarkeit von Parametern zur Charakterisierung des Endproduktes aus mechanischbiologischen Restmüllbehandlungsverfahren. *Schriften-reihe des Mnis-teriums für Umwelt, Jugend und Familie,* Austria.

Gerzabek, M.H., Danneberg, O. & Kandeler, E. 1993. Bestimmung des Humifizierungsgrades. In F. Schinner, R. Öhlinger, E. Kandeler & R. Margesin (eds), *Bodenbiologische Arbeitsmethoden*: 107–109. Springer Verlag.

Kononova, M.M. 1975. Humus of Virgin and Cultivated Soils. In E. Gieseking (ed), *Soil Components Vol. 1, Organic Components*: 475–526.

Naumann, D., Schultz, C.P. & Helm, D. 1996. What can infrared spectroscopy tell us about the structure and composition of intact bacteria cells? In H.H. Mantsch & D. Chapman (eds), *Infrared Spectroscopy of Biomolecules*: 279–310. New York: Wiley-Liss.

Sanchez-Monedero, M.A., Cegarra, J., Garcia, D. & Roig, A. 2002. Chemical and structural evolution of humic acids during organic waste composting. *Biodegradation* 13: 361–371.

Schachtschabel, P., Blume, H.P., Brümmer, G., Hartge, K.H. & Schwertmann, U. 1998. *Lehrbuch der Bodenkunde*. Stuttgart: Ferdinand Enke Verlag.

Smidt, E., Lechner, P., Schwanninger, M., Haberhauer, G. & Gerzabek, M.H. 2002. Characterization of waste organic matter by FT-IR spectroscopy: Application in waste science. *Applied Spectroscopy* 56 (9): 1170–1175.

Smith, B. 1999. *Infrared Spectral Interpretation. CRC Press.*

Tomati, U., Madejon, E. & Galli, E. 2000. Evolution of humic acid molecular weight as an index of compost stability. *Compost Science and Utilization* 8 (2): 108–115.

Ziechmann, W. 1994. *Humic substances*. BI Wissenschaftsverlag.

European Symposium on Environmental Biotechnology, ESEB 2004 - Verstraete (ed)

Efficiency and monitoring of methane oxidation in bio-covers on landfills

S. Roeder, M.H.-Humer & P. Lechner
Institute of Waste Management, BOKU-University of Natural Resources and Applied Life Sciences, Vienna, Austria

ABSTRACT: Over the last decade the mitigation of methane emissions from landfills by means of methane oxidation in bio-covers has become a low-cost alternative to conventional landfill top covers. Especially bio-covers made of compost provide optimal conditions for methanotrophic bacteria and ensure high oxidation rates. To prove the efficiency of such bio-covers two different methods were improved and tested. An open wind tunnel was constructed and tested at laboratory scale before it was used to measure surface fluxes from two landfill sites, revealing obvious differences in emission rates between landfills with active bio-covers and conventional top-covers. Soil column expermiments were carried out and their results were compared to results derived from a gas transport model, showing that such a model may be useful for designing and monitoring bio-covers.

1 INTRODUCTION

1.1 *Methane oxidation in landfill covers*

It has long been known that microbial methane oxidation is an important natural process which significantly reduces methane emissions from several natural systems such as swamps or wetlands. Under aerobic conditions, microorganisms convert methane to water, carbon dioxide and microbial biomass.

Enhancing this natural potential of microbial methane oxidation in suitable landfill covers is a simple and low-cost measure to mitigate methane emissions, in particular on older landfill sites, on landfills with mechanically-biologically pre-treated waste and also as a practical method for landfills in developing countries or as an additional measure to gas extraction systems on bio-reactor landfills.

In various Austrian field trials it was shown that bio-covers made of compost, which ensure optimal ambient conditions for methanotrophic bacteria, foster microbial activity and attain very high oxidation rates up to 95–100% (Humer et al. 1999, Humer et al. 2001).

These investigations have proven that ripe waste compost, such as sewage sludge compost and municipal solid waste compost, are suitable substrates for methane oxidation. The oxidation rates in these composts were clearly higher than in natural soils. Apart from a proper compost quality, the design of the cover layer, in particular the construction of a gas distribution layer for a homogeneous gas supply is crucial.

1.2 *Detection of surface fluxes*

Local authorities demand the quantification of remaining CH_4 fluxes from bio-covers in order to permit practical application of these covers instead of conventional top-covers. From the variety of existing techniques for measuring surface fluxes, especially static and dynamic chamber methods are suitable for the detection of local phenomena (such as emission peaks close to cracks in the landfill or in the top-cover) while micrometeorological and optical methods are widely used to derive whole-landfill flux rates (Roeder et al. 2000, unpubl.).

1.3 *Modelling of methane oxidation*

The gas transport in soils is governed by the main processes diffusion, advection and dilution and altered by sink processes like dissolution in soil water, degeneration (e.g. methane oxidation) and sorption to soil particles (Kjeldsen 1996). The simulation of gas transport processes in methane-oxidation-layers may be a useful tool for designing compost-covers with enhanced oxidation-efficiency. Existing models for gas transport in soils (Bogner et al. 1997, Fasolt et al. 1995, Moldrup et al. 1996, Poulsen et al. 2001, Stein et al. 2001) incorporate some or all of the processes mentioned above, ranging from simple 1D models that provide analytical solutions to complex 3D models with numerical solutions.

2 MATERIALS AND METHODS

2.1 *Open wind tunnel*

The search for a system that can be operated easily and that is capable of detecting local phenomena but avoiding the disadvantages of static (only diffusion driven emissions are detected) and dynamic chambers (over- or under-estimation of fluxes caused by inadequate flow rates) resulted in the modification of a dynamic chamber concept. Finally an "open wind tunnel" was constructed, where (instead of a pump) the natural wind is used to carry the air and the emitted gases through the tunnel (Roeder et al. 2001, unpubl.).

The tunnel (Fig. 1) has a length of 3.4 m, a semicircular cross section with a diameter of 1 m and therefore covers an area of 3.4 m^2. It is made of plexiglas and reinforced with steel frames. In the rear and the front section of the tunnel inlets for gas concentration measurements are located and connected via a switch to a portable flame ionization detector (used for determination of CH_4 concentrations) and a portable infrared analyzer (used for determination of CO_2 concentrations). An anemometer for wind speed measurements is located in the rear section of the tunnel.

Surface fluxes are derived using Equation 1 from the wind speed inside the tunnel and the difference between gas concentrations at the inlet and the outlet of the tunnel.

$$F = \frac{v \cdot A_c \cdot \rho}{A_s} \cdot (C_{out} - C_{in}) \qquad (1)$$

where F = gas flux from the surface; v = wind speed inside the tunnel; A_c = area of the tunnel's cross section; ρ = gas density; A_s = area of the soil covered with the tunnel; C_{out} = gas concentration at the outlet of the tunnel; and C_{in} = gas concentration at the inlet of the tunnel.

2.1.1 *Laboratory experiments*

A test cell was constructed and filled with expanded clay, under which a system of tubes for gas supply was located. The open wind tunnel was placed on the test cell and charged with methane through the gas supply system. To simulate adequate wind speeds, two fans were used. The laboratory experiments were carried out from August 2000 till March 2001 in order to improve the measurement system and to quantify the accuracy of the system.

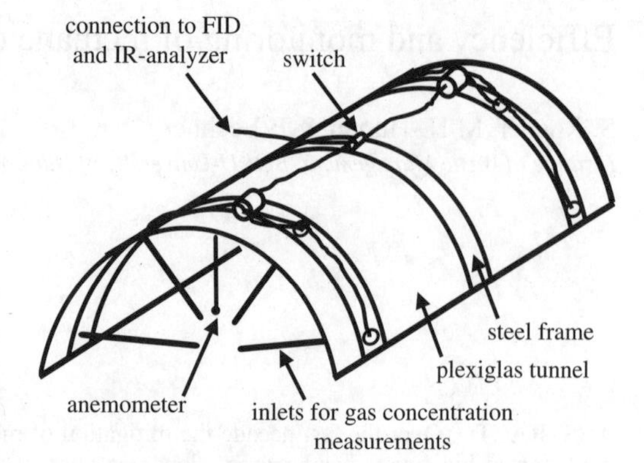

Figure 1. Open wind tunnel.

2.1.2 *Field experiments*

The first series of experiments took place from April till August 2001 at a landfill located in St. Pölten, Lower Austria, where five test cells (each of them covering an area of about 625 m^2) had been installed on the dumped waste. On four cells differently designed compost covers had been applied (some of them with others without a gas distribution layer made of coarse gravel), while the fifth cell had been left uncovered as a kind of reference. At several locations on each of the test cells probes had been installed, in order to record gas concentration profiles at different depths of the bio-cover (Humer et al. 2001).

The landfill's gas recovery system was in full operation while the experiments took place. The open wind tunnel was placed on different locations on each of the cells. After an adaptation time of five minutes wind speeds and gas concentrations were recorded over a period of at least ten minutes, in order to calculate gas fluxes from the landfill surface.

Applying the same procedure, a second series of field experiments was carried out at the main landfill of Vienna in September 2002. This landfill is equipped with an interim, conventional top-cover (about 0.3 m of compacted sandy loam) and a gas recovery system, which was in full operation during the experiments.

2.2 *Gas transport model*

In order to find a model reproducing the specific conditions in bio-covers properly and to minimize required input data, the question, if advection has a significant influence on the gas transport in bio-covers has to be answered. Advection can not only be caused by the pressure differences between ambient air and the landfill gas but also by a reduction of the gas volume (where 2 mol of oxygen plus 1 mol of methane gives 1 mol of carbon dioxide) occurring in the oxidation-layer, resulting in an encouraged uptake of N_2 from the ambient air (Kjeldsen 1996).

2.2.1 *Soil column experiments*

To find out, if this phenomena is of significant importance for the gas transport in compost-covers, a series of soil column experiments was carried out using Ar as a tracer.

Two plexiglas tubes with a diameter of 0.2 m and a height of 0.7 m were filled with sewage-sludge-compost up to a height of 0.6 m. Along the height of the columns (0.1 m apart from each other) probes were installed. To avoid the influence of changes in ambient air temperature, the columns were situated in an air conditioned chamber at a temperature of 25°C. CH_4 supply was connected to the bottom of each tube, while the headspace of each tube was connected to another gas supply, providing ambient air or a mixture of 40% Ar, 40% N_2 and 20% O_2. Acting as a kind of reference column the compost in one of the tubes was sterilized before the beginning of the experiments, inhibiting methane oxidation.

After supplying both columns with methane from the bottom and ambient air from the top for one week in order to reach stable conditions, the air supply for the headspace was switched to the tracer gas mixture. Then samples were taken through each probe and analyzed with a Shimadzu-Gas Chromatograph equipped with a HP Molecular Sieve 5A (10 m ×0.32 m I.D., 10 μm df.) for Ar, CH_4 and N_2. The interval for sampling at each probe was 50 minutes during the first three hours of the experiment. After leaving the columns unattended for the night, three further samples were taken from each probe, again at an interval of 50 minutes.

To provide soil property data for further model simulations, bulk density, particle density and moisture content were determined at six different depths in both soil columns after the completion of the column experiments.

2.2.2 *Model development*

Based on the results of the column experiments an existing model, that simulates the one dimensional transport of a mixture of four gases (CH_4, O_2, CO_2, N_2) by means of a finite difference scheme and incorporates advection, diffusion and methane oxidation (Stein et al. 2001) was selected and modified, adding Ar as a fifth gas in order to compare results from simulation and experiments. By means of this model estimations for gas concentrations, pressure and flux rates at different depths and times can be achieved.

3 RESULTS AND DISCUSSION

3.1 *Measurements with the open wind tunnel*

3.1.1 *Laboratory experiments*

The first series of experiments revealed a strong correlation between accuracy and wind speed. These results led to a lower operating limitation at the wind speed of 0.5 m/s which should be accessed in order to derive accurate flux rates. Also the angle between the wind direction and the axis of the tunnel should not access 30°. The measurements also showed, that in most cases the flux rates were overestimated which led to modifications of the inlets for gas concentration detection in the rear and front section of the tunnel. In the second series of laboratory experiments it could be shown, that the accuracy of the modified measurement system is about 30% at a detection limit of 1 g/m²d. These numbers are acceptable regarding the high variability of emission rates over space and time.

Table 1. Results of field experiments at the landfill of St. Pölten.

Section	CH_4 flux (g/m²d)	CO_2 flux (g/m²d)	CH_4:CO_2 flux ratio
Bio-cover	1 to 5	20 to 200	1:50 to 1:200
Uncovered	20 to 500	100 to 1000	1:5 to 1:10
Hot-spots	700 to 1000	1500 to 2500	1:2 to 1:5

3.1.2 *Field experiments*

The onsite measurements at St. Pölten revealed significant differences in flux rates of CH_4 between the cells with active bio-covers and the uncovered reference cell (see Table 1). It could also be shown, that the ratio of CH_4 and CO_2 fluxes decreased with the absence or the failure (hot-spots caused by cracks and fissures in the landfill surface) of the bio-cover indicating a reduced efficiency of methane oxidation.

The field measurements at the Viennese landfill showed CH_4 fluxes between 5 and 500 g/m²d and CO_2 fluxes between 40 and 800 g/m²d, resulting in flux ratios of 1:2 to 1:5. This shows, that due to the absence of adequate bio-covers methane oxidation does not play an important role on this landfill.

3.2 *Gas transport model*

3.2.1 *Soil column experiments*

Results from one of these experiments are shown in Figure 2.

The concetration profiles derived from the column experiments showed, that Ar concentrations were always at a higher level in the active column with undisturbed methane oxidation compared to the same depth in the sterilized reference column. This indicates, that the presumed additional advective flow caused by volume reduction through methane oxidation really takes place inside the oxidation layers.

3.2.2 *Model simulation*

The comparison of the data derived from model simulations to the results of the column experiments shows, that the model is capable of predicting gas

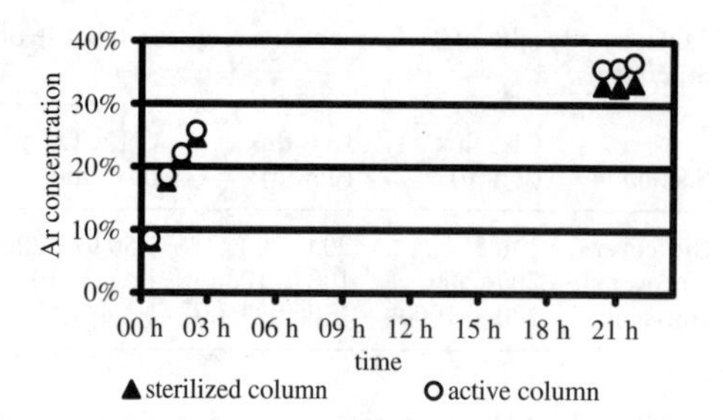

Figure 2. Ar concentration in soil columns at a depth of 0.1 m.

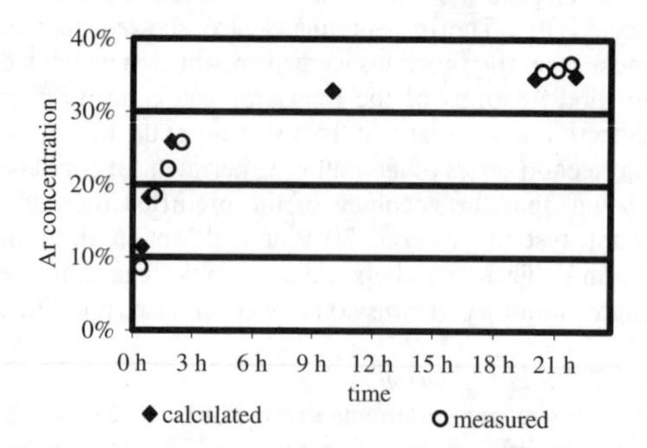

Figure 3. Comparison of calculated and measured Ar concentrations in the active column at a depth of 0.1 m.

concentration profiles. An example for such a comparison is shown in Figure 3.

The comparison of calculated gas concentration profiles with profiles measured during the field experiments in St. Pölten and of calculated fluxes with fluxes measured with the open wind tunnel shows, that it will be indispensable for a practical application of the model to record bulk density, particle density, moisture content and soil temperature properly. It will also be necessary to measure meteorological data that influences the gas transport such as atmospheric pressure and air temperature.

4 CONCLUSIONS

The measurement of gas fluxes from differently designed landfill surfaces by means of an open wind tunnel shows that the application of bio-covers can mitigate methane emission rates significantly. It also reveals that the tunnel acts as a useful device for measuring fluxes except of situations where the wind abates or the wind direction changes severly.

The application of a gas transport simulation model may be a useful, quick and cost-efficient tool for designing proper bio-covers and for predicting the influence of changes in environmental conditions on the efficiency of bio-covers in field. The proper practical use requires the recording of some metereological and soil data and supplementary, the results of such simulations must be verified by field measurements at least in form of an initial proof test. For that purpose the installation of probes for recording gas concentration profiles combined with the use of an open wind tunnel detecting the surface fluxes seems to be a promising approach.

REFERENCES

Bogner, J. et al. 1997. Kinetics of Methane Oxidation in a Landfill Cover Soil: Temporal Variations, a Whole-Landfill Oxidation Experiment, and Modelling of Net CH_4 Emissions. *Environmental Science & Technology* 31(9): 2504–2514.

Fasolt, B. 1995. Determination of the Gas Emission Flux from Landfills by Concentration Measurements of CH_4 and CO_2 in the Top Layer. In T.H. Christensen et al. (eds), *Proceedings Sardinia 995; Fifth International Landfill Symposium; Cagliari, 2–6 October 1995*. Cagliari: CISA, Environmental Sanitary Engineering Centre.

Humer, M. et al. 1999. Compost as a Landfill Cover Material for the Elimination of Methane Emissions. In W. Bidlingmaier et al. (eds), *Proceedings of the International Conference ORBIT 99 on Biological Treatment of Waste and the Environment Part II (Perspectives on Legislation and Policy, Product Quality and Use, Biodegradable Polymers, and Environmental and Health Impacts)*. Berlin: Rhombos-Verlag.

Humer, M. et al. 2001. Design of a Landfill Cover Layer to Enhance Methane Oxidation. Results From a Two Year Field Investigation. In T.H. Christensen et al. (eds), *Proceedings Sardinia 2001; Eighth International Landfill Symposium; Cagliari, 1–5 October 2001*. Cagliari: CISA, Environmental Sanitary Engineering Centre.

Kjeldsen, P. 1996. Landfill Gas Migration in Soil. In T.H. Christensen et al. (eds), *Landfilling of Waste: Biogas*. London: E & FN Spon.

Moldrup, P. et al. 1996. Modelling Diffusion and Reaction in Soils: A Diffusion and Reaction Corrected Finite Difference Calculation Scheme. *Soil Science* 161(6): 347–354.

Poulsen, T.G. et al. 2001. Modelling Lateral Gas Transport in Soil Adjancent to Old Landfill. *Journal of Environmental Engineering* 127(2): 145–153.

Roeder, S. 2000, unpublished. Quantifizierung von Methanemissionen bzw. der Aufnahme von Methan an Deponieoberflächen zum Nachweis der Wirksamkeit der mikrobiologischen Methanoxidation in Kompost-Abdeckschichten. Vienna: Institute of Waste Management, BOKU-University of Natural Resources and Applied Life Sciences.

Roeder, S. et al. 2001, unpublished. Untersuchungen zur Messung von Methanemissionen auf Deponien. Vienna: Institute of Waste Management, BOKU-University of Natural Resources and Applied Life Sciences.

Stein, V.B. et al. 2001. A Numerical Model for Biological Oxidation and Migration of Methane in Soils. *ASCE Practice Periodical of Hazardous, Toxic and Radioactive Waste Management* 5(4): 225–234.

European Symposium on Environmental Biotechnology, ESEB 2004 - Verstraete (ed)
© 2004 Taylor & Francis Group, London, ISBN 90 5809 653 X

Development of an integrated biowaste management infrastructure – lessons learned from the Netherlands

F.M.L.J. Oorthuys, E.H.M. van Zundert & A.J.F. Brinkmann
Grontmij Water and Waste Management, De Bilt, The Netherlands

ABSTRACT: This paper elaborates on a number of key issues related to the implementation and operation of the current infrastructure for household biowaste management in the Netherlands. It describes the historical background to the current system of separate collection, composting and compost re-use, as it was implemented in the early 1990s. It elaborates in particular on four elements of the system: biowaste recovery rates, treatment facilities, marketing of compost, and financial aspects. It also summarises a number of current developments that may impact upon the future features of the biowaste management infrastructure. The paper concludes with a number of lessons that can be learned from the operational experiences in the Netherlands.

1 INTRODUCTION

In the Netherlands, recycling of biowaste through composting has been an operational practice for a long time. In the first decades of the 20th century comingled municipal solid waste (MSW) was composted on relatively small scale. As organics were the main component of MSW in that period, and other components such as plastics were absent, the compost could without further treatment be applied as a soil conditioner on arable land. Composting of biowaste on a more industrial level started in 1929, with the private organisation VAM composting commingled MSW from the larger cities in the urbanised western part of the Netherlands. MSW was transported by rail to the VAM, located in the north-eastern part of the Netherlands, with the compost being used to further develop arable land in the same region.

Increased environmental awareness in the 1970s and 1980s caused the practices of application of mixed MSW compost to be politically reconsidered. The Waste Management Hierarchy established in 1979 favoured organic waste to be diverted from landfill, while at the same time the Soil Protection Act set strict standards for quality of soil conditioners and fertilisers to be applied on land. As a result, it was decided that a system would be set up for separate collection of household biowaste, which would then be converted into high-quality compost, as opposed to mixed MSW with significantly higher heavy metal and other polluting contents.

As per 1 January 1994, municipalities have been responsible for separate collection of household biowaste, also referred to as Vegetable, Fruit and Garden waste (VFG), with strict compost standards ensuring quality of the end products. Also in this period, landfill bans were imposed upon organic wastes, increasing further the need for alternative treatment. As a result, an entire new infrastructure was established in the early 1990s, comprising separate collection of household biowaste and composting at some 25 treatment facilities. Dedicated campaigns to promote public awareness have assisted to run the scheme successfully since.

This paper looks into some of the results of the Netherlands infrastructure for (household) biowaste management, in particular lessons that can be learned after approximately 10 years of operational experience.

2 MONITORING

As from the start of separate collection of household biowaste a system has been implemented to monitor the performance of the entire biowaste treatment chain. The Ministry of Spatial Planning, Housing and Environment (VROM) initiated this system which was continued from halfway the 1990s by the Dutch Waste Management Council (AOO), the platform in which different government layers in the Netherlands coordinate on waste policy.

The monitoring system is supplied with data from municipalities and waste companies, which are subsequently summarised in public reports. The monitoring process has ensured that operational experiences are used for regular evaluation of existing waste management plans on different governmental layers, as well as the evaluation of general regulations on biowaste management. As any system of separate biowaste collection relies on the voluntary cooperation of households, an important aim of the monitoring system in its first years was also to increase knowledge and support with the general public.

This paper highlights four main categories of issues that have been monitored since the introduction of the system:

- the biowaste recovery rate;
- biowaste treatment facilities;
- the marketing of biowaste compost produced;
- financial aspects.

2.1 *Recovery rate*

In the first years after 1994, the biowaste recovery rate quickly rose to approximately 60%, equaling approximately 100 kilogram biowaste per person per annum. In recent years, a slight decrease in recovery rate to approximately 90 kg per person can be noted, which is generally attributed to less environmental interest with the general public.

As can be seen from Table 1, significant differences do exist between rural and urban areas, which can be attributed to the presence of gardens and participation rates amongst the public. The table also illustrates the recovery targets, which have been specified per level of urbanisation.

Since the mid-1990s, the total amount of separately collected household biowaste has been at a constant level of approximately 1.5 Mtonnes per year, corresponding to 20% of the total household waste production.

Table 1. Recovery rate (AOO, 2003).

Level of urbanisation	Amount of separately collected Biowaste (kg/inhabitant) year 2000	year 2001	Target (kg/inhabitant)
>2,500 households/km^2	26	24	35
1,500–2,500 households/km^2	78	77	85
1,000–1,500 households/km^2	101	98	105
500–1,000 households/km^2	119	116	125
<500 households/km^2	136	129	140
National average	92	88	

Ensuring the quality of separately collected biowaste, i.e. minimising non-biodegradable components in the waste, is the responsibility of the municipality, which uses primarily education and awareness campaigns to acquire co-operation from households. Separately collected biowaste is inspected upon arrival at the treatment facility. If the quality does not meet the official waste acceptance criteria (generally <5% non-organic pollutants) the waste will not be accepted for composting, while the associated costs for transport and alternative treatment will be incurred to the municipality of origin.

Legally, municipalities can only abandon separate collection if, despite efforts, the recovery rate remains very low, the biowaste quality is very bad and, as a result, the separate collection costs are very high. In recent years, this legal clause has been used by a number of major municipalities to cease separate collection in city centres and town areas with a large percentage of high-rise buildings.

2.2 *Biowaste treatment facilities*

The available treatment capacity for household biowaste is approximately 1.7 Mtonnes per year, with a variety of technologies being applied in 25 full-scale facilities (refer to Figure 1 and Table 2).

As can be seen from Figure 1 and Table 2, indoor aerobic composting is the most applied household biowaste treatment technology (approx. 75% of total treatment capacity), where as outdoor composting and anaerobic digestion (AD) form a minor part. This can be attributed to the fact that at the time of implementing these facilities, indoor composting was considered a proven technology, where as AD was not.

2.3 *Marketing of biowaste compost*

The annual compost production amounts to 600–700 ktonnes (approximately 40% of separately collected fresh biowaste, after mass loss in the composting process), while almost all compost is marketed within the Netherlands in the following sectors:

- agriculture 50%
- horticulture 28%
- private markets 19%

The use of the produced compost as soil conditioner is regulated by two separate pieces of legislation.

The first is the so called Law on Fertilisers, to which all materials applied as fertiliser/soil conditioner have to comply. This regulation states maximum quantities of nitrogen en phosphorous (kg/ha/year) that can be applied on land, depending on the time of year, the cultivation and the soil characteristics.

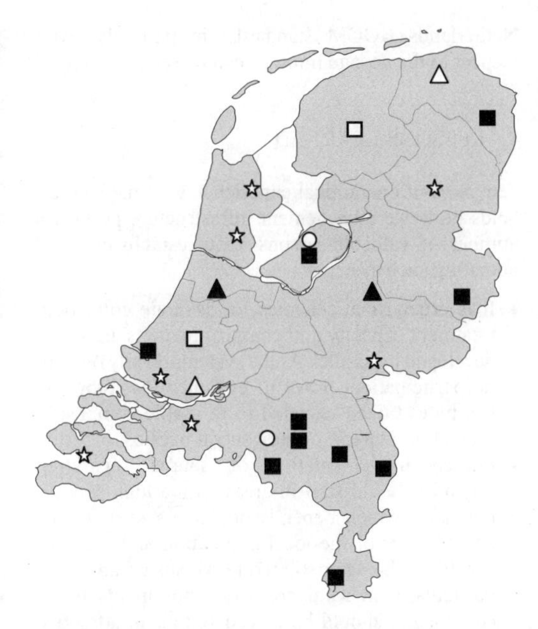

Figure 1. Distribution of treatment facilities in the Netherlands (Grontmij, 2003).

	Type
☆	Hangar-composting Bühler or VAM-concept
■	Tunnel-composting
▲	Open air composting VAR-concept
□	Hanger-composting PACOM-comcept
○	Anaerobic digestion BIOCEL or Valorga-concept
△	Other concepts

Table 2. Type of biowaste treatment in the Netherlands.

Type	Number	Licensed capacity (ktonnes/ year)	Relative capacity (%)
Hangar-composting Bühler or VAM concept	7	704	41.2
Tunnel-composting	10	564	33.0
Open air compost-ing VAR-concept	2	229	13.4
Hangar-composting PACOM-concept	2	76	4.5
Anaerobic digestion BIOCEL-concept	1	35	2.0
Other concepts	3	100	5.9
Total	25	1,708	100.0

The second is the statutory standard BOOM (Besluit kwaliteit en gebruik overige organische meststoffen). BOOM defines maximum contents of heavy metals in sewage sludge, compost and topsoil.

Table 3. Netherlands and (proposed) European compost standards.

	BOOM		EU working document	
Parameter (mg/kg dm)	Compost	High quality compost	Compost Class 1	Compost Class 2
Cd	1	0.7	0.7	1.5
Cr	50	50	100	150
Cu	60	25	100	150
Hg	0.3	0.2	0.5	1
Ni	20	10	50	75
Pb	100	65	100	150
Zn	200	75	200	400
Maximum dosage	6 ton ds/ha/yr	Best practice	Best practice	30 tonnes/ ds/ha/yr

For compost, two different classes are being distinguished, which are referred to as 'standard compost' and 'high quality compost' (see also Table 3). Where as high quality compost can be applied according to best agronomic practice, for standard compost a maximum dosage does apply, equalling 6 tonnes dry matter per hectare per year for arable land and 3 tonnes dry matter per hectare per year for grassland.

As standards for high quality compost are in practice seldom met, and as nutrient contents in compost are relatively low, the maximum dosage (tonnes solids/ha/annum) in the BOOM regulation is the limiting factor for compost application. However, as no such maximum dosage factor does exist for topsoil, compost producers have used this as an escape by marketing their product as topsoil. This legal gap is currently being evaluated.

Compost is generally marketed at a price close to zero (positive or negative), and does therefore not contribute significantly to the revenues of a composting plant (mainly gate fees).

In recent years, a number of private composting companies have started to produce more dedicated types of compost from household biowaste, e.g. by blending it with other (in)organic soil improvers. In particular the replacement of peat used for potting soils seems to be promising, which would cause a shift from agriculture as main outlet, to horticultural applications.

2.4 *Financial aspects*

Composting of biowaste is a more cost-effective treatment method than incineration: gate fees for composting facilities range between 35 and 55 Euro/tonne, while fees for incineration vary between 70–115 Euro/tonne. Average costs for comingled waste collection are approximately 60–70 Euro/tonne, where as costs for separate biowaste collection range from 100 to 170 Euro/tonne. Consequently, whether separate collection of biowaste followed by composting in total is more cost-effective than commingled collection followed by incineration depends on the extra costs for separate collection, compared to the difference in treatment facilities' gate fees. In general, it can be stated that in urban areas with low biowaste recovery rates and relatively low incineration gate fees, separate biowaste collection is not necessarily the most cost-effective option. However, in more rural areas with higher biowaste recovery rates this is more likely to be the case, in particular if incineration gate fees are higher.

Operational experience to date shows that composting is a more cost-effective biological treatment method for household biowaste than anaerobic digestion, mainly caused by the robustness, flexibility and less complicated technical nature of composting technologies compared to anaerobic digestion.

3 CURRENT DEVELOPMENTS

As indicated above, a number of developments on a national level currently determine the future features of the biowaste management structure, including:

- Slight decrease in public participation in the separate collection schemes;
- Municipalities ceasing separate collection operations in some urban areas, particularly in high-rise buildings and city centres;
- Diversification of compost products, and evaluation of compost quality standards.

In addition to these developments, there is increasing concern about the health risk for workers in composting facilities, in particular because of high bio-aerosol concentrations. Safe working standards and measures to be taken are currently under investigation.

On a European level, the recently adopted Animal By-Products Regulation (Regulation (EC) No 1774/2002) poses additional standards for composting facilities, in particular in relation to sanitation of biowaste and process/product control. Research in the past half year showed that composting facilities in the Netherlands have no major difficulties complying with the standards in the Regulation (Grontmij, 2003).

The EU Working Document on Biowaste (2nd Draft, February 2001) outlines the features of a future biowaste directive. The philosophy of the working document is very much in line with existing practices in the Netherlands, as are proposed standards with respect to collection, processing and products. The compost standards in the document are less strict than Netherlands BOOM standards, in particular with respect to the dosage limit, as can be seen in Table 3.

4 LESSONS LEARNED

Ten years of operational experience with the Netherlands biowaste management infrastructure provide a number of valuable lessons about establishing and operating such systems:

- It is crucial that schemes for separate collection, treatment capacity and compost markets are being developed in parallel. As the systems do rely on public participation, prove of evidence that biowaste has indeed been recycled to re-usable compost is crucial for support of the general public.
- Phased implementation of source separation schemes should start in areas with a high success potential, which depends on factors such as the demography of the local population and the geographical characteristics (rural versus urban).
- Particularly in rural areas (promotion of) home composting should be considered as an attractive addition to a system of source separation. In urbanised areas, particular in the larger cities, separate collection should be outweighed against alternatives such as mixed collection combined with various kinds of thermal treatment.
- A variety of full-scale proven technologies do exist for the biological treatment of household biowaste. Which technology is most suitable for a particular situation depends on many factors, including the required treatment capacity, operational flexibility, acceptable emission levels, and capital and operational costs of the facility.
- Potential end-users of compost should be directly involved in market development. This will enhance confidence in products which are waste derived, but as operational practice shows of high quality.

REFERENCES

Grontmij (2003) Separate collection and treatment of VFG waste. Description of he existing infrastructure. Report prepared for VVAV, April 2003.

AOO (2003) Monitoring Report 2001.

Regulation (EC) No 1774/2002 of the European Parliament and of the Council of 3 October 2002 laying down health rules concerning animal by-products not intended for human consumption.

European Commission. Working Document 'Biological Treatment of Biowaste' 2nd Draft. February 2001.

European Symposium on Environmental Biotechnology, ESEB 2004 - Verstraete (ed)

Microbial hydrogen metabolism and environmental biotechnology

K.L. Kovács, Z. Bagi, B. Bálint, J. Balogh, R. Dávid, B.D. Fodor, Gy. Csanádi, T. Hanczár, Á.T. Kovács, D. Latinovics, G. Maróti, L. Mészáros, K. Perei, A. Tóth & G. Rákhely
Department of Biotechnology, University of Szeged and Institute of Biophysics, Biological Research Centre, Hungarian Academy of Sciences, Szeged, Hungary

ABSTRACT: The purple sulfur phototrophic bacterium, *Thiocapsa roseopersicina* BBS contains several [NiFe] hydrogenases. These enzymes can be used e.g., as fuel cell H_2 *splitting* catalyst or in photoheterotrophic H_2 *production*. Methanotrophic bacteria utilize H_2 to supply reductant for their methane monooxygenase (MMO) enzyme systems. H_2 *driven enzyme activity* plays determining role in methane oxidation. This process is of great importance in decreasing the emission of the greenhouse gas methane, in bioremediation of halogenated hydrocarbons and related hazardous compounds, and in formation of the easily storable and transportable renewable energy carrier methanol. Microorganisms that *supply* H_2 *in situ* facilitate the biodegradation of organic material and concomitant biogas production. Fast, efficient, and economic treatment of organic waste, sludge, manure is achieved and generation of significant amount of renewable fuel from waste is intensified. The technology has been field tested under mesophilic and thermophilic conditions with positive results.

1 [NIFE] HYDROGENASES

Understanding the molecular fundamentals of hydrogen production and utilisation in biological systems is a goal of supreme importance for basic and applied research (Cammack et al. 2001). The key enzyme in biological H_2 metabolism is hydrogenase. This unique enzyme catalyses the formation and decomposition of the simplest molecule occurring in biology: H_2. The simple-looking task is solved by sophisticated macromolecular machinery. Hydrogenases are metalloenzymes harbouring Ni and Fe, or only Fe atoms, arranged in an exceptional structure. This study focuses on the hydrogenases with [NiFe] active centres. Like most redox metalloenzymes, hydrogenases are usually extremely sensitive to inactivation by oxygen, high temperature, CO, CN and various environmental factors. These properties are not favourable for most biotechnological applications, including biohydrogen production, water denitrification, bioconversion of biomass, and other bioremediation uses.

Their physiological function vary: they can serve as redox safety valves to dispose of excess reducing power, or generators of chemical energy by taking up and oxidising H_2, or maintaining a reducing environment for reactions of crucial importance, such as the fixation of atmospheric nitrogen. In some organisms, the numerous functions are performed by the same enzyme, but more frequently, a separate, specialised hydrogenase carries out each *in vivo* biochemical function.

In metal-containing biological catalysts, it is the protein matrix, surrounding the metal centres, which provides the unique environment for the Fe and Ni atoms and allows hydrogenases to function properly, selectively, and effectively. The [NiFe] hydrogenases are composed of at least two distinct (heterodimer) polypeptides, containing highly conserved metal binding domains. The large subunit harbours the active centre, fastened to the protein by 4 cysteine ligands. The Fe atom ligates 2 CN and 1 CO diatomic molecules and it is fixed to the Ni atom via sulphur bridges. The small subunit contains 2–3 Fe_4S_4 clusters, which are 15 angstroms apart, and thus, form a conducting wire inside the protein to facilitate the transport of electrons between the active centre and the protein surface. A major goal for hydrogenase basic research is to understand the intimate protein-metal interaction in this complex structure. The problem is not simple to address, as some of the methods for scientific investigation provide information on the metal atoms, without directly detecting the protein matrix around them. Other modern techniques reveal details of the protein core, but do not expose the metal centres within. A combination of the various molecular approaches is expected to uncover the fine molecular details of the catalytic action (Kovács & Bagyinka 1990, Szilágyi et al. 2002).

2 ASSEMBLY OF [NIFE] HYDROGENASES

A number of accessory gene-products govern the metal uptake, their attachment into the right place at the right time, the formation and ligation of the CN and CO groups, and the incorporation and fixation of this labile inorganic structure into the protein matrix (Paschos et al. 2001). Our present understanding suggests that the concerted action of, at least, 15–20 such accessory proteins is necessary for the formation of an active NiFe hydrogenase (Cammack et al. 2001). Some of the participating proteins are ***hy***drogenase ***p***leiotrop, called Hyp. They take part in the fabrication of every hydrogenase synthesised in the cell. Others specifically work on one type of [NiFe]. Sometimes, the accessory genes are neatly arranged around the structural genes, but most often, they are scattered in the genome (Fodor et al. 2001, Maróti et al. 2003, Fodor et al. 2003).

3 HYDROGENASES OF *THIOCAPSA ROSEOPERSICINA* BBS

T. roseopersicina is a phototrophic purple sulphur bacterium; the strain marked BBS has been isolated from the cold water of the North Sea. Its anaerobic photosynthesis uses reduced sulphur compounds (sulphide, thiosulfide, or elementary sulphur), but it can also grow on organic compounds (sugar, acetate) in the dark. The bacterium contains a nitrogenase enzyme complex, thus it is capable of fixing atmospheric N_2, a process accompanied by H_2 production (Vignais et al. 1995). Previous studies in our laboratory have revealed that *T. roseopersicina* contains at least two membrane-associated [NiFe] hydrogenases with remarkable similarities and differences. One of them (HydSL/HynSL (for recent nomenclature change see Vignais et al. 2001)) shows extraordinary stability: it is much more active at 80°C, than around 25–28°C. It is to be noted that *T. roseopersicina* cannot grow above 30°C. HynSL of *T. roseopersicina* is also reasonably resistant to oxygen inactivation and stays active after removal from the membrane. The other [NiFe] hydrogenase, HupSL, is very sensitive to all these environmental factors and thus it resembles the [NiFe] hydrogenases known from other microorganisms. The structural genes coding for these enzymes have been cloned and sequenced (Colbeau et al. 1994, Rákhely et al. 1998, Rákhely et al. 2003). The translated protein sequences indicate a significant sequence homology between the two [NiFe] hydrogenases. Despite the pronounced differences in stability, the two small subunits are identical in 46% of their amino acids and the two large subunits show 58% sequence identity. In order to understand the physiological roles of these hydrogenases, mutants lacking either or both of them have been generated in our laboratory by marker exchange mutagenesis. Much to our surprise, the hydrogenase-deleted mutants grew just as avidly as the wild type strain. The phenomenon was finally understood when two additional [NiFe] hydrogenases were discovered in the cytoplasm of the bacterium. According to our current understanding, there are four distinct [NiFe] hydrogenase molecular species in *T. roseopersicina*, representing all hydrogenase forms thus far described in various microorganisms, in a single cell (Kovács et al. 2002). This makes *T. roseopersicina* one of the best candidates for studies of [NiFe] hydrogenase structure-function relationships and assembly. The outstanding situation allows us to address specific questions concerning the assembly of each of these enzymes and the regulation of their biosynthesis. Answers to these questions will be of direct relevance in designing an optimal catalyst for biological hydrogen production and/or utilisation and to protein engineering of scientifically intriguing and biotechnologically important redox enzymes in general. From the fragmented information available, there is no clear answer as to why *T. roseopersicina* needs so many distinct hydrogenases. Our working hypothesis links this abundance of various [NiFe] hydrogenases to the fact that this bacterium should be able to perform various metabolic activities (photoautotrophic, photoheterotrophic, heterotrophic metabolism) in order to survive in its natural habitat (Imhoff 2001). Having numerous hydrogenases at hand increases the chances of survival for the bacterium and increases our chances to understand basic phenomena of hydrogenase catalysis.

4 METHANOTROPHIC HYDROGENASES

Methane-oxidizing bacteria (methanotrophs) have attracted considerable interest over the past twenty years because of their potential in producing bulk chemicals (e.g. propylene oxide) and single-cell protein and for use in biotransformation (Dalton et al. 1995). Methanotrophs are unique in that they only grow on methane, although some will also grow on methanol. Methanotrophs oxidize methane using the enzyme methane monooxygenase (MMO). The soluble enzyme complex (sMMO) is present in some but not all methanotrophs (Murrell & Dalton 1992). The sMMO is a remarkable enzyme in that it can also oxidize a large number of other substrates such as alkanes, alkenes and even aromatic compounds. The other form of MMO, found in all methanotrophs, is the membrane-bound or particulate form (pMMO) (Nguyen et al. 1998). It has narrower substrate specificity than sMMO.

Both MMO enzymes require reducing power for catalysis. The *in vivo* electron donor for the sMMO is NADH. Under physiological conditions, the reducing power is supplied by the further oxidation of the methanol (via formaldehyde and formate to CO_2) produced by the MMO. Since biodegradation processes using MMO are cooxidation processes, alternative ways of supplying reducing power are needed. A possible alternative is hydrogen.

Little is known about hydrogenases of methanotrophs. De Bont (1976) reported hydrogen-uptake activity in *Methylosinus trichosporium*, which was induced during N_2 fixation. The presence of an uptake hydrogenase was suggested since acetylene reduction by whole cells could be driven by hydrogen. Constitutive hydrogen-evolving activities (1 nmol min^{-1} (mg dry wt cell)$^{-1}$) from formate under anoxic conditions were reported for *Methylomicrobium album* BG8 and *Methylosinus trichosporium* OB3b (Kawamura et al. 1983). Chen & Yoch (1987) detected distinct constitutive and inducible hydrogen-uptake activities in *Methylosinus trichosporium* OB3b. Hydrogen-uptake activity in *Methylosinus trichosporium* OB3b was shown to be able to supply reducing power for both sMMO and pMMO activities (Shah et al. 1995). Hydrogen driven propylene oxidation by *Methylococcus capsulatus* (Bath) was demonstrated by Stanley & Dalton (1992), but the mechanism was not investigated in detail. There exist at least two [NiFe] hydrogenases in *M. capsulatus* (Bath) (Hanczár et al. 2002). The genes encoding a membrane-bound [NiFe] hydrogenase has been sequenced and characterized (Csáki et al. 2001).

4.1 *Hydrogen-driven MMO activities*

As only the soluble hydrogenase utilized NADH, *in vivo* assays could be applied to investigate this activity further. Hydrogen-driven MMO activities were measured to obtain information on the *in vivo* function of this hydrogenase. The apparent K_S for hydrogen was again 0.8 mM in both assays. Maximal rates of MMO activities were 140 nmol min^{-1} (mg cell protein)$^{-1}$ for the sMMO and 260 nmol min^{-1} (mg cell protein)$^{-1}$ for the pMMO. Positive control assays with 20 mM sodium formate as electron donor confirmed that maximum rates were not limited by NADH, but by the activity of the MMOs. Cells grown in the presence of 5% hydrogen or under nitrogen-fixing conditions (thus generating hydrogen *in situ* inside the cells) did not show any difference in V_{max} or K_S of hydrogen-driven MMO activities.

Raising the incubation temperature from 45 to 57°C did not bring about a pronounced increase of the hydrogen-driven pMMO activity. This preliminary observation indicated *in vivo* heat stability of the hydrogenase and pMMO activities. The temperature dependent difference in the solubility of hydrogen may also explain the small activity difference, particularly as similar results were obtained for the hydrogen-driven sMMO activity.

Most hydrogenases are sensitive to oxygen exposure. The interaction between O_2 and the functionally active hydrogenase could be studied only by indirect methods, such as the hydrogen-driven MMO activity assays, since direct hydrogenase assays (both hydrogen evolution and hydrogen uptake ones) require the complete absence of oxygen. Increasing the oxygen concentration to 10% (v/v) clearly had a positive effect. Further increase of the O_2 concentration to 15% still did not cause any drop in either of these sMMO or pMMO activities (Hanczár et al. 2002). Increased oxygen concentrations are likely to improve the rate of product formation through the methane monooxygenases rather than the hydrogenase(s). This suggests that the activity of the MMO itself is the rate-limiting factor in the combined (hydrogen-driven MMO) activity.

5 UTILISATION OF HYDROGEN METABOLISM IN BIOTECHNOLOGICAL APPLICATIONS

Hydrogen evolution by intact bacterial cells is frequently observed in nature. In microbial ecosystems the role of these microorganisms is creation and maintenance of anaerobic, reductive environment as well as supplying the universal reducing agent, molecular hydrogen. Gaseous hydrogen is usually not released from the natural ecosystems unless there is an excess of reductive power which needs to be disposed of in order to ensure the optimal metabolic and growth equilibrium in the population. H_2 generated *in vivo* by hydrogen forming bacteria is utilized by hydrogen consuming members of the microbiological community. Hydrogen is transferred to the recipient microorganism very effectively by interspecies hydrogen transfer. The molecular details of this process are not fully understood, but its significance in safeguarding the optimum performance of the entire ecosystem and the delicate regulatory mechanisms should be appreciated. In the mixed population bacterial systems presented here the advantages of interspecies hydrogen transfer are exploited (Kovács & Polyák 1991).

5.1 *Biogas*

Decomposition of wastes anaerobically to form biogas is one of the earliest applications of biotechnology. It is well known that three distinct microbe populations take part in the anaerobic digestion process. These microbe populations are the polymer degrading, so-called hydrolyzing bacteria, the acetogens and the methanogens. The first group, the polymer degraders attack the macromolecules using extracellular enzymes and producing intermediers. Because of its abundance in Nature, cellulose is the main substrate for hydrolyzing bacteria. The acetogens then use these sugars and oligosaccharides and produce organic acids, like acetate, succinate, formate, propionate, and carbondioxide. The third group is the methanogens. These microorganisms belong to the Archaebacteria and thus possess unique molecular and cellular properties. They produce methane using acetate, hydrogen and carbondioxide.

Our recent research results show that H_2 has an important role in the anaerobic fermentation.

Some other experiments at various scales and using distinct organic waste sources have been carried out. Volumes between 0.1 m^3 and 10,000 m^3 were used. In every experiment the methane production increased. Best results were obtained with pig manure. In this case the biogas production increased to 200%. The experiments thus proved that it was possible to increase the gas production using intensified microbiological biomass decomposition.

REFERENCES

Cammack, R., M. Frey & R. Robson, (eds) 2001. *Hydrogen as a fuel: Learning from Nature.* London: Taylor & Francis.

Chen, Y.P. & Yoch D.C. 1987. Regulation of two nickel-requiring (inducible and constitutive) hydrogenases and their coupling to nitrogenase in *Methylosinus trichosporium* OB3b. *J. Bacteriol.* 169: 4778–4783.

Colbeau, A., Kovács, K.L., Chabert, J. & Vignais, P.M. 1994. Cloning and sequencing of the structural (*hupSLC*) and accessory (*hupDHI*) genes for hydrogenase biosynthesis in *Thiocapsa roseopersicina. Gene* 140: 25–31.

Csáki, R., Bodrossy, L., Hanczár, T., Murrell, J.C. & Kovács, K.L. 2001. Molecular characterization of a membrane bound hydrogenase in the methanotroph *Methylococcus capsulatus* (Bath). *FEMS Microbiol. Lett.* 205: 203–207.

Dalton, H, Golding, B.T., Waters, B.W., Higgins, R. & Taylor, J.A. 1995. Oxidations of cyclopropane, methylcyclopropane, and arenes with the mono-oxygenase system from *Methylococcus capsulatus. J. Chem. Soc. Chem. Commun.* 1981: 482–483.

De Bont, J. & A.M. 1976. Hydrogenase activity in nitrogen-fixing methane-oxidizing bacteria. *Antonie Van Leeuwenhoek* 42: 255–259.

Fodor, B., Rákhely, G., Kovács, Á.T. & Kovács K.L. 2001. Transposon mutagenesis in purple sulfur photosynthetic bacteria: Identification of *hypF*, encoding a protein capable to process [NiFe] hydrogenases in α, β and γ subdivision of proteobacteria. *Appl. Environm. Microbiol.* 67: 2476–2483.

Fodor, B.D., Kovács, Á.T., Csáki, R., Hunyadi-Gulyás, É., Klement, É., Maróti, G., Mészáros, L.S., F.-Medzihradszky, K., Rákhely, G. & Kovács, K.L. 2003. Modular broad-host-range expression vectors for single protein and protein complex purification. *Appl. Environm. Microbiol.* (in press).

Hanczár, T., Bodrossy, L., Csáki, R., Murrell, J.C. & Kovács K.L. 2002. Hydrogen driven methane oxidation in *Methylococcus capsulatus* (Bath). *Arch. Microbiol.* 177: 167–172.

Imhoff, J.F. 2001. True marine and halophilic anoxygenic phototrophic bacteria. *Arch. Microbiol.* 176: 243–254.

Kawamura, S., O'Neil, J.G. & Wilkinson J.F. 1983. Hydrogen production by methylotrophs under anaerobic conditions. *J. Ferment. Technol.* 61: 151–156.

Kovács, K.L. & Bagyinka, Cs. 1990. Structural properties and functional states of hydrogenase from *Thiocapsa roseopersicina. FEMS Microbiol. Rev.* 87: 407–412.

Kovács, K.L. & Polyák, B. 1991. Hydrogenase reactions and utilization of hydrogen in biogas production and microbiological denitrification systems. *Proceedings of the 4th IGT Symposium*, Chapter 5, pp 1–16, Colorado Springs.

Kovács, K.L., Fodor, B.D., Kovács, Á.T., Csanádi, Gy., Maróti, G., Balogh, J., Arvani, S. & Rákhely, G. 2002. Hydrogenases, accessory genes and the regulation of [NiFe] hydrogenase biosynthesis in *Thiocapsa roseopersicina. Int. J. Hydrogen Energy* 27: 1463–1469.

Maróti, G., Fodor, B.D., Rákhely, G., Kovács, Á.T., Arvani, S. & Kovács, K.L. 2003. Accessory proteins functioning selectively and pleiotropically in the biosynthesis of [NiFe] hydrogenases in *Thiocapsa roseopersicina. Eur. J. Biochem.* 270 (10): 2218–2227.

Murrell, J.C. & Dalton, H. 1992. *The Methane and Methanol Utilizers*. Plenum Press, N.Y.

Nguyen, H.H.T., Elliott, S.J., Yip, J.H.K. & Chan, S.I. 1998. The particulate methane monooxygenase from *Methylococcus capsulatus* (Bath) is a novel copper-containing three-subunit enzyme – Isolation and characterization. *J. Biol. Chem.* 273: 7957–7966.

Paschos, A., Glass, R.S. & Böck, A. 2001. Carbamoyl-phosphate requirement for synthesis of the active center of [NiFe]-hydrogenases. *FEBS Lett.* 488: 9–12.

Rákhely, G., Colbeau, A., Garin, J., Vignais, P.M. & Kovács, K.L. 1998. Unusual gene organization of HydSL, the stable [NiFe] hydrogenase in the photosynthetic bacterium *Thiocapsa roseopersicina. J. Bacteriol.* 180: 1460–1465.

Rákhely, G., Kovács, Á.T., Maróti, G., Fodor, B.D., Csanádi, Gy., Latinovics, D. & Kovács, K.L. 2003. A cyanobacterial type, heteropentameric NAD^+ reducing [NiFe] hydrogenase in the purple sulfur photosynthetic bacterium, *Thiocapsa roseopersicina. Appl. Environm. Microbiol.*, (in press)

Shah, N.N., Hanna, M.L., Jackson, K.J. & Taylor, R.T. 1995. Batch cultivation of *Methylosinus trichosporium* OB3b. 4. Production of hydrogen-driven soluble or particulate methane monooxygenase activity. *Biotechnol. Bioeng.* 45: 229–238.

Stanley, S.H. & Dalton, H. 1992. The biotransformation of propylene to propylene oxide by *Methylococcus capsulatus* (Bath): 1. Optimization of rates. *Biocatalysis* 6: 163–175.

Szilágyi, A., Kovács, K.L., Rákhely, G. & Závodszky, P. 2002. Homology modelling reveals the structural background of the striking difference in thermal stability between two related [NiFe] hydrogenases. *J. Mol. Model.* 8: 58–64.

Vignais, P.M., Billoud, B. & Mayer, J. 2001. Classification and phylogeny of hydrogenases. *FEMS Microbiol. Rev.* 25: 455–501.

Vignais, P.M., Toussaint, B. & Colbeau, A. 1995. Regulation of hydrogenase gene expression. In R.E. Blankenship, M.T. Madigan & C.E. Bauer (eds), *Anoxygenic Photosynthetic Bacteria*, Chapter 55. pp 1175–1190. Kluwer.

Composition and function of biofilms and consortia

European Symposium on Environmental Biotechnology, ESEB 2004 - Verstraete (ed)
© 2004 Taylor & Francis Group, London, ISBN 90 5809 653 X

Biodiversity of phototrophic biofilms

G. Roeselers, M.C.M. van Loosdrecht & G. Muyzer
Department of Biotechnology, Delft University of Technology, Delft, The Netherlands

ABSTRACT: Phototrophic biofilms are microbial communities driven by light. Surface attached photosynthetic microbes derive energy from captured photons, reduce carbon dioxide, and provide organic compounds and oxygen that fuel the growth of heterotrophic microorganisms in the biofilms. In the present study the structure and species composition of phototrophic biofilms are determined. Biofilms recovered from a sedimentation tank of the wastewater treatment plant at Fiumicino Airport Rome (Italy), and cultivated in a special designed phototrophic biofilm incubator were analyzed. The gene sequences of the small subunit ribosomal RNA were used as a molecular marker for the identification of microorganisms. Denaturing gradient gel electrophoresis (DGGE) of PCR-amplified 16S RNA gene fragments provided an overview of the diversity and showed drastic populational changes.

1 INTRODUCTION

Much of what we know about microorganisms has been learned under laboratory conditions that are mostly not representative of how microorganisms are found in nature. Microorganisms predominantly exist in diverse communities where a variety of interactions exist. Mutualism, commensalism, antagonism, and saprophytism are but a few of the more common microbial interactions.

Considering the number of known microbes and the estimated number of those yet to be described, the complexity of microbial communities and interactions in nature can truly be appreciated.

Inclusion in a community brings about profound changes in microorganisms, which may therefore exhibit important differences from their planktonic counterparts, including physiological properties, susceptibility to antimicrobial agents, interaction with host tissues and immunological response.

However, the lack of methods for studying these communities in situ has hampered detailed analysis. Recently, several molecular and microscopy based techniques have been developed that enable the detailed study of microbial communities in situ. It has become clear that in nature a dominant part of microorganisms persists in sessile communities. These surface attached populations of microorganisms surrounded with a glycocalyx matrix are usually referred to as biofilms. Biofilm formation is implicated in a significant amount of pathogenic infections. Bacterial biofilms also cause fouling, equipment failure and product contamination. Antibiotic doses, which kill planktonic growing cells, need to be highly increased to kill cells in biofilms.

The structure, growth dynamics and physiology of heterotrophic biofilms have been extensively studied. But until recently phototrophic biofilms have received little attention. Phototrophic biofilms occur on contact surfaces in a range of terrestrial and aquatic environments. Phototrophic biofilms can best be described as surface attached microbial communities mainly driven by light as the energy source with a photosynthesizing component clearly present. Diatoms and cyanobacteria generate energy and reduce carbon dioxide, providing organic substrate and oxygen. The photosynthetic activity fuels processes and conversions in the total biofilm community, including the heterotrophic fraction (Guerrero et al., 2002). The microorganisms produce an extracellular matrix that holds the biofilm together (Sutherland, 2001) (Fig.1).

The potential use of phototrophic biofilms has not been exploited in detail. In contrast to heterotrophic biofilms, there are no numeric predictive models available for photo-autotrophic driven biofilms. Nevertheless, phototrophic biofilms have a variety of important applications. They cause biofouling of marine constructions and ship hulls, and they deteriorate building materials and monuments. The EU-financed project PHOBIA (PHOtrophic BIofilms and their potential Applications towards the development of a unifying concept) is a study of phototrophic biofilm growth that

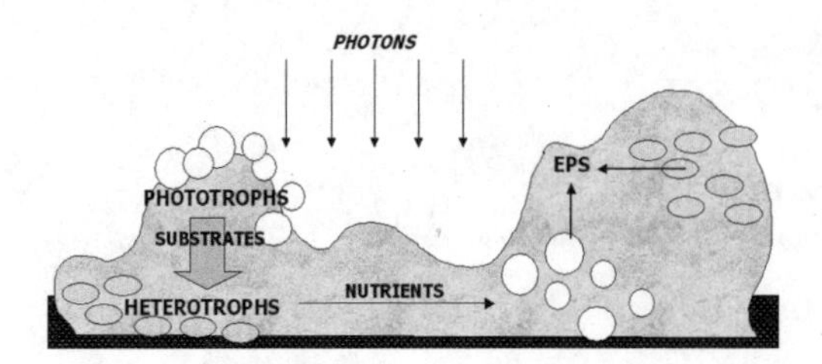

Figure 1. Photosynthetic microbes derive energy from captured photons, reduce carbon dioxide, and provide the organic substrates and oxygen that sustain the growth of heterotrophic microbes in the biofilms. A matrix of polymeric substances (EPS) provides the attachment of the microorganisms to the substratum, and holds the biofilm together.

integrates structural development and physiology. The project involves a total of six partners from different institutes and universities in Europe. Each partner is responsible for specific experiments in which the physiochemical microenvironment, the structure, the dynamics, the biodiversity and the relevant physiological processes of biofilms will be studied. With the data obtained, a unifying concept of phototrophic biofilm growth will be made via a model describing development and activity. Applied aspects of the biofilms will be explored, e.g., for testing new antifoulants, and for testing the capacity of phototrophic biofilms to remove nutrients and metals from the aquatic environment

One objective in this study is to assess the microbial diversity of aquatic phototrophic biofilms, and to monitor possible population changes. This report presents preliminary results obtained with biofilms from a sedimentation tank of the wastewater treatment plant (WWTP) at Fiumicino Airport (Rome). Bacterial diversity analysis was performed using denaturing gradient gel electrophoresis (DGGE) (Fischer and Lerman 1979) of PCR-amplified rRNA gene fragments (Muyzer, 1999) obtained from DNA extracted from WWTP biofilm samples and biofilms grown in a small incubator.

2 MATERIALS AND METHODS

2.1 *Sampling*

Samples of approximately 1.5 g (wet weight) were scraped off with a razorblade knife from the sedimentation tank wall and from the object slides in the incubator and transferred to 2 ml Eppendorf tubes. The biofilm samples were transported on dry ice to the laboratory and stored at −80°C until further use.

2.2 *Incubator*

A small incubator (designed by M. Staal, Centre for Estuarine and Marine Ecology, Yerseke, the Netherlands) that contained a single microscope slide was inoculated with a sample from the WWTP of Fiumicino Airport. Mineral medium with a pH to approximately 7.5 was pumped through the incubator with a velocity of 0,133 ml/s ($V = 1{,}04 \times 10^{-3}$ m/s). The medium was prepared according to Rippka et al. (1979). Vitamin B12 was added for the growth of particular cyanobacteria that require this.

The incubator was inoculated by adding 200 μl suspended biomass from the Fiumicino Airport sample into the medium container.

A light source was provided by placing a cathode lamp emitting a light spectrum close to daylight above the incubator. Object slides were harvested at 5 and 10 days after inoculation.

2.3 *DNA extraction*

Genomic DNA was extracted from the biofilm samples and the gradient column samples using the Ultra Clean Soil DNA Isolation Kit™ (Mobio Laboratories) according to manufacturer's protocol. The extracted DNA was run on 1% (w/v) agarose gels for 40 minutes at a constant voltage of 100 V. Gels were incubated in a 0.5 μg/μl ethidium bromide solution and analyzed under a UV illuminating cathode lamp. Genome size was checked relative to a λ/*Eco*RI + *Hin*dIII digest (MBI fermentas). DNA was stored at −20°C.

2.4 *PCR amplification of rDNA fragments*

To amplify the bacterial 16S rRNA encoding gene fragments 1 μl of the DNA dilutions was used as template DNA in 50 or 100 μl PCR reactions using forward primer 341F-GC and reverse primer 907R and PCR conditions as described by Schäfer and Muyzer (2001). Primer pair CYA359F+GC and CYA781R (Nübel et al., 1997) was used in separate PCR reactions for the specific amplification of the 16S rRNA gene fragments of oxygenic phototrophs. Reactions were cycled at the following parameters: 94°C for 5 min, followed by 34 cycles of 94°C for 1 min, 55°C for 1 min, and 72°C for 1 min, with a final extension step at 72°C for 10 min (Nübel et al., 1997).

2.5 *Denaturing gradient gel electrophoresis*

DGGE was applied according to Schäfer and Muyzer (2001). One mm thick 6% acryl amide gels with a urea-formamide (UF) gradient of 20–80% were used on top of the gradient gel. An acrylamide gel without UF was cast on top of the gradient gel to obtain good slots. From each PCR reaction 20 μl product, containing approximately 1 μg DNA was mixed with 6 μl of 10x gel loading solution and loaded onto the gel (Ferrari and Hollibaugh, 1999). Gels were run in TAE buffer (Tris, Acetic acid, and EDTA, pH 8) for 16h at

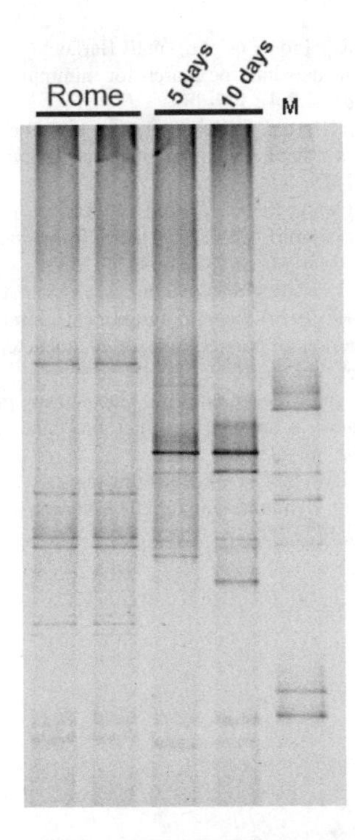

Figure 2. DGGE analysis of bacterial 16S rRNA gene fragments. The first two lanes (Rome) show a duplicate of the samples derived from the WWTP at Fiumico airport. The third and the fourth lane show biofilm samples that were grown in an incubator for respectively 5 and 10 days.

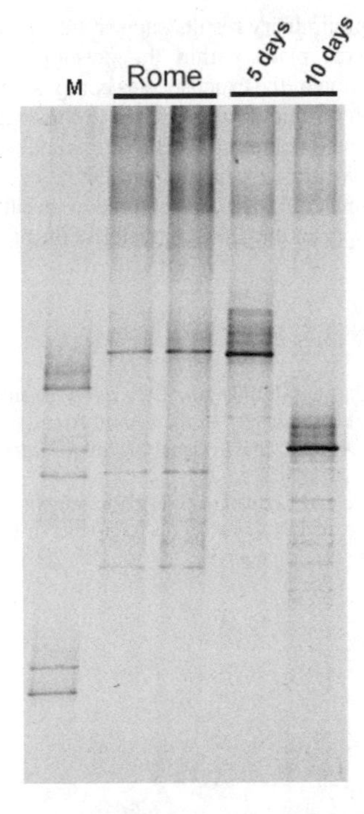

Figure 3. DGGE analysis of 16S rRNA gene fragments of oxygenic phototrophs. The first two lanes (Rome) show a duplicate of the samples derived from the WWTP at Fiumico airport. The third and the fourth lane show biofilm samples that were grown in an incubator for respectively 5 and 10 days.

100V and at a constant temperature of 60°C. Gels were incubated in an ethidium bromide solution and analyzed and photographed under a UV illuminating cathode lamp using a Fluor-S-Multimage (Bio-Rad).

3 RESULTS

An inventory was made of the microbial diversity in freshwater phototrophic biofilms. General bacterial 16S rDNA PCR products and 16S rDNA PCR products specific for oxygenic phototrophs were separated by DGGE. The resulting banding patterns were very different for the samples from the WWTP at Fiumicino Airport, the 5 days old biofilms from the incubator, and the 10 days old biofilms from the incubator. The PCR-DGGE of the general bacterial 16S rDNA fragments showed different banding patterns for the WWTP samples and the biofilms grown in the incubator. The 5 days old biofilm showed the same dominant band as the 10 days old biofilm, but the latest showed some additional bands that were absent in the 5 days old biofilm (Fig. 2)

The PCR-DGGE with the oxygenic phototrophic 16S rDNA fragments showed the same dominant band in the WWTP samples and the 5 days old biofilm. However, the 10 days old biofilm showed a completely different banding pattern, suggesting a strong population shift between 5 and 10 days after inoculation (Fig. 3).

4 CONCLUSIONS AND DISCUSSION

The DGGE analysis of the bacterial 16S rRNA and oxygenic phototrophs 16S rRNA gene fragments suggests a difference in community composition between the phototrophic biofilms that grow in the WWTP at Fiumicino airport and the phototrophic biofilms grown in the incubator. Although, initially the same species were present in the incubator the different conditions have led to a different community composition. As the developing biofilm increases in thickness the light distribution might change. The pioneer organisms that made the first attachment to the glass surface might become deprived from light.

These preliminary results suggest that a strong succession takes place within the developing biofilm. Determination of the exact species composition and the distribution of functional microbial groups will provide insight in the dynamics of niche differentiation in phototrophic biofilms. Extraction and DNA sequencing of the specific DGGE bands is required to reveal the identity of the species and fully characterise the biodiversity.

REFERENCES

Ferrari, V.C., and Hollibaugh, J.T. (1999). Distribution of microbial assemblages in the central Arctic ocean basin studied by PCR/DGGE: analysis of a larger data set. *Hydrobiologia* 401: 55–68

Fischer, S.G., and Lerman, L.S. (1979). Length-independent separation of DNA restriction fragments in two dimensional gel electrophoresis. *Cell* 16: 191–200

Guerrero, R., Piqueras, M., and Berlanga, M. (2002). Microbial mats and the search for minimal ecosystems. *Int Microbiol* 5(4): 177-88

Muyzer, G. (1999). DGGE/TGGE a method for identifying genes from natural ecosystems. *Curr Opin Microbiol* 2(3): 317-22

Nübel, U., Garcia-Pichel, F., and Muyzer, G. (1997) PCR primers to amplify 16S rRNA genes from cyanobacteria. *Appl. Environ. Microbiol*. 63: 3327–3332

Rippka, R., Deruelles, J., Waterbury, J., Herdman, M., and Stanier, R. (1979) Generic assignments, strain histories and properties of pure cultures of cyanobacteria. *J. Gen. Microbiol.* 111: 1–61

Schäfer, H., and Muyzer, G. (2001) Denaturing gradiënt gel electrophoresis in marine microbial ecology. *Meth. Microbiol.* 30: 425–468

Sutherland, I.W. (2001). The biofilm matrix – an immobilized but dynamic microbial environment. *Trends Microbiol* 9(5): 222–7

European Symposium on Environmental Biotechnology, ESEB 2004 - Verstraete (ed)
© 2004 Taylor & Francis Group, London, ISBN 90 5809 653 X

Sulfate reduction at low pH by a defined mixed culture of acidophilic bacteria

S. Kimura & D.B. Johnson
School of Biological Sciences, University of Wales, Bangor, U.K.

ABSTRACT: Sulfidogenesis in a bench-scale bioreactor containing a mixed culture of acid-tolerant sulfate reducing bacteria (aSRB) and acidophilic heterotrophic bacteria was investigated. A 2 l bioreactor was inoculated with *Desulfosporosinus* strain M1 and an *Acidocella*-like isolate (PFBC). PFBC was added to the reactor in order to remove acetic acid that was produced by the aSRB by incomplete oxidation of glycerol, and which is highly toxic at relatively low concentrations (often $<500\,\mu M$) to most acidophiles. The bioreactor pH was maintained at 4.0 and glycerol was supplied as energy- and carbon-source. Zinc sulfate was also added to the bioreactor to remove sulfide generated by the aSRB as insoluble ZnS. Changes in concentrations of sulfate, zinc, glycerol and acetic acid were monitored, together with total cell counts. The relative abundance of *Desulfosporosinus* M1 and *Acidocella* PFBC was estimated using fluorescent *in situ* hybridization. The oxidation of glycerol was shown to be coupled to the reduction of sulfate. At the pH that the culture was maintained (pH 4.0), zinc was selectively precipitated as zinc sulfide while iron remained in solution. With this mixed culture, the concentrations of acetic acid were maintained at low levels (<1 mM), presumably due to it being metabolized by the acetotrophic *Acidocella* PFBC. Previous attempts to demonstrate sulfidogenesis at low pH by pure cultures of SRB have proved unsuccessful. These results presented in this paper suggest that syntrophic mixed cultures of aSRB and acetotrophic microorganisms are effective at low pH, and may be used for selective removal of metals from acidic waste waters.

1 INTRODUCTION

Sulfate is often found in high concentrations in mineral leaching environments. Therefore dissimilatory sulfate reduction might be predicted to be commonplace in acidic anaerobic zones, and there have been several reports documenting sulfidogenesis in these environments (Tuttle et al. 1969, Gyure et al. 1990, Fortin et al. 2000). Although attempts have been made to isolate and cultivate acid-tolerant or acidophilic sulfate reducing bacteria from mine waters, this has mostly been unsuccessful. It appears that most mine water isolates are neutrophilic and are not active below pH 5.0 (Tuttle et al. 1969, Küsel et al. 2001). One of the reasons for the lack of success is the use of inappropriate media. For example, some commonly used substrates, such as lactate, are toxic to microbes at low pH. Solid media specifically designed to isolate SRB at low pH (3.5) were developed by Sen and Johnson (1999) and were used to isolate acid tolerant SRB (aSRB) successfully from acid mine drainage-impacted sites.

This paper describes sulfidogenesis in a bench-scale bioreactor containing a mixed culture of aSRB and acidophilic heterotrophic bacteria.

2 MATERIALS AND METHODS

2.1 *Bioreactor culture*

The growth medium for 2 l bioreactor contained 5 mM glycerol, 5 mM $ZnSO_4$, 5 mM K_2SO_4, 1 mM $FeSO_4$, vitamin mixture (Widdel & Pfenning 1981), trace elements and basal salts (Johnson 1995). The pH was maintained between 3.8 and 4.2 by adding 0.5 M NaOH or 0.1 M H_2SO_4. The amount of sulfuric acid used in pH maintenance was monitored carefully to allow accurate calculations to be made of net sulfate reduction. Oxygen-free nitrogen was continuously bubbled through the culture to maintain low dissolved oxygen concentrations and the temperature was maintained at 30°C. The bioreactor was inoculated with an *Acidocella*-like isolate, PFBC, and *esulfosporosinus* isolate M1. *Acidocella* PFBC was first isolated from a supposedly pure aSRB culture, and was found subsequently to be able to grow anaerobically with aSRB although other known *Acidocella*- like isolates have not been reported to grow in the absence of oxygen (Hallberg et al. 1999). *Desulfosporosinus* M1 was isolated from sediment collected from the White River, Montserrat, West Indies. Previous work (Sen 2001) had shown that this SRB was an "incomplete oxidizer" of glycerol (and ethanol), producing stoichiometric amounts of acetic acid, as equation [1]:

$$4C_3H_8O_3 + 3SO_4^{2-} + 6H^+ \rightarrow 4CH_3COOH + 4CO_2 + 3H_2S + 8H_2O \quad [1]$$

Incomplete oxidation of organic substrates is a common feature in *Desulfosporosinus* and *Desulfotomaculum* spp. (Castro et al. 2000).

2.2 *Analytical techniques*

A Thoma counting chamber (Weber Scientific International Ltd, England) was used for enumeration of total bacteria in a culture sample. Samples were filtered through 0.2 μm cellulose nitrate membranes (Whatman, UK) and concentrations of glycerol and acetic acid were analyzed using enzymatic assays (R-Biopharm GmbH, Germany). Sulfide production was estimated by measuring changes in soluble zinc in the culture:

$$Zn^{2+} + H_2S \rightarrow ZnS + 2H^+ \quad [2]$$

As shown in equation [2], the production of zinc sulfide generated protons, thereby contributing to net pH control in the reactor (as sulfidogenesis generates net alkalinity, equation [1]). Zinc concentrations were measured using an atomic absorption spectrophotometer (Pye Unicam, SP9-10) and concentrations of sulfate were determined turbidometrically as $BaSO_4$ (Hydrocheck, Cambridge, UK).

2.3 *Fluorescent in situ hybridization (FISH)*

To fix samples from the bioreactor culture, 1 ml sample solution was centrifuged and resuspended in 3% (v/v) paraformaldehyde in PBS. The sample was incubated at 4°C for 2 hours for fixation. The fixed samples were analyzed using a modified method of that described by Bond & Banfield (2001). Probe hybridization was carried out by adding both 25 ng of fluorescein-labeled eubacterial probe (EUB338Fl) (Amann et al. 1990) and 25 ng of Cy3-labeled ALF1B (Manz et al. 1992) which target *Acidocella* PFBC, or aSRB probe (Robert et al. unpubl.) which targeted *Desulfosporosinus* M1. Mounting medium

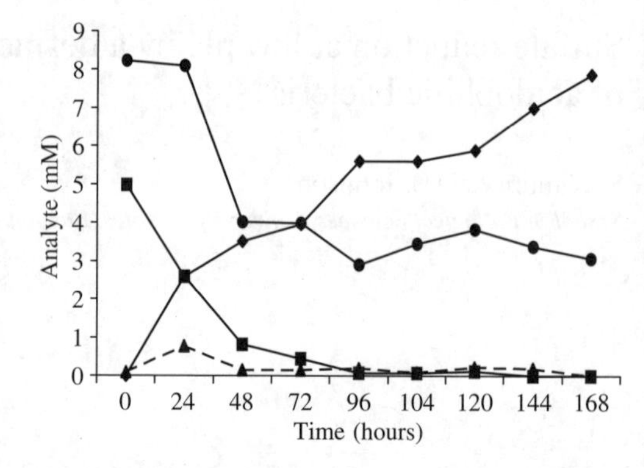

Figure 1. Sulfate reduction and removal of zinc. (▲), acetic acid; (■), glycerol; (◆), sulfate reduced; (●), soluble zinc.

(70% glycerol in 100 mM sodium tetraborate, pH 9.2, containing 3 mg/ml N-propyll gallate) was applied to reduce fading of the probe signal during enumeration. Various concentrations of formamide (0–50%, v/v) were tested to optimize specificity and maximize signal response for each probe, using pure cultures of target and related microorganisms. The optimum formamide concentrations of ALF1B and aSRB probes were 20 and 30% respectively. To enumerate *Acidocella* PFBC and *Desulfosporosinus* M1, Cy3-labeled cells were counted relative to those stained with the EUB338Fl probe using Nikon ECLIPSE E600 microscope.

3 RESULTS

3.1 *Sulfate reduction and removal of soluble zinc*

Figure 1 shows changes in concentrations of glycerol, acetic acid, sulfate and zinc in the bioreactor over 168 hours. Oxidation of glycerol was coupled with sulfate reduction, as measured either by changes in sulfate concentrations or those of soluble zinc, over the first 120 hours of culture incubation. By this time, 5.0 mM glycerol had been consumed and 5.2 mM zinc had been removed from the culture, though most of this had occurred during the first 48 hours of incubation. Beyond this, there was a second phase of sulfate reduction (sulfate concentration decreased by 2 mM) which was not coupled to glycerol oxidation (which was depleted by that time) or removal of soluble zinc. At no time was acetic acid observed to accumulate beyond a maximum concentration (0.8 mM) that was recorded after 24 hours incubation.

3.2 *Microbial population dynamics*

Changes in the microbial population in the bioreactor are shown in Figure 2. Total cell numbers increased

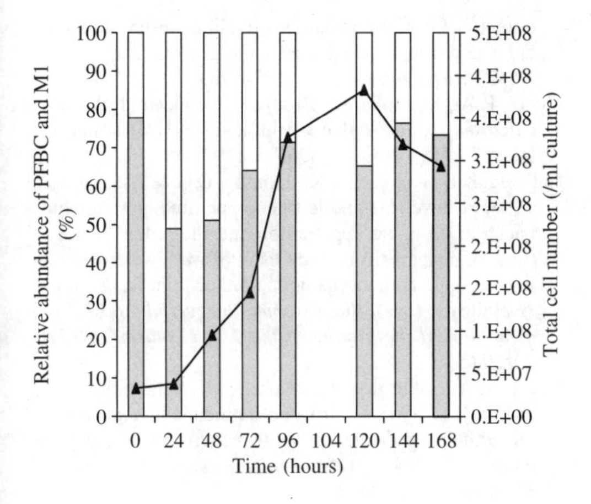

Figure 2. Microbial population dynamics in the bioreactor mixed culture. (▲), total number of cells (/ml culture); unshaded bar, *Desulfosporosinus* M1; shaded bar, *Acidocella* PFBC.

ten-fold during incubation. A gradual increase was observed at the beginning of incubation and the rate accelerated before reaching the highest number after 120 hours incubation. The data from population composition analysis by FISH are presented as the relative abundance of *Acidocella* PFBC or *Desulfosporosinus* M1, to total number of cells stained by the eubacterial probe. After the first 24 hours of incubation, the relative percentage of *Desulfosporosinus* M1 increased significantly and was maintained for a further 24 hours. After that time, relative numbers of *Acidocella* PFBC started to increase, and total cell numbers showed a simultaneous large increase. The *Acidocella* PFBC population appeared to decrease at 120 hours but recovered after a further 24 hours, and comprised more than 70% of bioreactor population (on a numerical basis) at the end of incubation. Population analysis showed a gradual change from *Desulfosporosinus* M1 to *Acidocella* PFBC dominance in the bioreactor over time.

4 DISCUSSION

A problem associated with the use of SRB in the treatment of acidic mine waters and metal recovery processes is that SRB used in these processes are neutrophiles and cannot tolerate lower pH. In these systems, exemplified by the Budelco zinc refinery in the Netherlands (Barnes et al. 1994), SRB are grown in a separate reactor to avoid direct contact with the acidic groundwater. Use of acid-tolerant SRB would simplify such systems as, without the necessity of shielding the bacteria from the water to be remediated, it would be possible to combine the sulfide producing reactor and a metal precipitating reactor into a single unit.

The results obtained from the bioreactor culture showed that *Desulfosporosinus* M1 could grow and produce sulfide at pH 4.0 in the presence of the heterotroph, *Acidocella* PFBC. Attempts at growing *Desulfosporosinus* M1 in pure culture in the bioreactor set at pH 4.0 to 6.0 were not successful, though previously this (and related aSRB) have been cultured in acidic liquid media, but without pH control, in anaerobic jars (data not shown). This suggests that successful growth of aSRB such as *Desulfosporosinus* M1 in consistently low pH media (such as the pH-statted bioreactor in the present study) may only be possible when they are grown in mixed culture with other acidophiles, such as *Acidocella* PFBC.

As a result of incomplete oxidation of glycerol by *Desulfosporosinus* M1, acetate is produced and excreted into the culture liquor. At pH 4.0, most of this acetate would become protonated, as the pK_a of acetic acid is 4.75. In its undissociated form, acetic acid (as many other small molecular weight acids) is toxic to acidophilic bacteria, even at low concentrations (Norris & Ingledew 1992). *Acidocella* PFBC is related (99% sequence similarity of their 16S rRNA gene) to *Acidocella* WJB3, which has been found not only to be more tolerant of acetic acid than most other acidophiles, but also to metabolize it in acidic (pH 2.5 to ~4.0) media (Gemmell & Knowles 2000). *Acidocella* PFBC has also been found to degrade acetic acid in pure culture, but only under aerobic conditions. However, the fact that this putative "obligate aerobe" was growing simultaneously in oxygen-free medium with the obligately anaerobic *Desulfosporosinus* M1 does indicate that it can metabolize in anaerobic conditions. Other work has confirmed that *Desulfosporosinus* M1 cannot degrade acetate, and that *Acidocella* PFBC cannot metabolize glycerol (S. Kimura, unpubl.).

The fact that acetic acid was detected, but never accumulated to >0.8 mM throughout the duration of the experiment, indicates that it was being removed efficiently by *Acidocella* PFBC. Oxidation of 5 mM glycerol exclusively by *Desulfosporosinus* M1 would have been predicted to produce 5 mM acetic acid (equation [1]). Anaerobic consortia of SRB and acetotrophic methanogenic archaea have also been described, in which the acetate produced by the SRB is utilized by the archaea (Steyer et al. 2000) However, there were no methanogens present in the acidophilic mixed culture used in the present work and, although moderately acidophilic methanogens have been isolated (Horn et al. 2003), no extreme acidophiles have been described.

The results from this experiment also showed the possibility of selective removal and recovery of metals from acidic liquors, using sulfidogenic biotechnology. Waters draining metal mines are frequently acidic, and contain elevated concentrations of heavy metals and metalloids (Johnson 2003). The culture liquor in the experiment described contained both ferrous iron and zinc, both of which form highly insoluble sulfides. However, the solubility product of iron sulfide ($4.0 \cdot 10^{-19}$ M^2) is significantly greater than that of zinc sulfide ($1.0 \cdot 10^{-23}$ M^2). At pH 4.0, the concentration of sulfide ions (S^{2-}) is great enough to allow formation of ZnS, but insufficient for FeS, which is why ferrous iron remained in solution while zinc was removed. This general principle could be applied for selective removal of other chalcophilic metals, such as copper and cadmium, from waste streams containing mixtures of metals.

5 CONCLUSION

This study demonstrated sulfidogenesis in a bioreactor maintained at pH 4.0, by a mixed culture of acid-tolerant SRB and an acetotrophic, non-sulfidogenic heterotroph. At this pH, sulfidogenesis promoted the segregation of soluble iron and zinc by selective precipitation of zinc sulfide. Sulfate-reducing biotechnology, conducted in acidic liquors, could have important applications in the bioremediation of metal-contaminated wastewaters.

REFERENCES

Amann, R.I., Krumholz, L. & Stahl, D.A. 1990. Fluorescent-oligonucleotide probing of whole cells for determinative, phylogenetic, and environmental studies in microbiology. *J. Bacteriol.* 172:762–770.

Barnes, L.J., Scheeren, P.J.M. & Buisman, C.J.N. 1994. Microbial Removal of Heavy Metals and Sulphate from Contaminated Groundwaters. In J.L. Means & R.E. Hinchee (eds), *Emerging technology for Bioremediation of Metals*. Boca Raton, Fl: Lewis Publishers.

Bond, P.L. & Banfield, J.F. 2001. Design and performance of rRNA targeted oligonucleotide probes for *in situ* detection and phylogenetic identification of microorganisms inhabiting acid mine drainage environments. *Microbial Ecol.* 41:149–161.

Castro, H.F.,Williams, N.H. & Ogram, A. 2000, Phylogeny of sulfate-reducing bacteria. *FEMS Microbiology Ecology* 31:1–9.

Fortin, D., Roy, M., Rioux, J.P. & Thibault, P.J. 2000. Occurrence of sulfate-reducing bacteria under a wide range of physico-chemical conditions in Au and Cu-Zn mine tailings. *FEMS Microbiol. Ecol.* 33:197–208.

Gemmell, R.T. & Knowles, C.J. 2000. Utilisation of aliphatic compounds by acidophilic heterotrophic bacteria. The potential for bioremediation of acidic wastewaters contaminated with toxic organic compounds and heavy metals. *FEMS Microbiol. Lett.* 192:185–190.

Gyure, R.A., Konopka, A., Brooks, A. & Doemel, W. 1990. Microbial sulfate reduction in acidic (pH 3) strip-mine lakes. *FEMS Microbiol. Ecol.* 73:193–202.

Hallberg, K.B., Kolmert, Å.K., Johnson, D.B. & Williams P.A. 1999. A novel metabolic phenotype among acidophilic bacteria: aromatic degradation and the potential use of these organisms for the treatment of wastewater containing organic and inorganic pollutants. In R. Amils & A. Ballester (eds), *Biohydrometallurgy and the Environment Toward the Mining of the 21st Century*: 719–728. Amsterdam: Elsevier.

Horn, M.A., Matthies, C., Kusel, K., Schramm, A. & Drake, H.L. 2003. Hydrogenotrophic methanogenesis by moderately acid-tolerant methanogens of a methane-emitting acidic peat. *Applied and Environmental Microbiology* 69:74–83.

Johnson, D.B. 1995. Selective solid media for isolating and enumerating acidophilic bacteria. *Journal of microbiological methods*. 23:205–218.

Johnson, D.B. 2003. Chemical and microbiological characteristics of mineral spoils and drainage waters at abandoned coal and metal mines. *Water, Air and Soil Pollution Focus* 3:47–66.

Küsel, K.A., Roth, U., Trinkwalter, T. & Peiffer, S. 2001. Effect of pH on the anaerobic microbial cycling of sulfur in mining – impacted freshwater lake sediments. *Environ. Exp. Bot.* 46:213–223.

Manz, W., Amann, R., Ludwig, W., Wagner, M. & Schleifer, K.H. 1992. Phylogenetic oligodeoxynucleotide probes for the major subclasses of Proteobacteria: problems and solutions. *Syst. Appl. Microbiol.* 15:593–600.

Norris, P.R. & Ingledew, W.J. 1992. Acidopilic bacteria: adaptations and application. In R.A. Herbert & R.J. Sharp (eds), *Molecular biology and biotechnology of extremophiles*: 121–131. Cambridge: Royal Society for Chemistry.

Sen, A.M. 2001. Acidophilic sulphate reducing bacteria: candidates for bioremediation of acid mine drainage pollution. School of Biological Sciences. University of Wales, Bangor.

Sen, A.M. & Johnson, D.B. 1999. Acidophilic sulphate-educing bacteria: candidates for bioremediation of acid mine drainage. In R. Amilis & A. Ballester (eds), *Biohydrometallurgy and the Environment Toward the Mining of the 21st Century*: 709–718. Amsterdam: Elsevier.

Steyer, J.P., Bernet, N., Lens, P.N.L. & Moletta, R. 2000. Anaerobic treatment of sulfate rich wastewaters: process modeling and control. In P.N.L. Lens & L.H. Pol, (eds), *Environmental technologies to treat sulfur pollution:* 207–228. London: IWA publishing.

Tuttle, J.H., Dugan, P.R., Macmillan, C.B. & Randles, C.I. 1969. Microbial dissimilatory sulfur cycle in acid mine water. *J. Bacteriol.* 97:594–602.

Widdel, F. & Pfenning, N. 1981. *Arch. Microbiol.* 129: 395–400.

European Symposium on Environmental Biotechnology, ESEB 2004 - Verstraete (ed)
© 2004 Taylor & Francis Group, London, ISBN 90 5809 653 X

A model for anoxic iron corrosion by *Sh. oneidensis* MR-1 based on cell density related H_2 consumption and Fe(II) precipitation

W. De Windt, J. Dick, N. Boon & W. Verstraete
Laboratory of Microbial Ecology and Technology (LabMET), Ghent University, Gent, Belgium

S. Siciliano
Department of Soil Science, University of Saskatchewan, Saskatoon, Saskatchewan, Canada

H. Gao & J. Zhou
Environmental Sciences Division, Oak Ridge National Laboratory, Oak Ridge, Tennessee

ABSTRACT: In the absence of oxygen, a protective H_2 film is formed around an Fe(0) surface, inhibiting the electron flow from this surface. Our study of anoxic corrosion of Fe(0) beads revealed that, in the presence of *Shewanella oneidensis* MR-1, H_2 removal and precipitation of Fe mineral particles on the cell surface are determining processes for corrosion. These two biologically mediated processes were governed by cell density. Addition of supernatant of a corrosion assay with high cell concentration induced metabolic activity in a corrosion assay with low cell concentration, resulting in increased H_2 consumption and Fe release from Fe(0) beads. Homoserine lactone-like molecules were detected in the supernatant by a bio-assay, suggesting the involvement of a quorum-sensing regulatory mechanism. The interaction of *Sh. oneidensis* MR-1 with mineral particles was further investigated by means of the mutant *Sh. oneidensis* COAG. The COAG mutant exhibited increased biomineralization (up to 50% increase) and the putative interactions between COAG cells and mineral precipitates were validated by means of transcriptional analysis and Proteomics.

1 INTRODUCTION

Shewanella oneidensis MR-1 is a facultatively aerobic Gram-negative bacterium with remarkably diverse respiratory capacities (Heidelberg et al., 2002). The respiratory versatility of this metal-ion reducing bacterium presents opportunities for bioremediation of both metal and organic pollutants under anaerobic conditions. It is also these properties that make *Shewanella oneidensis* an interesting species for studying corrosion and precipitation of corrosion products (Caccavo et al., 1997; Dubiel et al., 2002; Glasauer et al., 2002).

One of the processes involved in biocorrosion of metal surfaces by bacteria has been described to be the removal of a depolarizing H_2 layer. Under anoxic conditions a polarizing H_2 film is formed at the metal surface that protects Fe(0) from further corrosion (Hamilton and Lee, 1995). This hydrogen film may be consumed due to energy metabolism of a bacterial community consisting of facultative and obligate anaerobes and coupled to the reduction of nitrate (Kielemoes et al., 2000). In this study, *Shewanella oneidensis* MR-1 was chosen as a model organism to investigate the metabolic process of H_2 removal from Fe(0). We hypothesized that the H_2 layer formed at the Fe(0) surface under anoxic conditions is consumed by MR-1 and linked to denitrification.

Next to H_2 consumption, *Shewanella oneidensis* interacted in other ways with the Fe(0) corrosion process by specifically influencing the precipitation of corrosion products. Controlled mineral formation and deposition of corrosion products has been described for several species of *Shewanella* with Fe(II) precipitating as nanocrystals at the cell surface during iron respiration. It is not clear if the nanocrystals are formed *de novo* or adsorbed from the surrounding medium (Beveridge, 1989; Glasauer et al., 2001). However, recent studies indicated that certain outer membrane proteins may allow *Shewanella oneidensis* to recognize mineral surfaces and interact with them on a nanoscale (Lower et al., 2001). As a part of this research, we obtained a *Shewanella oneidensis* mutant with the running name COAG due to its auto-aggregating properties, which increased bioprecipitation of mineral metal particles. This strain exhibits overexpression of several outer membrane proteins, as well as some metabolic processes. Micro-array and Proteomics studies were used to characterize the COAG mutant.

Finally, we demonstrated the importance of quorum sensing in the H_2 consumption process by *Shewanella oneidensis* MR-1 (De Windt et al., 2003). We report for the first time a direct relationship between a quorum-sensing regulated metabolic action and increased corrosion.

2 RESULTS AND DISCUSSION

H_2 is produced under anoxic conditions at surfaces containing Fe(0) and in a moist atmosphere according to Schlegel (1990):

$$Fe(0) + 2H^+ \rightarrow Fe\ (II) + H_2$$

This reaction was effectively observed in the corrosion assay: H_2 escaped from the Fe(0) surface and accumulated to an equilibrium concentration in the headspace that amounted to 1000 ppmV in the control reactors. *Shewanella oneidensis* MR-1 was capable to utilize this hydrogen as an energy source. This was coupled to reduction of nitrate to nitrite and further to gaseous N-compounds (Fig. 1a,b).

At high cell densities of OD 1.0 and higher, H_2 consumption reached approximately 100% of the H_2 released in the headspace, indicating that metabolic activity of the cells at this density was high. These high densities of metabolically active *S. oneidensis* MR-1 cells supported increased Fe precipitation at their cell surface resulting in a significant fraction of Fe precipitates between 0.45 and 8.0 μm (Fig. 2a). The precipitate was identified by SEM-EDX as vivianite (Fig. 2b). Jorand et al. (2000) already indicated that the presence of *Shewanella putrefaciens* CIP 80-40 biomass probably delayed the availability of Fe(II). Liu et al. (2001) describe the significance of cell number in precipitation of Fe mineral phases on the surface of *Shewanella oneidensis* CN32 cells.

Both H_2 consumption and precipitation of corrosion residues influenced release of Fe(II) from Fe(0) beads and both processes were governed by cell density of *Shewanella*. The results of corrosion assays indicated that the sudden increase in the H_2 consumption profile (Fig. 1a) due to higher cell densities ultimately resulted in increased corrosion since Fe(II) release out of the Fe(0) beads increased from 153 ± 25 mg/L to 196 ± 7 mg/L after 20 hours. Thus, consumption of boundary H_2 was a driving force for corrosion of the Fe(0) beads (De Windt et al., 2003).

Supernatant from corrosion assays with high cell concentrations contained molecules that stimulated H_2 utilization and total Fe release by MR-1. The

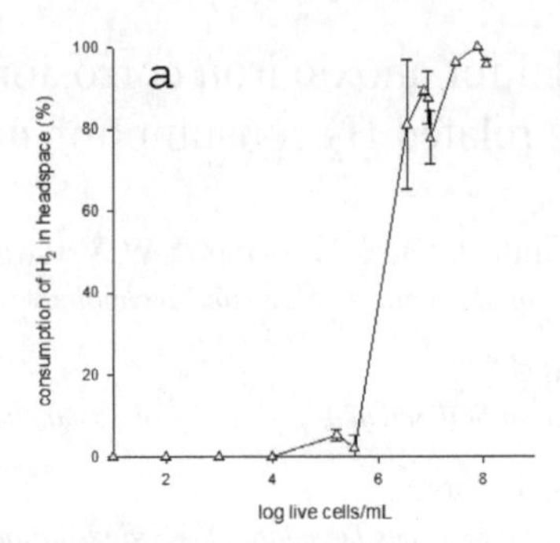

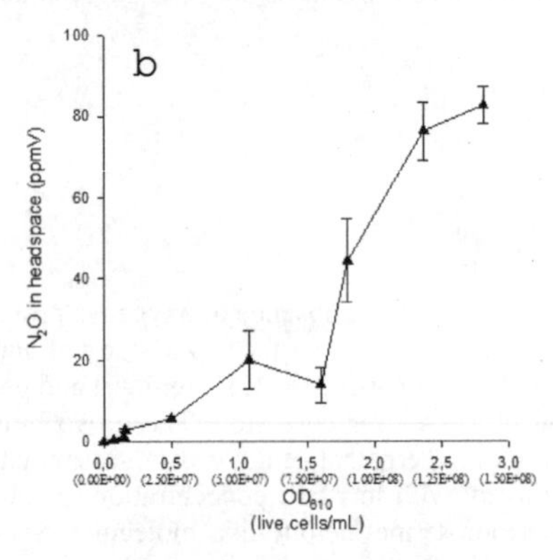

Figure 1. (a) H_2 consumption in the headspace of corrosion assays by *Shewanella oneidensis* MR-1 measured after 20 h incubation as a function of the initial amount of live cells/mL (Δ); Consumption is given as a percentage of the H_2 measured in the headspace of 'control' corrosion assays containing no bacteria. (b) N_2O production in the headspace of corrosion assays measured after 20 h incubation at several initial cell densities of MR-1 (▲).

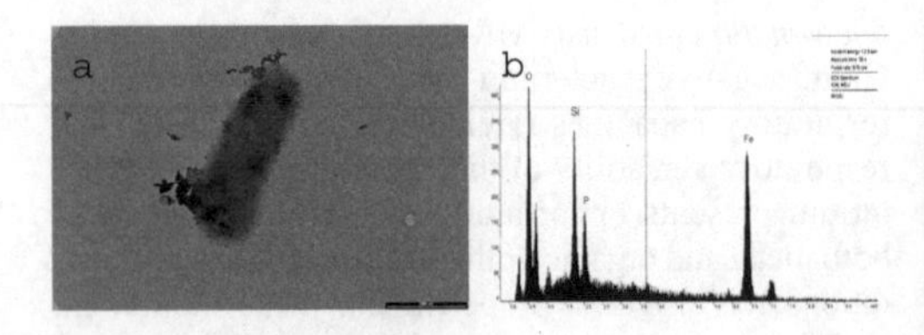

Figure 2. (a) Whole mount Electron Microscopy illustrating surface precipitation on a *Sh. oneidensis* MR-1 cell. (b) EDX spectrum of the electron dense material on the cell membrane of *Sh. oneidensis* MR-1, suggesting the presence of iron phosphate, probably vivianite.

response of a bioreporter strain to the presence of AHL molecules in the supernatant was determined by fluorometric response (Fig. 3). Supernatant of a corrosion assay with high cell number (OD 2.0) of MR-1 was, after filtration over a 0.22 μm Millipore filter, added to a corrosion assay with low cell number

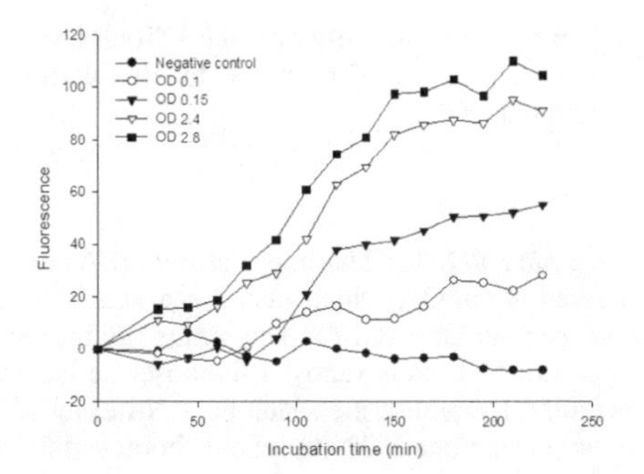

Figure 3. Effect of supernatant of corrosion assays with different initial cell concentrations on the fluorescence of the AHL reporter strain *E. coli* MT102 (pJBA130) (stable GFP).

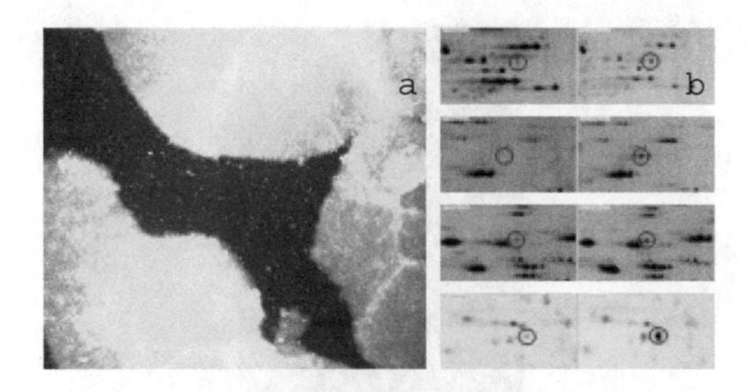

Figure 4. (a) Aggregates of *Shewanella oneidensis* COAG as opposed to cells in suspension in the liquid channels between the aggregates. (b) 2D gel electrophoresis of a part of the proteome of the COAG mutant consisting of envelope or outer membrane proteins of the COAG mutant (right) and MR-1 (left). Spots that showed overexpression with the mutant were encircled.

of MR-1 (OD 0.01). By adding the high-density supernatant, the metabolic activity of the MR-1 cells at low concentration increased.

This was measured as an increase of H_2 consumption, coupled to increased N_2O production as a standard for increased denitrification. N_2O increased from 0.5 ± 0.2 ppmV to 2.4 ± 0.4 ppmV in the headspace. H_2 consumption in headspace equaled $(13 \pm 2)\%$ of the concentration measured in the control corrosion assays after 20 h when no supernatant was added and increased to $(67 \pm 14)\%$ upon addition of supernatant. To further ascertain that this effect was due to quorum-sensing molecules in the supernatant, the pH of the high cell density supernatant (OD 1.5) was increased to 11 in order to hydrolyze the HSL molecules and afterwards adjusted back to neutral. Upon addition to corrosion assays with low cell density, this supernatant with hydrolyzed HSL molecules had no longer an effect on H_2 consumption, which remained insignificant at $4.3 \pm 2.2\%$, or on N_2O production, which remained at the insignificant level of 0.06 ± 0.04 ppmV. This is the first mention ever of a direct relationship between a quorum-sensing regulated metabolic action and increased corrosion.

The *Shewanella oneidensis* COAG mutant possesses several unique traits that make it an interesting strain for further research on biofilm formation and interaction with metal particles. Four phenotypical characteristics are attributed to this COAG strain: auto-aggregation (Fig. 4a), faster development of thicker biofilms, increased interaction with metal precipitates and overexpression of envelope proteins (Fig. 4b).

When *Sh. oneidensis* COAG was released in a corrosion assay at the same high cell density as the MR-1 strain, the effects on total Fe(II) release and precipitation of corrosion products were dramatic. With *Sh.* COAG, 90.74 ± 27.45 mg/l Fe(II) precipitates > 8.0 μm could be recovered and no smaller particles, whereas the wild type at the same OD yields 33.80 ± 9.77 mg/l Fe(II) precipitates between 0.45 and 8.0 μm and 24.50 ± 9.06 mg/l Fe(II) precipitate > 8.0 μm. These results suggest strong interactions between the *Sh.* COAG cells and metal particles that form larger complexes altogether.

To further investigate the *Sh. oneidensis* COAG mutant, transcription analysis of the COAG versus MR-1 wild type strains was done by means of microarray analysis. After analysis and normalization of the data (Imagene and Genespring) only statistically significant differences and fold-changes larger than 2.0 were retained for further study. From this analysis, it appeared that several membrane proteins such as cytochrome complexes and type IV pili were overexpressed with the COAG mutant and these could play an important role in biofilm formation and aggregation. Several operons and genes in relation to fermentation and anaerobic respiration were overregulated with the COAG mutant, although strictly aerobic conditions were maintained during growth of both MR-1 and COAG. A possible explanation for these results could be substratum- and oxygen-limitation within the large aggregates, causing anoxic conditions. Also, all genes in relation to Ni/Fe dehydrogenase activity were overexpressed with the mutant, giving rise to increased H_2 consumption by COAG. Several toxin metabolic genes, stress proteins and nucleotide salvage processes were overregulated, indicative for stress conditions within the cells due to the aggregation state. Finally, several electron and anion transport systems were overregulated, which could be expected from cells in an aggregation state. Due to the fact that whole operons were overregulated with the COAG strain in relation to other genes or operons, clusters of genes can be found with similar levels of overexpression. To illustrate this, the cluster analysis of all significantly overexpressed genes in the COAG strain compared to the MR-1 wild-type is shown in Fig. 5.

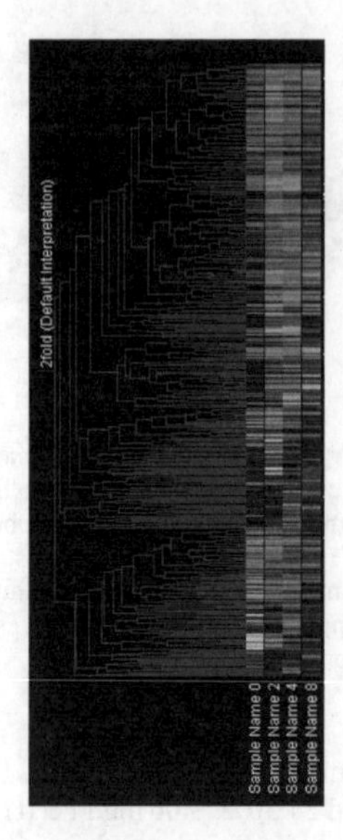

Figure 5. Cluster analysis of overexpressed genes in COAG in a micorarray study with competitive hybridization with MR-1.

3 MATERIALS AND METHODS

3.1 *Corrosion assays*

Sh. oneidensis MR-1 cells were harvested from an overnight LB culture in sterile 50 mL centrifuge tubes and washed three times with 50 mL M9 medium. After dilution to a defined cell concentration, the cells were suspended in 45 mL sterile M9 medium supplemented with sodium acetate and sodium nitrate to final concentrations of 50 mg/L acetate and 20 mg/L NO_3^--N. 15 mL of the cell suspension was injected in a 100 mL Schott bottle containing 15 iron beads of 57 mg Fe[0] each and closed hermetically by a rubber stopper and screw-cap. Each of these reactors was set up in triplicate. These reactors were flushed 10 times by 2 min of Ar overpressure followed by 2 min of underpressure. 10% (v/v) acetylene was added to inhibit further reduction of N_2O gas to N_2 in the headspace of the corrosion assays. Each assay was incubated for 20 hours by shaking at 150 rpm and at 28°C. The control corrosion assays were not inoculated. H_2 was measured in the headspace by means of a Hydrogen Monitor. N_2O was quantified with a gas chromatograph with N_2 as carrier gas in a stainless steel Porapack Q column with a total length of 3 meters and a mesh size of 80/100 and operated at a temperature of 35°C. The GC was equipped with an electron capture detector with a 10 mCi 36Ni radioactive source. Distinction between small Fe precipitate associated with culturable *Shewanella* cells and larger, micrometer-scale precipitate not associated with culturable cells, was partially based on filtration on Millipore filters. Total Fe in the corrosion assays was determined by dissolution in concentrated HCl.

3.2 *AHL detection*

Escherichia coli MT102 harbors plasmid pJBA130 that expresses *gfp*mut3*, which encodes a stable Green Fluorescent Protein (GFP). The strain fluoresces in response to N-3-oxohexanoyl-L-homoserine lactone, N-octanoyl-L-homoserine lactone, N-hexanoyl-L-homoserine lactone and N-butyryl-L-homoserine lactone (Andersen et al., 2001). To examine presence of homoserine lactones in the supernatant of corrosion assays a new protocol was used. Supernatant from corrosion assays was filtered through a 0.45 μm filter after centrifugation at 5000 × g for 10 min. 50 μL of the supernatant was mixed with 50 μL of LB broth in a PCR vial and the *E. coli* MT102 (pJBA1300) reporter strain was inoculated at an OD_{600} of 1.5. The PCR vials were incubated at 30°C in the 96-well block thermal cycling system of an ABI-Prism 7000 Sequence Detection System Real-Time PCR machine (Applied Biosystems, Foster City, USA) with online monitoring of the fluorometric response by the ABI-Prism 7000 SDS Software package (Applied Biosystems, Foster City, USA). The negative control consisted of a PCR vial with 50 μL milliQ water mixed with 50 μL LB broth, inoculated with the *E. coli* MT102 (pJBA130) reporter strain inoculated at an OD_{600} of 1.5.

ACKNOWLEDGEMENTS

This work was supported by a grant from the Institute for the Promotion of Innovation by Science and Technology in Flanders within scope of the STWW project titled "*Shewanella putrefaciens*: an omnipotent bacterium involved in microbial induced corrosion and in bioremediation of halogenated compounds".

REFERENCES

Andersen, J.B., Heydorn, A., Hentzer, M., Eberl, L., Geisen berger, O., Christensen, B.B. et al. 2001. *gfp*-based N-acyl homoserine-lactone sensor systems for detection of bacterial communication. *Appl Environ Microbiol* 67: 575–585.

Beveridge, T.J. 1989. Role of cellular design in bacterial metal accumulation and mineralization. *Annu Rev Micro biol* 43: 147–171.

Caccavo, F., Schamberger, P.C., Keiding, K., and Nielsen, P.H. 1997. Role of hydrophobicity in adhesion of the dissimilar tory Fe(III)-reducing bacterium *Shewanella alga* to amorphous Fe(III) oxide. *Appl Environ Microbiol* 63: 3837–3843.

De Windt et al. 2003. Cell density related H_2 consumption in relation to anoxic Fe(0) corrosion and precipitation of corrosion products by *Shewanella oneidensis* MR-1. *Environ Microbiol* 5: 1192–1202.

Dubiel, M., Hsu, C.H., Chien, C.C., Mansfeld, F., and Newman, D.K. 2002. Microbial iron respiration can protect steel from corrosion. *Appl Environ Microbiol* 68: 1440–1445.

Glasauer, S., Langley, S., and Beveridge, T.J. 2001. Sorption of Fe (hydr)oxides to the surface of *Shewanella putrefaciens*: cell-bound fine-grained minerals are not always formed *de novo*. *Appl Environ Microbiol* 67: 5544–5550.

Glasauer, S., Langley, S., and Beveridge, T.J. 2002. Intracellular iron minerals in a dissimilatory iron-reducing bacterium. *Science* 295: 117–119.

Hamilton, W.A., and Lee, W. 1995. Biocorrosion. In Sulfate Reducing Bacteria. Barton, L.L. (ed). New York, USA: Plenum Press, pp. 243–264.

Heidelberg, J.F., Paulsen, I.T., Nelson, K.E., Gaidos, E.J., Nelson, W.C., Read, T.D. et al. 2002. Genome sequence of the dissimilatory metal iron-reducing bacterium *Shewanella oneidensis*. *Nat Biotechnol* 20: 1118–1123.

Jorand, F., Appenzeller, B.M.R., Abdelmoula, M., Refait, P., Block, J.C., and Genin, J.M.R. 2000. Assessment of vivianite formation in *Shewanella putrefaciens* culture. *Environ Technol* 21: 1001–1005.

Kielemoes, J., De Boever, P., and Verstraete, W. 2000. Influence of denitrification on the corrosion of iron and stainless steel powder. *Environ Sci Technol* 34: 663–671.

Krause, B., and Nealson, K.H. 1997. Physiology and enzymology involved in denitrification by *Shewanella putrefaciens*. *Appl Environ Microbiol* 63: 2613–2618.

Liu, C., Zachara, J.M., Gorby, Y.A., Szecsody, J.E., and Brown, C.F. 2001. Microbial reduction of Fe(III) and sorption/precipitation of Fe(II) on *Shewanella putrefaciens* strain CN32. *Environ Sci Technol* 35: 1385–1393.

Lower, K.S., Hochella, M.F., and Beveridge, T.J. 2001. Bacterial recognition of mineral surfaces: nanoscale interactions between *Shewanella* and α- FeOOH. *Science* 292: 1360–1363.

European Symposium on Environmental Biotechnology, ESEB 2004 - Verstraete (ed)
© 2004 Taylor & Francis Group, London, ISBN 90 5809 653 X

Biological iron oxidation by *Thiobacillus ferrooxidans* and *Leptospirillum ferrooxidans* in a biofilm airlift reactor

S. Ebrahimi, R. Kleerebezem, J.J. Heijnen & M.C.M. van Loosdrecht
Department of Biotechnology, Delft University of Technology, Delft, The Netherlands

F.J. Fernández Morales
Department of Chemical Engineering, University of Castilla-La Mancha, Ciudad Real, Spain

ABSTRACT: In this study, the oxidation of ferrous iron by a mixed culture of *Thiobacillus ferrooxidans* and *Leptospirillum ferrooxidans* bacteria was investigated in a continuous biofilm airlift reactor. Small basalt particles (0.2 mm) were used as carrier material for biofilm formation. The reactor was operated at a constant temperature of 30°C and at pH values ranging from 1.3 to 1.7. The reactor was fed with a nutrient solution containing 5.6 g l^{-1} ferrous iron, at various loading rates. Feasibility and engineering aspect of biological ferrous oxidation in an air-lift reactor were investigated. Specific attention was paid to biofilm formation and competition between both types of bacteria. It was found that both types of acidophilic bacteria were capable of effective colonization of the carrier material, resulting in ferrous iron removal capacities of more than 0.05 mol l^{-1} hr^{-1}. Coexistence of *T. ferrooxidans* and *L. ferrooxidans* in the system was observed by fluorescent in situ hybridization (FISH) measurement.

1 INTRODUCTION

Recently, ferrous iron (FeII) oxidizing bacteria have been exploited for biological treatment of hydrogen sulphide (H_2S) containing gases. The removal of H_2S is achieved in a two-stage chemical-biological process. In the first stage H_2S is absorbed and oxidized to elemental sulfur (S^0) with ferric iron (FeIII). S^0 is removed from the solution by a separator, and FeIII is subsequently regenerated by biological oxidation of FeII in an aerobic bioreactor. FeIII formed can then be recycled to the H_2S absorber. To avoid precipitation of FeIII the overall process is conducted at pH-values between 1 and 2.

For biological oxidation of FeII with molecular oxygen (O_2) two types of mesophilic acidophilic bacteria are available: *Thiobacillus ferrooxidans* and *Leptospirillum ferrooxidans*. The principal difference between both types of bacteria is that *T. ferrooxidans* has the capacity to oxidize both FeII and reduced sulfur compounds (e.g. S^0), whereas *L. ferrooxidans* is only capable of FeII oxidation. Since S^0 oxidation results in a decrease of the final product (S^0) yield, this work is aimed at directing the competition between both types of bacteria towards *Leptospirillum*.

Biological oxidation of FeII is an important sub-process in the integrated chemical-biological process for H_2S removal. The application of biofilm type bioreactors for this purpose offers several advantages over suspended growth bioreactors: (i) Immobilization enables reactor operation at high biomass concentrations, leading to high volumetric conversion capacities, and (ii) in the integrated process the regenerated solution is recycled to the first stage (H_2S absorber). In case of operation with suspended biomass, a separator may be required to prevent the bacteria entering the first stage. In the first stage the suspended bacteria may be exposed to toxic H_2S. Application of a separator is not necessary if a biofilm type of reactor is employed.

There are several studies that show successful immobilization and high oxidation rates using packed bed or fluidized bed reactors with various inert carrier matrix materials such as activated carbon, glass beads or ion-exchange resin. (Grishin and Tuovinen, 1988). The reported values for iron oxidation rates in biofilm type reactors are up to one order of magnitude higher than suspended cell reactors. Halfmeier et al (1993) have carried out an extensive study on FeII oxidation by *T. ferrooxidans* in a 3 l fixed bed and an airlift reactor.

The result in the fixed bed reactor using sintered glass rings, as a carrier material was successful and a ferrous oxidation rate of 3.6 g l^{-1} h^{-1} was achieved, which was five times higher than a suspended cells bioreactor. However, no colonization by *T. ferrooxidans* of quartz sand particles in airlift reactor was observed (Halfmeier et al., 1993).

In the present work, we have studied the biological oxidation of FeII in a continuous internal loop airlift reactor using basalt as carrier material for biofilm formation. Basalt possesses a high specific surface area ($\sim$2000 $m^2 m^{-3}$) allowing a high oxygen and substrate flux to the biomass, resulting in a high volumetric treatment capacity. Application of airlift reactors is also favored because the precipitation of ferric iron complexes may cause major clogging problems in long-term operation of fixed bed reactors. Even at very low pH ($<$1.5) precipitation of FeIII in the form of jarosites cannot be completely avoided.

In the integrated process for H_2S removal, full separation of S^0 from the scrubbing liquid is normally not possible. Therefore some S^0 will enter the bioreactor. Some bacteria may furthermore enter the first stage. If the bioreactor is operated with *T. ferrooxidans*, partial oxidation of S^0 may occur therewith reducing the product yield. To prevent oxidation of elemental sulphur, *L. ferrooxidans* is therefore the preferred iron oxidizing bacterium. However, operation of a full-scale process with a pure culture is economically not feasible. Therefore we inoculated the bioreactor with a defined mixed culture consisting of both *T. ferrooxidans*, and *L. ferrooxidans*. The competition between both types of bacteria in the biofilm was investigated using FISH.

2 MATERIAL AND METHODS

2.1 *Reactor and operating conditions*

A laboratory-scale airlift reactor with a liquid volume of 3.0 l as described by Tijhuis, was used (Tijhuis, 1994). A schematic representation of the experimental setup is shown in Figure 1. The temperature was maintained at 30 $\pm$ 1°C by means of a thermostated water jacket. The pH of the reactor influent was maintained at a value of 1.3. The pH of the liquid in the reactor was measured, but not controlled. At all times the pH in the reactor was maintained at a value below 2.0. The airflow rate was set between 3.0 and 5.0 l/ min by means of a mass flow controller. Prior to entering the reactor, air was led through a humidifier in order to minimize evaporation. The air was sparged in the reactor by means of a sintered glass stone, attached to a thin glass tube. Small basalt was used as carrier material with a density of 3 kg/l and a settling velocity of 50 m/h. The diameter of the basalt particles was measured by image analysis as 0.2–0.3 mm (the mean diameter was 0.26 mm). The available carrier surface per reactor volume was 110 m^2/m^3. The initial carrier concentration was 80 g l^{-1} (5% volume). Under these conditions, the basalt was suspended homogeneously in the airlift. This carrier material has a rough surface, which results in a good potential for biofilm development. Before use, basalt particles was treated with 1 M H_2SO_4 solution for 2 h and washed until neutral to remove acid soluble salts.

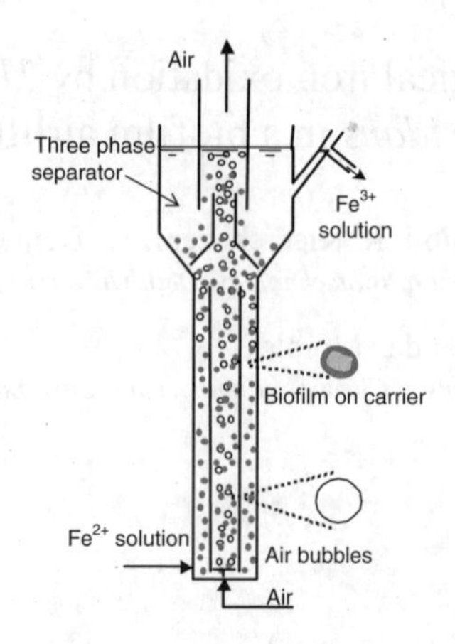

Figure 1. Schematic representation of the biofilm airlift reactor.

2.2 *Microorganisms and media*

A mixed culture of *T. ferrooxidans* and *L. ferrooxidans* was used as inoculum. Fluorescent in situ hybridization (FISH) measurements were regularly performed to check the distribution and abundance of *T. ferrooxidans* and *L. ferrooxidans* in the suspended biomass and the biofilm.

The composition of the medium for an influent Fe^{2+} concentration of 12 g l^{-1} were as follows (in g l^{-1}): $FeSO_4{\cdot}7H_2O$, 12; $(NH_4)_2SO_4$, 2; KCl, 0.1; K_2HPO_4, 1; $MgSO_4{\cdot}7H_2O$, 0.2; and trace elements (Vishniac and Santer 1957). The pH was set with 1 M H_2SO_4 solution to 1.3.

2.3 *Analytical procedures*

Total iron and FeII concentrations were determined with the colorimetric ortho-phenanthroline method, according to American Standard Methods (ASTM) D1068. Off line analysis of O_2 and CO_2 concentrations in the off-gas and on line redox-potential measurements were used to characterize the treatment performance of the reactor.

Regularly, dynamic respiration measurements in a biological oxygen monitor (BOM) were conducted

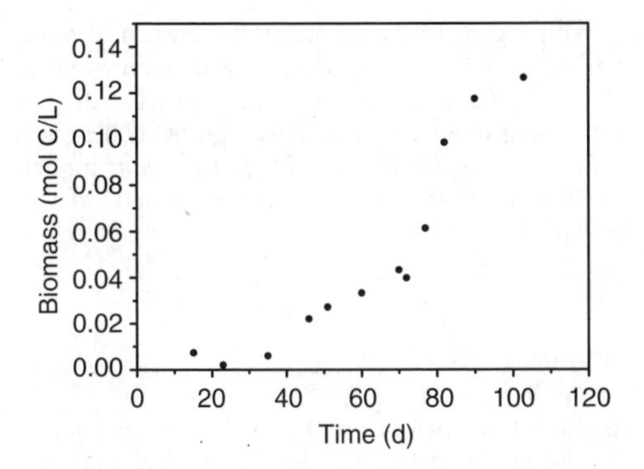

Figure 2. Estimated biomass concentration as a function of time.

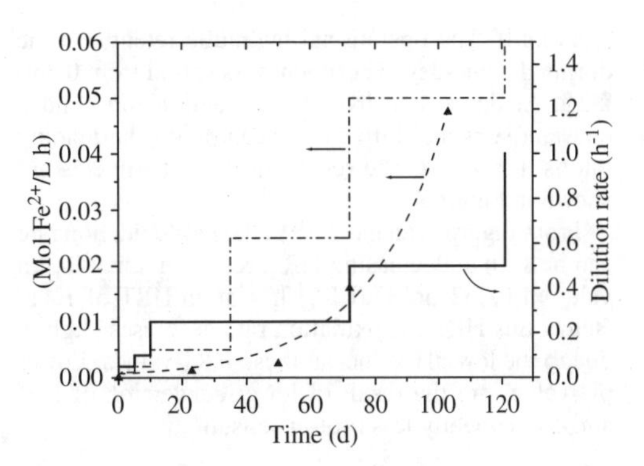

Figure 3. Volumetric ferrous iron loading rate (-··-), ferrous iron capacity (-▲-), dilution rate (—).

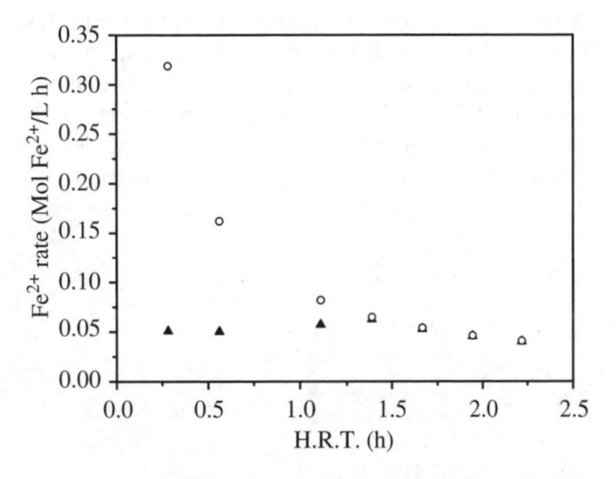

Figure 4. Relation between the hydraulic retention time and the volumetric loading rate (○) and the FeII removal capacity (▲).

with suspended biomass from the reactor. Herewith a distinction could be made between the actual contribution of the suspended and immobilized biomass to the treatment capacity.

The biomass was detached from the support by centrifugation and sonication for FISH analysis. Specific probes for *T. ferrooxidans* and *L. ferrooxidans*, as well as a eubacterial probes were used to analyse the microbial composition of the biomass.

3 RESULTS AND DISCUSSION

3.1 *Reactor start up – Biofilm formation*

Before startup the reactor was filled with a 0.1 M FeII solution and seeded with 240 g ($80\ g\ l^{-1}$) pumice carrier particles. The reactor was inoculated with a mixed culture of *T. ferrooxidans* and *L. ferrooxidans*, and operated batchwise. When 95% conversion of FeII was achieved (about 1 day), continuous substrate dosing was initiated at a dilution rate of $0.03\ h^{-1}$. After a few days, the dilution rate was increased to $0.1\ h^{-1}$, which is in the same order of magnitude as the critical dilution rate for *T. ferrooxidans* and *L. ferrooxidans* bacteria. The reactor was operated until a constant removal capacity (90% FeII conversion) was achieved. From the FeII oxidation capacity in absence of substrate limitation the actual biomass concentration in the reactor was calculated using a presumed maximum specific substrate conversion rate of 7.8 mol Fe^{2+} $molC^{-1} hr^{-1}$ and a maintenance coefficient of 0.35 mol Fe^{2+} $molC^{-1} hr^{-1}$.

The resulting biomass concentration as a function of time is shown in Figure 2. The actual FeII removal capacity as a function of time as well as the volumetric loading rate and the dilution rate during start-up of the reactor is shown in Figure 3.

The results demonstrate that an effective increase of the treatment capacity was achieved in time. BOM respiration measurements with suspended biomass furthermore demonstrated that more than 95% of the biological activity was located in immobilized biomass. This suggests that effective biofilm formation and solid retention was achieved. The biomass growth rate in absence of substrate limitation was estimated to be in the range of 0.06 to $0.08\ hr^{-1}$, which is in the same order of magnitude as the maximum growth rate of *T. ferrooxidans* and *L. ferrooxidans*. This confirms that basically complete biomass retention was achieved in the reactor. The biomass concentration in the reactor seemed to stabilize after approximately 100 days of operation. This suggests that either mass transfer limitations (gas-liquid or liquid-biofilm) or biomass retention limitations became predominant.

3.2 *Effect of the loading rate on the conversion rate*

To study the effect of the substrate-loading rate on the conversion rate, an experiment was carried out at a variable dilution rate and constant ferrous concentration

(0.1 mol/l). The operational hydraulic retention time during the one-day experiment was varied from 0.3 to 2.2 h. At decreasing hydraulic retention time and a constant substrate influent concentration the loading rate is increasing. The results of the experiments are shown in Figure 4.

From Figure 4, an increase in the FeII oxidation rate can be seen at decreasing HRT, reaching a maximum value of 0.062 mol FeII L^{-1} h^{-1} at an HRT of 1.4 h. Below this HRT the oxidation rate decreases slightly due to the low pH-values at these HRT-values. Lower pH-values are the result of lower conversion of FeII and consequently less proton consumption.

4 CONCLUSIONS

Airlift reactors are a promising concept for high-rate biological FeII oxidation at low pH-values. The high oxygen mass transfer that can be achieved combined with the highly effective biomass retention of the slow growing bacteria, high treatment capacities can be achieved at short hydraulic retention times.

With regard to the competition between *T. ferrooxidans* and *L. ferrooxidans*, a clear domination of *L. ferrooxidans* was observed using FISH at the end of the experimental period. This is probably the result of the much higher affinity of *L. ferrooxidans* for FeII resulting in preferred growth during periods of FeII limitation.

REFERENCES

Grishin, S.I. and Tuovinen, O.H., 1988. Fast kinetics of Fe^{2+} oxidation in packed bed reactors. Applied and Environmental Microbiology, 54(2): 3092–3100.

Halfmeier, H., Schafer-Treffeldt, W. and Reuss, M., 1993. Potential of Thiobacillus Ferrooxidans for waste gas Purification. Part 2. Increase in Continuous Iron Oxidation kinetics using Immoblized Cell. Appl. Environ. Microbiol, 40: 582–587.

Tijhuis, L., 1994. The biofilm airlift suspension reactor; biofilm formation, detachment and heterogeneity, TUDelft.

Vishniac, W. and Santer, M. 1957. The thiobacilli. Bacteriol. Rev. 21: 195–213.

European Symposium on Environmental Biotechnology, ESEB 2004 - Verstraete (ed)
© 2004 Taylor & Francis Group, London, ISBN 90 5809 653 X

PCR-TGGE and fluorescence microscopy in the study of a lab-scale contaminant plume

H.C. Rees, S.A. Banwart & D.N. Lerner
Groundwater Protection and Restoration Group, Department of Civil and Structural Engineering, University of Sheffield, Sheffield, UK

R.W. Pickup
CEH Windermere, The Ferry House, Far Sawrey, Cumbria, UK

ABSTRACT: A novel imaging system using the fluorescent oxygen tracer Ruthenium (II)-dichlorotris (1,10-phenanthroline) ($Ru(phen)_3Cl_2$) was developed to investigate the biodegradation of potassium acetate (KAc) by an *Acinetobacter* sp. and MW7, an environmental sample from a contaminated site in St Albans, England. A flow cell containing quartz sand (212–300 μm) covered with a biofilm was set up to allow a degrading plume of KAc to be formed. Oxygen distribution was monitored using UV excitation of $Ru(phen)_3Cl_2$ and was imaged using a CCD camera. KAc concentration was measured using ion chromatography analysis. Microbial analysis was carried out by destructively sampling across transects of the plume followed by fluorescence microscopy for total cell counts using DAPI, and active cell counts using the tetrazolium salt, CTC. In the MW7 experiments changes in the microbial population were monitored with PCR-TGGE.

1 INTRODUCTION

Natural attenuation in which natural processes reduce the concentration of pollutants at contaminated sites is a rapidly growing field of research. Although dispersion reduces the concentration of contaminants, it is the activity of the indigenous microbial populations (bioremediation), and chemical transformation, which will ultimately break them down.

Aquifers contaminated by pollutants such as BTEX, phenols, and polycyclic aromatic compounds can suffer long-term impact from contamination. However, microbial communities in the groundwater have the potential to break down a wide range of pollutants if they can survive the prevailing environmental conditions. The study of these living systems and the environmental processes that control them is an important tool in finding restoration strategies for many contaminated sites.

Natural attenuation is usually limited by the supply of available electron acceptors such as O_2 and NO_3^- (Oya and Valocchi, 1998; Cirpka et al. 1999). Biodegradation therefore occurs at the fringes of contaminant plumes where electron acceptors can be replenished by transverse mixing (Lerner et al. 2000). Here pollutants are also diluted giving rise to complex mixing processes which in turn affect the rate of biodegradation.

Measurements of transverse mixing at the plume fringe have been investigated by Huang et al. 2001, 2003. Measurements were obtained using a noninvasive imaging technique which used fluorescent dye tracers to look at transverse dispersion and biodegradation within a plume. Here we use this technology as before to look at biodegradation within a plume in terms of oxygen usage. However, this data will be related to the prevailing microbial community structure and its biodegradation activity.

PCR-TGGE, initially developed by Rosenbaum & Riesner (1987), is a technique which is now routinely used in microbiology labs to study microbial diversity, although to a lesser extent than its sister technique PCR-DGGE. This technique involves the PCR of environmental DNA followed by separation of the fragments to produce a fingerprint of the microbial population. Here we use this methodology to study the microbial population across a transect of a lab-scale contaminant plume.

To investigate biodegradation activity, microbiological and chemical analyses are carried out. Measurement of numbers of active cells versus total cells are made across a transect of the plume to illustrate where biodegradation is occurring. Chemical analysis are used to measure the amount of biodegradation down the flow of the plume and to look at biodegradation and dilution effects at the fringes of the plume.

Our goal is to use tools such as fluorescence imaging, microscopy and molecular techniques such as PCR-TGGE to allow us to better understand biodegradation at the lab-scale and therefore to improve our conceptualization of biodegradation processes and our confidence in natural attenuation at the field-scale.

2 MATERIALS AND METHODS

2.1 *Growth medium and inoculum*

The mineral medium consisted of: NaCl, 0.5 g/l; KH_2PO_4, 0.5 g/l; $MgSO_4$, 0.5 g/l; $NaNO_3$ 0.069 g/l dissolved in ultra high quality (UHQ) water adjusted to pH 7.0 and autoclaved at 121°C for 20 minutes. Either *Acinetobacter* sp. or MW7 water was inoculated into 100 ml mineral medium plus 0.3 ml 10% potassium acetate (KAc) and incubated overnight at room temperature with shaking (200 rpm).

High quality quartz sand (212–300 μm) was used as the porous matrix and was washed with distilled water and sterilized by autoclaving before use. A sterile glass column, was filled with the sterile quartz sand/mineral medium slurry before the addition of the cultured cells for biofilm production.

2.2 *Set up and operation of the flow cell*

The basic experiment has been described previously (Huang et al. 2003), modifications are described (see Figure 1 for experimental setup). The experiment was conducted in a Perspex flow cell with internal dimensions 200 × 100 × 5 mm. Inlet, outlet, and a filling pipe were located at the corners of the internal space. A point source injection pipe was placed centrally through the internal space of the flow cell, 20 mm from the top. The injection pipe had an internal diameter of 3 mm, was sealed at the bottom end, and was perforated in the flow direction. Septa ports for aqueous sampling were placed down the central line of the flow cell and across a transect of the cell.

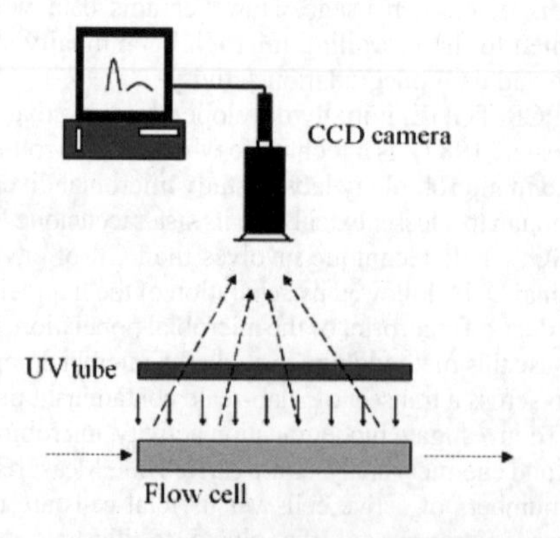

Figure 1. Schematic of experimental setup. Dotted arrows indicate the direction of flow, solid arrows the UV illumination from the UV tube, and hashed arrows the transmitted light from the fluorescent tracer dye in the flow cell. Images are recorded by a CCD camera and shown on a computer screen.

Before assembly the inside surfaces of the flow cell were sterilized by cleaning with absolute ethanol and rinsing with sterile UHQ water. The quartz sand/mineral media slurry covered in biofilm was injected into the flow cell using a sterile 50 ml syringe, a filter at the bottom of the flow cell prevented loss of quartz sand through the outlet pipes. The average porosity of the matrix was determined ($n = 0.53$) from the mass of quartz sand.

For the experiments the dye Ruthenium(II)-dichloritris(1,10-phenanthroline) ($Ru(phen)_3Cl_2$) was used at a concentration of 1.0×10^{-4} M in the mineral medium. Under ultra violet light the $Ru(phen)_3Cl_2$ is excited and emits fluorescent light which is quenched by molecular oxygen. The mineral medium was pulled through the flow cell by a fluid metering pump (Integrated Dispensing Systems) at a constant flow rate of 0.13 ml/min. To create the plume, the same mineral media containing fluorescent oxygen indicator and ~10 mM KAc was injected by a syringe pump (Harvard apparatus, model 2400003) at a constant flow rate of 8.6 μl/min, through the injection pipe.

2.3 *Imaging of plume formation*

The process was recorded by a CCD camera (Hitachi color KP-D581) with a UV long pass filter placed in the front of the camera. The camera was connected to a computer via a frame grabber board (Data Translation) and images visualized with the Global Lab Imaging software version 2 (Data Translation).

2.4 *Chemical and biological analysis*

Aqueous samples were collected via the septa ports at zero and 24 hours, filtered and analysed for acetate concentration using a Dionex DX-120 ion chromatograph (Dionex, Sunnyvale, CA, USA). Oxygen concentration of inflow and outflow solutions was measured using a hand held oxygen probe (WPA O20).

The porous medium was sampled by destructive sampling with five samples being taken across a transect of the plume under UV illumination. Total concentration of proteins was measured on 0.5 g samples of quartz sand using the Bradford assay (Bradford, 1976). Protein measurements were also carried out on sonicated samples to allow a percentage to be calculated for the number of cells knocked off the sand during the sonication treatment and thus allow cell counts to be adjusted correspondingly.

To determine whether the cells were metabolically active, the number of electron transport system active

(ETS-active) cells was measured using 5-cyano-2,3-ditolyl tetrazolium chloride (CTC) staining (Rodriguez et al. 1992). Cells were detached from the quartz sand using an ultrasonic bath (Decon) for 40 minutes. The sonicated samples were incubated overnight in the dark at room temperature in a 10 mM CTC solution. 10 μl aliquots were spotted onto microscope slides containing wells in triplicate, and mounted with Vecta Shield, containing DAPI for total counting (4′6-diamidino-2-phenylindole, Porter and Feig, 1980). At least 10 fields of view were counted per slide triplicate with a Zeiss Axioplan 2 Imaging microscope fitted for epifluorescence. CTC counts were also compared to killed controls where samples were heat treated at 80°C for 10 minutes before CTC was added.

3 RESULTS AND DISCUSSION

The plume formation was recorded with a CCD camera for 24 hours by which time a steady state had been reached. Elucidation of the plume was made possible by use of the oxygen tracer $Ru(phen)_3Cl_2$. A typical plume is shown in Figure 2 with bright fluorescence of the dye occurring where oxygen was used up by biodegradation processes. Darker areas indicate the presence of oxygen due to dye quenching and the change from light to dark illustrated that there was an oxygen gradient at the fringe of the plume. When oxygen concentrations were calculated from calibrations preformed prior to the experimental run a very steep oxygen gradient was found in the fringe region.

Aqueous samples were taken at time zero and time 24 hours for acetate measurement. Typical results are shown in Table 1. Acetate concentrations were seen to decrease down the gradient of the plume. The conservative tracer, sodium fluorescein was run under the same conditions as the oxygen tracer experiment (see Figure 2B), and as the concentration remained at steady state down the plume the decrease in concentration of acetate was therefore, due to biodegradation. However, the decrease seen from the core to the outside of the plume would likely be due both to concentration gradients arising from mixing and biodegradation as conservative tracer experiments also showing decreased concentrations from the core to the outside of the plume. The fringes create a large area for transverse mixing of acetate and oxygen and hence, concentration gradients (Huang et al. 2003), therefore allowing aerobic biodegradation to occur. Low acetate concentrations were found in the background samples 4 and 5 and also in the inflow samples where acetate levels should have been zero. This was probably due to the high dilution factor needed (20 × dilution) to be able to run the samples in triplicate. A high dilution meant that a small peak which would usually be read near detection limits is multiplied up and becomes a much larger value such as those found. This fact therefore, also suggests that samples 6 and 9 which were supposed to be in the fringe region were in fact outside the fringe of the plume as there is little acetate present and the levels recorded were at background levels.

Samples of the quartz sand were also taken across a transect of the plume and in the source region for total and active counts. The percentage of active cells

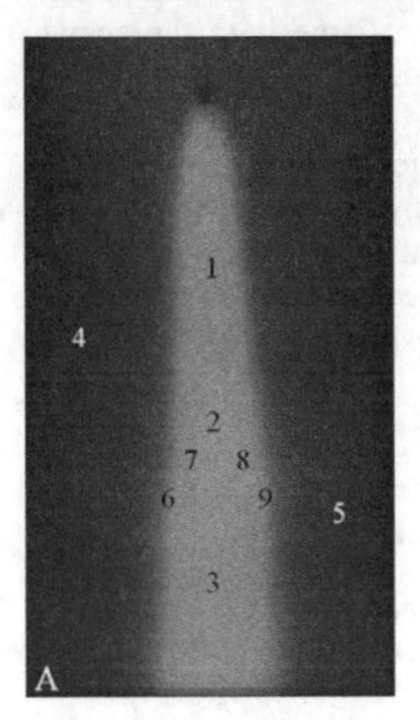

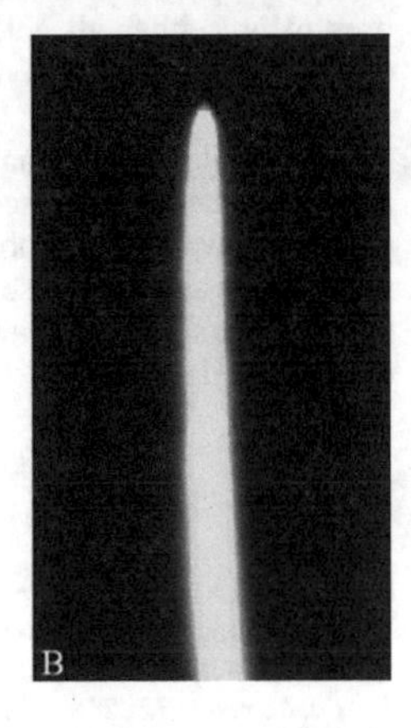

Figure 2. (A) Experimental result showing oxygen distribution in the porous medium after 24 hours. The figure also shows position of sample ports 1–9. (B) Experimental result using a conservative tracer, sodium fluorescein under the same conditions.

Table 1. Acetate concentrations before and after the experimental run. Time zero samples 2, from inside the flow cell and 10, from the acetate injection pipe were taken before the experimental run; samples 1–9 were taken from inside the flow cell (see Figure 2 for positions) at time 24 hours; time 24 inflow was taken from the media inlet pipe; time 24 outflow was taken from the outlet pipes connected to the fluid metering pump; and 10 time 24 was taken from the acetate injection pipe after the experimental run.

Sample	Acetate concentration (ppm)
2, time zero	14.0
10, time zero	595.0
1	351.0
2	287.0
3	280.9
4	14.8
5	12.5
6	14.5
7	54.3
8	73.5
9	14.8
Time 24 inflow	13.8
Time 24 outflow	35.4
10, time 24	569.0

Table 2. Percentage of active cells across a transect the plume (1 and 5, background; 2 and 4, fringe; 3 core) and in the carbon source area (6) at time zero hours and time 24 hours for *Acinetobacter* sp. and MW7 water cells.

Sample	*Acinetobacter* sp. t = 0	*Acinetobacter* sp. t = 24	MW7 t = 0	MW7 t = 24
1	12.5	9.4	2.2	1.9
2	11.6	22.4	3.2	5.1
3	16.5	3.9	3.2	1.9
4	22.4	25.3	2.3	2.6
5	13.6	8.8	1.8	1.8
6	14.2	28.9	3.2	2.0

was calculated from this data, typical results are shown in Table 2. Results show that in the single culture experiments with the *Acinetobacter* sp. there was a higher percentage of active cells in all samples than when the mixed culture from MW7 water was used. This could be due to differing abilities of cell types to take up CTC. Also evident from single culture experiments was a higher percentage of active cells in the fringe and source regions, along with a reduced percentage of active cells in the core of the plume compared to background levels. This trend was not as clearly defined in the mixed culture situation although the fringe regions (2 and 4) did have elevated numbers of active cells compared to background levels. As a higher percentage of active cells was found in the fringes of the plume it is reasonable to assume that aerobic biodegradation was occurring more intensively there as suggested by previous data (Huang et al. 2003).

To investigate whether or not there was an effect on the microbial population over the 24 hours of the experiment, PCR-TGGE analysis was performed on the mixed culture samples. Typical TGGE gels are shown in Figure 3. Results indicated that there was no obvious difference in the populations across the transect of the plume, although between zero and 24 hours there was the loss of one band across the transect. This showed that at least one species present was unable to cope with the prevailing chemical conditions supplied by the plume formation. Samples taken also differed from the growth culture used to create the biofilm and from the original MW7 water sample inoculated into the growth medium. The loss of bands and therefore species between the original inocula and the growth culture shows that not all of the indigenous species can use acetate for growth, or can not tolerate the concentrations used. The results after 24 hours show that the species within the mixed culture (with one exception) were able to cope with the prevailing chemical conditions. However, it is also possible that the experiment was not allowed to continue for long enough for there to be a significant change in the population.

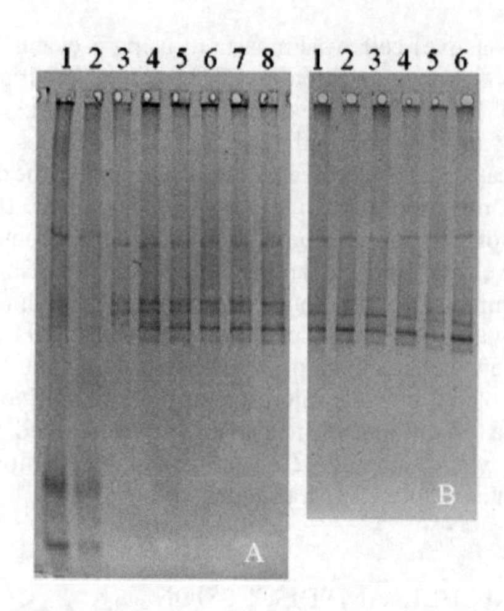

Figure 3. 12% TGGE gels (45–60°C for 3 hours). (A) Lanes 1 and 2, duplicates of MW7 water; lanes 3–8, samples across a transect of the plume at time zero (B) lanes 1–6, samples across a transect of the plume at time 24 hours.

4 CONCLUSIONS

The non-invasive technique using $Ru(phen)_3Cl_2$ has successfully been used to elucidate plume formation and to allow for aqueous and porous medium sample collection. Results have shown that compared to the background levels, active cells are more abundant in the fringes and source areas, and fewer active cells are found in the core of the plume due to a lack of oxygen. TGGE has illustrated population variation between inocula and plume samples but shows little variation across the plume. This implies that the populations present are either all able to cope with the acetate load, or that populations did not have sufficient time to develop.

The results indicate that techniques such as TGGE and fluorescence microscopy are useful tools in the study of prevailing microbial communities in this lab-scale plume and that these techniques will allow us to better understand and investigate biodegradation at the field scale.

REFERENCES

Bradford, M.M. 1976. A rapid and sensitive method for the quantitation of microgram quantities of protein utilizing the principle of protein-dye binding. *Analytical Biochemistry* 72: 248–254.

Cirpka, O.A.; Frind, E.O.; Helmig, R.J. 1999. Numerical simulation of biodegradation controlled by transverse mixing. *Journal of Contaminant Hydrology* 40: 159–182.

Huang, W.E.; Smith, C.C.; Lerner, D.N.; Thornton, S.F.; Oram, A. 2002. Physical modeling of solute transport in porous media: evaluation of an imaging technique using UV excited fluorescent dye. *Water Research* 36: 1843–1853.

Huang, W.E.; Oswald, S.E.; Lerner, D.N.; Smith, C.C.; Zheng, C. 2003. Dissolved oxygen imaging in a porous medium to investigate biodegradation in a plume with limited electron acceptor supply. *Environmental Science and Technology* 37: 1905–1911.

Oya, S.; Valocchi, A.J. 1998. Transport and biodegradation of solutes in stratified aquifers under enhanced in situ bioremediation conditions. *Journal of Water Resources Research* 34: 3323–3334.

Porter, K.G.; Feig, Y.S. 1980. The use of DAPI for identifying and counting aqauatic microflora. *Limnology and Oceanography* 25: 943–948.

Rodriguez, G.G.; Phipps, D.; Ishiguro, K.; Ridgway, H.F. 1992. Use of a fluorescent redox probe for direct visualization of actively respiring bacteria. *Applied and Environmental Microbiology* 58: 1801–1808.

Rosenbaum, V.; Riesner, D. 1987. Temperature gradient gel electrophoresis: thermodynamic analysis of nucleic acids and proteins in purified form and in cellular extracts. *Biophysical Chemistry* 26: 235–246.

European Symposium on Environmental Biotechnology, ESEB 2004 - Verstraete (ed)
© 2004 Taylor & Francis Group, London, ISBN 90 5809 653 X

Syntrophic sulfate-reducing cocultures

A.J.M. Stams
Laboratory of Microbiology, Wageningen University, Hesselink van Suchtelenweg; Wageningen, The Netherlands

ABSTRACT: Sulfate reduction and methanogenesis are two important anaerobic processes in nature. In addition, these processes are applied for environmental biotechnological purposes. Methanogenesis mainly occurs when organic matter is degraded in the absence of inorganic electron acceptors. It is negatively affected by the presence of e.g. sulfate. In that case, sulfate reducers compete with methanogenic consortia for common substrates. In the absence of sulfate sulfate reducers play an important role as hydrogen producers in syntrophy with methanogens, while they can take over the role of hydrogen scavengers from methanogens in the presence of sulfate. There are examples of sulfate reducers, which act as hydrogen producers or hydrogen consumers in syntrophic cultures, despite the fact that they can oxidize the same substrates with sulfate in pure culture. One example is growth of *Syntrophobacter* on propionate in coculture of a *Desulfovibrio* species. Another example is a thermophilic methanol-degrading *Desulfotomaculum* species, which uses hydrogen when cocultured with methanol-utilizing homoacetogens.

1 INTRODUCTION

Sulfate-reducing and methanogenic environments are characterised by their complex food-chain and the symbiotic and competitive relationships between the different microbial members of that food chain. Methanogens use only a limited range of substrates, and therefore fermentative and acetogenic microorganisms are required to degrade complex organic molecules to the substrates that methanogens can use (Schink & Stams, 2002). Hydrogen consumption by methanogens allows the degradation of reduced organic compounds (e.g. propionate, butyrate) by acetogenic bacteria. This results in an obligate syntrophic relationship between acetogenic bacteria and methanogens. Only a narrow range of hydrogen partial pressures allows growth of hydrogen-producing and hydrogen-consuming microorganisms. Sulfate reducers have a much broader substrate range than methanogens, and therefore syntrophic degradation seems to be less important than in sulfate-reducing environments than in methanogenic environments. Over the years methanogenic and sulfate-reducing bioreactors have been studied at Wageningen University. It has become clear that sulfate reducers can act as hydrogen-consuming and hydrogen-producing microorganisms depending on the presence or absence of sulfate.

2 SULFATE REDUCTION IN BIOREACTORS

For anaerobic wastewater purification the presence of sulfate is undesired. In many types of wastewater sulfate is present due to the use of sulfuric acid in the industrial process or the use of sulfite as bleaching agent. The presence of sulfate results in a competition between sulfate-reducing bacteria and methanogenic consortia for common substrates. There is direct competition between methanogens and sulfate reducers for hydrogen and acetate, while sulfate reducers also compete with acetogenic bacteria for compounds like propionate and butyrate. The occurrence of sulfate reduction in methanogenic bioreactors and the competitive interactions have been discussed elsewhere (Stams et al. 2003; Oude Elferink et al. 1994). It appears that it is almost impossible to fully suppress sulfate reduction. On the other hand it is also difficult to obtain a completely sulfate-reducing process. Both can be related to the easiness with which hydrogen-utilizing sulfate reducers outcompete methanogens and the difficulty that acetate-degrading sulfate reducers have to outcompete acetoclastic methanogens. Fully sulfate-reducing bioreactors are required in those cases where advantage is taken of the occurrence of sulfate reduction. This is the case for processes to remove heavy metals as metal sulfides from groundwater or wastewater and processes to remove sulfur

Table 1. Fermentative conversions by sulfate reducing bacteria in pure culture or in coculture with methanogens. Gibbs free energy changes according to Thauer et al. (1977).

Desulfovibrio	
Pyruvate $\rightarrow$ acetate + CO_2 + H_2	$\Delta G^{0'} = -47.3$ kJ
Lactate $\rightarrow$ acetate + CO_2 + $2H_2$	$\Delta G^{0'} = -4.2$ kJ
Ethanol $\rightarrow$ acetate + $2H_2$	$\Delta G^{0'} = +9.6$ kJ
Syntrophobacter	
Propionate $\rightarrow$ acetate + CO_2 + $3H_2$	$\Delta G^{0'} = +76.1$ kJ
Desulfobulbus	
3 lactate $\rightarrow$ 2 propionate + acetate + CO_2	$\Delta G^{0'} = -164.8$ kJ
3 ethanol + CO_2 $\rightarrow$ 2 propionate + acetate	$\Delta G^{0'} = -123.4$ kJ

from groundwater, wastewater or flue-gases. Removal of sulfur is achieved by conversion of organic and inorganic sulfur compounds to sulfide followed by partial oxidation of sulfide to elemental sulfur.

3 SULFATE REDUCERS AS ACETOGENS

Sulfate reducers have the ability to grow in the absence of sulfate or other electron acceptors. Sulfate reducers like *Desulfovibrio* and *Desulfomicrobium* can grow by oxidation of pyruvate to acetate, CO_2 and H_2 (Table 1). They are also able to oxidize lactate and ethanol to acetate, but only when the hydrogen produced is efficiently removed by methanogens. This was already shown by Bryant et al. (1977). Sulfate reducers even were the dominant acetogenic bacteria in a methanogenic reactor treating whey (Chartrain & Zeikus, 1986). *Syntrophobacter* species ferment propionate to acetate, CO_2 and H_2 in the presence of hydrogenotrophic methanogens. These acetogenic bacteria were shown to be true sulfate reducers able to grow in pure culture with propionate and sulfate (Schink & Stams, 2002). This is very remarkable because *Syntrophobacter* was obtained in a defined coculture with a *Desulfovibrio* species (Boone & Bryant, 1980). Thus, it seems that some sulfate-reducers prefer to grow syntrophically in the presence of sulfate, despite the fact that they can reduce sulfate themselves. By cocultivation with a *Desulfovibrio* species, *Syntrophobacter* species grow somewhat faster on propionate than in pure culture (Table 2).

Desulfobulbus species also grow with propionate and sulfate, but unlike *Syntrophobacter* species they cannot oxidize propionate to acetate in coculture with methanogens. However, in the absence of sulfate they can ferment lactate and ethanol ($+CO_2$) to acetate and propionate.

Growth and activity of sulfate reducing bacteria in the absence of sulfate may explain why such bacteria are always present in high numbers in seemingly fully

Table 2. Specific growth rates (day^{-1}) of fatty acid oxidizing bacteria and consortia. Data taken from Stams et al., 2003.

	Pure culture	Syntrophic culture
Butyrate		
Desulfoarculus baarsii	0.4	–
Desulfobacterium autotrophicum	0.7–1.1	–
Desulfococcus multivorans	0.17–0.23	–
Desulfotomaculum acetoxidans	1.2–1.3	–
Syntrophomonas wolfei	–	0.2–0.3
Syntrophomonas sapovorans	–	0.6
Syntrophospora bryantii	–	0.25
Propionate		
Desulfobulbus elongatus	1.39	–
Desulfobulbus propionicus	0.89–2.64	–
Desulfococcus multivorans	0.17–0.23	–
Syntrophobacter fumaroxidans	0.02	0.15–0.17
Syntrophobacter pfennigii	0.07	0.07
Syntrophobacter wolinii	0.06	0.02–0.21

methanogenic bioreactors. The presence of high numbers of sulfate reducers is of importance when considering the competition between sulfate reducers and methanogenic consortia in the presence of sulfate.

4 SULFATE LIMITED BIOREACTORS

Sulfate reducers can create lower hydrogen threshold values than methanogens, around 0.2 and 1 Pa, respectively. This is due to the fact that the Gibbs free energy change of sulfate reduction is higher than that of methanogenesis.

$$4H_2 + SO_4^{2-} + H^+ \rightarrow HS^- + 4H_2O \quad \Delta G^{0'} = -151.9\,kJ$$

$$4H_2 + HCO_3^- + H^+ \rightarrow CH_4 + 3H_2O \quad \Delta G^{0'} = -135.6\,kJ$$

Therefore, when sulfate is added to a methanogenic environment, sulfate reducers may take over the role of hydrogenotrophic methanogens in the syntrophic degradation of organic compounds like propionate and butyrate. On the other hand, sulfate reducers can oxidize propionate and butyrate directly without the involvement of interspecies electron transfer. Comparing data in the literature, it is obvious that isolated butyrate- and propionate-degrading sulfate reducers grow much faster than syntrophic methanogenic or sulfate-reducing consortia (Table 2). It is not clear yet under which conditions butyrate and propionate are degraded by sulfate-reducing bacteria directly or by sulfate-reducing consortia. Most likely this is related to the availability of sulfate. The competition of different sulfate reducers for sulfate might be important in this respect. Under sulfate-limiting conditions the different types of sulfate reducers have to compete for the available sulfate. Laanbroek et al. (1984) studied the competition of different sulfate reducers for sulfate in sulfate-limited chemostats. They found that *Desulfomicrobium baculatum* was the most successful competitor for limiting amounts of sulfate, followed by *Desulfobulbus propionicus* and then by *Desulfobacter postgatei*. It is thought that in natural environments these bacteria are important in the conversion of hydrogen and lactate (*D. baculum*), propionate (*D. propionicus*) and acetate (*D. postgatei*). If *Desulfomicrobium* and *Desulfovibrio* species indeed have the highest affinity for sulfate than syntrophic sulfate-reducing cocultures will be established under sulfate-limiting conditions. In this respect, it is worth to mention that in most wastewaters that are treated anaerobically, the amount of sulfate is mostly lower than required for complete sulfate reduction.

Interestingly, there are also examples where sulfate reducers use hydrogen in syntrophy with other anaerobic bacteria, while they can use that substrate in pure culture. One example was discovered when sulfate reduction with methanol was studied in our Laboratory. Methanol is an excellent substrate for methanogens and homoacetogens, while methanol is rarely reported to be a good electron donor for sulfate reduction. When an anaerobic bioreactor was operated at a low temperature, indeed very little methanol was used for sulfate reduction. However, in a bioreactor operated at 65°C, most of the methanol was used for sulfate reduction (Weijma, 2000). A thermophilic *Desulfotomaculum* strain was isolated from that sludge, but the amount of sulfide that was produced by the pure culture was much lower than the amount of sulfide produced in the bioreactor or in the enrichment cultures from which the sulfate reducer was isolated Remarkably, when the sulfate reducer was grown with hydrogen and sulfate, high sulfide concentrations were formed. We were able to isolate two non-sulfate reducing bacteria (*Moorella mulderi* and *Thermotoga lettingae*), which were present in the enrichment cultures (Balk et al., 2002; 2003). When a coculture of the three bacteria was made, again high sulfide concentrations were formed with methanol (unpublished results). We concluded from our observations that the sulfate reducer consumed the hydrogen, which was produced by the other two bacteria.

5 CONCLUSIONS

Sulfate reducers use a broad range substrate range for growth. This allows these bacteria to grow in a wide variety of anaerobic environments and under a wide variety of environmental conditions. Sulfate reducers both can function as hydrogen producers in coculture with methanogens or other hydrogen-utilizing anaerobes and as hydrogen-utilizers in coculture with hydrogen producers. In some cases sulfate reducers act unexpectedly as hydrogen producers or hydrogen consumer in sulfate reducing environments. The quantitative importance syntrophic sulfate-reducing cocultures needs to be investigated further. In any case, it is obvious that the occurrence of sulfate-reducing bacteria in an environment cannot be directly linked to their function in that environment. This should be taken into consideration when studying the ecology of sulfate reducing bacteria in anaerobic bioreactors and other environments.

REFERENCES

Balk, M., Weijma, J. & Stams, A.J.M. 2002 *Thermotoga lettingae*, sp. nov, a novel thermophilic, methanol-degrading bacterium isolated from a thermophilic bioreactor. *International Journal of Systematic and Evolutionary Microbiology* 52, 1361–1368

Balk, M., Weijma, J., Friedrich, M.W. & Stams, A.J.M. 2003 Methanol conversion by a novel thermophilic homoacetogenic bacterium *Moorella mulderi* sp.nov. isolated from a bioreactor. *Archives of Microbiology* 179, 315–320

Boone, D.B. & Bryant, M.P. 1980 Propionate-degrading bacterium, *Syntrophobacter wolinii* sp.nov. gen.nov., from methanogenic ecosystems. *Applied and Environmental Microbiology* 40, 626–632

Bryant, M.P., Campbell, L.L., Reddy, C.A. & Crabill, M.R. 1977 Growth of *Desulfovibrio* in lactate or ethanol media low in sulfate in association with H_2-utlizing methanogenic bacteria. *Applied and Environmental Microbiology* 33, 1162–1169

Chartrain, M. & Zeikus, J.G. 1986 Microbial ecophysiology of whey biomethanation: characterization of bacterial trophic populations and prevalent species in continuous culture. *Applied and Envionmental Microbiology* 51, 188–196

Laanbroek, H.J., Geerligs, H.J., Sijtsma, L. & Veldkamp, H. 1984 Competition for sulfate and ethanol among *Desulfobacter*, *Desulfobulbus* and *Desulfovibrio* species isolated from intertidal sediments. *Applied and Environmental Microbiology* 47, 329–334

Oude Elferink, S.J.W.H., Visser, A., Hulshoff Pol, L.W. & Stams, A.J.M. 1994 Sulfate reduction in methanogenic bioreactors. *FEMS Microbiology Reviews* 15, 119–136

Schink, B. & Stams, A.J.M. 2002 Syntrophism among prokaryotes. In: Dworkin, M., Schleifer, K.-H. & Stackebrandt, E. (Eds) *The Prokaryotes* (electronic third edition) Springer Verlag, New York. (without page numbers)

Stams, A.J.M., Oude Elferink, S.J.W.H. & Westermann, P. 2003 Metabolic interactions between methanogenic consortia and anaerobic respiring bacteria. *Advances in Biochemical Engineering and Biotechnology* 81, 31–45

Thauer, R.K., Jungerman, K. & Decker, K. 1977 Energy conservation in chemotrophic anaerobic bacteria. *Bacteriological Reviews* 41, 100–180

Weijma, J. 2000 Methanol as electron donor for thermophilic biological sulfate and sulfite reduction. *PhD thesis, Wageningen University,* Wageningen, The Netherlands

European Symposium on Environmental Biotechnology, ESEB 2004 - Verstraete (ed)

Occurrence of *Legionella pneumophila* in environmental samples

L. Devos, N. Boon & W. Verstraete
Laboratory of Microbial Ecology and Technology (LabMET), Faculty of Agricultural and Applied Biological Sciences, Ghent University, Ghent, Belgium

ABSTRACT: During the last 5 years, several outbreaks of Legionnaires' disease have been reported throughout Europe, resulting in hundreds of infections and several deaths. *Legionella pneumophila* has been shown to be the most infectious agent for Legionnaires' disease. Biofilms that are present throughout the water distribution system are probably the most important sources of *Legionella*-outbreaks. Although it is accepted that *Legionella* spp. exist in biofilms, only a limited number of studies have attempted to characterise the bacteria's survival and interaction within these complex ecosystems. The relationship between biofilm formation and *Legionella* multiplication has not yet been quantified and the prevention or abatement of biofilm formation has attracted limited attention as a(n additional) control measure. Preventing the spread of *Legionella* by limiting biofilm formation or the association of *Legionella* in the biofilm requires a systematic and stringent approach to ensure the biosafety of water and materials. However such approach may be a promising control measure to lower the risk for *Legionella* multiplication.

Several studies have tried to answer the question as to whether *L. pneumophila* can grow actively or survive in natural and man-made aquatic environments. The survival or growth however depends on the influence of its partners and opponents, since *L. pneumophila* mostly persists in multispecies biofilms. Some 42 environmental samples were analysed for the presence of *L. pneumophila* by the plate technique, a *Legionella*-specific agglutination test and PCR with *Legionella*-specific primers. Of these samples, 31% were positive for the three methods. The composition of the microbial communities was determined by means of universal primers. Correlations are sought between the occurrence of *L. pneumophila* and the presence of some other species. It is currently investigated if there is a link between microbial hotspots and the presence of *L. pneumophila*.

European Symposium on Environmental Biotechnology, ESEB 2004 - Verstraete (ed)
© 2004 Taylor & Francis Group, London, ISBN 90 5809 653 X

Filamentous bacterial structures: a unifying theory supported by experimental evidences

A.M.P. Martins[1,2], C. Picioreanu[1], J.J. Heijnen[1] & M.C.M van Loosdrecht[1]
[1]*Kluyver Laboratory for Biotechnology, Department of Biochemical Engineering, Delft University of Technology, Delft, The Netherlands*
[2] *Environmental Technologies Centre, ISQ, Av. Prof. Cavaco Silva, Porto Salvo Portugal*

ABSTRACT: This study presents a unifying theory about the morphology of bacterial structures, i.e. filamentous or compact and smooth structures. It is postulated that diffusion micro-gradients of substrate concentration inside biological flocs are the main factor responsible for the morphology of bacterial structures. At low bulk liquid substrate concentrations the micro-gradients of substrates in flocs are steep and filamentous bacterial structures, and especially filamentous bacteria due to their preferential unidirectional growth, give easier access to the substrate at the outside of the flocs, and thereby proliferate. At high bulk liquid substrate concentration the micro-gradients of substrates in flocs are lower and there is no substantial advantage for filamentous organisms to grow outside the floc. This hypothesis is supported by laboratory results, which are summarized in this paper.

1 INTRODUCTION

The development of filamentous bacterial structures is highly undesirable in activated sludge systems. Bulking sludge, a term used to describe the excessive growth of filamentous bacteria (Jenkins et al., 1993; Wanner, 1994; Eikelboom, 2000), is a typical example of such type of structures. Despite much research, bulking sludge seems to remain a continuing problem in operating wastewater treatment plants (Eikelboom et al., 1998; Lakay et al., 1999) and a comprehensive solution does not seem to be available. One reason for not finding a good general solution might be the absence of a consensus on the exact level at which the problem should be approached. The dominant approach found in the literature is by trying to identify the dominant filamentous bacterium in a bulking sludge (Jenkins et al., 1993; Eikelboom, 2000). By studying and understanding the ecophysiology of the filamentous bacterium it is hoped that a solution to avoid the occurrence of a specific filament can be found. Another approach is the recognition that the general characteristic is the bacteria morphology (Martins et al., 2003c). Realizing how the morphology affects the ecology of the bacteria could lead to a general solution independent of the species involved. In this approach the occurrence of a specific filamentous bacterium is a second order problem, only of relevance when filamentous bacteria occur and interfere with the settling characteristics of the sludge. Since filamentous bacterial structures are formed by a group of bacteria with a specific morphology, but not a specific physiology, we believe a generic approach would be feasible.

Recently, we hypothesized that substrate diffusion gradients inside biological flocs were a relevant factor responsible for the bacterial structures morphology, and, eventually, bulking sludge (Martins et al., 2003a, 2003b, 2003c). The objective of this paper is to show some of the experimental results supporting this hypothesis.

2 MATERIAL AND METHODS

The experiments were performed in well-controlled sequencing batch reactor (SBR) systems with a 2 L working volume (Applikon Fermenter). These systems allow a proper "scale-down" of the conditions bacteria experience in full-scale wastewater treatment plant containing a selector (Martins et al., 2003a, 2003b, 2003d).

The reactors were operated continuously for 75 to 90 days, in cycles of 4 or 6 h. The hydraulic residence time and the solids retention time (SRT) were 8 and 12 h, and 10 days, respectively. Acetate was given as the only carbon source.

Fully aerobic conditions and anoxic-aerobic and anaerobic-aerobic conditions, typical of biological nutrient removal systems (BNR), were applied. Using different concentrations of electron donor (acetate) and electron acceptor (e.g. dissolved oxygen and nitrate) and feeding periods allowed simulation of different substrate gradients (macro-gradients along the time dependent systems and micro-gradients in activated sludge flocs) and a variable relative size of the selector.

Details about the operational conditions (e.g. operating phases and length of the SBR cycles), synthetic wastewater composition, analysis performed (analytical work, microbiological analysis and methods used) and calculation procedures (e.g. oxygen uptake rate, specific substrate consumption rate, specific storage polymers production rate, specific denitrification rate, bio-P activity) are described elsewhere (Martins et al., 2003a, 2003b, 2003d).

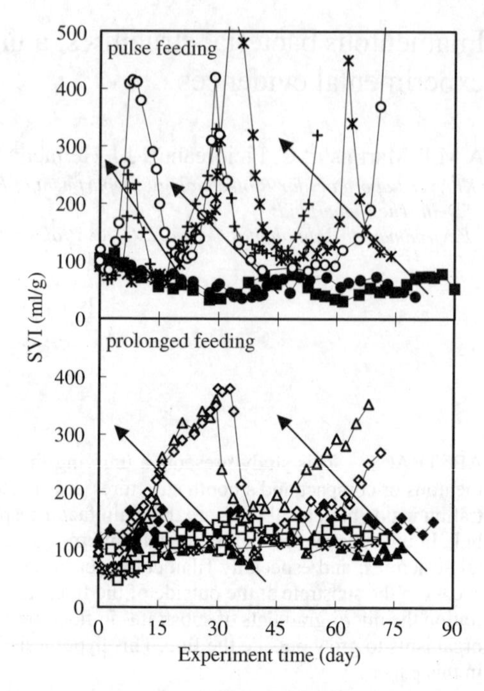

Figure 1. Effect of oxygen concentration in the feast phase on the sludge settleability, expressed as sludge volume index (SVI), during the whole operating period (in mg O_2/1): (■) 2.9; (✳) 1.1; (+) 0.2; (●) 2.9, (○) 0.7; (×) 2.5; (□) 0.9; (▲) 3.0; (△) 0.3; (◆) 2.9; (◇) 0.3. The arrows indicate the decrease in bulk liquid dissolved oxygen concentration during the feast phase (Martins et al., 2003b).

3 RESULTS AND DISCUSSION

3.1 *Activated sludge characteristics*

In full aerobic systems, strong macro-gradients over the cycle time in SBRs (or reactor length in continuously fed systems) of soluble readily available substrate concentration (e.g. organic substrate, dissolved oxygen), typical of systems with a plug-flow regime (or pulse feeding in SBRs), providing that all the nutrients are in sufficient amounts, led always to well settling sludge (sludge volume index, SVI, less than $100\,\mathrm{mlg^{-1}}$). Whenever acetate was added in a limiting rate, a condition in which the acetate concentration in the reactor was always very low, like occurring in continuously fed completely mixed systems (or prolonged feeding in SBRs) the sludge settleability decreased (SVI $> 150\,\mathrm{mlg^{-1}}$). Decreasing the oxygen concentration in the bulk liquid during the feast phase (which represents the contact time in a selector) to less than $1.1\,\mathrm{mg\ O_2L^{-1}}$ had a strong negative impact on sludge settleability (Figure 1). This negative effect was stronger at high acetate loading rate (pulse feeding). In this case, porous and irregularly shaped flocs with finger-type filamentous structures and with many filamentous bacteria (type 021N and *Thiotrix* sp.) extending from the floc surface, promoting bridging between the flocs, were commonly observed (Figure 2).

In anoxic-aerobic and anaerobic-aerobic conditions the sludge settleability did not change with the feeding pattern and was always very good (SVI $< 80\,\mathrm{mlg^{-1}}$). The only condition which led to a strongly increased SVI ($120\,\mathrm{mLg^{-1}}$) was the presence of dissolved oxygen at very low concentrations (not detectable in the bulk liquid; about $12\,\mathrm{mg\ O_2h^{-1}}$ were transferred through the liquid surface) in the anoxic stage of a denitrifying system. When strictly anoxic conditions were imposed the settleability of the sludge improved and after 10 days the SVI decreased to less than $80\,\mathrm{mLg^{-1}}$. The sludge settleability deteriorated again after returning to microaerophilic conditions.

A trend in the morphology of the flocs was observed during these experiments. Plug-flow systems without nutrient limitation (e.g. dissolved oxygen in aerobic systems or nitrate in denitrifying systems) contained many compact and smooth shaped flocs usually larger than 500 μm (Figure 3). Small, porous, and filamentous shaped flocs were dominant in substrate rate limiting systems.

3.2 *Kinetic aspects*

No significant difference was observed between maximum specific acetate uptake rate ($0.5\,\mathrm{Cmol{\cdot}Cmol{\cdot}h^{-1}}$ in full aerobic conditions) and maximum storage polymers (poly-β-hydroxybutyrate, PHB, in full aerobic

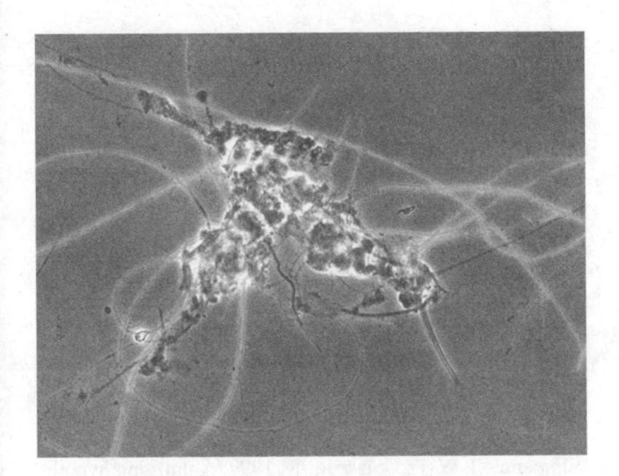

Figure 2. Filamentous bacteria *Thiothrix nivea* (identified by fluorescent *in situ* hybridization) protruding from the floc center of the activated sludge floc. SBR system with low dissolved oxygen concentration ($<0.2\,mg\ O_2/L$) and high substrate loading rate in the feast phase.

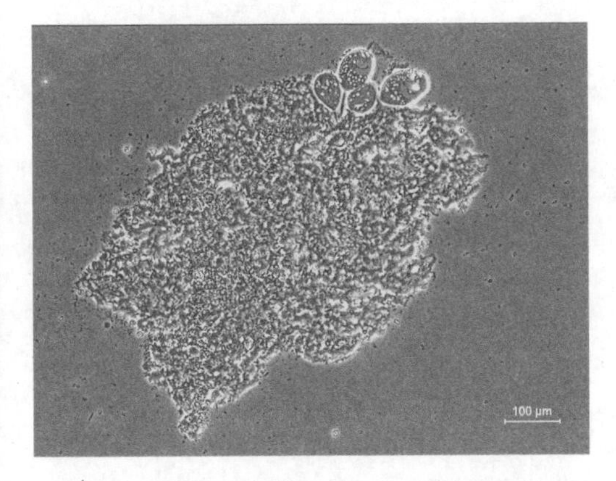

Figure 3. Phase-contrast photomicrograph of typical biological floc at steady-state in simulated plug-flow denitrifying systems, without nutrients limitation.

conditions) production rate ($0.3\,Cmol{\cdot}Cmol{\cdot}h^{-1}$) of well settling sludge and filamentous bulking sludge. This indicates that the traditional theory based on kinetic selection might not hold generally true. In denitrifying systems the rates were considerably lower (about three to five times) than in aerobic systems.

3.3 *Diffusion based hypothesis*

The transport of substrates (e.g. organic substrate, oxygen and nutrients) in microbial aggregates is expected to be mainly diffusional. If substrate diffusion limitation does not exist then more regularly shaped and compact flocs are formed (Wilén and Balmér, 1999, Martins et al., 2003a, 2003b, this study). In case diffusion limitation occurs filamentous bacterial structures predominate. These experimental results are in line with the diffusion-based hypothesis (Martins et al., 2003a, 2003b, 2003c) on the importance of diffusion process inside microbial aggregates. According this hypothesis the morphology of bacterial aggregates, i.e. filamentous or compact and smooth structures, originates from the presence of substrate gradients in sludge flocs. At low bulk liquid substrate concentration the micro-gradients of substrates in flocs are steep and filamentous bacterial structures, and especially filamentous bacteria due to their typical morphology (grow predominantly in one or two directions), get easier access to the substrate at the outside of the flocs, and thereby proliferate. A network of filamentous bacteria will then be formed, affecting the settleability and compaction of activated sludge. Other biological systems, like biofilms growing in substrate transported limited regimes, have shown similar finger and filamentous type structures (Ben-Jacob et al., 1994; van Loosdrecht et al., 1995; Wimpenny and Colasanti, 1997), as supported by modeling results (Picioreanu et al., 1998).

More experimental results are certainly needed to prove this theory but, as shown in this study, there are strong evidences indicating that a general principle – diffusion – together with the unidirectional growth of filamentous bacteria, probably dictate the morphology of activated sludge flocs, and, eventually, the settling characteristics of activated sludge. Modeling studies with preferential uni- or bi-directional growth of filamentous bacteria in dynamic floc structures, where gradients of substrate concentration are quantified, could be a valuable framework in the understanding of this complex phenomenon.

ACKNOWLEDGEMENTS

António Martins received financial support from the Portuguese State in the context of PRAXIS XXI by the Doctoral Scholarship BD/19538/99.

REFERENCES

Ben-Jacob, E.; Schochet, O.; Tenenbaum, A.; Cohen, I.; Czirók, A.; Vicsek, T. 1994. Generic modelling of cooperative growth patterns in bacterial colonies. *Nature* 368(3): 46–49.

Eikelboom, D. H.; Andreadakis, A.; Andreasen, K. 1998. Survey of filamentous populations in nutrient removal plants in four european countries. *Water Sci Technol* 37(4/5): 281–289.

Eikelboom, D. H. 2000. Process control of activated sludge plants by microscopic investigation. IWA Publishing, London, U-K.

Lakay, M. T.; Hulsman, A.; Ketley, D.; Warburton, C.; de Villiers, M.; Casey, T. G.; Wentzel, M. C.; Ekama, G. A. 1999. Filamentous organism bulking in nutrient removal activated sludge systems. Paper 7: Exploratory experimental investigations. *Water SA* 25(4): 383–396.

Jenkins, D.; Richard, M. G.; Daigger, G. T. 1993. Manual on the causes and control of activated sludge bulking and foaming, 2nd ed. Lewis Publishers, Inc., Boca Raton, Fla.

Martins, A. M. P.; van Loosdrecht, M. C. M.; Heijnen, J. J. 2003a. Effect of feeding pattern and storage on the sludge settleability under aerobic conditions. *Water Res* 37(11): 2555–2570.

Martins, A. M. P.; van Loosdrecht, M. C. M.; Heijnen, J. J. 2003b. Effect of dissolved oxygen concentration on the sludge settleability. *Appl Microbiol Biotechnol* 62(5–6): 586–593.

Martins, A. M. P.; Pagilla, K.; Van Loosdrecht, M. C. M.; Heijnen, J. J. 2003c. Filamentous bulking sludge – a critical review. *Water Res* (in press).

Martins, A. M. P., Van Loosdrecht, M. C. M., Heijnen, J. J. 2003d. Bulking sludge in biological nutrient removal systems. *Biotechnol Bioeng* (in press).

Picioreanu, C., Van Loosdrecht, M. C. M., Heijnen, J. J. 1998. Mathematical modelling of biofilm structure with a hybrid differential-discrete cellular automaton approach. *Biotechnol Bioeng* 58: 101–116.

Van Loosdrecht, M. C. M., Eikelboom, D., Gjaltema, A; Mulder, A.; Tijhuis, L.; Heijnen, J. J. 1995. Biofilm structures. *Water Sci Technol* 32(8): 35–43.

Wanner, J. 1994. Activated Sludge Bulking and Foaming Control. Technomic Publishing Co., Lancaster, PA.

Wilén, B.-M.; Balmér, P. 1999. The effect of dissolved oxygen concentration on the structure, size and size distribution of activated sludge flocs. *Water Res* 33(2): 391–400.

Wimpenny, J. W. T.; Colasanti, R. 1997. A unifying hypothesis for the structure of microbial biofilms based on cellular automaton models. *FEMS Microb Rev* 22: 1–16.

European Symposium on Environmental Biotechnology, ESEB 2004 - Verstraete (ed)
© 2004 Taylor & Francis Group, London, ISBN 90 5809 653 X

Physiological and proteomic analysis of EPS production and biofilm formation for a phenol degrading bacterium

A.L. Geng, C.J. Lim, A. E.W. Soh, B. Zhao & T.S. Leck
Institute of Environmental Science and Engineering, School of Civil & Environmental Engineering, Nanyang Technological University, Singapore

ABSTRACT: The objective of this study is to investigate the biofilm formation for a phenol-degrading bacterium. Strain EDP3, was isolated from municipal activated sludge using the enrichment isolation technique. The 16S rRNA gene sequence analysis revealed that EDP3 belonged to the gamma group of *Proteobacteria*, with an identity of 97.0% to the 16S rRNA gene sequences of *Acinetobacter calcoaceticus*. This strain could mineralize up to 1000 ppm phenol at room temperature. At higher phenol concentration, i.e. greater than 750 ppm, biofilms were developed by forming microbial granules in the culturing media. Scanning electron microscope (SEM) analysis indicated that the biofilm formation was due to the production of extracellular substances (EPS). In addition, at 1000 ppm phenol concentration, both bacteria growth and phenol degradation profiles showed two plateaus. This indicates that higher phenol concentration, or higher toxicity, has induced some phenotypic changes in the bacterium. Two-dimensional gel electrophoresis has been conducted in order to further investigate such toxicity responses.

1 INTRODUCTION

Phenols are commonly employed chemicals that are widely used in many industries, including petroleum refining, petrochemical, pharmaceutical and resin manufacturing plants. They are therefore major pollutants found in many industrial wastewaters. Phenol removal from wastewater is therefore catching researchers' attention. Biotreatment of phenol containing wastewater is generally preferred due to the lower costs and the possibility of complete mineralization. While phenol wastewaters are usually treated in continuous activated sludge processes, high phenol loading rates and fluctuations in phenol loading have been reported to cause the breakdown of the systems. These difficulties are attributed to the substrate inhibition effects resulted from phenol toxicity. Cell immobilization is one of the strategies to overcome substrate inhibition. Aerobic granulation, a new form of cell immobilization, represents another alternative, which has shown great potentials in biotreatment of high strength phenol containing wastewater (Jiang et al., 2002).

Phenol loading has significant effects on the activity and the structure of microorganisms. While over exposure to phenol usually results in a decrease, or complete loss, of enzyme activity and cell growth (Kibret et al., 2000), microorganisms are capable of a variety of physiological responses to increase their tolerance towards phenol toxicity. The objective of this study was to investigate the phenol loading responses in an *Acinetobacter sp.*, a granule formation phenol degrading bacterium. Phenol degradation kinetics, scanning electron microscopy and two-dimensional gel electrophoresis analysis have been conducted to understand the physiological and phenotypic changes in response to the variations of phenol toxicity.

2 MATERIALS AND METHODS

2.1 *Bacterial strain and culture conditions*

The organism used throughout this work was strain EDP3. It was isolated from activated sludge, and was selected for its high phenol degradation activity. Stock cultures of *Acinetobacter sp.* EDP3 were maintained by periodic subtransfer on phenol agar slants and stored at 4°C. Phenol agar slants were made from the mineral salt medium supplemented with 500 ppm phenol and 2% agar. All batch cultures were performed in 250 ml Erlenmeyer flasks with bug stoppers and 100 ml culturing medium. The mineral salt medium contained (gl^{-1}): K_2HPO_4, 0.65; KH_2PO_4, 0.19; $NaNO_3$, 0.5; $MgSO_4.7H_2O$, 0.1; $FeSO_4.7H_2O$,

0.00556; $(NH_4)_2SO_4$, 0.5; and 10 ml trace mineral solution per liter medium. The pH of the medium was adjusted to 6.8–7.0. The trace mineral solution contained (gl^{-1}): nitrilotriacetic acid, 1.5; $MnSO_4.H_2O$, 0.5; $CoCl_2.6H_2O$, 0.1; $CaCl_2$, 0.1; $ZnSO_4.7H_2O$, 0.1; $CuSO_4.5H_2O$, 0.01; H_3BO_3, 0.01; $Na_2MoO_4.2H_2O$, 0.01 and $AlK(SO_4)_2.12H_2O$, 0.01.

2.2 *Analytical methods*

Cell growth was monitored spectrophotometrically by measuring the absorbance at 600 nm. Phenol was analyzed using the 2690 Alliance® HPLC Systems from Waters™ (Milford, MA, USA). The chromatographic analysis was conducted on a Symmetry® C18 Column (4.6 × 150 mm, 5 μm particles size) obtained from Waters. The mobile phase was composed of 55% distilled water and 45% acetonitrile and the flow rate was 1 ml/min. The photodiode array detector was set at 270 nm with single channel detection.

2.3 *Scanning electron microscopy analysis*

For scanning electron microscopy preparation, cells were fixed for 4 hours in 2% (v/v) glutaraldehyde, washed 3 times with 0.10 M sodium cacodylate buffer, and dehydrated with T-butyl alcohol of increasing concentration (50, 70, 85, 95, 100%, v/v). Dehydrated cells were filtered through a 0.2 μm polycarbonate filter (Millipore, USA), dried with a freeze dryer (Bal-Tec CPC 030), sputter-coated with gold at 20 mA in a high vacuum (2.8×10^{-6} Torr) and low temperature (−170°C) cryo-chamber for 90 seconds, and then viewed with the scanning electron microscope at 20 kV.

2.4 *Preparation of cell extracts for two-dimensional gel electrophoresis*

Cultures of *Acinetobacter sp.* EDP3 grown at 400 ppm and 1000 ppm were both harvested at their late exponential growth phase, respectively, by centrifugation at 8,000 × g for 10 min at 4°C. The cells were washed twice by 40 mM Tris – HCl (pH 8.0) and centrifuged under the same conditions. Cell pellets were immediately used or stored at −20°C future analysis. The cells were re-suspended in 40 mM Tris – HCl (pH 8.0) to a cell density of 0.5 g wet weight per ml. The cell suspension was sonicated for 30 min using a fl inch probe on Sonicator® for 10 seconds with a 20 seconds cooling interval between each pulse. Samples were then treated with DNAase and RNAase (final concentrations of 1 mg/ml and 5 mg/ml, respectively) for 20 mins at room temperature, then centrifuged at 12,000 × g for 10 min at 4°C. The supernatant was collected in a 50 ml centrifuge tube and ice-cold methanol was added to a final volume of 40 ml. The tube was then placed at −80°C for an hour before centrifugation at 12,000 × g for 30 min at 4°C. The pellet was re-suspended in 0.5 ml lysis solution (8 M urea, 4% CHAPS, 40 mM Tris (Base)).

2.5 *Two-dimensional gel electrophoresis (2D-GE)*

Approximately 500 μg of proteins from each cell extract was analyzed by 2D-GE. A pH 4–7 immobilized pH gradient (13 cm) IPG strip gel (Amersham Biosciences) was rehydrated overnight with rehydration solution (8 M urea, 2% CHAPS, 0.002% Bromophenol blue) and the sample in a total volume of 250 μl. IEF was performed using an IPGphor IEF system (Amersham Biosciences) and conducted by stepwise increase of the voltage as follows: 500 V for 1 hr, 1000 V for 1 hr, 8000 V till the total volt-hours (Vh) reaching 24 KVh. After IEF separation, strips were equilibrated twice for 15 min each time with SDS equilibration buffer (50 mM Tri/HCl, pH 8.8; 6 M urea, 30% glycerol, 2%SDS, 0.0002% bromophenol blue). IPG strips were then placed over a 12.5% polyacrylamide gel (18 × 16 cm) and ran at 15 mA/gel for 15 min and then the current was increased to 30 mA/gel until the bromophenol blue had run off the end of the gel. The gels were stained using Coomassie blue R350. ImageMaster v 3.01 software (Amersham) was used for spots detection and detailed analysis.

3 RESULTS AND DISCUSSIONS

3.1 *Scanning electron microscope analysis of biofilms*

Acinetobacter sp. EDP3 formed granules in response to phenol toxicity at higher phenol concentration. Cells grown in 1000 ppm phenol were collected at the late exponential phase for SEM analysis. Figures 1 and 2

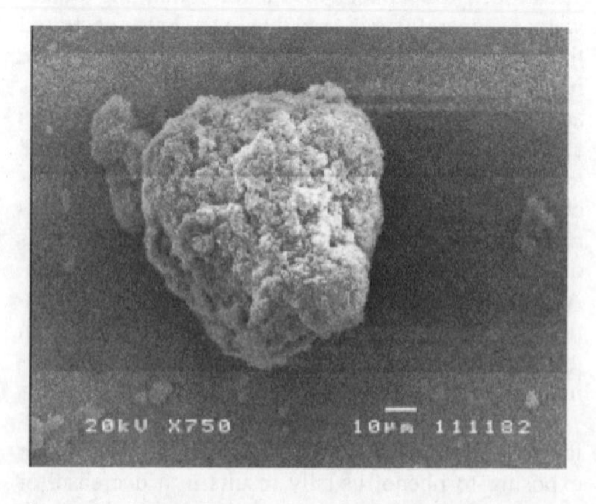

Figure 1. SEM picture of aerobic granules formed by *Acinetobacter sp.* EDP3.

give the results of the SEM pictures for the granules and the biofilms. It can be seen that strain EDP3 is coccus and its size is around 1 μm. In response to high phenol concentration, without any supporting medium, biofilms were developed by cell self-aggregation. Some extracellular polymeric substances in the granules interlinked the cells forming a porous spherical structure.

Figure 2. SEM picture of biofim formed by *Acinetobacter sp.* EDP3.

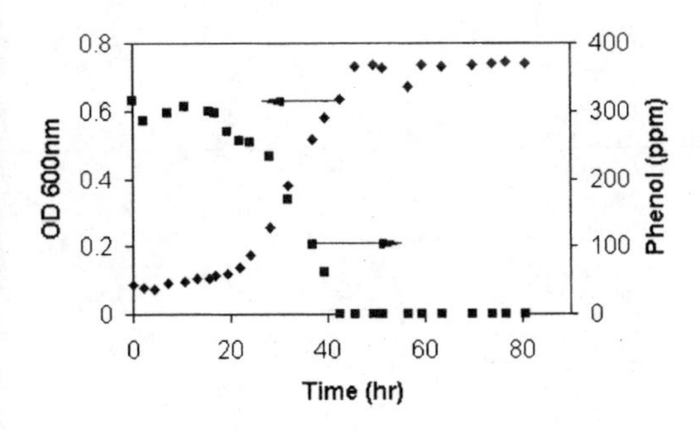

Figure 3. Cell growth and phenol degradation at 400 ppm phenol.

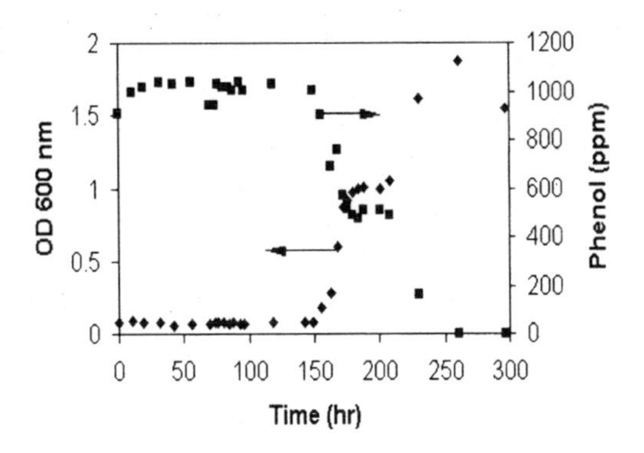

Figure 4. Cell growth and phenol degradation at 1000 ppm phenol.

3.2 *Bacteria growth and phenol degradation kinetics*

Experiments were conducted in duplicate. 400 ppm and 1000 ppm phenol were used in the experiments. Figures 3 and 4 show the cell growth and phenol degradation profiles under initial phenol concentration of 400 ppm and 1000 ppm, respectively. It can be seen that, at 400 ppm phenol initial concentration, the lag phase for bacteria to grow is around 20 hours, whereas that for 1000 ppm phenol initial concentration was around 150 hours. In addition, both cell growth and phenol degradation profiles for 400 ppm phenol culture showed one plateau, whereas those for

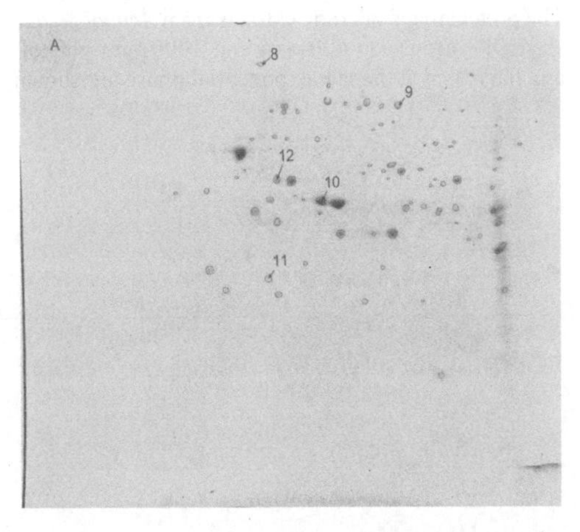

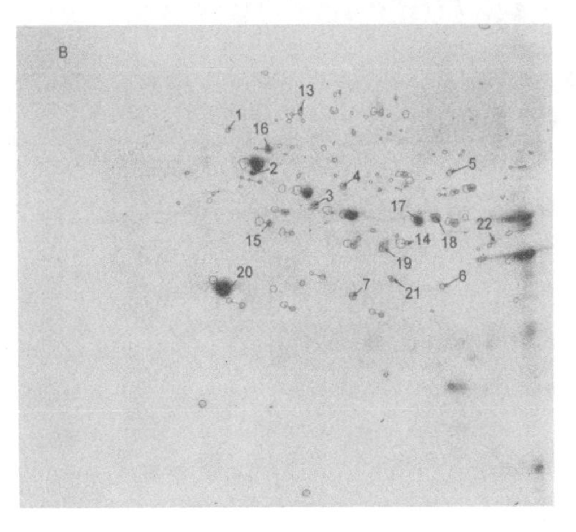

Figure 5. A&B. 2-DE maps from this strain grown on 400 ppm(A) and 1000 ppm(B) phenol.

1000 ppm gave two plateaus. Bacteria growth reached stationary phase when phenol is consumed for the case of 400 ppm phenol concentration. For the case of 1000 ppm phenol concentration, bacteria growth reached the first stationary phase at 150 hours. Phenol was partially degraded at that time. The first stationary phase lasted for around 30 hours, bacteria continued to grow and phenol was further degraded until completed used, when the second stationary phase was reached. Obviously, at higher phenol concentration, some phenotypic changes were going on in the bacteria to adapt to phenol toxicity. Another set of enzymes might have been induced while bacteria growth further proceeded to a second exponential phase. This could be further confirmed by the proteome mapping approach.

3.3 *Two-dimensional gel electrophoresis*

The protein maps of crude extracts from *Acinetobacter sp.* EDP3 grown on 400 ppm and 1000 ppm phenol\ and harvested at the late exponential phase are shown in Figure 5 A&B. It is clear that higher phenol concentration induced the production of more proteins. It is interesting to note that spots 8–12 are present only in map A, and spots 1–7 are present only in map B. For the remaining of the spots, a modulation of protein expression occurred: spots 13–15 were downregulated; spots 16–22 were overexpressed if cells were grown in 1000 ppm concentration of phenol. Further analysis will be done by identifying the protein spots and elucidate the shift of biodegradaton pathway in response to phenol toxicity variations.

REFERENCES

Jiang, H.L., Tay J.H., Tay, S.T.L. (2002). Aggregation of immobilized activated sludge cells into aerobically grown microbial granules for the aerobic biodegradation of phenol. *Lett. Appl. Microbiol.* 35: 439–445.

Kibret, M., Somitsch, W., Robra, K.H. (2000). Characterization of a phenol degrading mixed population by enzyme assay. *Water Res.* 34: 1127–1134.

European Symposium on Environmental Biotechnology, ESEB 2004 - Verstraete (ed)
© 2004 Taylor & Francis Group, London, ISBN 90 5809 653 X

Application of the anammox process in wastewater treatment: mathematical modeling

W.R.L. van der Star, C. Picioreanu & M.C.M. van Loosdrecht
Department of Biotechnology, Delft University of Technology, Delft, The Netherlands

X. Hao
R&D Center for Sustainable Environmental Biotechnology, Beijing Institute of Civil Engineering and Architecture, Beijing, China

ABSTRACT: Mathematical models of the behaviour of anammox in multispecies biofilm were reviewed and the effect of temperature, inflow variations, as well as the presence of COD were discussed. An extension of the current (one dimensional) models to two dimensional model will yield a more realistic description, especially with regard to the concentration of intermediates like nitrite.

1 INTRODUCTION

The discovery of anaerobic ammonium oxidation process (anammox) in the nineties has considerably changed our view of the biological nitrogen cycle. Until then, the global nitrogen cycle was viewed upon unidirectionally: ammonia is aerobically oxidized to nitrite/nitrate, which can be reduced dinitrogen gas, which can be fixed again to ammonia. The anammox process however enables direct conversion of ammonia to dinitrogen gas, with nitrite acting as the electron acceptor (Van de Graaf et al. 1995) and can thus be seen as a shortcut of the existing routes (Figure 1).

The anammox process is performed by a dedicated group in the order of the Planctomycetales (Strous et al. 1999a), and is characterized by an extremely slow, autotrophic growth (doubling time is approximately 11 days) and is inhibited by its own electron acceptor nitrite (Strous et al. 1998). Reducing equivalents for the carbon fixation during growth are provided by oxidation of part of the nitrite to nitrate. The overall stoichiometry is as follows:

$$1.32\,NO_2^- + 1NH_4^+ + 0.066\,CO_2 + 0.13H^+ \rightarrow$$
$$0.066\,CH_{1.8}O_{0.5}N_{0.15} + 1.02\,N_2 + 0.26\,NO_3^-$$
$$+ 0.26\,NO_3^- + 2.03H_2O$$

The unidirectional view of the N-cycle is also the basis for the conventional ammonium removal in wastewater treatment via nitrification–denitrification.

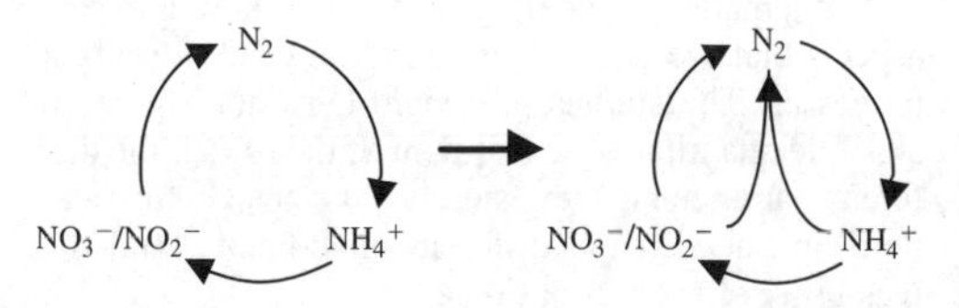

Figure 1. From a unidirectional to the present N-cycle representation.

The employment of anammox however has led to various innovative N-removal concepts in which the anammox process plays an important role. To address the feasibility of those processes and to identify the parameters that are crucial for prolonged operation, several modelling studies were performed. In this paper an overview is provided of the outcome of those studies.

2 COMPETITION FOR SUBSTRATES

Ammonium and nitrite are the substrates for anaerobic ammonium oxidation, and both processes involved in nitrification compete for one of those substrates. Aerobic ammonium oxidizers (like *Nitrosomonas* spp.) compete for ammonium, and convert it to nitrite, whereas nitrite oxidizers (like the *Nitrobacter* spp.) aerobically oxidize nitrite to nitrate. The competition between the anammox organisms and the organisms involved in nitrification are depicted in Figure 2.

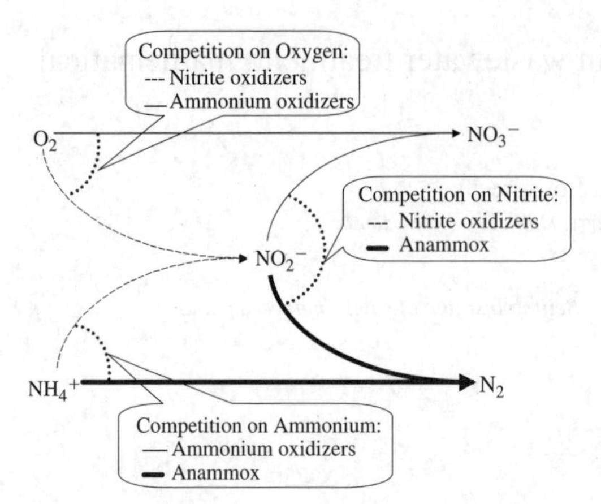

Figure 2. Competition of anammox, aerobic ammonium oxidizers and nitrite oxidizers.

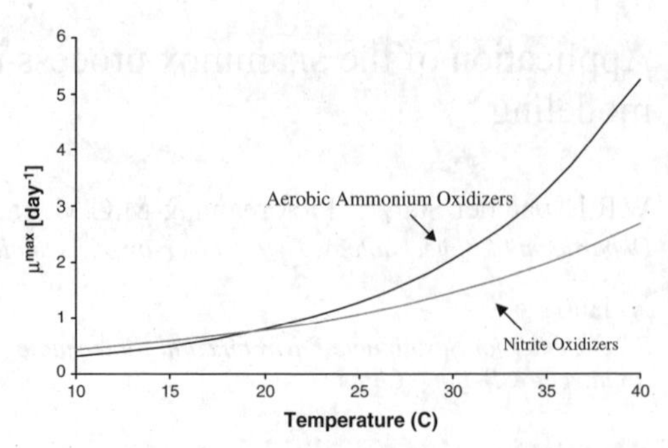

Figure 3. Growth rate of ammonium oxidizers and nitrite oxidizers at various temperatures.

The behavior of three different types of organisms for three different substrates (ammonium, nitrite, oxygen) makes it complex to validate the different process options. Furthermore, practical validation (on labscale as well as on a larger scale) is complicated by the slow growth of the anammox organisms.

Mathematical modeling can be a key into the aspects that are crucial for operation of the different processes. The studies presented here don't give an absolute quantitative assessment of the role of the different parameters; they merely give insight in relationship between the different phenomena, and the feasibility of process options.

3 NOVEL AMMONIUM REMOVAL PROCESSES

Since the anammox process requires ammonium and nitrite in a ratio 1:1.3, a considerable amount of the ammonia in wastewater has to be converted to nitrite by aerobic ammonium oxidizing bacteria. This partial nitrification process can physically be separated from the anammox process in a two-reactor process. It is also possible however, to complete the reaction in one reactor.

3.1 *Two reactor processes*

In the two-reactor concept, part of the WW-ammonia is oxidized to nitrite (the SHARON process), to provide the ideal feed mixture for the anammox process, which is performed in the subsequent reactor. In this paragraph both processes will be treated.

3.1.1 *SHARON process*

In conventional nitrification reactors, the conversion from ammonia to nitrite is always accompanied by nitrite oxidation. Changing of process parameters (like DO concentration, pH etc) in the process has never led to out-competition of the nitrite oxidizers in favour of the ammonium oxidizers over prolonged periods.

However, at slightly elevated temperatures (above 20°C) the higher specific growth rate of ammonium oxidizers (Figure 3) can be employed. This feature is used in the SHARON system (Hellinga et al. 1998); a nitrification reaction with no sludge retention. In this process the residence time is chosen such, that ammonium oxidizers can maintain. The residence time is so short however, that the nitrite oxidizers are washed out.

3.1.2 *Anammox reactor*

In contrast to the SHARON process, in the anammox reactor sludge retention is needed. With a hydrolic residence time of approximately one day the process has proven to be stable over prolonged periods on a labscale (Van Dongen et al. 2001), and the first anammox reactor in the world at the WWTP Dokhaven, Rotterdam is now operated since summer 2002.

3.2 *One reactor process*

To perform the anammox process as well as the partial nitrification process in one reactor, growth conditions should be satisfied for both processes. Due to the contradictory oxygen requirements (oxygen is needed for nitrification, but inhibits anammox), a homogenous reactor with both processes is impossible.

However, in the CANON (Completely Autotrophic Nitrogen removal Over Nitrite) process (Sliekers et al. 2002), particles are formed consisting of nitrite oxidizers and anaerobic ammonium oxidizers (see Figure 4).

On the outside of the particle ammonium oxidizers produce nitrite, which diffuses into the granule, where the anammox organisms combine with ammonium to form dinitrogen gas. Other similar- one-reactor processes include the OLAND (Philips et al. 2002) and the anaerobic/aerobic deammonification process (Hippen et al. 2001) process.

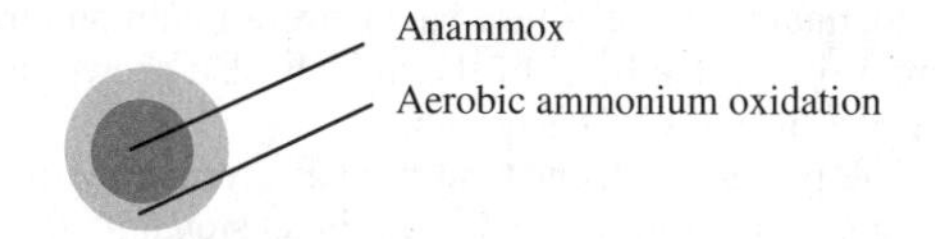

Figure 4. Schematic representation of a CANON particle.

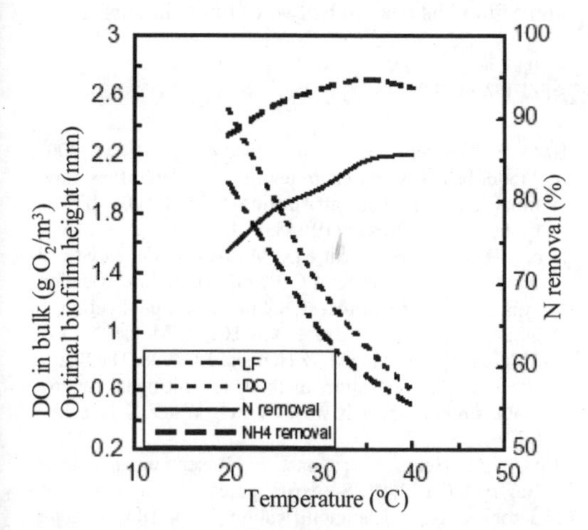

Figure 5. Effect of temperature on ammonium removal, biofilm size and optimal DO (Hao et al. 2002).

4 MODELING THE CANON PROCESS

Crucial to the CANON process are the concentrations of dissolved oxygen and ammonium. In the first place, oxygen must be consumed fully in the outer layer of the CANON particle, in order to maintain anoxic conditions for the anammox process at the core of the particle. Moreover, if oxygen is a limiting substrate, than the (undesired) nitrite oxidizers experience a competition on two substrates: nitrite (with anammox) and oxygen (with aerobic ammonium oxidizers). This double limitation is the key factor that determines the wash out of nitrite oxidizers.

4.1 *Influence of temperature*

The optimal temperature for the anammox is about 40°C (Strous et al. 1999b), and thus lower temperatures are suboptimal with respect to the conversions in the system. Also, at lower temperatures, the aerobic ammonium oxidation activity is lower, which leads to a higher penetration of oxygen into the biofilm. Both effects lead – in the temperature range from 20 to 30°C – to a better ammonium removal at higher temperatures. At temperatures above 30°C the effect is less pronounced, because the anammox biomass concentration is not limiting in the biofilm anymore (Hao et al. 2002).

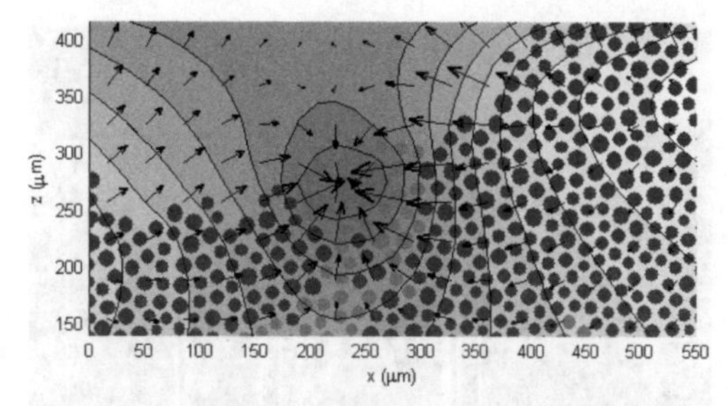

Figure 6. Lateral gradients of nitrite concentration (arrows) with aerobic ammonium oxidizers (dark) and anammox organisms (light) (Picioreanu et al. 2003).

4.2 *Influence of process variations*

Full scale processes will generally suffer from disturbances, in terms of flow, as well in the concentration of the nutrients (Hao et al. 2002). As soon as a steady state is reached however, for the SHARON-ANAMMOX process, the difference of short-term and long-term dynamics is not significant. The fact that the anammox organisms grow slowly (even with respect to the long term variations), makes the system less sensitive to changes. This in contrast to the behavior of processes with faster growing organisms.

4.3 *2-Dimensional models*

Most biofilm models are 1-dimensional, and thus cannot include heterogeneity – in terms of biomass composition as well as concentration gradients – parallel to the surface. The difference between 1D models, and two dimensional models – which can take this effect into account – were evaluated recently (Picioreanu et al. 2003).

Modeled in 2D, the nitrite profiles of the CANON process are markedly different. This is due to the double function nitrite has in the system: it functions both as a product and as a substrate. In two-dimensional models of the CANON process this leads to gradients parallel to the modelled surface (Figure 6).

By the nature of one dimensional mathematical models, lateral gradients can never be detected, and thus their influence cannot be established. The lateral gradients have also an influence on the biofilm composition and the ratio between aerobic ammonium oxidizers and anammox organisms present in the biofilm (Figure 7). Two-dimensional models therefore give a more realistic description of the expected biofilm morphology and composition.

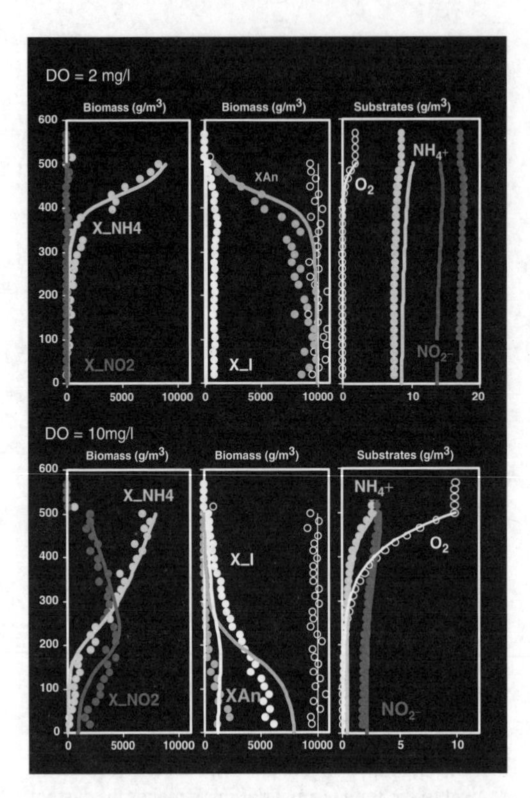

Figure 7. Biofilm composition at different bulk oxygen concentrations: differences between 1D (dots) and 2D models (lines) (Picioreanu et al. 2003).

4.4 *Nitrogen removal in the presence of COD*

Also for high COD waste streams the anammox process can be employed. Within the anoxic part of the biofilm however, more competition can be expected, due to the present of heterotrophic nitrite and nitrate reducing organisms. Here the temperature effect is even more important than in the absence of COD: at lower temperatures anammox activity must be present in thicker (or more compact) biofilms in order to have an acceptable ammonium removal (Hao & Van Loosdrecht 2003).

5 CONCLUSIONS

Mathematical models have proven to be efficiently applicable to evaluate processes and critical process conditions in ammonium removal processes, where the anammox process is employed.

The slow growth of the anammox organisms make the biofilm robust towards fluctuations; on a long as well as on a short timescale. The anammox fraction and growth rate are better in biofilms at elevated temperature, and lead to a better overall ammonium removal rate. On basis of the models also growth in the presence of COD is possible.

The present mathematical models are mainly one dimensional. However, only two dimensional models can detect gradients parallel to the surface as they develop for nitrite. This has a considerable impact on the biofilm composition. Therefore the future generation of models need to be two or three dimensional.

LITERATURE

Hao, X., Heijnen, J.J. & Van Loosdrecht, M.C.M. 2002. Model-based evaluation of temperature and inflow variations on a partial nitrification-ANAMMOX biofilm process. Water Res 36(19):4839–4849.

Hao, X.D. & Van Loosdrecht, M.C.M. 2003. Model-based evaluation of the influence of COD on a partial nitrification-Anammox biofilm process. submitted for publication.

Hellinga, C., Schellen, A.A.J.C., Mulder, J.W., Van Loosdrecht, M.C.M. & Heijnen, J.J. 1998. The Sharon process: an innovative method for nitrogen removal from ammonium-rich waste water. Water Sci Technol 37(9):135–142.

Hippen, A., Helmer, C., Kunst, S., Rosenwinkel, K.H. & Seyfried, C.F. 2001. Six years' practical experience with aerobic/anoxic deammonification in biofilm systems. Water Sci Technol 44(2–3):39–48.

Philips, S., Wyffels, S., Sprengers, R. & Verstraete, W. 2002. Oxygen-limited autotrophic nitrification/denitrification by ammonia oxidisers enables upward motion towards more favourable conditions. Appl Microbiol Biotechnol 59(4–5):557–566.

Picioreanu, C., Kreft, J.U. & Van Loosdrecht, M.C.M. 2003. A particle based multidimensional multispecies biofilm model. submitted for publication.

Sliekers, A.O., Derwort, N., Gomez, J.L.C., Strous, M., Kuenen, J.G. & Jetten, M.S.M. 2002. Completely autotrophic nitrogen removal over nitrite in one single reactor. Water Res 36(10):2475–2482.

Strous, M., Fuerst, J.A., Kramer, E.H.M., Logemann, S., Muyzer, G., Van de Pas-Schoonen, K.T., Webb, R., Kuenen, J.G. & Jetten, M.S.M. 1999a. Missing lithotroph identified as new planctomycete. Nature (London) 400(6743):446–449.

Strous, M., Heijnen, J.J., Kuenen, J.G. & Jetten, M.S.M. 1998. The sequencing batch reactor as a powerful tool for the study of slowly growing anaerobic ammoniumoxidizing microorganisms. Appl Microbiol Biotechnol 50(5):589–596.

Strous, M., Kuenen, J.G. & Jetten, M.S.M. 1999b. Key physiology of anaerobic ammonium oxidation. Appl Environ Microbiol 65(7):3248–3250.

Van de Graaf, A.A., Mulder, A., de Bruijn, P., Jetten, M.S.M., Robertson, L.A. & Kuenen, J.G. 1995. Anaerobic oxidation of ammonium is a biologically mediated process. Appl Environ Microbiol 61(4):1246–1251.

Van Dongen, U., Jetten, M.S.M. & Van Loosdrecht, M.C.M. 2001. The SHARON-Anammox process for treatment of ammonium rich wastewater. Water Sci Technol 44(1):153–160.

European Symposium on Environmental Biotechnology, ESEB 2004 - Verstraete (ed)
© 2004 Taylor & Francis Group, London, ISBN 90 5809 653 X

Screening, genetic classification and role of single bacterial strain in a MTBE degrading *consortium*

M. Camilli, N. Cimini & D. Scarabino
EniTecnologie, Biological Science Department, Rome, Italy

M. Molinari
ENI, R&M Division, Rome, Italy

ABSTRACT: MTBE bioremediation was approached by a preliminary research activity based on the sampling and chemical characterization of MTBE contaminated groundwater and wastes and the microbial screening of *consortia* adapted to the alkyl ether molecule. Two autochthonous microbial *consortia* able to degrade MTBE with a mean velocity of 100 mg/L culture broth/24 hs were selected from bacterial enrichment cultures (subculturing procedure) using 12 samples of waters coming from different and distant contaminated sites. The 16 S rDNA based classification indicated *Acidovorax sp; Variovorax sp.*; *Hydrogenophaga pseudoflava*; Sphingo*monas sp.*; *Ultramicrobacterium* and *Gordonia nitida* as the six species most similar to the components of one of the identified *consorta.* Only *Variovorax sp.* showed the capacity to degrade MTBE as a pure culture. The highest MTBE degradation rate (average 13.0 mg degraded g^{-1} cells, h) was observed in the co-culture *Acidovorax sp./Variovorax sp.*. A new process for the remediation of MTBE contaminated groundwaters based on the use of immobilized members of the selected *consortium* represents the final research purpose of this work.

1 INTRODUCTION

Alkyl ethers such as methyl *tert*-butyl ether (MTBE) have been used as oxygenate gasoline additives to boost the octane number and to reduce the vehicle emissions. The most common oxygenate MTBE is currently added to gasoline at concentrations up to 15% (v/v) and its use has increased every year since (Wilson 2001). Due to the widespread use of MTBE in reformulate fuels documented release sites are continuously being reported. A recent survey identified MTBE as the most common contaminant of urban aquifers in the USA and USEPA recommends a concentration of 20–40 ug/L as indicative Health Advisory level for drinking waters and 70 ug/L for nonprimary water sources (EPA 1994). The risk for the environment posed by MTBE is exacerbated by its high water solubility and mobility in groundwater systems. The relative bioremediation recalcitrance of MTBE combined with its high water solubility make it the single largest environmental threat to ground water quality. Traditional methods to remove gasoline constituents from groundwater do not work well for MTBE. Air stripping is very difficult because MTBE's low volatility requires large air/water ratios to make the treatment effective. In addition, MTBE low affinity for organic carbon prevents efficient removal by granular activated carbon (Fortin 2001). Albeit at slower rates than other gasoline constituents (BTEX), available research indicates that many naturally occurring microbial populations can biodegrade MTBE as well as *tert*-butyl alcohol (TBA, an MTBE metabolite of concern) (Fortin 2001, Garnier 1999, Wilson 2001). Initial studies suggested that MTBE is resistant to microbial degradation, but more recent research has shown that MTBE can be partially or fully biologically degraded by mixed cultures of bacteria (M100) (Salanitro 1994) containing several microbial species whose roles in MTBE degradation were not well defined. Moreover the cometabolic degradation of MTBE was shown to be an important biodegradation mechanism of certain pure strains growing on gaseous alkanes (Liu 2001). Most recently, pure cultures of bacteria capable of growth on MTBE as sole carbon source have been isolated (*Rubrivivax gelatinosus* strain PM1 (Hanson 1999), *Rhodococcus* sp., and *Hydrogenophaga flava* ENV735 (Steffan 2001, Hatzinger 2001). These strains could be useful in field for the remediation of MTBE contaminated sites. Strategies for enhancing *in situ* biodegradation of MTBE (i.e. accelerated natural attenuation and biobarrier bioaugmentation) (De Flaun. 2002) and *ex situ* biological treatment applications (biotrickling filters) are urgently needed. This scenario suggested, as preliminary research activity, the selection and eventually the manipulation of further effective MTBE-degrading bacterial populations.

2 MATERIALS AND METHODS

2.1 *Water samples*

Twelve groundwater samples were collected from different sites located in distant regions of the italian territory (northern and central Italy). Subsurface water samples were taken from gasoline service stations and fuel storage tanks areas chronically contaminated by MTBE.

2.2 *Analytical methods*

The concentrations of MTBE and BTEX in cell free diluted (1:1000) liquid samples were automatically analyzed with an HP 6890 gascromatograph (GC) equipped with purge & trap device (O-I Analytical 4560; autosampler 551-A) and with a flame ionization detector (FID). The GC was fitted with a J&W Scientific DB-5MS 95% metil 5% phenylsiloxane (30 m × 0.25 mm ID df 0,25 um) capillary column (EPA methods 8015 and 5030B). The injection was by pulsed splitless injection mode. The oven temperature was held at 36°C for 7 min. and was ramped to 70°C (10°C/min.), to 120°C (17°C/min.) and to 220°C (30°C/min.) The oven temperature was isothermally stopped at 70, 120 and 220°C for 1, 2 and 1 min. respectively. The MTBE was quantified by using standards of known concentrations. The MTBE daughter product TBA was detected by GC-MS (GC-MS DS Mod. MAT/90, Finnigan).

2.3 *Microbiological methods*

Total heterotrophic bacteria were enumerated in the water samples as CFU (Colony Forming Unit) by plate counting on TSA (Trypticase Soy Agar, Oxoid) as growth *medium* The plates were incubated at 30°C for a period of 24 hours.

Screening of MTBE-degraders was performed using series of 500 ml flasks where 50 ml of water samples have been added as *inoculum* to a same volume (50 ml) of Minimal Salt Medium (MSM, (Salanitro 1994), composition in g/L: KH_2PO_4 0.695; K_2HPO_4 0.854; $(NH_4)_2SO_4$ 1.234; $MgSO_4$ $7H_2O$ 0.460; $CaCl_2$ $2H_2O$ 0.176; $FeSO_4$ $7H_2O$ 0.01; pH 6.90). The micronutrient (5.0 mL/L MSM) solution composition was (mg/L): H_3BO_3 60; $CoCl_2$ $6H_2O$ 40; $ZnSO_4$ $7H_2O$ 20; $MnCl_2$ $7H_2O$ 6; $NaMoO_4$ $2H_2O$ 6; $NiCl_2$ $6H_2O$ 4; $CuCl_2$ $2H_2O$ 2.1. Concentrations of MTBE from 200 to 500 mg/L were further added acting as sole source of carbon and energy and the flasks sealed with screw caps and teflon septa were put in a rotatory shaker at room temperature against blaks (MTBE and MSM). Daily aliquots of microbial suspension (40 uL) were withdrawn from the systems, processed and GC analyzed. Every 15 days, an enrichment subculturing procedure (refreshment of half volume of running cultures with a half volume of new MSM) was applied. After 5 subculturing steps per sample, the grown microorganisms were transferred to both TSA or MSM agarized plates (including MTBE-wetted paper disks) for the isolation of the colonies (members) of the developed MTBE-degrading microbial *consortium*. Isolated colonies were collected and the portion of 16S rDNA of the extracted DNA was amplified by Polymerase Chain Reaction (PCR). The amplification products were sequenced and compared to bacterial rDNA 16S sequence (GenBank Database www.ncbi/nlm/nih.gov) for classification. Sequencing was performed according to ABI™ Dye Terminator Cycle Sequencing Ready Reaction Kit, Applied Biosystems.

The classified isolates were tested as single strains with regard to their specific degradation capabilities on MTBE (MSM solutions of alkyl ether at the optimal 6.5–7.5 range of pH values).

3 RESULTS

The main chemical and microbiological parameters of groundwater samples used in the present work are shown in Table 1.

All the samples appeared contaminated by the representative chemicals of a gasoline contamination. The associate microbial charge, except for CM samples, demonstrated the probable presence of MTBE adapted *microflora*. The samples characterized by high concentration of MTBE and low amounts of BTEX (CVT1, P3 and RV) seemed to be more suitable for the isolation of bacteria specifically degrading the alkyl ether. In three water samples: CVT1 (Fig. 1), P3 and P6 the presence of MTBE-degrading bacteria was clearly assessed.

In all the positive cases the time required (several days) to perform the complete alkyl ether degradation could be related to both toxic effect of MTBE metabolism daughter products and/or to the slow growing nature of the enriched *consortium*.

Due to the absence of TBA or other daughter metabolites in the GC-MS responses, the second option appeared to be more probable, according to previous literature observations (Hatzinger 2001). The mean MTBE degradation rate calculated in the actively degrading cultures (2nd refreshment) was of the order

Table 1. Chemical characteristics of sampled groundwaters.

Water samples	MTBE (μg/L)	Benzene (μg/L)	Toluene (μg/L)	*p*-Xylene (μg/L)	*o*-Xylene (μg/L)	Total Heterotroph C.F.U./mL
CVT1	57000	–	21,5	17	10,3	1.5×10^5
P3	378	15	–	–	10	5.0×10^3
P5	18000	2500	9000	9000	5000	5.0×10^3
P6	1300	341	1800	1300	909	4.0×10^3
RV	1100	–	18	15	21	3.0×10^4
RM	2800	–	3200	1530	830	3.0×10^5
RP	1600	–	84	351	158	1.4×10^4
CM7	9400	2100	2100	2100	393	3.0×10^4
CM8	8500	988	568	160	50	1.5×10^3
CM9	8000	937	565	152	49	4.0×10^2
CM10	7500	874	502	37	32	1.4×10^3
CM11	14000	2000	9100	1840	923	2.0×10^3

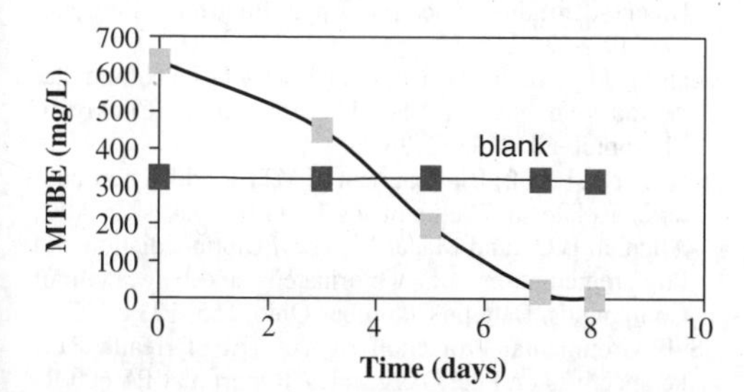

Figure 1. **MTBE degradation**. CVT1 microbial *consortium*.

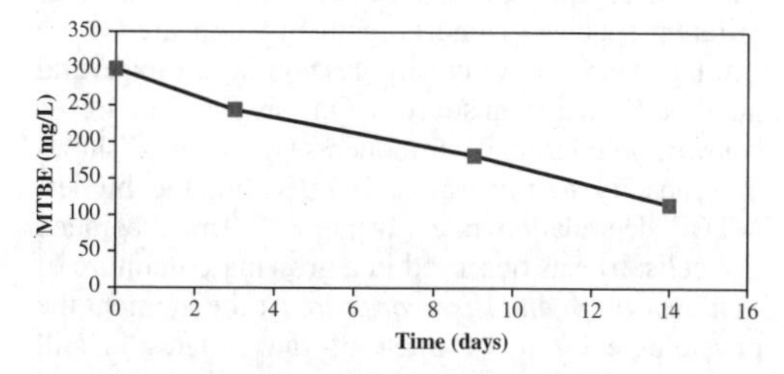

Figure 3. **MTBE Degradation** (*Variovorax*).

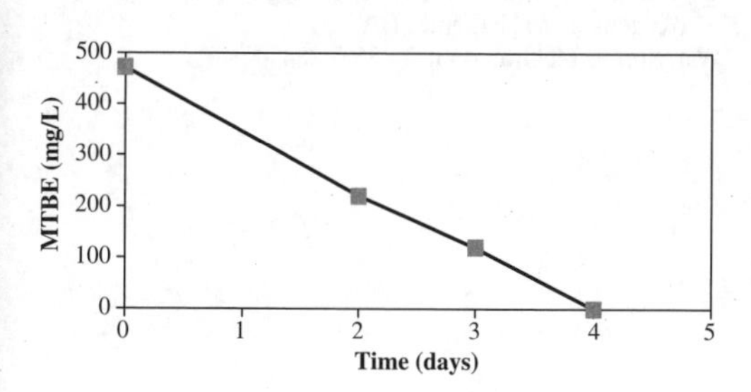

Figure 2. **MTBE Degradation**. *Acidovorax* and *Variovorax* co-culture.

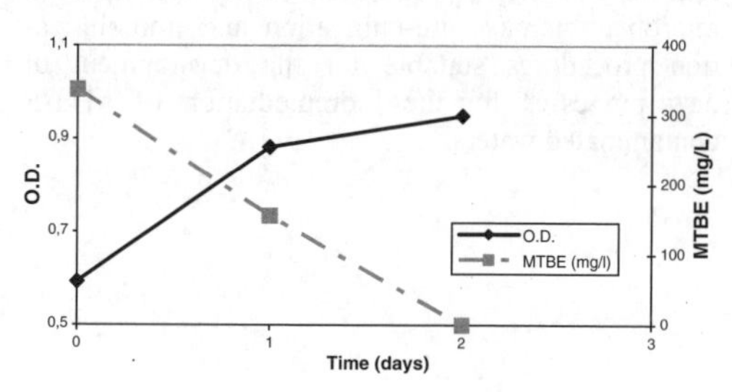

Figure 4. **MTBE degradation vs. OD**. *Acidovorax* and *Variovorax* growth.

of 100 mg/L/24 hs. The microbial constituents of CVT1 MTBE-degrading *consortium* after selection and isolation on TSA plates and classification by 16S rDNA nucleotide sequence resulted to be represented (99% data bank homology) by the following six single strains: *Acidovorax sp.*; *Variovorax sp.*; *Hydrogenophaga pseudoflava*; Sphingo*monas sp.*; *Ultramicrobacterium* and *Gordonia nitida*.

The identified A*cidovorax, Variovora* and *Hydrogenophaga genera* were of the same family (*Comamonadaceae*) of the already studied MTBE-degrading *Rubrivivax gelatinosus* and PM1 *bacterium*.

The isolate components of the P3 consortium appeared morphologically similar to those isolated from CVT1.

The experiments assessing the specific MTBE degrading potential of each bacterial strain indicated *Acidovorax/ Variovorax* as co-culture and *Variovorax* as single strain as able to mineralize MTBE without the metabolic contribution of other *consortium* members. Figures 2 and 3 show the performances on MTBE of the above mentioned isolates.

The MTBE degradation velocity of the pure culture was slower than the whole *consortium*, while the co-culture one was comparable (100 mg MTBE/L/24 hs). The co-culture itself appeared to be slow-growing (Fig.4). The measured MTBE degradation rate of the co-culture measured at the Optical Density (OD) maximum value (OD = 1) was 13.0 mg MTBE g $cell^{-1}$ h^{-1} average. The physiological role of the singular components of the microbial *consortium* is under investigation.

4 CONCLUSIONS

Bacterial enrichment cultures of groundwater samples coming from gasoline contaminated sites indicates the possibility to select autochthonous microbial *consortia* able to completely degrade MTBE. The bacterial composition of the selected *consortia* (six different species of microorganisms*)* appeared *to* be similar even from water samples coming from several hundred Kms distant sources. One single strain of the *consortium* microbial components (*Variovorax*) shows the capacity to mineralize MTBE, but the highest MTBE degradation rate (average 13.0 mg degraded g^{-1} cells, h) was observed in a growing co-culture of *Acidovorax sp. and Variovorax sp.* At the moment the physiological role of most of the isolates is still unknown. The activities of screening, isolation and characterization of MTBE-degrading bacteria represent the first step of the running studies on specific microbial biomass pre-cultivation and immobilization procedures suitable for the development of new processes for the bioremediation of MTBE contaminated waters.

REFERENCES

De Flaun M.F. & Stwffan R.J. (2002) Bioaugmentation p. 434–442 in:G. Bitton (ed), Encyclopedia of environmental microbiology. John Wiley & Sons, New York, NY

Fortin N.Y. *et al.* (2001) Methyl *tert*-butyl ether (MTBE) degradation by a microbial *consortium*. Environ. Microbiol. **3**: 407–416

Garnier P.M. *et al.* (1999) Cometabolic biodegradation of methyl tert-butyl ether by *Pseudomonas aeruginosa* grown on pentane. Appl. Microbiol. Biotechnol. **51**: 498–503

Hanson J.R. *et al.* (1999) Biodegradation of methyl *tert*-butyl ether by a bacterial pure culture Appl. Environ. Microbiol. **65**: 4788–4792

Hatzinger P.B. *et al.* (2001) Biodegradation of methyl *tert*-butyl ether by a pure bacterial culture. Appl. Environm. Microbiol. **67**: 5601–5607

Liu C.V. *et al.* (2001) Kinetics of methyl *tert*-butyl ether Cometabolism at low concentrations by pure cultures of Butane-degrading Bacteria. Appl. Environ. Microbiol. **67**: 2197–2201.

Salanitro J.P. *et al* (1994) Isolation of a bacterial culture that degrades methyl *tert*-butyl ether. Appl. Environm. Microbiol. **60**: 2593–2596

Steffan *et al.* (2000) Biotreatment of MTBE with a new bacterial isolate in Wickramanayake G.B., Gavaskar A.R., Alleman B.C., and Magar V.S (*eds*) Bioremediation and Phytoremediation of Chlorinated and Recalcitrant Compounds, Battelle Columbus Ohio, 165–173

US Environmental Protection Agency (EPA) Health Risk Perspectives on Fuel Oxygenates. Report n° EPA 600/R-94/217. US Environmental Protection Agency, Washington DC 1994.

Wilson G.J. *et al.*(2001) Aerobic biodegradation of gasoline oxygenates MTBE and TBA

Wat. Sci. & Technol. **43** n°2 : 277–284.

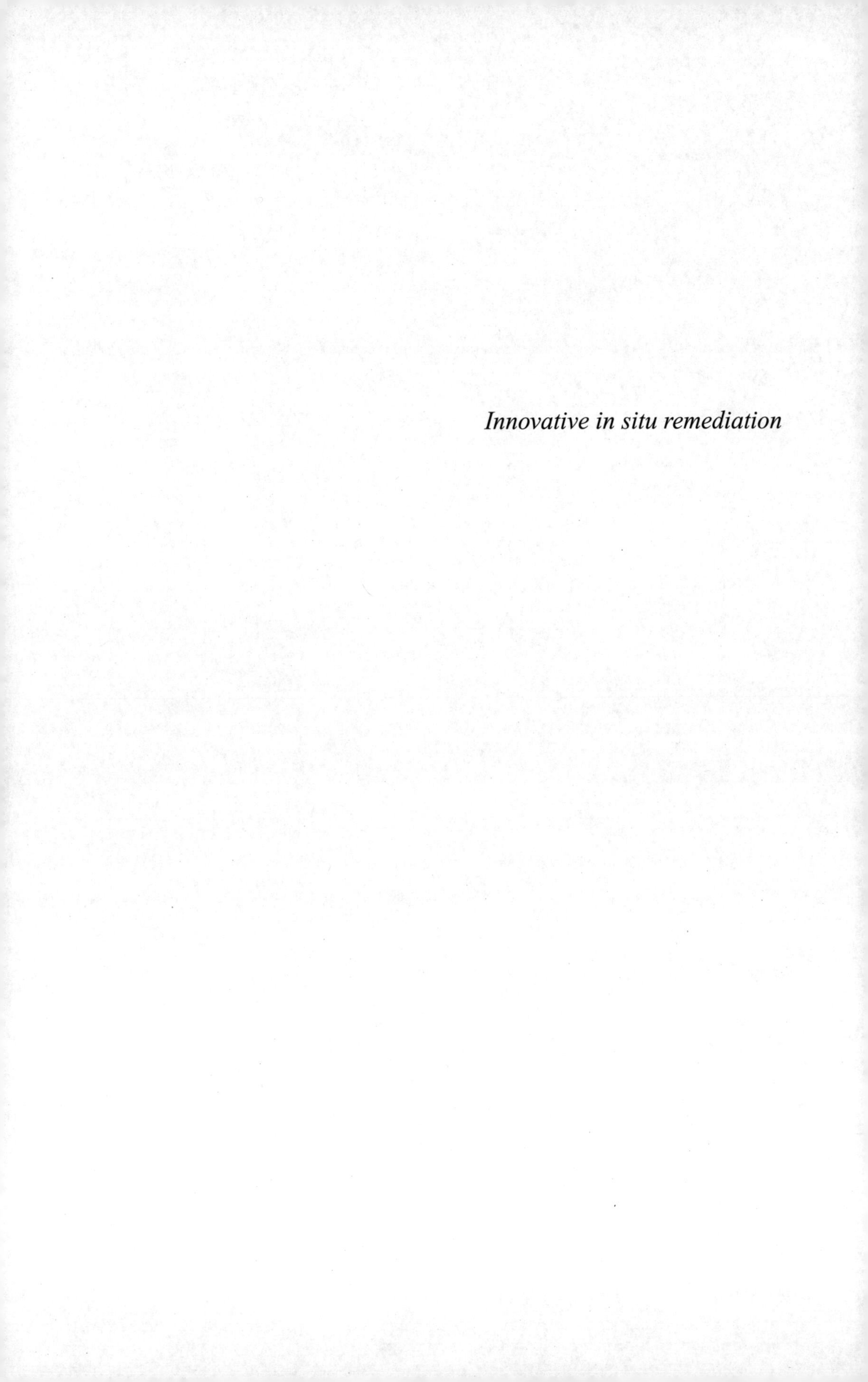

Innovative in situ remediation

European Symposium on Environmental Biotechnology, ESEB 2004 - Verstraete (ed)
© 2004 Taylor & Francis Group, London, ISBN 90 5809 653 X

Use of stable isotope fractionation for the assessment of bioattenuation of chlorinated ethenes

I. Nijenhuis, J. Andert, K. Beck, A. Vieth, M. Kästner & H.H. Richnow
Department of Bioremediation, UFZ Centre for Environmental Research Leipzig-Halle, Leipzig, Germany

G. Diekert
Institute for Microbiology, Friedrich-Schiller University, Jena, Germany

ABSTRACT: Estimation of the *in situ* biological degradation of pollutants at contaminated field sites is a difficult task. Assessment of biodegradation usually includes the application of both, culture dependent and molecular biological, culture independent, techniques. More recently, the application of stable isotope fractionation has been suggested as a tool to monitor and estimate biodegradation of pollutants at contaminated sites. Application of stable isotope techniques relies on the observation that during biological conversion of low molecular weight organic substrates, the lighter stable isotope (^{12}C or ^{1}H) is preferentially used over the heavier stable isotope (^{13}C or ^{2}D) resulting in an enrichment of the heavier isotope in the remaining substrate and relatively higher amounts of the lighter isotope in the products. Experimentally determined compound specific isotope fractionation factors can then be used to estimate the extent of degradation. Our laboratory investigated a deep and a shallow aquifer at a tetrachloroethene (PCE) contaminated site in the Leipzig area. In both aquifers a decrease in the concentration of PCE was observed but only in the shallow aquifer a significant enrichment of ^{13}C was observed in the residual amount of PCE indicating microbial degradation. Using compound specific isotope fractionation factors published in literature, the extent of degradation on a groundwater flow path down stream of a contamination source could be estimated. Fractionation factors for PCE are currently only available for a few mixed cultures but no pure culture data is available. Preliminary laboratory data indicated significant isotope fractionation of PCE by pure cultures of *Sulfurospirillum multivorans* (formerly *Dehalospirillum multivorans*), *Desulfitobacterium* sp. PCE-S and a mixed culture containing *Dehalococcoides ethenogenes* during reductive dechlorination. A more extensive database of compound specific stable isotope fractionation data for mixed and pure cultures will be required to validate the stable isotope approach for the quantification of biodegradation of environmental pollutants.

1 INTRODUCTION

The chlorinated solvents, tetrachloroethene (PCE) and trichloroethene (TCE) are among the most common groundwater pollutants and are considered to be carcinogenic agents (ASTDR 2003; WHO 2003). Because of their properties, these solvents were used on a large scale in the dry-cleaning industry as well as for the degreasing of metal parts. As a result of spillage, many sites exist in the USA and Europe with contamination of PCE or TCE (EPA 2003; WHO 2003). One cost efficient technology for the treatment of these contaminated sites is the use of natural attenuation or enhanced natural attenuation, using the potential of the natural microflora for the detoxification of pollutants (Lee *et al.* 1998).

While PCE persists under aerobic conditions, under strict anaerobic conditions, which are present in most contaminated aquifers, the chlorinated ethenes can be reductively dechlorinated by microorganisms to the harmless endproduct of ethene (Tandoi *et al.* 1994; Maymo-Gatell *et al.* 1997; Lovley 2001). Assessment of natural attenuation is a complicated task since concentrations of contaminants will depend on several factors including sorption, dilution and biodegradation. A decrease in contaminant concentration alone, therefore, does not provide evidence for biological degradation. Traditionally used methods for the assessment of *in situ* biodegradation potential involve the use of culture dependent methods including microcosms. These culture dependent techniques, while giving valuable information, are labor intensive and time consuming and may not reflect the *in situ* potential for biodegradation. More recently, molecular biological techniques for the detection of specific microorganisms have been included in the assessment strategy but the presence of a molecular marker for either the organism or the desired biodegrading enzyme does not necessarily result in actual *in situ* activity (van de Pas *et al.* 2001; Siebert *et al.* 2002). In recent years, the application of stable isotope techniques has been suggested for the assessment of natural attenuation *in situ*. These techniques depend on the fact that biological systems prefer substrates with the lighter stable isotope (^{12}C or ^{1}H) over the heavier stable isotope (^{13}C or ^{2}D). This preference results in the enrichment of the heavier stable isotope in the residual substrate which then can be used as a measure for the extent of biodegradation. The preference of the biological system for the lighter stable isotope over the heavier one can be expressed as the compound-specific stable isotope fractionation factor (αC, αH/D). This factor may depend on a multitude of factors including biodegrading microorganism, reaction mechanism, pathway of degradation, enzyme etc. To validate the use of stable isotope techniques for the assessment of biologically mediated natural attenuation, fractionation factors determined in laboratory studies are required. A few studies have reported stable isotope fractionation of PCE and TCE by mixed reductively dechlorinating cultures, several tested reductive dechlorination of TCE (Sherwood Lollar *et al.* 1999; Bloom *et al.* 2000; Slater *et al.* 2001), but only one study has been reported using PCE (Slater *et al.* 2001).

A variety of microorganisms is capable of the reductive dechlorination of PCE and TCE, many can partially dechlorinate to *cis*-DCE, while some organisms, belonging to the genus of *Dehalococcoides* are capable of complete dechlorination to ethene (Maymo-Gatell *et al.* 1997; Holliger *et al.* 1998; He *et al.* 2003).

In this study, *Sulfurospirillum multivorans* (formerly *Dehalospirillum multivorans*) and *Desulfitobacterium* sp. PCE-S cultures capable of reductive dechlorination PCE to *cis*-DCE and a mixed culture containing *Dehalococcoides ethenogenes* dechlorinating PCE to vinyl chloride and ethene were used to investigate compound-specific stable isotope fractionation of PCE.

2 RESULTS

2.1 *Field data*

Two PCE contaminated aquifers, a deep and a shallow one, at the site of a former dry-cleaning plant in Leipzig, Germany, were analysed for isotopic ratio and concentration of pollutant (Vieth *et al.* 2003). In both the shallow and deep aquifers, a decrease in the concentration of PCE was observed down the flow path. Only in the shallow aquifer, an enrichment in ^{13}C isotopes in the residual PCE was observed while in the deeper aquifer the isotope ratio remained constant along the groundwater flow path. Decrease in concentration of the pollutant may result from dilution, absorption and abiotic and biotic degradation. Absorption and dilution have not been observed to cause significant isotope fractionation while biological degradation has been found to induce an isotopic shift in the pollutant (Richnow *et al.* 2002). These results suggest microbial degradation of PCE in the shallow aquifer but not in the deep aquifer (Vieth *et al.* 2003).

2.2 *Laboratory data*

Pure cultures of *Sulfurospirillum multivorans* (Scholz-Muramatsu *et al.* 1995), *Desulfitobacterium* sp. strain PCE-S (Miller *et al.* 1997) and a mixed culture containing *Dehalococcoides ethenogenes* (Fennell *et al.* 1997) were used to determine compound specific stable isotope fractionation factors for PCE during reductive dechlorination. The concentration gradient and change in stable isotope composition of PCE were determined over time during growth of these cultures. The stable isotope fractionation factor, αC, was then determined using the Rayleigh-equation (Hoefs 1997):

$$\frac{R_t}{R_0} = \left(\frac{C_t}{C_0}\right)^{\left(\frac{1}{\alpha}-1\right)} \tag{1}$$

R_t and R_0 are the isotope composition of PCE at times t and zero and C_t and C_0 the concentration of PCE at time t and zero respectively.

The stable isotope fractionation factor (αC) for the reductive dechlorination of PCE by *Desulfitobacterium* sp. strain PCE-S (αC = 1.0045) and the mixed culture containing *Dehalococcoides ethenogenes* (αC = 1.0048) were in the same order of magnitude as the previously published value (αC = 1.0055) by Sherwood Lollar *et al.* (Sherwood Lollar *et al.* 2001). Interestingly, the stable isotope fractionation factor for PCE by *S. multivorans* (αC = 1.0005) was one order of magnitude lower compared to the other values.

3 DISCUSSION AND CONCLUSION

Stable isotope fractionation has been observed for PCE in the field and in microbial cultures in laboratory studies. We observed a ten fold difference in stable isotope fractionation factor comparing *S. multivorans* (α = 1.0005) to *Desulfitobacterium* sp. strain PCE-S and the *Dehalococcoides ethenogenes* containing mixed culture (α = 1.0045–1.0048). This difference could be due to several factors including the enzyme,

enzyme location, and reaction mechanism. The cultures used in this study all dechlorinate PCE via TCE and the reductive dehalogenases of these microorganisms are all thought to contain corrinoid co-factors in the reactive center (Neumann *et al.* 1996; Magnuson *et al.* 1998; Miller *et al.* 1998). We therefore assume that the reaction mechanism is similar for the tested cultures and would not cause a significant effect in isotope fractionation. Both the enzyme itself and its location could cause the difference in isotope fractionation. The reductive dehalogenases of *Desulfitobacterium* PCE-S and *Dehalococcoides ethenenogenes* strain 195 are thought to be located on the outside of the cytoplasmic membrane, while the PCE reductive dehalogenase of *S. multivorans* is thought to be located in the cytoplasm (Miller *et al.* 1998; Neumann *et al.* 1998; Magnuson *et al.* 2000). The low isotope fractionation by *S. multivorans* could therefore be due to the diffusion or transport of PCE through the membrane. Additionally, the enzyme structure itself could cause the observed differences.

Using compound specific α's, the extent of degradation *in situ* can be estimated along a groundwater flowpath using the following equation (Vieth *et al.* 2003):

$$B[\%] = \left[1 - \left(\frac{R_t}{R_0}\right)^{\left(\frac{1}{\frac{1}{\alpha}-1}\right)}\right] \times 100 \qquad (2)$$

Where B% is the percentage of degradation which value can be calculated applying α and the isotope compositions of the contaminant at the source (R_0) and during the reaction (R_t).

While the concentration of a contaminant is affected by various factors, including dilution and adsorption, the isotope composition of the contaminant is only significantly affected by microbial degradation in case of steady state environmental conditions.

We found a ten fold difference in α between the different bacterial cultures. The estimated percentage biodegradation, based on isotope composition, will depend on the used stable isotope fractionation factors. Additionally, the calculation requires that R_0 is representative for the plume. Using the field data as reported by Vieth *et al.* (Vieth *et al.* 2003) and equation 3, B% was calculated, as shown in Figure 1. The low αC (1.0005) results in a higher estimated value for B and full estimated biodegradation in the second well, while the higher αC (1.00055) results in a lower estimated value for B. B only reflects the relative amount of biodegradation of PCE, not a concentration, and is therefore not a quantitative measure.

Stable isotope techniques appear to be a valuable tool for the *in situ* assessment of bioattenuation at polluted sites but the example in Figure 1 illustrates

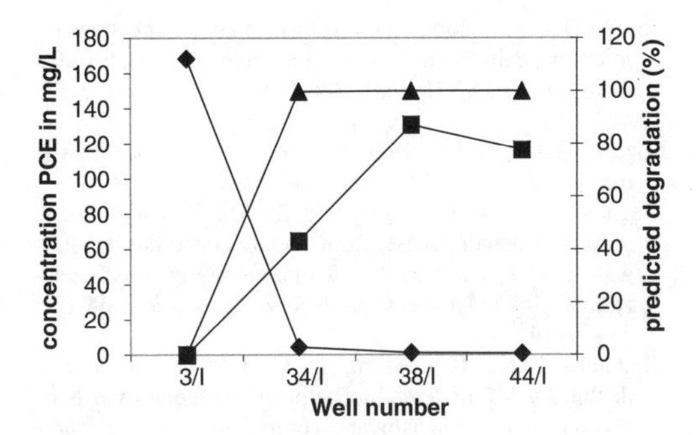

Figure 1. Estimation of the percentage of biodegradation (B%) using αC = 1.0005 (triangles) or αC = 1.0055 (squares) and the actual concentration of PCE (diamonds) in the wells in a flow path in the shallow aquifer as described by Vieth *et al.* (Vieth *et al.* 2003).

that use of stable isotope fractionation must be used with caution. Before application of these techniques, knowledge about the responsible *in situ* biodegrading microbial community must be obtained. Additionally, the relationships between biodegrading community or microorganism and stable isotope fractionation of the pollutant of interest should be studied.

ACKNOWLEDGEMENT

We would like to thank Dr. James Gossett, Cornell University, for providing the mixed culture containing *Dehalococcoides ethenogenes*. I.N. is funded through an EU Marie Curie Host Fellowship BIOISOTOPE Contract EVK1-CT-2000-56120.

REFERENCES

ASTDR (2003). Agency for toxic substances and drug registry, www.atsdr.cdc.gov.

Bloom, Y., R. Aravena, et al. (2000). "Carbon isotope fractionation during microbial dechlorination of trichloroethene, cis-dichloroethene, and vinyl chloride: implications for assessment of natural attenuation." *Environ. Sci. Technol.* **34**: 2768–2772.

EPA (2003). Environmental Protection Agency, www.epa.gov.

Fennell, D.E., J.M. Gossett, et al. (1997). "Comparison of butyric acid, ethanol, lactic acid, and propionic acid as hydrogen donors for the reductive dechlorination of tetrachloroethene." *Envir. Sci. Technol.* **31**: 918–926.

He, J., K.M. Ritalahti, et al. (2003). "Detoxification of vinyl chloride to ethene coupled to growth of an anaerobic bacterium." *Nature* **424**(6944): 62–65.

Hoefs, J. (1997). *Stable isotope geochemistry.* Berlin, Springer-Verlag.

Holliger, C., G. Wohlfarth, et al. (1998). "Reductive dechlorination in the energy metabolism of anaerobic bacteria." *FEMS Microbiology Reviews* **22**(5): 383–398.

Lee, M.D., J.M. Odom, et al. (1998). "New perspectives on microbial dehalogenation of chlorinated solvents: insights from the field." *Annual Review of Microbiology* **52**: 423–452.

Lovley, D.R. (2001). "Bioremediation. Anaerobes to the rescue." *Science* **293**(5534): 1444–1446.

Magnuson, J.K., M.F. Romine, et al. (2000). "Trichloroethene reductive dehalogenase from Dehalococcoides ethenogenes: sequence of tceA and substrate range characterization." *Applied and Environmental Microbiology* **66**(12): 5141–5147.

Magnuson, J.K., R.V. Stern, et al. (1998). "Reductive dechlorination of tetrachloroethene to ethene by a two-component enzyme pathway." *Applied and environmental Microbiology* **64**(4): 1270–1275.

Maymo-Gatell, X., Y. Chien, et al. (1997). "Isolation of a bacterium that reductively dechlorinates tetrachloroethene to ethene." *Science* **276**(5318): 1568–1571.

Miller, E., G. Wohlfarth, et al. (1997). "Comparative studies on tetrachloroethene reductive dechlorination mediated by *Desulfitobacterium* sp. strain PCE-S." *Archives of Microbiology* **168**(6): 513–519.

Miller, E., G. Wohlfarth, et al. (1998). "Purification and characterization of the tetrachloroethene reductive dehalogenase of strain PCE-S." *Archives of Microbiology* **169**(6): 497–502.

Neumann, A., G. Wohlfarth, et al. (1996). "Purification and characterization of tetrachloroethene reductive dehalogenase from Dehalospirillum multivorans." *The Journal of Biological Chemistry* **271**(28): 16515–16519.

Neumann, A., G. Wohlfarth, et al. (1998). "Tetra-chloroethene dehalogenase from *Dehalospirillum multivorans*: cloning, sequencing of the encoding genes, and expression of the pceA gene in Escherichia coli." *Journal of Bacteriology* **180**(16): 4140–4145.

Richnow, H.H., A. Vieth, et al. (2002). "Isotope fractionation of toluene: a perspective to characterise microbial *in situ* degradation." *The Scientific World Journal* **2**: 1227–1234.

Scholz-Muramatsu, H., A. Neumann, et al. (1995). "Isolation and characterization of *Dehalospirillum multivorans* gen nov, sp nov, a tetrachloroethene-utilizing, strictly anaerobic bacterium." *Arch. Microbiol.* **163**(1): 48–56.

Sherwood Lollar, B., G.F. Slater, et al. (1999). "Contrasting carbon isotope fractionation during biodegradation of trichloroethylene and toluene: Implications for intrinsic bioremediation." *Organic Geochemistry* **30**(1): 813–820.

Sherwood Lollar, B., G.F. Slater, et al. (2001). "Stable carbon isotope evidence for intrinsic bioremediation of tetrachloroethene and trichloroethene at area 6, Dover Air Force Base." *Environmental Science & Technology* **35**(2): 261–269.

Siebert, A., A. Neumann, et al. (2002). "A non-dechlorinating strain of *Dehalospirillum multivorans*: evidence for a key role of the corrinoid cofactor in the synthesis of an active tetrachloroethene dehalogenase." *Arch Microbiol* **178**(6): 443–9.

Slater, G.F., B.S. Lollar, et al. (2001). "Variability in carbon isotopic fractionation during biodegradation of chlorinated ethenes: implications for field applications." *Environmental Science & Technology* **35**(5): 901–907.

Tandoi, V., T.D. DiStefano, et al. (1994). "Reductive dehalogenation of chlorinated ethenes and halogenated ethanes by a high-rate anaerobic enrichment culture." *Environ. Sci. Technol.* **28**: 973–979.

van de Pas, B.A., H.J. Harmsen, et al. (2001). "A *Desulfitobacterium* strain isolated from human feces that does not dechlorinate chloroethenes or chlorophenols." *Arch. Microbiol.* **175**(6): 389–394.

Vieth, A., J. Müller, et al. (2003). "Characterisation of *in situ* biodegradation of tetrachloroethene (PCE) and trichloroethene (TCE) in contaminated aquifers monitored by stable isotope fractionation." *Isotopes in Environmental Health Studies* **39**(2): 113–124.

WHO (2003). Guidelines for drinking water quality, World Health Organization.

European Symposium on Environmental Biotechnology, ESEB 2004 - Verstraete (ed)
© 2004 Taylor & Francis Group, London, ISBN 90 5809 653 X

In-situ biodegradation of benzene and toluene in a contaminated aquifer monitored by stable isotope fractionation

A. Vieth[1], M. Kästner[1], M. Schirmer[2], H. Weiß[3], S. Gödeke[3], H.H. Richnow[1] & R.U. Meckenstock[4]
[1]*UFZ-Centre for Environmental Research, Department of Biormediation; Leipzig-Halle, Leipzig, Germany*
[2]*UFZ-Centre for Environmental Research, Department of Hydrogeology, Leipzig-Halle, Leipzig, Germany*
[3]*UFZ-Centre for Environmental Research Interdisciplinary, Department of Industrial and Mining Landscapes; Leipzig-Halle, Leipzig, Germany*
[4]*Eberhard-Karls University of Tübingen, Center for Applied Geosciences, Tübingen, Germany*

ABSTRACT: Intrinsic biodegradation of benzene and toluene in a heavily contaminated aquifer at the site of a former hydrogenation plant was investigated by means of isotope fractionation processes. The carbon isotope compositions of benzene and toluene were monitored in two campaigns during a time period of 12 months to assess the extent of the *in-situ* biodegradation and the stability of the plume over time. The spatial distribution of the carbon isotope composition of benzene suggested that *in-situ* biodegradation occurred at marginal zones of the plume where concentrations were lower than 30 mg l^{-1}. The investigation of the vertical structure of the benzene plume provided evidence for *in-situ* degradation processes at the upper and lower fringes of the plume. The Rayleigh-model, applied to calculate the extent of biodegradation and residual theoretical concentrations of toluene, showed that *in-situ* biodegradation was a relevant attenuation process. The biodegradation rate constant for toluene was estimated to be $k = -5.7 \pm 0.5\ \mu M\ d^{-1}$ on the ground water flow path downstream of the source area. The results show that isotope fractionation can be used to quantify microbial *in-situ* degradation in contaminated aquifers and to develop conceptual models for natural attenuation approaches.

European Symposium on Environmental Biotechnology, ESEB 2004 - Verstraete (ed)
© 2004 Taylor & Francis Group, London, ISBN 90 5809 653 X

Comparison of different MULTIBARRIER concepts designed for treatment of groundwater containing mixed pollutants

L. Bastiaens, J. Dries, J. Vos, Q. Simons, M. De Smet & L. Diels
Vito (Flemish Institute for Technological Research), Mol, Belgium

ABSTRACT: A study was set up to develop and evaluate the possibilities of multibarriers for the remediation of groundwater contaminated with a mixed pollution. A multibarrier is a multifunctional permeable reactive barrier in which biological and physicochemical pollutant removal processes can be combined. Different combinations of biodegradation, reductive dehalogenation using zerovalent iron and sorption were considered to treat a mixed pollution consisting of heavy metals (As, Zn), aromatics (Benzene, toluene, *m*-xylene) and volatile chlorinated compounds (PCE, TCE). Biodegradation of BTEX in batch-sytems was found to be negatively influenced by the presence of zero-valent iron and other pollutants. Starting from the same aquifer mixture, addition of different electron-acceptors selected for a different BTEX-degrading microbial population as determined by PCR-DGGE. Based on column experiments which simulated different multibarrier concepts, sequential as well as mixed (Fe(0)/bio) multibarriers proved to be suitable for the treatment of the tested pollutant mixture. Good results were obtained with O_2, NO_3^-, SO_4^{2-}, Fe^{3+} as terminal electron acceptor.

1 INTRODUCTION

In order to remediate contaminated groundwater, many techniques have been developed and applied. However, groundwater is often polluted with complex mixtures of different chemicals and most remediation techniques only deal with one or a few pollutant types. Sanitation of groundwater polluted with mixtures of hazardous compounds has received wrongly less attention and remains a problem. A combination of different existing technologies may be a solution, but one should consider (I) the impact of one remedition technique on another one and (II) the influence of co-contaminants on the removal efficiency of the processes.

This study focusses on MULTIBARRIERs, i.e. permeable reactive barriers in which different pollutant removal processes (biological and physicochemical processes) are combined to treat in situ groundwater containing mixed pollutants (Figure 1). Different MULTIBARRIER concepts to treat groundwater were designed, evaluated and compared. One of the objectives of the study was to answer the question whether the removal processes should be applied one after the other (sequential MULTIBARRIER), or whether a combination of different processes in one zone (mixed MULTIBARRIER) is also possible. The latter may require more optimisation but the installation is expected to be less complex and less expensive.

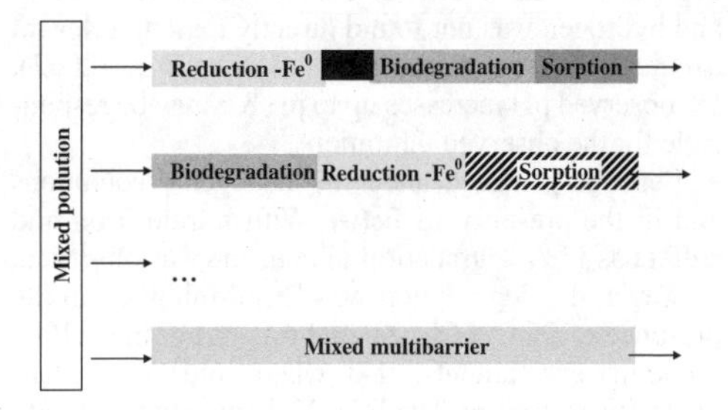

Figure 1. Different multibarrier concepts.

Another important question concerns the choice of the terminal electron acceptors (TEA) for the biological process.

2 METHODS AND RESULTS

2.1 *Batch enrichment experimental set-ups*

A mixture of 13 different BTEX, VOCls and/or heavy metal contaminated aquifer samples was used as starting material for the enrichment of benzene(B), toluene (T) and *m*-xylene (*m*X) degrading bacteria under denitrifying, sulfate reducing, iron reducing and methanogenic conditions. The enrichments were

Table 1. Summary of BTX degradation activity during the enrichment experiment: B/T/X, significant benzene, toluene or m-xylene degradation; 0, no degradation; N/D, not determined.

	Benzene only		BTX mixture‡		Mixed pollution*	
TEAP†	$-Fe^{0}$♂	$+Fe^{0}$♀	$-Fe^{0}$♂	$+Fe^{0}$♀	$-Fe^{0}$♂	$+Fe^{0}$♀
Denitrification	B	N/D	BTX	N/D	N/D	N/D
Iron reduction	B	0	BTX	TX#	TX	TX#
Sulfate reduction	B	0	TX	TX#	TX	TX#
Methanogenesis	N/D	N/D	TX	TX#	TX	TX#

†terminal electron acceptor process; ♂ 45 g mixed aquifer + 30 g sand + 150 ml medium; ♀ + 15 g Fe^{0} FeA4.
‡benzene, toluene and m-xylene at 2 mg/l each, *BTX at 2 mg/l each, PCE (2 mg/l), TCE (5 mg/l), Zn (5 mg/l) and As (0.2 mg/l).
#degradation halted due to pH increase.

performed in closed 250 ml vial containing 45 g mixed aquifer, 30 g filter sand and 150 ml medium. The effect of zero-valent iron (Fe(0)) and other pollutants (PCE, TCE, Zn, As) on the enrichment of anaerobic benzene and BTmX-degrading bacteria was evaluated.

Table 1 summarizes the observed BTEX degradation potential over 300 days of incubation under the different enrichment conditions. Toluene and *m*-xylene were rapidly degraded in all enrichment sets without Fe(0), and the degradation rates increased after repeated injection of new substrate. In the presence of Fe(0) the degradation halted after less than 100 days. As Fe(0) and hydrogen was not found directly toxic to relevant axenic BTEX-degrading strains (Dries et al. 2002), the observed pH increases up to pH 8,5 may be responsible for the observed inhibition.

Benzene persisted under methanogenic conditions and in the presence of Fe(0). With nitrate, iron and sulfate as TEA degradation of benzene was observed. However, the degradation was found inhibited in the presence of PCE, TCE, Zn and As, and in the sulfate reducing enrichments also when toluene and *m*-xylene were present. The BTmX-degradation capacity remained active after several (10%) transfers.

The evolution of the microbial population in the denitrifying, sulfate reducing and iron reducing enrichment cultures and the transfer cultures was followed by PCR-DGGE using general eubacterial primers amplifying a 450 fragment of the 16S rRNA gen (63F & GC-518R). Especially after a few transfers the complexity of the obtained DGGE DNA patterns decreased and a few dominant band appeared in the sulfate and iron reducing enrichments; On the other hand, the observed population remained rather constant and complex in the denitrifying transfers. The Obtained PCR-DGGE DNA patterns were different depending on the added TEA, indicating that the added TEA selected for a different population (results not shown). The dominant bands in the different DGGE-patterns have been cloned and are being sequenced.

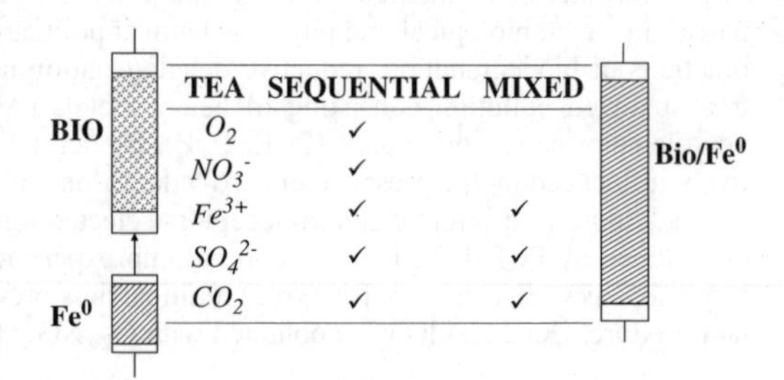

Figure 2. Potentially interesting multibarrier concepts.

2.2 *Simulation of different multibarrier concepts in laboratory scale column systems*

Potentially interesting sequential and mixed multibarrier concepts depending on the TEA are given in Figure 2. As oxygen and nitrate are known to be transformed rapidly by zero-valent iron and to have a negative influence on the performance of zero-valent iron, only sequential MULTIBARRIERS (Fe0 + BIO) were found potentially interesting with these TEAs.

Column tests were set-up to evaluate some of these sequential and mixed MULTIBARRIERS (Table 2). A model pollutant mixture was defined consisting of (i) the heavy metals zinc (5 mg/l) and arsenate (0.2 mg/l), (ii) the chlorinated ethenes PCE (2 mg/l) and TCE (5 mg/l), and (iii) the aromatic hydrocarbons benzene, toluene and *m*-xylene (BTX, 2 mg/l each). The artificial groundwater used in the tests containing 0.5 mM $NaHCO_3$, 0.5 mM $KHCO_3$, 0.5 mM $CaCl_2.6H_2O$, 0.5 mM $MgCl_2.6H_2O$. The investigated pollutant removal processes were reductive dehalogenation of the chlorinated ethenes (VOCls) with zero valent iron, sorption/reduction of the metals and biodegradation of BTmX and also some VOCls. In the biobarrier oxygen (system A), nitrate (system B), sulfate (system D) as well as iron (system E) were added as terminal electron acceptor (TEA). In all

Table 2. Characteristics of different MULTIBARRIER laboratory scale column systems.

Column trains	Column characteristics (TEA)	Acronym	Fe (0) (w%)	FS (w%)	Aquifer (w%)	HRT	Remarks
System A: Zerovalent iron wall followed by a micro-aerobic biobarrier							
A1	1-Fe(0)column	A1-1	100	–	–	13 h	Aquifer = aquifer from near landfill
	2-Bio column (O_2)	A1-2	–	80	20	13 h	TEA: Oxygen supplied as H_2O_2
System B: Zerovalent iron wall followed by a denitrifying biobarrier							Aquifer = MB-aquifer enriched
B1	1-Fe(0)column	B1-1	100	–	–	16 h	under denitrifying conditions
	2-Bio column (NO_3^-)	B1-2	–	80	20	16 h	TEA:10 mM NO_3^-
System C: Partially mixed Fe(0)/biowall followed by an anerobic biowall							Aquifer = MB-aquifer (mixture of 13
C	1-mixed zone	C1-1	20	60	20	1 w	contaminated aquifer materials)
	2-bio zone	C1-2	–	78	22	2 w	TEA: Not added
System D: sequential and mixed barriers under sulfate reducing conditions							
D1	1-Bio column (SO_4^{2-})	D1-1	–	80	20	22 h	
D2	1-Fe(0)column	D2-1	35	–	65	22 h	Aquifer = MB-aquifer enriched
	2-Bio column (SO_4^{2-})	D2-2	–	80	20	22 h	under sulphate reducing conditions
D4	1-Mixed column (SO_4^{2-})	D4-1	63	24	13	20 h	TEA: 0.5 mM SO_4^{2-}
System E: sequential and mixed barriers under iron reducing conditions							
E1	1-Bio column (Fe^{3+})	E1-1	–	80	20	19 h	
E2	1-Fe(0)column	E2-1	35	–	65	21 h	Aquifer = MB-aquifer enriched
	2-Bio column (Fe^{3+})	E2-2	–	80	20	19 h	under iron reducing conditions
E4	1-Mixed column (Fe^{3+})	E4-1	63	24	13	21 h	TEA: 2.5 mM Fe(III)EDTA

TEA: terminal electron acceptor; HRT: estimated hydraulic retention time; h: hours; w: weeks; FS: filter sand.

concepts, the sorption part was considered mainly as a polishing step. In one column set-up (system C) no TEA was added leaving the Fe(III) generated during corrosion of the zero-valent iron or CO_2 as only TEAs.

Table 3 summarises the removal efficiencies of the organic compounds. In the sequential multibarrier simulated in system A and B, the chlorinated compounds PCE and TCE were removed for more than 80% in the Fe(0) column, while the major part of the BTmX compounds were degraded in the biocolumn. In the micro-aerobic system the removal was improved by a higher dissolved oxygen concentration.

PCE and TCE were removed for more than 95% in the mixed zone of system C. 20 to 50% of the BTmX compounds were biodegraded as determined by comparing the results of the test column train with a formaldehyde killed set-up.

In the column trains of system D and E PCE and TCE were only removed when Fe(0) was present. With the mixed system higher removal efficiencies were obtained in comparison with the sequential set up. Under sulfate reducing conditions the degradation of BTmX was the less efficient, explicable by the observed pH raises.

No of the added heavy metals were detected in the effluents of the multibarrier systems, except in the columns where Fe(III)EDTA was added as TEA. This is in agreement with the results of other tests where a very high retention of Zn and As on the zero-valent iron and the used aquifer mixture was observed (results not shown).

3 CONCLUSION

The use of zero-valent iron for the remediation of groundwater polluted with certain chlorinated compounds is a proven technology at this moment (Gillham 1996; O'Hannesin and Gillham, 1998, Bastiaens et al., 2002a, 2002b). As it concerns an abiotic reductive removal process, not much attention has been given to the microbiology in zero-valent iron barriers during many years. However, recently it has been shown that bacteria are presence in and near iron barriers and that they may play an important role in the functioning of the barriers (Scherer et al, 2000; Gu et al., 2002; Van Eekert et al., 2003). Especially the hydrogen form during anaerobic corrosion of the iron may select for certain bacterial groups.

In a first part of the presented study experiments were performed to examine the compatibility between key-compounds in multibarriers, being zero-valent iron, BTmX-degrading bacteria, TEA and a mixed pollution. Zero-valent iron was found to have only an indirect negative effect on BTmX-degrading micro-organisms due to pH-effects. Further the presence of other pollutants reduced the benzene removal efficiency. Interesting BTmX-degrading cultures were

Table 3. Organic pollutant removal efficiencies (%) in the different MULTIBARRIER columns.

Column trains	Characteristics (TEA)	Acronym	PCE	TCE	Benzene	Toluene	*m*-xylene	Remarks
System A: Zerovalent iron wall followed by a micro-aerobic biobarrier[1]								
A1	1-Fe(0)column	A1-1	60–80	90–99	4–13	10–53	NA	
	2-Bio column (O_2)	A1-2	80–100	92–100	99–100	99–100	>90	
System B: Zerovalent iron wall followed by a denitrifying biobarrier[1]								
B1	1-Fe(0)column	B1-1	66–99	97–100	6–18	16–54	NA	
	2-Bio column (NO_3^-)	B1-2	80–93	97–100	89–100	99–100	>90	
System C: Partially mixed Fe(0)/biowall followed by anaerobic biowall[2,1]								
C1	1-mixed zone	C1-1	95–100[1]	95–100[1]	20[2]	30[2]	40[2]	pH increase
	2-bio zone	C1-2	95–100[1]	95–100[1]	20[2]	30[2]	50[2]	in mixed zone
System D: sequential and mixed barriers under sulfate reducing conditions[1]								
D1	1-Bio column (SO_4^{2-})	D1-1	0–11	0–15	11–44	97–100	NA	
D2	1-Fe(0)column	D2-1	55–71	88–98	4–54	0–36	NA	
	2-Bio column (SO_4^{2-})	D2-2	55–78	89–98	23–67	12–36	<25	With Fe(0)
D4	1-Mixed column (SO_4^{2-})	D4-1	76–89	99–100	18–2	7–23	>90	pH up to 8,5
System E: sequential and mixed barriers under iron reducing conditons[1]								
E1	1-bio column (Fe^{3+})	E1-1	0–11	9–18	24–29	100	NA	
E2	1-Fe(0)column	E2-1	53–67	85–92	50–68	72–91	NA	
	2-Bio column (Fe^{3+})	E2-2	54–67	86–92	52–71	77–90	>90	
E4	1-Mixed column (Fe^{3+})	E4-1	58–73	94–97	55–64	94–98	>90	

[1]Removal efficiencies determined based on the concentration in het effluent of a (sub) column and the concentration in the influent of the column train;[2] Removal efficiencies determined relative to the poisonned control.
NA: Not available (yet);

enriched from a aquifer mixture under denitrifying, sulfate and iron reducing conditions.

In the second part of the study, the enriched cultures were used as inoculum for column set ups simulating different multibarrier concepts. Both mixed and sequential MULTIBARRIER configurations showed to be suitable for sanitation of mixed groundwater pollution. Indications were found for an improved removal in mixed systems. In all tested MULTIBARRIER concepts chlorinated ethenes and heavy metal removal (except system D) was observed when zerovalent iron was present. Biodegradation of BTmX was observed under aerobic, denitrifying and iron reducing conditions, but to a much lesser extend when sulfate was present as TEA.

In a third stage of this work, a partially mixed MULTIBARRIER was installed on pilot scale in a container system (5 m × 2.4 m × 2.4 m) in which an aquifer is simulated. The tested MULTIBARRIER consists of a mixed Fe0 + BIO zone followed by an anaerobic BIO-zone and a sorption zone. Iron (III) originating from the corrosion of the zero-valent iron was selected as TEA. Besides monitoring of the chemical composition of the groundwater and the field parameters, in situ mesocosm socks and molecular techniques like PCR-DGGE are being used to monitor changes in the microbial population in the different zone.

ACKNOWLEDGEMENTS

This work was funded by a VITO Ph.D grant (to J.Dries) and by EC project n° QLRT-2000-00163 (MULTIBARRIER).

REFERENCES

Bastiaens, L., Maesen, M., Vos, J., Kinnaer, L., Diels, L., Nuyens, D., O'Hannesin, S., 2002a. Permeable reactive iron barrier for treatment of groundwater highly polluted with TCE. Monterey

Bastiaens, L., Vos, J., Maesen, M., Kinnaer, L., Diels, L., Weytingh, K., van de Velde, A., Berndsen, E., Peene, A., O'Hannesin, S., 2002b. Feasibility studie, design and implementation of a European permeable reactive iron barrier. Monterey

Dries, J., Bastiaens, L., Springael, D., Diels, L., Agathos, S.N., 2002. The effect of zero valent iron on the enrichment of anaerobic BTX degrading microbiota under mixed multibarrier conditions. Sixth International Symposium on Environmental Biotechnology and Fourth International SYmposium on cleaner bioprocesses and sustainable development, Veracruz, June 9–12.

Dries, J., Geuens, S., Bastiaens, L., Springael, D., Agathos, S.N., Diels, L. 2003. Multibarrier, a technology concept for in-situ remediation of mixed groundwater pollution. CONSOIL 2003. 1650–1656.

Gillham, R.W., 1996. In-situ Treatment of groundwater: Metal-enhanced degradation of chlorinated Organic Contaminants, In: Aral M.M., (ed.), Advances in groundwater Pollution Control and Remediation, Kluwer Academic Publishers, pp 249–274.
Gu, B., Watson, D.B., Wu, L., Phillips, D.H., White, D.C., Zhou. J., 2002. Microbiological characteristics in a zero-valent iron reactive barrier. Environmental Monitoring and Assessment 77: 293–309.
O'Hannesin, S.F., and Gillham, R.W., 1998. Long-term performance of an In-situ "Iron Wall" for remediation of VOCs. Ground Water, 36:164–170.
Scherer, M.E., Richter, S., Valentine, R.L., Alvarez, P.J.J., 2000. Chemisrty and microbiology of permeable reactive barriers for in-situ groundwater clean up. Critical Rev. in Environmental Science & technology. 30(3):363–411.
Van Eekert, M., Draaisma, C., Diels, L., Schraa, G., 2003. Treatment of multiple pollution in a multibarrier system. CONSOIL 2003, 1666–1674.

European Symposium on Environmental Biotechnology, ESEB 2004 - Verstraete (ed)
 ISBN 90 5809 653 X

Biodegradation of fuel oxygenates by MTBE-degrading microorganisms

F. Fayolle[1], D. Lyew[1,2], A. François[1], H. Mathis[1], S.R. Guiot[2], F. Monot[1]
[1]*Institut Français du Pétrole, Rueil-Malmaison, France*
[2]*Biotechnology Research Institute, NRC, Montreal, Canada*

ABSTRACT: Methyl *tert*-butyl ether (MTBE) and ethyl *tert*-butyl ether (ETBE) are widely used as additives to gasoline and are frequently detected in surface and ground waters. *Rhodococcus ruber* IFP 2007, isolated for its capacity to grow on ETBE owing to the presence of a cytochrome P450, is able to convert both ETBE and MTBE to *tert*-butyl alcohol (TBA). *Mycobacterium austroafricanum* IFP 2012 possesses a different oxidation system and is able to use MTBE as a carbon and energy source. The capacities of these strains to degrade fuel oxygenates have been investigated and compared. Their performances in lab-scale biofilters have been determined and the process is being optimized prior to its implementation in a pilot-scale biobarrier reactor.

1 INTRODUCTION

Methyl *tert*-butyl ether (MTBE) is used as an additive to gasoline to increase its octane index and to decrease release of CO and unburnt hydrocarbons in the exhaust gases. In France and in Spain, ethyl *tert*-butyl ether (ETBE) is also used. The presence of MTBE in ground- and surface waters was first detected in the USA, and several authors have also recently reported contamination of aquifers in Europe. The contamination of aquifers was caused by the release of MTBE-supplemented gasoline into the environment. The frequent presence of MTBE in groundwaters is mainly due to the high solubility of this compound and to its apparent recalcitrance to biodegradation under both aerobic and anaerobic conditions (Fayolle *et al.*, 2001). Its half-life in contaminated aquifers was estimated to be 2 to 3 years. However, the biodegradation of MTBE and of its degradation intermediate, *tert*-butyl alcohol (TBA) by cometabolism has been observed (Fayolle *et al.*, 2001). Recent studies have isolated bacterial strains capable of growing on these oxygenates. Hernandez-Perez *et al.* (2001) isolated a strain, *Rhodococcus ruber* IFP 2007, which was able to grow on ETBE. Three strains, *Rubrivirax gelatinosus* PM1 (Hanson *et al.*, 1999), *Hydrogenophaga flava* ENV735 (Hatzinger *et al.*, 2001) and *Mycobacterium austroafricanum* IFP 2012 (François *et al.*, 2002) have been isolated for their capacity to grow on MTBE as a sole carbon and energy source.

This paper is focused on the degradation of these fuel oxygenates by *R. ruber* IFP 2007 and by *M. austroafricanum* IFP 2012. The characterization of these strains would provide data that may be useful in their application in the remediation of contaminated sites.

2 GROWTH OF *R. RUBER* IFP 2007 AND *M. AUSTROFRICANUM* IFP 2012 ON OXYGENATES

Several strains have been isolated from different sources by enrichment on either fuel ethers or on their derived tertiary alcohols. These strains have been identified by 16S rDNA sequencing, and their degradation capacities have been studied (Table 1).

Table 1. Strains isolated on ethers or their derived alcohols.

Ether or alcohol degradation capacities	Reference	Genotypic identification
ETBE to TBA	IFP 2007	*Rhodococcus ruber*
	IFP 2005	*Rhodococcus zopfii*
	IFP 2009	*Mycobacterium* sp.
TBA to CO_2	IFP 2011	*Stenotrophomonas maltophilia*
	IFP 2003	*Burkholderia cepacia*
MTBE to CO_2 TBA to CO_2	IFP 2012	*Mycobacterium austroafricanum*

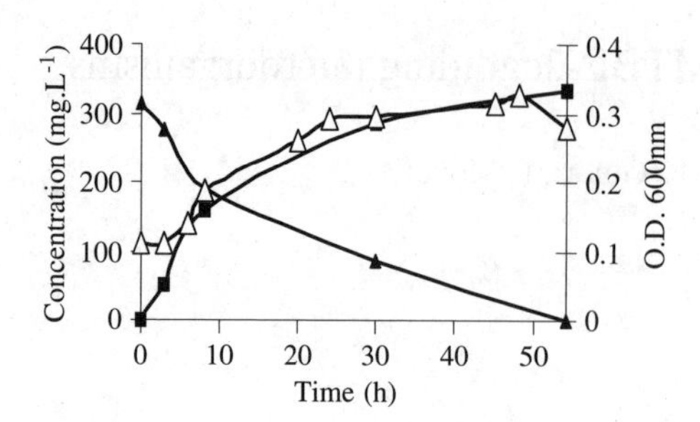

Figure 1. Growth of *R. ruber* IFP 2007 on ETBE (△-△: optical density at 600 nm ; ▲-▲: ETBE; ■-■: TBA).

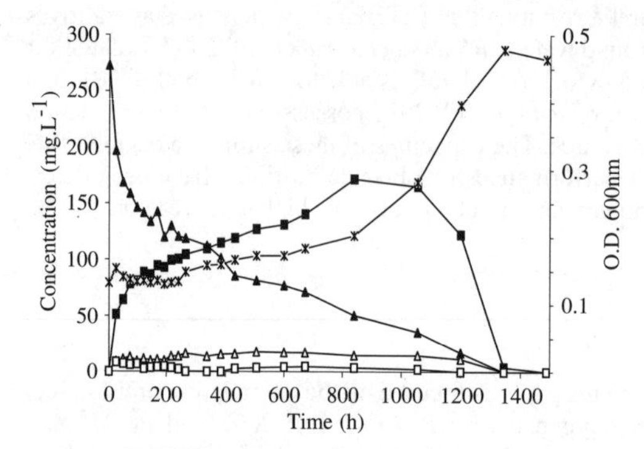

Figure 2. Growth of *M. austroafricanum* IFP 2012 on MTBE (✱-✱ : optical density at 600 nm ; ▲-▲: MTBE; ■-■: TBA; △-△: TBF; □-□: HIBA).

The growth of *R. ruber* IFP 2007 on ETBE is presented in Figure 1. ETBE was only partially utilized and *R. ruber* IFP 2007 grew on the C-2 compound liberated by the breakage of the ether bond, resulting in a stoichiometric accumulation of TBA in the culture medium. *R. ruber* IFP 2007 was not able to grow on either MTBE or TAME.

The growth of *M. austroafricanum* IFP 2012 on MTBE was slow and several degradation intermediates were detected in the culture medium: *tert*-butyl formate (TBF), TBA and 2-hydroxyisobutyric acid (HIBA), leading to the partial characterization of the MTBE degradation pathway (François *et al.*, 2002). The growth comprised two phases: (i) a stationary phase corresponding to the conversion of MTBE to TBF and TBA, and (ii) growth on the TBA formed (Figure 2).

The time required for growth of *M. austroafricanum* IFP 2012 on MTBE was 25 times higher than that for *R. ruber* IFP 2007 growing on a similar concentration of ETBE. However, *M. austroafricanum* IFP 2012 was able to grow more efficiently on TAME (result not shown) than on MTBE since growth on MTBE was 3 times longer than on TAME at similar ether concentrations. No growth of *M. austroafricanum* IFP 2012 was observed on ETBE.

Table 2. Efficiency of biomass production during growth on MTBE and its degradation intermediates.

Microorganism	Substrate	$Y^{exp}_{X/S}$ (*) (a)	$Y^{theo}_{X/S}$ (*) (b)	δ_e (c)
R. gelatinosus PM1	MTBE	15.8	92.1	5.8
H. flava ENV735	MTBE	35.2	92.1	2.6
M. austroafricanum	MTBE	33.4	92.1	2.8
IFP 2012	TBA	45.1	73.7	1.6
	HIBA	43.9	55.3	1.3

(*) g cell dry weight.mol^{-1}.
(a) $Y^{exp}_{X/S}$, experimental biomass yield.
(b) $Y^{theo}_{X/S}$, theoretical biomass yield (cell dry weight/mol substrate) $= 3.07 \times e$, where e is equal to the number of electrons available through hydrolysis of substrates:

$$\text{Substrate} + y\,H_2O \text{ -----> } z\,CO_2 + (e/2)H_2.$$

(c) δ_e, energy discrepancy index $= Y^{theo}_{X/S}/Y^{exp}_{X/S}$.

It has to be noted that MTBE is a poor growth substrate. The efficiency of biomass production was compared with that of *R. gelatinosus* PM1 and *H. flava* ENV735 and also to that obtained on two of the intermediates of MTBE degradation, TBA and HIBA (Table 2). The values of discrepancy index also show that MTBE is better utilized by *M. austroafricanum* IFP 2012 and *H. flava* ENV735 than it is by *R. gelatinosus* PM1.

3 DEGRADATION OF ETHERS BY RESTING CELLS OF *R. RUBER* IFP 2007 AND *M. AUSTROFRICANUM* IFP 2012

The capacity to degrade MTBE was tested using resting cells of *R. ruber* IFP 2007 harvested after growth on ETBE. MTBE was degraded to TBA and formate (Hernandez-Perez *et al.*, 2001). The strain was unable to use either TBA or formate as carbon and energy source. Since the ultimate goal is bioremediation of MTBE-contaminated sites using this strain, the use of an alternative carbon source had to be considered. Accordingly, we tested its capacity to degrade MTBE in the absence (Figure 3a) and in the presence (Figure 3b) of a co-substrate, isopropanol.

The MTBE degradation rates were 33.6 ± 0.7 and 31.1 ± 1.1 mg. g^{-1}.h^{-1} in the absence and in the presence of isopropanol, respectively, showing that the presence of isopropanol had no negative effect on the MTBE degradation rate. Furthermore, the ETBE/MTBE degrading enzymes are constitutively expressed by this strain (Chauvaux *et al.*, 2001).

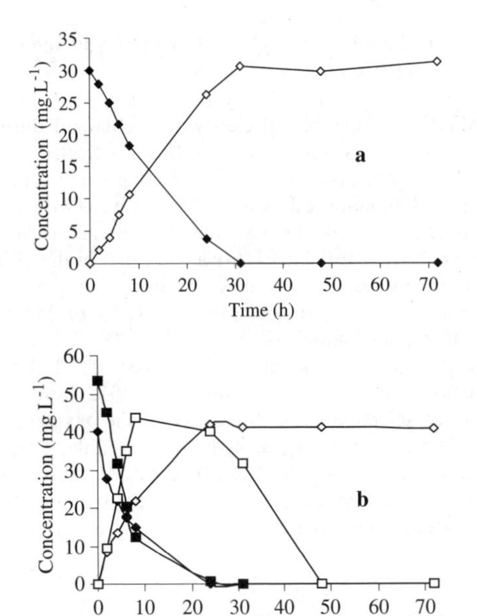

Figure 3. Degradation of MTBE by resting cells of *R. ruber* IFP 2007 in the absence (a) or in the presence (b) of isoprisopropanol. (◆-◆: MTBE; ◇-◇: TBA, ■-■: isopropanol; □-□: acetone).

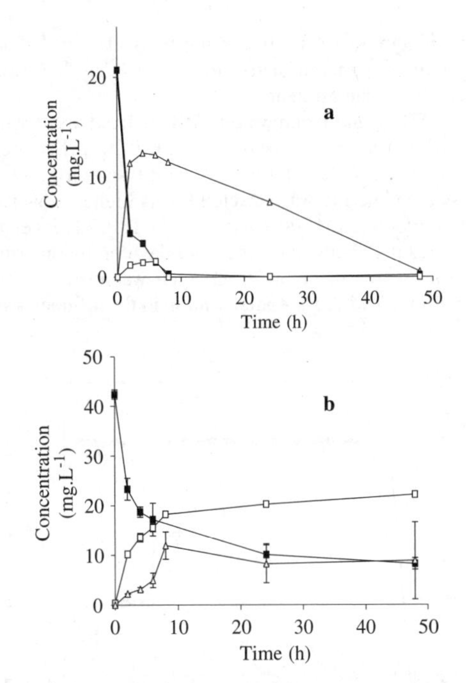

Figure 4. Degradation of MTBE by resting cells of *M. austrafricanum* IFP 2012 at $20\,mg.L^{-1}$ MTBE concentration (a) or at $40\,mg.L^{-1}$ MTBE concentration (b) (■-■: MTBE; □-□: TBF, Δ-Δ: TBA).

Therefore, an additional carbon source could allow for the use of *R. ruber* IFP 2007 to decontaminate MTBE-contaminated aquifers. Nevertheless, such a process would require working with a mixed culture combining *R. ruber* IFP 2007 with one of the strains isolated for their capacity to use TBA as a carbon and energy source (see Table1).

In contrast, MTBE was used as a carbon and energy source by *M. austroafricanum* IFP 2012 and the capacity of MTBE degradation was studied at different MTBE concentrations (Figure 4 a, b). The initial MTBE degradation rate was higher (Figure 4 a) at low MTBE concentrations. Moreover, MTBE degradation stopped when TBA and TBF accumulated in the conversion medium. These results show that limitations in MTBE degradation were mainly due to the negative effect of degradation intermediates (i.e., TBF and TBA). The effect of TBF on the initial MTBE degradation rate was shown to be especially strong even at a low TBF/MTBE concentration ratio (François *et al.*, 2003).

The enzymatic system responsible for the initial oxidation of ETBE and MTBE in *R. ruber* IFP 2007 is a cytochrome P450. The genes involved in this catalytic step have been located, cloned and sequenced (Chauvaux *et al.*, 2001). This cytochrome P450 cannot oxidize TBA, explaining why TBA accumulated in the medium when using this strain. The enzymatic system involved in the initial oxidation of MTBE in *M. austroafricanum* IFP 2012, although not yet elucidated, is not a cytochrome P450. It is not able to efficiently oxidize ETBE (results not shown). There are strong evidences that, in this strain, the initial oxidation of both MTBE and TBA is catalyzed by the same enzymatic system, which displays a higher affinity towards MTBE than towards TBA (François *et al.*, 2002).

4 ETBE AND MTBE DEGRADATION IN LAB-SCALE BIOFILTERS

ETBE degradation by a mixed culture composed of *R. ruber* IFP 2007 for ETBE degradation and *M. austroafricanum* IFP 2012 for TBA degradation was tested in a continuous biotrickling filter using pumice as a solid support. The volume of the bed was 4 L and oxygen was provided by aeration with air. Different ETBE concentrations were used (Figure 5). The elimination efficiency was 100% and the maximum elimination rate was about 6 mg ETBE.$L^{-1}.h^{-1}$. The system has remained active for 83 days. No TBA was detected in the effluent.

MTBE degradation was carried out in an immerged biofilter using *M. austroafricanum* IFP 2012. In a previous step, different types of solid supports were tested and perlite was selected for its higher capacity to immobilize *M. austroafricanum* IFP 2012 cells. The results obtained during a 40-day continuous run are presented in Figure 6. Oxygen was provided by aeration. The MTBE concentration in the influent was about 12 mg.L^{-1} throughout the experiment and various residence times were tested.

Beginning with a 2-day residence time, a 100% MTBE elimination efficiency was obtained during the first 5 days, after which MTBE was detected in the effluent. The residence time was increased to maintain high elimination efficiency. Eventually, with a residence time of about 4 days, the elimination efficiency was about 100%. No TBA was detected in the effluent. These results show the suitability of using both of these strains for remediation of ETBE- or MTBE-contaminated aquifers.

Present investigations are focused on MTBE degradation using *M. austroafricanum* IFP 2012, especially on the selection and optimization of the oxygen providing system (oxygen, hydrogen peroxide, oxygen release compound or ORC) (Lyew *et al.*, 2003) and on the scale-up using a 2-m^3 pilot reactor simulating an in-situ biobarrier system.

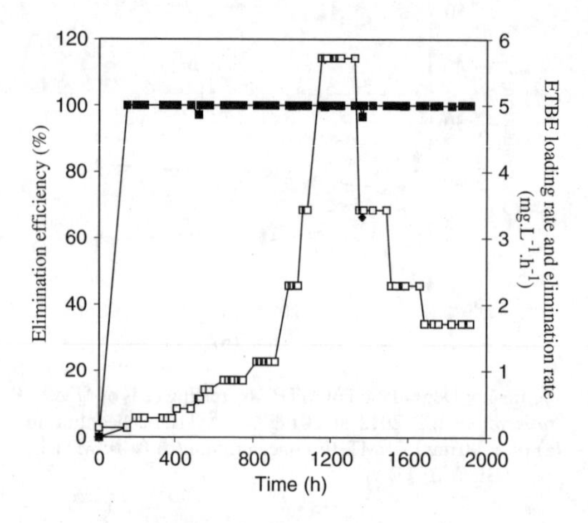

Figure 5. ETBE degradation capacity of *R. ruber* IFP 2007/*M. austroafricanum* IFP 2012 in a biotrickling filter (◆-◆: ETBE elimination rate; ■-■: elimination efficiency; □-□: ETBE loading rate).

5 CONCLUSIONS

Although ETBE and MTBE are persistent in the environment especially because of their low biodegradability, some microbial strains able to use these compounds as carbon sources have been isolated. Physiological studies aiming at elucidating the enzymes and genes involved in their catabolic pathways provide information on the main limitations to their degradation. Such studies also contribute to a rational approach of the implementation of an efficient, reliable and stable bioremediation process.

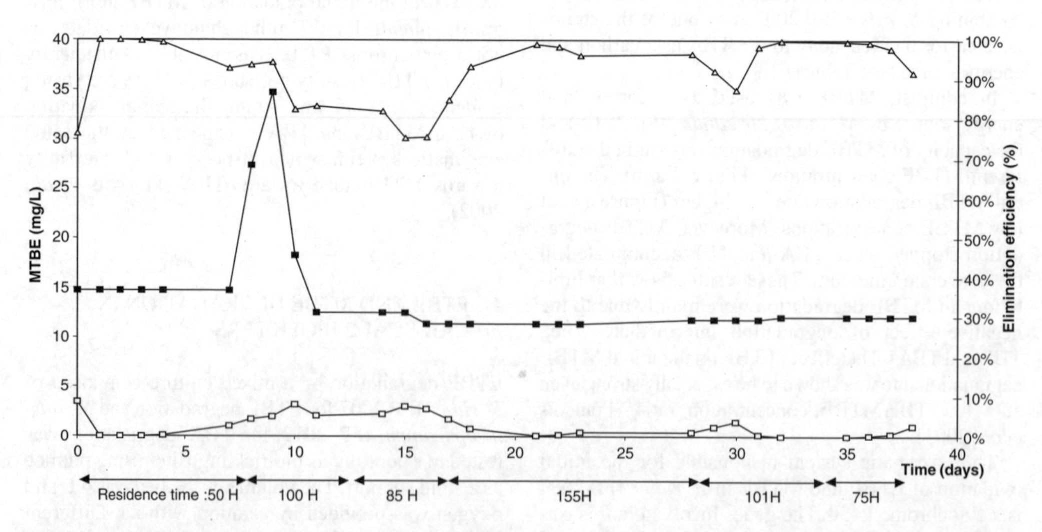

Figure 6. MTBE degradation capacity of *M. austroafricanum* IFP 2012 in an immerged biofilter (■-■: MTBE concentration in the inlet; □-□: MTBE concentration in the outlet, Δ-Δ: elimination efficiency).

REFERENCES

Chauvaux, S., Chevalier, F., Le Dantec, C., Fayolle, F., Miras, I., Kunst, F., Béguin, P. 2001. Cloning of a genetically unstable cytochrome P-450 gene cluster involved in the degradation of the pollutant ethyl *tert*-butyl ether by *Rhodococcus ruber. J. Bacteriol.* 183:6551–6557.

Fayolle, F., Vandecasteele, J.-P., Monot, F. 2001. Microbial degradation and fate in the environment of methyl *tert*-butyl ether and related fuel oxygenates. *Appl. Microbiol. Biotechnol.* 56:339–349.

Francois, A., Garnier, L., Mathis, H., Fayolle, F., Monot, F. 2003. Roles of *tert*-butyl formate, *tert*-butyl alcohol and acetone in the regulation of methyl *tert*-butyl ether degradation by *Mycobacterium austroafricanum* IFP 2012. Appl. *Microbiol. Biotechnol.*, 65:256–262.

François, A., Mathis, H., Godefroy, D., Piveteau, P., Fayolle, F., Monot, F. 2002. Biodegradation of methyl *tert*-butyl ether and other fuel oxygenates by a new strain, *Mycobacterium* austroafricanum IFP 2012. *Appl. Environ. Microbiol.* 68:2754–2762.

Hanson, J.R., Ackerman, C.E., Scow, K.M. 1999. Biodegradation of methyl *tert*-butyl ether by a bacterial pure culture. *Appl. Environ. Microbiol.* 65:4788–4792.

Hatzinger, P.B., McClay, K., Vainberg, S., Tugusheva, M., Condee, C.W., Steffan, R.J. 2001. Biodegradation of methyl *tert*-butyl ether by a pure bacterial culture. *Appl. Environ. Microbiol.* 67:5601–5607.

Hernandez-Perez, G., Fayolle, F., Vandecasteele, J.-P. 2001. Biodegradation of ethyl *t*-butyl ether (ETBE), methyl *t*-butyl ether (MTBE) and *t*-amyl methyl ether (TAME) by *Gordonia terrae. Appl. Microbiol. Biotechnol.* 55:117–121.

Lyew, D., Guiot, S., Fayolle, F., Monot, F. 2003.Comparison of different means of providing oxygen to a biobarrier to be used for the degradation of MTBE. In press.

European Symposium on Environmental Biotechnology, ESEB 2004 - Verstraete (ed)

Natural attenuation of landfill leachate in reaction to BTEX contamination

S. Botton & J.R. Parsons
University of Amsterdam, Department of Environmental and Toxicological Chemistry, Amsterdam, The Netherlands

W.F.M. Röling & M. Braster
Free University of Amsterdam, Department of Molecular Cell Phisiology, Amsterdam, The Netherlands

ABSTRACT: BTEX degradation under iron-reducing conditions is studied by means of microcosms inoculated with polluted and clean sediment or groundwater collected from the aquifer next to Banisveld landfill, Boxtel, the Netherlands. Degradation of benzene, toluene and m-xylene was observed in microcosms containing polluted sediment. In clean sediment microcosms no biological removal of BTEX could be detected. Second generation cultures enriched from toluene degrading microcosms were used for PLFA analysis with ^{13}C-labelled toluene in order to identify the bacteria responsible for toluene removal by tracing the ^{13}C-pollutant in the growing biomass. PLFA patterns were compared to the PLFA profile of *Geobacter metallireducens,* an indigenous iron reducing bacterium known to degrade toluene under anaerobic conditions. Similarities between the two profiles as well as the incorporation of labelled contaminant into the extracted PLFAs suggested the involvement of this bacterium in the removal of the pollutant.

1 INTRODUCTION

BTEX (benzene, toluene, ethylbenzene, xylene) represent a group of monoaromatic pollutants that are very often found in groundwater systems as a consequence of illegal dumping of petroleum waste in landfills. The environmental fate of such contaminants, characterized by relatively high solubility in water and proved toxicity for the environment and human health, needs to be monitored.

The resilience of subsurface environments in response to aromatic hydrocarbon pollution is mainly determined by the ability of the indigenous microbial communities to mineralize the pollutants. In some cases the removal of BTEX in such systems has been observed, but biodegradation processes seem to depend on the intrinsic characteristics of the investigated site (redox conditions, microbial communities, etc). Among the BTEX, benzene in particular was found to be the most recalcitrant to biological removal.

In this project, the biodegradation of BTEX in subsurface environments is investigated. In particular, this research aims at giving an insight of the biological processes that this group of contaminant undergoes within the plume of pollution originating from landfills.

2 MATERIALS AND METHODS

Groundwater and sediment samples collected from the Banisveld site were used as inoculum for the microcosms.

Five series of microcosms were started: crimp-cap bottles were filled with anaerobic medium, the headspace was replaced by a N_2/CO_2 (80/20) gas mixture and, after autoclaving, vitamins and Fe(III) were added. Bottles were then sluiced into an anaerobic glove bag and inoculated with groundwater or sediment and spiked with known amount of one of the contaminant. Each microcosm was prepared in duplicate.

Fe(III) was chosen as the electron acceptor because of the iron reducing conditions of the plume of pollution generating from the Banisveld landfill (Breukelen 2003). Amorphous Fe(III) oxyhydroxide was provided to the microcosms due to the high bioavailability of this iron form towards microbial reduction (Lovely 1986).

Each series of microcosms was monitored along with two controls: a sterile control, to detect abiotic removal of the contaminant, i.e. adsorption to the sediment; and a second control, without contaminant, to monitor variations in the concentration of Fe(III) not related to the oxidation of the BTEX.

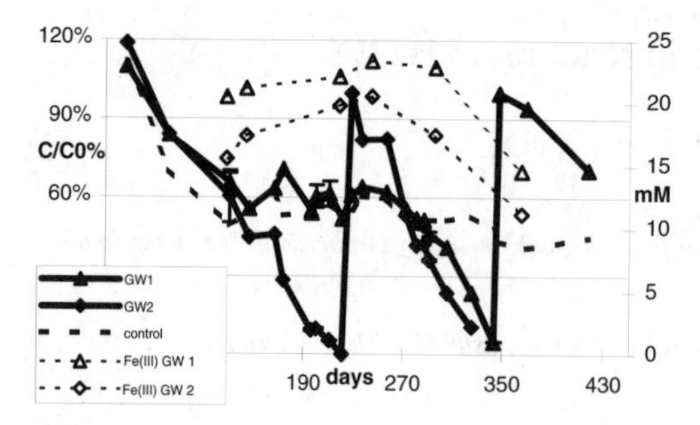

Figure 1. Residual concentration of benzene and Fe(III) in anaerobic microcosms inoculated with groundwater from Banisveld.

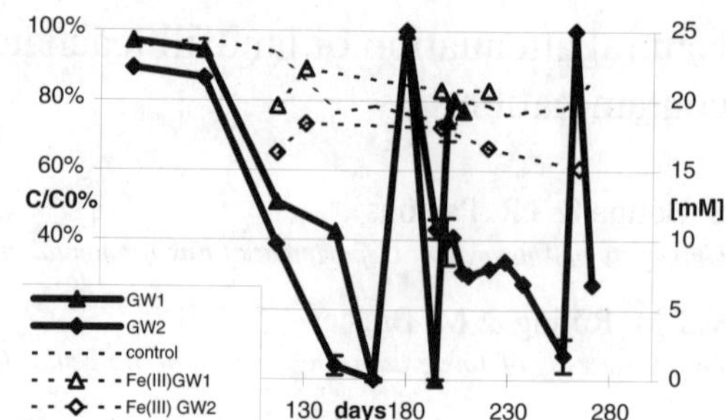

Figure 2. Residual concentration of toluene and Fe(III) in anaerobic microcosms inoculated with groundwater from Banisveld.

3 RESULTS AND DISCUSSION

3.1 *BTEX biodegradation*

A decrease in benzene and Fe(III) concentrations was observed in both replicates of groundwater microcosms, Figure 1.

Benzene was also removed from each of the microcosms inoculated with sediment collected from the polluted area. In these microcosms iron reduction was also detected.

The microcosms inoculated with the clean sediment did not show benzene removal, although a very high rate of iron reduction was observed, indicating that the microbial communities are still alive and possibly using different carbon sources.

A similar trend was observed in microcosms spiked with toluene. The concentration of toluene in both groundwater microcosms decreased and when it was not detectable any more microcosms were spiked again.

During the same period Fe(III) was reduced and consequently the concentration of Fe(II) increased, suggesting that toluene was removed by iron-reducing microorganisms, (Figure 2).

The microcosm GW 2 was used as inoculum for a series of enrichment cultures. The following step was to spike the cultures with higher concentrations of pollutant, provided as ^{13}C-toluene, in order to enrich the cultures of the toluene degrading bacteria as well as enable the tracing of the pollutant along the biodegradation pathway.

Toluene was also removed from sediment microcosms. The pollutant was added again when depleted and a decrease in toluene concentration as well as an increase in Fe(II) concentration were observed.

No significant variation in toluene concentration was measured in sediment microcosms inoculated with clean soil, although iron reduction was detected.

Decrease in m-xylene concentration was detected in sediment microcosms.

There are no indications of biological removal for ethylbenzene, p- and o-xylene sediment microcosms. The concentration of the pollutants did not vary significantly during almost 300 days of incubation. Fe(II) concentration increased in all the microcosms suggesting that microbial biomass is still active but using alternative carbon sources.

3.2 *PLFA analysis*

Second generation cultures enriched from toluene degrading microcosms were used for PLFA (Phospho-Lipidic Fatty Acid) analysis.

As previous studies reported (Röling 2001) the occurrence of *Geobacter* species at the Banisveld site seems to be related to the position of the plume of pollution, suggesting the involvement of this species in the biological processes occurring in the plume. In order to evaluate the role of *Geobacter* in toluene removal, a culture of this bacterium was cultivated under the same conditions of the toluene degrading enrichments and PLFA profiles were compared.

The PLFA profiles obtained were in good agreement: in both cases the PLFAs C16:1w9c, C16:0 and C18:1w11c represent 80% of the detected fatty acids, which is consistent with previous studies; (Lovley 1993; Ludvigsen 1999; Zhang 2003) and with the hypothesis that *Geobacter* is actively present in the enrichment cultures.

The extracted PLFAs were monitored in time and the concentration of each of fatty acid increased, Figure 3, indicating that they were extracted from the growing biomass.

Furthermore, the spectra of the single PLFAs were analysed, revealing that after the addition of labelled toluene the fragments in which the molecule is fractionated were enriched in heavier isotopes (m/z +1, +2). This finding proved the occurrence of ^{13}C in the growing biomass and therefore enables to link the

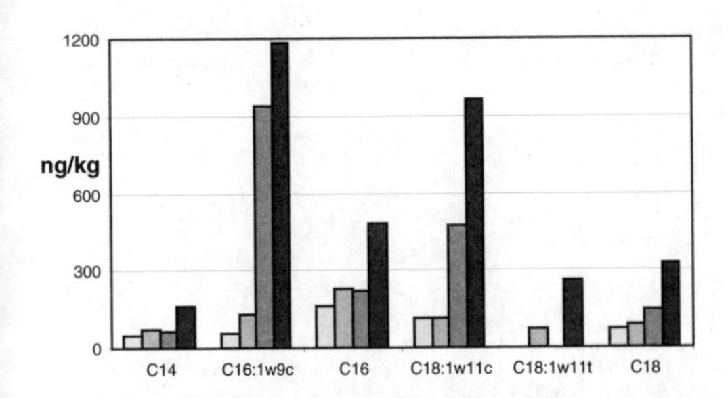

Figure 3. Variation in time of PLFAs extracted from toluene degrading enrichment cultures. Fatty acids are designated as A:BwC, where A is the total number of carbon atoms, B the number of double bonds, C the position of the double bond from the methyl-end of the fatty acids. The suffixes c and t stand for cis for trans respectively.

removal of toluene with a culture that shows high similarity in PLFA profiles with *Geobacter metallireducens*.

4 CONCLUSIONS AND FUTURE WORK

Benzene, toluene and m-xylene were degraded in microcosms inoculated with groundwater or sediment collected from the plume of pollution in Banisveld. The anaerobic removal of the contaminant occurred under iron-reducing conditions.

Analysis of the PLFA profiles of toluene degrading enrichment cultures suggested the involvement of *Geobacter* species. Further molecular analysis (DGGE) are being carried out in order to gain more information on the changes of microbial communities in reaction to BTEX pollution.

REFERENCES

Breukelen, B.M.v., 2003, Natural Attenuation of Landfill Leachate: a Combined Biogeochemical Process Analysis and Microbial Ecology Approach, Phd thesis.

Lovley, D., Giovannoni, R., Champine, W., Phillips, E.J.P., Gorby, and Goodwin, 1993, Geobacter Metallireducens gen. nov. sp. nov.,a microorganism capable of coupling the complete oxidation of organic compounds to the reduction of iron and other metals., *Archives of Microbiology* 159:336–344.

Lovley, D.R., and Phillips, E.J.P., 1986, Organic Matter Mineralization with Reduction of Ferric Iron in Anaerobic Sediments, *Appl.Environ.Microbiol.* 51(4):683–689.

Ludvigsen, L., Albrechtsen, H.-J., Ringelberg, D.B., Ekelund, F., and Christensen, T.H., 1999, Distribution and Composition of Microbial Populations in a Landfill Leachate Contaminated Aquifer (Grindsted, Denmark), *Microbial Ecology* 37:197–207.

Röling, W.F.M., Breukelen, B.M. v., Braster, M., Lin, B., and Verseveld1, H.W.v., 2001, Relationships between Microbial Community Structure and Hydrochemistry in a Landfill Leachate-Polluted Aquifer, *Appl.Environ. Microbiol.* 67(10):4619–4629.

Zhang, C.L., Li, Y., Ye, Q., Fong, J., Peacock, A.D., Blunt, E., Fang, J., Lovley, D.R., and White, D.C., 2003, Carbon isotope signatures of fatty acids in Geobacter metallireducens and Shewanella algae, *Chemical Geology* 195:17–28.

European Symposium on Environmental Biotechnology, ESEB 2004 - Verstraete (ed)
© 2004 Taylor & Francis Group, London, ISBN 90 5809 653 X

Molecular techniques and *in situ* mesocosm socks for monitoring *in situ* BTEX bioremediation

B. Hendrickx, W. Dejonghe, W. Boënne, L. Bastiaens & D. Springael
Flemish Institute for Technological Research (Vito), Mol, Belgium

T. Lederer & M. Cernik
Aquatest-Stavebni Geologie, Inc., Prague, Czech Republic

W. Verstraete
Laboratory of Microbial Ecology and Technology, Gent, Belgium

M. Bucheli, I. Rüegg & Th. Egli
Swiss Federal Institute for Environmental Science and Technology (EAWAG), Duebendorf, Switzerland

ABSTRACT: An *in situ* combined monitoring system was used to examine the effect of a BTEX contamination on the diversity of a soil community in a BTEX contaminated aquifer. Therefore, mesocosm socks were filled with non-sterile and sterile uncontaminated aquifer material (from part A0) and positioned under the groundwater table in a monitoring well located in the contaminated area (part A1) of the site. Sterile and non-sterile A0 aquifer material inserted in A1 evolved towards the community structure characteristic for location A1. In all three cases this structure seems to be dominated by the same *Pseudomonas* species as determined by sequencing of 16S rRNA genes. Also the BTEX composition of the BTEX catabolic genotypes seems to evolve similar in the sterile and non-sterile A0 samples inserted at location A1, both developing *tmoA*- and *xylE*-like genotypes. On the contrary, at location A0 no *xylE*-like genotypes were detected.

1 INTRODUCTION

Recently, there is a growing interest in *in situ* bioremediation and monitored natural attenuation (MNA) of sites containing contaminated groundwater plumes. MNA consists of monitoring the combination of natural biological, chemical and physical processes that reduce the mass, toxicity, mobility, volume or concentration of the contaminants (e.g., biodegradation, dispersion, dilution, sorption and volatilization) at such level that no human intervention is needed. To increase our understanding of the processes involved in *in situ* bioremediation or (M)NA, there is a strong need to understand better the indigenous microbial community behavior, their dynamics and activities under the implemented conditions at such sites. In batch- and column laboratory experiments, i.e., contained model ecosystems, *in situ* conditions are simulated as good as possible. However, ultimately, the relevance of laboratory microcosm data for the field must be compared to the data obtained on site and *in situ.* In this study an *in situ* mesocosm sock system was designed to follow up the microbial community dynamics in BTEX polluted aquifers. The system consists out of a polyamide membrane pocket filled with aquifer material, which is contained within a perforated holder of polypropylene, which allows to install the system in a monitoring well. After specified time intervals the socks are recovered from the wells to examine the microbial community structure by PCR-DGGE fingerprinting of 16S rDNA and the BTEX catabolic gene composition by PCR. The use of the designed *in situ* mesocosm sock system to examine the effect of a BTEX contamination on (i) the diversity of an oligotrophic soil community, (ii) the presence of specific BTEX catabolic genotypes within this community, and (iii) the community adaptation in a BTEX contaminated aquifer, was studied.

2 MATERIALS AND METHODS

2.1 *Mesocosm sock and* in situ *implementation*

A mesocosm sock consisted of a polyamide membrane pocket that can be filled with aquifer material, which is

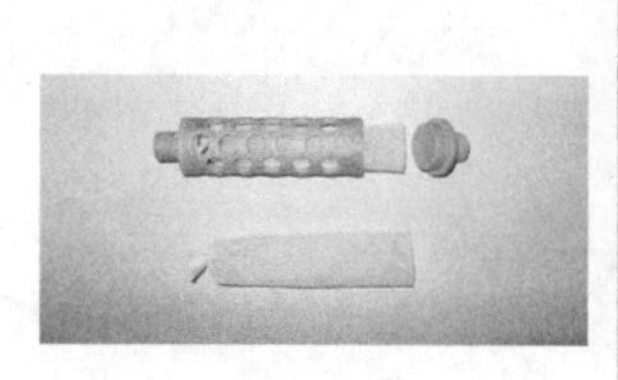

Figure 1. *In situ* mesocosm sock system and implementation system.

contained within a perforated holder of polypropylene, which allows to install the system in a monitoring well (Figure 1). The membranes were purchased from Solana N.V., Schoten, Belgium and had a mesh width of 6 micron and a percentage open surface of 5% (49 PA 6/5).

Aquifer inserted in the mesocosms was obtained from an oil-refinery site, situated in Northern Bohemia (Czech Republic), containing a BTEX (mainly benzene) contaminated groundwater plume. Figure 2 shows the location of the strongly contaminated part A1 versus the non-contaminated part A0 on this Czech site.

Mesocosm socks were filled with non-sterile and sterile uncontaminated A0 aquifer obtained from the saturated zone Z at the A0 area (A0Z) and were positioned under the groundwater table in monitoring wells located in the contaminated A1 and the non-contaminated A0 area.

2.2 *DNA extraction, PCR amplification, DGGE analysis and cloning*

DNA was extracted from soil as described previously (Hendrickx *et al.*, submitted for publication). In brief, 2 g of soil was suspended in 4 ml of Tris-glycerol-buffer together with glass beads to obtain a mechanical disruption of the bacterial cells. This was followed by an enzymatic (lysozyme and proteïnase K) and chemical (SDS) lysis of the cells. DNA was then purified by a phenol/chloroform/isoamylalcohol extraction, the addition of PVPP and finally by using the Wizard DNA Clean-Up System (Promega).

A 496 bp eubacterial 16S rRNA gene fragment was amplified using the forward primer GC-63F (with GC-clamp) and the reverse primer 518R as described by El-Fantroussi *et al.* (1999). The primer sets for the detection of the BTEX catabolic genotype *tmoA* and *xylE* were used as described by Hendrickx *et al.* (submitted for publication).

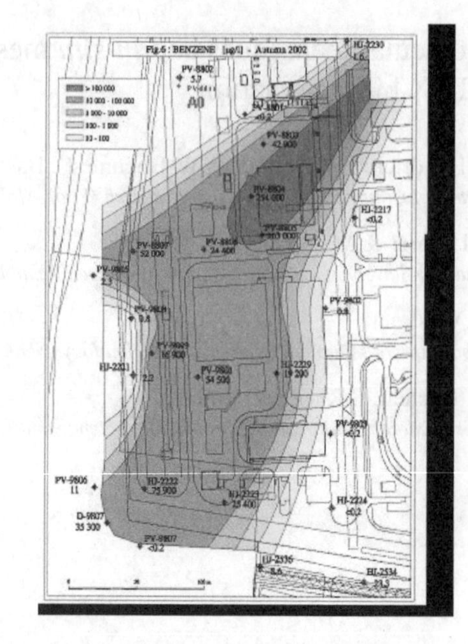

Figure 2. Map of the BTEX (mainly benzene) contaminated site in the Czech Republic showing the benzene contamination plume and indicating relevant sampling locations and monitoring wells.

To separate the eubacterial 16S rDNA PCR-products obtained with primer set GC-63F/518R, an 8% polyacrylamide gel with a denaturing gradient of 35% to 65% was used. For the PCR-products obtained with GC-TMOA-F/TMOA-R, a 6% polyacrylamide gel with a denaturing gradient of 40% to 70% was used. Electrophoresis was performed at a constant voltage of 120 V for 15 h in 1x TAE running buffer at 60°C on a DGGE-machine (INGENYphorU-2, INGENY International BV, The Netherlands).

PCR products were cloned into plasmid vector pCR®2.1-TOPO® using the TOPO TA Cloning® Kit (N.V. Invitrogen SA, Merelbeke, Belgium) as described by the manufacturer. Cloned inserts with different DGGE patterns were sequenced by the Westburg company (Westburg, Leusden, The Netherlands).

2.3 *BTEX degradation tests*

To study the BTEX degradation potential present in the aquifer material inserted for 10 months *in situ* in the mesocosm socks, batch degradation tests were performed in the laboratory. Ten gram of aquifermaterial retrieved from the socks was incubated into glass jars that contained 190 ml of Evian water, 20 ppm of benzene and 5 ppm of toluene. Degradation tests were performed under aerobic and nitrate (5 mM $NaNO_3$)

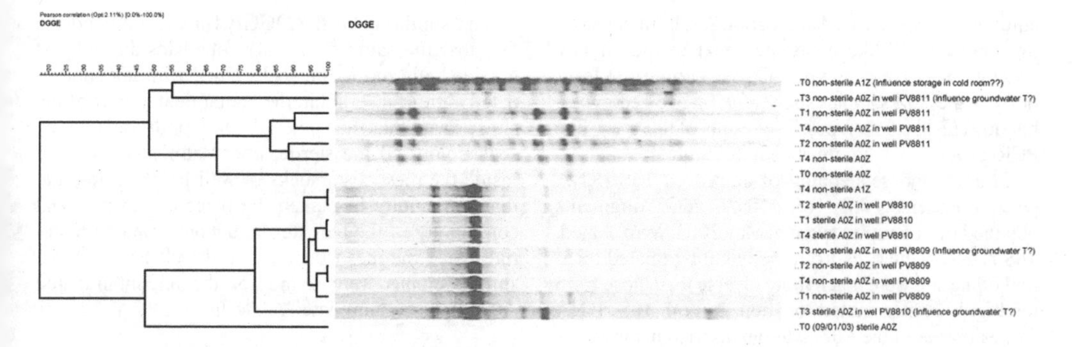

Figure 3. UPGMA clustering of eubacterial community 16S rRNA gene fingerprints obtained from the aquifer material from the mesocosm socks and the aquifer outside the socks at different points in time using the pearson moment-based similarity coefficient.

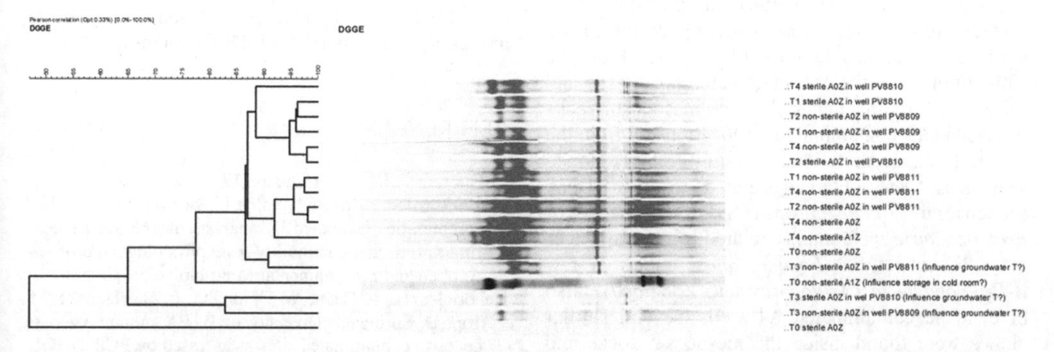

Figure 4. UPGMA clustering of the DGGE fingerprints of the PCR amplicons obtained with the primer set GC-TMOA-F/TMOA-R on DNA obtained from the aquifer material from the mesocosm socks and the aquifer outside the socks at different points in time using the pearson moment-based similarity coefficient.

reducing conditions. The concentration of BT was followed in time by headspace analysis on a gaschromatograph.

3 RESULTS AND DISCUSSION

Use of *in situ* mesocosm system to follow the effect of BTEX contamination on the structure of an oligotrophic microbial community and on the BTEX catabolic genotypes of an uncontaminated aquifer under *in situ* conditions. On January 9, 2003, mesocosm socks were filled with either non-sterile or sterile aquifer material sampled from the saturated zone of the uncontaminated area A0 (designated A0Z) and positioned in the groundwater in monitoring wells PV-8809 (insertion of non-sterile A0Z material) and PV-8810 (insertion of sterile A0Z material) located in the contaminated area A1. As a control, sterile and non-sterile A0Z aquifer material was also inserted into monitoring well PV-8811 in the non-contaminated location A0. Figure 3 shows the results of the eubacterial 16S rDNA PCR-DGGE analysis.

The original A0Z community changed both at location A0 (monitoring well PV8811) and location A1 (monitoring wells PV8809 and PV8810), but evolved into different communities. At location A0 (monitoring well PV8811), the bacterial community of the implemented non-sterile A0Z material in the socks evolved into a community with an identical profile of that of the A0Z aquifer. At location A1, the microbial community of both the sterile (monitoring well PV8810) and the non-sterile (monitoring well PV8809) A0Z material evolved into similar communities, indicating that the bacterial community present in the contaminated A1 aquifer rather colonized the A0Z material installed in the contaminated A1 part of the site than that the A0 community evolved into a new population.

PCR results addressing the composition of the BTEX catabolic genotypes indicated that the composition of the BTEX catabolic genotypes seemed to evolve similar in the sterile and non-sterile A0Z aquifer samples inserted at location A1, both developing mainly *xylE*-like genotypes next to the already present *tmoA*-like genotypes. On the contrary at location A0, the A0Z samples showed *tmoA*-like genotypes, but no *xylE*-like genotypes. The results of the *tmoA*-like PCR-DGGE analysis are shown in Figure 4.

Cloning and sequencing of amplified 16S rRNA gene fragments. All 16S rRNA gene fragments obtained from samples taken in May 2003, were cloned. 16S rDNA sequences from the uncontaminated location of the site were much more diverse than those from the highly contaminated location. In both 16S rDNA clone libraries of the A0Z material inserted in the mesocosm socks and the freshly retrieved aquifer material taken from outside the sock at the A0Z location, *Actinobacteria* (14 clones/27 clones) and *Proteobacteria* (13/27) seem to comprise the two main groups of resident bacteria. At the uncontaminated A0 well, several similar clones were found inside the mesocosm socks and outside in the A0Z aquifer. From the clone libraries of the uncontaminated non-sterile and sterile A0Z aquifer material in the mesocosm socks inserted in the contaminated A1 location and from the material outside these socks, mainly (22/23) cloned sequences were obtained related to 16S rDNA sequences of *Proteobacteria*. 16S rDNA sequences of *Proteobacteria* were mainly related to the sequences of γ-*Proteobacteria* (*Pseudomonas*) and of a few β-*Proteobacteria* (*Acidovorax* and *Zoogloea*). Also here at the contaminated A1 wells, several similar clones were found inside the mesocosm socks and outside in the A1Z aquifer.

The similarity of the DGGE fingerprints and the fact that the same clones are found inside and just outside both mesocosm socks filled with uncontaminated aquifer inserted in the contaminated part of the site, proved that the bacterial community outside the sock colonized the sterile, uncontaminated material inside the mesocosm socks of well PV8810 and that this community has taken the place of the bacterial community of the non-sterile, uncontaminated material inside the mesocosm socks of well PV8809. No similar clones were found for the uncontaminated A0 and the contaminated A1 wells.

4 CONCLUSIONS

Mesocosm socks can be used to study *in situ* the differences in the microbial community structure and the BTEX catabolic genotypes present in a polluted and non-polluted part of a BTEX contaminated site.

REFERENCES

El-Fantroussi, S., L. Verschuere, W. Verstraete & E.M. Top. 1999. Effect of phenylurea herbicides on soil microbial communities estimated by analysis of 16S rRNA gene fingerprints and community-level physiological profiles. Applied and Environmental Microbioly 65:982–988.

Hendrickx, B., F. Faber, W. Dejonghe, L. Bastiaens, E.M. Top, D. Springael. Diversity of BTEX monooxygenase genes in contaminated sites as assessed by PCR-DGGE. Submitted for publication.

European Symposium on Environmental Biotechnology, ESEB 2004 - Verstraete (ed)
© 2004 Taylor & Francis Group, London, ISBN 90 5809 653 X

Biotransformation of PCBs and their intermediates by plants and bacteria – study of possible metabolic connections in nature

T. Macek[1,2], K. Francova[1,2], J. Rezek[1,2], K. Demnerova[1,2] & M. Mackova[1,2]
[1]*Inst. of Organic Chemistry and Biochemistry CAS, Prague, Czech Republic*
[2]*Dept. of Biochemistry and Microbiology, ICT Prague, Prague, Czech*

L. Kochánková
Dept. of Environ. Chem., Fac. of Chem. Technol., ICT Prague, Prague, Czech Republic

M. Sylvestre
Institut National de la Recherche Scientifique, INRS-IAF, Pointe-Claire, Québec, Canada

ABSTRACT: PCBs are metabolised by bacteria to chlorobenzoic acids, which can be further transformed by organisms living in the contaminated environment (microorganisms, plants, insects etc). Plants used for transformation of PCBs different enzymes than bacteria and the main intermediates of plant metabolism are hydroxychlorobiphenyls. In order to follow the fate of products of bacterial and plant PCB metabolism in the environment, we studied further degradation of plant-formed hydroxychlorobiphenyls by bacterial enzymes of upper biphenyl pathway and on the other hand degradation of chlorobenzoic acids by plants.

1 INTRODUCTION

Polychlorinated biphenyls (PCBs) are toxic, recalcitrant synthetic organic compounds. These stable hydrophobic compounds are widely dispersed in the environment as a consequence of their manufacture and extensive use in diverse applications. They are omnipresent in sediments, soil and living organisms. This fact is a cause for concern because they are suspected of having adverse effects on the human reproductive, endocrine, neural and immune system. The high cost and public opposition to current physical remediation technologies have motivated the exploitation of biological destruction systems using microorganisms and recently also plants (Macek et al., 2002) to clean-up PCB-contaminated sites. In the environment, PCBs are co-metabolically transformed to chlorobenzoic acids by aerobic bacteria through the biphenyl catabolic pathway (Abramowicz, 1995). The initial reaction of the aerobic degradation pathway of biphenyl and PCBs is catalyzed by biphenyl dioxygenase (BPDO). The four enzymatic steps required to transform PCBs into corresponding benzoates/chlorobenzoates.

The enzymes of the bacterial biphenyl catabolic pathway are very versatile. Beside PCB they can also catalyze reactions using several hydroxybiphenyls and chlorohydroxybiphenyls as substrate (Sondossi et al., 1991). Although BPDO was shown to be able to oxygenate hydroxybiphenyls and hydroxychlorobiphenyls (Sondossi et al., 1991) there is still little information on the mode of transformation of the hydroxychlorobiphenyls by biphenyl dioxygenases. This information is timely in the context of the exploitation of the interaction between plants and microbes to degrade PCBs.

Plants metabolise recalcitrant organic xenobiotics in general in a different manner than microorganisms, the compounds are usually activated, conjugated and stored (Lee and Fletcher 1992, Macek et al., 2000). In case of PCB metabolization by plants, the situation is complicated by the wide range of PCB congeners differing in chlorine substitution, the presence of different enzyme systems available in plants for activation reactions, and the possible availability of bacterial PCB metabolites to plants and vice-versa.

Recent findings show that plants are major contributors to the recycling of organic matter. Thus the ability of rhizosphere bacteria to degrade organic pollutants was shown to be strongly influenced by plant exudates (Fletcher and Hedge, 1995). Furthermore, evidences suggested that plant enzymes, including oxygenases of the P450 family can metabolize PCBs to yield hydroxylated derivatives (Kucerova et al., 2000). The purpose of the present investigation was to evaluate the capacity of enzymes of the biphenyl/chlorobiphenyl pathway, especially BPDO of two well-characterized PCB degrading bacteria, *Burkholderia* sp. LB400 and *Comamonas testosteroni* B-356 to metabolize *ortho*-substituted hydroxybiphenyls carrying chlorine atoms on the hydroxyl-substituted ring. The chlorohydroxybiphenyls tested were identified in recent studies as products of 2-chlorobiphenyl, 3-chlorobiphenyl and 3,5-chlorobiphenyl conversion by plants (Chroma et al., 2002). Confirming the ability of bacterial enzymes to accept as substrates the intermediates of plant metabolism of PCBs is crucial for preparation of transgenic plants expressing bacterial enzyme for PCB-degradation. A second objective was to identify the metabolites generated by both catalytic systems in order to compare their mode of attack on these substrates.

To exploit the fate of bacterial intermediates (chlorobenzoic acids – CBAs) in plant cells the degradation of four CBAs by three different plant species was tested.

2 MATERIALS AND METHODS

2.1 *Bacterial strains and culture media*

The bacterial strains used in this study were *E. coli* M15 [pREP4](from QIAGEN Inc., Chatworth, California) harboring plasmids pQE31 carrying *bphAE* from LB400 or B-356 or carrying any of B-356 *bphF* (Hurtubise et al., 1995*), bphG,* (Hurtubise et al., 1995), *bphB, bphC* or *bphD.* The media used was Luria-Bertani (LB) broth supplemented with appropriate antibiotics.

2.2 *Plant material and estimation of CBAs*

Plant cells from Plant Tissue Culture Collection of the Dept. of Natural Products, IOCB, *Solanum nigrum* (black nightshade), line SNC-9O, *Medicago sativa* (alfalfa), strain ALF, *Nicotiana tabacum* (tobacco), WSC, and *Armoracia rusticana* (horseradish) K54 grown under aseptic conditions in liquid MS medium (as described for PCBs by Mackova et al., 1997) were used after 5 day precultivation for transformation of CBAs. The media samples after incubation with 200 mg/L 2-, 3-, 4-CBA and 2,5-diCBA were directly injected into HPLC apparatus Hewlett-Packard 1100 and their decrease was followed by DAD detection.

2.3 *Enzyme assays for BPDO and BPDO coupled to other pathway enzymes*

Enzyme assays for BPDO were performed at 37°C and the assay buffer used was 100 mM morpholineethansulphonic acid (MES; pH 5.5). The reaction mixture (200 μl total volume) contained 100 nmol of NADH, 50 nmol of the substrate, 1.2 nmol of $FeSO_4$, and 0.6 nmol of each BPDO component. The enzyme components were added individually as required. The reaction was initiated by adding the substrate dissolved in 2 μl acetone. Reaction mixture was incubated with shaking for 5 min at 37°C in Eppendorf tubes. Reaction products of BPDO were extracted with 3 ml of ethyl acetate; the organic layer was dehydrated with ammonium sulphate. The solvent was evaporated to 1 ml. To identify the obtained products by GC-MS, routine derivatization procedures were used to compare spectral features and retention times of trimethylsilyl or butylboronate derivatives of analyzed products with those of authentic samples.

3 RESULTS AND DISCUSSION

In a previous studies, strains B-356 and *Pseudomonas putida* KT2440 carrying the complete *bph* operon from strain B-356 were shown to transform 5-chloro-2-hydroxybiphenyl to 5-chloro-2-hydroxy-benzoate (Sondossi et al., 1991). In the present investigation, we have compared the abilities of His-tagged purified LB400 and B-356 BPDOs to catalyze the oxygenation of 2-hydroxy-3-chlorobiphenyl, 2-hydroxy-5-chlorobiphenyl and 2-hydroxy-3,5-dichlorobiphenyl. Activities of the purified His-tagged BPDOs when biphenyl was the substrate were 98 nmol/min/mg for LB400 BPDO and of 581 nmol/min/mg for B-356 BPDO.

Both enzyme preparations catalyzed the hydroxylation of the three chlorohydroxybiphenyls tested. The same patterns were obtained for both enzyme preparations when MSTFA-derived metabolites were analyzed by GC-MS, showing that both enzymes metabolized these compounds similarly. The metabolites and their GC/MS characteristics are listed in Table 1. When 2-hydroxy-5-chlorobiphenyl was the substrate, the trimethylsilyl (TMS) derivative of the major metabolite exhibited a mass spectral fragmentation pattern characterized by ion fragments at m/z 367 ($M-CH_3$), 332 ($M-CH_3-Cl$), 293 (M-TMSO), 277 (M-TMSO-O). Based on this fragmentation pattern and on the fact that 3,3′-dihydroxybiphenyl was found to produce an oxo-derivative as major metabolite (Sondossi et al., 1995), metabolite 3 has been tentatively identified as 6-hydroxy-5-(2-hydroxy-5-chlorophenyl)-3-cyclohexene-1-one (Fig. 1) which is likely to be generated from the rearrangement of *cis*-2,3-dihydro-2,3-dihydroxy-2′-hydroxy-5′-chlorobiphenyl (metabolite 1, Fig. 1) obtained by catalytic oxygenation of 2-hydroxy-5-chlorobiphenyl.

The reaction medium contained minor amounts of a metabolite showing the mass spectral features of a dihydroxychlorobiphenyl (Table 1). It was most probably the 2,2′-dihydroxy-5-chlorobiphenyl or the

Table 1. Spectral features of metabolites produced from catalytic oxygenation of hydroxychlorobiphenyls.

Substrate[a]	Metabolite no.:	Retention time (min.)	Spectral features M+	M − 15
3,5diCl2HP				
	1	23.1	488	473
	3	22.07	398	383
	4	22.05	416	401
3Cl2HBP				
	1	21.9	454	439
	2	21.8	452	437
	3	20.6	364	349
	4	20.8	382	367
5Cl2HBP				
	1	21.9	454	439
	2	21.5	452	437
	3	20.6	364	349
	4	20.8	382	367

[a] 3,5diCl2HBP, 2-hydroxy-3,5-dichlorobiphenyl; 3Cl2HBP, 2-hydroxy-3-chlorobiphenyl; 5Cl2HBP, 2-hydroxy-5-chlorobiphenyl.

Figure 1. Hydroxylated metabolites generated by the catalytic oxygenation of 2-hydroxychlorobiphenyl. Metabolite 1, *cis*-2,3-dihydro-2,3-dihydroxy-2′-hydroxychlorobiphenyl; metabolite 2, 2-hydroxychloro-2′,3′-dihydroxybiphenyl; metabolite 3, 5-(chloro-2-hydroxyphenyl)-6-hydroxy-3-cyclohexen-1-one; metabolite 4, 2,2′-dihydroxychlorobiphenyl or the 2,3′-dihydroxychlorobiphenyl (could not be distinguished).

2,3′-dihydroxy-5-chlorobiphenyl (metabolite 4, Fig. 1). These metabolites were also generated by the loss of OH and rearrangement of *cis*-2,3-dihydro-2,3-dihydroxy-2′-hydroxy-5′-chlorobiphenyl. Minor amounts of *cis*-2,3-dihydro-2,3-dihydroxy-2′-hydroxy-5′chlorobiphenyl (metabolite *1*, Fig. 1) were also found as well as 2-hydroxy-5-chloro-2′,3′-dihydroxybiphenyl (metabolite *2*, Fig.1), which was most likely generated by spontaneous rearrangement and dehydrogenation of *cis*-2,3-dihydro-2,3-dihydroxy-2′-hydroxy-5′-chlorobiphenyl. The presence of these metabolites suggests that the dihydrodiol metabolite is unstable under the reaction conditions used and readily transformed to more stable structures. When the reaction mixture was treated with butylboronate a very small peak of a metabolite identified as *cis*-2,3-dihydro-2,3-dihydroxy-2′-hydroxy-5′-chlorobiphenyl was detected. The size of the peak provides a further evidence to demonstrate that a large portion of this intermediate was lost rapidly after its formation in the reaction medium.

A similar pattern of metabolization was obtained when 2-hydroxy-3-chlorobiphenyl was the substrate. Thus the major metabolite was identified as 5-(chloro-2-hydroxyphenyl)-6-hydroxy-3-cyclohexen-1-one (metabolite *3*, Fig. 1), which was likely generated from the rearrangement of *cis*-2,3-dihydro-2,3-dihydroxy-2′-hydroxy-3′-chlorobiphenyl. The reaction medium also contained minor amounts of dihydroxybiphenyl (most probably the 2,2′-dihydroxy-3-chlorobiphenyl or the 2,3′-dihydroxy-3-chlorobiphenyl) (metabolite 4, Fig. 1), generated by the rearrangement of *cis*-2,3-dihydro-2,3-dihydroxy-2′-hydroxy-3′-chlorobiphenyl. Minor amounts of *cis*-2,3-dihydro-2,3-dihydroxy-2′-hydroxy-3′-chlorobiphenyl as well as 2-hydroxy-3-chloro-2′,3′-dihydroxybiphenyl were also found.

When the chlorohydroxylated compounds were used as substrates in coupled reactions of BPDO together with 2,3-dihydro-2,3-dihydroxybiphenyl-2,3-dehydrogenase (*BphB*) plus 2,3-dihydroxybiphenyl 1,2-dioxygenase (*BphC*), yellow *meta*-cleavage metabolites were produced for all three of them. The mass balance could not be determined because the molar extinction coefficients for these *meta*-cleavage metabolites are unknown. However, OD readings much lower than the observed values of 0.4 to 0.7 should have been observed if only trace amounts of *cis*-2,3-dihydro-2,3-dihydroxy-2′-hydroxychlorobiphenyl were converted to the *meta*-cleavage intermediate. This indicates that the dihydrodihydroxy metabolite of the hydroxychlorobiphenyl is stable for a period long enough to be converted by *BphB* to pursue the sequence of reactions leading to CB.

This assumption is also supported by the observation that except for some 2,2′,3′-trihydroxychlorobiphenyl, no other hydroxylated metabolites were detected in the reaction medium of the coupled reaction employing all pathway enzymes (*BphAEFGBCD*).

Chlorobenzoic acids are one of the end products of so-called upper aerobic bacterial degradation pathway of PCBs and they can be available to plants growing in consortia with PCB degrading microbes in the rhizosphere. To explain the fate of xenobiotics and their intermediates we also intended to find the answer, if plants are able to metabolise CBAs. As a model we choose four plant tissue cultures, namely *Solanum nigrum, Medicago sativa, Nicotiana tabacum* and *Armoracia rusticana*. 2-, 3-, 4-CBA and 2,5-diCBA were used as model compounds.

Plant cells of horseradish and black nightshade showed to be the most active cultures in respect of CBAs metabolism. Both species metabolised about 70% of 2-chloro and 2,5-dichlorobenzoic acids within one week. 2 CBA is probably the most suitable substrate due to its best solubility and bioavailability (Figure 2). Cells of alfalfa and tobacco plants did not degrade significant amounts of CBAs and all chosen CBAs proved much higher toxic effect on vitality and growth of these two species in comparison with horseradish and black nightshade cultures. Our study proved that certain plant species are able to degrade some chlorobenzoic acids entering the environment due to industrial and agricultural activities. From the literature it is known, that plants help the rhizospheric microflora to remove several organic and inorganic pollutants by supporting its activities and living conditions. In our study we showed that plant cells could help to remove CBAs also by their direct metabolisation and removal from the environment, and that plants can be partly responsible for the fate of bacterial PCB metabolites.

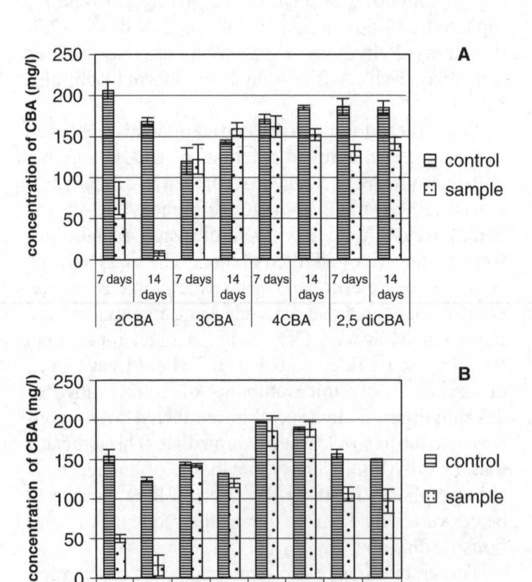

Figure 2. Graph shows the amount of the residual amounts of chlorobenzoic acids after 7 and 14 d cultivation. As controls cells killed by boiling for 20 min. were used. A: *Armoracia rusticana* K54. B: *Solanum nigrum* SNC 9O.

4 CONCLUSIONS

Data confirm that both B-356 BPDO and LB400 BPDO oxygenate mono-substituted hydroxy- and hydroxychlorobiphenyls on the non-substituted ring. This finding and proof of the ability of plants to transform CBAs are important from the point of better understanding of co-operation between plants and bacteria in the rhizosphere during phyto/rhizore-mediation of recalcitrant organic compounds (Macek et al., 2002). The plant metabolites and intermediates are available to soil microorganisms (e.g. from falling leaves, died roots, exudates etc.) and vice versa and thus question had to be answered, if the intermediates and products of both metabolisms can serve as substrates to enzymes of other organisms living in contaminated environment. The mechanism of transformation of chlorobenzoic acids, enzymes involved and products formed are not known yet and will be studied in further research.

ACKNOWLEDGEMENT

The work was supported by the grant of the GACR No. 526/01/1292, the EU 5FW grant No. QLK3-2001-00101 and grant of ICT 320080015.

REFERENCES

Abramowicz, D.A., 1990. Aerobic and anaerobic biodegradation of PCBs: a review. *Critical Reviews in Biotechnology* 10, 241–251.

Chroma, L., Mackova, M., Kucerova, P., in der Wiesche, C., Burkhard, J., Macek T., 2002. Enzymes in plant metabolism of PCBs and PAHs. *Acta Biotechnologica* 22, 34–41.

Chroma, L., Moeder, M., Kucerova, P., Macek, T., Mackova, M., 2003. Plant enzymes in metabolism of polychlorinated biphenyls. *Fresenius Environmental Bulletin* 12, 291–295.

Fletcher, J.S., Hegde, R.S., 1995. Release of phenols by perennial plant roots and their potential importance in bioremediation. *Chemosphere* 31, 3009–3016.

Francova, K., Mackova, M., Macek, T. & and Sylvestre, M. 2004. Ability of bacterial biphenyl dioxygenases from *Burkholderia sp.* LB400 and *C. testosteroni* B-356 to catalyse oxygenation of ortho-hydroxybiphenyls formed from PCBs by plants. *Environmental Pollution*: 127(1), 41–48.

Hurtubise, Y., Barriault, D., Powlowski, J., Sylvestre, M., 1995. Purification and characterization of the *Comamonas testosteroni* B-356 biphenyl dioxygenase components. *Journal of Bacteriology* 177, 6610–6618.

Hurtubise, Y., Barriault, D., Sylvestre, M., 1996. Characterization of active recombinant his-tagged oxygenase component of *Comamonas testosteroni* B-356 biphenyl dioxygenase. *Journal of Biological Chemistry* 271, 8152–8156.

Kucerova, P., Mackova, M., Chroma, L., Burkhard, J., Triska, J., Demnerova, K., Macek, T. 2000. Metabolism of PCBs by *Solanum nigrum* hairy root clone SNC-9O and analysis of transformation products. *Plant and Soil* 225, 109–115.

Macek, T., Mackova, M., Kas, J., 2000. Exploitation of plants for the removal of organics in environmental remediation. *Biotechnology Advances* 18(1), 23–35.

Macek, T., Macková, M., Kucerová, P., Chromá, L., Burk-hard, J. & Demnerová, K. 2002. Phytoremediation. In: *Focus on Biotechnology*, (Hofman M. and Anne J., eds.), Vol. 3, Kluwer Academic Publ., Dordrecht, pp. 115–137.

Mackova, M., Macek, T., Ocenaskova, J., Burkhard, J., Demnerova, K. & Pazlarova, J. 1997. Biodegradation of polychlorinated biphenyls by plant cells. *International Biodeterionation and Biodegradation* 39: 317–325.

Sondossi, M., Sylvestre, M., Ahmad, D., Massé, R., 1991. Metabolism of hydroxybiphenyl and chlorohydroxybiphenyl by biphenyl/chlorobiphenyl degrading *P. testosteroni*, strain B-356. *Journal of Industrial Microbiology* 7, 77–88.

Sondossi, M. Lloyd, B.A., Barriault, D., Sylvestre, M., Simard, M., 1995. Microbial transformation of a dihydroxybiphenyl. *Acta Crystallographica* Section C 51, 491–494.

Swanson, G.M., Ratcliffe, H. E., Fischer, L. J., 1995. Human exposure to polychlorinated biphenyls (PCBs): A critical assessment of the evidence for adverse health effects. *Regulatory Toxicology and Pharmacology* 21, 136–150.

European Symposium on Environmental Biotechnology, ESEB 2004 - Verstraete (ed)
© 2004 Taylor & Francis Group, London, ISBN 90 5809 653 X

In situ bioprecipitation for remediation of metal-contaminated groundwater

W. Ghyoot & K. Feyaerts
Umicore, Olen, Belgium

L. Diels & K. Vanbroekhoven
Vito, Mol, Belgium

X. De Clerck & W. Gevaerts
ARCADIS Gedas, Deurne, Belgium

E. Ten Brummeler & P. van den Broek
Bremcon NV, Zwijndrecht, Belgium

ABSTRACT: Metal-polluted groundwater co-contaminated by sulfates may be decontaminated by *in situ* immobilisation based on biological sulfate reduction. This remediation technique is based on the stimulation of sulfate reducing bacteria. Sulfates are reduced to sulfides and heavy metals precipitate as insoluble metal-sulfides. Laboratory tests successfully demonstrated the removal of soluble Zn, Ni, Co and sulfate. Different carbon sources were evaluated as well as the requirements for nutrient addition, pH and redox conditions. A metal-contaminated site was chosen for the pilot-test. Molasses was injected as a carbon source. Monitoring showed a shift of pH and redox conditions followed by a decrease of sulfate and metals concentrations. The results and requirements for *in situ* metal bioprecipitation are highlighted.

1 INTRODUCTION

Industrial processing of non-ferrous metals has resulted in historical contamination of soils and groundwaters with metals and salts. Pump and treat has been widely applied for groundwater remediation. Modelling of groundwater flow and contaminant extraction is applied for prediction of the remediation time. Often though it is found that the time required to achieve remediation goals is (much) longer than predicted. As an alternative for pump and treat *in situ* remediation techniques are being developped with potentially lower operational costs. These can be based on chemical immobilization (injection of reagents) or on biologically induced phenomena. If sulfates are present, *in situ* bioprecipitation based on biological sulfate reduction and metalsulfides precipitation can be feasible. This technique was studied on lab- and pilotscale to demonstrate its principles.

2 MATERIALS AND METHODS

2.1 *Chemical analyses and sequential extraction*

The pH, redox and conductivity were measured with portable instruments. All samples were filtered on 0.45 μm filters before chemical analysis. Metals, sugar, dissolved organic carbon (DOC) and fatty acids were determined according to Standard Methods (APHA 1998). Sequential extraction on aquifer samples was performed according to a modified procedure from Tessier et al. (1979).

2.2 *Batch tests*

Different carbon sources and additives were assessed for stimulation of sulfate reduction. Aquifer (80 g) and groundwater (186 ml) were added to 250 ml flasks. For batch experiment 3 the ratio of aquifer (150 g) and groundwater (90 ml) was different. The appropriate carbon source and additives were added. Samples were incubated at 12°C. Sampling was executed under anaerobic conditions.

2.3 *Pilot test*

The location of the injection and monitoring wells is indicated in Figure 1. The filters were placed 8.5 to 10 m below ground level. The geology up to 10 m below ground level consists of sand and clay-containing sand.

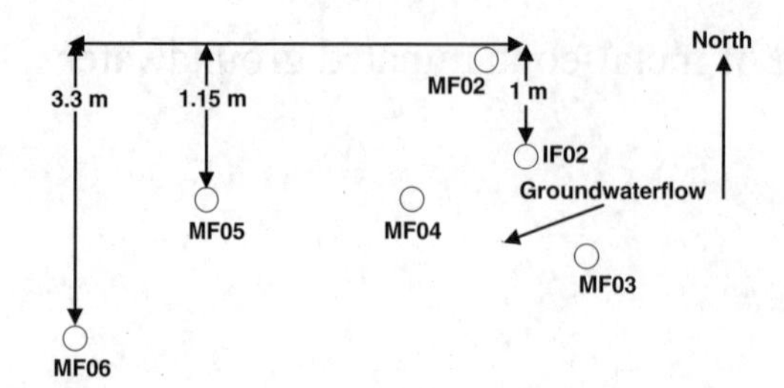

Figure 1. Location of injection (IF02) and monitoring wells (MF02, MF03, MF04, MF05, MF06) for the pilot test.

2.4 *Calculation*

Results for sulfate and metal removal are expressed as percentage remaining in solution:

$$\% \, remaining = \frac{C_i}{C_o} 100$$

where C_i = concentration at time i; and C_o = original concentration.

3 RESULTS AND DISCUSSION

3.1 *Batch tests*

In a first series of batch experiments some carbon sources were tested with and without nutrient addition (NH_4NO_3, KH_2PO_4). After 22 weeks a decrease of sulfate and metal concentrations was observed for the methanol addition (Table 1). No decrease was observed with acetate or molasses. Because previous tests were succesfull with molasses it was added to the control on week 22. Eight weeks later sulfate and metals removal was completely.

A significant reduction in soluble metal concentrations was noticed within weeks after the molasses addition though without a drop in redox or a decrease of sulfate concentration. It was hypothesized that molasses adsorbs to the aquifer and forms metal complexes thus reducing soluble concentrations. Upon molasses degradation sulfate is reduced and metals are precipitated. At week 30 methanol showed excellent and acetate only minor removal.

A sequential extraction was performed on samples from the molasses-supplemented and the methanol condition after 49 weeks. The exchangeable Zn fraction was reduced from 641 to <2 mg/kg and for Cd from 92 to <0.1 mg/kg. The results showed that Zn and Cd moved from the exchangeable to the Fe-Mn-oxide fraction.

Ni was an exception with the exchangeable fraction being reduced from 234 to 230 mg/kg (molasses) or from 234 to 82 mg/kg (methanol) only.

In a second series of tests acetate, lactate, ethanol, methanol, molasses and a hydrogen releasing component (HRC) were evaluated with and without Vitamine B12 addition. The latter is known as an activator of sulfate reducing bacteria (SRB). Samples were incubated during 16 weeks. Sulfate reduction was observed from week 6 on (first sampling date). Ranking the carbon sources according to a decreasing stimulation of SRB: HRC > molasses > ethanol > methanol = lactate > acetate. The influence of Vit B12 was minimal and it seemed not required. Cd was completely removed with all carbon sources at week 6. Zn was completely removed at week 6 with most carbon sources (HRC, lactate, ethanol, molasses, methanol) and almost completely with acetate at week 8. Co, Ni and Fe removal were also removed though incompletely. Best carbon sources were HRC, molasses and ethanol.

Table 1. Results of batch test 1 (% remaining).

Parameter	Unit	Start	Con + Mol*		MeOH	Ac
Time	Week	0	22	30	30	30
pH		5.9	5.9	5.6	6.3	6.0
Eh	mV	87	25	−185	−344	−121
SO_4^{2-}	%	100	108	37	6	0
Cd	%	100	39	0	0	86
Zn	%	100	81	0	0	70
Ni	%	100	79	19	0	70
Co	%	100	80	11	0	69
Fe	%	100	80	15	0	63

* No carbon source week 0–22, molasses added week 22; MeOH = methanol, Ac = acetate.

A third series of batch experiments was performed with molasses. In one condition extra nutrients (N, P) and a mineral solution were added to the molasses. In the control hardly any sulfate reduction was observed though some metal concentrations decreased. Addition of molasses clearly stimulated sulfate reduction and metal removal. The addition of nutrients did not result in extra removal though the kinetics of sulfate reduction were enhanced by the nutrients addition.

The order of metal removal was Cd > Zn > Co > Ni > Fe. Cd was the first metal to precipitate and to the highest extent. This corresponds to the solubility of the respective metalsulfides formed. Solubility products are for CdS: $pK_{sp} = 26.1$, for ZnS: $pK_{sp} = 23.8$, for CoS: $pK_{sp} = 20.4$, for NiS: $pK_{sp} = 18.5$ and for FeS: $pK_{sp} = 18.2$.

Ranking the carbon sources according to decreasing efficiency (where most efficient means yielding the fastest kinetics for sulfate reduction and the lowest final metal concentrations) results in: HRC > molasses > ethanol > lactate > methanol > acetate. The addition of nutrients (N, P) or the addition of Vit B12 was not required for activation of the SRB.

Table 2. Data on molasses injections.

No	Day	Q_E (m^3/h)	Vol. (m^3)	Q_I (m^3/h)	Molasses (kg)	KBr (kg)	Na_2CO_3 (kg)
1	0	0.5	4	2	60	1	0
2	55	NM	0.3	NM	3.7	0	0
3	139	NM	1	NM	17.5	1	2

Q_E = extraction flow rate; Q_I = injection flow rate; NM = not measured.

3.2 In situ *pilot test*

3.2.1 *Molasses injections*

Given the positive results and because it is a relatively cheap carbon source molasses was chosen for an *in situ* pilot test. No nutrients or minerals were added. Three infiltrations were performed (Table 2). For the first injection groundwater was extracted from IF02 under anaerobic conditions and collected in a recipient under N_2 atmosphere. Molasses and a tracer (KBr) were added. The mixture was infiltrated again in IF02 under gravity flow. A second injection was executed on day 55. Again water was extracted from IF02 and infiltrated after addition of molasses. The third injection was executed on day 139. The dilution water consisted in a 45/55 mixture of groundwater sampled previously from the monitoring filters and groundwater extracted from MF06.

3.2.2 *pH, redox and bromide*

Injection of molasses had a pronounced effect on the redox potential. It dropped below −200 mV (after 1 to 2 weeks), remained low during 10 weeks and then rapidly increased above 100 mV. MF05 was less influenced. About 4 weeks after the third injection the redox dropped under −200 mV during 2 weeks.

Starting from a value above 5.5 pH dropped to 4.5 after 10 weeks in all filters except for MF06. The pH dropped due to fatty acid formation. The pH 4.5 was maintained several weeks and then increased to 5. After the third injection the pH remained between 4.5 and 5 because carbonate buffer had been added.

Bromide was added as an inert tracer. It is supposed not to adsorb to the aquifer and indicates the distribution of liquid after injection. A peak bromide concentration was noticed after the first and third injection. MF03 was most rapidly influenced by the first injection while for the other filters more time was required. Due to this rapid response it might be that preferential flows occurred towards MF03.

3.2.3 *Sugar, fatty acid and DOC concentrations*

Molasses is not directly consumed by SRB but first fermented by fermentative bacteria. This results in the production of hydrogen and fatty acids. Hydrogen can be consumed by the hydrogen oxidizing SRB and

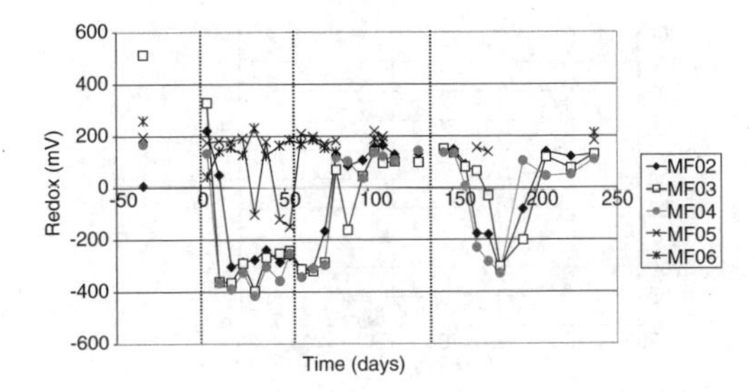

Figure 2. Redox potential during pilot test. Dotted lines indicate injection of molasses.

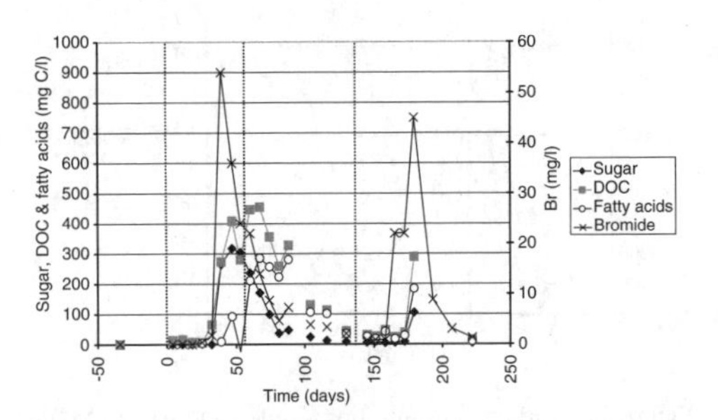

Figure 3. Carbon concentrations, redox and Br^- in MF02.

acetate by the acetate oxidizing SRB. Sugar and dissolved organic carbon (DOC) were analyzed to assess molasses distribution and fermentation. The difference between DOC and sugar concentrations was assumed to be due to fatty acids.

Molasses was first detected in MF03 followed by MF04 and MF02 corresponding to the tracer results. There was an overlap in the effect of the first and the second injection. In this way DOC concentration remained high enough in order to sustain sulfate reduction. After the third injection the DOC concentrations were significantly lower than after the first injection. The molasses concentration was similar for both injections (15 and 17.5 kg/m^3) though the load was different (60 and 17.5 kg). It was hypo-thesized that molasses adsorbed to the aquifer. The remain-ing part solubilized in the groundwater. When insufficient molasses is added (such as in the third injection) only adsorption can occur. Fatty acids were analyzed on day 108. A total concentration of 343 ppm was detected in MF04, dominated by acetic acid (276 ppm).

3.2.4 *Sulfate and metal concentrations*

Following the first and second injection increasing metal concentrations were observed in MF02. This was attributed to the pH drop caused by acidification. However, soon after the second injection a substantial decrease of sulfate and metal concentrations was observed. After the third injection metal concentrations

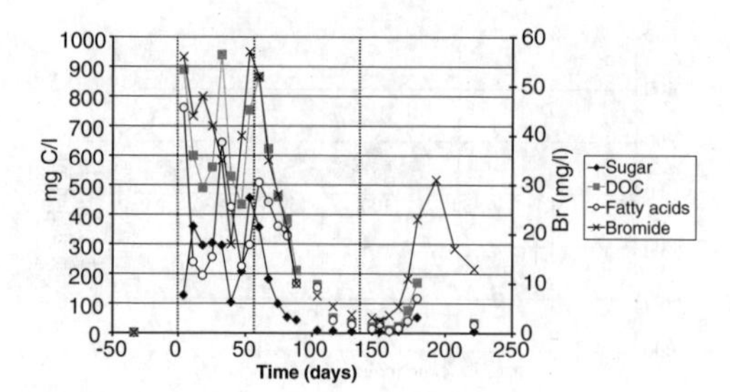

Figure 4. Carbon concentrations, redox and Br^- in MF03.

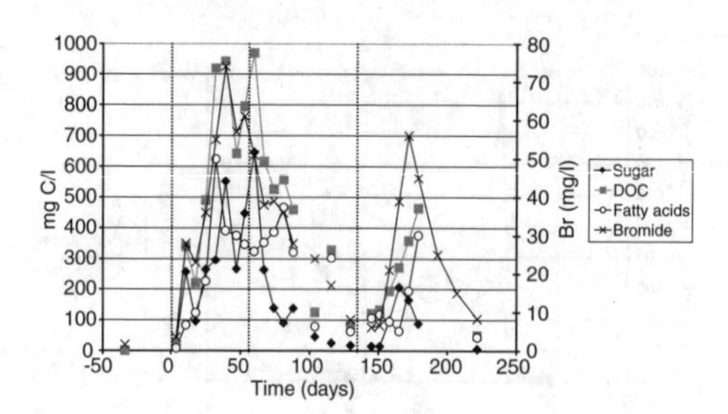

Figure 5. Carbon concentrations, redox and Br^- in MF04.

increased again. Most probably an influx of metals occurred with the injection water. Sulfate was not removed because insufficient molasses had been added.

Similar observations were made for MF03. For MF04 sulfate and metal concentrations decreased after injections one and two. Though the third injection resulted in increasing concentrations. It was much more pronounced than in MF02 because concentrations before the third injection were significantly lower in MF04. Sulfate and metal concentrations raised until they reached their original levels in both MF02 and MF04. Only cadmium remained below detection limit.

In MF05 a small decrease of the sulfate concentration was noticed but metal concentrations remained on the same level. MF06 was not influenced by the molasses injections.

4 CONCLUSIONS

- *In situ* metal bioprecipitation was induced after addition of molasses.
- Ranking metal removal yields Cd > Zn > Co > Ni. Cd precipitated first and to the highest extent.
- Ranking carbon sources yields HRC > molasses > ethanol > actate > methanol > acetate. HRC yields fastest kinetics and lowest metal concentrations.
- Addition of nutrients (N, P) or Vit B12 was not required for SRB stimulation though kinetics are favoured by their addition.

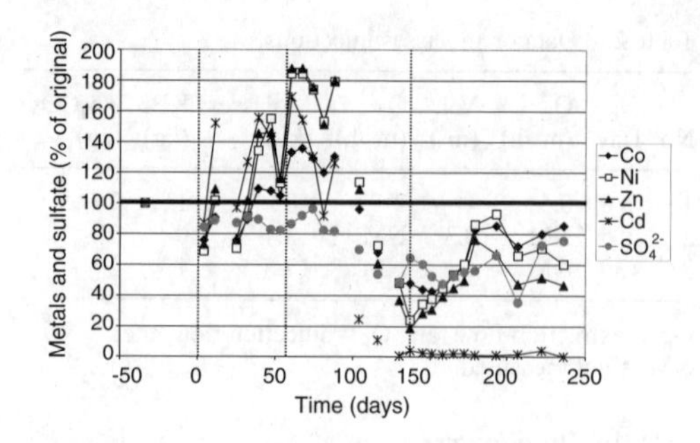

Figure 6. Metal and sulfate concentrations in MF02.

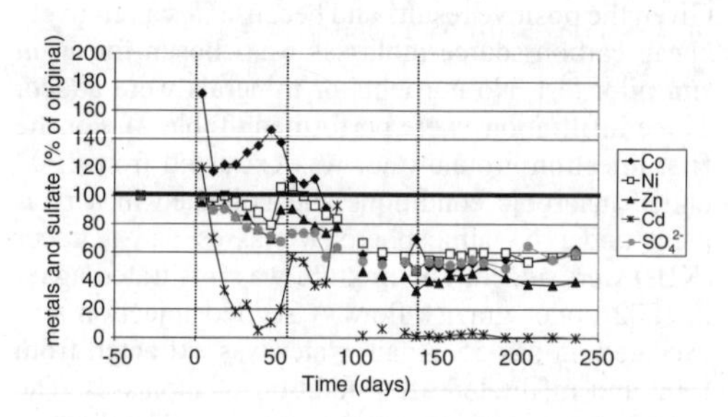

Figure 7. Metal and sulfate concentrations in MF03.

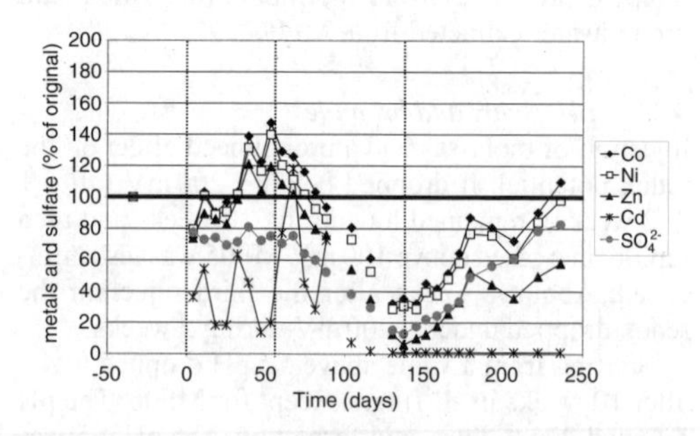

Figure 8. Metal and sulfate concentrations in MF04.

- Redox, pH and DOC are useful control parameters. DOC addition requires careful control.
- Full-scale implementation requires specific lay-out of injection and monitoring filters. Irreversibility of precipitation and final metal concentrations require further research.

REFERENCES

APHA, 2003. *Standard methods for examination of water and wastewater*. American Public Health Association Publications.

Tessier, A., Campbell, P.G.C. & Bisson M. 1979. Sequential extraction procedure for speciation of particulate trace metals. *Analytical Chemistry* 51: 844–851.

European Symposium on Environmental Biotechnology, ESEB 2004 - Verstraete (ed)
© 2004 Taylor & Francis Group, London, ISBN 90 5809 653 X

Metals remediation compound (MRC™): a new slow-release product for *in situ* metals remediation

Anna Willett[1], Jeremy G.A. Birnstingl[2] & Stephen S. Koenigsberg[1]
[1]*Regenesis, San Clemente, CA*
[2]*Regenesis, London, UK*

Contamination of groundwater by metals has not been widely addressed by engineered *in situ* remediation technologies, despite the documentation of metals contamination at greater than 50% of sites from the National Priorities List and at Department of Defense and Department of Energy locations. Metals remediation compound (MRC™) is a slow-release metals remediation product that removes dissolved metals from groundwater via *in situ* immobilization (precipitation and/or sorption to soil particles). The immobilized metals are stable under reducing conditions and may be stable under oxidizing conditions, depending on the identity of the metal.

MRC consists of an organosulfur compound esterified to a carbon backbone. This organosulfur ester is embedded in a polylactate matrix, making MRC a thick, viscous liquid. Upon injection into an aquifer, the organosulfur compound is slowly released when MRC's ester bonds are cleaved via hydrolysis by water and microbial enzymatic action. The organosulfur moiety interacts with metal ions, either to complex them or to reduce them and complex them sequentially. These complexes sorb strongly to soil, filter media, or other solid supports. MRC also slowly releases lactate, which acts as an electron donor and carbon source for naturally-occurring bacteria and creates the optimal conditions for metals immobilization by the organosulfur compound. For sites with mixed metal and chlorinated solvent contamination, MRC provides a substrate for accelerated reductive dechlorination and metals immobilization.

MRC's ability to remove dissolved metals, such as arsenic, copper, chromium, cadmium, mercury, and lead, from solution has been tested in the laboratory and verified in situ via injection into metals-contaminated aquifers. Results from these laboratory investigations and field applications will be presented.

European Symposium on Environmental Biotechnology, ESEB 2004 - Verstraete (ed)
© 2004 Taylor & Francis Group, London, ISBN 90 5809 653 X

Enrichment of MTBE degrading bacteria from gasoline contaminated subsurface soil

D. Moreels, L. Bastiaens & L. Diels
Vito (Flemish Institute for Technological Research), Mol, Belgium

D. Springael, F. Ollevier & R. Merckx
KU Leuven, Leuven, Belgium

ABSTRACT: Biodegradation of methyl *tert*-butyl ether (MTBE) in aquifer material was studied during slurry enrichments (batch-systems) simulating *in situ* conditions. Soil samples taken at 4 different depths at a gasoline-contaminated site were used and the microbial population of each soil sample was separately enriched. The MTBE-degradation kinetics and the community dynamics as a response to different implemented *in situ* important parameters (e.g. additions of nutrients, addition of extra oxygen and presence of co-contaminants) were determined and compared between the different depths. In all enrichments the MTBE-biodegradation capacity was active for more than 600 days. PCR-DGGE fingerprints revealed that addition of extra nutrients or benzene did not lead to significant changes in the eubacterial population. No differences could also be observed between the community composition of the enrichment conditions after 1 year of incubation at different oxygen concentrations (8, 9.5 and 11.5 mg/l DO). In contrast, adding propane did induce differences in band intensity of the fingerprint. A different eubacterial community was observed at the 4 different examined depths.

1 INTRODUCTION

Methyl *tert*-butyl ether (MTBE) is an anthropogenic chemical used as an octane enhancer in fuels from the late 1970s in concentrations ranging from 2 to 15% (v/v). More than 20 million tons of MTBE were consumed in 2000. Leaks of MTBE are due to point (e.g. leaking pipelines) and non-point sources (e.g. atmospheric fall-out) (Johnson et al, 2000; Pankow et al, 1997). MTBE has been detected in environmental compartments such as ambient air (Pankow et al, 1997) and groundwater (Zogorski et al, 1996) in concentrations of 1 μg/l to 200 mg/l respectively. MTBE is a persistent molecule with estimated half-lives of a few days in the atmosphere to more than several years in groundwater. Due to its relatively high water solubility (50 g/l) and low retardation compared to other gasoline components, MTBE will partition more to the groundwater and will migrate faster through the subsurface (Squillace et al, 1997). Furthermore, recent concerns about possible acute and chronic toxicity towards exposed humans have led to extensive research on its toxicity. From these studies it was concluded that MTBE is an animal carcinogen with the potential to cause cancer in humans (NSTC, 1997). Recently, several laboratory studies have shown that MTBE can be biodegraded by pure or mixed bacterial cultures (Deeb et al, 2000). Field studies have indicated that biodegradation could be an important natural attenuation mechanism (Borden et al, 1997). As such, biodegradation could present an economical and ecological valuable alternative to mostly expensive physical–chemical technologies for treatment of contaminated aquifers. Nonetheless, there is still a lack of information on (1) the current occurrence and distribution of MTBE degrading bacteria in the subsurface, (2) their metabolic needs and (3) the effect of subsurface conditions and implemented treatment conditions on MTBE degradation and on the bacterial populations. In several reports, it was found that adding extra nutrients (Streger et al, 2002), higher dissolved oxygen concentrations (Salanitro et al, 2000) or co-contaminants (e.g. benzene and propane) (Deeb et al, 2001; Steffan et al, 1997) to the groundwater stimulated MTBE biodegradation. Therefore, the effect of addition of these compounds under *in situ* simulating nutrient poor conditions on MTBE degradation and on the subsurface soil microbial community composition was followed in batch enrichments with molecular biological techniques, namely a culture-independent technique based on PCR amplification of 16S rRNA genes followed by denaturing gradient gel electrophoresis (DGGE).

2 MATERIALS AND METHODS

2.1 *Used soil*

The 4 used soil samples (P250, P300, P450, P600) were taken at depths of 250, 300, 450 and 600 cm below ground surface (bgs) at a gasoline contaminated site. Sample P250 was located in the vadose zone, P300 in the capillary zone and P450 and P600 in the saturated zone. All samples except P250 were initially contaminated with benzene, toluene, ethylbenzene and xylenes (BTEX) in concentrations ranging from 0.043 to 17 mg/kg ds. All samples were initially contaminated with MTBE, in concentrations ranging from 7 to 14 mg/kg ds. The soil samples had a near neutral pH and were of the sandy clay loam type in the case of P250 and of the loamy sand type in the case of the other samples. All soil samples contained a microbial population capable of degrading MTBE and benzene (Moreels et al).

2.2 *Enrichment cultures*

120 ml glass flasks (Air/liquid = 50/50) were filled with 6 g of soil and 54 ml of artificial groundwater (Fontes et al, 1991). Eight different culture conditions were started as described in Table 1. Nitrogen and phosphorus were added as NH_4NO_3 and $Na_2HPO_4 \cdot 2H_2O$ respectively until a C/N/P of 100/10/1 was obtained. Pure MTBE and benzene were added to the liquid phase using a microliter syringe. Pure propane and oxygen were added by injection in the headspace of the vial, the amount based on the Henry constant.

Table 1. Overview of the conditions implemented on the soil samples in the microcosms.

	Addition of					
Microcosm number	NP	MTBE* (mg/l)	Be* (mg/l)	Pr* (mg/l)	O_2* (mg/l)	$HgCl_2$* (mg/l)
C1	−	5	5	−	8	250
C2	−	5	−	−	8	−
C3	+	5	−	−	8	−
C4	+	5	−	−	9.5	−
C5	+	5	−	−	11.5	−
C6	+	5	5	−	11.5	−
C7	+	−	5	−	11.5	−
C8	+	5	−	5	11.5	−

*: Concentration in liquid phase; +: added; −: not added; NP: nutrients; Be: benzene; Pr: propane.

The flasks were stopped with Viton stoppers and incubated statically at 20°C.

Aqueous subsamples from the incubated microcosms were taken through the septum in function of time and poisoned with 500 mg/l $HgCl_2$ prior to analysis. Residual MTBE and benzene concentrations in the water phase were measured by headspace GC-MS. Dissolved oxygen concentrations were measured with a calibrated electrode oxygen meter (Strathkelvin Instruments Limited). When depleted, MTBE, benzene, propane and oxygen were re-added. After 400 days of incubation, added MTBE concentrations were raised in order to increase the enrichment process.

2.3 *DNA-analyses*

DNA was extracted as described by Hendrickx et al (submitted). The extracted DNA was purified with polyvinylpyrollidone (PVPP) and the DNA clean up Wizard[R] protocol. 16S rRNA gene fragments from soil microbial communities were amplified by PCR using primer set P63f and P518r, according to El Fantroussi et al (1999). DGGE was performed as described by Muyzer et al (1998).

3 RESULTS

3.1 *Biodegradation of MTBE in the enrichment cultures*

The evolution of the MTBE concentration in all the enrichments was followed in time and in all cases MTBE degradation was observed for more than 600 days. Figure 1 shows for example the results for the poisoned control and the enrichments with and without propane with soil P250. MTBE biodegradation was sustainable in both enrichment cultures. The results of C5 are representative for other enrichment cultures of the other soil samples. After 400 days of incubation MTBE concentrations were raised in order to attain more biomass. In the poisoned control minimal losses of MTBE were observed. After 600 days, MTBE was re-added in the poisoned control.

Table 2 depicts the first order rate constants k corresponding to the biodegradation of MTBE in relation to the different implied test conditions in the enrichment cultures of soil sample P250. Losses in the poisoned control were taking into account. At the start of the enrichment it was found that addition of propane and implementing high dissolved oxygen concentrations of 9.5 and 11.5 mg/l DO stimulated the MTBE biodegradation rate. Adding nutrients or benzene did have negative effects. Concerning the other soil samples, the effect of nutrients and dissolved oxygen concentrations depended on the soil sample studied, while addition of benzene and propane in general did promote MTBE biodegradation. The effect of the additives on longer term was evaluated by determining the MTBE-degradation rate after 655 days of incubation. Since biomass production was not taking into account, the presented first order rate constant k2 is only an approximation. It can be concluded from Table 2 that in most conditions the first order rate constant increased with at least a factor 4. Adding extra nutrients, implementing dissolved oxygen concentrations and adding benzene had no effect on the MTBE biodegradation rate after more than 600 days of incubation. In contrast, in the culture condition containing propane (C8), the first order rate decreased with a factor 14. The slower MTBE degradation rate can also be observed in figure 1. In this condition, a relative higher oxygen demand was observed compared to other culture conditions. At day 655, dissolved oxygen concentrations were decreased to a value of 3 mg/l. It was observed that each time dissolved oxygen concentrations decreased below 5 mg/l, MTBE degradation slowed down or even stopped.

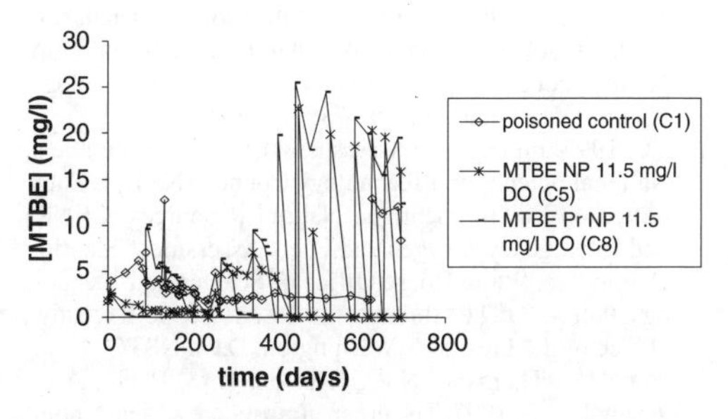

Figure 1. Biodegradation of MTBE in enrichment cultures of soil sample P250. Presented is the effect of adding propane on the MTBE biodegradation rate.

Table 2. First order rate constant k of MTBE degradation in enrichment cultures of soil sample P250.

Culture condition	First order rate constant k1 (/day)	First order rate constant k2 (/day)
C2 (MTBE, no additions, 8 mg/l DO)	0.060	0.2391
C3 (MTBE, NP, 8 mg/l DO)	0.0040	0.2361
C4 (MTBE, NP, 9.5 mg/l DO)	0.011	0.2346
C5 (MTBE, NP, 11.5 mg/l DO)	0.0080	0.2369
C6 (MTBE, Be, NP, 11.5 mg/l DO)	0.0040	0.2395
C8 (MTBE, Pr, NP, 11.5 mg/l DO)	0.056	0.0039

k1: first order rate constant at the start of incubation;
k2: first order rate constant after 655 days of incubation.

3.2 *DGGE-analysis of soil microbial communities*

DGGE analyses of 16S rRNA gene fragments amplified by PCR were performed from the different enrichment cultures of the 4 soil samples at the start and after 1 year of incubation. The results of soil sample P250 are given in Figure 2. The implementation of the different growth conditions during 1 year, affected the composition of the microbial population slightly, which might indicate that different populations are being enriched. Especially the fingerprint of the enrichment culture implemented with propane (C8), showed significant changes in band intensity compared to other fingerprints. It can be seen by investigating the fingerprints of the unsaturated soil sample P250 that a band appears in all MTBE containing enrichment cultures after an incubation period of 1 year, but is absent in the culture condition C7 (containing only benzene). This amplified DNA fragment might thus be representative for an MTBE degrading bacterium. In this regard, PCR amplification of DNA extracted from a number of the enrichment cultures with *Rubrivivax gelatinosus* PM-1 specific primers (Hristova et al, 2003) indicated that at least this MTBE degrading bacterium might be a member of the bacterial community of several soil samples of site P. DGGE-analysis showed that the microbial population composition of the 4 soil samples was different before the enrichment and that the composition of the microbial population remained different even after 1 year of incubation under similar *in situ* simulating incubation conditions (results not shown).

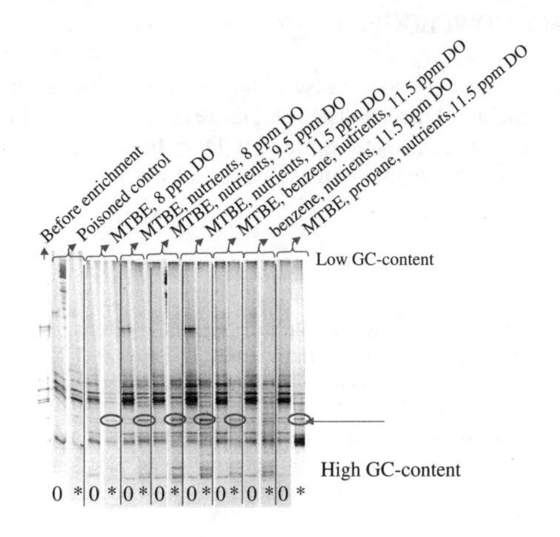

Figure 2. DGGE community fingerprint obtained for soil samples P250. All samples were analysed after 1 year of incubation. *After 1 year of incubation; 0, at the start of the incubation.

4 CONCLUSIONS

Enrichment studies with soil samples taken from 4 different depths of a contaminated site were carried out to investigate whether there was a difference in the rate of MTBE biodegradation and soil community composition between soil samples and between the different enrichment conditions. This study led to the following conclusions. (1) Sustainable MTBE biodegradation was observed to occur after more than 600 days in batch enrichments of soil samples originating from different depths of a gasoline contaminated site. In order to maintain suitable culture conditions for sustainable MTBE biodegradation, the parameter oxygen appeared to have a crucial role. (2) DGGE analysis of 16S rRNA gene fragments revealed that the microbial population composition for each depth sample was different before the enrichment and remained different even after 1 year of incubation under similar enrichment conditions. (3) 1 year of incubation with extra nutrients did not lead to differences in microbial population composition. Also implementing dissolved oxygen concentrations of 9.5 and 11.5 mg/l or adding benzene did not lead to differences in fingerprint pattern. In contrast, adding propane as co-contaminant of MTBE did lead to significant differences in the composition of the bacterial population. (4) Future research will concentrate on isolation and characterization of the MTBE degrading bacteria. Spread plating on agar media revealed several different morphotypes, which are being tested for their capacity to degrade MTBE. Further, DGGE fingerprint bands of several enrichment cultures are being cloned and will be sequenced in order to identify the bacteria present.

ACKNOWLEDGEMENTS

We thank Miranda Maesen and Queenie Simons for excellent technical assistance. This research was funded by a PhD-scholarship granted to D.M. by the Flemish Institute for Technological Research (Vito).

REFERENCES

Borden, R.C., Daniel, R.A., LeBrun, L.E., Davis, C.W. 1997. Intrinsic biodegradation of MTBE and BTEX in a gasoline-contaminated aquifer. Water Resources Research. 33, 1105–1115.

Deeb, R.A. Scow, K.M., Alvarez-Cohen, L. 2000. Aerobic MTBE biodegradation: an examination of past studies, current challenges and future research directions. Biodegradation. 11, 171–186.

Deeb, R.A., Hu, H-Y, Hanson, J.R., Scow, K.M., Alvarez-Cohen, L. 2001. Substrate interactions in BTEX and MTBE mixtures by an MTBE-degrading isolate. Environ. Sci. Technol. 35, 312–317.

El Fantroussi, S., Verschuere, L., Verstraete, W., Top, E.M. 1999. Effect of phenylurea herbicides on soil microbial communities estimated by analysis of 16S rRNA gene fingerprints and community-level physiological profiles. Appl. Environ. Microbiol. 65, 982–988.

Fontes, D.E., Mills, A.L., Hornberger, G.M., Herman, J.S. (1991). Physical and chemical factors influencing transport of microorganisms through porous media. Appl. Environ. Microbiol. 57, 2473–2481.

Hristova, K., Gebreyesus, B., Mackay, D., Scow, K.M. 2003. Naturally occurring bacteria similar to the methyl tert-butyl ether (MTBE)-degrading strain PM-1 are present in MTBE-contaminated groundwater. Appl. Environ. Microbiol. 69, 2616–2623.

Johnson, R., Pankow, J., Bender, D., Price, C., Zogorski, J. 2000. MTBE: To what extent will past releases contaminate community water supply wells? Environ. Sci. Technol/ News. May 1, 210A–217A.

Moreels, D., Bastiaens, L., Ollevier, F., Merckx, R., Diels, L., Springael, D. Evaluation of the intrinsic methyl *tert*-butyl ether (MTBE) biodegradation potential of hydrocarbon contaminated subsurface soils in batch microcosm systems. Submitted.

Muyzer, G., Smalla, K. 1998. Application of denaturing gradient gel electrophoresis (DGGE) and temperature gradient gel electrophoresis (TGGE) in microbial ecology. Antonie Van Leeuwenhoek. 73: 127–141.

NSTC. 1997. Interagency Assessment of Oxygenated Fuels. National Science and Technology Council (NSTC), Committee on Environment and Natural Resources (CENR) and Interagency Oxygenated Fuels Assessment Steering Committee. White House Office of Science and Technology Policy (OSTP) through the CENR of the Executive Office of the President. Washington, D.C.: NSTC.

Pankow, J.F., Thomson, N.R., Johnson, R.L., Baehr, A.L., Zogorski, J.S. 1997. The urban atmosphere as a non-point source for the transport of MTBE and other volatile organic compounds (VOCs) to shallow groundwater. Environ. Sci. Technol. 31, 2821–2828.

Salanitro, J.P., Johnson, P.C., Spinnler, G.E., Maner, P.M., Wisniewski, H.L., Bruce, C. 2000. Field-scale demonstration of enhanced MTBE bioremediation through aquifer bioaugmentation and oxygenation. Environ. Sci. Technol. 34, 4152–4162.

Squillace, P.J., Pankow, J.F., Korte, N.E., Zogorski, J.S. 1997. Review of the environmental behaviour and fate of methyl *tert*-butyl ether. Environ. Toxicol. Chem. 16, 1836–1844.

Steffan, R.J., McClay, K., Vainberg, S., Condee, C.W., Zhang, D. 1997. Biodegradation of the gasoline oxygenates methyl *tert*-butyl ether, ethyl *tert*-butyl ether, and *tert*-amyl methyl ether by propane-oxidizing bacteria. Appl. Environ. Microbiol. 63, 4216–4222.

Streger, S.H., Vainberg, S., Dong, H., Hatzinger, P.B. 2002. Enhancing transport of *Hydrogenophaga flava* ENV735 for bioaugmentation of aquifers contaminated with methyl *tert*-butyl ether. Appl. Environ. Microbiol. 68, 5571–5579.

Zogorski, J., Morduchowitz, A., Baehr, A., Bauman, B., Conrad, D., Drew, R., Korte, N., Lapham, W., Pankow, J., Washington, E. 1997. Fuel oxygenates and water quality coordinated by the interagency oxygenated fuel assessment, Office of Science and Technology Policy, Executive Office of the President, Washington, DC.

European Symposium on Environmental Biotechnology, ESEB 2004 - Verstraete (ed)
© 2004 Taylor & Francis Group, London, ISBN 90 5809 653 X

Reductive precipitation of the nuclear fuel cycle contaminant Tc(VII) by *Geobacter sulfurreducens*

J.C. Renshaw & J.R. Lloyd
Williamson Research Centre for Molecular Environmental Sciences, Dept. of Earth Sciences, University of Manchester, Manchester, UK

F.R. Livens & I. May
Centre for Radiochemistry Research, Dept. of Chemistry, University of Manchester, Manchester, UK

M.V. Coppi & D.R. Lovley
Dept. of Microbiology, University of Massachusetts, Amherst, USA

S. Glasauer
Dept. of Microbiology, University of Guelph, Guelph, ON, Canada

ABSTRACT: Technetium is a key radioactive pollutant in subsurface environments contaminated with radioactive waste. However, dissimilatory Fe(III)-reducing bacteria can control its mobility through the enzymatic reduction and precipitation of soluble Tc(VII) to insoluble Tc(IV). We have investigated the mechanism of electron transfer to Tc(VII) in the model dissimilatory Fe(III)-reducing bacterium *Geobacter sulfurreducens*. Reduction of Tc(VII) by *G. sulfurreducens* required hydrogen as the electron donor, suggesting the involvement of a hydrogenase, while CO profiling indicated that Ni and Fe are the metal cofactors. Enzymatic activity and the reduced Tc(IV) precipitate was localized in the periplasm using Cu(II) (a selective inhibitor for periplasmic hydrogenases) and transmission electron microscopy. Finally, a deletion mutant unable to synthesize the primary periplasmic NiFe hydrogenase of *G. sulfurreducens* was unable to reduce Tc(VII), confirming a role for this protein in electron transfer to the fission product.

1 THE NUCLEAR LEGACY

The nuclear weapons and nuclear energy generation programmes of the past 60 years have created a legacy of radioactive waste and contaminated land around world. In the UK, at sites such as Sellafield, Aldermaston and Dounreay there is soil and groundwater contamination with transuranium elements and fission products. Comparable (or worse) problems exist in the U.S. at Rocky Flats (plutonium contaminated oil leaks), Hanford (leaking high level waste tanks) and Savannah River (contaminated land and groundwater), and in the former Soviet Union at Kyshtym (HLW tank explosion), Chelyabinsk (leaks and HLW disposal to surface waters) and Chernobyl (reactor explosion). These few examples alone illustrate the widespread and challenging nature of the problem. The current estimated costs of decontamination and safe disposal of these wastes is colossal. In the UK, the recent government white paper on the nuclear legacy estimated the cost at £48 billion (Managing the nuclear legacy, 2002). In the USA, the estimated cost of dealing with military waste alone is $1 trillion.

One of the key radionuclides contaminating nuclear sites is technetium (Tc). This fission product of uranium is produced in kilogram quantities during nuclear reactions. The stable form of Tc in aerobic conditions is the pertechnetate anion (TcO_4^-), which is highly soluble and very mobile in the environment. It also has a long half-life and has high bioavailability as an analogue of sulphate (Wildung *et al.*, 1979). These factors combine to make Tc a significant radiological hazard in the environment.

2 MICROBES & RADIONUCLIDES

Microbes can influence the speciation and mobility of radionuclides in the environment in a number of ways. They can sorb radionuclides to the cell surface, or can produce a range of ligands that can complex and leach metal ions from minerals, leading to either an increase or decrease in mobility of the radionuclide. Microbes can also alter the oxidation state of metal ions, including radionuclides, and this can have significant effects on the mobility of the radionuclide in the environment. An understanding of the effect of microbial metabolism on radionuclide solubility and speciation will help in predicting their transport behaviour and in developing methods to limit migration of the radionuclides. The effect of microbes on radionuclides could also be exploited in bioremediation technologies to treat contaminated land and in processes to remove radionuclides from wastes and effluents prior to disposal. Many bacteria can reduce metal ions, including significant environmental pollutants such as uranium, chromium and mercury, either indirectly or through direct enzymatic mechanisms (Lovley, 1993, Lloyd *et al.*, 2002). Tc(VII), in the form of highly mobile TcO_4^-, can be reduced by microorganisms to insoluble Tc(IV); the ability to reduce Tc(VII) appears to be widespread amongst bacteria (Fredrickson *et al.*, 2000, Lloyd *et al.*, 1998, Lloyd *et al.*, 1999, Wildung *et al.*, 2000).

Dissimilatory Fe(III) reduction is the dominant respiratory process in many sedimentary environments, and here Fe(III)-reducing bacteria have the potential to reduce Tc(VII) *in situ*. However, little is known about the mechanism of enzymatic reduction of Tc(VII) by these organisms. Our aim in this study was to fully determine the mechanism of TcO_4^- reduction by the model organism *Geobacter sulfurreducens*. This organism was selected because it is a close relative to bacteria that dominate in the subsurface when Fe(III) is utilized as an electron acceptor (Snoeyenbos-West *et al.*, 2000). Other factors making this organism ideal for laboratory studies include the availability of a genomic sequence (at www.tigre.org) in combination with a genetic system for making knockout mutants required for biochemical studies (Coppi *et al.*, 2001).

3 DIRECT ENZYMATIC REDUCTION OF Tc(VII) BY *GEOBACTER SULFURREDUCENS*

3.1 G. sulfurreducens *can couple Tc(VII) reduction to the oxidation of* H_2

Resting cells of *G. sulfurreducens* can couple oxidation of hydrogen to the reduction of Tc(VII) (Lloyd *et al.*, 2000). In contrast, acetate and formate, which can be used as electron donors for the reduction of Fe(III) by the organism (Caccavo Jr *et al.*, 1994), do not support efficient reduction of Tc(VII); hydrogen is the only efficient electron donor for Tc(VII) reduction. This indicates that the mechanism of Tc(VII)

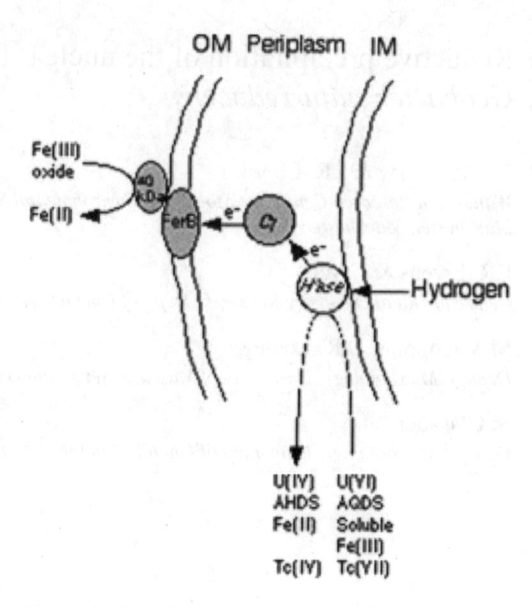

Figure 1. Proposed mechanism for Tc(VII) reduction in *G. sulfurreducens*.

reduction is distinct from that of Fe(III) reduction. Reduction of Fe(III) by *G. sulfurreducens* involves a cytochrome-mediated electron-transport system (Fig. 1; Caccavo Jr *et al.*, 1994, Gaspard *et al.*, 1998, Lloyd *et al.*, 1999, Leang *et al.*, 2003), whilst the absolute requirement of hydrogen for Tc(VII) reduction suggests the involvement of a hydrogenase in the reduction pathway. A number of other bacteria, including *E. coli, Shewanella putrefaciens* and sulfate-reducing bacteria can also couple the oxidation of H_2 to the reduction of Tc(VII), again suggesting the involvement of hydrogenase enzymes (Lloyd *et al.*, 1997, Lloyd *et al.*, 2001, Lloyd *et al.*, 1999, Wildung *et al.*, 2000, De Luca *et al.*, 2001).

3.2 *Localization and product of Tc(VII) reduction*

Transmission electron microcopy showed that the product of Tc(VII) reduction, Tc(IV) as TcO_2, is precipitated predominantly in the periplasm, with additional reduced Tc deposits associated with the outer and plasma membranes. Further investigation of reduction of Tc(VII) by resting cells of*G. sulfurreducens* confirmed that the enzyme responsible for Tc(VII) reduction was localized in the periplasm. Treatment of whole cells with α-chymotrypsin prior to incubation with TcO_4^- did not affect the ability of the cells to reduce Tc(VII), indicating that the reductase was not accessible to protease treatment on the surface of the cell, in contrast to the surface bound reductase for insoluble Fe(III) oxides (Lloyd *et al.*,

2002). However, treatment of the cells with Cu(II) did significantly inhibit reduction of Tc(VII). Cu(II) is a specific inhibitor of periplasmic hydrogenases, suggesting that the mechanism of Tc(VII) reduction involves a periplasmic hydrogenase.

Coppi *et al.* (2003) recently identified a key periplasmic NiFe hydrogenase in *G. sulfurreducens* (*Hyb*). This enzyme was required for hydrogen-dependent growth using Fe(III) as an electron acceptor, suggesting that *Hyb* is the critical respiratory hydrogenase in *G. sulfurreducens*. When the *Hyb* gene was knocked out, *G. sulfurreducens* lost its ability to reduce TcO_4^-, indicating that *Hyb* also plays a key role in the the Tc(VII) reduction pathway. These results, in combination with those of other studies on hydrogenase-dependent reduction in *Escherichia coli* (Lloyd *et al.*, 1997) and *Desulfovibrio fructosovorans* (De Luca *et al.*, 2001) suggest that enzymatic reduction of Tc(VII) is mediated via a periplasmic NiFe hydrogenase in *G. sulfurreducens*.

4 INDIRECT REDUCTION OF Tc(VII) BY *GEOBACTER SULFURREDUCENS*

G. sulfurreducens can also reduce Tc(VII) indirectly via an Fe(II)-mediated mechanism (Lloyd *et al.*, 2000). The organism can couple the oxidation of acetate to the reduction of Fe(III), and the microbially produced Fe(II) can then abiotically reduce Tc(VII). When Fe(III) was supplied as the citrate complex, Tc(VII) was reduced but not precipitated from solution. However, when Fe(III) was present as an insoluble Fe(III) oxide, Tc(VII) was reduced and formed insoluble TcO_2, although the rate of Fe(III) reduction was much slower for the insoluble oxide than for the citrate. The rate of Fe(III) oxide reduction was enhanced by the addition of anthraquinone disulfonate (AQDS), which acts as an electron shuttle to the oxide, and led to the formation of magnetite. The reduced Tc was almost exclusively associated with the magnetite surface. Although AQDS can also act as an electron shuttle to Tc(VII), the rate and extent of Tc(VII) reduction was much lower than observed in the presence of AQDS and Fe(III) oxide.

5 SUMMARY

G. sulfurreducens can reduce the environmentally significant radionuclide Tc from the highly soluble and mobile Tc(VII) (TcO_4^-) to insoluble Tc(IV) (TcO_2). Reduction can occur either through direct, enzymatic mechanisms or via indirect pathways. The enzymatic mechanism couples the oxidation of hydrogen to Tc(VII) reduction through the NiFe periplasmic hydrogenase *Hyb*. The indirect mechanism couples the oxidation of acetate to the reduction of Fe(III); the microbially synthesized Fe(II) can then reduce Tc(VII). The environmental relevance of these competing pathways is currently under investigation.

ACKNOWLEDGMENT

This work was supported by the Biological and Environmental Research Program (BER), U.S. Department of Energy, Grant No. DE-FG02-02ER63422.

REFERENCES

Caccavo Jr, F., Lonergan, D.J., Lovley, D.R., Davis, M., Stolz, J.F. & McInerney, M.J. 1994. *Geobacter sulfurreducens* sp. nov., a hydrogen and acetate-oxidizing dissimilatory metal reducing microorganism. Appl. Environ. Microbiol. 60: 3752–3759.

Coppi, M.V., Leang, C., Sandler, S.J. & Lovley, D.R. 2001. Development of a genetic system for *Geobacter sulfurreducens*. Appl. Environ. Microbiol. 67: 3180–3187.

Coppi, M.V., O'Neil, R.A. & Lovley, D.R. 2003. Identification of a critical uptake hydrogenase required for hydrogen-dependent reduction of Fe(III) and other electron acceptors by *Geobacter sulfurreducens* (in press).

De Luca, G., de Philip, P., Dermoun, Z., Rousset, M. & Vermeglio, A. 2001. Reduction of Tc(VII) by *Desulfovibrio fructososvorans* is mediated by the nickel-iron hydrogenase. Appl. Environ. Microbiol. 67: 4583–4587.

Fredrickson, J.K., Kostandarithes, H.M., Li, S.W., Plymale, A.E. & Daly, M.J. 2000. Reduction of Fe(III), Cr(VI), U(VI), and Tc(VII) by *Deinococcus radiodurans* R1. Appl. Environ. Microbiol. 66: 2006–2011.

Gaspard, S., Vazquez, F. & Holliger, C. 1998. Localization and solubilization of the iron(III) reductase of *Geobacter sulfurreducens*. Appl. Environ. Microbiol. 64: 3188–3194.

Leang, C., Coppi, M.V. & Lovley, D.R. 2003. OmcB, a c-type polyheme cytochrome, involved in Fe(III)-reduction in *Geobacter sulfurreducens*. J. Bacteriol. 185: 2096–2103.

Lloyd, J.R., Blunt-Harris, E.L. & Lovley, D.R. 1999. The periplasmic 9.6-kilodalton c-type cytochrome of *Geobacter sulferreducens* is not an electron shuttle to Fe(III). J. Bacteriol. 181: 7647–7649.

Lloyd, J.R., Chesnes, J., Glasauer, S., Bunker, D.J., Livens, F.R. & Lovely, D.R. 2002. Reduction of actinides and fission products by Fe(III)-reducing bacteria. Geomicrobiol. J. 19: 103–120.

Lloyd, J.R., Cole, J.A. & Macaskie, L.E. 1997. Reduction and removal of heptavalent technetium from solution by *Escherichia coli*. J. Bacteriol. 179: 2014–2021.

Lloyd, J.R., Mabbett, A.N., Williams, D.R. & Macaskie, L.E. 2001. Metal reduction by sulphate-reducing bacteria: physiological diversity and metal specificity. Hydrometallurgy 59: 327–337.

Lloyd, J.R., Nolting, H.-F., Sole, V.A., Bosecker, K. & Macaskie, L.E. 1998. Technetium reduction and precipitation by sulfate-reducing bacteria. Geomicrobiol. J. 15: 43–56.

Lloyd, J.R., Sole, V.A., Van Praagh, C.V.G. & Lovley, D.R. 2000. Direct and Fe(II)-mediated reduction of technetium by Fe(III)-reducing bacteria. Appl. Environ. Microbiol. 66: 3743–3749.

Lloyd, J.R., Thomas, G.H., Finlay, J.A., Cole, J.A. & Macaskie, L.E. 1999. Microbial reduction of technetium by *Escherichia coli* and *Desulfovibrio desulfuricans*: Enhancement via the use of high-activity strains and effect of process parameters. Biotechnol. Bioeng. 66: 122–130.

Lovley, D.R. 1993. Dissimilatory metal reduction. Annu. Rev. Microbiol. 47: 263–290.

Magnuson, T.S., Hodges-Myerson, A.L. & Lovley, D.R. 2000. Characterization of a membrane-bound NADH-dependent Fe(III) reductase from the dissimilatory Fe(III)-reducing bacterium *Geobacter sulfurreducens*. FEMS Microbiol Lett 185: 205–211.

Managing the nuclear legacy – a strategy for action. 2002. Department of Trade and Industry, U.K. Government.

Snoeyenbos-West, O., Nevin, K.P., Anderson, R.T. & Lovley, D.R. 2000. Enrichment of *Geobacter* species in response to stimulation of Fe(III) reduction in sandy aquifer sediments. Microbial. Ecol. 39: 153–167.

Wildung, R.E., Gorby, Y.A., Krupka, K.M., Hess, N.J., Li, S.W., Plymale, A.E., McKinley, J.P. & Frederickson, J.K. 2000. Effect of electron donor and solution chemistry on products of dissimilatory reduction of technetium by *Shewanella putrefaciens*. Appl. Environ. Microbiol. 66: 2451–2460.

Wildung, R.E., McFadden, K.M. & Garland, T.R. 1979. Technetium sources and behaviour in the environment. J. Environ. Qual. 8: 156–161.

European Symposium on Environmental Biotechnology, ESEB 2004 - Verstraete (ed)

Biobarrier stability lab-scale studies for treatment of hydrocarbon-contaminated groundwater

S.R. Guiot, R. Cimpoia & C. Rhofir
Environmental Bioengineering Group, Biotechnology Research Institute, National Research Council Canada, Montreal, Canada

A. Peisajovich
Environmental Affairs Section, Transport Canada, Dorval, Canada

G. Leclair
Intervention and Restoration Section, Environment Canada, Montreal, Canada

ABSTRACT: The stability of a laboratory-scale bioreactive barrier in the treatment of contaminated groundwater was evaluated over a 16 month operation. Simulated groundwater contained diesel as the source of total petroleum hydrocarbons (TPH), and selected metals, as found at a North-Quebec site. Two 5-L capacity barrier prototypes, packed with peat moss granules and inoculated, were operated at 10°C. The biobarrier showed a TPH removal efficiency varying between 70 and 97%, depending on the TPH load. Sorption of TPH onto peat contributed to the TPH removal; however microbial biodegradation was the ultimate mechanism of TPH removal as shown by the biomineralization of chemical models: over 62% for hexadecane and 88% for naphthalene. Biomass enriched relatively rapidly in hydrocarbon-degrading microbial populations.

1 INTRODUCTION

The biological barrier (biobarrier) consists of a permeable bioreactive zone placed across the path of a contaminated plume. The bioreactive zone is packed with a highly permeable material, which serves as the support medium for attached growth of indigenous and/or inoculated microorganisms. Thus as the contaminated groundwater moves under natural hydraulic gradient through the bioreactive zone, the contaminants are removed or biodegraded, leaving uncontaminated groundwater to emerge from the downgradient side. The biobarrier system completely contains the contaminated plume, preventing off-site migration of the dissolved-phase contamination. This technology has the potential to operate for a long period of time with low maintenance requirements and low operator intervention (Guerin *et al.* 2002).

The objectives of the present study were to evaluate the stability of laboratory-scale biobarriers in the treatment of contaminated groundwater after a long-term operation. A simulated groundwater is used as the influent during the continuous biobarrier operation. It contains diesel as the source of total petroleum hydrocarbons (TPH), as well as selected metals that were found to be present at the North-Quebec site. They included iron, aluminum, calcium, copper, cobalt, zinc, manganese, molybdenum, lead and silver. The influent of the biobarrier also contained sources of nitrogen and phosphorus to support microbial growth and activities inside the biobarrier. Two biobarriers were operated at 10°C. One biobarrier – called "R" – had a previous history of TPH removal, but had stopped operating for a period of six months. This biobarrier was not emptied or re-inoculated. Its continuous operation re-started simply by passing the influent through it. The second biobarrier – called "N" – used virgin peat moss as the packing material and was inoculated with an enrichment culture developed from the contaminated soil. The biobarriers operated with hydraulic retention time (HRT) of about two days, corresponding to a linear liquid velocity of 12.5 cm/d.

The biobarriers' stability was evaluated by following during one and half year of operation, process parameters such as the efficiency in the removal of TPH, the contribution of sorption and biodegradation to the removal of TPH, and the buildup of active biomass in the system as well as the enrichment levels in the populations necessary for the TPH biodegradation.

2 MATERIALS AND METHODS

2.1 *Biobarrier operation*

The biobarrier consisted of a rectangular packed-bed reactor (25 × 20 × 10 cm), and a total volume of 5 L (Yerushalmi and Guiot 2001). Granulated peat moss (Produits Recyclable Bioforêt, Québec, Canada) is used as the packing material. Granulated peat moss has shown promising properties in terms of its cost, availability, long-term durability, rigidity, sorption of contaminants (both organic contaminants and inorganic metals), and bacterial immobilization capacity. The peat moss granules had a diameter of 0.2 to 0.4 cm, particle density of 1.1 g/mL and porosity of 55%. The biobarrier content in peat granules was 0.78 kg (dry wt), leaving a void volume of 2 L. The setup for continuous operation is schematically presented in Figure 1. The influent of biobarriers had the following composition of minerals and metals (mg/L): KH_2PO_4 870; K_2HPO_4 2260; NH_4Cl 900; $MgSO_4 \cdot 7H_2O$ 97; $AlK(SO_4)_2 \cdot 12H_2O$ 42; $FeSO_4 \cdot 7H_2O$ 70; $Ca(NO_3)_2 \cdot 4H_2O$ 51; $Pb(NO_3)_2$ 0.032; Ag_2SO_4 0.0043; $Co(NO_3)_2 \cdot 6H_2O$ 0.29; $CuSO_4$ 0.16; $ZnSO_4 \cdot 7H_2O$ 0.29; $MnSO_4 \cdot H_2O$ 1.69; $Na_2MoO_4 \cdot 2H_2O$ 0.48. Diesel was added to the influent solution, simulating contaminated groundwater with an inlet TPH concentration ranging from 12 to 61 mg/L in both biobarriers. The influent solution was continuously agitated in the feed tank and occasionally sparged with sterile air in order to ensure an initial DO concentration of >10 mg/L in the influent. A four-channel peristaltic pump model Gilson minipuls 2 (Gilson Medical Electronics, WI, USA) was used to deliver the contaminated water to the four inlet ports on the side of the biobarriers.

The "N" biobarrier was initially operated in batch mode in order to promote the attachment of microbial cells to the support material and to develop a microbial biofilm. This was done by inoculating the packed biobarrier with the enrichment culture (10% v/v), followed by the addition of diesel to make a 50 mg/L TPH concentration, and continuous recirculation of liquid inside the biobarrier at a flow rate of 32 L/d.

The enrichment culture was developed from the contaminated soil sample, using diesel at a concentration of 100 mg/L. The cultures were enriched by regular transfers of 10% (v/v) inoculum into fresh minimal salts medium (MSM) every three weeks, as described in Yerushalmi *et al.* (2003).

2.2 *Analytical techniques*

After extraction and purification, the hydrocarbons were analyzed by a gas chromatograph coupled to a flame ionization detector (GC-FID) (Perkin-Elmer, Norwalk, CT, USA). The inlet and outlet concentrations of the TPH were analyzed on a demand basis, while the concentration gradient of TPH as well as

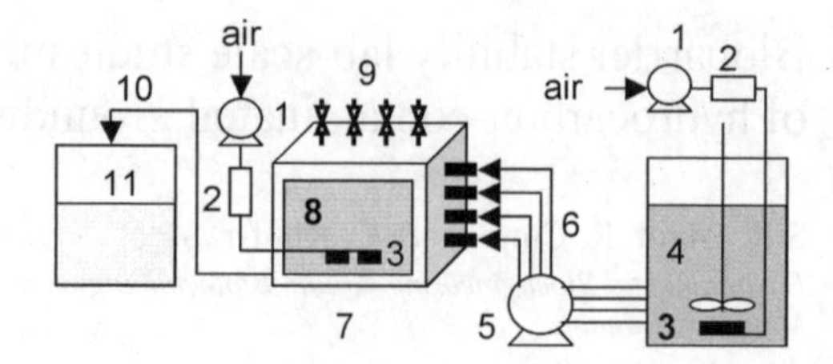

Figure 1. Schematic diagram of the biobarrier set-up during continuous operation. 1-pump, 2-air filter, 3-air diffusers, 4-feed tank with TPHs, 5-feed pump, 6-feed lines, 7-biobarrier, 8-packing, 9-top ports, 10-effluent line, 11-effluent tank.

biomass inside the biobarriers were determined three times during their continuous operation. Details of the methods used can be found in Yerushalmi *et al.* (2003). Suspended solids (SS) and volatile suspended solids (VSS) were processed according to Standard Methods (APHA 1995).

2.3 *Activity tests*

Respirometric activity was used as an indicator of the biomass relative content in active microbial populations. The test is based on the measurement of specific O_2 consumption in presence of glucose (initial concentration of 2 g/L), as a typical C-source readily usable by all bacteria. The tests were carried out at room temperature in 120 mL serum bottles and performed in duplicate. O_2 was measured in the gas phase by GC-FID. The O_2 consumption rate was calculated by a least-squares-based linear regression over 3 or 4 values within a maximal activity range. The specific activity was obtained by dividing the rate by the VSS content of the serum bottle.

Microcosm tests measure the extent of mineralization of ^{14}C-labeled compounds by the microbial culture. The CO_2 produced by dissimilation of the labeled substrate was subsequently measured by scintillation counting, and then converted to mg or % of substrate over a given period of time. Hexadecane, an aliphatic hydrocarbon, and naphthalene, an aromatic hydrocarbon, were used in separate bottles as representative hydrocarbons during the mineralization tests (initial concentration of 100 ppm (v/v) for both). Details are given in Yerushalmi *et al.* (2003).

3 RESULTS AND DISCUSSION

3.1 *Removal of hydrocarbons*

The biobarrier systems both supported removal of hydrocarbons with high efficiencies, as presented in Figure 2, which shows the dynamics of change in TPHs concentration in the inlet and outlet streams of both

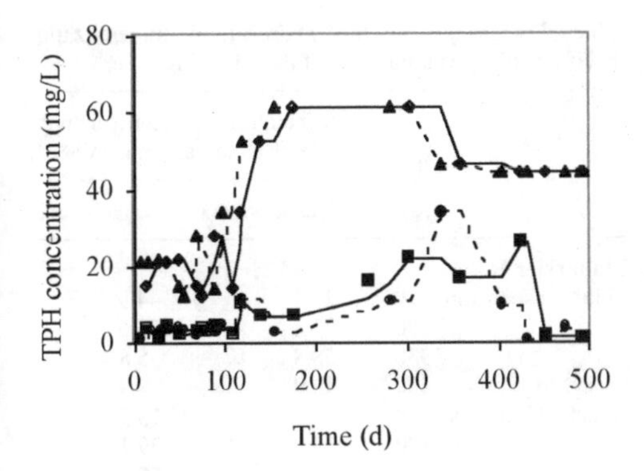

Figure 2. TPH concentrations at inlet and oulet of the biobarriers (—◆– R in; —■– R out; ...▲.. N in; ...●.. N out).

biobarriers. During the entire operation the liquid flow rate entering in the biobarriers ranged from 0.85 to 1.25 L/d, corresponding to linear velocities ranging from 10 to 15 cm/d, and HRT (based on the barrier void volume), from 2.5 to 1.7 d. During the first 100 days of operation the TPH inlet concentration averaged at 20 ± 7 mg/L. This corresponded to an average organic loading rate (OLR) of 4.5 ± 1.8 mg TPH/L of biobarrier (L_{bbr}).d. After a short initial adaptation period the removal efficiency of hydrocarbons reached 80 ± 9% in the "R" biobarrier and 85 ± 8% in the "N" biobarrier. The average concentration of effluent was respectively 4.0 ± 2.8 mg/L and 3.2 ± 1.1 mg/L. In a second phase of operation (days 100–400), the TPH inlet concentration was increased: on average 53 ± 8 and 55 ± 7 mg TPH/L for the "R" and "N" barriers respectively, which corresponded to OLR of 9.5 ± 4.5 and 11.5 ± 1.4 mg TPH/L_{bbr}.d. This led to a decline in the removal efficiency, which fluctuated around 70 ± 22 and 73 ± 48% for the "R" and "N" barriers, respectively. TPH outlet concentrations therefore varied around 15.8 ± 7.7 and 15 ± 13.3 mg/L, respectively. This indicates that at such OLRs, the removal capacity of the biobarriers was exceeded.

Consistently, accumulation of TPHs was observed at the first port of biobarrier, as shown by the concentration profiles of hydrocarbons along the length of biobarriers in both systems, measured after 300 days and compared to that measured after 100 days (Figure 3). However, the TPHs concentration declined with the distance from the inlet. As expected, since the peat moss is characterized by a high capacity for sorption of petroleum hydrocarbons (Cohen *et al.* 1991), the TPH measured on peat granules were high after one year of operation, and represented 1000 and 700 mg/kg dry peat on average, for the biobarriers "R" and "N", respectively, as compared to less than 100 mg/kg dry peat at the beginning. After the TPH inlet concentration was somewhat reduced (to 44 mg/L, day 320), the biobarriers progressively returned to superior per-

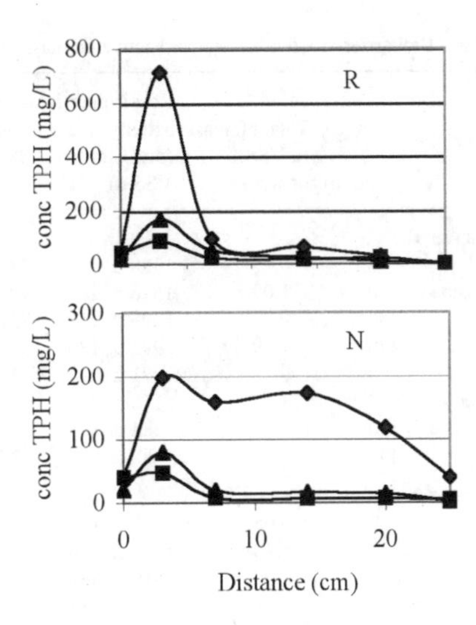

Figure 3. Gradients of TPH concentrations along the length of the biobarriers (–▲— day 100, –◆— day 300, –■— day 500).

formances, reaching values similar to or even higher than those at the beginning: removal efficiency of 97 ± 0.7 and 91 ± 5%, and residual TPH concentration of 1.7 ± 0.5 and 3.3 ± 4 mg/L, for the biobarriers "R" and "N", respectively. Consistently the TPH accumulation decreased significantly at the first port (Figure 3, day 500). This dynamics in the TPH concentrations indicate the TPH accumulation on peat and in biobarrier liquid was only transient, and the biobarrier was able to recover with time in response to a diminution in the organic load.

3.2 *Biomass retention*

Since the substratum used for biomass attachment, the peat moss, is a fully organic material, VSS measurements could not be used to evaluate the microbial biomass change in the biobarrier system. Hence overall respirometric activity tests on aliquots of overall and liquid biomass were performed. The enrichment culture used to inoculate the biobarrier "N" was assumed to be a 100% active microbial culture. The active biomass percentage in all biomass samples was then assessed in comparing their respirometric activity value to the one of the enrichment culture. VSS measurements and estimations of active biomass are detailed in Table 1. The results show the retention of a relatively dense suspension of microorganisms in the interstitial liquid i.e. ~1 g VSS/L, probably because of the high HRT in the system. With an active biomass portion making 4 to 6% of the substratum during the first year of operation, it can be estimated the biobarriers contained a total of 6 to 10 g VSS/L_{bbr} of attached active biomass, which is a normal range for fixed film

Table 1. Biobarriers content in total and active biomass.

	Time (d)	Total biomass (g VSS/L or wet kg)	Respirometric activity (mgO_2/g VSS.d)	Active biomass (% total VSS)
Biobarrier R				
Liquid suspension	0	0.75	1388 ± 61	72
	120	1.09	1516 ± 40	79
	330	0.73	1396 ± 15	72
	490	1.59	995 ± 13	52
Peat + attached biomass	0	199	71 ± 2	3.7
	120	173	110 ± 2	5.7
	330	164	84 ± 8	4.4
	490	180	42 ± 8	2.2
Biobarrier N				
Liquid suspension	0	0.65	1930 ± 16	100
	120	1.40	1153 ± 166	60
	330	1.00	819 ± 83	42
	490	1.01	652 ± 28	34
Peat+ attached biomass	0	187	45 ± 3	2.3
	120	199	94 ± 5	4.9
	330	187	82 ± 4	4.2
	490	188	50 ± 43	2.6

Table 2. Change in the hydrocarbon mineralization specific activity as a function of the operating time.

	Time (d)	Mineralization specific activity of active biomass (mg/gVSS.d) Hexadecane	Naphthalene
Biobarrier R			
Liquid suspension	0	28.1	44.9
	120	23.8	59.6
	330	58.3	5.8
	490	33.9	2.9
Attached biomass	0	9.5	13.9
	120	13.9	29.1
	330	25.3	50.5
	490	32.2	78.1
Biobarrier N			
Liquid suspension	0	24.6	29.7
	120	26.9	69.1
	330	33.9	2.1
	490	33.2	3.0
Attached biomass	0	1.0	0.7
	120	13.1	35.3
	330	23.5	47.1
	490	19.3	57.9

reactors. However at the end of the operation, the active portion of the attached biomass seemed to have decreased down to near 2%.

3.3 *Biomineralization extent and specific activity*

Sorption of TPH onto peat contributed to the removal of hydrocarbons in the biobarrier; however microbial biodegradation was the ultimate mechanism of removal of the hydrocarbons as shown by the high biomineralization potential of ^{14}C-labeled chemical models, using attached and free microbial culture withdrawn from the two biobarriers. Generally the mineralization extent was higher for naphthalene (over 83% in 5 cases out of 7 after one week of incubation) than for hexadecane (lower than 72% in 6 cases out of 7). As the non-mineralized balance represents the assimilated carbon, this difference is consistent with an higher growth yield on aliphatics than on aromatics.

The biodegradation specific activity was estimated by reporting the ^{14}C-labeled substrate mineralization rate to the active biomass values obtained from the respirometric tests, as detailed in Table 2. The activity values are used as an indicator of the relative density of the microorganisms exclusive to the catabolism of the substrate used in the test. Results showed a continuous enrichment of the active attached biomass in both hexadecane- and naphthalene-degrading bacteria (HDB and NDB, respectively). Overall the enrichment was sustained upon attachment and higher than in free suspension. As far as the attached growth was concerned, enrichment was higher for naphthalene than for hexadecane. While HDBs slightly increased their content in the liquid suspension, NDBs drastically decreased just after a few months. Typically heterotrophic bacteria degrade alkanes faster than aromatics (Geerdink *et al.* 1996). Possibly the selective pressure exerted by the TPH as the only C-source was more effective for NDB than for HDB, explaining both a higher attached growth of NDB and their faster washout from the liquid.

4 CONCLUSIONS

The biobarrier systems supported constant removal of hydrocarbons at high efficiencies, as long as the TPH loading did not exceed the biodegradation capacity of the system. TPH were essentially removed by sorption, but only transiently, as biomineralization contribution was proved. The biomass content of the barrier got stabilized, which minimizes the risks of clogging, while clearly enriched in the populations necessary for the TPH biodegradation. However saturation of support by sorbed TPH may occur on the long-term, if the TPH loading exceeds the biodegradation potential.

REFERENCES

APHA, AWWA, and WEF. 1995. Standard methods for the examination of water and wastewater, 19th Edn.. Washington, D.C.: American Public Health Association.

Cohen, A.D., Rollings, M.S., Zunic, W.M. & Durig, J.D. 1991. Effects of chemical and physical differences in peats and their ability to extract hydrocarbons from water. *Water Resour. Res*. 25: 1047–1060.
Geerdink, M.J., van Loosdrecht, M.C.M. & Luyben, K.C.A.M. 1996. Biodegradability of diesel oil. *Biodegradation*, 7: 73–81.
Guerin, T.F., Horner, S., McGovern, T. & Davey, B. 2002. An application of permeable reactive barrier technology to petroleum hydrocarbon contaminated groundwater. *Water Research*, 36: 15–24.
Yerushalmi, L. & Guiot, S.R. 2001. Biodegradation of benzene in a laboratory-scale biobarrier at low dissolved oxygen concentrations. *Bioremediation J.*, 5(1): 63–77.
Yerushalmi, L., Rocheleau, S., Cimpoia, R., Sarrazin, M., Sunahara, G., Peisajovich, A., Leclair G. & Guiot, S.R. 2003. Enhanced biodegradation of petroleum hydrocarbons in contaminated soil. *Bioremediation J.*, 7(1): 37–51.

Biodegradation and bioremediation

European Symposium on Environmental Biotechnology, ESEB 2004 - Verstraete (ed)
© 2004 Taylor & Francis Group, London, ISBN 90 5809 653 X

Microbial growth with mixtures of carbon substrates: what are its implications for the degradation of organic pollutants in particular and for microbial ecology in general?

T. Egli
Swiss Federal Institute for Environmental Science and Technology (EAWAG), Dübendorf, Switzerland

ABSTRACT: Microbial growth in ecosystems occurs typically in a dilute, carbon-energy-limited (oligotrophic) environment, in which a multiplicity of natural carbon compounds is present together with pollutants. There is now much experimental evidence that under such conditions microorganisms are utilizing many different carbon substrates simultaneously, a behavior referred to as "mixed substrate growth". From defined pure culture experiments one can deduce that mixed substrate growth confers a number of advantages to a microbial cell. This includes the ability to grow fast and more efficiently at low nutrient concentrations, which results in a competitive advantage. Also, the simultaneous consumption of carbon substrates improves a cell's metabolic flexibility towards changes in the environment, e.g. allows faster induction of enzymes necessary for new substrates. Furthermore, it might also reduce (or even abolish?) threshold concentrations for consumption of carbon compounds. The presently available experimental data suggest that in low polluted ecosystems simultaneous utilization of natural carbon substrates together with carbonaceous pollutants is probably the rule rather than the exception. How this affects pollutant degradation rates is still unclear, scenarios for both enhanced and reduced rates of degradation for individual pollutants can be constructed.

1 CARBON-ENERGY-LIMITED GROWTH AS A KEY FACTOR FOR MIXED SUBSTRATE GROWTH

In most ecosystems growth of heterotrophic microbes is taking place in a dilute environment where the availability of carbon/energy sources is severely restricted (Morita, 1988). A multitude of easily degradable carbon compounds of natural origin is usually present at very low concentrations, which are for individual compounds typically in the range of a few micrograms per liter or lower (Münster, 1993). In addition, in low polluted environments organic chemicals are present at similar or lower concentrations compared to those of individual natural carbon sources. (Systems heavily polluted with organic carbonaceous chemicals are not considered here because in such systems other factors limit biodegradation, e.g., the availability of terminal electron acceptors, of nitrogen, or of phosphorus.) This contrasts strongly with the experimental conditions used for physiological and kinetic investigations in the laboratory where most often pure microbial cultures are cultivated in batch cultures at high concentrations of a particular carbon source (or organic chemical) the serves as the only source of carbon and energy.

There is now convincing evidence that under such conditions microorganisms do not specialize on growth with a single particularly favorable carbon source, as predicted by the concept of carbon catabolite repression, leading to "diauxic growth" (Magasanik, 1976). Instead, they simultaneously take up and utilize for growth as many as possible of the carbon sources present. This behavior is usually referred to as "mixed substrate growth" (Harder & Dijkhuizen, 1976).

Only recently data has been reported from which general physiological and kinetic principles can be deduced concerning the consequences of mixed substrate growth for microbial growth in ecosystems. Also, little is still known with respect to the consequences of the interaction of natural substrates with pollutants, although it has always been assumed that their presence affects pollutant biodegradation in the environment (Alexander, 1994).

Experimental evidence for the simultaneous utilization of mixtures of carbon substrates by heterotrophs was obtained mainly from carbon-limited chemostat cultures (the literature is compiled in Harder & Dijkhuizen, 1976, and Egli, 1995). Mixed substrate growth with two or more carbon substrates under carbon-limited conditions in the chemostat has been demonstrated for a number of different combinations of substrate mixtures and microbes. Many of the combinations are known to provoke diauxic growth patterns when present at high concentrations in batch culture. Probably the most extensively studied case is that of the growth of *Escherichia coli* with mixtures of up to six different sugars (Lendenmann et al., 1996). Also the simultaneous utilization of pollutants together with alternative, easily degradable carbon sources has been demonstrated for a number of cases. Examples reported from our laboratory include growth of *Chelatobacter heintzii* with nitrilotriacetate (NTA) and glucose (Bally & Egli, 1996), *Comamonas testosteroni* with mixtures of acetate and p-toluenesulfonate (Tien, 1997), *Methylobacterium* DM4 with acetate plus dichloromethane (Tien, 1997), or *Ralstonia eutropha* with fructose and 2,4-D (Füchslin, 2002), or *Pseudomonas putida* F1 and *Ralstonia pickettii* PKO1 with mixtures of succinate and benzene (Bucheli-Witschel et al., 2003).

Generally, under carbon-limited growth conditions mixed substrate growth is the rule (i.e., the author is not aware of an example where this was not the case). However, also during growth in batch cultures at high carbon concentrations simultaneous growth with several carbon sources appears to occur more frequently than assumed, especially with carbon compounds that support medium to low growth rates (Egli, 1995). Although most investigations have been dealt with mixtures of carbon substrates, there is also experimental evidence that the behavior can be extended to growth with mixtures of nitrogen sources under nitrogen-limited conditions, or to the simultaneous use of electron acceptors (e.g., oxygen plus nitrate) under electron acceptor-limited growth conditions (see Egli, 1995).

The information presently available indicates that the ability to simultaneously utilize mixtures of carbon substrates confers at least three important advantages to a microbial cell:

1) a kinetic advantage in competition for substrates at low concentrations,
2) an improved ability to respond to changes in substrate availability, and
3) the ability to circumvent threshold concentrations for individual compounds for growth and utilization.

Below the experimental evidence available for this will be outlined.

2 KINETICS OF MIXED SUBSTRATE GROWTH

Virtually all models used to describe the kinetics of growth and substrate utilization are based on the assumption that a single compound is controlling the rate of growth (summarized in Kovarova & Egli 1998). However, already in the seventies experimental data suggested an effect of simultaneous utilization of multiple carbon substrates on growth kinetics (Law & Button, 1976).

An extensive kinetic study of Lendenmann and co-workers on mixed substrate growth of *E. coli* with defined mixtures of up to six sugars performed in carbon-limited chemostat finally demonstrated the principles and advantages of mixed substrate growth (Egli et al., 1993; Lendenmann et al., 1996; Lendenmann & Egli, 1998). With all mixtures tested the steady-state concentration of sugars at a set dilution (growth) rate was consistently lower during simultaneous utilization of mixtures compared to growth with single sugars. The steady-state concentration of individual sugars was approximately linearly dependent on their contribution to the total sugar consumption rate. This principle has now been confirmed for a number of combinations of structurally unrelated substrates feeding into different catabolic pathways, including easily degradable carbon sources plus pollutants (see e.g., Kovarova & Egli, 1998, Füchslin, 2002; Lovanh et al., 2002).

Extrapolating this kinetic principle to environmental conditions where cells can be assumed to grow with dozens of different substrates simultaneously, one can deduce that cells performing mixed substrate growth are able to grow relatively fast at minute concentrations of individual carbon sources. Hence, the puzzling fact that marine bacteria grow quickly at the extremely low carbon concentrations in the open ocean can be explained by growth with mixed substrates. It was demonstrated in chemostat cultures with dual substrate mixtures – note that this was before the kinetic principle was known!–that cells with the ability to use mixtures of substrates have a competitive kinetic advantage and are able to outcompeted mutants unable to use one of the two substrates offered (Dijkhuizen & Davis, 1980).

3 METABOLIC FLEXIBILITY AND MIXED SUBSTRATE UTILIZATION

Many of the enzyme systems responsible for the breakdown of organic pollutants are inducible. Hence, it is of interest to know how fast the proteins in such pathways can be synthesized once this compound becomes available in the environment. Studies on the dynamics of enzyme induction have been performed in our laboratory for the NTA-degrading bacterial strain *Chelatoacter heintzii* in continuous culture (Bally & Egli, 1996). The key enzyme of NTA catabolism, a monooxygenase (NTA-MO), is inducible in *C. heintzii*. In these experiments a glucose-limited chemostat culture of *C. heintzii* growing at a constant dilution rate was subjected to a medium shift where glucose was replaced by either NTA only, or by mixtures of NTA plus glucose as the carbon/energy source(s). After a shift from glucose to NTA only the cells needed more than 20 hours before they started to induce NTA-MO. Hence, during the first 20 hours of the shift the cells were unable to utilize NTA, as a result the complexing agent accumulated in the bioreactor and wash-out of the culture was observed. Only after induction of NTA-MO growth resumed and the culture recovered. Surprisingly, when the culture was shifted from glucose to a mixture of 1% glucose plus 99% NTA the time required for induction of NTA-MO was reduced from some 20 hours to 10 hours. This positive effect of glucose on the induction of NTA-degradation capacity was investigated also for mixtures containing higher proportions of glucose. The general result was that the more glucose the culture was left with the faster induction of NTA-MO and growth with NTA started. For example, leaving the cells with 90% glucose reduced the time for induction to less than 30 minutes. Under these conditions no significant accumulation of NTA was observed and growth with NTA started almost immediately. Just recently, we have obtained similar results for shifts of *Pseudomonas putida* F1 and *Ralstonia pickettii* PKO1 from succinate to either benzene alone or to mixtures of succinate plus benzene (Bucheli-Witschel et al., 2003).

These results suggests that under carbon-limited growth conditions the availability of alternative carbon/energy sources does not inhibit but rather support the induction of other catabolic enzymes, most likely by supplying energy and building blocks for the synthesis of proteins that have to be newly produced (see Egli, 2001). Taken this scenario one step further one comes to the conclusion that the simultaneous use of mixtures of carbon/energy sources gives the organism an increased degree of metabolic freedom and flexibility to synthesize proteins according to their need.

4 THRESHOLD CONCENTRATIONS AND MIXED SUBSTRATE GROWTH

Threshold concentrations for growth and utilization for carbon substrates (which is not necessarily the same) have frequently been reported (see e.g. Alexander, 1994, and the discussion in Egli, 1995). However, frequently the analytical limits of detection for compounds hamper careful investigations. Nevertheless, for pure cultures threshold concentrations for growth with single substrates typically appear to be in the range of 1–100 microgram per liter. It has been shown for a number of examples that utilization thresholds for particular compounds can be lowered by the simultaneous utilization of alternative carbon substrates. For example, *Pseudomonas aeruginosa* was able to grow with a mixture of 45 compounds, each present at a concentration of 1 microgram per liter, whereas the bacterium was unable to grow with the individual compounds present at this concentration. Or, a threshold for induction of 3-phenyl-propionate (3 ppa) in the range of 3 mg per liter was reported for *Escherichia coli* during growth with glucose. When 3 ppa was supplied in the medium at concentrations above 3 mg per liter induction of 3 ppa-degrading enzymes occurred and the compound was degraded in the presence of glucose down to concentrations far below the threshold concentration for induction (Kovar et al., 2002).

The presently available information gives a reasonable explanation for the observation that many compounds can be consumed by microbes even if they are present at the nanogram per liter range or lower (Alexander, 1994). However, it appears that a systematic experimental approach is needed to obtain a more complete picture as to why threshold concentrations for utilization (or growth, or induction) occur for some compounds but not for others. This particularly important to understand, set and justify acceptable residual levels of pollutants in the environment.

5 OUTLOOK

The ecological significance of mixed substrate utilization appears to be obvious for heterotrophic microorganisms: It allows fast growth at low concentrations, gives metabolic flexibility, and probably also allows to get around the threshold concentrations for utilization that are observed for growth with single substrates. All this will improve the competitiveness of a cell in an oligotrophic environment.

The consequences for biodegradation are not as clear-cut. Certainly, mixed substrate utilization kinetics will allow pollutants to be degraded to lower residual concentrations than during growth with the pollutant as a single substrate; also threshold concentrations might be lower or even non-existent. With respect to pollutant degradation rates, however, two opposite processes have to be considered. On one hand, mixed substrate growth will result in improved competitiveness and will allow faster selection of a pollutant-degrading bacterial population to a size that will result in significant rates of biodegradation. On the other hand, the simultaneous use of a pollutant together with alternative carbon substrates will reduce the consumption rate for the pollutant. Hence, we have to know, which of the two processes is more important for the total pollutant degradation capacity in a system (i.e., the product of number of competent cells times their rate of pollutant utilization). Only detailed investigation will give us answers.

REFERENCES

Alexander, M. 1994. *Biodegradation and Bioremediation*. San Diego: Academic Press.

Bally, M. & Egli, T. 1996. Dynamics of substrate and enzyme synthesis in *Chelatobacter heintzii* during growth in carbon-limited continuous culture with different mixtures of glucose and nitrilotriacetate. *Applied and Environmental Microbiology* 62: 133–140.

Bucheli-Witschel, M., Rüegg, I., Hafner, T., & Egli T. 2003. Dynamics of benzene degradation by *Pseudomonas putida* F1 and *Ralstonia pickettii* PKO1 in the presence of alternative substrate. 1st FEMS Congress of European Microbiologists. Ljubljana, Slovenia. Abstract book, p. 322. Delft (NL): Elsevier on behalf of FEMS.

Dykhuizen, D. & Davies, M. 1980. An experimental model: bacterial specialists and generalists competing in chemostats. *Ecology* 61: 1231–1227.

Egli, T. 1995. The ecological and physiological significance of microbial growth with mixtures of substrates. *Advances in Microbial Ecology* 14: 305–386.

Egli, T. 2001. Biodegradation of metal-complexing aminopolycarboxylic acids. *Journal of Bioscience and Bioengineering* 92: 89–97.

Egli, T., Snozzi, M., Lendenmann, U. 1993. Kinetics of Microbial Growth with Mixtures of Carbon Sources. *Antonie van Leeuwenhoek* 63: 289–298.

Füchslin, H.P. 2002. *Microbial Competition and Mixed Substrate Utilisation in the Laboratory: Towards a Better Understanding of Microbial Behaviour in the Environment.* PhD thesis No 14641. Swiss Federal Institute of Technology, Zürich, Switzerland.

Harder, W. & Dijkhuizen, L. 1976. Mixed substrate utilization. In A.C.R. Dean, D.C. Ellwood, C.G.T. Evans & I. Melling (eds), *Continuous culture 6. Applications and New Fields*: 297–314. Chichester, England: Ellis Horwood.

Kovar, K., Chaloupka, V. & Egli, T. 2002. A threshold substrate concentration is required to initiate the degradation of 3-phenylpropionic acid in *Escherichia coli*. *Acta Biotechnologica* 22: 285–298.

Kovarova-Kovar, K. & Egli, T. 1998. Growth kinetics of suspended microbial cells: from single-substrate-controlled growth to mixed-substrate kinetics. *Microbiology and Molecular Biology Reviews* 62; 646–666.

Law, A.T., & Button, D.K. 1976. Multiple-carbon-source-limited growth kinetics of a marine coryneform bacterium. *Journal of Bacteriology* 129: 115–123.

Lendenmann, U., Snozzi, M. & Egli, T. 1996. Kinetics of simultaneous utilization of sugar mixtures by *Escherichia coli* in continuous culture. *Applied and Environmental Microbiology* 62: 1493–1499.

Lendenmann, U., & Egli T. 1998. Kinetic models for the growth of *Escherichia coli* with mixtures of sugars under carbon-limited conditions. *Biotechnology and Bioengineering* 59: 99–107.

Lovanh N., Hunt, C.S., & Alvarez, P.J.J. 2002. Effect of ethanol on BTEX biodegradation kinetics: aerobic continuous culture experiments. *Water Research* 36: 3739–3746.

Magasanik, B. 1976. Classical and postclassical modes of regulation of the synthesis of degradative bacterial enzymes. *Progress in Nucleic Acids Research and Molecular Biology* 17: 99–115.

Morita, R.Y. 1988. Bioavailability of energy and its relationship to growth and starvation survival in nature. *Journal of Canadian Microbiology* 43: 436–441.

Münster, U. 1993. Concentrations and fluxes of organic carbon substrates in the aquatic environment. *Antonie van Leeuwenhoek* 63: 243–264.

Tien, A. 1997. *The Physiology of a Defined Four Membered Mixed Bacterial Culture During Continuous Cultivation with Mixtures of Three Pollutants in Synthetic Sewage.* PhD thesis No 11905. Swiss Federal Institute of Technology, Zürich, Switzerland.

van der Kooij, D., Oranje, J.P., & Hijnen, W.A.M. 1982. Growth of *Pseudomonas aeruginosa* in tap water in relation to utilization of substrates at concentrations of a few micrograms per liter. *Applied and Environmental Microbiology* 44: 1086–1095.

European Symposium on Environmental Biotechnology, ESEB 2004 - Verstraete (ed)
© 2004 Taylor & Francis Group, London, ISBN 90 5809 653 X

Simultaneous degradation of atrazine and phenol by *Pseudomonas* sp. ADP: effects of toxicity and adaptation

G. Neumann, N. Kabelitz & H.J. Heipieper
Department of Bioremediation, Centre for Environmental Research (UFZ) Leipzig-Halle, Permoserstr, Leipzig, Germany

R. Teras, L. Monson & M. Kivisaar
Department of Genetics, Institute of Molecular and Cell Biology, Tartu University and Estonian Biocentre, Tartu, Estonia

ABSTRACT: The strain *Pseudomonas* sp. ADP is only able to degrade atrazine as sole nitrogen-source and therefore needs a carbon- and energy-source for growth. In addition to the typical C-source for *Pseudomonas*, Na_2succinate, the strain can also grow with phenol as carbon source. Thereby, phenol is oxidized to catechol by a multi-component phenol hydroxylase. Catechol is degraded via the *ortho*-pathway using the catechol-1, 2-dioxygenase. It was possible to stimulate the strain in order to degrade very high concentrations of phenol (1000 mg/l) and atrazine (150 mg/l) simultaneously. With cyanuric acid, the major intermediate of atrazine degradation, as N-source both the growth rate and the phenol degradation rate were similar to those measured with ammonia as N-source. With atrazine as N-source growth rate and the phenol degradation rate were reduced to about 35% of those data obtained for cyanuric acid. This presents clear evidence that although the first three enzymes of the atrazine degradation pathway are constitutively present, either these enzymes or the uptake of atrazine are the bottleneck that diminishes the growth rate of *Pseudomonas* sp. ADP with atrazine as N-source. Whereas atrazine and cyanuric acid showed no significant toxic effect on the cells, phenol reduces growth and activates/induces typical membrane-adaptive responses known for the genus *Pseudomonas*. Therefore, *Pseudomonas* sp. ADP is an ideal bacterium to investigate the regulatory interactions between several catabolic genes and stress response mechanisms during the simultaneous degradation of toxic phenolic compounds and a xenobiotic N-source such as atrazine.

1 INTRODUCTION

Atrazine (2-chloro-4-(ethylamino)-6-(isopropyl-amino)-1,3,5-triazine) is a herbicide used for controlling broadleaf and grassy weeds and is relatively persistent in soils (Hayes & Laws 1991). Atrazine and its metabolites have been detected in ground and surface waters at levels exceeding the Environmental Protection Agency's maximum contaminant level of 3 ppb (Hayes & Laws 1991).

Pseudomonas sp. ADP was the first isolated bacterium capable of degrading the herbicide atrazine (Mandelbaum et al. 1995, Wackett et al. 2002). Since then, most of the understanding of the genes and enzymes involved in atrazine degradation derives from studies using this strain, in which the first three enzymatic steps of atrazine degradation have been defined (de Souza et al. 1998). The genes *atzA*, *atzB*, and *atzC*, which encode the enzymes atrazine chlorohydrolase (AtzA), hydroxyatrazine ethylaminohydrolase (AtzB), and N-isopropylammelide isopropylaminohydrolase (AtzC), convert atrazine sequentially to cyanuric acid (de Souza et al. 1998). Cyanuric acid is catabolized by *Pseudomonas* sp. ADP to carbon dioxide and ammonia (de Souza et al. 1998). These first three genes have been localized on an approximately 100 kb plasmid, pADP-1 (de Souza et al. 1998). Recently, pADP-1 was completely sequenced and was shown to contain the genes for the complete catabolism of cyanuric acid to CO_2 and NH_3 as well namely, *atzD*, *atzE*, and *atzF* (30) as well. Structural and functional studies showed that the genes encoding the initial reactions of atrazine catabolism are not organized in an operon, but are dispersed and flanked by transposase copies.

However, the strain is only able to use atrazine as sole nitrogen-source and therefore needs an additional

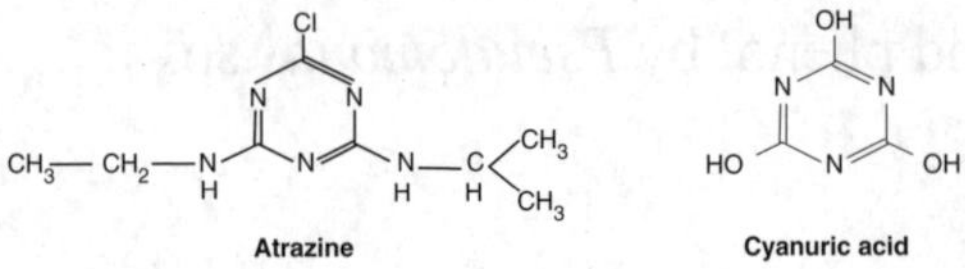

Figure 1. The investigated *s*-triazines.

carbon- and energy-source for growth. In the present, work we investigated the degradation of phenol in the presence of different N-sources including the two *s*-triazines atrazine and cyanuric acid (Fig. 1). During our experiments we studied the effect of simultaneous degradation of those compounds as well as membrane-adaptive responses of the cells to the toxic compounds phenol and atrazine.

2 RESULTS

Pseudomonas sp. ADP can grow with phenol as sole carbon source. It cleaves catechol via the *ortho*-pathway using the catechol-1,2-dioxygenase, but was not able to degrade chlorophenols.

2.1 *Degradation of atrazine*

In order to adapt the ADP strain we selected atrazine concentrations of 100 or 150 mg/l atrazine, 2 g/l cyanuric acid and the phenol concentration of 500 mg/l which caused about 50 percent growth inhibition and the best physiological membrane adaptation reactions. Cyanuric acid is quite hydrophilic and therefore non-toxic and an easily degradable N-source. Although atrazine has a higher logP value than phenol (2.71 to 1.45) it is not that toxic to ADP as phenol (data not shown) and its complete degradation (cf. Fig. 2) was proved via HPLC-analysis (data not shown).

2.2 *Degradation of phenol*

Completely adapted, the strain was able to degrade phenol in amounts of 1000 mg/l totally within 8 hours with cyanuric acid as N-source (cf. Fig. 3). At the same time the growth rates μ were up to 0.32 h-1 which corresponds to a doubling time (tD) of about 2 hours. The degradation rates for phenol were up to 150 mg/h. These growth rates and the cell yields are comparable to those obtained with medium containing ammonia as N-source. Furthermore, there exists a correlation between an increased atrazine concentration and an increasing cell yield, although it takes the cells much longer to reach the same optical density.

2.3 *Simultaneous degradation of atrazine and phenol*

With atrazine as N-source, phenol was completely degraded after about 30 hours, with decreased growth

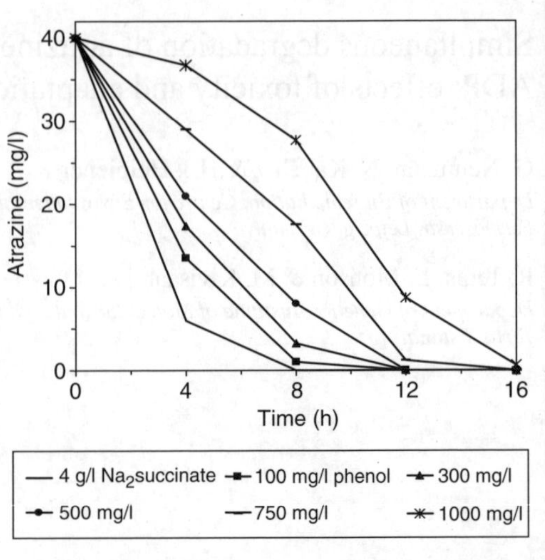

Figure 2. Degradation of atrazine by *Pseudomonas* sp. ADP with phenol as C-source. Due to the low water solubility of atrazine, the scale does not show the nominal concentration in the medium. Atrazine was applied in an amount of up to 150 mg/l by using a small reservoir made of an Eppendorf tube with a semi-permeable membrane. This guaranteed the constant presence of the maximum atrazine concentration which corresponds to the maximum solubility of this compound in water (30 mg/l). This application hindered a precipitation of atrazine in the medium that would have disturbed the measurement of the optical density (O.D.).

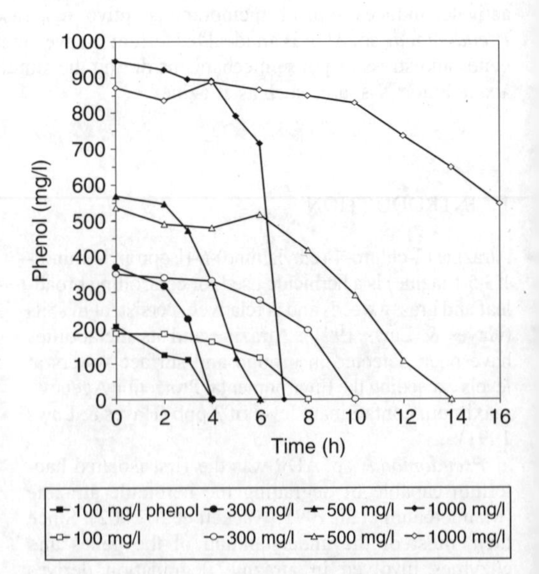

Figure 3. Comparison of the phenol degradation by *Pseudomonas* sp. ADP with cyanuric acid (■ closed symbols) or atrazine (□ open symbols) as sole N-source.

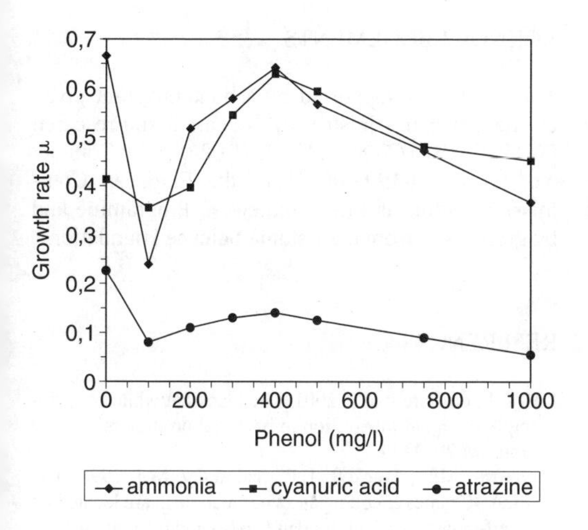

Figure 4. Comparison of the growth rates of *Pseudomonas* sp. ADP growing on NH_4^+, cyanuric acid or atrazine as N-source. The "zero-points" correspond to the control cultures grown on Na_2succinate as C-source.

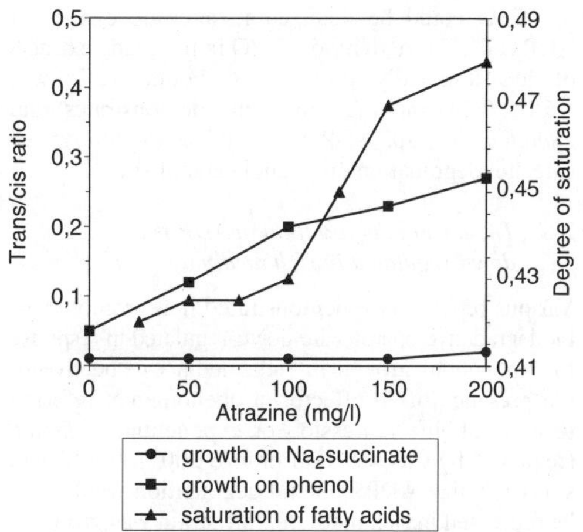

Figure 5. Effect of the atrazine concentration on changes in the membrane fatty acid composition of *Pseudomonas* sp. ADP.

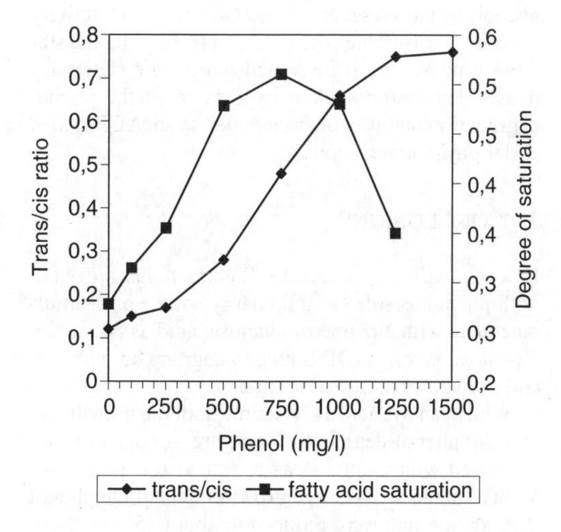

Figure 6. Effect of the phenol concentration on changes in the membrane fatty acid composition of *Pseudomonas* sp. ADP.

and degradation rates, respectively (μ: 0.12 h-1; tD: 5.75 h). It was possible to stimulate the strain to degrade very high concentrations of phenol and atrazine simultaneously and as shown in Fig. 4 at growth rates similar to those obtained in "normal" mineral medium only containing ammonia as N-source and salts supplemented with a standard vitamin solution.

2.4 *Physiological adaptation to atrazine and phenol*

As it is known that phenol as well as atrazine is a toxic compound, the adaptation of the strain was also investigated (Heipieper et al. 2003).

As it is shown in Fig. 5 the strain showed no increasing *trans/cis* ratio when cultivated on atrazine + Na_2succinate, which confirms the assumption of atrazine not being very toxic. Nevertheless, an increasing degree of saturation of the membrane fatty acids was observed when atrazine was added as a toxin within the logarithmic growth phase.

Figure 6 emphasizes the toxicity of phenol. The degree of saturation decreases significantly at 1000 and 1250 mg/l because the cells are no longer able to adapt to such high concentrations. As the *cis-trans* isomerase is in no need for energy or co-factors in order to function, the *trans/cis* ratio is still increasing at those concentrations.

2.5 Pseudomonas *sp. strain ADP utilizes phenol via ortho-pathway of catechol degradation*

Phenol is usually degraded via the catechol degradation pathway. Two ways for catechol ring fission are existing: the *meta*- and *ortho*-pathway. The majority of bacteria use the *meta*-pathway of catechol degradation, especially if bacteria have multicomponent phenol hydroxylases like *Pseudomonas* sp. CF600 (Powlowski & Shingler 1990). However, there is also evidence that the multicomponent phenol hydroxylase and the *ortho*-pathway of catechol degradation can coexist (Ehrt et al. 1995). Both *meta*- and *ortho*-pathways are distinguishable by measuring characteristic enzymes, catechol-2,3-dioxygenase (C23O) for the *meta*-pathway, and catechol-1,2-dioxygenase (C12O) for the *ortho*-pathway. The activities of both enzymes were measured in ADP cells grown in phenol minimal media. Activity of C12O but no evidence of C23O could be detected in a crude extract of ADP cells. The activity of C12O in the crude extracts of logarithmically growing (6 hours) cells was 1.66 ± 0.24 μmol/mg min. This demonstrates that *Pseudomonas* sp. ADP uses the *ortho*-pathway of catechol degradation on phenol catabolism.

2.6 *The phenol degradation in ADP is down-regulated in rich medium*

Various reports have demonstrated that promoters of biodegradative operons are down-regulated in response to exponential growth in rich media irrespective of the presence of an effector, a phenomenon referred to as catabolite repression or exponential silencing (reviewed by Cases & de Lorenzo 2001). In order to study whether ADP's phenol degradation would also be repressed in rich medium, this strain was grown in LB medium in the presence of phenol and the activity of the *ortho*-pathway enzyme C12O was measured (data not shown). No activity of C12O could be detected when bacteria were grown in LB in the absence of phenol. In the presence of phenol, the C12O activity was only measurable when cells were entering the stationary phase and it increased during the stationary phase, demonstrating that expression of the phenol degradation pathway in *Pseudomonas* sp. ADP is also under physiological control.

3 CONCLUSIONS

It was possible to cultivate *Pseudomonas* sp. ADP with phenol as sole C- and energy source and simultaneously with atrazine or cyanuric acid as N-source. *Pseudomonas* sp. ADP is able to degrade both phenol and *s*-triazines at the same time.

With cyanuric acid as N-source both the growth rate and the phenol degradation rate are similar to those measured with ammonia as N-source (cf. Fig. 3, 4). With atrazine as N-source growth rate and the phenol degradation rate were reduced to about 35% of those data obtained with cyanuric acid. Due to the slight toxicity of atrazine the maximum growth rate occurred at a lower phenol concentration than with cyanuric acid or ammonia.

Although the first three enzymes of the atrazine degradation are constitutively present, they are the bottleneck that diminishes the growth rate of *Pseudomonas* sp. ADP with atrazine as N-source.

Pseudomonas sp. ADP is an ideal bacterium to investigate the degradation of phenolic compounds, the degradation of atrazine and the influence of stress adaptation mechanisms.

Additionally, the strain offers great opportunities to study the regulatory interactions between different degradation pathways and stress response mechanisms.

ACKNOWLEDGEMENTS

This work was supported by a bilateral grant given by the German Ministry for Education and research (WTZ-BMBF) Contract No. EST 02/006, by Contract No. QLK3-CT-1999-00041 of the European Commission within its Fifth Framework Programme and by grant 4481 from the Estonia Science Foundation.

REFERENCES

Cases, I., de Lorenzo, V. 2001. The black cat/white cat principle of signal integration in bacterial promoters. *EMBO Journal* 20: 1–11.

de Souza, M.L., Wackett, L.P., Sadowsky, M.J. 1998. The *atz*ABC genes encoding atrazine catabolism are located on a self-transmissible plasmid in *Pseudomonas* sp. strain ADP. *Applied and Environmental Microbiology* 64: 2323–2326.

Ehrt, S., Schirmer, F., Hillen, W. 1995. Genetic organization, nucleotide sequence and regulation of expression of genes encoding phenol hydroxylase and catechol 1,2-dioxygenase in *Acinetobacter calcoaceticus* NCIB8250. *Molecular Microbiology* 18: 13–20.

Hayes jr., W.J., Laws jr., E.R. 1991. Classes of Pesticides, Handbook of Pesticide Toxicology, vol. 3. Academic Press.

Heipieper, H.J., Meinhardt, F., Segura, A. 2003 The *cis-trans* isomerase of unsaturated fatty acids in *Pseudomonas* and *Vibrio*: biochemistry, molecular biology and physiological function of an unique stress adaptive mechanism. *FEMS Microbiology Letters* 229: 1–7.

Isken, S., Heipieper, H.J. 2002. Toxicity of organic solvents to microorganisms. In G. Bitton (ed.), *Encyclopedia of Environmental Microbiology*. New York: John Wiley.

Kabelitz, N., Santos, P.M., Heipieper, H.J. 2003. Effect of aliphatic alcohols on growth and degree of saturation of membrane lipids in *Acinetobacter calcoaceticus*. *FEMS Microbioogy. Letters* 220: 223–227.

Mandelbaum, R.T., Allan, D.L., Wackett, L.P. 1995. Isolation and characterization of a *Pseudomonas* sp. that mineralizes the *s*-triazine herbicide atrazine. *Applied and Environmental Microbiology* 61: 1451–1457.

Martinez, B., Tomkins, J., Wackett, L.P., Wing, R., Sadowsky, M.J. 2001. Complete nucleotide sequence and organization of the atrazine catabolic plasmid pADP-1 from *Pseudomonas* sp. strain ADP. *Journal of Bacteriology* 183: 5684–5697.

Powlowski, J., Shingler, V. 1990. In vitro analysis of polypeptide requirements of multicomponent phenol hydroxylase from *Pseudomonas* sp. strain CF600. *Journal of Bacteriology* 172: 6834–6840.

von Wallbrunn, A., Richnow, H.H., Neumann, G., Meinhardt, F., Heipieper, H.J. 2003. The enzymatic mechanism of *cis/trans* isomerization of unsaturated fatty acids in *Pseudomonas putida*. *Journal of Bacteriology* 185: 1730–1733.

Wackett, L.P., Sadowsky, M.J., Martinez, B., Shapir, N. 2002. Biodegradation of atrazine and related *s*-triazine compounds: from enzymes to field studies. *Applied Microbiology and Biotechnology* 58: 39–45.

European Symposium on Environmental Biotechnology, ESEB 2004 - Verstraete (ed)

Removal of selected polar micropollutants in membrane bioreactors

H. De Wever, S. Van Roy, J. Vereecken, C. Dotremont & L. Diels
Vlaamse instelling voor technologisch onderzoek, Mol, Belgium

J. Müller & T. Knepper
ESWE-Institute for Water Research and Water Technology, Wiesbaden, Germany

S. Weiss & T. Reemtsma
Technical University Berlin, Berlin, Germany

ABSTRACT: This paper discusses laboratory-scale studies on the removal of polar persistent pollutants (P3). In a first test the removal of readily aerobically biodegradable linear alkylbenzene sulphonates (LAS) was compared in a sidestream membrane bioreactor (MBR) and a conventional activated sludge systems (CAS). LAS removal efficiencies over 97% were found, going hand in hand with the formation of sulfophenyl carboxylates, some of them being identified as P3 compounds. The effect of hydraulic retention time, hydrophobicity of the membrane used in the MBR, and LAS composition was negligible in the range tested. The removal of a technical mixture of naphthalenesulphonates was compared in a submerged MBR and a CAS. In this experiment, degradation of the monosulphonates started immediately. For the disulphonates, the order in which degradation started was similar for both reactor systems but degradation always started much earlier in the MBR. In both systems, removal efficiencies over 90% were achieved for 4 isomers. Degradation of the 1,3-isomer only started towards the end of the test period. The 1,5-isomer was never degraded.

1 INTRODUCTION

Due to particle retention by membrane separation, membrane bioreactors (MBRs) can be operated at much higher sludge concentrations than conventional activated sludge systems (CAS). This generally results in a higher effluent quality. The removal of dissolved organic compounds, measured as the chemical oxygen demand (COD) or biochemical oxygen demand, is well described. However, it is not clear at present whether MBRs have the potential to achieve a better mineralization of micropollutants. Polar micropollutants are generally too small to be retained by the micro- or ultrafiltration membranes commonly applied in MBRs. Hence, the only benefit MBRs may have compared to CAS systems for their removal, is the intensification of biological processes at high biomass concentrations.

In this paper we report on comparative tests between MBR and CAS for removal of persistent polar pollutants (P3). Two groups of P3 were studied: LAS which are known to be biodegradable, whereas some of their degradation products are persistent, and naphthalenesulphonates.

2 MATERIALS AND METHODS

2.1 *Removal of LAS*

A MBR and a CAS system were operated in parallel. The MBR consisted of an aerated compartment with an active volume of 25 l. Sludge separation was performed with an external tubular cross-flow ultrafiltration membrane (Storck, the Netherlands) having a mean pore size of 30 nm. The CAS system had an active volume of 8 l. The average sludge concentrations in MBR and CAS were 7.7 g l^{-1} and 3.9 g l^{-1}. Oxygen concentration in the aeration tank was kept between 1 and 2 mg l^{-1}, and pH between 7 and 8.

Both reactors were inoculated with activated sludge from a municipal wastewater treatment plant and were fed with a synthetic domestic wastewater to which increasing amounts of LAS detergent were added. Table 1 summarizes the experimental set-up.

At regular time intervals, sludge samples were monitored for their heterotrophic and LAS degradation capacity by oxygen uptake rate measurements. For the determination of LAS degradation, an experimental protocol was used, proposed by Ellis *et al.* (1996).

2.2 *Removal of naphthalene sulphonates*

A submerged MBR was operated in parallel with a CAS system. The MBR was equipped with Kubota membranes and had an active aerated volume of 21 l. Continuous aeration was provided to reduce membrane fouling. The CAS had an aerobic compartment of 4.2 l. The oxygen concentration was kept between 1 and 2 mg l^{-1}. Average sludge concentrations were 6.7 g l^{-1} and 3.8 g l^{-1} in MBR and CAS respectively.

The experimental set-up is summarized in Table 2. The technical mixture of naphthalene sulphonates used in the test, contained different amounts of the mono- and disulphonates and was added to a final concentration which was environmentally relevant and still detectable by high pressure liquid chromatography (HPLC)-fluorescence analysis. Because the 1,5-disulphonate was present in the lowest amount and was the least sensitively analysed, extra additions were made.

2.3 *Analytical methods*

The performance of the reactors was monitored by analysing influent, effluent and mixed liquour samples for selected parameters according to standard procedures (APHA 1995). Chemical Oxygen Demand (COD), nitrogen and phosphorus were determined with photometric cuvette tests (Dr. Lange).

For the LAS test, samples were taken, preserved with formaldehyde, filtered, and LAS and their degradation products quantified by liquid chromatography electrospray mass spectrometry (LC-ES-MS) as described by Eichhorn & Knepper (2002). For the naphthalenesulphonate test, samples were filtered over 0.45 μm and frozen pending analysis. The method for naphthalene sulphonate analysis was HPLC coupled to fluorescence detection.

Table 1. Overview of experimental conditions for LAS removal from synthetic domestic wastewater. HRT: hydraulic retention time. Muster 4 was from Henkel (Germany) and UFASAN from Unger (Norway).

	HRT (h)	LAS (mg/l)	LAS source	Membrane	Days
Period 1	20	1	'Muster 4'	PVDF	1–28
Period 2	10	1	'Muster 4'	PVDF	28–47
Period 3	10	1	UFASAN	PVDF	47–79
Period 4	10	5	UFASAN	PVDF	79–93
Period 5	10	10	UFASAN	PVDF	93–110
Period 6	10	10	UFASAN	Ceramic	110–130

Table 2. Overview of experimental conditions for naphthalene sulphonate removal from synthetic domestic wastewater. HRT: hydraulic retention time, SRT: sludge retention time.

	HRT (h)	SRT (D)	Concentration NS (μg/l)	Days
Period 1	20	Infinite	300	1–20
Period 2	20	Infinite	500	20–42
Period 3	10	Infinite	500	42–52
Period 4	10	Infinite	500*	52–57
Period 5	10	100	500*	57–80
Period 6	5	100	500*	80–101
Period 7	10	100	1000	101–115

* Extra addition of 1,5-naphthalene disulphonate.

3 RESULTS AND DISCUSSION

3.1 *Removal of LAS*

During the entire test period, the MBR effluent quality in terms of COD was always superior to the one of the CAS mainly as a result of a higher removal in suspended solids. In addition to the normal fluctuations, very high effluent COD values were observed in the CAS from day 100 to day 113. During this period, severe problems occurred with the pressurized air system in the building where the tests were running. Although this had implications on the oxygen supply to both reactors, it was visually observed from the color of the activated sludge, that anaerobic conditions had prevailed over a much longer period of time in the CAS than in the MBR.

Figure 1 shows that the LAS elimination efficiency was similar in MBR and CAS and was generally over 97%. This corresponds with literature data on LAS removal in full-scale CAS systems (Schröder *et al.* 1999, Huang *et al.* 2000, Eichhorn & Knepper 2002). When oxygen supply was interrupted (day 100 till day 113), a dramatic effect was visible in the CAS system. The removal efficiency dropped to zero but recovered fairly quickly. Simultaneously, high concentrations of SPC metabolites appeared in the effluent (not shown), implying that biodegradation was incomplete. This may have been due to the fact that aeration

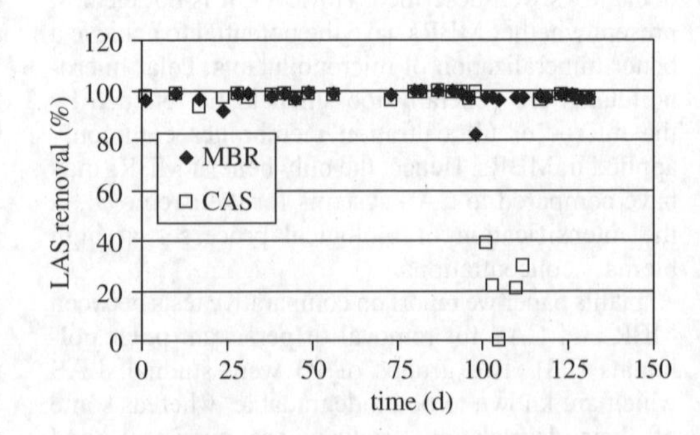

Figure 1. Comparison of LAS removal in MBR and CAS.

problems were more severe in the CAS than in the MBR, or it may indicate that MBRs show a more stable performance than CAS systems in case of perturbations.

With respect to the operational parameters tested, hydraulic retention time (HRT) did not influence LAS removal or SPC (sulfophenyl carboxylate) production (Table 3). Likewise, a change in LAS source and composition on day 47 did not affect the removal efficiencies. Upon changes in LAS concentration, the procentual LAS removal remained constant in both reactor systems. Hence, an increase in LAS and SPC effluent levels became apparent. This happened earlier or to a larger extent in the CAS than in the MBR. LAS effluent concentrations already increased in the CAS, when influent levels had been theoretically raised from 1 to 5 mg l^{-1} (in period 4). This was not the case for the MBR. The average values given for Period 5 in Table 3, are confounded by the fact that the problems with oxygen supply had started. When both the MBR and CAS system had recovered from this perturbation, similar effluent LAS levels of around 150 $\mu g\ l^{-1}$ were attained. However, SPC concentrations remained higher in the CAS than in the MBR. All these observations may indicate that MBRs adapt more quickly to changes in operational conditions and may prove to be more robust for micropollutant degradation.

It is likely that the residual LAS and SPC effluent concentrations could be further reduced by increasing the HRT. This was however not tested.

Overall, the ratio of effluent SPC relative to the influent LAS concentration, calculated using $\mu g\ l^{-1}$ instead of mM concentrations for simplicity, was systematically lower in the MBR than in the CAS, from period 3 onwards. Differences were low since the average ratio amounted to 0.04 in the MBR and 0.06 in the CAS.

If pronounced differences in degradation pathways existed between MBR and CAS, this would become visible in the isomer distribution patterns. From the available data, it could be concluded that the degradation pathways in MBR and CAS were similar and that no obvious relation between operational parameters and degradation pattern existed.

As expected, the ultrafiltration membrane of the MBR did not retain LAS or SPC. The fact that the measurements on the MBR supernatant are similar to the ones on the MBR effluent, indicates that accumulation did not take place. In addition, replacing the PVDF membrane with a more hydrophilic ceramic one, did not affect the removal efficiencies or the retention behaviour.

The respirometric response of the activated sludge to the addition of LAS at concentrations as low as 1 mg l^{-1} indicated that LAS removal was not only due to sorption phenomena, but that biodegradation was a major removal route.

3.2. *Removal of naphthalene sulphonates*

Like in the previous test, the COD of the MBR effluent was always lower than the one of the CAS.

Table 3. LAS concentrations (top) and SPC concentrations (bottom) in μg/l (average ± standard deviation) per test period.

		MBR		
Period	Influent	Effluent	Supernatant	CAS effluent
1	218 ± 82	10 ± 10	8 ± 2	7 ± 4
2	477 ± 307	7 ± 2	9 ± 3	9 ± 6
3	217 ± 61	4 ± 3	5*	6 ± 3
4	2787 ± 1076	5 ± 1	5*	24 ± 6
5	4624 ± 1266	245 ± 286	37*	1987 ± 2143
6	6198 ± 2489	156 ± 20	134 ± 47	1287 ± 2054

		MBR		
Period	Influent	Effluent	Supernatant	CAS effluent
1	8 ± 5	20 ± 16	15 ± 7	16 ± 9
2	3 ± 2	21 ± 4	14 ± 3	32 ± 10
3	11 ± 4	16 ± 8	24*	24 ± 7
4	130*	61 ± 19	88*	163 ± 67
5	131*	109 ± 95	106*	149 ± 72
6	144 ± 20	94 ± 56	188 ± 14	313 ± 198

The values given are averages of 4 samples, except when marked with*.

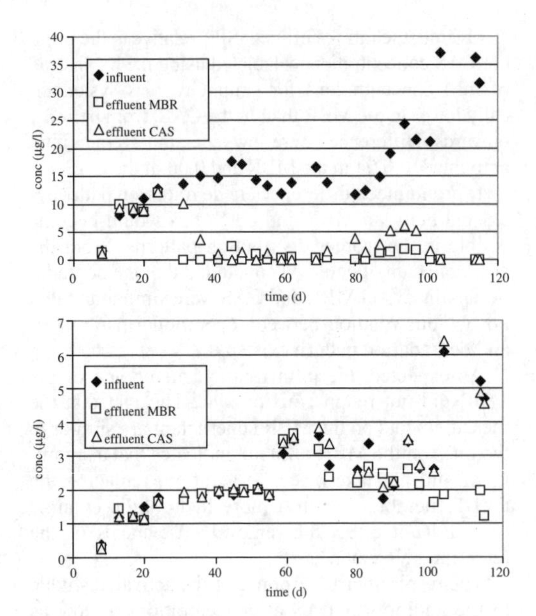

Figure 2. Comparison of degradation efficiency for 2 napthalene disulphonate substrates in MBR and CAS. Top: 1,6-isomer, bottom: 1,3-isomer.

Only when the HRT was reduced to 5 h (period 6), COD removal efficiencies temporarily decreased in the CAS.

The naphthalene monosulphonates were removed from the onset of the experimental period. For the disulphonates, an adaptation period of varying length was observed. Figure 2 shows the evolution in removal efficiency for 2 selected pollutants. Removal of the 1,6-isomer was complete after 30 d of operation in the MBR. In the CAS it took 10 more days to achieve complete elimination. From then on, degradation was stable. Only when the HRT was reduced to 5 h, a temporary increase in effluent concentrations was observed. Under the present conditions, it therefore seems advisable to keep the HRT around 10 h.

Figure 2 also indicates that degradation of the 1,3-naphthalene disulphonate only started towards the end of the test period and that it only occurred in the MBR. Because the test was stopped after 115 days of operation, it is not known whether this isomer would also be degraded in the CAS after some time lag.

Table 4 shows that the adaptation time needed to achieve 90% removal of a particular naphthalene sulphonate is systematically shorter in the MBR. Whether this is due to its higher sludge concentration or to the retention of slow growing or specific naphthalene sulphonate degrading bacteria by the membrane, is not clear at present. It is important to note however that the only difference between MBR and CAS seems to lie in the adaptation time needed.

Table 4. Onset of biodegradation for different naphthalene sulphonates. This implies that removal efficiencies on two consecutive sampling occasions were >90%.

	MBR	CAS
1-NSA	Day 0	Day 20
2-NSA	Day 0	Day 20
2,6-NDSA	Day 24	Day 41
1,6-NDSA	Day 31	Day 41
1,7-NDSA	Day 45	Day 104
2,7-NDSA	Day 52	Day 104
1,3-NDSA	Never	Never
1,5-NDSA	Never	Never

When one waits sufficiently long, the same overall removal efficiencies are attained. To confirm these observations, tests are now underway with real domestic wastewater.

In our tests, removal efficiencies close to 100% were achieved for naphthalene disulphonates. This does not completely correspond with full-scale experience on CAS treatment of sewage (Altenbach 1996) or MBR treatment of tannery wastewater (Reemtsma *et al.* 2002). The 1,6- and 2,6-isomer generally showed elimination efficiencies similar to our results. For the 1,7- and 2,7-isomers the results were variable. 1,5-Naphthalene disulphonate was the most stable isomer in all studies.

4 CONCLUSIONS

Our results do not provide an indication that MBRs work better than CAS in removing trace organics. However, the time needed to attain new degradation capacities was shorter.

ACKNOWLEDGEMENTS

This work was carried out in the framework of the European Project "Removal of Persistent Polar Pollutants Through Improved Treatment of Wastewater Effluents (P-THREE)" under contract EVK1-CT-2002-00116.

REFERENCES

APHA. 1995. *Standard Methods for the Examination of Water and Wastewater.* 19th Edition, American Public Health Association/American Water Works Association/ Water Environment Federation, Washington DC, USA.

Altenbach, B. 1996. *Determination of substituted benzene- and naphthalene sulfonates in wastewater and their behaviour in sewage treatment.* Ph.D. Thesis, EAWAG/ETH, Zürich, Switzerland.

Eichhorn, P. & Knepper, T.P. 2002. α,β-unsaturated sulfophenylcarboxylates as degradation intermediates of linear alkylbenzenesulfonates: evidence for Ω-oxygenation followed by β-oxidations by liquid chromatography-mass spectrometry. *Environmental Toxicology and Chemistry* 21(1): 1–8.
Ellis, T., Barbeau, D. S., Smets, B. F. & Grady, C. F. L. Jr. 1996. Respirometric technique for determination of extant biokinetic parameters describing biodegradation. *Water Environment Research* 68, 917–926.
Huang, X., Ellis, T.G. & Kaiser, S.K. 2000. *Extant biodegradation testing with linear alkylbenzene sulfonate in laboratory and field activated sludge systems.* CD-ROM Proceedings of the Water Environment Federation 72nd Annual Conference and Exposition. Water Environment Federation, Anaheim, California, October, 2000, http://www.public.iastate.edu/\tilde;tge/wc0054p3.pdf
Reemtsma, T., Zywicki, B., Stüber, M., Klöpfer, A. & Jekel, M. 2002. Removal of sulfur-organic polar micropollutants in a membrane bioreactor treating industrial wastewater. *Environmental Science and Technology* 36: 1102–1106.
Schröder, F.R., Schmitt, M. & Reichensperger, U. 1999. Effect of waste water treatment technology on the elimination of anionic surfactants. *Waste Management* 19: 125–131.

European Symposium on Environmental Biotechnology, ESEB 2004 - Verstraete (ed)
© 2004 Taylor & Francis Group, London, ISBN 90 5809 653 X

The reduction of coloured compounds using whole bacterial cells (*Shewanella* strain J18 143)

C.I. Pearce & J.T. Guthrie
Department of Colour Chemistry, University of Leeds, Leeds, UK

J.R. Lloyd
Williamson Research Centre for Molecular Environmental Science and Department of Earth Sciences, University of Manchester, Manchester, UK

ABSTRACT: The delivery of colour, in the form of dyes, onto textile fibres is an inefficient process. As a result, the wastewater produced by the textile industry is coloured. Colour pollution in aquatic environments is an escalating problem, despite substantial research into modifying the dyeing process to increase the level of delivery and uptake of the dyes. The recalcitrant nature of modern synthetic dyes has led to the imposition of strict environmental regulations. A cost-effective process to remove colour from wastewater that is produced by the textile industry is needed (Willmott et al. 1998). The research presented here concerns the use of a patented biocatalyst, isolated from soil contaminated with textile dyes, for the reduction of water-soluble azo dyes that are present in textile dyeing wastewater.

1 COLOUR IN DYE WASTEWATER

Coloured wastewater is a consequence of batch processes in both the dye manufacturing and the dye-consuming industries. In manufacture, 2% of dyes produced are discharged directly in aqueous wastewater. 10% of the dyes used are subsequently lost during the textile coloration process (Easton 1995). To give an indication of the scale of the problem, the annual market for dyes is more than 7×10^5 tonnes per year (Robinson 2001). The main reason for these losses is the incomplete exhaustion of dyes on to the fibre. Coloured wastewater is particularly identified with reactive azo dyes, used for dyeing cellulose fibres, which make up approximately 30% of the total dye market (Kamilaki 2000). The appealing properties of these dyes include the fact that a wide range of brilliant shades is available, across a number of application methods, providing a high wet fastness. The structure of Remazol Black B, shown in Figure 1, is characteristic of reactive azo dyes, with the chromophore containing azo groups, the reactive centres and the solubilising components.

Residual colour is a problem with reactive dyes because, in current dyeing processes, as much as 50% of the dye is lost to the wastewater. These losses are due to low levels of dye-fibre fixation and to the presence

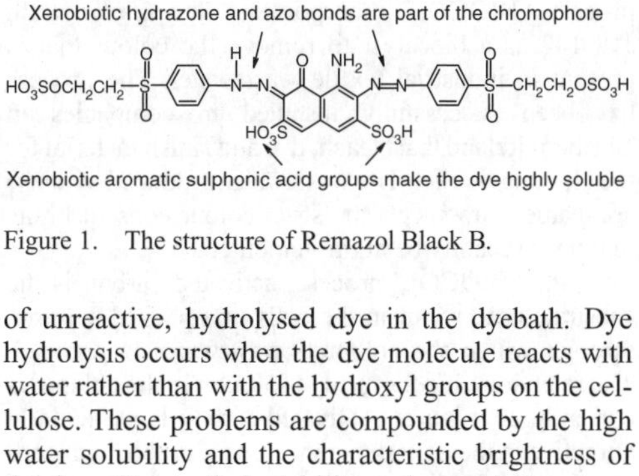

Xenobiotic hydrazone and azo bonds are part of the chromophore

Xenobiotic aromatic sulphonic acid groups make the dye highly soluble

Figure 1. The structure of Remazol Black B.

of unreactive, hydrolysed dye in the dyebath. Dye hydrolysis occurs when the dye molecule reacts with water rather than with the hydroxyl groups on the cellulose. These problems are compounded by the high water solubility and the characteristic brightness of the dyes. Due to their stability and xenobiotic nature, wastewaters containing azo dyes are not degraded by conventional treatment processes involving light, chemicals or activated sludge (Leisinger 1981).

The dyes are therefore released into the environment, in the form of coloured wastewater. This can lead to acute effects on exposed organisms due to toxicity of the dyes, abnormal coloration, and reduction in photosynthesis due to absorbance of light entering the water (Slokar 1998, Strickland 1995). Also, public perception of water quality is greatly influenced by colour. The presence of unnatural colours is aesthetically unpleasant and tends to be associated with contamination (Waters 1995). A system in the dyehouse that is dedicated to colour removal and gives the correct balance between cost effectiveness and environmental impact is necessary to solve these problems. Currently, the major methods of textile dye treatment involve physio-chemical processes such as coagulation, ozonation and absorption. Such methods are often very costly and, although the dyes are removed, accumulation of concentrated sludge creates a disposal problem. Alternative approaches that utilise microbial biocatalysts to reduce the azo dyes offer potential advantages over physio-chemical processes have been and are the focus of recent research (Willmott et al. 1998, Robinson 2001).

2 COLOUR REMOVAL USING A BIOCATALYST

2.1 *Reactive azo dye solutions*

An investigation into the apparent colour removal that naturally occurs in the regions that surround textile wastewater discharge pipes revealed a bacterium, *Shewanella* strain J18 143 that can couple growth to the biodegradation of reactive azo dyes.

A consortium of research partners (the Colour Chemistry Department at Leeds University, Questor Technologies Ltd., the British Textile Technology Group and the Earth Sciences Department at Manchester University) has developed a novel biotechnological process, "BIOCOL", incorporating *Shewanella* strain J18 143 as a biocatyst to remove the colour that is present in industrial textile wastewater. The process has been successfully installed in companies in Northern Ireland that bleach, dye and finish material for the clothing market. The wastewater produced by these companies now meets the strict colour consent limits that have recently been implemented.

In the "BIOCOL" process, activated carbon is the substrate used to retain the cells in a packed column, through which the coloured wastewater is pumped. The retention time for the dye solution in the column is currently 20 minutes. Activated carbon is particularly attractive in these circumstances, as it is able to absorb the dyes, providing a large surface area on which bioreduction of the dyes by *Shewanella* strain J18 143 can take place. Activated carbon can be expensive to regenerate. However, in the "BIOCOL" process, regeneration is only required once every 6 months and is carried out by Questor Technologies as part of the package provided when the "BIOCOL" process is installed.

It is thought that the colour removal process involves the transfer of electrons from the cell to the dye molecule, via a dye reductase. This results in the reduction of the chromophore, i.e. the azo bonds, to produce a significantly less coloured solution that contains amines (Figure 2).

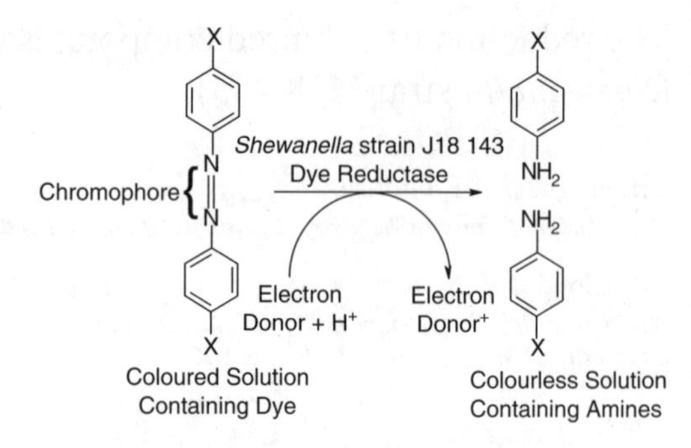

Figure 2. Biocatalysed reduction of an azo dye.

The enzyme system responsible for the destruction of azo dyes by bioreduction, and the performance of the biocatalyst has yet to be completely understood. This information is required if the colour removal process is to be optimised to its full potential.

Preliminary studies, involving growing and resting cells of *Shewanella* strain J18 143, were carried out to determine the growth conditions necessary to increase dye reductase activity. Growing cell assays demonstrated that the dye reductase for Remazol Black B was upregulated under anaerobic conditions. The type of electron donor used had a significant effect on the rate of dye reduction and no dye reduction was observed in the absence of an electron donor. Sodium formate produced the fastest rate of dye reduction indicating that the enzyme system responsible for the reduction was coupled to formate dehydrogenase activity. Temperature will have an important effect on the viability of the *Shewanella* strain J18 143 cells and their ability to reduce the dye. The enzyme system responsible for dye reduction was still active at 50 to 60°C, which was surprising, as the activity of most ambient enzyme systems tends to fall off after about 40°C. This property of the *Shewanella* strain J18 143 cells gives the "BIOCOL" process an advantage over other biological treatment systems as the cells will be able to survive fluctuations in the temperature of the wastewater. It was found that the dye reductase enzyme system in *Shewanella* strain J18 143 cells was stable in neutral to alkaline conditions. This pH range is advantageous to the "BIOCOL" process because reactive dye wastewater has an alkaline pH. The concentration of dye substrate can influence the efficiency of dye removal through a combination of factors including the toxicity of the dye at higher concentrations, and the ability of the enzyme to recognise the substrate efficiently at the very low concentrations that may be present in some wastewaters. The dye reductase enzyme functioning inside the *Shewanella* strain J18 143 cells in this system was able to reduce dye in very dilute solutions as well as more concentrated dye solutions.

As it is unlikely that the highly charged sulphonated azo dyes pass through the *Shewanella* cell membrane, dye reduction probably involves extra cellular reducing activity. This reducing activity is achieved using mediator compounds, such as anthraquinone-2,6-disulphonic acid (AQDS), to shuttle reduction equivalents from the cells and facilitate the reduction of the extracellular azo dye. The process of dye reduction using *Shewanella* strain J18 143 cells was more efficient with the addition of AQDS and it took place in the initial 30 seconds.

3 COLOUR REMOVAL FROM RELATIVELY INSOLUBLE COLOURED COMPOUNDS USING *SHEWANELLA* STRAIN J18 143

From the improved results that were obtained with the use of a redox mediator it is possible to surmise that, for the reduction to take place, the dye does not need to be inside the cell. Therefore, the *Shewanella* strain J18 143 cells should have to ability to reduce azo bonds in relatively insoluble coloured compounds such as keto-hydrazone/azo pigment dispersions and azo dyes when they are reactively dye onto cotton fabric.

3.1 *Keto-hydrazone/azo pigment dispersions*

A range of keto-hydrazone/azo pigment dispersions was assayed in the same way as the dye in the presence the *Shewanella* strain J18 143 cells. It was found that each pigment was reduced to a certain extent, but reduction took several days, the process being much slower than that which occurred with the equivalent dye solution.

The fastest rate of reduction was observed with HD Sperse Orange EX5 pigment dispersion. The structure of the orange pigment is shown in Figure 3. This pigment has a relatively simple structure compared to the other pigments studied and it is similar to many azo dyes. HD Sperse Orange EX5 pigment dispersions were used to carry out more work on the optimisation of the pigment reducing system. As with the reduction of Remazol Black B, addition of AQDS as a redox mediator increased the rate of pigment reduction by acting as an electron shuttle between the relatively insoluble pigment substrate and the cell bound reductase.

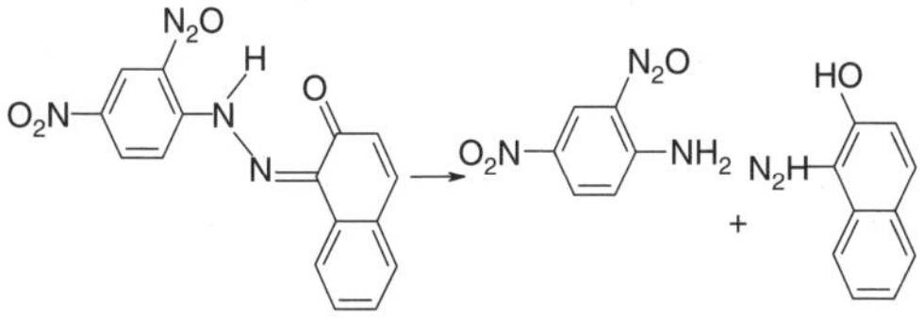

Figure 3. Reduction of HD Sperse Orange EX5 using *Shewanella* strain J18 143.

3.2 *Reactive azo dyes on cotton fabric*

When Remazol Black B dye is covalently bonded to cotton fabric in the reactive dyeing process, it can almost be considered as a pigmentary species. Any intramolecular rotation in the dye molecule is significantly restricted as it is covalently bonded to the cotton at both ends of the molecule. This is worth considering as the reduction mechanism for the dye in solution may rely on intramolecular rotation.

The reduction of dye on cotton fabric by *Shewanella* strain J18 143 cells, in the presence of AQDS, was monitored by the change in reflectance and transmittance of the dyed cotton samples. Figure 4 shows that exposing the dyed cotton to *Shewanella* strain J18 143 cells has a significant effect on the K/S value of the dyed cotton. The K/S value is related to the concentration of dye on the fabric.

Recent research has shown that the reduction of Remazol Black B dye on cotton fabric takes place over the initial 24 hours and, with optimization, the time required to remove the colour could be reduced further. The system could be used as a means of recycling dyed fabric.

4 CONCLUSION

Colour pollution has been recognised as an escalating problem, resulting in the implementation of strict colour consent limits for textile effluents. To comply with these consent limits, the biocatalyst *Shewanella* strain J18 143 can be used in an industrial process for the removal of colour from textile effluent. Currently this is a "black box" process and research is being undertaken to understand the mechanism of dye reduction by *Shewanella* strain J18 143, allowing optimisation and modeling of the process. *Shewanella* strain J18 143 is also capable of reducing certain relatively insoluble pigments. This reduction process is affected by pigment structure, pigment solubility and

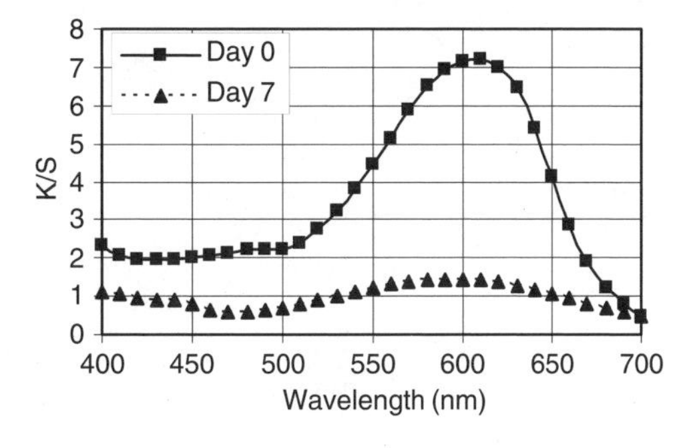

Figure 4. Change in K/S values for cotton fabric dyed with Remazol Black B (1.0% omf), treated with *Shewanella* strain J18 143.

the azo-hydrazone tautomeric equilibrium of the pigment. *Shewanella* strain J18 143 can be used to treat intensely coloured fabric and that the level of colour removal is great enough to possibly allow the fabric to be re-dyed and recycled. The fact that the *Shewanella* strain J18 143 can reduce solid substrates opens up a whole range of other industrial applications in which the reduction of compounds such as transition metal ions would be both environmentally and commercially beneficial.

REFERENCES

Easton, J. 1995. The dye maker's view, In Peter Cooper (ed.), *Colour in Dyehouse Effluent*. Bradford: Society of Dyers and Colourists.

Kamilaki, A. 2000. *The Removal of Reactive Dyes from Textile Effluents – A Bioreactor Approach Employing Whole Bacterial Cells*. Leeds: University of Leeds.

Leisinger, T. et al. 1981. *Microbial Degradation of Xenobiotics and Recalcitrant Compounds: FEMS Symposium no. 12*. London: Academic Press for the Swiss Academy of Sciences and the Swiss Society of Microbiology on behalf of the Federation of European Microbiological Societies.

Robinson, T. et al. 2001. Remediation of dyes in textile effluent: a critical review on current treatment technologies with a proposed alternative. *Bioresource Technology*. 77: 247–255.

Slokar, Y.M. & Le Marechal, A.M. 1998. Methods of Decoloration of Textile Wastewater. *Dyes and Pigments*. 37(4): 335–356.

Strickland, A.F. & Perkins, W.S. 1995. Decolorization of Continuous Dyeing Wastewater by Ozonation. *Textile Chemist and Colorist*. 27(5): 11–15.

Waters, B.D. 1995. The dye regulator's view. In Peter Cooper (ed.), *Colour in Dyehouse Effluent*. Bradford: Society of Dyers and Colourists.

Willmott, N.J. et al. 1998. The biotechnology approach to colour removal from textile effluent. *Journal of the Society of Dyers and Colourists*. 114: 38–41.

European Symposium on Environmental Biotechnology, ESEB 2004 - Verstraete (ed)

Chloroform cometabolism by bacterial strains grown on butane

S. Fedi[1], Y. Pii[1], D. Frascari[2], A. Zannoni[2], M. Nocentini[2], D. Zannoni[1]
[1]*Department of Biology, University of Bologna, via Irnerio, Bologna, Italy*
[2]*Department of Chemical and Mining Engineering and Environmental Technologies, University of Bologna, Bologna, Italy*

ABSTRACT: In this microcosm study we have monitored the performances of 12 butane-utilizing consortia during a long-term aerobic cometabolic degradation of chloroform (CF). After approximately 100 days of continuous CF degradation, a sudden improvement of the biodegradative performances (7-fold increase of r_c and 4-fold increase of r_c/r_s ratio) was observed in two microcosms in which the highest amount of depleted CF ($>70\,mg\ kg_{dry\ soil}^{-1}$) was observed. Two bacterial strains, isolated from the CF-degrading microcosms, were identified belonging to the genera *Rhodococcus* and *Stenotrophomonas*, respectively. A strain, named F, was further characterized to establish the involvement of the butane-monooxygenase in CF degradation.

1 INTRODUCTION

Bioremediation by microbial aerobic cometabolism is a promising biological approach for the remediation of sites contaminated with chlorinated aliphatic hydrocarbons (CAHs). Although chloroform (CF) is one of the most recalcitrant CAHs, some bacteria have been shown to dehalogenate it through an aerobic cometabolic process. For example, the methane-utilizing bacterium *Methylosinus trichosporium* OB3b is able to cometabolize several CAHs, including CF and trichloroethylene (TCE), using two types of methane monooxygenases (Oldenhuis et al. 1989, Tsies et al. 1989) while in recent years, McClay et al. (1996) have shown CF degradation to be present in several toluene-oxidizing strains. Further, Kim et al. (1997) reported that propane was an effective cometabolic substrate to drive the trasformation of CF through the use of mixed cultures in microcosms enriched from aquifer solids.

The design of an *in-situ* or *on-site* cometabolic bioremediation treatment is based on results of a preliminary lab-scale microcosm study in order to evaluate whether the indigenous biomass of the site has the capability to degrade all the contaminants present and to estimate the main parameters characterizing the process, namely: (i) the acclimatation time necessary for the onset of the cometabolic biodegradation process, (ii) the primary substrate utilization rate, iii. the contaminant biodegradation rate, (iv) the mass of substrate required to degrade a given amount of contaminants (transformation yield), and (v) the amount of contaminants degraded per unit of biomass inactivated by the degradation products (transformation capacity).

The first part of this study reports on the performances of several butane-utilizing consortia as monitored during a long-term aerobic CF cometabolic degradation process. Fifteen slurry microcosms, set up by variable combinations of two soils and three groundwaters, were exposed to periodical additions of butane and CF under different conditions for an experimental period of time between 100 and 370 days. In the second part of the present study, we have investigated the composition of the selected consortia deriving from the best-performing microcosms. Isolates from the CF-degrading microcosms were grouped into different operational taxonomic units (OTU) by means of amplified ribosomal DNA restriction analysis (ARDRA). Two main OTU were identified and the phylogenetic position of the strains was determined by 16S rDNA sequences. A strain, named F, was further characterized to esstablish the involvement of the butane monooxygenase in CF degradation. Competition between increasing concentrations of butane and CF, along with the inhibitory effect of acetylene were also evaluated.

2 MATERIALS AND METHODS

2.1 *Microcosms preparation and operation*

Twelve slurry microcosms were prepared using 155 mL Amber serum bottles with Teflon-lined rubber septa. Each microcosm contained 25 g of a sandy soil sampled in a non-polluted aquifer, 70 mL of non-contaminated groundwater and 75 mL of headspace air. To avoid bacterial contamination, bottles, caps and all the tools used for preparing the microcosms were autoclaved (121°C, 20 min). The groundwater had previously been amended with 100 mg/L of nitrate and 10 mg/L of phosphate, in order to prevent nutrients from being limiting factors in biomass growth. Microcosms were divided into three groups, corresponding to two real-case scenarios and to a third, intermediate situation. The operational conditions in each group were as following:

Group #1 (2 microcosms): the microcosms were initially spiked with butane (2.1 mg/L in the aqueous phase); CF (1 mg/L in the liquid phase) was the added after 7 ÷ 8 pulses of butane; this condition is representative of processes where the specialized biomass develops before the addition of the contaminant, such as *on-site* biological pump-and-treat processes or *in-situ* treatments with biologically active barriers.

Group #2 (5 microcosms): addition of butane (2.1 mg/L) after a prolonged (35 ÷ 69 days) exposition of the indigenous biomass to CF (0.8 ÷ 2.1 mg/L); this situation is representative of an *in-situ* treatment in which the growth substrate is introduced in the contaminated zone some time after the contamination has occurred;

Group #3 (5 microcosms): the microcosms were initially spiked with both butane (2.1 mg/L) and CF (0.9 ÷ 1.7 mg/L in the aqueous phase).

The microcosms were placed on an orbital shaker (150 rpm and room temperature: 20 ÷ 22°C) and operated for a period variable between 100 and 250 days. Upon completion of each pulse of butane or CF, a new spike was introduced. The initial concentration of butane in each spike was kept constant at 2.1 mg/L, whereas the initial CF concentration in each spike was time-varied in some of the microcosms, in order to evaluate the relationship between concentration and degradation rate and to detect possible toxic effects. More precisely, the initial CF concentrations of the pulses introduced in the 3 groups of microcosms were characterized by the following average values: group #1, 1.7 mg/L; group #2, 2.0 mg/L; group #3, 2.4 mg/L. The desired CF concentrations were obtained by spiking a CF aqueous solution (4 g/L) into the sealed microcosms. The fully aerobic condition ($O_2 > 8\%_{v/v}$) was periodically verified by headspace analysis, and was maintained by additions of pure oxygen. Microcosms were periodically opened (in the absence of CF and butane) and stripped with air to eliminate dissolved carbon dioxide and possible volatile products of CF degradation. Nutrients (N and P) were provided every time the nitrate concentration was lower than 20 mg/L.

2.2 *Analysis*

Butane and CF gas concentrations were determined by analyzing 40 μL headspace samples using a Hewlett Packard 6890 gas-chromatograph equipped with a column joined to both a flame ionization detector for butane analysis and an electron capture detector for CF analysis. The total mass and the liquid-phase concentration of butane and CF were calculated from the gas-phase concentrations by mass balances. Oxygen and carbon dioxide volumetric percentages in the microcosms headspace were determined using a Varian gas-chromatograph equipped with a thermal conductivity detector. Nitrate concentration in the liquid phase was measured by Ion Chromatography.

2.3 *Microbiological characterization of the consortia*

Bacterial counts: bacterial colonies grown on R_2A agar plates were grouped in different clusters and were counted on the basis of their different morphologies; the viable cells concentration was expressed as colony forming units per mL of suspension (CFU/mL). Genomic DNA was extracted, from colonies belonging to each cluster, to be used for PCR amplification of 16S rDNA. Amplified ribosomal DNA restriction analysis (ARDRA): a 2 μL sample of each amplification mixture was analyzed by agarose gel electrophoresis in Tris-acetate-EDTA buffer containing 0.5 μg/mL of ethidium bromide. Amplified 16S rDNA was purified from each PCR reaction mixture by using a QIAquick PCR purification and was sequenced. The 16S rDNA nucleotide sequences were aligned with the most similar sequences in the Ribosomal Database Project (RDP) database by using RDP utilities.

Growth of the isolated strains on different intermediates of the alkane metabolic pathways was tested by streaking out the culture on mineral medium (MM) plates containing 0.1% (v/v for liquid, W/v for solid) of different intermediates. Visible growth was observed within 48–72 h.

2.4 *Growth conditions and chloroform degradation assay*

Strains were cultured at 30° in a mineral medium and grown *n*-butane as the only reduced carbon source. Cell growth was monitored by removing a portion of the cultures and measuring the optical density at 660 nm (OD_{600}). Harvested cells had a typical OD_{600} of 0.50.

Chloroform degradation assay: cells were harvested by centrifugation (6.000 × g, 10 min), washed twice with the same buffer as in the growth medium, and then suspended to a constant cell density (based on OD_{600}). Assays were conducted in 11 ml serum vials sealed with Teflon-coated butyl rubber stoppers. Chloroform was added as a diluted aqueous solution which was made daily from chloroform-saturated solution at a room temperature. The amount of CF to be added was estimated from solubility tables (Schwille 1979). The concentrations of CF in the liquid phase were calculated from Henry's law constants (Gossett 1987). The reaction mixtures (1.20 ml) containing the same phosphate buffer as in the growth medium and CF solution were equilibrated at 30°C, with constant shaking for at least 30 min before the assay was started. The reactions were started by adding 250 μl of a concentrated cell suspension (approximately 0.30 mg of protein). For the time course assays, a sample (40 μl) of the gas phase was removed for analysis by gas-chromatography using a Hewlett Packard 6890 gas-chromatograph equipped with a column and an electron capture detector for CF analysis.

2.5 *Acetylene inactivation assay*

To determine whether or not acetylene inactivated CF degradation, cells were exposed to acetylene prior addition of CF. Cell suspensions (250 μl) were incubated for 10 min in sealed 11 ml vials which contained phosphate buffer, and 0.1% vol/total (vial vol) acetylene. Control cells were preincubated in the phosphate buffer. After preincubation, acetylene was removed from the vials by opening the cap and purging with air for 3 minutes. The vials were resealed and then the reactions were initiated by the addition of CF to the reaction mixture.

3 RESULTS

3.1 *Long-term trends of both butane consumption-and CF degradation-rates*

The microbial utilization of butane began after a lag-time equal to 4.7 ± 1.5 days (average ± 95% confidence interval) and proceeded with a maximum rate, relative to the first pulse, characterized by an overall 16 ± 6 mg/(L d) average and by a higher value in the microcosms where CF was initially absent [group #1: 30 mg/(L d)] with respect to those where the utilization of butane began in the presence of the chlorinated contaminant [group #2: 9 mg/(L d), group #3: 18 mg/(L d)]. The degradation of CF began immediately in the microcosms where a butane-utilizing biomass had already developed (group #1), after a 6.8-days lag-time in the microcosms characterized by a pre-exposition to CF and after a 15.9-days lag-time in the microcosms characterized by the contemporary introduction of butane and CF. CF maximum degradation rates during the first pulse were characterized by an overall 0.24 ± 0.07 mg/(L d) average (in correspondence of an average of 1.4 ± 0.3 mg/L in the aqueous phase) with negligible differences between the 3 groups of microcosms.

After two to three pulses of butane, all the microcosms reached a pseudo steady-state condition, characterized – in each microcosm – by moderate pulse-to-pulse fluctuations of the biomass concentration and substrate utilization rate around a quite constant value. CF concentrations up to 14 mg/L in the aqueous phase were degraded without any indication of toxicity due to CF.

Two microcosms (B14, belonging to group #3, and B15, belonging to group #1) showed, after 100-days (net time) of CF degradation (calculated as the time elapsed from the first exposition to contaminant minus the total time periods during which no CF was present in the microcosms), a sudden and remarkable increase of the CF biodegradation rate, accompanied by a less significant increase of the butane utilization rate. More precisely, in microcosm B15 (Figure 1) a 7.5-fold r_c increase (from 0.63 to 4.73 mg/(L d), in terms of average values), a 1.8-fold r_s increase [from 37 to 66 mg/(L d)], a 1.4-fold increase of the $(r_c/c_c)/(r_s/c_s)$ ratio (from 1.7 to 2.5%) and a 4-fold T_y increase (from 1.5 to 5.9%) were observed.

In order to investigate whether the observed increases in CF degradation rate could be induced in other microcosms by means of a bioaugmentation treatment, 0.2 mL of suspension sampled from microcosm B15 were introduced in microcosm B16 (group #3). Both butane and CF degradation rates immediately began to increase and, after 10 days, a pseudo-stationary situation was achieved (Figure 2), characterized – with respect to the situation previous to the inoculation – by a 10-fold r_c increase (from 0.44 to 4.4 mg/(L d):), a 3-fold r_s increase (from 35 to 106 mg/(L d)) and a 6-fold T_y increase (from 0.9 to 5.5%), whereas the $(r_c/c_c)/(r_s/c_s)$ ratio remained quite constant (from 1.3 to 1.5%). This result confirms that, as reported in previous studies (Frascari et al. 2002), the introduction of small amounts of selected CF-degrading biomasses in batch reactors (0.3% of the liquid volume in this experiment) determines an immediate change in the behavior of the inoculated microcosm, which quickly reaches degradation performances similar to those of the parent reactor.

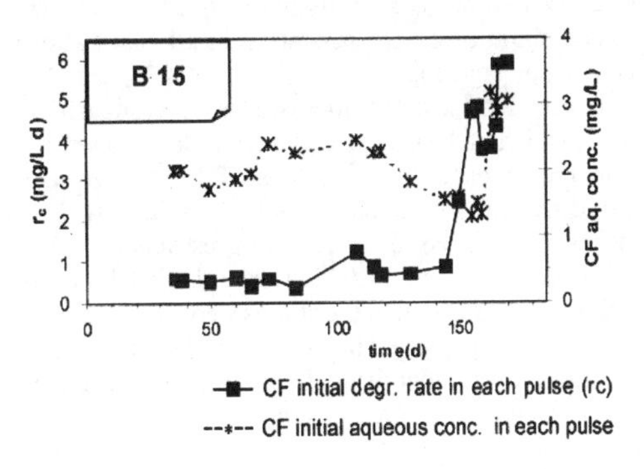

Figure 1. Microcosms B15: CF maximum degradation rate and CF initial aqueous concentration in each pulse.

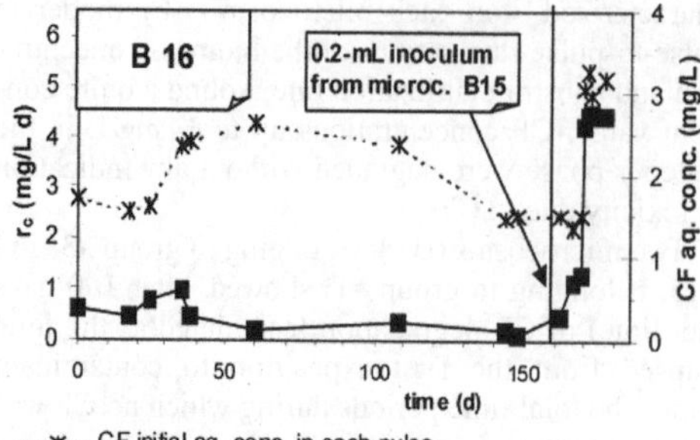

Figure 2. Microcosm B16: maximum degradation rate and CF initial aqueous concentration in each pulse before and after introduction of 0.2 mL aliquots from microcosm B16.

3.2 *Characterization of the high performing bacterial consortium obtained in microcosm B15*

To further investigate the characteristic of the biomass present in microcosm B15 after the sudden improvement of the degradation performances, CF depletion was monitored in the absence of butane until exhaustion of the degradation process and subsequent reintroduction of butane, and the results were compared with the data of similar experiments conducted on 2 microcosms in which no increase of the CF degradation rate had been observed. In the absence of butane microcosm B15, proved to be able to deplete 8 subsequent pulses of CF, was characterized by an average initial concentration of 3.3 mg/L, for a total degradation period of 60 days. Upon exhaustion of the CF degradation process, the biomass concentration was measured and butane was re-introduced in the 3 microcosms. A biomass concentration equal to half the level measured before the prolonged degradation of CF only was found in B15, whereas in the other 2 microcosms the degradation of CF until exhaustion determined a 17 to 34-fold decrease in biomass concentration. The consumption of butane re-started in all cases after a 1- to 3-day lag-time.

Based on these indications, we have investigated the composition of the selected consortia. Bacteria isolated from the CF degrading microcosm B15 were grouped into different operational taxonomic units (OTU) by means of amplified ribosomal DNA restriction analysis (ARDRA). Two main OTUs were identified and the phylogenetic position of the strains were determined by 16S rDNA sequences. The two isolated strains, named F and R, were identified belonging to the genera *Rhodococcus* and *Stenotrophomonas*, respectively. They could utilize a wide variety of carbon sources tested and most of the intermediates of the proposed propane and butane metabolic pathways (Table 1).

Table 1. Carbon sources and intermediates of the propane and butane pathways tested for growth of strains F and R.

	Strain tested	
Carbon source	Strain F	Strain R
1-Butanol	+	+
1-Propanol	+	+
Acetic acid	+	+
Butyric acid	+	+
Caproic acid	+	+
Heptane	+	+/–
Hexane	+	+/–
Pentane	+	+/–
Propionic acid	+	+
Valeric acid	+	+

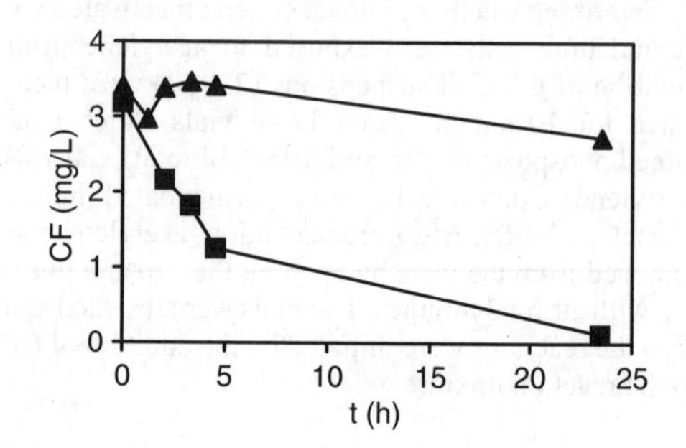

Figure 3. Time course of CF degradation by strain F. Cells were incubated in the presence of 3.5 mg/L of CF (▲), and CF plus 0.5% (vol/total vial vol) acetylene (■).

Rhodococcus strain F was further characterize to establish the involvement of the butane monooxygenase in CF degradation. Bacterial culture of *Rhodococcus* F were grown on butane and tested for CF degradation. After 5 hours the cells of strain F grown on butane had consumed about 70% of the CF initially present (Figure 3). Acetylene, an inhibitor of several monooxygenases was used to test its effect on CF degradation by strain F. When acetylene (0.5% vol/total vol vial) was added to the reaction mix, along with CF and cell suspension, the CF degradation was completely inhibited (Figure 3).

4 CONCLUSIONS

In summary, this study indicates that the prolonged cometabolic biodegradation of chlorinated solvents can lead to the selection of bacterial strains able to express remarkable high biodegradation rates and characterized by a strong resistance to the toxic degradation products. Further, the isolation of these strains and their inoculation in other reactors allowed to obtain, within a few days, the same degradation performances observed in the parent reactor.

REFERENCES

Frascari, D. Pinelli, D. & Nocentini, M. 2002. Aerobic cometabolic degradation of chloroform with butane: influence of system features on biomass adaptation. In A.R. Gavaskar & A.S.C. Chen (eds), *Proceedings of the Third International Conference on Remediation of Chlorinated and Recalcitrant Compounds*.

Gosset, M.J. 1987. Measurement of Henry's low constants for C1 and C2 chlorinated hydrocarbons. *Environmental Science Technology* 21: 202–208.

Kim, Y. Semprini, L. & Arp, D.J. 1997. Aerobic cometabolism of chloroform and 1,1,1-trichloroethane by butane-grown microorganisms. *Bioremediation Journal* 1(2): 135–148.

McClay, K. Fox, B.G. & Steffan, R.J. 1996. Chloroform mineralization by toluene-oxidizing bacteria. *Applied Environmental Microbiology* 62: 2716–2722.

Schwille, F. 1979. Dense chlorinated solvents in porous and fractured media. In R.A. Freeze & J.A. Cherry (eds), *Groundwater*: 131. Englewood Cliffs (N.J.): Prentice-Hall.

Tsien, H. Brusseau, G.A. Hanson, R.S. & Wackett, L.P. 1989. Biodegradation of trichloroethylene by *Methylosinus trichosporium* OB3b. *Applied Environmental Microbiology* 55: 3155–3161.

European Symposium on Environmental Biotechnology, ESEB 2004 - Verstraete (ed)
© 2004 Taylor & Francis Group, London, ISBN 90 5809 653 X

Development of an integrated approach for the removal of tributyltin from waterways and harbors

K. Pynaert & L. Speleers
Environmental Research Center, Hofstade-Aalst, Belgium

ABSTRACT: Tributyltin (TBT), an effective biocide frequently used (in the past) as a paint additive on ship hulls to prevent fouling, has accumulated into sediments of harbors and waterways. In the framework of a LIFE-Environment project, we are currently investigating the extent of TBT-contamination in the port of Antwerp (Belgium) and possible means of remediation. Average concentrations of 50–1000 μg/kg DW were found, however with peaks of more than 40000 μg/kg DW close to the ship repair site. In lab-scale experiments, resuspension of such sediment particles was shown to pose a risk of TBT-remobilization, particularly at a pH higher than 8.1. Sampling of the water phase during an on-site dredging operation confirmed these results, as TBT-concentrations of more than 500 ng/L were detected at a water pH of 8.29. Although not acute toxic, such high concentrations demand a cautious dredging approach and the need for sediment remediation (bio-, phyto-, electrochemical, thermal, …).

1 GENERAL INTRODUCTION

Tributyltin (TBT) is a tri-alkylated organometallic compound with strong biocidal properties towards a variety of bacteria, mollusks, fish, …, and has been used in paints to protect surfaces (e.g. ship hulls) from fouling and as preservative in the wood, textile and leather industry. It is also used as stabilizing material in PVC-production and as a catalyst in the chemical industry. Recent estimates have shown that the annual world production around the year 2000 was close to 50000 tons (Inoue et al. 2000). From all these sources, TBT has entered the environment, where it was shown to exhibit severe adverse effects on non-target organisms and was even considered by some experts to be the most toxic compound ever deliberately released in the environment by mankind (Fent 1996). That is why the International Maritime Organization (IMO) has passed a resolution in 2001 to ban the application of TBT-based antifouling paints on ships and boats as of 2003 and has also proposed to establish a mechanism to prevent the potential future use of other harmful substances in antifouling systems (Champ 2000).

Tributyltin that has entered the aquatic environment can have different fates. It can be degraded by abiotic and biotic factors, which usually involve sequential removal (de-alkylation) of alkyl groups from the tin atom and generally result in a toxicity reduction. UV light and chemical cleavage are the most important abiotic factors in aquatic and terrestrial ecosystems, but biotic processes have been demonstrated to be the most significant mechanisms for TBT-degradation both in soil and in fresh water, marine and estuarine environments (Dowson et al. 1996). The half-life in aerobic environments is reported to be between 14 days and several months (Champ & Seligman 1996). However, TBT has a high soil-water partitioning coefficient and bio-concentration factor, both varying between 100 and 100000 (Fent & Looser 1995; Ohtsubo 1999) depending on type of soil and organism respectively. This implies a high tendency to adsorb to particulate matter and/or bioaccumulate. Rate of TBT-degradation and the partitioning behavior are influenced by several factors, such as the nature and density of microbial populations, TBT-solubility, dissolved and/or suspended organic matter, pH, salinity, temperature and light (Blunden & Chapman 1982; Langston & Pope 1995). Several organisms are particularly sensitive for bioaccumulation, and the effects vary from imposex (females developing male characteristics) in dog whelks and mollusks at concentrations as low as 10 ng/L, to growth inhibition and shell deformation in fish and oysters from 100 to 700 ng/L. Acute toxic effects on fish are observed at concentrations higher than 700 ng TBT^+/L.

Harbors and marinas with heavy shipping traffic are particularly susceptible to TBT-accumulation, also due to the hull cleaning water discharge from ship repair sites. Like many other major ports, the port of Antwerp faces the issue of sediment TBT-contamination and has submitted a LIFE-Environment project entitled 'Development of an integrated approach for the removal of tributyltin from waterways and harbors'. The project was granted (LIFE02 ENV/B/000341) and is carried out by two dredging companies (Envisan and DEC), a consultant (APEC) and a research laboratory (ERC). The tasks of ERC within the project are to characterize the sediment, to make an evaluation of possible TBT-remobilization when contaminated sediment is resuspended (e.g. by dredging) and finally to conduct lab-scale experiments for the remediation of contaminated sediment and water. More information on the project can be found on the website http://www.portofantwerp.be/tbtclean.

2 MATERIALS AND METHODS

2.1 *Measurement of organotin compounds*

Organotins are 'leached' from approx. 0.5 g wet sediment sample with acetic acid/ethanol (solvent). After sonication, an acetate buffer is added (pH 5.3) and derivatization is performed with sodiumtetraethylborate. Water samples (5 mL) are directly buffered and derivatized. The volatile organotin compounds are sampled in the headspace with SPME (solid phase micro-extraction) and subsequently analyzed with GC-MS (6890N GC, 5973 MS quadropole, Agilent). Accurate quantitative analysis is made possible through use of deuterated organotins as internal standards (Devos et al. 2003).

2.2 *Location of the sampling points*

The port of Antwerp is located to the north of the city of Antwerp (Belgium), at the mouth of the river Schelde. Two sediment-sampling campaigns were carried out: (1) general screening of the whole harbor area, (2) detailed screening of the immediate surroundings of Antwerp Ship Repair. Samples were taken with a Van Veen grab, sampling the top layer of the sediment.

2.3 *Lab-scale resuspension experiments*

Sediment from the port of Antwerp was resuspended in harbor water (4L) in a glass reactor vessel equipped with an electrical mixer rotating at 150 rpm. The most important harbor water characteristics were: pH 8.29, <10 ng TBT^+/L, 2693 mg NaCl/L, 40 mg total COD/L, 9 mg suspended solids (SS)/L, 897 mg volatile solids/L and 7 mg total N/L. In the experiments, a pH controller was used (Type R305, Consort nv, Turnhout, Belgium), adding 0.1 N of either HCl or NaOH via two peristaltic pumps. Sediment from the highly contaminated area around Antwerp Ship Repair was used in these experiments, characteristics that are shown in the result section Table 1. Suspended solids concentration in the tests was 1 g SS/L, unless stated otherwise.

Table 1. Characteristics of the highest contaminated sediment sample from the Antwerp ship repair site. This sediment was also used for the resuspension experiments.

Parameter	Value
pH	7.74
TOC (%)	3.13
Grain size parameters (%)	
Fraction <2 μm	33.1
Fraction <63 μm	59.9
Fraction 63–125 μm	12.6
Fraction 125–250 μm	24.8
Fraction >250 μm	2.65
General parameters (mg/kg dry weight)	
Mineral oil	1020
Metals (Σ Zn, Cu, Ni, Hg, As, Pb, Cr, Cd)	1434
PAHS (Σ 16 of EPA)	5.10
EOX (as Cl)	65
Organotin compounds (μg/kg)	
Total tributyltin (TBT)	43013
Total dibutyltin (DBT)	1370
Total monobutyltin (MBT)	870

2.4 *On-site water sampling during dredging*

In order to obtain sediment for pilot-scale bioremediation experiments, a dredging campaign was carried out by DEC and the harbor authorities. During dredging, samples of the water phase were taken at two depths (4 and 8 m) with an Eyckelkamp peristaltic pump, immediately filtered over a 0.45 μm glass fiber filter and analyzed for organotin compounds.

3 RESULTS

3.1 *Screening for TBT in the harbor sediment*

The first measuring campaign revealed that TBT was indeed present in all sediment samples, in varying concentrations (Fig. 1).

Because the highest concentration (1064 μg/kg DW) was detected close to Antwerp Ship Repair, an area where ship hull cleaning water is discharged into the harbor, a more detailed sampling campaign was performed. As expected, much higher TBT concentrations were measured in these sediment samples (detail in Fig. 1), with a maximum of 43013 μg/kg DW. The latter sample was characterized in more detail and used for experiments on TBT-remobilization.

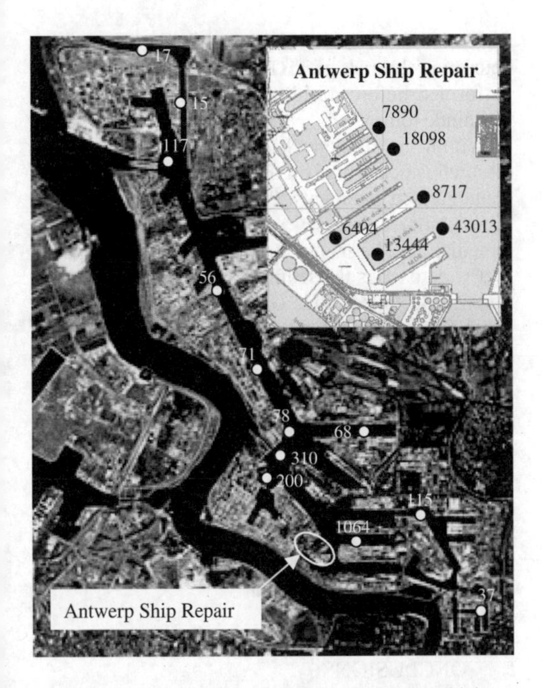

Figure 1. Air photograph of the port of Antwerp with location and TBT-concentration (μg/kg) of the sediment samples. A GIS-map of the ship repair site + sampling points is shown in more detail (copyright Gemeentelijk Havenbedrijf Antwerpen).

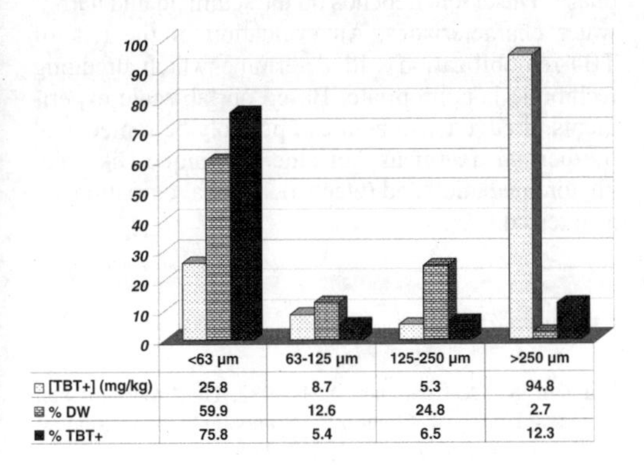

	<63 μm	63-125 μm	125-250 μm	>250 μm
□ [TBT+] (mg/kg)	25.8	8.7	5.3	94.8
▨ % DW	59.9	12.6	24.8	2.7
■ % TBT+	75.8	5.4	6.5	12.3

Figure 2. Fractionation of the highly TBT-contaminated sediment sample + TBT-distribution in the fractions.

3.2 *Characterization of highly TBT-contaminated sediment*

The sediment sample in which we detected 43013 μg TBT^+/kg DW was subsequently characterized in more detail. It was separated in four fractions by sieving (sieves of 63, 125 and 250 μm), and TBT was measured in each fraction.

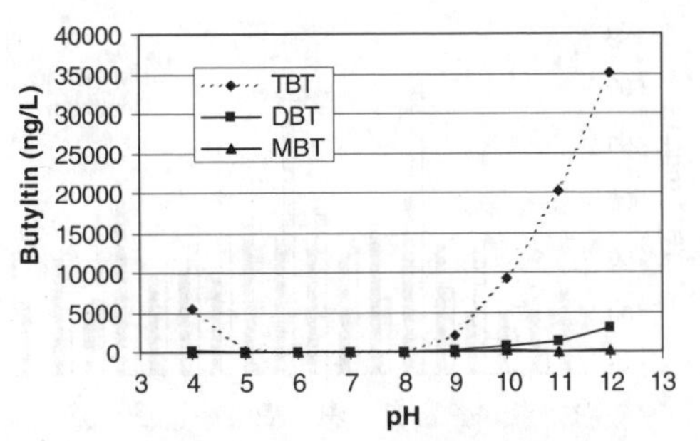

Figure 3. Butyltin (TBT, DBT and MBT) remobilization in harbor water as a function of pH.

The results of the analysis of the different fractions are shown in Figure 2. The fraction >250 μm was highly enriched in TBT, however only made up less than 3% of the dry weight. The biggest absolute amount of TBT was found in the fraction <63 μm, and no TBT-free fraction was present.

These results suggested that physical separation of a fraction by means of for example filtering or hydrocycloning was not going to be effective as a remediation strategy. Microscopic analysis of the fraction 250 μm revealed the presence of paint chippings, likely responsible for the high TBT concentration in this fraction. Within the fractions, TBT was relatively well correlated with metal concentration as well as with the organic carbon content (results not shown), confirming TBT's metal and hydrophobic features.

3.3 *Lab-scale resuspension experiments*

Highly contaminated sediment was homogeneously resuspended in 4L harbor water (1 g SS/L), and TBT release to the water phase was followed over time. Preliminary tests indicated that pH has a marked influence on TBT-remobilization. Therefore, a pH range was tested, the results of which are shown in Figure 3. Between pH 6 and 8, no TBT was desorbed from the sediment during 4 hours of mixing. Above pH 8 and below pH 6, increased amounts of TBT are remobilized.

More detailed pH experiments in the relevant range of harbor water pH (8.29) revealed that TBT desorption starts from pH 8.1. This would imply that resuspension of contaminated sediment particles in the port of Antwerp could indeed give rise to elevated TBT levels in the water column. To verify this, samples of the water column were taken during on-site dredging.

3.4 *On-site water sampling during dredging*

Dredging was carried out with a grab dredger, and caused significant sediment resuspension. Background sampling of the water column prior to dredging

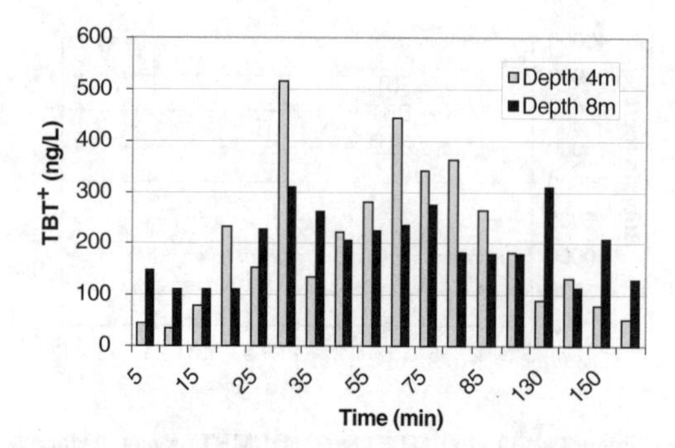

Figure 4. Tributyltin concentration in the water column at 4 and 8 m depth during (first 120 min) and after dredging.

showed that TBT was already present at concentrations lower than 50 ng/L between 4 and 10 m, but elevated concentrations were measured closer to the water surface (up to 150 ng/L). Results of the measurements during dredging are shown in Figure 4. Dredging lasted for 120 minutes, and additional samples of the water column were taken during sediment settling (for about 1 hour). Peaks of more than 500 ng TBT/L were detected at 4 m depth during dredging, but the concentration dropped to background levels once dredging had stopped.

4 DISCUSSION

Due to the adverse environmental effects of TBT and the stringent legislation for dumping TBT-contaminated sediments, the remediation of such sediments is highly relevant. In this paper, we have investigated the extent of TBT-contamination in the port of Antwerp, one of the largest European harbors. In most samples, TBT-concentration was <100 μg/kg, but nevertheless higher than the stringent Belgian legislative norm for sediment dumping (7 μg TBT/kg). Near the Antwerp ship repair site, where contaminated water is discharged in the harbor, very high TBT-concentrations of more than 43000 μg/kg were detected (Fig. 1). Characterization of such sediment samples showed that every fraction is contaminated (Fig. 2), and that physical fraction separation would not substantially reduce the TBT-load. Furthermore, dredging simulation experiments and on-site water sampling during dredging indicated that TBT desorbs from the sediment particles at relevant harbor water pH. These observed water TBT-levels were not acute toxic, but however high enough to advise the use of a dredging technique that minimizes particle resuspension. In general, the local environmental parameters determining the extent of TBT desorption (type of sediment, sediment and harbor water pH, TBT concentration, …) will differ for each harbor and preliminary (lab-scale) tests to determine the risk of TBT remobilization should be performed before deciding on the type of dredging technique to use.

Within the framework of the TBT-Clean project, research is performed on the remediation of the TBT-contaminated sediments. Preliminary results indicate a possibility to use several techniques for sediment remediation. Thermal treatment at temperatures above 170°C detoxifies the sediment by stepwise debutylation of TBT to less toxic DBT, MBT and finally inorganic Sn. Electrochemical oxidation of TBT is also shown feasible (Stichnothe et al. 2001). As an alternative, chemical oxidation with potassium permanganate is also tested, but results are not yet available. The success of this technique will mainly depend on the organic content of the sediment and related costs. Bioremediation of TBT-contaminated sediments can be a cost-effective treatment method. Lagooning and bioreactor experiments are performed and preliminary results are encouraging but need to be confirmed before they will be published.

5 CONCLUSIONS

Sediment in the port of Antwerp is contaminated with tributyltin. Locally, this contamination is very high and resuspension of such sediment particles (e.g. by dredging) was shown to remobilize TBT to the water phase. The extent depends on the sediment and harbor water characteristics. An evaluation of the risk of TBT-remobilization will determine which dredging technique is appropriate. Based on lab-scale experiments, dredged sediment can possibly be remediated by thermal treatment, but other techniques like bio/phytoremediation and (electro)chemical oxidation are also tested.

REFERENCES

Blunden, S.J. & Chapman, A.H. 1982. The Environmental Degradation of Organotin Compounds – a Review. *Environ Technol Letters* 3: 267–272.

Champ, M.A. 2000. A review of organotin regulatory strategies, pending actions, related costs and benefits. *Sci Total Environ* 258: 21–71.

Champ, M.A. & Seligman, P.F. 1996. *Organotin: Environmental Fate and Effects*. London: Chapman & Hall.

Devos, C., Vliegen, M., Moens, L. & Sandra, P. 2003. Automated headspace SPME and GC-MS in SIM-mode for the analysis of organotin compounds in water and sediment samples (in prep.).

Dowson, P.H., Bubb, J.M. & Lester, J.N. 1996. Persistence and degradation pathways of tributyltin in freshwater and estuarine sediments. *Estuarine Coastal and Shelf Science* 42: 551–562.

Fent, K. 1996. Ecotoxicology of organotin compounds. *Critical Reviews in Toxicology* 26: 1–117.
Fent, K. & Looser, P.W. 1995. Bioaccumulation and bioavailability of tributyltin chloride: influence of pH and humic acids. *Water Res* 29: 1631–1637.
Inoue, H., Takimura, O., Fuse, H., Murakami, K., Kamimura, K. & Yamaoka, K. 2000. Degradation of triphenyltin by a fluorescent pseudomonad. *Appl Environ Microbiol* 66: 3492–3498.
Langston, W.J. & Pope, N.D. 1995. Determinants of TBT Adsorption and Desorption in Estuarine Sediments. *Mar Pollut Bull* 31: 32–43.
Ohtsubo, M. 1999. Organotin compounds and their adsorption behavior on sediments. *Clay Science* 10: 519–539.
Stichnothe, H., Thoming, J. & Calmano, W. 2001. Detoxification of tributyltin contaminated sediments by an electrochemical process. *Sci Total Environ* 266: 265–271.

European Symposium on Environmental Biotechnology, ESEB 2004 - Verstraete (ed)
© 2004 Taylor & Francis Group, London, ISBN 90 5809 653 X

Functional genomic analysis of *Deinococcus radiodurans* resistance to ionizing radiation

Y. Liu & J. Zhou
Environmental Sciences Division, Oak Ridge National Laboratory, Oak Ridge, TN

ABSTRACT: *Deinococcus radiodurans* R1 is a remarkable bacterium for its extreme resistance to ionizing radiation. To define the repertoire of *D. radiodurans* genes responding to acute ionizing irradiation, transcriptome dynamics were examined in the recovering irradiated cells. The expression patterns of the majority of the induced genes resembled that of *recA* that showed up-regulation in the early phase and down-regulation in the late phase. The *RecA*-like induced genes include those involved in DNA metabolism, cell envelope formation, cellular transport, and transcription and translation process.

1 INTRODUCTION

Deinococcus radiodurans (DEIRA), an extremely radiation resistant bacterium, was discovered from the spoiled canned meat that had been radiation-sterilized, and is the most characterized member of the radiation resistant bacterial family *Deinococcaceae* that is distributed in a wide array of environments from elephant feces to the contaminated sediments beneath a radioactive waste tank at Hanford, WA. DEIRA is also remarkably resistant to a wide range of other DNA extreme conditions including exposure to desiccation, ultraviolet radiation, and oxidizing agents. DEIRA's ionizing radiation resistance is believed to be incidental, a consequence of this organism's adaptation to a long-term dehydration (Battista et al. 1999). In addition, DEIRA is vegetative, easily cultured and transformed, and has a unique gram-negative cell wall structure that shows a positive reaction to Gram's staining. It can survive both acute and chronic ionizing radiation exposures without lethality or induced mutation. Understanding the mechanisms by which DEIRA repairs DNA damages may help us in formulating new radiotherapy for cancer treatments (Rew 2003).

Whole genome sequence has revealed an unusual genetic composition of DEIRA: two chromosomes and two plasmids within a relative small genome. Comparative genomic and experimental analyses indicate that DEIRA's extreme radiation resistance phenotype is complex, likely determined collectively by an assortment of protection and DNA repair systems. Remarkably, the number of genes identified to be involved in DNA repair in DEIRA is less than that reported for *Escherichia coli* (Makarova et al. 2001). This suggests that the organism's extreme resistance phenotype may also be attributable to the functionally unknown genes and pathways. Despite these efforts, the molecular mechanisms underlying its resistance remain poorly understood. Thus, a comprehensive genome-wide examination of the genes and pathways would be helpful for a further understanding of how DEIRA responds to and recovers from irradiation. Here we report the analysis of genomic expression within cells recovering from 15 kGy using whole-genome DNA microarrays. We find that the hallmark components of DEIRA's recovery encompass differential regulation of systems involved in DNA metabolism, cell envelope formation, cellular transport, and transcription and translation processes.

2 DNA DAMAGE-SPECIFIC INDUCTION IN RESPONSE TO IRRADIATION

After acute radiation, DEIRA undergoes three typical recovery phases: early (0–3 h), middle (3–9 h), and late phase (9–24 h). In the first two phases, cell growth is inhibited; damages are being repaired, whereas in the late phase, the damages are repaired and the cell restarts to grow. For genes with statistically significant expression ratios, we found that about quarter of the genome was induced and about one to fifth was repressed during DEIRA recovery. Differential regulation in the early and mid phases of recovery illustrates that this time interval is an active period of coordinated gene expression. Within this large pool of significantly expressed

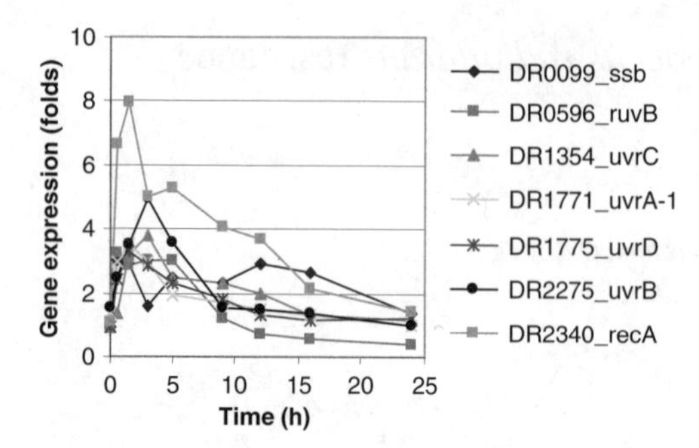

Figure 1. *RecA*-like expression patterns of seven SOS-like response genes in DEIRA.

genes, we identified a subgroup of *recA*-like expression patterns that showed an up-regulation in the early phase and down-regulation in the late phase. As DEIRA RecA is critical to genomic restoration after irradiation (Carroll et al. 1996), its induction is considered a dominant marker for the onset of an overall DNA repair.

Generally, no marked error-prone SOS response is observed in DEIRA (Moseley et al. 1983). However, there have been a few reports of possible existence of SOS response in DEIRA (Makarova et al. 2001). Among genes with *recA*-like expression patterns in DEIRA, several orthologs are known to be activated during the error-prone DNA repair (SOS) response of *E. coli* and *Bacillus subtilis*, including *recA*, *ssb*, *uvrABCD*, and *ruvB* (Fig. 1). Like other bacteria, DEIRA encodes the LexA repressor-autoprotease (DRA0344 and DRA0074), which in *E. coli* and *B. subtilis* controls the expression of the SOS regulon. However, we have not been able to identify LexA-binding sites (SOS-box) in the up-stream of the above-mentioned SOS-like response genes in DEIRA (unpublished data). Mutation of one of the *lexA* genes (DRA0344) did not increase RecA protein following gamma radiation (Narumi et al. 2001). In addition, DEIRA does not encode proteins of the DinP/UmuC DNA polymerases that play a critical role in translesion DNA synthesis and associated error-prone repair such as SOS repair in *E. coli*. These observations suggest that DEIRA may not possess a typical SOS response system that usually involves RecA-LexA interaction. Interestingly, however, mutation of RecA did diminish the SOS-like response to gamma radiation for the above-mentioned genes in DEIRA (unpublished data), suggesting that a possible novel RecA-involved mechanism regulates a specific damage-response (SOS-like), where an unknown element(s) may interact with RecA to coordinate the expression of a group of undefined genes (RecA regulon) in DEIRA. It would be interesting to determine the RecA regulon and LexA-like partner(s) that together with the RecA regulates the damaged-related expression.

DEIRA has repair pathways that include excision repair, mismatch repair, and recombinational repair. Not surprisingly, most genes involved in these repair pathways were induced to some extent following the acute radiation.

3 TRANSPORT AND REPAIR OF DAMAGED MOLECULES

Generally, the damaged DNA, if not repaired by direct damage reversal or modification, would be hydrolyzed to oligonucleiotides or nucleosides, purines, pyrimidines, and sugars, and be exported out of cell to prevent being re-integrated into newly synthesized DNA thus resulting in possible mutations. These products are found in the cytoplasm and also in the surrounding growth medium, suggesting that DEIRA exports the DNA degradation products once they are formed (Battista 1997). Not surprisingly, genes encoding enzymes like excinuclease (UvrABCD), exodeoxyribonuclease (DR0186, DR0353), extracellular nuclease (DRB0067), were up-regulated following radiation. Pyrophosphatases of the MutT/Nudix superfamily are thought to support one of the major house-cleaning systems of the cell (Fisher et al. 2002). Five genes of this family with higher specificity to deoxynucleotide triphosphates, including the MutT ortholog, were induced in the early phase. We also found early-phase activation of several genes involved in export systems including protein-export membrane protein SecG (DR1825) and MDR-type exporter (DR2098); and transporters of a major facilitator family with undefined activities. The overall gene expression profile showed that genes involved in base, nucleotide, and amino acid biosynthesis were not actively expressed although the transcription and translation machineries were implicated to be highly operative (Table 1). By contrast, many genes encoding the nucleotide, peptide, and carbohydrate transporters, and ribonucleoside-diphosphate reductase (DRB0107, DRB0108, DRB0109) were induced in the early recovery phase. Therefore, we hypothesize that during the initial phase of cellular cleansing, amino acids, nucleotides, nucleosides, sugars, and phosphate may be imported into the cell as precursors for both nucleic acid synthesis by ribonucleoside-diphosphate reductase and protein synthesis.

Furthermore, the ComEA system, which has been identified through its role in DNA transformation competence, might be involved in the export of damaged DNA in irradiated DEIRA cells. Two genes of this system were induced: DR0361 (metallobetalactamase family enzyme ComE) and DR0207 (secreted protein ComEA) that is among the top 10 most strongly induced genes. However, the role of these genes in the radiation resistance of DEIRA is unclear.

Table 1. Summary of significantly up- and down-regulated genes of cells at 3 h post irradiation in different function categories.

Function Group*	Total number	UP** (%)	Down (%)
Cell envelope	75	40.00	16.00
Transcription	26	38.46	19.23
Transport and ATP-binding proteins	185	31.89	11.89
Protein synthesis	111	30.63	22.52
DNA metabolism	79	26.58	18.99
Hypothetical proteins	950	25.26	18.21
Regulatory functions	122	22.95	17.21
Unknown function	170	22.94	17.06
Energy metabolism	196	21.43	26.53
Fatty acid and phospholipid	52	21.15	21.15
Central intermediary metabolism	150	20.00	21.33
Cofactors, prosthetic groups, and carriers	60	20.00	26.27
Conserved hypothetical proteins	487	18.27	21.35
Bases and nucleotides	51	17.65	19.61
Cellular process	88	17.05	29.55
Amino acid biosynthesis	79	16.46	31.65
Protein fate	83	15.66	39.76
Total or average	2972	23.90	22.28

* According to TIGR's annotation. ** Although total 3003 genes were arrayed, only 2972 data points whose average intensity value was over the reliability criteria, i.e. 200, have been analyzed.

4 POSSIBLE INVOLVEMENT OF CELL ENVELOPE IN RADIATION RESISTANCE

Thirty of 75 (40%) DEIRA genes implicated in cell envelope formation including integrated membrane proteins showed a *recA*-like expression pattern (Table 1). Two genes encoding an NlpA like lipoprotein are among the top 10 most up-regulated candidates. This extraordinary expression profile attracted our attention because the cell envelope of DEIRA has an unusual feature: stained like gram-positive, but the actual structure is more like gram-negative. A very thick complex cell wall separates the plasma and outer membranes. This 14- to 20 nm cell wall is composed of a peptidoglycan layer and numerous fine compartments. The chemical structure of the cell wall and membrane of DEIRA are considered unique to DEIRA (Makarova et al. 2001). Interestingly, the genome-wide proteome following radiation has shown an enhanced production of cell envelope proteins (Lipton et al. 2002).

Does this unique cell envelope structure play a role in protection from ionizing radiation-caused damage? The answer is probably not, because the cell suffers a large number of double-strand breaks of DNA molecules once acutely irradiated (Battista et al. 1999). Rather, this organism rapidly and accurately repairs DNA damage. However, we argue that the cell envelope may still play a role in this rapid and accurate repair machinery because many sensory proteins embedded in the membranes percept external and internal changes, thus signaling cellular responses including damage repair. The dose-dependent delay of the onset of cellular replication suggests the existence of a checkpoint where a sensory protein(s) monitors the extent of repair and signal the initiation of replicative DNA synthesis (Fig. 2). Damage-related signal transduction pathways are still mysterious in prokaryotes. The initial signal must be damage-related, even the damage itself (Fig. 2). The questions of how this signal is perceived and thereafter transduced, what the sensory proteins and other signaling elements involved, and where they locate, have not been addressed yet. Two recent papers (Hua et al. 2003, Earl et al. 2002) reported an undefined signaling element (DR0167) named PprI or IrrE, that locates upstream of RecA in the signal transduction pathway because mutation of this gene diminished RecA expression leading to a severe radiation sensitive phenotype like *recA* mutants. Also, many functional structure proteins involved in transport, respiration, and even damage protection enrich in the membrane.

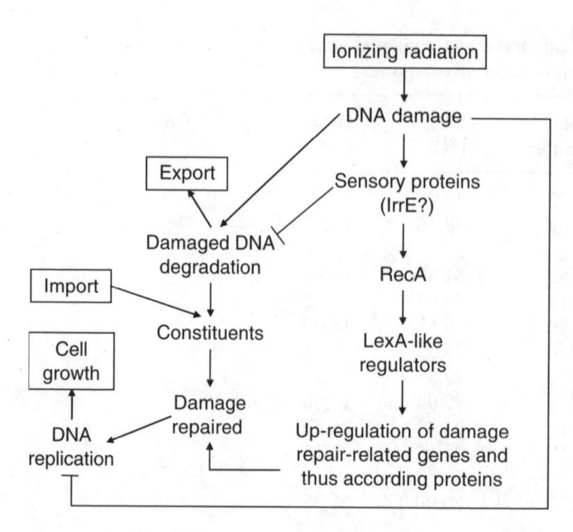

Figure 2. The regulatory network of DEIRA response to ionizing radiation where the radiation-caused damages are perceived by an unidentified sensory protein(s) that signals downstream to activate and over-express RecA cascading a wide array of damage-related reactions leading to cell recovery from the damage and restart to grow.

For instance, one of the three catalases in *Deinococcus radiophilus* is a membrane-associated constitutive enzyme (Yun et al. 2000). On the other hand, the specialized cell wall characterized as compartmentalized layers may, in our imagination, facilitate damage repair in the way that the exported damaged macromolecules are arrested, sorted, and even further processed so that reusable constituents would be imported back to the cell for the syntheses of proteins, RNAs, and DNAs, or the unusable parts be repelled outside otherwise.

5 PROTECTIVE MECHANISMS AGAINST OXIDATIVE DAMAGE

Previous evidences support that DEIRA protective mechanisms against oxidative damage may also be involved in its extreme radiation resistance. The lethal effect of ionizing radiation is known to be oxygen enhanced by generating reactive oxygen species (ROS), which damage cellular macromolecules. DEIRA shows characteristic reddish color due to the presence of carotenoids whose major components have been identified as deinoxanthin. Since these carotenoids act as effective antioxidants *in vitro*, the resistance of this bacterium to ionizing radiation appears to be partly attributed to the carotenoids. Consistent with the *in vitro* results, a mutant of DEIRA deficient in the carotenoids shows higher sensitivities than the wild type strain against ionizing radiation and hydrogen peroxide (Carbonneau et al. 1989). These carotenoids are destroyed by gamma radiation so that the irradiated cell becomes discolored. The cell survived the radiation gradually regains the red color during the recovery phases. This process well agrees with the discovery that the pigment gene (DR1790) expression and protein production were gradually recovered following acute ionizing radiation (Lipton et al. 2002).

DEIRA also produces relatively high levels of two kinds of superoxide dismutase (SodAC) encoded by three genes (DR1279, DR1546, and DRA0202), two kinds of catalase (KatAE) encoded by DR1998, DRA0146, and DRA0259 (Markillie et al. 1999). The activities of DEIRA catalase and superoxide dismutase are induced by oxidative stress. The high levels of catalase and superoxide dismutase produced by *Deinococcaceae* species were, in part, regulated by growth phase. A gradual increase in total catalase superoxide dismutase activity occurred during exponential and stationary phase, which quantitatively correlates with the amounts ROS generated by aerobic respiration. Mutants in either catalase (KatA) or/ and superoxide dismutase (SodA) were more sensitive to ionizing radiation than the wild type, suggesting that both catalase and superoxide dismutase play an important role in cell resistance to ionizing radiation (Markillie et al. 1999). The expression of all annotated *kat* and *sod* genes were immediately suppressed following ionizing radiation, then either gradually up-regulated like a 'cell growth-related' pattern, or quickly peaked at 3 h post irradiation then followed by a gradual decline to the normal levels as a delayed '*recA*-like' pattern. Apparently, as with that of pigment gene (DR1790), the down-regulation of these oxygen free radical destroyers immediately following ionizing radiation is possibly regulated in a cell growth-related manner, whereas the delayed '*recA*-like' response is, at least in part, a response to the accumulation of ROS. Interestingly, we found that the genes encoding the tricarboxylic acid (TCA) cycle were repressed in the early and mid phases, whereas genes encoding the glyoxylate shunt were activated in this interval so that the total NADH generated by the cycle would be reduced leading to less production of ROS by the oxidative phosphorylation. These results suggest that metabolically induced oxidative stress is repressed during the recovery.

6 CONCLUSION

The present analysis of the global gene expression of DEIRA in response to acute ionizing irradiation revealed a complex transcription regulatory network in which RecA together with other yet known proteins plays an essential role in damage repair that also involves cell envelope formation, cellular transport, and transcription and translation process.

REFERENCES

Battista, J.R. 1997. Against all odds: the survival strategies of *Deinococcus radiodurans*. *Annu Rev Microbiol.* 51: 203–24.

Battista, J.R., Earl, A.M. & Park, M.J. 1999. Why is *Deinococcus radiodurans* so resistant to ionizing radiation? *Trends Microbiol.* 7: 362–5.

Carbonneau, M.A., Melin, A.M., Perromat, A. & Clerc, M. 1989. The action of free radicals on *Deinococcus radiodurans* carotenoids. *Arch Biochem Biophys.* 275:244–51.

Carroll, J.D., Daly, M.J. & Minton, K.W. 1996. Expression of recA in *Deinococcus radiodurans*. J Bacteriol. 178: 130–5.

Earl, A.M., Mohundro, M.M., Mian, I.S. & Battista, J.R. 2002. The IrrE protein of *Deinococcus radiodurans* R1 is a novel regulator of recA expression. *J Bacteriol.* 184: 6216–24.

Fisher, D.I., Safrany, S.T., Strike, P., McLennan, A.G., Cartwright, J.L. 2002. Nudix hydrolases that degrade dinucleoside and diphosphoinositol polyphosphates also have 5-phosphoribosyl 1-pyrophosphate (PRPP) pyrophosphatase activity that generates the glycolytic activator ribose 1,5-bisphosphate. *J Biol Chem.* 277: 47313–7.

Hua, Y., Narumi, I., Gao, G., Tian, B., Satoh, K., Kitayama, S. & Shen, B. 2003. PprI: a general switch responsible for extreme radioresistance of *Deinococcus radiodurans*. *Biochem Biophys Res Commun.* 306: 354–60.

Lipton, M.S., Pasa-Tolic' L., Anderson, G.A., Anderson, D.J., Auberry, D.L., Battista, J.R., Daly, M.J., Fredrickson, J., Hixson, K.K., Kostandarithes, H., Masselon, C., Markillie, L.M., Moore, R.J., Romine, M.F., Shen, Y., Stritmatter, E., Tolic' N., Udseth, H.R., Venkateswaran, A., Wong, K.K., Zhao, R. & Smith, R.D. 2002. Global analysis of the *Deinococcus radiodurans* proteome by using accurate mass tags. *Proc Natl Acad Sci USA.* 99: 11049–54.

Makarova, K.S., Aravind, L., Wolf, Y.I., Tatusov, R.L., Minton, K.W., Koonin, E.V. & Daly, M.J. 2001. Genome of the extremely radiation-resistant bacterium *Deinococcus radiodurans* viewed from the perspective of comparative genomics. *Microbiol Mol Biol Rev.* 65: 44–79.

Markillie, L.M., Varnum, S.M., Hradecky, P. & Wong, K.K. 1999. Targeted mutagenesis by duplication insertion in the radioresistant bacterium *Deinococcus radiodurans*: radiation sensitivities of catalase (katA) and superoxide dismutase (sodA) mutants. *J Bacteriol.* 181: 666–9.

Moseley, B.E. & Evans, D.M. 1983. Isolation and properties of strains of *Micrococcus (Deinococcus) radiodurans* unable to excise ultraviolet light-induced pyrimidine dimers from DNA: evidence for two excision pathways. *J Gen Microbiol.* 129: 2437–45.

Narumi, I., Satoh, K., Kikuchi, M., Funayama, T., Yanagisawa, T., Kobayashi, Y., Watanabe, H. & Yamamoto, K. 2001. The LexA protein from *Deinococcus radiodurans* is not involved in RecA induction following gamma irradiation. *J Bacteriol.* 183: 6951–6.

Rew, D.A. 2003. *Deinococcus radiodurans*. *Eur J Surg Oncol.* 29: 557–8.

Yun, E.J. & Lee, Y.N. 2000. Production of two different catalase-peroxidases by *Deinococcus radiophilus*. *FEMS Microbiol Lett.* 184: 155–9.

European Symposium on Environmental Biotechnology, ESEB 2004 - Verstraete (ed)

Metabolic products of PCBs in bacteria and plants – comparison of their toxicity and genotoxicity

P. Lovecká, M. Macková & K. Demnerová
Institute of Chemical Technology, Department of Biochemistry and Microbiology, Prague, Czech Republic

ABSTRACT: Different systems for the measurement of ecotoxicity have been used for the estimation of the toxicity of intermediates of bacterial and plant PCB metabolism in comparison with primary polychlorinated biphenyls. Luminescent bacteria (*Vibrio fischeri*) and mammalian cells (keratinocytes) showed similar response reactions to the different toxicants examined. Chlorobenzoic acids (intermediates of bacterial PCB metabolism) exhibited the lowest toxicity. Products of plant metabolism (hydroxychlorobiphenyls) were the most toxic compounds; their toxicity exceeded that one of initial individual monochlorobiphenyls. Other method (Bioscreen and germination of seeds) following growth of bacteria of different species in presence of the toxicants, showed a different response of the selected model organisms based on their abilities to degrade PCBs. Ames test with *Salmonella typhimurium* was used for genotoxicity measurements. Chlorobenzoic acids exhibited lower toxicity then initial PCB, but in some cases still significant.

1 INTRODUCTION

Microbial and plant species may possess enzymes capable of metabolizing certain environmentally persistent xenobiotics that contaminate soil and water. It has been shown that some bacteria metabolize different organic molecules including highly persistent polychlorinated biphenyls (Abramowicz et al., 1990, Pazlarová et al., 1997, Nováková et al., 2002). Some plant species have been described to have also ability to metabolise PCBs (Wilken et al., 1995, Macková et al., 1997) and polyaromatic hydrocarbons (Kucerová et al., 2001). Generally, plant metabolism has been studied to a lesser extent than that in bacteria or mammals and little information is available. Not much is known about the intermediates in plants, their toxicity and the effect which such compounds may have on animals and other organisms (Macek et al., 2002). In plants, organic compounds are in the first phase transformed into more reactive ones, then conjugated with sugars, amino acids, etc. to less phytotoxic ones and deposited in vacuoles or lignified parts of the cell wall. Unfortunately this fact does not mean that metabolites or products have lower toxicity towards other living systems.

Different systems for ecotoxicity measurement were evaluated during last 15 years. Many of them are based on measurement of viability of different organisms and their ability to survive in presence of different toxicants. In our experiments we studied metabolism of PCBs in bacteria and plants (Pazlarová et al., 1997). These organisms are mainly involved in transformation of toxic compounds in nature and they are responsible for further fate of those xenobiotics and their intermediates in the environment. We use bacterial strains and several plant species that are able to transform PCBs. The decrease of the content of PCB congeners after biotransformation is detected by gas chromatography with EC detection. Toxicity of identified bacterial and plant products was studied using plant cells, microbial and mammalian cell system (e.g. Chu et al., 1997).

2 MATERIALS AND METHODS

2.1 *Polychlorinated biphenyls, products of bacterial and plant metabolism*

Commercially available individual congeners of monochlorinated PCBs were used as models for toxicity studies. Bacterial and plant products and intermediates of PCB metabolism were chosen according to previous results (see Table 1) (Kucerová et al., 2000).

2.2 *Toxicity assay using luminescent bacteria*

A working suspension of luminescent bacteria was prepared by reconstituting a vial of lyophilized cells

Table 1. Compounds used for the toxicity measurement.

PCB 1	2-chlorobiphenyl
PCB 2	3-chlorobiphenyl
PCB 3	4-chlorobiphenyl
RPM 3	3-chloro-4-hydroxybiphenyl
RPM 4	4-chloro-4-hydroxybiphenyl
RPM 6	2-chloro-5-hydroxybiphenyl
3 CBA	3-chlorobenzoic acid
2,3 di CBA	2,3-chlorobenzoic acid
2,5 di CBA	2,5-chlorobenzoic acid

of *Vibrio fisheri,* using 0.5 ml of 2% NaCl aqueous solution at 2–5°C. The bacterial suspension was added to 0.5 ml dilution series of toxicant (chlorobenzoic acids, congeners of PCB, hydroxychlorobiphenyls) in 2% NaCl. Luminescence was measured after 15 minutes of incubation. The results are expressed as EC50, the higher EC50 is, the less toxic is the tested toxicant.

2.3 *Toxicity assay using the Bioscreen test*

This test allows evaluation of the effect of various concentrations of toxicants on bacterial viability and growth in comparison with the controls cultivated without toxicants. Values of $OD_{400\,nm}$ and the time dependence of growth are monitored in parallel samples incubated with or without toxicants for 2 days. The effect of three different groups of concentrations of toxicants (PCB, chlorobenzoic acids and hydroxychlorobiphenyls) at three concentrations (10 mg/l, 20 mg/l and 40 mg/l) was followed using three different bacterial strains (*Pseudomonas sp.* R9 – isolated from the soil contaminated with polyaromatic hydrocarbons, *Pseudomonas sp.* P2 – isolated from the soil contaminated with PCB and *E. coli* from the Collection of the Department of Biochemistry and Microbiology). Minimal medium containing glycerol (5 g/l) as basic medium and toxicants as described above were used for the incubation. As control the growth of bacteria in minimal medium was monitored.

2.4 *Toxicity assay using germinating seeds*

This method is based on measuring of the germinating root length of *Lactuca sativa* in presence of toxicants. The temperature of incubation is 22°C without light for 4 days and referent medium consists of 18.5 g/l $CaCl.2H_2O$, 2.3 g/l KCl, 49.3 g/l $MgSO_4.7H_2O$, 25.9 g/l $NaHCO_3$ (2.5 ml for 1000 ml H_2O). Results are expressed as the coefficient of inhibition I (%), see Equation (1)

$$I = (Lc - Ls)/Lc * 100 \qquad (1)$$

where Lc = length of root of control
Ls = length of root of sample.
The sample is interpreted to be toxic, when the I - value is greater than 30%.

2.5 *Measurement of genotoxicity using Ames test*

The Ames test is used world-wide as an initial screen to determine the mutagenic potential of new chemicals and drugs (Mortelmans et al., 2000). First of all the toxicity of the substances to *Salmonella* strains was tested. The Ames test has a range of specific modifications and enables detection of a wide variety of mutagens. The detection system using *Salmonella typhimurium* His differs in mutations within histidine operon, compared with the original strain. Mutations in the histidine operon are induced by mutagens leading to reversion to protothrophy, e.g. the ability to synthesize histidine. To enhance the sensitivity of indicator strains differing in mutation type with the tested substance, markers are added. Finally, the results were compared using the parameter Rt/Rc, where Rt represents the total number of revertants of a particular concentration of a tested substance, where Rc is a total number of spontaneous revertants on control plates. The sample is mutagenic when Rt/Rc is higher than 2.

3 RESULTS AND DISCUSSION

3.1 *Measurement of the toxicity*

The main products of bacterial degradation of PCB are chlorobenzoic acids, which can be further degraded by other bacteria present in contaminated environment.

Analysis of the products of bacterial and plant metabolism has shown that plant products (hydroxychlorobiphenyls) of the first phase of PCB transformation are similar to those detected in mammalian cells. While some bacteria further degrade PCBs to chlorobenzoic acids, plants are generally unable to open the ring and degrade the chemical structure of biphenyl. To better understand the fate of PCBs (and all other toxicants) in the environment, the cooperation of plants and microorganisms living in the rhizosphere in contaminated soil should be evaluated (Macek et al., 2002). The ability of plant enzymes to further metabolise the products of bacterial PCB degradation (the chlorobenzoic acids) has been followed in our laboratories as well as the ability of bacterial enzymes to transform the plant-formed hydroxychlorobiphenyls (Francova et al., 2003, 2004).

Commercial PCBs were produced as mixtures of different congeners varying in their degree of chlorination. The simplest PCBs are monochlorobiphenyls which are usually not present in commercial mixtures, but they can be formed in the natural environment by microbial dechlorination or degradation of more highly chlorinated congeners. We used monochlorobiphenyls and their products identified in bacterial and/or plant cells as models for the toxicity measurements of compounds which can exist naturally in contaminated soils and increase the toxicity of polluted areas. As was previously described bacterial and plant intermediates of PCBs are structurally different. Toxicity of bacterial (chlorobenzoic acids) and plant intermediates (hydroxychlorobiphenyls) of PCB transformation compared to the toxicity of initial monochlorobiphenyls measured using luminescent bacteria is shown in Figure1.

From the results it can be concluded that hydroxychlorobiphenyls as primary products of PCB metabolism in plants are the most toxic (i.e. the lowest LD50), comparing to the other groups of tested compounds. Hydroxychlorobiphenyls are more soluble in water than the original PCB and thus are more available and toxic for living organisms. A similar response was previously obtained with mammalian cells (keratinocytes), which also showed the highest sensitivity to hydroxychlorobiphenyls. Comparing both systems for the ecotoxicity measurement luminescent, we found that bacteria are susceptible to lower concentrations of tested compounds. Toxicity of the same compounds was tested using an independent method – namely Bioscreen measurement. The Bioscreen test showed a different response of the selected model organisms. *Pseudomonas sp.* R9, which is not able to degrade PCBs, exhibited the strongest response to individual congeners of polychlorinated biphenyls, while hydroxychlorobiphenyls and chlorobenzoic acids were less toxic. The opposite effect was followed with *Pseudomonas sp.* P2, which is able to degrade PCBs and thus it was not susceptible to PCBs but to products of PCB degradation – chlorobenzoic acids. *E. coli* exhibited a similar response to all three groups of tested compounds, see Figure 2.

Results of measuring with seeds of *Lactuca sativa* are shown in Figure 3. Monochlorobiphenyls are not toxic for plant system, but chlorobenzoic acids and hydroxychlorobiphenyls are more toxic then PCB. It is probably due to their better solubility in water.

3.2 *Measurement of mutagenicity by Ames test*

Due to the toxicity of the tested substances for our tested model microorganisms it was necessary to establish an appropriate concentration, which did not kill the bacterium itself. The lowest concentration showing any mutagenic effect was chosen and a concentration gradient was prepared. Genotoxic effect was proved in case of chlorobenzoic acids. Original PCBs and hydroxychlorobiphenyls showed less genotoxicity than chlorobenzoic acids, only with TA 100 cells weak genotoxic effect was measured.

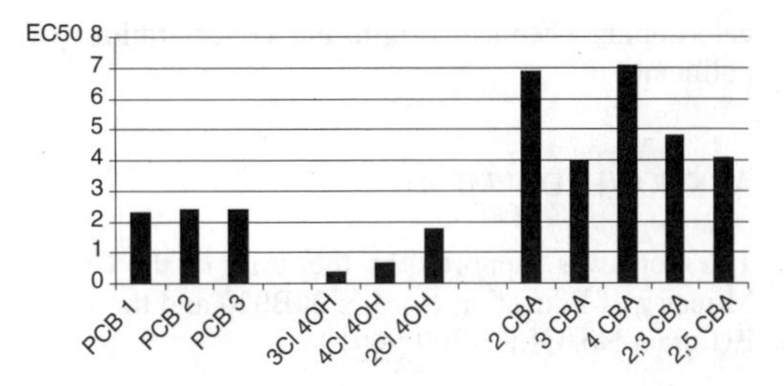

Figure 1. Toxicity of monochlorobiphenyls and products of bacterial and plant metabolism measured by luminescent bacteria *Vibrio fischeri*.

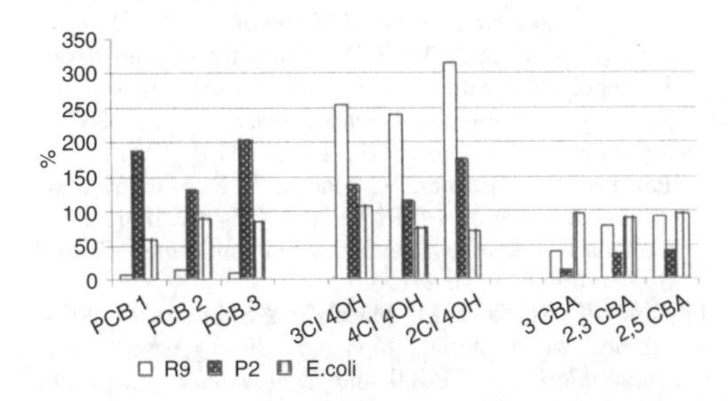

Figure 2. Evaluation of specific growth rates of three bacterial strains incubated with PCBs, chlorobenzoic acids and hydroxychlorobiphenyls at a concentration of 10 mg/l.

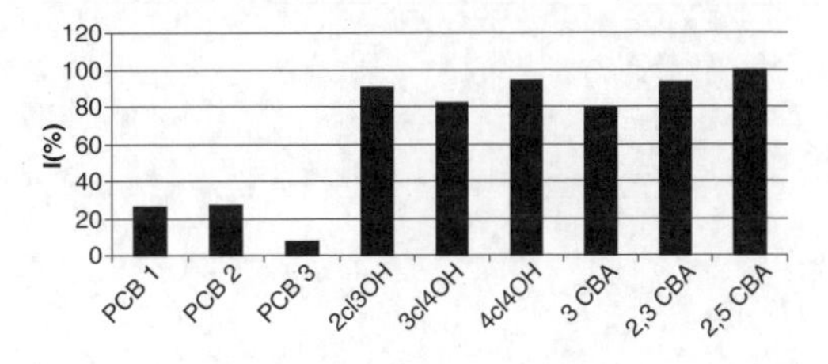

Figure 3. Toxicity of monochlorobiphenyls and products of bacterial and plant metabolism measured by seeds of *Lactuca sativa*.

4 CONCLUSIONS

In our study we showed different responses of organisms selected at two trophic levels to the presence of pollutants. Namely we tested the toxicity of some PCBs and their bacterial and plant-formed products. The data showed that the phenomenon which could appear with particular low soluble contaminants transformed by any living system unfortunately can not be generalised to properties of all contaminants polluting the environment. In each case basic data documenting behaviour of pollutants and their products should be evaluated using proper system for ecotoxicity analysis and compared with analytical results determining chemical origin and concentrations of pollutants.

ACKNOWLEDGEMENT

The work was supported by the grant of the Czech Ministry of Education No. LN 00B030 and the 5FW EU grant No. QLK3-2001-00101.

REFERENCES

Abramowicz, M. 1990. Aerobic and anaerobic biodegradation of PCBs: a review. *Crit. Rev. Biotechnol.* 10: 241–245.

Chu, S., He, Y. & Xu, X. 1997. Determination of acute toxicity of polychlorinated biphenyls to *Photobacterium phosphoreum. Bull. Environ. Contam. Toxicol.* 58: 263–267.

Demnerová, K., Stiborova, H., Leigh, M.B., Pieper, D., Pazlarová, J., Brenner, V., Macek, T. & Macková, M. 2003. Degradation of PCBs and CBs by indigenous bacteria isolated from contaminated soil. *Water Air and Soil Pollution.* 3 (3): 47–55.

Francova, K., Macková, M., Macek, T. & Sylvestre, M. 2004. Ability of bacterial biphenyl dioxygenases from *Burkholderia sp.* LB400 and *Comamonas testosteroni* B-356 to catalyse oxygenation of ortho-hydroxybiphenyls formed from PCBs by plants. *Environmental Pollution.* 127 (1): 41–48.

Kucerová, P., Macková, M., Chromá, L., Burkhard, J., Demnerová, K. & Macek, T. 2000. Metabolism of polychlorinated biphenyls by *Solanum nigrum* hairy root clone SNC-9O and analysis of transformation products. *Plant and Soil.* 225: 109–115.

Kucerová, P., in der Wiesche, C., Wolter, M., Macek, T., Zadrazil, F. & Macková, M. 2001. The ability of different plant species to remove polycyclic aromatic hydrocarbons and polychlorinated biphenyls from incubation media. *Biotechnology Letters.* 23: 1355–1359.

Macek, T., Macková, M., Kucerová, P., Chromá, L., Burkhard, J. & Demnerová, K. 2002. Phytoremediation. In: *Focus on Biotechnology*, (Hofman M. and Anne J., series eds.), Vol. 3, Kluwer Academic Publishers, Dordrecht, pp. 115–137.

Mortelmans, K. & Zeiger, E. 2000. The Ames *Salmonella*/ mikrosome mutagenicity assay. *Mutation Research.* 455: 29–60.

Nováková, H., Vosahlíková, M., Pazlarová, J., Macková, M., Burkhard, J. & Demnerová, K. 2002. Degradation of PCBs by *Pseudomonas* P2. *International Biodegradation and Biodeterioration.* 50 (1), 47–54.

Pazlarová, J., Demnerová, K., Macková, M. & Burkhard, J. 1997. Analysis of PCB-degrading bacteria: physiological aspects. *Lett. Appl. Microbiol.* 24: 334–336.

Wilken, A., Bock, M., Bokern, M. & Harms, H. 1995. Metabolism of different PCB congeners in plant cell culture. *Environ. Chem. Toxicol.* 17: 2017–2022.

European Symposium on Environmental Biotechnology, ESEB 2004 - Verstraete (ed)
 ISBN 90 5809 653 X

Evaluation of alkane biodegradation potential of environmental samples by competitive PCR

S. Heiss-Blanquet, S. Rochette & F. Monot
Institut Français du Pétrole, Département de Biotechnologie et Chimie de la Biomasse, Rueil-Malmaison, France

ABSTRACT: A molecular method for evaluation of the biodegradation potential of microflorae of polluted sites was tested focusing on the detection of the *alkB* gene, encoding alkane hydroxylase which catalyses the hydroxylation of *n*-alkanes. *AlkB*-related genes are found in many bacterial genera, and often have reduced sequence similarities. Therefore, four different primer pairs were designed, based on classification of available partial sequences of alkane hydroxylase genes. These primers allowed amplification of *alkB*-related genes in 14 out of 15 tested alkane-degrading strains or microcosms. In addition, competitive PER experiments showed that *alkB* genes could be correctly quantified in the tested strains. *AlkB* genes were also quantified in pristine or contaminated water samples and compared to the corresponding *n*-heptane degradation rate. Results show that *alkB* gene copy numbers follow the same tendency as degradation rates and suggest the suitability of this method for a rapid evaluation of hydrocarbon degradation capacities of environmental samples.

1 INTRODUCTION

Contamination by petroleum hydrocarbons often occurs in aquatic or terrestrial environments. Depending on the type of hydrocarbons present in the pollution source, i.e. long or short chain alkanes, aromatics, alicyclic or polyaromatic hydrocarbons, degradation by indigenous microorganisms will be more or less rapid and complete. For a better understanding and prediction of the biodegradation processes either by natural attenuation or enhanced by biostimulation or bioaugmentation, it is crucial to study the microbial activity at the contaminated site. This can be achieved by cultivation-based techniques, such as monitoring respiration, assessing enzyme activities, or by measuring the contaminant concentration (Knudsen, 2002; Morra, 2002). Molecular approaches, such as studying the microbial community by DGGE or RFLP of PCR amplified 16S rDNA fragments, can give information about composition changes within the population. However, these latter methods do not provide functional characteristics of the microcosm. In contrast, utilization of gene probes or primers specific for catabolic genes allow detection of organisms possessing relevant degradation pathways. This information could be a key element for the evaluation of remediation processes.

In the present work, we explored the possibility of evaluation of degradation capacities by detection of functional genes, using the *alkB* gene as a model.

This enzyme encodes the non-heme alkane hydroxylase, and it is a key enzyme found in bacterial alkane degradation pathways. As part of the alkane monooxygenase complex, consisting of the electron transfer peptides rubredoxin and rubredoxin reductase and the alkane hydroxylase itself, it catalyses the initial oxidation of alkanes to the corresponding alcohol. The *alkB* gene has first been characterized in *Pseudomonas putida* GPo1, where it is located on the OCT plasmid as part of the *alkBFGHJKL* operon (Kok et al., 1989; van Beilen et al., 1994). It has subsequently been detected in a large variety of gram-positive and gram-negative strains (Smits et al. 1999; Vomberg & Klinner 2000; van Beilen et al., 2002). Comparison of these alkane hydroxylase sequences shows reduced similarity, sometimes less than 40% at the amino acid level. For this reason, highly degenerated primers have been used for detection of *alkB* genes (Smits et al., 1999). In some cases, however, these primers led to unspecific amplifications or yielded no amplification product at all, especially in gram-positive strains (Smits et al., 1999; Heiss-Blanquet, unpubl.).

In order to improve detection of *alkB* genes in isolated strains and microcosms, we developed less degenerated primers based on groups of *alkB* genes with increased sequence similarity. Two of the four primer pairs targeting the major groups of alkane hydroxylases were then used for quantitative PCR experiments. After validation with model strains, *alkB* genes were quantified in water samples collected at a site contaminated with C7 to C15 hydrocarbons. The number of genes of each sample was compared to its degradation activity.

2 RESULTS

Several bacterial strains originating from gasoline polluted soils and degrading *n*-heptane or 2-methyl-hexane were isolated in our lab. For their genetic characterization, attempts to amplify the *alkB* gene with highly degenerated primers (Smits et al. 1999) were undertaken. In only four of eight strains, an amplification product could be obtained. In order to detect *alkB* genes more efficiently, more specific primers were designed, based on sequence similarities between distinct groups of *alkB* genes. Indeed, when aligning known *alkB* gene sequences four groups of more closely related genes can be observed (Figure 1).

The largest group in this dendrogram consists of *alkB* sequences from gram-positive strains, named "*Rhodococcus*" group. Another branch regroups *alkB* sequences from *Acinetobacter* strains, and two groups contain sequences from *Pseudomonas* and other gram-negative strains. It is interesting to note that strains in *Pseudomonas* 1 group can degrade *n*-alkanes from C6 to C12, whereas *Pseudomonas* aeruginosa strains from group *Pseudomonas* 2 degrade *n*-alkanes ranging from C12 to C16. Alkane hydroxylases from two *P. fluorescens* and *Burkholderia* strains which can grow on C12–C28 (*P. fluorescens* CHA0) and on C10–C16 *n*-alkanes, respectively, are quite closely related to the *P. aeruginosa alkB* sequences, but occupy an intermediate position between these and alkane hydroxylases from gram-positive strains (van Beilen et al., 2003).

Primers with a maximum degenerescence degree of 16 were designed for each group, and PCR amplification products could be obtained for all eight strains tested (Table 1). In addition, primer pairs are specific for the groups they were designed for, since only one primer pair lead to amplification products in each strain. *AlkB* fragments could also be amplified from six out of seven *n*-heptane grown microcosms with at least one of the primer pairs (not shown).

As the designed primers seem to efficiently amplify *alkB* genes, their use for quantitative PCR was tested. For these experiments, we focused on the *Rhodococcus* and *Pseudomonas* 1 primer pairs, since *alkB* genes belonging to the *Pseudomonas* 2 group should not be relevant for the *n*-heptane grown cultures which were used here (see above). As for *alkB* genes from *Acinetobacter* strains, several studies showed their limited role in alkane-degradation in different ecosystems (Whyte et al., 2002; Sei et al., 2003). These results were confirmed by the observation that amplification of *alkB* genes with the *Acinetobacter* primers was obtained in only one of the six populations that responded positively in the PCR experiments (Heiss-Blanquet, unpubl.).

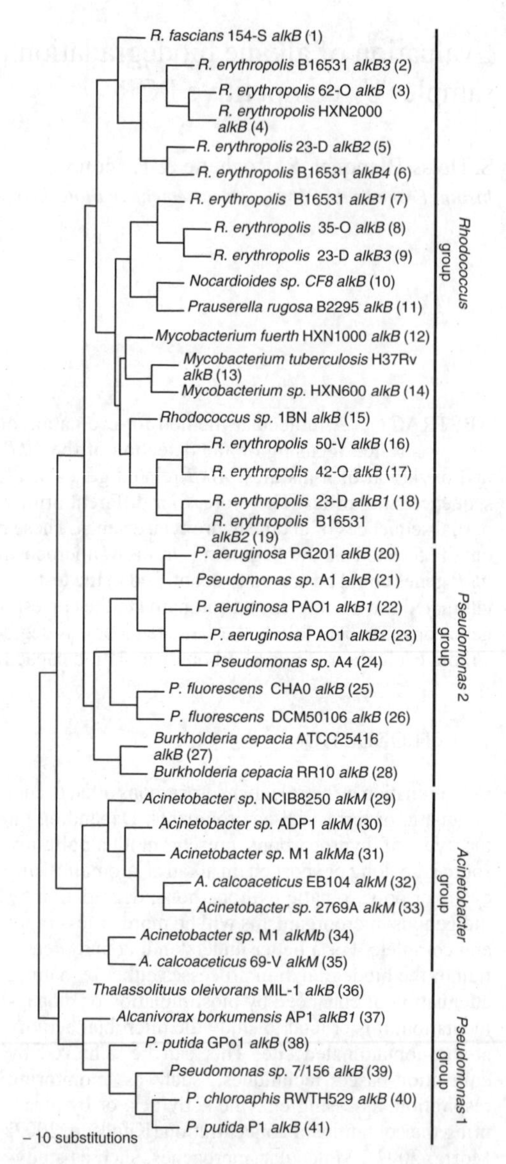

Figure 1. Dendrogram of partial *alkB* nucleic acid sequences. Alignment was done with the Clustal program and the tree was established with the Neighbour-joining and Jukes-Cantor algorithms. Accession numbers: (1) aj301873 (2) aj301876 (3) aj301874 (4) aj301875 (5) 301869 (6) aj301877 (7) aj009586 (8) aj301871 (9) aj301870 (10) af350429 (11) aj009587 (12) aj300339 (13) z95121 (14) aj300338 (15) aj401611 (16) aj301866 (17) aj301867 (18) aj301868 (19) aj297269 (20) ae004581 (21) aj311787 (22) aj009580 (23) aj009581 (24) aj311786 (25) aj009579 (26) af090329 (27) aj293344 (28) aj293306 (29) aj009584 (30) aj002316 (31) ab049410 (32) aj233398 (33) aj009585 (34) ab049411 (35) aj009582 (36) aj431700 (37) aj295164 (38) aj245436 (39) ay034587 (40) aj250560 (41) aj233397.

Competitor DNA fragments for quantitative PCR were constructed as described by Jin et al. (1994) and standard curves with serial dilutions of pCR2.1 TOPO plasmids containing *alkB* fragments from the *Rhodococcus* and the *Pseudomonas* 1 group were established. The results showed that amplification efficiency was the same for competitors and *alkB* gene fragments and that quantification was linear for at least 3 orders of magnitude.

For further validation of the method, quantification of *alkB* genes was checked using *P. putida* GPo1, *P. putida* SH41, *Gordonia sp.* SG11 and *Rhodococcus* sp. SH22 as model strains. For this purpose, the copy number of *alkB* genes per genome had to be known. *P. putida* GPo1 contains one copy of the *alkB* gene on the OCT plasmid, and 1–2 copies of the plasmid per strain (Staijen et al. 1997). Southern Blot hybridization with total DNA of the three other strains showed that there was one *alkB* gene copy in *P. putida* SH41 and *Gordonia sp.* SG11 and four copies in *Rhodococcus* sp. SH22. Indeed, multiple *alkB* genes in *Rhodococcus* strains have been frequently observed (van Beilen et al. 2002).

Table 1. PCR amplification of alkane hydroxylase fragments in seven alkane-degrading strains with the four primer pairs.

	Primer pair of group			
	Pseu 1	Pseu 2	Rho	Aciento
Rhodococcus rubber IV11	−	−	+	−
Mycobacterium ratisbonese FR13	−	−	+	−
Rhodococcus sp. SH22	−	−	+	−
Rhodococcus sp. PN12	−	−	+	−
Pseudomonas putida SH41	+	−	−	−
Rhodococcus opacus FO11	−	−	+	−
Gordonia sp. SG11	−	−	+	−
Pseudomonas aeruginosa GL1	−	+	−	−

− no amplification product obtained with the respective primer pair.
+ amplification product corresponding to an alkane hydroxylase fragment.

The four strains were grown on a minimal culture medium containing *n*-heptane as a carbon and energy source, and harvested during the logarithmic growth phase (optical density ~ 0,2–0,5). Cell numbers were determined by epifluorescence microscopy, DNA extracted and submitted to quantitative PCR experiments. DNA extraction and competitive PCR were also performed for dilutions of the original culture, in order to test the quantification of low amounts of *alkB* genes. A comparison of *alkB* gene copy numbers with cell numbers is shown in Table 2.

The results show that quantification of *alkB* genes is in the same range as cell numbers for each of the tested strains. For *Rhodococcus sp.* SH22, between 100 and 200% of the expected *alkB* genes are found, whereas for *Gordonia sp.* SG11, only 12–17% of the expected value are detected. The reason might be an incomplete DNA extraction, but the amount of *alkB* genes found is still in the right order of magnitude. For the two *P. putida* strains, 100–200% of the expected amount are detected, decreasing to 30% for the dilutions of the *P. putida* SH41 culture, probably because of loss of cells or DNA in the diluted solutions.

After validation of the competitive PCR tools for quantification of the two groups of *alkB* genes, the method was applied to three water samples originating

Table 2. Quantification of *alkB* gene copies by competitive PCR in models trains belonging to the groups of *Rhodococcus* and *Pseudomonas* 1. ND = not determined.

Primer belonging to group	Strain		Number of *alkB* copies	Cells counted	Copies per cell	% detected*
Rhodococcus	*Rhodococcus sp.* SH22	Undiluted culture	8.1. 10^8	1.1. 10^8	7.4	185
		1:10 dilution	4.8. 10^8		4.4	110
		1:100 dilution	5.3. 10^8		4.8	120
Rhodococcus	*Gordonia sp.* SG11	Undiluted culture	5.1. 10^7	4.2. 10^8	0.1	12
		1:10 dilution	7.1. 10^7		0.2	17
		1:100 dilution	ND			
Pseudomonas 1	*P. putida* SH41	Undiluted culture	1.8. 10^8	1.0. 10^8	1.8	180
		1:10 dilution	3.1. 10^7		0.3	31
		1:100 dilution	2.9. 10^7		0.3	29
Pseudomonas 1	*P. putida* GPo1	Undiluted culture	3.8. 10^7	3.4. 10^7	1.1	112
		1:10 dilution	ND			
		1:100 dilution	ND			

* percent of *alkB* gene copies per cell detected by competitive PCR in relation to the *alkB* copy number per genome determined by Southern Blot hybridization.

from a site in the Champagne region, which has been polluted by Mogaz, a motor fuel with a composition similar to that of gasoline. Sample D1 was collected upstream of the pollution source, the F4 sample was very close to the pollution source, whereas sample D6 was located 600 m downstream in the plume. To evaluate if *alkB* gene copy numbers are indicative of degradation capacities, the three samples were incubated with minimal medium containing either *n*-heptane or glucose. These conditions should result in cultures with largely different amounts of alkane-degrading strains, thus permitting analyses of very different samples.

Determination of degradation activity of *n*-heptane showed that the activity of the *n*-heptane-grown D6 and F4 samples was about 6–8 times higher than that of the glucose-grown cultures. The degradation activity of the glucose-grown D1 sample was very low, but increased more than 100 times, when this sample was enriched on *n*-heptane, thus resulting in the same activity as the *n*-heptane-grown F4 sample (about 55 μmoles $CO_2 h^{-1}$ 10^{10} bacteria^{-1}). Incubation on *n*-heptane also led to a significant increase in copy numbers of *alkB* genes from the *Rhodococcus* group in all three samples. Interestingly, *alkB* genes from the *Pseudomonas* 1 group did not increase in the D6 sample when incubated with *n*-heptane. They increased slightly in the F4 sample, but more than 1000 times (average of ~1 copy per cell) in the *n*-heptane-grown D1 sample.

These results show that, depending on the initial sample composition and on culture conditions, strains with *alkB* genes from different groups, either *Pseudomonas* 1 or *Rhodococcus*, might predominate after enrichment with *n*-heptane. While *alkB* genes from the *Pseudomonas* 1 group represent the majority in the D1 sample, 6–14 higher amounts of *alkB* genes belonging to the *Rhodococcus* group can be found in the D6 and F4 samples. The major contribution to the overall *n*-heptane degradation activity may thus originate from either of the two *alkB* gene groups studied.

3 CONCLUSIONS

In the present work, detection of the known diversity of *alkB* genes by PCR was optimized and the molecular tools for quantitative detection by competitive PCR were validated. The two most relevant groups of *alkB* genes for degradation of short chain alkanes, the *Rhodococcus* and *Pseudomonas* 1 groups, were quantified in water samples that had been incubated in glucose or *n*-heptane supplemented media. The results show that *n*-heptane degradation activity increased in parallel to the number of *alkB* genes in the alkane-grown samples. Although the ratio between the two groups of *alkB* genes depends on the sample studied, degradation activity can roughly be correlated to the number of *alkB* genes of the two groups taken together. This indicates that both groups have to be taken into account for evaluation studies, when short chain alkanes (<C11) are considered. The results obtained thus suggest that monitoring degradation genes allows following the evolution of degradation at a contaminated site thereby providing a molecular indicator for the efficiency of decontamination treatments or natural attenuation processes.

REFERENCES

Jin, C., Mata M. & Fink D.J. 1994. Rapid construction of deleted DNA fragments for use as internal standards in competitive PCR. *PCR Methods Appl.* 3: 252–255.

Knudsen, G.R. 2002. Quantifying the metabolic activity of soil- and plant-associated microbes. In C.J. Hurst, R.L. Crawford, G.R. Knudsen, M.J. McInerney & L.D. Stetzenbach (eds), *Manual of Environmental Microbiology:* 591–596. Washington: ASM Press.

Kok, M., Oldenhuis, R., van der Linden, M.P.G., Raatjes, P., Kingma, J., van Lelyveld, P.H. & Witholt, B. 1989. The *Pseudomonas oleovorans* alkane hydroxylase gene. Sequence and expression. J. Biol. Chem. 264: 5435–5441.

Morra, M.J. 2002. Assessment of extracellular enzymatic activity in soil. In C.J. Hurst, R.L. Crawford, G.R. Knudsen, M.J. McInerney & L.D. Stetzenbach (eds), *Manual of Environmental Microbiology:* 597–601. Washington: ASM Press.

Sei, K., Sugimoto, Y., Mori, K., Maki, H. & Kohno, T. 2003. Monitoring of alkane-degrading bacteria in a sea-water microcosm during crude oil degradation by polymerase chain reaction based on alkane-catabolic genes. *Environ. Microbiol.* 5(6): 517–522.

Smits, T.H.M., Röthlisberger, M., Witholt, B. & van Beilen, J.B. 1999. Molecular screening for alkane hydroxylase genes in gram-negative and gram-positive strains. *Environ. Microbiol.* 1(4): 307–317.

Staijen, I.E., Hatzimanikatis, V. & Witholt, B. 1997. The AlkB monooxygenase of *Pseudomonas oleovorans* – synthesis, stability and level in recombinant *Escherichia coli* and the native host. *Eur. J. Biochem.* 244: 462–470.

Van Beilen, J.B., Wubbolts, M.G. & Witholt, B. 1994. Genetics of alkane oxidation by *Pseudomonas oleovorans. Biodegradation* 5: 161–174.

Van Beilen, J.B., Smits, T.H.M., Whyte, L.G., Schorcht, S., Röthlisberger, M., Plaggemeier, T., Engesser, K.H. & Witholt, B. 2002. Alkane hydroxyalse homologues in gram-positive strains. *Environ. Microbiol.* 4(11): 676–682.

Van Beilen, J.B., Li, Z., Duetz, W.A., Smits, T.H.M. & Witholt, B. 2003. Diversity of alkane hydroxylase systems in the environment. *Oil & Gas Science and Technology* – Rev. *IFP* 58(4): 427–440.

Vomberg, A. & Klinner, U. 2000. Distribution of *alkB* genes within n-alkane-degrading bacteria. *J. Appl. Microbiol.* 89: 339–348.

Whyte, L.G., Schultz A., van Beilen, J.B., Luz, A.P., Pellizari, V., Labbé, D. & Greer, C.W. 2002. Prevalence of alkane monooxygenase genes in Arctic and Antarctic hydrocarbon-contaminated and pristine soils. *FEMS Microbiol. Ecol.* 41: 141–150.

European Symposium on Environmental Biotechnology, ESEB 2004 - Verstraete (ed)
© 2004 Taylor & Francis Group, London, ISBN 90 5809 653 X

Biodegradation of short and medium chain polychlorinated alkanes

E. Heath
"Jožef Stefan" Institute, Ljubljana, Slovenia

S.R. Jensen & W.A. Brown
McGill University, Montreal, QC, Canada

ABSTRACT: Polychlorinated alkanes (PCAs) have due to their chemical stability worldwide applications, however, they are also persistent, bioaccumulative and toxic, and their environmental fate is of public concern. In order to estimate biodegradation potential of polychlorinated alkanes, two inputs are required: chemical characterization of PCA mixtures and definition of dehalogenase enzyme activity on PCAs. To address the first input, the free radical chlorination process was modelled using a Monte Carlo simulation. Even though the reaction probabilities were based on available literature data, the model showed to capture gross characteristics of PCA industrial mixture. To define dehalogenase enzyme activity, the biodegradation of chlorinated alkanes by oxygenolytic dehalogenase of *Pseudomonas* sp. 273 was studied. We examined the effect of carbon chain length, intramolecular chlorine distribution and overall chlorine content using commercially available compounds and chloroalkanes synthesized in our laboratory. The outcome is a definition of the chemical structures that both support and limit dehalogenation. This data will contribute to a better understanding of the environmental fate of PCAs.

1 INTRODUCTION

Polychlorinated alkanes (PCAs) are a mixture of isomers of chlorinated alkanes with carbon chain length between C_{10} and C_{30} and 30–70% of chlorine content by weight. Due to their good heat stability, they are mainly used as flame-retardants and extreme pressure additives (Tomy *et al*, 1998, Tomy & Stern, 1999). Industrial production of PCA involves free radical chlorination of alkanes (Tomy *et al*, 1998). This substitution reaction results in a poorly defined mixture of positional isomers with varying degrees of chlorination. The annual world consumption of PCAs is close to 300 kt with a significant proportion of this ending up in the environment. Unfortunately, PCAs are persistent and bioaccumulative and are becoming ubiquitous being found in the world's waters, soils and biota. They are of special concern since they are potential carcinogens, toxins, mutagens, teratogens and endocrine disrupters.

Although, microorganisms have demonstrated the ability to degrade a variety of chlorinated compounds (Hardman, 1991, Slater *et al*, 1995, 1997, Wischnak & Muller, 2000, Yokota *et al*, 1986), previous studies mostly focus on the degradation of short chain chlorinated hydrocarbons, with the exception of mono and terminally dichlorinated alkanes (Armfield *et al*, 1995, Heath *et al*, 2002, Wischnak *et al*, 1998). The maximum degree to which PCAs can be degraded biologically depends upon a number of biotic and abiotic factors, including bioavailability, genetic expression, and genotype, to name a few (Bock & Muller, 1996, Maier, 2000). Ultimately, the maximum degree to which the PCA mixture can be degraded, depends upon substrate specificity of the dehalogenase enzymes for the many isomers present in the PCA mixture. The main goal of our research was to determine the affinity of an oxygenolytic dehalogenase to a number of isomers found in commercial PCA mixtures. Attempts were made to generalize the results by identifying intramolecular chlorination sequences that were recalcitrant to the enzyme under study (biophobes), and those that were not (biophores).

Because of the lack of commercially available pure chlorinated alkanes, we synthesized a variety of polychlorinated alkanes with chain lengths between C_6 and C_{16} in our laboratory, and used these along with purchased chloroalkanes to study their aerobic degradation by the oxygenolytic dehalogenase expressed by *Pseudomonas* sp. 273.

In addition to establishing the biophores and biophobes describing the interaction with the dehalogenase, the composition of the PCA mixture needs to be established. To date, no analytical technique exists that is capable of resolving the thousands of isomers associated with commercial PCA mixtures. To address this problem a Monte Carlo model was developed incorporating the experimental degradation data obtained using pure alkanes.

2 MATERIALS AND METHODS

2.1 *Hydrocarbon synthesis*

The reaction apparatus consisted of reaction vessel that contained an olefin in CCl_4. Chlorine gas was bubbled through the solution at room temperature and pressure. The reaction was monitored using gas chromatography and stopped once the all initial olefin was used up (1-decene, 5-decene, 1,9-decadiene, 1,5,9-decatriene, Sigma-Aldrich Canada Ltd, Mississauga, ON, Canada) or when lower chlorinated products became present. The composition of the reaction products was confirmed by mass spectrometry and nuclear mass resonance analyses (data not shown).

2.2 *Culture and culture conditions*

Biodegradation tests were performed in shake flasks containing 100 mL of mineral salt media (Heath *et al,* 2002). The flasks were incubated at 30°C at constant agitation (200 rpm). A volume of 100 μL of chlorinated alkane was added to the mineral media, unless otherwise stated. Between tested chlorinated substrates, compounds synthesised in our laboratory (1,2-dichlorodecane, 5,6-dichlorodecane, 1,2,9,10-tetrachlorodecane, 1,2,5,6,9,10-hexachlorodecane) as well as commercially available chlorinated alkanes were included (1,6-dichlorohexane, 1-chlorodecane, 1,10-dichlorodecane, 1-chlorohexadecane, Sigma-Aldrich Canada Ltd, Mississauga, ON, Canada).

2.3 *Hydrocarbon analysis*

At the end of the incubation period, selected samples were extracted with chloroform. 1,12-dichlorododecane (DCDD) was used as internal standard. The analyses of the chlorinated alkanes were made using capillary gas chromatograph with flame ionization detection.

2.4 *Analysis of chloride release*

Chloride release of chlorinated compounds was quantified using ion chromatography (IC) with an electrical conductivity detector.

3 RESULTS

To formulate the upper limit for aerobic degradation of PCA mixtures, the biophores/biophobes associated with the dehalogenase need to be elucidated, and a complete characterization of the PCA mixture is required.

3.1 *Modelling the free radical chlorination process*

Industrial mixtures of PCAs are produced by the free radical chlorination of n-alkanes in which chlorine radicals replace hydrogen atoms. There is currently no analytical technique available to resolve thousands of isomers present in PCA industrial mixtures (Muller & Schmidt, 1984, Tomy *et al*, 2000). In an attempt to estimate the distribution of PCA isomers, a Monte Carlo simulation was developed. The main input into the model is a set of rules describing the discrimination with which the chlorine radicals replace the various hydrogen atoms. Assuming that the reactivity of all hydrogen atoms were equal a relatively broad distribution was predicted (Figure 1, "simple rules").

From the chemistry literature, the rate of replacement of a hydrogen atom is related to the stability of the resulting carbon free radical intermediate. A more "realistic" set of rules was derived from this body of work, resulting in a much narrower distribution (Figure 1, "realistic rules"). Although a complete chemical characterization is not possible, researchers have been able to quantify PCAs with respect to the gross degree of chlorination (Tomy *et al*, 2000). These results are shown in Figure 1 (group abundance profile) and are in good agreement with the results in which the literature rules are applied ("realistic" rules).

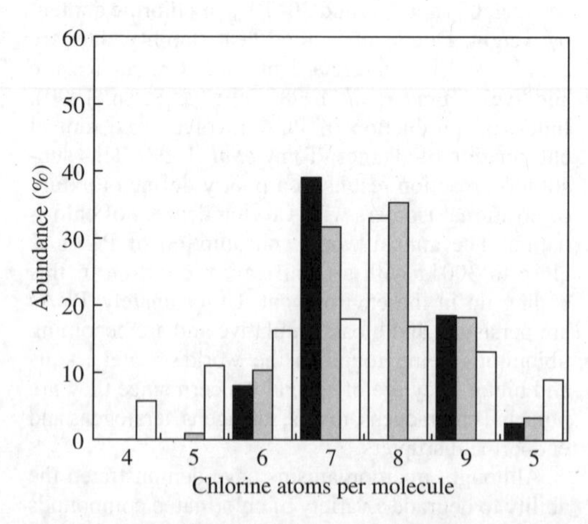

Figure 1. Comparison of experimental (white bars: "simple" and grey bars: "realistic" rules) and group abundance profiles (black bars: Tomy *et al*, 2000) of the C_{11} fraction of S-PCA 60.

3.2 *Aerobic biodegradation of pure PCA compounds*

To determine dehalogenase enzyme substrate specificy, aerobic biodegradation of pure chloroalkanes was studied. The effect of carbon chain length (C_6–C_{16}), halogen position (terminal, vicinal and internal), and overall carbon chlorine content (14–61%, wt/wt) were examined.

Monochloroalkanes. Based on the experimental results, all tested monochlorodecanes (*1-chlorodecane* and *1-chlorohexadecane*) were completely dehalogenated when volatilization was eliminated as a competing mechanism. All compounds tested were dehalogenated by greater than 97%.

Dichloroalkanes. Results for dichloroalkanes tested varied significantly with carbon chain length and with the position of the chlorine atoms. In all cases, 1,10-dichlorodecane was completely dehalogenated. As such, it was used as control to establish the enzyme activity in all experiments. In comparison, only up to 56% of 1,2-dichlorodecane was dehalogenated.

There is evidence in the literature (Jimenez & Bartha, 1996) that adding a solvent, which increases the affinity of the cell for organic/aqueous interface, increases the bioavailability of a substrate. To test this possibility, 1,10-DCD was mixed with 1,2-DCD. In this case, 1,10-DCD served not only as a solvent, but also to induce the dehalogenating enzymes and as a co-substrate. Based on the results, the addition of 1,10-DCD did not significantly augment the yield or rates of biological dehalogenation of 1,2-DCD (data not shown). However, GC analysis of the chloroform extract revealed an unknown peak. Attempts to identify the compound by GC-MS were unsuccessful.

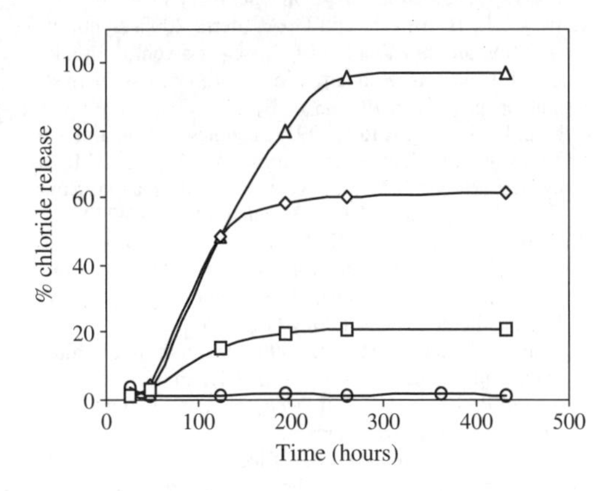

Figure 2. Biological dehalogenation of 50 μL of pure 5.6-dichlorodecane (○), 50 μL pure 1,10-DCD (△), a mixture containing 25 μL of each of 5,6-dichlorodecane and 1,10-DCD (◇), and a mixture containing 25 μL each of 1,2,9,10-TCD and 1,10-DCD (□).

Similar experiments were performed with *5,6-dichloroalkane*. Again, the substrate was incubated alone and together with 1,10-DCD. In all cases, no or very little dehalogenation of 5,6-DCD was obtained, while 1,10-DCD was completely dehalogenated. However, after extraction with chloroform, GC revealed no trace of 5,6-DCD in 50% of the flasks containing only 5,6-DCD as a carbon source. GC analysis however did reveal an unknown peak with a retention time close to that found in the flasks containing 1,2-DCD. In the remaining flasks, all the 5,6-DCD was recovered.

1,6-dichlorohexane (DCH) was the shortest chloroalkane tested. This compound was chosen in order to assess the ability of the oxygenase, to dehalogenate short chain hydrocarbons. This compound was added to a series of shake flasks in concentrations ranging from 0.25 to 1 g/L. In total, up to 5% of chlorine was removed from this substrate.

Polychloroalkanes. Tetrachlorodecane (TCD) is a viscous oil at room temperature and does not mix well with growth medium. When incubated at 30°C and 200 rpm, no chloride was released into the media, while complete dehalogenation of the control 1,10-DCD was achieved. To improve mass transfer and to increase the affinity of the cell for the organic/aqueous interface, TCD was mixed with 1,10-DCD (25 μL each) prior to incubation. After 9 days of incubation 21% of chloride release was observed with no additional chloride release after 35 days. Extraction of the flask content resulted in 100% recovery of TCD and 60–85% of 1,10-DCD.

1,2-5,6-9,10-hexachlorodecane (HCD) is a solid at the optimal growth temperature for *Pseudomonas* sp. 273 (30°C). When added in the solid state, no chloride was released into the medium. The solid HCD was then dissolved in 1,10-DCD (1:10). In this case, the amount of chloride released was stoichiometric with the number of moles of 1,10-DCD added (data not shown). This suggests that none of the HCD was degraded.

Our study shows that the position and amount of halogenation influences the total amount of dehalogenation. The highest dehalogenation rates were associated with alkanes chlorinated on the terminal positions (mono and dichlorinated alkanes with carbon chain length above 10). Not all dichlorinated alkanes were degraded to the same extent. 1,10-DCD was a preferred dichlorinated substrate of the organism. In this case, the two chlorine atoms did not seem to influence each other, and the dehalogenase interacted with the compound in the same way as with its monochlorinated analogues. Short chain dichlorinated hydrocarbon 1,6-dichlorohexane gave low degrees of dehalogenation, which suggests a lack of substrate affinity of dehalogenase enzymes.

When chlorine atoms are present in vicinal arrangement (1,2-DCD, TCD, HCD), dechlorination occurred to a much lesser extent. Both TCD and HCD were not degraded even when mixed with 1,10-DCD (enzyme inducer, co-substrate and solvent). These results show that vicinal carbons present at the end of the carbon chain are not suitable for oxygenolytic dehalogenase studied. This hypothesis was supported by the fact that no chloride was released during 5,6-DCD biodegradation. Also, after extraction, in most flasks no trace of 5,6-DCD was observed, while an oily substance was visible. This indicates partial degradation of 5,6-DCD. β-oxidation of 5,6-DCD was most likely initiated and proceeded until the vicinal chlorine atoms terminated the process. Future studies are planned to identify these compounds.

A similar hypothesis to that above was postulated for 1,2-DCD where approximately 50% of chloride was released during dehalogenation. As suggested by the results of 5,6-DCD, β-oxidation likely commences at the unchlorinated end of the molecule. In this case, β-oxidation would continue until a 5,6-dichlorohexanoic acid or 3,4-dichlorobutyric acid is produced. This compound could than spontaneously form a lactone releasing chloride ions in the process (Wiberg & Roy, 1991). Thus the release of chloride resulting from degradation of 1,2-DCD is due to an abiotic mechanism rather than the action of the dehalogenase enzyme. Again, future studies are planned to identify the metabolic intermediate of this degradation process.

4 CONCLUSIONS

This study provides important information on aerobic degradation and dehalogenation of oxygenolytic enzyme on PCA. It was shown that single terminally substituted chlorine atoms are easily removed. It is also shown that vicinal chlorine atoms terminate oxygenolytic dehalogenation. In this case, β-oxidation can take place when a terminal alkyl group is present, resulting in the production of short chain chlorinated acids. Depending on the acids produced, some may undergo cyclisation and abiotic dechlorination.

To draw an integral picture of enzyme specificity associated with dehalogenation of *Pseudomonas* sp. 273 on short chain chlorinated alkanes (C_{10}–C_{14}), dehalogenation experiments of multiple monochlorinated secondary carbons are needed. These experiments, including the synthesis of compounds not commercially available, are currently underway in our laboratory. Once the study is completed, the data will be used to formulate a set of rules for the biodegradation potential of the individual species in the PCA mixtures. Using the distribution generated by Monte Carlo model, the rules will be then applied to assess the biodegradation potential of the mixture.

REFERENCES

Armfield, S. J., P. J. Sallis, P. B. Baker, A. T. Bull, and D. J. Hardman. 1995. Dehalogenation of haloalkanes by *Rhodococcus erythropolis* Y2. Biodegradation **6:**237–246.

Bock, C., and R. Muller. 1996. Biodegradation of α, ω-dichlorinated alkanes by *Pseudomonas* sp. 273. DECHEMA monographs **133:**721–726.

Hardman, D. J. 1991. Biotransformations of halogenated compounds. Critical Reviews in Biotechnology **11:**1–40.

Heath, E., M. Bratty, and W. Brown. 2002. Biodegradation of chlorinated alkanes by *Pseudomonas* sp. 273, p. 42–51. *In* P. Glavic and D. B. Voncina (ed.), Zbornik referatov s posvetovanja Slovenski kemijski dnevi 2002, vol. 1. FKKT, Maribor, Slovenia.

Jimenez, I. Y., and R. Bartha. 1996. Solvent-augmented mineralization of pyrene by a *Mycobacterium* sp. Applied and Environmental Microbiology **62:**2311–2316.

Maier, R. M. 2000. Microorganisms and organic pollutants, p. 363–402. *In* R. M. Maier, I. L. Pepper, and C. P. Gerba (ed.), Environmental microbiology. Academic Press, San Diego, California, USA.

Muller, M. D., and P. P. Schmid. 1984. GC/MS analysis of chlorinated paraffins with negative ion chemical ionization. Journal of High Resolution Chromatography & Chromatography Communications **7:**33–37.

Slater, J. H., A. T. Bull, and D. J. Hardman. 1995. Microbial dehalogenation. Biodegradation **6:**181–189.

Slater, J. H., A. T. Bull, and D. J. Hardman. 1997. Microbial dehalogenation of halogenated alkanoic acids, alcohols and alkanes. Advances in Microbial Physiology **38:**133–176.

Tomy, G. T., A. T. Fisk, J. B. Westmore, and D. C. G. Muir. 1998. Environmental chemistry and toxicology of polychlorinated n-alkanes. Reviews in Environmental Contamination and Toxicology **158:**53–128.

Tomy, G. T., and G. A. Stern. 1999. Analysis of C_{14}–C_{17} polychloro-n-alkanes in environmental matrixes by accelerated solvent extraction-high-resolution gas chromatography/electron capture negative ion high-resolution mass spectrometry. Analytical Chemistry **71:**4860–4865.

Tomy, G. T., B. Billeck, and G. A. Stern. 2000. Synthesis, isolation and purification of C_{10}–C_{13} polychloro-n-alkanes for use as standards in environmental analysis. Chemosphere **40:**679–683.

Wiberg, K. B. a. W., F. Roy, 1991. Lactones. 2. Enthalpies of Hydrolysis, Reduction, and Formation of C4-C13 Monocyclic Lactones, Strain Energies and Conformations. Journal of the American Chemical Society **113:**7697–7705.

Wischnak, C., F. E. Loffler, J. Li, J. W. Urbance, and R. Muller. 1998. *Pseudomonas* sp. strain 273, an aerobic α,ω – dichloroalkane-degrading bacterium. Applied and Environmental Microbiology **64:**3507–3511.

Wischnak, C., and R. Muller. 2000. Degradation of chlorinated compounds, p. 241–271. *In* J. Klein (ed.), Environmental Processes II – Soil Decontamination, 2nd ed, vol. 11b. Wiley-Vch, Weinheim.

Wyatt, I., C. T. Coutts, and C. R. Elcombe. 1993. The effect of chlorinated paraffins on hepatic enzymes and thyroid hormones. Toxicology **77:**81–90.

Yokota, T., H. Fuse, T. Omori, and Y. Minoda. 1986. Microbial dehalogenation of haloalkanes mediated by oxygenase or halidohydrolase. Agricultural Biology and Chemistry **50:**453–460.

Soil clean-up

European Symposium on Environmental Biotechnology, ESEB 2004 - Verstraete (ed)

Examining the ecological niche of polycyclic aromatic hydrocarbon (PAH) degrading *Mycobacterium* spp. in PAH contaminated soil

M. Uyttebroek, P. Breugelmans & D. Springael
Catholic University of Leuven, Leuven, Belgium

J.-J. Ortega-Calvo
Instituto de Recursos Naturales y Agrobiología de Sevilla, Seville, Spain

A. Ryngaert
Flemish Institute for Technological Research, Mol, Belgium

ABSTRACT: Recent studies indicate that some PAH degrading bacteria may have adapted to the low bioavailability of sorbed polycyclic aromatic hydrocarbons (PAHs). This could play an important role in the selection of PAH degrading bacteria in PAH contaminated soil. This study suggests that *Mycobacterium* spp. are better adapted to degradation of sorbed PAHs than PAH degrading bacteria from other genera, such as *Sphingomonas* and *Pseudomonas*. Moreover, *Mycobacterium* spp. were especially found in the clay fraction of a PAH contaminated soil, where PAH concentration is high and bioavailability of PAHs is low.

1 INTRODUCTION

Polycyclic aromatic hydrocarbons (PAHs) form a group of toxic hydrophobic organic contaminants which are difficult to degrade by microbial attack due to their low water solubility and sorption to soil particles. In enrichments, PAH degrading bacterial strains isolated from contaminated soils are often *Mycobacterium* spp. Moreover, *Mycobacterium* strains are preferentially selected over other PAH degrading bacteria when the PAH is provided in the enrichments sorbed to hydrophobic carriers, indicating that *Mycobacterium* is better adapted to degradation of sorbed PAHs than other PAH degrading bacteria such as *Sphingomonas* and *Pseudomonas* (Bastiaens et al. 2000, Grosser et al. 2000). Contact between *Mycobacterium* cells and the source of contamination might play an important role in efficient degradation of the PAH molecule, suggesting that in nature, these bacteria are closely connected to soil particles containing PAHs (Wick et al. 2002a). Other bacteria produce biosurfactants to enhance PAH degradation. This indicates that different bacteria may exploit different niches in PAH contaminated soils. This work focuses on examining the ecological niche of PAH degrading bacteria with emphasis on *Mycobacterium* spp. in PAH contaminated soil. Therefore, we compared the degradation of ^{14}C-labeled phenanthrene, sorbed to different porous synthetic amberlite sorbents, by different PAH degrading species. In addition, the distribution of indigenous *Mycobacterium* cells over the different fractions of a PAH contaminated soil was examined using PCR-DGGE.

2 MATERIALS AND METHODS

2.1 *Bacterial strains and culture conditions*

Mycobacterium gilvum VM552, *Sphingomonas* sp. LH162 and *Pseudomonas aeruginosa* 19SJ were grown on minimal P-buffered medium (pH 7.5) with phenanthrene crystals as sole carbon source (0.2% wt/vol). The P-buffered medium contained (per liter distilled water): 0.875 g $Na_2HPO_4.2H_2O$; 0.1 g KH_2PO_4; 1 g $(NH_4)_2SO_4$; 0.2 g $MgCl_2.6H_2O$; 0.1 g $Ca(NO_3)_2.4H_2O$ and 5 ml of a trace element solution ([per liter] 800 mg Na_2-EDTA; 300 mg $FeCl_2$; 10 mg $MnCl_2.4H_2O$; 4 mg $CoCl_2.6H_2O$; 1 mg $CuSO_4$; 3 mg $Na_2MoO_4.2H_2O$; 2 mg $ZnCl_2$; 0.5 mg LiCl; 0.5 mg $SnCl_2.2H_2O$; 1 mg H_3BO_3; 2 mg KBr; 2 mg KI; 0.5 mg $BaCl_2$). The bacteria were grown in 250 ml erlenmeyer flasks, containing 100 ml of liquid P-buffered medium, at 25°C on a shaker at 125 rpm.

2.2 *Model sorbents*

The phenanthrene log K_D ($l\,kg^{-1}$) for the porous synthetic amberlite sorbents are respectively 3.0 (IRC-50), 3.5 (XAD-7HP) and 4.2 (XAD-2), based on Grosser et al. (2000). The sorbents were washed according to the method of Cornelissen et al. (1998) and dried overnight at 70°C. In glass Pyrex tubes, 100 mg of these sorbents were pre-equilibrated for 7 days (20°C) with 4.5 ml of ^{14}C-labeled phenanthrene solution by shaking them at an angle of 45°. This solution contained 0.5 μg total phenanthrene per ml phosphate-buffer (0.006 μCi ^{14}C-phenanthrene ml^{-1} P-buffer). The 0.1 M P-buffer (pH 5.8) was prepared by diluting 8.5 ml of 1 M K_2HPO_4 and 91.5 ml of 1 M KH_2PO_4 to 1 liter with distilled water.

2.3 *Culture preparation and phenanthrene degradation in the presence of model sorbents*

After growth, the bacterial cultures were filtered over a ROBU glasfilter nr. 2 (40–100 μm) and reincubated at the same conditions without the phenanthrene crystals for 2.5 days. The bacteria were centrifuged, washed twice with liquid P-buffered medium and diluted in P-buffered medium based on the optical density (660 nm). To enumerate the bacteria in the cultures, viable bacterial counts were done by spreading serial 10-fold dilutions in 10^{-2}M $MgSO_4$ on solid agar plates (15 g agar per liter P-buffered medium with 0.2% glucose). The plates were incubated at 28°C for 2 weeks.

0.5 ml of the prepared bacterial culture was added to the Pyrex tubes (no change in pH) after the sorption-step. Mineralization of the ^{14}C-labeled phenanthrene was followed as production of $^{14}CO_2$ by means of liquid scintillation counting. The trapping agent was 1 ml of 0.5 M NaOH. The experiments were done in triplicate.

2.4 *Soil and fractionation procedure*

The soil was contaminated with PAHs. The fractionation procedure was based on the method of Stemmer et al. (1998). Low-energy sonication (200 J/g ovendry soil) of field moist samples (<2 mm) in combination with wet sieving and repeated centrifugation gave 4 particle size fractions: 2000–200 μm (coarse sand); 200–53 μm (fine sand); 53–2 μm (loam) and 2–0.1 μm (clay). The fractions were kept an overnight at 4°C before DNA extraction.

2.5 *DNA extraction, PCR amplification, agarose gel electrophoresis and denaturing gradient gel electrophoresis (DGGE)*

Total DNA was extracted from the different soil fractions, using a modified protocol of El Fantroussi et al. (1997). Eubacterial 16S rRNA gene fragments were amplified by PCR using the eubacterial primer set 63F & 518R (Leys et al., in prep.). *Mycobacterium* 16S rRNA gene fragments were amplified by PCR using

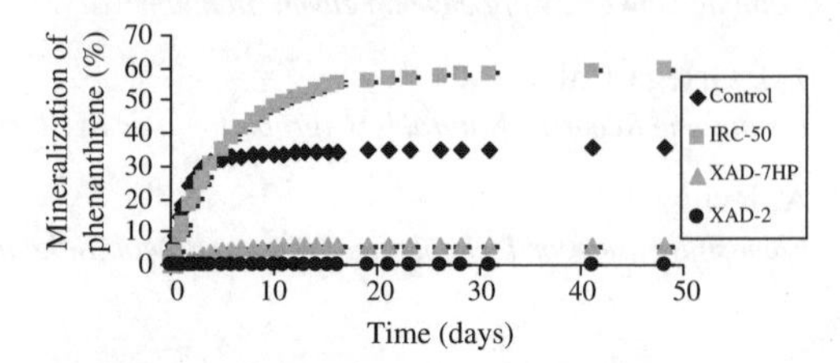

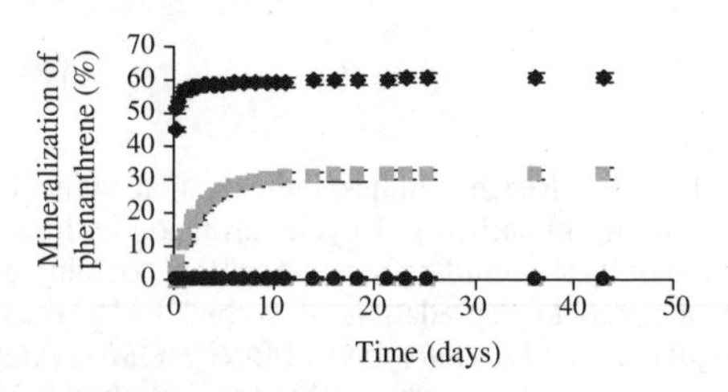

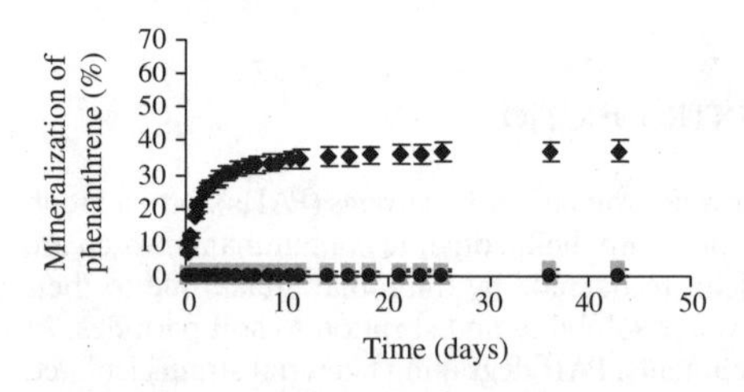

Figure 1. Mineralization of phenanthrene (as $^{14}CO_2$) with control (no sorbent) and the amberlite sorbents by (a) *Mycobacterium gilvum* VM552, (b) *Sphingomonas* sp. LH162 and (c) *Pseudomonas aeruginosa* 19SJ. The error bars are standard errors of triplicate measurements.

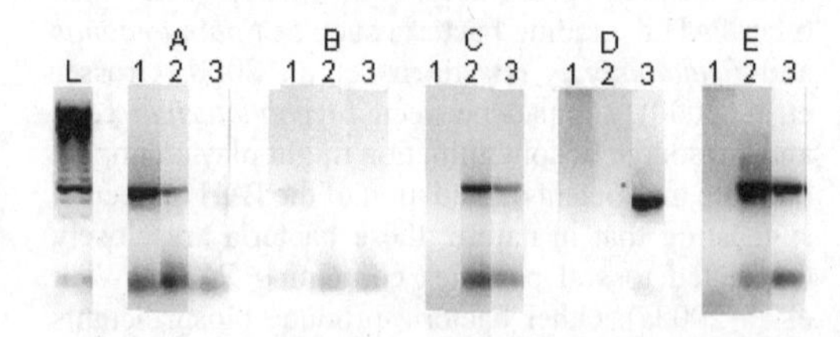

Figure 2. PCR detection of *Mycobacterium* spp. in the different soil fractions of a PAH contaminated soil. Soil DNA was not diluted (1), 10-fold (2) and 100-fold (3) diluted. Lanes: (A) coarse sand (2000–200 μm); (B) fine sand (200–53 μm); (C) loam (53–2 μm); (D) clay (2–0.1 μm); (E) total soil (<2000 μm); (L) 100 bp DNA ladder.

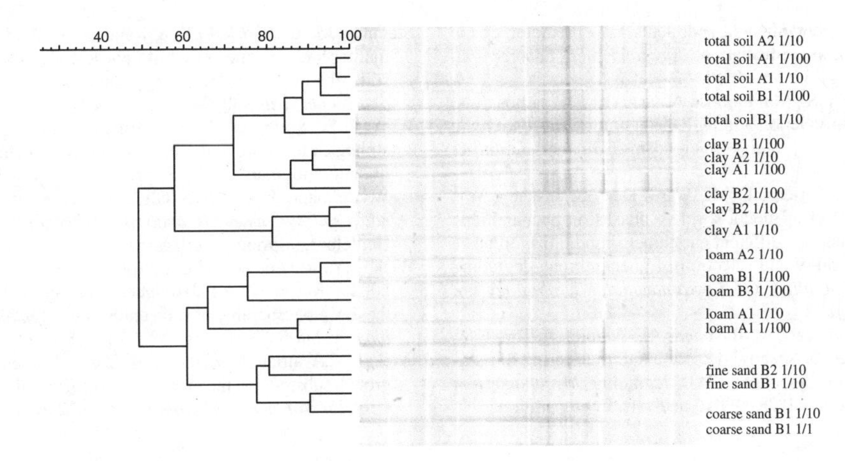

Figure 3. Dendrogram of 16S rRNA gene based PCR-DGGE community profiles of Eubacteria in the different soil fractions of a PAH contaminated soil. Replicates of soil fractions are indicated by A and B. Replicates of DNA extraction are indicated by 1, 2 and 3. Soil DNA was not diluted (1/1), 10-fold (1/10) and 100-fold (1/100) diluted.

the *Mycobacterium* specific Myco-primer set (Leys et al., in prep.). Because of PCR inhibition by unknown soil constituents, DNA extracts were diluted 10- and 100-fold. The amplification product was visualised by agarose (1.5%) gel electrophoresis. DGGE analysis was performed as described by Leys et al. (in prep.).

3 RESULTS AND DISCUSSION

In the first part of the study, we compared the degradation of ^{14}C-labeled phenanthrene, sorbed to different porous synthetic amberlite sorbents, by different PAH degrading species: *Mycobacterium gilvum* VM552 (Fig. 1a), *Sphingomonas* sp. LH162 (Fig. 1b) and *Pseudomonas aeruginosa* 19SJ (Fig. 1c). The amount of bacteria used was 1.11×10^7 CFU/ml, 6.55×10^7 CFU/ml and 8.10×10^5 CFU/ml, respectively. *Mycobacterium* sp. VM552 degraded phenanthrene, sorbed to IRC-50 at a rate (53 ng ml^{-1} day^{-1}) comparable to the rate observed with the control without sorbents (120 ng ml^{-1} day^{-1}). VM552 also showed minor but significant degradation of phenanthrene sorbed to XAD-7HP. *Sphingomonas* LH162 showed a much higher degradation rate of phenanthrene with the control (1080 ng ml^{-1} day^{-1}), compared to the rate observed with IRC-50 (55 ng ml^{-1} day^{-1}). *Pseudomonas* sp. 19SJ showed no significant degradation of phenanthrene sorbed to the amberlite sorbents. Therefore, *Mycobacterium* spp. seem to be better adapted to degradation of sorbed PAHs than the other PAH degrading bacteria. This can possibly be explained by adhesion of the *Mycobacterium* cells to the sorbents, due to the presence of mycolic acids in their outer cell wall (Bastiaens et al. 2000, Wick et al. 2002a, b). This leads to reduction of the mean distance between the pollutant and the bacterial cell, which may enhance mass transfer by an increased diffusion gradient (Wick et al. 2001).

In the second part of the study, the distribution of indigenous *Mycobacterium* cells over the different fractions of a PAH contaminated soil was examined using PCR. The amount of *Mycobacterium* cells seems to be highest in the clay fraction of the PAH contaminated soil since the clay fraction gives the best signal for the 100-fold dilution, compared to the other soil fractions (Fig. 2). Different soil fractions contained different eubacterial communities with the community of the clay fraction representing the major soil community. This was shown by DGGE of eubacterial 16S rRNA gene PCR amplicons (Fig. 3). However, *Mycobacterium* diversity between the different soil fractions was not significantly different (data not shown). The clay fraction also contained the highest amount of total carbon of all fractions (data not shown) and probably also the highest PAH concentration and therefore bioavailability of PAHs might be low in that fraction. This could play an important role in the selection of PAH degrading bacteria in PAH contaminated soil.

REFERENCES

Bastiaens, L., Springael, D., Wattiau, P., Harms, H., De Wachter, R., Verachtert, H. & Diels, L. 2000. Isolation of adherent polycyclic aromatic hydrocarbon (PAH)-degrading bacteria using PAH-sorbing carriers. *Applied and Environmental Microbiology* 66: 1834–1843.

Cornelissen, G., Rigterink, H., Ferdinandy, M.M.A. & Van Noort, P.C.M. 1998. Rapidly desorbing fractions of

PAHs in contaminated sediments as a predictor of the extent of bioremediation. *Environmental Science and Technology* 32: 966–970.
El Fantroussi, S., Mahillon, J., Naveau, H. & Agathos, S.N. 1997. Introduction and PCR detection of *Desulfomonile tiedjei* in soil slurry microcosms. *Biodegradation* 8: 125–133.
Grosser, R.J., Friedrich, M., Ward, D.M. & Inskeep, W.P. 2000. Effect of model sorptive phases on phenanthrene biodegradation: different enrichment conditions influence bioavailability and selection of phenanthrene-degrading isolates. *Applied and Environmental Microbiology* 66: 2695–2702.
Leys, N., Ryngaert, A., Bastiaens, L., Wattiau, P., Top, E. & Springael, D. Specific detection and monitoring of natural and/or inoculated PAH-degrading *Mycobacterium* spp. in soil by PCR-DGGE analysis. In preparation.
Stemmer, M., Gerzabek, M.H. & Kandeler, E. 1998. Organic matter and enzyme activity in particle-size fractions of soils obtained after low-energy sonication. *Soil Biology and Biochemistry* 30: 9–17.
Wick, L.Y., Springael, D. & Harms, H. 2001. Bacterial strategies to improve the bioavailability of hydrophobic organic pollutants. In R. Stegmann, G. Brunner, W. Calmano & G. Matz (eds.), *Treatment of contaminated soil: fundamentals, analysis, applications*: 203–217. Heidelberg: Springer-Verlag.
Wick, L.Y., Ruiz de Munain, A., Springael, D. & Harms, H. 2002a. Responses of *Mycobacterium* sp. LB501T to the low bioavailability of solid anthracene. *Applied Microbiology and Biotechnology* 58: 378–385.
Wick, L.Y., Wattiau, P. & Harms, H. 2002b. Influence of the growth substrate on the mycolic acid profiles of mycobacteria. *Environmental Microbiology* 4: 612–616.

European Symposium on Environmental Biotechnology, ESEB 2004 - Verstraete (ed)
© 2004 Taylor & Francis Group, London, ISBN 90 5809 653 X

In situ remediation of PAH contaminated soil by combination of biodegradation and Fenton reaction

J.H. Langwaldt, M.R.T. Palmroth, T.A. Aunola & T.A. Tuhkanen
Tampere University of Technology, Institute of Environmental Engineering and Biotechnology, Tampere, Finland

ABSTRACT: The *in situ* remediation of PAHs in soil from a creosote oil contaminated site was studied. Treatment of the contaminated unsaturated soil combined the advanced oxidation process of Fenton-like treatment with biodegradation of the contaminants. The soil was contaminated with up to 19 mg PAHs/g and free phase of creosote oil was present at the site. Over two weeks, hydrogen peroxide (8–9%, 900 L) was injected in 4 to 5 m depth. Prior to hydrogen peroxide injection the total numbers of bacteria ranged from 4×10^8 to 2×10^9 cells/g horizontally and from 3×10^8 to 2×10^9 cells/g vertically. On average 53/62% (horizontally and vertically) of the bacteria were viable. After termination of the injection the numbers of bacteria were at the same order of magnitude as prior to the treatment. Indigenous soil bacteria utilized sole PAH compounds of up to 100 mg/L in bioassays. The *in situ* biodegradation of PAHs will be further monitored.

1 INTRODUCTION

Creosote oil contaminated environments are widespread and expensive to remediate. Treatment of creosote oil contaminated soil is often carried out by on-site stabilization, excavation and off-site incineration or disposal. *In situ* remediation of contaminated soils is considered more cost-efficient than on-site and off-site treatment (EPA, 1998). The low bioavailability of hydrophobic contaminants, such as polyaromatic hydrocarbons (PAHs) in creosote oil, is limiting their *in situ* biodegradation. Advanced oxidation processes (AOP) enhance the bioavailability of contaminants, because the partially oxidized metabolites formed by the AOP are more water-soluble. Furthermore, the formation of oxygen by decomposition of hydrogen peroxide facilitates aerobic biodegradation of contaminants. Thus a short-term AOP, such as the Fenton reaction, and subsequent biodegradation may enhance contaminant removal more than natural processes and may reduce remediation costs.

Earlier studies have shown that Fenton reaction can successfully decrease the mass of contaminants (EPA, 1998). The survival of bacteria during Fenton reaction has been shown both in water phase (Howsawkeng et al., 2001) and in soil (Palmroth et al., 2003a). In our previous work, the combination of Fenton reaction like oxidation and the biodegradation of PAHs has been studied in laboratory scale columns (Aunola et al., 2002, Palmroth et al., 2003a). The preliminary studies showed that the combination of AOP and biodegradation removed up to 55% of the PAHs (Aunola et al., unpublished). Further, survival of bacteria and their ability of PAH-degradation was shown (Palmroth et al., submitted).

2 MATERIAL AND METHODS

2.1 *The studied site*

The studied creosote oil contaminated site is at a wood-preservation facility and has been described earlier (Haapea and Tuhkanen, 2002). It has been estimated that contamination of the subsurface by creosote oil occurred 20 to 30 years ago. The site is located in a groundwater area classified as important groundwater resource. The site is characterized by an esker of sand and gravel. The bedrock is in a depth of 5 m.

2.2 *Soil sampling and installation of injection and monitoring wells*

Soil samples were taken at 11 points from the surface down to the bedrock in one meter intervals. Soil samples were stored in plastics bags at 4–5°C in the dark. Four steel injection wells (1.5″ OD) were installed with 1×1 m spacing. Further, small diameter wells (7) were installed for monitoring of pore water quality (Fig. 1). Liquid samples were recovered with a peristaltic pump.

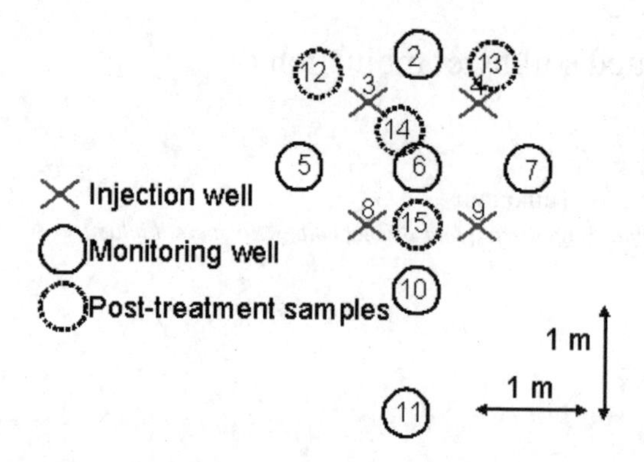

Figure 1. Location of the injection and monitoring wells.

After termination of the Fenton-like treatment additional soil samples were obtained from 4 points.

2.3 *The Fenton-like treatment*

Fenton-like treatment was carried out by injection of hydrogen peroxide into the unsaturated zone. The injected hydrogen peroxide solution (8–9%) was made up by dilution of a 50% H_2O_2 solution with uncontaminated, iron-free humic groundwater from the area. During two weeks 900 L of H_2O_2 solution were injected at depth of 4–5 m, i.e. directly above the bedrock.

2.4 *Analyses*

The samples were dried with anhydrous sodium sulphate (Na_2SO_4) prior to extraction. Soil samples (6 to 9 g) were extracted with dichloromethane (60 ml) in a Soxhlet apparatus for 16 h. The volume of the extract was recorded after extraction. The extracts were filtered (0.2 μm) and diluted with dichloromethane prior to analysis. Internal standard was added into run vials. The extracts were analyzed with a HP 6890 gas chromatograph equipped with a HP 5972A mass-selective detector in selected ion monitoring (SIM) mode. The gas chromatograph was installed with a Supelco Equity-5 capillary column.

Total iron concentration were determined from the pore water and soil samples after extraction with 5 M HCl by atomic absorption spectroscopy.

2.5 *Viability and activity of indigenous bacteria*

BacLight™live/dead assay with SYTO 9 and propidium iodide stains (Viability Kit L-7012, Molecular Probes) was used to determine the viability based on cell wall integrity of the cells in treated and untreated soil (Palmroth et al., 2003b).

Pyrene, anthracene, acenaphtene, fluorene and phenanthrene, 1 g/L, were dissolved in n-hexane/n-pentane and added to the 96-well plates to yield following final volume concentrations: 5, 25, 50 and 100 mg/L. The same volumes of pure solvent were added to control wells to estimate growth on solvents and/or natural organic carbon in the soil suspension. The PAH solutions and hexane were allowed to evaporate under filtered N_2 stream in a desiccator before the plates were used.

Solution containing 1-methoxy-5-methyl-phenazinium methyl sulphate and 2-(4-iodophyl)-3-(4-nitrophenyl)-5-(2,4-disulfonyl)-2-H-tetrazolium monosodium salt was prepared in autoclaved deionised water. The solution was added to the wells of the plate wells, except the corner well to minimise edge effects. Soil (1 g) was added to 50 ml of mineral medium. Soil suspension (190 μl) was added to each well. Utilization of PAHs is linked to in a redox reaction, which causes an absorbance increase at 450 nm.

3 RESULTS AND DISCUSSION

3.1 *PAH concentrations*

The depth profiles of total PAH concentrations in three sampling points are presented in Figure 2. The concentration of PAHs was highest at a depth of 1–2 m with up to 19 mg/g in (Fig. 2). The unsaturated soil contained droplets of creosote oil. A free phase of creosote oil was present as on top of the bedrock.

The PAH-composition in the sampled soil and the free phase of the dense non-aqueous phase liquid (DNAPL) was similar to a German creosote oil product used for wood-preservation (Fig. 3). Currently the post-treatment soil samples are analyzed for their PAH-content.

3.2 *Fate of H_2O_2 in the subsurface*

The concentrations of H_2O_2 in the monitoring wells ranged from 0 to 5% during the injection of H_2O_2. Two days after termination of the H_2O_2 injection, the highest concentration of residual H_2O_2 in pore water was 1.4%.

3.3 *Iron in the subsurface*

In the soil the total iron concentrations ranged from 3.7 to 17.2 mg/g. In pore water from a down gradient of the injection zone located monitoring well, dissolved iron concentration increased during the Fenton-like treatment. Leaching of iron from the solids into the pore water has been earlier observed in laboratory experiments under similar conditions (unpublished results). In general, Finnish soils contain elevated concentrations of iron, which is favourable for Fenton-like treatment of contaminated soil.

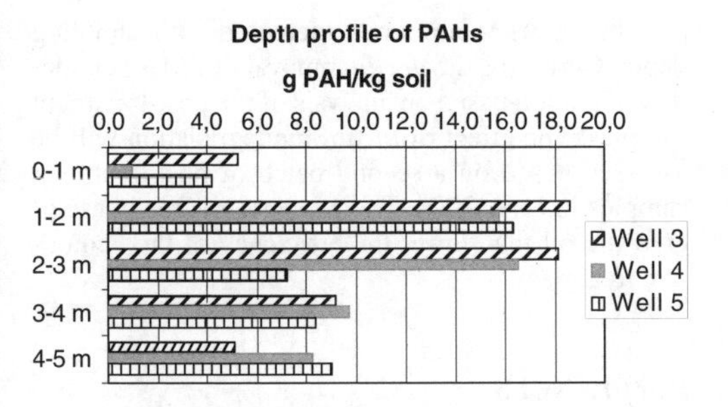

Figure 2. PAH concentration with depth.

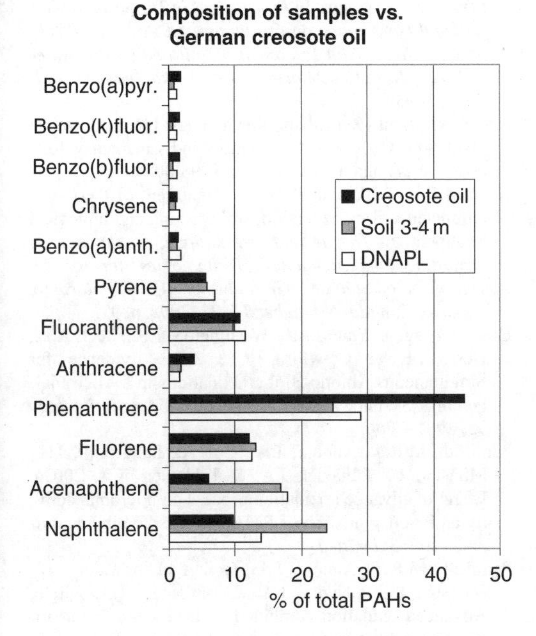

Figure 3. Composition of PAHs in sampled soil (3–4 m depth), extracted DNAPL and creosote oil.

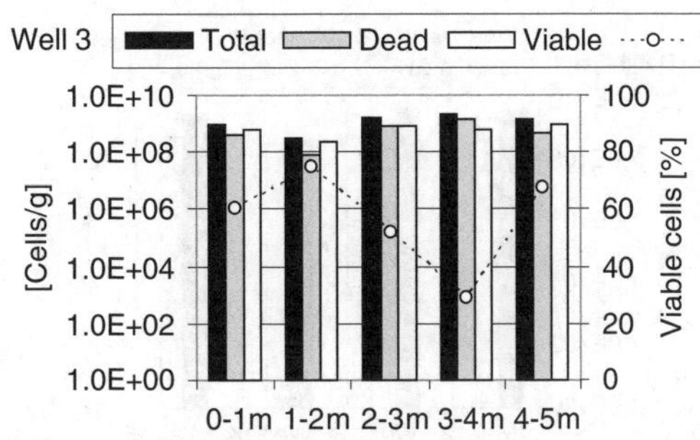

Figure 4. Vertical profile of numbers of bacteria in soil samples from well 3 prior to H_2O_2-injection.

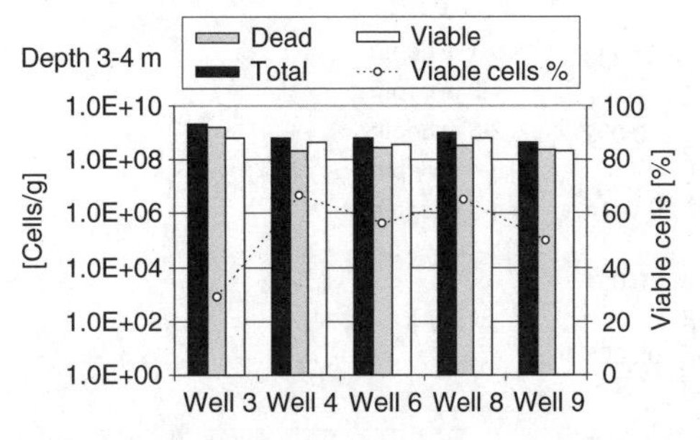

Figure 5. Horizontal profile of numbers of bacteria at a depth of 3–4 m prior to H_2O_2-injection.

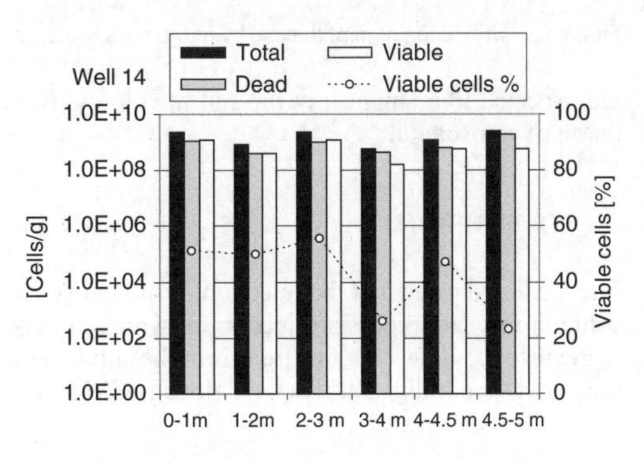

Figure 6. Vertical profile of numbers of bacteria in soil samples from well 14 after H_2O_2-injection.

3.4 *Abundance of bacteria*

Prior to the injection of hydrogen peroxide, the total numbers of bacteria ranged from 3×10^8 to 2×10^9 cells/g vertically (Fig. 4) and from 4×10^8 to 2×10^9 cells/g horizontally (Fig. 5). On average 53/62% (horizontally and vertically) of the bacteria were viable (Figs 4 and 5). After termination of the injection the numbers of bacteria were at the same order of magnitude as prior to the injection (Fig. 6 and 7).

3.5 *Biodegradation of PAHs*

The indigenous soil bacteria degraded naphthalene, fluorene, phenanthrene and anthracene at concentrations of up to 100 mg/L. The utilization of PAHs increased with sampling depth (Fig. 8). Bacteria sampled from 0–4 m, above the present free phase, showed inhibition by high concentrations (100 mg/L) of sole PAH compounds (Fig. 8). Earlier results showed no biodegradation of naphthalene by bacteria extracted from aged soil samples without naphthalene (Palmroth et al., 2003). Thus, the indigenous may have to be exposed to naphthalene to remain their ability to degrade the compound. Naphthalene concentrations in the vapours

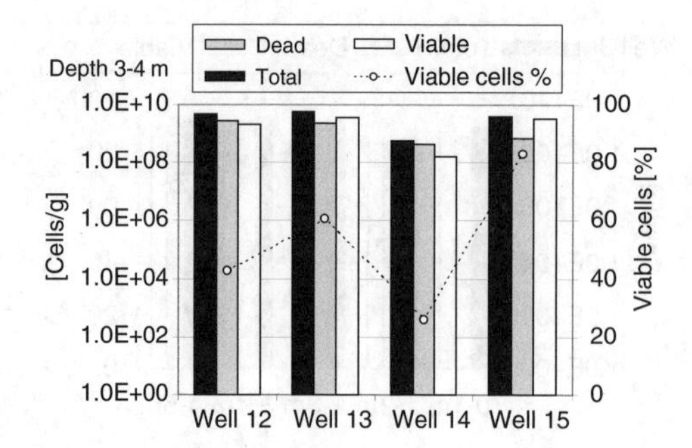

Figure 7. Horizontal profile of numbers of bacteria at a depth of 3–4 m after H_2O_2-injection.

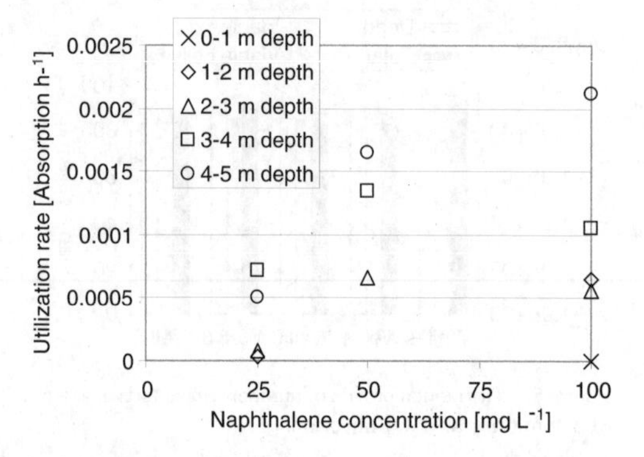

Figure 8. Utilization of naphthalene by indigenous bacteria.

are expected to be highest in the soil next to the free phase of creosote oil.

4 CONCLUSIONS

The studies showed that the injection of diluted H_2O_2 solution is a safe process. Injection of hydrogen peroxide resulted in leaching of iron from the solids into the pore water. Indigenous bacteria biodegraded PAHs and the biodegradation rate increased with sampling depth. Currently, the *in situ* removal of PAHs is under determination based on analysis of the post-treatment samples. The effect of *in situ* biodegradation will be assessed based on a second batch of post-treatment samples in early 2004. The presence of free phase of creosote oil may lower the efficiency of the Fenton-like treatment.

REFERENCES

Aunola, T., Palmroth, M., Goi, A. & Tuhkanen, T., 2002. Treatment of PAH contaminated soil by Fenton and Fenton-like reactions. In *Proceedings of The Second International Conference on Oxidation and Reduction Technologies for In-Situ Treatment of Soil and Groundwater (ORT-2). Toronto, Ontario, Canada. November 17–21, 2002*, 38–39.

EPA. 1998. In Situ Remediation Technology: In Situ Chemical Oxidation. Office of Solid Waste and Emergency Response. Washington DC. EPA 542-R-98-008.

Haapea, P. & Tuhkanen, T. (2002) Treatment of PAH contaminated soil by ozonation, soil washing and biological treatment. In *The Second International Conference on Oxidation and Reduction Technologies for In-Situ Treatment of Soil and Groundwater (ORT-2). Toronto, Ontario, Canada. November 17–21, 2002*, p. 97.

Howsawkeng, J., Watts, R.J., Washington, D.L., Teel, A.L., Hess, T.F. & Crawford, R.L. 2001. Evidence for Simultaneous Abiotic-Biotic Oxidations in a Microbial-Fenton's System. *Environmental Science & Technology* 35, 2961–2966.

Palmroth, M.R.T., Aunola, T.A., Goi, A., Langwaldt, J.H., Münster, U., Puhakka, J.A. & Tuhkanen, T.A. 2003a. Effect of advanced oxidation process on microbial activity and biodegradation of PAHs in creosote oil contaminated soil. *Submitted*

Palmroth, M.R.T., Aunola, T.A., Goi, A., Langwaldt, J.H., Münster, U., Puhakka, J.A. & Tuhkanen, T.A. 2003b. Advanced oxidation combined with biodegradation for in situ remediation of creosote oil contaminated soil. In *Proceedings of the Second European Bioremediation Conference. Chania, Greece, p. 63.66, June 30-July 4, 2003*, p. 63–66.

European Symposium on Environmental Biotechnology, ESEB 2004 - Verstraete (ed)
© 2004 Taylor & Francis Group, London, ISBN 90 5809 653 X

Fate and metabolism of [^{15}N]2,4,6-trinitrotoluene (TNT) in soil

M. Weiß, R. Russow, H.H. Richnow, M. Kästner
Centre for Environmental Research Leipzig-Halle UFZ, Department of Bioremediation, Permoserstraße, Leipzig, Germany

R. Geyer
Center for Biomarker Analysis/Mass Spectroscopy Lab., University of Tennessee, Knoxville, USA

ABSTRACT: The fate of the label from [^{14}C] and [^{15}N]TNT was analysed in bioreactors under aerobic conditions in soil and in soil treated by a fungal remediation process with *Stropharia rugosoannulata*. The experiments show a different fate of ^{15}N in comparison to ^{14}C from TNT of up to 18%. Three N-mineralisation processes were identified in detailed experiments with [^{15}N]TNT. About 2% of the ^{15}N label was found as NO_3^- and NH_4^+ showing simultaneous processes of direct TNT denitration (I) and reduction with cleavage of the amino groups (II). The ^{15}N-enrichment of NO_2^-/NO_3^- up to 7.5% indicates the formation of Meisenheimer complexes with a denitration of TNT. N_2O and N_2 contained 1.4% of the label. The isotopic composition of the N_2O (38 at.%) demonstrated that both N atoms were generated from the labelled TNT and clearly indicate a novel process (III). We propose the formation by cleavage of N_2O from condensed azoxy-metabolites. Additionally, 1.7% of the ^{15}N label was detected as biogenic amino acids in the straw overgrown by the fungus. Overall, 60–85% of the applied [^{15}N]TNT were degraded and 52–64% were found as non-extractable residues in the soil matrix. 3% were detected as 2-amino-4,6-dinitrotoluene and 4-amino-2,6-dinitrotoluene.

European Symposium on Environmental Biotechnology, ESEB 2004 - Verstraete (ed)

Biodegradation of atrazine in the presence of soil models

P. Besse, M. Sancelme & A.M. Delort
Laboratoire de Synthèse et Etude de Systèmes à Intérêt Biologique, Université Blaise Pascal, Aubière Cedex, France

T. Alekseeva, C. Taviot-Guého & C. Forano
Laboratoire des Matériaux Inorganiques, Université Blaise Pascal, Aubière Cedex, France

ABSTRACT: Atrazine, still a widely used herbicide for control of broad leaf weeds in major crops (corn, sugarcane), is a relatively common contaminant of groundwater. Before entering this compartment, atrazine is first in contact with soil where degradation by microorganisms occurs. Within the framework of a CNRS programme, we investigated the sorption and biodegradation of atrazine by *Pseudomonas* sp. strain ADP in the presence of different soil models. Humic acids were chosen as a model of the organic part of soil, and anionic (Layered Double Hydroxide = LDH) and cationic (Bentonites) clays were chosen as models for the mineral part. No atrazine adsorption on LDH occured, but the presence of this matrix in the incubation medium increased dramatically the rate of its biodegradation. With Ca-bentonites, the effects observed are completely reversed: strong adsorption of atrazine and decrease of the rate of biodegradation. The use of humic acid-clay complexes lead to a strong increase in the sorption capacities of atrazine.

1 INTRODUCTION

The present study is part of an interdisciplinary research program, entitled TRANZAT (Programme Environnement, Vie et Sociétés of the CNRS, France), addressing a major and actual environmental concern related to agriculture, that is the diffuse contamination of soils and freshwater ecosystems with xenobiotic chemicals used for crop protection.

Atrazine is still nowadays one of the most widely used herbicides in agriculture, for control of broad leaf weeds in major crops such as corn or sugarcane, despite its prohibition since last years in several countries of Western Europe, first in Germany and from July 2003 in France. Consequently, it is the most frequently detected herbicide in surface and groundwater and its concentration often exceeds at least 10 to 100 fold the standard quality (0.1 μg/L). Obviously, such agrochemical pollutants have to pass through the soil compartment before entering the adjacent rivers and streams.

The part of atrazine reaching the soil undergoes a series of chemical and biological transformations. More often, the fate of herbicides in the environment is largely dependent on metabolism by microorganisms. However, with respect to their great abundance and diversity in soils, only a small part of soil microflora has the capability of mineralizing completely the molecule, while the greater part transforms it partially. A number of bacterial strains, able to use atrazine as the sole source of nitrogen and/or carbon, have been isolated: some were able to mineralize atrazine, others catalyze a hydrolytic dechlorination reaction, producing hydroxyatrazine, or N-dealkylations (Wackett et al 2002). In this study, the bacterium *Pseudomonas* sp. strain ADP (Mandelbaum et al 1995) was chosen as a model as its atrazine metabolic pathway, as well as the different genes involved, is well characterized (Sadowsky & Wackett 2000).

However, in soils, the potential activity of herbicide degrading bacteria can often be greatly modified, depending on the physico-chemical properties of the soil (pH, composition …), but also on the interactions herbicide – soil components (sorption, retention, formation of complexes), or on the interactions bacteria – soil components (Huang et al 2002). Moreover, the bioavailability of sorbed compounds remains an opened question, one of the difficulties to answer being due to the complexity of soils. According to the authors, soil-sorbed organic compounds have been considered unavailable for biodegradation without prior desorption (Ogram et al 1985, Crocker et al 1995) or still degradable by microorganisms. In this last case, the use of biosurfactants to facilitate the desorption, the production of extracellular enzymes (Singh et al 2003) or the attachment of cells to soil particles containing concentrations in contaminants supporting high mineralization rates (Park et al 2003) constitute some of the ways used by microorganisms to get the contaminant bioavailable. The use of biosurfactants was not clearly established yet (Calvillo & Alexander 1996). Only the positive effect of synthetic surfactants on bioavailability of hydrophobic contaminants was described (Brown et al 1999, Steffan et al 2002).

In this work, the biodegradation of atrazine (AT) by a reference bacterium, *Pseudomonas* sp. strain ADP, in the presence of solid matrixes was studied. As soils are very complex materials with various compositions, we have chosen to work with different soil models: commercially available humic acids (Aldrich) as model of the organic part of the soil, and cationic (bentonite) or synthetic anionic (Layered Double Hydroxides = LDH) clays as models for the mineral part. To mimic soils in a more appropriate way, anionic and cationic clay-humic acids complexes were also tested. Two processes involved in the fate of pesticide in soils have been studied separately: sorption of atrazine on the solid matrix and its biodegradation.

2 ATRAZINE ADSORPTION ON A SOLID MATRIX

2.1 *Syntheses of the different sorbents*

Cationic clay was Ca-form of industrial Aldrich bentonite, obtained by ion exchange without additional purification. Anionic clay (LDH) used for the experiments was $Mg_3Al(OH)_6NO_3 . H_2O$ synthesized by coprecipitation method (Figure 1) (De Roy et al 1992).

Complexes with various percentages of adsorbed humic acid (HA) were prepared with both anionic and cationic clays. Characterization of these soil models was realized by RX diffraction, MEB and BET. The adsorption of HA on clays lead to a great modification of the textural properties (specific surface, aggregates ...) without any fundamental change in the structure.

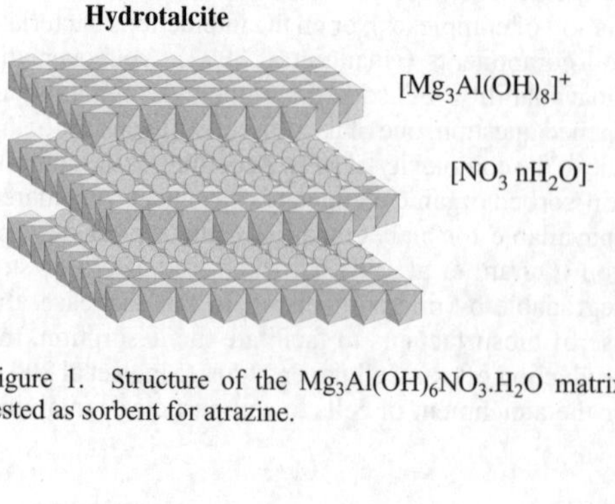

Hydrotalcite

$[Mg_3Al(OH)_8]^+$

$[NO_3\ nH_2O]^-$

Figure 1. Structure of the $Mg_3Al(OH)_6NO_3.H_2O$ matrix tested as sorbent for atrazine.

2.2 *Atrazine adsorption*

Adsorption of AT was carried out using a batch-equilibration technique (ratio solid/liquid 1/10) in darkness at 25°C. So various AT concentrations in Volvic water were stirred with 1 g of solid. At pH = 7, AT is under its neutral form.

No AT adsorption was observed on LDH which confirms that hydrotalcites are not good sorbents for hydrophobic pesticides. For all other cases, adsorption isotherms fitted the Freundlich equation. The adsorption capacity (Kf) values varied between 0.82 and 45.5, being the smallest for LDH with 2% HA and highest for Ca-bentonite with 10% HA (3.80 for Ca-bentonite alone). These data indicate that HA was the main factor controlling atrazine adsorption.

3 BIODEGRADATION OF ATRAZINE IN THE PRESENCE OF SOIL MODELS

As most of the matrixes used were potential sorbents of atrazine, a preliminary step consisting of a 24 h pre-incubation of the solid matrix with the 0.1 mM atrazine solution, was carried out. After this step, the experiments were initiated by inoculation with resting cells of *Pseudomonas* sp. strain ADP. Negative controls – incubation without cells in the presence or in the absence of the matrix – indicated that atrazine is chemically stable in water and that its biosorption is negligible.

3.1 *In the presence of humic acids*

Representative AT degradation time courses in the presence of different concentrations in humic acids (Aldrich) are shown on Figure 2.

No significant adsorption of atrazine was observed during the pre-incubation step, even at higher HA concentrations. These results are in agreement with those reported by Meredith & Radosevich (1998),

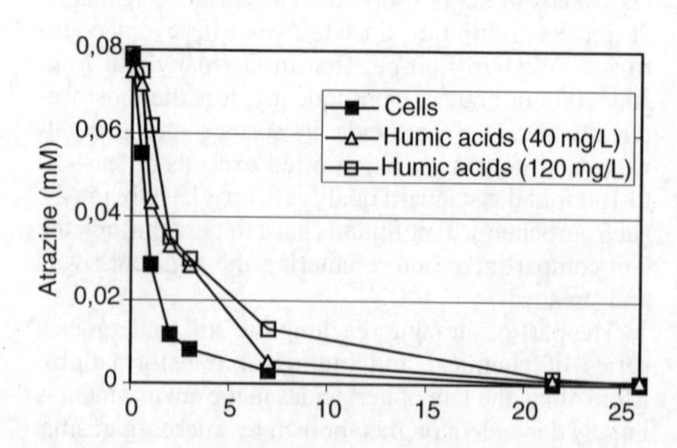

Figure 2. Time courses of the concentrations of atrazine during its incubation with resting cells of *Pseudomonas* sp. strain ADP in the absence or in the presence of humic acids.

who estimated that less than 1% of atrazine was associated with HA under similar conditions.

The effect of HA on biodegradation, whatever its concentration, was also very small: a slight decrease in the initial rate was observed with low HA concentrations whereas an increase was noted with high concentrations (800 mg/L). Under our conditions, HA do not affect the biodegradation of AT.

3.2 *In the presence of clays*

The same type of experiments was carried out with various quantities of LDH or Ca-bentonite (ratio solid/liquid 0.1–1/10).

The presence of LDH in the incubation medium increased dramatically the biodegradation rate of AT (Figure 3). This effect is LDH dependent.

No adsorption was observed during the pre-incubation step as previously shown by the isotherms. Interactions AT-solid matrix do not exist; so the accelerating effect of LDH on AT biodegradation has to come from interactions between the bacterium and LDH. Characterizations of such interactions are in progress.

In the case of cationic clays, all the phenomena are reversed. Ca-bentonite strongly adsorbs AT. The more important the quantity of bentonite was, the more AT was adsorbed (around 50% for a ratio solid/liquid 1/10) during the pre-incubation step.

Concerning the biodegradation, a decrease in the initial rate was noticed, and a plateau appeared at the beginning of the incubation. The length of the plateau was a function of the quantity of cationic clay (data not shown).

As adsorption of AT on the solid matrix occured, the question was to know if what was analyzed in the supernatants during the incubation corresponded really to biodegradation, or also to disappearance of AT due to its affinity for the matrix. Extraction of the solid matrix by a mixture Water/Methanol 1/4 was carried out on each sample of the incubation. The supernatant and the methanolic extract were analyzed by HPLC in parallel. We have shown that (i) by adding the content in AT in the supernatant and in the methanolic extract of the samples taken during the pre-incubation period, we found the initial AT concentration, indicating that the interactions Ca-bentonite – AT are weak; (ii) the plateau observed at the beginning of the incubation was the result of an equilibrium between biodegradation and desorption of AT in the liquid phase. As the concentration of the "free" AT in the medium is decreasing - because of the biodegradation process – sorbed AT was released and degraded.

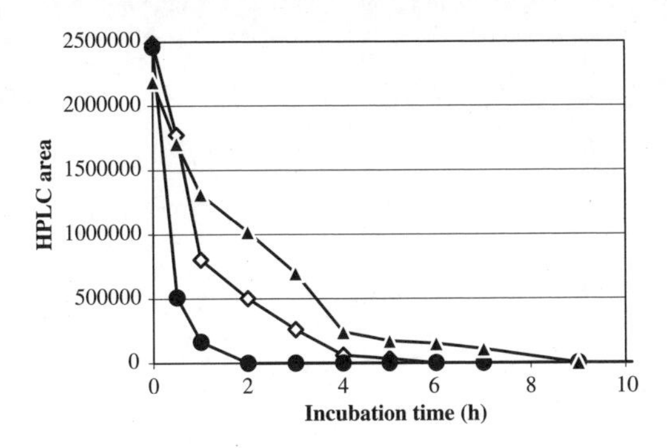

Figure 3. Biodegradation of AT by *Pseudomonas* sp. strain ADP in the absence (▲) or in the presence of 200 mg (◇) or 500 mg (●) LDH (Total volume: 10 mL of a 0.1 mM AT solution).

3.3 *In the presence of humic acid – clay complexes*

Synthetic HA – clay complexes with anionic and cationic clays were prepared with a content of 2 and 10% HA, and characterized. Similar experiments as those described previously were carried out.

The pre-incubation step showed a strong increase in the AT adsorption, as demonstrated previously with the isotherms, getting to 30% with LDH-10% HA and 75% in the presence of Ca-bentonite with 10% HA (ratio solid/liquid 1/10).

When LDH-HA complexes are present in the incubation medium, the accelerative effect of LDH was reduced, as much as the HA content in the complex was high. This effect was almost completely lost with a high quantity of LDH with 10% HA.

Although AT adsorption was higher on HA – cationic clay complexes than with Ca-bentonite alone, the initial rate of AT biodegradation was less affected. If a slow-down of the biodegradation rate was observed compared to the blank corresponding to the cells alone, no plateau was clearly evidenced. The presence of HA decreased the inhibitor effect of Ca-bentonite on the biodegradation rate.

4 CONCLUSION

The fate of organic contaminants in the environment, in particular in soils, depends on many abiotic and biotic processes: first of all adsorption, but also chemical and biological transformations. Adsorption/desorption are probably the most important mode of interactions between soil components and pollutant that controls its availability towards microbial degradation, retention and mobility through the soil profile.

In this work, different behaviors of the soil models chosen towards atrazine have been observed, from the absence of adsorption (LDH) to quite high adsorption with Ca-bentonite – 10% HA. With all the models tested, only weak interactions between the solid matrix and AT were involved, as the simple extraction of the solid with an organic solvent allowed to find again the initial AT concentration.

This adsorption process could not be correlated directly to the biodegradation rate of AT by *Pseudomonas* sp. strain ADP: for examples, (i) no adsorption was detected with LDH and a dramatic increase of the biodegradation rate was observed compared to cells alone; (ii) a higher AT adsorption of Ca-bentonite – 10% HA complex than Ca-bentonite alone was measured, but the effect on the slow-down of the biodegradation was lower with the first model. Many complex mechanisms are involved in the environmental fate and impact of pesticides, their (bio) availability being one of the key factors, particularly for the sorbed-compounds: interactions contaminants – soil components depending on the properties of both actors, but also interactions between bacteria and the soil components (as shown by the assay with LDH). The characteristics of the bacteria [production of biosurfactants, surface charge, chemotaxis towards the contaminant…] are important factors that must also be taken into account in such approach (Park et al 2003). The determination of these characteristics is in progress.

The authors acknowledge greatly Dr. Hans-Peter Buser from Syngenta Crop Protection AG (Basel, Switzerland) for having providing them atrazine, and CNRS for financial support.

REFERENCES

Brown, D.G., Guha, S. & Jaffe, P.R. 1999. Surfactant-enhanced biodegradation of a PAH in soil slurry reactors. *Bioremediation J.* 3: 269–283.

Calvillo, Y.M. & Alexander, M. 1996. Mechanism of microbial utilization of biphenyl sorbed to polyacrylic beads. *Appl. Microbiol. Biotechnol.* 45: 383–390.

Crocker, F.H., Guerin, W.F. & Boyd, S.A. 1995. Bioavailability of naphthalene sorbed to cationic surfactant-modified smectite clay. *Environ. Sci. Technol.* 29: 2953–2958.

De Roy, A., Forano, C., El-Malki, M. & Besse, J.P. 1992. Anionic clays: trends in pillaring chemistry. In M.L. Occeli & H. Robson Van Nostrand (eds), *Synthesis of microporous materials*: 108–169. New York: Reinhold.

Huang, P.M., Bollag, J.M. & Senesi, N. 2002. Interactions between soil particles and microorganisms. Impact on the terrestrial ecosystems. P.M. Huang, J.M. Bollag, N. Senesi (eds), IUPAC Series on Analytical and Physical Chemistry of Environmental Systems, Vol. 8, John Wiley & Sons, England.

Mandelbaum, R.T., Allan, D.L. & Wackett, L.P. 1995. Isolation and characterization of a *Pseudomonas* sp. that mineralizes the *s*-triazine herbicide atrazine. *Appl. Environ. Microbiol.* 61: 1451–1457.

Meredith, C.E. & Radosevich, M. 1998. Bacterial degradation of homo- and heterocyclic aromatic compounds in the presence of soluble/colloidal humic acid. *J. Environ. Sci. Health* B33: 17–36.

Ogram, A.V., Jessup, R.E., Ou, L.T. & Rao, P.S.C. 1985. Effects of sorption on biological degradation rates of (2,4-dichlorophenoxy) acetic acid in soils. *Appl. Environ. Microbiol.* 49: 582–587.

Park, J.-H., Feng Y., Ji, P., Voice, T.C. & Boyd, S.A. Assessment of bioavailability of soil-sorbed atrazine. *Appl. Environ. Microbiol.* 69: 3288–3298.

Sadowsky, M.J. & Wackett, L.P. 2000. Genetics of atrazine and *s*-triazine degradation by *Pseudomonas* sp. strain ADP and other bacteria. In J.C. Hall, R.E. Hoagland & R.M. Zablotowicz (eds), *Pesticide biotransformations in plants and microorganisms*. ACS Symp Ser 777: 268–282. Oxford University Press, Oxford.

Singh, N., Megharaj, M., Gates, W.P., Churchman, G.J., Anderson, J., Kookana, R.S., Naidu, R., Chen, Z., Slade, P.G. & Sethunathan, N. 2003. Bioavailability of an organophosphorus pesticide, Fenamiphos, sorbed on an organo clay. *J. Agric. Food Chem.* 51: 2653–2658.

Steffan, S., Tantucci, P., Bardi, L. & Marzona, M. 2002. Effects of cyclodextrins on dodecane biodegradation. *J. Inclusion Macrocyclic Chem.* 44: 407–411.

Wackett, L.P., Sadowsky, M.J., Martinez, B. & Shapir, N. 2002. Biodegradation of atrazine and related *s*-triazine compounds: from enzymes to field studies. *Appl. Environ. Biotechnol.* 58: 39–45.

European Symposium on Environmental Biotechnology, ESEB 2004 - Verstraete (ed)
© 2004 Taylor & Francis Group, London, ISBN 90 5809 653 X

Comparison of the degradation characteristics of poly(3-hydroxybutyrate-co-3-hydroxyvalerate) in water and soil by isolated soil microorganisms

B.I. Sang
Water Environment & Remediation Research center, Korea Institute of Science and Technology, Seoul, Korea

K. Hori & H. Unno
Department of Biotechnology, Tokyo Institute of Technology, Yokohama, Japan

ABSTRACT: Microbial degradation of P(3HB-co-3HV) films in water and soil by isolated soil microorganism was conducted to investigate degradation characteristics. While colony growth and P(3HB-co-3HV) degradation capability of bacteria and actinomycetes were affected by nutrient condition, fungi were not largely affected and their colony growth rates were much faster. The P(3HB-co-3HV) degradation rate was ordered as follows; fungi > actinomycetes ⩾ bacteria in water and fungi > actinomycetes > bacteria in soil. The degradation by bacteria occurred mainly by bacterial colonization in water and soil and the depolymerase took an important role on P(3HB-co-3HV) degradation by actinomycetes and fungi as comparison with bacteria. The actinomycetous and fungal hyphae grew with penetrating the films and this growth characteristic could be an factor to enhance degradation rate.

1 INTRODUCTION

In near future, it is expected to increase the practical application of biodegradable polymer in many areas including agriculture. The discharged biodegradable polymer could be accumulated in agricultural fields if the large amounts of biodegradable polymer are discarded and the degradation rate is slow than expected. The accumulated polymer could change the soil characteristics like water permeability and content capacity and follow the decrease of the agricultural production. In most case, the microbial degradation of biodegradable polymer in environment occurs in water and soil with the important role of bacteria, actinomycetes and fungi. However there is little investigation of the degradation rates by each microorganism type and the degradation mechanisms by them in water and soil. This information could be useful for sound usage of biodegradable polymer in agricultural area and for sound design and research of biodegradable polymer manufacture. For this aim, the degradation mechanisms of poly(3-hydroxybutyrate-co-3-hydroxyvalerate) [P(3HB-co-3HV)], most famous and already commercialized biodegradable plastic, in water and soil by isolated P(3HB-co-3HV)-degrading microorganisms were investigated.

2 MATERIALS AND METHODS

2.1 *Polymer studied*

P(3HB-co-3HV) containing 12% 3HV was obtained from Aldrich as powder. For degradation experiments, the films of P(3HB-co-3HV) were prepared by conventional solvent casting techniques with 2% (w/v) solution of the P(3HB-co-3HV) in chloroform using glass petri dishes as casting surfaces.

2.2 *Microorganisms*

Three bacterial, *Acidovorax delafieldii* B7-7, *Acidovorax delafieldii* N7-21, and *Acidovorax delafieldii* N7-28, 2 actinomycetous, *Streptomyces* sp. B2-21 and *Streptomyces griseus* N2-10 and 2 fungal strains, *Fusarium oxysporium* E1-3 and *Paecilomyces lilacinus* N4-5, isolated from soil by our laboratory were used for colony and clear-zone growth on solid medium and degradation experiments in water and soil.

2.3 *Measurement of colony and clear-zone growth rate on solid medium*

To measure the colony and clear-zone growth rates of P(3HB-co-3HV)-degrading microorganisms, two agar plates containing 0.25% (w/v) of P(3HB-co-3HV) overplayed plates, one poor in nutrients (yeast extract 0.5 g/l in basal mineral medium; BM plate) and one rich in nutrients (nutrient broth 8.0 g/l, Difco Laboratories, Detroit, U.S.A.; NB plate), were prepared. Clear-zone radii were measured with a ruler at least once a day. Distinct zones could be estimated to fractions to millimeters.

2.4 *Degradation experiments in water and soil*

Three film pieces (1 × 3 cm × 0.05 mm) of P(3HB-co-3HV) disinfected by washing in 70% ethanol for 1 hour and rinsed twice with sterile distilled water were used for degradation experiments in 200 ml flask containing 10 ml of basal mineral medium with 0.1 g/l of yeast extract at 28°C and 100 r.p.m.. For degradation experiments in soil, 2 film pieces (1 × 3 cm × 0.05 mm) of P(3HB-co-3HV) disinfected by washing in 70% ethanol for 1 hour and rinsed twice with sterile distilled water were buried with inoculation of each microorganism in each soil contained in the petri-dishes and incubated at a constant temperature of 30°C, and a constant water content (35%). Film pieces were removed aseptically, washed in distilled water and cleaned by sonication. The films were dried to constant weight in an oven for 1 hour at 80°C and overnight in the desiccator at room temperature. The artificial soil (2 mm length × 1 mm dia.), ISOLITE™ from Isolite Industry Co. Ltd., Japan, was used for the P(3HB-co-3HV) degradation experiments in soil under the unified physicochemical conditions.

2.5 *Analytical method for SEM observation*

After degradation experiments, P(3HB-co-3HV) films were post-treated as mentioned above. Surface appearances of the films were observed using a scanning electron microscope (SEM) with 15 kV acceleration after Pt-Pd coating of the sample.

3 RESULTS AND DISCUSSION

3.1 *Colony and clear-zone growth rate in solid medium*

On solid medium, the colony and clear-zone growth of bacteria were affected by the nutrient condition. While the colony in rich nutrient condition (NB plates) grew faster than in poor nutrient condition (BM plates), the size of clear-zone was shown to be smaller in rich condition. Growth of actinomycetes was also affected by the nutrient condition. Especially, *Streptomyces* sp. B2-21 nearly showed the clear-zone in rich condition. Both colony and clear-zone growth of fungi showed to be faster in rich nutrient condition than in poor condition and the growth rates of fungal colony and clear-zone were much faster than those of bacteria and actinomycetes as shown in Table 1. It indicates that fungi could take a relatively important role on the microbial degradation of biodegradable polymer in soil that usually contains high or low organic compounds in natural conditions.

3.2 *P(3HB-co-3HV) degradation mechanisms in water and soil*

P(3HB-co-3HV) films were degraded in water after approximate 200 hours of inoculation completely by used bacteria strains except *Acidovorax delafieldii* N7-28 and the degradation in soil by bacteria occurred more slowly as shown in Figs 1 and 2. From the staining of the film surface incubated for 3 days in water with Violet Red solution and surface observation of degraded film surface in water and soil, while bacteria in water started to colonize over all surface of the films from the beginning of degradation, the bacterial colonization in soil occurred from the contact points of soil particles to the film surface. Hemispherical holes by bacterial colonization were observed on the

Table 1. Colony and clear-zone growth rate of isolated microorganisms.

	Nutrient condition	Bacteria	Actinomycetes	Fungi
Colony growth rate (mm^2/hr)	Poor	0.03–0.10	0.20–0.68	14.51–36.53
	Rich	0.14–0.40	0.29–0.51	9.26–37.68
Clear-zone growth rate (mm^2/hr)	Poor	0.98–1.85	3.70–4.91	15.04–22.21
	Rich	0.40–1.31	0.14–3.80	3.69–46.81

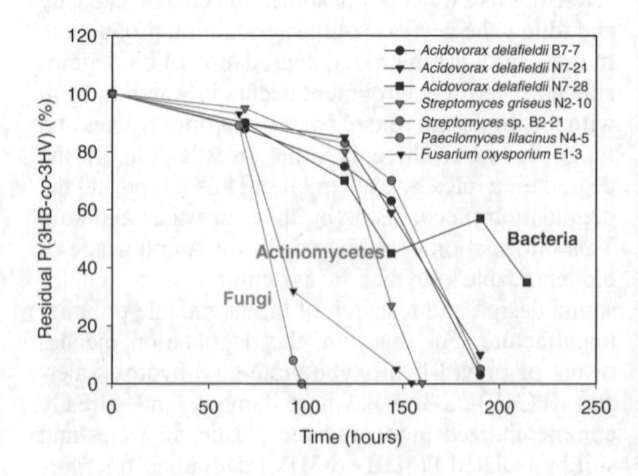

Figure 1. Degradation rate in water.

film surfaces in soil and from these holes the degradation was progressed to the entire surface of the films. Bacterial colonization also occurred in water on the film surface but the big hemispherical holes found on the degraded films in soil could not be found. It is assumed that the degradation in water occurred by the bacterial colonization and the depolymerase produced by attached and suspended bacteria simultaneously and the degraded small fragments by bacterial colonization and depolymerase washed out continuously into the medium. On the ground of low depolymerase activity in medium as shown in Fig. 3, it suggested that the degradation in water mainly occurred by bacterial colonization and the role of the suspended bacteria on the degradation was minor.

In Figs 1 and 2, actinomycetes could degrade P(3HB-co-3HV) films within 160–200 and 380–450 hours in water and soil, respectively. The degradation rates by actinomycetes showed to be similar in water with bacteria but be faster in soil than bacteria. Especially the degradation rate in water and soil was slower at the early stage than bacteria but after then

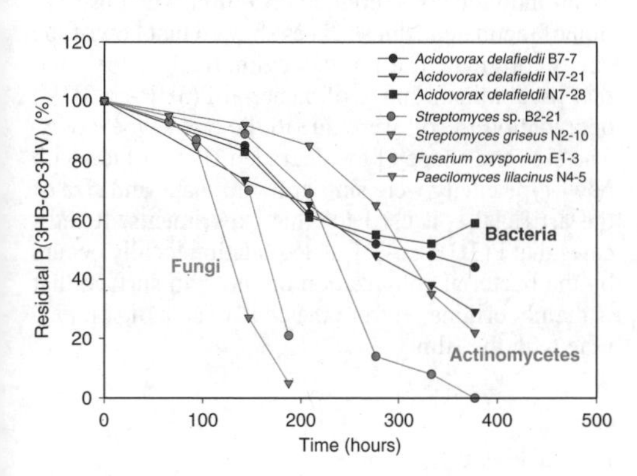

Figure 2. Degradation rate in soil.

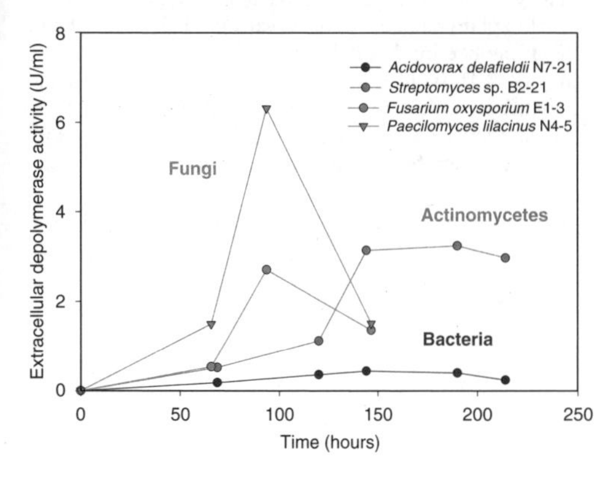

Figure 3. Extracellular depolymerase activity in water.

suddenly increased. Fig. 3 gives the cue why the degradation rate increased fast. In water, the activity of extracellular depolymerase excreted by attached and suspended actinomycetes was 6 times higher than that from bacteria. And it is thought that the depolymerase enhanced the degradation in water and the steep weight loss of P(3HB-co-3HV) films in soil from mid-stage also was caused by increase of the depolymerase excretion. The growth characteristic on the film surface, actinomycetous hyphae growing with penetrating the film, was observed in the surface analysis of degraded films using SEM and penetrating hyphae made the cylindrical holes giving increased surface area to be attacked by microorganisms and depolymerase and made the film into small fragments finally. Besides the role of depolymerase on the enhancing degradation rate, this growth characteristic of actinomycetes should be considered.

Fungi showed very fast degradation rate as comparison with bacteria and actinomycetes as shown in Figs 1 and 2. Fast degradation rate by fungi is caused by several factors as follows. As shown in Fig. 3, the depolymerase activity of fungi was approximately 2 and 12 times higher than actinomycetes and bacteria, respectively. On solid medium, the colony and clearzone growth rate were much faster than bacteria and actinomycetes as shown in Table 1. This fast growth rate could give the advantageous situation in water, and especially in soil. From the surface observation of degraded films, it was found that fungal hyphae also grew with penetrating the films in water and soil, and the many hemispherical holes made by depolymerase were observed on the film surfaces in water. In soil, fungi grew with high dense hyphae around P(3HB-co-3HV) films as similar with actinomycetes as shown in Fig. 4. At early phase the degradation in soil was occurred by separated fungal hyphae and several lines etched by depolymerase from fungal hyphae were

Figure 4. Actinomycetes growth around P(3HB-co-3HV) films after 6 (A) and 11 days (B); fungi growth after 8 days (C) and soil aggregate by fungi around the films (D).

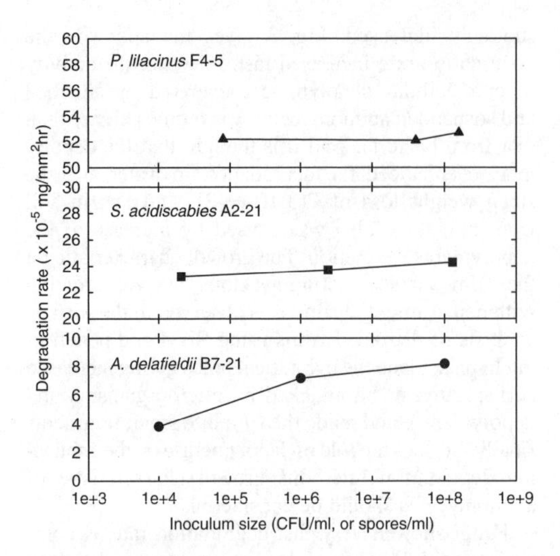

Figure 5. Degradation rates of each type of PDMs in soil with different inoculum sizes.

observed on the film surfaces. However after fungi grew densely on the film surfaces the hemispherical holes were also observed in soil. It indicates that the extracellular depolymerases from attached microorganisms take an important role on P(3HB-co-3HV) degradation in soil.

Figure 5 showed the degradation rates of each type of P(3HB-co-3HV)-degrading microorganisms (PDMs) in soil with different inoculum size. Increase of inoculum size reduced the lag time for microbial degradation of P(3HB-co-3HV) film but the degradation rates by fungi appeared not to be influenced by the different inoculum size. Bacterial degradation rate increased with the inoculum size and actinomycetous degradation rate showed slight dependence on the inoculum size. Fungal PDM, *P. lilacinus* F4-5, showed 7.2 and 2.2 times of degradation rates by bacterial PDM, *A. delafieldii* B7-21, and actinomycetous PDM, *S. acidiscabies* A2-21, respectively.

P(3HB-co-3HV) film degradation in water by bacteria originated in the bacterial colonization that made cavities on the entire film surface. Each cavity enlarged and merged into a larger one continuously with time. P(3HB-co-3HV) film was degraded from the surface and thinned. In water, the large and typical cavity on the P(3HB-co-3HV) film was observed only when bacterial colonization occurred on the virginal surface of P(3HB-co-3HV) film at the early stage of the degradation experiments. P(3HB-co-3HV) film degradation by bacteria in soil also occurred with bacterial colonization mainly forming in the contact area between a soil particle and P(3HB-co-3HV) film. The various sizes of cavities formed by bacterial colonization were observed and these cavities enlarged and merged each other into another larger one continuously with time to the complete degradation. Two types of cavity, the shapes of disk and bowl, were observed on the P(3HB-co-3HV) film surface. The cavity of disk type appeared mainly in the contact area between a soil particle and P(3HB-co-3HV) film. And, the cavity of bowl type formed mainly in the rest area of the contact area between a soil particle and P(3HB-co-3HV) film, and inside of the cavity of disk type. The cavity of bowl type appeared to be formed by an individual bacterial colonization. The observation of incubated film surfaces showed that bowl-type cavity originated from the cylindrical shape, and that predominant mode of bacterial P(3HB-co-3HV) degradation in soil were due to the bowl-type cavity. The size and shape of the degraded film surface with bowl-type cavity were similar to the shape and size of the artificial soil used for this experiments. It indicates that P(3HB-co-3HV) degradation locally occurs by the bacterial colonization on the film surface that is mainly originated from the contact area of soil particle with the film.

REFERENCES

Sang, B-I., Hori, K., Tanji, Y. & Unno, H. 2001. A kinetic analysis of the fungal degradation process of poly(3-hydroxybutyrate-co-3-hydroxyvalerate) in soil. Biochemical Engineering Journal 9: 175–184.

Sang, B-I., Hori, K., Tanji, Y. & Unno, H. 2002. Fungal contribution to in situ biodegradation of poly(3-hydroxybutyrate-co-3-hydroxyvalerate) film in soil. Applied Microbiology and Biotechnology 58: 241–247.

European Symposium on Environmental Biotechnology, ESEB 2004 - Verstraete (ed)
© 2004 Taylor & Francis Group, London, ISBN 90 5809 653 X

In-situ removal of atrazine using synergy between plant roots and rhizospheric organisms

Gupta, S. & Schulin, R
Swiss Federal Research Station for Agroecology and Agriculture (FAL), Reckenholzstrasse, Zürich

Kanekar, P. & Pakniker, K.M.
Agharkar Research Institute (ARI), Pune, India

Schwitzguébel, J-P.
Laboratory of Biotechnology, EPFL, Lausanne, Switzerland

Raghu, K.
Jai Research Foundation (JRF), VAPI, Gujarat, Indiaew Institute

Kathrin, W.
Institute of Terrestrial Ecology (ITO)-ETHZ, Grabenstrasse, Schlieren (ZH), Switzerland

ABSTRACT: This paper presents results of an Integrated Research Project financed by Swiss and Indian Government to develop a new gentle, appropriate and efficient atrazine remediation technique, based on the synergetic use of rhizospheric microorganisms and roots of plants such as weeds, shrubs and wild legumes and agronomic crops either to extract or to degrade or to stabilise in sites heavily impacted by atrazine.

Agriculture crops such as corn and sugarcane or non-agricultural plants such as Gliricidia, Vetiver have shown resistance to atrazine. In most of our later studies, we used Vetiver plants because these plants shows optimum uptake and degradation potential for Atrazine. The degradation and uptake potential and resistant of Vetiver plants to atrazine are investigated by two independent measurement methods such as Biochemical and radiotracer methods. The technique has been developed with the use of growth experiments in laboratory, nutrient solution, pot experiment and mini plot field experiment. These results are still need to be validated in large scale field experiment.

1 INTRODUCTION

Clean soil and groundwater are the basis for healthy food and for the preservation of human health. Due to rapid economic development, urbanisation and enhanced consumption during last years, a steady increase in concentration and in number of agricultural sites, ground waters, surface waters, agricultural products where the Maximum contaminant Levels Goals (MCLG) with respect to pesticides are recorded and documented world wide.

Phytoremediation techniques which are complementary to classical bioremediation techniques are increasingly used to remove contain or render inactive the Persistent Organic Pollutants (POP). Growing plants can be accomplished at a cost ranging from 2 to 4 orders of magnitude less than the current engineering cost of excavation and reburial.

The main objective of this Integrated Research Project financed by funds made available by the Indo-Swiss Colloboration in Biotechnology is to develop better understanding of soil–plant interaction either to extract or to degrade or to stabilise atrazine and their metabolites in sites exceeding Maximum contaminant Levels Goals (MCLG). There are no systematic studies on soil–plant Interaction exist in literature.

The main objective of this research is the development and evaluation of these new, gentle, appropriate and efficient biological processes, based on the use of plants such as weeds, shrubs and wild legumes and agronomic crops either to extract or to degrade or to stabilise atrazine in sites heavily impacted by atrazine. Following steps as shown Figure 1 are used in the developing process. Some of the critical points will be presented and discussed in the presentation.

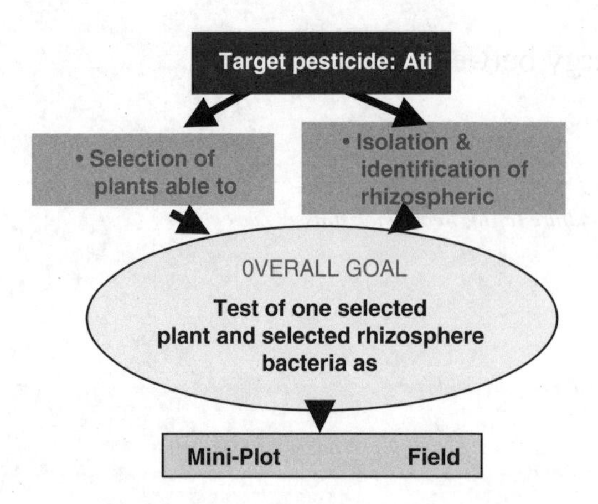

Figure 1. Development of field applicable phytoremediation technique for two pesticides.

1. To select site specific non-agricultural and agricultural crops for their ability to either degrade or to extract or to stabilise pesticides under investigation in contaminated soils.
2. To delineate pathways employed in the uptake and metabolism of atrazine by site specific plants and identification of metabolites.
3. To isolate readily identified soil microbial consortia present in the rhizosphere of the selected specific plants, which are involved in the metabolism of target pesticides.
4. To test one or more selected plants and their rhizosphere co-culture or consortium augmented in soil for transformation and reduction of target pesticides and their metabolites in greenhouse studies.

2 MATERIALS AND METHODS

In order to establish the importance of soil-plant interaction in field condition, greenhouse studies on degradation and transformation of lindane and atrazine and their metabolites are planned with the selected plants and their rhizospheric co-culture or consortium. Two different analytical techniques bio-chemical and radio tracer are used for these studies.

3 RESULTS AND DISCUSSION

3.1 *Investigation to select plants can survive at 2 ppm of atrazine conc in soil*

In one pot culture experiment agricultural and non-agricultural plant species for resistance to atrazine. are investigated. Seeds/seedlings of around 5 agricultural and 10 non-agricultural plant species were tested for growth in 2 ppm atrazine spiked black cotton soil

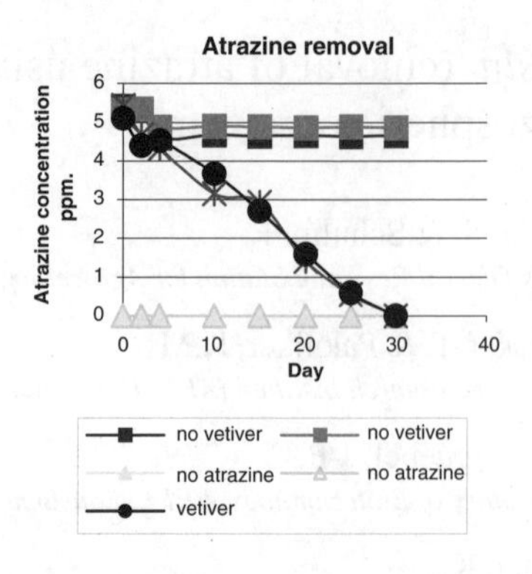

Figure 2.

Table 1.

Treatment	Soil + Vetiver (Set 1)	Sterile soil(Set 2)	Unsterile soil (Set 3)	Test (Set 4)
Atrazine ppm. 4	0	4.4	2.8	0

in porcelain pots. Two non-agricultural and two agricultural plant species e.g. Gliricidia and Vetiver, corn and sugarcane.

3.2 *Investigation to assess the atrazine degrading capability of corn plants in black cotton soil*

For soil experiments under green house condition 10 kg black cotton soil was used in porcelain pots. Soil was spiked with 2 ppm atrazine. Seeds were used for experiment and growth of the plant was monitored from germination upto one month. Residual atrazine was extracted from soil with methanol and analyzed by HPLC. Vetiveria plants having 30 g weight were planted in each pot. One pot was not spiked with atrazine, which served as positive control. The experiment was run in triplicate with appropriate controls (soil + atrazine, no plants, soil + plants, no atrazine) kept in green house throughout the experiment. The corn plant showed 70% degradation of atrazine from soil during one month of growth. Microorganisms have been isolated from corn rhizosphere and tested for degradation of atrazine.

3.3 *Investigation to assess the atrazine degrading capability of Vetiver plants in hydroponics study*

After 30 days of incubation period complete removal of atrazine from hydroponics system with Vetiveria plant was observed. In hydroponics system without Vetiveria plants, 14% removal was observed. There was no difference in biomass also. No change was

Table 2. Percent distribution of ^{14}C atrazine residues in 15 days old *Vetiveria* plants.

	Percent ^{14}C residues in						
Growth conditions	Hoagland solution	Shoot	Root	Bottle washings	Root washings	Foam (Volatilization)	Total
Unsterilized conditions	56.75	22.42	11.63	0.07	1.72	0.02	92.61
Plant exposed for 15 days	74.41	19.97	8.79	0.1	1.26	0.02	104.55

observed in pH of the hydroponics solution during the experiment. The experiment indicated uptake of atrazine by Vetiver plant system.

3.4 *Investigation to assess the atrazine removal capability of Vetiver plants in black cotton soil at 4 ppm*

An experiment was conducted with glass bottles filled with 100 g black cotton soil spiked with 4 ppm atrazine. Following experimental sets were established in triplicate.

After 15 days of atrazine exposure, atrazine was extracted from all experimental sets with dichloromethane. HPLC analysis was done to determine residual atrazine concentration.

3.5 *Investigation to assess the max tolerable atrazine conc capcity of Vetiver plants*

Black cotton soil obtained from field having no atrazine exposure history was spiked with gradient of atrazine. Commercially available atrazine was mixed in water and mixed thoroughly in 10 kg soil so as to attain active ingredient concentration in range of 50 ppm to 15 000 ppm. 10 porcelain pots were filled with 10 kg soil spiked with gradient of atrazine concentration. Vetiveria plants having 30 g weight were planted in each pot. One pot was not spiked with atrazine, which served as positive control.

All the Vetiver plants were not exposed to atrazine before the experiment. Effect of atrazine was noted on the basis of wilting of leaves, retardation of growth, new offshoots etc.

Results showed that growth of Vetiver was not adversely affected by application of atrazine up to 10 000 ppm atrazine. It showed some wilting effect and retardation of growth of Vetiver at 15 000 ppm application. Thus Vetiver can be used at atrazine accidental spill sites for phytoremediation. Removal of atrazine from soil with the help of Vetiver at high concentration will prove its potential for phytoremediation.

Evapotranspiration losses were same in exposed and unexposed plants. There was no retardation of growth of Vetiveria exposed to atrazine as compared to control Vetiveria plants. There was no difference in biomass also. No change was observed in pH of the hydroponics solution during the experiment. The experiment indicated uptake of atrazine by Vetiver plant.

3.6 *Uptake of ^{14}C atrazine by Vetiver plants in nutrient solution*

In Vetiver plants grown for 15 days under unsterilized conditions there was considerable uptake of ^{14}C residue both in shoot (22.42%) and shoot (11.63%). When the plants were allowed to grow up to 30th day the uptake in shoots and roots were 24.59% and 9.02% respectively. In Vetiver plants grown for 15 days under sterilized conditions the uptake of ^{14}C residues in shoot and root were 19.97% and 8.79% respectively. The mass balance with plants grown in sterilized conditions was good. (Table 2 and Figure 3).

3.7 *Plant roots (Vetiver)–microbial interaction for remediation of atrazine contaminated soil (Figure 3)*

In this study Vetiver plants which are resistant to atrazine and capable of removing appreciable amount of atrazine are grown in soil with different variations. Bacterial culture R2 (*Arthrobacter* sp.) capable of atrazine degradation was isolated from rhizosphere of Vetiver plant. It has shown capacity to remove atrazine from contaminated soil. Interactive potential of Vetiver and bacterial culture isolated from its rhizosphere was tested for atrazine removal from soil. Experimental set up. Vetiver–microbial interaction for removal of atrazine from soil. (1) soil + 25 ppm atrazine + Vetiver; (2) soil + Vetiver (without atrazine); (3) soil + 25 ppm atrazine + Vetiver + culture; (4) sterile soil + 25 ppm atrazine; (5) sterile soil + 25 ppm atrazine + culture; (6) soil + 25 ppm atrazine + culture; (7) soil + 25 ppm atrazine.

1. After 4 days of atrazine exposure, the soil planted with Vetiver along with bacterial culture showed complete removal of atrazine.
2. Vetiver plants alone showed 31% atrazine removal from soil.

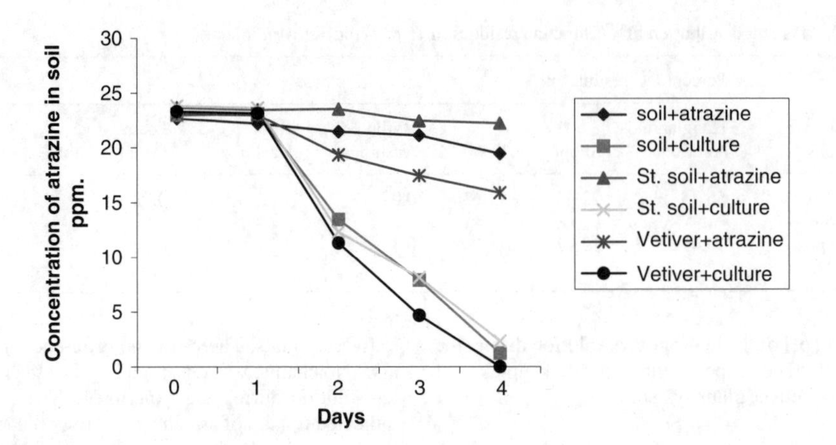

Figure 3.

3. Bacterial culture was able to remove 94% atrazine from unsterile soil.
4. Plant–microbial interaction showed rapid atrazine removal when compared with culture and plants alone. When comparing the results of the 3rd day, culture and plants alone showed 65% and 24% atrazine removal respectively, plant–microbial interaction showed 80% atrazine removal.
5. Sterile soil spiked with atrazine showed 6% atrazine removal after 5 days of exposure, which could be accounted for adsorbtion and other physical factors.
6. Normal flora of soil contributes to 14% atrazine removal.

Thus plant–microbial interaction has enhanced effect on atrazine removal. This plant–microbial remediation model can be explored for remediation of atrazine contaminated soil.

REFERENCES

Chaudhry, Q., Blom-Zandstra, M., Gupta, S. and Joner, E., 2004. Utilising the Synergy between Plants and Microbes to Enhance Breakdown of Organic Pollutants in the Environment Environmental Science & Pollution Research (in preparation).

Hanson, J.E., Stoltenberg, D.E., Lowery, B. and Binning, L.K., 1997. Influence of Application Rate on Atrazine Fate in a Silt Loam Soil, J. Environ. Qual. 26: 829–835.

Mirgain, G., Green, A. and Monteil, H., 1993. Degradation of Atrazine in Laboratory Microcosms Isolation and Identification of the Biodegrading Bacteria." Environmental Toxicology and Chemistry, Vol.12, p. 1627–1634.

Wackett, L.P., Sadowsky, M.J., Martinez, B. and Shapir N., 2002. Biodegradation of atrazine and related s-triazine compounds: from enzymes to field studies. Appl Microbiol Biotechnol. Jan; 58(1): 39–45.

Gas treatment

European Symposium on Environmental Biotechnology, ESEB 2004 - Verstraete (ed)
© 2004 Taylor & Francis Group, London, ISBN 90 5809 653 X

Fungal biodegradation of toluene in gas-phase biofilters

E. Estévez, M.C. Veiga & C. Kennes
University of La Coruña, Chemical Engineering Laboratory, Faculty of Sciences, Campus da Zapateira, La Coruña, Spain

ABSTRACT: Two new fungal strains, *Exophiala oligosperma* and *Paecilomyces variotii*, were found to use toluene as sole source of carbon and energy. In batch assays, toluene was mineralized without detection of any intermediate metabolite. According to mass balance calculations more than two third of the substrate was converted to carbon dioxide and water. Each strain was used individually as pure biocatalyst in two identical biofilters fed toluene polluted air for several months. All media were sterilized on start-up and an antibiotic was added to the nutritive solution in order to make sure that the pollutant was degraded exclusively by the fungi. The absence of any contaminant strain was checked by observations under optic and scanning electron microscopes as well as by basic microbiological studies. Toluene and carbon dioxide concentrations were measured both at the inlet and outlet of the biofilters. Elimination capacities of almost 80 and 55 $g/m^3 \cdot h$ with more than 99% removal efficiencies, were reached, respectively, with *Exophiala oligosperma* and *Paecilomyces variotii*, with complete mineralization of the pollutant.

1 INTRODUCTION

Toluene is a major component of gasoline and is widely used in industry. It is therefore the most frequently encountered monoalkylbenzene in industrial waste gases released from different sources, including paint manufacturing, plastics production, printing ink production and printing processes, production of pharmaceuticals, petroleum refining, surface coating, etc. (Pope *et al.*, 1989). Although it is not carcinogenic according to the World Health Organization (WHO, 1985), it is toxic and is considered as being a priority air pollutant.

Several conventional technologies are available for air pollution control, based on either mass transfer or oxidation processes (Kennes & Veiga, 2001). However, over the past few years biological treatment technologies have gained much interest mainly as a result of their high efficiency and relatively low cost (Kennes & Veiga, 2001). Bioprocesses represent a suitable alternative for the removal of pollutants at concentrations below approximately 5 g/m^3 and flow rates up to 200,000 m^3/h (Thalasso *et al.*, 2003), provided the target compounds are biodegradable.

The biodegradation of toluene by bacteria was discovered several decades ago and has been widely documented (Smith, 1990; Singleton, 1994). Different bacterial populations have been used in biofilters for the treatment of toluene polluted air (Jorio *et al.*, 1998; Mendoza *et al.*, 2004; Prado *et al.*, 2002; Veiga *et al.*, 1999; Wu *et al.*, 1999; Zilli *et al.*, 2000). Although relatively high elimination capacities were obtained in those recent studies compared to earlier reports on biofiltration of alkylbenzenes with compost or soil biofilters (Miller & Canter, 1991), active studies are being undertaken in several research group in order to further increase the performance of reactors used for waste gas treatment. Therefore, different new types of bioreactors and packing materials have recently been developed or improved, and tested either at lab-scale or industrial-scale (Kennes & Veiga, 2001, 2002). Another possibility consists in searching for more performant biocatalysts. It was recently suggested that the use of specialized fungi might allow improving the performance of gas-phase biofilters by enhancing the rate of mass transfer of pollutants from the air to the biocatalyst (van Groenestijn & Hesselink, 1993; Kennes & Veiga, 2001). Besides, fungi are more tolerant to extreme conditions compared to bacteria. However, the complete biodegradation of monoalkylbenzenes by fungi was reported for the first time only quite recently (Weber *et al.*, 1995).

In the present work, two new fungal strains were shown to use toluene as sole source of carbon and energy. The strains were inoculated in biofilters, avoiding external microbial contamination, and optimal performance of the bioreactors was evaluated over several months.

2 MATERIALS AND METHODS

2.1 *Microbial strains*

The fungal strains used in this study, namely *Exophiala oligosperma* and *Paecilomyces variotii*, were grown on a mineral medium (Mendoza *et al.*, 2003; Prado *et al.*, 2002) in 250 ml bottles. They were originally found in a biofilter which had reached a very high performance with elimination capacities exceeding 120 g/m^3 · h and removal efficiencies of 99.9% (Veiga & Kennes, 2001).

2.2 *Design and performance of the biofilters*

The biofiltration of toluene was performed in two identical glass biofilters packed with perlite and operated under the same conditions but each inoculated with a different pure culture of either *Exophiala oligosperma* or *Paecilomyces variotii*. A clean air flow was mixed with a toluene contaminated air stream fed to the biofilters according to the scheme shown in Figure 1 (Prado *et al.*, 2002). The combined humidified air stream flew in a downflow mode through the filter beds. Several sampling ports allowed for the analysis of biofilm samples and air samples at different depths. The reactors were operated at room temperature. Other characteristics of the experimental set-up were as described elsewhere (Estévez *et al.*, 2004).

2.3 *Analytical methods*

Toluene concentrations in the gas-phase were measured on a HP-5890 series II gas chromatograph (GC) equipped with a Flame Ionization Detector and a 30 m HP-5 capillary column according to previously specified conditions (Prado *et al.*, 2002). The amount of carbon dioxide produced was detected with a gas-sensor.

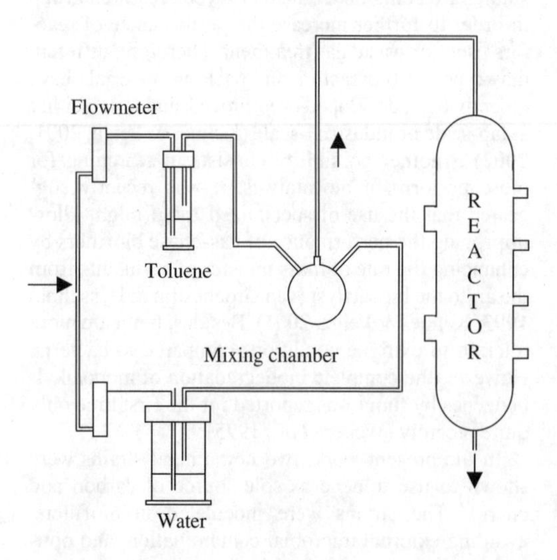

Figure 1. Biofilter fed toluene polluted air. The arrows indicate the air flow.

2.4 *Biofilter performance*

The performance of the biofilters was usually checked on a daily basis. The following parameters were used to evaluate and compare such performance:

$$TL = \frac{C_{in}.Q}{V}(g/m^3.h)$$

$$EC = Q.\frac{(C_{in} - C_{out})}{V}(g/m^3.h) \quad (1)$$

$$RE = \frac{(C_{in} - C_{out})}{C_{in}}.100(\%)$$

where TL = toluene load; EC = elimination capacity; RE = removal efficiency; C_{in} = inlet concentration; C_{out} = outlet concentration; Q = air flow rate; V = volume of the filter bed.

3 RESULTS AND DISCUSSION

3.1 *Fungal toluene biodegradation in batch assays*

The fungi *Exophiala oligosperma* and *Paecilomyces variotii* are, respectively, a new species and a new strain. They biodegrade toluene in a slightly acidic mineral medium with an optimum activity around 30°C. When grown in batch cultures on toluene as single carbon and energy source, complete substrate removal is observed with recoveries of about two third of the carbon source as carbon dioxide with both strains. Accumulation of potential intermediate metabolites in the culture broth was not detected. To our knowledge only two other *Exophiala* strains have been identified and isolated on substituted benzenes. *Exophiala lecanii-corni* was isolated on toluene (Woertz *et al.*, 2001), and *Exophiala jeanselmei* was isolated earlier on styrene, although that strain cannot use toluene as carbon and energy source (Cox *et al.*, 1997). No data have been published earlier on the growth of *Paecilomyces* strains on monoalkylbenzenes. Contrary to what had been suggested several years ago, the ability to degrade specific pollutants can apparently not be used for taxonomic purposes in fungi since such characteristic seems to be strain-specific rather than an ability of given genera or species (Nyns *et al.*, 1968; Oudot *et al.*, 1993).

3.2 *Biofiltration of toluene with the fungal strains*

3.2.1 *Inoculation and start-up*

Both cultures were grown on toluene in sterilized mineral medium. *Exophiala oligosperma* presented a typical filamentous growth, while *Paecilomyces variotii* grew in the form of pellets. Because of these two different forms of growth, the homogenous inoculation of the packing material was somewhat more difficult in case of the *Paecilomyces* strain than with the *Exophiala* strain. However, after a few days heavy biomass growth and biofilm development was observed in both biofilters.

3.2.2 *Long term biofilter operation*

A relatively constant toluene load was maintained during approximately the first two weeks after inoculation. Afterwards, on the third week of operation, the load was gradually increased from about $10\,g/m^3 \cdot h$ to more than $100\,g/m^3 \cdot h$. The *Exophiala* strain appeared to perform better than *Paecilomyces variotii*. The results, compared in Figure 2, show that the black yeast reached a maximum elimination capacity of almost $80\,g/m^3 \cdot h$ for a removal efficiency close to 100%, while this value was about twenty to twenty five percent lower in case of the other strain. Both strains were able to grow at higher loads even above $100\,g/m^3 \cdot h$. However, the removal efficiency gradually decreased under such conditions. At such high loads, some inhibitory effect was observed for the biofilter inoculated with the *Paecilomyces* strain, contrary to the other biocatalyst.

As in batch assays, analysis of carbon dioxide in the air leaving both biofilters allowed to ensure that most of the pollutant was degrade to carbon dioxide and water. The absence of any intermediate product confirmed the previous assumption.

The presence of the inoculated strains as single dominant organisms in the filter beds was regularly checked. The addition of an antibiotic in the culture broth used for the inoculation of both reactors as well as during the start-up phase, when supplying the nutritive solution, contributed to avoid bacterial contamination. Once heavy biofilm growth was reached, microbial contamination was much less probable than in the early stages of operation of the biofilters. No other strains were found in any of the biofilters during the almost six months operation. A typical Scanning Electron Micrograph (SEM) of a filter bed sample taken from the biofilter inoculated with *Exophiala oligosperma* after four months operation is shown in Figure 3. Abundant growth is clearly observed as well as the formation of a characteristic filamentous network. No bacteria were detected at all under the microscopes in any of the several filter bed samples taken from both biofilters. Bacteria were not detected either by basic microbiological techniques.

Very few other papers have been published on the biofiltration of benzene compounds by fungal strains. In most cases no special care was taken to avoid microbial contamination and other strains were present in the filter beds, often at non negligible concentrations (Cox *et al.*, 1997; García *et al.*, 2001; Woertz *et al.*, 2001). The present results as well as data on reactor performance reported by other groups are compared in Table 1. In some cases, elimination capacities of $>200\,g/m^3 \cdot h$ were reached. However, such results were so far only observed for short periods or under unstable conditions. Optimization work should allow maintaining such encouraging results for longer periods of time.

4 CONCLUSIONS AND FUTURE RESEARCH PERSPECTIVES

Although the mineralization of toluene was discovered quite recently in fungi, these organisms allow to

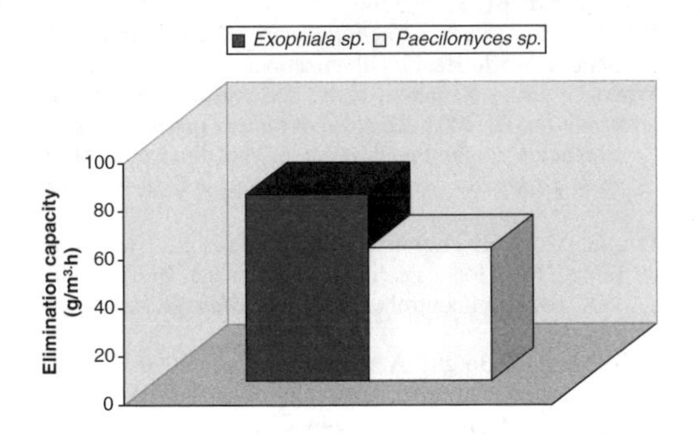

Figure 2. Comparison of the maximum biofilter performance reached with almost 100% removal efficiency with *Exophiala oligosperma* and *Paecilomyces variotii*.

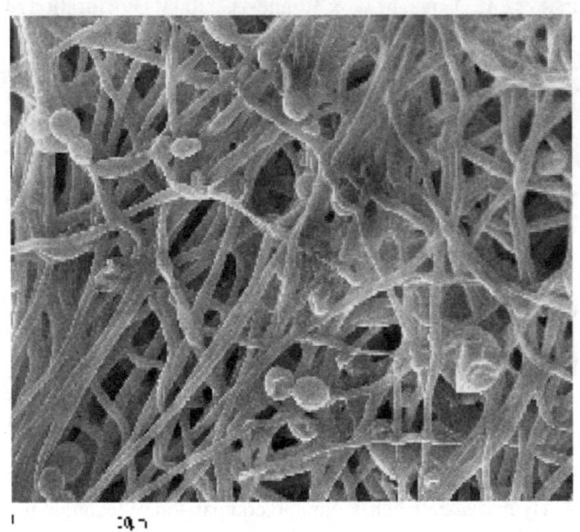

Figure 3. SEM photograph of a filter bed sample withdrawn from the biofilter inoculated with the *Exophiala* strain.

Table 1. Maximum toluene elimination capacities reached in biofilters with different fungi.

Dominant strain	Elimination capacity ($g/m^3 \cdot h$)	References
E. oligosperma	≈80	Present study
P. variotii	55	Present study
E. lecanii-corni	≈80–90**	Woertz *et al.*, 2001
S. apiospermum	≈90–100**	García *et al.*, 2001
*E. jeanselmei**	62	Cox *et al.*, 1997

* With styrene as pollutant, instead of toluene.
** Elimination capacities >200 $g/m^3 \cdot h$ were obtained in short term experiments or under unsteady-state conditions.

reach a high reactor performance when present in biofilters used for waste gas treatment. Most of the isolated strains tested in biofiltration studies use toluene as sole source of carbon and energy and do basically completely mineralize the pollutant. Present and future research on the optimization of such systems should allow reaching stable elimination capacities exceeding 100 $g/m^3 \cdot h$ in long-term operation.

ACKNOWLEDGMENT

The present research was financed by the Spanish Ministry of Science and Technology.

REFERENCES

Cox, H.H.J., Moerman, R.E., van Baalen, S., van Heiningen, W.N.M., Doddema, H.J. & Harder, W. 1997. Performance of a styrene-degrading biofilter containing the yeast *Exophiala jeanselmei*. *Biotechnol. Bioeng.* 53: 259–266.

Estévez, E., Veiga, M.C. & Kennes, C. 2004. Biofiltration of waste gases with the fungi *Exophiala oligosperma* and *Paecilomyces variotii*. (in press).

García-Peña, E.I., Hernández, S., Favela-Torres, E., Auria, R. & Revah, S. 2001. Toluene biofiltration by the fungus *Scedosporium apiospermum* TB1. *Biotechnol. Bioeng.* 76: 61–69.

Jorio, H., Kiared, K., Brzezinski, R., Leroux, A., Viel, G. & Heitz, M. 1998. Treatment of air polluted with high concentrations of toluene and xylene in a pilot-scale biofilter. *J. Chem. Technol. Biotechnol.* 73: 183–196.

Kennes, C. & Veiga, M.C. 2001. *Bioreactors for Waste Gas Treatment*. Dordrecht: Kluwer Academic Publishers.

Kennes, C. & Veiga, M.C. 2002. Inert filter media for the biofiltration of waste gases – characteristics and biomass control. *Re/Views in Environmental Science & Bio/Technology*. 1: 201–214.

Mendoza, J.A., Prado, O.J., Veiga, M.C. & Kennes, C. 2004. Hydrodynamic behaviour and comparison of technologies for the removal of excess biomass in gas-phase biofilters. *Water Res.* 38: (in press).

Mendoza, J.A., Veiga, M.C. & Kennes, C. 2003. Biofiltration of waste gases in a reactor with a split-feed. *J. Chem. Technol. Biotechnol.* 78: 703–708.

Miller, D.E. & Canter, L.W. 1991. Control of aromatic waste air streams by soil bioreactors. *Environ. Prog.* 10: 300–306.

Nyns, E.J., Auquiere, J.P. & Wiaux, A.L. 1968. Taxonomic value of the property of fungi to assimilate hydrocarbons. *Antonie van Leuwenhoek*. 34: 441–447.

Oudot, J., Dupont, J., Haloui, S. & Roquebert, M.F. 1993. Biodegradation potential of hydrocarbon-assimilating tropical fungi. *Soil Biol. Biochem.* 25: 1167–1173.

Pope, A.A., Brooks, G., Moody, T., Most, C. & Patterson, G. 1989. *Toxic air pollutants/source crosswalk: a screening tool for locating possible sources emitting toxic air pollutants*. US-EPA. Report EPA-450/2-89-017.

Prado, O.J., Mendoza, J.A., Veiga, M.C. & Kennes, C. 2002. Optimization of nutrient supply in a downflow gas-phase biofilter packed with an inert carrier. *Appl. Microbiol. Biotechnol.* 59: 567–573.

Singleton, I. 1994. Microbial metabolism of xenobiotics: Fundamental and applied research. *J. Chem. Technol. Biotechnol.* 59: 9–23.

Smith, M.R. 1990. The biodegradation of aromatic hydrocarbons by bacteria. *Biodegradation*. 1: 191–206.

Thalasso, F., Veiga, M.C. & Kennes, C. 2003. Biofiltration for Waste Gas Handling. In S.N. Agathos & W. Reineke (eds), *Biotechnology for the Environment: Wastewater Treatment and Modeling, Waste Gas Handling*: pp. 239–258. Dordrecht: Kluwer Academic Publishers.

Van Groenestijn, J.W. & Hesselink, P.G.M. 1993. Biotechniques for air pollution control. *Biodegradation*. 4: 283–301.

Veiga, M.C., Fraga, M., Amor, L. & Kennes, C. 1999. Biofilter performance and characterization of a biocatalyst degrading alkylbenzene gases. *Biodegradation*. 10: 169–176.

Veiga, M.C. & Kennes, C. 2001. Parameters affecting performance and modeling of biofilters treating alkylbenzene-polluted air. *Appl. Microbiol. Biotechnol.* 55: 254–258.

Weber, F.J., Hage, K.C. & de Bont, J.A.M. 1995. Growth of the fungus *Cladosporium sphaerospermum* with toluene as the sole carbon and energy source. *Appl. Environ. Microbiol.* 61: 3562–3566.

WHO, 1985. *Toluene. Environmental Health Criteria 52*. Geneva: World Health Organization.

Woertz, J.R., Kinney, K.A., McIntosh, N.D.P. & Szaniszlo, P.J. 2001. Removal of toluene in a vapor-phase bioreactor containing a strain of the dimorphic black yeast *Exophiala lecanii-corni*. *Biotechnol. Bioeng.* 75: 550–558.

Wu, G., Conti, B., Leroux, A., Brzezinski, R., Viel, G. & Heitz, M. 1999. A high performance biofilter for VOC emission control. *J. Air Waste Manage. Assoc.* 49: 185–192.

Zilli, M., Del Borghi, A. & Converti, A. 2000. Toluene vapour removal in a laboratory-scale biofilter. *Appl. Microbiol. Biotechnol.* 54: 248–254.

European Symposium on Environmental Biotechnology, ESEB 2004 - Verstraete (ed)
© 2004 Taylor & Francis Group, London, ISBN 90 5809 653 X

BioDeNOx: novel process for NO_x removal from flue gases based on chemically enhanced biological NO and iron reduction

Peter Van der Maas, Bram Klapwijk & Piet Lens
Sub-Department of Environmental Technology, Wageningen University, Bomenweg 2, Wageningen, The Netherlands

ABSTRACT: A novel process for NO_x removal from flue gases, called BioDeNOx, combines the principles of wet absorption of NO into a aqueous Fe(II)EDTA solution with biological reduction of NO in a bioreactor. This research investigates the core processes of the biological regeneration of Fe(II)EDTA: reduction of NO and EDTA chelated Fe(III). The reduction of NO to N_2 was found to be biologically catalyzed. The NO reduction kinetics follow first order with the NO/nitrosyl concentration. Besides absorbent, Fe(II)EDTA serves as electron donor for NO reduction. This implies that redox cycling of FeEDTA plays an important role in the biological denitrification process. However, continuous reactor experiments demonstrated that not the denitrification capacity, but the iron reduction capacity was the limiting the load of the bioreactor. When treating flue gas containing 3,3% O_2 and 500 ppm NO, approximately 90% of the electron flow was used for Fe(III)EDTA reduction and only 10% for NO reduction. Batch experiments strongly suggest that the reduction of EDTA chelated Fe(III) is not a direct enzymatic conversion, but an indirect nonenzymatic reaction. The redox couple S^0/S^{2-} plays an important role in the electron transfer between the bacteria and Fe(III)EDTA. The redox couple sulfide/elemental accelerates the electron transfer between the bacteria and Fe(III)EDTA. Thus, Fe(III)EDTA reduction is most likely a nonenzymatic conversion.

1 INTRODUCTION

The emission of nitrogen oxides (NO_x) to the atmosphere causes serious environmental problems, e.g. acid rain and depletion of the ozone layer *(Harding, 1996)*. Selective catalytic reduction (SCR) processes, which are generally applied for NO_x emission abatement, require high temperatures and expensive catalysts *(Davis, 1992)*. Therefore, biological techniques using denitrification may represent promising alternatives for the conventional SCR techniques, because denitrification occurs at ambient temperatures with the use of cheap microbial inocula (e.g. soil or activated sludges). One major drawback of biological techniques like trickling filtration is a rather low treatment efficiency at economic retention times *(Du Plessis, 1998)* because of the rather slow transfer of NO from the gas to the liquid phase. When using aqueous solutions of Fe(II)EDTA as scrubber liquor, the mass transfer can be accelerated. These Fe(II)EDTA solutions have the ability to form stable complexes with NO, and therefore provides high absorption efficiencies for gaseous NO *(Demmink, 1997)*. In an instantaneous reaction, the nitrosyl-complex is formed according to the reactions 1 and 2:

$$NO\ (g) \leftrightarrow NO\ (aq) \quad (1)$$

$$NO\ (aq) + Fe(II)EDTA^{2-} \leftrightarrow Fe(II)EDTA\text{-}NO^{2-} \quad (2)$$

The principles of wet absorption of NO with biological reduction of NO in a bioreactor are combined in the so called BioDeNOx process *(Buisman, 1999)*, as schematically represented in Figure 1.

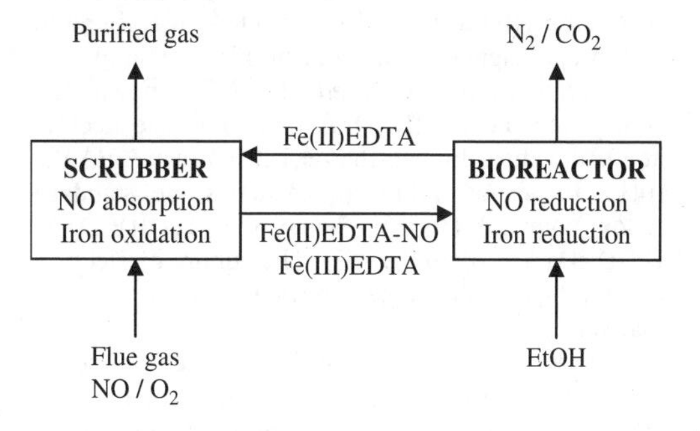

Figure 1. Schematic principle of BioDeNOx.

In this process, the biological reduction of NO to di-nitrogen gas (N_2) takes place under thermophilic conditions, at around 50–55°C, which is the adiabatic temperature of scrubber liquors. When ethanol is used as electron donor, the denitrification reaction occurs according to the overall reaction 3 *(Buisman, 1999)*:

$$6\ Fe(II)EDTA\text{-}NO^{2-} + C_2H_5OH \rightarrow 6\ Fe(II)EDTA^{2-} + 3\ N_2 + CO_2 + 3\ H_2O \quad (3)$$

Since industrial flue gasses generally contain 2–8% oxygen, part of the Fe(II)EDTA is oxidized to Fe(III)EDTA according to:

$$4\ Fe(II)EDTA^{2-} + O_2\ (aq) \rightarrow 4\ Fe(III)EDTA^{-} + 2\ H_2O \quad (4)$$

To regenerate the absorption liquor, the Fe(III)EDTA that is formed by reaction 4 has to be reduced back to Fe(II)EDTA. Thus, besides NO reduction, reduction of EDTA chelated Fe(III) is a core reaction within the regeneration pathway of the BioDeNOx process.

This research aims to elucidate the processes taking place in the bioreactor and to optimize the biological regeneration of Fe(II)EDTA. Here we summarize our main results obtained so far.

2 CONTINUOUS REACTOR

The technical feasibility of the BioDeNOx concept was demonstrated in a bench scale installation with a continuous flue gas flow of 650 l/h *(Van der Maas, submitted)*. Ethanol was supplied as the electron donor for the biological regeneration process. Figure 2A shows that the NO removal efficiency was 70–80% at an influent concentration of 500 ppm NO. Note that this removal efficiency is the result of the limited scrubber/absorption efficiency and that all scrubbed NO was fully converted to N_2.

The ORP in the bioreactor, which indicates the ratio between Fe(III)EDTA and Fe(II)EDTA, was found to be a proper signal for the control of the ethanol supply *(Van der Maas, submitted)*. The NO removal efficiency and the Fe(III)EDTA reduction rate decline at ORP values higher than −140 mV vs Ag/AgCl (pH 7.0), which value approximates the standard redox potential (E_0) of the system Fe(III)EDTA/Fe(II)EDTA *(Kolthoff, 1957)*. The supply of electron donor (ethanol) should be controlled at an ORP below that value.

Not the NO reduction capacity, but the iron reduction capacity was the limiting the load of the bioreactor *(Van der Maas, submitted)*. Inoculation of the bioreactor with sludge with good properties for reduction of EDTA chelated Fe(III), resulted in a stable operation with a flue gas containing 3,3% O_2 and 500 ppm NO. Under these conditions, app. 90% of the electron flow is used for Fe (III) EDTA reduction and only 10% for NO reduction (Fig. 2B).

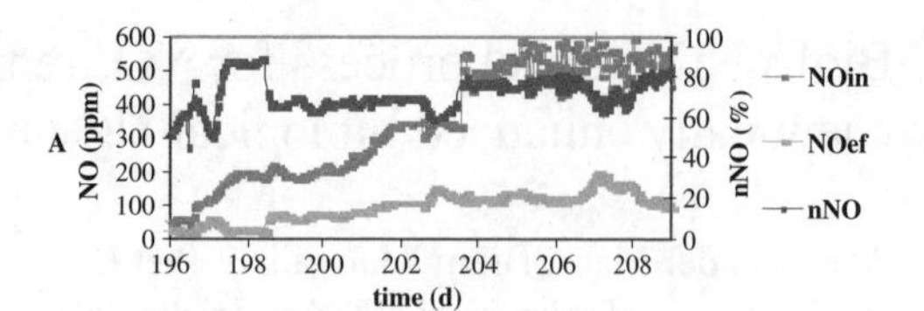

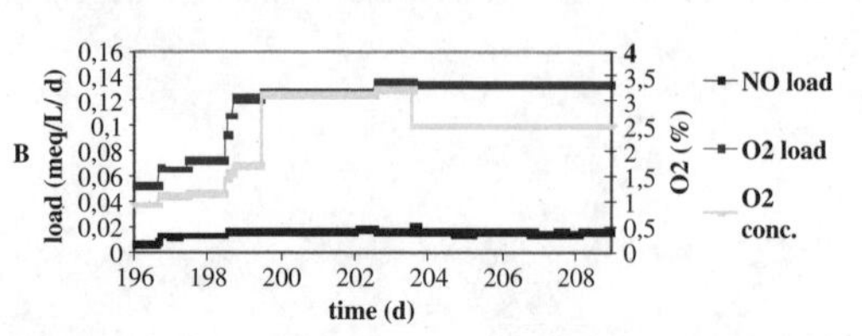

Figure 2. NO removal efficiency (A) and bioreactor load (B). The load is expressed as molar electron equivalents per liter per day ($meq \cdot l^{-1} \cdot d^{-1}$).

3 NO REDUCTION

Various sludges from full scale denitrifying and anaerobic reactors, as well as several electron donors were screened on their capability to catalyze NO reduction under thermophilic conditions (55°C, pH 7.2) *(Van der Maas, 2003)*. Reduction of NO to N_2 showed to be biologically catalyzed with nitrous oxide (N_2O) as an intermediate. Fe(II)EDTA turned out to be a suitable electron donor for NO reduction according to reaction 5:

$$Fe(II)EDTA\text{-}NO^{2-} + 2H^+ \rightarrow N_2O + H_2O + 2\ Fe(III)EDTA^{-} \quad (5)$$

Similar as found for NO reduction, the presence of ethanol did not affect the rate of N_2O reduction to N_2 *(Van der Maas, 2003)*. This indicates that electrons provided by Fe(II) also served N_2O reduction, according to reaction 6:

$$N_2O + 2\ Fe(II)EDTA^{2-} + 2\ H^+ \rightarrow N_2 + H_2O + 2\ Fe(III)EDTA^{-} \quad (6)$$

Figure 3 shows that the NO reduction rate follows first order kinetics with the NO/nitrosyl concentration. The NO reduction rate was independent of the presence of any external e-donor like ethanol, acetate or hydrogen. This possibly indicates that Fe(II)EDTA

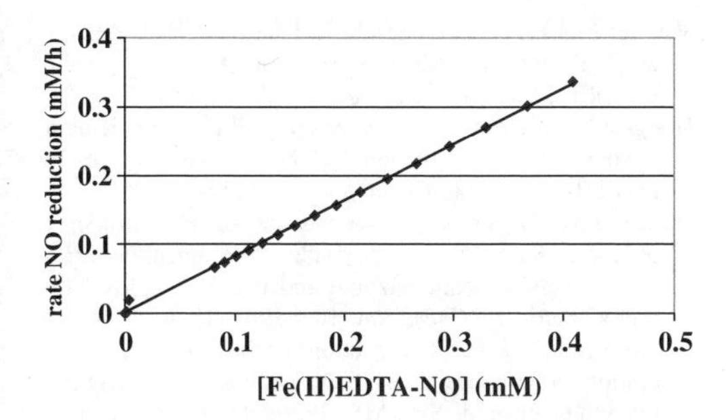

Figure 3. NO reduction rate as function of the NO concentration.

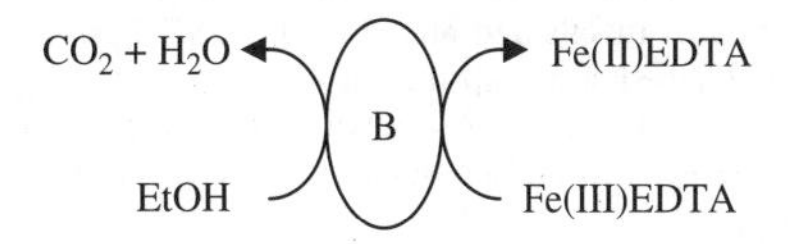

CO$_2$ + H$_2$O ◄ (B) ► EM$_{red}$ ► Fe(II)EDTA

EtOH — EM$_{ox}$ ◄ — Fe(III)EDTA

Figure 4. Direct enzymatic versus indirect nonenzymatic Fe-reduction. EM = electron mediator, B = bacterium.

is the preferred electron donor for NO reduction. That hypothesis seems to be supported by literature *(Hodges, 1974)*: Fe(II)EDTA (E_0 = 0.10 V) chemically reduces Ferri-cytochrome c (E_0 = 0.26 V) which is an electron carrier for bacterial NO reductase *(Wasser, 2002)*.

4 IRON REDUCTION

As mentioned above, not NO reduction but iron reduction was the limiting process of the bio-regeneration. To optimize this process, it is important to know the principle of the reduction of EDTA chelated Fe(III). Batch experiments showed that reduction of EDTA chelated Fe(III) was catalyzed by all the four inocula tested *(Van de Sandt, 2002)*. On the other hand it is known that Fe(III)EDTA in high concentrations (e.g. 25 mM) is a poor electron acceptor for most dissimilatory Fe-reducing bacteria *(Finneran, 2002)*. This rises the question whether, in the application of BioDeNOx, the reduction of Fe(III)EDTA is a direct enzymatic conversion (catalyzed by dissimilatory Fe-reducing bacteria) or an indirect nonenzymatic reduction with the use of one or more electron shuttling compounds, i.e. sulfur, quinones, riboflavin. Both options are schematically represented by Figure 4.

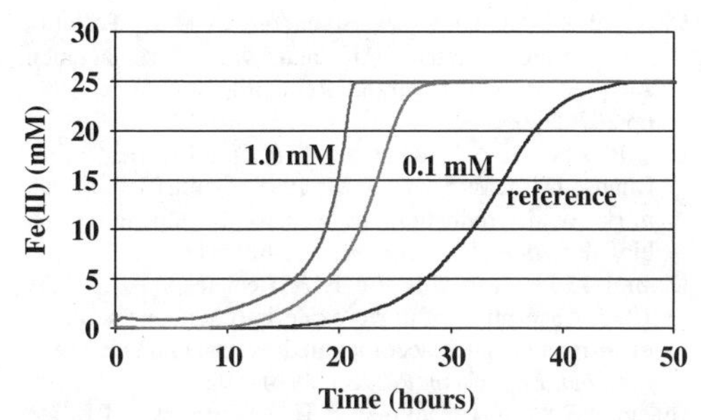

Figure 5. Acceleration of Fe(III)EDTA reduction by low concentrations of S^0/S^{2-}.

Figure 5 shows that low concentrations of sulfide/sulfur accelerate the reduction of EDTA chelated Fe(III) by approximately a factor 5. Sulfide chemically reduces Fe(II)EDTA with the concomitant formation of elemental sulfur *(Neumann, 1984)*. Figure 5 strongly indicates that, in the bio-regeneration of BioDeNOx, the reduction of EDTA chelated Fe(III) is not a direct enzymatic conversion, but an indirect nonenzymatic reaction. The redox couple S^0/S^{2-} plays an important role in the electron transfer between the bacteria and Fe(III)EDTA.

5 CONCLUSION

NO_x removal by chemical absorption to aqueous Fe(II)EDTA solutions, combined with the biological regeneration of that scrubber liquor, is a promising alternative for conventional chemical and biological treatment techniques. Biological reduction of both NO and Fe(III)EDTA are the crucial conversions during the regeneration of Fe(II)EDTA. Redox cycling of FeEDTA plays an important role in the biological denitrification process. The redox couple sulfide/elemental accelerates the electron transfer between the bacteria and Fe(III)EDTA. Thus, Fe(III)EDTA reduction is most likely a nonenzymatic conversion.

ACKNOWLEDGMENT

This research was financially supported by the Dutch Foundation for Applied Sciences (STW-NWO), grant STW WMK 4963, and Biostar Development CV.

REFERENCES

Harding, A.W., Brown, S.D. & Thomas, K.M. 1996, Release of NO from the combustion of coal chars. *Combustion and Flame*. 107: 336–350.

Davis, W.T., Pakrasi, A. & Buonicore, A.J. Air Pollution Engineering Manual. Air and Waste Management Association, Van Nostrand Reinhold, New York. 1992, pp 207–262.

Du Plessis, C.A., Kinney, K.A., Schroeder, E.D., Chang, D.P.Y. & Scow, K.M. 1998, Denitrification and nitric oxide reduction in a aerobic toluene-treating biofilter. *Biotechnol. Bioeng.* 58: 408–415.

Demmink, J.F., Van Gils, I.C.F. & Beenackers, A.A.C.M. 1997, Absorption of nitric oxide into aqueous solutions of ferrous chelates accompanied by instantaneous reaction. *Ind. Eng. Chem Res.* 36: 4914–4927.

Buisman, C.J.N., Dijkman, H., Verbraak, P.L. & Den Hartog, A.J. 1999, Process for purifying flue gas containing nitrogen oxides. United States Patent US5891408.

Van der Maas, P., Van den Bosch, P., Klapwijk, B. & Lens, P. *Submitted.* NOx removal from flue gas by an integrated technique using chemical absorption and biological conversion.

Kolthoff, I.M. & Auerbach, C. 1952, Studies on the system Iron-Ethylenediamine Tetraacetate. *J. Am. Chem. Soc.* 74: 1452–1456.

Van der Maas, P., Van de Sandt, T., Klapwijk, B. & Lens, P. 2003, Biological Reduction of Nitric Oxide in Aqueous Fe(II)EDTA Solutions. *Biotechnol. Prog.* 19: 1323–1328.

Hodges, H.L., Holwerda, R.A. & Gray, H.B. 1974, Kinetic studies of the reduction of Ferricytochrome c by $Fe(EDTA)^{2-}$. *J. Am. Chem. Soc.* 96: 3132–3137.

Wasser, I.M., De Vries, S., Moënne-Loccoz, P., Schröder, I. & Karlin, K.D. 2002, Nitric oxide in biological denitrification: Fe/Cu metalloenzyme and metal complex NOx redox chemistry. *Chem. Rev.* 102: 1201–1234.

Van de Sandt, T. 2002, Key reactions of the biological regeneration within the BioDeNOx concept using various inocula and electron donors. MSc-thesis 03–05, Wageningen University.

Finneran, K.T., Forbush, H.M., VanPraagh, C.G. & Lovley, D.R. 2002, *Desulfitobacterium metallireducens* sp. nov., an anaerobic bacterium that couples growth to the reduction of metals and humic acids. *IJSEM papers in press*, published online april 2002.

Neumann, D. W. & Lynn, S. 1984, Oxidative absorption of H2S and O2 by iron chelate solutions. *AIChE Journal.* 30: 62–69.

European Symposium on Environmental Biotechnology, ESEB 2004 - Verstraete (ed)
© 2004 Taylor & Francis Group, London, ISBN 90 5809 653 X

Application of haloalkaliphilic sulfur-oxidizing bacteria for the removal of H_2S from gas streams

H. Banciu[1], R. Kleerebezem[1], G. Muyzer[1], J.G. Kuenen[1] & D.Y. Sorokin[1,2]
[1] *Dept. of Biotechnology, Delft University of Technology, Delft, The Netherlands*
[2] *Institute of Microbiology, Russian Academy of Sciences, Moscow, Russian Federation*

ABSTRACT: Since H_2S absorption from gases is enhanced under haloalkaline conditions, the organisms capable of living at these conditions and metabolizing the sulfur compounds could be an attractive option for biotechnology. Recently, two new haloalkaliphilic genera have been described. *Thioalkalimicrobium* spp. and *Thioalkalivibrio* spp. are chemolithoautotrophic sulfur-oxidizing bacteria actively involved in the natural sulfur cycle in alkaline soda lakes. The representatives of the first genus have a high growth rate, low biomass yield and low salt tolerance, whereas the second group is characterized by a low growth rate, high biomass yield and high salt tolerance. The energy sources for these extremophiles are inorganic sulfur compounds like thiosulfate, sulfide, polysulfide, elemental sulfur, tetrathionate or thiocyanate. Thiosulfate, sulfide and polysulfide are oxidized at relatively high rates at low to moderate salt concentration (0.6 to 2 M of total Na^+). The capacity of these organisms to convert reduced sulfur compounds in haloalkaline conditions makes them highly attractive for application in biotechnological processes for hydrogen sulfide removal from waste streams.

1 INTRODUCTION

Several environmental problems are caused by sulfur compounds like sulfate (pollution of surface water, acid mine drainage), SO_2 (acid rain), H_2S (odor problems, high toxicity, acid rain) and methylated sulfur compounds (odor problems, toxicity, climate change). The aim of sulfur biotechnology is to prevent loss of sulfur compounds to the atmosphere and to avoid complete oxidation of sulfur compounds to sulfate. Current research is therefore focused on the production of sulfur compounds that can be easily separated from the waste streams, stored and re-used for other purposes. One of the successful processes is the production of elemental sulfur from H_2S-containing gas streams by sulfur-oxidizing bacteria in the Thiopaq® process (Paques BV, Balk, The Netherlands) (Fig. 1). In this system gasses can be treated by the absorption of H_2S in a scrubber unit, subsequent biological oxidation of sulfide to elemental sulfur at neutral pH and separation of the sulfur and recycling of the percolation water to the scrubber (Janssen et al. 2001). A variety of gas streams (e.g. pressurized natural gas, synthesis gas, biogas and refinery gas) can be treated with this two-steps process. Points for major innovation of this process are the enhancement of the stripping efficiency of H_2S in the scrubber (by elevating the pH) and the reduction

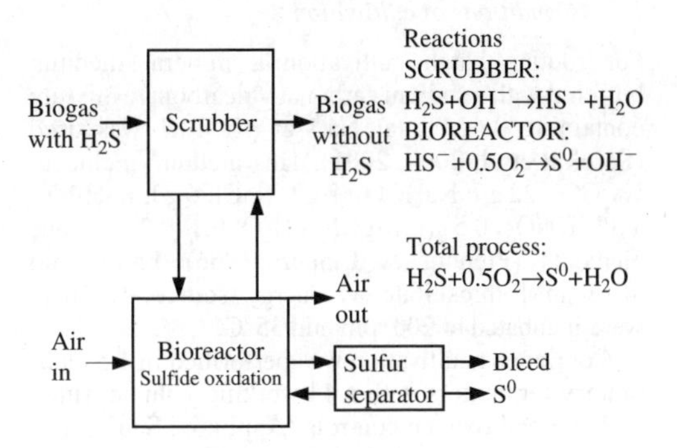

Figure 1. Block process diagram of the Thiopaq®-bioprocess and reaction mechanisms involved.

of the bleed stream of the aerobicreactor (by maintaining high salt conditions). Moreover, since high CO_2 content is usual for H_2S-containing industrial gases, use of alkaline carbonates in the scrubber instead of organic or inorganic alkali (NaOH) is beneficial for selectivity of H_2S absorption.

In our laboratory, various sulfur-oxidizing bacteria have been isolated under conditions of high salt and high pH (Sorokin et al. 2000, 2001, 2002). These alkaliphilic sulfur-oxidizing bacteria that originate from soda lakes of Siberia (Russia) and Kenya, can tolerate very high pH (up to 10.6–11) and salt concentrations (1–4 M Na^+) making them attractive for biotechnological sulfide removal. The aim of the present research was to characterize the physiology and growth kinetics of the newly isolated haloalkaliphilic chemolithotrophic sulfur-oxidizing bacteria with respect to their potential use for sulfur production from H_2S-containing gas streams at haloalkaline conditions.

2 MATERIALS AND METHODS

2.1 *Biological material*

Pure cultures of obligately chemolithoautotrophic, haloalkaliphilic, sulfur-oxidizing bacteria of the genus *Thioalkalimicrobium* and *Thioalkalivibrio* have been used in this work. The organisms have been isolated from soda lakes in Kenya and Central Asia (Sorokin et al., 2000, 2001). The type strains *Thioalkalimicrobium aerophilum* (strain AL 3) and *Thioalkalivibrio versutus* (strain AL 2), originally isolated from a Siberian soda lake and some strains from the Kenyan soda lakes, routinely maintained in our laboratory, were used for the ecophysiological experiments. *Tv. versutus*, strain ALJ 15 was used as model organism to study the effect of varying sodium concentration on the growth kinetics.

2.2 *Conditions of cultivation*

For routine batch cultivation a mineral medium buffered with a sodium carbonate/bicarbonate mixture containing 0.6 M total Na^+ at pH 10.1 was used (Sorokin et al. 2000, 2001). This medium included: Na_2CO_3, 22 g/l; $NaHCO_3$, 8 g/l; NaCl, 6 g/l; K_2HPO_4, 1 g/l; KNO_3, 0.5 g/l; $MgCl_2 \cdot 6H_2O$, 0.1–0.2 g/l; trace elements (Pfennig & Lippert 1966) 1 ml/l, and 40–80 mM thiosulfate as energy source. Cultures were incubated at 200 rpm and 35°C.

Continuous cultivation was performed in 1.5 l laboratory fermentors with a 1 l working volume, fitted with pH and oxygen controls (Applikon, Schiedam). The pH was controlled by automatic titration with 2 M NaOH and HCl. The dissolved oxygen concentration was controlled at the level of minimum 50% air saturation by the stirring speed. The temperature was controlled at 35°C. For chemical stability, the medium was supplied from two reservoirs, containing acidic and alkaline solutions in double strength. The acidic solution contained: KH_2PO_4 – 1 g/l, $MgCl_2 \cdot 6H_2O$ – 0.2 g/l, 2 ml/l trace elements mixture and 1 ml/20 l silicone antifoam. The alkaline base included KNO_3 – 1 g/l and $Na_2CO_3/NaHCO_3$ two times concentrated. Thiosulfate was sterilized separately as 2 M solution and added to the alkaline solution to give a final concentration of 40 mM.

2.3 *Respiration measurements*

Cells were harvested by centrifugation, washed and resuspended in buffers containing 0.6, 2 or 4 M total Na^+ at a pH corresponding to that of the growth medium. The respiration rates were measured at 35°C in a 5 ml cell mounted on a magnetic stirrer and fitted with a dissolved oxygen probe (Yellow Spring Instruments, Ohio). Thiosulfate, sulfide, polysulfide and elemental sulfur (as acetone solution) were added at final concentrations of 50 μM and 34 μM. The buffers consisted of a mixture of carbonate and bicarbonate, and 50 mM KCl, pH 10, with sodium concentration between 0.1 and 4 M. The rate values represent average results obtained from 3 to 5 independent measurements with standard deviations less than 10%.

2.4 *Chemical analysis*

Thiosulfate concentrations were determined by cyanolytic procedures (Kelly et al. 1969) or by standard iodimetric titration after neutralization of the medium with 50% acetic acid. Elemental sulfur was detected by cyanolysis (Sörbo 1957). Sulfate was determined by a modified turbidimetric method (Kolmert et al. 2000) Cell protein was measured by Lowry method (Lowry et al. 1951) using bovine serum albumin (BSA) as standard. When elemental sulfur was present, it was extracted with acetone from the biomass pellet before hydrolysis to avoid interference with Lowry assay. For accurate measurement of the dry weight, cells were washed with 0.6 M or 2 M NaCl solutions to avoid cell lysis. The data represent average values obtained from 3 independent measurements.

2.5 *Kinetics analysis*

The kinetic constants ($qO_{2\ max}$) and apparent K_s were estimated from respiration experiments using washed cells of the cultures grown at different salt concentration. To increase the sensitivity of the K_s measurements at 1–5 μM substrate concentration, the respiration experiments were run at 10% air saturation. The K_s values were calculated from the VO_2-S plots based on three independent measurements. Y_{max} values were determined graphically from q_s-D ($q_s = \mu/Y$) and 1/Y-1/D plots, respectively, on the basis of the Pirt modification of the Monod growth model. For each dilution rate at least three steady state biomass concentrations were measured with an interval of 1 volume change. Each determination was done in triplicate; the data represent average values with standard deviation less than 10%. The maximum specific growth rate for each sodium concentration was determined experimentally as the dilution rate at which wash-out of the biomass and accumulation of thiosulfate started.

3 RESULTS

3.1 *Growth kinetics and survival strategy*

The kinetic parameters and some eco-physiological features of the representatives of *Thioalkalimicrobium* and *Thioalkalivibrio* grown in continuous culture under energy limitation are presented in Table 1.

The representatives of *Thioalkalimicrobium* group are characterized by higher growth rate and lower biomass yield than the organisms belonging to *Thioalkalivibrio* group. The rates of inorganic sulfur oxidation are much higher in the *Thioalkalimicrobium* cells compared with *Thioalkalivibrio*. First group is incapable of oxidizing elemental sulfur, a property that is present in *Thioalkalivibrio*.

The genotypic and phenotypic diversity of *Thioalkalimicrobium* is low, with only three species being isolated to date. The genus *Thioalkalivibrio* (7 species described to date) is very heterogeneous comprising several metabolic variants (e.g. denitrifyiers and SCN^--oxidizing species) (Sorokin et al. 2002) and moderate to extreme salt tolerants species, as opposed to the low salt tolerant *Thioalkalimicrobium*.

The competition and starvation experiments at low Na^+ concentration (0.6 M) clearly demonstrated the ability of *Thioalkalimicrobium* to outcompete *Thioalkalivibrio* at high dilution rates ($D > 0.02\,h^{-1}$). At low dilution rates ($D < 0.02\,h^{-1}$) the cells of *Thioalkalivibrio* were dominant in a mixed culture of both groups (Fig. 2). *Thioalkalivibrio* cells also survived much longer upon complete starvation.

3.2 *Influence of sodium concentration on growth parameters and sulfur oxidation rates*

Thioalkalivibrio versutus strain ALJ 15 showed an optimal growth at pH 10 and 0.6 to 2 M Na^+. Strain ALJ 15 was able to grow up to 4 M Na^+, up to pH 10.6 and a temperature up to 45°C, being a good example of triple extremophily (thermotolerant haloalkaliphilic bacteria). When grown in batch culture at alkaline pH and different sodium concentration, cells of strain ALJ 15 were capable of adjusting the optimum Na^+ concentration for inorganic sulfur oxidation) to values close to which they were cultivated. (Fig. 3).

The high salt tolerance of *Thioalkalivibrio versutus* strain ALJ 15 made it suitable for the investigation of the effect Na^+ concentration on the growth parameters in continuous culture, under energy limitation and at pH 10.

The influence of Na^+ concentration on maximum specific growth rate (μ_{max}), maximum biomass yield (Y_{max}) and maximum capacity for oxygen consumption ($qO_{2\,max}$) with different inorganic sulfur compound are presented in Table 2. The increase of sodium concentration from 0.6 M to 2 and 4 M led to a two to three-fold decrease in the values of kinetic parameters. The

Table 1. Kinetic and eco-physiological properties of the two groups of chemolithoautotrophic sulfur bacteria.

Parameter	*Thioalkalimicrobium*	*Thioalkalivibrio*
μ_{max} (h^{-1})	0.33	<0.2
Y_{max} (g protein/ mol $S_2O_3^{2-}$)	2–3	4–6
$qO_{2\,max}$ (mmole O_2/g prot min^{-1})		
$S_2O_3^{2-}$, HS^-	2.5–5	0.3–0.8
S_8^{2-}	1.5–2.5	0.2–0.8
S^0	0–0.05	0.2–0.6
K_s (μM)	3–5	1–3
Sulfur production*	No	Yes
Group diversity	Homogenous	Heterogenous
Salt tolerance (Na^+)	0.2–1.2 M	0.2–4 M
Resistance to starvation	Low	High
Survival strategy	R-strategists	K-strategists

* At high aeration.

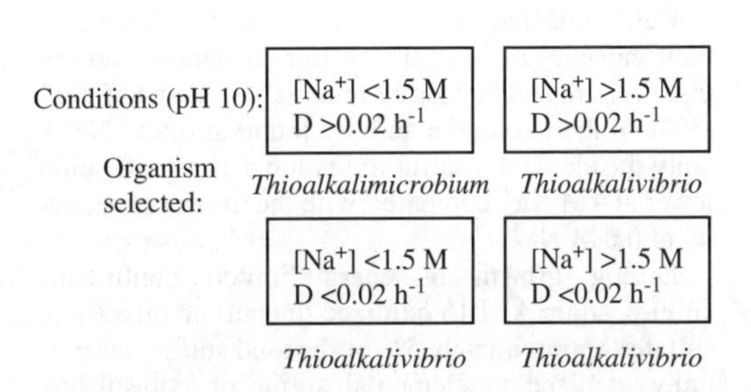

Figure 2. Selection of haloalkaliphilic sulfur-oxidizing bacteria grown in mixed culture at different Na^+ concentration, dilution rates and at pH 10.

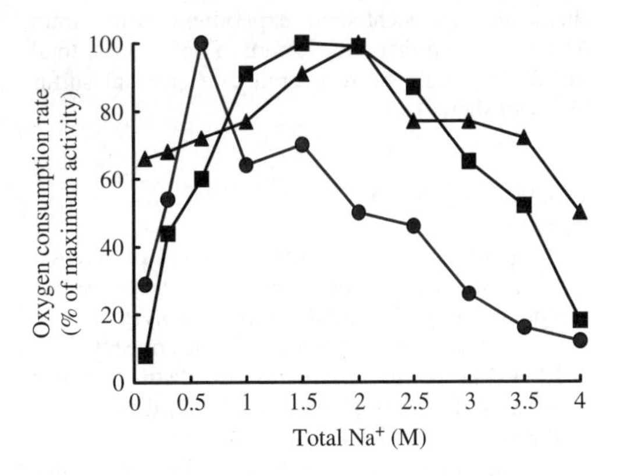

Figure 3. Influence of Na^+ on the oxygen consumption rate with $S_2O_3^{2-}$ in the washed cells of *Thioalkalivibrio versutus* strain ALJ 15 grown at different salt concentrations and at pH 10. Symbols: circles, 0.6 M Na^+-grown cells; squares, 2 M Na^+-grown cells; triangles, 4 M Na^+-grown cells.

Table 2. Kinetic parameters of *Thioalkalivibrio versutus* strain ALJ 15 grown in thiosulfate-limited chemostat, at pH 10 and at different salt concentrations.

Parameter	Salt concentration ($M\ Na^+$) 0.6	2	4
μ_{max} (h^{-1})	0.29	0.21	0.11
Y_{max} (g protein/mol $S_2O_3^{2-}$)	7.9	6.0	4.0
$qO_{2\ max}$ (mmole O_2/g prot min^{-1})			
$S_2O_3^{-2}$	0.96	0.65	0.36
HS^-	0.61	0.40	0.32
S_8^{2-}	0.90	0.63	0.40
S^0	0.50	0.20	0.05

apparent affinity constant (K_s) for thiosulfate did not change significantly with the increase of salt concentration but was rather growth rate-dependent (data not shown).

Cells collected from the effluent at 4°C, tested for their capacity of inorganic sulfur oxidation, showed that they are able to respire relatively well at $4\,M\ Na^+$ (30% of the maximum activity found at $0.6\,M\ Na^+$). Only the elemental sulfur oxidation was significantly lower at $4\,M\ Na^+$ compared with the maximum activity at $0.6\,M\ Na^+$.

During growth in energy-limited continuous culture, strain ALJ 15 oxidized thiosulfate directly to sulfate. Approximately 5% of the total sulfur was partially oxidized to elemental sulfur or polysulfide, which was subsequently oxidized to sulfate.

An attempt was made to grow the strain ALJ 15 in sulfide-limited chemostat at high salt concentration ($2\,M\ Na^+$) and alkaline pH. During this experiment a low redox potential (−200 mV) was achieved in the chemostat. Transient-state experiment with strain ALJ 15 showed that a maximum of 35% of the total sulfide [S] could be recovered as elemental sulfur (data not shown).

4 CONCLUSIONS

Our experiments showed that biological production of elemental sulfur from reduced inorganic sulfur compounds (e.g. thiosulfate, sulfide, polysulfide) is possible at extreme conditions of salt concentration and pH. The representatives of two distinct groups of chemolithoautotrophic, haloalkaliphilic, sulfur-oxidizing bacteria, *Thioalkalimicrobium* and *Thioalkalivibrio*, are able to oxidize reduced sulfur compounds at a sodium concentration higher than 0.5 M (sea water salinity) and at alkaline pH. Some strains are extremely versatile being capable of growing well within a range of $0.6-4\,M\ Na^+$ and at pH 10. Their growth kinetic parameters are comparable to those found in neutrophilic sulfur-oxidizing bacteria.

The end product of substrate oxidation by haloalkaliphilic sulfur-oxidizing bacteria was sulfate. Elemental sulfur appeared in many cases as intermediate product of inorganic sulfur compound oxidation. Substrate over-loading and low redox potential in the system may enhance elemental sulfur production starting from sulfide and polysulfide as will be the subject of future work.

ACKNOWLEDGEMENT

This work was financially supported by the STW projects DST.4653 and WCB.5939.

REFERENCES

Kelly, D.P., Chambers, L.A. & Trudinger, P.A. 1969. Cyanolysis and spectrophotometric estimation of trithionate in mixture with thiosulfate and tetrathionate. *Anal Chem* 41:898–901

Kolmert, A., Wikstrom, P. & Hallberg, K.B. 2000. A fast and simple turbidimetric method for the determination of sulfate in sulfate-reducing bacterial cultures. *J Microbiol Methods* 41:179–184

Janssen, A.J.H., Ruitenberg, R. & Buisman, C.J.N. 2001. Industrial applications of new sulphur biotechnology. *Water Sci Technol* 44(8):85–90

Lowry, O.H., Rosebrough, N.J., Farr, A.L. & Randall, R.J. 1951. Protein measurement with the Folin phenol reagents. *J Biol Chem* 193:265–275

Pfennig, N. & Lippert, K.D. 1966. Über das vitamin B12-bedürfnis phototropher schwefelbacterien. *Arch Microbiol* 55:245–256

Sörbo, B. 1957. A colorimetric determination of thiosulfate. *Biochim Biophys Acta* 23:412–416

Sorokin, D.Y., Lysenko, A.M., Mityushina, L.L., Tourova, T.P., Jones, B.E., Rainey, F.A., Robertson, L.A. & Kuenen, J.G. 2001. *Thioalkalimicrobium aerophilum* gen. nov., sp. nov. and *Thioalkalimicrobium sibericum* sp. nov., and *Thioalkalivibrio versutus* gen. nov., sp. nov., *Thioalkalivibrio nitratis* sp.nov., novel and *Thioalkalivibrio denitrificans* sp. nov., novel obligately alkaliphilic and obligately chemolithoautotrophic sulfur-oxidizing bacteria from soda lakes. *Int J Syst Evol Microbiol* 51:565–580

Sorokin, D.Y., Robertson, L.A. & Kuenen, J.G. 2000. Isolation and characterization of alkaliphilic, chemolithoautotrophic, sulphur-oxidizing bacteria. *Antonie van Leeuwenhoek* 77:251–262

Sorokin, D.Y., Tourova, T.P., Lysenko, A.M., Mityushina, L.L. & Kuenen, J.G. 2002. *Thioalkalivibrio thiocyanoxidans* sp. nov. and *Thioalkalivibrio paradoxus* sp. nov., novel alkaliphilic, obligately autotrophic, sulfur-oxidizing bacteria from soda lakes capable of growth on thiocyanate. *Int J Syst Evol Microbiol* 52:657–664.

European Symposium on Environmental Biotechnology, ESEB 2004 - Verstraete (ed)
© 2004 Taylor & Francis Group, London, ISBN 90 5809 653 X

Solvent-tolerant bacterial community for biofiltration of very high concentration gaseous solvent streams

M. Leethochawalit, J.A.S. Goodwin & M.T. Bustard
Chemical Engineering, School of Engineering and Physical Sciences, Heriot-Watt University, Edinburgh, United Kingdom

P.C. Wright & H. Radianingtyas
Department of Chemical and Process Engineering, University of Sheffield, Sheffield, United Kingdom

ABSTRACT: The aerobic biodegradation of high concentrations of gaseous acetone streams by a previously enriched solvent-tolerant bacterial consortium within a 1.9 litre vapour-phase fixed bed biofilter was investigated. Acetone biodegradation rate was measured and the characteristics of the bacterial community were investigated. 16S rRNA sequencing was used subsequently to identify these strains.

Successful gas-phase biofiltration of solvent vapour at loadings of up to $360\,g\,m^{-3}\,h^{-1}$ was implemented; this is higher than any other reported elimination capacity for acetone in the recognised literature. The mixed solvent-tolerant bacterial consortium was immobilized on sintered glass rings as a bioprocess intensification strategy. Acetone was tracked as the sole carbon source within a minimal salts medium. Removal efficiencies of up to 100% acetone were successfully demonstrated by the biofiltration system. The long-term effect on biofilter performance was also investigated. It showed that the consortium was able to adapt to high concentrations of acetone vapour ($17\,g\,m^{-3}$) over an extended period with removal efficiency above 80%. This result demonstrates that the biofilter could deal with high concentrations of acetone efficiently. It also shows that the consortium had a significant ability to adapt to high-concentration feed streams, which may occur under industrial conditions.

1 INTRODUCTION

Acetone is an important chemical widely used in many industrial processes e.g. semi-conductor manufacture. The worldwide use of acetone, its high vapour pressure and high water solubility, lead to a significant effect on both public health and the environment.

Since the last decade, biofiltration has become a very attractive option for treating waste gases because of its low running costs (Groenestijn and Hesselink, 1993; Leethochawalit et al., 2001). However, one major problem is the ability of such systems to handle fluctuating pollutant and high concentration loads (Kennes and Thalasso, 1998). This study was an investigation of solvent-tolerant bacterial community and the possibility that it can be used in vapour-phase bioreactors (biofilters) to treat high concentrations of gaseous acetone stream.

2 EXPERIMENTAL SECTION

2.1 *Bioreactor*

The biofilter column used in this study consisted of a glass tube with an inner diameter of 0.088 m and a height of 0.7 m (Figure 1). The microbial immobilisation matrix was Siporax® sintered glass pall rings (15 mm o.d., 10 mm i.d., 15 mm length and $1.7 \times 10^{-6}\,m^3$ volume). The packed-bed volume was $1.91 \times 10^{-3}\,m^3$, and the total matrix surface area was $797\,m^2$. The bioreactor was designed that it could be operated in either gas or liquid continuous mode.

2.2 *Inoculation of bioreactor*

Firstly, the solvent-tolerant microorganisms, obtained from an outdoor oil waste sump at Heriot-Watt University, Edinburgh, UK, were enriched and cultivated as described previously (Bustard et al., 2000). The enrichment cultures were grown in 4 $\times$250 cm^3 shake flasks, each containing around 100 ml of culture, to a cell density of 9 $\times 10^8$ cell cm^{-3}, and subsequently added to the biofilter column with 1000 cm^3 of Minimal Salts Medium (MSM) (Angelidaki et al., 1990) containing 20 cm^3 acetone, thereby giving inoculum and acetone concentrations of 3 $\times 10^8$ cells cm^{-3} and 7.9 $\times 10^3$ g m^{-3} respectively.

This liquid mixture was then re-circulated through the bed at 0.006 $m^3 h^{-1}$ using a peristaltic pump, with the biofilter operated in submerged liquid mode.The circulating liquid was replaced with fresh MSM containing 7.9 $\times$ 10^3 g m^{-3} IPA once per week. In order to build a strong biofilm onto support matrices, the biofilter was operated in submerged liquid mode for three months before changing to gas phase operation.

2.3 *Gas-phase biofilter operation*

After the biofilm formation period, gas phase experiments were performed in the same column by draining the liquid. Air was bubbled through liquid acetone to produce an acetone-saturated air stream, which was then diluted with air using balancing valves and rotameters, before being supplied to the bioreactor base via a plenum chamber and perforated plate distributor at a rate of 0.036 $m^{-3} h^{-1}$. Every 7 days, a batch of 2000 cm^3 MSM was washed slowly through the bed to prevent drying out. Samples were withdrawn by syringe from the gas inlet and outlet for gas chromatographic analysis.

2.4 *Biodegradation of acetone by a mixed microbial consortium*

Shake flasks (500 ml) were set up in triplicate and contained 1% (v/v) acetone in 196 ml MSM and 2 ml of inoculum (3 $\times$ 10^7 cells ml^{-1}) added. Control experiments were set up. Flasks were stoppered with foam bungs and placed within an IKA KS 250 orbital shaker at 150 rpm for the duration of the experiment. Samples were taken at 24 h periods for analysis. The experiment was conducted at 18°C.

2.5 *Solvent concentration determination*

Acetone concentrations were determined using a Shimadzu GC-17A gas chromatograph equipped with a Carbowax BP20 column (length = 15, 1 μm film). Helium was used as the carrier gas at 11.4 ml min^{-1} and the temperatures of the FID system and injection port were 300°C and 250°C respectively. During the analysis, the oven temperature was ramped from 70°C to 150°C at a heating rate of 50°C min^{-1}.

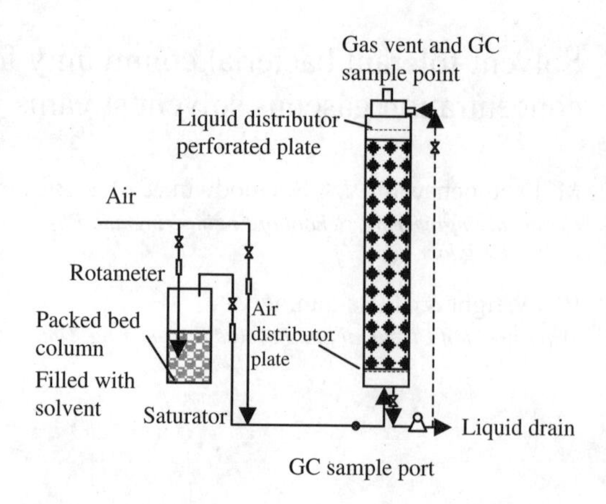

Figure 1. Schematic of experimental set up of the three phase fixed bed bioreactor.

2.6 *Bacterial identification*

Samples from the enrichment consortium were plated out onto 1.5% (w/v) nutrient bacteriological agar (Oxoid, UK) for 3 days at 18°C. Morphologically different colonies were selected and sub-cultured onto the same medium until uniform colonies were achieved. Preliminary identification was based on colony morphologies and Gram staining. Oxidation tests were carried out using 1% tetramethyl-p-phenylenediamine dihydrochloride and catalase tests were carried out using hydrogen peroxide reagent.

2.7 *16S rRNA gene sequencing*

Genomic DNA from individual bacterial strains was extracted using a Qiagen DNEasy Tissue Kit (Qiagen, Crawley, W. Sussex, UK) according to manufacturer's instructions. The 16S rDNA was selectively amplified from genomic DNA by using PCR with oligonucleotide universal primers Eubac 27F and Eubac 1492R (DeLong et al., 1993). PCR amplification was undertaken with a DNA Thermal Cycler model 2700 (Applied Biosystems, UK) under the following conditions: 100–150 ng of template DNA, 1x reaction buffer (Applied Biosystems, UK), 2.5 mM $MgCl_2$ (Applied Biosystems, UK), 1.25 U of Taq DNA polymerase (Applied Biosystems, UK), 0.5 μM upstream primer (Operon, Germany), 0.5 μM downstream primer (Operon, Germany), 200 μM of each dNTP (Applied Biosystems, UK), and PCR H_2O (Sigma-Aldrich, UK) combined in a total volume of 50 μl. The tubes were incubated at 95°C for 3 min and then subjected to the following thermal cycling programme: denaturation at 95°C for 90 sec, primer annealing at 59°C for 90 sec, and chain extension at 72°C for 90 sec with an additional extension time of 10 min on the final cycle, for a total of 40 cycles. The amplified DNA was purified using a Qiaquick PCR Purification Kit (Qiagen, Crawley, W. Sussex, UK) and subjected to sequencing reaction using ABI PRISM® BigDye™ Terminators v3.0 Cycle Sequencing Kit (Applied Biosystems, UK) according to manufacturer's instructions with bacteria universal primers (Eubac 27F, Eubac 1492R (DeLong et al., 1993), Eubac 357F, Eubac 519R, Eubac 803F, Eubac 909R, Eubac 1114F, and Eubac 1221R (Colquhoun et al., 1998), Operon, Germany) prior to sequencing in an ABI Prism 310 (Applied Biosystems, UK). Fragments of 16S rRNA (1200–1500 bp) were compared to the most similar sequences in the GenBank and EMBL nucleotide sequence databases based on percent similarities.

3 RESULTS AND DISCUSSION

3.1 *Characterisation and identification of mixed-consortium*

Broad identification of the solvent-tolerant mixed bacteria was carried out and the results are summarised in Table 1. Strains S-1, S-2, S-4 and S-6 were rod shaped, while S-3 and S-5 were cocci. Two thirds of the cultures were gram-negative and catalase-positive, namely strains S-2, S-4, S-5 and S-6. Strains S-2, S-4 and S-6 were also found to be oxidase-positive. The fact that five of the strains were catalase-positive suggests that the majority of the strains had an oxidoreductase enzyme and were capable of aerobic respiration. Half of the mixed-consortium bacteria exhibited cytochrome oxidase activity, which is indicated by the oxidation reaction and the formation of water as oxidation product (Salle, 1973). The identification results of the isolates by 16S rRNA sequence analysis are shown in Table 2. Sequencing of the 16S rRNA of the strain S-1 showed a close relationship to *Staphylococcus* sp. The sequencing of strain S-2 and S-4 showed equal similarities to two different species; *Defluvibacter lusatiae* and *Aquamicrobium defluvium*. Both of these species are gram-negative, rod-shaped. *Defluvibacter lusatiae* was isolated and identified for the first time from activated sludge from an industrial waste water plant in Germany (Fritshe et al., 1999). *Aquamicrobium defluvium* was isolated and reported from activated sewage sludge in Germany as a thiophene-2-carboxylate-metabolizing bacterium in 1998 (Bambauer et al., 1998). The strain S-3 was identified to be *Kocuria* sp. The genus *Kocuria* is reported for the first time in this paper as solvent-tolerant and suitable for biofiltration purposes. The strain S-6 displayed similarities to various species but all of them were in the genus of *Flavobacterium,* which is reported to be organic-solvent-tolerant (Sardessai and Bhosle, 2002). However, the genus *Flavobacterium* is catalase and oxidase negative. Subsequently, strain S-6 will require further identification.

Table 1. Characteristics of mixed microbial consortium isolated from oil-contaminated soil.

	Strains					
Characteristics	S-1	S-2	S-3	S-4	S-5	S-6
Gram Stain	+	−	+	−	−	−
Shape	rod	rod	cocci	rod	cocci	rod
Oxidase	−	+	−	+	−	+
Catalase	−	+	+	+	+	+

Table 2. Tentative 16S rRNA sequence similarities.

Strain No.	Strain name	% similarity
S-1	*Staphylococcus* sp.	99
S-2	*Defluvibacter lusatiae*	98
	Aquamicrobium defluvium	98
S-3	*Kocuria* sp.	99
S-4	*Defluvibacter lusatiae*	98
	Aquamicrobium defluvium	98
S-5	*Hydrogenophaga* sp.	95
S-6	*Flavobacterium* sp.	94

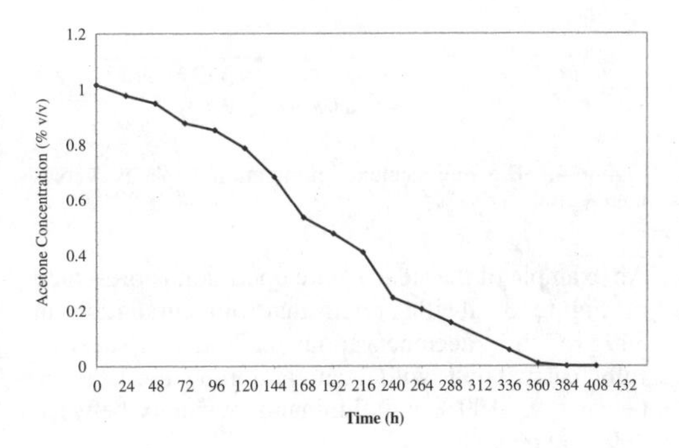

Figure 2. Microbial biodegradation of 1% (v/v) acetone.

3.2 *Biodegradation of acetone by a mixed microbial consortium*

Biodegradation of acetone by a free-cell mixed consortium was investigated. Figure 2 shows that this mixed microbial consortium had 100% biodegradation efficiency for 1% (v/v) acetone solution after 360 h. The average rate of microbial biodegradation was calculated to be 2.29×10^{-3}% (v/v) h^{-1}.

3.3 *Acetone biodegradation during steady state*

A very high solvent concentration was applied to the biofilter to assess the microbial metabolic capability.

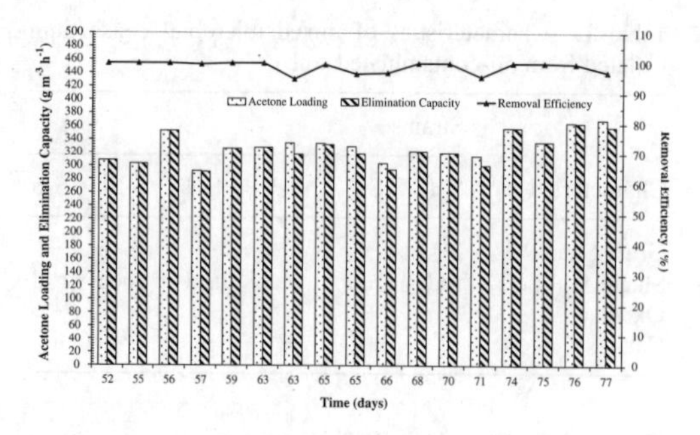

Figure 3. Mass loading, elimination capacity and removal efficiency of acetone vapour in the biofilter.

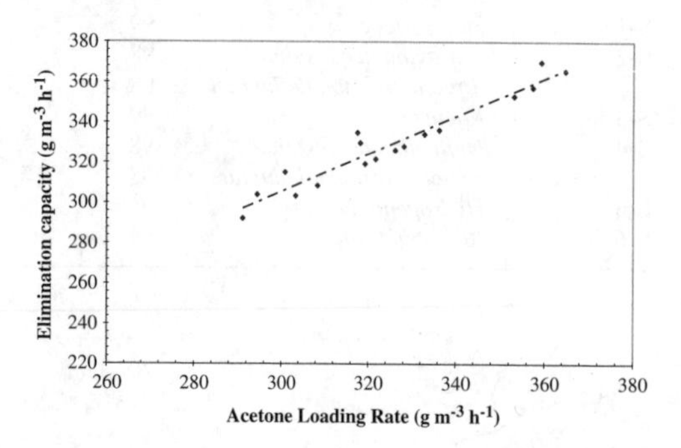

Figure 4. Biofilter acetone elimination capacity versus loading rate.

An example of the steady state operation is presented in Figure 3. It illustrated that biodegradation at 340 g m^{-3}h^{-1} acetone vapour loading was successfully carried out with acetone removal efficiency between 95–100% and elimination capacity between 300–360 g m^{-3}h^{-1}.

This is higher than any other reported elimination capacity for acetone in the recognised literature (Chang and Lu, 2003; Hwang et al., 1997). Furthermore, the high mass loading was achievable with very high inlet concentration (over 17 g m^{-3}). Figure 4 illustrates the linear relationship between acetone elimination capacity and the mass loading of the biofilter. It can be seen that the elimination capacity matches the mass loading, which means that the biofilter is operating under-capacity, despite the elevated mass loadings. These results demonstrated that the solvent-tolerant microbial consortium in the biofilter had the ability to metabolise very high loadings of acetone.

3.4 *Long term effect of acetone biodegradation*

After nine months of gas phase operation, biofilter performance was investigated for 21 days. For the final period of operation, the loading of the biofilter was reduced to 230–320 gm^{-3}h^{-1}. Results from this period are presented in Figure 5. It can be observed that the average removal efficiency was above 80%. The overall results show that the consortium was able to adapt to high concentrations of acetone vapour over an extended time period of ten months.

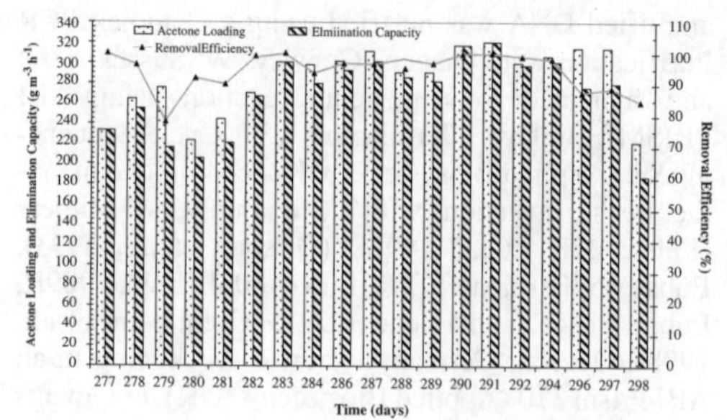

Figure 5. Long term effect on mass loading, elimination capacity and removal efficiency of acetone vapour.

4 CONCLUSIONS

The biodegradation of acetone at high concentration by a solvent-tolerant bacterial consortium was achieved with acetone removal efficiency of 95–100% at mass loading rates up to 360 g m^{-3}h^{-1}. The solvent-tolerant bacteria were identified by 16S rRNA sequencing analysis with similarity percentage up to 99%. The results also showed that the consortium was able to adapt to high concentrations of acetone vapour over an extended period with removal efficiency above 80%. This demonstrates the feasibility of a high performance gas phase biofilter for treatment of very high concentrations of solvent vapour over a long term period.

REFERENCES

Angelidaki, I., Petersen, S. P., and Ahring, B. K., 1990, Effects of lipids on thermophilic anaerobic digestion and reaction of lipid inhibition upon addition of bentonite, *Applied microbiology and biotechnology* **33**:469–472.

Bambauer, A., Rainey, F. A., Stackebrandt, E., and Winter, J., 1998, Characterization of *Aquamicrobium defluvii* gen. nov. sp. nov., a thiophene-2-carboxylate-metabolizing bacterium from activated sludge, *Archive in Microbiology* **169**:293–302.

Bustard, M. T., McEvoy, E. M., Goodwin, J. A. S., Burgess, J. G., and Wright, P. C., 2000, Biodegradation of propanol and isopropanol by mixed microbial consortium, *Applied Microbiology and Biotechnology* **54**:424–431.

Chang, K., and Lu, C., 2003, Biofiltration of isopropyl alcohol and acetone mixtures by a tricke-bed air biofilter, *Process Biochemistry* **in press**:1–9.

Colquhoun, J., Mexson, J., Goodfellow, M., Ward, A., Horikoshi, K., and Bull, A., 1998, Novel rhodococci and other mycolate actinomycetes from the deep sea., *Antonie Leeuwenhoek* **74:**27–40.

DeLong, F., Franks, D., and Alldredge, A., 1993, Phylogenetic diversity of aggregate-attached vs. free-living marine bacterial assemblages *Limnol, Ocean* **38:**924–934.

Fritshe, K., Auling, G., Andreesen, J. R., and Lechner, U., 1999, *Defluvibacter lusatiae* gen. nov., sp. nov., a new chlorophenol-degrading member of the α-2 subgroup of Proteobacteria, *Systematic and Applied Microbiology* **22:**197–204.

Groenestijn, J. W. V., and Hesselink, P. G. M., 1993, Biotechniques for air pollution control, *Biodegradation* **4:**283–301.

Hwang, S.-J., Tang, H.-M., and Wang, W.-C., 1997, Modeling of Acetone Biofiltration Process, *Environmental Progress* **16(3):**187–193.

Kennes, C., and Thalasso, F., 1998, Waste Gas Biotreatment Technology, *Journal of Chemical Technology & Biotechnology* **72:**303–319.

Leethochawalit, M., Bustard, M. T., and Wright, P. C., 2001, Novel vapor-phase biofiltration and catalytic combustion of volatile organic compounds, *Industrial Engineering Chemistry Research* **40:**5334–5341.

Salle, A. J., 1973, Fundamental Principles of Bacteriology, McGraw-Hill, Inc., U.S.A.

Sardessai, Y., and Bhosle, S., 2002, Tolerance of bacteria to organic solvents, *Reseach in Microbiology* **153:**263–268.

POSTERS

Microbial diversity

European Symposium on Environmental Biotechnology, ESEB 2004 - Verstraete (ed)
© 2004 Taylor & Francis Group, London, ISBN 90 5809 653 X

16S-rRNA analysis for characterization of rhizosphere Zn-resistant bacteria isolated from mining area under semiarid climate

R. Bennisse, F. Chamkh & A.I. Qatibi
Microbiological Engineering group, Sciences & Techniques Faculty, Cady Ayyad University, Marrakech, Morocco

C. Joulian
Microbiology Laboratory IRD, Universités de Provence et de la Méditerranée, Marseille, France

M. Labat
Post Harvest Microbial Biotechnology, Microbiology Laboratory, IFR-BAIM, Universités de Provence et de la Méditerranée, Marseille, France

ABSTRACT: In order to exploit the ability of the rhizosphere bacterial communities in the remediation of heavy-metal-contaminated soils, we have isolated and identified, by classical techniques, 32 strains associated with *Salsola vermiculata* and *Peganum harmala*, two heavy metal tolerant plants grown in a semiarid polluted sandy soils contaminated mainly with Zn. Three Zn-resistant and prevailing strains in the rhizosphere were identified using 16S-rRNA gene sequence analysis. Phylogenetic dendogram based on 16S-rRNA gene sequence indicating that two from the three Zn-resistant isolates are related to pathogenic strains *Bacillus anthracis* (similarity average of 99.9%) and *Enterococcus durans* (similarity average of 99.5%). The third strain is related to *Arthrobacter chlorophenolicus* (similarity average of 99.4%), a non-pathogenic bacterium.

1 INTRODUCTION

The use of genes encoding 16S-rRNA as molecular markers has become a routine technique for the identification of bacteria present in metal polluted soils. However, direct analysis of microbial communities based on this gene is not appropriate to distinguish between the sensitive and the resistant heavy metal-bacteria present in contaminated soil. Many studies have previously shown that not all bacteria present in metal polluted soils can develop resistance to heavy metals (Sabry *et al.* 1997, Konopka *et al.* 1999). Furthermore, due to natural variability in soil types, soil organic matter contents and soil management, it is difficult to say that a shift in the structure of bacterial communities represent the effect of an increased concentration of heavy metals on soil micro-organisms and which ones are not influenced by other factors. A more direct method of studying the effect of heavy metal pollution on micro-organisms might be to estimate the number of tolerant micro-organisms. Increases in the numbers of metal-tolerant bacteria have been seen as the results of heavy metal pollution by the traditional plat count technique (Huysman *et al.* 1994, Bennisse *et al.* 2003). In the present study we analysed the rhizosphere bacterial communities associated with two heavy metal tolerant plants identified as *Salsola vermiculata* and *Peganum harmala* (Bennisse *et al.* 2003) that are found abundantly in one of the major polymetallic (Pb-Zn-Cu) occurrence in the Moroccan Hercynian province (Draâ Sfar, Marrakesh, Morocco), and presenting long-term contamination mainly with Zn and Pb from mining operations. The first step was to determine the bacterial community composition by the traditional identification of pure cultures isolated on laboratory media and to isolate the culturable Zn-resistant bacteria. The isolation of Zn-resistant bacteria was assessed by plate culture as no method exists to test heavy-metal resistance directly without purification of bacteria. The second step was to select isolates that are highly tolerant to Zn and prevailing in the rhizosphere of two metallophyte plants, to identify them by using 16S-rRNA gene sequence analysis.

2 MATERIALS AND METHODS

2.1 *Enumeration of bacteria*

Culturable heterotrophic aerobic bacteria were enumerated by plating appropriate dilutions on Luria-broth agar plates supplemented with cycloheximide as an antifungal agent (50 $\mu g.l^{-1}$ final concentration). Colony forming units (CFU) were counted after 5 days incubation at 30°C.

2.2 *Measurements of Zn-resistant bacteria*

Zinc-resistant bacteria were enumerated by plating diluted soil samples on sterile LB agar plates supplemented with filter-sterilised stock solution of $ZnSO_4$ to give final concentration of 10 mM. The plates were incubated at 30°C, and CFU were counted after 5 days.

2.3 *Identification and diversity analysis*

In order to identify the isolated bacteria, biochemical tests were conducted according to the methods described by Smibert & Krieg (1981) and/or using the strips API20NE and API20E (Bennisse *et al.* 2003). Diversity analysis was carried out by using the Shannon species diversity index (Barkey *et al.* 1985, Dean-Ross 1990).

2.4 *DNA extraction from Zn-resistant bacteria and PCR amplification*

DNA was extracted from the isolates using the Wizard Genomic DNA Purification kit, according to the manufacturer's protocol (Promega, Charbionnière, France). The 16S-rRNA gene was amplified with the forward primer Fd1 (5′ AGA GTT TGA TCC TGG CTC AG 3′) and the reverse primer Rd1 (5′ AAG GAG GTG ATC CAG CC 3′) for strains R31and R51 or the reverse primer R6 (5′ TAC GGT TAC CTT GTT ACG AC 3′) for strain BR42. PCR reactions were performed in the GeneAmp PCR System 2400 (Perkin-Elmer, Applied Biosystems, Norwalk, Conn.) using the following protocol: 1 min at 94°C, 30 or 40 cycles of 30 s at 94°C, 1 min at 55°C, 2 min at 72°C, and a final extension of 5 min at 72°C. PCR products were purified with the Nucleo Spin Extract kit (Macherey Nagel, Düren, Germany).

2.5 *Cloning of 16S-rRNA genes*

Purified PCR products of strains R31 and R51 were ligated into pGEMT-easy plasmids and transformed into JM109 chemically competent *Escherichia coli* cells, according to the manufacturer's protocol (Promega, Charbionnière, France). Clone libraries were screened by direct PCR amplification from colonies with the vector specific primers SP6 (5′ ATT TAG GTG ACA CTA TAG TT 3′) and T7 (5′ TAA TAC GAC TCA CTA TAG GG 3′), and the following protocol: 2 min at 96°C, 40 cycles of 30 s at 94°C, 1 min at 55°C, 3 min at 72°C, and a final extension of 10 min at 72°C. Plasmids containing the right length insert ($\approx$1.5 kb) were purified using the Wizard Plus SV Minipreps DNA Purification System, according to the manufacturer's protocol (Promega, Charbionnière, France).

2.6 *Sequencing and phylogenetic analysis*

Purified plasmids (strains R31 and R51) and PCR products (strain BR42) were sent for sequencing to Genome Express (Grenoble, France). Partial 16S-rRNA sequences obtained in our study and reference sequences obtained from the Ribosomal Database Project II (Maidak *et al.*, 2001) and from Genbank database (Benson *et al.*, 1999), were manually aligned with the sequence editor BioEdit (Hall, 1999). Only unambiguously aligned sequence positions were exported to Treecon (Van de Peer & De Wachter, 1994) for phylogenetic analyses. Pairwise evolutionary distances were computed by the method of Jukes & Cantor (1969). Dendrograms were constructed by the neighbor-joining method (Saitou & Nei, 1987). Confidence in the tree topology was determined by bootstrap analysis using 100 resamplings of the sequences (Felsenstein, 1985).

3 RESULTS AND DISCUSSION

As observed in Table 1, the number of cultivable heterotrophic bacteria of the two heavy metal stressed soils were lower than the control soil.

Diversity indices (DI) show extremely low bacterial diversities (0.38 to 1.65) and high frequency of Zn-resistance bacteria (7 to 34.5%) at the polluted soils compared with the unpolluted soil. The low DI indicates that heavy metals induce stress that shifts community composition to lower diversity. However, several factors and not only the toxic metal concentration could contribute to the observed decrease. The fact that these soils were low in organic content in addition to their sandy nature (Bennisse *et al.* 2000), may have contributed to the decrease of the bacterial diversity. The more direct method of studying the effect of heavy metal toxicity may be to estimate the number of tolerant bacteria. High frequencies of metal resistance are immediate consequences of adaptation to heavy metal pollution (Huysman *et al.* 1994).

Amongst the 32 strains isolated in our study, three isolates designed R31, BR42 and R51, exhibited high Zn-resistance and were prevailing in the rhizosphere

Table 1. Enumeration of the aerobic heterotrophic bacteria, generic composition and diversity of the bacterial population in rhizosphere soils.

Z	Rhizosphere Soils	CFU · g^{-1}	Isolates	RA (%)	DI
A	*Salsola vermiculata*	5.8 10^6	*Bacillus* sp **R31**(+)	11	1.08
			Xanthomonas sp AR31(−)	9.5	
			Pseudomonas sp AR32(−)	67	
			Flavobacterium sp AR33(+)	7	
			Bacille G+ sp AR34(−)	5	
			Pseudomonas sp AR35(−)	0.5	
	Perganum harmala	6.1 10^5	*Brevundimonas* sp R41(+)	3	0.38
			Flavobacterium sp AR41(+)	3	
			Flavobacterium sp AR42(−)	91	
			Bacille G + sp AR43(−)	3	
B	*Salsola vermiculata*	1.1 10^6	*Enterobacteriaceae* sp BR31(−)	26.5	1.65
			Pseudomonas sp BR32(−)	22	
			Flavobacterium sp BR33(−)	1.2	
			Enterococcus sp **R51**(+)	30.5	
			Bacille G+ sp R52(−)	9	
			Bacille G+ sp R53(−)	4	
			Enterobacteriaceae sp R54(−)	4.8	
			Bacille G+ sp R55(−)	2	
	Perganum harmala	4.4 10^4	Bacille G+ sp BR41(−)	27	1.07
			Arthrobacter sp **BR42**(+)	34.5	
			Acinetobacter sp BR43(−)	38.5	
C	*Perganum harmala*	3.4 10^6	*Pseudomonas* spTR41(−)	29	2.08
			Flavobacterium sp TR42(−)	7.5	
			Pseudomonas sp TR43(−)	3	
			Xanthomonas sp TR44(−)	3	
			Bacille G+ sp TR45(−)	7.5	
			Pseudomonas sp TR46(−)	6.5	
			Bacille G+ sp TR47(−)	14.5	
			Alcaligens sp TR48(−)	15.5	
			Alcaligens sp TR49(−)	3.5	
			Bacille G+ sp TR401(−)	2.5	
			Xanthomonas sp TR402(−)	7.5	

Z, zones; CFU·g^{-1}, total culturable heterotrophic bacterial numbers (n = 3); RA, relative abundance based on the number of characterised strains in the total communities; ID, diversity index; G+, Gram-positive; C, zone control; (−), Zn-sensible strains; (+) Zn-resistant strains at 10 mM. No *S. vermiculata* was found in the control.

(Table 1). Based on phylogenetic analyses of partial 16S-rRNA gene sequences, it was apparent that the three isolates were respectively closely related to *Bacillus anthracis* (similarity average of 99.9%; Fig. 1), *Arthrobacter polychromogenes* and *A. oxydans* (similarity average of 99.4%; Fig. 3) and *Enterococcus durans* (similarity average of 99.5%; Fig. 2).

Since individual species amongst these three genera exhibit very high phylogenetic identities on 16S-rRNA gene sequence, isolates R31, BR42 and R51 could represent new species of the genera *Bacillus*, *Arthrobacter* and *Enterococcus*, respectively. Their phylogenetic position at the species level can however not be inferred based solely on their 16S-rRNA gene sequence.

The phylogenetic analysis reveals the presence of two strains closely related to pathogenic bacteria, *B. anthracis* and *E. durans*. However, because non-pathogenicity is an important criterion since potential pathogens would cause safety problems in large-scale operation and in the disposal of used cell mass, only strain BR42, closely related to the non-pathogenic bacterium *Arthrobacter chlorophenolicus*, was selected for ongoing studies to examine its implication in association with *Perganum Harmala* in the phytoremediation of heavy-metal-contaminated soils. Furthermore, additional experiments are ongoing in our laboratory to understand the mechanisms behind the association between this bacterium and the plant,

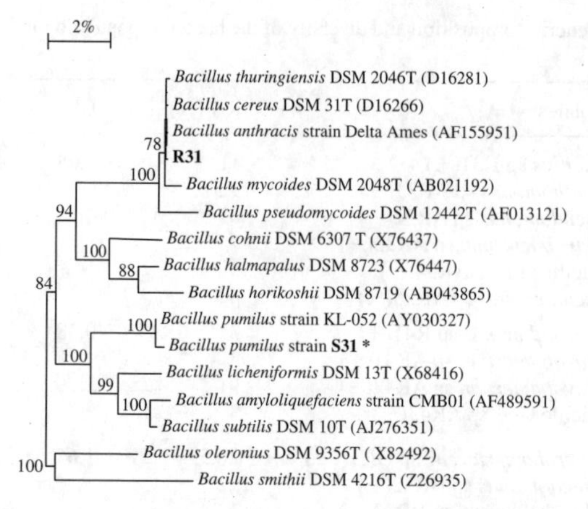

Figure 1. Phylogenetic dendrogram based on 16S-rRNA sequence analyses (1443 unambiguous nucleotides) indicating the position of strain R31, a Zn-resistant and prevailing strain isolated from the rhizosphere of *S. vermiculata*. *Strain S31 is a Zn-resistant strain isolated in the non-rhizosphere of *S. Vermiculata*. Scale bar, 2 nucleotide substitutions per 100 nucleotides. Only bootstrap values above 50%, expressed as a percentage of 100 replications, are shown at branching points.

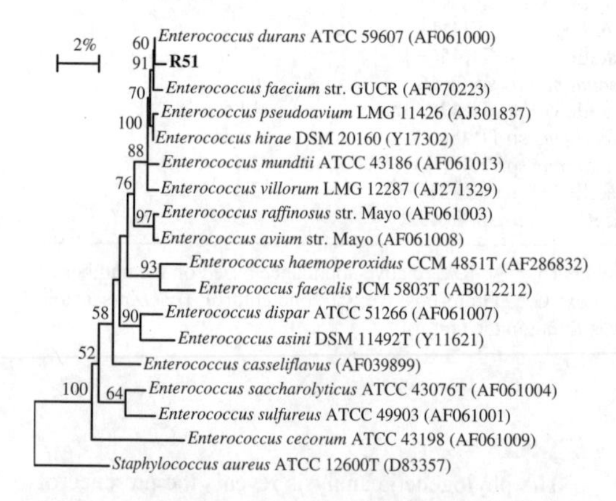

Figure 2. Phylogenetic dendrogram based on 16S-rRNA sequence analyses (1440 unambiguous nucleotides) indicating the position of strain R51, a Zn-resistant and prevailing strain isolated in the rhizosphere of *S. vermiculata*. Scale bar, 2 nucleotide substitutions per 100 nucleotides. Bootstrap values above 50%, expressed as a percentage of 100 replications, are shown at branching points.

since such an understanding could facilitate the management of these soil bacteria for a restoration and/or bioremediation program (Baker *et al.* 1994, Salt *et al.* 1995).

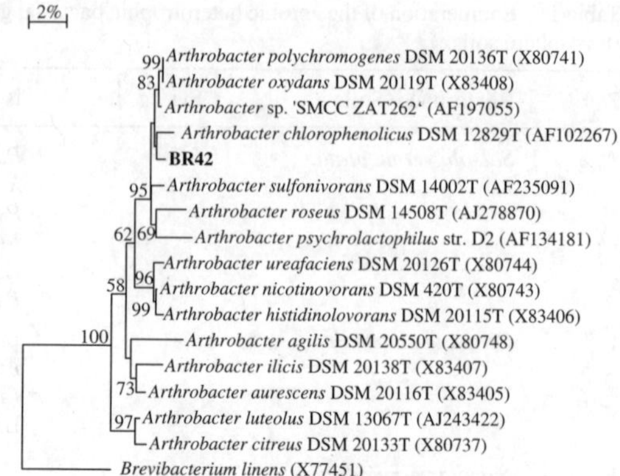

Figure 3. Phylogenetic dendrogram based on 16S-rRNA sequence analyses (1174 unambiguous nucleotides) indicating the position of strain BR 42, a Zn-resistant and prevailing strain isolated in the rhizosphere of *P. harmala*. Scale bar, 2 nucleotide substitutions per 100 nucleotides. Bootstrap values above 50%, expressed as a percentage of 100 replications, are shown at branching points.

ACKNOWLEDGMENTS

The authors are indebted to A. El Asli of Al Akhawayn University, Ifrane, Morocco for improving the manuscript.

REFERENCES

Baker, A.J.M., McGrath, S.P., Sidoli, C.M.D. & Reeves, R.D. 1994. The possibility of in situ heavy metal decontamination of polluted soils using crops of metal-accumulating plants. *Res Conserv Recycl* 11: 41–49.

Barkey, T., Tripp, S.C. & Olson, B.H. 1985. Effect of metal-rich sewage sludge application on the bacterial communities of grasslands. *Appl Environ Microbiol* 49: 333–337.

Bennisse, R., Qatibi, A.I. & Chraibi, K. 2000. Rhizosphere and non-rhizosphere bacterial communities in a Zinc polluted soil. *Proceeding of Third International Symposium Environment, Catalysis and Process Engineering ECGP'3* Fès (Maroc) 13–14 Nov 2000.

Bennisse, R., Qatibi, A.I., Labat, M., ElAsli, A.G. & Berhada, F. 2003. Rhizosphere bacterial diversity in heavy metals contaminated soils. *World J Microbiol Biotechnol* Ref. WIBI 4066: (in press).

Benson, D.A., Boguski, M.S., Lipman, D.J., Ostell, J., Ouellette, B.F., Rapp, B.A. & Wheeler, D.L. 1999. Gen Bank. *Nucleic Acids Res* 27: 12–17.

Dean-Ross, D. 1990. Response of attached bacteria to zinc in artificial streams. *Can J Microbiol* 36: 561–566.

Felsenstein, J. 1985. Confidence limits on phylogenies: an approach using the bootstrap. *Evolution* 39: 783–791.

Hall, T.A. 1999. BioEdit: a user-friendly biological sequence alignment editor and analysis program for Windows 95/98/NT. *Nucleic acids Symp Ser* 41: 95–98.

Huysman, F., Verstraete, W., & Brookes, P.C. 1994 Effect of manuring practices and increased copper concentrations on soil microbial populations. *Soil Biol Biochem* 26: 103–110.
Jukes, T.H. & Cantor, C.R. 1969. Evolution of protein molecules. In *Mammalian protein metabolism*, pp. 21–132. Edited by H.N. Munro. New York: Academic Press.
Konopka, A., Zakharova, T., Bischoff, M., Oliver, L., Nakatsu, C. & Turco, R.F. 1999. Microbial biomass and activity in lead-contaminated soil. *Appl Environ Microbiol* 65: 2256–2259.
Maidak, B.L., Cole, J.R., Lilburn, T.G., Parker, C.T. Jr, Farris, R.J., Garrity, G.M., Olsen, G.J., Schmidt, T.M. & Tiedje, J.M. 2001. The RDP II (Ribosomal database project). *Nucleic Acids Res* 29: 173–174.
Sabry, S.A., Ghozlan, H.A. & Abou-Zeid, D.M. 1997. Metal tolerance and antibiotic resistance pattern of bacterial population isolated from sea water. *J Appl Microbiol* 82: 245–252.
Saitou, N. & Nei, M. 1987. The neighbor-joining method: a new method for reconstructing phylogenetic trees. *Mol Biol Evol* 4: 405–425.
Salt, D.E., Blaylock, M., Kumar, N., Dushenkov, V., Ensley, B.D., Chet, I. & Raskin, I. 1995. Phytoremediation : A novel strategy for the removal of toxic metals from the environment using plants. *Biotechnol* 13: 468–475.
Smibert, R.M. & Krieg, N.R. 1981. General characterisation. In *Manual of methods for microbiology*. American society for microbiology, Washington, DC pp. 409–443.
Van de Peer, Y. & De Wachter, R. 1994. TREECON for Windows: a software package for the construction and drawing of evolutionary trees for the Microsoft Windows environment. *Computer Appl Biosc* 10: 569–570.

European Symposium on Environmental Biotechnology, ESEB 2004 - Verstraete (ed)
© 2004 Taylor & Francis Group, London, ISBN 90 5809 653 X

Diversity of bacteria isolated from waste crude oil samples. Production of bioemulsifiers and removal polycyclic aromatic hydrocarbons in a land farming treatment inoculated with a microbial consortium

C. Calvo[1,2], F.L. Toledo[1], B. Rodelas[2], M.V. Martínez Toledo[1,2] & J. González López [1,2]
Group of Environmental Microbiology,[1] Institute of Water Research,[2] Department of Microbiology, University of Granada, Spain

C. García Fandiño
Repsol YPF, Puertollano, Ciudad Real. Spain

ABSTRACT: Bacterial strains isolated from waste crude oil were characterized. Most of the strains belonged to genus Bacillus and only two of them to Gram-negative bacteria. The capacity of these microorganisms to grow and remove phenanthrene and pyrene was evidenced in the majority of Bacillus strains. Further, the ability to synthesized extracellular bioemulsifiers was detected, in *B. subtilis* 28-15, *A. faecalis* 212-2 and *Enterobacter* sp 214-6. Considering these properties it has been constructed a consortium composed of *B. pumilus* 28-11, *A. faecalis* 212-2, *M. luteus* 212-4 and *Enterobacter* 214-6 to be used as inoculant in a land farming treatment.

1 INTRODUCTION

Bioremediation of contaminated aquatic and soil environments has arisen as an effective biotechnology, with a range of advantages compared to more traditional technologies. Polycyclic aromatic hydrocarbons (PAHs) extensively occur as soil and water pollutants, and are important environmental contaminants because of their recalcitrance (Desai and Banat, 1997). Bacteria naturally inhabiting contaminated sites are of interest as potential agents for PAHs bioremediation (Aitken et al., 1998; Barathi and Vasudevan, 2001). In this paper we have characterized bacteria isolated from waste oil with capacity of removing PAHs under different experimental conditions.

2 MATERIALS AND METHODS

2.1 *Microorganisms*

We have studied 15 bacterial strains isolated in our laboratory from waste crude oil samples (Calvo et al., 2002). All strains were selected according to their ability to grow on minimal solid media with PAHs.

2.2 *Phylogenetic characterization*

These strains were characterized by a genetic approach, inferring their phylogeny from the 16S rDNA sequence.

2.3 *Culture of PAHs degrading-microorganism*

Growth in the presence of PAHs was confirmed in minimal liquid media (BH) with phenanthrene or pyrene (0.1% w/w) as sole carbon source.

2.4 *Extraction of exopolysaccharide (EPS)*

EPS were obtained by precipitation with ethanol (Quesada et al., 1993). The influence of hydrocarbon substances was studied by addition to the media of the following hydrocarbons (1% w/vol): n-octane, toluene, xilene, mineral light oil, mineral heavy oil or crude oil (kirkuk).

2.5 *Chemical composition of EPS and emulsification activity*

Carbohydrates and proteins were determined by colorimetric assays (Dubois et al., 1956; Bradford, 1976). Emulsification assays were performed according to Cooper and Goldenberg (1987).

2.6 *Land farming assays*

These assays were performed in soil plots of 25 m^2 with hydrocarbons (Repsol YPF, Puertollano, Spain). One of these plots was inoculated with a consortium composed of four bacteria: *B. pumilus* 28-11, *A. faecalis* 212-2, *M. luteus* 212-4 and *Enterobacter* 214-6.

3 RESULTS AND DISCUSSION

Alignment of the sequences of nearly full-length 16S rDNA of the 15 strains and EMBL database sequences demonstrated the affiliation of most strains to the genus *Bacillus*. Eight strains clustering together with strains of the *B. pumilus* species, and two strains affiliated to *B. subtilis*.

These results are in agreement with several reports that describe *Bacillus* as one of the major bacterial groups that effectively degrade hydrocarbons (Zhuang et al., 2002; Shimura et al., 1999). Other Gram-positive

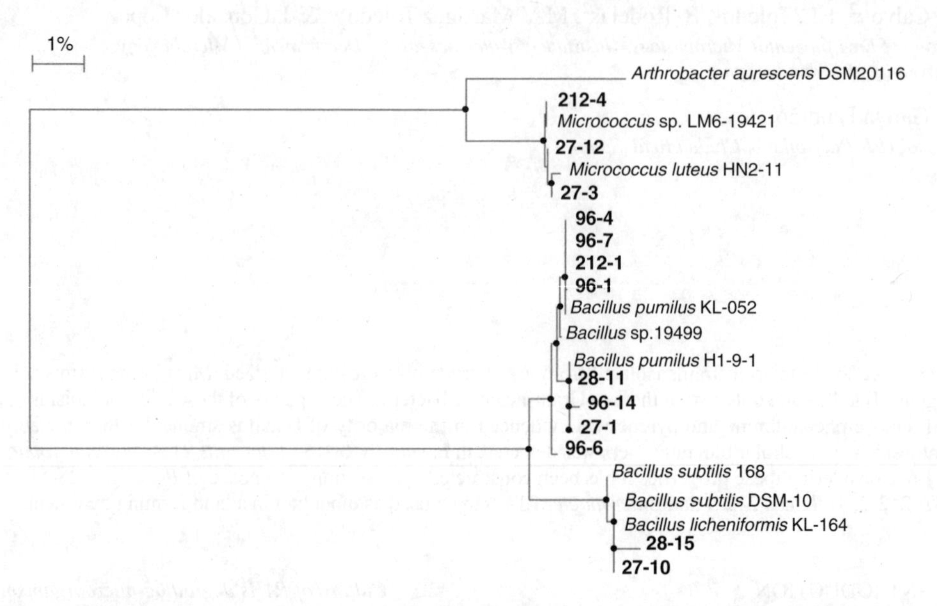

Figure 1. Neighbour-joining phylogenetic tree based on the nearly full length of the 16S rDNA gene of the 13 Gram-positive strains described in this paper and sequences from EMBL that gave the highest scores in similarity searches. Nodes highlighted with circles indicate more than 50% of bootstrap value.

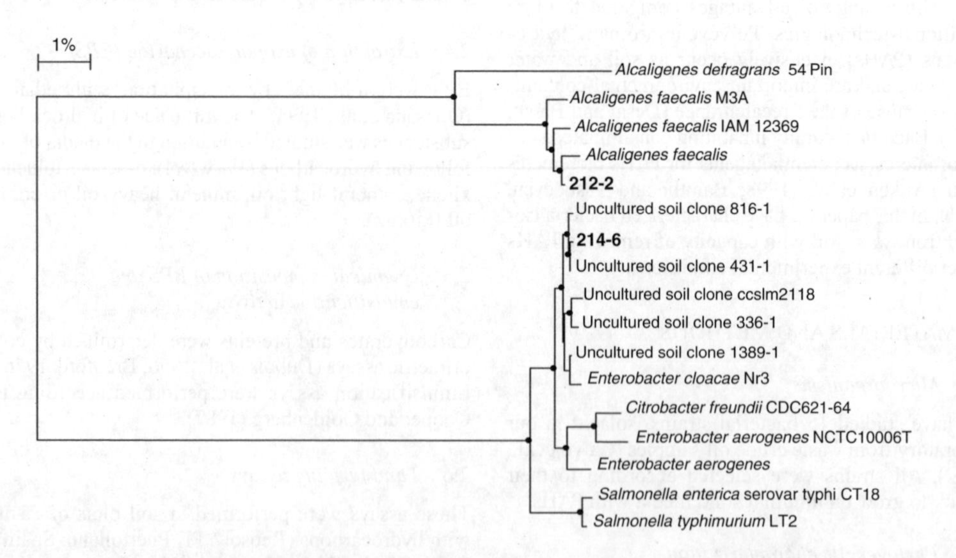

Figure 2. Neighbour-joining phylogenetic tree based on the nearly full length of the 16S rDNA gene of the 2 Gram-negative strains described in this paper and sequences from EMBL that gave the highest scores in similarity searches. Nodes highlighted with circles indicate more than 50% of bootstrap value.

bacteria were affiliated to the *Micrococcus luteus* species (three strains) (Fig. 1).

16S rDNA from only two Gram-negative strains was amplified and sequenced; one of them was affiliated to *Alcaligenes faecalis*, while 16S rDNA of strain 214-6 rendered the highest identity scores in homology searches with the sequence of several unidentified non-cultivated soil bacteria and members of the *Enterobacteriaceae*. The phylogenetic tree inferred from these sequences shows that strain 214-6 clusters in the periphery of the *Enterobacter* cloacae species (Fig. 2).

Growth of these strains in minimal liquid media with phenanthrene and pyrene, have shown that *Bacillus* strains were the most efficient. Viable counts indicated that strains *Bacillus subtilis* 28-15 and 212-1 and *Bacillus pumilus* 28-11 used phenanthrene as only carbon source in minimal medium (Fig. 3), while they were not able to use pyrene. However, *Bacillus pumilus* strains 27-1, 96-1 and 96-7 utilized only pyrene (Fig. 4).

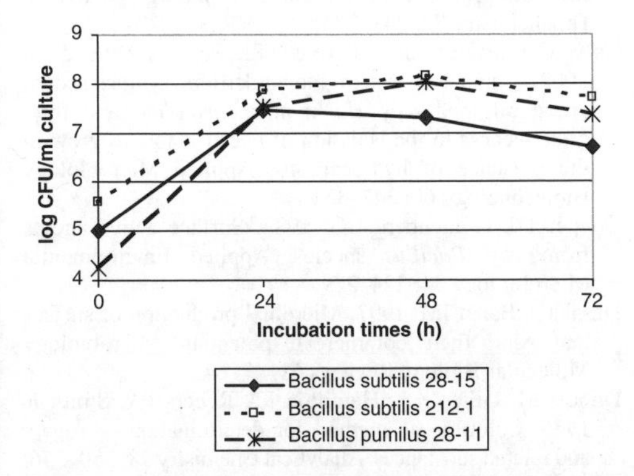

Figure 3. Growth of *Bacillus subtilis* strains 28-15 and 212-1, and *Bacillus pumilus* strain 28-11 in the minimal BH liquid medium added of phenanthrene at 0.1% w/v.

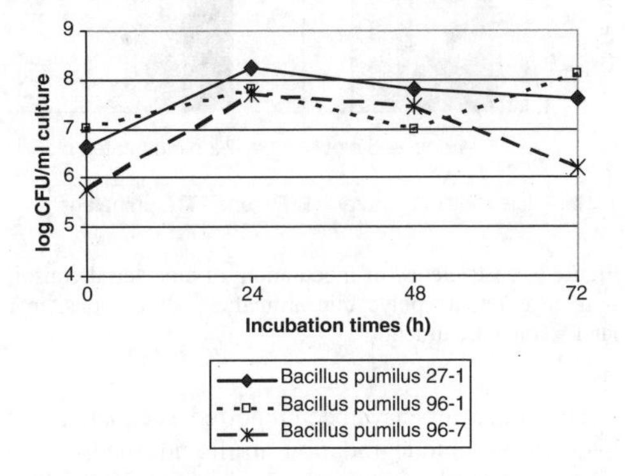

Figure 4. Growth of *Bacillus pumilus* strains 27-1, 96-1 and 96-7 in the minimal BH liquid medium added of pyrene at 0.1% w/v.

Table 1. Emulsifying activity (%) of EPS synthesized by *B. subtilis* 28-15, *A. faecalis* 212-1 and *Enterobacter* sp. 214-6 on different hydrocarbons.

EPS	Octane	Toluene	Xilene	Light oil	Heavy oil	Crude oil
B. subtilis 28-15	21.4	40.4	35.7	52.4	0	85.7
A. faecalis 212-2	47.6	57.1	57.1	57.1	23.8	73.8
Entero-bacter sp.214-6	11.9	45.2	57.1	23.8	11.9	81.0

Table 2. Yield and chemical composition of exopolysaccharides (EPS) produced by *Bacillus subtilis* strain 28-11, *Alcaligenes faecalis* strain 212-2 and *Enterobacter* sp. strain 214-6 in culture media supplemented with hydrocarbons.

Strain	Substrate	Yield production (g/l)	Carbohydrate (%)	Proteins (%)
B. subtilis 28-15	Glucose	0.4	33.9	14.1
	Toluene	0.3	49.1	15.9
	Light oil	0.1	28.5	13.5
	Heavy oil	0.4	25.1	16.4
A. faecalis 212-2	Glucose	0.5	50.9	22.1
	Octane	0.5	70.3	10.4
	Light oil	0.4	32.3	15.4
	Heavy oil	0.5	29.7	15.2
	Crude oil	0.4	39.6	14.1
Enterobacter sp. 214-6	Glucose	1.0	41.8	24.6
	Octane	0.9	50.5	9.1
	Light oil	0.9	45.9	9.2
	Heavy oil	1.1	54.2	9.0
	Crude oil	0.6	56.6	8.6

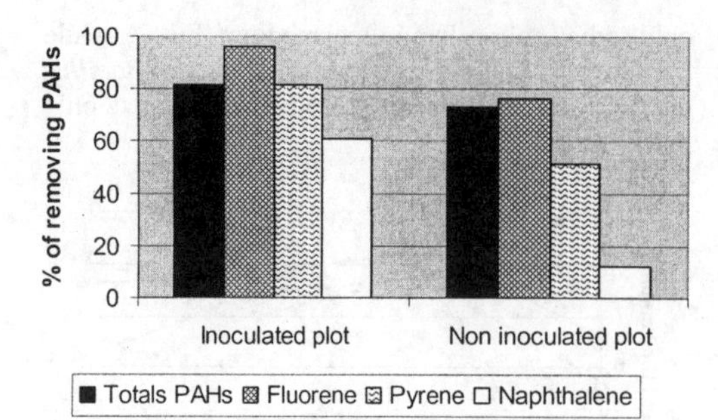

Figure 5. Efficiency of inoculation, of a bacterial consortium to eliminate polycyclic aromatic hydrocarbons, in a land farming treatment.

Bioemulsifiers have been reported as enhancers of hydrocarbon biodegradation in liquid media, soil slurries and soil microcosms (Sutherland, 2001). Production of biosurfactant compounds by these strains was also studied. *B. subtilis* 28-15, *A. faecalis* 212-2 and *Enterobacter* 214-6 produced EPS in media with different hydrocarbons. These bioemulsifiers were able to produce stable oil/water emulsions with various hydrocarbons (Table 1).

Polysaccharide production and chemical composition were influenced by media composition (Table 2).

Next and according to these results, we constructed a bacterial consortium: *B. pumilus* 28-11, *A. faecalis* 212-2, *M. luteus* 212-4, *Enterobacter* 214-6 for using as inoculum in an experimental land farming treatment. Assays carried out have shown a PAH removal of nearly 90 % in plot amended with the consortium, in contrast with a 73% of elimination in control plot without inoculation (Fig. 5).

3 CONCLUSIONS

We conclude that the samples of waste crude were efficient sources for the isolation of bacteria with degrading and/or emulsifying abilities.

ACKNOWLEDGEMENTS

This research was supported by Programa Nacional de I + D, Ministerio de Ciencia y Tecnología (MCYT), Spain (REN2000-0384-P4-02). Work by B. Rodelas was granted by Programa Ramón y Cajal (MCYT, Spain).

REFERENCES

Aitken MD, Stringfellow WT, Nagel RD, Kazunga C, Chen SH. 1998. Characteristics of phenanthrene-degrading bacteria isolated from soils contaminated with polycyclic aromatic hydrocarbons. Canadian Journal Microbiology. 44: 743–752.

Barathi S, Vasudevan N. 2001. Utilization of petroleum hydrocarbons by *Pseudomonas fluorescens* isolated from petroleum contaminated soil. Environmental International. 26: 413–416.

Bradford MM. 1976. A rapid and sensitive method for the quantification of microgram quantities of protein utilising the principle of protein-dye binding. Analytical Biochemistry 72: 248–254.

Calvo C, Martínez-Checa F, Toledo FL, Porcel J, Quesada E. 2002. Characteristics of bioemulsifiers synthesised in crude oil media by *Halomonas eurihalina* and their effectiveness in the isolation of bacteria able to grow in the presence of hydrocarbons. Applied Microbiology Biotechnology 60: 347–351.

Cooper DJ, Goldenberg BG. 1987. Surface active agents from two *Bacillus* species. Applied Environmental Microbiology 54: 224–229.

Desai JD, Banat IM. 1997. Microbial production of surfactants and their commercial potential. Microbiology Molecular Biology Reviews 61: 47–64.

Dubois M, Gilles KA, Hamilton KJ, Rebers PA, Smith F. 1956. Colorimetric method for determination of sugars and related substances. Analytical Chemistry 28: 350–356.

Quesada E, Béjar V, Calvo C. 1993. Exopolysaccharide production by *Volcaniella eurihalina*. Experientia 49: 1037–1041.

Shimura M, Mukerjee-Dhar G, Kimbara K, Nagato H, Kiyohara H, Hatta T. 1999. Isolation and characterization of a thermophilic *Bacillus* sp JF8 capable of degrading polychlorinated byphenyls and naphthalene. FEMS Microbiology Letters 178: 87–93.

Sutherland IW. 2001. Microbial polysaccharides from Gram-negative bacteria. International Dairy Journal 11: 663–674.

Zhuang WQ, Tay JH, Maszenan AM, Tay STL. 2002. *Bacillus naphthovorans* sp. Nov. from oil-contaminated tropical marine sediments and its role in naphthalene biodegradation. Applied Microbiology Biotechnology 58: 547–553.

European Symposium on Environmental Biotechnology, ESEB 2004 - Verstraete (ed)
© 2004 Taylor & Francis Group, London, ISBN 90 5809 653 X

Effects of exogenous co-planar polychlorinated biphenyls (PCBs) on the microbial reductive dechlorination of PCBs pre-existing in an anaerobic sediment of Venice Lagoon

G. Zanaroli, F. Fava & L. Marchetti
DICASM, Faculty of Engineering, University of Bologna, Bologna, Italy

J.R. Pérez-Jiménez & L.Y. Young
Biotechnology Center for Agriculture and the Environment, Cook College, Rutgers, The State University of New Jersey, New Brunswick, USA

ABSTRACT: The occurrence of reductive dechlorination processes towards pre-existing PCBs and the possibility of enhancing them by the addition of exogenous PCBs were investigated in a contaminated sediment of Porto Marghera (Venice Lagoon, Italy) suspended, under strictly anaerobic conditions, in water collected from the same site. After a five-month lag phase, several pre-exiting hexa-, penta- and tetra-chlorinated biphenyls were slowly bioconverted into tri- and di-, *ortho*-substituted PCBs. The spiked co-planar, dioxin-like 3,3′,4,4′-tetrachlorobiphenyl, 3,3′,4,4′,5- and 2,3′,4,4′,5-pentachlorobiphenyls, 3,3′,4,4′,5,5′- and 2,3,3′,4,4′,5-hexachlorobiphenyls were extensively transformed to lower chlorinated, mostly *ortho*-substituted congeners, during the 16-months experiment. The reductive dechlorination of the exogenous PCBs did not influence significantly the biotransformation onset of the sediment-carried PCBs. PCB dechlorination initiated when sulfate was completely depleted and methanogenesis started to take place, thus suggesting that sulfate-reducing bacteria started to use PCBs as electron acceptors only when sulfate was completely depleted or a possible involvement of methanogenic bacteria in the process.

1 INTRODUCTION

PCBs are contaminants of great environmental concern as they are poorly biodegradable, toxic and highly hydrophobic compounds that tend to strongly accumulate in anoxic freshwater and marine sediments (Brown & Wagner 1990). PCBs occurring in freshwater sediments have been proven to be dechlorinated in anaerobic slurry microcosms developed in the presence of defined mineral media (Bedard & Quensen 1995, Wiegel & Wu 2000).

With few exceptions, the dechlorination activity is in general directed to the *meta-* and *para-* chlorines of the biphenyl molecule, resulting in the bioconversion of highly chlorinated PCBs into low-chlorinated *ortho*-substituted congeners (Bedard & Quensen 1995, Wiegel & Wu 2000). A little is still known about the occurrence of the same processes towards aged PCBs in marine sediments (Alder et al. 1993, Palekar et al. 2003); to our knowledge, the reductive dechlorination of pre-existing PCBs was documented only once in the literature (New Bedford Harbor sediments under methanogenic conditions; Alder et al. 1993). Some other studies have documented the reductive dechlorination of PCBs in sediment slurries developed with salt rich media; however, spiked standard mixtures of PCBs and/or synthetic media were employed in these studies (Lake et al. 1992; Øfjord et al. 1994; Berkaw et al. 1996), and this markedly limited the practical relevance of the information provided.

Dechlorination of weathered PCBs was recently detected in 3 marine contaminated sediments of the Brentella Canal (Porto Marghera) of the Venice Lagoon, when suspended either in a synthetic marine medium and in water coming from the site (i.e. under geochemical conditions closer than the previous case to those occurring *in situ*). The dechlorination occurred concurrently with sulfate-reduction and methanogenesis under both conditions, but it was more extensive in the presence of the site water than with the mineral medium (Fava et al. 2003a). Very recently, the occurrence of the same processes was documented in a fourth historically contaminated sediment of the Venice Lagoon suspended in water of the same site;

by using specific microbial inhibitors (i.e. BES, molybdate, antibiotics, etc.) it was observed that (a) pre-existing PCB dechlorination was mediated by sulfate-reducing and spore forming indigenous bacteria, (b) the dechlorination was selective towards the *meta* and *para* positions of PCB molecules, and (c) the processes were not "primed" by the addition of exogenous 2,3,4,5,6-pentachlorobiphenyl (Fava et al. 2003b). Therefore, more information on the biological fate of weathered PCBs occurring in marine sediments and in particular in those of the Porto Marghera area (Venice Lagoon), is desirable. In this study we analyzed the occurrence of these processes in another historically PCB-contaminated sediment of the Venice Lagoon, by also studying the possibility of enhancing them through their supplementation with different exogenous PCB congeners.

2 EXPERIMENTAL APPROACH AND METHODS

The sediment employed was black, silty mud and it contained approximately 1.6 mg/kg (on dry wt basis) of a mixture of PCBs which could be partially ascribed to PCBs of Aroclor 1242 and Aroclor 1254.

A set of 8 slurry-phase anaerobic microcosms consisting of a PCB contaminated sediment of Porto Marghera area suspended at 25% (v/v) in water of the same contaminated area was developed under N_2:CO_2 (70:30) atmosphere according to Fava et al. (2003a). Four of them (2 biologically active and 2 autoclave-sterilized) were used to investigate the occurrence of reductive dechlorination processes towards sediment-carried PCBs whereas the other 4 microcosms (2 biologically active and 2 autoclave-sterilized) were spiked with 3,3′,4,4′-tetrachlorobiphenyl, 3,3′,4,4′,5-pentachlorobiphenyl, 2,3′,4,4′,5-pentachlorobiphenyl, 3,3′,4,4′,5,5′-hexachlorobiphenyl and 2,3,3′,4,4′,5-hexachlorobiphenyl (all from Ultra Scientific, Rhode Island, USA) in order to investigate: a) the possibility of "priming" (i.e. stimulating) the dechlorination of PCBs pre-existing in the sediment, and b) the potential biological fate of target co-planar dioxin-like PCBs under the geochemical conditions created in the microcosms.

Each exogenous PCB was individually added (as solutions at 10,000 mg/l prepared in acetone) to a final concentration of 100 mg/kg of dry sediment; a volume of pure acetone identical of that employed to provide PCBs in the spiked microcosms was added to the parallel non spiked microcosms. Sterile microcosms were prepared through autoclave-sterilization performed at 121°C for 1 h in three consecutive days; in the case of spiked control microcosms, exogenous PCBs were added after autoclave sterilization.

The developed microcosms were then incubated stationary at 25 ± 1°C in the dark for 16 months, during which they were periodically sampled and analyzed to determine the volume and the composition of the headspace gas, as well as the concentration of pre-existing PCBs and of $SO_4^=$ according to Fava et al. (2003a).

3 RESULTS AND DISCUSSION

A total amount of sediment carried PCBs corresponding to 1.60 ± 0.13 mg/kg of dry sediment was found to occur in the non-spiked sterile and biologically active microcosms after 7 days of microcosm incubation. The total amount of pre-existing PCBs in the spiked microcosms was expected to be the same; however, we could not determine it, as the added PCBs interfered with GC-ECD congeners estimation.

No transformation of sediment-carried PCBs was detected in the sterile microcosms until the end of the experiment. On the contrary, a significant change in PCB profile was found to occur in the corresponding non-spiked biologically active microcosms, as compared to the sterile ones, starting from the 5th month of incubation. At the end of the experiment (after 16 months), several hexa-, penta- and tetra-chlorinated congeners were found to be bioconverted on a molar basis into less chlorinated PCBs, such as 2,2′,5/2,2′,4/4,4′-chlorobiphenyl and 2,4/2,5-chlorobiphenyl. 4-monochlorobiphenyl also accumulated in the biologically active microcosms at the end of the experiment, whereas 2-monochlorobiphenyl was depleted (Figure 1).

A little sulfate consumption was observed in the active microcosms since the first month of incubation. Sulfate was then quickly depleted, becoming 8.4% of the initial concentration (1.95 ± 0.03 g/l) after 2 months of incubation and undetectable at the end of the 3rd month of experiment (Table 1). No significant biogas production was observed in the active microcosms until sulfate was not completely depleted. A large amount of biogas (27.5 ± 16.5 ml) consisting of more than 47% of methane was detected in the same microcosms between the 3rd and the 5th month of incubation, i.e. immediately after complete sulfate depletion and before PCB dechlorination started. Methane production was detected at a lower rate all over the experiment (up to the 16th month) (Table 1).

Very similar trends, both in terms of sulfate consumption and methane production, were observed in the parallel biologically active spiked microcosms, where the overall amount of produced methane was about 60% of that detected in the non spiked ones (Table 1). The biotransformation of several sediment-carried PCBs could not be quantified in the spiked microcosms, as some of them were produced from the spiked congeners dechlorination. However, the fate of about 30% of the GC-ECD peaks ascribed to pre-existing PCBs could be monitored and compared

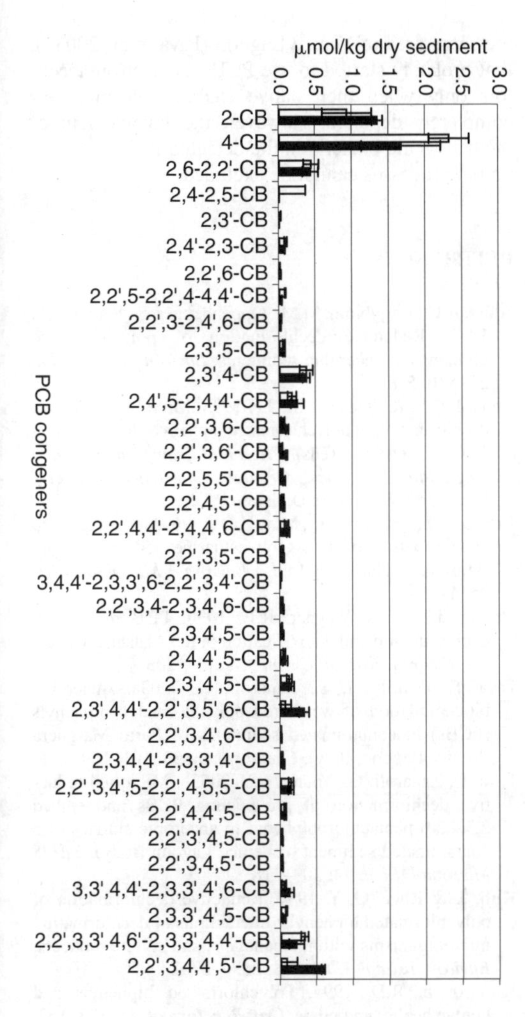

Figure 1. Average concentration of each sediment-carried PCB congener (±standard deviation as error bar) in the non spiked sterile (black bar) and biologically active (white bar) microcosms after 16 months of incubation.

Table 1. Overall sulfate consumption, gas and methane production (±standard deviation) in the biologically active microcosms after 16 months of incubation. Initial sulfate concentration in all microcosms was 1.95 ± 0.03 g/l

	Non spiked microcosms	Spiked microcosms
SO_4^- consumption (%)	100	100
Biogas production (ml)	34.7 ± 20.4	20.4 ± 4.9
CH_4 production (ml)	15.0 ± 8.5	8.6 ± 2.9

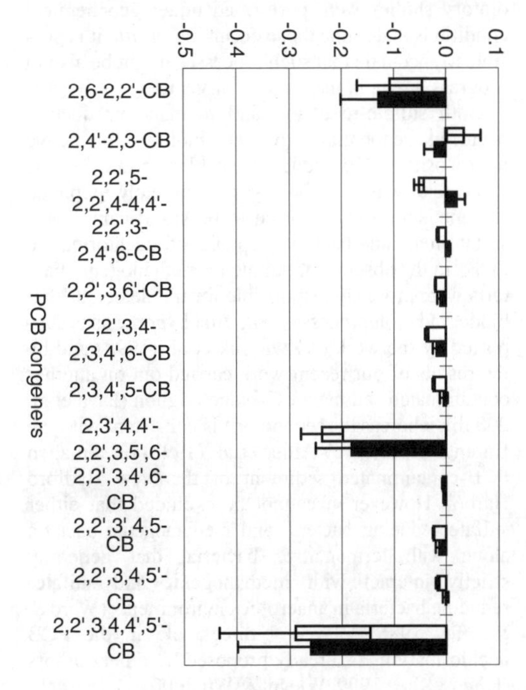

Figure 2. Changes in the PCB concentration in the biologically active non spiked (white bar) and spiked (black bar) microcosms (all estimated vs. the sterile ones) at the end of the 16th month of incubation.

with that observed in the non spiked microcosms (Figure 2). A significant transformation of such pre-existing PCBs and of the exogenous PCBs was detected starting from the 5th month of incubation. However, the dechlorination of the exogenous PCBs only slightly influenced the bioconversion extent and pattern of pre-existing PCBs (Figure 2), suggesting that the dechlorination of the latter was not significantly primed by the addition of the 5 exogenous coplanar congeners.

The exogenous PCBs were markedly bioconverted into less chlorinated congeners, such as 3,3′,5,5′-/2,3′,4,4′-, 2,3′,4′,5- and 2,4,4′,5-tetrachlorobiphenyl, 2,4,4′-, 2,3′,4- and 2,3′5-trichlorobiphenyl and 3,4- and 3,4′-dichlorobiphenyl between the 5th and the 8th month of incubation, and were found to be depleted by more than 80% at the end of the experiment (data not shown). This finding is of great relevance, as indicates that indigenous microbial consortia selected in the spiked primary microcosms were able to rapidly and extensively dechlorinate some dioxin-like coplanar PCBs that are regarded as some of the most toxic PCBs reported in the literature (Kimbrough 1995). The extensive dechlorination of the spiked PCBs only slightly intensify rate and extent of the dechlorination of the PCBs pre-existing in the sediment. The lack of significant priming effects in aged PCB contaminated marine sediments has also been observed in a different contaminated sediment of the Venice Lagoon spiked with 2,3,4,5,6-pentachlorobiphenyl (Fava et al. 2003b) and in a sediment of the LCP Chemicals Superfund site in coastal Georgia (USA) (Palekar et al. 2003).

Taken together, these data suggest that PCB pre-existing in the contaminated sediment of Porto Marghera employed in this study can undergo microbial reductive dechlorination. This finding supports previous observations on the occurrence of dechlorination processes in contaminated sediments of the Venice Lagoon (Fava et al. 2003a, b). As all these laboratory studies were performed under geochemical conditions that mime those occurring *in situ,* it is possible to speculate that such processes might be also in progress *in situ* in the Lagoon of Venice.

Both sulfate-reduction and methane production occurred sequentially in the biologically active microcosms; PCB reductive dechlorination became detectable when sulfate was completely depleted and methanogenesis started to be significant. Thus, sulfate-reducing bacteria capable of dechlorinating PCBs in the absence of sulfate or methanogenic bacteria were probably responsible for the detected PCB biodegradation processes. The first hypothesis is supported by the work of Zwiernik et al. (1998) and by the results of our recent work carried out on another contaminated sediment of Venice Lagoon (Fava et al. 2003b), whereas the second one is consistent with the findings reported by Alder et al. (1993) on the aged PCB-contaminated sediment of the New Bedford Harbor. However, it cannot be excluded that either sulfate reducing bacteria and methanogenic bacteria along with fermentative bacteria, that generally strictly interact with methanogenic and sulfate-reducing bacteria in anaerobic environments (Ward & Wrinfey 1985), played a direct role in the PCB dechlorination, as already proposed by other authors (Kim & Rhee 1999, Wiegel & Wu 2000).

4 CONCLUSIONS

The occurrence of microbial-mediated, reductive dechlorination processes towards weathered PCBs and spiked 3,3′,4,4′-tetrachlorobiphenyl, 3,3′,4,4′,5- and 2, 3′,4,4′,5-pentachlorobiphenyls, 3,3′,4,4′,5,5′- and 2,3,3′,4,4′,5-hexachlorobiphenyls has been shown in a contaminated sediment of Porto Marghera (Venice Lagoon, Italy. The dechlorination of pre-existing PCBs was slow and only slightly intensified by the addition of the exogenous dioxin-like co-planar congeners, that were rapidly and extensively dechlorinated. PCB dechlorination seemed to be mediated by sulfate-reducing bacteria, previously reported as responsible for PCB dechlorination in sediments of the same area of Venice Lagoon (Fava et al. 2003b), that probably started to use PCBs as electron acceptors only when their native electron acceptor was completely depleted. However, the involvement of methanogenic bacteria in the detected PCB-dechlorination processes cannot be excluded.

REFERENCES

Alder, A.C., Häggblom, M.M., Oppenheimer, S., Young, L.Y. 1993. Reductive dechlorination of polychlorinated biphenyls in anaerobic sediments. *Environ. Sci. Technol.* 27: 530–538.

Bedard, D.L. & Quensen III, J.F. 1995. Microbial reductive dechlorination of polychlorinated biphenyls. In L.Y. Young & C.E. Cerniglia (eds), *Microbial Transformation and Degradation of Toxic Organic Chemicals*: 127–216. New York: Wiley-Liss Division, Wiley.

Berkaw, M., Sowers, K.R., May, H.D. 1996. Anaerobic *ortho* dechlorination of PCBs by estuarine sediments from Baltimore Harbor. *Appl. Environ. Microbiol.* 62: 2534–2539.

Brown, J.F. Jr. & Wagner, R.E. 1990. PCB movement, dechlorination and detoxification in the Acushnet estuarine. *Environ. Toxicol. Chem.* 9: 1215–1233.

Fava, F., Gentilucci, S., Zanaroli, G. 2003a. Anaerobic biodegradation of weathered polychlorinated biphenyls (PCBs) in contaminated sediments of Porto Marghera (Venice Lagoon, Italy). *Chemosphere* 53: 101–109.

Fava, F., Zanaroli, G., Young, L.Y. 2003b. Microbial reductive dechlorination of pre-existing PCBs and spiked 2,3,4,5,6-pentachlorobiphenyl in anaerobic slurries of a contaminated sediment of Venice Lagoon (Italy). *FEMS Microbiol. Ecol.* 44: 309–318.

Kim, J. & Rhee, G.-Y. 1999. Reductive dechlorination of polychlorinated biphenyls: interactions of dechlorinating microorganisms with methanogens and sulfate reducers. *Environ. Toxicol. Chem.* 18: 2696–2702.

Kimbrough, R.D. 1995. Polychlorinated biphenyls and human health: an update. *Crit. Rev. Toxicol.* 25: 133–163.

Lake, J.L., Pruell, R.J., Osterman, F.A. 1992. An examination of dechlorination process and pathways in new Bedford Harbor sediments. *Mar. Environ. Res.* 33: 31–47.

Øfjord, G.D., Puhakka, J.A., Ferguson, J.F. 1994. Reductive dechlorination of Aroclor 1254 by marine sediment cultures. *Environ. Sci. Technol.* 28: 2286–2294.

Palekar, L.D., Maruya, K.A., Kostka, J.E., Wiegel, J. 2003. Dehalogenation of 2,6-dibromobiphenyl and 2,3,4,5,6-pentachlorobiphenyl in contaminated estuarine sediment, *Chemosphere* 53: 593–600.

Ward, D.M. & Wrinfey, M.R. 1985. Interactions between methanogenic and sulfate-reducing bacteria in sediments. *Adv. Microbiol. Ecol.* 3: 141–175.

Wiegel, J. & Wu, Q. 2000. Microbial reductive dehalogenation of polychlorinated biphenyls. *FEMS Microbiol. Ecol.* 32: 1–15.

Zwiernik, M.J., Quensen III, J.F., Boyd, S.A. 1998. $FeSO_4$ amendments stimulate extensive anaerobic PCB dechlorination. *Environ. Sci. Technol.* 32: 3360–3365.

European Symposium on Environmental Biotechnology, ESEB 2004 - Verstraete (ed)
© 2004 Taylor & Francis Group, London, ISBN 90 5809 653 X

Characterization of the microbial communities within and in the vicinity of an iron barrier

Thomas Van Nooten[1,2], Leen Bastiaens[1], Dirk Springael[2], Brigitte Borremans[1] & L. Diels[1]
[1]*Flemish Institute for Technological Research, Mol, Belgium*
[2]*Katholieke Universiteit Leuven, Leuven, Belgium*

The use of zero valent iron in reactive barriers has been shown to be very effective for passive, long-term applications of groundwater remediation. Because the contaminants are removed by abiotic processes, little is known about the microbial activity and characteristics within and in the vicinity of the Fe^0-barrier matrix. Major uncertainties need to be resolved with respect to the adaptation of indigenous microorganisms to the strongly reducing Fe^0 environment, changes in the microbial community composition, and their beneficial or detrimental effects on the longevity and long-term efficiency of the Fe^0 barriers. On the one hand, the accumulation of biomass, the production of gas bubbles, and the formation of mineral precipitates can have a negative impact on the reactivity of the barrier by blocking reactive sites. On the other hand, microorganisms can positively affect the performance of the barrier by contributing to the degradation of contaminants, by consuming abiotically produced gas bubbles, and by contributing to mineral dissolution.

The depletion of dissolved oxygen and the production of cathodic H_2 by Fe^0 corrosion provide a reducing environment favorable to a wide variety of hydrogen consuming anaerobic microorganisms. These include sulfate- and metal-reducing bacteria, methanogens, and denitrifying bacteria within and downgradient of the barrier.

The objective of this study is to examine the microbial population in reactive iron barriers and to get a clearer view on the different groups of microorganisms that are present within and in the vicinity of the Fe^0-barrier matrix. Molecular methods including PCR and PCR-DGGE are used for this purpose. Different sets of specific PCR-primers are being tested and will be applied at first on samples of lab scale experiments. Later, samples from pilot-scale set-ups and *in situ* reactive iron barriers will be examined. Available results will be presented.

European Symposium on Environmental Biotechnology, ESEB 2004 - Verstraete (ed)
© 2004 Taylor & Francis Group, London, ISBN 90 5809 653 X

Microbial activity and characterization during hydrocarbon phytoremediation

M.R.T. Palmroth, P.E.P. Koskinen, U. Münster & J.A. Puhakka
Institute of Environmental Engineering and Biotechnology, Tampere University of Technology, Tampere, Finland

J. Pichtel
Natural Resources and Environmental Management, Ball State University, Muncie, Indiana, USA

ABSTRACT: Microbial communities in soil contaminated by hydrocarbons (HCs) were characterized during phytoremediation to determine the effects of plants on indigenous microbial activity. In the laboratory, diesel fuel was readily removed from artificially contaminated soil by phytoremediation, while weathered HCs at field site were less bioavailable. In contaminated soil the utilisation of the volatile fraction of diesel fuel was higher in the MT2 assay that of uncontaminated soil. Diesel fuel utilisation in the MT2 assay was highest in contaminated soil vegetated with trees for both laboratory and field site communities. Potential maximum rates of microbial extracellular enzyme activities were more dependent on the soil type than contamination. The abundance of alkane hydroxylase and naphthalene dioxygenase genes was low in the field site but the addition of fresh HCs increased the abundance of these genes. In summary, hydrocarbon contamination did not affect microbial activity in soil. However, the increased bioavailability of the contaminants improved their removal in phytoremediation.

1 INTRODUCTION

Phytoremediation uses plants to remove, contain or render harmless environmental contaminants. Plants or plant-associated microflora convert many pollutants to non-toxic forms (Cunningham & Berti 1993; Cunningham et al. 1995). In addition, plants exude organic and inorganic substances to the soil environment during normal metabolism (Anderson et al. 1993) and the root exudates may favour microbial activity and biodegradation in the root zone. Thus phytoremediation can be used to enhance bioremediation of contaminants, such as hydrocarbons.

Microbial communities in soil contaminated by hydrocarbons were characterized during laboratory- and field-scale phytoremediation of hydrocarbons (HCs) to determine the effects of plants on indigenous microbial activity. The purpose of the study was to assess the effect of plants on the removal of hydrocarbons and on the microbial activities in both weathered and artificially contaminated soil.

2 MATERIALS AND METHODS

2.1 *Setup of experiments*

Experiments were made both in the laboratory with soil spiked with diesel fuel and at field-scale with weathered HC contaminated soil.

In the laboratory, 0.5% (w/w) diesel fuel (Neste Futura) was mixed with the soil. Plant treatments included Scots pine (*Pinus sylvestris*); Poplar (*Populus deltoides x Wettsteinii*); a grass mixture (*Festuca rubra, Poa pratensis, Lolium perenne*); and a legume mixture (White clover *Trifolium repens* and Green Pea *Pisum sativum*). Control treatments consisted of pots of each plant treatment without diesel. The effect of recontamination of soil with 2% diesel fuel was studied during the following summer as a short-term, one-month experiment for the trees only (Palmroth et al. 2002).

Field site contained soil contaminated with weathered HCs deriving from diesel fuel and lubrication oil and consisted of four treatment sectors. Two of the sectors contained composted biowaste. One sector was fertilized with NPK-fertilizer and one was unfertilized. White clover and grass mixture was sown to the entire plot. Scots pines and poplars were planted to three sectors. Second composted biowaste sector contained no trees. The study lasted for four growing seasons. The definitions of sample types collected from the field site for microbial community analyses were as shown in Table 1.

Mineral oil concentration was determined from field site samples with ISO committee draft method (ISO, 2000), which involves extraction with acetone/heptane and washing with water and Florisil. The extraction step was repeated twice and the cleaned combined extract was analysed with GC/MS and external

Table 1. Sample types collected from the field site for microbial community analyses.

Sample	Definition
No grass	Unfertilized soil, no vegetation containing spot
NA pine	Unfertilized soil, pine rhizosphere
NA poplar	Unfertilized soil, poplar rhizosphere
KP pine	Compost amended soil, pine rhizosphere
KP poplar	Compost amended soil, poplar rhizosphere

Table 2. Removal of diesel fuel (%) in the different plant treatments in the laboratory experiment.

Plant treatment	30 d	90 d	120 d	180 d	330 d	14 d	28 d
Poplar	60	90			96	69	86
Pine	50	89			93	77	60
Legume	52	96	98	97			
Grass	37	81	81	83			
No vegetation	52	86	84	73			

calibration method. Diesel fuel concentration in spiked soil was determined with twice repeated hexane extraction and the extract was analysed with GC/MS using external calibration (Palmroth et al. 2002).

2.2 *Carbon source utilisation patterns of soil microbiota*

Carbon source utilisation of soil microbial communities were assessed using Biolog ECO (Biolog, 2000) and MT2 plates (Biolog Inc., Hayward). A 90 μl aliquot of 1:100 soil suspension was injected into each of the 96 wells of the plates.

In this study, the utilisation of the volatile fraction of diesel fuel was used as an indicator of diesel fuel degradation (Palmroth et al. 2003). Diesel fuel was allowed to evaporate into a desiccator atmosphere and the Biolog MT2 plates without lid were incubated in the closed desiccator, thus exposing the plates to the maximum vapour pressure of diesel fuel volatile compounds. Absorbance was measured at 590 nm with a Victor™ Multilabel reader (Wallac, Turku, Finland) or Multiscan Ascent (Thermo Labsystems, Vantaa, Finland). The data was analysed according to Palmroth et al. (2003).

2.3 *Extracellular enzymatic activities*

Potential maximum rates of microbial extracellular enzyme (MEE) activity were determined in a multiwell assay with fluorogenic enzyme substrates. The activity of 14 enzymes taking part in the hydrolysis of C, N, and P compounds was determined in a 96-well assay (Palmroth et al. 2003). The stock and working solutions of enzyme substrates and the model fluorogenic molecules, 4-methylumbelliferone (MUF) and 7-amino-4-methylcoumarin (AMC), were prepared in dimethyl sulfoxide. Standards of MUF and AMC (0.001 to 50 μM) were prepared in duplicate in soil suspensions i.e. standard curves were prepared separately for each sample. The 1:100 soil suspension (90 μl) was transferred to each of the wells in a 96-well microtiter plate. The fluorescence of the hydrolysed model substrates and standards was measured at excitation 355 nm/emission 460 nm with a fluorometer (Fluoroskan Ascent FL, Thermo Labsystems), immediately after addition of soil suspension and at least once during incubation.

The rates of hydrolysed enzyme substrate [μM MUF/h/g] or [μM AMC/h/g] at 3 h incubation time were calculated for dry soil to enable comparison of different soil samples. The potential maximum rate of microbial extracellular enzyme hydrolysis (v_{max}) is referred in text as v_{max}. The term potential is used in this context due to the interference of the soil matrix to the assay. The v_{max} and K_m values were calculated from Michaelis-Menten plots with SigmaPlot™ Enzyme Kinetics Package 1.1.

2.4 *Genetic characterization*

The abundance of alkane hydroxylase (*alk*B) and naphthalene dioxygenase (*nah*Ac) genes were screened from tree rhizospheres in field as described earlier (Koskinen et al. 2003). In addition, *alk*B and *nah*Ac genes were studied from artificially contaminated soil vegetated with pines or poplars after 3 months of phytoremediation in laboratory.

3 RESULTS AND DISCUSSION

3.1 *Hydrocarbon removal*

Hydrocarbon removal in the plant treatments during the first and second diesel fuel application in the laboratory-scale were as shown in Table 2. Diesel fuel was removed more rapidly in the legume treatment than in other plant treatments. The presence of poplar and pine enhanced removal of diesel fuel, but removal under grass was similar to that with no vegetation. In the second diesel fuel application to the trees, diesel was removed considerably faster than in the first year. When pine and poplar treatments were compared, poplar treatment clearly enhanced diesel removal more than pine treatment.

Figure 1 shows the trend of HC concentrations at different sectors in the field site throughout the experiment. With the exception of the initial decrease of HC concentrations, no decrease in HC concentrations was observed and treatment objectives were not met within the over 3-year field study. Nonetheless, soil with no amendments had the highest concentrations of HCs over the whole research period of the

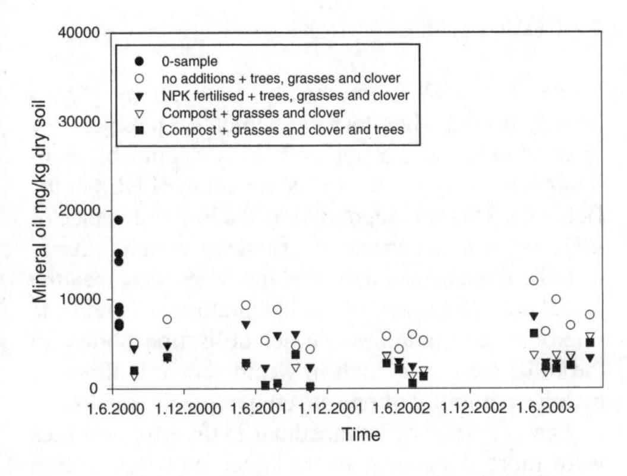

Figure 1. Hydrocarbon concentrations at the field site.

different soil treatments. This sector also had the lowest density of vegetation (result not shown). The treatments with composted biowaste had the lowest HC concentrations, but part of this is explained by the diluting effect of the biowaste supplementation. The results suggest that plants and/or soil fertilization slightly enhanced HC degradation in the field site.

3.2 *Carbon source utilisation patterns of soil microbiota*

Carbon source utilization patterns indicated minor differences in microbiota in soil vegetated with pine compared to microbiota in soil vegetated with poplar in artificially contaminated soil. The microbiota in all soil samples, contaminated and uncontaminated, was metabolically diverse as indicated by the ECO plate tests. This was also the case for field site communities. Furthermore, tree rhizosphere and compost affected the carbon source utilization profiles of field communities (results not shown).

Biolog MT2 plates with diesel fuel as the carbon source were used to study diesel fuel utilisation by soil microbes. In artificially contaminated soil, the utilisation of the volatile fraction of diesel fuel increased compared to the utilisation in uncontaminated soil. Community diesel fuel utilisation was highest in contaminated soil vegetated with trees (Table 3). The increased HC utilization by trees was also seen with field communities. Communities in tree rhizospheres had higher growth rates with volatile fraction of diesel in the MT2 plates than those in non-vegetated soil. In artificially spiked soil the results obtained with diesel fuel utilisation assay were in accordance with actual diesel removal (Table 2). The growth rates were higher with field samples than with laboratory samples, indicating that the bacteria in weathered soil had maintained their ability to degrade diesel fuel, although GC/MS results indicated that short-chained hydrocarbons had been removed from soil (data not shown).

Table 3. Maximum specific growth rates (μ_m) of microbial communities grown with diesel in the MT2 plates.

Community	Type of contamination	μ_m(1/d)
No Grass field	Weathered HCs	0.010 ± 0.001
NA pine field	Weathered HCs	0.121 ± 0.010
NA poplar field	Weathered HCs	0.110 ± 0.035
KP pine field	Weathered HCs	0.079 ± 0.011
KP poplar field	Weathered HCs	0.130 ± 0.011
Unvegetated uncontaminated soil laboratory	None	X
Unvegetated soil laboratory	2% fresh diesel 14 days ago	0.006 ± 0.005
Pine soil laboratory	2% fresh diesel 14 days ago	0.007 ± 0.002
Poplar soil laboratory	2% fresh diesel 14 days ago	0.013 ± 0.007

X = Absorbance increase not statistically significant during incubation time.

3.3 *Extracellular enzymatic activities*

Microbial extracellular enzyme activities were assessed during phytoremediation laboratory experiment in soils contaminated with diesel fuel compounds and in control soils to determine how plants affect microbial activity. Phosphomonoesterase v_{max} was 200 μM MUF/h/g dry soil or lower, which could indicate that the growth environment was not P limited. V_{max} of aminopeptidases and esterases increased after diesel contamination, but started to decrease with time (Palmroth et al. 2003). The potential maximum rates of aminopeptidase activity were at the scale of 10 to 10^2 μM AMC/h/g dry soil at other sampling times than 14 days after second diesel addition, where the rates were at the scale of 10^3 μM AMC/h/g dry soil. The presence of plants did not influence the activity of esterases and their v_{max} were at the scale of 10^3 to 10^4 μM MUF/h/g dry soil. The activity of other studied hydrolytic enzymes, glucosidases, chitinase and xylosidase, remained below detection limit.

Microbial extracellular enzyme activities of field site communities were assessed to study the effects of plants and soil type on microbial activity. Extracellular enzymatic activities of esterases increased as a result of compost addition up to 10^5 μM MUF/h/g dry soil, but during growing season they were similar in all treatments. Phosphomonoesterase activities and aminopeptidase activities were high, up to 10^3 and 10^4 μM MUF/h/g dry soil, respectively, in fertilised soil. In the beginning of growing season fertilisation caused activity of glucosidases to increase from undetected to 10^3 MUF/g/h and above.

In summary, the activities of extracellular hydrolytic enzymes were more dependent on soil type than hydrocarbon contamination.

Table 4. Screening of *alk*B and *nah*Ac genes from bacterial communities in field site and utilizing HCs in MT2 plates (2 parallel samples) and in laboratory phytoremediation experiment (3 parallel samples).

Communities	*Alk*B detection	*Nah*Ac detection
2 Parallel Samples:		
No Grass field	− −	− −
NA pine field	− −	− −
NA poplar field	− −	+ −
KP pine field	− −	− −
KP poplar field	− −	− −
No Grass MT2 diesel	+ −	− −
NA pine MT2 diesel	+ +	+ +
NA poplar MT2 diesel	+ −	− −
KP pine MT2 diesel	+ −	− −
KP poplar MT2 diesel	− −	+ −
3 Parallel Samples:		
Pine laboratory	− − −	+ − −
Poplar laboratory	− − −	+ + −

+ = positive signal, − = negative signal.

3.4 *Genetic characterization*

The screening of *alk*B and *nah*Ac genes revealed that their abundance in the field site communities was low (Table 4). This suggests that the alkane and naphthalene degradation pathways were not very active in the field site communities. Results are in agreement with the HC analyses from the field site. However, the abundance of both *alk*B and *nah*Ac genes increased when communities were grown with diesel in the MT2 plates. This indicates that *alk*B and *nah*Ac genes were re-activated after the addition of fresh HCs. Furthermore, *nah*Ac genes were detected from artificially spiked soil after 3 months of treatment. A*lk*B genes were neither detected from communities during phytoremediation of field site nor artificially contaminated soil. This is likely due to the lack of short chained alkanes in the soils. *Alk*B primers mainly target on the genes involved in the degradation alkanes ranging from C5 to C12 (Whyte et al. 1997), which are volatile and easily degradable. Therefore, *nah*Ac is suggested to be more relevant marker gene for remediation studies than *alk*B. In summary, results suggest that the low abundance of *alk*B and *nah*Ac genes in the prevailing field communities were due to low bioavailability and/or lack of easily degradable contaminants because of weathering of soil.

4 CONCLUSIONS

Diesel fuel HCs were readily removed by phytoremediation in laboratory experiments. However, low bioavailability and/or lack of easily degradable contaminants likely limited the degradation of HCs in the field site. This was supported by the low abundance of *alk*B and *nah*Ac genes in the field site communities.

MT2 plate assays indicated that trees increased the diesel utilization rates of both laboratory and field site microbial communities. Diesel utilisation ability in the MT2 assay was high in weathered soil, although the bioavailability of present contaminants was low.

The activities of extracellular hydrolytic enzymes were more dependent on soil type and fertilisation than hydrocarbon contamination. Carbon source utilization patterns indicated only minor differences. In summary, hydrocarbon contamination did not affect microbial activity in soil. However, the increased bioavailability of the contaminants improved their removal in phytoremediation.

REFERENCES

Anderson, T.A., Guthrie, E.A. & Walton, B.T. 1993. Bioremediation in the Rhizosphere. Environmental Science and Technology 27: 2630–2636.

Cunningham, S.D. & Berti, W.R. 1993. The Remediation of Contaminated Soils with Green Plants: An Overview. In Vitro Cell. Dev. Biol. 29P: 207–212.

Cunningham, S.D., Berti, W.R. & Huang, J.W. 1995. Phytoremediation of Contaminated Soils. Trends in Biotechnology 13: 393–397.

Koskinen, P.E.P., Palmroth, M.R.T., Münster, U. & Puhakka, J.A. 2003. Characterization of microbial communities during phytoremediation of weathered hydrocarbon contaminated soil. *Proceedings of the Second European Bioremediation Conference, Chania, Crete, Greece June 30th–July 4th, 2003,* pp. 429–432.

Palmroth, M.R.T., Pichtel, J. & Puhakka, J.A. 2002. Phytoremediation of diesel fuel contaminated subarctic soil. Bioresource Technology 84(3): 221–228.

Palmroth, M.R.T., Münster, U., Pichtel, J. & Puhakka, J.A. 2003. Assessment of diesel fuel utilisation and microbial activity during phytoremediation. Submitted to Biodegradation.

Whyte, L.G., Bourbonnière, L. & Greer, C.W. 1997. Biodegradation of petroleum hydrocarbons by psychrotrophic *Pseudomonas* strains possessing both alkane (*alk*) and naphthalene (*nah*) catabolic pathways. Applied and Environmental Microbiology 63(9): 3719–3723.

European Symposium on Environmental Biotechnology, ESEB 2004 - Verstraete (ed)
© 2004 Taylor & Francis Group, London, ISBN 90 5809 653 X

Molecular microbial diversity and technological features of a packed-bed biofilm digestor capable of olive mill wastewater anaerobic bioremediation and valorization

M.C. Colao & M. Ruzzi
DABAC, University of Tuscia, Viterbo, Italy

L. Bertin & F. Fava
DICASM, Faculty of Engineering, Bologna, Italy

ABSTRACT: The stability, bioremediation efficiency and methane productivity along with the biology of a granular activated carbon (GAC) packed-bed biofilm reactor capable of olive mill wastewater (OMWs) anaerobic digestion were investigated. The GAC-reactor was found to be an effective, reproducible and stable OMW digestor. We observed temporal shifts in the microbial community composition, based on 16S rDNA sequence diversity, when the biofilm community structure was compared with that of the inoculum after 9 months of operation. In particular, specific microbial populations were enriched during the course of the study and they mainly consisted of *Proteobacteria*, *Flexibacter-Cytophaga-Bacteroides*, and sulphate-reducing bacteria. The dominant sequence among *Archaea* (70% of clones) was closely related to *Methanobacterium formicicum*.

1 INTRODUCTION

Olive mill wastewaters (OMWs) are toxic effluents that constitute a problem of environmental concern for many Countries of the Mediterranean sea area (Hamdi 1996). They can be advantageously disposed through anaerobic digestion, as this technique offers the opportunity to couple OMW-decontamination with their valorisation through methane production (Fiestas Ros de Ursinos & Borja-Padilla 1996). However, this process is often unsatisfactory, as unable to completely remove phenolic compounds, that are the most toxic and recalcitrant constituents of OMWs (Beccari et al. 2000). The possibility of intensifying OMW-phenols removal using a specialized culture immobilized on granular activated carbon (GAC) in a packed-bed anaerobic reactor has been preliminary demonstrated in a recent work (Fava et al. 2003). In the present study, we further investigated the GAC packed-bed digestor by testing its stability, biodegradation efficiency and methane productivity under a large range of high OMW loads. For a more complete assessment of the system, we also elucidated the structure and spatial distribution of the microbial community within the reactor, using a combination of Terminal-Restriction Fragment Length Polymorphism (T-RFLP), sequence, and phylogenetic analysis of 16S rRNA genes.

2 EXPERIMENTAL APPROACH

Two OMWs, i.e., OMW1 and OMW2 with about 20 and 35 g/l of COD, and 1.5 and 2.0 g/l of total phenolic compounds, respectively, were employed. Two experimental amended OMWs, AOMW1 and AOMW2, were prepared by diluting (1:1) OMW1 and OMW2, respectively, with water; diluted wastewaters were then amended with $Ca(OH)_2$ (to adjust their pH to 6.5), urea (0.45 g/l) and NaOH (to adjust their pH to 7.8 ± 0.2). Both AOMWs were vigorously mixed and purged with 0.22 μm filter-sterilized O_2-free N_2 at room temperature for 3 h, before being employed in the experiments. The GAC packed-bed anaerobic reactor employed in this study, that had a configuration identical to that of the reactor employed previously (Fava et al. 2003), consisted of an hermetically closed glass column (diameter: 8 cm; height: 45 cm) thermostated at 35°C, equipped with an influent line (placed at the bottom) and an effluent line (placed at the top) that moved the digested wastewater and the biogas produced to a closed reservoir

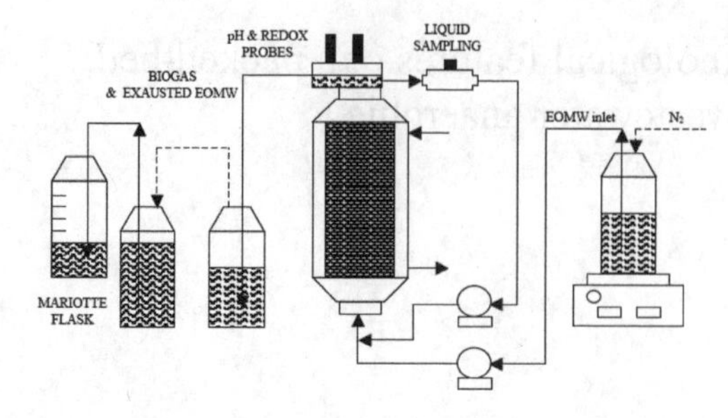

Figure 1. Scheme of the anaerobic packed-bed loop reactors developed and employed in the study.

hydraulically connected to a "Mariotte" bottle (Fig.1). The reactor was also equipped with a recycle line, along which a redox and a pH probe were placed. The reactor system, that had an empty volume of 2.4 l, was sterilized and then filled with 1.190 kg (dry weight) of GAC previously sterilised in autoclave. Then it was purged with filter-sterilized O_2-free N_2 and then filled with a deoxygenated suspension of the microbial inoculum [provided by Beccari's team (2000); 20.1 mg (on dry-weight basis) of biomass in a l of deoxygenated AOMW2]. The reactor medium was then recycled (upflow) for two weeks. To sustain biofilm formation, the reactor medium was completely replaced with fresh deoxygenated AOMW2 that was then recycled for two more weeks. Then, the reactor was forced to operate in continuous mode, by feeding it with either AOMW1 (COD: 10000 mg/l; phenols: ~600 mg/l) or AOMW2 (COD: 15000 mg/l; phenols: ~800 mg/l) at increasing dilution rates (D, expressed as the ratio between AOMW influent flow rate and the reactor reaction volume) corresponding to 0.415, 0.692, 1.038, 1.385, 2.077, 2.769 and 3.462 d^{-1}. The recycle rate was increased proportionally with D to have a reactor recycle ratio (defined as the ratio of the returned flow rate to the influent flow rate) of 77 and identical for all experiments. Steady state conditions were attained when COD and phenolic compound concentrations in the reactor effluent remained constant for at least one week long period. Six ml samples of reaction medium were taken daily from a sampling port placed along the recycle line of the reactor (Fig. 1); they were filtered and then analysed for COD, the concentration of total phenolic compounds, volatile fatty acids (VFAs) and sulfate as detailed below. Biogas resulting from the AOMW digestion was quantified through the "Mariotte" bottle system (Fig. 1), while its methane content was determined through gas-chromatographic analysis of biogas samples collected at the reactor headspace.

The distribution and structure of the microbial communities in the anaerobic reactor were evaluated using two molecular approaches. Samples from the mobile phase and biofilm, collected at different heights of the reactor column (5, 18 and 36 cm), as well as from inlet, outlet and inoculum, were treated to extract total DNA for molecular analysis. Thereafter, small-subunit rRNA genes were amplified by PCR, using universally conserved primers specific for either *Bacteria* (Osborn et al. 2000) or *Archaea* (Godon et al. 1997). The labelled products were digested with restriction enzymes *Hha*I and *Rsa*I, and the T-RFs were separated by capillary electrophoresis and detected by laser-induced fluorescence on an automated gene sequencer. At the same time partial clone libraries of 16S rRNA genes were generated from community samples.

3 RESULTS AND DISCUSSION

The bioremediation efficiency and the biogas productivity of the GAC-biofilm reactor were preliminary investigated through seven sequential 3-week-long experiments (experiments No. 1–7, Fig. 2) performed at different and increasing organic loads (calculated by multiplying COD or phenolic compound content of the employed AOMW by the D at which the reactor was forced to operate). In general, the pollutant removal (expressed as COD depletion yields, which were calculated by dividing the amount of pollutant removed in the reactor under steady state conditions by the amount of pollutant occurring in the reactor influent) increased with the organic load, and slightly decreased by increasing this parameter, in the case of total phenols. Notably, methane production (expressed as l of CH_4 produced per g of COD removed) increased sharply from experiment No. 2 to that No. 3 after which it remained constant (Fig. 2C). Taken together, these findings indicate that the GAC-process is characterized by a good versatility and tolerance towards high and variable OMW organic loads. The yields of pollutants removal observed through experiments No. 1–3 were very similar to those achieved at comparable organic loads with the previously developed GAC-system (Fig. 2; Fava et al. 2003); this evidence suggests that GAC-reactor is also a quite reproducible biotechnological process.

To investigate the stability of the GAC-system, it was forced to operate at a relatively high organic load [33 and 1.8 g/(l × d) of COD and phenolic compounds, respectively] for a 2-month period (experiment No. 8; Fig. 2). During this experiment, pollutant depletion yields similar to those obtained in the 3-weeks experiment No. 5 (carried out with comparable organic loads), along with a methane production higher than that achieved in the latter case (Fig. 2), were stably observed all over the 2 months. These findings suggest that the GAC system is characterized by a remarkable stability. Finally, it is interesting

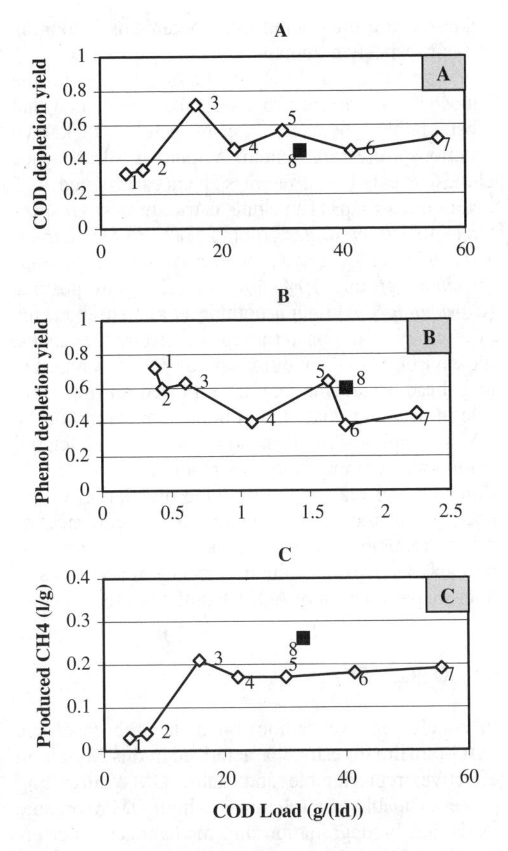

Figure 2. COD depletion yields (A), phenolic compound depletion yields (B) and methane production (C) of the GAC-reactor as a function of the organic load achieved, under steady state conditions, during the 7 sequential 3 weeks-experiments (experiments No. 1–7) and the final 2 months-long experiment (experiment No. 8).

to note that the volumetric productivity (in terms of pollutant removal and methane production, expressed as removed pollutant/produced methane per day per reaction volume) of GAC-system achieved throughout experiment No. 8 was significantly higher (by about 100% and 300% in terms of removal of COD and phenolic compounds, respectively, and by about 70% in terms of methane production) than those obtained with other up-flow packed-bed biofilm OMW digestors already described in the literature (Rozzi et al. 1989, Morelli et al. 1990, Marquez 2001) and with the improved contact digestor developed (with the same microbial inoculum employed in this study) by Beccari et al. (2000).

A large array of volatile fatty acids (VFAs) occurred in the effluents, where they were responsible for 30 to 60% of effluent COD. Generally, acetate prevailed on propionic acid that in turn was more abundant than all the other detected acids (i.e., iso-butyric acid, butyric acid and valerianic acid). The marked VFA production resulted in the production of effluents more acidic (pH 5.2 ± 0.2) that the influents (pH 7.8 ± 0.2). Sulphate, that occurred in AOMW2 and AOMW1 pumped into the reactor at 101.14 ± 25.90 and 62.12 ± 4.91 mg/l, respectively, was almost completely depleted during AOMW digestion. The amount of total immobilized biomass available in the GAC-system was estimated to be 44.50 g (on dry weight basis).

The distribution and structure of the microbial community in the GAC packed-bed bioreactor were evaluated by PCR analysis of 16S rRNA genes.

PCR performed with *Bacteria*-specific primers gave products of the expected size with all samples analysed, and data from *Hha*I and *Rsa*I generated T-RFs patterns were combined to achieve a more accurate characterization of the microbial communities into the reactor. *Hha*I digestion of amplicons generated with universal primers for *Bacteria* gave a total of 10 peaks (Fig. 3A). Differences in profiles, as well as changes in absolute numbers of peaks discernible, could be seen among the various samples taken through the reactor column packed-bed. Peaks with the highest intensity in T-RFs profiles had a molecular size of 493 bp (mobile phase), 335 bp (biofilm 5 cm) and 62 bp (biofilm 18 cm and 36 cm), respectively. Differences in the prominent peaks were also observed in the analysis of *Rsa*I digestion products. In the GAC-reactor considered in this study, a larger number of taxons were observed in the mobile phase than in biofilm samples, in particular from bottom and medium region of the reactor. Marked differences in the *Bacteria* fingerprints were present among inoculum and samples taken through the reactor, indicating that many members of the starter community were lost during the reactor operations. Our data are in agreement with findings of different authors suggesting that inocula generally play a minor role in the establishment of the final microbial community operating in anaerobic digestors treating complex and simple organic matter. Therefore, most of microbial inhabitants were introduced with OMW inlet or alternatively, were derived from minor and undetectable forms originally occurring in the inoculum.

A fragment of the expected size was generated with *Archaea*-specific primers only with DNA extracted from samples of biofilm collected at 18 cm, mobile phase and outlet of the GAC-reactor. None of the archaeal T-RFs detected in the reactor apparently derived from the inoculum that did not harbor any detectable taxons belonging to this domain. The T-RFs pattern was less complex than the one generated with eubacterial primers and showed the presence of a major peak whose size was 332 (*Hha*I) and 80 bp (*Rsa*I), respectively, indicating the presence of a

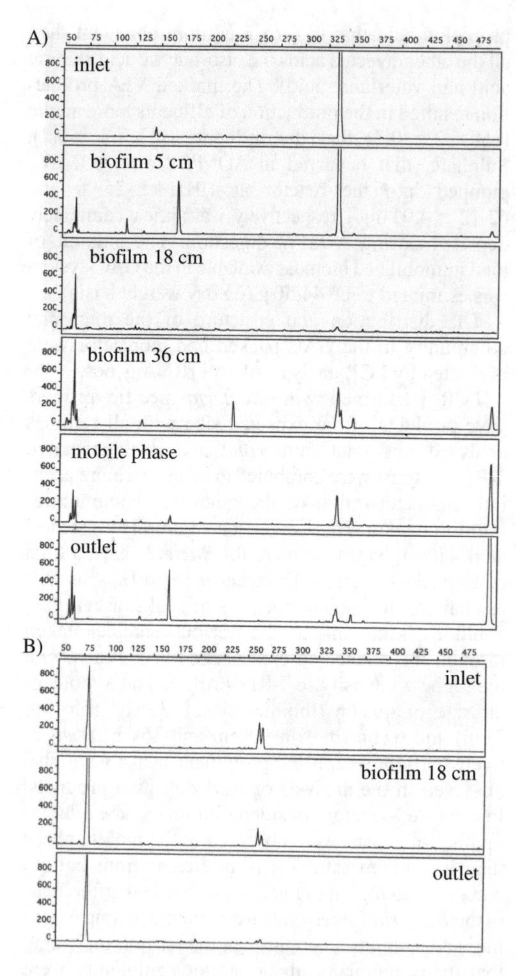

Figure 3. Electropherogram of the 5′ T-RFs derived from PCR amplified *Bacteria* (*Hha*I digestion, A) and *Archaea* (*Rsa*I digestion, B) communities 16S rDNA.

dominant specie (Fig. 3B). Diversity in the whole *Bacteria* community was considerably higher than among *Archaea*, as it would be expected because of the higher diversity detected in this domain. The dominant sequence among *Archaea* community (70% of the clones) is 100% similar to the 16S rDNA of *Methanobacterium formicicum* (T-RF of 80 bp), a hydrogenophylic methanogenic member of this domain.

To investigate the bacterial diversity in more detail and identify prominent bands in the T-RFs patterns, partial clonal libraries of 16S rRNA genes were constructed from biofilm, inlet and outlet samples. Sequencing and BLAST searching of the bacterial and archaeal clones resulted mainly in matches with unknown and uncultured microorganisms belonging to their respective domain.

Comparative analysis of the sequences to the Ribosomal Database Project database indicated that most of the genera or taxon expected with known functions were represented. A total of 30 sequence clones revealed 7 different sequences grouped in 5 taxonomic groups. The clone distribution was respectively 40% *Bacteroides* group, 25% δ-*Proteobacteria,* 10% γ-*Proteobacteria*, 15% *Anaerobaculum thermoterrenum* group, 10% low G + C gram-positive (*Clostridium*). Although nothing is known about the physiology of the bacteria represented by the clones, the environmental conditions of the reactor suggested that these organisms may have played an important role in decontamination and biomethanization of the OMW employed in this study. A similar bacterial community composition was reported to occur in a fluidized-bed reactor fed by wine distillation waste; notably, in the present study, the occurrence of sulfate-reducing bacteria was also demonstrated probably as a result of the occurrence of processes of microbial reduction of AOMW sulfate in the reactor.

4 CONCLUSIONS

In conclusion, we demonstrated that the anaerobic GAC-biofilm digestor characterized in this work is an effective, reproducible and stable OMW digesting process capable of a tolerance to high OMW organic loads and biodegradation and methanogenic performances higher than those exhibited by alternative bench scale digestors described so far. The results of this study, also, highlight the importance of using the analysis of the microbial community structure in combination with the analysis of the main chemical and physical parameters in the assessment of new biotechnological processes specifically developed for the disposal and valorization of difficult-to-manage agro-industrial wastewaters.

REFERENCES

Beccari, M., Majone, M., Petrangeli Papini, M. & Torrisi, L. 2000. Enhancement of anaerobic treatability of olive oil mill effluents by addition of $Ca(OH)_2$ and bentonite without intermediate solid/liquid separation. In *Proc. 1st World Congress of the "International Water Association", Paris, 3–4 July 2000.*

Fava, F., Bertin, L., Colao, M.C., Berselli, S., Majone, M. & Ruzzi, M. 2003. Use of packed-bed biofilm reactors in the anaerobic digestion of olive mill wastewater. In *Proc. of the "Second Bioremediation Conference", Chania, 30 June–4 July 2003.*

Fiestas Ros de Ursinos, J.A. & Borja-Padilla, R. 1996. Biomethanization. *Int. Biodeter. Biodeg.* 38: 145–153.

Godon, J.-J., Zumstein, E., Dabert, P., Habouzit, F. & Moletta, R. 1997. Molecular microbial diversity of an anaerobic digestor as determined by small-subunit rDNA sequence analysis. *Appl. Environ. Microbiol.* 63: 2802–2813.

Marques, I.P. 2001. Anaerobic digestion of olive mill wastewater for effluent re-use in irrigation. *Desalination* 137: 233–239.

Morelli, A., Rindone, B., Andreoni, V., Villa, M., Sorlini, C. & Balice, V. 1990. Fatty acids monitoring in the anaerobic depuration of olive oil mill wastewater. *Biol. Wastes* 32: 253–263.

Osborn, A.M., Moore, E.R.B. & Timmis, K.N. 2000. An evaluation of terminal-restriction fragment length polymorphism (T-RFLP) analysis for the study of microbial community structure and dynamics. *Environ. Microbiol.* 2: 39–50.

Rozzi, A., Passino, R. & Limoni, M. 1989. Anaerobic treatment of olive mill effluents in polyurethane foam bed reactors. *Process Biochem.* 26: 68–74.

European Symposium on Environmental Biotechnology, ESEB 2004 - Verstraete (ed)
© 2004 Taylor & Francis Group, London, ISBN 90 5809 653 X

Characterization of microbial communities removing NOx from flue gas in BioDeNOx reactors

R. Kumaraswamy, G. Muyzer, M.C.M. van Loosdrecht & J.G. Kuenen
Department of Biotechnology, Julianalaan Delft, the Netherlands

ABSTRACT: BioDeNOx is a novel integrated physico-chemical and biological process for the removal of nitrogen oxides (NOx) from flue gas. Due to the high temperature of flue gas the process is performed at a temperature between 50–55°C. Flue gas containing CO_2, oxygen, SO_2 and NOx, is purged through a $Fe(II)EDTA^{2-}$ containing liquid. The $Fe(II)EDTA^{2-}$ complex effectively binds the NOx; the bound NOx is converted into N_2 in a complex reaction sequence. It is evident that though the process looks simple, due to the large number of parallel potential reactions and serial microbial conversions, it is much more complex. There is a need for a detailed investigation in order to properly understand and optimise the process.

1 INTRODUCTION

Based on our present hypothesis in the BioDeNOx process, NO removal from flue gas consists of four reactions (Fig. 1), i.e., two chemical reactions within the scrubber, and two microbial reactions within the reactor (Buisman et al. 1999).

1. Wet absorption and complexation of NOx by $Fe(II)EDTA^{2-}$. This reaction rate is very fast, which leads to an increased mass transfer rate of NOx. However the design of the scrubber also influences the mass transfer and reaction rate.

   $$Fe(II)EDTA^{2-} + NO \rightarrow Fe(II)EDTA\text{-}NO^{2-}$$

2. Oxidation of $Fe(II)EDTA^{2-}$ to $Fe(III)EDTA^{-}$ by oxygen in the flue gas. This reaction is an unwanted chemical reaction as it lowers the availability of $Fe(II)EDTA^{2-}$ for NO absorption. It consumes most of the oxygen input in the scrubber creating a microaerophilic environment.

   $$2Fe(II)EDTA^{2-} + 1/2O_2 + 2H^+ \rightarrow 2Fe(III)EDTA^{-} + H_2O.$$

3. Biological reduction of NOx to N_2 by denitrifying bacteria, with ethanol as electron donor.

   $$6Fe(II)EDTA\text{-}NO^{2-} + C2H5OH^{+} \rightarrow 2HCO_3^{-} + 2H^+ + 3N_2 + H_2O + 6Fe(II)EDTA^{2-}$$

4. Regeneration of $Fe(II)EDTA^{2-}$ by reduction of $Fe(III)EDTA^{-}$ with ethanol as electron donor.

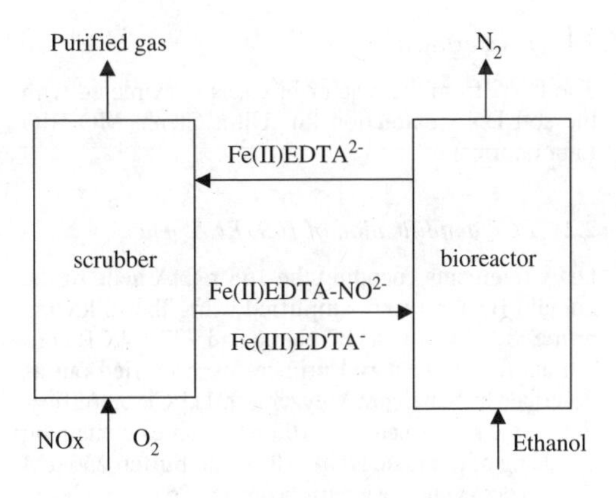

Figure 1. Schematic drawing of BioDeNOx process.

$$12Fe(III)EDTA^{-} + C_2H_5OH + 5H_2O \rightarrow 2HCO_3^{-} + 12Fe(II)EDTA^{2-} + 14H^+$$

Out of the two microbial reactions mentioned above, though NO reduction to N_2 by denitrification decides the effluent gas quality, microbial $Fe(III)EDTA^{-}$ reduction is very essential to get back the active $Fe(II)EDTA^{2-}$ complex for NO absorption. It also competes with the denitrification in the electron flow from electron donor. When the conditions in the bioreactor are evaluated by listing possible electron donors and acceptors, it becomes clear that more microbial reactions might be possible besides the postulated reactions 3 and 4.

1.1 *Goal of the project*

Analyzing the possible potential reactions gave a complete overview of the microbiology of BioDeNOx. The overview raised the questions below about microbiology of the process.

1. Which organisms are present and what are their roles?
2. How can the identified microbial community with known physiology be related to the process?
3. What is the measurable microbial diversity of the BioDeNOx reactor?
4. How does microbial population dynamics relate to the performance and stability of the reactor?

To answer these questions, the research is conducted by combining traditional microbiological methods with modern molecular ecological tools in a multiphasic approach. In this paper we will see a simple example of PCR-DGGE is used to study the microbial diversity and to identify the unculturable microorganisms in the BioDeNOx reactor (Muyzer et al. 1999).

2 MATERIALS AND METHODS

2.1 *DNA extraction*

The DNA from the reactor biomass is extracted with the soil DNA extraction kit (Ultra Clean, MO BIO Laboratories).

2.2 *PCR amplification of 16S rRNA gene*

DNA fragments encoding the 16S rRNA gene of the domain Bacteria were amplified using the following primers: 341 F with a GC-clamp and 907R. PCR conditions for the bacterial primers were carried out as described by Schäfer & Muyzer (2001).PCR amplifications were performed with 100-μl volumes containing 5–100 ng of template DNA, 10 × Taq buffer, 250 μM of each deoxynucleoside triphosphate, 25 pmol of each primer and 2.5 U of Taq DNA polymerase (Sigma). PCR products were analyzed by electrophoresis in 1% w/v agarose gels, which is then stained with ethidium bromide and destained with water.

2.3 *DGGE analysis of PCR products*

DGGE was performed as described by Schäfer & Muyzer (2001) using the D-code system (Bio-Rad Laboratories) with a 1 mm gel. PCR products were applied directly onto 6% w/v polyacrylamide gels with denaturing gradients from 20 to 70% (100% denaturant is 7 M urea and 40% v/v formamide). DGGE bands were excised from the gels, and reamplified using primer sets 341 F with GC-clamp and 907R. After PCR, products of the second amplification were electrophoresed again in a denaturing gel to check the purity of the bands, the PCR amplifications were purified with QIAquick Gel extraction kit (QIAGEN laboratories).

3 RESULTS AND DISCUSSION

3.1 *Species composition of BioDeNOx reactor*

The microbial complexity of the bioreactors from Paques and Wageningen has been studied by PCR-DGGE analysis of 16S rDNA gene fragments (Fig 2). The following results were obtained

1. All the reactors showed a low measurable population complexity with a maximum of three bands.
2. The microbial communities changed over time.
3. The measurable microbial community of BioDeNOx reactor from Wageningen is slightly different from that of Paques reactor based on the position of bands in DGGE gel (Fig. 2).

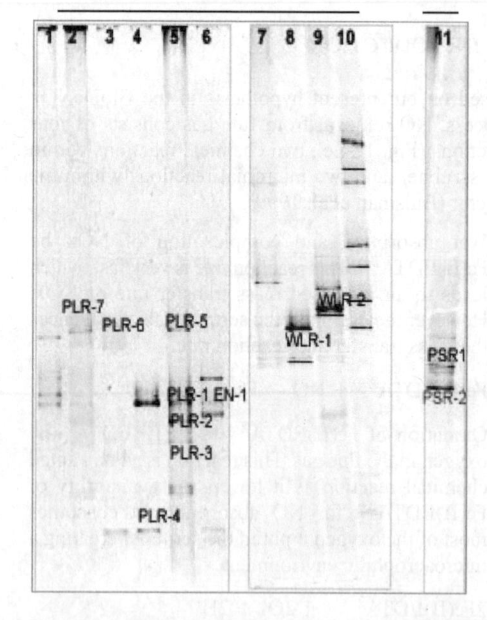

Figure 2. DGGE analysis of 16S rRNA gene fragments obtained from biomass samples taken at different time points from the BioDeNOx reactors. The bands labeled were taken for sequencing. Lanes: 1-Paques reactor (2001 Aug, 55°C), 2-Paques reactor (2002 Feb, 50°C), 3-Paques reactor (2001 Oct, 55°C), 4-Paques reactor (2002 April, 40°C), 5-Paques reactor (2002 May 40°C), 6-Enrichment (50°C) (2002 April, 55°C), 8-Wageningen lab scale reactor (2002 Jan), Wageningen lab scale reactor (2002 April), 10-DNA marker, 11-Paques pilot scale reactor (2002 April, 50°C).

The cell morphology does not show any variation over time, nor with temperature or other parameters, rod-shaped bacteria dominate the reactors at all time points. The change in the DGGE banding pattern clearly indicates the advantage of this technique over conventional microscopic observations. This can be explained as different populations dominating at different conditions such as temperature, pH, and redox potential of the reactor and also the composition of flue gas.

It is debatable whether the same reactions (i.e. NO reduction and Fe(III) reduction) are performed by different microbial population or different reactions (i.e. incomplete ethanol oxidation, sulfur reduction) by different populations. Difference in the source of inoculum of the reactors or the difference in operational characteristics may also be a reason for the differences in the microbial community of both the reactors. Despite these differences the first question to answer is whether a common group of bacteria was enriched.

Therefore sequence analysis of the microbial community members were performed by excising DGGE bands (Fig. 2).

3.2 *Sequence analysis of microbial community members*

The dominant DNA fragments were excised and amplified from the denaturing gradient gel and amplified using universal 16S rRNA primers. The partial 16S rRNA sequences of DGGE bands (400–500 bp) were obtained and compared to sequences stored in Genbank using BLAST program to find similar sequences. Subsequently the sequences were imported into the software program ARB and a distance tree was generated showing the phylogenetic affiliation of BioDeNOx community members (Fig. 3). The sequencing results of DGGE bands showed a close relationship of the organism represented by DGGE band with *Deferribacter*

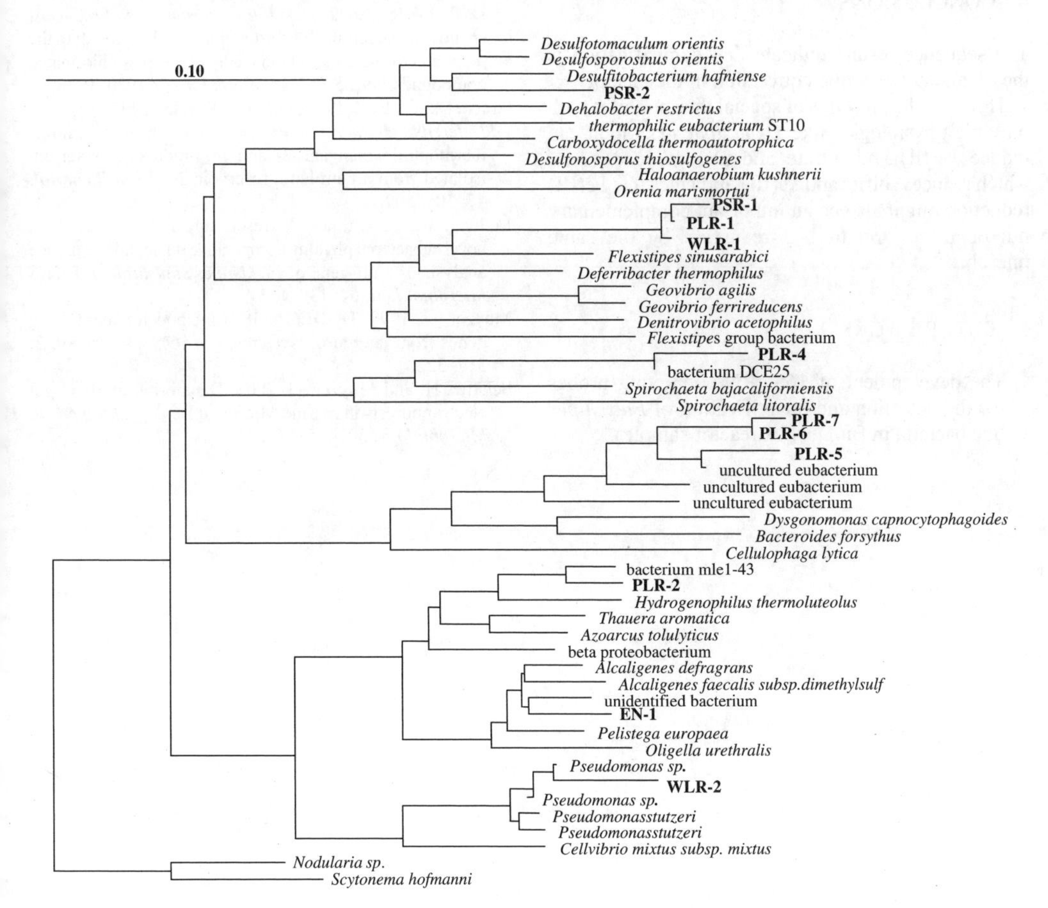

Figure 3. Evolutionary tree based on partial 16S rDNA sequences showing the phylogenetic affiliation of different bacterial community members identified by PCR-DGGE in the BioDeNOx reactors. The scale bar represents an estimated 10% sequence divergence. Names in bold refers to the sequences obtained from DGGE band (Fig. 2).

thermophilus (Greene et al. 1997) a thermophilic dissimilatory iron reducing and nitrate reducing bacterium isolated from a petroleum reservoir and *Flexistipes sinusarabici* a thermophilic bacterium isolated from hot brines water of red sea (Ludwig et al. 1991). Another DGGE band is closely related to *Desulfitobacterium* a thermophilic sulfite reducing bacterium (Finneran et al. 2002).

The lab-scale reactors of Wageningen and Paques Biosytems and the pilot scale reactor at United States showed the presence of microorganisms closely related to *Deferribacter sp.* and *Flexistipes sp.* despite the different inocula used. Apart from the bacteria related to *Flexistipes*, there were also bacteria found that were closely related to more common denitrifying bacteria in wastewater treatment such as *Alcaligenes* and *Pseudomonas*.

4 CONCLUSIONS

The sequence results indicate that denitrification is the dominant metabolic conversion in the reactor.

However the presence of special thermophiles with multiple physiologies like *Deferribacter*, which can reduce Fe(III) and nitrate and *Desulfitobacterium*, which reduces sulfite and Fe(III), indicates that Fe(III) reduction might also be an important complementary microbial reaction to NO reduction by the same microbes.

5 FUTURE WORK

1. The development of 16S rRNA targeting probes for the detection and quantification of *Flexistipes* like bacteria in BioDeNOx reactor samples.
2. Isolation and ecophysiological characterization of bacterial community members involved in the BioDeNOx process. The knowledge of growth kinetics (yield, specific growth rate and affinity constant [Ks]), mixed substrate uptake and basic characteristics (doubling time, optimal temperature and pH) of a representative or dominant isolate can be used to optimize the process characteristics (like threshold temperature, alternative electron donor for the process, etc).
3. The study of relationship between structure (species composition and diversity in the bioreactor) and functionality (activity of the bioreactor).

REFERENCES

Buisman 1999. United States Patent US5891408.

Finneran, K. T., H. M. Forbush, C. V. VanPraagh, and D. R. Lovley. 2002. *Desulfitobacterium metallireducens* sp. nov., an anaerobic bacterium that couples growth to the reduction of metals and humic acids as well as chlorinated compounds. *Int.J.Syst.Evol.Microbiol.* 52: 1929–1935.

Greene, A. C., B. K. Patel, and A. J. Sheehy. 1997. *Deferribacter thermophilus* gen. nov., sp. nov., a novel thermophilic manganese- and iron-reducing bacterium isolated from a petroleum reservoir. *Int.J.Syst.Bacteriol.* 47: 505–509.

Ludwig, W., G. Wallner, A. Tesch, and F. Klink. 1991. A novel eubacterial phylum: comparative nucleotide sequence analysis of a tuf-gene of *Flexistipes sinusarabici*. *FEMS Microbiol.Lett.* 62: 139–143.

Muyzer, G. 1999. DGGE/TGGE a method for identifying genes from natural ecosystems. *Curr.Opin.Microbiol.* 2: 317–322.

Schäfer, H. and Muyzer, G. 2001. Denaturing gradient gel electrophoresis in marine Microbial Ecology. *Methods in Microbiology*. 30: 425–468.

European Symposium on Environmental Biotechnology, ESEB 2004 - Verstraete (ed)

Analysis of the diversity of ammonia oxidizing and denitrifying bacteria in submerged filter biofilms for the treatment of urban wastewater by temperature gradient gel electrophoresis

B. Gómez-Villalba, B. Rodelas, C. Calvo & J. González-López
Grupo de Microbiología Ambiental, Departamento de Microbiología, Facultad de Farmacia, e Instituto del Agua, Universidad de Granada, Granada, Spain

ABSTRACT: Wastewater treatment by means of fixed biofilm technology is one of the major biotechnological processes used worldwide. In this study, nitrogen removal from wastewater is achieved by a combination of nitrification and denitrification in a submerged filter biofilm. The system was very efficient for the removal of organic matter, and significantly reduced N in the effluent water. The spatial and temporal diversity of ammonia oxidizing and denitrifying bacteria in biofilms of the treatment plant was estimated by a temperature-gradient gel electrophoresis (TGGE) approach.

1 INTRODUCTION

Wastewater treatment is one of the major biotechnological processes used worldwide (Wagner et al. 2002). Nutrient disposal to sensitive areas, particularly nitrogen and phosphorus from wastewater treatment plants, causes eutrophization, strongly reducing water quality. These effluents are regulated by the European Union Directive CEE 271/91.

Fixed film technology is widely used for the removal of organic matter and nitrogen through the biological process of nitrification–denitrification (Gómez et al. 2000, Hem et al. 1994, Lynga & Balmer 1992). The high cost of first installation and running operation of a conventional urban wastewater treatment system makes us pose this pilot-scale investigation, developing a biofilm treatment system that allows getting purified water to be poured into sensitive zones. Amongst other advantages compared to activated sludge technology, submerged filters provide an easier control, reduce room, lower cost and minimize odours and noise. Submerged filters have been used under different working conditions in order to remove suspended solids, organic matter, pathogens and nitrogen (Gálvez 2001).

Physiology and biochemistry of both nitrifying and denitrifying bacteria are extensively known, however, the information about the ecology of these organisms is still scarce (Bothe et al. 2002), The recent introduction of molecular techniques in microbial ecology has significantly implemented the knowledge of these organisms in their habitats, as it makes possible the study of non-culturable bacteria. In 1993, Muyzer et al. introduced the denaturing gradient gel electrophoresis (DGGE) technique in the field of microbial ecology. This method and its homologous, temperature gradient gel electrophoresis (TGGE), allow the separation of DNA fragments of similar size depending on their nucleotide sequence. DGGE has been often applied to the study of the population dynamics of microorganisms in natural habitats (Muyzer & Smalla 1998, Muyzer 1999), and in recent years it has considerably helped to broaden the knowledge of the microbial communities of biological systems for the depuration of wastewater, particularly the ammonia oxidizers (Purkhold 2000). TGGE has the advantage of providing more reproducible diversity profiles and the drawback of a much higher cost of the needed equipment.

In this study, the TGGE approach was used to monitor the spatial and temporal distribution of nitrifying and denitrifying microbial communities in a pilot-scale plant for the treatment of urban wastewater. The system consisted of two submerged filters working in pre-denitrification mode. The first one worked under anoxic conditions, keeping the denitrifying activity by the use of inlet urban wastewater as carbon source, while the second one worked aerobically oxidizing organic nitrogen and ammonia to nitrate.

2 MATERIALS AND METHODS

2.1 *Lab scale wastewater treatment plant*

The pilot-scale plant used for these experiments is depicted in Fig. 1. It consisted of two methacrylate cylindrical columns (2 m high, 6 cm inner diameter each), both packed with clayey schists of 5–7 cm average size, and 1.75 g/cm^3 density, up to 1 m high.

The columns are connected with a valve that allows a separate cleaning of the biofilters. In order to avoid filter clogging, cleaning cycles of the biofilters were carried out every 3 days. Hydraulic charge was kept at 0.354 m^3/m^2/h, and the nitrifying column was supplied with an airflow of 4.2 m^3/m^2/h. Water flow was regulated by a peristaltic pump (Watson Marlow 505S).

The system was operated in pre-denitrification mode, with a recycle rate of 1000 ml/h (100%) or 2000 ml/h (200%). The biofilters operate downflow (denitrifying column, anoxic) and upflow (nitrifying column, oxic), fed with urban wastewater coming from primary treatment of the wastewater treatment plant "EDAR Churriana" (EMASAGRA, Granada, Spain). Water was collected every 3 days and stored at room temperature. Its average composition was: BOD5, 135 mg/l; COD, 450 mg/l; Total nitrogen, 50 mg/l; and suspended solids, 120 mg/l.

Samples taken from the treatment plant were analysed according to the Standard Methods for Examination of Water and Wastewater (APHA, 1995). The main analysis for the purposes of the present study concerned COD and total nitrogen.

2.2 *DNA extraction and TGGE*

Total DNA was periodically extracted from biofilm samples, recovered from four locations of the reactor (25DN, 75DN, 25N and 75N, as shown in Fig. 1). Biofilm recovery and DNA extraction were achieved following a slight modification of previously described methods (Watanabe et al. 1998, Gálvez 2001).

Two functional genes were used as target for the study of bacterial diversity: the ammonia monooxygenase *amoA* gene for β-subclass ammonia-oxidizing bacteria, and the nitrous oxide reductase *nosZ* gene for denitrifying bacteria. Fragments of both genes of a size adequate for TGGE separation (ca. 475 bp for *amoA* and 250 bp for *nosZ*) were amplified by PCR, using pairs of universal primers previously described (Nicolaisen & Ramsing 2002, Scala & Kerkhof 1998).

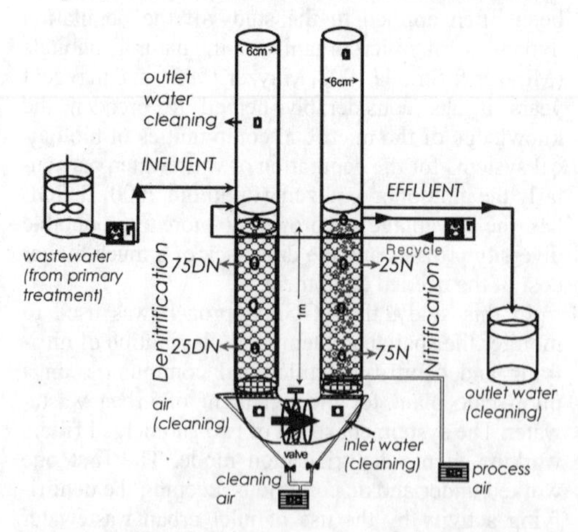

Figure 1. Diagram of the pilot-scale plant used in the study, showing the sampling points throughout the system (25DN, 75DN, 25N and 75N).

3 RESULTS

3.1 *Organic matter and nitrogen removal by denitrifying–nitrifying submerged filters*

The values of COD and total N in the submerged filter treatment plant for a 25 days period are shown in Figs. 2 and 3.

Both COD and total N were significantly reduced under the two recycle rates tested. COD in the effluent water was reduced to 78 mg/l with a recycle rate of 1000 ml/h, while a higher reduction (only 56 mg/l in

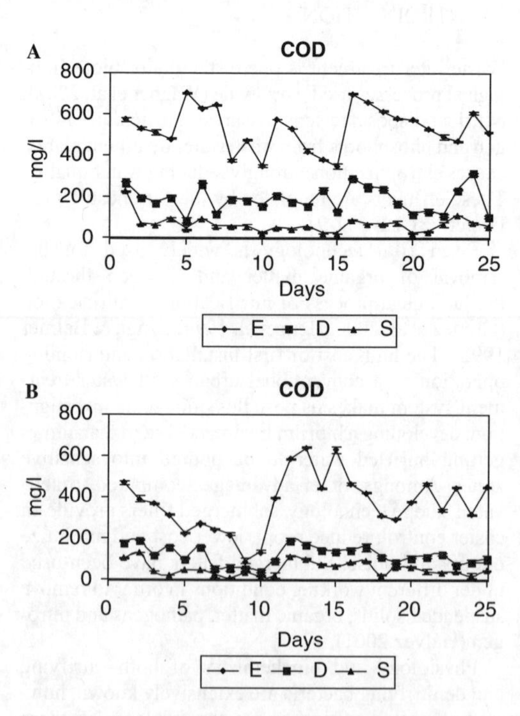

Figure 2. COD (mg/l) in the inlet water (E), final of denitrifying biofilm (D) and effluent water (S), for a recycle rate of 1000 ml/h (A) and 2000 ml/h (B).

the effluent water) was possible when the recycle rate was raised to 2000 ml/h (Fig. 2). However, the recycle rate of 2000 ml/h allowed for a total nitrogen removal of 43%, while at the 1000 ml/h recycle rate the elimination of N was more efficient and improved to a 60% (Fig. 3).

3.2 *Spatial and temporal distribution of functional microbial groups*

TGGE profiles showed the presence of nitrifying and denitrifying bacteria in both the anoxic and aerobic filter. The community profiles of denitrifying bacteria remained stable over space and time, while profiles of ammonia oxidizers greatly varied at different sampling times and locations in the reactor. Several bands separated by TGGE were reamplified and sequenced, in order to analyse the composition of these microbial communities in the biofilm.

Phylogeny inferred from *amoA*/AmoA revealed the prevalence of *Nitrosomonas* species, with 5 sequences affiliated to *N. oligotropha*, 6 sequences affiliated to *N. europaea*, and 3 sequences that showed only a 75.7–76.1% similarity of the DNA sequence with the closest described species (*N. nitrosa*). This value is indicative of previously undiscovered species (Purkhold et al. 2000).

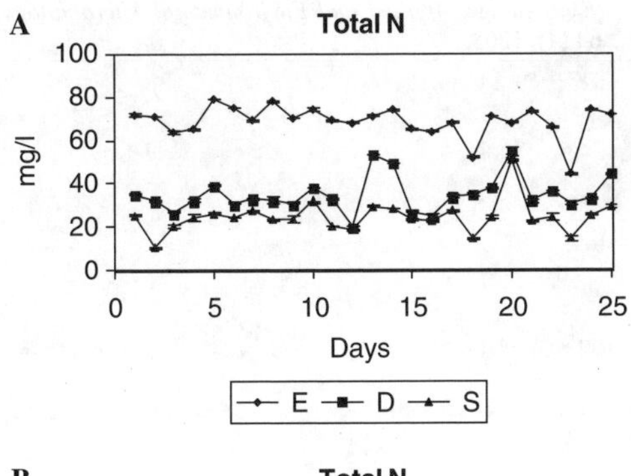

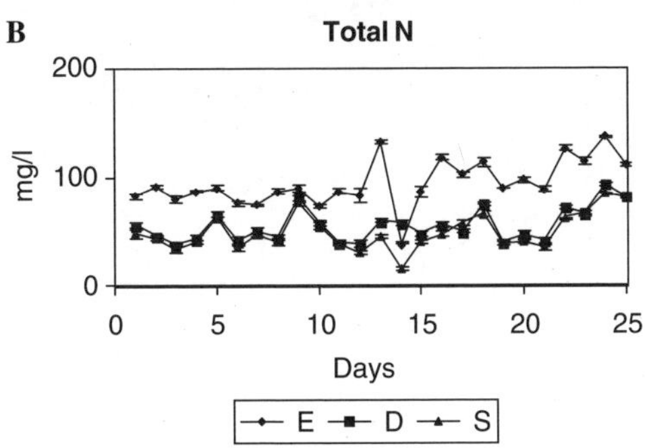

Figure 3. Total N (mg/l) in the inlet water (E), final of denitrifying biofilm (D) and effluent water (S), for a recycle rate of 1000 ml/h (A) and 2000 ml/h (B).

18 *nosZ* sequences were obtained from TGGE bands, although many redundancies were found, most probably due to the use of degenerated primers for amplification. Most of the new sequences were related to *nosZ* of Gamma-proteobacteria (*Pseudomonas*), while the rest clustered in the periphery of previously known denitrifying Alpha-proteobacteria (*Bradyrhizobium* and *Azospirillum*).

4 DISCUSSIONS

Biological treatment of wastewater by the submerged filter technology needs to be performed under the pre-denitrification mode, in order to be economically advantageous. Under the working conditions tested here, the removal of organic matter was over 75%, and the COD in effluent water was never over 125 mg/l, as required by the EU Directive 271/91. These results are in agreement with previous work on similar systems (Gálvez 2001).

Total nitrogen was reduced by a 60% with the 1000 ml/h recycle rate. This level is near the maximum that can be achieved for the low COD/N ratio in the influent water (Van Loodsdrech & Jetten, 1998).

Recent studies on the microbial diversity of nitrifying and denitrifying bacteria on biological wastewater treatment plants are based on molecular methods, mainly FISH and DGGE. In this study, the use of the TGGE method is used for the first time for the profiling and characterisation of these communities in a submerged filter system. The predominance of ammonia oxidizers of the genus *Nitrosomonas* over *Nitrosospira* in the system has also been reported in previous studies in plants based on the activated sludge technology (Purkhold et al. 2000).

Copies of the *nosZ* gene isolated from the submerged filter biofilm were related to those of well known denitrifyers, mostly falling within the *Pseudomonas* group. Similar results were obtained in a previous study using an equivalent denitrification-nitrification system (Sakano et al. 2002). However, copies of *nosZ* recovered from natural samples (continental shelf sediments) are commonly unrelated to any previously known sequences (Scala & Kerkhof 1999).

ACKNOWLEDGEMENTS

This research was supported by Programa Nacional de I + D, Ministerio de Ciencia y Tecnología (MCYT), Spain (AMB99-0666-C02-01). Work by B. Gómez-Villaba was supported by an PhD grant (Plan FPI, MCYT, Spain). Work by B. Rodelas was funded by Programa Ramón y Cajal (MCYT, Spain).

REFERENCES

APHA. 1995. Standard Methods for the Examination of Water and Wastewater, 19th Ed. American Public Health Association, Washington DC.

Bothe, H., Jost, G., Schloter, M., Ward, B.B. & Wizel, K.P. 2000. Molecular analysis of ammonia oxidation and denitrification in natural environments. *FEMS Microbiology Reviews* 24 (5): 673–690.

Gálvez, J. 2001. Eliminación de la material orgánica y del nitrógeno en el agua residual urbana mediante lechos inundados. PhD Thesis, University of Granada, Spain.

Gómez, M.A., Gónzalez-López, J. & Hontoria, E. 2000. Influence of carbon source on nitrate removal contaminated groundwater in a denitrifying submerged filter. *Journal of Hazardous Materials* 80 (1–3): 69–80.

Hem, L.J., Rusten, B. & Ødegaard, H. 1994. Nitrification in a moving bed biofilm reactor. *Water Science and Technology* 28 (6): 1425–1433.

Lynga, A. & Balmer, P. 1992. Denitrification in a non-nitrifying activated sludge system. *Water Science and Technology* 26 (5–6): 1097–1104.

Muyzer, G., de Waal, E.C. & Uitterlinden, A.G. 1993. Profiling of complex microbial populations by denaturing gradient gel electrophoresis analysis of polymerase chain reaction amplified genes coding for 16s RNA. *Applied and Environmental Microbiology* 59 (3): 695–700.

Muyzer, G. 1999. DGGE/TGGE, a method for identifying genes from natural ecoystems. *Current Opinion in Microbiology* 2 (3): 317–322.

Muyzer, G. & Smalla, K. 1998. Application of denaturing gradient gel electrophoresis (DGGE) and temperature gradient gel electrophoresis (TGGE) in microbial ecology. *Antonie van Leewenhoek* 73 (1): 127–141.

Nicolaisen, M.H. & Ramsing, N.B. 2002. Denaturing gradient gel electrophoresis (DGGE) approaches to study the diversity of ammonia-oxidizing bacteria. *Journal of Microbiological Methods* 50 (2): 189–203.

Purkhold, U., Pommerening-Roser, A., Juretschko, S., Schmid, M., Koops, H. & Wagner, M. 2000. Phylogeny of all recognized species of amonia oxidizers based on comparative 16S rRNA and AmoA sequence analysis: implications for molecular diversity surveys. *Applied and Environmental Microbiology* 66 (12): 5382–5386.

Sakano, Y., Pickering, K., Ström, P., & Kerkhof, L. 2002. Spatial distribution of total, ammonia-oxidizing, and denitrifying bacteria in biological wastewater treatment reactors for bioregenerative life support. *Applied and Environmental Microbiology* 68 (5): 2285–2293.

Scala, D.J. & Kerkhof, L.J. 1998. Nitrous oxide reductase (*nosZ*) gene-specific PCR primers for detection of denitrifiers and three *nosZ* genes from marine sediments. *FEMS Microbiology Letters* 162 (1): 61–68.

Scala, D.J. & Kerkhof, L.J. 1999. Diversity of nitrous oxide reductase (*nosZ*) genes in continental shelf sediments. *Applied and Enivronmental Microbiology* 65 (4): 1681–1687.

Van Loosdretch, M. & Jetten, M.S.M. 1998. Microbiological conversion in nitrogen removal. *Water Science and Technology* 38 (1): 1–7.

Watanabe, K., Yamamoto, S., Hino, S. & Harayama, S. 1998. Population dynamics of phenol degrading bacteria in activated sludge. *Applied and Environmental Microbiology* 64 (4): 1203.

European Symposium on Environmental Biotechnology, ESEB 2004 - Verstraete (ed)
© 2004 Taylor & Francis Group, London, ISBN 90 5809 653 X

Bacterial chemotaxis towards phenanthrene in two PAH contaminated soils

L. Fredslund, K. Bijdekerke & D. Springael
Laboratory for Soil and Water Management, Catholic University of Leuven, Kasteelpark, Heverlee, Belgium

R. De Mot
Centre for Microbial and Plant Genetics, Catholic University of Leuven, Kasteelpark, Heverlee, Belgium

C.S. Jacobsen
Geological Survey of Denmark and Greenland, Copenhagen, Denmark

ABSTRACT: Chemotactic responses towards organic pollutants have recently been described for a number of aromatic-degrading soil bacteria. This study represents a new approach for isolation of chemotactic xenobiotic degrading bacterial strains from hydrocarbon-contaminated soil environments. The presence of soil bacteria showing chemotaxis towards phenanthrene was examined by this method in two Polycyclic Aromatic Hydrocarbon (PAH) polluted soils. Eubacterial communities isolated for their chemotactic features or by conventional phenanthrene enrichment culturing were compared by 16S rRNA gene PCR analysis and subsequent DGGE-fingerprinting for the two soils. The isolated strains were then tested for chemotaxis towards phenanthrene by using different chemotactic assays.

1 INTRODUCTION

Biological removal of xenobiotics in soil is dependent on the presence of pollutant-degrading bacteria. Recent studies indicate that direct contact between a degrading bacterium and the contaminant source could be of major importance for utilization of the contaminants. This means that the bacteria must migrate to and seek contact with the substrate source as a first step in colonisation of the pollutant source. As such, migration or translocation of bacteria on a micro- and macro-scale towards the pollutant source could be an important feature for biodegradation of organic pollutants in soil.

Chemotactic responses based on the detection of and active migration along a gradient towards pollutants like naphthalene and toluene has been described for a number of aromatic-degrading bacteria (Marx and Aitken 1999, Parales et al. 2000, Parales and Harwood 2002), and in *Ralstonia eutropha* JMP134 (pJP4), a gene involved in chemotaxis was found to be associated with genes for degradation of the herbicide 2,4-D (Hawkins and Harwood 2002).

This study represents a new approach for isolation of novel xenobiotic degrading bacterial strains from hydrocarbon-polluted soil environments. For two very different PAH polluted soils strains were isolated based on selection for their pollutant directed chemotactic responses, using a Lutrol F127 assay, which include immobilization of the chemotactic bacteria in Petri dishes and subsequent growth on the contaminating source in the same dishes. Individual strains were subsequently analysed in detail for their chemotactic and phenanthrene degradative features. The communities isolated for their chemotactic features were compared by means of 16S rRNA gene PCR-DGGE-fingerprinting with phenanthrene degrading isolates obtained from the same soils using enrichment culturing.

2 MATERIALS AND METHODS

2.1 *Soils*

The two PAH polluted soils were collected in 2000. Soil D originated from the gas factory of Barlocher near Munich, Germany. Soil E was sampled from a former railway road site in Andujar, Spain. Soil D is characterised by a large fraction of coarse particles and stones, whereas soil E has a high content of clay. Both soils have a high content of phenanthrene; soil D of ~240 ppm and soil E of ~1100 ppm.

2.2 *Lutrol F127 assay*

The gelling agent used for this work was the co-polymer Lutrol F127, consisting of 73% of polyethylene glycol and 27% polypropylene glycol with an average molecular weight of 12,000. A solution of Lutrol of *e.g.* 25% w/v in a minimal medium remains liquid at room temperature yet solidifies at higher temperatures (Gardener and Jones 1985). The liquefaction/solidification process is reversible by changing incubation temperatures. Thus it is possible to have the medium in a liquid state, allow a bacterial inoculum to move towards a chemoattractant, raise the temperature and trap the cells by solidification. This means that the bacteria can no longer move but can divide to form colonies using the chemoattractant as sole carbon and energy source (Thomson et al. 2001).

2.3 *Phenanthrene enrichment culturing*

We determined the number of culturable phenanthrene degraders in the two soils by conventional MPN-enumeration (triplicate 10-fold dilution series) in 5 ml Wicks minimal media (WMM) with phenanthrene in 10 ml glass tubes on a rotary shaker (200 rpm). Six of these tubes (10^{-2} and 10^{-6} dilutions) then served as starting points of parallel series of enrichment culturing of phenanthrene degraders from the soils. We made five subsequent reinoculations for each tube, extracting the DNA of the cultures by each reinoculation. The culturable phenanthrene degrading strains were isolated from each tube on phenanthrene spray plates of WMM at the first and last transfer.

2.4 *Chemotactic tests*

2.4.1 *Modified capillary assay*

The assay is developed from the classical capillary assay of Adler (1973) and modified by Grimm and Harwood (1997). The assay was performed as described by Ortega-Calvo et al. (2003).

2.4.2 *Soft-agar swarm plate assay*

The soft-agar swarm plate assay was modified after Parales and Harwood (2002). Soft-agar swarm plates were prepared as 0.3% agar in WMM covered with a thin film of phenanthrene applied in acetone. 10 µl of bacterial suspension was pipetted in the center of the plates, and the diameter of the ring of growth measured at 24 h and 48 h.

2.4.3 *Agarose plug assay*

The agarose plug assay was performed as described by Yu and Alam (1997), with plugs of 2% agarose in phenanthrene saturated WMM solution, and a 10^7 CFU ml^{-1} bacterial cell suspension. Negative controls were performed with plugs of 2% agarose in WMM. A positive response is obtained if a chemotactic band appears around the agarose plug after 5 to 30 minutes, as visualised by dark-field microscopy.

2.5 *DNA extraction, PCR amplification, and denaturing gradient gel electrophoresis (DGGE)*

Total DNA was extracted from the different soil fractions, using a modified protocol of El Fantroussi et al. (1997). The Eubacterial 16S rRNA gene was amplified by PCR with the eubacterial primerset 63F & 518R (El Fantroussi et al. 1999). Because of PCR inhibition, DNA extracts were diluted 10-fold. DGGE analysis was performed as described by El Fantroussi et al. (1999).

3 RESULTS AND DISCUSSIONS

Total CFUs counted by plating on WMM was $6{,}6 \times 10^7$ CFUs per g of soil D, and $4{,}2 \times 10^7$ CFUs per g of soil E. Total numbers of CFUs proved lower when counted on 1/10 Tryptic Soy Broth or Tapwater plates (data not shown).

Total numbers of culturable phenanthrene degraders were estimated to $8{,}8 \times 10^5$ bacteria for both soils D and E by MPN counting.

For each soil approximately 10 different strains were isolated by their chemotaxis towards phenanthrene with the Lutrol F127 assay (Fig. 1). For each soil we isolated around 15 different phenanthrene degrading strains from the enrichment culturing.

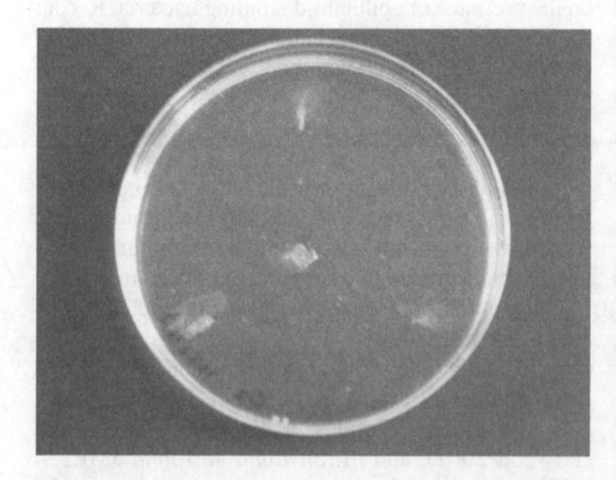

Figure 1. 10 µl of a 10^{-2} soil suspension was added at the edge of a liquid Lutrol F127 plate in 3 spots surrounding a phenanthrene-covered agar-plug in the centre. The bacteria were then allowed to move for 48 hours at room temperature before the plate was solidified by raising the temperature to 29°C. The bacteria subsequently grew slowly from the diffusing phenanthrene.

A total number of approximately 50 strains are currently tested by the four chemotaxis assays described above to describe their chemotaxis abilities. Furthermore a partial sequence of their 16S rRNA gene will be cloned and sequenced to determine the phylogenetic affinity of the different strains. We expect the results to give indications as to whether chemotaxis is a widespread feature of phenanthrene degrading soil bacteria, but also to state if isolation based on chemotaxis leads to novel or different phenanthrene degrading soil bacterial strains compared to isolation by conventional enrichment culturing.

ACKNOWLEDGMENTS

This work was supported by the FWO research project contract no. G.0254.03, The Danish Agricultural and Veterinary Research Council, and Novozymes A/S, Denmark.

REFERENCES

Adler, J. 1973. A method measuring chemotaxis and use of the method to determine optimum conditions for chemotaxis by *Escherichia coli*. J. Gen. Microbiol. 74: 77–91.

El Fantroussi, S., J. Mahillon, H. Naveau, & Agathos, S.N. 1997. Introduction and PCR detection of *Desulfomonile tiedjei* in soil slurry microcosms. *Biodegradation* 8: 125–133.

El Fantroussi, S., L. Verschuere, W. Verstraete, & Topp, E.M. 1999. Effect of phenylurea herbicides on soil microbial communities estimated by analysis of 16S rRNA gene fingerprints and community-level physiological profiles. Appl. Environ. Microbiol. 65(3): 982–988.

Gardener, S., & Jones, J.G. 1985. A new solidifying agent for culture media which liquefies on cooling. J. Gen. Microbiol. 139: 731–733.

Grimm, A.C., & Harwood, C.S. 1997. Chemotaxis of *Pseudomonas putida* to the polyaromatic hydrocarbon naphtalene. Appl. Environ. Microbiol. 63(10): 4111–4115.

Hawkins, A.C., & Harwood, C.S. 2002. Chemotaxis of *Ralsonia eutropha* JMP134(pJP4) to the herbicide 2,4-Dichlorophenoxyacetate. Appl. Environ. Microbiol. 68(2): 968–972.

Marx, R.B., & Aitken, M.B. 1999. Quantification of chemotaxis to naphtalene by *Pseudomonas putida* G7. Appl. Environ. Microbiol. 65(7): 2847–2852.

Ortega-Calvo, J.J., A.I. Marchenko, A.V. Vorobyov, & Borovick, R.V. 2003. Chemotaxis in polycyclic aromatic hydrocarbon-degrading bacteria isolated from coal-tar- and oil-polluted rhizospheres. FEMS Microbiol. Ecol. 44(3): 381.

Parales, R.E., J.L. Ditty, & Harwood, C.S. 2000. Toluene-degrading bacteria are chemotactic to the environmental pollutants benzene, toluene, and trichloroethylene. Appl. Environ. Microbiol. 66(9): 4098–4104.

Parales, R.E., & Harwood, C.S. 2002. Bacterial chemotaxis to pollutants and plant-derived aromatic molecules. Curr. Opin. Microbiol. 5: 266–273.

Thomson, R., R. Pickup, & Porter, J. 2001. A novel method for the isolation of motile bacteria using gradient culture systems. Jour. Microbiol. Methods. 46: 141–147.

Yu, H.S., & Alam, M. 1997. An agarose-in-plug bridge method to study chemotaxis in the Archeon *Halobacterium salinarum*. FEMS Microbiol. Letters. 156: 265–269.

European Symposium on Environmental Biotechnology, ESEB 2004 - Verstraete (ed)
© 2004 Taylor & Francis Group, London, ISBN 90 5809 653 X

Microbial ecology of an acid polycyclic aromatic hydrocarbon contaminated soil from a former gaswork manufacturing plant

S. Vermeir, M. Uyttebroek & D. Springael
Laboratory for Soil and Water Management, Catholic University of Leuven, Leuven, Belgium

ABSTRACT: Two PAH contaminated soils, one with pH 7 and one with pH 2, sampled from a former gaswork plant were analyzed regarding microbial ecology and the presence of PAH utilizing bacteria. The soils were inhabited by different eubacterial communities based on DGGE 16S rRNA gene fingerprinting. From both soils, enrichment cultures were obtained utilizing pyrene or phenanthrene as source of carbon and energy at different pH. However, no pure PAH degrading strains could be recovered from these cultures except from one phenanthrene utilizing culture obtained at pH 7 from the neutral pH soil. Based on eubacterial DGGE fingerprinting, the cultures utilizing pyrene and phenanthrene from the low pH soil consisted of one bacterial strain and this for all pHs.

1 INTRODUCTION

Sites localizing former gaswork manufacturing plants often contain locations contaminated with high concentrations of polycyclic aromatic hydrocarbons (PAHs) and displaying acid pH. Other acid environments in which PAH or other hydrocarbons occur are soils in the vicinity of coal pile storage basins and acid tar basins. In spite of the wide range of PAH contaminated environments, most studies on PAH biodegradation have focused on micro-organisms that can grow on common laboratory media at neutral pH. Data on the microbial ecology of acid soils and on the existence of bacteria able to degrade PAH under acid conditions are as such scarce. Stapleton et al. (1998) reported on the degradation of hydrocarbons such as naphthalene and toluene by a microbial community inhabiting downstream areas of a long-term coal pile storage basin.

In this paper, we report on a study investigating the microbial ecology of an acid PAH contaminated soil sampled from a former gaswork manufacturing plant. The eubacterial community was studied by means of Denaturing Gradient Gel Electrophoresis (DGGE) fingerprinting of eubacterial 16S rRNA genes amplified by PCR from total soil DNA extracts, counting of Colony Forming Units (CFU) on relevant media and by setting-up enrichment cultures for isolating PAH-degrading bacteria. The results were compared with results obtained from a PAH contaminated soil derived from a nearby location at the same site but with neutral pH.

2 MATERIALS AND METHODS

2.1 *Culture conditions*

The medium used to enrich and count PAH degrading bacteria was the medium described by Harrison (1981) and contained per liter 2 g $(NH_4)_2SO_4$, 0.1 g KCl, 0.5 g K_2HPO_4, 0.5 g $MgSO_4 \cdot 7H_2O$, 0.1 g Tripticase Soy Broth (TSB) and crystals of the appropriate PAHs or glucose 0.1% (w/v) as sole carbon source. The pH of the medium was adapted to the right pH by addition of H_2SO_4 1N. PAHs were added to the medium after sterilization and inoculation. Solid media were prepared by adding 2.5% (w/v) agar for media with pH 7, 5 and 3 and by adding agarose 0.7% (w/v) for media with pH 2 as described by Johnson (1995). Medium 869 and 0.1X869 was prepared as described (Mergeay et al., 1985). In all cases, cycloheximide at 100 mg/l was added as anti-fungal agent.

2.2 *CFU counting and enrichment of PAH-degrading bacteria*

CFU counting was performed by spreading serial dilutions of a water extract from the soil on the solid media described above. A soil water extract was obtained by suspending 2 g of soil in 18 ml of $MgSO_4$-solution (10^{-2} M) by mixing for 2 hours on a reciprocal shaker. After mixing, the soil suspension was left standing for 2 hours after which the water phase was used for preparation of the dilution series. CFU were counted on 869 medium and glucose medium after two weeks of incubation at 25°C and on PAH medium after 4 to 8 weeks of incubation.

Enrichment of PAH degrading bacteria was performed in 15 ml glass tubes containing the media described above and pyrene or phenanthrene as C-source. The cultures were inoculated with 100 μl of the aqueous soil extract prepared from the soils as described above. The tubes were incubated while shaking at 25°C and regularly checked for turbidity development. In case of growth, 500 μl of the culture was transferred into a new tube containing the same medium. In all cases, control tubes were inoculated containing the same medium without added PAHs.

2.3 *DNA extraction, PCR and DGGE analysis*

Total DNA from enrichment cultures and soil was extracted using a modified protocol as described by El Fantroussi et al. (1999). Eubacterial 16S rRNA gene fragments were amplified by PCR using the eubacterial primerset 63F & 518R (El-Fantroussi et al., 1999). If necessary, DNA extracts were diluted 10- and 100-fold prior to PCR amplification. DGGE analysis was performed as described by El-Fantroussi et al. (1999).

3 RESULTS AND CONCLUSIONS

From a former gaswork manufacturing plant, soil samples from two different PAH contaminated locations were analysed regarding microbial ecology and the presence of acid pH resistant PAH utilising bacteria. The soil at one location (B3) displayed a neutral pH while the soil at the second location displayed a pH of two (B7). Both soils contained total PAH concentrations of around 500 mg/kg. The two soils demonstrated complete different eubacterial communities based on DGGE 16S rRNA gene fingerprinting with the low pH soil showing less diversity than the neutral pH soil (data not shown). On media with pH 7, the number of heterotrophic and of glucose utilising CFU obtained for soil B7 were 2 to 3 log lower than the corresponding CFU numbers obtained for the soil F3. For both soils, CFU numbers on media with lower pH (pH 5, 3 and 2) decreased with CFU numbers below the detection limit at pH 3. No CFU were obtained on PAH containing medium.

Both from the neutral pH (B3) and low pH (B7) soil, enrichment cultures were obtained utilizing PAH (pyrene and phenanthrene) as sole source of carbon and energy and this at different pH (Table 1 and Fig. 1). Interestingly, from the neutral pH soil B3, pyrene and/or phenanthrene utilizing cultures were obtained at pH 5 and 7 while from the low pH soil, enrichment cultures growing on phenanthrene and/or pyrene were obtained at pH 5, 3 and 2. However, from

Table 1. Overview of the results obtained for the enrichment of PAH degrading bacteria from soils B7 and B3.

Soil	pH	Phenanthrene	Pyrene	Without PAH
B7	2	−	+	−
	3	+	+	−
	5	+	+	−
	7	+	+	−
B3	2	−	−	−
	3	−	−	−
	5	+	+	−
	7	+	+	−

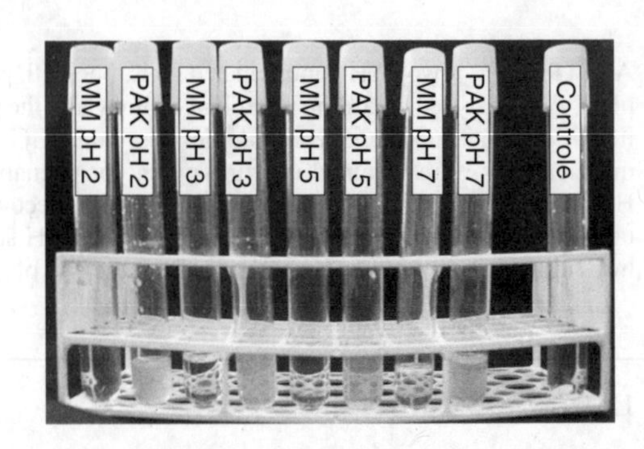

Figure 1. Examples of enrichment cultures (PAK) growing on pyrene at different pH and appropriate inoculated controls (MM) derived from the acid soil B7. The "Controle" vial indicate a non-inoculated vial but containing pyrene.

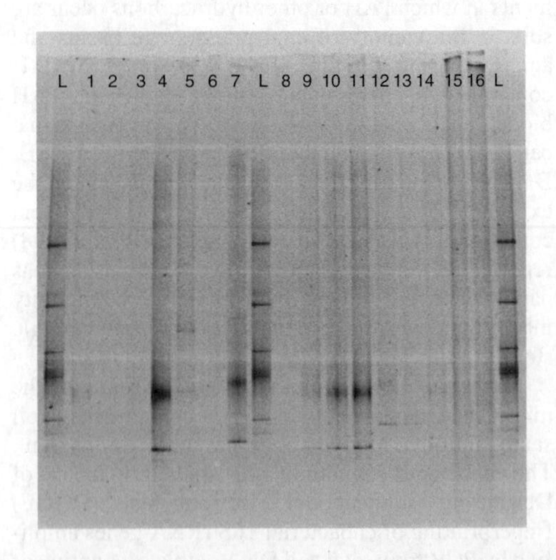

Figure 2. DGGE eubacterial community fingerprints of cultures growing on phenanthrene and pyrene, enriched from acid soil B7. Lanes 1–4 and 8–11: Cultures growing on pyrene at pH 2 (lane 1 & 8), pH 3 (lane 2 & 9), pH 5 (lane 3 & 10) and pH 7 (lane 4 & 11); Lanes 5–7 and 12–14: Cultures growing on phenanthrene at pH 3 (lane 5 & 12), pH 5 (lane 6 & 13) and pH 7 (lane 7 & 14). Lane L: DGGE marker.

none of these cultures, pure PAH degrading strains could be recovered on solid agar plates containing appropriate selective media except from the phenanthrene utilizing cultures obtained at pH 7 from the neutral pH soil.

Using DGGE eubacterial 16S rDNA fingerprinting, it was shown that the enrichment cultures utilizing pyrene and phenanthrene from soil B7 demonstrated single bands indicating that they consisted of one bacterial strain and this for all pHs (Fig. 2). Interestingly, the pyrene and phenanthrene utilizing cultures consisted of clearly different strains. The pyrene utilizing cultures showed the same DGGE pattern for cultures enriched on pH 3, 5 and 7. The phenanthrene utilizing enrichment cultures showed the same pattern for cultures enriched on pH 3 and 5, but a different pattern for this enriched on pH 7. Possibly, the B7 soil contains a pyrene degrading population able to grow at pH 3, 5 and 7 and at least two phenanthrene degrading populations, one able to grow at pH 7 and one at lower pH. To our knowledge, these cultures are the first described acid resistant bacterial cultures growing on and degrading 3- and 4-ring PAHs at low pH.

REFERENCES

El Fantroussi, S., Verschuere, L., Verstraete, W. and Top, E.M. 1999. Effect of phenylurea herbicides on soil microbial communities estimated by analysis of 16S rRNA gene fingerprints and community-level physiological profiles. *Appl. Environ. Microbiol.* 65: 982–988.

Harrison, A.P.J.R. 1981. *Acidifilium cryptum* gen. nov., sp. nov., heterotrophic bacterium from acidic mineral environments. *Int. J. System. Bacteriol.* 31: 327–332.

Johnson, D.B. 1995. Selective solid media for isolating and enumerating acidophilic bacteria. *J. Microbiol. Methods.* 23: 205–218.

Mergeay, M., Nies, D., Schlegel, H.G., Gerits, J., Charles, P., & Van Gijsegem, F. 1985. *Alcaligenes eutrophus* CH34 is a facultative chemolitotroph with plasmid-bound resistance to heavy metals. *J. Bacteriol.* 162: 328–334.

Stapleton, R.D., Savage, D.C., Sayler, G.S., & Stacey, G. 1998. Biodegradation of aromatic hydrocarbons in an extremely acidic environment. *Appl. Environ. Microbiol.* 64: 4180–4184.

European Symposium on Environmental Biotechnology, ESEB 2004 - Verstraete (ed)
© 2004 Taylor & Francis Group, London, ISBN 90 5809 653 X

Effects of chromate on culturable and non-culturable soil bacterial community

C. Viti, F. Decorosi, A. Mini, G. Rosi & L. Giovannetti
Dipartimento di Biotecnologie Agrarie, Università di Firenze, Florence, Italy

ABSTRACT: In this research the short and middle term effects of chromate exposure on soil bacterial community were estimated. We monitored the number and the tolerance to Cr(VI) of culturable heterotrophic bacteria by plate count technique and the genomic diversity of total bacteria in soil microcosms artificially polluted with three different concentrations of K_2CrO_4 (50, 250, and 1000 mg Kg^{-1} of soil) by Terminal Restriction Fragment Length Polymorphism (T-RFLP). The percentages of Cr(VI)-tolerant bacteria increased with the level of soil contamination. Analysis of T-RFLP profiles showed that the bacterial communities of the microcosm polluted with 50 mg Kg^{-1} of K_2CrO_4 and the control microcosm were related and both differed from the bacterial communities of microcosms amended with 250 and 1000 mg Kg^{-1} of K_2CrO_4.

1 INTRODUCTION

Chromium is an element that has many industrial uses: alloys, refractory materials, leather tanning, textile dyeing, wood preservatives, batteries and electroplating, only to cite a few (Cervantes et al. 2001). This extremely wide range of applications has resulted in large volumes of chromium waste to be discharged into the environment. In soil chromium occurs mainly with two different oxidation states: Cr(III) and Cr(VI), which have opposite chemical and physical characteristics (Bartlett & James 1996). Cr(VI) compounds are more dangerous than those of Cr(III) because the former is a highly soluble and mobile ion, able to cross cell membranes, whereas the latter is a less mobile, relatively non-toxic ion, stable at the pH range normally common in soil systems. These two different oxidation states can inter-convert and, generally, the reduction of Cr(VI) to Cr(III) is favoured, but high concentrations of Cr(VI) may overcome the reducing capability of the environment and thus Cr(VI) may persist as a pollutant (Cervantes et al. 2001); moreover, a part of Cr(III) can be transformed in Cr(VI) in Bartlett-positive soils (Bartlett et al. 1979).

Heavy metals have ecological effects on soil microbial community but most of our knowledge comes from data obtained studying the impact of metal such as Cu and Zn, whereas little is known regarding the effects of Cr upon microbial community structure and diversity (Shi et al. 2002, Viti & Giovannetti 2001). Therefore, the aim of this study was to determine the short and middle term effects of chromate upon soil microbial community. We monitored the number of cultivable heterotrophic bacteria and their tolerance to Cr(VI) through direct count plates method. T-RFLP technique was selected to detect any shift in the genetic diversity of the bacterial community caused by chromate.

2 MATERIALS AND METHODS

2.1 *Preparation of microcosms*

Soil used for microcosms preparation comes from an agricultural soil near Cerbaia (Florence, Italy) where no pesticides have never been used. The soil characteristics are reported in Table 1. On the basis of texture, the soil could be defined as a silt-clay loam (U.S. Department of Agriculture & Natural Resources Conservation Service, 1998). Samples were collected from up to a 20 cm depth. Right after collection, soil was

Table 1. Soil chemical and physical characteristics.

Soil pH	Organic C ($g\ Kg^{-1}$)	N Kjeldahl ($g\ Kg^{-1}$)	Avail. P ($g\ Kg^{-1}$)	Particle size distribution (%)		
				Sand	Silt	Clay
7.5	20.35	1.05	0.044	13	52	35

Table 2. K_2CrO_4 concentrations used for the microcosm preparation.

	K_2CrO_4 concentration
Microcosm C	No Cr(VI)
Microcosm 1	50 mg Cr(VI) Kg^{-1} of soil
Microcosm 2	250 mg Cr(VI) Kg^{-1} of soil
Microcosm 3	1000 mg Cr(VI) Kg^{-1} of soil

sifted with a sterilised sieve (pore diameter 2 mm) to remove gravel and plant residues and then it was stored at 4°C.

Soil was divided in four parts, each of 3.5 kg. One was used as control, the other three were amended with different K_2CrO_4 concentrations (Table 2). During the 120 days of the experiment, microcosms were kept at room temperature in a greenhouse and moistened periodically with 15 mL of sterile distilled water.

2.2 *Microcosms sampling*

Samplings were made after 7, 30 and 120 days from microcosms preparation. Two containers from each microcosm type were randomly chosen, pooled and mixed. The samples obtained were used for the following experiments.

2.3 *Enumeration of culturable heterotrophic bacteria and Cr(VI)-tolerant bacteria*

Plate count method was used to monitor the number of culturable heterotrophic bacteria population and Cr(VI)-tolerant bacteria. 10 g of soil were diluted in 90 mL of sterile saline solution (8.5 gL^{-1}), mixed thoroughly on a magnetic stirrer at 120 rpm for 120′ and then let rest for 60′ to allow settling of deposit. Standard serial dilutions followed and 100 μL aliquots of each dilution were spread in triplicate on plates with R2A medium (Oxoid, Basingstoke, Hampshire, England). Plates were incubated at 25°C in the dark for 7 days.

The number of Cr(VI)-tolerant bacteria was assessed using the same procedure described above, but using R2A medium plus 0.25 mM, 1 mM and 3 mM of K_2CrO_4 as suggested by Mergeay (1995).

2.4 *DNA extraction from soil*

DNA was extracted using the FastDNA SPIN Kit for Soil and the FastPrep Instrument (Qbiogene, Montreal, Canada). One additional step was added to the standard protocol: samples were incubated at 90°C in order to inactivate nucleases. Moreover, samples were put in a dryer for 30′ before final elution because we wanted to remove any ethanol trace and prevent any possible future inhibition in PCR reaction.

2.5 *Amplification condition*

PCR was performed using primers 63F (5′-CAGGCCTAACACATGCAAGTC-3′) and 1389R (5′-ACGGGCGGTGTGTACAAG 3′) (Liu et al. 1997) (MWG-Biotech AG, Ebersberg, Germany) labeled at the 5′ end with 6-FAM and HEX, respectively. 4 μL template DNA was used for 50 μL reactions, final concentrations of PCR reagents were: PCR buffer (1X), dNTP (0.2 mM), primers (0.3 μM), $MgCl_2$ (3.75 mM), BSA (Sigma-Aldrich, St. Louis, MO., USA) (0.04 mg mL^{-1}), *Taq* DNA polymerase (Polymed, Florence, Italy) (0.02 U μL^{-1}) and sterilised distilled water up to the final volume. Reactions were carried out in the Thermal Cycler GeneAmp PCR System 9600 (Applied Biosystems, Foster City, CA. USA) as described: initial denaturation step at 94°C for 5′, 35 cycles of 94°C for 45″, 60°C for 30″, 72°C for 90″, final extension step at 72°C for 6′. Amplification products were visualised and quantified on 1% agarose gels prior to further analysis.

2.6 *Amplified products purification and restriction analysis*

Removal of labeled primers and nucleotides that were not incorporated during PCR reaction was carried out with the QIAquick PCR Purification Kit (Qiagen N.V., The Netherlands). Purified amplicons (about 10 μg) were digested in individual reactions with *Alu*I and *Hha*I restriction enzymes according to the supplier (New England Biolabs, Beverly, MA. USA).

2.7 *Capillary electrophoresis and data analysis*

Aliquots of the digests were mixed with 1 μL of the GeneScan-500 TAMRA (Applied Biosystems, Foster City, CA. USA) used as internal size standard and loaded in the ABI Prism 310 automatic sequencer (Applied Biosystems, Foster City, CA. USA). The samples were run under standard denaturing electrophoresis conditions for 30′. The data were analysed using the GeneScan 3.1 software program (Applied Biosystems, Foster City, CA. USA). Peaks less than 100 fluorescence units were omitted from the analysis.

Terminal restriction fragments from each primer were aligned and converted to binary data (presence or absence of a peak). Data derived from the two different digestions were assembled to obtain a combined restriction profile for each bacterial community.

3 RESULTS AND DISCUSSION

The number of total CFU in microcosms C and 1 were similar and no statistically significant change occured ($P < 0.05\%$) through the 120 day-experiment (Table 3). Microcosms 2 and 3, with respect to microcosm C

Table 3. Number of CFU g^{-1} of dry soil (mean of three replicates ± standard deviation) on R2A medium.

	7 days	30 days	120 days
Microcosm C	$8.7 \pm 0.4 \times 10^6$	$1.0 \pm 0.1 \times 10^7$	$4.3 \pm 0.5 \times 10^6$
Microcosm 1	$7.7 \pm 0.4 \times 10^6$	$8.6 \pm 0.1 \times 10^6$	$4.6 \pm 0.4 \times 10^6$
Microcosm 2	$2.3 \pm 0.1 \times 10^6$	$1.1 \pm 0.1 \times 10^7$	$1.5 \pm 0.1 \times 10^7$
Microcosm 3	$6.3 \pm 0.1 \times 10^5$	$1.1 \pm 0.1 \times 10^7$	$2.8 \pm 0.2 \times 10^7$

Table 4. Percentages of Cr(VI)-tolerant bacteria of the four microcosms grown on R2A agar plates amended with three different concentrations of Cr(VI) after 7 and 120 days.

	0.25 mM*		1 mM*		3 mM*	
	7**	120**	7**	120**	7**	120**
Microcosm C	27***	37	12	10	8	7
Microcosm 1	85	52	44	12	23	6
Microcosm 2	78	88	65	78	40	68
Microcosm 3	100	98	78	85	66	77

* K_2CrO_4; ** days; *** percentages of tolerant bacteria with respect to the control (100% plate without K_2CrO_4).

and microcosm 1, displayed a consistently low number of cultivable heterotrophic bacteria at the beginning of the experiment, but the number of CFU increased over time and was about ten times greater than the control at the end of the experiment (120 days). This trend is presumably due to the leaching, by cellular lysis of Cr(VI)-sensitive microorganisms, of nutrients which could be used for Cr(VI)-tolerant heterotrophs growth.

Percentages of Cr(VI)-tolerant bacteria were estimated (with respect to the number of total heterotrophic culturable bacteria grown on R2A chromate-free plates) using three different concentrations of K_2CrO_4. Bacteria able to grow in presence of the Cr(VI) concentrations used were present in all microcosms, but a clear relationship existed between the level of chromate pollution of microcosms and the frequency of Cr(VI)-tolerant bacteria (Table 4). This is in accord with Viti and Giovannetti (2001) who found that the frequency of heterotrophic Cr(VI)-tolerant bacteria in soil is related to Cr contamination level, and with the results of Choi and Young (1995) which showed that the presence of Cr(VI) in soil may influence the natural selection for resistant character.

Using forward and reverse restriction fragments derived from amplicons digestions, distinctive bacterial community restriction profiles were obtained. Therefore, T-RFLPs were further analysed in order to compare the genetic diversity of bacterial communities of the four microcosms. The degree of similarity among the communities was quantified by numerically

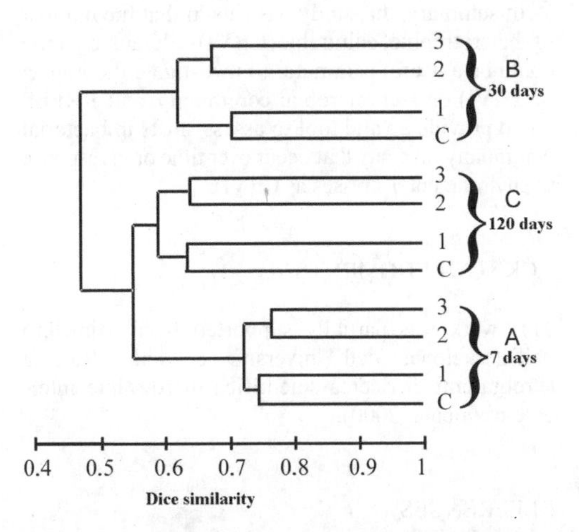

Figure 1. Dendrogram based on the UPGMA clustering of Dice similarity coefficients of combined T-RFLP profiles of bacterial communities (C = microcosm C; 1 = microcosm 1; 2 = microcosm 2; 3 = microcosm 3) obtained with the enzymes *Alu*I and *Hha*I.

analysing the fragment profiles of T-RFLP. Each community was represented with a restriction fragment profile obtained by combining four single T-RFLP profiles made up of two differently labelled primes and two restriction enzyme digestions (*Alu*I and *Hha*I). The similarity between each pair of combined profiles was calculated by the Dice similarity coefficient, and subsequently the UPGMA (unweighted pair group method using arithmetic averages) clustering algorithm was applied to the value matrix. The analysis of the dendrogram (Fig.1) shows that three main clusters were obtained: cluster A, cluster B and cluster C, which housed T-RFLP profiles of the bacterial communities of the four microcosms at 7, 30, 120 days from the beginning of experiment, respectively. Moreover in each cluster it is possible to identify two clearly separated sub-groups, one constituted by the microbial communities of microcosm C (unpolluted) and microcosm 1 (polluted with 50 mg Cr(VI) kg^{-1} of soil) and the other composed by microbial communities of microcosms 2 and 3, polluted with 250 and 1000 mg Cr(VI) kg^{-1} of soil, respectively. These data, if confirmed by other experiments in progress, mainly focused to define the reproducibility of the T-RFLP approach, point out that (i) shifts in the composition of all microbial communities, due to abiotic and biotic factors, occurred over time; (ii) different concentrations of chromate in soil produce different changes in the composition of the soil bacterial community; (iii) even the lowest level of chromate used (50 mg Cr(VI) kg^{-1} of soil) to pollute the microcosms affects the microbial community diversity.

In summary, this study has shown that the number of heterotrophic culturable Cr(VI)-tolerant bacteria could be a useful parameter to investigate the impact of Cr(VI) on soil microbial community, and T-RFLP could provide a valid tool to assess shifts in bacterial community diversity that occur over time or in response to environmental stresses as Cr(VI).

ACKNOWLEDGMENTS

This work was partially supported from Ministero dell'Istruzione, dell'Università e della Ricerca (Programmi di ricerca scientifica di rilevante interesse nazionale, 2000).

REFERENCES

Bartlett, B.R. & James, B.R. 1979. Behavior of chromium in soils: III. Oxidation. J. Environ. Qual. (8): 31–35.

Bartlett, R.J. & James, B.R. 1996. Chromium. In Sparks, D.L. (ed.), *Methods of soil analysis. Part 3. Chemical methods. Book Series no.5.* ASA & SSSA Madison: USA.

Cervantes, C. Campos-Garcia, J. Devars, S. Gutierrez-Corona, F. Loza-Tavera, H. Torres-Guzman, J.C. & Moren-Sanchez, R. 2001. Interactions of chromium with microorganisms and plants. FEMS Microb. Rev. (25): 335–347.

Choi, S.C. & Young, L.Y. 1995. Presented at the 95th General Meeting of the American Society for Microbiology, May 21–25. American Society for Microbiology, Washington, D.C.

Liu, W. Marsh, T.L. Cheng, H. & Forney, L.J. 1997. Characterization of microbial diversity by determining terminal restriction fragment length polymorphisms of genes encoding 16S rRNA. Appl. Env. Microb. (63): 4516–4522.

Mergeay, M. 1995. Heavy metal resistances in microbial ecosystems. In Antoon Akkermans, Jan Dirk Van Elsas, Frans J. De Bruin (eds.), Molecular microbial ecology manual. Chapter 6.1.7. The Netherlands: Kluwer Academic.

Shi, W. Becker, J. Bischoff, M. Turco, R.F. & Konopka, A.E. 2002. Association of microbial community composition and activity with lead, chromium and hydrocarbon contamination. Appl. Env. Microb. (68): 3859–3866.

Viti, C. & Giovannetti, L. 2001. The impact of chromium contamiantion on soil heterotrophic and photosynthetic microorganisms. Ann. Microb. (51): 201–213.

European Symposium on Environmental Biotechnology, ESEB 2004 - Verstraete (ed)
© 2004 Taylor & Francis Group, London, ISBN 90 5809 653 X

Emission and distribution of microorganisms from municipal waste dumps

W. Barabasz
Department of Microbiology, Agricultural University, Cracow, Poland

1 INTRODUCTION

Waste dumps and other municipal objects such as composting plants, sewage purification plants, sewage sludge etc., besides their positive role in environment protection, may also negatively affect the neighborhood, including atmospheric air. Chemical compounds, microorganisms (aerosols), and odors are main factors polluting the air. All waste dumps are considered as necessary evil and the place for their localization is more and more difficult to find. It results from the fact that they are the source of ground water pollution, the habitat of insects transferring bacteria, rodents and disease-forming microorganisms. Municipal waste dumps are one of the major source of threat for human, and its size first of all depends on the amount and the type of wastes, as well as the way of dump organization, its exploitation and current meteorological and natural conditions. Waste dumps react towards surrounding soil, surface and ground water, they pollute atmosphere and indirectly, distant agricultural, municipal and recreation areas [Barabasz *et al.* 1998, Borrello *et al.* 1999, Colakoglu 1996. Kuratowska 1997, Wyllie *et al.* 1990].

Air is a natural environment where microorganisms may occur. However, conditions, the air creates, is not suitable for development of specific microflora due to the lack of nutrients, and due to unfavorable physical and chemical properties. It is the way for microorganism transport rather than the habitat for their living. Nevertheless, various microorganisms can remain in the air even for very long time [Gonzales *et al.* 2000, Hendry *et al.* 1993].

The aim of studies was to compare the level of atmospheric air pollution with microorganisms around three municipal waste dumps differing with their exploitation time.

2 MATERIAL AND METHODS

Three municipal waste dumps differing with the exploitation time were chosen for microbial studies. The oldest dump – Barycz in Cracow of 37 ha area – exploited since 1974, received 198 000 tons of wastes; waste dump Krzyż near Tarnów of 12.68 ha and exploited since 1985, received 46 000 tons of wastes; waste dump in Bolesław near Olkusz of 13.09 ha and exploited since 1997, received 283 000 tons of wastes in 2000.

Following solid mediums were used in detailed microbial tests for quantitative and qualitative determination of microorganisms:

1. Nutrient agar – agar MPA (total bacteria number and diagnostics);
2. Agar SS – bacteria from *Salmonella* and *Shigella* genera;
3. Chapman's medium – disease-forming *staphylococcus* determination;
4. Endo agar – *Escherichia coli* bacteria;
5. Gaus' medium and Pochon's medium (total number of actinomycetes);
6. Malt agar and Czapek's medium (total number of fungi and diagnostics);
7. King's medium – medium for *Pseudomonas fluorescens*.

Number of microorganism colonies (so-called "CFU" – colony-forming units) of bacteria, actinomycetes and fungi grown on mediums applied was calculated in detailed studies, and then, using tables, the results were recalculated onto the number of microorganisms per 1 m^3. All microbial, analytical, taxonomic and diagnostic tests were performed according to generally accepted norms and needs applying standard methods of microbial techniques.

3 RESULTS

Detailed results of quantitative and qualitative studies along with the evaluation of the level of atmosphere pollution with microorganisms (bacteria, actinomycetes, fungi) in chosen municipal waste dumps in Barycz, Krzyż and Bolesław, as well as at their closest neighborhood and control sites pointed out that microbiocenotic composition of microorganisms was very different referring to both the amount and quality. There were places with multiple exceeding of norms and occurrence of great number of microorganisms around dumps, as well as such with very small population of all studied groups of microorganisms.

Studies revealed that the amount of microorganisms occurring in the atmospheric air around 3 analyzed waste dumps greatly differed to one another. The largest number of microorganisms was observed in municipal waste dump in Barycz and actinomycetes were the group of microorganisms that mostly exceeded permissible norms. Number of norm exceeding by bacteria and fungi was similar and amounted to about 10%. Quantitative studies carried out in waste dump in Krzyż and Bolesław showed similar tendencies – always actinomycetes dominated referring to number of norm exceeding.

Qualitative data upon microbiocenotic composition of microorganisms occurring in studied waste dumps revealed that their largest species differentiation was observed around the oldest one, i.e. Barycz in Cracow. The 56 bacteria species, including 4 disease-forming ones, 38 actinomycetes and 49 fungi species, including 15 toxin-producing fungi, were isolated from the atmospheric air.

Detailed diagnostic tests upon the microorganisms occurring in atmospheric air point out that there is natural saprophytic microflora on studied area, typical for rural environment originating from surrounding fields, orchards, and farm buildings. However, there are also microorganisms characteristic for polluted air and typical for urban area and large cities.

Among isolated saprophytic bacteria, the following ones occurred the most often: *Bacillus megaterium, Bacillus mycoides, Bacillus brevis, Bacillus sp., Staphylococcus sp., Micrococcus sp., Sarcina maxima*.

Staphylococcus aureus was the most often met disease-forming bacteria. However, it should be underlined that not every strain of that species invokes the disease. Most of the bacteria found were typical ubiquitous ones (so-called "ubiquists") that occur in different environments.

Microbiocenotic composition of actinomycetes was similar; 38 species were isolated. Among them, not-disease-forming ones were found, thus they are not major threat for habitants of the neighborhood of waste dumps in Barycz, Krzyż and Bolesław.

Qualitative study results point out that the frequency occurrence of particular fungi species was variable. Following species were the most often met: *Cladosporium herbarum, Aspergillus niger, Alternaria alternata, Alternaria geophila, Rhizpous nigricans, Verticillium celulosae* General tendency was proven, from which it follows that the farther from every waste dump, the poorer microbiocenotic composition of microorganisms. The greatest species differentiation of microorganisms was always observed in the active zone of the dump at its adjacent neighborhood.

4 DISCUSSION

The increase of Pole's consciousness referring to pro-health prophylaxis requires the improvement of technology in various branches of human's economic activity. More and more needs in relation to ecological conditions or product's microbial purity are also put. Therefore, learning the air microflora is of particular importance for evaluation of hygienic conditions occurring in living quarters, factories, and public compartments. Great number of microorganisms occurring in the air is often the indicator of bad sanitary state of the environment that surrounds large people communities [Gonzalez *et al.* 2000]. Because air may be the way for transfer the microbial pollution from polluted areas to pure ones, ways of air flow should be monitored in details and the sources of pollution should be eliminated.

Microorganisms, reflux and gases are the main elements that have very negative influence on sanitary and hygienic state of the environment around large municipal waste dumps. Recently, more and more research upon air analysis methods has been published, as well as searches and determinations of indicators for air bacterial pollution have been carried out. These are bacterial species or genera selected as representatives of microflora originating from given pollution sources, i.e. soil, surface waters, humans and animals. Waste dumps and inconveniences associated with them have lately become well visible problem of environment protection. Waste management, particularly industrial and municipal ones, should be considered as very significant, but at the same time very difficult problems that have not been solved for a large scale in Poland yet.

Results of many authors such as: Cronholm 1980, Cvetenić *et al.* 1997, Kuratowska 1997, Horner *et al.* 1995, Wyllie *et al.* 1990, involved in municipal wastes and disease-forming microorganisms occurring in the air, revealed that municipal waste dumps, particularly in large city agglomerations, may be the source of infectious bio-aerosol that may, under some weather conditions, negatively affect the human's health due to spreading of disease-forming microorganisms, in particular their preserved forms (spores, conidia and sporules).

From quantitative data it is clear that municipal waste dumps Barycz, Krzyż and Bolesław may have the effects on sanitary state of atmospheric air around the closest neighborhood. It was found that dumps were not the only source of air pollution with microorganisms at other study points, particularly those distant from waste dumps even several kilometers, which was proved by microbiocenotic composition of microorganisms in analyzed air.

5 CONCLUSIONS

1. Atmospheric air studied in municipal waste dumps Barycz, Krzyż and Bolesław, around, near and away from them, as well as at control points contained

many microorganisms that occurred in various number and variable microbiocenotic composition.

2. Following items were isolated from studied atmospheric air and classified:
   - 56 bacteria including 4 disease-forming ones;
   - 38 actinomycetes;
   - 49 fungi including 15 toxin-producing fungi;
3. Microbial tests, particularly comparison of species composition of isolated microorganisms revealed that municipal waste dumps Barycz, Krzyż and Bolesław were not the only sources of atmospheric air pollution with microorganisms in a zone of studied waste dumps influence.

LITERATURE CITED

1. Barabasz W., Marcinowska K., Bis H., Chmiel M., Galus A., Paśmionka I., Grzyb J., Frączek K., Opalińska-Piskorz J., Pawlak K., Flakowa K., Kornaś G., Kultys H., Król T.: Mikrobiologiczne skażenie powietrza atmosferycznego wokół wysypiska odpadów komunalnych w Baryczy k/Krakowa. /w/ Air Protection in Theory & Application. PAN, Instytut Podstaw Inżynierii Środowiska, Prace i Studia, 48, III, 145–157, 1998
2. Borrello P., Gucci P.M., Musmeci L., Pirrera A.: The microbiological characterization of the bioaerosol and leachate from an urban solid refuse dump: preliminary data. Ann.Ist.Super Sanita. 35,3,467–471, 1999
3. Colakoglu G.: Fungal spore concentrations in the atmosphere at the Anatolia quarter of Istanbul, Turkey. J Basic Microbiol. 36,3,155–162, 1996
4. Cronholm L.S.: Potential health hazards from microbial aerosole in densely populated urban region. Appl. & Environm. Microbiol., 39,6–12, 1980
5. Cvetenić Z., Pepeljnjak S.: Distribution and mycotoxin-producing ability of some fungal isolates from the air. Atmospheric Environment. 30,3,491–495, 1997
6. Gonzalez C.A., Kogevinas M., Gadea E., Huici A., Bosch A., Papke O.: Biomonitoring study of people living near or working at a municipal solid-waste incinerator before and after two years of operation. Arch.Environ. Health. 55,259–267, 2000
7. Hendry K.M., Cole E.C.: A review of mycotoxins in indoor air. J.Toxicol.Environ.Health. 38,2,183–198, 1993
8. Horner W.E., Helbling A., Salvaggio J.E., Lehrer S.B.: Fungal allergens. Clinical Microbiology Reviews. 8,2,161–179, 1995
9. Kuratowska A.: Rezerwuary chorobotwórczych czynników biologicznych w aerosferze, hydrosferze i litosferze. /w/ Ekologia – jej związki z różnymi dziedzinami wiedzy. PWN, Warszawa-Łódź, 1997
10. Wyllie T., Morehouse L.G.: Mycotoxic Fungi, Mycotoxins, Mycotoxicoses. Marcel Dekker Inc. New York and Basel, 1990

European Symposium on Environmental Biotechnology, ESEB 2004 - Verstraete (ed)

Microbial population changes and dynamics during bioremediation of an oily-sludge contaminated soil

K.I. Chalkou, C.K. Meintanis & A.D. Karagouni
University of Athens, Department of Biology, Section of Botany, Microbiology Lab., Athens, Greece

ABSTRACT: Indigenous bacterial populations in contaminated soil can degrade a wide range of target constituents of the oily sludge disposed to the environment by oil refineries, qualifying bioremediation as a potentially useful tool in the cleansing and treatment of petroleum contaminated soils. In this work we aimed to analyse the microbial community which is the protagonist in an oily-sludge contaminated soil. More than 200 bacterial strains that can use crude oil and naphthalene as the sole carbon and energy source were isolated through enrichments from three different refinery sludge deposition sites in central Greece. Far-range distribution genes encoding for key enzymes of both aromatic and aliphatic hydrocarbon degradation pathways were detected by polymerase chain reaction and DNA hybridisation. In addition, DGGE analysis of *in situ* microbial community structures was used to estimate population changes during bioremediation progress.

Oil residues containing high molecular mass hydrocarbons, rich in aromatic compounds are frequent end products of crude oil processing and are poorly degradable, while their disposal poses an environmental problem. Indigenous bacteria in contaminated soil can degrade a wide range of target constituents of the oily sludge disposed to the environment by oil refineries. Most of the microbial catabolic pathways responsible for the degradation, including the *alk* (*n*-alkanes) and *nah* (naphthalene) pathways have been extensively characterised and are generally located on large, transmissible plasmids usually found in *Pseudomonas* spp.

In the present work bacterial strains isolated through enrichment from oil – residuals contaminated soil, able to grow on crude oil as the sole carbon source were examined for their efficiency to degrade petroleum hydrocarbons. The fastest growing strains were divided into four consortia, representing populations from deposition sites with different contamination level. Genes encoding key enzymes of both aromatic and aliphatic hydrocarbon degradation pathways were detected in the above strains by PCR and DNA hybridization using specific primers, while the presence of the studied genes in each consortium was related to its efficiency to degrade crude oil. Additionally, biodiversity and population changes were correlated to the concentration of petroleum hydrocarbons and the contamination level of the soil.

MATERIALS AND METHODS

Bacterial strains. *Pseudomonas putida* DSMZ4476 carrying the NAH plasmid and *Escherichia coli* DSMZ8830 carrying the OCT plasmid were used as reference strains to determine the specificity of the primers designed for the detection of 2,3 catechol – dioxygenase (*nah*H) and aliphatic alcohol dehydrogonase (*alk*J) genes in the isolates.

Sampling sites. Soil samples were collected from 3 sites used as an oily sludge disposal area by the refinery of MOTOROIL (Hellas) S.A., and from a non-contaminated control area.

Strain isolation and culture conditions. Bacterial strains were isolated through enrichment procedures on minimal salts medium supplemented with crude oil 5% (w/v) or naphthalene 1% (w/v) as the sole carbon source. Cultures were incubated with agitation at 28°C.

Chemical analysis. Biodegradation activity of each consortium was determined after extraction of the culture with n-hexane, centrifugation and GC-MS (Saturn 4D ion trap, Hewlett Packard 5840A/6890) analysis of the hydrocarbon phase. For the determination of hydrocarbon concentration in soil samples, EPA extraction method 1664 was used, followed by GC-MS analysis.

Primer design. Oligonucleotide primers to amplify regions of *nah*H and *alk*J genes were designed on the

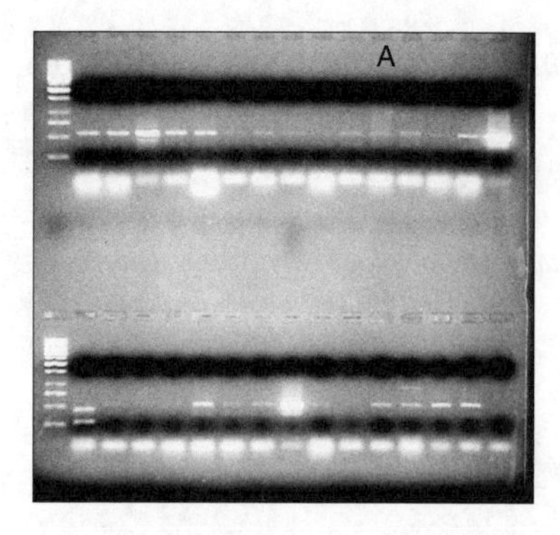

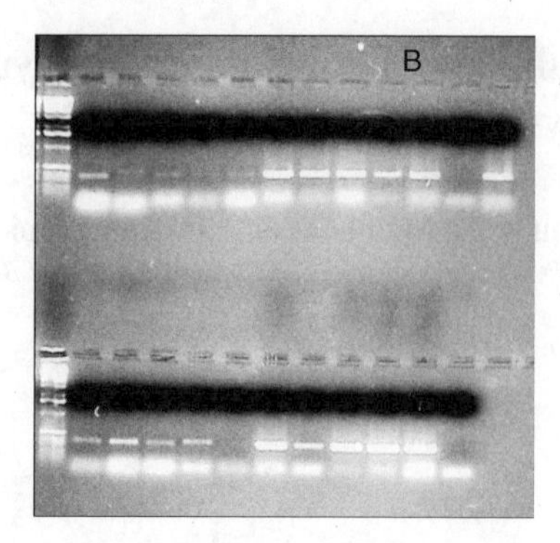

Figure 1. Detection of *nah*H (A) and *alk*J (B) genes.

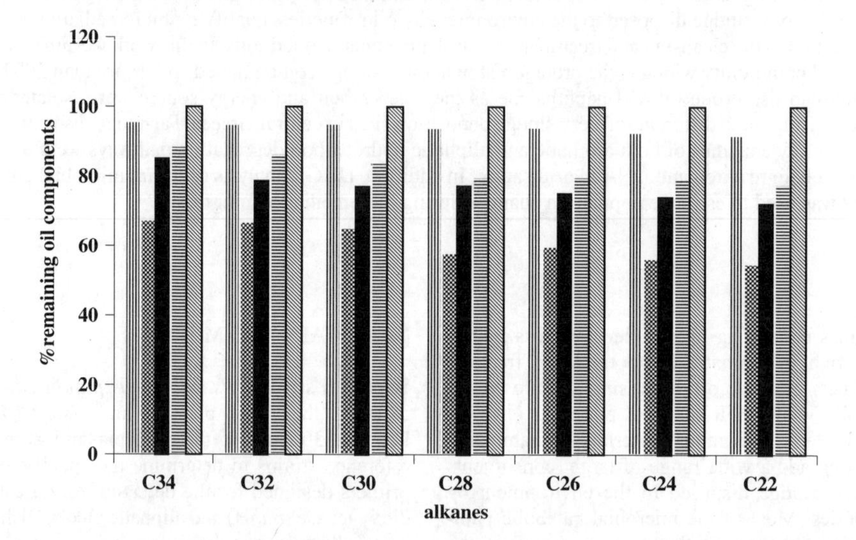

Figure 2. Percentage remaining of high molecular weight alkanes from crude oil after 10 days culture (▥) no inoculum added), (▩) consortium A, (■) consortium B, (▤) consortium C, (▧) consortium D.

Table 1. More than 200 bacteria able to grow on crude oil and/or naphthalene as sole carbon source were isolated from four different habitats. The faster growing isolates of each habitat were grown in consortium cultures.

Area	Total hydrocarbons (μg/g soil)	Total isolates	Isolates carrying *nah*H	Isolates carrying *alk*J
A	26609,87	13	3	3
B	21831,52	84	30	59
C	15118,44	102	29	42
D (control)	4075,26	5	0	1

basis of published sequences by using the Hitachi Software DNASIS. PCR amplification of a 476 bp fragment of the *nah*H gene was performed using the forward primer *nah*HF 5′ – gcagacaaggaata(ct)actgg – 3′ and the reverse *nah*HR 5′ – tcgaatcgttcgttgagc – 3′. For the detection of *alk*J gene a 352 bp region was amplified using the forward primer *alk*JF 5′ – tcggcc(ct)aatttgcagtttc – 3′ and the reverse *alk*JR 5′ – tttacccat(ca)ctacaagtacc – 3′.

RESULTS AND DISCUSSIONS

- More than 200 bacterial strains able to grow on crude oil and/or naphthalene as the sole carbon source were isolated from four oil-sludge deposition sites with different contamination level.
- The number of isolates was different for the studied habitats. Only few dominant species were isolated from the heavily contaminated soil with deposition time 6 months before sampling. Biodiversity

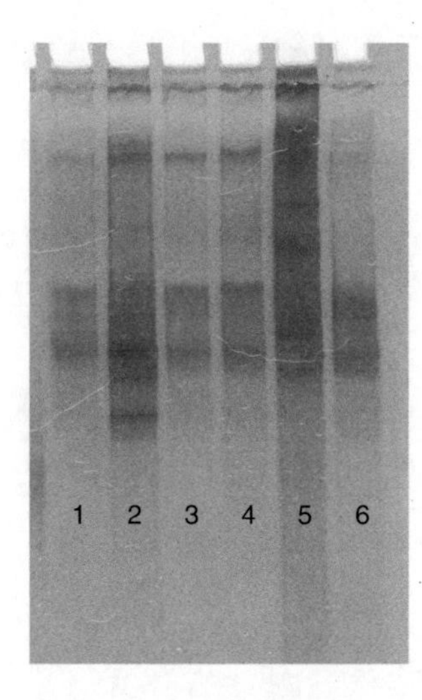

Figure 3. DGGE analysis of total community DNA from the sampling sites, using primers F984GC: 5′ – AACGC GAAGAACCTTAC-3′ R1387: 5′ – CGGTGTGTACAAG GCCCGGGAACG-3′. The GC-rich sequence, is attached at 5′ end of forward primer (F984) to prevent complete melting during separation in the denaturing gradient. 5: (D), 2: (C), 4,6: (B) and 1,3: (A)
Gel composition: Polyacrylamide 6% (wt/vol)
Denaturing Gradient: 40–60% (formamide and urea).
100% denaturants corresponds to 7 M urea and 40% formamide (vol/vol).

increases as the soil texture improves in the habitats with last deposition time 1.5 and 3 years before sampling.

- The ability of the isolates to degrade crude oil is far better for bacteria of consortium A, as they are adapted in high H/C concentrations, while consortium D exhibited poor degradation activity, as it comes from a non contaminated habitats; similar results have been reported.
- The *nah*H gene is more common among the isolates probably because it is mainly located on transmissible plasmids rather than the bacterial chromosome.
- In addition, there seems to be a relation between the percentage of indigenous population carrying the *alk*J gene and the biodegradation activity of the bacteria isolated; in the habitats where the *alk*J gene is rarely found, such as the control area and the 3 years old deposition site, the biodegradation of crude oil was low. Yet, all four consortia follow the same biodegradation pattern, with maximum activity on shorter-length chain alkanes.

REFERENCES

MacNaughton SJ, Stephen JR, Venosa AD, Davis GA, Chang YJ, White DC. (1999) Microbial population changes during bioremediation of an experimental oil spill. Applied and Environmental Microbiology 65: 3566–74.

Milcic-Terzic J., Lopez-Vidal Y., Vrivc M.M. and Saval S. (2001) Detection of catabolic genes in indigenous microbial consortia isolated from a diesel-contaminated soil. Bioresource Technology 78: 47–54.

Mishra S., Jyot J., Kuhad R.C. and Lal B. (2001) In situ bioremedation potential of an oily sludge-degrading bacterial consortium. Current Microbiology 43: 328–335.

Verstraete W. and Top E.M. (1999) Soil clean up: lessons to remember. International Biodeterioration and Biodegradation 43: 147–153.

Vinas M., Grifoll M., Sabate J. and Solanas A.M. (2002) Biodegradation of a crude oil by three microbial consortia of different origins and metabolic capabilities. Journal of Industrial Microbiology and Biotechnology 28: 252–260.

European Symposium on Environmental Biotechnology, ESEB 2004 - Verstraete (ed)
© 2004 Taylor & Francis Group, London, ISBN 90 5809 653 X

New softeners influence on the activated sludge process

E. Grabińska-Sota & J. Kalka
Environmental Biotechnology Department Silesian University of Technology, Gliwice, Poland

ABSTRACT: The objective of this study was investigation of the influence of new generation fabric softeners on activated sludge processes: degradation of organic compounds, nitrification rate and settlement properties. It was shown, that preparations were biodegradable under conditions simulating wastewater treatment plant. Efficiency of disulphine blue active substances removal was dependent on the structure of preparation: the length of hydrophobic chain and structure of hydrophilic part. It has ranged within the scope 95.3 to 97.7%.

1 INTRODUCTION

One of the most important groups of washing agents is group of fabric softeners with softening and antielectrostatic action (Bahadur & Chand 1997). Softeners introduced to the last washing cycle eliminate the harsh feel produced by modern laundering process and prevent the build up of antielectrostatic charges (Poźniak 2001). After use through the laundry process fabric softeners active ingredients are predominantly discharged to the sewer as part of household wastewaters. Consumption of softeners in Europe is estimated during a year on 6 kg per capita (Puchta et al. 1993) and concentration of those compounds appearing in sewage is 4.5 mg/dm^3 (Huber 1984). This sewage is treated at municipal treatment plants, where can influence on the activated sludge process. Until recently the most commonly used for delivering fabric-conditioning benefits are imidazolium alyldimethylammonium salts (Puchta et al. 1993). Producers of those compounds still developing new substances characterizing with good useful properties and safe for the environment. The aim of presented experiment was to evaluate the influence of 3 new preparations with antielectrostatic properties on activated sludge process.

2 MATERIALS AND METHODS

2.1 *Compounds*

Investigated preparations were:

- 1-decyl-3-cyclohexyloxymethylimidazolium chloride (A11), with molecular weight 358 and active substance content 94.7%,
- 1-deyl-3-hexyloxymethylimidazolium chloride (A12) with molecular weight 357 and active substance content 97.9%,
- 1,8-bis[(2,7-dooxaoktamethylene)dimethyl-oktyl] ammonium chloride (A15), with molecular weight 528 and active substance content 96–98%.

Preparations differed in both hydrophobic and hydrophilic parts. They have possessed not only antielectrostatic properties, but also have showed fungicidal actions (Brycki 1998, Urbanik 1998). Samples were prepared by professor Pernak's group from Poznań University of Technology.

2.2 *Biodegradation tests*

Biodegradation tests were conducted according to OECD guidelines for testing of chemicals – coupled unit test – Method 303A (OECD 1993), in conditions simulating biological wastewater treatment plant. Four models of activated sludge were prepared – one for each preparation and the fourth was taken as a control (fig.1).

The small activated sludge units consisted of storage vessel (A) of capacity at least 24 dm^3 contained synthetic sewage and dosing pump (B) introduced the sewage into the aeration vessel (C), which was about 4 dm^3 volume. The liquor passes into the separator (D) and treated effluent leaved the apparatus to be collected in vessel (F). Sludge was returned from the bottom of the separator to the aeration vessel by the means of an air-lift pump (E). Aeration of the mixed liquor was effected by the use of aeration cube (s).

The control model was supplied only with synthetic municipal sewage. During the experimental investigation preparations were in a concentration of 10 mg/dm^3.

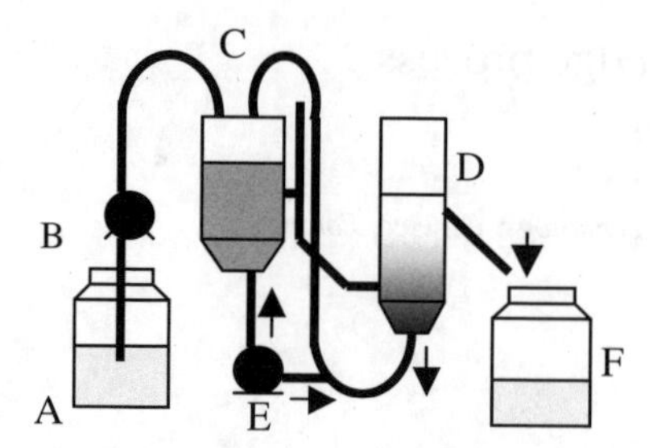

Figure 1. Suitable apparatus for biodegradability testing.

Activated sludge for the tests was supplied from municipal wastewater treatment plant. Activated sludge processes were controlled by measurement:

- in aeration chamber: pH and oxygen concentration;
- in average daily samples of sewage: COD removal (potassium dichromate method according to Polish Standards (1974)), disulphine blue active substance removal (Osburn 1982, Waters 1994), mineral forms of nitrogen;
- in samples of activated sludge: concentration of suspended solids, Mohlmann's index.

Nitrification rate was calculated according to the following formula (Surmacz-Górska 2000):

$$V_x = \frac{L_{dN-NO_x}}{V_N \times X_N} \; [\mathrm{gN/g_{asdm}d}] \qquad (1)$$

where L_{dN-NOx} = daily load of oxidased forms of nitrogen occurring in aeration chamber ($N-NO_2$ and $N-NO_3$); V_N = volume of aeration chamber; X_N = concentration of activated sludge.

3 RESULTS

During the experiment conventional operational conditions were maintained. Experiments were conducted with 3 h hydraulic retention time, the concentration of activated sludge had differed between 3.0–3.8 g_{asdm}/dm^3 and substrate loading was from the range 0.22 to 0.37 $g_{COD}/g_{asdm}d$ (table 1).

Range of pH of influent (7.0–7.1) was adequate for aerobic microorganisms degrading organic compounds. Changes of reaction on slightly alkaline have proved that acidic degradation products were not created. The highest value of Mohlmann's index was observed for control sludge (table1). Investigated preparations did not cause increase of Mohlmann's index; improvement of settling properties of activated sludge with preparations A11, A12 and A15 was observed.

Table 1. Parameters of biodegradation.

Investigated activated sludge with preparation	Substrate loading ($g_{COD}/g_{asdm}d$)	Concentration of suspended solids (g_{asdm}/dm^3)	Mohlmann's index (average value) (cm^3/g)
A11	0.37	3.0	128
A12	0.29	3.2	71
A15	0.33	3.8	105
Control	0.22	3.1	181

Table 2. Biodegradation of CSA (medium values).

Activated sludge with	% removal of DBAS	% removal of COD
A11	97.7	82.6
A12	94.7	76.6
A15	96.7	82.4
Control	–	85.2

Table 3. Chemical characteristic of raw and treated sewage.

Activated sludge with		(pH)	COD (mg/dm^3)	DBAS (mg/dm^3)
A11	Raw sewage	7.08	240	10
	Treated sewage	8.00	42	0.23
A12	Raw sewage	7.10	252	10
	Treated sewage	7.80	60.2	0.47
A15	Raw sewage	7.09	248	10
	Treated sewage	7.91	48	§0.39
Control	Raw sewage	7.10	236	–
	Treated sewage	8.20	35	–

This phenomenon is extremely important for correct operation of wastewater treatment plant.

Efficiency of disulphine blue active substances removal was dependent on the structure of preparation: the length of hydrophobic chain and structure of hydrophilic part. It has ranged within the scope 95.3 to 97.7% (table 2).

Chemical characteristic of raw and treated sewage was presented in table 3. Decomposition rate of organic substances was decreased when preparation A12 was added to sewage. It could be provoked by bactericidal properties of this preparation (Pernak & Skrzypczak 1996, Pernak et al. 1997).

Bactericidal properties of A12 could also influence the value of factor BOD/COD and nitrification rate (table 4, fig. 2).

It could be supposed that A12 inhibited nitrification as all sludge parameters was maintained at the same level.

Table 4. Average nitrification rates obtained in the study.

Investigated process	Nitrification ratio ($gN/g_{asdm}d$)
Activated sludge with A11	0.060 ± 0.014
Activated sludge with A12	0.009 ± 0.002
Activated sludge with A15	0.067 ± 0.011
Control	0.1287 ± 0.08

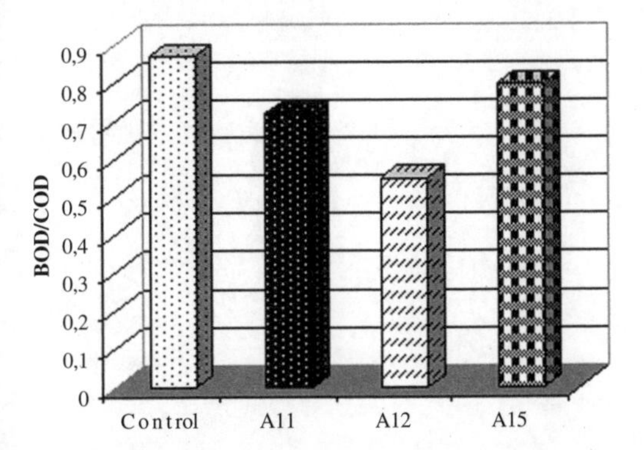

Figure 2. BOD/COD ratio determinated for raw sewage amended with different preparations.

4 CONCLUSIONS

At the base of obtained results it could be concluded, that according to OECD guideline all preparations were biodegradable, but introduced cyclohexan ring in preparation A12 decreased COD and DBAS removal. It influenced also nitrification rate.

REFERENCES

Bahadur, P. & Chand, M. 1997. Studies on dodecylammoniumchloride in the presence of additives. *Tenside Surf. Det.* 34: 347–356.

Brycki, B. 1998. Aktywność przeciwdrobnoustrojowa oraz inne właściwości aplikacyjne czwartorzędowych soli amoniowych. *Proceedings of III Symposium Quaternary Ammonium Salts and the Fields of Application, Ponzań*

Huber, L.H. 1984. Ecological behavior of cationic surfactants from fabric softeners in the aquatic environment. *JAOCS* 61(2): 377–382.

OECD. 1993. Guideline for testing of chemicals. Paris.

Osburn, Q.W. 1982. Analytical methods for cationic fabric softners in waters and wastes. *JAOCS* 59(10): 453–457.

Pernak, J. & Skrzypczak, A. 1996. 3-alkylthiomethyl-1-ethylimidazolium chlorides. Correlation between critical micele concentration and minimum inhibitory concentrations. *Eur. J. Med. Chem.* 31: 901–903.

Pernak, J. 1997. Synthesis and microbial activity of new 1-benzyl benzimidazolium chlorices. *Arch. Pharm. Med. Chem.* 330: 253–258.

Polish Standards PN-74/C-04578 Arkusz 03: Woda i ścieki. Badanie zapotrzebowania tlenu i zawartości węgla organicznego. Oznaczanie chemicznego zapotrzebowania tlenu (ChZT) metoda dwuchromianową.

Pożeniak, R. 2001. Czwartorzędowe sole amoniowe jako środki antystatyczne. In Zieliński R. (ed), *Quaternary Ammonium Salts and the Fields of Application*, Poznań: Wydawnictwo Instytut Technologii Drewna

Puchta et al. 1993. A new generation softeners. *Tenside Surf. Det.* 30: 186–191.

Surmacz-Górska, J. 2000. Usuwanie zanieczyszczeń organicznych oraz azotu z odcieków powstających w wysypiskach odpadów komunalnych. Gliwice: Academic Press.

Waters, J. 1994. Analysis of low concentrations of cationic surfactants in laboratory test liquars and environmental sample. In Cross, J. & Singer, E. (eds), Cationic Surfactants. Analytical and Biological Evaluation. New York, Basel: M. Dekker.

Water purification

European Symposium on Environmental Biotechnology, ESEB 2004 - Verstraete (ed)
© 2004 Taylor & Francis Group, London, ISBN 90 5809 653 X

Bioprotection of the structure and function of activated sludge microbial communities against chloroaniline shock loads

N. Boon & W. Verstraete
Laboratory of Microbial Ecology and Technology (LabMET), Faculty of Agricultural and Applied Biological Sciences; Ghent University, Ghent, Belgium

E.M. Top
Department of Biological Sciences, University of Idaho, Moscow

S.D. Siciliano
Department of Soil Science, University of Saskatchewan, Saskatoon, Saskatchewan, Canada

ABSTRACT: Bioaugmentation of bioreactors generally focuses on the removal of specific xenobiotics. Often, less attention is hereby paid to the recovery of other reactor functions such as ammonium–nitrogen removal. This work evaluated the effects on activated sludge reactor functions of a 3-chloroaniline (3-CA) pulse and the protective bioaugmentation by inoculation with the 3-CA degrading strain *Comamonas testosteroni* I2*gfp*. Changes in functions such as nitrification, removal and sludge compaction were monitored with the sludge community structure, in particular the nitrifying populations. Molecular techniques characterized and enumerated the ammonia oxidizing microbial community immediately after a 3-CA shock load. Two days after the 3-CA shock, ammonium accumulated and the nitrification activity did not recover over a 12-day period in the non-bioaugmented reactors. In contrast, nitrification in the bioaugmented reactor started to recover from day 4 on. The molecular analysis showed that the ammonia oxidizing microbial community of the bioaugmented reactor recovered both in structure, activity and abundance, while in the non-bioaugmented reactor the number of ribosomes of the ammonia oxidizers decreased drastically and the community composition changed and did not recover. This study demonstrates that bioaugmentation of wastewater reactors to accelerate the degradation of toxic chlorinated organics such as 3-chloroaniline protected the nitrifying bacterial community, thereby allowing faster recovery from toxic shocks.

1 INTRODUCTION

The main purpose of a wastewater treatment plant is to convert the organics present in an incoming wastewater stream to N_2 and CO_2 as well as to decrease the amount of suspended solids entering the environment. The biological treatment of industrial wastewaters by mixed microbial communities is however often disrupted by organic and inorganic chemicals present in the wastewater stream, resulting in inhibited nitrification, decreased carbon removal and modification of sludge compaction properties (Bitton, 1994). Little is known about the composition of mixed microbial communities in reactors when biological processes are disrupted by, or are recovering from xenobiotic shocks. Most investigators focus on xenobiotic degradation but in terms of the day to day functioning of the reactors, the restoration of biological activity such as ammonia removal, is of primary importance.

The restoration of activity in wastewater treatment plants is a time-consuming and costly process. Many wastewater treatment plants contain buffering tanks or specialized microbial strains to protect the plant from a variety of chemical shocks (Eichner et al., 1999) but there is no known process to specifically protect reactors from chloroaniline shocks. The effect of chloroaniline on activated sludge microbial communities is not known but the non-chlorinated compound aniline inhibits nitrification (McCarthy, 1999). In our previous work, a 3-CA degrading strain *Comamonas testosteroni* I2 was isolated from activated sludge and used in a Semi Continuous Activated Sludge system to biodegrade a continuous input of 3-CA contaminated wastewater. For more than 14 days, strain I2*gfp* was capable to augment the removal of all 3-CA (Boon et al., 2002a; Boon et al., 2000).

Few reports have investigated the recovery of disrupted reactor functions after bioaugmentation. Therefore, the aim of this work was to evaluate the short-term effects of a 3-CA shock load and bioaugmentation with a 3-CA degrading inoculant on reactor performance parameters such as nitrification, carbon removal, and sludge settling.

2 MATERIAL AND METHODS

2.1 *Experimental setup*

The experiments were conducted with sludge freshly collected from a domestic wastewater treatment plant (Bourgoyen-Ossemeersen, Ghent, Belgium), and subjected to a SCAS (Semi-Continuous Activated Sludge) procedure, as described previously (Boon et al., 2003).

Before a shock dose was applied, the reactors were operated for 6 days to stabilize. At day 0, Reactors C (control) ($n = 2$) continued to receive only milk powder and were control reactors. In addition to the milk powder, reactors S (stressed) ($n = 2$) and SB (stressed and bioaugmented) ($n = 2$) both received at day 0 a shock load of 250 mg 3-CA L^{-1} in the reactor mixed liquor. The bioaugmentation experiment was performed in reactors 3, in which *Comamonas testosteroni* I2gfp was inoculated to a final concentration of $(5.4 \pm 0.37) \times 10^8$ cells/ml. This strain, which has been chromosomally marked with gfp (Green Fluorescent Protein), mineralizes 3-CA, fluoresces green under UV-light and is rifampin as well as kanamycin resistant (Boon et al., 2000).

Ammonium, nitrite, nitrate, chemical oxygen demand (COD), sludge volume index (SVI) and 3-CA were analyzed as describe previously (Boon et al., 2003).

2.2 *Molecular analysis*

DNA and RNA were extracted from the different reactors, representing respectively the active bacteria present. Analysis of the composition of the ammonium oxidizing bacteria (AOB) was performed with Denaturing Gradient Gel Electrophoresis (DGGE), using primers specific for the AOB 16S rRNA (Boon et al., 2002b). Bands of interest were cut out and sequenced.

Quantification of the number and the activity of AOB was performed with Fluorescent *in situ* Hybridization and Quantitative PCR, respectively (Boon et al., 2003).

3 RESULTS

3.1 *Reactors performance*

The first pair of reactors (type C) was used as a control with no inoculation or shock load applied. The second pair consisted of the 3-CA treated reactors (stressed reactors); they received a shock load treatment of 3-CA at day 0 (type S). The third pair of reactors were the bioaugmented reactors, which received the same 3-CA pulse at day 0, but were in addition, inoculated with *C. testosteroni* I2*gfp* cells, able to degrade 3-CA (type SB).

In the non-bioaugmented reactors S, nitrate concentration remained very low and ammonium accumulated for the duration of the experiment. In the bioaugmented reactors SB from day 4 on, nitrification recovered resulting in the accumulation of nitrate. The nitrite was present in reactors SB until day 6 and at day 8, all the nitrite was converted to nitrate. The restored nitrification activity resulted in the complete removal of ammonium from day 6 on.

Also other reactor parameters were affected by the 3-CA pulse. In all reactors that received the 3-CA pulse the sludge volume index increased, indicating that the settlement of the activated sludge was poor (results not shown). In contrast to nitrification, the SVI did not recover in the bioaugmented reactors during the first 10 days, however the sludge compactability improved on day 12. Only the non-bioaugmented shock loaded reactors type S accumulated high amounts of COD (Figure 1B). However from day 4 on the COD removal increased again. The bioprotection of the reactors type SB was very effective and no differences in COD removal compared to control reactors C were observed.

3.2 *Analysis of ammonia oxidizing bacteria*

Since the 3-CA shock and the bioprotection had a major influence on the nitrification activity in reactors S and SB, a more detailed analysis of the ammonia oxidizing bacteria (AOB) was performed. At day 0 (good nitrification activity before the shock), DGGE patterns obtained for AOB are very similar, even between the RNA and DNA samples (results not shown). A visual comparison of the RNA-DGGE patterns showed that two bands seemed to play a critical role in the nitrification. The upper band showed the highest similarity with an uncultured *Nitrosomonas*, while the lower matched with a *Nitrosococcus/Nitrosomonas*.

Two days after the shock load, nitrification was inhibited in reactors S as well as SB, the RNA-DGGE patterns of the AOB community structure differed between these two reactors (Figure 2). On day 2 in reactor S, the *Nitrosomonas* band was hardly visible, and after 8 days both bands had almost disappeared.

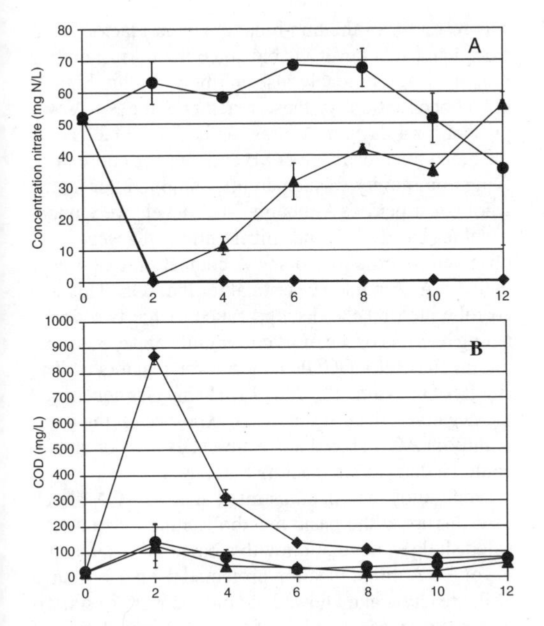

Figure 1. Performance of the reactors in reference to the stabilization period (day 0). Reactors type C were not stressed nor bioaugmented (●), reactors type S were stressed on day 0 with a shock load of 250 mg L^{-1} of 3-CA (♦), reactors type SB were on day 0 stressed as reactors type S but also bioaugmented by inoculating *C. testosteroni* I2*gfp* (▲). Panel (A) nitrate concentration in the effluent; (B) COD in the effluent; Values represent the mean ± standard error of two reactors (n = 2).

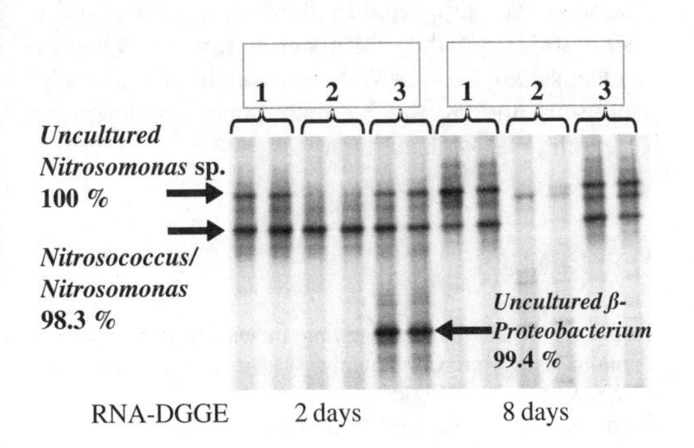

Figure 2. Analysis of the DGGE profiles at days 2 and 8 of the different reactors, using partial AOB 16S rRNA gene fragments, based on RNA. 1, reactors type C without inoculant and without 3-CA shock; 2, non-bioaugmented reactors type S with 3-CA shock load; and 3, the bioaugmented reactors type SB with 3-CA shock load.

In the bioaugmented reactors SB however, both bands were still clearly detectable at day 2. Interestingly, a third potential ammonia oxidizer, identified as an uncultured β-Proteobacterium, became more dominant.

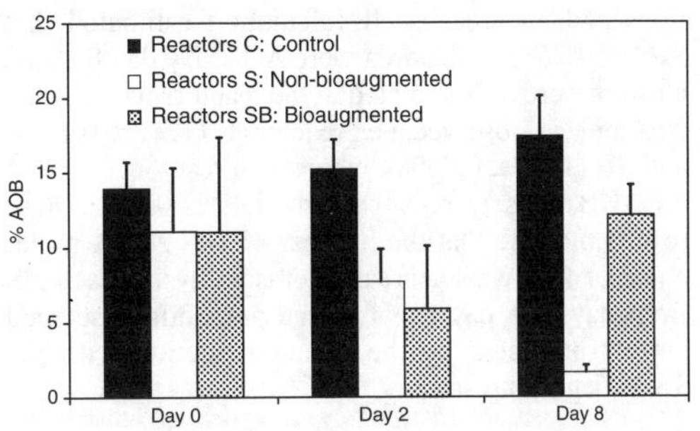

Figure 3. Relative abundance of cDNA of ammonia oxidizers in activated sludge reactors, as determined by Quantitative PCR. The relative amounts of AOB was estimated by dividing the number of AOB rRNA copies by the total number of bacterial rRNA copies Values represent the mean ± standard error (n = 3).

The DNA sequence of this fragment was identical to a sequence from a PCR fragment in freshwater sediments. The first known sequence of an ammonia oxidizers similar to this band is *Nitrosococcus mobilis*. By day 8, when nitrification was restored in the bioaugmented reactors SB, the RNA-DGGE profiles were similar to their original pattern and the additional band had disappeared.

3.3 *Quantification of the ammonia oxidizing bacteria*

The DGGE analysis showed that shifts occurred in the AOB community. Since it is not possible to obtain quantitative data from the DGGE patterns, Quantitative PCR and FISH were used to examine the activity and prevalence of the AOB in the activated sludge. From days 0, 2 and 8, the amount of rRNA molecules, present in the reactors, was estimated by amplifying the 16S rRNA cDNA of the bacteria and the ammonia oxidizers from the different reactors and measuring the increasing amounts of amplification products. (Figure 3).

The data showed that the relative amounts of rRNA copies of the ammonia oxidizers decreased after two days, both in the stressed bioaugmented and in the non-bioaugmented reactors. Eight days after the shock load, the number of rRNA copies of the ammonia oxidizers decreased further in the non-bioaugmented reactors, while in the bioprotected reactors the number of rRNA copies of the ammonia oxidizers recovered.

To confirm the trend observed in the real-time PCR data, FISH analysis was also applied. From the different reactors at days 0 and 8, the area of all the bacteria and of the AOB was measured and the ratio of AOB and bacteria was calculated. At day 0, clusters of AOB could be observed in the activated sludge flocs and an area based calculation estimated that 2.23 ± 0.76% of the area were AOB. By day 8, AOB clusters were only detected in the reactors where nitrification was observed, i.e., reactors C (1.73 ± 0.76%) and SB (1.60 ± 0.05%), whereas in reactor S, almost no AOB cells were observed (0.47 ± 0.59). These results indicate that the number of 16S rRNA molecules of the AOB decreased drastically in reactor S from day 0 to day 8, while the populations seemed largely protected by the bioaugmentation with the 3-CA degrading strain.

4 DISCUSSION

We investigated the effects of a 3-chloroaniline (3-CA) shock load on the basic functions of a wastewater treatment reactor and focused on the recovery of some functional activities. Bioaugmentation by the inoculation of a 3-CA degrading strain was used to investigate whether rapid 3-CA removal could decrease the recovery period. The most drastic effect of the 3-CA shock load observed was on the nitrification activity. Initially, nitrification was totally inhibited in reactors S and SB where a 3-CA shock load was applied. The non-substituted form of 3-CA, i.e. aniline is known to inhibit nitrification, and, like most inhibitors, aniline inhibits nitrification activity by acting as a suicide substrate for the ammonium monooxygenase (McCarthy, 1999). It is likely that 3-CA inhibition is based on the same mechanism (D.J. Arp, personal communication). As a consequence, the reestablishment of the nitrification activity in the activated sludge after the removal of the inhibitor through degradation or wash-out requires *de novo* synthesis of the enzymes. After two days in the bioaugmented reactors SB, 3-CA was degraded completely, allowing the slow recovery of nitrification activity, visible from day 4 on (Fig. 1).

In the RNA-DGGE profiles of the AOB community, an additional very intense band appeared at day 2 in the bioaugmented reactors 3. These data suggest that this bacterial population became a highly active member of the community. This population may have had a selective advantage over the other two populations, represented by the initial AOB populations. Work of Suwa et al. (1994) showed that various *Nitrosomonas* sp. strains isolated from activated sludge could be either sensitive or insensitive to ammonium. The short-term inhibition of nitrification by 3-CA led to an accumulation of ammonium up to 65 mg NH_3-N L^{-1}, which may have inhibited the initial AOB. The new band at day 2 may represent an ammonia oxidizer that is insensitive to higher ammonium concentrations and which then became more active. From day 4 on, the ammonium concentration decreased again, such that the nitrification activity by the original AOB was restored. In the non-bioaugmented reactor 2 this extra band was never visible since the corresponding organism was probably also inhibited by the high 3-CA concentration. In these reactors S the 3-CA was present for 4 days, in contrast with less than 2 days in the bioaugmented reactors SB and thus the entire nitrifying community was apparently inhibited during the whole, test period. Although 3-CA levels were undetectable by day 6, no nitrification recovery was observed in these reactors S. A contact time of 4 days with 3-CA seemed to be critical for the AOB. This was corroborated by the disappearance of the two dominating bacteria of the AOB community at day 8, as visible in the RNA-DGGE patterns for reactors S. The results of real-time PCR and FISH analyses confirmed the negative effect of the 3-CA exposure on the community of AOB since hardly any AOB cells or cluster could be detected in reactors S at day 8.

This study has demonstrated that a 3-CA shock load disrupted the basic metabolic functions of activated sludge. Inoculation with a bacterial strain capable of 3-CA mineralization protected the performance of the reactors and allowed a rapid recovery of nitrification. Our work clearly indicates that inoculation of wastewater treatment systems subject to toxic shock loads, with a specific degrader of the toxic compound, can result in faster recovery of essential metabolic functions.

ACKNOWLEDGMENT

This work was supported by the project grant G.O.A. (1997–2002) of the "Ministerie van de Vlaamse Gemeenschap, Bestuur Wetenschappelijk Onderzoek" (Belgium) and by the Flemish Fund for Scientific Research (FWO-Vlaanderen).

REFERENCES

Bitton, G., 1994. Toxicity testing in wastewater treatment plants using microorganisms, Wastewater microbiology. Wiley-Liss, New York.

Boon, N., De Gelder, L., Lievens, H., Siciliano, S.D., Top, E.M. and Verstraete, W., 2002a. Bioaugmenting bioreactors for the continuous removal of 3-chloroaniline by a slow release approach. Environ. Sci. Technol., 36: 4698–4704.

Boon, N., De Windt, W., Verstraete, W. and Top, E.M., 2002b. Evaluation of nested PCR-DGGE (denaturing gradient gel electrophoresis) with group-specific 16S rRNA primers for the analysis of bacterial communities from different wastewater treatment plants. FEMS Microbiol. Ecol., 39: 101–112.

Boon, N., Goris, J., De Vos, P., Verstraete, W. and Top, E.M., 2000. Bioaugmentation of activated sludge by an indigenous 3-chloroaniline degrading *Comamonas testosteroni* strain, I2*gfp*. Appl. Environ. Microbiol., 66(7): 2906–2913.

Boon, N., Top, E.M., Verstraete, W. and Siciliano, S.D., 2003. Bioaugmentation as a tool to protect the structure and function of an activated sludge microbial community against a 3-chloroaniline shock load. Appl. Environ. Microbiol., 69: 1511–1520.

Eichner, C.A., Erb, R.W., Timmis, K.N. and Wagner-Döbler, I., 1999. Thermal gradient gel electrophoresis analysis of bioprotection from pollutant shocks in the activated sludge microbial community. Appl. Environ. Microbiol., 65: 102–109.

McCarthy, G.W., 1999. Modes of action of nitrification inhibitors. Biol. Fertil. Soils, 29: 1–9.

Suwa, Y., Imamura, Y., Suzuki, T., Tashiro, T. and Urushigawa, Y., 1994. Ammonia-oxidizing bacteria with different sensitivities to $(NH_4)_2SO_4$ in activated sludges. Water Res., 28(7): 1523–1532.

European Symposium on Environmental Biotechnology, ESEB 2004 - Verstraete (ed)
© 2004 Taylor & Francis Group, London, ISBN 90 5809 653 X

Bioremediation of Lead (Pb^{2+}) from water effluents by *Rhizobium sp.* (BJVr-12) Extracellular Polysaccharides (EPS) immobilized on modified coconut husk

E.T. Paner & M.L.D. Udan
National Institute of Molecular Biology and Biotechnology, University of the Philippines Los Banos College, Laguna, Philippines

ABSTRACT: This study attempted to establish the capacity of *Rhizobium sp.* (BJVr-12) extracellular polysaccharides (EPS) to adsorb lead and compare the efficiency of biotrap columns consisting of 20-g acid-modified coconut husks with EPS (biotrap) and without EPS (control biotrap) in reducing lead content of water effluents. Results showed that the biotrap can remove 99.6% lead ions at optimum flowrate of 1 ml/min after 15 minutes elution. Control biotrap removed a maximum of 94.6% lead ions. Recovery of adsorbed lead ions was done by acid desorption process using 6 M HCl. Recycled biotraps with EPS could still be effective in adsorbing a maximum of 99.5% lead ions at 1 ml/min flowrate after 15 minutes. The linear result of the Langmuir model indicates that the biosorption of lead ions by immobilized EPS on modified coconut husk column involves a monolayer adsorption and constant energy of adsorption under low lead ion concentrations.

1 OBJECTIVES

In general, the study attempted to establish the capacity of *Rhizobium sp.* (BJVr-12) EPS to adsorb lead and compare the efficiency of the acid- modified coconut husk with and without EPS in reducing lead content of waste water.

Specifically, it aimed to:

1) evaluate the capacity of the modified coconut husk to remove heavy metal ions from the aqueous solution.
2) determine the optimum conditions for the maximum reduction of lead in the effluents using column biotrap with acid- modified coconut husk and EPS.
3) establish Langmuir and Freundlich constant for the biosorption process of modified coconut husk with and without EPS.
4) recover the adsorbed lead from the biotrap by means of acid desorption process, and
5) recycle the biotrap.

2 INTRODUCTION

Lead, which is used in the manufacture of paints, glass, batteries and gasoline combustion (McGraw Hill Encyclopedia, 1992) is now believed to be a threat for our health because of its accumulation in the bodies of water. According to our Department of Environment and Natural Resources (DENR), the maximum limit of Pb^{2+}(in mg/L) for the protection of public health are 0.5 for marine waters (class SC) and 0.3 for inland waters (class C).

Because the Philippines is the world's largest coconut producer, waste disposal of husks is a problem. Utilization of husks in biotraps may reduce this problem.

Ion exchange capacity of husks is increased by acid treatment (Randal, 1977); regeneration and recovery of metals can be done by acid elution (Mattuschka & Straube, 1993).

2.1 *Adsorption isotherm*

Different weights of modified coconut husk (MCH) with and without EPS (0.5 g, 1.0 g and 2.5 g) were used to determine the extent of its adsorption capacity. Fifty ml of 50 ppm lead solution was combined in Erlenmeyer flasks and shaken (120 rpm) at different time intervals. Eluents were collected after 0 min, 15 min, 30 min, 1 hr, 2 hrs, 4 hrs, 8 hrs, 12 hrs, 16 hrs, 20 hrs, and 24 hrs of continuous shaking; each experiment was replicated three times and eluents analyzed by Atomic Absorption Spectrophotometer for lead.

2.2 *Optimization of flowrate and weight of the biotraps used*

Four liters of 50 ppm lead solution were allowed to pass through prepared biotraps (using 50 ml plastic syringe as columns) at different weights and flowrates:

10 g biotrap using 1 ml/min flowrate
20 g biotrap using 1 ml/min flowrate
10 g biotrap using 2 ml/min flowrate
20 g biotrap using 2 ml/min flowrate

Two types of biotraps were used:

- Acid-modified coconut husk with EPS;
- Acid-modified coconut husk without EPS. Eluents were sampled at 0 min, 15 min, 30 min, 1 hr, 2 hrs, 4 hrs, and 8 hrs. after flowing through each column (biotraps). This experiment was replicated six times for each biotrap and flowrate condition (1 ml/min and 2 ml/min rates).

3 RESULTS AND DISCUSSION

3.1 *Biosorption of lead in the biotrap*

A 50 ppm lead solution was passed through the biotrap for eight (8) hours and eluents were collected at a specified time interval. Results showed that the modified coconut husk with EPS (MCH + EPS, 20 g) gave a much higher capacity of removing lead from the aqueous solutions (99.6% lead removed) as compared to the column material with coconut husk alone (MCH, 20 g resulted to 94.6% lead removed). The greater removal of lead is attributed to the anionic composition of the EPS surface ensuring greater binding site for metal cations. The coconut husk alone can sequester metal cations because of their cellulosic materials which contain an appreciable amount of carboxylic groups. Moreover, coconut husk had an increased ion exchange capacity upon treatment with sulfuric acid, forming the modified coconut husk (Volesky, 1993; Baes & Okada, 1997).

3.2 *Optimization of flowrate*

A flowrate of 1 ml/min was proven to be the optimum flowrate by which a maximum amount of lead (99.6%) was trapped in the column material. A slower flowrate allowed the lead and the column material (MCH + EPS) to attain equilibrium by reaching a greater amount of exchanging sites. However, no significant difference was observed in the results using 1 ml/min and 2 ml/min flowrates.

3.3 *Langmuir and Freundlich adsorption isotherm models*

The Langmuir and Freundlich adsorption isotherm models were applied to the biosorption of lead by EPS on modified coconut husk.

The Freundlich equation, defined as

$$\ln Q = \ln K + (1/n) \ln C_{eq}$$

where Q is the lead uptake in mg/g cell, C_{eq} is the concentration of Pb^{+2} in liquid phase and K and n are the Freundlich constants, deals with the heterogeneous surface adsorption. Langmuir model, on the other hand, assumes the equation:

$$C_{eq}/Q = 1/bQ_o + C_{eq}/Q_o$$

where b is the constant related to the energy or net enthalpy of adsorption, Q_o is the number of moles of solute adsorbed per unit weight of adsorbent in forming complete monolayer on the surface, and C_{eq} and Q are defined previously in Freundlich equation. This model assumes monolayer adsorption and constant energy of adsorption. It also assumes that there is no migration of sorbent molecules on the surface plane. Q_o and b can be calculated using the linear regression formula with Q_o as the slope and b as the y-intercept. A high Q_o value signifies a high number of available sites for the solute molecules on the EPS surface. On the other hand, a high b shows a high affinity of the EPS for the solute molecules. The summary of the experimental results are shown on Table 1. Comparison of the Langmuir constant, Q_o and Freundlich constant, K, for the two set-ups shows that EPS immobilized on modified coconut husk gave a fairly higher value as compared to the set-up of modified coconut husk alone. This result indicates the presence of more anionic groups in the former that sequester lead ions in solution. The difference of the affinity constants, b and n, on the other hand, indicates a relative difference for Pb^{2+} ions. (See Table 1).

The results showed that biosorption of lead ions by MCH + EPS followed the Langmuir model for a monolayer adsorption and constant energy of adsorption under low Pb^{2+} ion concentration.

Table 1. Langmuir and Freundlich Constants for 2.5 g of material.

	Husk + EPS	Husk alone
Langmuir constants		
Q_o	0.001	0.0009
b	−3436.07	−4198
r	1	1
Freundlich constants		
K	0.7938	0.4914
n	1.7289	−2.2849
r	0.9182	−0.9182

3.4 *Desorption of the biosorbed lead*

The acid desorption process used 6 M HCl and 6 M citric acid. HCl (hydrochloric acid) proved to be a better desorbing agent, removing 71.15% lead ions from a 20 g column material of MCH + EPS, as compared to 14.18% lead ions removed using citric acid as desorbing agent.

3.5 *On recycling of biotraps*

The recycled biotrap (MCH + EPS column material) removed a maximum of 99.5% lead while the unused biotrap (MCH + EPS column material) removed a maximum of 99.6% lead. The difference in the % removal is insignificant; the decrease maybe attributed to the incomplete removal of lead ions from the biotrap during desorption process. Nevertheless, recycled biotraps can still be effective.

REFERENCES

Department of environment and Natural Resources. Environmental Management Bureau. 1990. DENR Administrative Order No. 35. Diliman, Quezon City, Philippines.

Kunin, R. 1960. Elements of Ion Exchange. Rohm & Haas, Co., Philadelphia.

Mattuschka, B. & G. Straube. 1993. Biosorption of Metals by a Waste Biomass. J of Chemical Technology. 58: 57–83.

McGraw Hill Encyclopedia of Chemistry. 1992. 585–587.

Philippine Coconut Authority. The Philippine Coconut Industry. Quezon City: Domestic Trade Division

Randal, J.M. 1977. Variation in effectiveness of Barks as Scavengers for Heavy Metal Ions. Forest Product Journal. 27(11): 51–55.

Sarkanen, K.V. & C.H. Ludwig. 1971. Lignin Occurrence, Structure and Reaction. Weighly Interscience, N.Y.

Volesky, B. et al. 1993. Removal of Lead from Aqueous Solutions by Penicillium Biomass. Biotechnology & Bioengineering. 42: 785–787.

Volesky, B. & I. Prasetyo. 1993. Cadmium removal in a Biosorption Column. Biotechnology & Bioengineering. 42: 1010–1015.

Volesky, B. & Z.R. Holen. 1993. Biosorption of Lead and Nickel by Biomass of Marine Algae. Biotechnology & Bioengineering. 43: 1001–1009.

Wood, J.M. & H.K. Wang. 1983. Microbial resistance to heavy metals. Environmental Science Technology. 17: 582–590A.

European Symposium on Environmental Biotechnology, ESEB 2004 - Verstraete (ed)
 ISBN 90 5809 653 X

Influence of hydraulic retention time and sulfide toxicity on sulfidogenic FBR

A.H. Kaksonen & J.A. Puhakka
Institute of Environmental Engineering and Biotechnology, Tampere University of Technology, Tampere, Finland

P.D. Franzmann
CSIRO, Land and Water, Underwood Avenue, Floreat, Australia

ABSTRACT: The influence of hydraulic retention time (HRT) and sulfide toxicity on ethanol and acetate oxidation were studied in a sulfate-reducing fluidized-bed reactor (FBR) treating acidic metal-containing wastewater. Incomplete acetate oxidation restricted the operation of the FBR at HRTs below 6.5 h, since the alkalinity produced by acetate oxidation is necessary for wastewater neutralization. At a HRT of 6.5 h, Zn and Fe precipitation rates were over 600 mg $l^{-1}d^{-1}$ and 300 mg $l^{-1}d^{-1}$, respectively, and the alkalinity produced by substrate utilization increased the wastewater pH from 3 to 7.9–8.0. Both uncompetitive and noncompetitive inhibition models described well the sulfide inhibition of the sulfate-reducing culture. Dissolved sulfide inhibition constants (K_i) for ethanol and acetate oxidation were 225–248 mg S l^{-1} and 338–356 mg S l^{-1}, respectively, and the corresponding K_i values for H_2S were 76–84 mg S l^{-1} and 118–124 mg S l^{-1}.

1 INTRODUCTION

Sulfate-reducing bioreactors for treating wastewaters from mining and mineral processing are becoming an alternative to conventional chemical treatment. The bioprocess is based on biological hydrogen sulfide production (Equation 1) by sulfate-reducing bacteria (SRB), followed by metal sulfide precipitation (Equation 2) and neutralization of the water by the alkalinity produced by the microbial oxidation of the electron donor (Equation 3) (Christensen et al., 1996; Dvorak et al., 1992):

$$2\,CH_2O + SO_4^{2-} \rightarrow H_2S + 2\,HCO_3^- \quad (1)$$

where CH_2O = electron donor

$$H_2S + M^{2+} \rightarrow MS(s) + 2\,H^+ \quad (2)$$

where M^{2+} = metal, such as Zn^{2+}

$$HCO_3^- + H^+ \rightarrow CO_2\,(g) + H_2O \quad (3)$$

In a previous study, we showed the potential of an ethanol-fed sulfate-reducing fluidized-bed reactor (FBR) to precipitate Zn and Fe from acidic sulfate-containing wastewater (Kaksonen et al., 2003a). The aim of this study was to evaluate the effect of hydraulic retention time (HRT) on substrate utilization in the FBR and the concomitant wastewater treatment performance. Further, batch FBR kinetic experiments were conducted to determine the effects of dissolved sulfide (DS) and H_2S on the sulfidogenic oxidation of ethanol (Equation 4) and its major degradation intermediate, acetate (Equation 5) (Oude Elferink et al., 1994):

$$2CH_3CH_2OH + SO_4^{2-} \rightarrow 2CH_3COO^- + HS^- + H^+ + 2H_2O \quad (4)$$

$$CH_3COO^- + SO_4^{2-} \rightarrow 2HCO_3^- + HS^- \quad (5)$$

2 MATERIALS AND METHODS

2.1 *Bioreactor*

A laboratory-scale FBR (500 ml) was used at 35°C with silicate mineral (∅ 0.5–1 mm, Filtralite, Norway) as biomass carrier material. The carrier was fluidized by FBR recycle flow maintaining a 20% bed expansion and an empty bed volume of 350 ml. The empty bed volume was used in calculations of HRT and loading rates. High recycle flow rate (700 ml min^{-1}; upflow velocity 29 m h^{-1}) resulted in recycle ratios of 730–2900, which ensured completely mixed conditions in the FBR. The original inoculum of the FBR was from methanogenic granular sludge and sediments from Outokumpu's Pyhäsalmi mine, Finland, and was enriched and maintained for over two years in a sulfate reducing FBR treating synthetic acidic wastewater supplemented with ethanol as described by Kaksonen et al. (2003a and 2003b). The enrichment consists of a diverse sulfate-reducing community (Kaksonen et al., in press).

2.2 *Continuous flow experiments*

In the continuous flow experiments, synthetic wastewater (Table 1) containing sulfate, Zn and Fe was fed to the FBR. The wastewater was supplemented with ethanol maintaining the ethanol/sulfate ratio at stoichiometrical concentrations (Equations 4 and 5) to allow complete ethanol oxidation through sulfate reduction. The HRT was gradually decreased (from 20.7 to 6.1 h) until process failures occurred. Every 1 to 5 days, the feed solution was sampled, and 1 to 5 days integrated samples were collected from the FBR effluent into containers placed in a refrigerator. Sulfate, ethanol and acetate, soluble Zn and Fe, pH and alkalinity were analyzed in each sample. Samples for DS were taken directly from a sampling port in the recycle line.

2.3 *Batch kinetic experiments*

Batch kinetic experiments were conducted in the FBR to study the effects of DS and H_2S on ethanol and acetate oxidation. Before and between individual batch experiments, the FBR was continually fed (HRT 24 h) with ethanol supplemented synthetic wastewater (Table 1) to maintain a constant biomass content in the FBR and to wash out the excess sulfide. The pH of the feed was 4 and the resulting pH in the FBR and effluent was 6.9–7.3. Sulfate was added in excess, i.e. three times as much as stoichiometrically required for complete oxidation of the electron donor.

For each batch experiment, the feed flow was discontinued and the FBR was operated in recycle mode. A portion (50 ml) of the FBR liquid was replaced through the sampling valve placed in the recycling line with fresh anaerobic solution (pH 7) containing Na_2SO_4, nutrients, ethanol or acetate, and NaS_2. During the batch experiments, samples of the FBR liquid were collected from the recycling line at regular intervals and analysed for ethanol and acetate. In addition, pH, sulfate, and DS were determined in the first and last sample of each batch experiment. The amount of biomass in the FBR carrier material (mg dw g^{-1}) and FBR liquid (mg dw l^{-1}) were estimated as volatile solids (VS).

Table 1. Composition (mg l^{-1} except for pH) of the fluidized-bed reactor feeds.

	Continuous flow experiments	Batch kinetic experiments*
Ethanol	690	160
KH_2PO_4	56	28
NH_4Cl	110	55
Ascorbic acid	11	5.5
Thioglycolic acid	11	5.5
Na_2SO_4	1400	1820
$MgSO_4 \cdot 7H_2O$	2300	580
$FeSO_4 \cdot 7H_2O$	500	140
$ZnCl_2$	420	15
pH	3.0–4.1	4

* Feed to the reactor between individual batch experiments was continuous.

2.4 *Analytical methods*

For sulfate, ethanol, acetate, soluble metal and DS analyses, the samples were filtered through 0.45 μm polyethersulfone membrane syringe filters. Sulfate was determined by ion chromatography (Dionex DX-120, USA). DS was analyzed spectrometrically (Shimadzu UV-1601, Japan for HRT experiments or Helios Epsilon, 9423 UVE 100E, USA for sulfide inhibition experiments) by the colorimetric method described by Cord-Ruwisch (1985). Soluble Zn and Fe concentrations were measured with an atomic absorption spectrophotometer (Perkin Elmer 1100B, USA) according to the Finnish standards SFS 3044 (SFS, 1980a) and SFS 3047 (SFS, 1980b). Ethanol and acetate were determined by gas chromatography using a flame ionisation detector (Hewlett Packard 5890A or 5890 Series II, USA). pH was determined in unfiltered liquid samples using a WTW SenTix41 pH electrode (Germany) in the HRT experiments, and using an Orion Research pH/millivolt meter 811 (USA) or HANNA instruments pH meter Piccolo 2 (Italy) in the sulfide inhibition experiments. Total alkalinity, total suspended solids (TSS), volatile suspended solids (VSS), and volatile solids (VS) were determined as previously described (Kaksonen et al., 2003b).

2.5 *Kinetic calculations*

The ethanol and acetate oxidation rates were standardized to the total amount of biomass in the FBR. Uncompetetive (Equation 6) and noncompetetive (Equation 7) inhibition models were used to obtain the inhibition constants (K_i) (Maillacheruvu & Parkin, 1996) for DS and H_2S.

$$v = \frac{V_{\max} \cdot S}{K_m + S \cdot \left(1 + \frac{I}{K_i}\right)} \quad (6)$$

$$v = \frac{V_{\max} \cdot S}{(K_m + S) \cdot \left(1 + \frac{I}{K_i}\right)} \quad (7)$$

where v = oxidation velocity (mg gVS^{-1} min^{-1}), V_{max} = maximum oxidation velocity (mg gVS^{-1} min^{-1}), S = initial substrate concentration (mg l^{-1}), K_m = Michaelis-Menten constant (mg l^{-1}), I = inhibitor concentration (mg l^{-1}), and K_i = inhibition constant (mg l^{-1}). The models were fitted to the data using the least squares method. K_m values were determined in our previous study (Kaksonen et al., 2003a). The concentration of the undissociated sulfide [H_2S] in the batch kinetic experiments was calculated from the total dissolved sulfide concentration [DS] using the first dissociation constant of H_2S (K_{a1}) and equation (8):

$$[H_2S] = \frac{[DS]}{1 + 10^{pH - pKa1}} \quad (8)$$

where pK_{a1} is $-\log K_{a1}$. The concentration of S^{-2} was insignificant, since the second pK_{a2} value of H_2S is 12.9 (Dean, 1999), and reactor pH was 6.9–7.3 during the batch experiments. The first dissociation constant for H_2S at 35°C was calculated using van't Hoff equation (Zumdahl, 1998) and the enthalpies and the dissociation constant for H_2S at 25°C obtained from Dean (1999).

3 RESULTS

3.1 *The effects of HRT on FBR operation*

The effect of decreasing the HRT on the ethanol and acetate oxidation in fluidized-bed treatment of acidic Zn and Fe containing wastewater was studied in continuous flow experiments in a sulfate-reducing FBR. The data obtained, were plotted against HRT (Figs 1–3) to visualize the effect of HRT on the FBR efficiency. As the HRT was reduced from 20.7 to 6.5 h, the sulfate reduction rate increased to 4.3 g l^{-1} d^{-1}, ethanol oxidation rate to 2.6 g l^{-1} d^{-1}, and acetate oxidation rate to 2.2 g l^{-1} d^{-1} (Fig. 1). Maximum Zn and Fe precipitation rates increased to over 600 m l^{-1} d^{-1} and 300 mg l^{-1} d^{-1}, respectively (Fig. 1).

At the lowest HRT (6.5 h) that allowed stable FBR performance, percent ethanol oxidation, and metal

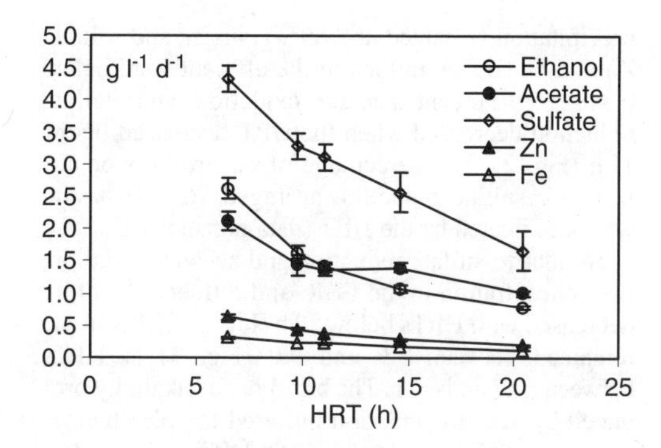

Figure 1. The effects of hydraulic retention (HRT) time on quantitative oxidation of ethanol and acetate, reduction of sulfate, and precipitation of Zn and Fe in the fluidized-bed reactor during the continuous flow experiment.

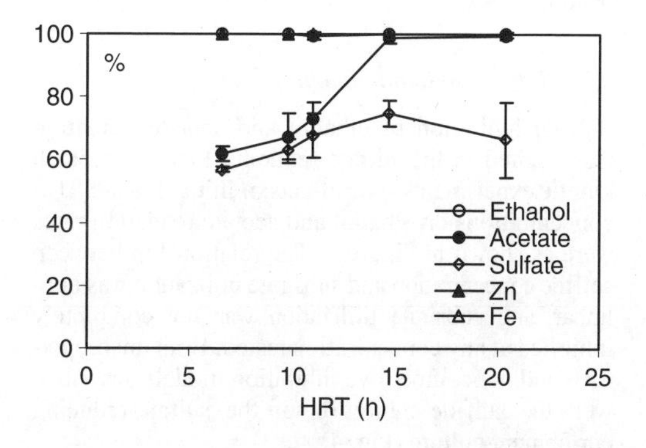

Figure 2. The effects of hydraulic retention time (HRT) on percent oxidation of ethanol and acetate, reduction of sulfate, and precipitation of Zn and Fe in the fluidized-bed reactor during the continuous flow experiment.

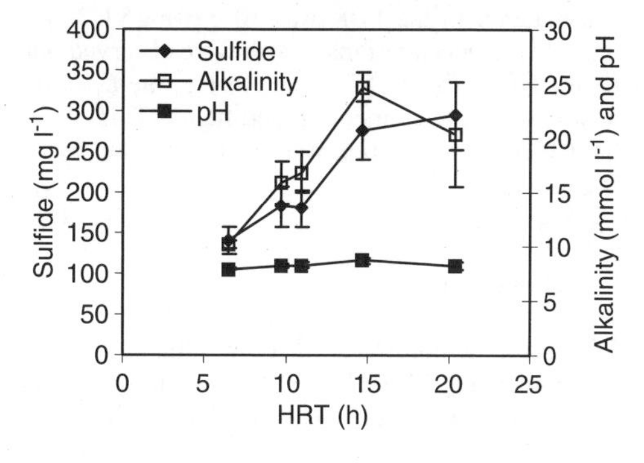

Figure 3. The effects of hydraulic retention time (HRT) on effluent dissolved sulfide, alkalinity and pH during the continuous flow fluidized-bed reactor experiment.

precipitation remained at 99.9% (Fig. 2), and soluble Zn and Fe concentrations in the effluent were below $0.8\,mg\,l^{-1}$. Percent acetate oxidation and sulfate reduction decreased when the HRT decreased below 12 h (Fig. 2). The percentage of electron donor utilized for sulfate reduction averaged $76 \pm 10\%$ and was not affected by the HRT (data not shown).

Similar to sulfate reduction and acetate oxidation, DS concentration in the FBR, and effluent alkalinity decreased with HRTs below 12 h (Fig. 3). Effluent pH remained between 7.5 and 9.0 (Fig. 3) at HRTs between 20.7 and 6.5 h. The bicarbonate alkalinity produced by acetate oxidation buffered the pH changes until process failure occurred at a HRT of 6.1 h (data not shown). Effluent TSS and VSS concentrations fluctuated between 7–2300 and $0–745\,mg\,l^{-1}$, respectively. The amount of TSS and VSS in the effluent was not significantly affected by the HRT (data not shown). Thus, decreased HRT did not result in biomass washout from the FBR.

3.2 *Sulfide inhibition kinetics*

Sulfide inhibition of ethanol and acetate oxidation was studied in the sulfate-reducing FBR with batch kinetic experiments. The effects of initial DS and H_2S concentrations on ethanol and acetate oxidation rates were as shown in Figure 4. The relationship between sulfide concentration and substrate utilization was nonlinear, and substrate utilization was not completely inhibited at any concentration tested. Both uncompetitive and noncompetitive inhibition models described well the sulfide inhibition of the sulfate-reducing enrichment culture (Fig. 4).

DS inhibition constants (K_i) for ethanol and acetate oxidation were $225–248\,mg\,S\,l^{-1}$ and $338–356\,mg\,S\,l^{-1}$, respectively, and the corresponding K_i values for H_2S were $76–84\,mg\,S\,l^{-1}$ and $118–124\,mg\,S\,l^{-1}$, for ethanol and acetate oxidation, respectively. During the continuous flow experiments with varying HRT, the DS concentration in the FBR was $50–370\,mg\,S\,l^{-1}$. The highest DS concentrations, which were observed during the HRTs above 15 h, were close to the K_i values obtained in batch kinetic experiments. This shows that the ethanol oxidation was partly inhibited by the sulfide.

4 CONCLUSIONS

This work demonstrates that an ethanol-utilizing sulfate-reducing FBR efficiently precipitates Zn and Fe from wastewater (pH 3) at a HRT of 6.5 h with an ethanol loading of $2.6\,g\,d^{-1}l^{-1}$. Under these conditions, ethanol oxidation is incomplete and acetate accumulates in the FBR. Incomplete acetate oxidation restricts the operation of the FBR at shorter HRT, since the alkalinity produced by acetate oxidation is necessary for wastewater neutralization. Uncompetetive and noncompetitive inhibition models describe the sulfide product inhibition of the sulfate-reducing FBR enrichment. Ethanol oxidation is more affected by sulfide toxicity than is acetate oxidation.

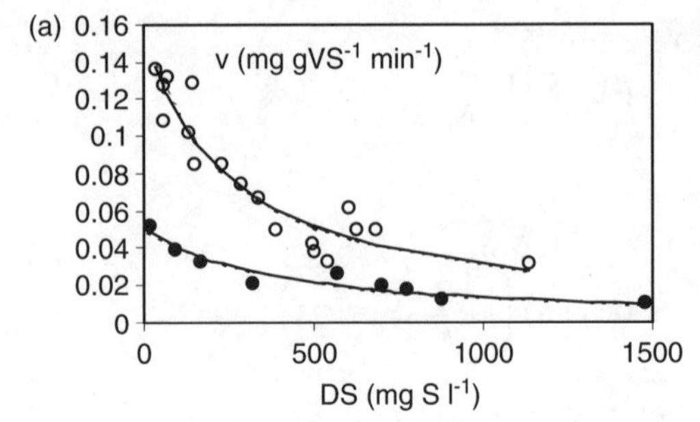

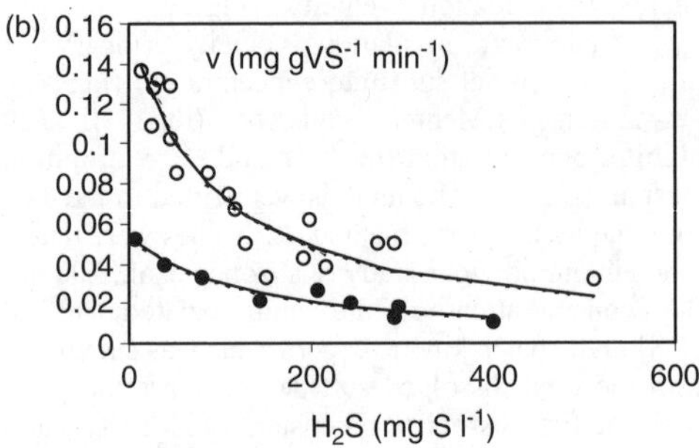

Figure 4. Effects of initial (a) dissolved sulfide (DS) and (b) hydrogen sulfide (H_2S) concentration on ethanol (unfilled symbols) and acetate (filled symbols) oxidation rates (v) in the batch kinetic experiments. Solid and dashed lines represent uncompetetive and noncompetetive inhibition model fits, respectively.

ACKNOWLEDGEMENTS

This work was supported by the National Technology Agency of Finland, Outokumpu Oyj, Finland, Finnish Graduate School in Environmental Science and Technology and Academy of Finland. We thank Ms. Sook Leng Thong for technical assistance.

REFERENCES

Christensen, B., Laake, M. & Lien, T. 1996. Treatment of acid mine water by sulfate-reducing bacteria: results from a bench scale experiment. Water Research 30: 1617–1624.

Cord-Ruwisch, R. 1985. A quick method for the determination of dissolved and precipitated sulfides in cultures of sulfate-reducing bacteria. Journal of Microbiological Methods 4: 33–36.

Dean, J.A. 1999. Langes's handbook of chemistry. 15th edition. McGraw-Hill, Inc. USA

Dvorak, D.H., Hedin, R.S., Edenborn, H.M. & McIntire, P.E. 1992. Treatment of metal-contaminated water using bacterial sulfate-reduction: results from pilot-scale reactors. Biotechnology and Bioengineering 40: 609–616.

Kaksonen, A.H., Franzmann, P.D. & Puhakka, J.A. 2003a. Performance and ethanol utilization kinetics of a sulfate-reducing fluidized-bed reactor treating acidic metal-containing wastewater. Biodegradation 14: 207–217.

Kaksonen, A.H., Riekkola-Vanhanen, M.-L. & Puhakka, J.A. 2003b. Optimization of metal sulfide precipitation in fluidized-bed treatment of acidic wastewater. Water Research 37: 255–266

Kaksonen, A.H., Plumb, J.J., Franzmann, P.D. & Puhakka, J.A. Simple organic electron donors support diverse sulfate-reducing communities in fluidized-bed reactors treating acidic metal-containing wastewater. FEMS Microbiology Ecology. In press.

Maillacheruvu, K.Y. & Parkin, G.F. 1996. Kinetics of growth, substrate utilization and sulfide toxicity for propionate, acetate, and hydrogen utilizers in anaerobic systems. Water Environment Research 68: 1099–1106.

Oude Elferink, S.J.W.H., Visser, A., Hulshoff Pol, L.W. & Stams, A.J.M. 1994. Sulfate reduction in methanogenic bioreactors. FEMS Microbiology Reviews 15: 119–136.

SFS. 1980a. SFS 3044: Metal content of water, sludge and sediment determined by atomic adsorption spectroscopy, atomisation in flame. General principles and guidelines. Finnish Standards Association, SFS. 8 p.

SFS. 1980b. SFS 3047: Metal content of water, sludge and sediment determined by atomic adsorption spectroscopy, atomisation in flame. Special guidelines for lead, iron, cadmium, cobalt, copper, nickel and zinc. Finnish Standards Association, SFS. 6 p.

Zumdahl, S.S. 1998. Chemical principles. 3rd edition. Boston, New York: Houghton Mifflin Company.

European Symposium on Environmental Biotechnology, ESEB 2004 - Verstraete (ed)
© 2004 Taylor & Francis Group, London, ISBN 90 5809 653 X

Biotechnology to concentrate heavy metals from polluted waters

M. Petre
Faculty of Ecology, Ecological University of Bucharest, Romania

F. Cutas & S. Litescu
National Institute for Biological Sciences Bucharest, Romania

ABSTRACT: The present work reports on the screening and assessment of some bacterial and fungal species to be used as heavy metals binders in wastewaters treatments. The laboratory experiments were performed to establish the most efficient biotechnology to concentrate the heavy metals from polluted waters by using bacterial and fungal biomass as efficient biosorbents. Among the microbial species screened for Cu (II) and Pb uptake, *Bacillus subtilis* exhibited the highest binding potential from bacteria, as well as *Pleurotus ostreatus,* from fungal species. The experiments concerning the application of biotechnology based on the metal binding potential of these microorganisms have been made by immobilization in poly-acrylamide (PAA) and collagen-poly-acrylamide (CPAA) and using a continuous flow column bioreactors. The results proved that the immobilized microorganisms can accumulate heavy metals at low concentrations ranging mainly from 0.3 to 0.9 mg/ml.

1 INTRODUCTION

Growth in human populations has generally been matched by a contaminant formation of a wide range of waste products, many of which cause serious environmental pollution effects if they are allowed to accumulate in the ecosystems. In this respect, environmental biotechnology could be the efficient way to solve this problem by the application of microorganisms and processes in waste treatment and management (Verstraete & Top 1992; Zarnea 1994).

Taking into consideration that microorganisms can actively take up the heavy metals by various ways from dilute solutions they should be used as metal bio-accumulators in wastewater treatments in order to extract these toxic metals from industrial effluents and reduce subsequent environmental poisoning (Smith 1996).

Considering the immobilized microorganisms as biocatalysts of any biotechnological process, which involves their metabolic activity it should be necessary to emphasize the importance of their application in wastewater treatments to concentrate heavy metals, like Cu, Cd, Hg, Pb, Zn (Levinson et al. 1994).

2 MATERIALS AND METHODS

2.1 *Materials*

The microorganisms used in these experiments have been screened from several species which occur in natural microbiota of the environment to be used as heavy metals binders in wastewaters treatments.

Bacterial species from *Bacillus* genus are aerobic or facultative anaerobic microorganisms with wide diversity of physiological ability concerning heat, pH and salinity. *Bacillus subtilis* has rod-shaped straight cells arranged in pairs or chains with rounded or squared ends.

From fungi, *Pleurotus ostreatus* species is characterized by hyphae with a dolipore septum and parenthesomes with basidiospores forming directly a mycelium and clavate basidium. This kind of species has lamellate hymenophore and monomictic or dimictic hyphal system.

A continuous flow bioreactor has been used to achieve the experiments concerning the screening of microorganisms which have shown the best potential to concentrate heavy metals from dilute solutions.

The laboratory experiments regarding the cellular binding of Cu, Zn and Pb ions were achieved by using heavy metal solutions, as it is shown in Table 1.

2.2 *Methods*

For the time being, most of the available immobilization methods has already been applied especially to filamentous fungi and, in fact, several biotechnological processes have been developed using the metabolic potential of these microorganisms to accumulate heavy metals from polluted effluents (Carlile & Watkinson

Table 1. Concentrations of the stock and prepared solutions of heavy metal compounds.

Heavy metal salts	Stock solutions (mg/ml)	Prepared solutions (μg/ml)
$CuSO_4 \times 5\ H_2O$	0.8	9
$ZnSO_4 \times 7\ H_2O$	0.6	8
$PbSO_4$	0.7	7
$Pb(NO_3)_2$	0.5	7

1996). To achieve an optimal immobilization of microorganism cells, the inner structure of the immobilization matrix is essential (Wainwright 1992). In most of cases the spherical shape of the matrix has the advantage of combining the fast production with the control of the final pellet diameter (Chen & Humphrey 1988). The preparative procedure of immobilization matrices, used in these experiments, was based on the radiopolymerization of acrylamide and collagen-acrylamide monomers in different aqueous solutions (10%, 15% and 20%).

The radiopolymerization of these acrylamide monomers in aqueous solutions was achieved by using a ^{60}Co radioactive source of 5–7 KGy debit. By using this procedure, there were made poly-acrylamide (PAA) and collagen-poly-acrylamide (CPAA) hydrogels, which were steamsterilized and hydrated in nutritive culture solutions in order to be available for microbial growth (Petre et al. 2001).

The immobilization of bacterial species has been made by adsorption on the surface of hydrogels. Although this process is essentially mild and allows a good retention of cell viability, however, desorption can occur rapidly under certain circumstances (Busscher et al. 1995). The adsorption was mainly due to electrostatic interactions between microorganism cell membrane and hydrogel surface. This kind of matrices as well as the immobilization method by adsorption minimize the loss of cellular activity in order to preserve the intact viable microorganisms and achieve the highest cell density per unit volume (Thomas & White 1992; Urrutia & Beveridge 1995).

In the case of fungal cells, the immobilization has been achieved by natural adherence and surface biofilm growth. In this way, the fungi adhere firmly and resist to removal by rising, being protected by irregularities of the support and their strong attachment on the substratum (Costerton et al. 1995; Rijnaarts et al. 1996). At the same time the fungal cells secrete extracellular polysaccharides with adhesive properties and are involved in cell adherence on different substrata (Rosevear 1990; Warren et al. 1992).

The immobilization of both bacterial and fungal cells was made following some important steps, such as: the aseptic inoculation of PAA and CPAA hydrogels which need to be previously steamsterilized and hydrated by nutritive culture media, the maintaining of these microbial cells immobilized on hydrogels in an incubation chamber for 24–72 h and the aseptic placement of these immobilized microbial cells into the culture vessel of the bioreactor (Petre et al. 2001; Petre et al. 2002).

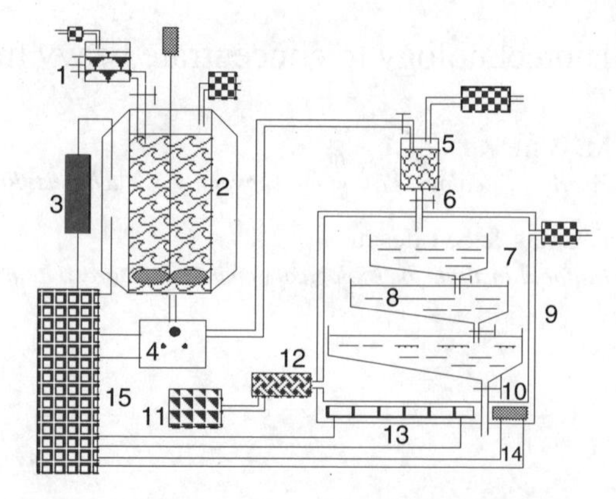

Figure 1. Schematic figure of continuous flow bioreactor 1 – collector of dilute heavy metal solutions; 2 – steam sterilization reservoir; 3 – water heater; 4 – peristaltic pump; 5 – inoculum reservoir; 6 – inoculation pipe; 7 – culture chamber; 8 – deposable multi-flatted beds; 9 – incubation room; 10 – effluent drain; 11 – air pump; 12 – Millipore air filter; 13 – electric heater; 14 – refrigerant; 15 – automation panel.

After immobilization in radiopolymerized PAA and CPAA gels, the microbial cells were placed into the inoculum reservoir of bioreactor, as it is shown in Figure 1.

3 RESULTS AND DISCUSSION

3.1 *Cultivation of immobilized bacterial and fungal cells in continuous bioreactor to concentrate heavy metals*

Any biotechnological process requires a suitable environment for growth of pure cultures of certain microorganisms (Vournakis & Runstadler 1989; Alexander 1996). The design of such bioreactors must incorporate a device for maintaining the constant temperature, inoculum reservoir, sterile air supply in aerobic processes, incubation room as well as an automation panel for bioprocess monitoring and management. The microorganisms, such as bacteria and fungi can actively take up the metals by various mechanisms and have a potential use in extracting rare metals from dilute solutions. In a similar way, microorganisms are being used to extract toxic metals from industrial effluents (Smith 1996).

In this respect, the experiments concerning the concentration of heavy metals such as Cu and Pb from dilute solutions by using immobilized bacterial and fungal cells have been achieved in a flow bioreactor designed for continuous cultures (see Fig. 1). The results of experiments concerning microbial biomass growth by using such laboratory bioreactors were registered during successive repetitions and the mean values are shown for *Bacillus subtilis* in Table 2, as well as for *Pleurotus ostreatus*, in Table 3.

After the immobilization of microbial cultures on PAA and CPAA hydrogels, during a period of time of 20 h for bacterial cells and 60 h for fungal cells the beginning of heavy metal concentration activity for Cu and Pb ions has been detected by electrochemical analysis, using for this purpose the ASV procedure. During a time period of continuous culture cycles of 250 h for *Bacillus subtilis* cells as well as 370 h for *Pleurotus ostreatus* have been registered the results of microbial biomass growth and the correspondent values of electrochemical investigations.

3.2 *Electrochemical analysis of heavy metal binding potential*

The electro-analytical assessment of heavy metal binding potential of immobilized microbial cells used in these biotechnological experiments were achieved by anodic stripping voltametery (ASV), using for that purpose an electrochemical unit at 303 A, with three electrodes, respectively, the indicator electrode – mercury electrode with suspended droplet, the reference electrode –Ag, $AgCl/Cl^-$ electrode and the auxiliary electrode – electrode with platinum wire. The applied potential domain has been tested from la −1.2 to +0.1 V and the ordinary electrolyte used in analysis was: HNO_3 0,1 M.

There were tested different periods of time for heavy metals concentration (10, 20, 40, 60, 120 sec), and I was established the best duration at 60 sec. The optimal potential for the electrochemical analysis was determined at −1.3 V. There were tested several scanning speeds for the electrochemical analysis (10, 15, 25, 50, 90, 150, 250, 500 mV s^{-1}) and finally, the best speed was established at 100 mV s^{-1}.

The electrochemical analysis have proved the presence of complex forms of copper (II) ions in all tested variants, as well as the migration of peak potentials at higher values of 30 mA being a significant argument for these results, as it is shown in F igure 2 and for lead ions at lower values of 0.140 mA, in Figure 3.

The samples for electrochemical analysis has been calibrated for $Zn^{2+}(ZnSO_4)$, $Pb^{2+}(PbSO_4)$, $Cu(CuSO_4)$, in both oxidation stages, for low concentration domains from 0.2 to 3.33 ppm (Petre et al. in press).

Table 2. The limits of binding and sensitivity thresholds of immobilized *Bacillus subtilis* cells, during the concentration of Cu and Pb from heavy metal solutions.

Bacterial biomass (g% d.w.)	Binding threshold Cu (mg/ml)	Binding threshold Pb (mg/ml)	Sensitivity threshold Cu (mg/l)	Sensitivity threshold Pb (mg/l)
5	0.2	0.1	3	1
7	0.4	0.3	5	2
10	0.7	0.5	7	4
15	0.8	0.6	8	6
20	0.9	0.7	10	7

Table 3. The limits of binding and sensitivity thresholds of immobilized *Pleurotus ostreatus* cells, during the concentration of Cu and Pb from heavy metal solutions.

Funga biomass (g% d.w.)	Binding threshold Cu (mg/ml)	Binding threshold Pb (mg/ml)	Sensitivity threshold Cu (mg/l)	Sensitivity threshold Pb (mg/l)
5	0.3	0.2	5	3
7	0.5	0.3	7	4
10	0.7	0.5	8	6
15	0.9	0.7	10	7
20	1.1	0.9	12	9

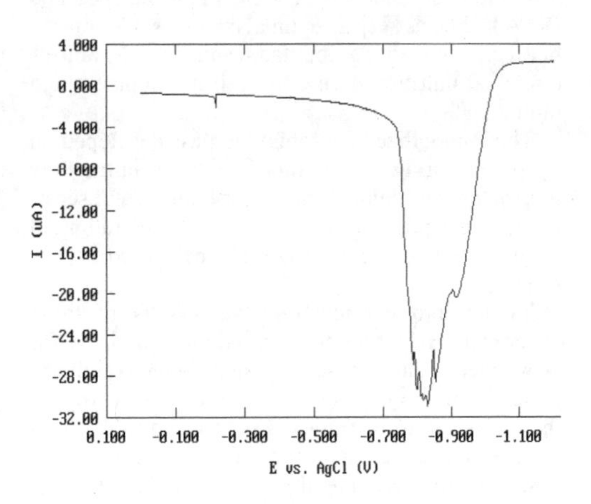

Figure 2. Diagram of electrochemical analysis by ASV, for copper ions (II) bounded in dilute solutions by using immobilized cells of *Bacillus subtilis*.

4 CONCLUSIONS

The procedures used in these experiments to immobilize bacterial and fungal species, such as *Bacillus subtilis* as well as *Pleurotus ostreatus*, involve non-destructive operations to preserve both the integrity and viability of these microbial cells.

The advantage of using such immobilization methods consists in high metabolic activity of microbial

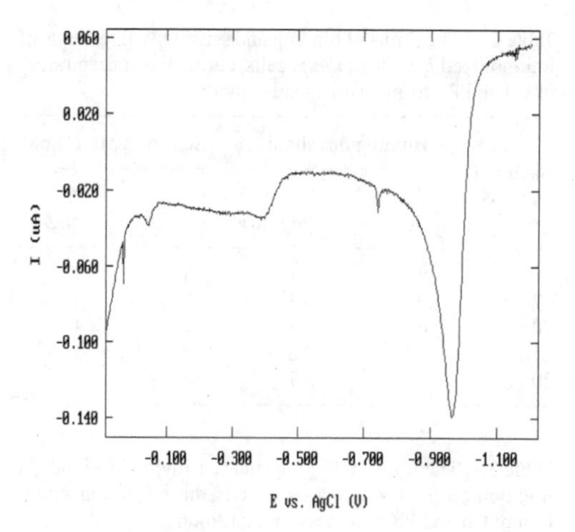

Figure 3. Diagram of electrochemical analysis by ASV, for lead ions bounded in dilute solutions by using immobilized cells of *Pleurotus ostreatus.*

cells, grown in continuous cultures, to concentrate Cu and Pb ions from dilute mineral solutions.

The biotechnology applied to concentrate heavy metal ions by using immobilized bacterial and fungal cells in continuous cultures is significantly influenced by the type of immobilization support. The achieved experiments have shown that the PAA and CPPA hydrogels keep an optimal microcosm for microbial cells concerning cellular respiration as well as microbial nutrition during the cell development and multiplication.

The immobilized microbial biomass developed on hydrogels during the continuous cultures in the flow bioreactor have significant higher binding and sensitivity thresholds for heavy metal concentration in comparison with the free microbial cells used as control samples.

Taking into consideration the results of these experiments regarding the applied biotechnology for heavy metal concentration by using bacterial and fungal cells immobilized on PAA and CPAA hydrogels, the next stage to be achieved is that one to design a pilot-scale installation to create the optimal microenvironment by increasing the chamber culture volume and debit of heavy metal solutions in correlation with the enhancement of microbial biomass development and its binding potential as biosorbent.

REFERENCES

Alexander, M. 1996. Bioremediation technologies. In M. Alexander (ed) Biodegradation and bioremediation: 248–270. London: Academic Press.

Busscher, H.J., Bos, R. & Vandermei, H.C. 1995. Initial microbial adhesion is a determinant for the strength of biofilm adhesion. *FEMS Microbiol. Lett.* 128: 229–234.

Carlile, M.J. & Watkinson, S.C. 1996. Fungi and biotechnology. In M.J. Carlile & S.C. Watkinson (eds) *The Fungi* 387–393. London: Academic Press.

Chen, T.L. & Humphrey, A.E. 1988. Estimation of critical particles diameter for optimal respiration of gel entrapped and palletized microbial cells. *Biotechnol. Lett.* 10: 699–702.

Costerton, J.W., Lewandowski, Z., Caldwell, D.E., Korber, D.R. & Lappin-Scott, H.M. 1995. Microbial biofilms. *Annu. Rev. Microbiol.* 49: 711–745.

Levinson, W.E., Stormo, C.E., Tao, H.L. & Crawford, R.L. 1994. Hazardous waste clean-up and treatment with encapsuled or entrapped microorganisms. In G.R. Chaudhry (ed) *Biological degradation and bioremediation of toxic chemicals* 455–470. London: Chapman and Hall.

Petre, M., Zarnea, G., Adrian, P., Gheorghiu, E & Sularia, M. 2001. Biocontrol of cellulose wastes pollution using immobilized fungi on complex polyhydrogels. In M. Healy, D.L. Wise & M. Moo-Young (eds) *Environmental monitoring and biodiagnostics of hazardous contaminants*: 227–243. Dordrecht/Boston/s London: Kluwer Academic Publishers.

Petre, M., Buleandra, M., Radu, G.L., Gheordunescu, V, 2002. Use of immobilized microbial sorbents to remove bioavailable heavy metals (Cu, Zn, Pb) from polluted waters. *Romanian Journal of Biochemistry* 1: 72–73.

Petre, M., Radu, G.L., Cutas, F., Adrian, P., Gheorghiu, E. 2003. Concentration of heavy metal ions (Cu, Pb, Zn) by using microbial species immobilized on radiopolymerized hydrogels. In: G.L. Radu (ed) *Progress in bioanalysis*. Bucharest: "Ars Docendi" Publishing House (in press).

Rijnaarts, H.H.M., Norde, W., Bouwer, E.J., Lyklema, J. & Zehnder, A.J.B. 1996. Bacterial deposition in porous media: effects of cell-coating, substratum hydrophobicity and electrolyte concentration. *Environm. Sci. Technol.* 30: 2877–2883.

Rosevear, A., 1990, Immobilized biocatalysts. In J.M. Walker & Gingold, E.B. (eds) Molecular Biology & Botechnology: 235–258. Cambridge: Cambridge University Press.

Smith, J.E. 1996. Environmental biotechnology. In J.E. Smith (ed) *Biotechnology*: 140–145. Cambridge: Cambridge University Press.

Thomas, O.R.T. & White, G.F. 1992. Immobilization of the surfactant degrading bacterium *Pseudomonas* C12B in polyacrylamide gel III: biodegradation specificity for raw surfactants and industrial wastes. *Enzyme Microb. Technol.* 13: 338–343.

Urrutia, M.M. & Beveridge, T.J. 1995. Formation of short-range ordered aluminosilicates in the presence of a bacterial surface (*Bacillus subtilis*) and organic ligands. *Geoderma* 65: 149–165.

Verstraete, W. & Top, E. 1992. Holistic environmental biotechnology: 1–18. Cambridge: Cambridge University Press

Vournakis, J.N. & Runstadler, P.W. 1989. Microenvironment: the key to improve cell culture product. *Biotechnology* 7: 143–145.

Wainwright, M. 1992. Fungal cell immobilization. In M. Wainwright (ed) *An introduction to fungal biotechnology* 5–60. Chichester: Wiley.

Warren, T.M., Williams, V. & Fletcher, M. 1992. Influence of solid surface adhesive ability and inoculum size on

bacterial colonization in microcosm studies. *Appl. Environm. Microbiol*. 58: 2954–2959.
Williams, V. & Fletcher, M. 1996. *Pseudomonas fluorescens* adhesion and transport through porous media are affected by lipo-polysaccharide composition. *Appl. Environm. Microbiol*. 62: 100–104.
White, G.F. & Thomas, O.R.T. 1990. Immobilization of the surfactant degrading bacterium *Pseudomonas* C12B in polyacrylamide gel beads I: effect of immobilization on the primary and ultimate biodegradation of SDS, and the redistribution of bacteria within beads during use. *Enzyme Microb. Technol*. 12: 697–705.
Zarnea, G. 1994. Theoretical basis of microbial ecology. In G. Zarnea (ed) *Treatise of the general microbiology* 154–163. Bucharest: Romanian Academy Publishing House.

European Symposium on Environmental Biotechnology, ESEB 2004 - Verstraete (ed)

Acidogenesis in two-phase anaerobic treatment of dairy wastewater

Burak Demirel & Orhan Yenigun
Institute of Environmental Sciences, Bogazici University, Bebek, Istanbul, Turkey

ABSTRACT: This study investigated the effects of variations in hydraulic retention time (HRT) on volatile fatty acid (VFA) production and on the behavior of the microbial ecology in the anaerobic reactor during acidogenesis of a dairy wastewater. A laboratory-scale, continuous flow-completely mixed anaerobic reactor, coupled with a conventional gravity settling tank, was operated at a HRT range between 24 and 12 hours. The degree of acidification and the rate of acid production gradually increased, proportionally to the organic loading rate, with respect to decreases in HRT. The highest degree of acidification and the rate of acid production were achieved at 12 hours. The numbers of total bacterial community and autofluorescent methanogens decreased significantly, as HRT was decreased. Due to continuous recycling and no pH control, the methanogens were not completely washed out from the reactor, however, their activity was suppressed by monitoring of HRT.

1 INTRODUCTION

Two-phase anaerobic digestion process offer substantial advantages, in comparison to conventional anaerobic design applications, such as increased process stability and control, higher specific activity of methane formers, higher organic loading rates and optimization of environmental and operational conditions required for the both reactor systems (Massey & Pohland 1978). Acid phase digestion is a key microbial step during two-phase anaerobic stabilization of wastes, thus, the performance of the acid phase reactor is of paramount importance, since it should provide the most appropriate feed for the subsequent methane phase reactor. Consequently, the optimum environmental and operational conditions for providing satisfactory acid reactor performance should be determined for the each particular waste type in question. Hydraulic retention time (HRT) is an important operational variable during acid-phase digestion, and the effects of HRT on anaerobic acidogenesis have previously been investigated (Massey & Pohland 1978, Dinopoulou et al. 1988, Cha & Noike 1997, Hwang & Hansen 1998, Yilmazer &Yenigun 1999, Fang & Yu 2000). However, contradictory results have been reported. Moreover, the microbial ecology of the acid phase reactors have occasionally been investigated (Cha & Noike 1997, Ince & Ince 2000), particularly covering the start-up (Anderson et al. 1994, Liu et al. 2002). The primary objective of this experimental work was to investigate the effects of variations in HRT on VFA production and distribution and on the behavior of the microbial ecology in the laboratory-scale mesophilic anaerobic reactor during acidogenesis of a dairy wastewater.

2 MATERIALS AND METHODS

2.1 *Experimental set-up*

The experimental set-up consisted of a feeding tank, a feeding pump, an anaerobic reactor, a wet test meter, a gravity settling tank, an effluent tank and a recycling pump. The continuous flow-completely mixed anaerobic reactor had a working volume of 5.2 litres. The substrate was kept in the feeding tank, and feeding was carried out using Masterflex C/L Cole-Parmer and Watson Marlow pumps. Influent feeding port and gas outlet were located on the top of the reactor. Gas outlet was connected to the wet test meter, in order to measure the amount of biogas produced daily. HRT was adjusted volumetrically, through controlling the flow rate of the influent feed. Temperature was maintained within a range of 35 ± 1°C, using an aquarium heater located in the water jacket around the reactor. No pH control was exerted on the system. Mixing of the anaerobic reactor and feeding tank were provided using magnetic stirrers. The settled sludge was continuously recycled to the reactor using a Welp pump.

Table 1. Characteristics of dairy wastewater.

Parameter	Unit	Range
pH		5.8–11.4
Total COD	mg/l	1155–9185
Soluble COD	mg/l	550–5960
TKN	mg/l	14–272
Total phosphorus	mg/l	8–68
Alkalinity	mg $CaCO_3$/l	316–972
Lipid	mg/l	7–62

2.2 *Seed sludge*

The seed sludge was obtained from the upflow anaerobic sludge blanket (UASB) reactor of the wastewater treatment plant of a local alcohol distillery facility. Total solids (TS) and total volatile solids (TVS) concentrations of the seed sludge were determined to be 90225 and 82550 mg/l, respectively, prior to inoculation.

2.3 *Feed*

The feeding solution was prepared daily, using dairy wastewater, micro and macro nutrients and distilled water. Characteristics of dairy wastewater are given in Table 1. Lactose was also added to the feeding solution, in order to increase the influent feed strength whenever required. Alkalinity was provided using sodium hydrogen carbonate ($NaHCO_3$). Nitrogen (N) was seldom added to the feeding solution, in the form of ammonium chloride (NH_4Cl), and phosphorus (P) was not used, since dairy wastewater contained proper amounts of both nutrients. Stock solutions of 100 mg/l $NiSO_4 \cdot 6H_2O$, 10 mg/l $(NH_4)_6Mo_7O_{24} \cdot 4H_2O$ and 100 mg/l $CoCl_2 \cdot 6H_2O$ were also prepared and used in feed, in order to provide molybdenum (Mo), nickel (Ni) and cobalt (Co) as micro nutrients. All of the chemicals were reagent grade, obtained from commercial sources.

2.4 *Analytical methods*

A Hach COD Reactor Model 45600 (digestion at 150°C for 2 hours) and a Hach DR/2010 spectrophotometer were used for COD analyses. Monitoring of pH was carried out using a WTW pH 330 pH-meter with a WTW SenTix probe. Influent and effluent VFA concentrations were determined using a HP 5890 Series II Gas Chromotograph with a flame ionization detector (FID) and a HP-FFAP column. The lipid content of the samples were determined using partition-gravimetric method as outlined in Standard Methods (APHA 1992). TKN, total phosphorus, COD, alkalinity, total solids (TS), total volatile solids (TVS), mixed liquor suspended solids (MLSS) and mixed liquor volatile suspended solids (MLVSS) analyses were also performed according to methods outlined in Standard Methods (APHA 1992). Samples for microbiological analyses were taken from the reactor at each steady-state condition. A method based on Pike et al. (1972) was used for homogenization of samples. Enumeration of the total bacterial community and autofluorescent methanogens was carried out using an Olympus BX-50 Model Epifluorescence Microscope fitted with a 100 W high-pressure mercury lamp. Magnification of 600 was used with Olympus ×60 water immersion lenses having a ×60 eyepiece. Differentiation between the methanogenic and non-methanogenic bacteria was achieved by determining the number of autofluorescent cells when irradiated with UV light. Morphological changes within the fluorescent methanogenic population were determined by subdividing the total population into six morphologically distinct groups: cocci, short rods (0.2–0.5 × 3 μm), medium rods (0.3–0.6 × 6 μm), long rods (0.3–0.6 ×10 μm), filaments and saccina (Morgan et al. 1991).

3 RESULTS AND DISCUSSION

3.1 *Reactor performance*

The reactor was operated at a HRT of 24 hours and in an organic loading rate (OLR) range between 1 and 5 kg COD/m/d during start-up. After the reactor achieved steady-state conditions at 24 hours, HRT was decreased, to 22, 20, 18, 16 and 12 hours, respectively, in order to provide VFA production in the system. The performance of the acid-phase digestion can be evaluated using the expressions of the degree of acidification and the rate of product formation (Dinopoulou et al. 1988), respectively. The degree of acidification can be quantified using the percentage of the initial substrate concentration converted to VFAs. The inital substrate concentration (S_i) was measured in mg COD/l and the quantity of fermentation products (VFAs) were converted to the theoretical equivalent in mg COD/l (S_p). The formula is as follows (Dinopoulou et al. 1988):

$$\text{Acidification (\%)} = (S_p / S_i) \times 100$$

The rate of product formation (r_p) per unit of reactor volume can be expressed as follows (Dinopoulou et al. 1988).

$$r_p = (P - P_o)/(HRT) \times 24$$

where P_o and P are the influent and effluent product concentrations in g/l, respectively, and HRT is the hydraulic retention time in hours. There was no net VFA production (Net VFA production is the difference between influent and effluent VFA samples at

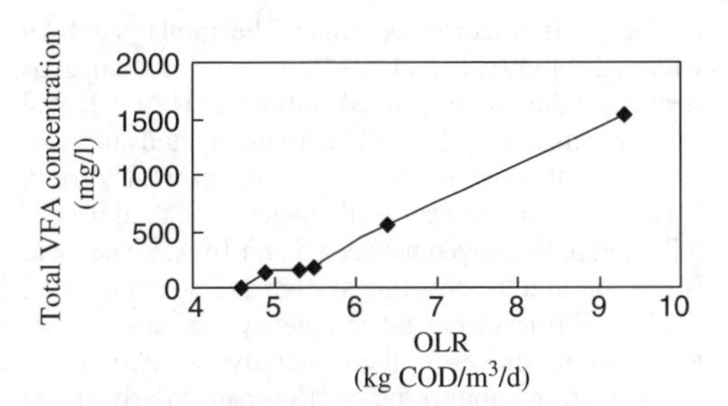

Figure 1. Total VFA concentration versus OLR.

Table 2. HRT, OLR, the degree of acidification and the rate of product formation (r_P).

HRT (hour)	OLR (kg COD/m/d)	Degree of acidification (%)	r_P (g/l/d)
24	4.6	0	0
22	4.9	6	0.1
20	5.3	6	0.2
18	5.5	7	0.2
16	6.4	23	0.8
12	9.3	56	3.1

steady-state conditions) in the system at 24 hours of HRT, thus, both the degree of acidification and the rate of product formation were zero. As HRT was decreased, total VFA production increased proportionally to the OLR (Figure 1). Consequently, both the degree of acidification and the rate of product formation increased gradually, with respect to increase in the OLR (Table 2). Acetic, propionic, butyric and valeric acids were commonly produced during mesophilic acidogenesis of dairy wastewater. HRT also seemed to determine the degree of acidification in this work, as reported previously by Dinopoulou et al. (1988). Poor degrees of acidification between 24 and 18 hours could be attributed to continuous recycling of biomass and high pH, since pH was not controlled.

3.2 *Bacterial enumeration studies*

The ratio between the number of autofluorescent methanogens and the number of total bacterial community was 18% in the seed sludge, prior to inoculation. This ratio was reported to be between 0.01 and 1% by Anderson et al. (1994) for start-up during pre-acidification of dairy wastewater, at a HRT of 12 hours and a pH range of from 5.0 to 5.5. Lower pH, HRT and a different reactor configuration (a CSTR), in comparison to this work, provided this lower ratio. Medium rods and *Methanococcus-like* species were observed to be the dominant methanogens in the seed. The numbers of total bacterial community and

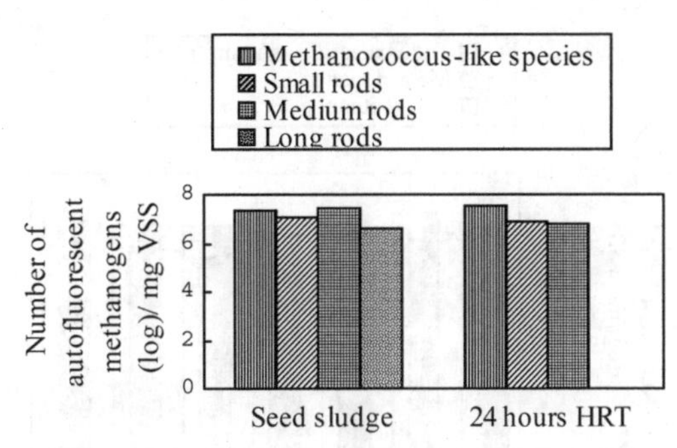

Figure 2. Variations in the numbers of autofluorescent methanogens at the end of start-up.

autofluorescent methanogens both decreased during start-up. This decrease could be attributed to operation at a relatively low HRT. However, the methane formers were active, because no net VFA production could be observed at 24 hours. A high pH of about 7.5 (on average) seemed to provide active methanogens in the system during this particular run. Variations in the numbers of autofluorescent methanogens during start-up is shown in Figure 2. Long rods could not be retained within the system. Changes within the morphology of the autofluorescent methanogens during start-up could have resulted due to the type of substrate used and/or the variations in the OLR. Moreover, the granulated structure of the seed sludge disintegrated soon after the reactor started operation. This might have resulted in high reactor effluent mixed liquor suspended solids (MLSS) and mixed liquor volatile suspended solids (MLVSS) concentrations during the rest of this study (data not shown), because it was previously reported that biogranules settled better, due to their larger sizes than the suspended sludge in the reactor, and, thus, had a less tendency of being washed out (Fang, 2000).

Changes in the numbers of the total bacterial community, non-methanogens and methanogens in the anaerobic reactor for the entire operation are shown in Figure 3. The number of non-methanogenic bacteria was determined by subtracting the number of autofluorescent methanogens from the total bacterial community (Anderson et al. 1994). The numbers of both total bacterial community and autofluorescent methanogens firstly decreased between 24 and 20 hours, in comparison to the seed, due to reactor operation at lower HRTs. The numbers of total bacterial community and autofluorescent methanogens then increased between 18 and 12 hours. This might have resulted from acclimation of the microbial ecology to reactor operating conditions. Cha and Noike (1997) also reported that the bacterial populations in the mesophilic acidogenic reactor decreased in a HRT range between 48 and 6 hours. Since there was a continuous recycling system and no pH control, complete

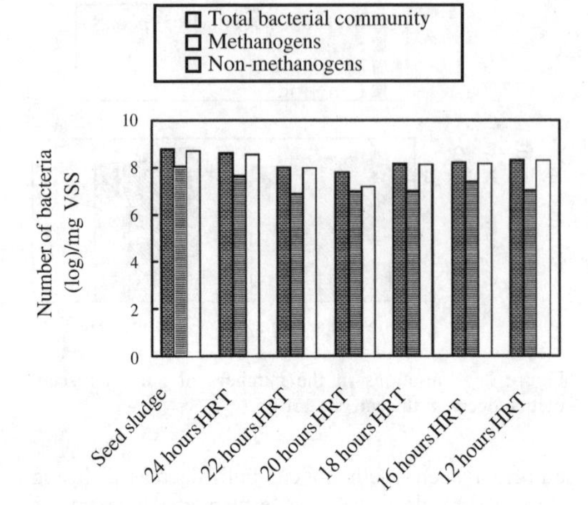

Figure 3. Variations in the numbers of total bacterial community, methanogens and non-methanogens.

wash-out of methanogens did not occur, however, the activity of the methane formers seemed to be suppressed by monitoring of HRT, particularly at low levels. Medium rods and *Methanococcus-like* species were generally observed to be the most abundant species in the system during the entire operation, in spite of variations in OLR, HRT and pH. The ratio between the numbers of autofluorescent methanogens and total bacterial community ranged between 5 and 16%. Ince and Ince (2000) reported this ratio to be between 0.01 and 3%, in a CSTR, at a HRT and a pH of 12 hours and 5.5 to 6.0, respectively, during pre-acidification of a dairy wastewater. This lower ratio resulted from a different reactor configuration and a lower operating pH. The lowest ratio of 5%, which was achieved at 12 hours of HRT, was also accompanied with the maximum rate of acid production.

4 CONCLUSIONS

The degree of acidification and the rate of product formation increased proportionally to the OLR, with respect to decrease in HRT from 24 to 12 hours. The maximum degree of acidification and the rate of product formation were both achieved at 12 hours of HRT. Poor rates of acid production between 24 and 18 hours could be attributed to continuous recycling of biomass and a high pH. Acetic, propionic, butyric and valeric acids were commonly produced during acidogenesis of dairy wastewater. The numbers of total bacterial community and autofluorescent methanogens decreased during start-up. Variations in HRT affected the numbers of total bacterial community and autofluorescent methanogens during the operation. The ratio between autofluorescent methanogens and total bacterial community ranged between 5 and 16%. Since there was a continuous recycling system and no pH control, methane formers were not completely washed-out from the system, however, their activity seemed to be suppressed by monitoring of HRT, particularly at low levels, for the range investigated.

REFERENCES

Anderson, G. K., Kasapgil, B., Ince, O. 1994. Microbiological study of two-stage anaerobic digestion during start-up. *Water Research* 28: 2383–2392.

APHA, AWWA, WPCF. 1992. Standard Methods for the Examination of Water and Wastewater, 18th ed. Washington: American Public Health Association.

Cha, G. C., Noike, T. 1997. Effect of rapid temperature change and HRT on anaerobic acidogenesis. *Water Science and Technology* 36(6–7): 247–253.

Dinopoulou, G., Sterritt, R. M., Lester, J. N. 1988. Anaerobic acidogenesis of a complex wastewater: 1. The influence of operational parameters on reactor performance. *Biotechnology and Bioengineering* 31: 958–968.

Fang, H. H. P. 2000. Microbial distribution in UASB granules and its resulting effects. *Water Science and Technology* 42(12): 201–208.

Fang, H. H. P., Yu, H. Q. 2000. Effect of HRT on mesophilic acidogenesis of dairy wastewater. *Journal of Environmental Engineering* 126: 1145–1148.

Hwang, S., Hansen, C. L. 1998. Characterization of and bio-production of short-chain organic acids from mixed dairy processing wastewater. *Transactions of the American Society of Agricultural Engineers*. 41: 795–802.

Ince, B., Ince, O. 2000. Changes to bacterial community make-up in a two-phase anaerobic digestion system. *Journal of Chemical Technology and Biotechnology* 75: 500–508.

Liu, W. T., Chan, O. C., Fang, H. H. P. 2002. Microbial community dynamics during tart-up of acidogenic reactors. *Water Research* 36: 3203–3210.

Massey, M. L., Pohland, F. G. 1978. Phase separation of anaerobic stabilization by kinetic controls. *Journal of Water Pollution Control Federation* 50(9): 2204–2222.

Pike, E. B., Carrington, E. G., Ashburner, P. A. 1972. An evaluation of procedures for enumeration bacteria in activated sludge. *Journal of Applied Bacteriology* 35: 309–321.

Yilmazer, G., Yenigun, O. 1999. Two-phase anaerobic treatment of cheese whey. *Water Science and Technology* 40: 289–208.

European Symposium on Environmental Biotechnology, ESEB 2004 - Verstraete (ed)
© 2004 Taylor & Francis Group, London, ISBN 90 5809 653 X

Optimisation of biological textile wastewater treatment

H. De Wever, B. Lemmens, S. Van Roy & L. Diels
Vlaamse instelling voor technologisch onderzoek, Mol, Belgium

ABSTRACT: Both a conventional activated sludge (CAS) and a membrane bioreactor (MBR) configuration were optimised for biological degradation of textile wastewater. At laboratory-scale, the MBR effluent quality was superior to that of the conventional system for the parameters COD, nitrogen and colour. This was not only due to an improved biodegradation efficiency, but was also related to COD and dyestuff retention by the ultrafiltration membrane of the MBR and accumulation of the retained compounds in the reactor. Addition of activated carbon did not improve the biological degradation of the recalcitrant compounds. On-site comparative tests at pilot-scale confirmed that effluent quality can be improved by the implementation of a MBR concept, although the differences were small at this scale. With respect to further treatment of the effluent to obtain reuse quality, MBR effluent could be directly fed to a reverse osmosis step, whereas CAS effluent needed a pretreatment by ultrafiltration.

1 INTRODUCTION

A biological treatment is often considered the most sustainable approach to remove a large fraction of Chemical Oxygen Demand (COD) from textile wastewaters. Although many textile companies have a biological treatment plant, the effluent still contains residual COD and is visually polluted. Therefore, attempts were made to optimize the biological treatment. This may not only improve the water quality for discharge, but may also open perspectives for water reuse, after further polishing steps. In this respect, the upgrade of a conventional activated sludge system (CAS) to a membrane bioreactor (MBR) configuration has proven to be successful for many other types of wastewaters (Lozier & Fernandez 2000, Rozzi *et al.* 2000, Andersen *et al.* 2002). MBRs are compact and intensified biological treatment systems which often show a superior biodegradation efficiency. They generate an effluent of a high quality and free of suspended solids, which can be fed to sensitive post-treatment technologies such as reverse osmosis, without the need for a pretreatment step. Consequently, MBRs are preferable to CAS in conditions where discharge norms are stringent or water reuse is aimed for.

In this paper, we investigated the optimisation of biological wastewater treatment of textile wastewaters both in CAS and MBR. Because degradation of azodyes does not occur in aerobic conditions but has been demonstrated in anaerobic-aerobic sequences (Terras *et al.* 1999, Matteoli *et al.* 2002), we tested the latter configuration. Comparative tests were performed both at laboratory- and pilot-scale. Reactor operation and performance was evaluated for the removal of COD, nutrients and dyestuffs.

2 MATERIALS AND METHODS

2.1 *Laboratory-scale experiments*

A MBR and a CAS were operated in parallel on the same textile wastewater from a company producing cotton towels. Both consisted of an anoxic and an aerobic biological compartment. Mixed liquour was recycled from the aerobic to the anoxic compartment. Oxygen concentrations were always above $2\,mg \cdot l^{-1}$ in the aerobic compartments, pH was kept neutral. The MBR was a sidestream one, equipped with a tubular membrane (WFF4385, Stork, Netherlands). Both reactors were inoculated with adapted sludge from a full-scale textile wastewater treatment plant.

As shown in Table 1, two different batches of wastewater were fed to both reactors in the test period. On day 34, activated carbon was dosed to both reactors to evaluate the impact on COD removal efficiencies. In the MBR, the carbon was retained in the reactor by the membrane, in the CAS it was gradually washed out.

Membrane performance was evaluated by pressure and flux measurements.

2.2 *Pilot-scale experiments*

The pilot-scale MBR consisted of three hydraulically connected cylindrical compartments, which were kept respectively in anaerobic, anoxic and aerobic conditions. Mixed liquour recirculation from the aerobic to the anoxic and from the anoxic to the anaerobic compartment was provided. Sludge separation was achieved by continuous recirculation over an external MBR with the same specifications as the one used in the laboratory-scale tests. The MBR was fed with textile wastewater taken from the buffer tank of the full-scale installation.

The full-scale installation was designed according to the Unitank® concept and included a denitrification step.

2.3 *Analytical methods*

Analyses of COD, nutrients and sludge concentrations were performed according to standard methods (APHA 1995). The dye removal efficiency was evaluated by absorbance measurements at 3 selected wavelengths.

Table 1. Experimental conditions during the comparative MBR and CAS tests at laboratory-scale tests.

Time (d)	Feed	Activated carbon	HRT MBR (d)	HRT CAS (d)
0–34	Batch 1	Absent	4.8–6.5	3–6
34–40	Batch 1	Present	6.5	6
40–68	Batch 1	Present	8–10	6
68–94	Batch 1	Present	7–10	6
94–115	None	Present	Infinite	Infinite
115–149	Batch 2	Present	5.5–5.6	5–6

* HRT: hydraulic retention time.

3 RESULTS AND DISCUSSION

One of the most difficult aspects of textile wastewater treatment is the removal of visual pollution. Many dyestuffs are hardly biodegradable but can be partly removed by sorption on the sludge flocs. For azodyes, mineralization has been reported to occur in two steps: reduction of the azobond in strictly anaerobic conditions and a further degradation of the products in subsequent aerobic conditions (Matteoli *et al.* 2002). Our own preliminary experiments with synthetic wastewater containing azodyes indeed showed that biodegradation can be achieved when aerobic treatment is preceded by an anaerobic treatment. Addition of an extra carbon source improved the colour removal.

In a next phase, dye removal was evaluated for real textile wastewater containing both azodyes and anthraquinone dyes. Tests were performed at laboratory- and pilot-scale.

3.1 *Laboratory-scale experiments*

As shown in Figure 1, COD removal was consistently better in the MBR than in the CAS and fluctuations were smaller. This was in part due to the fact that the MBR membrane retains suspended solids. However, because the COD of the filtered CAS effluent samples was still higher than the COD of the MBR, either the biodegradation efficiency was better or high molecular-weight soluble compounds were retained by the ultrafiltration membrane. Apparently, both phenomena occurred, since COD accumulation in the MBR supernatant was observed.

As a result, maximum COD removal efficiencies amounted to 89% in the MBR and 56% in the CAS.

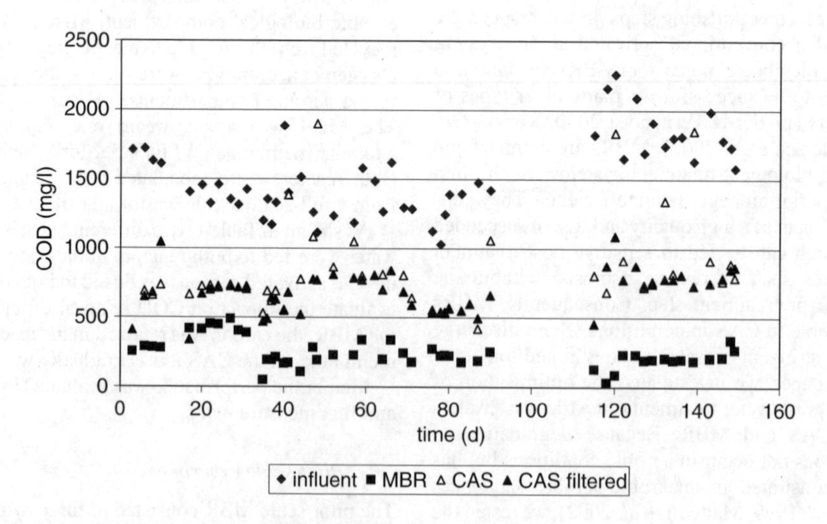

Figure 1. Evolution in COD removal for MBR and CAS at laboratory-scale.

On average, the COD removal efficiency was 30–35% higher in the MBR.

On day 34, powdered activated carbon was added to both MBR and CAS. Figure 1 indicates that this had an immediate effect on the effluent COD of both reactors. In the MBR, the effect was more pronounced because the effluent COD stabilised at around half of the value measured before carbon addition. In the CAS, a temporary increase in effluent COD was observed due to a pH increase to >8 which was not immediately corrected for. It is not clear whether the activated carbon actually improves the degradation of recalcitrant compounds. The fact that soluble COD accumulation occurred in the MBR supernatant seems to indicate that the sorption capacity had been reached and that no substantial biological regeneration of the activated carbon took place.

As shown in Figure 2, dye removal measured as a reduction in absorbance was generally better in the MBR than in the CAS. Removal efficiencies varied between 48 and 81% in the MBR and between 21 and 55% in the CAS depending on the wavelength used for the absorbance measurements. When activated carbon was dosed, dye removal efficiencies improved substantially, although the effect was limited in time. Analysis of the experimental data further indicated that strict anaerobic conditions were required for dye removal. When anoxic conditions prevailed at high nitrate concentrations, a reduction in colour removal was observed. Anthraquinone dyes were not decolorized, but removed to a limited extent by sorption phenomena.

As mentioned before, COD concentrations in the MBR supernatant increased in time. Around day 90, influent supply was stopped for about 2 weeks. During that period, the COD in the reactor did not decrease. This indicates that it consisted of high molecular-weight compounds which were recalcitrant.

3.2 *Pilot-scale experiments*

In a next step, the experimental observations were evaluated at larger scale. A pilot-scale external MBR was compared with a full-scale Unitank system. The MBR was operated at sludge loading rates of between 0.1 and 0.15 kg COD.kg sludge$^{-1} \cdot d^{-1}$ · Sludge concentrations varied between 8 and 17 g.l^{-1}.

COD removal efficiencies were around 80% in the MBR (see Figure 3). They were slightly higher than in the Unitank® system. The difference in effluent COD for MBR and Unitank® could completely be attributed to the retention of soluble high molecular weight COD by the ultrafiltration membranes. Hence, the

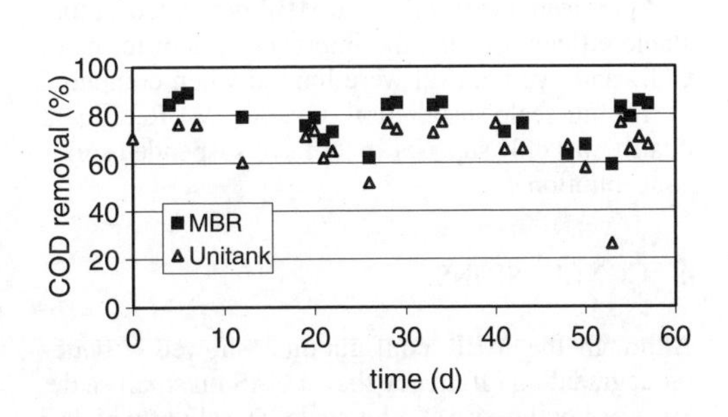

Figure 3. Evolution in COD removal for MBR and Unitank® at pilot- or full-scale respectively.

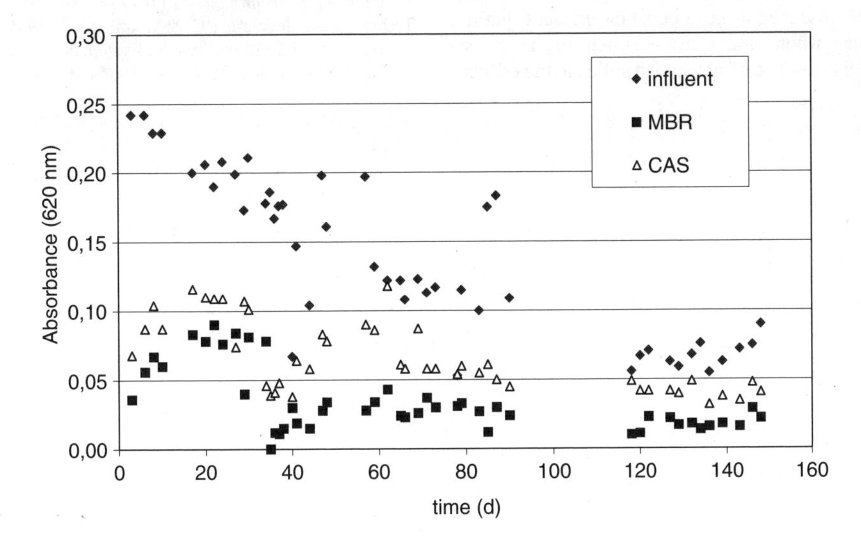

Figure 2. Evolution in dye removal for MBR and CAS at laboratory-scale.

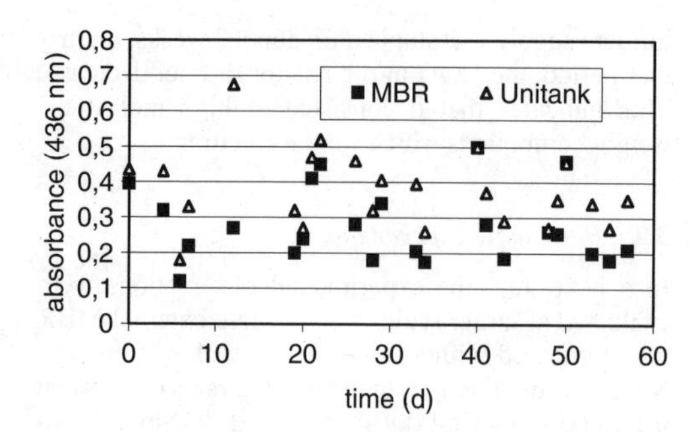

Figure 4. Evolution in dye removal for MBR and Unitank® at pilot- or full-scale respectively.

MBR does not show an improved biodegradation activity or efficiency.

As shown in Figure 4, dye removal was always 20–30% higher in the MBR than in the Unitank®.

Apart from the fact that the MBR generated a more stable effluent quality, the improvements in terms of COD and dye removal were limited when compared to the full-scale installation. Overall, MBR effluent quality was only superior in terms of suspended solids concentration.

4 CONCLUSIONS

Although the MBR configuration showed a better biodegradation efficiency than a CAS on small scale, this was not the case at pilot-scale. The slightly higher COD removal efficiencies observed in the pilot tests were due to physical retention of solutes by the membranes rather than biological elimination. Still, the MBR produced a more constant effluent quality. Particularly when water reuse is aimed for, its robustness and the absence of suspended solids in the effluent may prove to be advantageous for direct feeding to further polishing steps.

ACKNOWLEDGMENTS

This work was performed in the frame of IWT project AUT/980171 'Optimisation of biological purification of textile wastewater in view of water reuse' in collaboration with the company Santens nv and the former Seghers Better Technology for Water.

REFERENCES

Andersen, M., Kristensen, G.H., Brynjolf, M. & Grüttner, H. 2002. Pilot-scale testing membrane bioreactor for wastewater reclamation in industrial laundry. *Water Science and Technology* 46 (4–5): 67–76.

APHA 1995. *Standard methods for the examination of water and wastewater*. 19th Edition. American Public Health Association/American Water Works Association/Water Environment Federation, washington DC, USA.

Lozier, J. & Fernandez, A. 2000. Using a membrane bioreactor/reverse osmosis system for indirect potable reuse. In *Proceedings of the Conference on membranes in Drinking and Industrial Water Production.* Volume 2. Desalination Publications, L'Aquila, Italy.

Matteoli, D., Malpei, F., Bortone, G. & Rozzi, A. 2002. Water minimisation and reuse in the textile industry. In Lens, P., Hulshoff Pol, L., Wilderer, P. & Asano, T. (eds.), *Water recycling and resource recovery in industry*, IWA Publishing.

Rozzi, A., Malpei, F., Bianchi, R. & Matteoli, D. 2000. Pilot-scale membrane bioreactor and reverse osmosis studies for direct reuse of secondary textile effluents. *Water Science and Technology* 41 (10–11): 189–195.

Terras, C., Vandevivere, P.C. & Verstraete, W. 1999. Optimal treatment and rational reuse of water in textile *industry. Water Science and Technology* 39 (5): 81–88.

European Symposium on Environmental Biotechnology, ESEB 2004 - Verstraete (ed)
© 2004 Taylor & Francis Group, London, ISBN 90 5809 653 X

Nitrite production in a granular sequencing batch reactor

B. Arrojo, A. Mosquera-Corral, J.L. Campos, J.M. Garrido & R. Méndez
Department of Chemical Engineering, School of Engineering, University of Santiago de Compostela
Rua Lope Gómez de Marzoa s/n, Santiago de Compostela, Spain

ABSTRACT: The main objective of this research was focussed on the production of an effluent containing nitrite in an aerobic granular sequencing batch reactor (SBR) for heterotrophic carbon removal and combined nitrification-denitrification processes. For this purpose a granular SBR was operated at different carbon to nitrogen (COD/N) ratios in the feeding. The feeding flow was a synthetic medium, which contained acetate as carbon source (0.5 g COD/L) and ammonium as N source (100–200 mg N-NH_4^+/L). Different COD/N ratios of 15, 5 and 2.5 g/g in the feeding were tested.

The COD removal percentage was around 90% during the whole operational period. When, the reactor was operated with a COD/N ratio of 15 g/g no nitrite production was observed in the effluent and nitrate was the only oxidized compound. During the periods fed with COD/N ratios of 5 and 2.5 g/g, nitrite was measured in the effluent at values of 7–20 mg NO_2^--N/L. Changes on the COD/N ratio provoked the presence of different concentrations of nitrogen compounds in the effluent. The N removal percentages obtained in the reactor decreased with the increase of NH_4^+ concentration in the feeding from 80% to 40%.

1 INTRODUCTION

Conventional removal of ammonium requires usually large amounts of energy for aeration during nitrification and organic carbon for denitrification. Nitrification consists of the oxidation of ammonia to nitrite and a further oxidation of nitrite to nitrate in aerobic conditions. After this the nitrate generated is denitrified using an organic carbon source as electron donor to N_2 via nitrite. A process in which ammonium is restricted to the oxidation to nitrite using less energy in aeration and with low or even without a COD demand is a very attractive option for making the whole treatment process more sustainable (van Dongen et al., 2001). This can be achieved in biofilm systems by a partial oxidation of ammonium to nitrite, after then the nitrite produced can be converted into nitrogen gas with saving of oxygen and carbon source, respectively (Garrido et al.,1997).

The need to treat the generated wastewater at the same site where it is produced has led to investigation and development of more compact reactors with very high volumetric conversion capacities and with low surface requirements for the installation. The biofilm airlift suspension (BAS) are examples of this new systems, where the biomass grows as a biofilm on small suspended basalt particles. The processes of COD removal and both nitrification and denitrification can occur simultaneously in this type of reactor (Tijhuis et al., 1994; Van Benthum et al., 1997). These are continuously fed systems. Recently aerobic granular sludge was developed in a sequencing batch airlift reactor (SBAR) (Beun et al., 1999) where the removal of COD as well as nitrification and denitrification processes occurred (Beun et al., 2001).

In general, the population distribution in biofilms is the result of the difference in growth rate. The slow-growing nitrifiers are located inside the biofilms, the fast-growing heterotrophs are located more in the outer layer of the biofilms (Tijhuis et al., 1995; van Loosdrecht et al., 1995). In a discontinuously fed system, the acetate penetrates completely into the biofilms because of the temporarily high concentration in the liquid. O_2 is present only in the outer layer of the biofilms. Consequently, it can be expected that the nitrifiers are located in the outer, aerobic layer of the biofilms, and acetated is stored anoxically as PHB by the heterotrophs inside the biofilms and heterotrophs were present in the center of the granules where they grew with acetate during the feast period, using nitrate as the electron acceptor. They were also present in the outer layers of the granules where they could use oxygen as electron acceptor.

Nitrogen removal from wastewater via biological processes of nitrification–denitrification is dependent on the COD/N ratio of the wastewater treated (Sánchez et al., 2000). The effect of the COD/N ratio on the nitrification and denitrification processes using flocculent sludge has been widely studied elsewhere (Buys et al., 2000).

In the present work the conditions to produce nitrite in an aerobic granular sequential batch reactor using different COD/N ratios in the fed was evaluated. The effects of the COD/N ratio of the fed on the overall nitrogen removal efficiency were tested.

2 MATERIALS AND METHODS

2.1 *Experimental set-up*

A sequencing batch reactor (SBR) with a total volume of 2.5 L and a working volume of 1.5 L was used. Dimensions of the unit were: height of 465 mm and inner diameter of 85 mm, the height to the diameter ratio (H/D) being 5.5. The maximum level of the liquid was 264 mm, and the minimum level of 132 mm after effluent withdrawal. Oxygen was supplied to both reactors by using spargers to promote the formation of small air bubbles. A set of two peristaltic pumps was used to feed and to discharge the effluent, respectively. The influent was introduced in the system through ports located at the top of the reactors. The effluent was discharged through the sampling port placed at middle height of the column reactor (Fig. 1). A programmable logic controller (PLC) Siemens model S7-224CPU controlled the actuations over the pumps and valves, and thus the length of every operational period in the SBR. The reactor was operated at room temperature (15–20°C) and at oxygen concentration between 0 and 8 mg O_2/L and without pH control, which varied between 7.4 and 8.5.

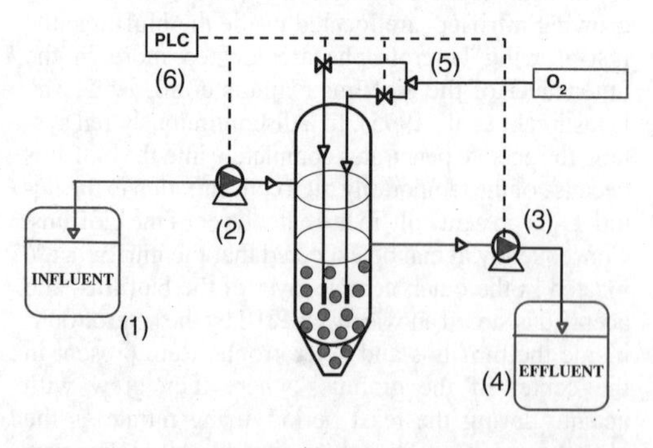

Figure 1. Experimental set up: (1) Feeding tank; (2) Feeding pump; (3) Effluent pump; (4) Effluent tank; (5) Air valves; (6) PLC.

2.2 *Strategy of operation*

The SBR was fed with a synthetic with the composition described in Table 1 according to Beun et al. (1999) and the trace solution to Smolders et al. (1995). The system was inoculated with a granular sludge collected from a SBR.

The operational conditions of the reactor are summarized in Table 2. Different COD/N ratios of 15, 5 and 2.5 g/g in the feeding were tested.

The system were operated in cycles of 3 hours with an exchange volume of 50%. Every cycle comprehended a feeding period of 3 min, a reaction period under aerobic conditions of 171 min, a settling period of 1 min, an effluent withdrawal period comprehended 3 min and an idle period of 2 min.

2.3 *Analytical methods*

The pH, nitrate, nitrite, ammonia, volatile suspended solids (VSS) and SVI were determined accordingly to Standard Methods (APHA, 1985). Concentrations of Chemical Oxygen Demand (COD) were determined by a modified method from the Standard Methods (Soto *et al.*, 1989).

3 RESULTS AND DISCUSSION

Applied OLR to the SBR during the experiments was constant around 2 kg COD/m^3 with COD removal percentages of 90%. Different COD/N ratios of 15, 5 and 2.5 g/g in the feeding were tested.

Table 1. Composition of the synthetic wastewater used to feed the SBR.

Parameters	Concentration (mg/L)
COD	500
NH4-N	100–200
Acetate	632
Phosphate	23

Table 2. Main operational stages of the SBR which were performed during the study.

Days	COD/N (g/g)	VSS (g/L)	Average Diameter (mm)
30	15.0	4–6	1.40
70	5.0	4–5	1.79
90	2.5	8–9	3.18
30	2.5*	4–5	3.09

* 50% VSS concentration.

(i) A COD/N ratio of 15 g/g was tested in an SBR operated during 30 days. No nitrite accumulation was observed in the effluent and nitrate was the only nitrogen oxidized compound. The percentages of N removal were around 85%.
(ii) The reactor was operated with a COD/N ratio of 5 during 60 days (Fig.2). The ammonia concentration in the influent was 100 mg N/L and the NLR of 0.4 kg N/m^3·d. After 20 days of operation stable conditions were reached and the percentage of ammonia oxidized was around 80%. At this point the concentrations of inorganic nitrogen compounds in the effluent were around 12 mg NO_3^--N/L and 17 mg NO_2^--N /L and denitrification percentage was 70%. The VSS concentration was between 4 and 5 g VSS/L with a SVI of 50–70 ml/g-VSS.
(iii) During the next step 90 days the COD/N ratio applied was 2.5 g/g. The ammonia concentration in the influent was increased to 200 mg N/L and after the first 30 days stable conditions were reached and ammonia oxidizing percentage was 63%. Denitrification percentage was of 55%. Concentrations in the effluent were 30 mg NO_3^--N/L and 0 mg NO_2^--N /L. The biomass concentration increased from 5 to 9 g VSS/L and the SVI decreased from 70 to 30 ml/g-VSS.

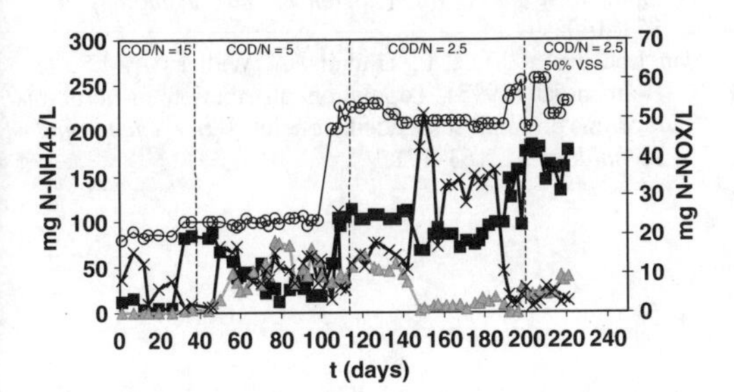

Figure 2. Ammonia concentration in the influent (○) and in the effluent (■), nitrate (×) and nitrite (▲) concentration in the effluent.

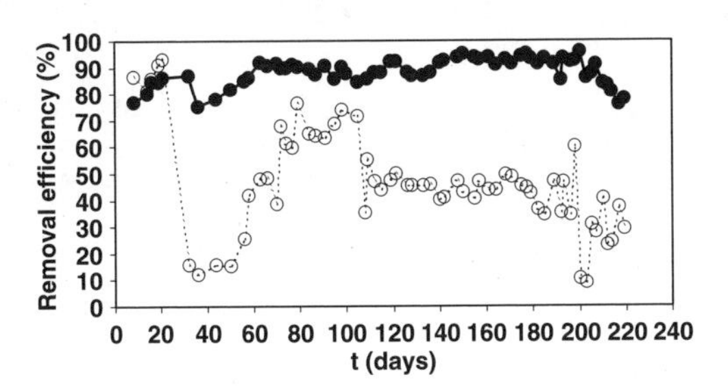

Figure 3. N removal efficiency (○) and COD removal efficiency (●).

(iv) The next period of 30 days was performed with a COD/N ratio of 2.5 g/g but with half of the amount of biomass from stage (iii). The percentage of ammonia oxidation was of 30% and the denitrification of 25%. (Fig. 3) Concentrations in the effluent were 7 mg NO_3^--N/L and 7 mg NO_2^--N /L.

The main processes for nitrogen removal in the SBR were nitrogen assimilation for biomass growth and nitrification–denitrification of ammonia. The fraction of nitrogen, which is removed by each mechanism, depended on the COD/N ratio of the influent in such a way that biomass assimilation can account for the removal of a large fraction of nitrogen when COD/N ratio in the influent to the SBR is high (Garrido et al., 2001). This is the case of the tested COD/N ratio of 15 g/g. On the other hand, with a low COD/N ratio nitrification–denitrification are the main mechanisms of nitrogen removal.

The different COD/N ratio tested in the reactor originated effluents with different composition regarding to nitrogen compounds. N removal percentages decrease with the increase of NH_4^+ concentration in the feeding from 80% to 40%.

It demonstrated that the nitrogen removal efficiency depended upon the concentration of COD and N and was also dependent upon the biomass concentration and of the diameter of the granules. Better nitrogen removal efficiencies were achieved for COD/N ratios up to 5 g/g.

The fact that the granules exhibited different denitrifying activities depends on the COD/N ratio. As occurred in biofilms (Tijhuis et al., 1994; van Loosdrecht et al., 1995) O_2 is present only in the outer layers and the inner layers, which were maintined under anoxic conditions, received the carbon source and nitrate to supor the denitrification process. If the carbon source is not enough (COD/N ratios low) the denitrification process is not completed and nitrite is produced.

Nitrate is consumed via denitrification while biodegradable compounds might be partly aerobically oxidized, partly used as electron donor for denitrifcation and partly stored in the biomass.

Biomass concentration increases during stage (iii) from 5 to 9 and produces a consequent increase on the nitrogen removal.

4 CONCLUSIONS

The production of nitrite in a granular SBR was achieved by using different COD/N ratios.

The COD removal percentage was around 95% during the whole operational period. When, the reactor was operated with a COD/N ratio of 15 g/g no nitrite production was observed in the effluent and nitrate was the only oxidized compound. During the periods fed with COD/N ratios of 5 and 2.5 g/g, nitrite was measured in the effluent at values of 7–20 mg N-NO_2^-/L. The amount of nitrogen removed by denitrification was different for each period.

Changes in the COD/N ratio originated effluent with different composition regarding to nitrogen compounds. N removal percentages decrease with the increase of NH_4^+ in the feeding from 80% to 40%.

ACKNOWLEDGMENTS

This work was funded by the European Commission through the ICON project (Project EKV-CT-2000-0054), Xunta de Galicia (PGIDT10XJ 120904 PM) and the Spanish CICYT which funded this research through the Oxanamon project (PPQ-2002-00771).

REFERENCES

APHA-AWWA-WPCF. (1985) Standard Methods for examination of water and wastewater. 16th Ed. Washington.

Beun J.J., Hendriks A., Van Loosdrecht M.C.M., Morgenroth E., Wilderer P.A. & Heijnen J.J. (1999). Aerobic granulation in a sequencing batch reactor. *Water Research* **33** (10), 2283–2290.

Beun J.J., Heijnen J.J. & van Loosdrecht (2001). N-removal in a granular sludge sequencing batch airlift reactor. *Biotechnology and Bioengineering* **75** (1), 82–92.

Garrido J.M., Van Benthum W.A.J., Van Loosdrecht M.C.M. & Heijnen J.J. (1997). Influence of Dissolved Oxygen Concentratin on Nitrite Accumulation in a Biofilm Airlift Suspension Reactor. *Biotechnology and Bioengineering*, **53**, 168–178.

Garrido J.M., Omil F., Arrojo B., Méndez R. & Lema J.M. (2001). Carbon and nitrogen removal from a wastewater of an industrial dairy laboratory with a coupled anaerobic filter-sequencing batch reactor system. *Water Science and Technology* **43**, 315–321.

Sánchez M., Mosquera-Corral, Méndez R. & Lema J.M. (2000). Simple methods for the determination of the denitrifying activity of sludges. *Bioresource Technology* **75**, 1–6.

Buys B.R., Mosquera-Corral A., Sánchez M. & Méndez R. (2000). Development and application of a denitrification test based on gas production. *Wat. Sci. & Tech.*, **41**(12), 113–120.

Smolders G.J.F., Klop J., van Loosdrecht M.C.M. & Heijnen J.J. (1995). A metabolic model of the biological phosphorus removal process. Effect of the sludge retention time. *Biotechnology and Bioen*gineering **48**, 222–233.

Soto M., Veiga M.C., Méndez R. & Lema J.M. (1989). Semi-micro COD determination method for high salinity wastewater. Environmental Technology Letters **10**(5), 541–548.

Tijhuis L., van Loosdrecht M.C.M. & Heijenen J.J. (1994). Formation and growth of heterotrophic aerobic biofilms on small suspended particles in airlift reactors. *Biotech nology and Bioen*gineering **44**, 595–608.

van Benthum W.A.J., van Loosdrecht M.C.M & Heijnen J.J. (1997). Control of heterotrophic layer formation on nitrifying biofilms in a biofilm airlift suspension reactor. *Biotechnology and Bioengineering* **53**, 397–405.

van Dongen U., Jetten M.S.M. & van Loosdrecht (2001). He SHARON – Anammox process for treatment of ammonium wastewater. *Water Science and Technology* **44**, 153–160.

van Loosdrecht M.C.M., Tijuhuis L., Widieks A.M.S. & Heijnen J.J (1995). Population distribution in aerobic biofilms on small suspended particles. *Water Science and Technology* **31**, 163–171.

European Symposium on Environmental Biotechnology, ESEB 2004 - Verstraete (ed)

Aerobic granulation in a sequencing batch reactor fed with industrial wastewater

B. Arrojo, A. Mosquera-Corral, J.M. Garrido & R. Méndez
Department of Chemical Engineering, School of Engineering, University of Santiago de Compostela
Rua Lope Gómez de Marzoa s/n, Santiago de Compostela, Spain

ABSTRACT: The main objective of this research was focused on the enhancement of the settling properties of the sludge of an industrial scale Sequencing Batch Reactor (SBR) by promoting granules formation. For this purpose, two laboratory scale SBRs, R1 and R2, were inoculated and fed with the same sludge and wastewater used in the industrial scale SBR in which a sludge with poor settling properties is normally generated. Both reactors were operated under similar conditions during most of the experimental period with exception of an anoxic phase of 10 and 30 min which was included at the beginning of every cycle of operation of R1. Organic and nitrogen loading rates (OLR and NLR) applied to both systems were as high as 7 g COD/(L·d) and 0.7 g N/(L·d), respectively. Nitrogen removal efficiency was 80% in both units even considering that R2 was operated always under aerobic conditions. Granules with similar morphology were developed in both systems. Size distribution was comprehended between 0.25 and 4.0 mm in both cases.

1 INTRODUCTION

The performance of the aerobic activated sludge process depends on the settling characteristics of the sludge generated in the reactor. The flocs in the sludge should be easily separated from the treated wastewater in the settler and recirculated to the reactor. However, in some cases flocs with poor settling properties are developed in the system, which might result in partial sludge washout and a decrease of the quality of the effluent. Most of the times this washout is caused by an excessive growth of filamentous microorganisms or development of flocs with a fluffy structure, which have a negative influence on the settling properties of the sludge. The control of the causes responsible for the production of sludge with poor settling properties is sometimes difficult. In some cases a change in either the reactor configuration or the operating strategy has influenced the characteristics of the developed sludge in these units. However, for some wastewater the formation of flocs with good settling properties is generally difficult. For these reasons, the research on the production of granular biomass as an alternative to promote a better biomass retention becomes interesting. Granules have good settling properties, due to the achievement of biomass aggregates with a high biomass density. Besides, the good settling characteristics of the granular sludge improves the separation of biomass from the treated wastewater, which is reflected either in a lower area requirement for settling, in continuous units, or allowing longer time for biological purification, in discontinuous units as SBRs that used granular biomass.

The formation of granular sludge was obtained in aerobic sequencing batch reactors (SBR) fed with synthetic wastewater (Tay et al., 2002a; Etterer & Widerer, 2001). In SBRs, the wastewater is treated in successive cycles of a few hours. At the end of every cycle, settling of the biomass takes place before the effluent is withdrawn, to keep the biomass in the reactor. There is evidence that the basis of granulation is the continuous selection of sludge particles that occurs inside the reactors. The part of the biomass, which does not settle fast enough, will be washed out with the effluent (Beun et al., 1999). Thus, the selection of the granules from a biomass mixture in an SBR can be easily performed using the difference in settling velocity between the granules (fast settling biomass) and the flocs (slow settling biomass). Because the settling velocity is an important selection criteria, utilisation of either relatively large column height diameter (H/D) ratio or short settling and water withdrawal periods is advantageous. As compared with conventional activated sludge flocs, heterotrophic granular sludge has regular, dense and strong microbial structures, good settling properties, high biomass retention, and the ability to withstand shock loads (Tay et al., 2002b).

Granulation could be a solution for operating some SBRs in which flocculent sludge with poor settling properties is developed. During a previous research, it was found that the settling properties of the sludge generated in an industrial scale SBR, with high COD and nitrogen removal efficiencies, were poor as the solids volume index (SVI) was never lower than 100 mL/g-VSS (Garrido et al., 2001) and the zone settling velocity (ZSV) was around 0.3 m/h. In order to study the feasibility of obtain granules in the industrial scale SBR, two laboratory scale SBRs were seeded with the sludge collected from the industrial scale SBR, fed with the same influent used in this unit and operated with several conditions to promote granules formation. An additional objective was to study the nitrogen removal in these granular systems and to investigate the causes that could influence the suspended solids content in the effluent. The formation of granules under different conditions was studied and diameter, density and final settling velocity were examined.

2 MATERIALS AND METHODS

2.1 *Experimental set-up*

Two sequencing batch reactors (SBR) with a total volume of 2.5 L and a working volume of 1.5 L were used. Dimensions of the units were: height of 465 mm and inner diameter of 85 mm, the height to the diameter ratio (H/D) being 5.5. The maximum level of the liquid was 264 mm, and the minimum level 132 mm after effluent withdrawal. Oxygen was supplied to both reactors by using spargers to promote the formation of small air bubbles. In case of R1 air was replaced temporarily by nitrogen gas during an anoxic period comprehended between 10 and 30 min. The flow of nitrogen and air was controlled by means of two electrovalves. A set of two peristaltic pumps was used to feed and to discharge the effluent, respectively, in both reactors. The influent was introduced in both systems through ports located at the top of the reactors. The effluent was discharged through the sampling port placed at middle height of the column reactor. A programmable logic controller (PLC) Siemens model S7-224CPU controlled the actuations over the pumps and valves, and thus the length of every operational period in the SBRs. The reactor was operated at room temperature (15–20°C) and at oxygen concentration between 0 and 8 mg O_2/L and without pH control, which varied between 7.4 and 8.5.

2.2 *Strategy of operation*

Two different feedings were used during the studies with both SBRs: a synthetic wastewater and an industrial wastewater from a laboratory for analysis of dairy products located near A Coruña, Spain (Table 1). The composition of the synthetic wastewater (SW) fed to reactor R1 was according to Beun et al. (1999). The industrial wastewater was the same as that used to feed the industrial scale SBR of 28 m^3 and was previously treated in an anaerobic filter in order to reduce the organic matter fraction. Additional information about the generation of the industrial wastewater and of the treatment plant can be found elsewhere (Garrido et al., 2001).

On day 27 the synthetic medium was replaced stepwise with a fraction of the industrial wastewater, maintaining the soluble COD concentration nearly constant (Table 2). After day 48, it was fed exclusively with the industrial wastewater. Operation of R2 started 50 days later and was directly fed with the industrial wastewater. The strategy of operation was similar in both systems from day 50 on. Table 2 shows some of the main operational stages and changes in the strategy of operation.

Both systems were inoculated with sludge collected from the industrial SBR and were operated in cycles of 3 hours with an exchange volume of 50%. Every cycle comprehended a feeding period of 3 min, a reaction period under anoxic and aerobic conditions of 171 min, a settling period of 1 min, an effluent

Table 1. Composition of wastewater of the laboratory for analysis of dairy products (IW) and the synthetic wastewater (SW) used to feed the SBRs.

Parameters	Concentration (mg/L) IW	Concentration (mg/L) SW
CODt	500–3000	500
CODs	300–1500	500
Total nitrogen	50–200	25
TSS	200–1200	0
VSS	100–1000	0
VFA	100–500	632*
Phosphate	20–60	23

* Acetate concentration in the SW.

Table 2. Main operational stages of the two SBRs which were performed during the study.

Day	R1	R2
1	Start up SW OLR = 1–2 g COD/L·d	
27	0.75 : 0.25*	
34	0.50 : 0.50*	
41	0.25 : 0.75*	
48	IW	
50		Start up IW
83	Anoxic period = 10 min	
133	Anoxic period = 30 min	
188–320	OLR = 4–7 g COD/L·d	

SW: synthetic wastewater; IW: industrial wastewater.
* Fraction of SW and IW which were fed to R1.

withdrawal period comprehended of 3 min and an idle period of 2.0. On day 83 an anoxic period of 10 min was included in the cycle of R1 and from day 133 on, this increased to 30 min until the end of operation of the reactor (day 320).

2.3 *Analytical methods*

The pH, nitrate, ammonia, volatile suspended solids (VSS), total suspended solids (TSS) and SVI were determined accordingly to Standard Methods (APHA, 1985). Concentrations of Chemical Oxygen Demand (COD) were determined by a modified method from the Standard Methods (Soto et al., 1989).

3 RESULTS AND DISCUSSION

Applied OLR and NLR were very high, up to 7 kg COD/m^3·d and 0.7 kg N/m^3·d, respectively. COD concentration in the influent was between 0.5 and 3 kg/m^3 and $NH4^+$ concentration was between 30 and 200 mg N/L. Overall COD removal efficiencies were most of the period between 85 and 95%. Lower values around 65% were measured in punctual periods. The concentration of ammonia in the effluent varied between 0 and 20 mg N/L and nitrate concentration between 0 and 40 mg N/L. Nitrogen removal efficiencies were up 80% (on day 188) in both reactors. These results are similar to those obtained by Garrido (Garrido et al., 2001) treating the same industrial wastewater in an SBR of 28 m^3 but operated with flocculent sludge. However, with granular SBR it was possible to operate at much higher OLR and NLR, around five fold higher, obtaining an effluent with similar characteristics.

The operation strategy in both SBRs was similar to those used by Beun (Beun et al., 1999) for promoting granular sludge in SBRs. The sludge used in both systems as inoculum was the typical flocculent activated sludge with a fluffy, irregular and loose morphology and relative abundance of filamentous microorganisms (Figure 1A). Settling properties of this sludge were: SVI of 200 mL/g VSS, and ZSV of 0.3 m/h. During the first seven experimental days an almost complete washout of the suspended biomass in both systems was observed. This was a result of the operation strategy of the systems, in which a very short settling and a fast effluent withdrawal period were applied to both reactors. Thus, either flocs or aggregates of biomass with settling velocity slower than 9 m/h were removed from the system as a result of mentioned conditions. Three weeks after the start up of the reactors the formation of small aggregates with an average diameter of 2.3 mm was observed in both systems. Suspended flocs gradually disappeared from the reactor and settling properties of the obtained aggregates were very good, SVI 60 mL/g and ZSV of 20 m/h. Microscopic examination of the sludge showed that the morphology of the granular biomass was completely different from the flocculent sludge that was used as inoculum. The shape of the granules was round with a cauliflower like aspect and very clear outline (Figure 1B).

The granular size distribution along the operational time increased gradually with time of operation (Figure 2). These results indicated that the formation of aerobic granules was a gradual process from the flocculent seeded sludge to compact aggregates, further to granular sludge, of 2.3 mm after three weeks of operation, and finally to mature granules of 3.5 mm of averaged diameter (day 220).

Biomass concentration was around 0.2 g TSS/L at the beginning of the experiments (Figure 3). It increased up to 3 g TSS/L after 50 days, and then fluctuated at this level until it reached a stable value around 5–6 g TSS/L. The percentage of ashes of the solids from the reactors ranged from 5 to 13%. The biomass concentrations obtained during this study were similar to those obtained in the industrial SBR, between 3 and 9 g TSS/L, the VSS/TSS ratio being lower, from 0.6 to 0.8 g/g, although the biomass settling properties of

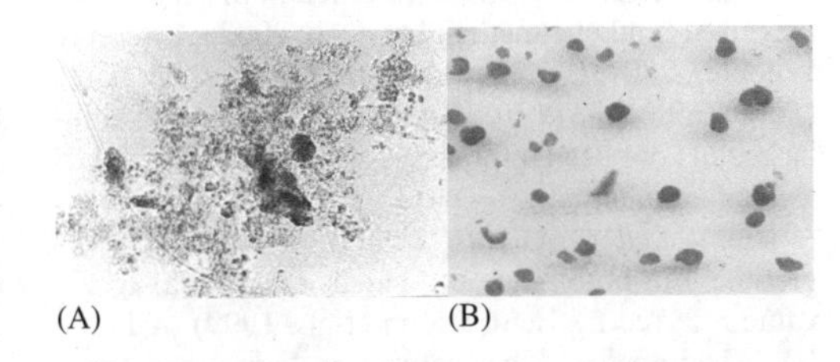

Figure 1. Pictures of the seeding sludge (A) and the aerobic granules at day 23 (B).

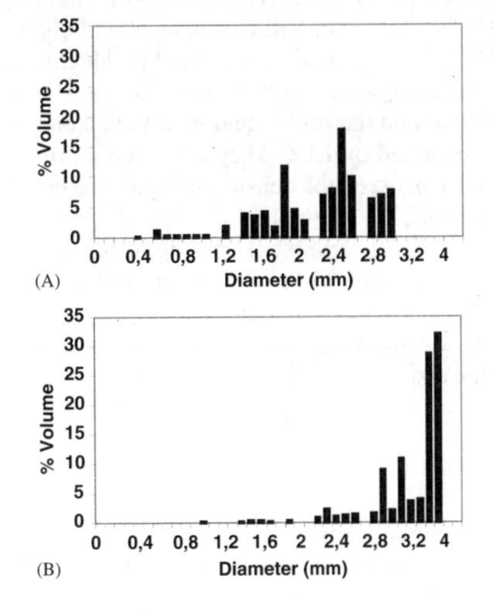

Figure 2. Comparison of the size distribution between granular sludge at the operating day 23 (A) and at day 220 (B).

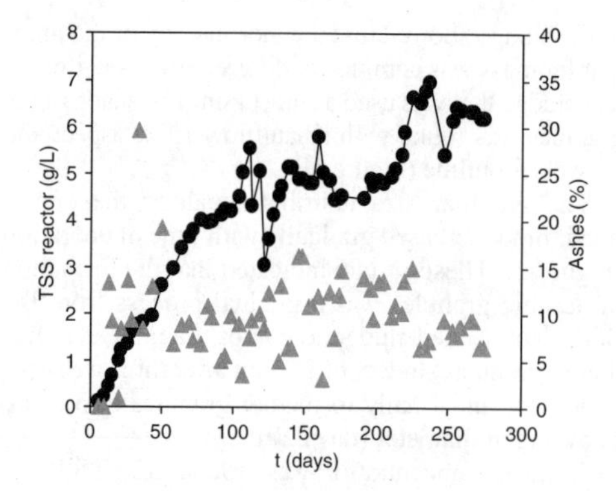

Figure 3. Concentration of TSS and % of ashes in R2.

the granules were much better. Experimental results showed that the utilisation of an industrial wastewater apparently had no significant effect on the development of the granules and biomass accumulation in the system.

Other authors obtained granules with similar physical characteristics to those obtained in this study, but they used synthetic media (Tay et al., 2002a; Etterer & Wilderer, 2001; Beun et al., 1999). On the other hand, Morgenroth (Morgenroth et al., 1997) obtained similar results but used a molasses solution free of solids as feeding medium.

Biomass density was around 10–15 g VSS/(L-granules) in both reactors, which was similar to the value referred by Beun (Beun et al., 1999) of 11.9 g VSS/(L-granules). This value was lower than those reported for granules or biofilms formed in airlift reactors of 20–30 g VSS/(L-granules) (Kwok et al., 1996) and 15–20 g VSS/(L-granules) (Tijhuis et al., 1994). Furthermore, Villaseñor et al. (2000) postulated that the degree of reduction of the degraded carbon source or the maximum specific growth rate of the biomass on the substrate used affected the density of the obtained biofilm. They observed an important difference between the densities of biofilms developed with formate, 20–30 g VSS/(L-granules), formaldehyde, 25–35 g VSS/(L-granules), or methanol 100–120 g VSS/(L-granules). In the present study the morphology of the granules obtained in both reactors fed with industrial wastewater were similar to those obtained with the synthetic feeding.

4 CONCLUSIONS

The formation of granules in two SBRs was achieved by using an industrial wastewater coming from a dairy analysis laboratory as influent. The operational strategy of the SBR allowed the formation of granules with better settling properties compared to the sludge collected from the industrial scale reactor. Granules with similar morphology were developed in both systems. Granules with good settling properties were obtained, SVI of 60 mL/g VSS, and ZSV of 20 m/h. This made feasible to operate the system with high exchange volume and thus organic and nitrogen loading rates applied to both systems were high, up to 7 g COD/(L·d) and 0.7 g $NH4^+$-N/(L·d). Nitrogen removal efficiency was similar in both units, even considering that R2 was operated always under aerobic conditions.

ACKNOWLEDGMENTS

This work was funded by the Spanish CICYT through the Oxanamon project (PPQ-2002-00771) and Xunta de Galicia (PGIDT10XJ 120904 PM).

REFERENCES

APHA-AWWA-WPCF. (1985). Standard Methods for examination of water and wastewater. 16th Ed. Washington.

Beun J.J., Hendriks A., Van Loosdrecht M.C.M., Morgenroth E., Wilderer P.A. & Heijnen J.J. (1999). Aerobic granulation in a sequencing batch reactor. *Water Research* **33** (10), 2283–2290.

Etterer T. & Wilderer P.A. (2001). Generation and properties of aerobic granular sludge. *Water Science and Technology* **43** (3), 19–26.

Garrido J.M., Omil F., Arrojo B., Méndez R. & Lema J.M. (2001). Carbon and nitrogen removal from a wastewater of an industrial dairy laboratory with a coupled anaerobic filter-sequencing batch reactor system. *Water Science and Technology* **43**, 315–321.

Kwok W.K., van Loosdrecht, M.C.M. & Heijnen J.J. (1996). Application of a biofilm airlift suspension reactor for acetic acid removal. Internal report, Delft University of Technology, Delft, The Netherlands.

Morgenroth E., Sherden T., Van Loosdrecht M.C.M., Heijnen J.J. & Wilderer P.A. (1997). Aerobic granular sludge in a sequencing batch reactor. *Water Research,* **31** (12), 3191–3194.

Soto M., Veiga M.C., Méndez R. & Lema J.M. (1989). Semi-micro COD determination method for high salinity wastewater. *Environmental Technology Letters* **10** (5), 541–548.

Tay J-H., Liu Q-S. & Liu Y. (2002a). Aerobic granulation in sequential sludge blanket reactor. *Water Science and Technology* **46** (4–5), 13–18.

Tay J-H., Liu Q-S. & Liu Y. (2002b). Hydraulic slection pressure-induced nitrifying granulation in sequencing batch reactors. *Applied Microbiology Biotechnology* **59**, 332–337.

Tijhuis L., van Loosdrecht M.C.M. & Heijenen J.J. (1994). Formation and growth of heterotrophic aerobic biofilms on small suspended particles in airlift reactors. *Biotechnology and Bioengineering* **44**, 595–608.

Villaseñor J.C., van Loosdrecht M.C.M., Picioreanu C. & Heijnen J.J. (2000). Influence of different substrates on the formation of biofilms in a biofilm airlift suspension reactor. *Water Science and Technology* **41** (4–5), 323–330.

European Symposium on Environmental Biotechnology, ESEB 2004 - Verstraete (ed)
© 2004 Taylor & Francis Group, London, ISBN 90 5809 653 X

Water recycling in Vojvodina Province – necessity and possibility

A. Belic & S. Belic
University of Novi Sad, Faculty of Agriculture, Department for Water Management, Serbia and Montenegro

M. Jarak
University of Novi Sad, Faculty of Agriculture, Department for Field Crops and Vegetables, Serbia and Montenegro

ABSTRACT: The reuse of wastewater in irrigation represents a rational option for additional water supply in the regions of water shortage, and in the other regions too, firstly because it represents an ecologically satisfactory solution. Modern management on pig breeding farms in Vojvodina Province, with liquid discharge technology, conditioned the accumulation of wastewater into lagoon. At the same time, contemporary agriculture means including also irrigation of growing plants. The investigation results have indicated the need and the possibility of farm wastewater utilization for irrigation. In this way it would be possible to ensure crop supply with nutrients, decrease of moisture shortage in the soil, environmental protection, and achieve significant economic effects. The changes in shallow groundwater quality as a consequence of using pig raising farm wastewater for irrigation are not more pronounced than those observed when applying mineral fertilizers.

1 INTRODUCTION

The reuse of wastewater is an important measure in the management of water resources. Irrigation with wastewater is a possible solution to the problem of water shortage in arid and semi-arid regions, as well as an ecologically acceptable solution to the problem of accumulated wastewater in other regions, too. The use of wastewater is both agriculturally and economically justified. However, it is necessary to minimize its possible undesirable effects on the environment. In irrigation, the most commonly used wastewater is wastewater from households, agriculture, and food industry.

The success in the use of wastewater in plant production depends on the coordination of the adopted strategies which all aim at the optimum plant production (both quantity and quality), preservation of the productivity of soil and protection of the environment. The choice of the irrigation method is a measure in health protection, along with other measures such as plant selection, wastewater treatments and the control of the exposure of workers to the direct contact with wastewater. Evaluation of certain irrigation methods from the aspect of use of the treated wastewater, based on the analysis of Kandiah (1990), is mentioned by Pescod (1992).

Modern management on pig breeding farms in Vojvodina introduction of liquid discharge technology, conditioned the accumulation of wastewater into lagoon which could be used for irrigation for growing plants. By this land reclamation measure is possible to close production on the farms, having in mind some useful characteristics of this kind wastewater.

The applicability of these waters is determined by their characteristics, amounts, the natural characteristics of the region, and the possibility of growing particular crops (de Hean, 1987). The choice of crops depends upon the local pedological, climatic, and market conditions, and especially on the degree of crops sensitivity to the contents of macro- and micro-nutrients, as well as to some hazardous components present in these wastewaters.

There are around seventy pig breeding farms in Vojvodina where the daily production of wastewater amounts to 1200 m^3. This paper shows the results of the analysis of wastewater from these farms and possibility of its reuse in irrigation analyzed on lysimetric station Rimski Sancevi near Novi Sad.

2 METHODOLOGY

To determine the characteristics of the wastewater, samples have been taken four times a year during three-year period. Of all the analyzed parameters, the subject of our present concern are those (pH, EC, TDS, Ca, Mg,

Na, Ca/sum cations, SO_4, Cl, NO_3, NH_4) which, on the one hand, characterize the effluent (increased concentrations as a consequence of the production process) and, on the other hand, in the case of irrigation influence the quality of groundwater, which is of great importance from the point of view of agricultural production and water supply. Only part of obtained results is presented here.

In the frame of a complex research project, at the lysimetric station Rimski Sancevi the effect of irrigation with diluted wastewater from the pig-raising farm on shallow groundwater quality has been studied (lysimeters IV and VII). Lysimetric station – drainage type with continuous levels consists of ten pots. The crops grown in the lysimeters were irrigated with the farm wastewater from Ilandza lagoon. The irrigating norms were determined on the basis of bioclimatic coefficient. During the experiments the water table was maintained at a level of 120 cm. In the lysimeters, a soil monolith was formed of the czernozem corresponding to the pedological profile of the surrounding soil. Besides, the experiment encompassed some additional investigations concerning the control of microbiological activity of the soil and shallow groundwater, chemical composition of the soil, as well as the observation of phenologicalstages of the crops (Belic et al, 1994).

The changes in mineralization of the shallow groundwater during the investigation period were monitored under the different production conditions (Abramov, 1985). In the lysimeters, the following crops were grown: maize, sugar beet, and sunflower. The wastewater for irrigation in the production process was used either concentrated or diluted with 6 parts of pure water. In the case irrigation was carried out with diluted wastewater its use was practiced during the growing season, whereas undiluted wastewater was used for irrigation before the crops sowing. In both cases, the irrigation with wastewater was treated as addition of nutrients, so that its amounts were determined by the N_{min} method. The amounts of nutrients were different in dependence of the growing variant applied, representing the values characteristic for the conditions of the production process.

3 RESULTS AND DISCUSSION

It is known that the wastewaters from pig-raising farm can exhibit substantial variations in their chemical composition, physical characteristics, and the amounts involved. The main factors causing such diversity are the production systems involved at the farm, the systems of collecting and disposing these wastewaters, as well as the characteristics of the raw water entering the production processes. In the lagoons too, various processes can take place, yielding transformation of complex organic compounds into less complex organic and/or inorganic forms, the scope of these processes being dependent of the residence time of the wastewaters in the lagoon.

For successful irrigation, it is necessary to meet the following basic conditions (Vermes & Kutera, 1984; Westcot, 1997):

- application of the optimum amount of wastewater
- acceptable quality of wastewater
- proper timing for irrigation
- appropriate method of irrigation
- prevention of the accumulation of salts around the root of the plant
- prevention by drainage of the rising of the level of groundwater
- maintenance of the optimum level of nutritious substances necessary for the plant

The success in the use of wastewater in plant production depends on the coordination of the adopted strategies which all aim at the optimum plant production (both quantity and quality), preservation of the productivity of soil and protection of the environment (Belic et al, 1996).

There are different methods of irrigation. Which method will be chosen depends in the first place on finances, but the possible harmful effects of the method have to be taken into account, too (Arceivala, 1981).

The choice of the irrigation method depends on the following factors (Ayers & Westcot, 1985; Chang et al, 1995):

- choice of the plants
- part of the plant, which is in direct contact with wastewater
- distribution of water, salts and pollutants in the soil
- water potential of the soil
- efficiency of the application of the method
- its possible undesirable effects on people and environment

The results show that the wastewater from pig breeding farms is low alkaline and highly mineralized, with a significant amount of organic matter. The wastewater also contains inorganic matter such as chlorides, sodium and sulphates. According to Kovcin (1993), the presence of all macro and micro components in food enables a successful and economical raising of pigs. Therefore, Ca, Na, Cl and micro components (mostly sulphates) are added into the food. This kind of wastewater also has a high concentration of ammonia ions. This parameter is influenced by the way of disposal of the wastewater as well as by the duration of the wastewater retention in lagoons.

For the majority of parameters highest seasonal values were registered during the spring. Also, it is characteristic that the majority of parameters attained their lowest seasonal values in the autumn period.

This points out to the importance of climatic factors, primarily of temperature and insolation, on the process of stabilization of the wastewater in the lagoon. This observation is in agreement with the considerations of Loehr (1974), Milojevic (1987) and Jahic (1990), concerning the effect of climatic factors on the processes of stabilization of wastewaters in the lagoon.

Taking into account the above characteristics of the wastewaters from the pig-breeding farm stored in the lagoon, a question arises as to the reuse of these waters. The contents of nitrogen, phosphorus and potassium may be a significant source of macronutrients for the crops. Also, a high content of organic matter in these wastewaters, being a factor of soil fertility, can contribute significantly to the increase in crop yields, which may be one of the main goals of their utilization. The wastewaters from a pig-breeding farm may contain different pathogenic microorganisms that may be hazardous in its utilization. However, these problems have not been encompassed by the present investigations.

During the investigation period the wastewater used for irrigation was characterized by a pH value in the range 7.1–7.8. However, the shallow groundwater in the lysimeters, irrespective of the production conditions and the quality of nutrients added exhibited significantly higher pH values, being in the range of 7.6–8.8. The lowest pH values were measured in the filtered water of the control vessels in which a grass mixture was grown.

Elecrical conductivity of the well water (EC_w) was 634 μS/cm. In contrast to this relatively low value, the shallow groundwaters were characterized by significantly higher average values, ranging from 817 to 925 μS/cm. It can be noticed that in the experiment variants involving higher doses of nutrients the values of electrical conductivity of filtered water showed an increase. By analyzing total dissolved solids (TDS) a similar conclusion may be drawn as from the analysis of the EC_w values, whereby it should be mentioned that in the variant of the vessels IV and VII, where largest quantities of wastewater were added in the beginning of the irrigation cycle, the TDS parameter had the highest values.

Content of calcium in shallow groundwater was the range of 0.97–5.38 meq, whereas this parameter for the well water was in the range of 3.63–4.48 meq. The highest average value (3.72 meq) was measured in the vessels I and X, as in the case of the previous parameters, and the lowest values in the vessels V and VI. In contrast to calcium, the analysis of magnesium content in the shallow groundwater did not show a tendency to be related to the production conditions. Magnesium content was in the range of 1.42–6.61 meq, whereas the average values were in the range of 3.05–4.03 meq. In the well water used for irrigation magnesium content was lower, the average value being 1.99 meq.

Sodium content in the shallow groundwater, with the average values being in the range of 2.51–3.85 meq was higher than that in the well water (1.01 meq). Although the measured extreme value amounted to 6 meq, it may be concluded that the value of sodium adsorbtion ratio (SAR) was below its critical value.

According to the approach that has been presented by a group of authors (Loyd & Heathcote, 1985; Kemmers, 1986), the ratio of contents of calcium and the sum of cations (Ca^{2+}/Σ_{cat}) in shallow groundwaters may be used as a hydrochemical indicator of its environmental status. By its presence Ca^{2+} ion is also the predominant cations in the exchangeable complex, and its way competes with the present H^+ ion, so that becomes a regulator of soil acidity. In an indirect way Ca^{2+} ion disturbs the pH-dependent macrobiological processes, and its influence increases with the retention time of groundwaters. Previous works (Belic, A et al., 1993) indicated a strong relationship (with a coefficient of correlation $r = 0.81$–0.96) between the Ca^{2+} content and pH value, i.e. the existence of competition between H^+ and Ca^{2+} ions, and an indirect effect of calcium ion the status of nutrients in the analyzed agro-ecosystem. The higher values indicate an environment of higher quality. By analyzing the average values of the Ca^{2+}/Σ_{cat} ratio in the filtered lysimeter water it is possible to draw the conclusion that in the variants VI and VII, in which highest amounts of wastewater were used for irrigation, the average value of this ratio was highest ($Ca^{2+}/\Sigma_{cat} = 0.38$). On the other hand, the lowest value ($Ca^{2+}/\Sigma_{cat} = 0.27$) was obtained for the control vessels IV and V, in which a mixture of grasses was grown. The average value of this ratio determined for the well water was $Ca^{2+}/\Sigma_{cat} = 0.57$.

In the shallow groundwater, chloride content was in the range of 0.14–2.68 meq, whereas the average value of this parameter for well water was 0.70 meq, the range being 0.37–0.82 meq. If mean values of chloride content are considered in relation to the experimental variant it can be noticed that the highest chloride contents were measured for the vessels IV and VII.

Sulphates content in the filtered water from the lysimeters were in the range of 0.94–7.35 meq, the highest mean values of 3.01 meq being registered in the experimental variants in the vessels I and X, and the lowest (1.51 meq) in the lysimeters with grass mixture. Content of sulphates was much lower in the well water, and the average value was 0.04 meq.

4 CONCLUSION

The obtained results point out to the possibility of closing the production circle on the farm without significant disadvantages to the quality of groundwater resources. The usage of wastewater for irrigation, under the controlled conditions of lysimetric station, indicates that pollution of shallow groundwater thus induced is not more significant than that caused by the unused (by plants) rationally introduced mineral nutrients. A rich microbiological activity in the soil, which can be considered a consequence of wastewater application for irrigation, as well as the economic effect lead to the serious consideration of introducing into a wider production such way of farming.

By analyzing the change in the quality of shallow groundwater in the lysimeters under conditions of growing maize and sunflower using the characteristic doses of fertilizers and wastewater for irrigation it can be concluded that the mode of soil utilization can have a significant effect on the changes of the quality of shallow groundwater. A decisive role in this respect has the soil characteristics. The changes in shallow groundwater quality as a consequence of using pig raising farm wastewater, either in concentrated or diluted state, for irrigation purpose are not more pronounced than those observed when applying mineral fertilizers. This conclusion indicates the possibility that such a use of wastewater may be considered as a solution acceptable from the environmental point of view, as far as the protection of shallow groundwater quality in agricultural areas is concerned. Besides, significant savings are thus achieved, as substantial amounts of nutrients, needed for crop growth and development, are introduced with the wastewater into the soil.

REFERENCES

Abramov, A.F., Peredkova, L.J., Mihailova, G.G. (1985) Orosenie bitovim stokami i hemiceskij sostav gruntovih vod, *Gidrotehnika i melioracija,* No. 9, p. 26–28, Moskva.

Arceivala, S.J. (1981) *Wastewater Treatment and Disposal,* Marcel Dekker, INC, New York.

Ayers, R.S. and Westcot, D.W. (1985) Water Quality for Agriculture, *FAO Irrigation and Drainage Paper* No. 29, Rome.

Belic, A., Jarak, M., Belic, S. (1993) Uticaj korišćenja mineralnih đubriva i otpadne vode sa svinjogojske farme na mikrobiolotку aktivnost i kvalitet voda prve izdani, *Voda i sanitarna tehnika*, No. 1–2, str. 57–63, Beograd.

Belic, S., Maksimovic, L., Dragovic, S. (1994) Evapotranspiracija i prinos kukuruza na podrucju sa plitkim nivoom prve izdani, str. 185–196, U: *Uređenje, korišćenje i zaлtita voda Vojvodine*, Institut za uređenje voda, Novi Sad.

Belic, S., Savic, R., Belic, A. (1996) Klasifikacija za ocenu upotrebljivosti voda za navodnjavanje, monografija *"Upotrebljivost voda Vojvodine za navodnjavanje"*, str. 5–36, Poljoprivredni fakultet, Institut za uredjenje voda, Novi Sad.

Chang, A.C., Page, A.L., Asano, T. (1995) *Developing Human Health-Related Chemical Guidelines for Reclaimed Wastewater and Sewage Sludge Applications in Agriculture,* WHO, Geneva.

de Hean, F.A.M. (1987) Pollution of soil and groundwater as the result of high manure applications, *lecture notes,* IHE, Delft.

Jahic, M. (1990) *Prečišćavanje zagadjenih voda*, Poljoprivredni fakulet, Institut za uredjenje voda, Novi Sad.

Kemmers, R.H. (1986) Calcium as hydro chemical characteristic for ecological states, *ICW Technical bulletins*, No. 47, Wageningen.

Kovcin, S. (1993) *Ishrana svinja,* Poljoprivredni fakultet, Novi Sad.

Lloyd, J.W., Heathcote, J.A. (1985) *Natural Inorganic Hydrochemistry in Relation to Groundwater*, Clarendon press, Oxford, England.

Loehr, C.R. (1974) *Agricultural Waste Management*, Academic press, New York.

Milojevic, M. (1987) *Snabdevanje vodom i kanalisanje naselja*, Naučna knjiga, Beograd.

Pescod, M.B. (1992) Wastewater Treatment and Use in Agriculture, *FAO Irrigation and Drainage Paper* No. 47, Rome.

Vermesh, L., Kutera, J. (1984) Waste water disposal and utilization in agriculture in Poland and Hungary, *Effluent and water treatment Journal,* December 4, pp. 465–469.

Westcot, D.W. (1997) Quality Control of Wastewater for Irrigated Crop Production, *FAO Water Reports* 10, Rome.

European Symposium on Environmental Biotechnology, ESEB 2004 - Verstraete (ed)
© 2004 Taylor & Francis Group, London, ISBN 90 5809 653 X

Metal bioremoval from industrial waste waters with an EPS-producing cyanobacterium

R. De Philippis, R. Paperi & M. Vincenzini
Department of Agricultural Biotechnology, University of Florence, Italy

ABSTRACT: The metal removal capability of the exopolysaccharide (EPS)-producing cyanobacterium *Cyanospira capsulata* was tested in laboratory solutions with two multimetal systems, composed by Cu, Ca and Mg or by Cu, Zn and Ni. In both cases, the cyanobacterial biomass removed significant amounts of the metals, maintaining a good efficiency in comparison with the single metal systems. The biomass was then tested with an industrial waste water containing six metals: Ba, Cr, Mn, Fe, Ni and Al. After one hour, about 60% of the initial amount of each metal was removed; a second and a third cycle, carried out with fresh biomass, removed about 85% and 97% of the initial metal amount. With another industrial waste water, mainly containing Fe, Al, Cu, Zn, Mn and Pb, the cyanobacterial biomass removed, after one cycle, about 85% of the initial amount of metals, showing a promising behaviour for industrial applications.

1 INTRODUCTION

The use of microorganisms for the removal of toxic heavy metals from polluted waters has been intensively investigated, in recent years, as an alternative to the conventional physicochemical methods, that are considered as less effective or too expensive if heavy metals are dissolved in huge volumes of water at relatively low concentrations (Volesky 2001). The most interesting potential advantages of this technique are (i) the possibility to treat waters containing low concentration of metals, in the order of few micromoles per litre; (ii) the capability of the microbial biomass to be active even in multimetal systems; (iii) the rapid kinetics of metal removal and (iv) the possibility to utilize naturally abundant renewable biomaterials (Wilde & Benemann 1993, Inthorn 2001). Microorganisms can remove metals from the surrounding environment with various mechanisms, including metabolically mediated processes, passive adsorption on cell envelopes, chelation of heavy metals by charged exocellular polysaccharides, either bound to cell surface or released during the growth (Volesky 2001). In particular, the good efficiency in metal chelation showed by many of the microorganisms studied has been related to the presence of a high number of negatively charged groups on the cell envelope (Volesky 1994). In this respect, the utilization of exopolysaccharide (EPS)-producing cyanobacteria seems to be quite promising, owing to the anionic nature showed by most of the cyanobacterial RPSs (De Philippis & Vincenzini 1998, De Philippis et al. 2001). Indeed, a recent investigation aimed to assess the copper removal capability of two filamentous EPS-producing cyanobacteria, *Cyanospira capsulata* and *Nostoc* PCC 7936, pointed out the very promising performances of the former cyanobacterium (De Philippis et al. 2003).

In literature, a large number of studies on the metal sorption carried out by microbial biomasses operating in single metal systems is available (see, for instance, the reviews published by Inthorn 2001; Volesky 1994, 2001; Wilde & Benemann 1993). On the other hand, only a reduced number of studies has been dedicated to the competitive interactions occurring among the metals when the microbial biomasses operate in multimetal systems (Echeverria et al. 1998; Singh et al. 1998; Prasad & Pandey 2000; Pradhan & Rai 2000, 2001) even if these systems are very similar to what occurs in many kind of industrial wastes. Finally, only very few studies have been dedicated to the metal bioremoval from real industrial waste waters (Lei et al. 2000; Volesky 2001).

Thus, the aim of this study was to investigate on the metal removal capability of *C. capsulata* operating in multimetal systems, either in pure laboratory solutions and in industrial waste waters, in order to assess the suitability of the exploitation of this cyanobacterium for the removal of heavy metals from polluted waters.

2 MATERIALS AND METHODS

The filamentous, heterocystous cyanobacterium *Cyanospira capsulata* utilized in this study was grown under laboratory conditions, as previously reported (Vincenzini et al. 1990). The experiments with pure metal solutions were carried out with 50 to 150 ml aliquots of whole cultures placed into small dialysis tubings (12 kDa cut-off) that were dipped for 24 hours into tap water and then into 250 mL of metal solutions (Cu, Zn or Ni, at concentration of 10 ppm at pH 5.5). All the experiments were done at least in triplicate and the data are reported as mean values ± standard deviation. The amount of metal removed was calculated from the difference of its concentration in solution, determined with an Atomic Absorption Spectrophotometer (Perkin Elmer, USA), before and after the contact with the cyanobacterial biomass. The concentration of biomass was routinely determined as amount of protein, according to the Lowry method (Herbert et al. 1971); the analysis of a large number of *C. capsulata* cultures showed that the protein content always ranged from 40 to 42% of the cell dry weight. The experiments carried out for assessing the interfence of the competing ions present in hard waters were done by adding $CaCl_2$ and/or $MgCl_2$ to final concentrations of 4 mM, when separately present, or of 2 mM each one, when contemporaneously present. The experiments on metal bioremoval from industrial waste waters were carried out with samples coming from the percolate of two different industrial wastes (defined as A and B; see Tabs 3–4 for their-metal composition) stored outdoors for several months and characterized by COD values of about 12–15 000 mg L^{-1}. Various amounts of fresh *C. capsulata* biomass, obtained from the centrifugation of cultures carried out under laboratory conditions, were directly suspended into the waste water at pH 5.5. After 1 hour of contact, the suspension was centrifuged at 5000 x *g* for 10 min at 20°C and the cell-free supernatant was analysed for its metal content by Atomic Absorption Spectrometry. Controls were always carried out by putting aliquots of pure metal solutions or of waste waters under the same conditions used for the experiments with the biomass.

3 RESULTS AND DISCUSSION

3.1 *Metal bioremoval in multimetal systems*

The removal capability of *C. capsulata* in multimetal systems was first tested in laboratory solutions containing Cu, Ca and Mg, in order to simulate the removal of metals from hard waters. The presence of competing ions such as Ca and Mg reduced the amount of copper removed by *C. capsulata* biomass to about 60–70% of the values observed in the absence of the two ions (Table 1). However, this reduction, previously observed also with *Oscillatoria anguistissima* (Ahuja et al. 1997) and *Phormidium laminosum* (Sampedro et al. 1995) biomasses, was rather limited, thus suggesting the effectiveness of a process operating with *C. capsulata* biomass for the bioremoval of heavy metals, even if solubilized in hard waters.

Table 1. Maximum values of specific copper removal (q_{max}, expressed as mg of Cu adsorbed per mass of adsorbent, measured as protein concentration) by *C. capsulata* biomass operating in single metal system or in the presence of Ca and/or Mg.

	q_{max}	
	[mg Cu (g protein)$^{-1}$]	(%)*
Cu	115 ± 5	100.0
Cu (+Ca)	83 ± 3	72.2
Cu (+Mg)	88 ± 2	76.5
Cu (+Ca + Mg)	78 ± 3	67.8

* data referred to as percentage of the q_{max} determined in the single component system.

Table 2. Maximum values of specific metal removal (q_{max}, expressed as mg of metal adsorbed per mass of adsorbent, measured as protein concentration) for Cu, Zn and Ni removed by *C. capsulata* biomass operating in single metal systems or in multicomponent system containing the three metals.

	q_{max}	
	[mg Cu (g protein)$^{-1}$]	(%)*
Cu	115.0 ± 5	100.0
Cu (+Zn + Ni)	71.0 ± 4	61.7
Zn	41.5 ± 1	100.0
Zn (+Cu + Ni)	12.4 ± 2	29.9
Ni	55.4 ± 2	100.0
Ni (+Cu + Zn)	26.6 ± 2	48.0

* data referred to as percentage of the q_{max} determined in the single component systems.

The capability of *C. capsulata* to act as a biosorbent for bivalent metals was then tested by utilizing copper, zinc and nickel as model metals, operating in single and in three component systems. In single metal systems, *C. capsulata* biomass showed the highest affinity towards copper, being the value of q_{max} (maximum specific metal removal, expressed as mg of metal removed per amount of biomass, determined as gram of protein) found for Cu significantly higher than those found for Ni and Zn (Table 2). The simultaneous presence of Cu, Zn and Ni in a multicomponent system strongly affected their q_{max}, the new values decreasing to about 62%, 48% and 30% for Cu, Ni and Zn, respectively. However, the smaller reduction found for the q_{max} of Cu in the three metal system, in comparison with the reduction in the q_{max} of Ni and

Table 3. Metal composition of the waste water A and amount of metals removed after the treatment carried out with *C. capsulata* fresh biomass.

Waste water composition		Metal removed (%)*		
Metal	Concentration (mg L^{-1})	1st cycle**	2nd cycle	3rd cycle
Ba	1.87	41	79	95
Cr	1.60	59	85	97
Mn	6.54	60	85	97
Fe	44.00	65	89	97
Ni	0.61	62	77	97
Al	8.99	66	78	97

* amount expressed as percentage of the metal concentration in the waste water before the treatment.
** each cycle was carried out with 140 mg L^{-1} dry weight of fresh biomass.

Zn, points out that *C. capsulata* biomass has the highest affinity for this metal, as it was previously observed for *Microcystis* sp.(Pradan & Rai 2000). These data seem to support the hypothesis that a great number of binding sites on cyanobacterial cells are specific for Cu (Echeverria et al. 1998). Indeed, recently it was demonstrated that in a multimetal systems there is one metal preferentially adsorbed by the microbial biomass, reducing the efficiency in the removal of the other metals (Pradhan & Rai 2001).

3.2 *Metal bioremoval from industrial waste waters*

The metal removal capability of *C. capsulata* biomass was tested with two industrial waste waters. *C. capsulata* biomass was suspended, at a final concentration of 140 mg dry weight per litre, into the waste water A, that contained six metals, namely Ba, Cr, Mn, Fe, Ni and Al, at a global concentration of 64 mg L^{-1}. After one hour, about 60% of the initial amount of each metal was removed, with the exception of Ba, that only showed a 40% removal (Table 3). Increasing, up to 190 and 280 mg dry weight L^{-1}, the amount of biomass utilized, the global amount of metals removed increased respectively up to about 70 and 78% of the initial metal concentration, showing a not linear correlation between the increase in the amount of biosorbent and the increase in the metal removal. An explanation of this behaviour may reside in the increase of the competitive interactions among the binding sites due to their increased proximity, as previously suggested for similar results obtained with other microorganisms (Ahuja et al. 1997; Vegliò et al. 1997).

For this reason, in the following experiments aimed to the improvement of the metal removal efficiency of the system utilized, it was tested the possibility to use a treatment constituted by consecutive cycles carried out with small amounts of fresh cyanobacterial biomass operating on the same water sample. After the second cycle, the amount of metals removed reached values of about 80–90% of the amount present before the biological treatment (Table 3). After the third cycle, the removal reached values corresponding to about 95–97% of the initial metal concentrations, thus demonstrating the good efficiency of *C. capsulata* in the biosorption of metals contained in the waste water A. Among the metals present, the highest value of specific metal uptake was observed for iron (q_{max} = 20.4 mg of Fe removed per g of biomass dry wt).

Table 4. Metal composition of the waste water B and amount of metals removed after the treatment carried out with *C. capsulata* fresh biomass.

Waste water composition		Metal removed (%)*	
Metal	Concentration (mg L^{-1})	1st cycle**	2nd cycle
Ba	1.39	85	99
Cr	3.84	83	97
Mn	6.32	85	97
Fe	57.10	85	97
Ni	0.80	85	98
Al	27.40	87	97
Cu	9.39	86	97
Pb	6.49	83	98
Zn	6.54	82	97
Hg	5.20	95	98

* amount expressed as percentage of the metal concentration the waste water before the treatment.
** each cycle was carried out with 140 mg L^{-1} dry weight of fresh biomass.

The second set of experiments was carried out with the industrial waste water B, containing ten metals, with the quantitative prevalence of Fe, Al, Cu, Zn, Mn and Pb (Table 4). In this case, a good efficiency in the metal removal was already observed after the first cycle, at the end of which the amount of metals removed always reached values higher than 80% of the initial metal concentration. After the second cycle, the removal attained values higher than 97%, thus demonstrating that the system was very efficient, in spite of the presence of a large number of different metals. Even in such a complex system, the highest value of specific metal uptake was again found for iron, being the q_{max} value 24.8 mg Fe (g biomass dry wt)$^{-1}$.

4 CONCLUSIONS

The results obtained with the multicomponent systems, and in particular with the two different industrial waste waters, point out the great potential of the use of *Cyanospira capsulata* for the bioremoval of heavy metals from polluted water bodies, suggesting further investigations on various kind of industrial waste waters with different characteristics.

ACKNOWLEDGMENTS

This research was partially supported by MIUR (Italian Ministry for the University and the Research); "*Fondi per la ricerca scientifica di Ateneo*", year 2002.

REFERENCES

Ahuja, P., Gupta, R. & Saxena, R.K. 1997. *Oscillatoria anguistissima*: a promising Cu^{2+} biosorbent. *Current Microbiology* 35: 151–154.

De Philippis, R. & Vincenzini, M. 1998. Exocellular polysaccharides from cyanobacteria and their possible applications. *FEMS Microbiology Reviews* 22: 151–175.

De Philippis, R., Paperi, R., Sili, C. & Vincenzini, M. 2003. Assessment of the metal removal capability of two capsulated cyanobacteria, *Cyanospira capsulata* and *Nostoc* PCC7936. *Journal of Applied Phycology* 15: 155–161.

De Philippis, R., Sili, C., Paperi, R. & Vincenzini, M. 2001. Exopolysaccharide-producing cyanobacteria and their possible exploitation: a review. *Journal of Applied Phycology* 13: 293–299.

Echeverria, J.C., Morera, M.T., Mazkiaran, C. & Garrido, J.J. 1998. Competitive sorption of heavy metals by soils. Isotherms and fractional experiments. *Environmental Pollution* 101: 275–284.

Herbert, D., Phipps, P.J. & Strange, R.E. 1971. Chemical analysis of microbial cells. In J.R. Norris & D.W. Ribbons (eds), *Methods in Microbiology*: Vol. 5B: 209–344. London: Academic Press.

Inthorn, D. 2001. Removal of heavy metal by using microalgae. In H. Kojima & Y.K. Lee (eds), *Photosynthetic Microorganisms in Environmental Biotechnology*: 111–135. Hong Kong: Springer-Verlag.

Lei, W., Chua, H., Lo, W.H., Yu, P.H.F., Zhao, Y.G. & Wong, P.K. 2000. A novel magnetite – immobilized cell process of heavy metal removal from industrial effluent. *Applied Biochemistry and Biotechnology* 84–86: 1113–1126.

Pradhan, S. & Rai, L.C. 2000. Optimization of flow rate, initial metal ion concentration and biomass density for maximum removal of Cu^{2+} by immobilized *Microcystis*. *World Journal of Microbiology and Biotechnology* 16: 579–584.

Pradhan, S. & Rai, L.C. 2001. Biotechnological potential of *Mycrocystis* sp. in Cu, Zn and Cd biosorption from single and multimetallic systems. *BioMetals* 14: 67–74.

Prasad, B.B. & Pandey, U.C. 2000. Separation and preconcentration of copper ions from multielemental solutions using *Nostoc muscorum* – based biosorbents. *World Journal of Microbiology and Biotechnology.* 16: 819–827.

Sampedro, M.A., Blanco, A., Llama, M.J. & Serra, J.L. 1995. Sorption of heavy metals to *Phormidium laminosum* biomass. *Biotechnology and Appied. Biochemistry* 22: 355–366.

Singh, S., Pradhan, S. & Rai, L.C. 1998. Comparative assessment of Fe^{3+} and Cu^{2+} biosorption by field and laboratory – grown *Microcystis. Process Biochemistry* 33: 495–504.

Gegliò, F., Beolchini, F. & Gasbarro, A. 1997. Biosorption of toxic metals: an equilibrium study using free cells of *Arthrobacter* sp. *Process Biochemistry* 32: 99–105.

Vincenzini, M., De Philippis, R., Sili, C. & Materassi, R. 1990. Studies on exopolysaccharide release by diazotrophic batch cultures of *Cyanospira capsulata. Applied Microbiology and Biotechnology* 34: 392–396.

Volesky, B. 1994. Advances in biosorption of metals: selection of biomass types. *FEMS Microbiology Reviews* 14: 291–302.

Volesky, B. 2001. Detoxification of metal – bearing effluents: biosorption for the next century. *Hydrometallurgy* 59: 203–216.

Wilde, E.W. & Benemann, J.R. 1993. Bioremoval of heavy metals by the use of microalgae. *Biotechnology Advances* 11: 781–812.

European Symposium on Environmental Biotechnology, ESEB 2004 - Verstraete (ed)
 ISBN 90 5809 653 X

Mutagenic activity of chlorine by-products of water microcontaminants – model research

T.M. Traczewska & K. Piekarska
Wrocław University of Technology, Institute of Environmental Protection Engineering, Wrocław, Poland

ABSTRACT: Potential mutagenic and carcinogenic properties of model water disinfected by chlorine, chlorine dioxide and mix of this compounds were examined on the basis of Ames tests. Additionally it was determined the most effective Amberlites resin to concentration of disinfection by-products of water microcontaminants. The most positive results of Ames test was received using the chlorine to disinfection process of model water. Research results indicate that chlorine of water generate mutagenicity disinfection by-products. When the chlorine dioxide was used to disinfection of model water there was not positive results of Ames tests. Using of the series system of three columns with XAD16, XAD7 and XAD2 resins seems to be the most optimal to concentration of organic microcontaminats from water.

1 INTRODUCTION

The increasing contamination of surface waters and consequently ground as well as underground waters necessitates the application of complicated and multi-stage purification methods (Namieśnik & Jaśkowski 1995). One of the processes generating noxious substances organic and inorganic is the process of disinfection. That is why a lot of water purification stations replace chlorine disinfection with chlorine dioxide or a mixture of those compounds in order to reduce the threat posed by secondary products dangerous health (Meier 1998, Fielding & Horth 1986, Świderska-Bróż 2001). The process of disinfection produces the whole variety of halogen compounds with characteristic mutagenic and carcinogenic qualities as well as substances whose toxicodynamic and genotoxic qualities have not been yet identified.

The assessment of full biological activity of water after disinfection can be made only with the use of bioindicative tests. Due to the complete assessment of biological activity of the whole spectrum of pollutants the tests allow for the determination of actual threat to humans. Ames test – recommended by WHO and EPA – was selected from the pool of bioindicative tests (Traczewska 2002).

The objective of research was the determination of potential mutagenicity of model water caused by the presence of non-volatile micro-contaminants from chlorine by-products in water during its disinfection with chlorine, chlorine dioxide and the mixture of these compounds. The tests were supposed to improve the procedure of preparation of extracts from disinfected water for Ames test.

2 MATERIALS AND METHODS

The model water was prepared from sewage water purified on a filter with activated carbon and mixed with water rich in dissolved humus compounds from the "Batorów" peat-bog. The water was disinfected with chlorine, chlorine dioxide and their mixture. The contact time was determined for 16 hours so it would match the estimated time the water is in the sewage system and the disinfectants would be completely used. Next the water was acidified to pH 2 and 1.0 dm^3 of it was run through a set of two filters with XAD resins. The methodology of isolating and concentrating of organic compounds from the water was based on Amberlite XAD resins which enable the recovery of organic compounds from large quantities of water and require small amounts of extractants. The resins used in the research had various polarity, adsorbing compounds of bigger and bigger molecular weight. XAD2 resin adsorbs hydrophobic organic compounds of MW up to 20 000, XAD16 similar compounds up to MW 40 000, XAD7 compounds of moderate polarity of MW up to 60 000. The extraction of organic compounds from deposits was performed with ethyl acetate, methyl alcohol and acetone in order to select the most efficient extractant.

The organic compounds from the first set of filters with XAD resins were physically and chemically analyzed, and from the other set underwent Ames test.

The physical and chemical analyses covered: OWO markers, colours in 350 nm, absorbance in 254 nm, total contents of chloroorganic compounds (TOX), GC/MS chromatographs, as well as basic analysis of the water contents.

The assessment of mutagenic and carcinogenic activity of the organic fraction extracted from the model water was made with the use of Ames test following the procedure recommended by the author (Maron & Ames 1983). Ames test covered the compounds present in the water sample condensed to the concentration rate of: 1000, 800, 500, 400, 300, 200, 100, 50, 25 and 10.

The bioindicative tests of the contamination of the model water disinfected with chlorine, chlorine dioxide and their mixture were performed with the use of two strains of tested bacteria *Salmonella typhimurium* TA98 and TA100. Before the beginning of the proper research genetic markers of those strains, their spontaneous reversion and their reaction to 100% mutagens were checked each time. Each test was performed both with and without the addition a metabolic activator, i.e. microsomal fraction from homogenate of the rat liver (S9 mix.), used in the transformation of promutagens into mutagens. Each case when the mutagenicity rate MR was greater than or equal to 2 (i.e. when the induced reversion was at least twice bigger than the spontaneous reversion) was considered a positive result for the particular model water sample.

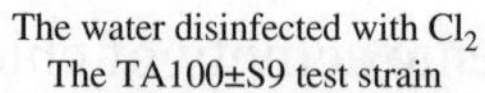

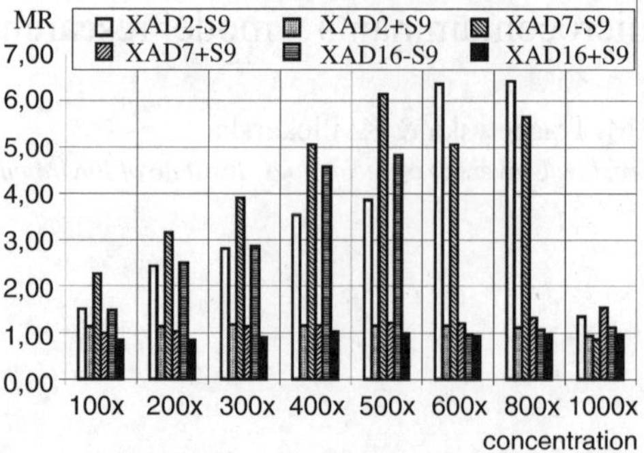

The water disinfected with Cl_2
The TA98±S9 test strain

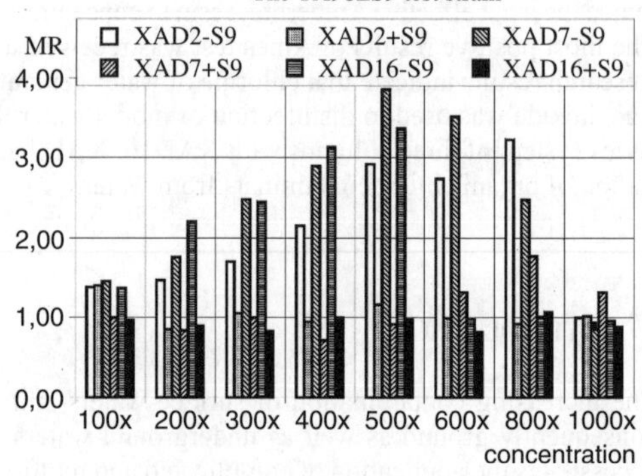

Figure 1 and 1a. The mutagenicity rate of microcontaminants of water disinfected with Cl_2.

3 RESEARCH FINDINGS

The model water colour was 150 gPt/m^3, Abs_{254} 98 m^{-1}, OWO 20.4 gC/m^3, N_{org} 0.6 gN/m^3, N_{NH4} 0.3 gN/m^3. The absorption capacities of resins were: XAD 2–0.48 mgC/g, XAD 7–2.16 mgC/g, XAD 16–3.09 mgC/g. The best extractant was acetone which enabled the recovery of 74% of OWO from resin XAD 2, 94% from XAD 7 and 79% from XAD 16. Methyl alcohol and ethyl acetate demonstrated 10–20% lower extraction efficiency.

Optimal doses of disinfectants were determined on the basis of their demand curves dependant on the contact time. Water of high colour and OWO required the application of doses in the range of 2–25 gCl_2/m^3 and 4–45 g ClO_2/m^3. Optimal dose of chlorine was 11 gCl_2/m^3 and 25 gClO_2/m^3 of chlorine dioxide. In the samples in which the disinfection was made with the use of both disinfectants the demand curve was made in such a way that first 10 gClO_2/m^3 was added and after 2 hours increasing doses of chlorine were added. The amount of 15 gCl_2/m^3 was considered optimal. The total amount of disinfectants was 10 gClO_2/m^3 + 15 gCl_2/m^3.

Chlorination of the model waters caused a decrease of absorption in 254 nm by about 40%, colour by 60%. OWO which was 20.4 gC/m^3 decreased slightly after chlorination. The concentration of TOX in the disinfected water was 2.7 gCl/m^3, and in extracts of XAD 2; 7; 16 respectively 2.1; 2.5; 2.4 gCl_2/m^3.

GC/MS chromatograms indicated several unusual peaks which cannot be identified with the use of the mass spectra library.

The results of Ames tests performed on organic extracts from the model waters disinfected with chlorine, chlorine dioxide and their mixture, produced with the use of three XAD resins (XAD2, XAD7 and XAD16) with different adsorption properties are presented in Figures 1–3.

In case of research done without the addition of microsomal fraction on the extract of the model water disinfected with chlorine (Figure 1 and 1a) an increase of the mutagenicity rate related with an increase of its

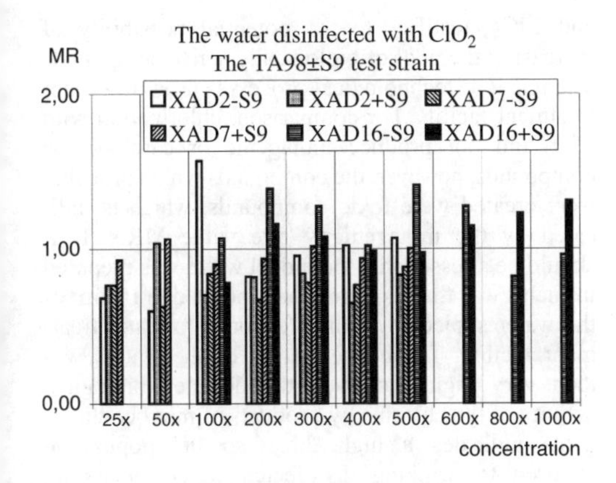

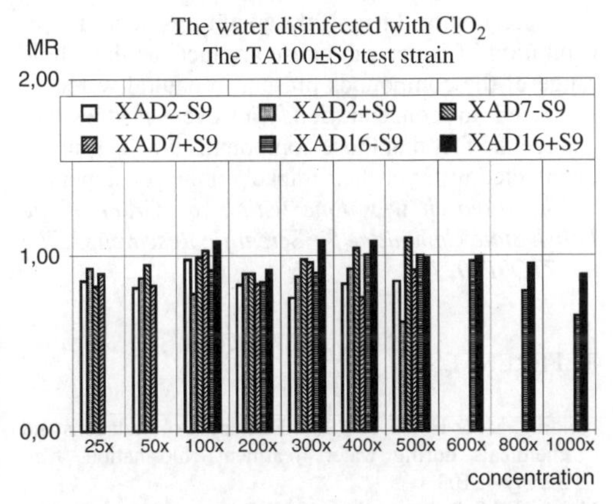

Figure 2 and 2a. The mutagenicity rate of microcontaminants of water disinfected with ClO_2.

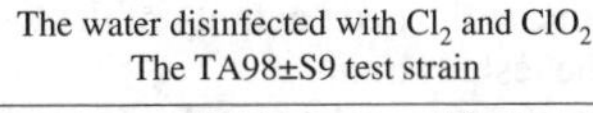

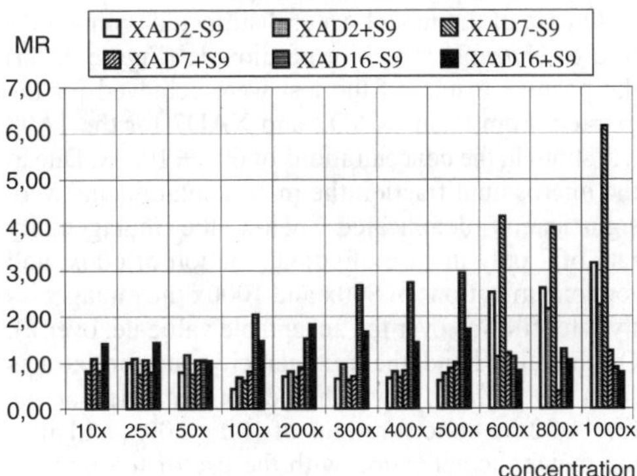

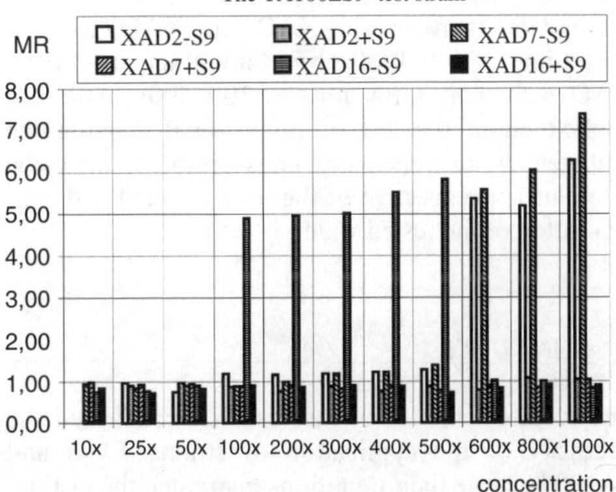

Figure 3 and 3a. The mutagenicity rate of microcontaminants of water disinfected with Cl_2 and ClO_2.

concentration rate was observed. This was observed from the concentration rate of 100x to 500x for the test strain TA98 and from the concentration rate of 100x to 800x for the TA100 test strain. For the more concentrated water the mutagenicity rate was almost one. In case of extracts produced with the use of XAD16 resin, MR = 1 for the tested samples was observed already with the water concentration of 600. In the research for the concentration rate of 200, 300, 400 and 500 the positive results of Ames test were achieved on the TA100 test strain, irrespective of the resin applied. For resins XAD2 and XAD7 the positive results were also achieved in case when the concentration rate was 600 and 800. In case of the TA98 test strain the positive results for all resins were achieved for the concentrations of 400 and 500 of the tested water, and for resins XAD2 and XAD7 for the concentrations of 600 and 800.

In the tests performed in the presence of the microsomal fraction from the rat liver a mutagenicity rate over or equal to 2 was observed in none of the tested samples. Monoxigenases present in the microsomal fraction then were able to detoxicate the mutagenic contaminants present in the model water samples disinfected with chlorine because in the tests with the use of the mixture of S9 a number of revertants induced at the level of spontaneous reversion was achieved.

At present the use of ClO_2 is more and more common in the disinfection of water (Figure 2 and 2a) because the application of this disinfectant results in fewer bi-products which pose a serious threat to health. That is why Ames test was also performed on the model water samples disinfected with chlorine dioxide in analogical conditions to those in case of disinfection with chlorine. No mutagenic contaminants were detected in the extracts of the water prepared that way in case of both test strains of *Salmonella typhimurim,* both in the tests done with and without the addition of fraction S9 mix. In almost all cases the value of mutagenicity rate was at the level of one or lower, which would indicate the toxic activity of the tested samples against the tested strains.

In case of the model water disinfected with a mixture of chlorine and chlorine dioxide (Figure 3, 3a) the positive results of the test were achieved for the extracts from resins XAD2 and XAD7 for the TA98 test strain in the concentrations of 600 ÷ 1000x. Due to the microsomal fraction the microcontaminants were significantly detoxicated (below the mutagenicity rate of 2 only in case of contamination of 600x, and for contaminations of 800x and 1000x the mutagenicity rate still was over the acceptable value i.e. over 2). On the other hand, the mutagenicity rate for extracts from resin XAD16 was higher or remained at the level of 2 for concentrations of 100 ÷ 500x, and after the metabolic activation with the use of the fraction from the rat liver it fell between 1 ÷ 2.

In case of the TA100 test strain the results achieved were also positive for the tests done without fraction S9 for extracts from resins XAD2 and XAD7 for concentrations of 600–1000x, and for extracts from resin XAD16 for concentrat ions of 100–500x. After the application in the test of microsomal fraction the mutagenicity rate decreased considerably to the value of about 1, irrespective of the resin applied and concentration of the tested water.

4 SUMMARY

The results achieved in the research confirmed the necessity of the application of strains TA98 and TA100 because their mutations guarantee the evaluation of the potential mutagenicity of a broad range of chlorine by-products of microcontaminants. It is important to use in the research the resins of various affinity to the adsorbed contaminants, which results in a wide range of those compounds. The sequence of resins XAD16, XAD7, XAD2 in three columns seems optimal.

A high mutagenicity rate of the microcontaminants which was achieved in the water disinfected with Cl_2, as well as in case of the disinfection of water with Cl_2 and ClO_2, testifies to the potential possibility of increasing the pool of biologically active compounds as a result of technological processes used in a water treatment facility. For comparison, chlorination with ClO_2 did not produce mutagenic or carcinogenic compounds, however, the compounds which probably were created were toxic compounds, which is indicated by the mutagenicity rate value $MR < 1$. It should be stressed that the model water was prepared in such a way that its 1 dm^3 corresponded to 10 dm^3 of the water subjected to disinfection in a water treatment facility. The concentration of the waters was then very high. However, the detected biological activity of the chlorine by-products of microcontaminants indicates a high threat to the population exposed to drinking disinfected water containing their precursors. In real-life conditions such a concentration of water can cause artefacts as the whole range of the compounds present in natural waters is subjected to concentration. However, it does not change the fact that those compounds do occur in the disinfected water or their impact on the consumers.

The research was done within the Grant of the Polish State Committee for Scientific Research (KBN) no. 7TO9D 02721

REFERENCES

Fielding, M. & Horth, H. 1986. Formation of mutagens and chemicals during water treatment chlorination. *Wat. Supply* 4: 103.

Maron, D.M. & Ames, B.N. 1983. Revised methods for the Salmonella mutagenicity test. *Mutation Res.* 113: 173–215.

Meier, J.R. 1988. Genotoxic activity of organic chemicals in drinking water. *Mutation Res.* 96: 211.

Namieśnik, J. & Jaśkowski, J. 1995. *Toxicology Outline* (polish). Eko-Phama (eds), Gdańsk.

Świderska-Bróż, M. 2001. Adverse changes of water quality during its purification and distribution (polish). *Inżynieria i Ochrona Środowiska* 4(3–4): 283–300.

Traczewska, T.M. 2002. *Biomonitoring of mutagenicity of microcontaminants of drinking water* (polish). Oficyna Wydawnicza Politechniki Wrocławskiej (eds), Wrocław.

European Symposium on Environmental Biotechnology, ESEB 2004 - Verstraete (ed)
 ISBN 90 5809 653 X

Determination of the degradation potential of food and industrial biowastes under thermophilic anaerobic conditions

T.M. Sivonen & R. Lepistö
Institute of Environmental Engineering and Biotechnology, Tampere University of Technology, Finland

ABSTRACT: Determination of suitable feed ratio for thermophilic codigestion of different biowastes is a crucial step, especially when biowastes with nitrogenous compounds or lipids of inhibitory concentrations are being degraded. In this study, a series of batch experiments were run at 55°C to determine the methane production potential of poultry slaughterhouse waste (PSHW), potato starch waste (PSW), and mixtures of these wastes. The PSW degraded easily (0.502 mL CH_4/h), but the PSHW constituted the limiting factor (0.286 mL CH_4/h). Based on the results, a ratio of 1:4 for PSHW to PSW was selected for co-digestion of the wastes in laboratory and pilot-scale digesters.

1 INTRODUCTION

In Finland, the annual production of organic waste amounts to 34 millions of tons (Finnish Ministry of Environment, 2003). No officially approved methods exist yet in Finland to manage biowastes, though new national legislation and EU directives require that wastes should be processed before disposal. Biowastes are usually rich in nutrients and thus amenable to anaerobic digestion.

Anaerobic digestion is a cost effective technology with major benefits such as high solids destruction, and thus less residual biomass, generation of rich products for land application, and production of methane gas, a valuable energy source for heating through co-generation.

Anaerobic thermophilic digestion has additional benefits over mesophilic digestion such as higher reaction rates (Moen et al., 2001), enhanced solubility of polymeric substrates and improved bioavailability of slightly biodegradable and insoluble compounds, thus promoting the bioconversion of the substrates, that is, higher digester efficiency (Hartmann & Ahring, 2003, Zinder, 1986).

Anaerobic digesters have been commonly operated by digesting sludge and manure of single origin. It has been, nevertheless, shown, that the biogas production can be enhanced by adding different types of biowastes to the feed (Yoneyama & Takeno, 2002, Angelidaki & Ahring, 1997), supplementing a better variety of substrates for microorganisms. Codigestion can also help in lowering the effects of toxic compounds, such as long-chained fatty acids (LCFAs) and ammonia (Angelidaki & Ahring, 1997).

Anaerobic degradation of slaughterhouse wastes with mesophilic digesters is characterized by long hydraulic reterntion time and low loading rates (Salminen, 2000). Slaughterhouse wastes usually have high concentrations of proteins and lipids, which, under anaerobic condition degrade into, among others, ammonia and LCFAs. At higher temperatures, the toxic effects of these compounds are more pronounced, but the formation of lipid scum layer in the digester is minimized. Starch, on the other hand, has been found to be readily degradable in mesophilic batch assays (Sanders et al., 2000).

The overall objective is to provide an efficient and economical codigestion treatment for sewage sludge, poultry slaughterhouse waste (PSHW) and potato starch waste (PSW). As a preliminary step, we focused on determining a suitable feed mixture for efficient start-up of laboratory and pilot digesters. To achieve our goal, we ran a series of batch assays on individual wastes and waste mixtures to determine the methane potential of the wastes.

2 METHODS

2.1 *Wastes*

Modified PSW was from Oy Kationi Ab, Lapua, Finland, producing modified potato starch for coating

Table 1. Properties of the PSHW and PSW.

	PSHW	PSW
pH	6.1	4.0
TS (g/kg)	239	430
VS (g/kg)	212	406
TCOD (g/kg)	360	429
SCOD (g/kg)	Na	26.7
Kj-N (g/kg)	14.6	0.8

Na = Not available; TCOD = Total COD;
SCOD = Soluble COD.

Table 2. The loads of the batch assays.

PSW	kg VS/m^3	PSHW	kg VS/m^3
5%	35.0	1%	14.5
10%	55.5	5%	23.1
15%	75.9	10%	33.9
33%	134		
50%	203		
Mixture 1: 5% PSW + 1% PSHW			37.9
Mixture 2: 5% PSW 1 3% PSHW			58.7

in the paper industry, and the PSHW (trimmings and offal) was from HK Broilertalo Oy, Eura, Finland. The wastes were homogenized before use in the batch assays. The waste characteristics are presented in Table 1.

2.2 *Seed sludge*

Seed used in the batch experiments originated from a thermophilic laboratory digester (55°C, 12-d SRT) inoculated with anaerobically digested mesophilic sludge and fed with a thickened mixture of primary and secondary sludge (60 and 40%, respectively) from Hämeenlinna municipal treatment plant, Finland.

2.3 *Batch assays*

The batch assays were conducted in duplicates using 50 mL serum vials with a headspace of 30 mL. 10 mL of seed sludge was added to the vials with the following concentrations (w/v) of wastes; 1–10% PSHW, 5–50% PSW, and two mixtures of the wastes that contained 5% PSW and 1 or 3% of PSHW. Media solution of macro and trace nutrients was used to dilute the wastes (Lepistö & Rintala, 1995). Vials were prepared under reduced atmosphere by flushing them with nitrogen, and sealed with butyl stoppers and aluminum crimps. Finally, to remove the residual oxygen, $NA_2S \cdot 9H_2O$ was added to achieve a final concentration of 0.25 g/L. The vials were then incubated in static cultures at 55°C. The loads of the batch assays are presented in Table 2.

2.4 *Analyses*

Total and soluble chemical oxygen demands (TCOD and SCOD), and total and volatile solids (TS, VS) were measured according to the Standard Methods (APHA, 1998). pH was measured with a calibrated Knick Portamess 751 meter equipped with a Mettler Toledo Inlab 412 electrode. Nitrogen was measured according to the Kjeldahl-method (Finnish Standards Association SFS, 1988) and methane production was determined with Shimadzu GC-14B gas chromatograph equipped with a thermal conductivity detector and a Porapak Q stainless steel column (mesh 80/100, 2 m × 0.32 cm).

2.5 *Calculations*

The methane production rate (V_{CH_4}) and initial specific methanogenic activity (ISMA) were calculated from the initial methane production by linear regression.

3 RESULTS AND DISCUSSION

3.1 *PSW batch assay*

The methane production started in vials with 5–15% waste additions in less than 2 hours of incubation, while 33 and 50% assays were largely inhibited (Figure 1). The methane production from 5–15% waste additions was similar to that of the sludge control for the first 10 hours, indicating that the microorganisms needed acclimation to the substrate. After 24–28 hours, the methane production of 5–15% assays started to decline, due to the formation and accumulation of CO_2, diluting the methane concentrations in the headspace. CO_2 also contributed to low pH (4.9–5.5) in the vials, which in turn partially inhibited methanogenesis (Winter & Zellner, 1990).

The V_{CH_4} was highest, 0.502 mL/h, for the 5% series (Table 3) and decreased with increasing PSW concentration (10 and 15%), indicating that the methanogens were inhibited by the higher loads. When high concentrations of carbohydrates are degraded, the interspecies hydrogen transfer capacity is exceeded and reduced compounds are produced (Sørensen et al., 1991). The average ISMAs of the methane production for the 5, 10 and 15% were 1.716, 1.345 and 0.933 mL CH_4/g VS_{added} h, respectively. The ISMA observed here are comparable to specific methanogenic activities (SMA) reported in the literature for carbohydrates using mesophilic UASB sludge (Fernández et al., 2001).

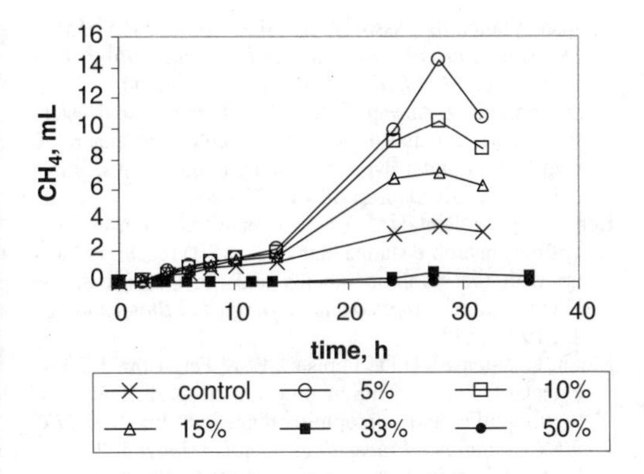

Figure 1. Methane production from PSW assays. Values are averages of the duplicates.

Table 3. The observed methane production rates (V_{CH_4}) and ISMAs of the batch assays.

Assays	V_{CH_4} (mL CH_4/h)	ISMA (mL CH_4/g VS_{added}* h)
PSW		
Control	0.133	0.455
5%	0.502	1.716
10%	0.396	1.354
15%	0.273	0.933
33%	Nd	Nd
50%	Nd	Nd
PSHW		
Control	0.107	0.446
1%	0.286	1.192
5%	0.109	0.454
10%	0.058	0.242
Mixtures		
Control	0.122	0.417
5% PSW + 1% PSHW	0.485	1.658
5% PSW + 3% PSHW	0.297	1.015

Nd = not determined; * = VS_{added} is the seed sludge VS (g).

3.2 *PSHW batch assays*

Methane production from PSHW started in all vials in less than 2 hours of incubation (Figure 2). Furthermore, lower values of V_{CH_4} and ISMA were measured for PSHW than for PSW (Table 3). The V_{CH_4} were of the same magnitude (5%) or less (10%) than the sludge control, while the best rates were achieved with 1% PSHW addition. The low bioavailability of lipids (Petruy & Lettinga, 1997) and high organic loads (Table 2) contributed to the low methane production. The results here compare favorably with values reported in literature. Lag period of up to 10 days has been reported for mesophilic anaerobic digestion of

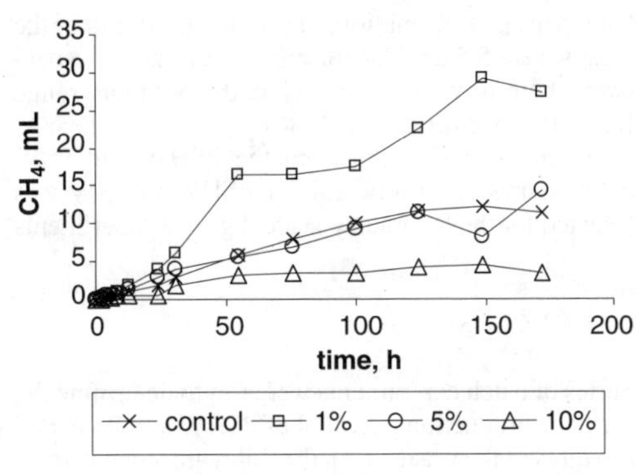

Figure 2. Methane production from PSHW assays. Values are averages of the duplicates.

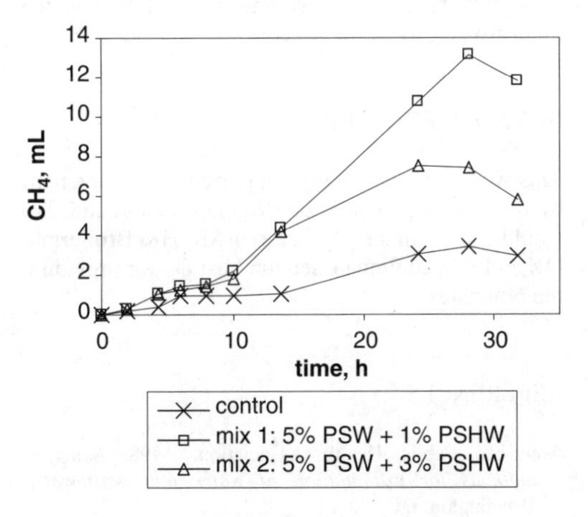

Figure 3. Methane production from two PSHW and PSW mixtures. Values are averages of the two duplicates.

poultry offal (Salminen, 2000). The longer lag periods were attributed to high concentration of LCFAs in the offal. Others reported that the acclimation of sludge to protein-rich feed reduces the lag phase and increases the methane yield (Perle et al., 1995).

3.3 *Waste mixture batch assays*

Based on the best observed methane production rates in PSHW and PSW batch assays, two different PSHW and PSW mixtures were assayed (mixture 1: 5% PSW + 1% PSHW and mixture 2: 5% PSW + 3% PSHW). Methane production started in all the assays in less than two hours from the start-up (Figure 3). The V_{CH_4} of both mixtures was higher (0.485 and 0.297 mL CH_4/h for mixtures 1 and 2, respectively) than that of the sludge control (0.122 mL CH_4/h). The lower V_{CH_4} and ISMA values for mixture 2 were probably due to higher percentage of PSHW in the mixture.

Due to the CO_2 formation, pH values at the end of the assays were 5.5 and 5.4 for mixtures 1 and 2, respectively, both of which were below the optimum range for methanogens (Zinder, 1986).

Based on the results from the PSHW and PSW batch assays, a ratio of 1:4 for PSHW to PSW was selected for the laboratory-scale digester experiments

4 CONCLUSIONS

Series of batch experiments were run to determine the methane production potential of PSHW and PSW and mixtures of the wastes with the following conculsions:

The PSW degraded easily (0.502 mL CH_4/h), but the poultry slaughterhouse waste constituted the limiting factor (0.286 mL CH_4/h).

A mixture of 5% PSW and 1% PSHW had the highest V_{CH_4}, 0.485 mL CH_4/h.

ACKNOWLEDGEMENTS

This work was financially supported (Tarja Sivonen) by the National Technology Agency of Finland. We would like to thank Oy Kationi Ab, HK Broilertalo Oy, and Hämeenlinnan Seudun Vesi Oy for providing the biowastes.

REFERENCES

American Public Health Association. 1998. *Standard methods for examination of water and wastewater.* Washington, DC, USA.

Angelidaki, I. & Ahring, B.K. 1997. Codigestion of olive oil mill wastewaters with manure, household waste and sewage sludge. *Biodegradation* 8: 221–226.

Fernández, B., Porrier, P. & Chamy, R. 2001. Effect of inoculum-substrate ratio on the start-up of solid waste anaerobic digesters. *Water Science and Technology* 44(4): 103–108.

Finnish Ministry of Environment. 2003. (in finnish) Biojätestrategiatyöryhmän ehdotus kansalliseksi biojätestrategiaksi.

Finnish Standards Association SFS. 1988. *SFS 5505; Determination of inorganic and organic nitrogen in wastewater. Modified Kjeldahl method.* Finland.

Hartmann, H. & Ahring, B.K. 2003. Phthalic acid esters found in municipal organic waste: enhanced anaerobic degradation under hyper-thermophilic conditions. *Water Science and Technology* 48(4): 175–183.

Lepistö., R. & Rintala, J. 1995. Acetate treatment in 70°C upflow anaerobic sludge blanket (UASB) reactors: Start-up with thermophilic inocula and the kinetics of the UASB sludges. *Applied microbiology and Biotechnology* 43: 1001–1005.

Moen, G., Stensel, H.D., Lepistö., R. & Ferguson, J. 2001. Effect of solids retention time on the performance of thermophilic and mesophilic digestion. In: *WEFTEC 2001 Conference Proceedings; October 2001*: 1–20.

Perle, M., Kimchie, S. & Shelef, G. 1995. Some biochemical aspects of the anaerobic degradation of dairy wastewater. *Water Research* 29(6): 1549–1554.

Petruy, R. & Lettinga, G. 1997. Digestion of milk-fat emulsion. *Bioresource Technology* 42: 17–26.

Salminen, E. 2000. *Anaerobic digestion of solid poultry slaughterhouse by-products and wastes.* Ph.D. Thesis. Jyväskylä Studies in biological and environmental science, University of Jyväskylä. Jyväskylä, Finland.

Sanders, W.T., Geerink, M., Zeeman, G. & Lettinga, G. 2000., Anaerobic hydrolysis kinetics of particulate substrates. *Water Science and Tehcnology* 41(3): 17–24.

Sørensen, A.H., Winther-Nielsen, M. & Ahring, B.K. 1991. Kinetics of lactate, acetate and propionate in unadapted and lactate-adapted thermophilic, anaerobic sewage sludge: the influence of sludge adaptation for start-up of thermophilic UASB-reactors. *Applied Microbiology and Biotechnology* 34: 823–827.

Winter, J. & Zellner, G. 1990. Adaptive properties of thermophiles: thermophilic anaerobic degradation of carbohydrates – metabolic properties of microorganisms from the different phases. *FEMS Microbiology Reviews* 75: 139–154.

Yoneyama, Y. &Takeno, K. 2002. Co-digestion of domestic kitchen waste and night soil sludge in full-scale sludge treatment plant. *Water Science and Technology* 45(10): 281–286.

Zinder, S.H. 1986. Thermophilic waste treatment systems. In: Brock, T.D. (ed.), *Thermophiles: General, Molecular, and Applied Microbiology*. Wiley Series in Ecological and Applied Microbiology, a Wiley Interscience Publication, John Wiley & Sons, New York, USA. 257–277.

European Symposium on Environmental Biotechnology, ESEB 2004 - Verstraete (ed)

Application of aquatic weeds as carbon sink in wastewater treatment

A. Mulder
Amecon Environmental Consultancy, Delft, The Netherlands

ABSTRACT: Greenhouse gas emissions and climate change will have major consequences for human life. Therefore it is relevant to develop technologies to mitigate the CO_2 emissions. Photosynthesis is a basic and relatively cheap process to sequester CO_2. In this paper the application of appropriate aquatic weeds as potential carbon sink is discussed. Based on an annual nitrogen excretion through sewage of 4.75 kg N per capita and a CO_2 fixation of 20 kg CO_2 per kg N the potential CO_2 sequestration rate is 100 kg CO_2 per capita per year. This value is about 2% of the total anthropogenic carbon dioxide emissions. This shows that wastewater treatment with aquatic weeds can have a modest contribution in the development of sustainable treatment processes.

1 INTRODUCTION

1.1 *Human impact on the carbon cycle*

There is serious concern for the major emissions of carbon dioxide which originate from the combustion of fossil fuels (Schlesinger, 1991 and Smil, 1996). The total emission is about 7 Gt C annual which equals 26 Gt CO_2 annual. Related to the global population the averaged annual emission rate is more than 4 t CO_2 per capita. This rate will differ considerably between the individual countries.

In the enormous reservoir of carbon in the oceans the anthropogenic emissions of carbon dioxide are sequestered partially (Figure 1 and Figure 2).

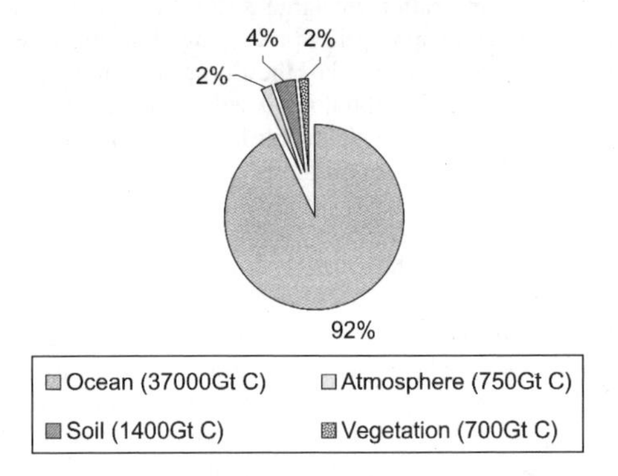

Figure 1. Distribution of the carbon reservoirs in the biosphere (based on data from Paustian et al., 1997 and Smil, 1996).

However in spite of this CO_2 uptake in the oceans the CO_2 concentration in the atmosphere rises and does play a major role in the observed climate change. Therefore it relevant to look for potential carbon sinks.

1.2 *Potential carbon sinks*

For mitigation of the anthropogenic CO_2 emissions a wide variety of actions have been proposed (Paustian et al., 1997). The most effective action is of course the reduction of the application of fossil fuel. However this is not fully realistic and therefore alternative actions are required as well. A promising process is the biological photosynthesis where with sunlight CO_2 and water are converted into biomass (equation 1).

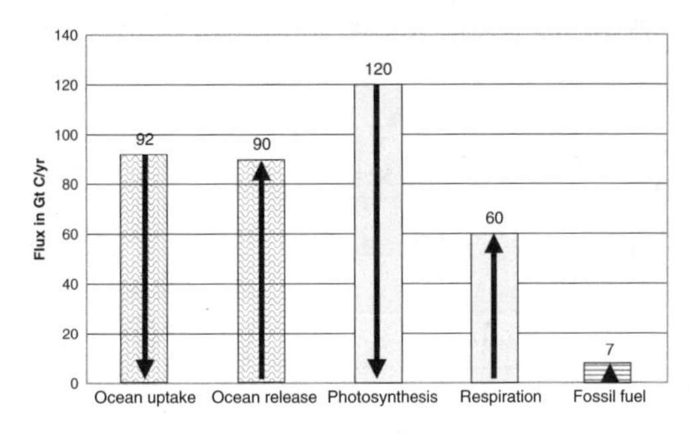

Figure 2. Simplified summary of major carbon fluxes in the biosphere (based on data from Schlesinger, 1991 and Smill, 1996).

$$CO_2 + H_2O \rightarrow CH_2O \quad (1)$$

In it relevant that the C which is converted into the biomass remains fixed and does not return by aerobic respiration as CO_2 or even worse as CH_4 by anaerobic fermentation. In the past vast amounts of biomass carbon were converted ultimately into graphite starting with formation of humus (Baas Becking, 1934 and Stumm, 1970 and equation 2).

$$CH_2O \rightarrow C(s) + H_2O \quad (2)$$

In this paper the feasibility of the application of aquatic weeds as carbon sink in wastewater treatment is discussed.

2 AQUATIC WEEDS AS CARBON SINK IN WASTEWATER TREATMENT

2.1 *Description of constructed wetlands*

Constructed wetlands are defined as a wetland with aquatic weeds designed for the purpose of wastewater treatment. Constructed wetlands are used for the treatment of anaerobic pre-treated wastewater from housing complexes and for tertiary treatment to effluents from conventional sewage treatment. The aquatic weeds to be applied are generally native plant species which grow locally and have rapid growth rates. Often applied are reed species *Phragmites australis* and *Phragmites communis* (Arceivala, 1998).

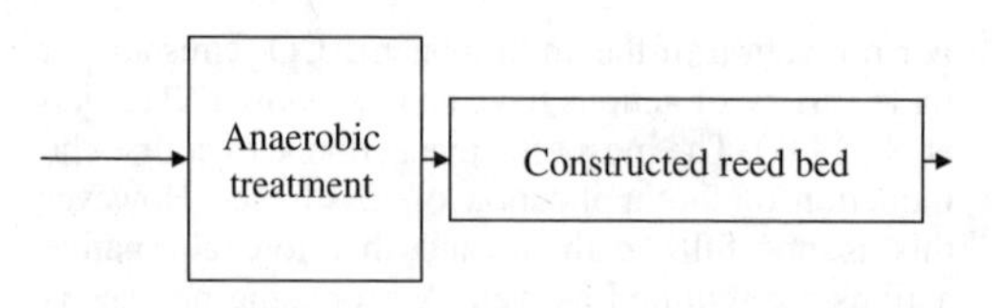

Figure 3. Typical layout for combined anaerobic pre-treatment and constructed wetland as carbon sink.

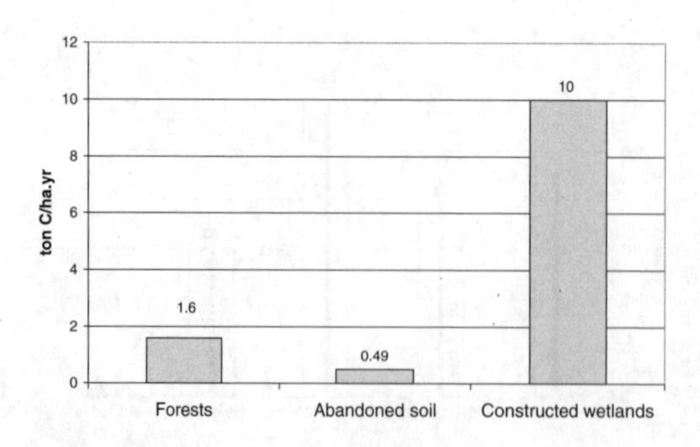

Figure 4. Comparison of the C sequestration rates in forests, abandoned land and constructed wetlands (data from Smil, 1996, Paustian et al., 1997 and Mandi et al., 1996).

For the sustainability it relevant to remove organic pollutants prior to the treatment in the constructed wetland in order to prevent there any uncontrolled release of methane. The methane produced in the anaerobic pre-treatment can be use as green energy. In the constructed wetlands high specific conversion rates are reported (Figure 4, Etherington, 1983 and Mandi et al., 1996). However the reported nutrient removal rates does not always meet the strong effluent standards. Therefore an additional option is to use the soil-vegetation biosystem has been developed a couple of decades ago but which may of interest now in view of CO_2 sequestration (Sopper, 1976).

A prerequisite for the application of constructed wetlands is the presence of enough sunlight which means that the process can be applied only in the European summer period from May–September. The application of this process seems appropriate for e.g. major tourist sites in southern Europe (e.g. French, Italy, Greece and Spain) where during summer many tourists remain. Besides the sustainability other important advantages of the proposed process are the technology is simple, investments are low, fast start-up and low maintenance during summer season.

Based on an annual nitrogen excretion through sewage of 4.75 kg N per capita and a CO_2 fixation of 20 kg CO_2 per kg N the potential CO_2 sequestration rate is 100 kg CO_2 per capita per year. This value is about 2% of the total annual anthropogenic carbon dioxide emissions of about 4000 kg CO_2 per capita.

2.2 *Feasibility of constructed wetlands as carbon sink*

Comparison of the effectiveness of several methods for mitigation of the carbon dioxide emission show that the constructed wetlands are quite effective (Figure 4). However application on large scale will require the development of appropriate processing of the produced reed (Polprasert, 1996). For the European scientists in environmental biotechnology it will be a challenge to optimize processes which contribute in the abatement of the carbon dioxide emissions.

3 CONCLUSIONS

Based on an annual nitrogen excretion through sewage of 4.75 kg N per capita and a CO_2 fixation of 20 kg CO_2 per kg N the potential CO_2 sequestration rate is 100 kg CO_2 per capita per year. This value is about 2% of the total anthropogenic carbon dioxide emissions. This shows that wastewater treatment with aquatic weeds can have a modest contribution in the development of sustainable treatment processes.

REFERENCES

Arceivala, S.J. 1998. Wastewater treatment for pollution control. 2nd ed. Tata McGraw-Hill Publish. Comp. Ltd., New Delhi.

Baas Becking, L.G.M. 1934. Geobiologie of Inleiding tot de Milieukunde. Van Stockum & Zoon N.V. Den Haag.

Etherington, J.R. 1983. Wetland ecology. The Institute of Biology's Studies in Biology no. 154. Edward Arnold Ltd., London.

Goldman, J.C. 1979. Outdoor algal mass cultures-I. Applications. Wat. Res. 13: 1–19.

Laegreid, M., Bockman, O.C. & Kaarstad, O. 1999. Agriculture, fertilizers & the environment. CABI Publish., Wallingford.

Mandi, L., Houhoum, B., Asmama, S. & Schwartzbrod. 1996. Wastewater treatment by reed beds an experimental approach. Wat. Res. 30: 2009–2016.

Mulder, A. 2003. The quest for sustainable nitrogen removal technologies. Wat. Sci. & Technol. 48 (1): 67–75.

Paustian, K., Andrén, O., Janzen, H.H., Lal, R., Smith, P., Tian, G., Tiessen, H., Van Noordwijk, M. & Woomer, P.L. 1997. Agricultural soils as sink to mitigate CO_2 emissions. Soil Use and Management 13: 230–244.

Polprasert, C. 1996. Organic waste recycling. John Wiley & Sons Ltd., Chichester, England.

Schlesinger, W.H. 1991. Biogeochemistry An analysis of global change. Academic Press, Inc. San Diego.

Sopper, W.E. 1976. Use of the soil-vegetation bio system for wastewater recycling. In: R.L. Sanks & T. Asano (eds), *Land treatment and disposal of municipal and industrial wastewater*: 17–43. Ann Arbor: Ann Arbor Science.

Smil, V. 1996. Cycles of life. Civilization and the biosphere. Scientific American Library, New York.

Stumm, W. & Morgan, J.J. 1970. Aquatic chemistry. An introduction emphasizing chemical equilibria in natural waters. Wiley-Interscience, New York.

European Symposium on Environmental Biotechnology, ESEB 2004 - Verstraete (ed)
© 2004 Taylor & Francis Group, London, ISBN 90 5809 653 X

Biomethanation of cheese whey using an upflow anaerobic filter

H. Gannoun, H. Bouallagui, Y. Touhami & M. Hamdi
UR-Procédés Microbiologiques et Alimentaires. Institut National des Sciences Appliquées et de Technologie (INSAT), Tunisie

ABSTRACT: The feasibility of cheese whey treatment by an upflow anaerobic filter (UAF) was tested. The effect of hydraulic retention time (HRT) (1, 2, 3, 4 and 5 days) and the feed concentration (5, 10, 15 and 20 g COD/l) on the extent of the degradation of this wastewater was examined. Varying the HRT from 5 to 1 day at 35°C had no effect on the fermentation stability and pH remained between 6.83 to 7.34. The performance of the reactor was depressed by changing the feed concentration from 15 to 20 g COD/l. The average total COD removals achieved was 80–90%. Physico-chemical pre-treatment using floculation and decantation were used to remove the fats and the suspended solids through the decanter and then to reduce the organic matter concentration. 50% of COD concentration was removed and the pre-treated effluent was used to feed the UAF reactor. *Lactobacillus paracasei ssp paracasei* 1 which was isolated from cheese whey showed higher acidification efficiency and can be used to enhance the acidification step in a separated reactor.

1 INTRODUCTION

In the last years, anaerobic treatment technology has developed remarkably for the treatment of specific industrial wastewaters (Garcia *et al.*, 1998; Lettinga *et al.*, 1999; Plumb *et al.*, 2001). In many developing countries, which are confronting rapidly increasing water pollutions problems, this technology becomes even more favourable and promising (Agrawal *et al.*, 1997; Shigeki *et al.*, 1999). In Tunisia, a large number of dairies dispose of their waste, especially cheese whey, into the environment in enormous quantities. Since cheese whey has a high COD of above 50 g/l, its disposal remains major problem. Anaerobic digestion of cheese whey offers an excellent solution in terms of both energy production and pollution control.

Since growth rates of anaerobic bacteria are very low, a long detention time in an anaerobic reactor is required for sufficient degradation of organic matter. Therefore much research has been directed to the development of techniques for maintaining a high concentration of biomass in anaerobic reactors (Jewell *et al.*, 1981; Kuba *et al.*, 1990). The upflow anaerobic filters (UAF) offer important advantages on other types of anaerobic reactors (Patel *et al.*, 1995). Among these, are ideal for soluble residues, they do not need solids or effluent recirculation, and accumulation of high concentrations of active solids (Jewell *et al.*, 1981). Furthermore, there is a little sludge production, with effluents substantially free of suspended solids (Veiga *et al.*, 1994).

On this basis, a study was conducted to asses the feasibility of pre-treated cheese whey by anaerobic treatment using an up flow anaerobic bed filter. The main objectives of this study were to investigate the effect of loading rate and the hydraulic retention time (TRH) on the reactor performance.

2 MATERIAL AND METHODS

2.1 *The up flow anaerobic bed filter (UAF)*

The schematic diagram of the experimental digester is shown in Fig.1. The reactor is consisted of glass column of 60 cm in height and 10 cm in diameter. The total volume of the reactor is 2 l. The digester was packed with the flocor as a support for the growth of microorganisms and inoculated initially with a mixture of cow dung slurry and an inoculum obtained from an active biogas digester of green wastewater treatment (Bouallagui *et al.*, 2003). After that, the digester was loaded with pre-treated cheese whey and operated at optimal mesophilic temperature range (35 ± 1°C). The UAF reactor was loaded with a dilute influent at different concentrations of COD

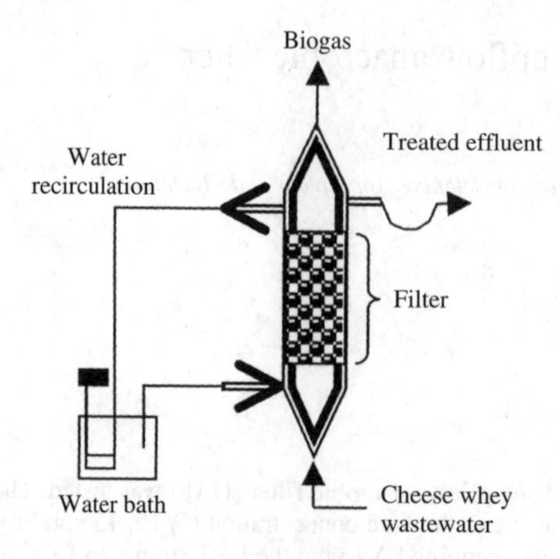

Figure 1. Schematic of the experimental set-up.

(5, 10, 15, 20 g/l) and the HRT was fixed at 5 days and further reduced at an HRT of 4, 3, 2, 1 day by changing the initial fed flow rate.

2.2 *Technical analysis*

The analysis of the samples taken from the digesters effluents were carried out when the steady-state was established.

The biogas produced was measured daily by a gas-meter, and the biogas samples were analysed using an ORSAT apparatus. Total solids (TS), total suspended solids (TSS), biological oxygen demand (BOD), carbon oxygen demand (COD) and pH were determined according to the standard methods (APHA, 1992).

3 RESULTS AND DISCUSSION

3.1 *Substrate characteristics and physicochemical pre-treatment*

The wastewater was obtained from a local factory for milk processing. It was analysed and stored at −20°C. The effluent is pre-treated for suspended solids (SS) and total greases removal. pH of the effluent was adjusted to 7.0 by addition of lime and the laboratory digesters are fed with the buffered wastewater. The physico-chemical characteristics of the wastewater before and after primary treatment are given in Table1.

The pre-treatment eliminated more than 50% of the total COD and about 60% of the total suspended solids content (TSS). The BOD/COD ratio is increased from 0.5 to 0.7 making cheese whey wastewater more suitable for a biological treatment.

Table 1. Cheese whey wastewater characteristics before and after physicochemical pre-treatment.

Parameters	Before pre-treatment	After pre-treatment
pH	4.46–4.6	7–7.2
Acidity (°D)	47	32
BOD_5 ($g\,O_2$/l)	40	18–22
COD (g/l)	50–60	25–30
TS (g/l)	59.09	42
TSS (g/l)	1.3	0.6

3.2 *Reactor performance*

The start up of digester is always a critical step because of the risk of strong acidification. However, in the case of cheese whey effluent, the start up occurred in 4 weeks without any specific problems. This is mainly due to the liberation of ammonia by protein degradation and a synergic interaction between the acidogenic and the methanogenic bacteria.

3.2.1 *Biogas production and quality*

The effect of varying feed concentration and HRT on the UAF reactor performance is shown in Fig. 2. The biogas production rate from anaerobic digestion of organic matter of cheese whey wastewater was improved by the increase of the feed concentration from 5 g COD/l to 15 g COD/l. However, there was a significant decrease in conversion of the substrate into biogas when the feed concentration increased from 15 g COD/l to 20 g COD/l. The methanogenesis was inhibited due to a pH decrease from 7.2 to 5.9 because of the rapidly degradation of substrate to volatile fatty acids (VFA). The highest biogas production rate of 3.2 l/d was obtained with 15 g COD/l feed concentration.

The volume of methane produced per kg of removal COD decreased by increasing the feed concentration and the highest conversion of wastewater to methane was obtained at low organic loading rate (1 g COD/l.d).

As the HRT was decreased, a gradual increase in the amount of biogas production rate (litre/day) was observed. Varying the HRT from 4 to 2 days had no effect on the fermentation stability. However, the methane production yield decreased from 0.32 l CH_4/g removal COD to 0.12 l CH_4/g removal COD by decreasing the HRT to 1 day. The failure of reactor performance at low HRT is due to the washout of substrate without submitting efficient biodegradation.

3.2.2 *COD removal and pH variation*

The total COD destruction and the pH variation at different feed concentrations are shown in Fig. 3. For the first three applied loading rate, the COD removal

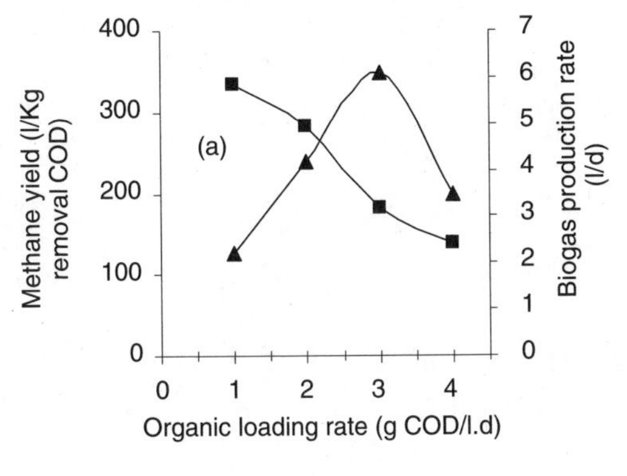

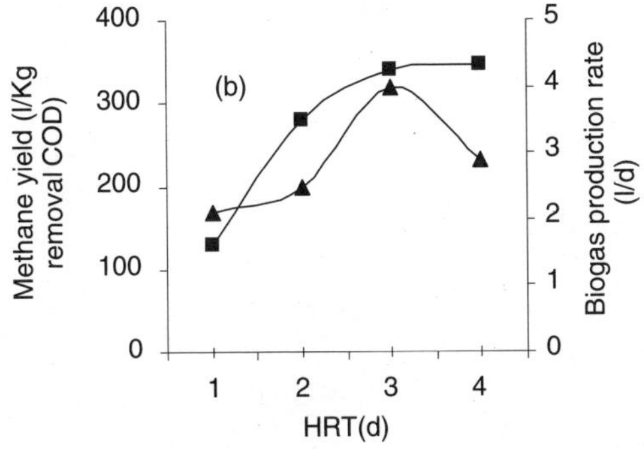

Figure 2. Biogas production rate (▲) and methane yield (■) variation at different organic loading rate (a) and different HRT (b).

efficiency was higher than 80%. It reaches its maximum at the weak loads ($L_R < 1.3$ Kg $COD/m^3 \cdot j$) and it was associated with the higher specific methane production per removal COD and a lower content of BOD in the digested effluent.

The pH was monitored continuously in the reactor. Under optimal conditions it remained to its neutral value (7 to 7.6) due to the process stability. The optimal activity of methanogenic bacteria and the volatile fatty acids were rapidly degraded to methane and carbon dioxide. However under 20 g COD/l feed concentration conditions, the pH decreased rapidly and the conversion of substrate to biogas was inhibited due to the high loading rate and the inhibition of methanogenic bacteria especially by increased VFA and pH decrease.

4 CONCLUSION

The results of this work prove that the upflow anaerobic filter reactor can be utilized for cheese whey wastewater management and energy recovery at relatively short time (2 days). It achieved good treatment efficiency up to 90% of COD removal and was operated stably at high loading rate.

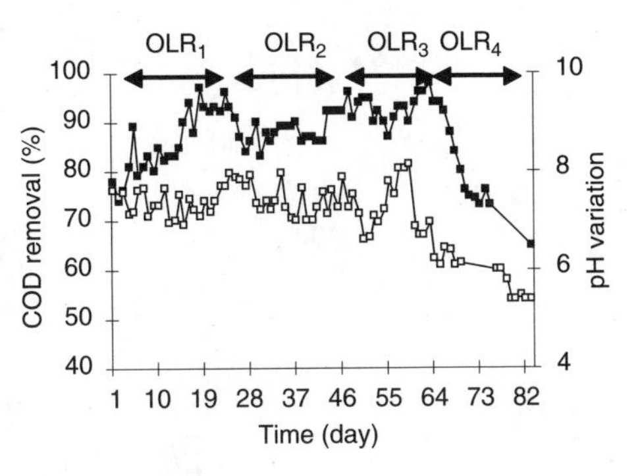

Figure 3. COD removal (■) and pH variation (□) at different loading rates.

REFERENCES

Agrawal, Lalit, K., Harada, H. and Okui, H. (1997). Treatment of Dilute in UASB Reactor at a Moderate Temperature: Performance Aspects. Journal of Fermentation and Bioengineering. Vol. 83, No. 2, 179–184.

Bouallagui, H., BenCheikh, R., Marouani, L. and Hamdi, M. (2003). Mesophilic biogas production from fruit and vegetable waste in tubular digester. Bioresource technology, Vol. 86, pp. 85–89.

Garcia, C.D., Buffiere, P., Moletta, R. and Elmaleh, S. Anaerobic digestion of wine distillery wastewater in down-flow fluidized bed. Water Research, Vol. 32, pp. 3593–3600.

Jewell, W.J., Switzenboum, M.S. and Morris, J.W. (1981). Municipal wastewater treatment with the anaerobic attached microbial film expanded bed process. J. Water. Poll, Vol. 53, pp. 482–490.

Kuba, T., Furumai, H. and Kusuda, T. A kinetic study on methanogenesis dy attached biomass in a fluidized bed. Water Research, Vol. 24, pp. 1365–1372.

Lettinga, G., Rebac, S., Parshina, S., Nozhevnikova, A.N., Van Lier, J.B. and Stams, A.J.M. (1999). High-rate anaerobic treatment of wastewater at low temperatures. Applied and Environmental Microbiology, Vol. 65, n°4. pp. 1969–1702.

Priti Patel, Manik Desai and Datta Madamwar, (1995). Biomethanation of Cheese Whey Using Anaerobic Upflow Fixed Film Reactor. Journal of Fermentation and Bioengineering. Vol. 79, No. 4, 398–399.

Plumb, J.J., Bell, J. and Stukey, D.C. (2001). Microbial Populations Associated with Treatment of an Industrial Dye Effluent in an Anaerobic Baffled Reactor. Applied and Environmental Microbiology. Vol. 67, 7, pp. 3226–3235.

Standard Methods for the Examination of Water and Wastewater (APHA,AWWA and WEF), 18th Edition (1992).

Shigeki, U. and Hideki, H.1999. Treatment of sewage by a UASB reactor under moderate to low temperature conditions. Bioresource Technology. Vol. 72, No. 2000, 275–282.

Veiga, M.C., Mendez, R. and Lema, J.M. Anaerobic filter and DSFF reactors in anaerobic reatment of tuna processing wastewater. Water Sci. Technol., 1994, Vol. 30, pp. 425–432.

European Symposium on Environmental Biotechnology, ESEB 2004 - Verstraete (ed)
© 2004 Taylor & Francis Group, London, ISBN 90 5809 653 X

High storage of PHB by mixed microbial cultures under aerobic dynamic feeding conditions

L.S. Serafim[1], P.C. Lemos[1,2], R. Oliveira[1], A.M. Ramos[1] & M.A.M. Reis[1]
[1]*CQFB/REQUIMTE, Chemistry Department, FCT/UNL, Caparica, Portugal*
[2]*Instituto de Tecnologia Química e Biológica (ITQB), UNL, Oeiras, Portugal*

ABSTRACT: Activated sludge submitted to aerobic dynamic feeding conditions showed a high and stable capacity to store polyhydroxybutyrate (PHB). The influence of carbon and nitrogen concentrations on the PHB accumulation yield was studied in a range of 30 Cmmol/l to 180 Cmmol/l for acetate and between 0 Nmmol/l and 2.8 Nmmol/l for ammonia. Low ammonia concentrations favored PHB accumulation. The maximum PHB content, 67.5%, was obtained for 180 Cmmol/l of acetate supplied in one pulse but with a slower PHB storage rate. In order to avoid substrate inhibition, 180 Cmmol/l of acetate were supplied in two different ways: continuously fed and in three pulses of 60 Cmmol/l each. In both cases the specific PHB storage rate increased and the PHB content obtained were, 56.2% and 78.5%, respectively. Addition of acetate by pulses controlled by the oxygen concentration was kept for 16 days. The PHB content was always above 70% of cell dry weight.

1 INTRODUCTION

Polyhydroxyalkanoates (PHAs) share identical properties with polypropylene, but with the advantage of being biodegradable. PHAs are industrially produced by pure cultures, like *Ralstonia eutropha*, using as carbon substrates glucose or a mixture of glucose and propionic acid. In industrial PHA production by pure cultures the polymer account for 85% of the cell dry weight. The major expenses in industrial PHA production deal with the substrate cost and the extraction of the polymer from inside the cells. The use of renewable sources obtained from waste organic carbon and mixed cultures (activated sludge) can significantly decrease the price of PHA. Utilization of open mixed microbial cultures facilitates the use of complex substrates since microbial population can adapt continuously to changes in substrate. Consequently, the need for sterilization and sterile fermentation systems is prevented, which contributes for the reduction of the final PHA price (Serafim et al. 2001).

Storage of polyhydroxyalkanoates (PHAs) by activated sludge can be particularly important in wastewater treatment systems, if sludge is submitted to consecutive periods of external substrate accessibility (feast) and unavailability (famine), generating a so-called unbalanced growth. Under these dynamic conditions, during excess of external carbon substrate, the uptake is driven to simultaneous growth of biomass and polymer storage and after substrate exhaustion stored polymer can be used as energy and carbon source. The storage phenomenon is usually dominant (70%) over growth, but under conditions in which substrate is present for a long time, physiological adaptation occurs and growth becomes more important (Van Aalst-Van Leeuwen et al. 1997; Dionisi et al. 2001). Bacteria use storage capacity to provide substrate for growth when it is depleted from the external medium, getting a strong competitive advantage over microorganisms without this ability (Majone et al. 1996).

The PHA yield in activated sludge is lower than the maximum value reported for industrial production by pure cultures. In fact, PHA accumulated inside the biomass in activated sludge systems can reach up to 62% of cell dry weight (Satoh et al. 1998). In order to turn the process with mixed cultures competitive with the based in pure cultures, optimization of PHA production is required.

In this work, activated sludge was submitted to aerobic dynamic substrate feeding (ADF) for 18 months. The influence of carbon and nitrogen concentrations and the reactor feeding regime were studied. A strategy of reactor operation based on oxygen concentration oscillations was implemented.

2 MATERIALS AND METHODS

Activated sludge was inoculated into a reactor where sequential cycles of 12 h were imposed, which comprised 10.5 h of aerobiosis, with feeding during the first 15 min, 1 h of settling and after that 30 min for withdraw half of liquid phase. The mean cell retention time for the reactor was 10 days. The mineral medium composition used was the same described in Lemos et al. (1998) supplemented with allylthiourea (10 mg/L) in order to avoid nitrification. Organic acids, suspended solids (SS), volatile suspended solids (VSS) and PHA determination were done according to Lemos et al. (1998). Oxygen uptake rate (OUR) was measured in a respirometer connected to the fermenter by a high frequency recirculating pump. The decrease in oxygen concentration along the time was evaluated during 2 min after the pump was stopped. Ammonia was determined using an ammonia gas sensing combination electrode ThermoOrion 9512.

The SBR with acetate addition controlled by dissolved oxygen concentration had similar operating parameters except the duration of aerobiosis; 10 h of starvation period were imposed after carbon source depletion. The SBR operation cycles were implemented using a rule-based supervisory controller. The acetate pulses were implemented with a DO-based feed controller.

3 RESULTS AND DISCUSSION

3.1 *System performance*

The biomass showed a quick adaptation to the ADF conditions by storing around 57% of PHB by cell cry weight in the first day (Figure 1).

The system showed to be stable during almost 18 months of continuous operation. However, cell PHA content was low (30–40%) when compared with values obtained for mixed cultures submitted to transient carbon supply in short term reactor operation (around 50%).

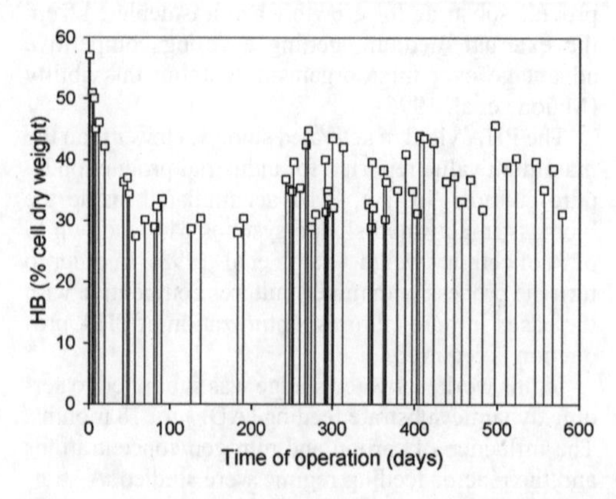

Figure 1. Evolution of PHB production in a SBR operated under ADF conditions fed with acetate.

3.2 *Effect of concentration of carbon and nitrogen sources*

Several experiments with different concentrations of HAc were performed in order to evaluate its impact on PHA production. Four concentrations of HAc were tested: 30 Cmmol/L (the same concentration added to the SBR), 60 Cmmol/L, 90 Cmmol/l and 180 Cmmol/l. The storage capacity of the biomass, the storage yield ($Y_{P/S}$, ratio between the PHB produced and the acetate consumed) and the PHB specific storage rate (q_P) for the different experiments can be observed in Figure 2.

The amount of PHB stored increased with acetate concentration, being 38.6% the lowest value obtained with 30 Cmmol/l and the highest value, 67.2% with 180 Cmmol/l of acetate supplied. This last value of PHB stored is higher than the usual values obtained with activated sludge.

The PHB storage rate increased with the carbon concentration in the range of 30 Cmmol/l to 90 Cmmol/l. However when 180 Cmmol/l of acetate were supplied the PHB production rate decreased sharply. This value of acetate concentration seemed to inhibit the PHB production. The storage yield was not significantly affected by acetate concentration.

The effect of nitrogen source was also analyzed. Four different concentrations of ammonia were tested: 0, 0.7, 1.4 and 2.8 Nmmol/l. The maximum amount of PHB stored, storage yield and PHB storage specific rate are shown in Figure 3.

The highest amount of PHB stored was obtained for 0.7 Nmmol/l (38.6%) and the lowest for 2.8 Nmmol/l (25.4%). The storage yield decreased inversely with ammonia concentration. When no ammonia was

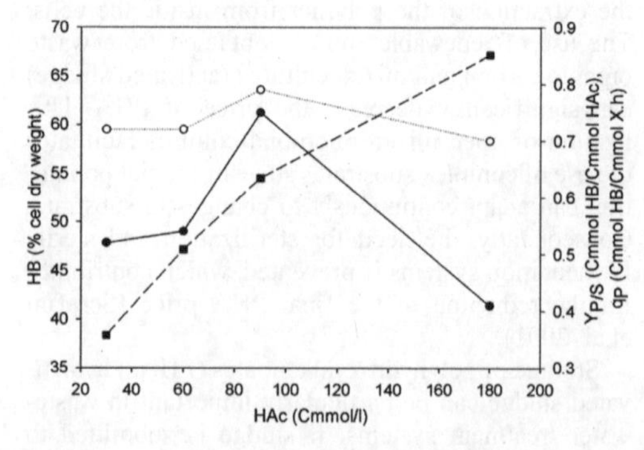

Figure 2. Comparison of PHB production in the experiments with different concentrations of acetate (■, cell PHB content; ○, storage yield; ●, PHB specific storage rate).

supplied, the storage yield (0.83 Cmmol PHB/Cmmol HAc) was more than the double of that obtained for 2.8 Nmmol/l (0.37 Cmmol HB/Cmmol HAc). Except for the case where no ammonia was supplied, the specific storage rate decreased inversely with the increase of ammonia concentration. The results showed that the higher the amount of ammonia used, the more carbon was deviated for growth and less important was the storage process.

3.3 *Effect of feeding regime*

As previously shown, high acetate concentration (180 Cmmol/l) favored PHB accumulation but the PHB specific storage rate decreased due to substrate inhibition. In order to overcome inhibition, the same amount of carbon substrate was fed to the reactor in two different ways: continuously and split in three pulses of 60 Cmmol/l each.

In the continuous experiment, the rate of substrate addition to the reactor was selected based on the value of substrate consumption rate previously determined for 60 Cmmol/l, in order to compare results for this experiment with the one in which the substrate was supplied in three pulses. In the later experiment, since acetate was not analyzed on-line, the exact time to add a new pulse was determined by the sudden increase in DO concentration caused by substrate exhaustion. The results in terms of maximum amount of PHB stored, storage yield and PHB storage specific rate are shown in Figure 4. In the experiment with three pulses average results of the three pulses are shown.

In the continuous experiment, the amount of PHB stored per cell dry weight (56.2%) as well as the storage yield (0.59 Cmmol HB/Cmmol HAc) and the storage rate (0.30 Cmmol HB/Cmmol X.h) were significantly lower than those obtained for the one pulse experiment (180 Cmmol/l). In the continuous experiment, concentration of substrate inside the reactor was always close to zero and then substrate limitation was likely the main factor responsible for the low PHB storage rate.

Splitting the substrate in three pulses led to a very high cell PHB content, 78.5%, at the end of the "feast" phase. This was the highest value of PHB content so far obtained by activated sludge and was in the range of PHB yields obtained with the pure cultures used in industrial processes, 80% (Choi et al. 1997).

The storage yield on substrate and the PHB storage rate were always very high, between 0.75 to 0.78 Cmmol HAc/Cmmol HB and between 0.68 to 0.82 Cmmol HB/Cmmol X.h, respectively and comparable to the values obtained in the experiments with 60 to 90 Cmmol/l.

3.4 *Reactor operated with acetate addition controlled by oxygen concentration*

The stability of the reactor fed with three pulses of substrate was determined along 16 days by on line controlling the frequency of substrate feed by the oxygen fluctuation. According to Figure 5, the high

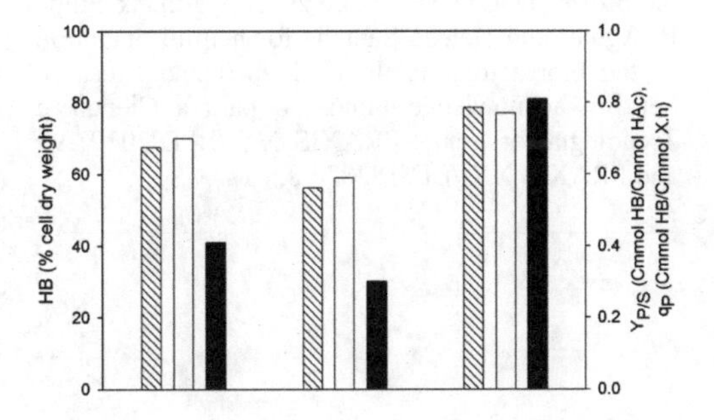

Figure 4. Effect of feeding regimen in PHB production (▧, cell PHB content; □, storage yield; ■, PHB specific storage rate).

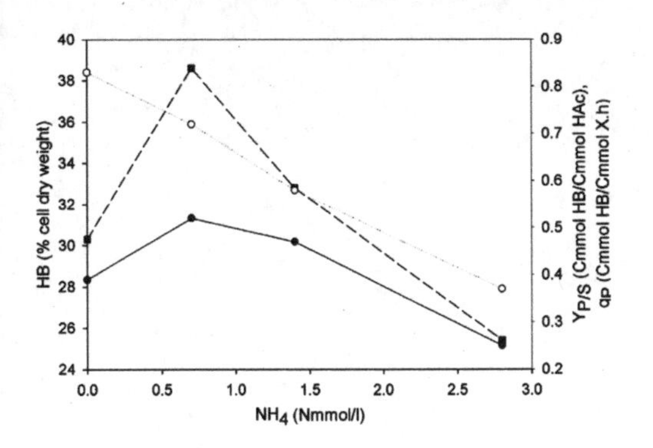

Figure 3. Comparison of PHB production in the experiments with different concentrations of ammonia (■, cell PHB content; ○, storage yield; ●, PHB specific storage rate).

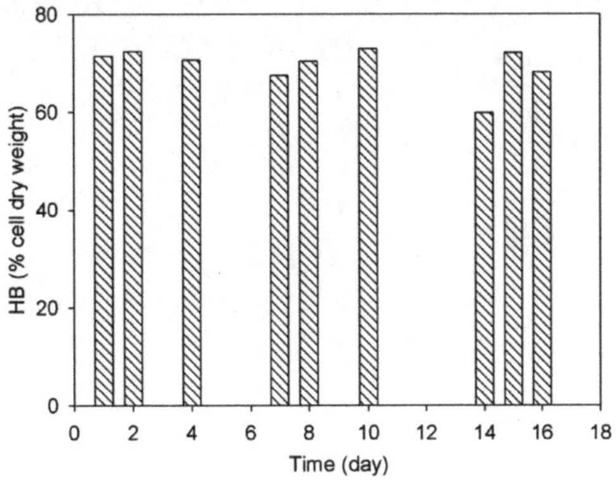

Figure 5. Performance of system with pulses of acetate controlled by oxygen concentration.

capacity of PHB storage by the sludge was maintained along the reactor operation.

The amount of PHB stored by the cells reached similar values to those previously obtained in the batch experiment performed with three carbon pulses and the high capacity of PHB storage by the sludge was maintained along the period of the SBR operation.

4 CONCLUSIONS

Operation of the SBR for almost one year under aerobic dynamic feeding, selected a population with a high storage capacity. It was shown that sludge submitted to aerobic dynamic feeding could accumulate high amounts of PHA by manipulating feeding concentrations and the reactor operating parameters.

ACKNOWLEDGEMENTS

The authors acknowledge the financial support of the Fundação para a Ciência e Tecnologia (FCT) through the project POCTI/35675/Bio/2002. We thank Filipe R. Aguiar and Helena Santana for helpful operation of the bioreactors. Paulo C. Lemos and Luisa S. Serafim acknowledge Fundação para a Ciência e Tecnologia for grants PRAXIS XXI/BPD/20197/99 and PRAXIS XXI/BD/18287/98.

REFERENCES

Dionisi, D., Majone, M., Tandoi, V., Beccari, M. 2001. Sequencing batch reactor: influence of periodic operation on performance of activated sludges in biological wastewater treatment. *Ind. Eng. Chem. Res.* 40: 5110–5119.

Lemos, P.C., Viana, C., Salgueiro, E.N., Ramos, A.M., Crespo, J.P.S.G., Reis, M.A.M. 1998. Effect of carbon source on the formation of polyhydroxyalkanoates (PHA) by a phosphate accumulating mixed culture. *Enzyme and Microbial Technology* 22: 662–671.

Majone, M., Massanisso, P., Carucci, A., Lindrea, K., Tandoi, V. 1996. Influence of storage on kinetic selection to control aerobic filamentous bulking. *Water Science and Technology* 34:223–232.

Satoh, H., Iwamoto, Y., Mino, T., Matsuo, T. 1998. Activated sludge as a possible source of biodegradable plastic. *Water Science Technology* 38(2): 103–109.

Serafim, L.S., Lemos, P.C., Crespo, J.G., Ramos, A.M., Reis, M.A.M. 2001. Polyhydroxyalkanoates production by activated sludge. In Emo Chiellini, Helena Gil, Gerhart Braunegg, Joahanna Buchert, Paul Gatenholm, Maarten van der Zee (eds.), Biorelated Polymers: Sustainable Polymer Science and Technology. Dordrecht: Kluwer Academic Plenum Publishing.

Van Aalst-van Leeuwen, M.A., Pot, M.A., Van Loosdrecht, M.C. M., Heijnen, J.J. 1997. Kinetic modelling of poly(β-hydroxybutyrate) production and consumption by *Paracocus pantotrophus* under dynamic substrate supply. *Biotechnology Bioengineering* 55(5): 773–782.

European Symposium on Environmental Biotechnology, ESEB 2004 - Verstraete (ed)
© 2004 Taylor & Francis Group, London, ISBN 90 5809 653 X

pH and organic leaching effect of the algal *Sargassum filipendula* in aqueous solutions

M.T. Veit, C.R.G. Tavares & R.M.M. Falleiro
Universidade Estadual de Maringá, Departamento de Engenharia Química – DEQ, Av. Colombo, Campus Universitário, Maringá-Paraná, Brasil

E.A. Silva

Universidade Estadual do Oeste do Paraná, Departamento de Engenharia Química – DEQ, Rua da Faculdade, Jardim La Salle, Toledo-Paraná, Brasil

ABSTRACT: The biosorption in marine algae has been associated to its cell wall biochemical properties. The pH and the organic leaching effect in different pH for the native and pre-treated (24, 48 and 72 hours) biomass of *Sargassum filipendula* were investigated. In different pH it was observed that smaller the pH, larger the leaching degree presented by the native and pre-treated biomass. The chemical and thermal treatment of the biomass contributed for the decrease of the leaching. The times of chemical treatment for the biomass showed similar values of the leaching. The ability of the pre-treated biomass to sequester the heavy-metal ions chromium(III) and/or nickel(II) from different solutions were compared. The competition effect between ions was studied in relation to the removal capacity, presenting the biosorbent higher affinity for chromium in solution. The pre-treated seaweed showed to be an efficient and low cost alternative to be considered in the acid-wastewaters-treatment.

1 INTRODUCTION

Many industrial operations contain a variety of metallic species present in its effluents, which must be removed previously for the discharge in the environment. The studies have shown that microorganisms, such as, bacteria, fungi, yeasts and algae can be used as biosorbents of low cost at passive removal of diverse heavy metals presents in concentrations of less than 100 mg/L in industrial effluents (Matheickal & Yu (1999)). In brown seaweeds (Phaeophyta), the biosorption of metallic ions phenomenon is attributed to the biochemists constituents of the cell wall, that is basically composed by three biopolymers types: alginate, fucoidan and cellulose. According to Davis et al. 2003, the heavy metal removal capacity from aqueous solutions by the brown seaweed is owed mainly to the presence of carboxylic groups in the biopolymer alginate. The alginate leaching process from the biomass structure for the solution generally accompanies the biosorption. To reduce the organic leaching an alternative indicated is the chemical and thermal pretreatment of the biomass (Matheickal et al. 1999, Matheickal & Yu (1999)). Several mechanisms are involved in the biosorption process, such as: ionic exchange, complexation, physical and/or chemical adsorption, coordination, quelation and inorganic microprecipitation (Sãg & Kutsal (1996)). The ionic exchange is considered the main mechanism of metallic ions removal by the algal biosorbents. Thus, the present work had for objectives: (i) to evaluate the effect of the contact time (24, 48 and 72 hours) in the pre-treatment of the native biomass using calcium chloride solution; (ii) to evaluate, in batch system, the effects of the pre-treatment in the pH, in the sorption capacity of the metallic ions by biomass and the resultant organic leaching from the exposition of the biosorbents samples in acid media.

2 MATERIALS AND METHODS

2.1 *Biomass and chemical and thermal pre-treatment*

The biomass used was the brown seaweed *Sargassum filipendula*, commonly found along of the littoral coast of Brazil. The algae were washed in current water, rinsed in distilled water and dried at 60°C for a period of 24 hours. Pre-treatment of this biomass was carried out as the following: 10 g of this native biomass was treated with 400 mL of 0.2 M $CaCl_2$ solution in three different contact times (24, 48 and 72 hours) under slow agitation and at room temperature (≅25°C). The initial pH of the $CaCl_2$ solution was adjusted at 5.0 using 0.1 N HCl. Every 24 hours of treatment, the biomass was submitted to successive washes with deionized water and placed again in contact with 400 mL of 0.2 M $CaCl_2$ solution. At the end of each one of the three treatment times, the pre-treated biomass was dried at 60°C for 24 hours and later was stored.

2.2 *Metal solution*

The salts used in the preparation of the metallic solutions were $CaCl_2.2H_2O$, $CrCl_3.6H_2O$ and $NiCl_2.6H_2O$. The solution concentrations of chromium(III) and nickel(II) used in the sorption experiments were both of 2 and 7 meq/L. The pH values were adjusted using 0.1 M NaOH, 0.1 N and 1.0 N HCl. All experiments utilized deionized water and were carried out in duplicate.

2.3 *pH, leaching, sorption and released of metallic ions in aqueous acid medium*

Experiments were carried out with deionized water, or without correction of the initial pH or with correction of the initial pH (1.5, 2, 3, 4, 5 and 6) for the *Sargassum filipendula* biomass in the forms native and pre-treated (1st, 2nd and 3rd treatment) in the following conditions: biomass concentration of 3 g/L (dry base), initial volume of 100 mL, temperature control of 30°C and rotation speed of 150 rpm. The tests were conducted along 100 hours, and the samples were filtered through a 0,45 μm membrane filter (Millipore).

Sorption experiments of the metallic ions chromium(III), nickel(II) and the combination of its mixtures were carried out in pH of 3 only for the pre-treated biomass (1st treatment) using the same experimental conditions described above. The pH was monitored in pre-established time intervals by means of pHmetro with temperature compensation (Digimed). The ions chromium and nickel concentrations in solution, and the ions calcium, magnesium, sodium and potassium released from the biomasses structure were determined by atomic absorption spectroscopy (Varian Spectr AA – 10 Plus). Solubility analyses were carried out by the mensuration of total organic carbon (TOC) present in the aqueous phase (Shimadzu TOC-5000A – Total Organic Carbon Analyzer).

3 RESULTS AND DISCUSSION

3.1 *Effect of the pH in the organic leaching in aqueous medium by the native and pre-treated biomass*

The effects of pH variation were evaluated for the biomasses contact with deionized water without and with pH correction. The results showed that as at correct pH of 4, 5 and 6 as at uncorrected pH, both the biomass, native and pre-treated presented the final pH of the aqueous medium around 7–7.5. The pH of 1.5, 2 and 3 presented values for the final pH around 2, 3 and 5.5, respectively, demonstrating the smallest deviation between the initial and final pH values from the solutions so for the native biomass as the pre-treated biomass. The pH variation with the time for the native and pre-treated biomass showed that as larger the pH from aqueous solution, smaller the amount of ions released, and consequently smaller the leaching degree.

In agreement with the total organic carbon results obtained (Figure 1), it can be observed that the pre-treated biomasses were inferior in relation to native biomass in all the pH values, indicating leaching reduction. With exception at pH of 1.5 and 2, the pre-treated biomasses presented organic leaching values similar between itself around 23 mg/L. Also, no significant effect in the leaching was observed for the different chemical treatment times of the biomass. The total organic carbon variation observed among the pre-treated biomasses in each studied pH was smaller of 12%. According to Matheickal & Yu (1999), the decrease of organic leaching is a fact attributed to the stability of the polymers alginate, obtained by the chemical and thermal treatment process.

The results of the metallic ions concentrations released from the biomass in solution during the leaching experiments showed that the native seaweed presented a significant alteration of all ions concentration

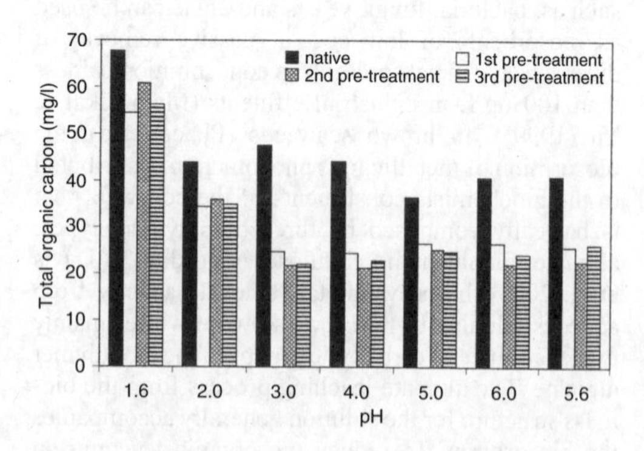

Figure 1. Organic leaching from native and pre-treated biomass at different pH values.

(calcium, magnesium, sodium and potassium), while the pre-treated biomasses presented a significant alteration only of the calcium ion concentration. At Figure 2 the ions concentrations released from the biomasses are presented.

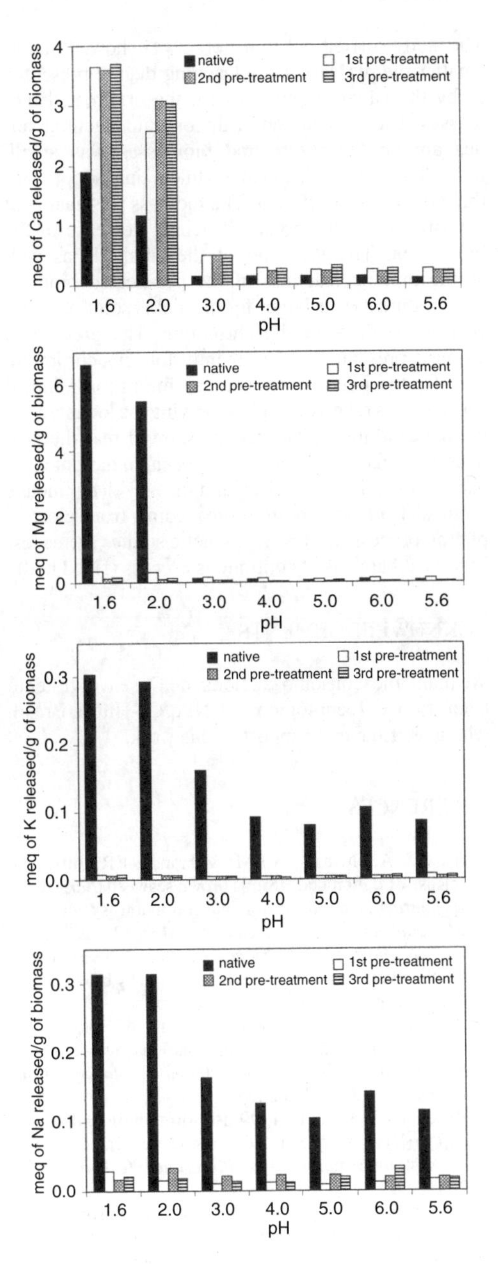

Figure 2. Released of calcium, magnesium, potassium and sodium ions from the native and pre-treated biomass.

3.2 *Effect of the pH in the organic leaching in different metallic solutions*

Organic leaching was studied only for the pre-treated (1st treatment) biomass of *Sargassum filipendula* in the presence of different initial concentrations of ions chromium(III), nickel(II) and the combination of its mixtures (50 mL of each concentration).

The realization of these experiments with metallic solutions occurred in initial pH of 3. This pH selection is due to the studied metals are present in their free ionic form (Cr^{3+} and Ni^{2+}). Moreover, another reason is the future application of this algal biosorbent in the effluent treatment from the electroplating process, which pH of its waters is between 2.5–3.0.

The pH results in relation with the time are presented in Figure 3. The initial pH value in relation to the final pH of the solution varied between 0.5–1.5. The smallest pH increase from the solution was verified for the simple and binary systems of high chromium concentration (7 meq/L).

At Figure 4 we can observe that for these same systems (Cr-7 meq/L, CrNi-7 meq/L and CrNi-7 and

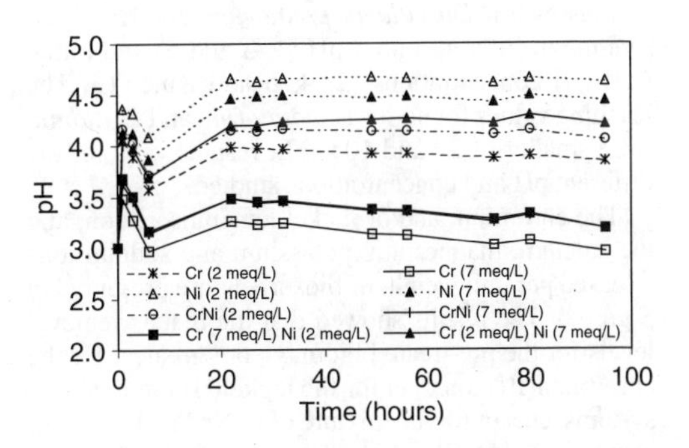

Figure 3. Temporal pH evolution caused by the pre-treated biomass contact at different initial metal concentrations.

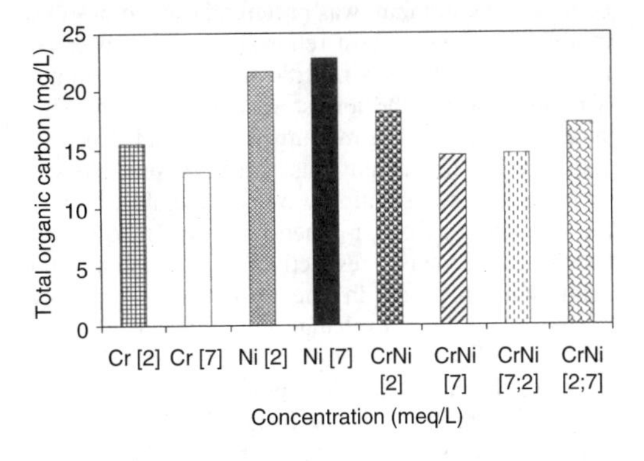

Figure 4. Organic leaching from pre-treated biomass at different initial metal concentrations.

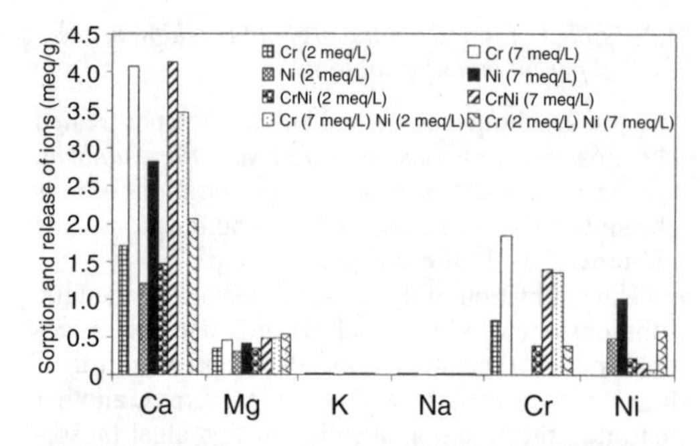

Figure 5. Sorption and metallic ions released by the pre-treated biomass of *Sargassum filipendula*.

2 meq/L), it was also obtained the smallest total organic carbon values, on average 14 mg/L. This value corresponds on organic leaching reduction between 18–38% in comparison the nickel concentrations (2 e 7 meq/L) and its mixtures (CrNi-2 meq/L and CrNi-2 and 7 meq/L). Matheickal & Yu (1999) also carried out experiments with the pre-treated ($CaCl_2$) seaweed biomasses of *Durvillaea potatorum* and *Ecklonia radiata* at different initial pH (1, 3 and 5) and initial lead (II) concentrations (2, 4, 6 and 8 meq/L). The leaching values found for *D. potatorum* and *E. radiata* were smaller of 20 and 10 mg/L, respectively, for the different pH and concentrations studied.

The chromium and/or nickel ions biosorption, and the calcium, magnesium, potassium and sodium ions released per unit weight of biosorbent are presented in Figure 5. The results showed that the metals removal levels for the pre-treated biomass, are greatest for the chromium(III) ion than for the nickel(II) ion in all the systems, except for the mixture of CrNi 2 and 7 meq/L, respectively. At other binary systems (CrNi-2 meq/L, CrNi-7 meq/L and CrNi-7 and 2 meq/L) it was observed that when in competition between the metallic ions, the chromium was preferentially removed by the biomass. The highest removal capacity observed for the biosorbent was of 1.84 meq/g for chromium (Cr-7 meq/L) and the lowest was of 0.07 meq/g for system containing the mixture CrNi-7 and 2 meq/L, respectively. The calcium ions concentrations released by the biomass in solution were considerable, mainly for the solutions of Cr-7 meq/L, CrNi-7 meq/L and CrNi-7 and 2 meq/L, respectively. The calcium and magnesium released in the systems, is probably related to the ionic exchange properties of the biosorbent during the chromium(III) and/or nickel(II) removal. The monovalent ions presence (Na^+ and K^+) in the different solutions was insignificant. According to Costa et al. 2001 that is due to the fact of these ions be present in smaller amounts in the biomass. This phenomenon also occurred in virtue of the chemical treatment has propitiated that the majority of the site was filled by the calcium ion.

4 CONCLUSIONS

The study carried out in different pH showed that as smaller the pH, larger the leaching degree presented as by the native biomass as for the pre-treated biomasses. The comparison of the organic leaching values among the pre-treated biomasses was small, indicating that the chemical treatment time adopted for the biomass is indifferent. The biomass pre-treatment contributed for the organic leaching decrease of the biosorbent. The variation of the final pH as well as the organic leaching of the pre-treated biomass (1st treatment) was larger for the nickel solutions concentrations than for the chromium. The pre-treated biomass presented higher affinity for ions chromium than for the ions nickel. The chromium ion removal capacity was relatively high, supplying the lowest residual concentrations. The results showed that the pre-treated biomass of *Sargassum filipendula* has potential use in the chromium(III) and/or nickel(II) metals removal from wastewaters proceeding from electroplating process. In this way, kinetic studies are necessary as at batch as at continuous process (fixed bed).

ACKNOWLEDGEMENTS

We thank the Conselho Nacional de Desenvolvimento Científico e Tecnológico – CNPq/CT-Hidro–Brasil, who gave financial support to this work.

REFERENCES

Costa, A.C.A., Tavares, A.P.M. & França, F.P. 2001. The release of light metals from a brown seaweed (*Sargassum* sp.) during zinc biosorption in a continuous system. *EJB Eletronic Journal of Biotechnology* 4(3): 125–129.

Davis, T.A., Volesky, B. & Mucci, A. 2003. A review of the biochemistry of heavy metal biosorption by brown algae. *Water Research* 37: 4311–4330.

Matheickal, J.T., Yu, Q. & Woodburn, G.M. 1999. Biosorption of cadmium(II) from aqueous solutions by pre-treated biomass of marine alga *Durvillaea potatorum. Water Research* 33(2): 335–342.

Matheickal, J.T. & Yu, Q. 1999. Biosorption of lead(II) and copper(II) from aqueous solutions by pre-treated biomass of Australian marine alga. *Bioresouce Technology* 69: 223–229.

Sãg, Y. & Kutsal, T. 1996. The selective biosorption of chromium (VI) and copper (II) ions from binary metal mixtures by *Rhizopus arrhizus* biomass. *Process Biochemistry* 31(6): 561–572.

European Symposium on Environmental Biotechnology, ESEB 2004 - Verstraete (ed)
© 2004 Taylor & Francis Group, London, ISBN 90 5809 653 X

Biological filtration for BOM and particle removal from drinking water supplies

D. Pak & D. Kim
Korea Institute of Science and Technology, Seoul, Korea

ABSTRACT: In this study, biological filtration was operated as a pretreatment of drinking water supplies. Total organic carbon level was monitored to show that biological filtration can remove natural organic matter. The effect of empty bed contact time on the removal of organic matter, color, turbidity and chlorophyll-a was investigated in two down-flow biofilters containing ceramic media. The empty bed contact time ranged from 2 hours to 6 hours. TOC removal was about 40% when EBCT was 4 and 6 hours. When EBCT was reduced to 2 hours, TOC removal reduced to 25%. After biological filtration, 90% of color was removed even though EBCT was reduced to 2 hours. Turbidity of raw water was varied from 5.5 NTU to 15.2 NTU. The effluent turbidity was maintained at less than 0.5 NTU.

1 INTRODUCTION

One of the problems with using surface water as drinking water source is often the presence of natural organic matter. This gives the water a high color and total organic carbon levels. Removal of natural organic matter is required since colored water is unattractive to consumers. It can cause odor and taste. Natural organic matter also leads to the formation of disinfection byproducts when water is chlorinated.

Biological filtration has attracted increased attention as an important process step for the production of aesthetically pleasing drinking water. In drinking water treatment it can achieve biodegradable organic matter (BOM) removal and particle removal within the same filter unit. Therefore benefits of biological filtration may include decrease of the potential for bacterial regrowth, reduction of chlorinated disinfection byproducts formed during secondary disinfection, reduction of chlorine demand, and decrease of corrosion potential. Additionally biological filtration has the potential to control taste and odor causing compounds and other micropollutants of health and aesthetic concern (Lundren et al. 1988, Manem & Rittmann 1992, Melin 1999, Melin et al. 2000, Rittmann 1995, Rittmann et al. 1995, Urfer et al. 1997).

In this study, biological filtration was operated as a pretreatment of drinking water supplies. Total organic carbon level was monitored to show that biological filtration can remove natural organic matter. Hydraulic retention time, the key engineering variable, was varied to investigate its effect on biological performance (BOM removal) and filtration performance.

2 MATERIALS AND METHODS

2.1 *Biological filter system*

Biological filter was operated as shown in Figure 1. The biofilter is 0.14 m in diameter and 1.5 m high. The filter media of ceramic beads with effective diameter of 3 to 5 mm were used to charge the filter to bed height of 0.7 m. The effective reactor volume was 10.8 liter. The biofilter was fed with raw water. The raw water used in this study was obtained from Paldang lake near Seoul and its characteristics are shown in Table 1.

2.2 *Analysis*

During the operation of biofilter samples were withdrawn at regular interval to analyze TOC, color, turbidity, and chlorophyll-a in the influent and effluent. Color, turbidity and chlorophyll-a were measured according to Standard Methods. TOC was analyzed by using TOC analyzer (5000A).

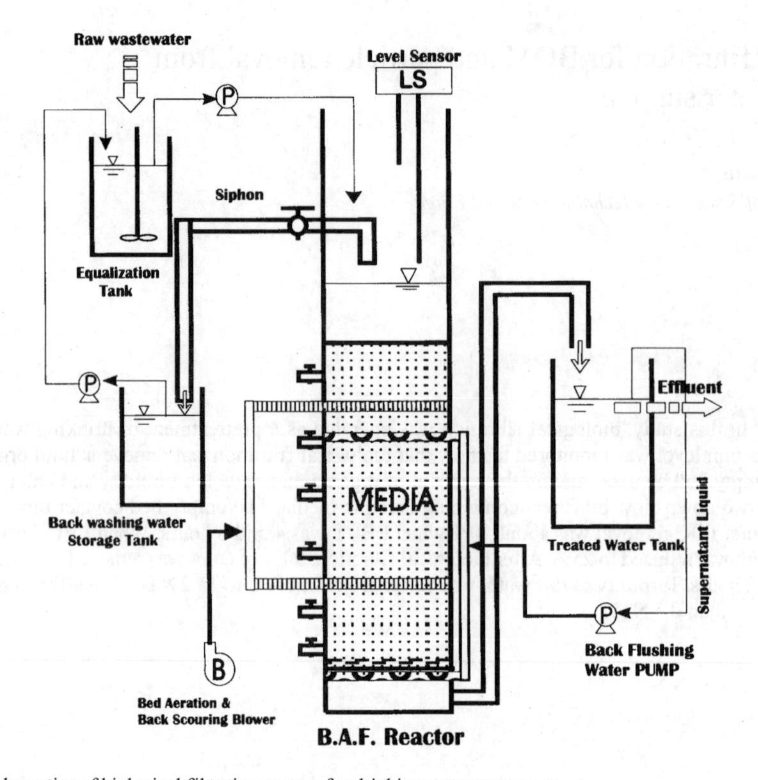

Figure 1. Schematics of biological filtration system for drinking water treatment.

Table 1. Characteristics of raw water from Paldang lake.

	Range	Average
pH	6.8–8.8	7.6
BOD	1.0–3.5	2.0
CODmn	2.8–14.4	3.2
TOC	1.6–5.2	2.7
T-N	1.9–4.5	2.6
NH3-N	0.08–0.65	0.11
NO3-N	1.6–3.1	2.2
T-P	0.02–0.22	0.08
SS	2.5–120	15
Turbidity	2.7–220	7.8
Alkalinity	22–84	44
Color	4–320	13
ABS	0–0.05	0.02
Hardness	25–75	57

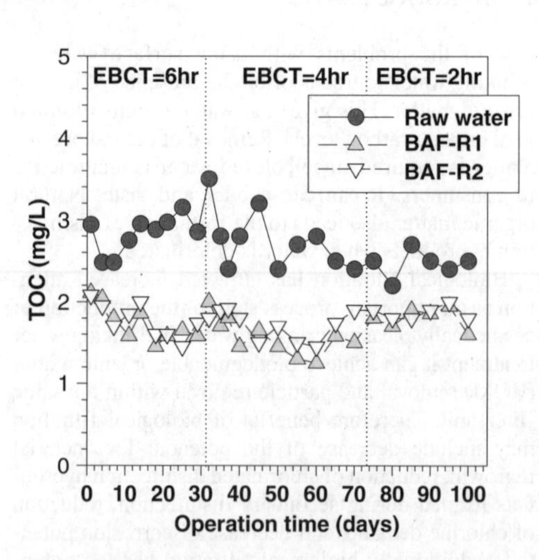

Figure 2. TOC concentration in influent and effluent from biological filtration system.

3 RESULTS AND DISCUSSION

3.1 *Removal of natural organic matters*

Figure 2 shows TOC concentration in influent and effluent from biofilters. TOC concentration in the influent was changed from 2.14 mg/l to 3.85 mg/l. About 40% of TOC was biologically degraded and mineralized to carbon dioxide and water. A number of researchers have shown that contact time significantly influences BOM removal within biological filter. Contact time, usually expressed as EBCT, is therefore

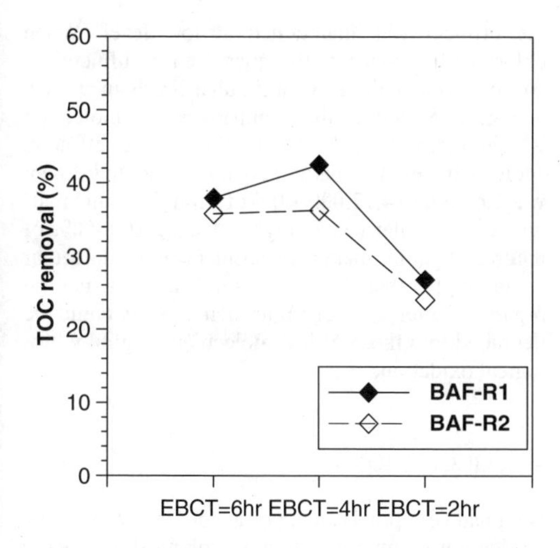

Figure 3. TOC removal rate in biological filter operated at different empty bed contact time.

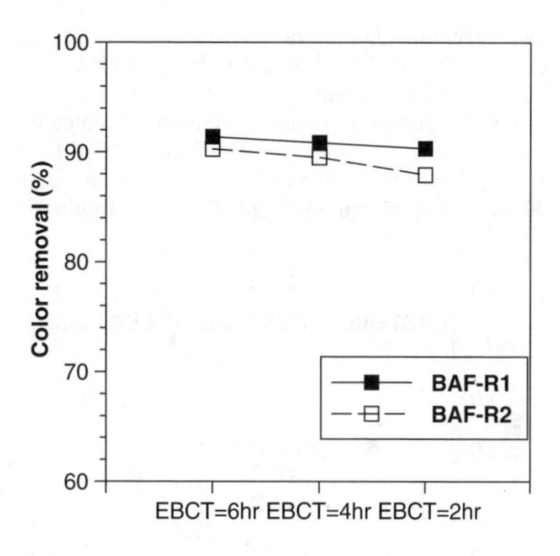

Figure 5. Color removal rate in biological filter operated at different empty bed contact time.

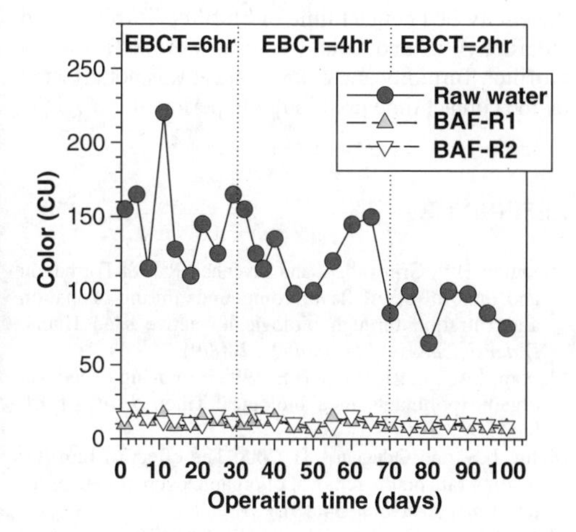

Figure 4. Color in influent and effluent from biological filtration system.

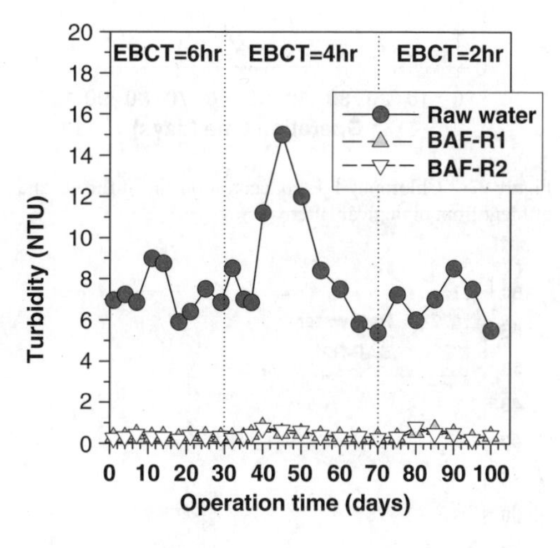

Figure 6. Turbidity in influent and effluent from biological filtration system.

a key design and operating variable. EBCT was varies from 2 hours to 6 hours. When EBCT was 4 hours or 6 hours, TOC removal was not affected by the contact time. However when EBCT was reduced from 4 hours to 2 hours, TOC removal was significantly influenced by the contact time as shown in Figure 3.

Natural organic matter gives the water a high color so that it has to be removed since colored water is unattractive to consumers. Figure 4 shows color removal by biological filtration. Color of raw water was varied from 75 CU to 220 CU. After biological filtration, 90% of color was removed. Even though EBCT was changed from 6 hours to 2 hours, color removal rate was affected by the contact time as shown in Figure 5.

3.2 *Particle removal*

Figure 6 shows turbidity removal by biological filtration. Influent turbidity was varied from 5.5 NTU to 15.2 NTU. The effluent turbidity was maintained at lower than 0.5 NTU. Average effluent turbidity was about 0.4 NTU. Biofilter has shown excellent turbidity removal. Even though EBCT was reduced from 6 hours to 2 hours, turbidity removal was not affected by contact time. During rainy season from 40 days to 60 days of operation, influent turbidity increased to 16 NTU, effluent turbidity was not increased.

Figure 7 shows chlorophyll-a removal by biological filtration. Chlorophyll-a in influent was varied from 13.5 μg/l to 61 ug/l, which was high fluctuation. Until 50 days of operation, chlorophyll-a concentration in the effluent was maintained at low level. When chlorophyll-a concentration increased in influent, its concentration in the effluent simultaneously increased.

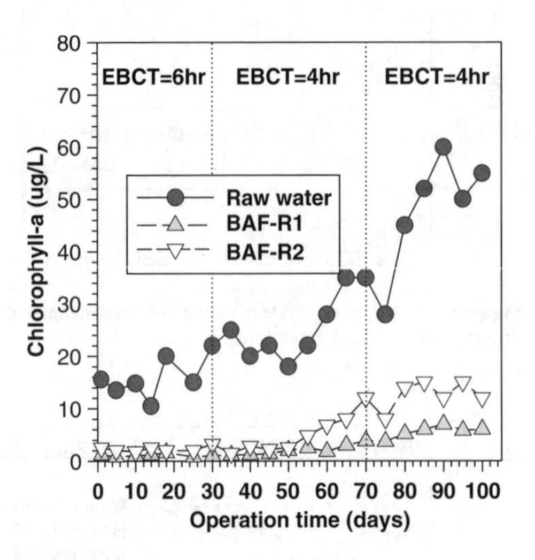

Figure 7. Chlorophyll-a concentration in influent and effluent from biological filter.

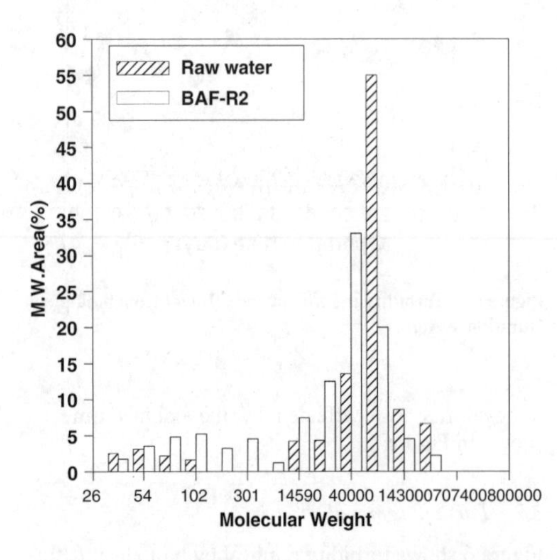

Figure 8. Comparison of molecular weight distribution before and after biological filtration.

Figure 8 shows the comparison of molecular weight distribution between influent and effluent. Molecular weight of natural organic matter in influent was from 40 to 4,12,000. 50% of natural organic matter has molecular weight higher than 72,800. 50% of natural organic matter in effluent has molecular weight higher than 40,000. This indicates that natural organic matter of high molecular weight could be degraded into those of low molecular weight by biological oxidation.

4 CONCLUSIONS

Biolgical filtration is an important process step for the production of microbially and aesthetically pleasing drinking water. Biodegradable organic matter in natural organic matters was removed by biofilm in filter. Biodegradable organic matter removal was affected by the empty bed contact time of biofilter. Turbidity and chlorophyll-a, and color removal were effective in biofilter. Turbidity and color removal was not affected by the contact time used in this experiment.

REFERENCES

Lundgren, B.V., Grimvall, A. and Savenhed, R. 1988 Formation and removal of off flavor compounds during ozonation and filtration through biologically active sand filters. *Water Science and Technology*, **20**(8/9).

Manem, J.A. and Rittmann, B.E. 1992 Removing trace level organic pollutants in a biological filter. *J. of AWWA*, **84**(4).

Melin, E.S. and Odegaard, H. 2000 The effect of biofilter loading rate on the removal of organic ozonation byproduct. *Water Research*, **34**(18).

Melin, E.S. and Odegaard, H. 1999 Biofiltration of ozonated humic water in expanded clay aggregate filters. *Water Science and Technology*, **40**(9).

Rittmann, B.E. 1995 Transformation of organic micropollutants by biological process. Quality and Treatment of Drinking Water, *The Handbook of Environmental Chemistry*.

Rittmann, B.E., Gantzer, C.J. and Montel, A. 1995 A biological treatment to control taste and odor compounds in drinking water treatment. *Advances in the Control of Taste and Odor in Drinking Water*. AWWA, Denver.

Urfer, D., Huck, P.M., Booth, S.D.J. and Coffey, B.M. 1997 Biological filtration for BOM and particle removal: a critical review. *J of AWWA*, **89**(12).

European Symposium on Environmental Biotechnology, ESEB 2004 - Verstraete (ed)
© 2004 Taylor & Francis Group, London, ISBN 90 5809 653 X

Towards aerobic stabilization of old landfills: simulation in lysimeters

L. Krzystek, A. Zieleniewska & S. Ledakowicz
Department of Bioprocess Engineering, Faculty of Process and Environmental Engineering, Technical University of Lodz, Lodz, Poland

H.-J. Kahle
Lausitzer Naturkundliche Akademie LANAKA e.V., Cottbus, Germany

ABSTRACT: The impact of aeration on the reduction of landfill gas emissions and leachates, and biodegradation of deposited material in different stages of conversion of organic substances was investigated. The simulation of aerobic landfill processes was carried out in the bench-scale lysimeters with a fixed bed of household solid waste stabilized under anaerobic conditions. Experimental studies showed that the aerobic waste stabilization was a very quick process. During a month the bed was managed to stabilize, reaching a significant reduction of both nitrogen compounds (N-NH_4^+ by around 70%), and easily biodegradable organic substances (ca. 90%). The aeration that started at the methanogenic phase was found to show better bed degradation in a shorter time than started in the acidogenic phase. The reduction of methanogenic potential of the landfill was even faster. The composition of gas at the outlet from the lysimeter changed and after one day already its content was similar to atmospheric air.

1 INTRODUCTION

The technology of MSW landfilling causes a long-term environmental impact. For modern landfill strategies the sustainability of the landfill represents the main goal to achieve. From the point of view of minimization of hazardous impact of old landfills on the environment caused by leakage of leachates and landfill gas emission, the aerobic stabilization of biodegradable substances and components containing nitrogen enables achieving of their sustainability. Experimental simulation of the landfill processes in lysimeters provides knowledge of the main processes that take place in the aerated landfill (Cossu et al. 2003).

In this paper the impact of aeration on the reduction of the landfill gas emissions and leachates, and biodegradation of deposited material in different stages of conversion of organic substances carried out in laboratory lysimeters was compared.

2 MATERIALS AND METHODS

2.1 *Substrate*

In landfill process simulation the lysimeters were filled with alternately laid layers of a municipal solid waste and compost mixture. The mixture of household waste contained organic wastes (vegetable and fruit – 10.9%; potatoes – 21.2%; bread – 2.3%; others – 3.6%) – 38%, paper and cardboard – 25%, plastics – 17%, textiles – 5%, other wastes – 15% (Ledakowicz & Kaczorek 2002). The waste was shredded to the size of 20–50 mm.

The bed was initially saturated with water (due to wetting the bed with tap water). A month after charging, around 200 ml of stabilized fermented sewage sludge was added to the lysimeters in order to initiate anaerobic digestion.

2.2 *Lysimeters and process parameters*

Experiments were carried out in four (L1–L4) laboratory lysimeters of working capacity of 15 dm^3. The lysimeters consisted of a glass cylinder of inner diameter 150 mm and height 850 mm, closed on top and bottom with stainless steel covers, equipped with pipes for leachate recirculation, taking samples for analysis, supply and removal of gases.

Initially, the processes of anaerobic stabilization took place in the lysimeters. Then, after reaching an appropriate stage of conversion of organic substances in the lysimeters (L2 – acidogenic phase, L3 – initial methanogenic phase, L4 – final methanogenic phase), the investigation of the process of aerobic stabilization started by supplying air to the lysimeters in a continuous way at the volumetric flow rate 6–10 dm^3/h. For control, in lysimeter L1 anaerobic degradation was carried out. Leachates taken from the lysimeters were recirculated once a day. Leachates from aerated lysimeters were also subjected to advanced oxidation processes, i.e. ozonation and UV radiation with the addition of H_2O_2.

Processes in the lysimeters were carried out at room temperature (around 20°C).

2.3 *Analytical methods*

In the leachates taken from lysimeters, pH, redox potential, BOD_5 (by the dilution method, APHA Standard Methods: 1989), COD (by the dichromate method, APHA Standard Method 1989), the content of volatile fatty acids – VFA (according to the Polish Standard PN-75C-04616 using Büchi – Distillation Unit B-324), $N\text{-}NH_4^+$ (by the method of distillation in the Büchi device), total N (by Kjeldahl method according to the Polish Standard PN-75-C-04576-17 in the Büchi device), total organic carbon – TOC (in Coulomat 702 Li/C, Strohlein Instruments, Germany) were analyzed. The composition of gas being formed in this way (CH_4, CO_2 and O_2 content) was regularly controlled (gas analyzer, LMS GasData).

3 RESULTS AND DISCUSSION

Changes in the basic indices of organic load (BOD_5, COD, VFA, content of N_{total}, $N\text{-}NH_4^+$), pH, redox potential and changes in biogas composition in time were controlled during the aerobic stabilization performance.

Values of these indices at the beginning and end of the process for each lysimeter are given in Table 1.

3.1 *pH*

As compared to the control lysimeter (L1) in which anaerobic degradation took place, in the leachates from aerated lysimeters (L-3 – initial methanogenic phase and L-4 – final methanogenic phase) an increase of pH to 8.5–9.0 was observed. In lysimeter L-2, in which aeration started in the acidogenic phase, pH was maintained on the level of around 5.5, like in the control lysimeter. Figure 1 illustrates changes observed in pH.

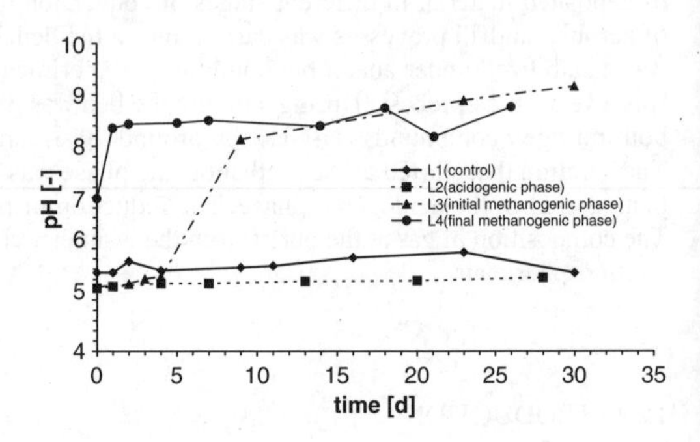

Figure 1. Changes of pH in leachates from lysimeters.

Table 1. Load indices of leachates from lysimeters at the beginning and end of processes tested.

	L1 (control)		L2 (acidogenic phase)		L3 (initial methanogenic phase)		L4 (final methanogenic phase)	
Index	Initial	Final	Initial	Final	Initial	Final	Initial	Final
BOD_5 [mg O_2/dm^3]	31.000	24.300	34.000	28.000	31.000	300	128	16.5
COD [mg O_2/dm^3]	52.500	40.300	62.800	55.400	58.100	5490	1228	1158
N_{total} [mg N/dm^3]	772	483	885	844	677	211	59.2	49.6
$N\text{-}NH_4^+$ [mg N/dm^3]	515	399	509	537	498	22.0	11.4	4.17
VFA [mg CH_3COOH/dm^3]	28.100	17.200	30.200	30.900	16.600	762	244	158
pH [-]	5.54	5.52	5.23	5.43	5.25	9.14	6.98	8.76
Redox [mV]	−23	−37	−8	193	−37	54	−51	188

3.2 BOD_5

In the first 10 days of aeration BOD_5 decreased abruptly in lysimeters L3 and L4. In the case of lysimeters L1 and L2 the values of this index decreased only slightly. Changes in BOD_5 are shown in Figure 2.

The highest degree of BOD_5 reduction was obtained in lysimeter L3. It was equal to 99.0%, while the lowest one was 17.6% for L2.

3.3 *COD*

Only in the case of one lysimeter L3, in the first 10 days a sudden decrease of this index was observed. In other lysimeters L1, L2 and L4 there was a stepwise slight decrease of COD. Figure 3 presents the obtained results.

The highest level of COD reduction was attained in lysimeter L3. It was 90.6%, and the lowest one – 5.7% was in L4.

3.4 *VFA*

It was found in the investigation that VFA content in lysimeter L2 oscillated on the same level, i.e. 28,000–33,000 mg CH_3COOH/dm^3, in L1 VFA content was slightly reduced (38.7%), in L3 it decreased below 10,000 mg CH_3COOH/dm^3. In the lysimeter L4 also a drop of VFA by around 35% in the first 10 days of the process was observed. These changes are shown in Figure 4.

The highest degree of VFA reduction equal to 95.4% was obtained in the lysimeter L3 and the lowest – 2.4% in L2.

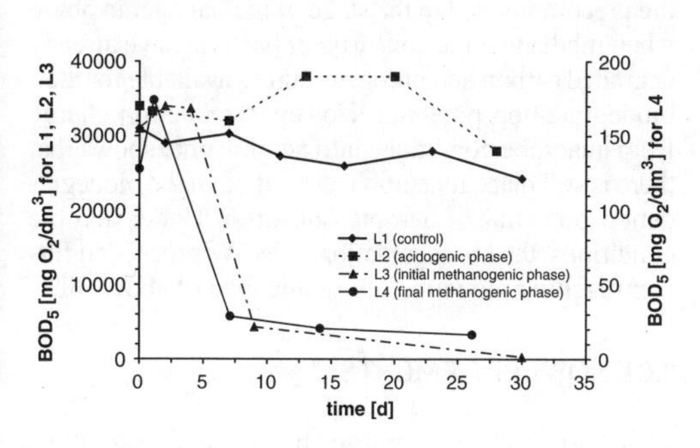

Figure 2. Changes of BOD_5 in leachates from lysimeters.

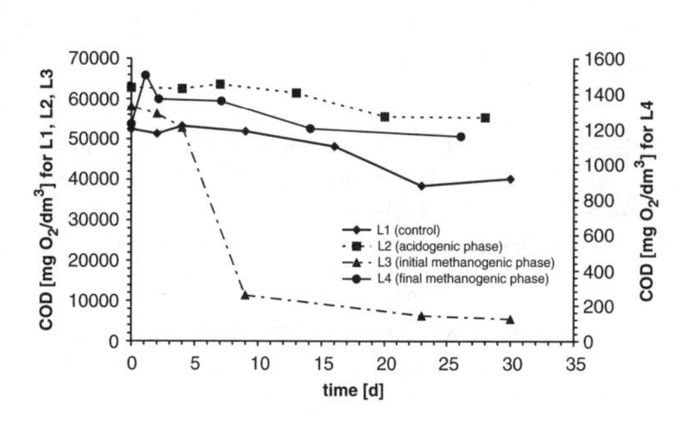

Figure 3. Changes of COD in leachates from lysimeters.

3.5 *Total nitrogen*

The initial concentrations of total nitrogen in the lysimeters L1, L2 and L3 were 700–900 mg N/dm^3, and for the lysimeter L4 60 mg N/dm^3. In the first 10 days of aeration, the total nitrogen decreased rapidly only in the lysimeter L3. In other lysimeters L1, L2 and L4 the concentrations decreased only slightly (Figure 5).

The highest reduction of total nitrogen concentrations was attained in the lysimeter L3 – it amounted to 68.8%, while the lowest 4.6% was obtained in L2. In L4 the reduction was 16%.

3.6 *Ammonium nitrogen*

A rapid decrease of ammonium nitrogen concentration was reported in the lysimeter L4 in the first 2 days of the process, while in L3 during the first 10 days. In the lysimeters L1 and L2 no big changes were observed which is shown in Figure 6.

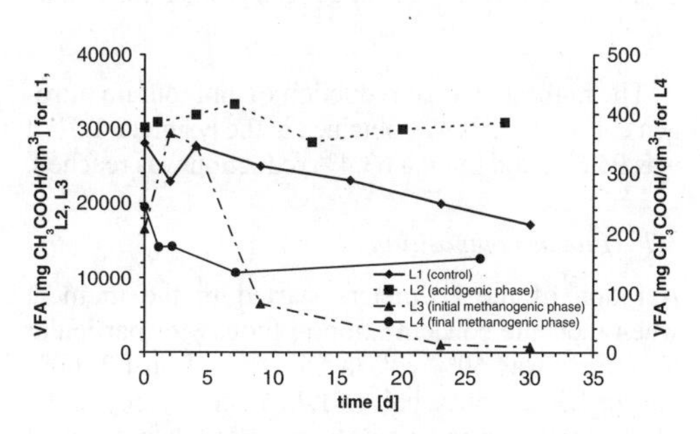

Figure 4. Changes of VFA in leachates from lysimeters.

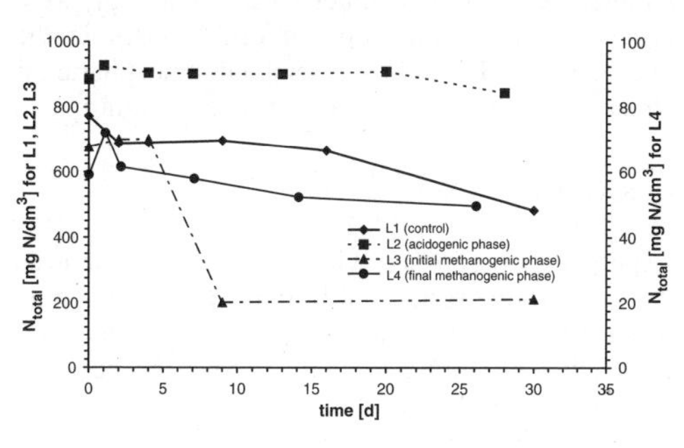

Figure 5. Changes of total nitrogen in leachates from lysimeters.

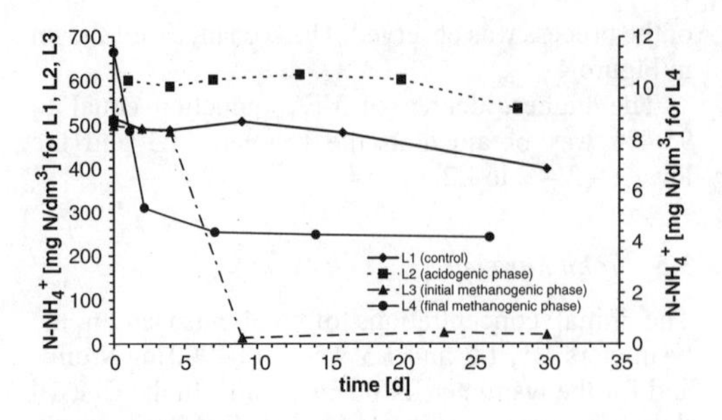

Figure 6. Changes of ammonia concentration in leachates from lysimeters.

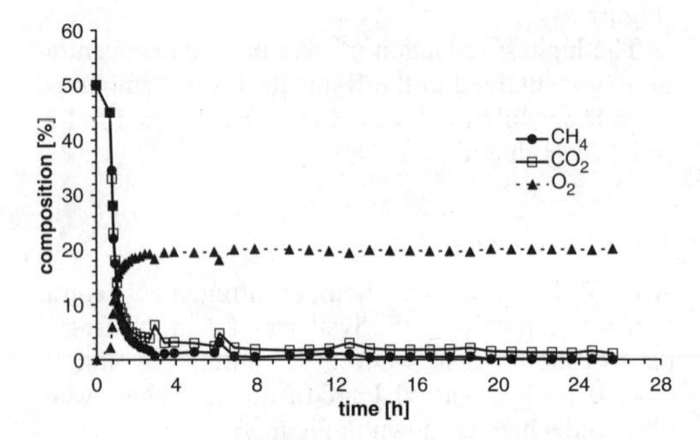

Figure 7. Changes of outlet gas composition for lysimeter L4.

The highest level of reduction of ammonium nitrogen concentrations was obtained in the lysimeter L3 – it was 95.6%, and in L4 a 63.4% reduction was reached.

3.7 *Biogas composition*

Aeration of the lysimeters started in the moment when methane concentration in biogas for particular lysimeters was 60% vol. in L1, 0% vol. in L2, 60% vol. in L3 and 50% vol. in L4. Since the beginning of the aerobic process a decrease of methane content (L3 and L4) and carbon dioxide (L2, L3 and L4), and an increase of oxygen content (L2, L3 and L4) were observed. The composition of outlet gases in the lysimeters L2, L3 and L4 resembled atmospheric air already after the first day of the process (Figure 7).

4 CONCLUSIONS

Experiments performed within this research showed that the aerobic waste stabilization was efficient only when aeration started at initial and final methanogenic phases. During around 30 days the bed was managed to stabilize, reaching a significant reduction of organic load indices. The aeration of lysimeters caused a quick reduction particularly of easily degradable organic substances (i.e. BOD_5 – in

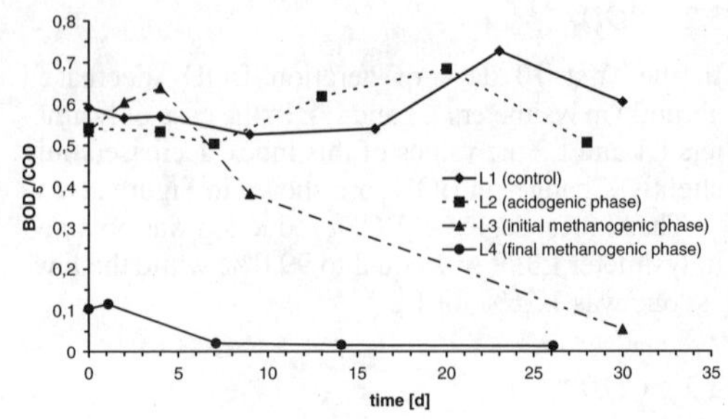

Figure 8. Changes of biodegradability in leachates from lysimeters.

L3 99% and in L4 87%) and $N\text{-}NH_4^+$ (in L3 96% and in L4 63%) and volatile fatty acids (in L3 95%). The reduction of a landfill methanogenic potential was even faster. Biodegradability of leachates in the lysimeters L3 and L4 (represented as a BOD_5 to COD ratio) initially increased, and next decreased significantly to the level of around 0.05 (Figure 8), which means that these leachates are very resistant to biodegradation. The composition of gas at the lysimeter outlet changed and after the first day already resembled that of the atmospheric air.

In the case when aeration started in the acidogenic phase no acceleration of the bed stabilization was achieved. Implementation of aeration into lysimeters is then recommended in the stage of methanogenic phase when methanogenic consortia of bacteria have already degraded carbon and nitrogen sources available for their biodegradation potential. However, reduction charge from anaerobe conditions into aerobic one, shows that there is still place for aerobic utilization of the biodegradation potential of aerobic consortia. Under aerobic conditions, the biodegradation of MSW proceeded further reaching reduction of organic load of above 90%.

ACKNOWLEDGEMENTS

This work has been supported by the State Committee for Scientific Research grant No. 4 T09C 01425.

REFERENCES

APHA-AWWA-WPCF Standard Methods for the Examination of Water and Wastewater. 17th edition, APHA, Washington, D.C., 1989.

Cossu, R., Raga, R. & Rossetti, D. 2003. The PAF model: an integrated approach for landfill sustainability. *Waste Management* 23 (2003): 37–44.

Ledakowicz, S. & Kaczorek, K. 2002. Laboratory simulation of anaerobic digestion of municipal leachate biodegradation in lysimeters. *Solid waste ISWA 2002 World Environmental Congress; Proceedings*: 1139–1146. Istanbul 2002.

European Symposium on Environmental Biotechnology, ESEB 2004 - Verstraete (ed)
© 2004 Taylor & Francis Group, London, ISBN 90 5809 653 X

Nitrile conversion by *M. imperiale CBS 498-74* resting cells in batch and UF-membrane bioreactors

M. Cantarella, A. Gallifuoco, A. Spera
Department of Chemistry, Chemical Engineering and Materials, University of L'Aquila, Monteluco di Roio, L'Aquila, Italy

L. Cantarella
Department of Industrial Engineering, University of Cassino, via Di Biasio, Cassino (FR), Italy

ABSTRACT: The bio-hydratation of acrylonitrile, propionitrile and benzonitrile catalysed by the NHase activity contained in resting cells of *Microbacterium imperiale CBS 498-74* was operated at 5°C, 10°C, and 20°C in laboratory scale batch or membrane bioreactors. The bioreactions were conducted in the presence of buffered media (50 mM Na_2HPO_4/NaH_2PO_4, pH 7.0) and in the presence of distilled water or tap-water, to simulate a possible end-pipe biotreatment process. Finally, the integral bioreactor performances were studied varying cell loading, from 0.1 to 16 mg_{DCW}/reactor in order to realize near 100% bioconversion of acrylonitrile, propionitrile and benzonitrile without consistent loss of NHase activity.

1 INTRODUCTION

Nitriles are an important group of compounds which appear in the environment *via* natural or industrial syntheses. They are toxic and carcinogen for living cells once released in the environment. However, the ability to degrade nitriles is quite common among microorganisms. Three different groups of enzymes are involved in the microbial hydrolysis of nitriles. Nitrilases (EC 3.5.5.1 and 3.5.5.7) catalyse the hydrolysis of nitriles in one step reaction to the corresponding carboxylic acids, forming ammonia. Nitrile hydratases, NHases, (EC 4.2.1.84) and amidases (EC 3.5.1.4) sequentially hydrolyse nitriles to the corresponding carboxylic acid, in a two step reaction with the amide as intermediate product.

NHases are useful biocatalysts that found industrial applications in the production of acrylamide (Yamada & Kobayashi 1986), nicotinamide and of interesting intermediate compounds (Kobayashi & Shimizu 1998). These enzymes also possess a great potential for environmental bioremediation. Their capability to transform highly toxic nitriles into the corresponding amides, which are more friendly compounds for the environment, outlines their biotechnological relevance (Jallageas et al. 1980, Schmid et al. 2002). Furthermore, the enzymatic processes present advantages over the traditional chemical ones because they operate under mild reaction conditions without toxic materials, realizing at the same time a reduction of emissions and wastes.

In the present work use was made of resting cells of *Microbacterium imperiale CBS 498-74*, a strain that allows the bioconversion of acrylonitrile into the corresponding amide with high conversion yield. A two-step degradation pathway of nitriles, which involves NHase and amidase, is adopted by the strain. The appropriate choice of operational conditions makes the amidase activity negligible as compared with NHase activity, and theoretical 100% conversion yields into amide are possible (Alfani et al. 2001, Cantarella et al. 1998a,b, 2002).

This report gives evidence on the total conversion of simulated streams of acrylonitrile, propionitrile and benzonitrile in continuous and batch laboratory-scale bioreactors either in the presence of appropriate buffered media or in the presence of distilled or tap-water. The study suggests the possibility to adopt, in the current industrial production of nitrile compounds, an end-pipe biotreatment process that would reduce the environmental impact of the waste streams.

2 MATERIALS AND METHODS

2.1 *Catalysed reactions*

$$R-C\equiv N \xrightarrow[+H_2O]{NHase} R-CO-NH_2 \xrightarrow[+2H_2O]{amidase} R-COOH+NH_4OH$$

R being : $CH_2=CH-$ or CH_3-CH_2- or C_6H_5-

Figure 1. NHase and amidase catalysed reactions.

2.2 *Microorganism culture conditions*

M. imperiale CBS 498-74 was obtained at the optimum initial glucose concentration for NHase production, $5\,g\,l^{-1}$, in a shake flask (500 ml) at 28°C for 24 h incubation. All fermentations were carried out in a rotary shaker G25-KC from New Brunswick Scientific (USA) with 220 rev min^{-1} orbital shaking. The composition of the culture medium (YMP-medium) was as follows: $3\,g\,l^{-1}$ yeast extract (Oxoid, England), $3\,g\,l^{-1}$ malt extract (Oxoid), $5\,g\,l^{-1}$ bacteriological-peptone (Oxoid). The medium was prepared in 50 mM Na_2HPO_4/NaH_2PO_4 buffer, pH 7.0 and sterilized by autoclaving at 121°C for 20 min. Specific activities in the cell were found to be 34.4 U mg_{DCW}^{-1} (Cantarella et al. 2002). One Unit (U) of NHase activity was defined as the amount of resting cells that catalyses the formation of 1 μmole of acrylamide per min under the adopted conditions. Dry cell weight (DCW) was determined by drying to constant weight a solution with a known optical density (OD). An average value of 0.26 $mg_{DCW}\,ml^{-1}$ per unit of OD was obtained. All runs were replicated at least twice and averaged values are reported.

2.3 *Chemicals*

Substrates and products were supplied by Aldrich (Germany). Reagent-grade, commercially available, compounds were used in all the experiments.

2.4 *Ultrafiltration membrane bioreactor*

An ultrafiltration kit (Amicon Model 52), (Grace, USA), of stirred type was employed as a membrane bioreactor. A fluoro-polymer membrane FS81PP produced by Dow Liquid Separations (England) with a molecular weight cut-off of 10,000 was used. The stirring, by means of a magnetic stirrer set at 250 rpm, was provided in all the runs to limit deposition onto the membrane due to concentration polarisation. The buffered substrate solutions were fed to the reactor using a peristaltic pump (Gilson Minipuls, France) assuring a constant flow-rate of $12 \pm 0.7\,ml\,h^{-1}$. The filtrates were collected using a fraction collector (LKB Instruments, Sweden) and aliquots were analysed for product determination. Reaction temperature was controlled within ±0.1°C. In bioreactor experiments, membranes totally retained the resting cells and no fouling was detected, under stirring conditions, within the explored cell concentrations. The membrane resistance to chemicals was fair and no rejection of solutes was determined.

2.5 *Batch bioreactor*

A reactor vessel of 100 ml volume was used. The stirring was assured by a magnetic bar and temperature was maintained by circulating water through the jacket from the constant temperature water bath.

2.6 *Analytical determinations*

Propionamide was identified by its retention time in a Hewlett Packard 5890 series II gas chromatograph, equipped with a HP 3396 series II integrator, and a RT-QPLOT-restek Cat 19716 capillary column (30 m × 0.53 mm i.d.), (USA). A base deactivated guard column Restek Cat 10002 (5 m × 0.53 mm i.d.) was also used. Helium carrier gas flow rate was 1.6 ml min^{-1}. The column, injector and flame ionization detector were held at 220°C.

The concentrations of acrylonitrile, benzonitrile and their amides were detected at 220 nm by HPLC with a Perkin Helmer Series 2 system equipped with UV detector and a Merck C18 reverse phase column operating at 30°C and at 0.5 ml/min mobile phase (acetonitrile and KH_2PO_4/H_3PO_4 buffer, 10 mM, pH 2.8; volumetric ratio 1–10). The nitrile and amide retention time being quite different, the absorbance at 220 nm allowed their quantitative determination.

3 RESULTS AND DISCUSSION

For a few years we have been investigating acrylamide production using resting cells from the same strain. The kinetic parameters and the operational conditions such as temperature, substrate and biocatalyst concentration to reach high acrylonitrile conversion, have been assessed in buffered media and reported elsewhere (Alfani et al. 2001, Cantarella et al. 1998a,b, 2002).

Figure 2 illustrates the effect of resting cell concentration on the conversion of 100 mM acrylonitrile in a continuous stirred UF-membrane reactor. The bioreactor was operated in 50 mM phosphate buffer, pH 7.0, at 10°C to reduce the thermal inactivation of NHase. With low amounts of resting cells the bioreactor is unable to operate as an integral one and only when cell concentration is as high as 16 mg_{DCW}/reactor the bioreactor converts 88% of the substrate into acrylamide.

A end-pipe treatment of industrial nitrile streams would be of great potential interest if their bioconversion into the less toxic amides or carboxylic acids could

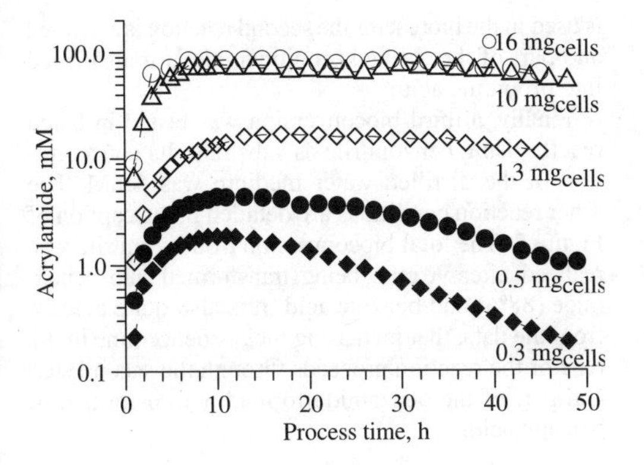

Figure 2. Continuous acrylonitrile conversion in UF-membrane bioreactor as a function of biocatalyst concentration. Acrylonitrile: 100 mM in Na-phosphate buffer, pH 7.0; T: 10°C; magnetic stirring: 250 rpm.

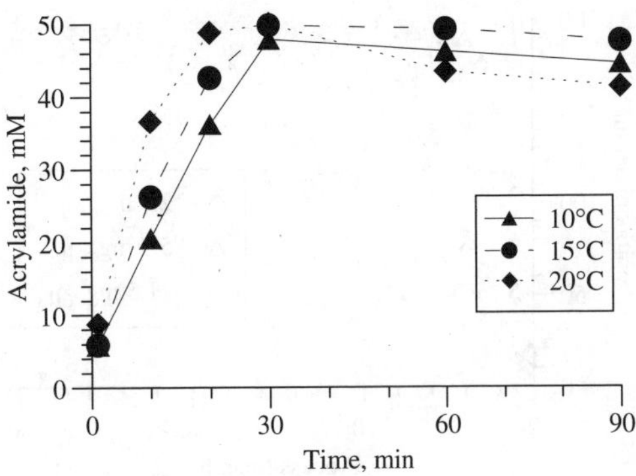

Figure 4. Time course of acrylamide bioproduction in batch reactor at different temperatures in distilled water. Reaction volume: 50 ml; acrylonitrile: 50 mM; resting cell: 10 mg_{DCW}; T: 20°C; magnetic stirring: 250 rpm.

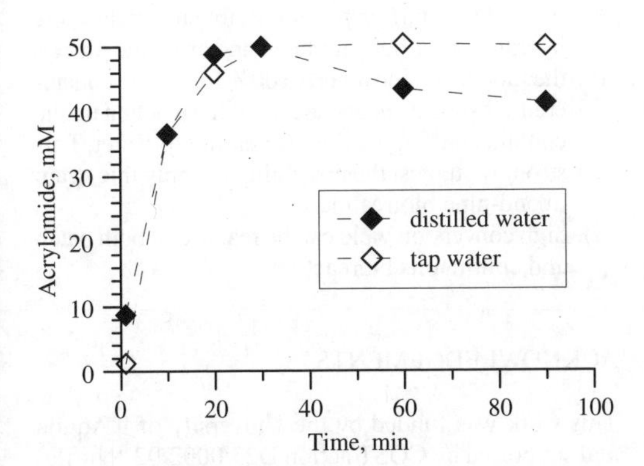

Figure 3. Time course of acrylamide bioproduction in batch reactor. Reaction volume: 50 ml; acrylonitrile: 50 mM; resting cell: 10 mg_{DCW}; T: 20°C; magnetic stirring: 250 rpm.

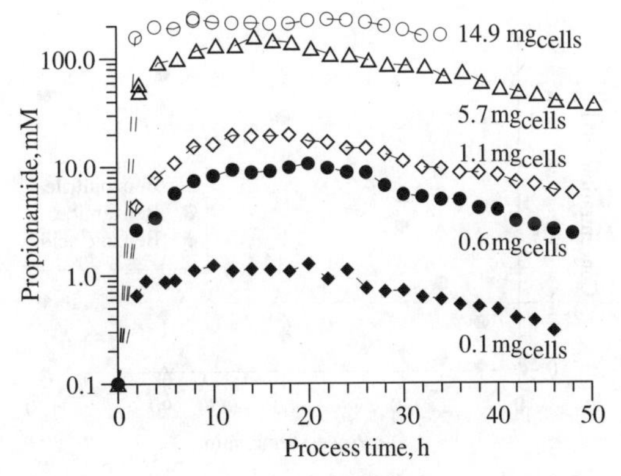

Figure 5. Continuous propionitrile conversion in UF-membrane bioreactor as a function of biocatalyst concentration. Propionitrile: 200 mM in Na-phosphate buffer, pH 7.0; T: 10°C; magnetic stirring: 250 rpm.

be performed in non buffered media. Following this reasoning the acrylonitrile bioconversion was studied in a first series of runs performed in batch reactor in either distilled or tap-water.

The reaction mixture contained 10 mg_{DCW} of cells and a lower concentration of acrylonitrile, since the economical balance of the main process calls for a high substrate conversion that implies a very low concentration in the waste streams. As Figure 3 shows, within roughly 30 min, both media reach 100% bioconversion, and probably a longer contact period in distilled water, activates the amidase activity that converts the amide into the corresponding acid.

This reaction was also conducted at three different temperature as illustrate in Figure 4. 20–30 min are necessary to accomplish the total conversion of acrylonitrile stream in the presence of distilled water. Apparently 10°C and 15°C too could be chosen to perform the bioreaction even though the reaction rate would be lower.

The bioreaction in distilled water was also performed in continuous UF-membrane reactor operating in differential mode at 4°C and 10°C (data not shown). The NHase activity decay rate, when compared with that obtained in a parallel experiment carried out in buffered media, was roughly the same at 4°C (half-life being 40 h) while at 10°C the half-life in distilled water resulted slightly higher (15 h) than that in buffered media (12 h).

The effect of resting cell concentration on the conversion of a feed stream of propionitrile 200 mM is shown in Figure 5. The 100% bioconversion was realized for 30 h of continuous performance with 14.9 mg_{DCW}/reactor. However, after 30 h of operation

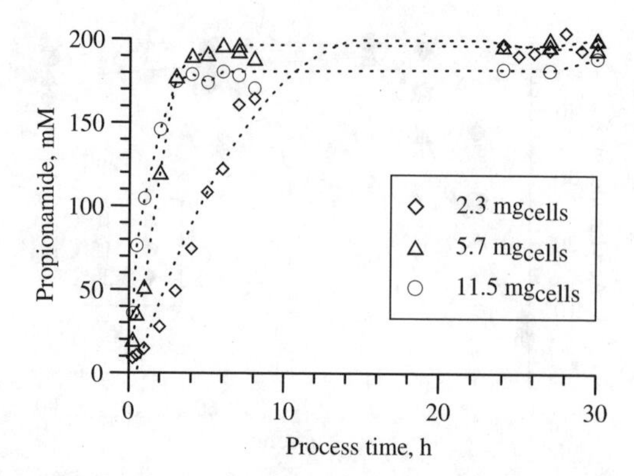

Figure 6. Time course of propionamide bioproduction in batch reactor. Reaction volume: 70 ml; propionitrile: 200 mM in Na-phosphate buffer, pH 7.0; T: 10°C; magnetic stirring: 250 rpm.

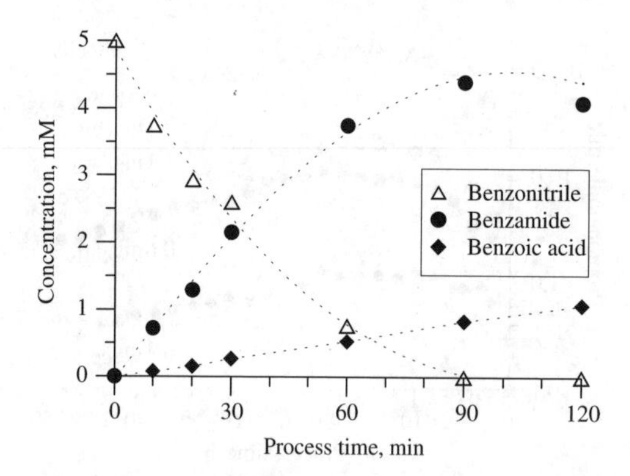

Figure 7. Time course of benzonitrile bioconversion in batch reactor. Reaction volume: 50 ml; 5 mM substrate in distilled water; resting cell: 2 mg_{DCW}; T: 20°C; magnetic stirring: 250 rpm.

the enzyme inactivation became evident and the bioconversion decreased. Of course, in order to keep the reaction going on a 100% conversion basis, an adequate policy of cells refresh should be pursued (Cantarella et al. 2004).

In a second series of experiments, the investigation on the influence of different amounts of resting cells on propionitrile bioconversion was also performed in batch reactor in buffered media. The results are shown in Figure 6 (see legend and caption for details). The curves indicate that, within the first 8 h of operation time, a conversion yield of 77, 88.5 and 85.5% is reached when the reactor is operated with 2.3, 5.7 and 11.5 mg_{DCW} respectively. To attain 90–93% conversion 24 h are necessary while the total conversion is obtained after 28 h operation. However, when 11. 5 mg of cells is used in the bioreactor the second reaction is activated and part of the formed propionamide is transformed into propionic acid.

Finally, a third bioconversion was tested in batch reactor using benzonitrile as substrate. Its concentration in the distilled water medium was 5 mM. The other reaction conditions are detailed in the caption of Figure 7. The total bioconversion of benzonitrile was realized after 90 min, being transformed into benzamide (88%) and benzoic acid. It is also quite evident from the data, that increasing the residence time in the reactor the reaction proceeds through the second step and part of the benzamide formed is transformed in benzoic acid.

4 CONCLUSIONS

The results of this study allow us to conclude:

i) the NHase of *M. imperiale* is able to catalyse the hydratation of both aliphatic and aromatic nitriles
ii) the bioconversion attains 100% for all the considered substrates, independently if the reaction media contains buffer, distilled water, or tap-water. This strongly suggest the possibility to apply this strain in end-pipe biotreatment
iii) high conversion yield can be reached in both batch and continuous bioreactors.

ACKNOWLEDGEMENTS

This work was funded by the University of L'Aquila and supported by COST action D25/0002/02 "Nitrile- and amide-hydrolyzing enzymes as tools in organic chemistry".

REFERENCES

Alfani, F., Cantarella, M., Spera, A. & Viparelli, P. 2001. Operational stability of *Brevibacterium imperialis CBS 489-74* nitrile hydratase. *J. Molecular Catal. B: Enzymatic* 11:687–697.

Cantarella, M., Cantarella, L., Gallifuoco, A., Frezzini, R., Spera, A. & Alfani, F. 2004. A study in UF-membrane reactor on activity and stability of nitrile hydratase from *Microbacterium imperiale CBS 498-74* resting cells for propionamide production. *J. Molecular Catal. B: Enzymatic* (in the press).

Cantarella, M., Spera, A. & Alfani, F. 1998a. Characterization in UF-membrane reactors of nitrile hydratase from *Brevibacterium imperialis CBS 489-74* resting cells. *Ann. N.Y. Acad. Sci.* 864:224–227.

Cantarella, M., Spera, A., Cantarella, L. & Alfani, F. 1998b. Acrylamide production in an ultrafiltration-membrane bioreactor using cells of *Brevibacterium imperialis CBS 489-74*. *J. Memb. Sci.* 147:279–290.

Cantarella, M., Spera, A., Leonetti, P. & Alfani, F. 2002. Influence of initial glucose concentration nitrile hydratase production in *Brevibacterium imperialis CBS 498-74. J. Molecular Catal. B: Enzymatic* 19:405–414.

Jallageas, J.C., Arnaud, A. & Galzy, P. 1980. Bioconversion of nitriles and their applications. *Adv. Biochem. Eng.* 14:1–32.

Kobayashi, M. & Shimizu, S. 1998. Metalloenzyme nitrile hydratase-structure, regulation, and application to biotechnology. *Nat. Biotechnol.* 16:733–736.

Schmid, A., Hollmann, F., Park, J.B. & Bühler, B. 2002. The use of enzymes in the chemical industry in Europe. *Current Opinion in Biotechnology* 13:359–366.

Yanenko, O., Astaurova, T., Pogorelova & Ryabchenko, L. 1996. Nitrile metabolism in *Rhodococcus:* the trends in improving of biocatalyst. *Proc. 10th International Biotechnology Symposium. Sydney, 95.*

Yamada, H. & Kobayashi, M. 1996. Nitrile hydratase and its application to industrial production of acrylamide. *Biosci. Biotechnol. Biochem.* 60:1391–1400.

European Symposium on Environmental Biotechnology, ESEB 2004 - Verstraete (ed)
© 2004 Taylor & Francis Group, London, ISBN 90 5809 653 X

Decolorization of azo and anthraquinone dyes by fungi

A. Anastasi, G.C. Varese, L. Casieri, E. Cane & V. Filipello Marchisio
Department of Plant Biology, University of Turin, Italy

ABSTRACT: Fungi are excellent candidates for employment in the bioremediation of coloured effluents through both the degradation of dyes and their removal by absorption onto the mycelium. This screening study has shown that numerous basidiomycete species, many of them never described in the literature, are very effective degraders of azo and anthraquinone dyes. Analysis of two of the more promising isolates, a *Cyathus stercoreus* strain and an unidentified basidiomycete, illustrated their efficient and fast decolorising potential and suggested that an independent manganese peroxidase was involved in degradation by the basidiomycete. Parallel investigation of an *Aspergillus* showed that it was capable of removing 98% of an azo dye by biosorption in a few days.

1 INTRODUCTION

Synthetic organic (azo, anthraquinone, indigo, etc.) dyes are used for several purposes in the textile, cosmetic, papermaking, food and pharmaceutical industries. Their complex aromatic molecular structure is suited to secure low-cost stability to light, water and oxidants, but also makes them toxic compounds that persist for a long time in the environment and are a threat to ecosystems and human health. Coloured effluents in waste water treatment systems are mainly eliminated by means of physical and chemical procedures, e.g. adsorption, concentration, chemical transformation and incineration, that are both very expensive and not free from drawbacks (Moreira et al. 2000). Bioremediation by micro-organisms is an environment-friendly and cost-competitive alternative (Novotny et al. 2001).

Bioremediation of coloured effluents is achieved by both degradation, i.e. rupture of the dye molecule and its subsequent mineralisation, and biosorption, i.e. removal of a dye by absorption on the part of a biomass (Banat et al. 1996). The poor information nowadays available shows that fungi could be excellent candidates for employment as both degraders and absorbers in the bioremediation of coloured effluents (Fu & Viraraghavan 2001).

Ligninolytic fungi, principally basidiomycetes such as *Phanerochaete chrysosporium, Bjerkandera adusta* and *Trametes versicolor*, have been used in most dye biodegradation studies, since the low specificity of their lignin-degrading enzymes, such as laccase and peroxidase, enables them to degrade numerous highly recalcitrant pollutants such as dyes (Soares et al. 2002). Production of these enzymes is greatly influenced by the composition of the culture medium and especially by the C:N ratio (Knapp et al. 2001).

Biosorption is mainly investigated with zygomycetes and mitosporic fungi, such as *Aspergillus*, using the active or inactivated mycelium, sometimes pretreated to modify the composition and/or charge of the fungal wall (Chu & Chen 2002, Fu & Viraraghavan 2002). Biosorption offers undeniable advantages, such as faster effluent treatment and recovery of the dye absorbed for future use. In addition, if conducted with an inactivated fungal biomass, it does not require the checking of restrictive parameter values to maintain the optimum conditions for the growth and activity of the living organism (Fu & Viraraghavan 2001).

This paper presents the results of:

- a screening of the efficiency of numerous basidiomycetes held in the University of Turin fungal collection (*Mycotheca Universitatis Taurinensis*, MUT) in the decolorization of an azo dye and two anthraquinone dyes. The best culture conditions for induction of their degradative capabilities were also investigated;
- a closer study in liquid cultures of three demonstrably active species to quantify their degradation potential (including identification of any enzymes involved), or their absorption potential.

2 MATERIALS AND METHODS

2.1 *Decolorization in solid cultures*

The screening was conducted on 151 isolates from 79 basidiomycete species in culture media supplemented with 0.02% w/v of the azo dye, Poly S-119, and the two anthraquinone dyes, Poly R-478 and RBBR, all from Sigma.

The four media had different C:N ratios: MEA (20 g malt extract, 20 g glucose, 2 g peptone, 18 g agar, 1 l deionised H_2O; C:N 20); PBA (40 g sorbose, 0.22 g ammonium tartrate, 0.6 g KH_2PO_4, 0.5 g $MgSO_4 \cdot 7H_2O$, 0.4 g K_2HPO_4, 74 mg $CaCl_2 \cdot 2H_2O$, 12 mg ferric citrate, 7 mg $ZnSO_4 \cdot 7H_2O$, 5 g $MgSO_4 \cdot 4H_2O$, 1 mg $CoCl_2 \cdot 6H_2O$, 0.1 mg thiamine HCl, 18 g agar, 1 l deionised H_2O; C:N 181); GLN (2 g glucose, 0.44 g ammonium tartrate, 0.01 g yeast extract, 18 g agar and trace elements, 1 l deionised H_2O; C:N 4.5); GHN (2 g glucose, 4.42 g ammonium tartrate, 0.01 g yeast extract, 18 g agar and trace elements, 1 l deionised H_2O; C:N 0.4). Trace elements were added to a final concentration, per litre of deionised H_2O, of 2 g KH_2PO_4, 0.5 g $MgSO_4$, 0.1 g $CaCl_2$, 5 mg $MnSO_4$, 10 mg NaCl, 1 mg $FeSO_4$, 1 mg $CoCl_2$, 1 mg $ZnSO_4$, 0.1 mg $CuSO_4$, 0.1 mg $AlK(SO_4)_2$, 0.1 mg H_3BO_3, 0.1 mg $NaMoO_4$, 0.001 mg thiamine, 0.001 mg biotine. The inocula, made in triplicate, were 5 mm mycelium disks cut with a sterile cork borer from the edge of colonies growing on MEA, placed in the centre of ∅ 50 mm capsules containing 10 ml of dyed nutritive medium. The decolorization halos were measured after 15 days' incubation in the dark at 24°C.

2.2 *Decolorization in liquid cultures*

Cyathus stercoreus (Schweinitz) de Toni and an unidentified basidiomycete DBB+ (UB), were selected on account of their ability to degrade Poly S-119 and Poly R-478 respectively, as demonstrated during the screening. *Aspergillus flavus* (Link) var. *flavus* was selected from another series of tests not described in this paper on account of its ability to absorb Poly S-119.

Liquid cultures were prepared in 500 ml Erlenmeyer flasks containing 30 ml of the medium found most suitable in the previous assays and the dyes at 0.02% w/v: GLN for *C. stercoreus* and *A. flavus* and GHN for UB, both buffered to pH 5.5 and 4.5 respectively with 20 mM sodium 2.2-dimethyl succinate. The inocula were 3 mycelium disks per flask prepared as described above. Cultures were done in triplicate and incubated statically at 24°C in the dark. Uninoculated and dyeless cultures were included for all fungi. Decrease of absorbance (expressed as percentage of decolorization) at the maximum visible wavelength of the dyes (514 nm for Poly R-478 and 475 nm for Poly S-119) was measured with a UV-VIS spectrophotometer (Pharmacia Biotech Ultraspec 3000) every two days until attainment of a plateau of values on a 0.1 ml aliquot of medium diluted 10-fold in water.

2.3 *Phenol-oxidase activity in liquid cultures*

The activities of manganese peroxidase (MnP), manganese-independent peroxidase (MiP) and laccase (Lcc) were assayed spectrophotometrically according to Tanaka et al. (1999). Oxidation of 2,2′-azino-bis-(3-ethylbenzothiazoline-6-sulphonic acid (ABTS) was monitored at 420 nm in 100 mM sodium acetate buffer pH 5.0 at 22°C. For peroxidase activities, the reaction was initiated by the addition of H_2O_2 to a final concentration of 100 μM. For MnP, the mixture also contained 100 μM $MnSO_4$. MiP activity was calculated by subtracting the Lcc activity measured in the absence of H_2O_2 from the total activity measured in its presence and absence of $MnSO_4$. MnP activity was calculated by subtracting the total activity measured in the presence of H_2O_2 and absence of $MnSO_4$ from that measured in their joint presence. Activities were monitored in the absence and presence of the dye and expressed as units per litre (U/l), where one unit is defined as the amount of enzyme catalysing 1 μmole of substrate per minute (ε of ABTS is $36{,}000\ M^{-1}cm^{-1}$).

3 RESULTS AND DISCUSSION

3.1 *Decolorization in solid cultures*

Most of the isolates (86%) decolorised one or more dyes. Poly S-119 was the most resistant (degraded by 34% of the isolates), whereas Poly R-478 and RBBR were degraded by 73% and 80% respectively. The higher recalcitrance of azo dyes respect to anthraquinonic dyes has already been observed by other workers (Jarosz-Wilkolazka et al. 2002, Abadulla et al. 2000). According to Jarosz-Wilkolazka et al. (2002) the high recalcitrance of sulphonated azo dyes such as Poly S-119 is due to the simultaneous presence of sulphone and azo groups. These, in fact, are not habitually present in nature and hence poorly biodegradable.

The highest degradative capabilities (expressed as the production of ∅ > 40 mm decolorization halos) were displayed by 27 isolates from 19 species. The best performance (complete degradation in less than 14 days) was obtained with 15 isolates from: *Ceriporia metamorphosa* (Fuckel) Stalpers, *Chondrostereum purpureum* (Persoon: Fries) Pouzar, *Cyathus stercoreus* (Schweinitz) de Toni (2 isolates), *Ganoderma applanatum* (Persoon) Patouillard, *Hypholoma sublateritium* (Fries) Quelet, *Lentinus conchatus* (Bulliard) J. Schroeter, *Lenzites betulinus* (Linnaeus: Fries) Fries, *Lopharia spadicea* (Persoon) Boidin, *Pholiota squarrosa* (Weigel: Fries) Kummer, *Polyporus ciliatus* Fries, *Polyporus squamosus* (Hudson) Fries, *Trametes pubescens* (Schumacher) Pilat (2 isolates), and UB. Only the ability of *Ceriporia metamorphosa* and *Ganoderma applanatum* to degrade anthraquinone dyes has been illustrated (Jarosz-Wilkolazka et al. 2002,

Novotny et al. 2001), whereas their degradation of azo dyes and the degradation of any dye by the other species has not been reported.

Many isolates decolorised the dyes on one medium only: 51% decolorised Poly S-119, 31% Poly R-478 and 26% RBBR. This finding shows that the greater the recalcitrance of the dye (Poly S-119), the more restrictive allowing its degradation become. PBA proved the best medium for the azo dye, MEA for the anthraquinone dyes. These results corroborate the suggestion by Martins et al. (2001) that a particular lack of nitrogen is needed to induce the ligninolytic enzymes involved in the degradation of azo compounds.

3.2 *Decolorization and enzyme assay in liquid cultures*

Cyathus stercoreus decolorised about 93% of the Poly S-119 in 12 days (Fig. 1). The mycelium was not stained and there was no decrease in absorbance in the uninoculated control, but the maximum enzyme activities were low: Lcc 5.8 U/l, MnP 12.0 U/l and MiP 9.2 U/l. This was confirmed in the dyeless tests where there was only a slight increase in MiP activity to 36.4 U/l. This weak correspondence between decolorization and activity could be the outcome of the adoption of non-optimal conditions of enzyme kinetics, poor enzyme diffusion in the cultures, or employment by the fungus of degradative mechanisms other than those being investigated.

UB decolorised about 63% of the Poly R-478 in 11 days (Fig. 2); here, too, the mycelium was not stained and there was no decrease in absorbance in the uninoculated control. MiP displayed the highest activity (384 U/l on the 16th day) and its pattern was close to that of the decolorization, whereas the maxima displayed by MnP and Lcc over the course of the three weeks were 119 and 60 U/l respectively. The dyeless test gave a maximum of 425 U/l for MiP at the end of the test, coupled with a more irregular pattern. The maxima for MnP and Lcc over the course of the three weeks were 135 and 57 U/l respectively. These results point to the constitutive nature of the three enzymes and underscore the importance of MiP in the decolorization of polymers, as also illustrated for *Pleurotus* and *Bjerkandera* species (Martinez 2002).

Aspergillus flavus reduced the Poly S-119 concentration by about 98% in 96 hours (Fig. 3) and there was no decrease in absorbance in the uninoculated control. The mycelium was intensely stained at the end of the test and treatment with 2 M NaOH was needed to release the acid dye. No phenol-oxidase activity was detected in the media. These two findings indicate that decolorization is not the work of degradative systems, but the result of biosorption by the mycelium.

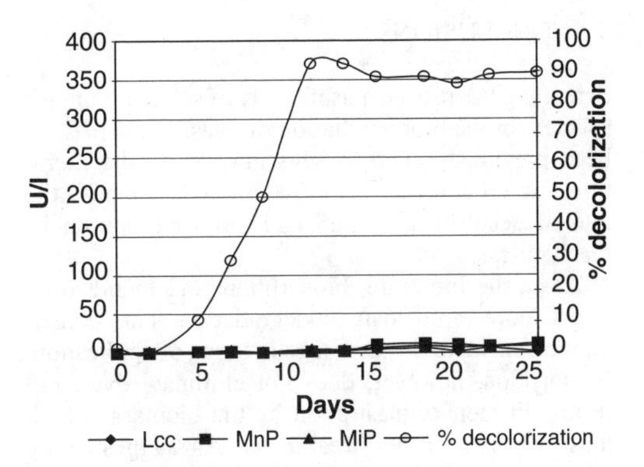

Figure 1. Per cent (%) decolorization of Poly S-119 by *Cyathus stercoreus* and enzyme activity values (U/l) for laccase (Lcc), manganese peroxidase (MnP) and manganese-independent peroxidase (MiP).

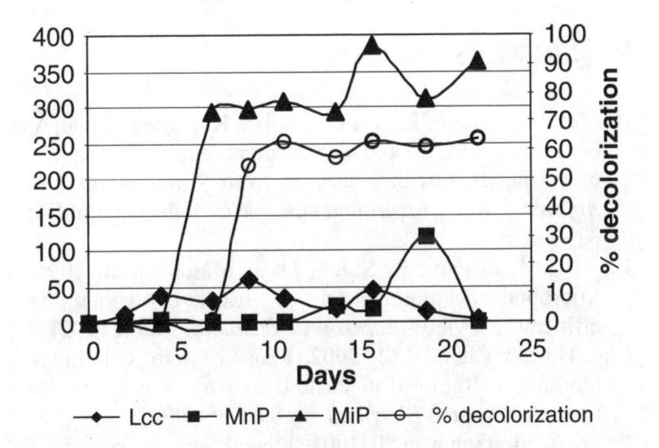

Figure 2. Per cent (%) decolorization of Poly R-478 by unidentified basidiomycete (UB) and enzyme activity values (U/l) for laccase (Lcc), manganese peroxidase (MnP) and manganese-independent peroxidase (MiP).

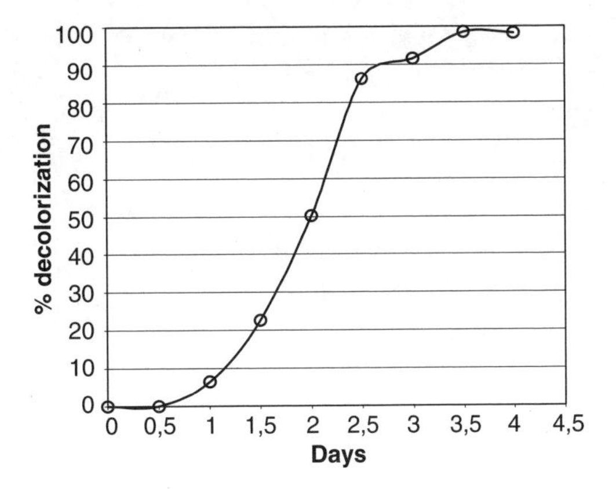

Figure 3. Per cent (%) decolorization of Poly S-119 by *Aspergillus flavus*.

4 CONCLUSIONS

Screening has proved a useful way of selecting fungus species for the bioremediation of dyes. Many had not been previously cited for this purpose in the literature. Liquid culture tests must follow to quantify the rate of decolorization and gain informations on its mechanisms.

As in the literature, biosorption was found to be much more rapid than biodegradation. This constitutes a considerable advantage in terms of application. Biosorption, however, does not eliminate dyes, but results in their sequestration by the biomass, which thus requires further treatments. As suggested by other workers (Robinson et al. 2001), thought could be given to combining degradation and absorption systems to secure the rapid and efficient bioremediation of coloured effluents.

REFERENCES

Abadulla, E., Tzanov, T., Costa, S., Robra, K., Cavaco-Paulo, A. & Gübitz, G.M. 2000. Decolorization and detoxification of textile dyes with a laccase from *Trametes hirsuta*. *Applied and Environmental Microbiology* 66(8): 3357–3362.

Banat, I.M., Nigam, P., Singh, D. & Marchant, R. 1996. Microbial decolourization of textile-dye-containing effluents: a review. *Bioresource Technology* 58: 217–227.

Chu, H.C. & Chen, K.M. 2002. Reuse of activated sludge biomass: I. Removal of basic dyes from wastewater by biomass. *Process Biochemistry* 37: 595–600.

Fu, Y. & Viraraghavan, T. 2001. Fungal decolourization of dye wastewaters: a review. *Bioresource technology* 79: 251–262.

Fu, Y. & Viraraghavan, T. 2002. Removal of Congo Red from an acqueous solution by fungus *Aspergillus niger. Advances in Environmental Research* 7: 239–247.

Jarosz-Wilkolazka, A., Kochmanska-Rdest, Y., Malarczyk, E., Wardas, W. & Leonowicz, A. 2002. Fungi and their ability to decolourize azo and anthraquinonic dyes. *Enzyme and Microbial Technology* 30: 566–572.

Knapp, J.S., Vantoc-Wood, E.J. & Zhang, F. 2001. Use of wood-rotting fungi for the decolorization of dyes and industrial effluents. In G.M. Gadd (ed.), *Fungi in Bioremediation*: 242–304. Cambridge: University Press.

Martinez, A.T. 2002. Molecular biology and structure-function of lignin-degrading heme peroxidase. *Enzyme and Microbial Technology* 30: 425–444.

Martins, M.A.M., Ferreira, I.C., Santos, I.M. & Queiroz, M.J. 2001. Biodegradation of bioaccessible textile azo dyes by *Phanerochaete chrysosporium. Journal of Biotechnology* 89: 91–98.

Moreira, M.T., Mielgo, I., Feijoo, G. & Lema, J.M. 2000. Evaluation of different fungal strains in the decolorization of synthetic dyes. *Biotechnology Letters* 22(18): 1499–1503.

Novotny, C., Rawal, B., Bhatt, M., Patel, M., Sasek, V. & Molitoris, H.P. 2001. Capacity of *Irpex lacteus* and *Pleurotus ostreatus* for decolorization of chemically different dyes. *Journal of Biotechnology* 89: 113–122.

Robinson, T., McMullan, G., Marchant, R. & Nigam, P. 2001. Remediation of dyes in textile effluent: a critical review on current treatment technologies with a proposed alternative. *Bioresource Technology* 77: 247–255.

Soares, G.M.B., Pessoa Amorim, M.T., Hrdina, R. & Costa-Ferreira, M. 2002. Studies on the biotransformation of novel disazo dyes by laccase. *Process Biochemistry* 37: 581–587.

Tanaka, H., Itakura, S. & Enoki, A. 1999. Hydroxyl radical generation by an extracellular low-molecular-weight substance and phenol oxidase activity during wood degradation by the white-rot basidiomycete Trametes versicolor. *Journal of Biotechnology* 75: 57–70.

European Symposium on Environmental Biotechnology, ESEB 2004 - Verstraete (ed)
© 2004 Taylor & Francis Group, London, ISBN 90 5809 653 X

Waste water purification to small treatment plants in regional areas and effluent recycling to feasible uses

P.S. Kollias
Dr. Civil – Sanitary engineer, Athens, Greece

V. Kollias
Dr. Physicist Researcher, University of Athens, Athens, Greece

T.C. Koliopoulos
Dr. Environmental engineer, University of Stratchclyde, Athens, Greece

S. Kollias
Dipl. Mathematician, University of Athens, Athens, Greece

ABSTRACT: Water is a precious good entirely necessary for the maintenance of our life. The pollution that had followed water used for domestic activities and the water scarcity, lead to take measures for waste water treatment, from spreaded settlements in regional areas. Investigations show the predomination of small treatment plants, against common treatment plants, to a centrobaric place. It is considered necessary the sustainable water use, of the treated used water with the tendency for recirculation to feasible uses.

1 INTRODUCTION

The ascertainment in many places of coming water shortage makes necessary the application of a suitable water management policy[1]. This makes essential the creation of small treatment plants placed to regional areas instead of common units, in centrobaric places. The treatment technology to small treatment plants is realized with conventional methods as: natural and mechanical aeration lagoons, underground disposal of waste, activated sludge with extended aeration, high load trickling filters and others. Treated effluent could be discharged to surface water aquifers, that must be protected from surface pollution.

2 THE CREATION OF SMALL TREATMENT PLANTS IN REGIONAL AREAS

Towns created in regional areas are found in great distances between them and their connection with sever pipes to a common centrobaric treatment plant is very expensive. For that reason it is necessary the realization of investigations about the costs of independent small treatment plants and common units, according to the following methodology.[3]

2.1 *Costs of independent small treatment plants ($S^i{}_1CIi$)*

This is found as follows:

- Construction costs of small treatment plants

$$S^i{}_1CIi = S^i{}_1CIprimi + S^i{}_1CIbioli + S^i{}_1disenfi + S^i{}_1\,sludgei.$$

Where $S^i{}_1CIprimi$ is the primary treatment cost, $S^i{}_1CIbioli$ the cost of biological treatment, $S^i{}_1disenfi$ the cost of disinfection and $S^i{}_1sludgei$ the sludge treatment.
- Operation costs and costs of relevant connections, between the different technical operations of the i treatment plants. This equals to $S^i{}_1CIoperi$.
- Costs to face environmental impacts $S^i{}_1Eii$. These include: landscape works, increase of effluent standards for adaptation to a water receiver. Total small treatment plants cost equals to

$$SCI = S^i{}_1CIi + S^i{}_1CIoperi + S^i{}_1Eii$$

2.2 *Costs of a common treatment plant (CC)*

This depends from the followings

- Construction costs of the common treatment plant (CCC)

CCC = Cprim + Cbiol + Cdisinf + Csludge

They are included the construction costs of primary, biological, disinfection and sludge units for the realization of the treatment of transferred wastes.

- Costs of external sewers for the transfer of wastes from settlements Cs
- Operation costs and costs for connections and different technical operations Co
- Costs to face environmental impacts and likely benefits of hosting the land site community Ce
- Total cost of the common treatment plant

SC = CCC + Cs + Co + Ce

2.3 *Investigations and comparisons*

It is examined the projected difference of costs D that equals to

D = SCI − SC

When D < 0 and SCI < SC it is selected the construction of the purification to small treatment plants covering regional areas.

3 THE TECHNOLOGY OF APPLIED TO SMALL TREATMENT PLANT METHODS

The treatment technology to small treatment plants referred to populations 200–5000 hab is realized with conventional methods including the following units as: a storm flow tank, screen, grit chamber, trickling filters, activated sludge process, disinfection and others.

3.1 *Methods suggested for treatment of wastes to small treatment plants [2]*

More feasible methods are the followings:

- Natural aeration lagoons
  They include a grit chamber and three lagoons that communicate through pipes. The water height varies from 1 to 1.2 m and the ratio between length/wide from 3/5. The slopes to the sides are ≤1/2. They can be used for population of 50–2000 hab. It is examined the efficiency of the reduction after treatment, to BOD_5, COD and SS. The advantages of this kind of treatment are: economy to the repairs, good efficiency to causing desease germs and others.
- Mechanical aeration lagoons
  They are composed from an aeration lagoon, followed from a decantation lagoon, that are separated through a dike, with a 3 m minimum width. It is attended an efficiency of BOD_5 30–35 mg/l COD 100–120 mg/l and SS 30 mg/l. They can be used for populations until 5000 hab. The main advantages are: Light repairs, stable quality of treated water and good efficiency to carried loads.
- Underground disposal of waste
  They are composed from an imhoff tank or a septic tank that is followed from a filtration bed with drains of suitable diameter. The system is possible to be applied for populations to about 300 hab. The considered advantages are: Light energy consumption and good tolerance to strong load variations.
- Spreading of wastes to ditches
  They are composed from a screen, a selected pretreatment and enfluent decharge to open trenches. The advantages include: the small cost, the almost worthless energy consumption and the good support of strong load variations. They can be used at about 150 hours a year.
- Activated sludge with extended aeration
  After a primary treatment (screen, grit chamber, sedimentation tank), follows aeration tank and final sedimentation tank, with activated sludge recirculation. The quality of produced effluent gives the following efficiency to parameters examined. BOD < 30 mg/l, COD < 90 mg/l and SS < 30 mg/l. It is necessary: the regular maintenance and the use of personal qualified. The main advantage is the very good quality of cleared used water.
- Activated sludge and decantation lagoons
  After influent pretreatment, it is decharged to an aeration tank and from there to a sedimentation lagoon. The expected quality of effluent to pollution parameters, varied as followings:
  BOD < 30 mg/l, COD < 120 mg/l and SS < 40 mg/l.
  The investment cost is relative high. The exploitation time enough long and the sludge volume important. An important advantage of the system is the facility of his integration to the regional environment.
- High load trickling filters
  It follows a conventional system with: pretreatment, primary tank, high load trickling filter, sedimentation tank and disinfection. The quality of treated waste is limited and the investment costs are high. The system is fiable and has low maintenance and electric consumption. It needs the evacuation of digested sludge once per six months.
- Decantation lagoons and rotating biological discs
  It consists at first from a decantation basin that is followed from biological discs.

  Then the effluent is directed to a secondary decantation basin. The main advantages are light maintenance and electric consumption.

4 THE PROBLEM OF EFFLUENT DISPOSAL AND THE RECYCLING TARGET

The treated water could be discharged to surface water aquifers, that must be protected for their planned use that could be classified as follows: water supply, swimming, fishing irrigation, general use. In order to be possible the fulfillment of the recycling target, water receiver quality standards must satisfy the following requirements:

- Water quality for domestic use
  The parameters examined according to EU directive standards include: hardness, PH, Total disolved solids, Iron, Manganese, Silica, Sodium, Alkalinity, Acidity, Chloride, Fluoride, Nitrate, Sulphate and others. Also algae material could produce toxic effects.
- Water quality in agriculture
  The ions exchange alters the physical characteristics of the soil. When clay has a sufficient lime or magnesium concentration, it is easily cultivated and has a good permeability. The control is realized with Sodium Absorption Ratio =

$$SAR = \frac{Na^{+}}{\sqrt{\frac{Ca^{++} + Mg^{++}}{2}}}$$

- Water quality for animal drinking purposes
  It is necessary the elements Al, As, Be, Bo, Cd, Cr, Co, Cu, F, Fe, Pb, Li, Mn, Hg, Ni, Se, V not surpass the foreseen standards.
- Water quality for industrial purposes
  Quality standards for industrial use include BOD_5, SO_4, Cl, NH_4, P, Colour, Smell, Oils and grease, $KmnO_4$, Organic and ammoniacal nitrogen, detergents, radioactive materials, COD, zinc and iron.

Also there must be examined the entrofication problems to the water bodies.

The quality of nitrogen, phosphorus and organic substances if will be carried out with treated effluents will produce entrofication problems to water bodies (disolve oxygen decrease and predominance of anoxic conditions).

5 TREATED WATER SECONDARY USES

They could be realized the following uses: green irrigation, recreational land, agricultural irrigation, industry, cooling, lakes for swimming or pisciculture, stock farming gardens viniculture. The following treatment is required.

Irrigation: Pretreatment, sedimentation, disinfection.

Filtration towards the ground water aquifers: Natural treatment in proportion with kind of soil or primary and secondary treatment.

Recreation land: Complete biological treatment – Chemical coagulation – flocculation followed by sand filtration – Disinfection.

6 GROUND WATER AQUIFER REPLENISHMENT WITH TREATED WATER IN COASTAL AREAS, TO FACE SALINE WATER INTRUSION

Small treatment plants that serve populations near coastal areas can be used for groundwater aquifer replenishment in order to face saline water intrusion. We notice the followings. They exist a separation surface between saline and fresh water. The subsurface movement of the boundary is influenced from the fresh water load namely from the height H, in reference to the sea surface. If $h\theta$ is the height from the reference level, p the fresh water density and $p\theta$ the sea water density, the equation of Chyben–Herzberg gives

$$h\theta = \frac{p \cdot H}{p\theta - p}$$

For $p = 1\,\text{gr/cm}^3$ and $p\theta = 1{,}025\,\text{gr/m}^3$ we have $h\theta = 40\,H$.

We see that for every 1 m lowering of fresh water there exists an ascending of 40 m to the brackish front. So it is considered very efficient, the replenishment with suitable quality treated used water, in order to face saline water intrusion.

7 SANITARY DANGERS AND THEIR INFLUENCE TO HUMAN HEALTH FROM THE USE OF TREATED WATERS

Sanitary dangers are created from the use of treated waters owed to toxic chemicals, possibly included to them or to causing disease organisms. The chemical toxicity results from chemical compounds (heavy metals) that could be concentrated to cultivations and transferred to the consumption circuit. To the treated waters they can be developed different microbes and it is necessary the realization of indicated controls. When we use sprinkled irrigation, the irrigation area must be far from dwellings (more than 1 km).

8 CONCLUSIONS

The considerable water consumption and water shortage created the necessity of waste water treatment and effluent recycling to feasible uses.

REFERENCES

1. Clarke, R., 1993. Water the international crisis.
2. Documentation technique. FNDE No 5. Ministere de l' Agriculture France. 1986. Les stations d' epuration adaptees aux petites collectivites.
3. Kollias, P.S., 2000. Sewerage and sewage treatment. p.608.

Biosafety

European Symposium on Environmental Biotechnology, ESEB 2004 - Verstraete (ed)
© 2004 Taylor & Francis Group, London, ISBN 90 5809 653 X

Bioregenerative life support in space missions: degradation efficiency and microbial community stability in an anaerobic liquefying compartment

H. De Wever, B. Borremans & L. Diels
Vlaamse instelling voor technologisch onderzoek, Mol, Belgium

D. Demey
EPAS, Gent, Belgium

C. Lasseur
European Space Agency, Noordwijk, Netherlands

ABSTRACT: The Micro-Ecological Life Support System Alternative (MELiSSA) is a bioregenerative system which basically transforms organic wastes into oxygen and food. In the first of five compartments, insoluble organic matter is liquefied in anaerobic thermophilic conditions. Because volatile fatty acids are needed for biotransformations in downstream compartments, methanogenesis is inhibited by acidification. In the present paper we evaluated the degradation efficiencies in an acidified membrane anaerobic bioreactor which was operated on fecal material. The stability and composition of the acidifying communities was evaluated by DGGE fingerprinting and compared with the community of a neutral anaerobic bioreactor operated on the same substrate. Cluster analysis indicated that the samples grouped per reactor.

Space life support systems are meant to provide astronauts with oxygen, food and water and have to remove waste, carbon dioxide, etc. Most air and water purification systems used at present, are based on physicochemical methods such as ion exchange and sorption, which use a lot of supplies and generate extra waste. For long-term space missions, this is unacceptable and more sustainable approaches need to be considered. The European Space Agency therefore investigates the potential of a bioregenerative system. MELiSSA or the Micro-Ecological Life Support System Alternative is inspired by natural lake ecosystems and aims to recover edible biomass and oxygen from organic wastes, carbon dioxide and minerals. MELiSSA consists of five interconnected compartments each colonised by specific microorganisms (Lobo & Lasseur, 2003). This is shown in Figure 1.

The liquefying compartment is the first step in the cycle. It is responsible for the biotransformation of human fecal material and other waste generated by the crew like non-edible plant material and toilet paper, into a soluble form. This occurs through a partial anaerobic digestion of the organic matter to volatile fatty acids. A complete anaerobic digestion is considered to consist of three steps with each step involving its own unique consortium of bacteria:

- Hydrolysis or liquefaction in which higher molecular mass organic compounds are transformed into lower molecular weight fermentation products.
- Acidogenesis involves the further conversion of the fermentation products into acetate, hydrogen and formate.
- Methanogenesis yielding the energy-rich end-product methane and carbon dioxide.

Hence, a complete anaerobic conversion leads to the production of stable gaseous end-products. The production of methane has almost no interest for Life Support, but could have for propulsion. Intermediate fermentation products have more interest for Life Support because they can be used in downstream compartments to produce edible biomass. Liquefaction and acidogenesis can still occur at lower pH, whereas methanogenesis does not. To avoid methane production, the first compartment is therefore slightly acidified.

The volatile fatty acids, minerals and ammonia produced during anaerobic fermentation are then fed into the second compartment. This is inoculated with photoheterotrophic anoxygenic *Rhodospirillum rubrum*, which consumes the volatile fatty acids with the concomitant production of edible organic biomass.

In the third nitrifying compartment ammonia is oxidized to nitrate.

Nitrate and the remaining minerals are then fed to the fourth compartment which is divided into a photoautotrophic bacteria compartment inoculated with *Arthrospira platensis*, and a higher plant compartment in which a variety of vegetables is cultivated.

From the above it is clear that the biotransformation products need to be cycled through the different compartments of the closed loop. On the other hand, the transfer of bacteria needs to be prevented, both for safety reasons and to avoid contamination of downstream compartments. This can be achieved by providing a membrane filtration step. For the first compartment, preference is given to the integration of the bioreactor and membrane filtration step into a membrane bioreactor concept.

The first compartment is crucial in the MELiSSA concept because it determines the fraction of organic waste that can be recycled in the loop. Bioconversion efficiencies therefore need to be optimized. Furthermore, to be able to control reactor operation in future space mission scenario's, it is essential to evaluate the stability of the mixed bacterial culture. As a first step to long-term monitoring of a pilot-scale anaerobic membrane bioreactor, we wanted to evaluate the degree of organic matter degradation that can be achieved in acidifying conditions and we wanted to compare microbial community composition of the acidified compartment with a non-acidified reactor. These results can be found in this paper. In a next step, similar studies will be performed on an acidified reactor operated at pilot-scale.

1 MATERIALS AND METHODS

1.1 *Reactor operation*

Two demonstration reactors with respective active volumes of 1.5 l and 1 l were operated at a temperature of 55°C. In the neutral reactor (1 l), pH was not corrected. In the acidified one (1.5 l), pH was adjusted to 6–6.5 to inhibit methanogenesis. Both reactors were continuously stirred. They were fed with fecal material collected from 8 different persons between age 24 and 40. The organic load was equal to 0.9 g organic matter.$l^{-1}.d^{-1}$. The hydraulic residence time was approximately 20 d. Three times a week, 150 ml of waste material was fed into the reactors. The same volume of liquid was removed from the neutral reactor. In the acidified reactor, 90 ml was removed directly from the reactor, 60 ml was removed via the ultrafiltration unit connected to it.

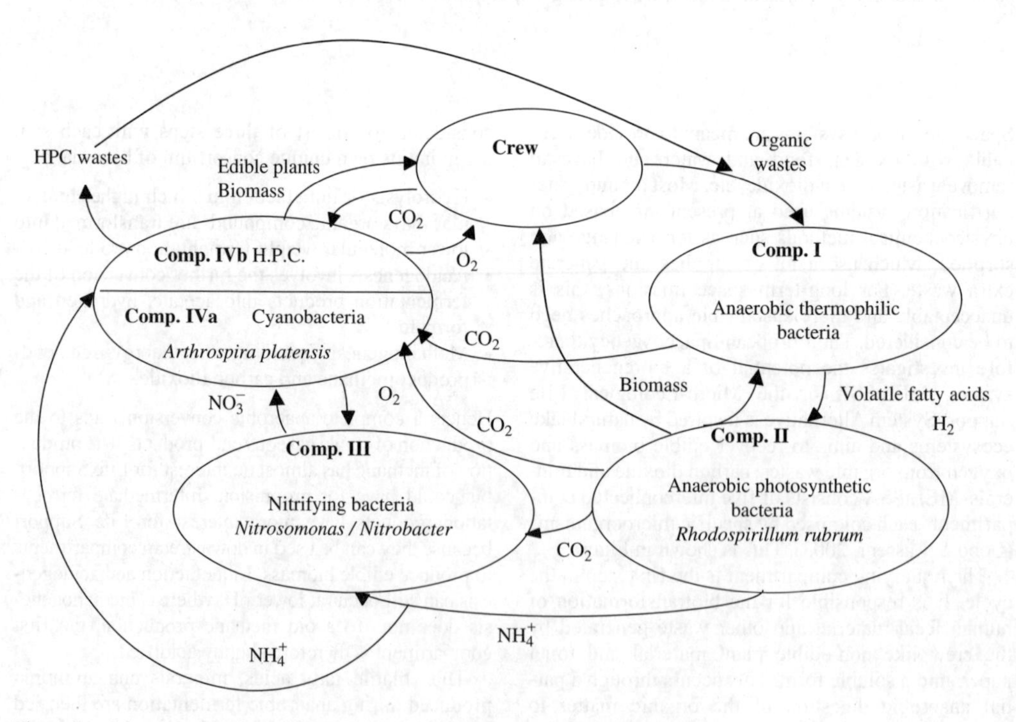

Figure 1. Schematic overview of the MELiSSA loop. HPC: Higher Plant Compartment.

1.2 *Reactor performance*

In calculating the total anaerobic conversion efficiency, it was assumed that organic polymers such as proteins and fibres are converted into volatile fatty acids (VFA) and biogas consisting of methane and carbon dioxide. In the acidified reactor, no methane was measured in the biogas. The production of other fermentation products such as lactate and alcoholic compounds was not considered.

The organic matter conversion efficiency was calculated as the difference between the organic matter fed to the reactor and the organic matter removed from it.

The membrane connected to the bioreactor has to separate the VFA and ammonium in the liquid phase from the organic material which has not been converted yet. Therefore, VFA and ammonium concentrations were measured in the reactor and in the permeate.

1.3 *Microbial community analysis*

At regular time intervals, 5-ml sludge samples from both bioreactors were taken, centrifuged, resuspended in 2 ml of a solution containing 15% glycerol and 0.85% NaCl, and then stored frozen pending analysis.

Total DNA was isolated from the samples using the protocol reported previously by El Fantroussi *et al.* (1999) and modified by Dr. F. Faber, University of Groningen, The Netherlands (personal communication).

A 455 bp eubacterial 16S rDNA from the extracted DNA was amplified by PCR as described by Marchesi *et al.* (1998) using the forward primer 63F (Marchesi *et al.* 1998) and the reverse primer 518R (Felske *et al.* 1996) in order to study the bacterial community composition.

The 16S rDNA amplicons were separated by denaturing gradient gel electrophoresis (DGGE) (Muyzer *et al.* 1993) with a 35 to 65% denaturant gradient on a 8% acrylamide gel. Resolved PCR products were visualised by UV transillumination. The digitised images were analysed with Bionumerics software. Cluster analysis was performed with Jaccard.

2 RESULTS AND DISCUSSION

2.1 *Reactor performance*

As shown in Figure 2, the total conversion efficiencies in the acidified reactor amounted to 28%. This was low compared to the fermentor where a complete anaerobic conversion down to methane took place. This is partly due to the accumulation of the fermentation products VFA which inhibit the hydrolysis of the original substrate. A continuous removal of the VFA from the reactor is therefore essential to reduce auto-inhibition. The high lignin content of the feed (19% of the dry weight) is mainly responsible for the low degradation rate. According to Chandler *et al.* (1980) and based on the lignin content, 30.5% of the MELiSSA substrate is estimated to be biodegradable.

The conversion efficiencies differed depending on the type of organic matter considered. Protein conversion rates were stable over the entire test period and amounted to 35%. Fibres were converted at an average percentage of 14%. Because the latter is problematic, complementary technologies are now investigated to improve fibre degradation, e.g. the use of ligninolytic fungi.

The connection of a filtration unit to the bioreactor seemed to improve the organic matter and total conversion efficiency to some extent. This may be due to the higher removal of liquefied fermentation products and a reduction in active biomass loss. The filtration unit forms a barrier for biomass (and non-degraded organic matter) which is recycled to the reactor.

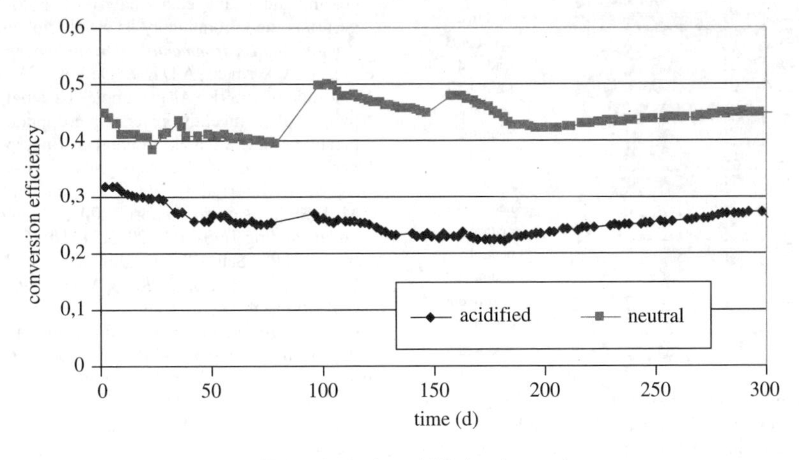

Figure 2. Biodegradation and conversion efficiencies in the acidified and neutral reactor.

2.2 *Membrane performance*

The amount of VFA and ammonium found in the reactor was equal to the amount in the permeate. VFA concentrations up to $3000\,mg.l^{-1}$ and NH_4^+-N concentrations up to $300\,mg.l^{-1}$ were found in the permeate and can hence be used downstream in the MELiSSA loop for the growth of plant material or edible biomass.

Organic matter was effectively retained by the membrane tested. Whereas the reactor contained 2% organic matter, the permeate only contained 0.4%. This increases the retention time of the slowly degradable compounds and may improve overall conversion efficiencies. On the other hand, if degradation is too slow, organic matter will accumulate and may have to be drained out of the reactor.

2.3 *Microbial community analysis*

Because universal primers were used for the DNA amplification, the fingerprints shown in Figure 3 only reflect the composition of the acidifying bacterial populations and these clearly change in time. However, some bands remain present throughout the experimental period and may relate to dominating populations.

Some bands are present in both reactors, while others are specific for either the acidified reactor or the neutral one. Both the species corresponding to common and specific bands are worth further investigations. For future space missions, the behaviour of the microbial community must be well understood to be able to predict organic matter conversion efficiencies and to allow model-based control of the overall reactor operation. Studies are therefore underway to identify and characterize some of the species involved.

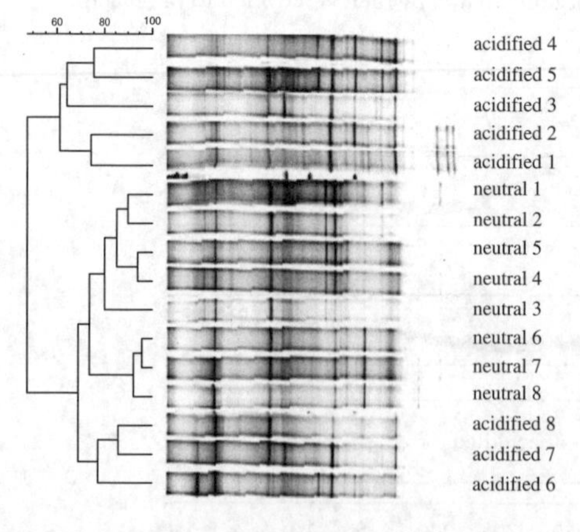

Figure 3. Dendrogram of the DGGE reactor fingerprints clustered with commercial software. The samples are numbered according to the sampling time and were taken over a period of 7 months.

Clustering analysis (Figure 3), leads to the following observations:

- there are 2 clusters: a top one consisting of 5 samples from the acidified reactor, and a bottom one with the other samples
- the first samples from the acidified and neutral reactor are expected to be more related to each other than samples taken at a later time point. However, the distance between these 2 samples and hence their degree of similarity is low. As a function of time, both reactor communities seem to converge to each other. E.g. later samples (e.g. number 8) show a higher similarity than the first sample
- samples are generally grouped per reactor. The environmental conditions (acidification or not) seem to determine the composition of the acidifying community. Whether this is also true for the methanogenic community is currently under investigation.

3 CONCLUSIONS

The present study shows that the fermentation products of organic wastes accumulated can be separated in a membrane bioreactor configuration. Acidification of the reactor to inhibit methanogenesis changes the acidogenic microbial community involved in the liquefaction processes.

REFERENCES

Chandler, J.A., Jewell, W.J., Gossett, J.M., Van Soest, P.J. & Robertson, J.B. 1980. Predicting methane fermentation biodegradability. Biotechnology and Bioengineering Symposium No.10: 93–107.

El Fantroussi, S., Verschuere, L., Verstraete, W. & Top, E.M. 1999. Effect of phenylurea herbicides on soil microbial communities estimated by analysis of 16S rRNA gene fingerprints and community-level physiological profiles. *Applied and Environmental Microbiology* 65: 982–988.

Felske, A., Akkermans, A.D.L. & De Vos, W.M. 1998. Quantification of 16S rRNAs in complex bacterial communities by multiple competitive reverse transcription-PCR in temperature gradient gel electrophoresis fingerprints. *Applied and Environmental Microbiology* 64: 4581–4587.

Lobo, M. & Lasseur, C. 2003. Leaving and living with MELiSSA. Annual report 2002, Memorandum of understanding TOS-MCT/2002/3161/In/CL.

Marchesi, J.R., Sato, T., Weightman, A.J., Martin, T.A., Fry, J.C., Hiom, S.J. & Wade, W.G. 1998. Design and evaluation of useful bacterium-specific PCR primers that amplify genes coding for bacterial 16S rRNA. *Applied and Environmental Microbiology* 64: 795–799.

Muyzer, G., de Waal, E.C. & Uitterlinden, A.G. 1993. Profiling of complex microbial populations by denaturing gradient gel electrophoresis analysis of polymerase chain reaction-amplified genes coding for 16S rRNA. *Applied and Environmental Microbiology* 59: 695–700.

European Symposium on Environmental Biotechnology, ESEB 2004 - Verstraete (ed)
© 2004 Taylor & Francis Group, London, ISBN 90 5809 653 X

Light as an ambient control factor in the systems of microbiological cultivation and biodestruction

A.Ye. Kouznetsov & S.V. Kalyonov
D. Mendeleyev University of Chemical Technology of Russia Miussqaya, Moscow, Russia

V.I. Soldatov & A.M. Seregin
State Unitary Enterprise "NPO Astrophizika", Moscow, Russia

M.G. Strakhovskaya,
Lomonosov Moscow State University, Biological faculty, Moscow, Russia

D.A. Skladnev
Institute for Genetics and Selection of Industrial Microorganisms, Moscow, Russia

ABSTRACT: To study some properties of light irradiation for regulation of microbial growth and biosynthetical processes of yeast, heterotrophic bacteria and halobacteria a complex including non-monochromatic sources of light, lasers with changing wavelength of irradiation from UV to visible range and systems for cultivation of microorganisms during irradiation has been designed. Possible ways of application of the complex to study biosynthesis of various substances and processes of biological waste water treatment are discussed.

Light-sensitive processes can be observed for various phototrophic and heterotrophic microorganisms and their enzymes, therefore light as an ambient factor can be applied for improvement of cultivation of microbial producers: bacteria, yeasts, fungi and for optimization of the biosynthesis, biotransformation and biodestruction processes. For today its influence is rather well investigated at the molecular and genetic levels. Key light parameters conditioning efficiency and molecular mechanism of the photoinduced processes are light wavelength and intensity.

Among the mechanisms of sensitivity of cells to light elucidated best of all is photoreparation by photolyase after UV exposure. The last one functionates together with other ways of response to stress factors. For anti-stress systems of reparation some intersection reactions are possible as well. Photostimulation of growth of microorganisms is observed under light in UV, blue or red ranges. The possibility of growth photostimulation is determined by the presence of cellular photoactive chromophores – pterins and pyridoxal (UV), flavins (blue light), components of a respiratory chain (blue and red light), phytochrome (red light). Depending on the wavelength used the photoinduced change of a proton gradient and activation of ion channels are observed in plasma membranes, that results in pH shift, redistribution of calcium ions and metabolism and cell division stimulation. Soft UV-radiation of UVA and UVB-ranges influences on receptors to UV-light located on cell surface. The primary chemical responses initiated by this light are accompanied by generation of free radicals, which participate in regulation of physiological state of cells and initiate the processes of oxidation of biosubstrates. This can stimulate cell growth.

In general, that one who is going to study the influence of visible and UV-light on microorganisms has to take into account processes of photoinduction, photoinhibition, photoreactivation, photoreparation, phototaxis, photodynamic effect etc. Physiological conditions of microbial population and the level of dissolved oxygen have substantial importance as well. For example our preliminary experiments on the influence of low-intensive laser light on some microorganisms have shown that to study this rather speculative field of investigations it is necessary to be very accurate from methodical point of view, because such factors as light-spectrum, background level of day or artificial light during manipulations in laboratory, oxygen concentration in microbial suspension, effects

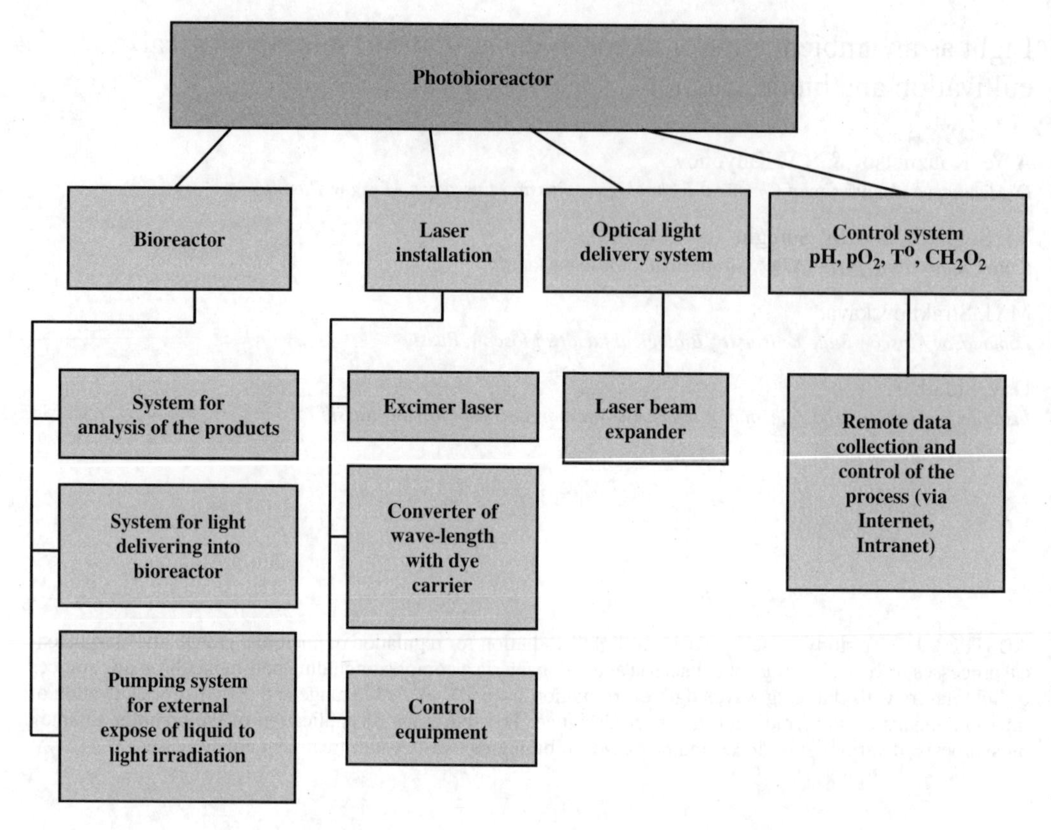

Figure 1. Block diagram of the photobiotechnological system for cultivation of the microbiological objects.

of synchronization of cells can influence on the results of the experiments.

To elucidate the utility of embedding of light as ambient factor into practice of the control of growth of heterotrophic microorganisms and their life circle development and to study the light-sensitive processes at microorganisms we have designed a bench-scale complex installation including (Figs 1, 2):

- lasers with changing wavelength of irradiation from UV- to visible range of spectrum, with variable output power and opportunity of frequency modulations;
- the system for cultivation of microorganisms with registration and automatic regulation of parameters conventional for the fermentation processes (T°, pH, pO_2, Eh, nutrient dosage, flow rate, air dosage) in periodic, fed-batch and flow conditions under irradiation of monochromatic and non-monochromatic light;
- the system for irradiation of microorganisms in microplates, shaker flasks, bioreactors by laser radiation in visible and ultra-violet ranges of spectrum with various wavelengths and by usual sources of non-monochromatic light in visible and UV-ranges;
- software based on LabView shell with a possibility of remote control and governing cultivation process through Internet.

It is supposed also to equip the complex with a system for registration of variation in absorption and fluorescence spectra of cells of microorganisms, luminometer (for ATP registration), sensor for detection of H_2O_2, a system of the morphological analysis.

Relying on the existing data it is possible to expect positive results of the application of the combined action of visible and UV-light at carrying out and governing fermentation processes, selection of more productive and more active strains. In particular, our experience of working with yeast of genus *Candida* has shown that one of the ways to improve the efficiency of the yeast cultivation could be through the simultaneous influence of visible and UV-light (Fig. 3).

Another way to stimulate microbial growth and activity is the use of laser irradiation. It has to be mentioned, that both low intensive laser radiation with a

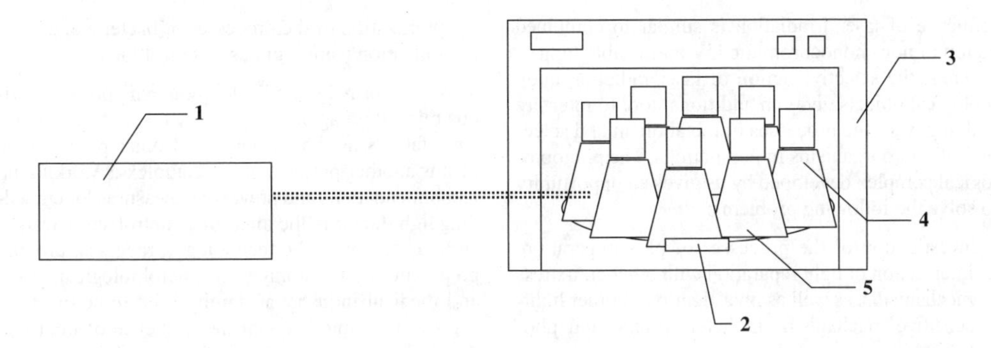

Figure 2. Scheme of the unit for cultivation of microorganisms in shaker flasks with laser irradiation. 1 – laser; 2 – shaker; 3 – shaker thermostat; 4, 5 – flasks with and without lighttight shields.

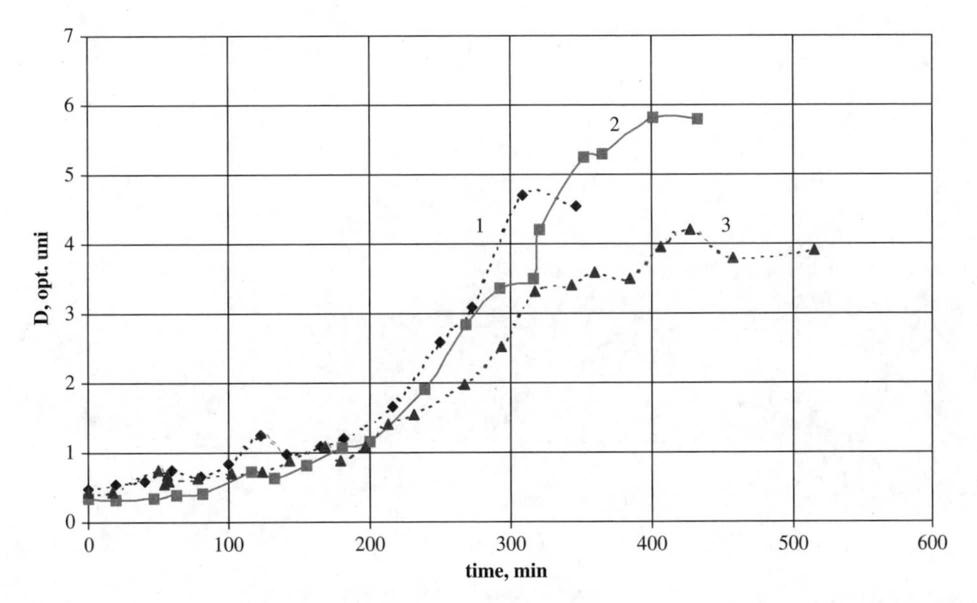

Figure 3. Growth of yeast *Candida tropicalis* under irradiation of UV- and visible light. 1 – control without UV-irradiation; 2 – with UV- and visible light irradiation; 3 – with UV-irradiation in the dark.

light density of 0,1–100 mW/cm^2 (that is comparable with solar radiation on the Earth surface), and high intensive laser radiation (with a light density of 1–100 W/cm^2) are biologically effective.

What are advantages of the laser application for biological system of cultivation and destruction?

Firstly, unlike traditional light sources the laser radiation allows to influence more purposefully and selectively on the intracellular processes, to regulate processes of biosynthesis, as well to act on the substrates, assimilated by microorganisms. For instance, the light adsorption of pigment melanin is the most pronounced in violet range, porphyrin and its derivatives – in red one. Laser light of red and infra-red ranges is absorbed by components of respiratory chain that stimulates cellular metabolism, including increase of potential of mitochondrial membranes, ATP synthesis and oxygen consumption. Red light of He-Ne laser should be absorbed strongly by green algae, destroying their structure. On the contrary, the green laser light should be absorbed poorly by chlorophyll, but actively by cells containing red pigments, and rapidly damage them. The system of reparation with photolyase functioning in response to UV irradiation has a maximal sensitivity to light at two wavelength at visible and UV ranges of spectrum. Our preliminary results obtained in the experiments with yeast and other microorganisms give also a possibility to suppose the influence of laser irradiation is similar to combined action of non-monochromatic UV and visible light.

Thus, the selective action of laser radiation upon biological objects gives an additional tool to interfere and to govern the processes of metabolism and selection of microorganisms in populations. The photobiological complex developed by us gives an opportunity to solve the following problems:

- investigation of the processes of photoreparation, intersection of light reparation with other anti-stress mechanisms, as well as investigation of other light-sensitive mechanisms in heterotrophic and phototrophic microorganisms;
- investigation of the genetic alterations in microorganisms under action of laser light of mutagenic and reparation ranges;
- optimization of the processes of bacteria and yeast cultivation using light as a control factor.

As, in principle, visible light can protect various microflora against such stressful for microorganisms factors as disinfectants and other prophylactic means another purpose of the complex is working up recommendations and preventive measures for embedding light factor in the microbial control, certification and attestation of the technology processes, product preparation and storage in microbiological, food and medical industry, in family life. Some practical ways to take into account the influence of light and diminish the damage of harmful microflora are being discussed.

European Symposium on Environmental Biotechnology, ESEB 2004 - Verstraete (ed)

Occurrences fecal coliforms in the water of the high Pirapó river basin

T.M. de Oliveira, J.D. Peruço, C.R.G. Tavares, P.A. Arroyo & E.S. Cossich
Departamento de Engenharia Química/UEM. Av. Colombo, Maringá (PR)

E.E. de Souza Filho & M.L. dos Santos
Departamento de Geografia/UEM. Av. Colombo, Maringá (PR)

ABSTRACT: Rivers are linear and not isolated systems that drain the water on the continetal mass superficially going to the oceans. These oceans reflect all the activities occurring in the area of these hydrological basins. The urban development and the creation of domestic and industrial effluents in the adjacent regions have compromised the natural conditions of the water quality. Therefore, this study aims to evaluating their impact on the receiving water, analyzing the presence of fecal coliforms in the water of both the high Pirapó river basin and its tributary at the left shore, the Maringá brook. The data collecting was performed in fixed places and in regular time intervals. It was verified a high number of fecal coliforms, compared to the legislation of the river framing. These organisms are not pathogenic, but they offer a satisfactory indication of some water contamination by human or animal feces.

1 INTRODUCTION

The 20th century urban occupation has been characterized by the lack of the soil use planning. As aconsequence, we observe an increasing concern with this disordered occupation, since the urbanization progress can compromise their surrounding environmental factors. The cities, while growing without a planning, prejudice not only the soil where they have been established, but also the water and the air.

As the cities grow several kinds of environmental impacts are observed: water polluted domestic and industrial sewerage; soil and valleys bottom degradation by allotments, paving and proofing of the soil making the infiltration difficult and changing the hydrologic cycle. Finally, the problems are even more worse, making the planning and the search for solutions to control the urban areas degradation difficult (JABUR, 2002).

These anthropic impacts can be observed in Maringá city in the Northwest region of the State of Paraná (Brazil) having almost 300,000 inhabitants (IBGE Census, 2000), situated at the 23°25′ South latitude 51°57′ West longitude, with 596 m of altitude.

Initially this city has been planned by the *Melhoramentos Norte do Paraná* Company. However, due to political conflicts between this company and the city hall, irregular allotments appeared which disfigured the idealization of a promising city, disguising it with a disorganizing expression, but, in the name of growing (FILETT, 2002).

Pirapó river, supplier of Maringá's water, is a tributary of Paraná river right shore with a 5,023 km^2 drainage area. According to LOPES (2002), quoting ZHADI (2002), this river is agonizing and the critical point is in Maringá city. The problem is not related to the amount of Pirapó's water, but to its quality.

According to Oliveira (1976), the water pollution is a result of the introduction of residues, in the form of matter and energy, making it prejudicial to men and other kinds of life, or even improper for a certain use established by it.

The document from SUREHMA (SUPERINTENDÊNCIA DE RECURSOS HÍDRICOS E MEIO AMBIENTE DO PARANÁ), number 004 from March 21st, 1991, according to the CONAMA resolution (CONSELHO NACIONAL DO MEIO AMBIENTE n°20 – BRAZIL), classifies the Pirapó river as a class II river. The class II rivers are designated to the domestic supply after the conventional treatment; to the amusement of primary contact and to the intensive creation destined to human feeding. Therefore, this study aims at identifying the organic indicators of Pirapó river basin water quality.

2 MATERIALS AND METHODS

With the purpose of evaluating the impacts of the urbanization on Maringá brook, water samples have been monthly collected from March to October, 2003.

The samples have been collected in seven different and fixed places, along the river and in regular time intervals, in the urban and agricultural zones of this river, to obtain data that could allow us to know the actual conditions of the water and its evolution.

Points 1 and 2, Maringá brook and Romeira stream, are near their riverheads and have urban and combined agricultural and stock raising characteristics. The point 3, Mandacaru stream, is a more urbanized river. Points 4 and 5, Maringá brook, can be observed after the entry of treated domestic effluents from one of the sewerage treatment stations of Maringá city, and before its confluence with the high Pirapó river basin, respectively. Points 6 and 7 are in Pirapó river, upstream and downstream the confluence with Maringá river.

The total *fecais* coliforms had been quantified by means of the plates of counting of *Escherichia coli* and coliforms, of 3 M Petrifilm, according to methods NMKL (147.1993) and AOAC (991.14), respectively. This is a ready-made culture medium system which contains Violet red Bile (VRB) nutrients, a cold-water-soluble gelling agent, an indicator o glucuronidase activity (BCIG), and a tetrazolium indicator that facilitates colony enumeration of *E. coli* and coliform.

3 RESULTS AND DISCUSSIONS

The population growing and the economical development of a region adjacent to a hydrological basin direct or indirectly interfere in its natural conditions. For example, the urban sewerage which entry in the hydric resources can deteriorate their water, causing the contamination by bacteria, partly pathogenic to men, and with organic substances degradable by microorganisms. This contamination was verified in high Pirapó river basin and its tributary, Maringá brook.

The bacteria of the coliforms group are indicators of fecal pollution, since they occur in a great number in human and other hot blood animals intestinal microbiotics, being eliminated in a great number by the feces. The presence of these coliforms in the presents a potential risk of pathogenic organisms, since they are more resistant in the water than the pathogenic bacteria of intestinal origin.

The water samples analysis indicated the presence of fecal coliforms above the standard considered ideal for the superficial water of this hydric resource of 1,000 UFC/100 mL (UFC: Unidade Formadora Colonias). The findings obtained in the five points of data collecting are shown in Table 1.

Table 1. Levels of fecais coliforms in the set of 7 points.

Points	Average rates of fecal coliforms (UFC/100 mL)
1	5,100
2	2,300
3	5,600
4	120,200
5	10,400
6	35,000
7	38,000

The differences in the average rates of fecal coliforms can be explained by the use and occupation of the soil. This is the case of point 2 that, as an agricultural region, showed a smaller rate of fecal coliforms. On the other hand, points 1 and 3 in urban and combined agriculture and stock raising areas have more probability to "naturally" receive sewerage.

The point 4 was the most critical one. It is downstream of the sewerage treatment station of Maringá city. Values between 82,300 and 351,000 UFC/100 mL were found.

The exorbitant increasing of fecal coliforms can be probably due to the great amount of treated domestic effluents that have probably been entering into the receiving water. This can indicate that there are some failures in the sewerage treatment. Other possibility is that the sewerage station has been planned to attend a certain demand of effluent, but due to the population growing it has become insufficient to attend the increase of residue creation.

A decrease in the average rate of fecal coliforms in point 5 was observed, when compared to point 4. This can be explained by the geographic formation of this region, which shows a predominance of a strongly waved relief where there are little plain surfaces, mainly in the small areas along the water course.

This declivity favors the self depuration process, that is, the natural degradation of the organic matter by the organism existent in the water. This can cause the destruction of the fecal coliforms. Other factor that can contribute to this destruction is its natural environment change. Therefore, the fecal coliforms groups can not feed and reproduce themselves under the water, otherwise they die, victim of several physical, chemical and biological factors.

The fecal coliforms findings detected in points 6 and 7, show an important pollution in Pirapó river that comes from other sources not originated from the influent studied, since the point 6, localized before the confluence of Maringá brook with this river, showed very high values of fecal coliforms.

4 CONCLUSION

The high rates of fecal coliforms observed in the receiving water studied, are due to the inadequate use and occupation of the region where it is localized, among other factors.

The contribution of Maringá's urban area to be organic pollution detected in Maringá brook water is mainly due to the domestic effluents entry, little and/or not treated.

Therefore, we can conclude that the superficial water of Maringá brook is inadequate to the population amusement, to fishing and to impounding water for supply.

REFERENCES

Filetti, C.R.G., *Ação dos Agentes na Expansão Irregular e Maringá*, Departamento de Engenharia Civil/ UEM, Maringá, 2002.

Jabur, A.S., Aspectos Qualitativos do escoamento Superfici al na Microbacia Hidrográfica do *Córrego Moscados no Município de Maringá.* Mestre em Geografia. UEM, Maringá- PR, 2002.

Oliveira, W.E., Técnicas de Abastecimento e *Tratamento de Água*, 2ª edição, CETESB, São Paulo, 1976.

Zahdi, A,*O Pirapó Agoniza*. Revista Mais, Maringá, p 18–22, julho de 2002.

European Symposium on Environmental Biotechnology, ESEB 2004 - Verstraete (ed)
© 2004 Taylor & Francis Group, London, ISBN 90 5809 653 X

Identification of human fecal pollution: detection and quantification of the human specific HF183 *Bacteroides* 16S rRNA genetic marker with real-time PCR

S. Seurinck, T. Defoirt & W. Verstraete
Laboratory of Microbial Ecology and Technology (LabMET), Ghent University, Ghent, Belgium

S. Siciliano
Department of Soil Science, University of Saskatchewan, Saskatoon, Saskatchewan, Canada

ABSTRACT: A real-time PCR assay for the detection and quantification in fecal and environmental samples of the human specific HF183 *Bacteroides* 16S rRNA genetic marker previously found by Bernhard & Field (2000a), was developed and proved to be more sensitive than the conventional HF183 PCR. The limit of detection in canal water was 1.94×10^2 copies per μl DNA extract and the limit of quantification was 5.83×10^2 copies per μl DNA extract. The intra- and intervariability coefficients of the real-time assay were less than 1% and 3%, respectively. The amount of marker copy number present per g of wet human feces varied from 10^2 to 10^8 between different individuals. The amount of marker copy number present in wastewater treatment plant influent varied from 10^8 to 10^9 per liter. The marker persisted upto 6 days at 28°C. Whereas, the limit of detection was not reached after 24 days of incubation at 4 and 12°C.

1 INTRODUCTION

Fecal contamination of water environments poses a threat to recreational water quality, because of the possible presence of pathogenic microorganisms associated with fecal material. Point source discharges such as raw sewage, stormwater, combined sewer overflows, effluents from wastewater treatment plants and industrial sources are considered to be the major contributors to fecal pollution (Griffin et al. 2001). Recently, policy makers have come to recognize the importance of non-point sources, such as agricultural run-off, dogs, horses, birds, and pleasure boats (Jagals et al. 1995, Alderisio et al. 1999). An effective means of microbial source tracking (MST) would be an important tool in water quality management.

A variety of methods have been proposed to identify fecal sources in water. Several MST-methods are based on differences in phenotypic and genotypic characteristics among populations of microorganisms, e.g. *Escherichia coli* and *Enterococci* in different fecal sources. These methods require a library of fingerprints of e.g. *E. coli* isolates from different fecal sources. The fingerprints of environmental isolates are then compared to the library, which would indicate if the fecal pollution in the environment is derived from a particular host group represented in the library. Phenotypic library-based methods, such as antibiotic resistance analysis (Wiggins 1996), and carbon source utilization analysis (Souza et al. 1999), as well as genotypic library-based methods, such as ribotyping (Carson et al. 2001), pulse-field gel electrophoresis (Parveen et al. 2001), and repetitive extragenic palindromic-PCR (Dombek et al. 2000, Seurinck et al. 2003), have been developed. Unlike these methods, which require culturing of target organisms, detection of host-specific markers holds promise as a rapid method for identifying sources of fecal contamination. Bernhard & Field (2000a) identified a human specific *Bacteroides* 16S rRNA genetic marker. They developed a conventional PCR assay to detect this marker in fecal samples and environmental water samples. The detection limit of the assay was approx. 7.1×10^{10} copies per g feces, corresponding to approx. 1.4 μg of dry feces per liter (Bernhard & Field 2000b).

The purpose of this study was to verify if the human specific HF183 Bacteroides 16S rRNA genetic marker can be used to detect human fecal pollution in Belgium and to develop a real-time PCR assay to quantify human fecal pollution at lower fecal contamination levels than could be detected by the conventional PCR assay of Bernhard & Field (2000b).

2 METHODS AND MATERIALS

2.1 *Sample collection and DNA extraction*

Human and animal fecal samples, and wastewater treatment plant influent were collected as described in Seurinck et al. (2003). DNA was extracted from feces with the QIAamp DNA Stool Mini Kit (QIAGEN, West Sussex, United Kingdom) according to the manufacturer's instructions. Water samples were filtered through a 0.22 μm Millipore filter. DNA extraction on the filters was performed as described by Boon et al. (2000).

2.2 *Real-time PCR assay*

The human specific HF183 *Bacteroides* 16S rRNA genetic marker was amplified in the real-time assay by using the human specific HF183 forward primer (5′ATCATGAGTTCACATGTCCG 3′) as developed by Bernhard & Field (2000b) and a newly developed reverse primer (5′TACCCCGCCTACTATCTAATG 3′). The real-time PCR was based on the principle of Heid et al. (1996). For quantification of the HF183 marker by real-time PCR, amplification was performed in 25-μl reaction mixtures by using buffers supplied with the qPCR™ core kit for Sybr® Green I (Eurogentec, Liège, Belgium). The PCR mixture contained 0.25 μM of each primer, 200 μM of each deoxynucleoside triphosphate with dUTP, 2.0 mM $MgCl_2$, 10 μl of real-time PCR 10X Buffer ($MgCl_2$-free), 2.5 U of Hot GoldStar DNA Polymerase, 3 μl Sybr® Green I 1/10, and DNase and RNase free filter sterile water (Sigma-Aldrich Chemie) to a final volume of 100 μl. The reactions were performed in MicroAmp Optical 96-well reaction plates with optical caps (PE Applied Biosystems, Nieuwerkerk a/d Ijssel, The Netherlands). The PCR temperature program was as follows: 50°C for 2 min and 95°C for 5 min, followed by 40 cycles of 95°C for 15 s, 53°C for 30 s, and 60°C for 1 min. The template DNA in the reaction mixtures was amplified and monitored with an ABI Prism SDS 7000 instrument (PE Applied Biosystems). Within each PCR-run all amplification reactions were performed in triplicate. The calibration curve was obtained by amplifying a tenfold dilution serie (1.4×10^7 to 14 copies per μl DNA extract) of plasmid extract. The plasmid contains the human specific HF183 *Bacteroides* 16S rRNA genetic marker.

2.3 *Conventional PCR assay*

All samples were also analyzed with the conventional PCR assay developed by Bernhard & Field (2000b).

2.4 *Assessment of the inter -and intravariability coefficient of the real-time PCR assay*

The inter- and intravariability coefficients were determined by means of the calibration curve analyzed in each real-time PCR assay. The intervariability coefficient was calculated based on the Ct data of four consecutive runs. Within each real-time PCR assay all samples were analyzed in triplicate.

The intravariability coefficient was calculated based on the Ct data within one run. Every sample within that run was analyzed six times.

2.5 *Assessment of the limit of detection (LOD) and limit of quantification (LOQ) in canal water*

Samples of the freshwater canal were taken with a sterile recipient, kept at 4°C and used within 24 hours after collection. Fresh human fecal material was suspended in 1 l canal water (Coupure, Gent, Belgium) to a final concentration of 1 g per l water. A tenfold dilution serie (10^{-1} to 10^{-9}) of the suspension was made in the same canal water. This procedure was repeated three times. The three series ware designated, Coupure 1, 2 and 3. Both, conventional PCR and real-time PCR were performed. The canal water as such did not contain the marker (checked with real-time PCR).

The theoretical blanc of the dilutionserie was used to calculate the LOD and LOQ. Therefore, it was assumed that the amount of copies per μl DNA extract from the 1 g per l suspension determined with the real-time PCR assay was correct. The theoretical blanc was then defined as the dilution that could not contain the marker anymore. The LOD and LOQ were calculated as follows:

$$LOD = blanc_{theor} + 3.3\ stdev(blanc_{theor})$$

$$LOQ = 3\ LOD$$

where $blanc_{theor}$ is the amount of copies per μl DNA extract in the theoretical blanc.

2.6 *Persistance of the human specific HF183* Bacteroides *16S rRNA genetic marker in canal water*

Fresh septic waste (100 ml) was suspended in 1 l of canal water (Coupure, Gent, Belgium). The suspension was partially incubated at 4, 12 and 28°C in aerobic conditions. Three series were set-up per incubation temperature. The samples at 4 and 12°C were incubated upto 24 days, and the samples at 28°C upto 8 days. The amount of copies per µl DNA extract was determined with the real-time PCR assay.

3 RESULTS AND DISCUSSION

3.1 *Specificity of the human specific HF183* Bacteroides *16S rRNA gene marker*

Different human and animal fecal samples, and wastewater treatment plant influent (wwtp-influent) were analyzed with the conventional and real-time PCR assay (Table 1).

The results in Table 1 can be considered as following a binomial distribution. With the conventional PCR, we obtained a 83% chance of the marker being present given a human source, and a 0% chance of the marker being present given a nonhuman source. Using the real-time PCR assay an even higher chance of the marker being present given a human source was obtained, namely 92%. Bernhard & Field (2000b) obtained a 88% chance, however they worked at a higher confidence level, since more samples were analyzed (16 human and 46 animal samples).

3.2 *Assessment of the inter- and intravariability coefficient of the real-time PCR assay*

Table 2 presents the results for the inter- and intravariability coefficient of the real-time PCR assay for the copy number range of 1.4×10^7 to 14 copies per µl DNA extract.

The intra- and intervariability coefficient of the real-time PCR assay was less than 1% and 3%, respectively.

3.3 *Determination of the copy number of the marker in human individuals and wastewater treatment plant influent in Belgium*

Table 3 presents the quantification results of the human specific HF183 *Bacteroides* 16S rRNA genetic marker in seven human individuals and four wwtp-influent samples, collected at the same wwtp but on four consecutive days. From the copy number data, the amount of bacteria can be calculated by considering the total DNA extract volume, the amount of feces used, and an average of five 16S rDNA operons per *Bacteroides* cell.

With the conventional PCR assay the genetic marker could not be detected in human 4 and 6. With the real-time PCR assay, human 6 also contained the marker. Sequencing of the amplicon of human 6 showed that indeed the correct markersequence was being amplified.

Table 1. The number of positive detection results for all fecal and wastewater treatment plant influent samples with the conventional PCR and real-time PCR assay.

Source	Number of samples	Conv. PCR	Real-time PCR
Human	7	5	6
Wwtp-influent	5	5	5
Pig	5	0	0
Cow	5	0	0
Horse	5	0	0
Chicken	5	0	0
Dog	5	0	0

3.4 *Assessment of the limit of detection (LOD) and limit of quantification (LOQ) of the real-time PCR assay in canal water*

With the conventional PCR, the human specific HF183 *Bacteroides* 16S rRNA genetic marker could be detected upto dilution 10^{-4} in Coupure 1, and upto dilution 10^{-3} in Coupure 2 and 3, corresponding to 100 µg of wet feces per l and 1 mg of wet feces per l.

Table 2. The inter- and intravariability coefficient for the real-time PCR assay within the range of 1.4×10^7 to 14 copies per µl DNA extract.

Quantity (copy number /µl DNA extract)	Inter CV (%)	Intra CV (%)
1.4×10^7	0.78	0.85
1.4×10^6	1.12	0.76
1.4×10^5	1.67	0.98
1.4×10^4	1.88	0.85
1.4×10^3	2.22	0.68
1.4×10^2	2.09	0.83
1.4×10^1	1.67	0.92

Table 3. Quantification results with the real-time PCR assay of the human specific HF183 *Bacteroides* 16S rRNA genetic marker in human individuals and wwtp-influent samples, collected at the same wwtp on four consecutive days.

	Copy number/µl DNA extract	Markerbacteria/g feces or/l influent
Sample	Average ± Stdev	Average ± Stdev
Human 1	$1.16 \pm 0.13 \times 10^4$	$1.16 \pm 0.13 \times 10^6$
Human 2	$2.95 \pm 0.28 \times 10^4$	$2.95 \pm 0.28 \times 10^6$
Human 3	$9.86 \pm 1.10 \times 10^3$	$9.86 \pm 1.10 \times 10^5$
Human 4	nd	nd
Human 5	$7.21 \pm 1.09 \times 10^6$	$7.21 \pm 1.09 \times 10^8$
Human 6	$8.37 \pm 0.03 \times 10^2$	$8.37 \pm 0.03 \times 10^4$
Human 7	$1.97 \pm 0.15 \times 10^6$	$1.97 \pm 0.15 \times 10^8$
Wwtp-infl 7/14	$5.40 \pm 0.97 \times 10^3$	$6.75 \pm 1.22 \times 10^8$
Wwtp-infl 7/15	$4.68 \pm 0.54 \times 10^3$	$5.85 \pm 0.68 \times 10^8$
Wwtp-infl 7/16	$2.00 \pm 0.25 \times 10^4$	$2.50 \pm 0.31 \times 10^9$
Wwtp-infl 7/17	$2.49 \pm 0.27 \times 10^4$	$3.11 \pm 0.34 \times 10^3$

nd: not detected.

Bernhard & Field (2000a) set up serial dilutions of raw sewage in filter-sterilized bay water. They could detect upto 10 μg of wet feces per l. This discrepancy is probably due to the higher amount of gene copies present in the raw sewage used by Bernhard & Field (2000a) for their dilution experiment. A LOD in terms of g of feces per l is of no use, since the amount of copies per g of feces varies between individuals.

With the real-time PCR, LOD and LOQ can be determined in terms of copies per μl DNA extract of the human specific HF183 *Bacteroides* 16S rRNA gene marker. The LOD and LOQ values for the three dilution series are reported in Table 4.

As expected the LOD and LOQ differ between the environmental samples. We report for the real-time assay for canal water, a LOD of 1.94×10^2 copy number/μl DNA extract and a LOQ of 5.83×10^2 copy number/μl DNA extract.

3.5 *Persistance of the human specific HF183* Bacteroides *16S rRNA genetic marker in canal water incubated at 4, 12, and 28°C*

The results of the persistance study are presented in Figure 1. The LOD of 1.94×10^2 copy number/μl DNA extract is also indicated in Figure 1.

Table 4. LOD and LOQ values for the different dilution series of fecal material in canal water.

Dilution series	LOD (copy number/μl DNA extract)	LOQ (copy number/μl DNA extract
Coupure 1	4.43×10^1	1.30×10^2
Coupure 2	2.14×10^1	6.43×10^1
Coupure 3	1.94×10^2	5.83×10^2

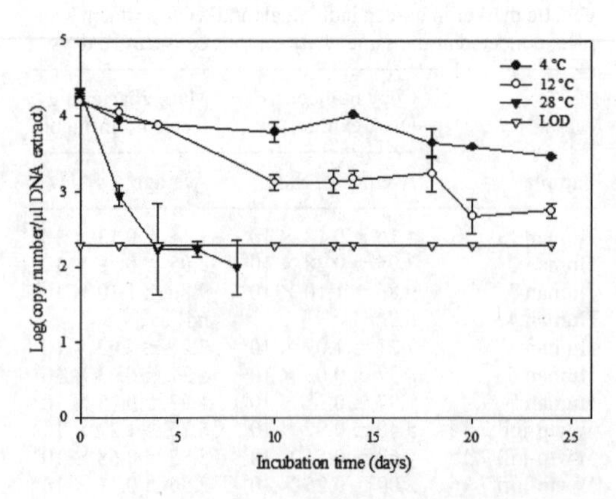

Figure 1. The persistance of the human specific HF183 *Bacteroides* 16S rRNA genetic marker in canal water during incubation at 4, 12, and 28°C under aerobic conditions.

As expected the copy number was already below the LOD after 6 days at 28°C. After 10 days of incubation at 12°C, the copy number decreased approximately one log-unit, but the LOD was not yet reached after 24 days of incubation. At 4°C, the copy number stayed constant throughout the incubation period. The water temperature of the canal water varies from 4 to 20°C during the year. Because of the long persistance of the human specific HF183 *Bacteroides* 16S rRNA genetic marker at these temperatures, we will not be able to make a distinction between recent and historical human fecal pollution.

REFERENCES

Alderisio, K.A., and N. DeLuca. 1999. Seasonal Enumeration of Fecal Coliform Bacteria from the Feces of Ring-Billed Gulls (*Larus delawarensis*) and Canada Geese (*Branta canadensis*). *Applied & Environmental Microbiology* 65:5628–5630.

Bernhard, A.E., and K.G. Field. 2000a. Identification of nonpoint sources of fecal pollution in coastal waters by using host-specific 16S ribosomal DNA genetic markers from fecal anaerobes. *Applied & Environmental Microbiology* 66:1587–1594.

Bernhard, A.E., and K.G. Field. 2000b. A PCR assay to discriminate human and ruminant feces on the basis of host differences in Bacteroides-Prevotella genes encoding 16S rRNA. *Applied & Environmental Microbiology* 66:4571–4574.

Boon, N., J. Goris, P. De Vos, W. Verstraete, and Top, E.M. (2000). Bioaugmentation of activated sludge by an indigenous 3-chloroaniline-degrading Comamonas testosteroni strain, I2gfp. *Applied & Environmental Microbiology* 66(7), 2906–2913.

Carson, C.A., B.L. Shear, M.R. Ellersieck, and A. Asfaw. 2001. Identification of Fecal *Escherichia coli* from Humans and Animals by Ribotyping. *Applied & Environmental Microbiology* 67:1503–1507.

Dombek, P.E., L.K. Johnson, S.T. Zimmerley, and M.J. Sadowsky. 2000. Use of repetitive DNA sequences and the PCR to differentiate *Escherichia coli* isolates from human and animal sources. *Applied & Environmental Microbiology* 66:2572–2577.

Griffin, D.W., E.K. Lipp, M.R. McLaughlin, and J.B. Rose. 2001. Marine Recreation and Public Health Microbiology: Quest for the Ideal Indicator. *BioScience* 51: 817–825.

Heid, C.A., J. Stevens, K.J. Livak, and P.M. Williams. 1996. Real time quantitative PCR. *Genome Research* 6(10): 986–994.

Jagals, P., W.O.K. Grabow, and J.C. Devilliers. 1995. Evaluation of indicators for assessment of human and animal fecal pollution of surface run-off. *Water Science & Technology* 31:235–241.

Parveen, S., N.C. Hodge, R.E. Stall, S.R. Farrah, and M.L. Tamplin. 2001. Phenotypic and genotypic characterization of human and non-human *Escherichia coli*. *Water Research* 35:379–386.

Seurinck, S., W. Verstraete, and S. Siciliano. 2003. The use of 16S–23S rRNA intergenic spacer region-PCR and repetitive extragenic palindromic-PCR analyses of *Escherichia coli* isolates to identify non-point fecal sources. *Applied & Environmental Microbiology* 69:4942–4950.

Souza, V., M. Rocha, A. Valera, and L.E. Eguiarte. 1999. Genetic structure of natural populations of *Escherichia coli* in wild hosts on different continents. *Applied & Environmental Microbiology* 65:3373–3385.

Wiggins, B.A., 1996. Discriminant analysis of antibiotic resistance patterns in fecal streptococci, a method to differentiate human and animal sources of fecal pollution in natural waters. *Applied & Environmental Microbiology* 62:3997–4002.

Solid waste, waste recycling and biofuels

European Symposium on Environmental Biotechnology, ESEB 2004 - Verstraete (ed)

Improved ethanol production from organic waste by wet oxidation pre-treatment

G. Lissens & W. Verstraete
Laboratory of Microbial Ecology and Technology, Gent University, Gent, Belgium

H. Klinke & A.B. Thomsen
Risø National Lab, Roskilde, Denmark

B.K. Ahring
Technical University of Denmark, Lyngby, Denmark

ABSTRACT: The feasibility of efficient ethanol production from the cellulose fraction of municipal solid waste (MSW) and woody yard waste by means of simultaneous saccharification and fermentation (SSF) by *Saccharomyces cerevisae* was investigated after (thermal) wet oxidation pre-treatment. The effects of varying wet oxidation parameters (e.g. temperature (185–200°C), pH and oxygen pressure (3–12 bar)) on the enzymatic cellulose and hemicellulose degradation of the organic waste were evaluated.

The SSF procedure at 10% dry matter (DM) revealed cellulose to ethanol conversion efficiencies ranging from 50–70% for MSW and 40–79% for yard waste at cellulase loadings varying from 5–25 FPU (filter paper units)/g DM, corresponding to 22 and 24 g/l ethanol for the highest enzyme loading. At moderate enzyme loadings (15 FPU/g DM), the ethanol yield was 65% and 69% of the theoretical yield for MSW and yard waste respectively. The wet oxidized filtrates did not exhibit any toxicity to the yeast.

1 INTRODUCTION

The production of organic waste such as municipal solid waste (MSW) increases worldwide as well as the concern for its effect on the environment (Liu et al. 2002). Apart from current disposal practices such as landfilling and incineration, organic waste is increasingly regarded as a valuable resource for material and fuel recovery (De Baere, 2000).

Organic waste (i.e. yard waste and kitchen waste) is mostly source-separated collected and is rich in lignocellulose, the main building block of plant material (Gellens et al. 1995). The most currently applied practice for energy recovery from these lignocellulosic wastes is the biological conversion into biogas, typically consisting of 55–65% of energy-rich methane gas (De Baere, 2000). However, the cellulose and hemicellulose contained in the waste can also be fermented into a competitively higher-value fuel, namely bio-ethanol.

Due to the inherent low biodegradability of lignocellulose and the low hydrolytic power of most ethanol producing organisms, a chemical and/or biological pre-treatment is needed prior to fermentation to break the intense bonds between the holocellulose and the shielding lignin. Although biological pre-treatments can be highly effective, a chemical treatment is mostly required to increase the ethanol yield during subsequent saccharification and fermentation (SSF) (Bjerre et al. 1996).

Thermal treatments involving steam and thermal hydrolysis are the most investigated processes for the pre-treatment of MSW prior to biofuel production. These studies however repetitively reported high concentrations of toxic compounds (i.e. furfural) formed out of dehydrated sugars during thermal treatment (Bjerre et al. 1996) and low lignin removal from the treated biomass. Alternatively, wet oxidation has been reported to permit fast lignin removal under alkaline conditions for pure biomasses (i.e. corn stover) (Varga et al. 2003) and low formation of fermentation inhibitors. Hence, the purpose of this study was to apply wet oxidation under varying process conditions to two mixed wastes, namely MSW and woody yard waste, to evaluate the feasibility of ethanol production from mixed waste.

2 MATERIALS AND METHODS

2.1 *Raw waste*

MSW was collected from a municipal waste plant in Denmark during wintertime. The waste was collected source-sorted and consisted mainly of kitchen waste. After collection, the waste was enriched with wheat straw to a final concentration of 8% DM (dry matter). Woody yard waste consisted of small branches from different trees and was also collected during wintertime. Both wastes were cut to a particle size smaller than 3 mm and dried to a DM content of 95%.

2.2 *Wet oxidation (WO) apparatus*

WO experiments were carried out in a high-pressure autoclave with a tubular loop and an impeller constructed at Risø National Lab (Bjerre et al. 1996). Four WO conditions were tested by varying the following parameters: T = 185–195°C, t = 10–15 min, O_2 pressure = 3–12 bar and Na_2CO_3 concentration = 0–2 g/l (Table 1). All experiments were performed batch-wise at 6% DM. After WO, solids and liquid were separated and solids were washed and dried prior to analysis.

2.3 *Chemical analysis*

The solid fractions derived from WO were shredded and hammer milled to 1 mm size prior to analysis. The wet oxidized solids and raw waste were analyzed for glucan, xylan, arabinan, Klason lignin and ash after strong and dilute acid hydrolysis with H_2SO_4 (Gilbert et al. 1952). The WO filtrates were analysed for their (monomeric and polymeric) sugar content following dilute acid hydrolysis and for their furan derivatives content. Analytical procedures (e.g. HPLC) were performed according to Klinke et al. (2002).

2.4 *Enzymatic assay*

A modified enzymatic assay (Varga et al. 2003) based on commercial cellulase and β-glucosidase

Table 1. Wet oxidation conditions.

Condition	A	B	C	D
MSW				
Temperature (°C)	185	185	195	195
Time (min)	10	10	10	10
O_2 pressure (bar)	3	12	3	12
Na_2CO_3 (g/l)	0	2	2	0
Yard waste				
Temperature (°C)	185	185	200	200
Time (min)	15	15	15	15
O_2 pressure (bar)	3	12	3	12
Na_2CO_3 (g/l)	0	2	2	0

(Celluclast®, Novozymes A/S, Denmark) was used to compare the enzymatic breakdown of the waste after WO. The assay was carried out both in the presence and absence (replacement by 0.2 M acetate buffer) of the WO filtrates to evaluate possible toxic effects of the filtrates during fermentation. The enzyme loading was at a constant value of 25 FPU (filter paper units)/g DM at a hydrolysis time of 48 h for all triplicate test tubes. The celluclast enzymatic activity was previously determined to be 67 FPU per ml (Thygesen et al. 2003). Incubations were carried out at 50°C on a shaker at 150 rpm (rounds per min).

2.5 *Simultaneous saccharification and fermentation*

A SSF procedure was developed to evaluate optimal WO conditions in function of the ethanol yield and the enzymatic degradation for each condition. The method consisted of an enzymatic pre-hydrolysis by Celluclast® (Novozymes) at 5 FPU/g DM at 50°C to convert part of the cellulose to glucose. The subsequent fermentation into ethanol was performed by bakers yeast at 32°C while Celluclast® was added at loadings varying between 0–20 FPU/g DM. The SSF was performed at 10% DM because of the relatively low carbohydrate content of the wastes employed. The total enzyme loading during SSF varied from 5–25 FPU/g DM. The cellulose to ethanol conversion was monitored gravimetrically by CO_2 loss.

3 RESULTS AND DISCUSSION

3.1 *Wet oxidation treatment*

The wet oxidation treatment considerably altered the composition of both raw wastes in terms of cellulose, hemicellulose and lignin. The composition of the raw MSW was as follows: 28% water solubles (e.g. salts and protein), 4.3% pectin, 15.6% resins/fats/waxes, 20.1% glucose, 7.2% xylose, 0.9% arabinose, 21.8% lignin and 1.9% total ash. The woody yard waste consisted of 5% water solubles, 7% pectin, 27.4% resins/fats/waxes, 24.8% glucose, 11.5% xylose, 2.2% arabinose, 22% lignin and 0.1% total ash.

For all WO conditions tested, the relative cellulose content of the wet oxidized solids increased due to the solubilisation of hemicellulose and lignin during wet oxidation. The relative cellulose content of the solids of both wastes was approximately doubled for WO condition B and condition D. Furthermore, the solubilisation of lignin (up to 67%) was also highest for condition B and D. However, the total carbohydrate recoveries were found to be lowest for the most severe WO condition D (highest oxygen pressure and temperature, Table 1).

These findings show that under the given WO conditions, the highly branched hemicellulose and lignin preferentially react (as measured by solubilisation into the liquid phase) whereas the effect of WO on cellulose is rather restricted to a decrease in crystallinity (Bjerre et al. 1996). Previous studies (Klinke et al. 2002; Lissens, unpubl.) have shown that wet oxidation temperature is one of the most important parameters with regard to oxidative sugar losses during WO. In this regard, WO condition D resulted in the highest production of sugar degradation products (e.g. carboxylic acids and fermentation inhibitors such as furfural) and lowest total carbohydrate recoveries.

Oxidative lignin conversion into carboxylic acids mainly was highest for condition B for both wastes. Oxygen pressure and initial pH were the most decisive WO parameters for delignification of both wastes during WO, a finding which has been reported previously for other biomasses such as wheat straw and poplar wood (Bjerre et al. 1996, Chang et al. 2001).

3.2 *Enzymatic convertibility of the WO wastes*

Figure 1 shows the conversion yields for the cellulose fraction of the raw and wet oxidized waste after incubation at 48 h with Celluclast®. While the conversion yields were relatively similar for all conditions tested with the MSW, the enzymatic cellulose conversion was significantly higher for condition B with the yard waste. Overall, from the cellulose up to 69% (condition B) for the MSW and up to 72% (condition D) for the yard waste could be recovered under the form of glucose (Figure 1). Despite the 8% higher conversion yield for condition D compared to condition B in the acetate assay, the highest overall enzymatic degradation was reached for condition B due to the comparatively lower carbohydrate losses. For both wastes, the enzymatic conversion efficiency of the raw waste was far lower compared to the wet oxidized waste (Figure 1). This difference in conversion yield was most pronounced for the woody yard waste, which has a much lower initial biodegradability compared to the MSW.

The high delignification degree and the higher enzymatic conversion yields found for condition B matched very well. The high pH and oxygen pressure assured high lignin destruction into carboxylic acids (decrease in pH with 3–4 units) and hence increase in enzymatic conversion (Klinke et al. 2002, Varga et al. 2003). The same tendency was observed for the enzymatic assay including the WO filtrates. The cellulose conversion yields of the MSW in the filtrate assay were in the same range while the ones of the yard waste were 5–15% lower.

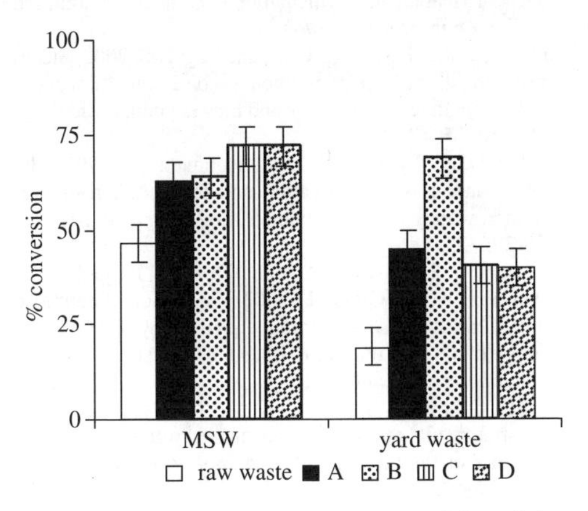

Figure 1. Enzymatic conversion yield (%) of the cellulose fraction of the raw and WO solids (condition A–D) in acetate buffer. Enzyme loading: 25 FPU (filter paper units) per g DM.

3.3 *Ethanol yield at various enzyme loadings*

Different cellulase loadings (5–25 FPU/g DM) were applied during SSF to determine the ethanol yield in function of the enzyme loading. As a whole, final ethanol concentrations in the range of 16.5–22 g/l and 11.7–24.4 g/l could be reached for MSW and yard waste, respectively. This corresponded to a maximum ethanol yield (at 25 FPU/g DM) of 70% for the MSW (Figure 2) and 79% for the yard waste (Figure 3). Even at moderate enzyme loadings (10–15 FPU/g DM), 60–70% of the original cellulose present in the waste could be converted into ethanol. For both wastes, 80% of the final ethanol yield was already reached after 48 h of fermentation for all enzyme loadings tested. However, extra nutrients under the form of yeast extract and casein extract had to be added to the woody waste in order to permit a rapid fermentation. The final glucose concentrations of the slurries after SSF were in all flasks lower than 0.15 g/l, indicating that the yeast fermented virtually all solubilized cellulose.

Due to the presence of fermentation inhibitors in the filtrates formed during other chemical pre-treatments (e.g. steam explosion), the WO liquid fraction is mostly

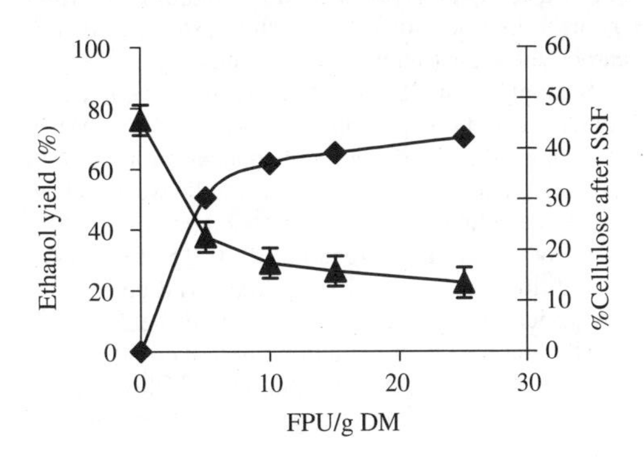

Figure 2. Ethanol yield from MSW during SSF for WO condition B. Key: ◆, ethanol yield; ▲, cellulose content.

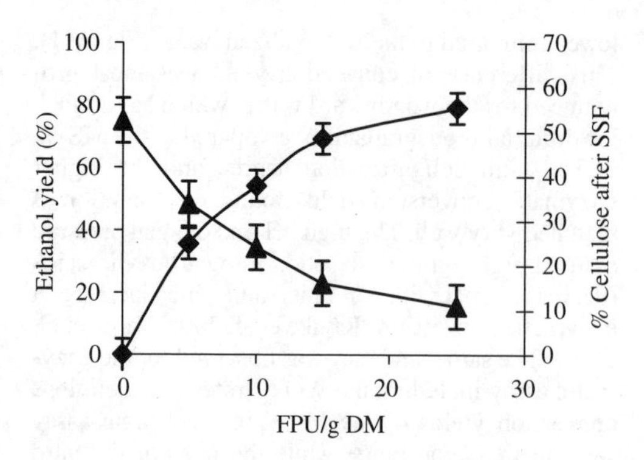

Figure 3. Ethanol yield from woody yard waste during SSF for WO condition B. Key: ◆, ethanol yield; ▲, cellulose content.

omitted from the SSF assay (Spindler et al. 1991). However, this study shows that the presence of the WO filtrates in the SSF assay did not exhibit any inhibition or toxicity towards the yeast, even at a high solids content of 10% DM. In fact, the conversion yields found in this study which included the WO filtrates were similar to the yields found in the absence of the filtrates. Moreover, in this study, only 0.2–0.8% DM of the original raw solids were converted during WO into possible fermentation inhibitors such as furfural.

4 CONCLUSIONS

This work shows that the presented wet oxidation method (T: 185–195°C, oxygen pressure: 3–12 bar, time: 10–1 5 min, alkaline pH) is a promising pretreatment for the simultaneous saccharification and fermentation (SSF) of organic waste (MSW and yard waste) into ethanol. The effect of high oxygen pressure under alkaline conditions showed to be decisive parameters for extensive delignification of MSW and yard waste (up to 67%). Solubilized lignin was further oxidized into non-toxic degradation products such as carboxylic acids (mainly acetate) and carbon dioxide.

By applying an SSF procedure with *Saccharomyces cerevisiae* and commercial cellulases at 10% DM, it was shown that a final ethanol concentration of up to 24 g/l of ethanol can be reached from the wet oxidized waste at a cellulose conversion efficiency of 60–70% at a moderate enzyme loading of 10–15 FPU/g DM. In addition, the wet oxidized filtrates did not exhibit any toxicity towards the yeast during fermentation.

The presented wet oxidation process could be particularly attractive for the treatment of fibrous side-fractions generated through separation and fermentation of grey MSW for the co-production of bio-ethanol. Alternatively, ethanol could be primarily produced from wet oxidized waste from the high-calorific fraction (carbohydrate-rich) while the energy contained in the low-calorific residue after fermentation could be recovered under the form of biogas by anaerobic digestion.

REFERENCES

Bjerre A.B., Olesen A.B., Fernqvist T., Ploger A. and Schmidt A.S. 1996. Pre-treatment of wheat straw using combined wet oxidation and alkaline hydrolysis resulting in convertible cellulose and hemicellulose. *Biotechnol. Bioeng.* 49: 568–577.

Chang V.S., Nagwani M., Kim C.H. and Holtzapple M.T. 2001. Oxidative lime pre-treatment of high-lignin biomass – Poplar wood and newspaper. *Appl. Biochem. Biotech.* 94: 1–28.

De Baere L. 2000. Anaerobic digestion of solid waste: state-of-the-art. *Water Sci. Technol.* 41: 283–290.

Gellens V., Boelens J. and Verstraete W. 1995. Source separation, selective collection and in-reactor digestion of biowaste. *Anton. Leeuw. Int. J.G.* 67: 79–89.

Gilbert N., Hobbs, I.A. and Levine J.D. 1952. Hydrolysis of wood using dilute sulfuric acid. *Industrial and Engineering Chemistry*, 44: 1712–1720.

Klinke H.B., Ahring B.K., Schmidt A.S. and Thomsen A.B. 2002. Characterization of degradation products from alkaline wet oxidation of wheat straw. *Bioresource Technol.* 82: 15–26.

Lissens G., Klinke H., Verstraete W., Ahring B. and Thomsen A.B. Wet oxidation pre-treatment of woody yard waste: parameter optimisation and enzymatic digestibility for ethanol production. Submitted to *J Chem Tech Biotech.*

Lissens G., Klinke H., Verstraete W., Ahring B. and Thomsen A.B. Wet oxidation treatment of organic household waste enriched with wheat straw for simultaneous saccharification and fermentation into ethanol. Submitted to *Environmental Technology*.

Liu H.W., Walter H.K., Vogt G.M. and Vogt H.S. 2002. Steam pressure disruption of municipal solid waste enhances anaerobic digestion kinetics and biogas yield. *Biotechnol. Bioeng.* 77: 121–130.

Spindler D.D., Wyman C.E. and Grohmann K. 1991. The simultaneous saccharification and fermentation of pretreated woody crops to ethanol. *Appl. Biochem. Biotech.* 28/29: 773–786.

Thygesen A., Thomsen A.B., Schmidt A.S., Jørgensen H., Ahring B.K. and Olsson L. 2003. Production of cellulose and hemicellulose-degrading enzymes by filamentous fungi cultivated on wet-oxidized wheat straw. *Enzyme Microbe Tech.* 32: 606–615.

Varga E., Schmidt A.S., Réczey K. and Thomsen A.B. 2003. Pre-treatment of corn stover using wet oxidation to enhance enzymatic digestibility. *Appl. Biochem. Biotech.* 104: 37–50.

European Symposium on Environmental Biotechnology, ESEB 2004 - Verstraete (ed)

Towards an efficient and integrated biogas technology

Z. Bagi, K. Perei & K.L.Kovács
Department of Biotechnology, University of Szeged, and Institute of Biophysics, Biological Research Center, Hungarian Academy of Sciences, Szeged, Hungary

ABSTRACT: Among the significant recent advances in understanding the ecology of anaerobic biodegradation of organic wastes is the recognition of the close syntropic relationship among the three distinct microbe populations and the importance of H_2 in process control. The regulatory roles of hydrogen levels and interspecies hydrogen transfer optimize the concerted action of the entire population. The concentration of either acetate or hydrogen, or both together, can be reduced sufficiently to provide a favorable free-energy change for propionate oxidation. We have shown that under these circumstances addition of hydrogen producers to the system and thereby shifting the population balance brings about advantageous effects for the entire methanogenic cascade. Proper management of the bacterial population is expected to facilitate the start-up of the fermentation. In order to reduce the costs of this treatment supplemented bacteria are grown in diluted industrial wastewater.

1 INTRODUCTION

Among the significant recent advances in understanding the ecology of anaerobic biodegradation of organic wastes is the recognition of the close syntropic relationship among the microbe populations which have important role in the biogas formation. Three distinct microbe populations take part in the anaerobic digestion process. These microbe populations are the polymer degrading, so-called hydrolyzing bacteria, the acetogens and the methanogens. The first group, the polymer degraders attack the macromolecules using extracellular enzymes and producing intermediers. Because of its abundance in Nature, cellulose is the main substrate for hydrolyzing bacteria. The acetogens then use these sugars and oligosaccharides and produce organic acids, like acetate, succinate, formate, propionate, and carbondioxide. The third group is the methanogens. These microorganisms belong to the Archaebacteria and thus possess unique molecular and cellular properties. They produce methane using acetate, hydrogen and carbondioxide. Methane is the main component of the biogas.

2 RESULTS AND DISCUSSION

Our results show that H_2 has an important role in the biogas fermentation. Although it is a major product of an intermediate step, only traces of hydrogen is found in the final product, biogas. This suggests that hydrogen may be a rate limiting substrate for methanogens. In order to check this hypothesis, the biogas forming natural consortium of microorganisms has been inoculated with a suitably selected hydrogen producing strain. Theoretically, there are three possibilities for the outcome of such experiment as follows:

1. The presence of the hydrogen producing bacterium has no effect on biogas formation. This would indicate that the hypothesis was wrong and biogas formation by methanogens is not limited by the amount of hydrogen available.
2. Hydrogen accumulates in the head space of the anaerobic fermentor.

   This would suggest that although the hypothesis was wrong and biogas formation is not limited by hydrogen, the caloric value of the biogas formed could be increased via the hydrogen component of the enriched biogas.
3. There is no hydrogen appearing in the final product but the amount of biogas formed increases.

   This would be interpreted as a proof for hydrogen being the rate limiting step in biogas formation, indeed.

During the anaerobic biodegradation hydrogen concentration is reduced to a much lower level than that of acetate. In addition, the hydrogen partial pressure can change rapidly within a few minutes. We have shown that under these circumstances addition of

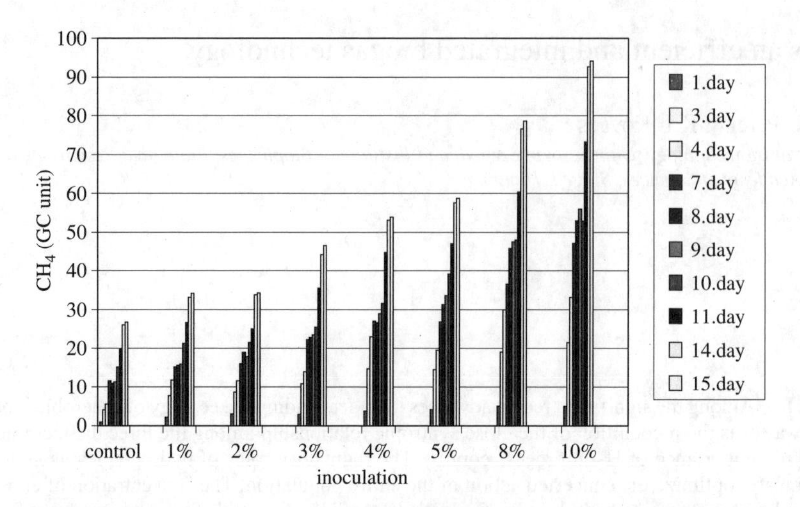

Figure 1. Gas production from a mixture of bagasse and manure using intensified microbiological biomass decomposition.

hydrogen producers to the system brings about advantageous effects for the entire microbiological methanogenic cascade. The decomposition rate of the organicsubstrate, which was animal manure in our first experiments, increases and both the acetogenic and methanogenic activities are amplified. In laboratory experiments some 2.6-fold intensification of biogas productivity has been measured.

It is to be noted that similar mechanisms have been suggested earlier and tests were carried out using hydrogen added externally from a gas cylinder. All those experiments failed and showed a strong inhibitory effect of hydrogen on methanogenesis. Indeed, too much hydrogen inhibits the metabolism of methanogens. Supplying the reducing power using the help of a bacterium, however, balances the microbial system and brings about the beneficial effect of additional hydrogen. Interspecies hydrogen transfer between the hydrogen producing and consuming microbial partners plays a determining role in the effectiveness of the biogas intensification process.

Some other experiments at various scales and using distinct organic waste sources have been carried out. Volumes between 50 ml and 10,000 m^3 were used. In every experiment the methane production increased. We got the best results with pig manure. We measured the biogas production from pig manure in a 100 litre fermentor. The manure was inoculated with a vigorously hydrogen producing bacterium strain two times, first at the beginning of the fermentation, and seven months later. Methane production increased after the inoculation and the accumulated biogas production increased significantly. In this case the biogas production increased to about 250%. The experiments thus proved that it was possible to increase the gas production using intensified microbiological biomass decomposition.

The pig slurry contains only about 5% dry weight, which can be elevated by adding biomass from energy plants. The energy plants' biomass content is about 35%, so it is possible to increase the solid content of the biomass input using energy plants. The most suitable energy plants in moderate continental climate areas, such as Hungary, are sweet sorghum, Jerusalem artichoke, which accumulate sugar as storage material and bagasse as a high cellulose content energy grass. Bagasse was used as a substrate in the experiment where we measured the gas production from energy plants.

The experiments were carried out at laboratory scale in batch fermenters.

The results substantiate the original idea and provide a proof of concept for the efficient intensification of biogas production through microbiological means. Field experiments are in progress.

REFERENCES

Claassen, P.A.M., van Lier, J.B., Lopez Contreras, A.M., van Niel, E.W.J., Sijitsma, L., Stams, A.J.M., deVries, S.S. & Weusthuis R.A. 1999. Utilisation of biomass for the supply of energy carriers. *Appl Microbiol Biotechnol* 52: 741–755.

Tiwari, G.N., Rawat, D.K. & Chandra, A. 1988. A simple analysis of conventional biogas plant. *Energy Conversion and Management* 28(1): 1–4.

Singh, S., Kumar, S., Jain, M.C. & Kumar D. 2001. Increased biogas production using microbial stimulants. *Bioresource Technology* 78(3): 313–316.

European Symposium on Environmental Biotechnology, ESEB 2004 - Verstraete (ed)
© 2004 Taylor & Francis Group, London, ISBN 90 5809 653 X

Effect of cell growth phase on hydrogen production by *Chlamydomonas reinhardtti*

D. Pak, J. Kim
Korea Institute of Science and Technology, Seoul, Korea

ABSTRACT: Effect of cell growth phase on hydrogen production was investigated with *Chlamydomonas reinhardtti*. Cells were continuously cultured in photobioreactor and taken at different growth phase to produce hydrogen. Hydrogen production by cells of different growth phase was compared under dark fermentation and sulfur deprivation. Hydrogen production by cells of mid or late exponential phase was higher than cells of early exponential phase or stationary phase in dark fermentation. Under sulfur deprivation, cell growth phase did not affect hydrogen production rate.

1 INTRODUCTION

The ability of green algae to photosynthetically produce molecular hydrogen has captivated the interest of scientific community (Akkerman 2002, Ghirardi 2000, Happe 2002). Photosynthetic activity of the green alga hydrogenase was only transient in nature due to the fact that photosynthesis and H_2O-oxidation entail the release of molecular O_2. Oxygen is a positive suppressor of hydrogenase gene expression and inhibitor of hydrogenase. One approach to overcoming the problem is to spatially separate oxygen evolution from hydrogen evolution. In this separated system, photosynthetic accumulation of reductant is followed by dark-period, nitrogen-limited H_2 evolution.

Recently, hydrogen production process was reported based on the green alga *C. reinhardtti* which differed from the system described above mainly by elimination of the dark fermentative stage and by using sulfur as the limiting nutrient (Horner 2002, Laurinavichene 2002, Tsygankov 2002). Deprivation of sulfur in green algae causes a reversible inhibition in the activity of oxygenic photosynthesis. The rate of photosynthetic oxygen revolution drop below those of oxygen consumption by respiration. In consequence, sealed cultures of green algae become anaerobic in the light. Following anaerobiosis, they produce hydrogen.

In this study, the hydrogen productions were compared between dark-period, nitrogen-limited phase and sulfur-limited phase. In the continuous algal cell culture, algal cells were harvested from the reactor by withdrawing part of medium and replacing with fresh medium. The effect of cell growth phase on starch accumulation and hydrogen production was investigated.

2 MATERIALS AND METHODS

Chlamydomonas reinhardtii UTEX 90 was obtained from UTEX, The Culture Collection of Algae at the University of Texas at Austin.

In pre-culture, *C. reinhardtii* was grown at 25°C in 250 ml Erlenmeyer flask containing 100 ml of culture medium. For agar culture, 1.5% agar was added to the culture medium. TAP (Tris-acetate-phosphate) was used as a culture medium. Fluorescent lamps were used for light supply. Light intensity for *Chlamydomonas* cultures was measured in terms of photosynthetically active radiation (PAR). Fluorescent lamps provide roughly 100 $\mu E/m^2$ sec. Light intensity is measured by Quantum sensor connected to LI-250 light meter.

In main-culture, *C. reinhardtii* was grown at 25°C in photo-bubble bioreactors shown in Figure 1. The cultures were continuously sparged with air containing 3% CO_2 for the purpose of agitation and CO_2 supply at a flow rate of about 100 ml/min. The mixing gas used in this experiment was made mixing air and CO_2 by gas mixer. Algal cells were harvested from the reactor by withdrawing part of medium and replacing with fresh medium.

Dark fermentation was performed in different growth phase. The cells were transferred to fermentation bottle of 140 ml fitted with a rubber stopper. The bottle was bubbled for 30 minutes with O_2-free N_2 gas (99.99%). After that capped the rubber stopper and

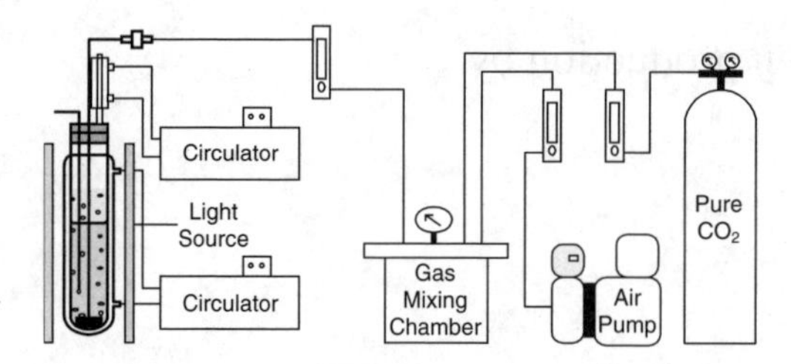

Figure 1. Schematic diagram of photobioreactor used to culture *C. reinhardtti*.

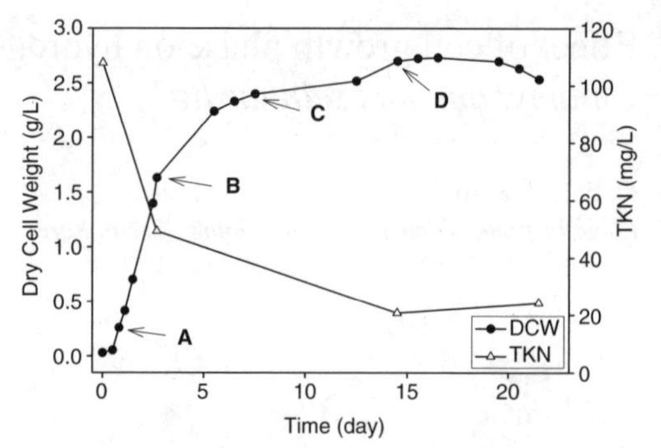

Figure 2. Sample point of *C. reinhardtti* during growth under illumination (A: early exponential phase, B: mid exponential phase, C: early stationary phase, D: mid stationary phase).

sealed fermentation bottle with aluminum foil to prevent light transmission. When this dark anaerobic condition was settled, algal cells were incubated at 25°C on a shaking incubator operating at 120 rpm. Sulfur deprivation was also performed in different growth phase. The cells were transferred to Erlmyer flask with rubber stopper.

Turbidometric measurements standardized (optical density) to a cell count curve are satisfactory for estimate of cell density. Cell growth was monitored by measuring the optical density (OD) at 660 nm with a spectrophotometer (HITACHI, U-2000). Starch is measured by iodo-starch reaction method (see Figure 6). Ethanol and acetate were assayed by a gas chromatography (GC) using FID detector of which temperature was 230°C. And also, column packing material is Porapak Q. Hydrogen was assayed by a gas chromatography (GC) using TCD detector.

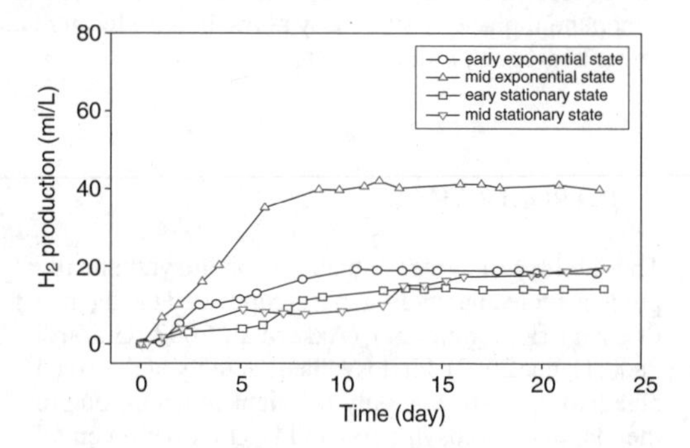

Figure 3. Hydrogen production by *C. reinhardtti* of different growth phase under dark fermentation.

3 RESULTS AND DISCUSSION

3.1 *Hydrogen production during dark fermentation*

Figure 2 shows that *C. reinhardtti* was grown under illumination to 2.5 g/l and nitrogen was taken up as a nutrient. In order to investigate the effect of cell growth phase on hydrogen production during dark fermentation, cells were taken at 4 different growth phase shown in Figure 2 and transferred to dark fermentation bottle.

Figure 3 shows hydrogen production accumulated during dark fermentation by *C. reinhardtti* taken at different growth phase. 40 ml/l of hydrogen production was obtained from cells of mid exponential phase. It was greater than those of cells of other growth phases.

Figure 4 shows the specific hydrogen production of *C. reinhardtti* of different growth phase. Specific hydrogen production of cell of early exponential growth phase is greater than cells of other gowth phase. However the accumulated starch in the cell of mid exponential phase was greater than that accumulated in the cell of early exponential phase. This indicates higher bioactivity of the cell of early exponential phase.

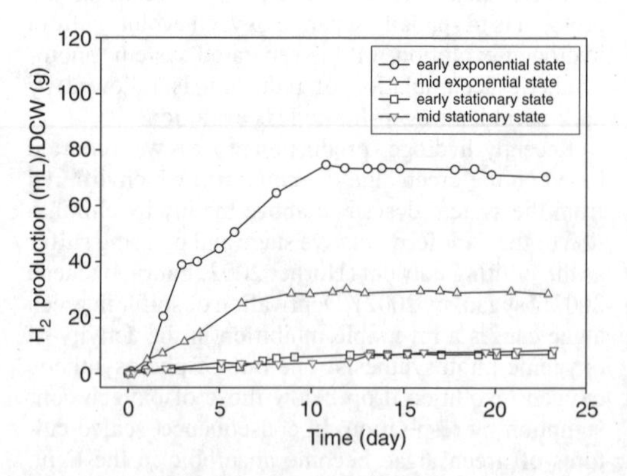

Figure 4. Specific hydrogen production of *C. reinhardtti* of different growth phase.

Hydrogen production rate was highest in cells of mid exponential phase shown in Table 1 and was 5.55 ml/l/d. This is twice of that obtained from cells of early exponential phase.

Table 1. Hydrogen production rate of *C. reinhardtti* under dark fermentation at different growth phase.

	Early exponential phase	Mid exponential phase	Early stationary phase	Mid stationary phase
Hydrogen production (ml/l/d)	2.47	5.55	1.28	0.72

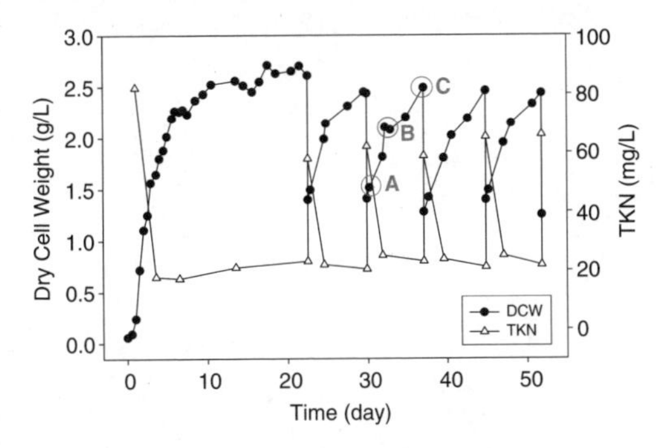

Figure 5. Continuous culture of *C. reinhardtti* in photobioreactor.

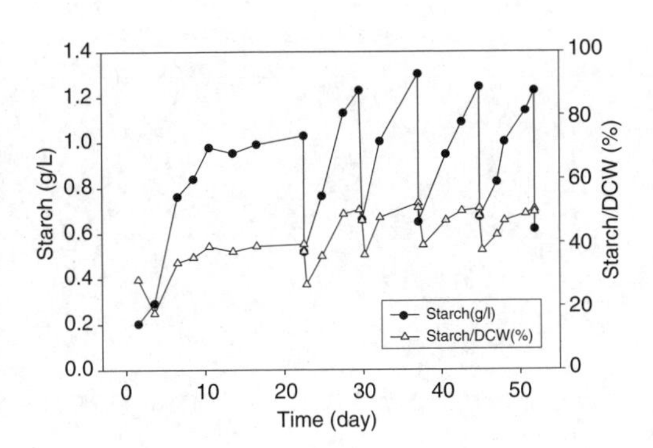

Figure 6. Starch accumulation in algae cell during growth under illumination.

3.2 *Hydrogen production of* C. reinhardtti *under sulfur deprivation*

C. reinhardtti was continuously cultured under illumination. After cells were withdrawn and the fresh medium was added to the photobioreactor, cells grew back to the level of 2.5 g/l in 8 hours shown in Figure 5. As cells grew, starch was accumulated in alga cell. Figure 6 shows a pattern of starch accumulation in the cell and specific starch production. During continuous culture, cells were taken at different growth phase and incubated under sulfur deprivation.

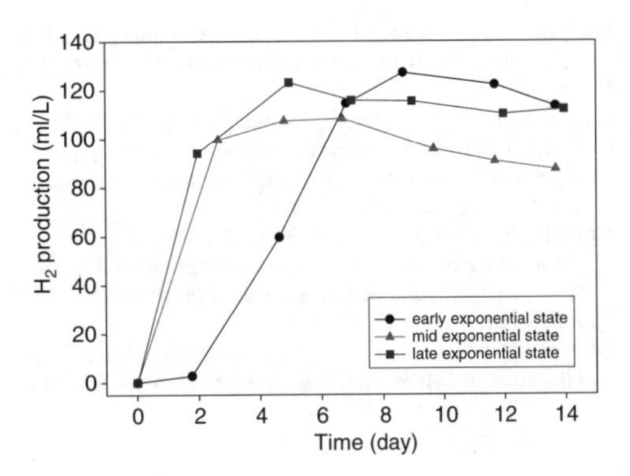

Figure 7. Hydrogen production of *C. reinhardtti* at different growth phase under sulfur deprivation.

Table 2. Hydrogen production rate of *C. reinhardtti* under sulfur deprivation at different growth phase.

	Early exponential phase	Mid exponential phase	Late exponential phase
Hydrogen production rate (ml/l/d)	22.04	22.88	23.50

Cells under sulfur deprivation produced 120 ml/l of culture. Initial biomass concentrations were different from each other as shown in Figure 7. Total hydrogen production was similar to each other. However, there was lag period at the beginning of hydrogen production by cells of early exponential phase.

Under sulfur deprivation, hydrogen production rate of cell was similar each other. This is about 4 times that obtained from cells under dark fermentation.

4 CONCLUSIONS

Chlamydomonas reinhardtti continuously grown in photobioreactor was used to produce hydrogen under dark fermentation or sulfur deprivation. Hydrogen productions of cells of different growth phase were compared each other. Hydrogen production of cells of mid or late exponential phase was higher than cells of other growth phases. Hydrogen production rate under sulfur deprivation was more than 4 times of that under dark fermentation.

REFERENCES

Akkerman, I., Janssen, M., Rocha, J. & Wijffels, R.H. 2002. Photobiological hydrogen production: photochemical efficiency and bioreactor design. *International J. of Hydrogen Energy*. 27, 1195–1208

Ghirardi, M.L., Zhang, L., Lee, J.W. & Flynn, T. 2000. Microalgae: a green source of renewable H_2. *TIBTECH*, 18, 506–511

Happe, T., Hemschemeier, A., Winkler, M. & Kaminski, A. 2002. Hydrogenase in green algae: do they save the algae's life and solve our energy problem. TREND in Plant Science, 7(6), 246–250

Horner, J.K. & Wolinsky, M.A. 2002. A power-law sensitivity analysis of the hydrogen-producing metabolic pathway in Chlamydomonas reinhardtti. *International J. of Hydrogen Energy*. 27, 1251–1255

Laurinavichene, T.V., Toistygina, I.V., Galiulina, R.R., Ghirardi, M.L., Seibert, M. & Tsygankov, A.A. 2002. Dilution methods to deprive Chlamydomonas reinhardtti culture of sulfur for subsequent hydrogen photoproduction. *International J. of Hydrogen Energy*. 27, 1245–1249

Tsygankov, A., Kosourov, S., Seibert, M. & Ghirardi, M. L. 2002. Hydrogen photoproduction under continuous illumination by sulfur-deprived synchronous Chlamydomonas reinhardtti culture. *International J. of Hydrogen Energy*. 27, 1239–1244

Winkler, M., Hemschemeier, A., Gotor, C., Melis, A. & Happe, T. 2002. Fe-hydrogenase in green algae: photofermentation and hydrogen evolution under sulfur deprivation. *International J. of Hydrogen Energy*. 27, 1431–1439

Composition and function of biofilms and consortia

European Symposium on Environmental Biotechnology, ESEB 2004 - Verstraete (ed)
© 2004 Taylor & Francis Group, London, ISBN 90 5809 653 X

Quantification of *Legionella pneumophila* by competitive polymerase chain reaction

P. Declerck[1], E. Lammertyn[2], I. Lebeau[2], J. Anné[2] and F. Ollevier[1]
[1]*Laboratory of Aquatic Ecology, Zoological Institute, Katholieke Universiteit Leuven, Charles De Beriotstraat Leuven, Belgium*
[2]*Laboratory of Bacteriology, Rega institute, Katholieke Universiteit, Minderbroedersstraat Leuven, Belgium*

A range of techniques is available for direct and indirect detection and identification of bacteria. Among the DNA based methods, Polymerase Chain Reaction (PCR) offers the highest specificity and sensitivity. During the last few years, many efforts have been made to convert PCR to a quantitative method. Hereto there are two approaches, namely competitive PCR (cPCR) and real-time quantitative PCR (RT-PCR). RT-PCR is an expensive method because specific and expensive equipment is needed, therefore until now the technique is only used in specialized laboratories. cPCR on the other hand, is a less expensive alternative and is one of the most common methods of quantitative PCR. This technique is based on a co-amplification of the target DNA (sample) with a homologous or heterologous DNA standard (competitor), which competes with the sample template DNA for the same set of DNA primers[1]. Since the competitor is added to the reaction mixture in known amounts, it is possible to calculate the amount of target DNA from the ratio of amplified products of sample and competitor DNA[2].

In this research a homologous competitor was developed in order to quantify the human pathogen *Legionella pneumophila* in water samples. This bacterium is the etiological agent of Legionnaires' disease, a severe kind of nosocomial and community acquired pneumonia, particularly among people with impaired host defenses. The competitor DNA fragment of 384 bp was obtained by inserting a 216 bp *Sma*I-*Pvu*II fragment of the pUC19 plasmid into the blunted *Sph*I site of a *Legionella pneumophila mip*-specific PCR fragment (168 bp) cloned in pBluescript. Determination of the concentration of the standard was done by OD_{260} measurement. When added to a test sample, both the *mip* gene and the *mip*-derived competitor DNA can be amplified using the primers PT69 and PT70[3]. Results and applications of quantitative PCR using the developed competitor will be discussed.

[1] Orlando *et al.* 1998;
[2] Rupf *et al.* 2001;
[3] Declerck *et al.* 2003

European Symposium on Environmental Biotechnology, ESEB 2004 - Verstraete (ed)
© 2004 Taylor & Francis Group, London, ISBN 90 5809 653 X

Mathematical modelling of biofilm dynamics

J. Dueck
Department of Environmental Process Engineering & Recycling, University Erlangen-Nuremberg, Erlangen Germany

S. Pylnik & L. Minkov
Faculty for Physics and Engineering, Tomsk State University, Tomsk, Russia

ABSTRACT: The mathematical simulation of the biofilm process is presented. The dynamics of biofilm is taken into account. The mechanisms responsible for increase or decrease of the biofilm thickness are suggested. The term for death-rate of microorganisms is derived. The dynamics of biofilm for different values of erosion coefficient r is illustrated by calculations. Analysis predicts that for sufficiently high values of r the thickness of the biofilm can only decrease.

The productivity of a water clarifying reactor depends on many factors. In particular, on its reaction surface, which can be especially large in case of a porous loading. On the other hand, the finer the pores are, the smaller is the water throughput.

The optimisation of the purification process by the experiments is usually time-consuming. Therefore, the efforts by modelling the purification process with application of the numeric methods are important.

A biofilm scheme is shown in Figure 1.

A particularly interesting target value withcomputations is the substrate flux biofilm, where by the entire performance of the reactor can be characterised.

The bio-kinetics of substrate consumption usually depends very strongly on the transport of the substrate into the film (Bruce et al. 2000, Cristian et al. 1998, Horn & Hempel, 2000, Saez & Rittmann, 1992). Therefore, the diffusion equation must be used for the description of the biofilm process.

In this work the model is presented, which describes the increase and the destruction of the biofilm.

The mathematical model contains the following equations:

Balance of the biomass:

$$\frac{\partial X_f}{\partial t} = Yq\frac{S_f}{K+S_f}X_f - bX_f^a, \qquad (1)$$

transport and consumption of the substrate:

$$\frac{\partial S_f}{\partial t} = D_f\frac{\partial^2 S_f}{\partial z^2} - q\frac{S_f}{K+S_f}X_f, \qquad (2)$$

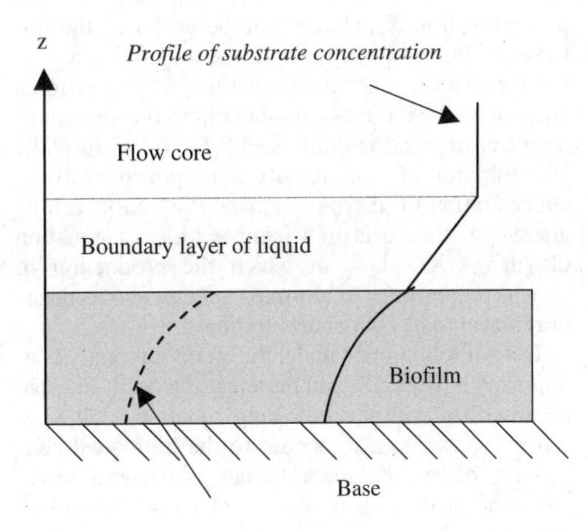

Figure 1. Scheme of a biofilm.

dynamics of the film thickness:

$$\frac{dL_f}{dt} = \frac{1}{\rho_B}\int_0^{L_f} Yq\frac{S_f}{K+S_f}X_f dz - rL_f \qquad (3)$$

with the initial:

$$L_f(0) = L_{f0}, X_f(0,z) = X_{f0}, S_f(0,z) = 0$$

and boundary conditions:

$$z = 0 : \frac{\partial S_f}{\partial z} = 0;$$

$$z = L_f : B\left(S_l - S_f(t, L_f(t))\right) = -D_f \frac{\partial S_f}{\partial z}$$

Some commentaries to the used equations. The first term in Equation (3) describes the increase of film due to reproduction of micro-organisms. It is assumed that the mass of new born bio-mass leads to a swelling of biofilm such as

$$\int_0^{L_f} \frac{dX_f}{dt} dz = \rho_B \frac{dL_f}{dt},$$

where the density of bio-mass ρ_B is assumed to be constant.

The analysis shows that the linear law for the death-rate of the bio-organisms leads to dissatisfactory results. The death-rate has been derived to be proportional to X_f^2. Hereby can be proposed the following scheme which leads to the magnitude $a = 2$. Let the death-rate of micro-organism be proportional not only to the bio-mass amount but to the amount of products of vital activity $\sim -bP_fX_f$. In its turn the concentration of vital activity is proportional to the concentration of micro-organisms $P_f \sim X_f$. It can be argued on the basis of a simple kinetics equation $dP_f/dt = k_1X_f - k_2P_f$ in which the production of toxic is proportional to bio-mass amount and its decay or removal to its own concentration.

Here of follows that under the quasi-stationary conditions $P_f = (k_1/k_2)X_f$ and therefore the death-rate can be described with the following equation $-bP_fX_f = -b(k_1/k_2)X_f^2$. Thus, in accord to the biological conception follows the quadratic law of bacteria death-rate. This derivation is made under the assumption that the intermediate products exert an influence on bio-organisms.

The rate of destroying of biofilm due to flow stresses is formulated to be proportional to the biofilm thickness. The second term in the right part of Equation (3) describes the destroying of the film, which is caused by the stress of ambient flow.

The mechanism of destroying is up to now far not clear (Bruce et al. 2000, Chang et al. 1991, Cristian et al. 1998, Gavrilescu & Macoveanu, 2000, Rittmann, B.E. 1982). In the literature is distinguished the erosion (entrainment slice by slice) by tangential stresses of flow and the sudden shedding of relatively large pieces of film. The last seems to be a rather random process. The character of film destroying depend on the flow

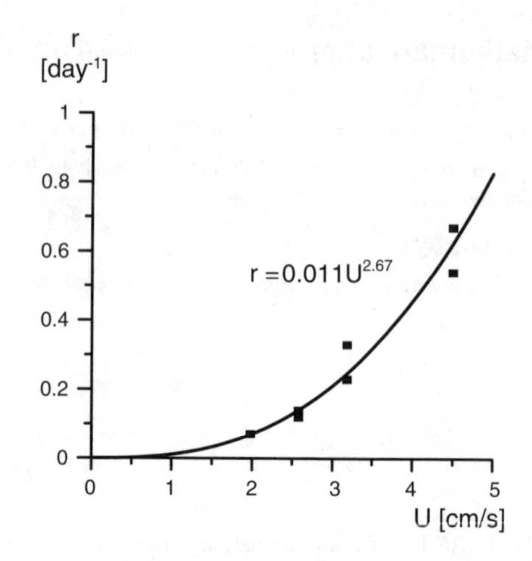

Figure 2. Erosion coefficient as a function of flow velocity.

property, particularly on the surface smoothness. Without detailed theory of film destroying the low of linear change of their removal will be used.

The process of slice removal is stochastic. Therefore, the proportionality of destroying rate and the film thickness L_f with an erosion coefficient r, which depends on the flow rate and the porous structure can be suggested. According to the literature, the empirical coefficient r of removal can lay between 0.1 and 10 day^{-1}.

From the rheological point of view, the bio-polymers are elasto-viscous liquids. The deformations are proportional to the strains from flow $\tau_{zy} \sim \sqrt{U}$ and to the film thickness L_f. If the value of deformation ΔL_f exeeds the crutial magnitude $L_f > \tau_{zy}/\sigma \sim \sqrt{U}$ then the layer of film can not be fixed and will be removed by the flow. Hereby U is the velocity of the liquid flow and σ the stability characteristic of film material.

Our own experiments confirm the law $dL_f = -rL_f\,dt$ and gives following expression $r = 0.01U^{2.7}$, which is illustrated in Figure 2.

We decide to consider as already known all parameters for the consumption kinetics of the substrate S_f, for the increase and death-rate of the micro-organisms X_f, mass exchange on the liquid-film interface B, diffusion coefficient of substrate D_f and initial concentration of the substrate in the water, the bio-mass in the biofilm and the thickness of the biofilm L_f.

The following has to be computed:

- the dynamics of the substrate concentration on the surface,
- the thickness of the biofilm L_f,
- the full bio-mass $Bio = \int_0^{L_f(t)} X_f\,(t,z)\,dz$ and
- the substrate flux into the film, depending on time $J = B(S_l - S_f(t, L_f))$.

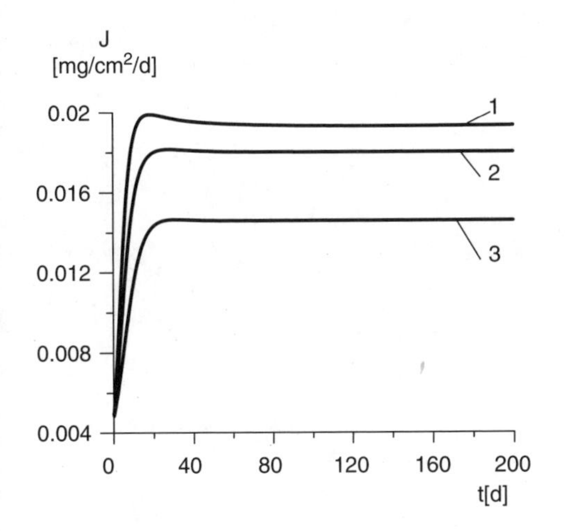

Figure 3. Dynamics of substrate flux into the biofilm.

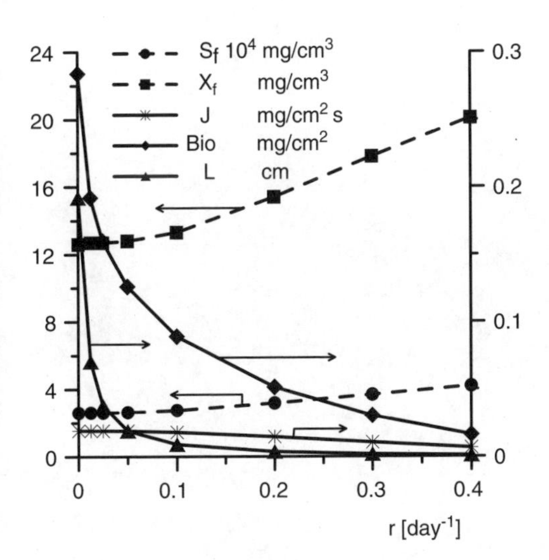

Figure 4. Stationary characteristics of biofilm in dependency on the erosion coefficient: r = 0 (1), 0.1 (2), 0.2 (3) d^{-1}.

For example, in Figure 3 the dynamics of the substrate flux in the biofilm is shown for different values of the erosion coefficient.

Contrary, the stationary value of the substrate flux depends strongly on the erosion coefficient (Figure 4).

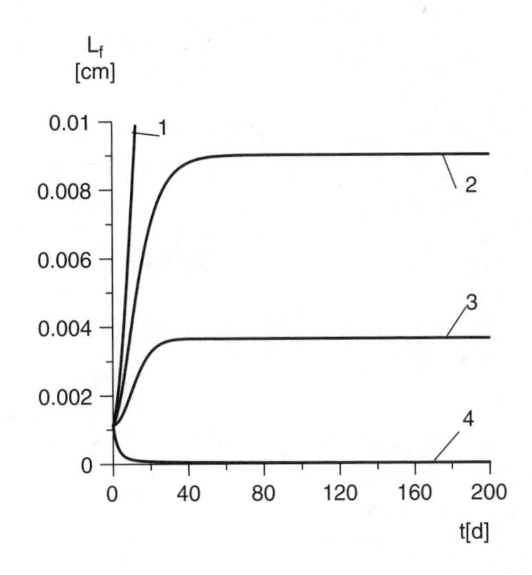

Figure 5. Dynamics of biofilm thickness: r = 0 (1), 0.1 (2), 0.2 (3), 0.57 d^{-1}.

Analysis predicts that for sufficiently high values of the erosion coefficient the thickness of the biofilm can only decrease.

Figure 5 presents the dynamics of L_f.

For $r \geqslant 0.56$ the biofilm is destroyed by water flow.

REFERENCES

Bruce, E., Rittmann, B., Perry L. McCarty, 2000. *Enviromental Biotechnology: Principles and Applications*, McGrawHill, Boston.

Chang, H.T., Rittmann, B.T., Amar, D., Ehlinger, O., Lesty, Y. 1991. Biofilm detachment mechanisms in liquid-fluidized bed, *Biotechnol. Bioengr.* 38: 499–506.

Cristian, P., Hollesan, L., Harremoes, P. 1998. Liquid film diffusion on reaction rate in submerged biofilters, *Water Res.*, 29: 377–388.

Gavrilescu, M., Macoveanu, M. 2000. Attached-growth process engineering in wastewater treatment, *Bioprocess Engineering*, 23: 95–106.

Horn, H., Hempel, D.C. 2000. Simulationsrechnungen zur Beschreibung von Biofilmsystemen, *Chemie Ingenieur Technik*, 72 (10): 1234–1237.

Rittmann, B.E. 1982. The effect of shear stress on loss rate, *Biotechnol. Bioengr.*, 24, pp. 1341–1370.

Saez, P.B., Rittmann, B.E. 1992. Accurate pseudo-analytical solution for steady-state biofilm, *Bioechnolol. Bioengr.*, 39: 379–385.

European Symposium on Environmental Biotechnology, ESEB 2004 - Verstraete (ed)

Ecological and biotechnological aspects of biofilm formation in waste water treatment

A.S. Sirotkin, G.I. Shaginurova, K.G. Ippolitov & V.M. Emelianov
Kazan State Technological University, Kazan, Russia

ABSTRACT: The influence of ecological factors on the development and properties of biofilms was analyzed. The conclusion about ecological succession in to the different age biofilms under the long-term hydrodynamic stress was drawn. Some effects in a "biofilm-adsorbent" system, such as accumulation of different substrata on the adsorbent surface and theirs accessibility to biodegradation, and also biological regeneration of the adsorbent were described.

1 INTRODUCTION

One of the most universal methods of sewage treatment intensification is biosorption. The efficiency of the biosorption method is provided by a synergetic action of biological and adsorptional systems. The application of adsorptional materials in biological treatment systems supposes a complex of the biological, physical and chemical processes, such as intensive exopolymer production, bioregeneration of adsorbents, accumulation of pollutants from water and the further biodegradation, electrostatic interaction with an adsorbent surface etc.

According to a traditional representation, a biofilm consists of live cells and dead cells, as well as of cellular fragments in matrix from the extracellular polymers fixed on a firm surface (Bishop et al. 1995). Biofilms also contain varying mixture of populations. On the surface of the biofilm there exists an active cell reproduction because the concentration of a substratum is higher. The substratum becomes limiting closer to the carrier surface, which is frequently the reason of biofilm destruction. The significant part of the biofilm is lost because of erosion of its surface under shear stress conditions or tearing off from the carrier surface as a result of its stratification. The biofilms are objects with a complex structure, which is characterized by a nonlinear change of density depending on the thickness of its layer. Studying of formation laws and factors influencing biofilms growth is very important for effective realization of waste water treatment processes using immobilised microorganisms. The purpose of this research was to investigate the biofilms growth and development, depending on various ecological factors: the availability of a substratum, waste water flow speed, microbial cultures associative interaction in biofilms biocoenosis.

2 MATERIALS AND METHODS

The biofilm development in continuous conditions of a nutrient medium flow was studied on an example of pure culture *Sphingomonas sp. L138* (Deutsche Sammlung von Mikroorganismen und Zellkulturen GmbH, Braunschweig, Germany, DSM 3376). The cells were immobilised on a glass surface in the model flowing channel. Glucose was used as the source of carbon. To research the microorganisms using confocal laser scanning microscopy (LSM), the cells were marked with transpozone – a protein fluorescing in a wave band of green color (GFB). It allowed to distinguish cells from the extracellular polymeric substances, which remained invisible in the given wave range. Extracellular polymeric substances (EPS) were detected by preliminary colouring tests of a biofilm by specific dye – concovalin A. The flow velocity of the nutrient medium varied from 0.0001 m^3/h up to 0.02 m^3/h. Received on the CLSM, series of images were processed by the Leica QUIN software (Stoerkel et al. 1994). This system is intended for the improvement of color and binary image processing and their qualitative and statistical analysis.

To research the process of the waste water treatment by the biofilm obtained from activated sludge, modelling solutions with phenol and the industrial sewage containing oil products or sulfur compounds were used. The concentration of these substances in modelling solutions corresponded to their concentration in industrial sewage in normal operation conditions and discharges. Activated carbon (PAC), ash of thermal power stations and zeolite containing rock (ZCR) were investigated as the biofilm carrier. The materials differed in a formation source, properties and cost. Researches were carried out in pilot-scale airtank using continuous mode of treatment. The adsorbent doze in the reactor was 0. 5 kg/m^3. Hydraulic retention time of 8–12 hours was performed.

3 RESULTS AND DISCUSSION

3.1 *Pure culture experiments*

Dependences of biofilm growth and extracellular polymers formation on height of the biofilm versus the cultivation time and the flow velocity in the channel were received. It was shown, that the biofilm develops most actively not only at small, but also at high flow velocities (Figs 1, 2). It is possible to explain the successful biofilm growth at low speeds by the absence of significant shear stress. High flow velocities stimulate the biofilm growth due to an increase of diffusion speed of the substratum and oxygen. At high speeds an adequate increase of the extracellular polymeric substances, filling the growing biofilm, is also provided. As a result of the analysis of images of extracellular polymers, secreted by the biofilm, a conclusion was made, that its development is practically proportional to the growth of microbial cells. Besides, it is revealed, that extracellular polymers are distributed in regular intervals in intercellular space at their maximal formation in case of low and high flow speeds. Under a hydrodynamical stress, there was a selection of the cells with the best adhesive ability. Thus, an ecological succession in biofilmes of various age was developed.

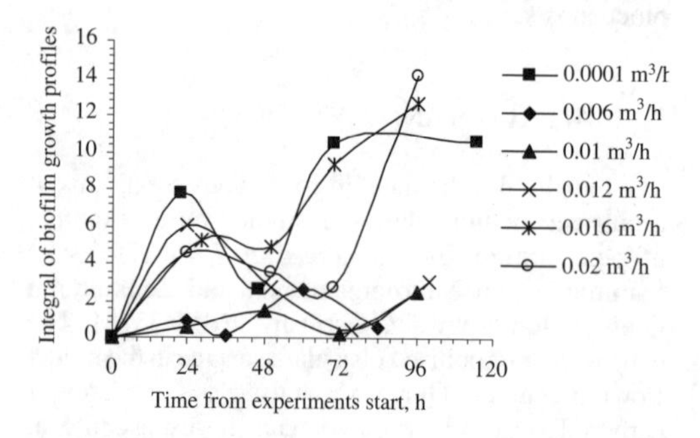

Figure 1. Biofilm growth cinetics at the different flow velocities.

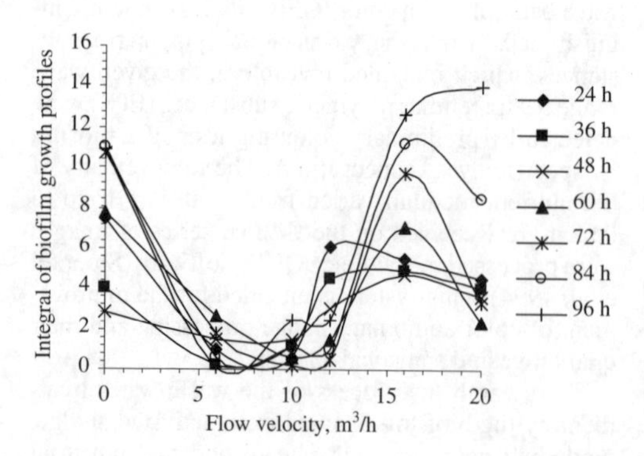

Figure 2. Biofilms growth at the different flow velocities.

The submitted data can be used for creating necessary hydrodynamical conditions in biosorption waste water treatment systems.

3.2 *Activated sludge biofilm experiments*

The efficiency of adsorbent biofilm systems was analyzed on removal COD, oil products, ammonium nitrogen in comparison with biological treatment by activated sludge. According to the results of researches, the efficiency of the organic substratum removal (on COD) in biosorptional systems, in comparison with activated sludge, was marked. In biosorptional systems, nitrification processes proceeded better (at the initial ammonium concentration 15 mg per liter, the elimination effect was 70%, 92% and 95% for the activated sludge system, ZCR biofilm and PAC biofilm systems, respectively). The best efficiency of nitrification process in systems of biosorption was provided with an adsorption of an organic substratum on PAC and ZCR due to what the ingibition of microorganisms – nitrificants was reduced. In case of ZCR system, the nitrification process was stimulated by an additional accumulation of NH^{4+} ions on the ZCR surface as a result of an ionic exchange.

In cases of discharged loadings (concentration of oil products up to 40 mg per liter) in adsorbent biofilm systems, the significant efficiency of oil products removal from the waste water was marked. Concentrations of oil products in the purified water were the following: for the biologic system – 5, for the ZCR biofilm system – 2 and for the PAC biofilm system – 2.2 mg per liter. Presumably, in case of hydrocarbons absorption by ZCR inside pores is supplemented with adsorption on the rock inclusions clay crystals surface, such as montmorillonit. As a result, there is more favorable "microclimate" for biofilm microorganisms in system with ZCR due to accumulation on the surface and availability to biooxidation of hydrocarbons molecules and ammonium nitrogen ions. Besides, on the ZCR surface as large pores material, more cells was accumulated in comparison with micro-and mesopored PAC particles.

Biological regeneration of adsorbents was estimated in the activated carbon - biofilm system in phenol-containing solutions. The peak efficiency of bioregeneration was provided in the first 7 days – approximately 33%. Aromatic hydrocarbons, in particular, phenol, are adsorbed in micropores, and their desorbtion was difficult – no more than on 45%. With the regard of micropores inaccessibility for the cells and theirs exoenzymes, it was marked that, under their influence, only mesoporous and external (mucroporous) activated carbon surfaces clearing can be carried out. At the continuous presence of microorganisms in the system with adsorbent, big losses of adsorptional capacities, connected to filling large pores by products of a metabolism, cells lysis and blocking of micropores, can emerge. The specified reasons define the bioregeneration process efficiency and limit prospects of its industrial application. By results of bioregenerations research of the activated carbon, loaded with pollutions from modelling and industrial waste waters, it was shown that the major factors determining efficiency of adsorbents bioregeneration are the following: the type and adsorptional properties of adsorbents structure; physical and chemical properties of adsorbed substances, such as adsorption affinity, the size of molecules and others; place and time of molecules presence in pores, convertibility of adsorption; competitive adsorption from a mixture with molecular and ionic pressure of substances from a solution; substances biodegradation availability, microbial activity etc (Sirotkin et al. 2000).

Table 1. Properties of the adsorbents in the biofilm – adsorbent systems.

Adsorbent	Interaction with microorganisms	Effect
ZCR	adsorption on the surface and in large pores, electrostatic interaction	more effective nitrification, stable performance at discharge of pollutants
PAC	adsorption on the surface	effective organics elimination, stable performance at discharge of pollutants
Ash	adsorption on the surface, electrostatic interaction	improvement of the sedimentation properties, stable performance at discharge of pollutants

To increase the treatment efficiency for sulfur-containing waste water, the biosorptional method with the thermal power station ashes airtank loading was recommended. In the ashes – biofilm system, the pollutants removing efficiency (on COD) was 20% higher than that in the airtank with the activated sludge. The sedimentation velocity increased 23–25%, and the sludge volume index in ash – biofilm system decreased 50%, as compared with the activated sludge. Such parameters is the result of the ash particles aggregation by activated sludge flocs. Owing to significant amounts Fe^{3+} and Al^{3+}, the ash surface usually has cationic properties. This material is not only the carrier of the biofilm, but also serves for the coagulation of activated sludge flocs and provides a good aggregates sedimentation. Each adsorbent tested demonstrated specific properties in relation to sewage components, as well as in interaction with activated sludge microorganisms (Table 1).

4 CONCLUSIONS

The data, described the influence of hydrolic shear stress on the biofilm growth can be used for creating necessary hydrodynamical conditions in biosorption waste water treatment systems.

On the basis of the received results, recommendations for application of carbon or mineral sorbents in traditional systems of biological clearing can be made depending on special industrial biological clearing problems.

The biosorptional technology of sewage treatment on the basis of existing airtank, approved in pilot and industrial scale, provided: (1) an effective COD, oil products, ammonium nitrogen and other pollution sewage treatment; (2) the achievement of high speeds of organic substances biooxidation and the efficiency of nitrification process; (3) the stability and reliability of clearing process in normal and discharge operation modes.

REFERENCES

Bishop, P.L. & Zhang, T.C. 1995. Effects of biofilm structure, microbial distributions and mass transport on biodegradation processes. *Water Scince Technology*. 31(1): 143–152.

Sirotkin, A.S. et al. 2000. Bioregeneration of contaminated adsorbents containing hazardous wastes. In *Bioremediation of soil contaminated with hazardous wastes*: 45–56. NewYork: Marcel Dekker.

Stoerkel, S. & Schneider, H.M. 1994. Proliferationsmessungen mit Hilfe des Leica-Bildanalysesystems QUANTIMET 500. *Mitteilungen fuer Wissenschaft und Technik Bd. X.* 12(8): 263–267.

European Symposium on Environmental Biotechnology, ESEB 2004 - Verstraete (ed)

Characterisation of three aerobic bacteria isolated from NPEO-contaminated sludges

D. Di Gioia, M. Pierini, F. Fava
DICASM, Facoltà di Ingegneria, Viale Risorgimento, Bologna

E. Coppini
GIDA S.p.A., via Baciacavallo, Prato

C. Barberio
Dipartimento di Biologia Animale e Genetica (DBAG), via Romana, Firenze

ABSTRACT: In this work, three aerobic bacteria were isolated from activated sludges contaminated by nonylphenol polyethoxylates and characterised. Two of them, namely BCaL1 and BCaL2 were Gram negative rods, the third, VA160, was a Gram positive bacterium with a rhizoid colony shape. Analysis of the 16S ribosomal DNAs showed that the three strains could be assigned to *Acinetobacter* (BCaL1) *Stenotrophomonas* (BCaL2) and *Bacillus* (VA160) genera. Their degradation ability towards commercial mixtures of nonylphenol ethoxylates, in particular towards a mixture having an average ethoxylation degree of five, was investigated in aerobic batch culture conditions. Results showed that the *Acinetobacter* and *Stenotrophomonas* strains were both able to degrade some components of the surfactant mixture. On the contrary *Bacillus* VA160 did not exhibit any detectable biodegradation activity, but, when co-cultured with the other two strains, it was capable of positively affecting their degradation ability.

1 INTRODUCTION

Nonylphenol polyethoxylates (NPEOs) are synthetic non ionic surfactants belonging to the group of alkylphenol ethoxylates. Commercial NPEOs are mixtures of isomeric forms with different length of the ethoxylic chain and branching of the nonyl group. Due to these features, NPEOs are not easily biodegradable compounds. Three main intermediates are produced from the aerobic biodegradation of NPEOs: nonylphenol (NP) and mono- and di-ethoxylated nonylphenols (Kravetz 1981). These intermediates, which are less water-soluble than the parental molecules, generally accumulate in aquatic environment where they can exert toxic effects towards animals (Yoshimura 1986, Patoczka & Pulliam 1990, Ekelund *et al.* 1990, Soto *et al.* 1991, Thibaut *et al.* 1999) and plants (Bokern *et al.* 1998). Short chain NPEOs and NP have been included among environmental estrogens. Despite their well-known toxicity, NPEOs are still widely used, particularly in some industries like textile, leather, or pulp, because of their cheapness combined with excellent detergent, wetting and solubilization properties (Balson & Felix 1995).

Information concerning the microbial degradation of nonylphenol polyethoxylates is still limited. Some members of the *Pseudomonas* (Maki *et al.* 1994, Frassinetti *et al.* 1996, John & White 1998) and *Xanthomonas* (Frassinetti *et al.* 1996) genera have been reported to mediate the biodegradation of such pollutants. Bacteria belonging to *Sphingomonas* (Fujii *et al.* 2000, Tanghe *et al.* 1999), *Alcaligenes* (Tanghe *et al.* 1999) and *Pseudomonas* (Fujii *et al.* 2000, Tanghe *et al.* 1999) genera along with a *Candida* (Frassinetti *et al.* 2001) strain have been reported to be able to degrade nonylphenol with branched or linear alkyl chain.

No specific remediation technologies for NPEO-contaminated wastes have been developed up to now; the possibility of employing microorganisms having specific degradation capabilities for such a purpose has to be explored. With the aim of obtaining bacteria capable of degrading two NPEO mixtures mainly employed in textile industries, several aerobic bacteria strains were isolated from activated sludge coming from conventional treatment plants receiving NPEO-contaminated wastewater (Barberio & Fani 1998, Barberio *et al.* 2001). In this work, three of these bacteria have been studied from a morphological and molecular point of view and their involvement in the degradation of NPEO mixtures has been investigated.

2 MATERIAL AND METHODS

2.1 *Media, culture conditions and strain isolation*

Growth media were Luria-Bertani (LB) medium, Nutrient Broth medium (NB) or HPW mineral medium (Barberio & Fani 1998), containing different amounts of surfactants as the carbon and energy source. HPW medium was amended with 0.2% (w/v) sodium acetate (HPWNA) for cometabolic experiments. Agar plates were also prepared from each of these media. Bacteria were grown aerobically at 30°C. When surfactants were present, growth experiments were carried out in the dark to avoid pollutant photooxidation. The surfactants employed were: industrial mixtures having an average of nine (NP9EO) and six (NP6EO) ethoxyl units (Hüls AG, Marl, Germany) and Igepal CO-520, having an average of five ethoxyl units (Sigma-Aldrich, Milan, Italy). BCaL1 and BCaL2 strains were obtained from NPEO-polluted sludge after one cycle of enrichment in HPW medium spiked with NP6EO and NP9EO at 15 mg/l each. Enrichment cultures were seeded on plates of the same medium, which was also employed for strain purification. Strain VA160 was obtained by directly plating the sludge on LB medium.

2.2 *Phenotypic characterization*

Strain cellular morphology was examined using a Nikon phase contrast microscope Alphaphot YS. Gram staining was performed according to Murray *et al.* (1994).

2.3 *Amplification and analysis of 16S rDNA*

Nearly all of the 16S rDNA gene of each strain was obtained, by amplification from crude cell lysates using primers 27f (5-GAGAGTTTGATCCTG-GCTCAG) and 1495r (5-CTACGGCTACCTTGT-TACGA), and analysed, by agarose gel electrophoresis, as described previously (Barberio *et al.* 2001). Amplified 16S rDNA was purified by the PCR Clean up kit (Roche) and sequenced using the enzymatic method (Sanger *et al.* 1997) with a Perkin Elmer automatic sequencer ABI prism 310. The oligonucleotides used were the two specific primers 27f and 1495r, and different groups of specific primers, depending on the strain. The annealing temperature (Ta) was 50°C. The 16S rDNA nucleotide sequences obtained were aligned with the most similar ones of GenBank using the BLAST program (Altschul *et al.* 1990).

2.4 *Evaluation of surfactant degradation*

Two methods were used to check the strain ability to degrade NPEOs: the two-phase Tetrakis (4-fluoro phenyl) borate (TAS), sodium salt (Omnia Reserch s.r.l., Milan, Italy) titration method (Tsuboichi *et al.* 1985) and High Pressure Liquid Chromatography (HPLC) (see 2.5 section).

When TAS titration method was used, cells from overnight LB cultures were harvested, washed in HPW medium, counted and re-suspended in the same medium at 10^9 cell/ml. A final concentration of 5×10^6 cell/ml was obtained by dilution in HPWNA medium. Cultures were incubated in a rotatory shaker (100 rpm). The bacterial growth was followed spectrophotometrically at 540 nm. At different times, 50 ml were collected from the cultures, acidified with 25 μl of 96% H_2SO_4 and stored at 4°C until assayed.

Biodegradation experiments, monitored via HPLC analyses, were performed in 100-ml pyrex bottles containing 20 ml of HPW medium to which 100 mg/l of Igepal CO-520 were added. For each degradation experiment, 7 bottles were prepared and extracted as follows: one after 30 min incubation and the other, in couples, after 7, 14 and 25 days. Incubation was performed at 30°C and 150 rpm. Inocula were prepared in HPWNA medium; cells were then harvested, washed in 50 mM phosphate buffer (pH 7), re-suspended in the same buffer in order to have 10^9 cell/ml and inoculated at 2% (v/v) in each bottle. At the pre-established times, the bottle content was acidified to pH 3 with 2N trichloroacetic acid and extracted twice with 5 ml diethyl ether. The organic fractions were collected, evaporated to dryness under a N_2 flux, re-suspended in 5 ml of hexane and injected into the HPLC system.

2.5 *Analytical methods*

HPLC analyses were performed using a Beckman Coulter apparatus. The separation of the components of Igepal CO-520 was obtained with a Supelcosil LC-Diol column (5 μm, 250 × 4 mm, Supelco Park Bellefonte, PA-USA), at 35°C and by employing a gradient elution of two phases: phase A, composed of hexane: methylene chloride 95:5, and phase B, composed of hexane: methylene chloride: methanol 50:40:10. The elution started with 70% phase A for 5 min, which was then changed to 40% in 10 min at a flow rate of 1 ml/min. The injection volume was 20 μl. The diode array detector was set at 235 and 280 nm. For quantification, calibration curves were obtained by using Igepal CO-520 at the concentration of 130 mg/l, 210 mg/l and 400 mg/l.

3 RESULTS AND DISCUSSION

3.1 *Isolation and phenotypic characterization*

BCaL1 and BCaL2 showed morphotypes barely distinguishable on HPW plates, but clearly different on LB plates. In this medium, the colonies of strain BCaL1 were pearly-white and convex, whereas BCaL2 colonies were flat and pale yellow. Both BCaL1 and BCaL2 colonies had smooth ends. Both strains were Gram-negative. Cells of strain BCaL1 were coccus-like of nearly 1 μm diameter, those of strain BCaL2 were small rods nearly 1.5 μm long × 1 μm diameter.

Strain VA160 did not grow on plates containing NP6EO and NP9EO as the sole carbon source. Colony shape was of rhizoid type on NB agar medium, but not on LB. Microscopic observation of the cells showed that they were non motile spore-forming rods (2.5 length × 0.7 μm diameter) mainly organized in long convoluted chains. After 40-hour growth on liquid LB, cell chains tended to form aggregates.

3.2 *Taxonomic analysis*

The 16S rDNA was successfully amplified from the three strains. The 16S rDNAs from the three isolates were sequenced and analysed (see 2.3 section). Results showed that strains BCaL1 and BCaL2 belonged to the γ subdivision of the *Proteobacteria* and could be assigned to the *Acinetobacter* and *Stenotrophomonas* genera, respectively. Strain VA160 clustered with group 1 *Bacilli* (Ash *et al.* 1991), *i.e.* the *Bacillus cereus* subgroup.

3.3 *Characterization of degradation ability by colorimetric assay*

The involvement of the three strains in NPEO degradation was preliminary checked by the TAS method (Tsuboichi *et al.* 1985), which is routinely used in treatment plants to measure degradation of NPEOs with ethoxylic units longer than four. The assays were carried out after growing the strains in the presence of a mixture of NP9EO and NP6EO (15 mg/l each) under cometabolic conditions (HPWNA medium). Results showed that, after 24 and 48 hours incubation, *Acinetobacter* BCaL1 and *Stenotrophomonas* BCaL2 strains degraded the initial NPEO by 5% and 8%, respectively (not shown). No surfactant degradation was observed with *Bacillus* VA160 strain. Furthermore, the growth of this strain on acetate was inhibited (by about 75%) by the addition of the NP9EO-NP6EO mixture.

3.4 *Characterization of degradation ability by HPLC analysis*

To better characterize the degradation capabilities of *Acinetobacter* and *Stenotrophomonas* strains, experiments were performed in the presence of 100 mg/l of Igepal CO-520 as the only source of carbon and energy, and NPEO degradation was monitored via HPLC at different incubation times.

HPLC analysis of Igepal CO-520 showed the presence of eight peaks, whose retention time increased

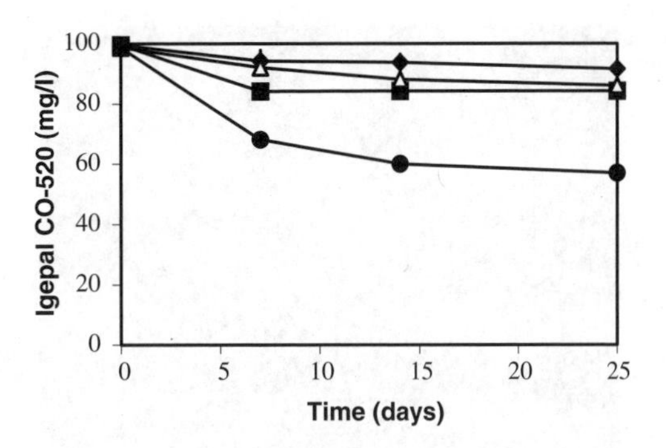

Figure 1. Degradation of Igepal CO-520 (initial concentration = 100 mg/l) by BCaL1 (◆), BCaL2 (△), a co-culture of BCaL1 and BCaL2 (■), a co-culture of BCaL1, BCaL2 and VA160 (●).

with the length of the ethoxyl chain, in agreement with data from Zhou *et al.* (1990).

The results of the biodegradation experiments showed that *Acinetobacter* BCaL1 was capable of degrading 7.9 mg/l (of the initial 100 mg/l) of Igepal CO-520 at the end of the 25-day incubation (calculated as the sum of the area of each peak attributed to components of the mixture), whereas *Stenotrophomonas* BCaL2 degraded 15 mg/l (Fig. 1). The degradation activity was mainly addressed towards the higher ethoxylated components of Igepal CO-520, with the accumulation of lower ethoxylated molecules (not shown). A co-culture of the two strains (cell ratio of 1:1), tested on the same surfactant, showed degradation values similar to those obtained with *Stenotrophomonas* BCaL2 (Fig. 1). To test if the presence of the *Bacillus* VA160 strain could affect the degradation performance of the *Acinetobacter*-*Stenotrophomonas* co-culture, a parallel biodegradation experiment was performed in the presence of all the three strains (cell ratio of 1:1:1). In this case, the overall removal of pollutants was of 42.3 mg/l (Fig. 1), i.e., 3–4 times higher than in the presence of single cultures of each degrading member. Also with the 3 membered co-culture, the specific biodegradation of the higher ethoxylated compounds along with the accumulation of low ethoxylated metabolites were observed. The *Bacillus* strain, as axenic culture, could not degrade Igepal CO-520 (not shown).

Microscopic observation, periodically performed on the cultures, showed the formation of cell aggregates since the second day of incubation which persisted up to the second week of incubation only in the three-membered co-culture (Fig. 2). VA160 cells occurring in this co-culture were well visible as partially empty sheaths protruding from the aggregates. Taken together, these data suggest that *Bacillus* VA160 may facilitate NPEO degradation carried out by *Acinetobacter* BCaL1 and *Stenotrophomonas* BCaL2 strains by stimulating the

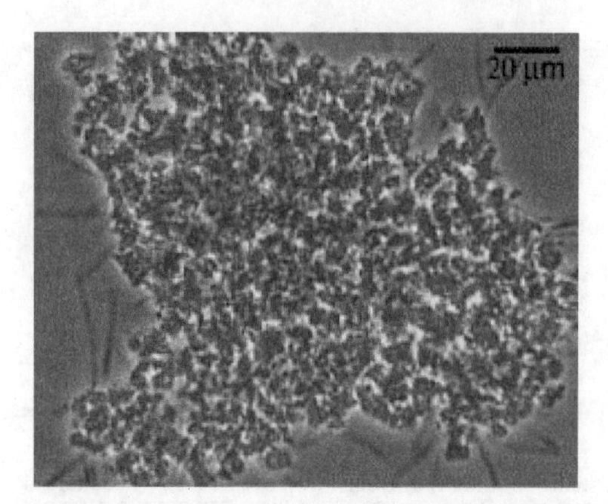

Figure 2. Example of an aggregate developed during growth of the co-culture of the three strains in MMH medium additioned of Igepal CO-520.

formation of cell aggregates within which the surfactant uptake was probably favored.

4 CONCLUSIONS

The overall results show that members of the genera *Acinetobacter* and *Stenotrophomonas* are involved in the degradation of NPEOs, confirming that *Acinetobacter* strains may degrade these surfactants (Barberio & Fani 1998, Barberio *et al.* 2001). There are no reports in the literature regarding alkylphenol ethoxylate degradation by *Stenotrophomonas* strains, although they are involved in the degradation of polypropylene glycol (Kawai 2002). The present data also show that *Bacillus* VA160 facilitates NPEO degradation carried out by *Acinetobacter* BCaL1 and *Stenotrophomonas* BCaL2 strains by probably inducing the formation of cell aggregates of the three strains, thus enhancing the pollutant bioavailability and uptake.

REFERENCES

Altschul S.F., Gish W., Miller W., Myers E.W., Lipman D.J. 1990. Basic local alignment search tool. *J. Mol. Biol.* 215: 403–410.

Ash C., Farrow J.A.E., Wallbanks S., Collins M.D. 1991. Phylogenetic heterogeneity of the genus *Bacillus* revealed by comparative analysis of small subunit ribosomal RNA sequences. *Lett. Appl. Microbiol.* 13: 202–206.

Balson T. & Felix M.S. 1995. In: Biodegradability of Surfactants, Karsa D.R. and Porter M.R. Eds., Blackie Academic & P,GB.

Barberio C. & Fani R. 1998. Biodiversity of an *Acinetobacter* population isolated from activated sludge. *Res. Microbiol.* 149: 665–673.

Barberio C., Pagliai L., Cavalieri D., Fani R. 2001. Biodiversity and horizontal gene transfer in culturable bacteria from activated sludge enriched in nonylphenol ethoxylates. *Res. Microbiol.* 152: 105–112.

Bokern M., Raid P., Harms H. 1998. Toxicity, uptake and metabolism of 4-n-nonylphenol in root cultures and intact plants under septic and aseptic conditions. *Environ. Sci. and Pollut. Res.* 5: 21–27.

Ekelund R., Bergman Å., Granmo Å., Berggren M. 1990. Bioaccumulation of 4- nonylphenol in marine animals. A Re-evaluation. *Environ. Pollut.* 64: 107–120.

Frassinetti S., Isoppo A., Corti A.,Vallini G. 1996. Bacterial attack of non ionic aromatic surfactants: comparison of degradative capabilities of new isolates from nonylphenol polyethoxylate polluted wastewaters. *Environ. Technol.* 17: 199–205.

Frassinetti S., Vallini G., D'Andrea F., Catelani G., Agnolucci M. 2001. Biodegradation of 4-(1-nonyl)phenol by axenic cultures of the yeast *Candida aquaetextoris*: identification of microbial breakdown products and proposal of a possible metabolic pathway. *Int. Biodeter. Biodeg.* 47: 133–140.

Fujii K., Urano N., Ushio H., Satomi M., Iida H., Ushio-Sata N., Kimura S. 2000. Profile of a nonylphenol degrading microflora and its potential for bioremedial applications. *J. Biochem.* Tokio. 128: 909–916.

John D.M. & White G.F. 1998. Mechanism for biotransformation of nonylphenol polyethoxylates to xenoestrogens in *Pseudomonas putida*. *J. Bacteriol.* 180: 4332–4338.

Kravetz L. 1981. Biodegradation of non ionic ethoxylates. *J. Am. Oil Chem. Soc.* 58: 58A–65A.

Maki H., Masnuda N., Fujiwara Y., Ike M., Fujita M. 1994. Degradation of alkylphenols ethoxylates by *Pseudomonas* sp. strain TR01. *Appl. Environ. Microbiol.* 60: 2265–2271.

Murray R.G.E., Doetsch R.N., Robinow C.F. 1994 Determinative and cytological light microscopy. In: Gerhardt P. (ed) *Methods for general and molecular bacteriology* ASM, Washington DC.

Patoczka J. & Pulliam G.W. 1990. Biodegradation and secondary effluent toxicity of ethoxylated surfactants. *Water Res.* 24: 965–972.

Sanger F., Nicklen S., Coulson A.R. 1977. DNA sequencing with chain-terminating inhibitors, *Proc. Natl. Acad. Sci.* USA. 74: 5463–5467.

Soto A.M., Justicia H., Wray J.W., Sonnenschein C. 1991. p-Nonylphenol: an estrogenic xenobiotic released from "modified" polystirene. *Environ. Health Perspect.* 92: 167–173.

Tanghe T., Dhooge W., Verstraete W. 1999. Isolation of a bacterial strain able to degrade branched nonylphenol. *Appl. Environ. Microbiol.* 65: 746–751.

Thibaut R., Debrauwer L., Rao D., Cravedi J.P. 1999. Urinary metabolites of 4-*n*-nonylphenol in rainbow trout (*Oncorhyncus mykiss*). *Sci. Tot. Environ.* 233: 193–200.

Tsuboichi M., Yamasaki N., Yanasisawa L. 1985. Two-phase titration of poly(oxyethylene)nonionic surfactants with Tetrakis(4-fluorophenyl)borate. *Analyt. Chem.* 57: 783–784.

Yoshimura, K. 1986. Biodegradation and fish toxicity of non ionic surfactants. *J. Am. Oil Chem. Soc.* 63: 1590–1596.

Zhou C., Bahr A., Schwedt G. 1990. Separation and determination of non-ionic surfactants of nonylphenol polyglycol ether type by liquid chromatography. *Analyt. Chim. Acta* 236: 273–280.

European Symposium on Environmental Biotechnology, ESEB 2004 - Verstraete (ed)
 ISBN 90 5809 653 X

Degradation of the endocrine – disrupting dimethyl phthalate and dimethyl isophthalate by mangrove microorganisms

J.-D. Gu
Laboratory of Environmental Toxicology, Department of Ecology & Biodiversity, The University of Hong Kong, and The Swire Institute of Marine Science, The University of Hong Kong, Shek O, Cape d'Aguilar, Hong Kong SAR, P.R. China

Y. Wang
Laboratory of Environmental Toxicology, Department of Ecology & Biodiversity, The University of Hong Kong, Hong Kong SAR, P.R. China

J. Li
Laboratory of Environmental and Molecular Microbiology, South China Sea Institute of Oceanography, Guangzhou P.R. China

ABSTRACT: Degradation of *ortho*-dimethyl phthalate (*o*-DMP) and dimethyl isophthalate (DMI) was investigated using microorganisms isolated from mangrove sediment in Hong Kong. One enrichment culture was capable of utilizing *o*-DMP as the sole source of carbon and energy, but none of the bacteria in the enrichment culture was capable of degrading *o*-DMP alone. In co-culture of two bacteria, degradation was observed and one bacterium in the culture was a new species unkown before. Biochemical utilization profile of this bacterium is limited to malate, acetate and phenyl acetate only. Complete degradation of *o*-DMP took place in less than 4 days at 400 mg/L. Degradation proceeds through monomethyl phthalate ester and phthalic acid before the aromatice ring opening. Using DMI as the sole carbon and energy source, *Klebsiella oxytoca* Sc and *Methylobacterium mesophilicum* Sr were isolated for enrichment culture. DMI was found to be degraded through the biochemical cooperation between the two species and the initial hydrolytic reaction of the ester bond by *K. oxytoca* Sc and the next step of transformation was by *M. mesophilicum* Sr, and IPA was degraded by both of them. Our data suggest that the plasticizer phthalate esters can be mineralized by consortia of bacteria and biochemical cooperation between bacteria plays an important role in the ecological community of the natural environment.

1 INTRODUCTION

Phthalate esters (PE) are widely used as plasticizers and additives in plastics manufacturing to improve mechanical properties of the plastic resin, particularly flexibility and softness (Giam *et al.*, 1978; Krauskopf, 1992; Sommer, 1985). It is typically used in cellulose ester-based plastics, such as cellulose acetate and cellulose butyrate (Staples *et al.*, 1997) and polyvinyl chloride (Edenbaum, 1992; Giam *et al.*, 1984). Additionally, dimethyl phthalate esters (DMP) are also used as a component of solvents, repellents, floatation reagents, and in the production of cosmetics, lubricants, carpeting, decorative cloths, and other products (Baikova *et al.*, 1999).

Due to the global utilization of plasticized polymers at large quantities, phthalates and their esters have been detected in all environmental samples (Bauer & Herrmann, 1997; Giam *et al.*, 1984). High concentrations were documented in landfill leachate, 300 μg/L DMP, 18,900 μg/L phthalic acid (PA), 6 μg/L monomethyl phthalate (MMP) (Mersiowsky, 2002). DMP is suspected to be responsible for functional disturbances in the nervous system and liver of animals. Meanwhile, it may also promote chromosome injuries in human leucocytes. Therefore, the US Environmental Protection Agency has listed them as priority pollutants (US EPA, 1992). Known for endocrine-disrupting activity, PEs also interfere with the reproductive system and normal development of animals and humans (Jobling *et al.*, 1995; Gray *et al.*, 1999).

Microorganisms are responsible for the complete destruction of PA and phthalate esters in the environments (Eaton, 2001; Fan *et al.*, 2004; Juneson *et al.*, 2001; Kim *et al.*, 2002; Wang *et al.*, 2003a, b). Degradation of this class of chemicals has been investigated under aerated conditions (Stahl and Pessen 1953; Sugatt *et al.*, 1984; Roslev *et al.*, 1998; Wang *et al.*, 2003a, b) and anoxic conditions (Kleerebezem *et al.*, 1999). Degradation of phthalate esters have not been reported on the microorganisms involved nor the biochemical processes. Selective bacteria have been know for their degradation of phthalate acid, but the phthalate esters are more complex in structures and possibly the degradative processes. However, degradation of DMP by pure species of microorganisms has been largely neglected, and few studies have been focused on the application of immobilized cells in degradation of DMP for potential bioremediation applications.

2 MATERIALS AND METHODS

2.1 *Microorganisms and culturing conditions*

The initial bacterial culture was established by adding 1.0 g of wet mangrove sediment taken from Mai Po Nature Reserve of Hong Kong into 100 ml of a mineral salt medium (MSM) in a 250 ml Erlenmeyer flask with *ortho*-DMP or DMI (starting concentration 100 mg/l) as the sole source of carbon and energy. Similar approach was reported elsewhere (Fan *et al.*, 2003; Wang *et al.*, 2003b, c). The MSM consisted of the following chemicals (mg/l): $(NH_4)_2SO_4$ 1,000, KH_2PO_4 800, K_2HPO_4 200, $MgSO_4 \cdot 7H_2O$ 500, $FeSO_4$ 10, $CaCl_2$ 50, $NiSO_4$ 32, $Na_2BO_7 \cdot H_2O$ 7.2, $(NH_4)_6Mo_7O_{24} \cdot H_2O$ 14.4, $ZnCl_2$ 23, $CoCl_2 \cdot H_2O$ 21, $CuCl_2 \cdot 2H_2O$ 10 and $MnCl_2 \cdot 4H_2O$ 30, and the initial pH of the culture medium was adjusted with HCl or NaOH to 7.0 ± 0.1.

The Erlenmeyer flasks were incubated in an INNOVA 4340 Incubator Shaker (New Brunswick Scientific, New Jersey, USA) kept at 150 rpm and 30.0 ± 0.5°C in the dark. The initial DMP- or DMI-degrading enrichment was transferred approximately once a week on the basis of depletion (more than 85%) of the substrate. At each transfer, 1.0 ml of the established culture was transferred to a new flask containing 90 ml of freshly made MSM with gradual increase of DMP or DMI concentrations (from 100 to 500 mg/l). The established cultures were further transferred 8 times prior to be used in the subsequently isolation of bacteria as described earlier (Wang *et al.*, 2003a, b).

2.2 *Degradation conditions*

All biodegradation experiments were carried out using the consortium cultures, pure isolates, and re-constituted consortia of the isolates from either of the substrate at 30°C in the dark. DMP or DMI biodegradation experiments were conducted in 250 ml flasks with 100 ml MSM (as described above), different concentrations of DMP or DMI and the re-constituted consortium. Tests were conducted in triplicate. Periodically culture aliquot (2 ml) from each flask were taken aseptically and stored frozen (−20°C) in a glass vial until analyzed. Sterile controls were prepared by autoclaving for 20 min before introduction of the substrate which had passed through 0.2-μm-pore-size membrane filter on a syringe (Pall Gelman Laboratory, Ann Arbor, Michigan).

2.3 *Identification of bacteria*

Bacteria in the enrichment cultures were spread on agar plates containing the appropriate DMP or DMI as the sole source of carbon and energy. After incubation, the colonies developed on the agar plates were further streaked on fresh agar plates for purification. When pure isolates were obtained, they were subjected to Gram staining and then API identification based on biochemical reaction for Gram negative isolates.

2.4 *Chemical analysis by HPLC*

In preparation for HPLC analysis, thawed culture samples from DMP or DMI degradation experiments were centrifuged (12,000 × *g*) and filtered through PVDF or Nylon Acrodisc Minispike syringe filters (0.2-μm-pore-size) (Pall Gelman Laboratory, Ann Arbor, Michigan). The first five drops were discard-ed to avoid the potential influence of phthalates or metabolites adsorption onto membranes. DMP or DMI in samples was separated and quantified on an Agilent 1100 series HPLC system (Agilent Techno-logies, Hewlett-Packard, California) consisting of a quaternary low-pressure degasser, a quaternary high-pressure pump, a model 7725*i* manual sample injector with a 20 μl sample loop, and diode array and multiple wavelength detectors. Separation of DMP and metabolites was accomplished by using a 4.6 × 150 mm Eclipse 5-μm XDB-C8 reversed-phase liquid chromatography column (Agilent Technologies). Methanol-water (50:50, v/v) delivered at a flow rate of 1.0 ml/min was used as the mobile phase in the HPLC analysis of DMP and mono-methyl phthalate (MMP). PA, MMP and DMPE were quantified by the external standards method at wavelength of 280 nm.

The mobile phase for DMI and related metabolites quantification consisted of (A) H_2O containing 0.12 mol/L ammonium acetate and (B) methanol. Gradients were as follows: 0–6 min B was held at 20%, 6–10 min B increased from 20% to 60%, B was held at 60% for 10 min, and B decreased to 20% in 6 min.

The calibration curves were linear for these compounds in the range from 10 to 1000 mg/l. The

UV-visible spectra were recorded at identical retention time for a particular chemical to confirm the identification of the compound.

2.5 *Microbial biomass*

Bacterial biomass in the culture aliquot was measured spectrophotometrically using at 600 nm.

3 RESULTS AND DISCUSSION

3.1 *Bacterial growth on DMP*

Mai Po Nature Reserve of Hong Kong is the largest intertidal wetland with mangrove and soft mud flat. This area is important to the conservation of local diversity of flora and fauna, and between 60,000 to 80,000 migratory and water birds use this site for staging and wintering. Due to the intensified industrialization and residential building development around the wetland, wastewater and industrial water have contributed to elevated level of metals and persistent organic compounds.

The bacteria used in this study was initially enriched and isolated from enrichment culture, and then re-constituted from pure species of microorganisms of the cultures showing capability of degrading *o*-DMP. Since none of the dominant bacterial species was capable of degrading DMP alone and re-constitution of them as a co-culture was attempted to assess the ability of consortium in degrading DMP. The re-constituted consortium was able to degrade the substrate and the consortium consisted of a *Rhodococcus* species and a new species currently under characterization. It is shown in Fig. 1 that 400 mg/l DMP were completely degraded by the consortium within 4 days. The concentrations of DMP in sterilized controls were constant during the experiments (data not shown), indicating that possible chemical hydrolysis and photolysis on degradation of DMP were negligible under current systems. At the same time, microbial biomass showed increased as substrate was utilized.

3.2 *Bacterial utilization of DMI*

Degradation of DMI was also observed in the enrichment culture using DMI as the sole source of carbon and energy and mangrove sediment as the source of microorganisms. From the enrichment culture after at least 5 enrichment transfers, two species of bacteria were isolated and identified as *Klebsiella oxytoca* Sc and *Methylobacterium mesophilicum* Sr. Similar to DMP degradation, DMI could not be degraded by either of the two isolates alone to the completion of mineralization. *K. oxytoca* Sc was able to transform DMI to mono-methyl isophthalte and the intermediate accumulated in the culture medium without further decrease over extended period of incubation. On the other hand, *M. mesophilicum* Sr was able to degraded the inter-mediate quickly to isophthalate. Furthermore, both species were shown to metabolize IPA at similar rate.

It appears that completed degradation of the phthalate esters require biochemical cooperation between different species of bacteria from community of the natural environment.

3.3 *Biochemical pathways of degradation*

Two major intermediates of biochemical transformation of DMP were identified as MMI and PA, respectively, by a combination of methods. Analysis of the culture aliquot on HPLC revealed that DMI was completely hydrolyzed with subsequent formation of MMI and IPA sequentially, which can be mineralized to CO_2 and H_2O. Transformation of DMI requires the participation of *K. oxytoca* Sc first forming monomethyl isophthalate and then *M. mesophilicum* Sr forming

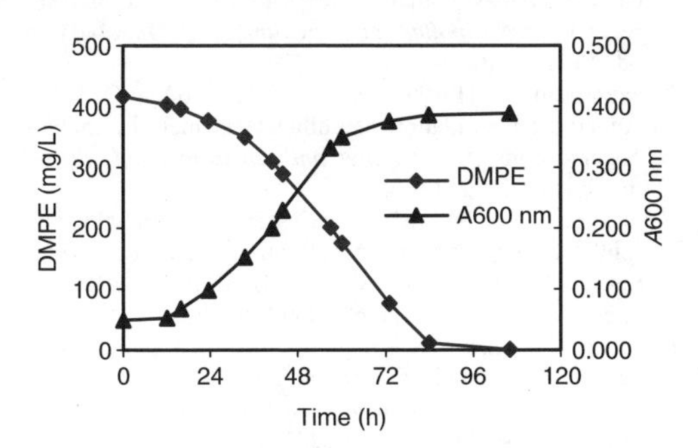

Figure 1. Degradation of dimethyl phthalte ester by a mixed culture of bacteria enriched from mangrove sediment and increase of microbial population as measured by optical density.

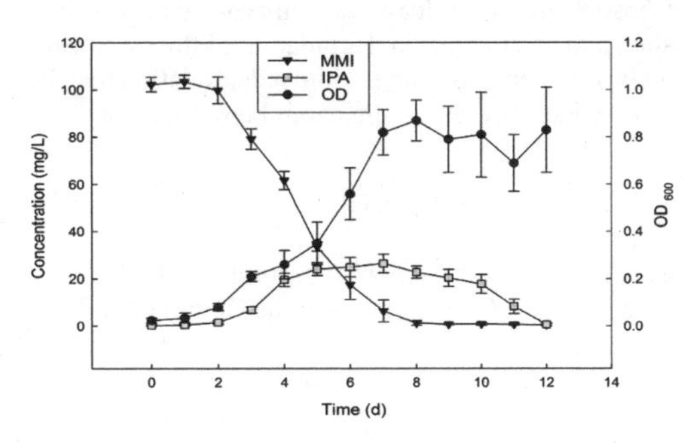

Figure 2. Microbial degradation of dimethyl isophthalte ester by a co-culture consisting of *Klebsiella oxytoca* Sc and *Methylobacterium mesophilicum* Sr isolated from mangrove sediment and simultaneous increase of their population.

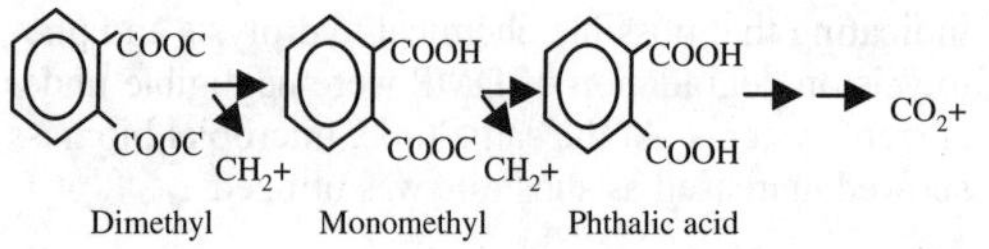

$COOC$ / $COOC$ → $COOH$ / $COOC$ + CH_2+ → $COOH$ / $COOH$ + CH_2+ → → CO_2+

Dimethyl Monomethyl Phthalic acid

Figure 3. Biochemical degradation pathway of ortho-dimethyl phthalte ester.

isophthalate acid (IPA). Interestingly, both species were equally capable of utilizing IPA. Similar biochemical process was also confirmed early using an enrichment culture obtained from activated sludge (Niazi *et al.*, 2001; Wang *et al.*, 2003a). When pure cultures of bacteria were isolated, they were not capable of degrading the phthalate ester by individual species, a combina-tion of them was observed to completely degrade the substrate (Wang *et al.*, 2003b). Phthalate acid is degradable by single species and *Comamonas acidovoran* fy-1 mineralized this substrate at concentration as high as 20,000 mg/l (Fan *et al.*, 2004).

The observed biochemical cooperation between two species of bacteria may have important ecological and evolutionary significance because the substrate concentration is generally low in natural environment and multiple species may interact closely to achieve the fully utilization of a particular carbon source. In addition, esterase involved in the initial steps of hydrolysis seem to be selective for the two different microorganisms implying that the two steps involve different esterases for DMI and MMI. It will be of great interest to investigate the esterases involved and distinguish the difference between the two isozyme and possible the evolutionary relationship between them.

4 CONCLUSIONS

Two enrichment cultures were made on DMP and DMI, but the individual bacterium from either of the cultures was not able to metabolize the respective substrate. Consortium of at least two micro-organisms was shown to be effective in degradation of the two chemicals. Biochemical cooperation in degradation may be more widely present in the natural environment.

ACKNOWLEDGEMENTS

This research was supported partially by ITS/260/00 of Hong Kong Government and the Chinese Academy of Sciences.

REFERENCES

Baikova, S.V., Samsonova, A.S., Aleshchenkova, Z.M. & Shcherbina, A.N. 1999. The intensification of dimethyl-phthalate destruction in soil. *Eurasian Soil Science* 32: 701–704.

Bauer, M.J. & Herrmann, R. 1997. Estimation of the environmental contamination by phthalic acid ester leaching from household wastes. *Science of the Total Environment* 208: 49–57.

Eaton, R.W. 2001. Plasmid-encoded phthalate catabolic pathway in *Arthrobacter keyseri* 12B. *Journal of Bacteriology* 183: 3689–3703.

Edenbaum, J. 1992. Polyvinyl chloride resins and flexible-compound formulating. In: J. Edenbaum (ed.), *Plastics Additives and Modifiers Handbook*: 17–41. New York: van Nostrand Reinhold.

Fan, Y., Wang, Y., Qian, P.-Y. & Gu, J.-D. 2004. Optimization of phthalatic acid batch biodegradation and the use of modified Richards model for modelling degradation. *International Biodeterioration & Biodegradation* 53: 57–63.

Giam, C.S., Chah, H.S. & Neff, G.S. 1978. Phthalate ester plasticizers: a new class of marine pollutants. *Science* 199: 419–421.

Gray, L.E., Wolf, C., Lambright, C., Mann, P., Price, M., Cooper, R.L. & Ostby, J. 1999. Administration of potentially antiandrogenic pesticides (procymidone, linuron, iprodione, chlozolinate, *p, p'*-DDE, and ketonazole) and toxic substances (dibutyl- and diethylhexyl phthalate, PCB 169, and ethane dimethane sulphonate) during sexual differentiation produces diverse profiles of reproductive malformations in the male rat. *Toxicology and Industrial Health* 15: 94–118.

Jobling, S., Reynolds, T., White, R., Parker, M.G. & Sumpter, J.P. 1995. A variety of environmentally persistent chemicals, including some phthalate plasticizers, are weakly estrogenic. *Environmental Health Perspectives* 103 (Suppl. 7): 582–587.

Juneson, C., Ward, O.P. & Singh, A. 2001. Biodegradation of bis(2-ethylhexyl)phthalate in a soil slurry-sequencing batch reactor. *Process Biochemistry* 37: 305–313.

Kawai, F. & Enokibara, S. 1996. Symbiotic degradation of polyethylene glycol (PEG) 20,000-phthalate polyester by phthalate ester- and PEG 20,000-utilizing bacteria. *Journal of Fermentation and Bioengineering* 82: 575–579.

Kim, Y-H., Lee, J., Ahn, J-Y., Gu, M.B. & Moon, S-H. 2002. Enhanced degradation of an endocrine-disrupting chemical, butyl benzyl phthalate, by *Fusarium oxysporum* f. sp. *pisi* Cutinase. *Applied and Environmental Microbiology* 68: 4684–4688.

Kleerebezem, R., Hulshoff Pol, L.W. & Lettinga, G. 1999. Anaerobic degradation of phthalate isomers by methanogenic consortia. *Applied and Environmental Microbiology* 65: 1152–1160.

Krauskopf, L.G. 1992. Monomerics for polyvinyl chloride (phthalate, adipate and trimellitakes). In: J. Edenbaum (ed.), *Plastics Additives and Modifiers Handbook*: 359–378. New York: van Nostrand Reinhold.

Niazi, J.H., Prasad, D.T. & Karegoudar T.B. 2001. Initial degradation of dimethylphthalate by esterases from *Bacillus* species. *FEMS Microbiology Letters* 196: 201–205.

Roslev, P., Madsen, P.L., Thyme, J.B. & Henriksen, K. 1998. Degradation of phthalate and di-(2-ethylhexyl) phthalate by indigenous and inoculated microorganisms in sludge-amended soil. *Applied and Environmental Microbiology* 64: 4711–4719.

Sommer, W. 1985. Plasticizers. In: R. Gächter & H. Müller (eds), *Plastics Additives Handbook* 251–296. New York: Hanser Publishers.

Stahl, W.H. & Pessen, H. 1953. The microbial degradation of plasticizers. *Applied Microbiology* 1: 30–35.

Staples, C.A., Peterson, D.R., Parkerton, T.F. & Adams, W.J. 1997. The environmental fate of phthalate esters: a literature review. *Chemosphere* 35: 667–749.

Sugatt, R.H., O'Grady, D.P., Banerjee, S., Howard, P.H. & Gledhill, W.E. 1984. Shake flask biodegradation of 14 commercial phthalate esters. *Applied and Environmental Microbiology* 47: 601–606.

US EPA. (1992 and update). *Code of Federal Regulations*, 40 CFR, Part 136, Washington DC.

Wang, Y., Fan, Y. & Gu, J.-D. 2003. Aerobic degradation of phthalic acid by *Comamonas acidovoran* fy-1 and dimethyl phthalate ester by two reconstituted consortia from sewage sludge at high concentrations. *World Journal of Microbiology & Biotechnology* 19: 811–815.

Wang, Y., Fan, Y. & Gu, J.-D. 2003. Microbial degradation of the endocrine-disrupting chemicals phthalic acid and dimethyl phthalate ester under aerobic conditions. *Bulletin of Environmental Contamination and Toxicology* 71: 810–818.

European Symposium on Environmental Biotechnology, ESEB 2004 - Verstraete (ed)
© 2004 Taylor & Francis Group, London, ISBN 90 5809 653 X

Quantification of microbial diversity in biofilms using denaturing gradient gel electrophoresis profile

H.H.P. Fang & T. Zhang
Environmental Biotechnology Lab, Center for Environmental Engineering,
The University of Hong Kong, Pokfulam Road, Hong Kong SAR, China

ABSTRACT: Microbial diversity affects thickness, structural complexity and ecological stability of biofilms. This study was conducted to demonstrate that diversity of biofilms might be quantified by either of two indices, Shannon-Weaver and Simpson, based on the DGGE (denaturing gradient gel electrophoresis) profiles of DNA extracted from the communities. Using such a method, the diversity indices of anaerobic marine biofilms were estimated, and compared with those of the corresponding microbial communities in suspension. The DGGE profiles of DNA revealed the presence of 22 species in both communities. The Shannon–Weaver and Simpson indices of the biofilms averaged 2.69 and 0.19, respectively. The former was about 9% lower and the latter was about 16% higher than the corresponding indices of the suspended communities. Results in both cases show that the suspended communities were more diverse than the biofilms. Details of the biofilm and suspended communities, derivations of both indices and implications of these results will be discussed.

1 INTRODUCTION

The microbial species diversity means the different types of microorganism species and their relative abundance in an ecosystem. Biodiversity will affected the stability of the biofilm microecosystem (Atlas and Bartha, 1998), biofilm microstructure, biofilm thickness (Christensen and Characklis, 1990) and will change following the successional shift of the biofilm community. Measures of species diversity might be simplified into the single indices using species relative abundance.

Previously the method to obtain the information on species relative abundance is based on culture, isolation and physiological/biochemical tests. This time-consuming method has biases due to the culture process. Recently, the molecular identification methods based on 16S rDNA have been extensively applied in the microbial community studies and revealed a lot of new species not been found using the culture depend methods.

In this study, microbial species diversity of biofilm and the suspended community was investigated using DGGE and cloning-sequencing method. The biodiversity index was calculated based on the DGGE profile and cloning–screening results. The difference between the biodiversity of biofilm and suspended community was discussed.

2 MATERIALS AND METHODS

2.1 *Culture of biofilm*

The biofilm was cultured on the mild steel coupons inside the anaerobic glass tank at the room temperature (20–22°C) as described elsewhere (Zhang and Fang, 2000). Samples of the microbial seed described above, biofilm on the coupons in the test and the suspended bacteria in the same test tank, were collected for microbial community analysis.

2.2 *DNA extraction, PCR and DGGE analysis*

Genomic DNA was extracted from samples of the microbial seed (MS) described above, biofilm on the coupons in the test (B_{5d}, B_{10d}, B_{20d}, B_{40d}, B_{60d}, B_{90d}) and the suspended bacteria (S_{20d} and S_{90d}) in the same test tank as described previously (Zhang and Fang, 2000).

Extracted genomic DNA was amplified using PCR primer set of the *Eubacteria* domain specific forward primer, 968F(968–983) with a 40-bp GC clamp, and the universal reverse primer 1392R using GenAmp® PCR system 9700 (Perkin Elmer Ltd., Foster City, USA).

DGGE was performed following the method of Muyzer et al. (1993) using a denaturant gradient of 40%–60% for 5 hr at 200 V and 60°C. The gel was

Table 1. Relative height of peaks on DGGE profile curve.

OTUs	MS	S_{20d}	S_{90d}	B_{5d}	B_{10d}	B_{20d}	B_{40d}	B_{60d}	B_{90d}
1						4	5	2	
2	3		5						
3	5	7							
4			8		3				
5	4								
6			14				5		
7		2	2				6	2	4
8	6	8	2	7	4	8	7	10	9
9		8							
10	18			8	3	8		9	4
11				8			6		
12	13	12	12	6	9	10	7	7	8
13	5		9					4	8
14	22	24	19	24	17	15	15	7	7
15		8							
16	26	6	19	20	20	19	23	21	24
17									2
18				3	3				
19	16								
20			2		3			2	
21	2			3	3			2	2
22		4							
Σhi	120	79	92	79	65	64	74	66	68

stained with silver nitrate. The band positions and intensities were recorded by an image analyzer (Q600S, Leica) with a digital CCD camera. The band positions and intensities were analyzed using a software (Quantimet Q600) provided by the manufacturer (Leica). The bands, or peaks, positions, in the DGGE profile corresponded to individual microbial identities, called OTUs, and the band intensities, i.e. peak heights, represented their relative quantities.

2.3 *Biodiversity indices*

The Shannon–Weaver index of general diversity, H (Shannon and Weaver, 1963) might be calculated using the following function:

$$H = -Pi \times [\log_2 (Pi)]$$
$$= -(hi/\Sigma hi) \times [\log_2 (hi/\Sigma hi)]$$

where Pi is the abundance of one species in a community, hi is the height of one peak (i) in the profile curve. Bigger value of H indicates higher species diversity.

The Simpson index, S (Simpson, 1949) was calculated using the following function:

$$S = \Sigma Pi^2$$

Lower value of S indicates higher species diversity.

Shannon–Wiener's diversity index reflects the general proportional abundance of the different species among the communities while Simpson's index indicates the general pattern of species abundance (from the most to the least dominant) in the communities.

3 RESULTS AND DISCUSSIONS

3.1 *The DGGE profile and cloning–screening results*

According the DGGE image, the relative height of the OTUs revealed on DGGE curve was shown in the Table 1. Totally there are 22 OTUs from all the samples, some shared by all the samples and the others appeared in one or more samples.

3.2 *Species diversity index*

The species diversity indexes calculated using DGGE profile were shown in Table 2. Based on T test, the average H value of the suspended cultures is significantly higher than that of the biofilm communities ($\alpha = 0.1$) and the average S value of the suspended cultures is remarkable lower than that of the biofilm communities ($\alpha = 0.01$). This indicated that the suspended cultures have higher species diversity than the biofilm communities.

Theoretically, the diversity should increase from low diversity of the pioneer populations to the higher and stable diversity of the climax community (Atlas and Bartha, 1998). However, there is no significant change in the species diversities of the samples from 5 day to

Table 2. Microbial species diversity index.

Suspended culture					Biofilm						
Index	MS	S_{20d}	S_{90d}	Mean	B_{5d}	B_{10d}	B_{20d}	B_{40d}	B_{60d}	B_{90d}	Mean
H	3.07	2.88	2.96	2.92 ± 0.09	2.64	2.69	2.43	2.75	2.87	2.76	2.69 ± 0.15
S	0.14	0.16	0.15	0.16 ± 0.012	0.19	0.20	0.20	0.18	0.17	0.19	0.19 ± 0.011

90 day. Such result might be due to two reasons: relative simple substrate and the biofilm still in growing. In other words, the general rule does not cover the ecosystem using simple substrate and the biofilm has not reached climax status in 90-day incubation.

It should be pointed out that the DGGE method might underestimate the species diversity. Some weak band cannot be viewed or detected. The single band in the gel might contain two very close fragments depending on the gel separation capacity although it was reported that DGGE might differentiate two fragments with difference of one base pair.

REFERENCE

Atlas R.M. and Bartha R. 1998. *Microbial ecology: fundamentals and applications*. 4th ed. Calif, Menlo Park: Benjamin/Cummings.

Christensen B.E. and Characklis W.G. 1990. Physical and chemical properties of biofilms. In: Characklis W.G. & Marshall K.C., *Biofilms*, New York: Wiley, pp. 93–129.

Muyzer G., de Waal E.C. and Uitterlinden A.G. 1993. Profiling of complex microbial population by DGGE analysis of polymerase chain reaction amplified genes encoding for 16S rRNA. *Appl. Environ. Microbiol.* 62: 2676–2680.

Simpson E.H. 1949. Measurement of diversity. *Nature.* 163: 688.

Zhang T. and Fang H.H.P. 2000. Digitization of DGGE (denaturing gradient gel electrophoresis) profile and cluster analysis of microbial communities. *Biotechnol. Let.* 22(5): 399–405.

Innovative in situ remediation

European Symposium on Environmental Biotechnology, ESEB 2004 - Verstraete (ed)
© 2004 Taylor & Francis Group, London, ISBN 90 5809 653 X

Monitoring sulfate-reducing bacteria (SRB) using molecular tools during *in situ* immobilization of heavy metals

J. Geets, B. Borremans, K. Vanbroeckhoven & L. Diels
Vito (Vlaamse Instelling voor Technologisch Onderzoek), Mol, Belgium

D. Van Der Lelie
Brookhaven National Laboratory, Upton, NY, USA
Vito (Vlaamse Instelling voor Technologisch Onderzoek), Mol, Belgium

J. Vangronsveld
Limburgs Universitair Centrum, Diepenbeek, Belgium

ABSTRACT: *In situ* bioprecipitation (ISBP) of heavy metals as metal sulfides by the activity of sulfate-reducing bacteria (SRB) is a promising strategy for sustaining groundwater quality. The feasibility of the ISBP process for a Zn, Cd, Ni and Co-polluted industrial site was studied in column experiments, using lactate, molasses or HRC® as supplemented electron donor and C-source. In addition, the effect of extra N/P was evaluated. In order to link the results of analytical analyzes to SRB-community composition and activity, the SRB population was monitored by molecular tools such as SRB-subgroup or – genus specific PCR and DGGE of the 16S rRNA- and *dsr* (dissimilatory sulfite reductase) gene.

1 INTRODUCTION

Sulfate-reducing bacteria (SRB) are known for their capacity to reduce and precipitate heavy metals (HM) as metal sulfides, which form stable precipitates due to their low solubility product. The activity of SRBs has potential for the creation of a bioreactive zone or barrier for the *in situ* precipitation of heavy metals as a remediation strategy for HM contaminated groundwater. Before going into the pilot scale application of *in situ* bioprecipitation (ISBP) at a HM contaminated site, the feasibility of the process was evaluated in column experiments. However, in order to optimize the ISBP process, an insight is needed in the composition and activity of the SRB-populations, as well as information on the way they are effected by process conditions, such as the added type of C-source/electron donor, or by other prokaryotes (e.g. fermenting bacteria, methane producing *Archaea*, acetogens). This can be done by combining analytical analyzes with molecular tools such as SRB-specific PCR, DGGE (Denaturing Gradient Gel Electrophoresis), and cloning and sequencing, based on either the 16S rRNA-gene or the *dsr* (dissimilatory sulfite reductase)-gene. The *dsr*-gene encodes for the DSR-enzyme, which catalyzes the reduction of sulfite to sulfide during sulfate respiration. Up to date, it was demonstrated that the ISBP process was dependent on the presence and activity of strains belonging to the SRB-genus *Desulfosporosinus*. However, the *dsr*-approach also indicates a role for strains belonging to the SRB-genus *Desulfomicrobium*.

2 MATERIALS AND METHODS

2.1 *Set-up of column experiments*

The study site is at the precincts of a non-ferrous industry in Olen (Belgium). Its groundwater is contaminated with the heavy metals Zn (444 mg/l), Cd (53.5 mg/l), Ni (144 mg/l) and Co (31.2 mg/l), and has a naturally high Fe-concentration (34.5 mg/l). Sulfate was present in a concentration of 1400 mg/l. Groundwater's pH was rather acidic (pH = 3,9).

Plexiglas columns (4 cm diameter, 50 cm height) were filled with aquifer under anaerobic conditions. Groundwater was mixed with nutrient solutions before being pumped into the columns. Inwards flow was 25 ml/day, resulting in a hydraulic retention time of 12.5 days. Added nutrients were lactate, lactate plus N/P, molasses and HRC® (Hydrogen Release Compound®,

Table 1. 16S rRNA-gene targeting PCR primers for SRB-subgroups and – genera.

Primer	Annealing temp. (°C)	Sequence 5′-3′	Specificity
DSV II 230F*	63	GAGYCCGCGTYYCATTAGC	*Desulfovibrio* sp., *Desulfomicrobium* sp.
DSV II 838R*	63	CCGACAYCTARYATCCATC	*Desulfovibrio* sp., *Desulfomicrobium* sp.
DSM172F**	64	AATACCGGATAGTCTGGCT	*Desulfomicrobium* sp.
DSM1469R**	64	CAATTACCAGCCCTACCG	*Desulfomicrobium* sp.
DFM 140F***	58	TAGMCYGGGATAACRSYKG	*Desulfotomaculum* sp., *Desulfosporosinus* sp.
DFM 842R***	58	ATACCCSCWWCWCCTAGCAC	*Desulfotomaculum* sp., *Desulfosporosinus* sp.
DSP140F****	60	AAAKCCGGGACAACCCTTG	*Desulfosporosinus* sp.
DSP1107R*****	60	CTAAAYACAGGGGTTGCG	*Desulfosporosinus* sp.
DSB 127F***	62	GATAATCTGCCTTCAAGCCTGG	*Desulfobacter* sp.
DSB II 1273R*	62	CYYTTTGCRRAGTCGCTGCCCT	*Desulfobacter* sp.
DBM 169F***	64	CTAATRCCGGATRAAGTCAG	*Desulfobacterium* sp.
DBM 1006R***	64	ATTCTCARGATGTCAAGTCTG	*Desulfobacterium* sp.
DBB II 121F*	66	CGCGTAGATAACCTGTCTTCATG	*Desulfobulbus* sp.
DBB II 1237R*	66	GTAGTACGTGTGTAGCCCTGGTC	*Desulfobulbus* sp.
DCC 140F****	65	CTRCCCYYGGATYSGGGATAAC	*Desulfococcus* sp., *Desulfonema* sp., and *Desulfosarcina* sp.
DCC 1273R****	65	CTYRCTCTCGCGAGYTCGCTACCCT	*Desulfococcus* sp., *Desulfonema* sp., and *Desulfosarcina* sp.

* Modified from Daly [Daly, 2000 #19]; ** [Loy, 2002 #47]; *** [Daly, 2000 #19]; **** This study; ***** Based on probe DFMII1107 [Loy, 2002 #47].

Regenesis). All electron donors were fed to columns with a final concentration of 0.02% C (w/v). For the estimation of the N/P ratio, the C : N : P ratio in the cell wall of anaerobic bacteria was taken into account (500 : 10 : 1 mol ratio). NH_4Cl was added as N-source and KH_2PO_4 as P-source. One column didn't receive any nutrients in order to investigate if HM precipitation took place without external stimulation. In addition, an abiotic control was created by poisoning the microbial population with formaldehyde.

2.2 *Analytical and molecular analyzes*

Samples of column in-and effluent were collected in function of time and analyzed for pH, DOC, DC, DIC, sulfate concentration, and concentrations of soluble Zn, Ni, Cd, Co and Fe. In addition, formation of methane and hydrogen gas in the column's effluent was followed.

Aquifer samples of columns were collected at different time points at 10 cm and 30 cm from the column's inlet. After extraction of total community's DNA, a rapid screening of the SRB population was performed using SRB-subgroup and – genus specific PCR primers for the 16S rRNA gene, which were either modified from Daly (Daly et al., 2000), or adapted from Loy (Loy et al., 2002), or newly designed in this study (Table 1).

During the column experiments, changes in bacterial community structure and diversity were monitored by DGGE fingerprinting of 16S rRNA gene fragments which were PCR-amplified using universal eubacterial primer pair GC40-63F and 518R (El-Fantroussi et al., 1999). In this way, both the SRB and other, non-sulfate reducing bacterial groups were analyzed. In addition, DGGE-analysis of the β-subunit of the *dsr*-gene enabled specific follow-up of the SRB community. For this purpose, a new forward PCR primer was designed, based on an internal *dsr* sequencing primer described by Pérez-Jiménez (Pérez-Jiménez et al., 2001), namely DSR p2060F (5′-CAACATCGTYCAYACCCAGGG-3′), to which a 40 bp GC-clamp was ligated to the 5′ end. By combining this forward primer with the DSR 4R reverse primer (Wagner et al., 1998) an approximately 400 bp *dsrB*-gene fragment was obtained for DGGE-analysis. DGGE fingerprint patterns of different column conditions were compared with each other and in function of time by cluster analysis, using the Pearson correlation coefficient and UPGMA clustering algorithm of the Bionumerics program (Bionumerics Version 1.01, Applied Maths, Belgium). Excised 16S rDNA DGGE-bands and *dsr*-PCR products were cloned into the PCR®2.1-TOPO® plasmid vector and *Escherichia coli* TOP10 cells using TOPO-TA cloning vector kit (Invitrogen). Clones containing the appropriate insert were selected for sequence analysis which was performed by Westburg Genomics (Westburg, the Netherlands). The obtained DNA-sequences were submitted to GenBank for preliminary identification using the program BLASTN

2.2.4 (http://www.ncbi.nlm.nih.gov/BLAST) (Altschul et al., 1990) of the Ribosomal Database Project to identify putative close phylogenetic relatives (Maidak et al., 1994).

3 RESULTS AND DISCUSSION

3.1 *Sulfate- and metal removal*

As an example of heavy metal-removal in function of time, Ni-removal efficiencies for the different column set-ups are given (Fig. 1). An overview of sulfate-removal efficiencies in is given in Fig. 2.

Overall, analytical results led to the conclusion that sulfate-consumption and heavy metal-removal were most efficient for the columns where lactate or lactate plus N/P was added. Removal efficiencies of these set-ups increased with a comparable slope and reached more or less the same percentages. For Zn and Co, removal efficiency remained constant as soon as it has reached a value of >95%. For Ni, the removal efficiency was between 85% and 90% at the end of the experiment, but this value had an increasing trend. The decrease in HM-concentration was accompanied by a decrease in sulfate concentration.

However, during the first weeks of the experiment, it seemed as if the set-up with molasses was going to be the most optimal condition for ISBP since HM removal efficiencies and sulfate-consumption rates were higher and increased faster than for the other set-ups. But after 13 weeks, the sulfate- and heavy metal removal-efficiency suddenly dropped. Moreover, heavy metals seemed to be released: after 15 weeks, the concentration of Zn, Co and Ni in the column's effluent was higher than in the influent. The same observation was made for the column fed with HRC®. After 10 weeks, the column reached removal efficiencies for Zn and Co of 95%, and 98% for Ni. At the end of the experiment (i.e. after 26 weeks), there was a sudden decrease in both metal removal efficiency and sulfate-removal.

The reason for this sudden ISBP insufficiency is not clear yet. It was first thought to be a consequence of competition between methanogenic prokaryotes (MP) and sulfate reducers. Indeed, the formation of methane was demonstrated, and the presence of *Archaea* was shown by PCR using the archaeal-specific primer pair ARC 344F-ARC 915R (Casamayor et al., 2002). However, there was also methane production in the columns fed with lactate, but these columns maintained their HM- and sulfate-removal efficiencies throughout the experiment. This suggests that the population of SRB and the population of MP exist together in a syntrophic way, e.g. the SRB oxidize lactate into acetate and/or hydrogen, which is in turn consumed by MP. An alternative explanation arose from the results of molecular analyzes (see section 3.2): the conditions

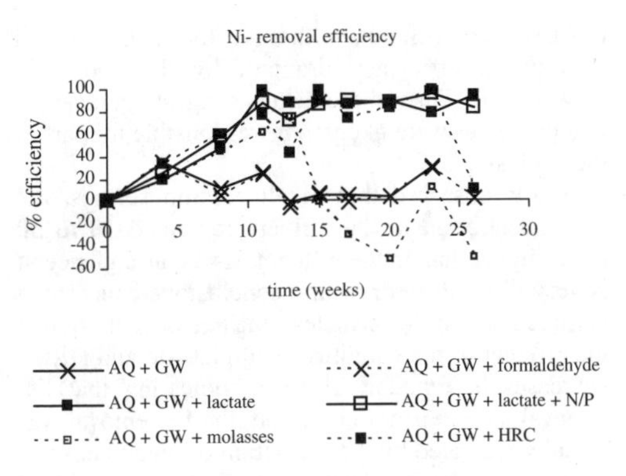

Figure 1. Ni-removal efficiency in function of time for different column conditions.

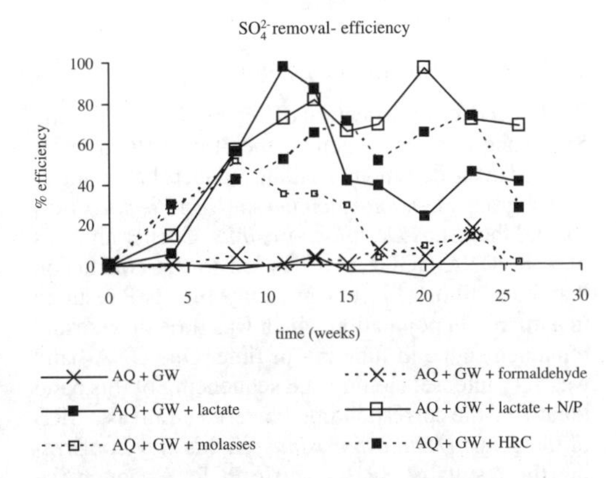

Figure 2. Sulfate-removal efficiency in function of time for different column conditions.

with molasses and HRC® enriched strains which could not be detected in the set-ups with lactate. The presence of these bacteria might have resulted in the conversion of C-sources and/or electron donors into products that cannot be metabolized by the bacteria involved in ISBP, or perhaps they competed for C-sources. A final explanation is a problem of inhibition of the SRB-population by the formed metal sulfides. This phenomenon was previously observed and studied in an anaerobic bioreactor treating acid mine drainage, by Utgikar (Utgikar et al., 2002).

A remarkable finding was that, after 4 weeks, Cd-concentration dropped from an average of 53.5 mg/l to a concentration which is below the detection limit of 5 µg/l, except for the two control tests (aquifer + groundwater, and aquifer + groundwater + formaldehyde). However, after 10 weeks, Cd-concentration also began to decrease in the column with no added nutrients. In addition, the Cd-concentration didn't increase in the effluent of the molasses-fed column, although the Zn, Ni and Co-removal efficiency dramatically dropped after 13 weeks. Together, these results suggest that adsorption processes were likely to be responsible for part of the Cd loss.

It was observed that in all column set-ups, the Fe-concentration in the effluent remained 10 to 50 times higher than in the influent. As a consequence, it is very difficult to draw any conclusions concerning Fe-precipitation. Nevertheless, the Fe-concentration in the effluent of the conditions with lactate and HRC® decreases in function of time, indicating that Fe-removal takes place. In addition, the Fe-removal was visually indicated by the formation of black FeS-precipitates on these columns' aquifer. For the set-up with molasses, the Fe-concentration in the effluent re-increases after 13 weeks, suggesting that Fe-removal efficiency has dropped.

3.2 *SRB-community composition and changes*

SRB-specific amplification of 16S rRNA genes using SRB-subgroup or – genus specific PCR primers resulted in the detection of sulfate reducers belonging to the subgroup *Desulfotomaculum* sp./*Desulfosporosinus* sp. and the genus *Desulfosporosinus*. Cluster analysis of 16S rDNA-DGGE profiles led to the conclusion that the addition of lactate or lactate plus N/P resulted in a microbial population which was little diverse and remained stable in function of time. One DNA-band was very intense; cloning and sequencing of this band identified the corresponding bacterial strain as a SRB of the genus *Desulfosporosinus*, in this way confirming the results of SRB-specific PCR. Although the same DNA band was present in the DGGE profiles of the columns where molasses or HRC® were added, this band became less intense when the heavy metal- and sulfate-removal efficiencies decreased, and population shifts resulted in the enrichment of other DNA bands. These bands are presently being identified by the cloning and sequencing approach.

Although the feasibility of DGGE-analysis of the *dsr*-gene is still under investigation, preliminary results show that it is possible to obtain a different DGGE-pattern for different SRB genera (unpublished results) and also for mixed SRB populations isolated from previous ISBP batch-experiments. As for the column experiments, so far the pool of *dsr*-PCR products has been cloned and sequenced, and results confirm the presence of Gram positive SRB (*Desulfotomaculum* sp., *Desulfosporosinus* sp.) but also demonstrate the presence of SRB of the genus *Desulfomicrobium*. This suggests that the analysis of the SRB-population based on the *dsr*-gene will reveal a greater diversity than does the 16S rRNA-gene approach. However, since the *dsr*-gene has been subject of lateral gene transfer events (Klein et al., 2001), it has to be evaluated how this lateral gene transfer influences the interpretation of *dsr*-DGGE fingerprints.

REFERENCES

Altschul, S.F., W. Gish, W. Miller, E.W. Myers, and D.J. Lipman. 1990. Basic local alignment search tool, *J. Mol. Biol.* 215: 403–410.

Casamayor, E.O., R. Massana, S. Benlloch, L. Ovreas, B. Díez, V.J. Goddard, J.M. Gasol, I. Joint, F. Rodríguez-Valera, and C. Pedrós-Alió. 2002. Changes in archaeal, bacterial and eukaryal assemblages along a salinity gradient by comparison of genetic fingerprinting methods in a multi-pond solar saltern. *Environ. Microbiol.* 4: 338–348.

Daly, K., R.J. Sharp, and A.J. McCarthy. 2000. Development of oligonucleotide probes and PCR primers for detecting phylogenetic subgroups of sulfate-reducing bacteria. *Microbiol. Ecol.* 146: 1693–1705.

El-Fantroussi, S., W. Verstraete, and E.M. Top. 2001. Enrichment and molecular characterization of a bacterial culture that degrades methoxy-methyl urea herbicides and their aniline derivatives. *Appl. Environ. Microbiol.* 67: 4943.

Kaksonen, A.H., M.L. Riekkola-Vanhanen, and J.A. Puhakka. 2003. Optimization of metal sulphide precipitation in fluidized-bed treatment of acidic wastewater. *Water Res.* 37: 255–266.

Klein, M., M. Friedrich, A.J. Roger, P. Hugenholtz, S. Fishbain, H. Abicht, L.L. Blackall, D.A. Stahl, and M. Wagner. 2001. Multiple lateral transfers of dissimilatory sulfite reductase genes between major lineages of sulfate-reducing prokaryotes. *J. Bacteriol.* 183: 6028–6035.

Loy, A., A. Lehner, N. Lee, J. Adamczyk, H. Meier, J. Ernst, K.H. Schleifer, and M. Wagner. 2002. Oligonucleotide microarray for 16S rRNA gene-based detection of all recognized lineages of sulfate-reducing prokaryotes in the environment. *Appl. Environ. Microbiol.* 68: 5064–5081.

Maidak, B.L., N. Larsen, M.J. McCaughey, R. Overbeek, G.J. Olsen, K. Fogel, J. Blandy, and C.R. Woese. 1994. The Ribosomal Database Project. *Nucleic Acid Res.* 22: 3485–3487.

Pérez-Jiménez, J.R., L.Y. Young, and L.J. Kerkhof. 2001. Molecular characterization of sulfate-reducing bacteria in anaerobic hydrocarbon-degrading consortia and pure cultures using the dissimilatory sulfite reductase (dsrAB) genes. *FEMS Microbiol. Ecol.*35: 145–150.

Utgikar, V.P., S.M. Harmon, N. Chaudhary, H.H. Tabak, R. Govind, and J.R. Haines. 2002. Inhibition of sulfate-reducing bacteria by metal sulfide formation in bioremediation of acid mine drainage. *Environ. Toxicol.* 17: 40–48.

Wagner, M., A.J. Roger, J.L. Flax, G.A. Brusseau, and D.A. Stahl. 1998. Phylogeny of dissimilatory sulfite reductases supports an early origin of sulfate respiration. *J. Bacteriol.* 180: 2975–2982.

European Symposium on Environmental Biotechnology, ESEB 2004 - Verstraete (ed)
© 2004 Taylor & Francis Group, London, ISBN 90 5809 653 X

Effect of ORC-injections on *in-situ* BTEX-biodegradation and on the endogenous micro-organisms

L. Bastiaens, Q. Simons, J. Vos, B. Hendrickx, R. Lookman, A. Ryngaert, M. Maesen & L. Diels
Vito (Flemish Institute for Technological Research), Mol, Belgium

ABSTRACT: A demonstration pilot-scale study is being conducted at a BTEX-contaminated site to evaluate the use of ORC® (Regenesis) for enhanced bioremediation. More specifically, the influence of ORC-injections on the chemical composition of the groundwater and on the endogenous micro-organisms at the tested site are being studied. Based on lab scale feasibility tests it was found that a limited oxygen availability reduced the degradation rate, which suggests that an increase of the oxygen concentration may enhance the BTEX-degradation. The ORC has been applied to the site in January 2003 and the evolution of chemical and microbial parameters is being monitored since. Monitoring of the *in-situ* and some chemical parameters revealed that ORC-injections led to increased dissolved oxygen concentration (up to 45 mg/L) and pH-increases up to pH 10.5. After the ORC-injection a part of the monitoring wells was found to give misleading results. The effect of the ORC-injection on the endogenous microbial population present in the aquifer at the site, is followed by microbial (platings) and molecular techniques (PCR-DGGE) using *in-situ* mesocosms systems. Analyses of aquifer samples in time indicated the presence and activity of BTEX-degrading micro-organisms at the site.

1 INTRODUCTION

At a petrol gas station in Belgium, an old mineral oil and BTEX-contamination is present in the groundwater. The pre-study revealed a relatively small-size contamination (about 12 m × 6 m) with a high concentration zone (Pz13: 5000–19000 ppb benzene, 3000–40000 ppb toluene, 600–2400 ppb ethylbenzene, 4700–14000 ppb xylenes and up to 23000 ppb mineral oil) and a small contamination plume. During the last years no significant concentration changes have been observed, indicating that the pollution will not be removed by natural attenuation.

The site was chosen as a test site in a demonstration project for 'remediation by enhanced microbial degradation using Oxygen Release Compound (ORC® regenesis) as oxygen supplier'. Although ORC has been used in several demonstration projects (Bianchi-Mosquera et al., 1994; Chapman et al., 1997; Borden et al., 1997), the ORC-experience in Belgium is rather limited and not much information is available concerning the effect of ORC on the pH in the groundwater and on the microbial population.

The goal of the project was to evaluate the effect of ORC on (I) a BTEX-pollution in Belgian subsoils, and (II) on the endogenous micro-organisms in the aquifer material. The demonstration project had special attention for site characterization, lab scale feasibility testing and detailed monitoring of chemicals as well as the micro-organisms.

2 MATERIALS AND METHODS

2.1 *Lab scale BTEX-degradation experiments*

BTEX-degradation experiments were set up in batch systems (10% aquifer material MIP4, groundwater Pz13) under aerobic, micro-aerophilic and anaerobic conditions as described by Bastiaens et al. (2003). Residual BTEX-concentrations were determined by high resolution GC-MS with headspace injection. Oxygen concentrations in the liquid phase of the vials were measured with a micro-electrode in the oxygen cell (Strathkelvin Instruments Limited) after sampling with a gastight syringe.

2.2 *Microbial analyses*

Viable counts in aquifer material were determined by microbial platings on rich medium 869 (Bastiaens et al., 2002). Growth on BTEX-compounds was determined by platings on phosphate buffered minimal medium

Wick (Van Herwijnen et al., 2003) with the BTEX-component, supplied in the gas phase, as sole carbon source.

DNA was extracted as described by Faber (personal communication) and purified using 0,1 g PVPP and the DNA clean up Wizard^R protocol. The microbial diversity was studied by (I) PCR-DGGE using general eubacterial 16S rRNA gene primer sets (forward primer 63F, reverse primer 518R, Marchesi et al., 1998), and (II) by PCR and agarose gel separation using 8 specific primer sets designed for amplifying genes involved in different aerobic BTEX-degradation pathways, being *tbm*D, *tmo*A, *xyl*A and *tod*C1 encoding mono-oxygenase or dioxygenase enzymes involved in the initial attack of the BTEX-compound, and *xyl*E, *cdo*, *tbu*E and *tod*E encoding dioxygenase enzymes involved in the lower pathway (B. Hendrickx, in preparation).

3 RESULTS

3.1 *Lab scale feasibility testing*

Aerobic microbial platings revealed that, although the number of countable bacteria was rather low (3.0 10^4 cfu/g soil), BTEX-degraders were found present (10^3–10^4 cfu/g soil). BTEX-degradation experiments were set up in order (I) to check the BTEX-degradation capacity of the endogenous bacteria in the aquifer material and (II) to test the effect of different oxygen concentration and of nutrient addition on the BTEX-degradation rate. Under aerobic and micro-aerophilic conditions all BTEX-compounds were degraded within 4 weeks, except o-xylene. Toluene, ethylbenzene and m-and p-xylene were degraded the most easily, followed by benzene and finally o-xylene. Addition of nutrients had only a slightly enhancing effect on o-xylene degradation. Reduction of the oxygen concentration from 8 mg/l to 2 mg/l extended the lag phase and reduced the BTEX-degradation rate. Under even more oxygen-limited conditions (nearly anaerobic conditions), only toluene and ethylbenzene, and also a small amount of m- and p-xylene, were degraded (results not shown, Bastiaens et al., 2003).

3.2 *ORC-injection*

In January 2003, a total of 480 kg of ORC were applied at the site, divided over 13 injections (2.5–5.5 m gbs), using Geoprobe direct-push equipment with lateral injection. The distance between ORC-injections points was 3 to 5 meters (Figure 1). The ORC-injection plan was based on data from chemical analyses on groundwater samples and MIP-probing. As the highest BTEX-concentrations were found in Pz13 and Pz32, more ORC per injection (42 kg ORC/injection versus 32 kg ORC/injection) was applied near these points. The demonstration project however is predominantly focused on Pz13.

3.3 *Monitoring results*

The monitoring plan included measurements of the *in-situ* parameters (pH, ORP, dissolved oxygen, temperature and conductivity) every 2 to 8 weeks, and of some chemical parameters (e.g. BTEX, nutrients, Mg) every 2 to 3 months. The effect of the ORC-injection on the endogenous microbial population present in the aquifer at the site, is followed by microbial (platings) and molecular techniques (PCR-DGGE using general eubacterial primers and BTEX-specific primers). For this purpose, *in-situ* mesocosms, containing aliquots of aquifer material, were developed and 16 of these mesocosms were installed in 4 different monitoring wells (meso1, meso2, meso3 and meso4) at the site.

Measurements in the field showed that the dissolved oxygen concentration clearly increased, and values up to 45 mg/l were detected (Figure 2). At locations with high oxygen concentrations, strongly elevated pH-values up to pH 10.5 were obtained, which is not optimal for biological activity. At the studied site, two spots (Pz13 and Pz31) with high pH values and high dissolved oxygen concentration can be seen, although ORC was also supplied to the intermediate region. In the intermediate region the magnesium concentration was also much lower in comparison with the regions around Pz13 and Pz31 (results not shown). In time, the oxygen is distributed over a larger area. Near 2 high pH monitoring wells direct measurements of pH and oxygen in the subsurface gave much lower values, indicating that probably a considerable amount of ORC was injected in the packing near the filter of the well causing misleading results.

Figure 3 shows the evolution of the BTEX-concentration at the site. The concentrations fluctuated but no clear trend is observed yet. At the time of ORC-injection, pure phase product has been observed in wells Pz32 and meso3. This free phase product, which

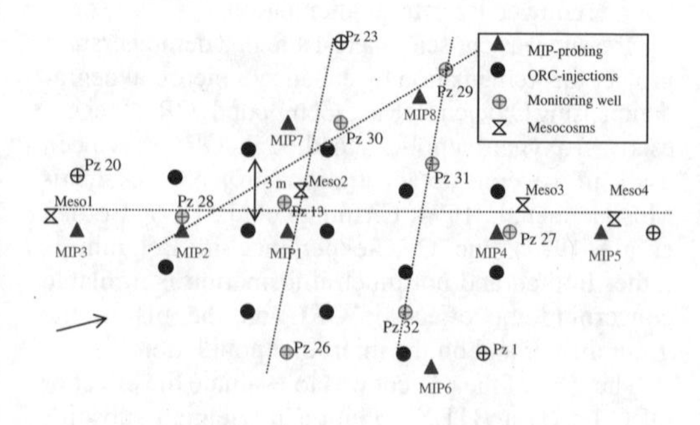

Figure 1. Schematic overview of the testsite.

was not expected at the site, may have been mobilized by water table changes or by pressure gradients induced by the ORC-injections.

Two and five months after the ORC-injection, the first mesocosms were harvested from meso1, meso2, meso3 and meso4. DNA was extracted from the aquifer material and PCR-DGGE analyses were performed. The eubacterial population was found to be different at the 4 locations, and a higher diversity was detected in time (results not shown).

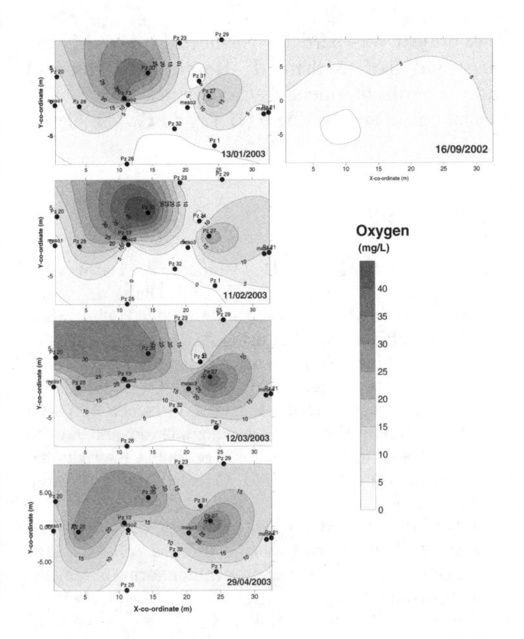

Figure 2. Evolution of dissolved oxygen concentration before and after injection of ORC.

Using BTEX-specific primer sets, some catabolic genes were detected (Table 1). Initially predominantly genes encoding for enzymes involved in the initial attack of BTEX-compounds have been detected. After 2 and even more pronounced after 5 months also genes encoding enzymes involved in the lower pathway were found present. There seems to be a relation between the pollution level of the groundwater

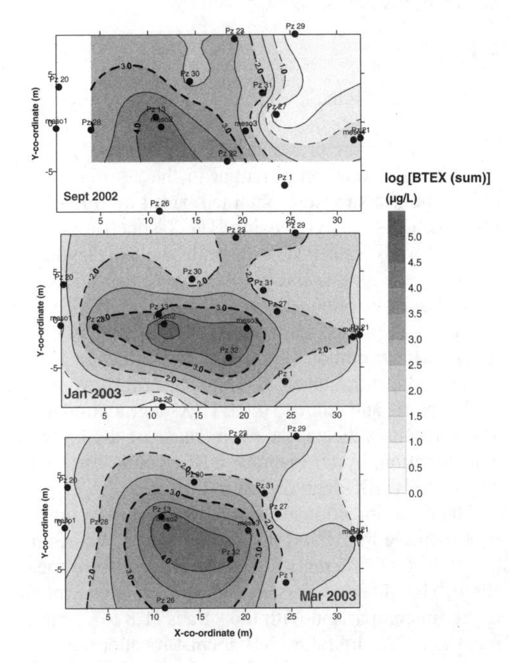

Figure 3. Evolution of BTEX-concentration before and after injection of ORC.

Table 1. BTEX-catabolic genes detected in the aquifer before ORC-injection (T0), and 2 (T1) and 5 months (T2) after ORC-injection.

	Initial attack				Lower pathway				
Catabolic genes	*tbm*D/ ...	*tmo*A/*bmo*A/ *tou*A/*tbu*A1/...	*xyl*A	*tod*C1/ *bed*C1	*xyl*EX (*Pseudomonas*)	*xyl*E (*Sphingomonas*)	*cdo*	*tbu*E	*tod*E
MESO 1 T0	−	+	−	−	−	−	−	−	−
MESO 2 T0	+	+	−	+	−	−	−	−	−
MESO 3 T0	−	+	−	+	−	−	−	−	−
MESO 4 T0	−	−	−	−	−	−	−	−	−
MESO 1 T1	−	+	−	−	−	−	−	−	−
MESO 2 T1	−	+	+	+	+	−	(+)	−	−
MESO 3 T1	+	+	−	+	−	−	−	−	−
MESO 4 T1	−	+	−	−	−	−	−	−	−
MESO 1 T2	−	−	−	−	−	−		−	−
MESO 2 T2	+	+	−	−	+	+		−	−
MESO 3 T2	+	+	−	−	−	+		−	−
MESO 4 T2	−	−	−	−	−	−		−	−

and the diversity of detected catabolic genes. In the aquifer material from locations meso2 and meso3, which are located in high-contaminated groundwater, 2 to 3 different groups of catabolic enzymes could be detected. Meso1 and meso4 contains only low BTEX-concentrations and only one or no positive signal was obtained with the specific BTEX-primer sets. RT-PCR on RNA-extracts from the aquifer samples revealed that *tmo*A-like genes were not only present but also active at the site.

4 CONCLUSIONS

Indications for the presence of an oxygen dependent microbial degradation potential in the aquifer material of the test site were obtained based on microbial platings, DNA-analyses using BTEX-specific primer sets and a lab scale BTEX-degradation test. Toluene, ethylbenzene and m-and p-xylene were degraded the most easily, followed by benzene and finally o-xylene. BTEX-degradation was not significantly influenced by addition of nutrients, but reduction of the oxygen concentration from 8 mg/l to 2 mg/l extended the lag phase and reduced the BTEX-degradation rate. These results indicate that an increase of the oxygen concentration, by for instance ORC, could lead to an improved BTEX-degradation in the field.

The ORC-injection at the site resulted in a significant increase in pH and dissolved oxygen. Values up to pH 10.5 and 45 mg/l dissolved oxygen were measured. The obtained oxygen concentrations are much higher in comparison with the values (1.8 to 20 mg/l) mentioned in literature (Bianchi-Mosquera et al., 1994; Chapman et al., 1997; Borden et al., 1997). It was found that the very high values were due monitoring wells that were affected by the ORC-injection, resulting in misleading measurements.

After the ORC-injection a more divers eubacterial population was detected in time. Based on molecular techniques BTEX-degrading bacteria were found present at the site, especially at the most contaminated spots. Genes encoding for enzymes involved in different pathways were detected.

In time there seems to be a shift from genes encoding for enzymes involved in the upper pathway to genes encoding for enzymes involved in the lower pathway. The effect of the ORC-injection and also of the pH-raise on the endogenous microbial population present in the aquifer at the site, will be further examined by harvesting more mesocosms from the field.

Up to now, the effect of the ORC on the BTEX-concentration is not clear, as free product appeared after ORC-injection. Although the site has been characterized by MIP-probing and chemical analyses of groundwater samples, the presence of free phase product was not expected. Possibly the free phase product was mobilized by pressure gradients induced during ORC-injection.

ACKNOWLEDGEMENTS

This project was partially financed by OVAM, and we would like to thank F. Boven (AIB-vincotte) and the projectpartners S. Camerlynck (DEC NV), F. Vanhoucke and B. Loete (JET) and G. Van Gestel (OVAM).

REFERENCES

Bastiaens, L., Springael, D., Wattiau, P., Harms, H., deWachter, R., Verachtert, H. and Diels, L. 2000. "Isolation of Adherent Polycyclic Aromatic Hydrocarbon (PAH)-degrading Bacteria using PAH-sorbing Carriers." Applied and Environmental Microbiology. 66:1834–1843.

Bastiaens, L., Vos, J., Maesen, M., Simons, Q., Lookman, R., Hendrickx, B. and Diels, L. 2003. Influence of ORC-injection on in-situ BTEX-biodegradation and an the endogenous micro-organisms. Seventh International Symposium in In situ and On-site Bioremediation, Orlando.

Bianchi-Mosquera, G.C. 1994. Enhanced degradation of dissolved benzene and toluene using a solid oxygen-releasing compound. Groundwater monitoring & remediation, Winter, 120–128.

Borden, R.C., Goin, R.T. and Kao, C.-M. 1997. Control of BTEX migration using a biological enhanced permeable berrier. Groundwater monitoring & remediation, Winter, 70–80.

Chapman, S.W., Byerley, B.T., Smyth, D.J.A. and Mackay, D.M. 1997. A pilot test of passive oxygen release for enhancement of in situ bioremediation of BTEX-contaminated ground water. Ground Water Monitoring & Remediation, 17:93–105.

Marchesi, J.R., Sato, T., Weightman, A.J., Martin, T.A., Fry, J.C., Hiom, S.J. and William G. Wade. 1998. "Design and Evaluation of Useful Bacterium-Specific PCR Primers That Amplify Genes Coding for Bacterial 16S rRNA." Applied and Environmental Microbiology. 64: 795–799.

Van Harwijnen, R., Springael D., Slot P., Govers H.A.J, and Parson J.R. 2003. Degradation of anthracene by *Mycobacterium* sp. Strain LB501T proceeds via a novel pathway, through o-phthalic acid. Appl. Environ. Microbiol., 69:186–190.

European Symposium on Environmental Biotechnology, ESEB 2004 - Verstraete (ed)
© 2004 Taylor & Francis Group, London, ISBN 90 5809 653 X

Involvement of the microbial community of the groundwater-surface water-interface in bioremediation

W. Dejonghe, R. Lookman, B. Borremans, A. Ryngaert, S. Springael, J. Bronders & L. Diels
Flemish Institute for Technological Research (Vito), Mol, Belgium

The main goal of the research presented in this abstract is to demonstrate and quantify the effect of the catabolic activity of the microbial community present in the interface between groundwater and surface water on the degradation of pollutants passing through this interface. The interface is a dynamic ecotone where active exchanges of water and dissolved material between the stream and groundwater in many porous sand- and gravel-bed rivers occur. Interfaces are important storage zones for organic carbon and are generally characterised by sharp physical and chemical gradients, thus enabling a broad spectrum of metabolic processes to occur within small spatial scales. As a consequence the interfaces are often hot spots in productivity and diversity of organisms and may contribute substantially to the carbon, nutrient and energy flow through the river system.

The main aim of the research is to characterise and quantify respectively the structure and the catabolic activity of the microbial community present in this interface by using molecular and classical isolation techniques. The genetic information, both at the DNA and RNA level, will be investigated in slices of undisturbed samples taken in and outside of the interface. The structure of the microbial community at different positions in the interface will be examined by Reverse Transcriptase-Polymerase Chain Reaction (RT-PCR) of extracted 16S rRNA, followed by Denaturing Gradient Gel Electrophoresis (DGGE). The presence but more importantly the expression of different catabolic genes in different positions in the interface will be studied by (RT)-PCR and quantified by real-time PCR. In addition, bacteria will be isolated from the different positions in the interface by classical agar plating and enrichment under different redox conditions. The genetic and catabolic diversity of these isolates and their degradation potentials will be studied. Overall, information will be obtained about the effect of the changing conditions in the interface on the presence, but more importantly, also the (catabolic) activity of bacteria and be coupled to the contribution of this interface to the overall cleaning of polluted sites.

European Symposium on Environmental Biotechnology, ESEB 2004 - Verstraete (ed)
 ISBN 90 5809 653 X

Isolation and characterization of bacteria with possible use for groundwater decontamination in multibarrier system

K. Demnerova, V. Spěváková, I. Melenová, P. Lovecká & M. Macková
Dept. of Biochemistry and Microbiology, Faculty of Food and Biochemical Technology, Institute of Chem. Technol. Prague, Technicka, Prague, Czech Republic

ABSTRACT: The suitability of relevant pollutants degrading microorganisms for MULTIBARRIER purposes was tested, their phenotypic and genotypic properties (DGGE, RFLP) were evaluated in order to select specific well-performing populations suitable for the MULTIBARRIER technology. Ecotoxicity analysis of landfill leachates was performed by testing of kinetic parameters of bacteria *Pseudomonas* sp. P2 by Bioscreen® apparatus. The toxicity of individual pollutants found in leachates and groundwater (PCE, TCE, benzene, toluene, xylene zinc and arsenic) or their mixtures at various concentrations was tested with biological system of luminescent bacteria *Vibrio fischeri* (Microtox®, Lumac®), and by Bioscreen®. For the same samples Ames *Salmonella typhimurium* His – test of mutagenicity was used.

1 INTRODUCTION

Landfill leachates are heavily polluted with organic and inorganic substances and contain a lot of toxic micropollutants such as BTEX (benzene, toluene, ethylbenzene, xylene), aromatic hydrocarbons, pesticides, chloroaromatic and aliphatic compounds and heavy metals. It was shown that from 90%–95% of the organic materials in municipal landfill leachates are of unknown composition.

There exist many studies concerning mutagenicity of groundwater, biogenotoxicity of organic xenobiotics in aquatic sediment materials or drinking water supplies (Haider et al., 2002; Guzella et al., 2002). Epidemiological studies have shown a correlation between genotoxicity of drinking water and increase of cancer risks (Koivusalo et al., 1997). Therefore, short-term genotoxicity tests are useful to identify potential carcinogens prior to costly and time-consuming long-term animal studies. In most publications on genotoxic effects of water samples only one system has been used and almost all data come from the *Salmonella*/ microsome assay (Ames test) (Haider et al., 2002).

Nowadays many technologies based on biological processes (Sayler, 1990, Sergeev et al., 1996, Macek et al., 2000) for groundwater protection and pollution treatment were described. So called permeable barriers constructed in a contaminated groundwater system allow groundwater to flow through, degrade or remove soluble pollutants. The use of permeable, subsurface reactive barriers of groundwater pollution and contamination is becoming widespread in the United States and Canada (Amos and Younger, 2003; US environmental Protection Agency, 1995).

The aim of the study is to develop such permeable multibarrier system which will be able to remove various pollutants from the groundwater on the principles described above (Amos and Younger, 2003). It is necessary to test interactions among barrier material, microorganisms, going to be used for the function of multibarrier, and required nutrients for purposes of the well performing MULTIBARRIER set up. As the major objective of this work is to contribute on development and study of feasibility of a MULTIBARRIER concept for treatment of groundwater contaminated by mixed pollution with the emphasis on the well-performance of the microbial biofilm involved and its compatibility and synergy with the coarse material, enrichment and selection of specific well-performing populations suitable for the MULTIBARRIER technology. Degradation products and degradation pathways of the pollutants during their chemical and microbial-based breakdown are identified with respect to the surface of the filling materials. Several methods are used for the evaluation of the toxicity and genotoxicity analysis of landfill leachates, individual substances mentioned above and their mixture.

The model groundwater contained a mixture of different pollutants representing different chemical families like AOX (e.g. PER, TCE, PCE, VC), BTEX, heavy metals and PCBs. An artificial contaminant mixture (so called MULTIBARRIER pollution

Table 1. Composition of landfill leachate.

	Landfill leachate P02231	Regular limit
pH	8.7	
Conductivity	27700 (μS/cm)	
Anions (mg/l)		
Sulphate	65.4	
Nitrate	<0.886	
Metals (μg/l)		
As	256	20
Cd	<5	5
Cr	557	50 (Cr(III))
Fe	7010	
Cu	19	100
Hg	<0.25	1
Pb	73	20
Mn	134	40
Ni	192	100
Zn	429	
AOX μg/l	2490	
BTEX (μg/l)	52.0	
Styrene (μg/l)	2.0	
135trimethylbenzene	2.0	
124trimethylbenzene	10	
p-isopropyltoluene	22	
14-dichlorobenzene	3.5	
Naphthalene	15	

mixture) was determined in respect with the incidence of the pollutants in the environment. It consists of benzene (B), toluene (T), m-xylene (mX) of concentration of 2 mg/l, tetrachloroethylene (PCE) (2 mg/l), trichloroethylene (TCE) (5 mg/l), 1,1-dichloroethane (11DCA) (5 mg/l), dichloromethane (DCM) (5 mg/l), arsenate (As) (0.2 mg As/l) and zinc (Zn) (5 mg/l). The landfill material was sampled from contaminated site in Belgium. This aquifer was historically polluted with VOCls, BTEX and metals (see Table 1).

2 RESULTS AND DISCUSSION

2.1 *Toxicity of MULTIBARRIER pollution mixture*

Toxicity of the defined mixture as well as the pollutants individually was measured by two different systems, i.e. bioluminescence test with *Vibrio fischerii* and determining growth curves by Bioscreen for PCB-degrading strain *Pseudomonas* sp. P2. The results show that the single organic compounds exhibit no toxic effect at the defined concentrations either on *Vibrio fischeri* or *Pseudomonas* sp. P2.

The addition of Zn and As to the mixture of the organics increases the toxicity considerably, though. The mixture of organic solvents (no Zn, As) had no effect on growth of *Pseudomonas* sp. but had a toxic effect on *Vibrio fischeri*. Zinc was found to be more toxic than arsenic (EC50(15,15) of As is 9.22 mg/l, EC50(15,15) of Zn 1.4 mg/l, respectively). On the whole, samples dissolved in DMSO were more toxic probably due to the better access of toxicants through bacterial wall.

2.2 *Culture estimation of the aquifer*

Total number of culturable heterotrophic and specific pollutant degrading microorganisms in the leachate material was counted by the conventional method of counting CFU on R_2A agar plates (supplemented with cycloheximide (50 mg/l) to inhibit fungus growth) after 24 hours of cultivation at 25°C aerobically. It contained 10^4 of microorganisms per 1 gram of the moist material, which is rather a small number. However, this result comprises the culturable clones only. Soil samples were previously found to be composed of a very large number of genetically separate clones. Such heterogeneity of the community may be determined by isolation of the total DNA from the material and subjecting it to PCR-DGGE protocol using universal primers.

2.3 *Primer selection and specificities*

There are several sets of primers designed to target different regions of the small subunit of ribosomal RNA (16S rRNA). rRNA genes of all living organisms contain a total of nine variable regions, V1 to V9, scattered in the molecule, which, for bacteria, is approximately 1,520 bp long. The length of PCR product is limiting factor of DGGE/TGGE: only partial sequences up to about 500 bp are separated well. Universal primers hybridize to evolutionarily conserved flanking regions. A high microbial diversity and dramatic pattern changes may be an effect of the reduction of the only one or two microorganisms. Choosing variable regions with less interspecies diversity can then reduce the effect of operon diversities.

At first, we were testing a universal bacterial primer set GC907f-1406r corresponding to *E. coli* positions 907–926 and 1392–1406. This primer set spans the highly variable region from V5 to V9 of 16S rRNA. The pure cultures of control strains from different phylogenetic groups were analyzed by PCR-DGGE using this set of primers. This analysis resulted with a total number of 4 bands for individual strains. It is a question whether the additional bands were the result of metastable conformers or different sequences (operons). However, these findings are in agreement with the results of SSCP analysis of V6 to V8 region, which gave 4 bands for pure culture of *Pseudomonas* sp., too. Such effects hamper the quantitative interpretation of community profiles such as DGGE, though. According to this study, it is the region of V2 and V3 that gives the less number of bands for microorganisms of our interest (*Pseudomonas* sp., *Agrobacterium* sp.). Thus

GC338f-518r set of primers was tested spanning the V3 region. In addition to that, this pair comprises the region of 200 bp only limiting the operons heterogeneity even further. On the other hand, such a short PCR product is not sufficient enough to identify the particular strain by excision of the individual band from the DGGE gel and sequencing. This primer pair gives only one band for all the strains with the exception of *Achromobacter xyloxidans* A19 giving 4 bands (not shown). Therefore these primers were chosen for further studies.

2.4 *Enrichment of indigenous soil microbial associations*

Batch culture experiments were employed in this study to isolate aerobic catabolic community able to utilize BTmX, PCE, and TCE in the mixture and individually as a sole carbon source from the leachate material. The principal selection pressure was a rising concentration of pollutant (from 50 ppm to 300 ppm of BTmX, and PCE, TCE, respectively). Toluene and m-xylene as well as the mixture were degraded easily under aerobic conditions at 25°C judging from the evident opalescence of culture media. Their degradation potential was successfully transferred six times to a fresh minimal medium and remained visible. However, the degradation of TCE and PCE did not occur, despite its low toxicity, as it was revealed during the third transfer (no more growth visible in the flasks and no chloride concentration in the medium was detected). Benzene seemed to be persistent under studied conditions at first. After 8 weeks of cultivation, the sixth transfer was plated out on the minimal medium agar plates with T, mX and the organic mixture as a sole carbon source and R2A plates in order to isolate individual colonies. Several different colonies were yielded – a mixture of Gram-positive and gram-negative rods characterized the associations. 4 days old culture on minimal medium plates was sprayed with 100 mM pyrocatechol in order to pick up the colony bearing catechol 2,3-dioxygenase (forming a visible yellowish zone of pyrocatechol cleavage within a few minutes). Catechol 2,3-dioxygenase plays a crucial role in degradation of a wide range of pollutants. Such colonies were then tooth picked and transferred onto a fresh minimal medium plate and LB plate (to check purity of the isolated strain). The plates were cultivated for one week at room temperature. The purity of isolated strains was checked then by PCR-DGGE. The pure strain was then inoculated back in the liquid minimal medium with TmX or organic mixture to find out whether the isolated strain is actually the BTmX degrader. In a case of growth on the minimal media, the culture was again subjected to PCR-DGGE protocol to have information on the composition of the association. It took rather a long time (4 months) of repeated isolations to obtain pure cultures of 7 strains. Gram-negative

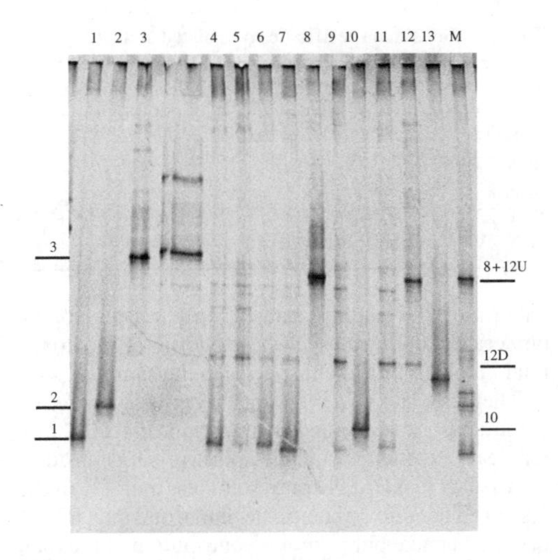

Figure 1. DGGE profile of the isolated and purified strains. Lane 1: BEN 1V; lane 2: BEN 1 Z BEN; lane 3: BEN 2; lane 4: XYL 22; lane 5: XYL 22V; lane 6: XYL 22M; lane 7: MIX 2M; lane 8: MIX 2V; lane 9: MIX 2 from PCE/TCE; lane 10: MIX 31V; lane 11: MIX 31M; lane 12: MIX 32; lane 13: UH133; lane 14: DGGE ladder (from the top to the bottom: *Pseudom.* sp. PS20, *Comamonas testosteronii* B356, *Ralst. eutropha* H850, *Enterob. agglomerans*).

rods seemed to be accompanied by Gram-negative cocci (small, barely visible colonies on MM agar plates), which actually were encouraging the growth on TmX and mixture as a sole carbon source. As it was found out later on, the mixture of gram-negative rods and Gram-negative coci has the ability to cometabolize TCE in the presence of toluene. Therefore, the presence of coci was vital for TmX, mixture degradation. After a long adaptation period, the pure cultures of strains were isolated and mixture of two strains was obtained.

2.5 *Evaluation of DGGE patterns*

The individual steps in molecular methods of community structure studies have their pitfalls. However, since all the samples were treated identically, we consider these pitfalls to be the same for all the samples allowing us to compare them. In a DGGE gel, the number and precise position of bands in a gel track were visually inspected to give an estimate of number and relative abundance of numerically dominant ribotypes in samples. The DGGE profile of the aquifer sample (see Fig. 1, lane 2,3) shows its diverse population. Only two identical bands were detected in the DNA sample obtained from the 24-hour's culture of the aquifer in R_2A medium meaning that only these two clones were culturable in R_2A. On the other hand, R_2A medium detected two more bands, which were at the original sample present only in a low copy number (under detection limit). The diversity of the DGGE

Table 2. Substrate specificities of isolated strains.

Strain	Ben	Tol	Xyl	Mix
BEN1V	+	–	–	–
BEN 1 BEN	+	–	–	–
BEN 2	++	++	+++	+++
XYL 22	–	++	+++	++
MIX 2V	–	+++	+++	+++

pattern of the aquifer sample only relates to the numerically dominant species and definitely not to the total number of different species in the sample.

There is a lack of a relationship between microbial community structure and degradation of BTEX in a real leachate plume, which is not surprising considering the fact that BTEX contributes less than 1% of the dissolved organic carbon in the plume and thus BTEX degraders make only a minor contribution to the total microbial community. However, under selection pressure *in vitro*, the BTEX degraders were selected already during the third enrichment on minimal media and dominate the association. As in the case of PCE and TCE enrichments, during the first and second enrichment, the detected bacterial community survives most probably only thanks to the remains of organic nutrients from the aquifer.

Six new endogenous aerobic isolates were obtained from landfill material from Belgium plant site for their ability to grow on benzene, toluene and m-xylene. Their potential use within the MULTIBARRIER project was tested. Out of these, two strains are gram-positive, all of them with growth temperature optimum at 20°C. According to the preliminary RFLP restriction profile of 16S rDNA, the diversity of these strains was revealed and confirmed by DGGE analysis of these strains. In order to taxonomically identify them, their 16S rDNA gene was amplified and the PCR product was directly sequenced with internal primers 16F357 and 16F1069. The results of the analysis of the nucleotide sequence using NCBI database, Basic Alignment Search Tool are shown in Table 3. Concerning their degradation abilities, *Pseudomonas* sp. BEN 2 seems to possess the widest degradation potential comprising all studied compounds. Toluene and m-xylene are readily degraded in 3 days at 18°C excreting in the presence of toluene a yellow compound into the liquid medium. The highest concentration of toluene and m-xylene degradable is 600 ppm so far. The other xylene isomers (o- and p-xylenes) were tested, too, revealing that o-xylene was not used as a growth substrate by any of the strains.

In order to find out whether the sequencing and identifying of the DNA fragments (200 bp) originating from the DGGE gel was sufficient to identify the isolate, the numbered bands were excised from the gel, eluted overnight and reamplified using GC free primers 338f-518r and sequenced using the same primers. Samples

Table 3. Identities of DGGE fragments related to the bands as determined by partial sequencing of 16S rDNA.

Strain	Closest relative in GenBank (access. no.)	% similarity	Phylogen. group
1	*Arthrobacter* sp. (X93356)	95	actinobacteria
2	*Agrobac.*sp. (AF508099)	98	α-proteobacteria
3	*Pseudom.*sp. (AF094745)	92	γ-proteobacteria
12D	*Micrococcus* sp. (AF218240)	100	actinobacteria
8	*Pseudom.* sp. (AF534198)	99	γ-proteobacteria
10	*Sinorhizo.* sp. (AF452129)	99	α-proteobacteria

[A] *E. coli* positions 357 to 432 (only 75 bp) of the forward sequence were sequenced.

number 1 and 12D were not determined probably due to the DNA loops extending over V2–V3 region of 16S rDNA. These partial sequences were then compared with the sequenced 712 bp 16S rDNA fragments.

ACKNOWLEDGEMENT

The work was sponsored by the grants 5FW EU No. QLK 3-CT-2000-00163, MSMT LN 00B030.

REFERENCES

Amos, P.W., Younger, P. 2003. Substrate characterisation for a subsurface reactive barrier to treat colliery spoil leachate. *Water Research* 37, 108–120.

Guzzella, L., Feretti, D., Monarca, S., 2002. Advanced oxidation and adsorption technologies for organic micropollutant removal from lake water used as drinking-water supply. *Water Research* 36, 4307–4318.

Haider, T., Sommer, R., Knasmuller, S., Eckl, P., Pribil, W., Cabaj, A., Kundi, M., 2002. Genotoxic response of Austrian groundwater samples treated under standardized UV (254 nm) – disinfections conditions in a combination of three different bioassays. *Water Research* 36, 25–32.

Koivusalo, M. and Vartiainen, T., 1997. Drinking water chlorination by-products and cancer. *Review of Environment and Health* 12, 81–90.

Macek, T., Macková, M., Káš, J. 2000. Exploitation of plants for the removal of organics in environmental remediation. *Biotechnol. Advances* 18 (1), 23–35.

Sayler, G.S. and Fox, R., 1990. *Environmental Biotechnology and waste treatment*, Plenum Press, New York.

Sergeev, V.I., Shimko, T.G., Kuleshova, M.L., Maximovich, N.G., 1996. Groundwater protection against pollution by heavy metals at waste disposal sites. *Water Science and Technology* 34 (7–8), 383–387.

European Symposium on Environmental Biotechnology, ESEB 2004 - Verstraete (ed)
© 2004 Taylor & Francis Group, London, ISBN 90 5809 653 X

Rapid biological treatment of residual DNAPL with slow release electron donor HRC-X™

Stephen S. Koenigsberg[1], Jeremy G.A. Birnstingl[2] and Anna Willett[1]
[1]*Regenesis, Calle Sombra, San Clemente, CA, USA*
[2]*Regenesis, Suite Greenwood, Princes Way, London*

The use of *in situ* bioremediation to stimulate the rapid dissolution, desorption, and biodegradation of residual DNAPL has been demonstrated in the laboratory and in well-documented field studies. Biodegradation of dissolved-phase contaminants increases the partitioning and subsequent biodegradation of residual DNAPL to the aqueous phase by (1) increasing the concentration gradient and driving force for dissolution and desorption and (2) increasing the overall solubility of the DNAPL by production of hydrophilic daughter products.

Specifically, the application of the slow release electron donor substrate, Hydrogen Release Compound-Extended Release (HRC-X™), has been successful in remediating high concentrations (>100 mg/L) of chlorinated ethenes, like PCE and TCE in residual DNAPL environments. *In situ* bioremediation with HRC-X is a low-cost method for residual DNAPL removal and avoids the costly and lengthy assessment associated with defining the exact location of the dispersed residual DNAPL.

HRC-X is a highly concentrated electron donor for bioremediation and has a field longevity of at least 3 years, as verified by field measurements of lactate and its derivative organic acids. Injection of HRC-X directly into the general residual DNAPL area of a contaminated aquifer results in the continuous release of lactic acid and fermentation of the lactic acid to hydrogen in and downgradient of the injection area. Hydrogen from HRC-X is used as an electron donor for reductive dechlorination, which results in dissolution of residual DNAPL and desorption of sorbed contaminants.

This presentation includes a description of HRC-X, as well as the mechanisms by which chlorinated ethene contaminants are dissolved, desorbed, and degraded. Case histories describing successful field applications of HRC-X and total project cost will be presented.

European Symposium on Environmental Biotechnology, ESEB 2004 - Verstraete (ed)
© 2004 Taylor & Francis Group, London, ISBN 90 5809 653 X

The efficacy of Oxygen Release Compound (ORC®): a nine year review

Stephen S. Koenigsberg,[1] Jeremy G.A. Birnstingl[2] & Anna Willett[1]
[1] *Regenesis, Calle Sombra, San Clemente, CA*
[2] *Regenesis, Suite Greenwood, Princes Way, London*

Oxygen Release Compound (ORC®) is proprietary formulation of intercalated magnesium peroxide that releases oxygen slowly and facilitates the aerobic bioremediation of a range of environmental contaminants, including petroleum hydrocarbons, certain chlorinated hydrocarbons, ether oxygenates (e.g. methyl *tert*-butyl ether [MTBE]), ammonia, certain herbicides, and arsenic. The history of ORC's introduction and acceptance represents a model for the evolution of an innovative technology. This statement comes by virtue of the fact that, since 1994, ORC has been applied 8,900 times worldwide and has been the subject of an extensive body of literature. This technology, known as a "time release electron acceptor", has now been clearly established as a sensible strategy for engineering accelerated bioattenuation on sites where design, capital, and management intensive options are either undesirable or contraindicated. ORC can be configured as a permeable reactive barrier, applied as a broader plume treatment, and emplaced post-excavation as part of the backfill. This presentation will summarize results from numerous field applications of ORC.

Some guidelines for using ORC have also emerged. It is contraindicated at sites where the BOD/COD load, seasonal or otherwise, is excessive or poorly understood, i.e., the technology is best applied to dissolved phase plumes and moderate levels of residual NAPL, once the majority of the source is removed. With regard to the range of compounds that can be addressed, ORC was first used for the remediation of BTEX and TPH groundwater contamination, and other applications have since been made for an array of other aerobically degradable compounds such as vinyl chloride, pentachlorophenol, polycyclic aromatic hydrocarbons and MTBE. With respect to MTBE, as early as 1996, consultants using ORC noticed that MTBE concentrations decreased at a higher than expected rate. Working on this foundation, in concert with published evidence that ethers are aerobically biodegradable, additional field experiments demonstrated that oxygen can indeed enhance the remediation of MTBE; a concept that has since been verified in other quarters.

European Symposium on Environmental Biotechnology, ESEB 2004 - Verstraete (ed)
© 2004 Taylor & Francis Group, London, ISBN 90 5809 653 X

Applications of Hydrogen Release Compound (HRC®) for accelerated bioremediation

Anna Willett[1], Jeremy G.A. Birnstingl[2] & Stephen S. Koenigsberg[1]
[1]*Regenesis, Calle Sombra, San Clemente, CA*
[2]*Regenesis, Suite Greenwood, Princes Way, London*

Hydrogen Release Compound (HRC®) is a food-grade, polylactate ester that, upon being deposited into an aquifer, slowly releases lactic acid for 12 to 18 months, creating the anaerobic conditions necessary for biodegradation of chlorinated solvents and many other contaminants. Fermentation of lactic acid from HRC by native bacteria produces a series of organic acids and results in the production of molecular hydrogen. Molecular hydrogen is an extremely efficient electron donor for a wide range of reductive biodegradation processes. The material is applied to the aquifer by push-point injection or backfill-auguring and can be applied in grid, barrier, or excavation formats. HRC is typically recommended for treatment of chlorinated solvent contamination found in the dissolved phase or sorbed to saturated soil. However, bioremediation can facilitate removal of residual source or DNAPL material at some sites. In addition to chlorinated solvent biodegradation, HRC has been used for bioremediation of explosives, perchlorate, chlorinated pesticides, and nitrate. HRC has been applied 474 times since 1997, making it the most widely-used electron donor for accelerating bioremediation. This presentation will give case histories, including European applications, and discussions of lessons learned.

European Symposium on Environmental Biotechnology, ESEB 2004 - Verstraete (ed)
© 2004 Taylor & Francis Group, London, ISBN 90 5809 653 X

New generation *in situ* bioremediation technologies

Merja Itävaara[1], Reetta Piskonen[1], Anu Kapanen[1], Terhi Kling[2], Eila Lehmus[2],
Janusz Sadowski[3] & Juhani Korkealaakso[2]

[1]*VTT Biotechnology, VTT, Finland*
[2]*VTT Building and transport, VTT, Finland*
[3]*VTT Information Technology, VTT, Finland*

In the area of remediation of contaminated soil and groundwater, there is a breakthrough going on: the traditional methods are replaced by *in situ* treatments, where the contaminants are degraded or extracted without removing the soil matrix.

The complex behavior of the pollutants in heterogeneous soil systems makes localization, distribution and remediation as well as monitoring and controlling the remediation processes a challenge. The major problems in *in situ* remediation are the inefficiency of monitoring the behavior and distribution of contaminants and their natural or enhanced biodegradation. The lack of methods to evaluate feasibility of the emerging treatment technologies makes *in situ* remediation a black box. Therefore, a multidisciplinary approach is needed for real practically oriented achievements.

Information is required on parameters, such as geophysical properties of soil structure, ground water flow, chemical composition of the contaminants, biodegradation potential and limiting factors as well as climatic conditions. A new integrated control and monitoring system will be developed for *in situ* remediation processes. The latest trend in the area of *in situ* remediation has been to integrate physical, chemical and biological treatments to get more efficient and cost-effective treatment systems, which can be managed as controllable processes. The main weakness in the *in situ* treatments is the inefficiency in monitoring the distribution, behavior and biodegradation of the contaminants. Without this information it is impossible to control and optimize the remediation processes.

Novel microbiological monitoring tools will be developed and integrated into the *in situ* treatment system. Monitoring of bioremediation has in the past based mainly on undirect methods. However, the latest developments in molecular biology have resulted in novel methods that allow direct monitoring of biodegradation processes.

The integrated monitoring and controlling system described would be able to determine spatial and temporal distributions of the contaminants in the soil and the effects of the remediation treatments to the mass transfer and bioavailability. Without this information it is impossible to control and optimize the remediation processes.

European Symposium on Environmental Biotechnology, ESEB 2004 - Verstraete (ed)
© 2004 Taylor & Francis Group, London, ISBN 90 5809 653 X

Analysis of structural properties of a mining soil in a process of phytostabilisation

L. Rizzi, G. Petruzzelli, G. Poggio & G. Vigna Guidi
CNR, Institute of Ecosystem Study, Unit of Soil Chemistry, Pisa, Italy

ABSTRACT: Particle size distribution, performed with a laser granulometer, and total porosity together with pore size and shape distribution, determined in soil thin sections by means of image analysis, were used to quantitatively evaluate modifications of soil structure of a sandy mining soil during a process of phytostabilisation. Experimental trials were carried out at a mesocosm scale. *Lolium italicum* and *Vetiveria zinanioides* were used for the setting up of mesocosms which were filled with soil alone or a mixture of soil (70%) and compost (30%) (v/v). A general lack of clear effects on soil structure ascribable to plants was found. The addition of compost caused a small yet evident increase of soil aggregation. Moreover, compost also raised total porosity by making greater the amount of pores in the size range of 0.05–0.5 mm.

1 INTRODUCTION

Phytostabilisation is an innovative technique with a low cost and very limited environmental impact which enables us to limit and prevent the diffusion of pollutants in contaminated soils or sites caused by leaching and/or erosion (Wong 2003). This result is obtained both by chemical–physical processes which occur at the level of the root apparatus of the plants, and also by means of a stabilisation of the soil operated by the roots and vegetation cover. Great importance is placed on the choice of plants to use (Bleeker 2002). They must develop an extensive root system and produce a large amount of biomass in soils which are often arid and with a low fertility. Furthermore, the plants should also be capable of tolerating high concentrations of heavy metals and be able to translocate into the epigeal part the minimum amount of pollutants. In land reclamation projects carried out on soils polluted by heavy metals the various aspects of soil structure, that represent one of the main factors influencing plant development, have not been sufficiently studied. In fact, the structure determines characteristics of drainage, erodibility and ease of soil tillage because it regulate the soil volume that can be explored by the roots, the availability of water for the plants and the movement of water, air, nutrients and pollutants.

The aim of this work is to investigate the variations in the physical structure of a mining soil that occur during an experimental process of phytostabilisation carried out at a mesocosm scale in greenhouse.

2 MATERIALS AND METHODS

2.1 *Soils*

The study areas are part of the mining district of Montevecchio, located in the south-west of the island of Sardinia (Italy) and which had been exploited for metal extraction, mainly Pb and Zn. Two soil samples were taken directly at the accumulation front of debris, high about 10 m, produced as a residue after separation of the useable part of the mineral. The first sample (A) was taken near the top of the heap, while the second (B) was sampled near the bottom.

2.2 *Plants*

Lolium italicum and *Vetiveria zizanioides* were used for the setting up of mesocosms in greenhouse after preliminary tests carried out with other plants.

2.3 *Mesocosms*

Mesocoms were prepared by filling pots of suitable size with 3 kg of air dried and carefully homogenised soil alone (hereafter referred to as A, and B) or a mixture of soil (70%) and compost (30%), on a volume basis, (hereafter referred to as A30%, and B30%).

An automatic irrigation system was used throughout the growing season of plants. The experiment was interrupted after 60 days when both plants had reached their maximum growth and had become to show signs of decline, also for the long period of unusually high temperature in late spring and early summer.

2.4 *Soil physical parameters*

The physical parameters of the soil structure taken into consideration are reported below.

2.4.1 *Particle size distribution*

The size distribution of primary components and aggregates were analysed by using a laser scattering technique working in the range 0.1–700 μm. Determinations were performed on soil–water suspensions passed through a 710 μm sieve (Pini et al. 1994).

2.4.2 *Porosity*

Thin sections were made, according to standard procedures (Pini et al. 1999), from undisturbed soil samples taken from superficial layers of mesocosms at the end of the experiment. Total porosity, pore shape, and size distribution of pores greater than 50 μm were determined on photographs of the sections by means of image analysis. Pores were divided into: (i) rounded, irregular, and elongated by using a shape factor; (ii) size classes according to their equivalent pore diameter (rounded and irregular pores) or to their width (elongated pores) (Pagliai et al. 1988).

3 RESULTS AND DISCUSSION

3.1 *General considerations*

The similar characteristics exhibited by both soils used in this experiment and listed in Table 1 (i.e. low pH, very little or none organic matter, high content of sand fraction and total Pb and Zn) gave evidence for a particularly harsh environment for plants to grow in. Such situation is even worsened by the severe shortage of rain typical of Mediterranean climates.

A set of plants were therefore chosen for their known capability to grow well in such infertile soils and to accumulate the least amount of Pb and Zn in their stems and leaves (EPA 2000). *Lolium italicum* and *Vetiveria zizanioides* were finally selected after having evaluated the results of preliminary laboratory trials carried out on microcosms with other plants (i.e. *Festuca arundinacea, Cynodon dactylon,* and *Poa pratensis*).

Compost was added to check how much a product rich in organic matter, easily available, and at a reasonable price could help plants to grow in such problematic soils, especially in their early crucial stages of development. The above characteristics make the use of compost also possible for planning a cost effective rehabilitation of mining sites (Bradshaw 2000). According to the present Italian legislation, composts with characteristics similar to those of the compost used in this experiment can be purchased and employed without any restriction for their low content of pollutants and pathogens.

Table 1. Selected characteristics of the investigated soils.

Parameter	Soil A	Soil B
pH (H_2O)	3.4	3.3
Sand (g kg^{-1})	960	991
OM (g kg^{-1})*	2.0	n.d.
EC (dS m^{-1})**	0.85	2.22
Pb-tot (mg kg^{-1})	1811	1623
Zn-tot (mg kg^{-1})	3510	2497

*OM: organic matter, **EC: electrical conductivity.

3.2 *Chemical determinations*

The Translocation Factor (TF), being TF = (metal content in shoots)/(metal content in roots), is a parameter commonly used in phytoremediation studies (Tassi et al. 2003, Tu et al. 2003) since it gives a clear picture on the capability of a given plant to distribute a metal extracted from the soil between its parts laying above and below the ground. TF should be as low as possible in phytostabilisation trials.

TFs (Table 2) are not greater than 0.23 for plants grown in mesocoms prepared with either soil A or B alone. This result means that, in the worst case mentioned before (i.e. Pb in *Lolium italicum* grown in soil B), the content of the metal in stems and leaves is about five times less than the content in roots. A sharp decrease of TF is observed whenever compost is added to soils. With regard to this finding, the behaviour of *Lolium italicum* is of particular interest, being the content of Pb below the detection limits in stems and leaves.

3.3 *Physical determinations*

Many parameters are commonly used for a quantitative evaluation of soil structure. Among them, particle size

Table 2. Translocation Factor of *Lolium italicum* and *Vetiveria zizanioides.*

	Translocation factor			
	Lolium italicum		*Vetiveria zizanioides*	
Soil sample	Pb	Zn	Pb	Zn
A	0.05	0.13	0.05	0.18
A30%	<0.01	0.08	0.02	0.14
B	0.23	0.11	0.07	0.12
B30%	<0.01	0.05	0.03	0.09

distribution performed with laser granulometers, and total porosity together with pore size and shape distribution determined on soil thin sections by means of image analysis techniques are reliable, widely used in soil physics and have been already found sensitive enough to make evident also slight modifications of soil structural characteristics (Pini et al. 1994, Pini et al. 1999). While samples used for particle size distribution are those commonly employed for chemical analyses, undisturbed soil samples are needed in order to get reliable information on the actual situation of the soil pore system.

Size distributions of primary components and aggregates of soil samples taken from mesocosms are shown in Figure 1. Actually, the laser granulometer was set to divide all particles present in the soil/water suspension, and falling in the analytical range of 0.1–700 μm, into 50 dimensional classes. However, to make comparisons between samples easier, the 50 classes were further grouped into four (<1 μm, 1–10 μm, 10–100 μm, and >100 μm).

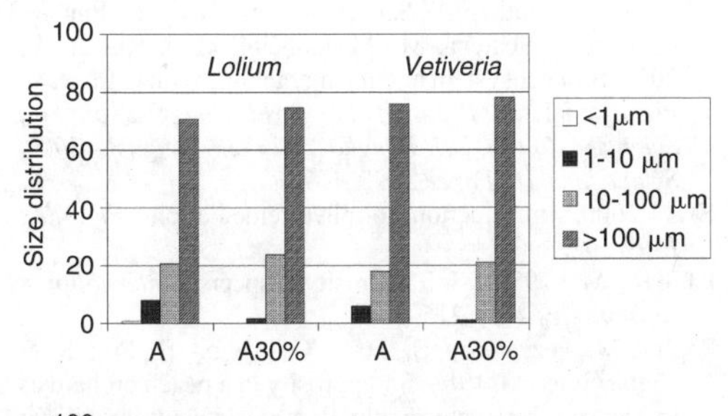

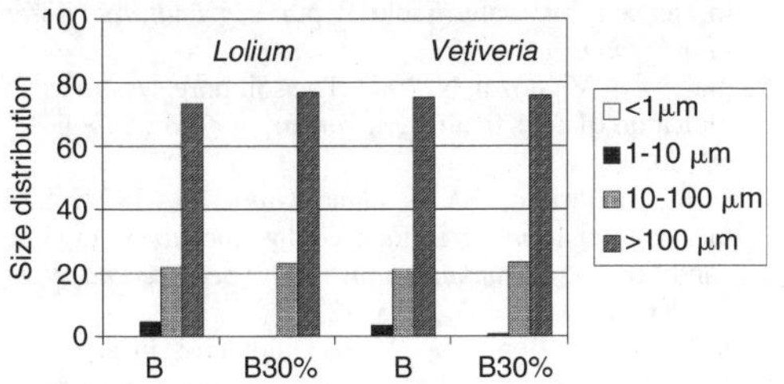

Figure 1. Particle size distribution in soils.

Table 3. Total porosity and shape of pores found in soil thin sections. All figures are expressed in per cent of the area actually analyzed.

Soil sample	Total porosity	Pore shape		
		Rounded	Irregular	Elongated
A	7.9	1.3	1.9	4.7
A30%	14.3	3.0	6.0	5.3
B	5.8	3.0	1.8	1.0
B30%	10.1	3.4	3.6	3.1

Granulometric compositions of both soils are very similar and they are particularly rich in sand-like material. In fact, particles belong mainly to the largest dimensional class (>100 μm) and to a lesser extent to that immediately smaller in size (10–100 μm). The impressive predominance of large particles, likely constituted more by primary components than by aggregates, can explain the lack of striking modifications caused by plants or by compost, since the difficulty of establishing strong linkages between sand particles to form stable aggregates is well known. However, the small content of particles <10 μm present in samples taken from mesocosms prepared with soils A and B alone, practically disappear after the addition of compost. This finding gives evidence that, even in these highly sandy soils, the structure can be improved through the ability of the added organic matter to promote the formation of some water stable aggregates, though limited to this small fraction of soil particles.

Values of total porosity, expressed as per cent of the total area of thin sections occupied by pores >50 m, and the part of this area taken up by pores with different shape are reported in Table 3, while related pore size distributions are shown in Figure 2.

No relevant differences were found between thin sections prepared from undisturbed soil samples taken from mesocosms planted with *Lolium italicum* or *Vetiveria zizanioides*. For this reason only figures referring to *Lolium italicum* are discussed.

Concerning total porosity, the micromorphometric method ranks soils in: very dense (total porosity lower than 5%), dense (total porosity in the range 5–10%), moderately porous (total porosity in the

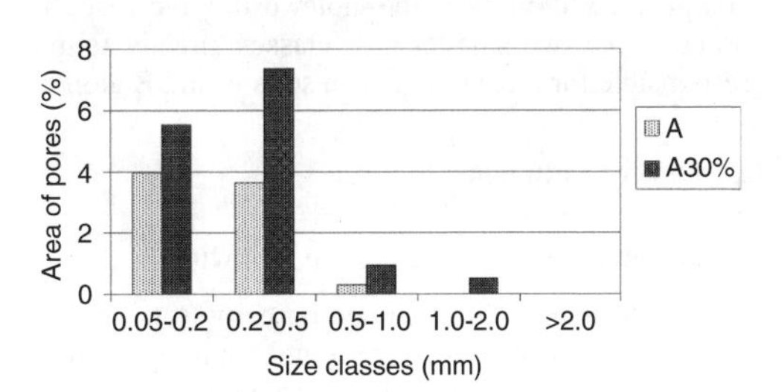

Figure 2. Pore size distribution in soils.

range 10–25%), highly porous (total porosity in the range 25–40%), and extremely porous (total porosity greater than 40%) (Pagliai 1988). As expected, total porosity exhibited a noteworthy increase when compost was added to both A and B soils allowing them to pass from dense to moderately porous. Similar results were found also in other experiments where total porosity was positively related to the amount of organic matter present in (D'Acqui 2002) or added to soil (Pagliai & Vignozzi 2002).

Pore shape and pore size distribution must be considered in addition to total porosity because they help to better relate the whole soil pore system, which is a crucial aspect of soil structure, to plant needs. In fact, many of the most important phenomena directly linked to plant growth, such as ease of root penetration, storage and movement of water and gases, depend on shape and size of pores. For soils A and B alone, pores were fairly well distributed among the three shape groups (rounded, irregular, and elongated) though elongated pores were more abundant in soil A and the rounded ones in soil B. After addition of compost the pore shape distribution did not change dramatically since pores of all shape classes increased in a more or less similar way (Table 1). Trends of pore size distributions were quite similar for soils A and B (Fig. 1) and most pores belonged to only two dimensional classes (0.05–0.2 mm, and 0.2–0.5 mm). Pores falling in the above ranges are commonly known for their capability to regulate transmission of water and exchange of gases and to allow the growth of most plant feeding roots. Compost had no effect on the size of pores, but for soil A30% where a very little percentage of pores in the size class 1.0–2.0 mm appeared after the treatment with the organic material. The striking result of compost addition was the noteworthy increase of pores in the two smaller size classes already found responsible for most porosity in soils A and B alone.

4 CONCLUSIONS

Main concluding remarks are the following:

- particle size distribution and porosity (total, shape and size distribution) were found appropriate for a quantitative evaluation of soil structure in phytostabilization trials carried out in mesocosms;
- a general lack of clear effects on soil structure ascribable to either plant employed was found throughout the experiment;
- despite the high presence of sand-like materials in both soils, the addition of compost caused a small yet evident increase of soil aggregation and an increment of total porosity, mainly due to pores in the size range 0.05–0.5 mm;
- *Lolium italicum* and *Vetiveria zizanioides* showed the needed characteristics for their possible utilization in phytostabilization processes in this mining environment.

ACKNOWLEDGEMENTS

CNR funded this work within the GNDRCIE project; we thank Mr. B. Castorina for providing plants of *Vetiveria zizanioides*, and Messrs.M. La Marca and M. Scatena for technical assistance.

REFERENCES

Bleeker, P.M., Assuncao, A.G.L., Teiga, P.M., de Koe, T. & Verkleij, J.A.C. 2002. Revegetation of the acidic, As contaminated mine spoil tips using a combination of spoil amendment and tolerant grasses. *The Science of the Total Environment* 300: 1–13.

Bradshaw, A. 2000. The use of natural processes in reclamation-advantages and difficulties. *Landscape and Urban Planning* 51: 89–100.

D'Acqui, L.P., Dodero, A., Santi, C.A., Pezzarossa, B., Pini, R., Petacco, F., Scatena, M., Mazzoncini, M. & Risaliti, R. 2002. Soil ecosystem and its interactions with C fluxes in Pianosa island. *Proc. of 17 World Congress on Soil Science, (on CD) Bangkok, 14–20 August 2002, Symposium 10, Paper* 2095: 1–7.

EPA, 2000. Introduction to phytoremediation. *EPA/600/R-99/107*.

Pagliai, M. 1988. Soil porosity aspects. *International Agrophysics* 4(3): 215–232.

Pagliai, M., Pezzarossa, B., Zerbi, G., Alvino, B., Pini, R. & Vigna Guidi, G. 1988. Soil porosity in a peach orchard as influenced by water table depth. *Agricultural Water Management* 16: 63–73.

Pagliai, M. & Vignozzi, N. 2002. The soil pore system as an indicator of soil quality. *Advances in GeoEcology* 35: 69–80.

Pini, R., Canarutto, S. & Vigna Guidi, G. 1994. Soil microaggregation as influenced by uncharged organic conditioners. *Communications in Soil Science and Plant Analysis* 25(11 & 12): 2215–2229.

Pini, R., Paris, P., Benetti. A., Vigna Guidi, G. & Pisanelli, A. 1999. Soil physical characteristics and understory management in a walnut (*Juglans regia* L.) plantation in central Italy. *Agroforestry Systems* 46: 96–105.

Tassi, E., Mastretta, C., Rizzi, L. & Barbafieri, M. 2003. PHYTODEC: European project for the use of vegetation in the management of heavy metal polluted sites. Enhancing metal bioavailability for phytoremedation strategy. In, *ConSoil 2003; Proc. intern. conf. (on CD), Gent, 12–16 May 2003*: 2682–2690. Leipzig: F&U confirm.

Tu, C., Ma, L.Q. & Bondada, B. 2003. Arsenic accumulation in the hyperaccumulator Chinese brake and its utilization potential for phytoremediation. *Journal of Environmental Quality* 31(5): 1671–1675.

Wong, M.H. 2003. Ecological restoration of mine degraded soils, with emphasis on metal contaminated soils. *Chemosphere* 50: 775–780.

Environmental genomics

European Symposium on Environmental Biotechnology, ESEB 2004 - Verstraete (ed)

Development and validation of a microarray for the identification of microorganisms responsible for bulking and foaming in WWTPs

Sofie Dobbelaere, Nico Boon, Han Vervaeren, Jorg Matthys, Kurt Sys & Willy Verstraete

Lab. Microbial Ecology and Technology (LabMET), Fac. Agric. and Appl. Biological Sciences, Ghent University, Coupure, Gent, Belgium

Vincent Denef & Syed Hashsham

Department of Civil and Environmental Engineering and the Center for Microbial Ecology, Michigan State University, Research Complex-Engineering, East Lansing, MI

ABSTRACT: A prototype microarray was developed, composed of 16S rRNA-targeted oligonucleotide probes covering all species relevant to foaming or bulking, including group-specific probes for the taxonomic levels of genus, family and higher. The probes were *in situ* synthesized on a microfluidic biochip or spotted on a glass-based array. Pure cultures were used to validate the system. Two labeling techniques (random and 16S rRNA specific) were compared and annealing and melting profiles were recorded. The first results presented here demonstrate that the technique is able to discriminate between match and mismatch probes. However, further optimization is needed to increase specificity.

1 INTRODUCTION

Microbial communities constitute the core component of a biological waste water treatment plant (WWTP) and are responsible for the overall conversions taking place in this system. Monitoring these communities may provide more powerful information on the structure and thus the quality of the WWTP microbial community. This information could then be used to reveal factors influencing the efficiency and stability of biological WWTPs and to develop strategies to improve system performance.

Up to recently, the characterisation of the microbiology of activated sludge was based on techniques such as plate counts, light microscopic observation, BIOLOG patterns, FISH mapping of specific strains and outlining of populations by means of their 16S rRNA genes (Eikelboom, 1975, Wagner et al., 1992, Jenkins et al., 1993, van der Merwe et al., 2003). These methods have served to gain insight in the composition of the microbial ecology of activated sludge. Yet, all these methods are quite time-consuming and thus cannot be implemented in high-throughput configurations.

Nucleic acid microarrays, which have recently been introduced for bacterial identification in microbial ecology (Guschin et al., 1997, Liu et al., 2001, Small et al., 2001), provide a powerful tool for parallel detection of hundreds or thousands of genes per single experiment, allowing for the simultaneous detection of a great variety of different microorganisms in a single sample and thus might be particularly useful for analysis of microbial community structure. However, most microarrays developed so far for bacterial identification consist of a limited number of probes and environmental applications are still in the early stages of evaluation. Up till now, substantial information on the performance of microarrays with complex environmental samples is still lacking.

Our aim was to establish the methodology, feasibility, and applicability of this technique for the analysis of microbial community structure in a complex environment, in particular domestic WWTPs, and to establish current detection limits. Therefore we constructed a first prototype microarray composed 16S rRNA-targeted oligonucleotide probes covering all species relevant to foaming or bulking.

Two common problems of sludge settling in wastewater treatment are the occurrence of foaming and bulking sludge. These give rise to poor separation of sludge and effluent. Bulking and foaming are complex phenomena that involve the overgrowth of filamentous bacteria. The relation of the growth of these filaments to the chemical composition of the influent and the operating parameters are largely unknown (Jenkins, 1992, Wanner, 1994). Diagnostic methods are available to identify the problem when occurring, e.g. by SVI determination, light microscopy, etc. However, these methods have serious drawbacks. Filamentous organisms can be morphologically non distinguishable, but genetic divers and the inverse is also true. In many cases, inaccurate methods to enhance the settlement of the foaming or bulking sludge are often applied, causing lower water treatment efficiency, high sludge volumes and high costs. Moreover, these methods do not allow prediction of the phenomena and their evolution in the early stages of development. With this prototype microarray we want to investigate whether this technique allows monitoring of bulking and foaming sludge events by early detection of the causing organisms. Here we present the results of the initial validation of the microarray methodology.

2 MATERIALS AND METHODS

2.1 *Microbial strains*

Gordonia amarae DSM 43392, *Tsukamurella paurometabola* DSM 20162, *Trichococcus flocculiformis* DSM 2094, *Acidovorax delafieldii* DSM 64 and *Nocardia farcinica* Trevisan LMG 4079 were used as sources of nucleic acids for these experiments.

2.2 *Probe design*

Oligonucleotide probes targeting the 16S rRNA of 91 organisms involved in bulking and foaming in WWTPs were designed using the Probe_Design tool of the ARB software package (Strunk et al., 1998). In addition, group-specific probes for the taxonomic levels of genus, family, and higher were designed using the Calculate_Multi-Probe tool of the ARB software package. Figure 1 illustrates the phylogeny of these probes. The probes were designed to have a G + C content of about 52%, a length of 18 nucleotides, a Tm of 58°C and as many centrally located mismatches with the target sites on the 16S rRNA genes of non-target organisms as possible. In addition, a universal bacterial probe, a NONEUB probe and an Arabidopsis probe were included in the list for hybridization control purposes. In the case of the Xeotron microarrays, for each probe a mismatch probe was designed by randomly changing 1 base in the center of the probe. The Match_Probes-tool of the ARB software package, the PROBEmer program (Emrich et al., 2003) and BLAST search (Altschul et al., 1990) were used to evaluate probe specificity.

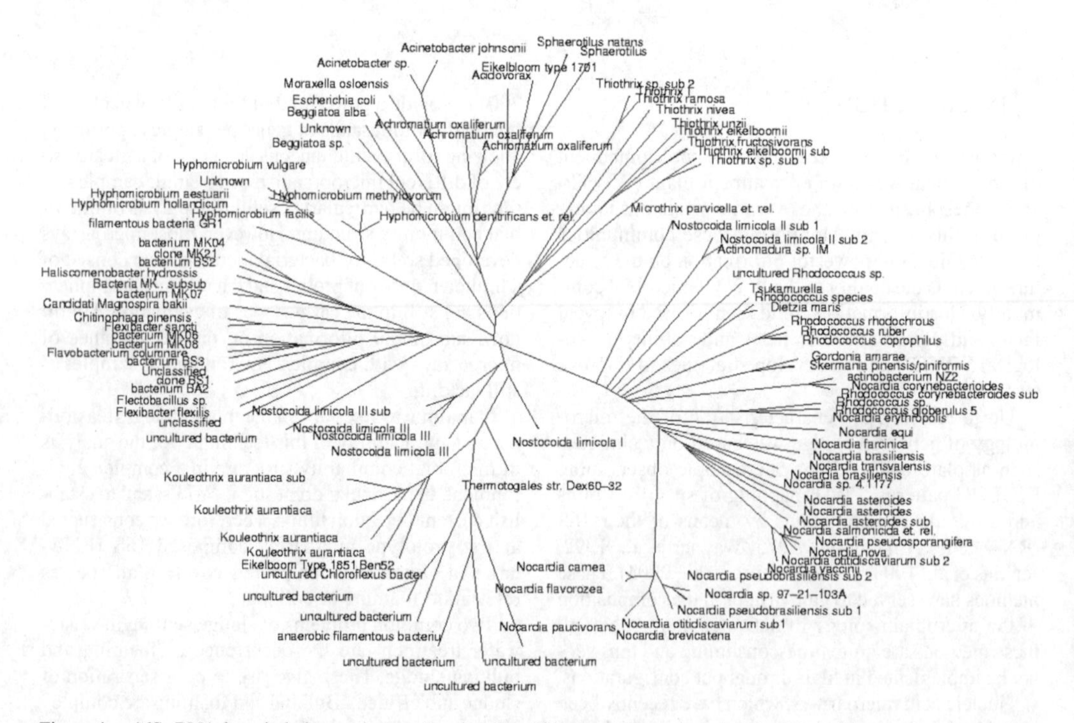

Figure 1. 16S rRNA-based phylogenetic tree showing the affiliation of organisms involved in bulking or foaming and for which probes were designed.

2.3 *Microarray construction*

Two different types of microarrays were used. The first one was a nano-chamber microarray biochip (XeoChip)® produced by Xeotron combining PGR chemistry, digital photolithography, and parallel microfluidics for *in situ* synthesis of probes. The probes were attached by the 5′ end with a 12-mer linker. The second type was a glass-based microarray produced by Eurogentec. For this type of array the probes contained a 5′-C6-amino linker used to immobilize the oligonucleotide probes to the array and to increase the on-chip accessibility of spotted probes to target DNA and were spotted at a concentration of 40 pmol/μl. Oligos were also synthesized by Eurogentec.

2.4 *RNA extraction*

Total cellular RNA was isolated by 'low pH hot phenol' extraction. Cells of 30 ml of an exponentially grown culture were collected and resuspended in 0.8 ml of a 50 mM sodiumacetate, 10 mM EDTA solution (pH 4.2). 0.5 g glass beads and 700 ml phenol (pH 4.2) were added and the cells were mechanically disrupted by bead-beating during 2 min. at max. speed. Samples were then incubated at 60°C for 10 min. and subsequently centrifuged at 5000 rpm for 5 min. The aqueous and organic phase were transferred to a new eppendorf tube. The bead-beating step was repeated for 1 min. by adding again 200 μl buffer to the beads. Both supernatants were combined and centrifuged for 10 min. at maximum speed. The aqueous phase was transferred to a new tube, washed with 800 μl phenol (pH 4.2) and 200 μl chloroform isoamylalcohol and centrifuged for 10 min. at maximum speed. This step was repeated twice. The aqueous phase was then washed with chloroform isoamylalcohol and the RNA was collected by precipitation overnight in 1 ml isopropanol and rinsed with 70% ethanol. If necessary, a DNAse treatment was performed on the samples.

2.5 *Preparation of fluorescently labeled cDNA*

A direct labeling procedure was employed to fluorescently label the cDNA. To 10 μg of purified RNA 30 μg random hexamer primers or 50 μg 16S specific primers were added. The mixture was denatured during 10 min. at 70°C and chilled on ice. To this, 1 × RT buffer; 500 μM dATP, dGTP and dTTP; 200 μM dCTP; 100 μM Cy3- or Cy5-dCTP (Amersham Pharmacia Biotech); 20U RNase inhibitor and 250U MML RT were added (final concentration, total volume 50 μl). The mixture was incubated at 42°C for 2 hours. Then 25 μl NaOH (1 M) was added and further incubated for 10 min. at 65°C. After incubation, 25 μl HCl (1 N) and 10 μl NaOAC (3M pH 5.2–5.5) were added and the mixture was purified using a Qiaquick column (Qiagen). Recovered cDNA was eluted with 30 μl of water.

2.6 *Microarray hybridization*

Two kind of hybridization experiments were performed. Annealing profiles (Xeochips) were recorded by hybridising the arrays at decreasing temperature, starting at 50°C and decreasing till 20°C, each time followed by a washing step and scanning using Genepix-software. Melting profiles (glass-based arrays) were recorded by hybridizing the microarrays overnight at 20°C, followed by different wash steps at increasing temperature, up to 60°C. After each wash step, the arrays were scanned. Hybridisation buffer contained 6 × SSPE pH 6.6, 1 mg/ml BSA, 20% formamide and 1–3 μg labeled cDNA. Following hybridisation, the arrays were washed at 30°C (or at increasing temperatures in the case of melting profiles) with 3 separate filter sterilized wash buffers: 2 × SSC and 0.1% SDS; 1 × SSC; 0.1 × SSC. Data were further analyzed in R (Ihaka & Gentleman, 1996).

3 RESULTS

The median intensities of the spots were used as this is less influenced by outliers than the mean. After excluding bad spots, the background was calculated using an interpolation procedure and signal/background ratios were calculated. Figure 2 shows an example of an annealing profile obtained upon hybridisation of a Xeotron microarray with a pure culture of *T. flocculiformis*. From this it can be seen that in the case of specific hybridisation a high signal is obtained for the match probe. In the case of non-specific hybridisation, no difference in signal could be observed between match and mismatch probe.

Based on these annealing and melting profiles, two temperatures were selected and the percentage of specific ('target'-probe) and non-specific hybridization ('non-target'-probe) at each temperature was determined. The results are shown in Table 1. In the case of specific hybridization, it is expected that all species- and group-specific probes will give signal (=100%). This is not the case. For the Xeochips, on average good signals were obtained for the group-specific probes, while the species-specific probes failed to give a signal in most cases. In addition, non-specific binding to non-target probes occurred. From this table it is also clear that by hybridizing at higher temperature, the non-specific binding can be reduced. However, also the specific hybridization was reduced in most cases.

About the same results were obtained for the glass-based microarrays. However, based on these melting profiles it is not possible to get information on the influence of temperature on the specificity of the signal.

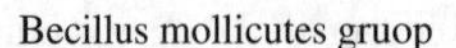

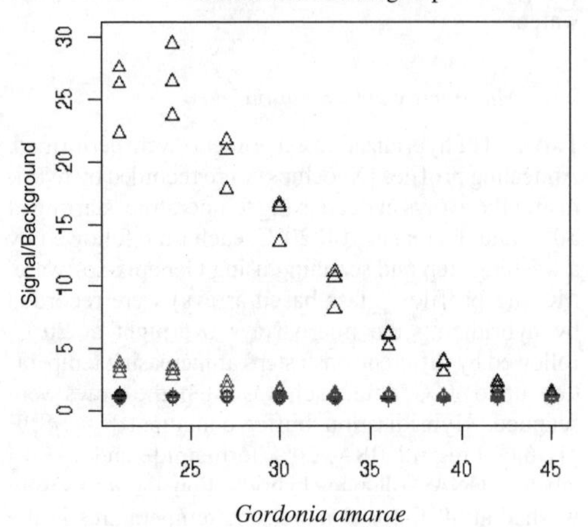

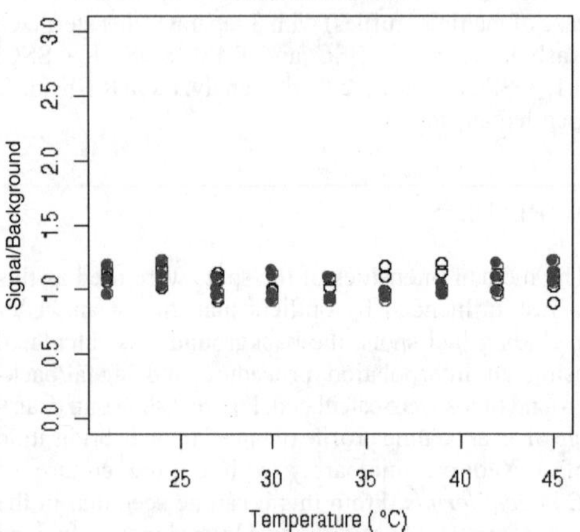

Figure 2. Annealing curve for a group-specific (B. mollicutes group) and a non-specific probe (*G. amarae*) after hybridisation with fluorescently labeled 16S rDNA of *T. flocculiformis*. Open symbols represent match probes, closed symbols represent mismatch probes.

4 DISCUSSION

The probe design of phylogenetic microarrays is very complex and not yet applied. Although in our study some specific probes gave a signal, the specificity certainly needs to be improved. Even at the lowest hybridization temperature a maximum of only 50% specific hybridization was obtained. Up till now the knowledge of (microarray) probe thermodynamics is limited. Although probes were designed taking into account generally accepted rules, it may be that they do not work properly due to some (yet) unknown factors. More fundamental knowledge of probe behaviour is necessary to overcome these problems. Recently, more information has become available on probe design

Table 1. Overview of the results obtained upon hybridization of the 2 microarray types (3 replicates) at 2 different hybridisation temperatures.

	% Specific hybridisation		% Non-specific hybridisation	
Organism	20°C	40°C	20°C	40°C
Xeochips				
T. flocculiformis	40	33	2.4	0.5
G. amarae	50	28	3.3	0.7
T. paurometabola	14	14	1.5	0.0
A. delafieldii	30	0.0	2.0	0.0
Glass-based arrays				
T. flocculiformis	20	20	2.5	2.5
G. amarae	33	33	3.1	3.1
N. farcinica	40	40	8.1	8.1

and hybridisation dynamics (Csontos & Bodrossy, 2003). This information will be used to further check the probes and eventually optimise or redesign them.

The use of well-defined primer sets targeting the 16S rRNA was expected to increase specificity and eliminate a lot of the non-specific hybridization observed so far, as this should result in a higher target concentration in the hybridization mix. However, the experiments performed on the glass-based arrays did not support this hypothesis. Especially the temperature seems to be a critical factor in increasing specificity. Alternatively, a PCR-step can be used to amplify the target. This should also increase the target concentration and as such the signal intensity.

In addition, the signal intensity can be improved by fragmentation (enzymatic or chemical) of the target. If the target is too long, sterical hindrance and the formation of secondary structures might hamper efficient binding to the probe. The use of shorter targets should improve homogeneity and intensity of the hybridisation signal.

REFERENCES

Altschul, S.F., Gish, W., Miller, W., Myers, E.W. & Lipman, D.J. 1990. Basic local alignment search tool. *J. Mol. Biol.* 215: 403–410.

Csontos, J. & Bodrossy, L. 2003. CalcOligo – a software for phylogenetic probe design [WWW document] URL www.diagnostic-arrays.com/calcoligo/index.htm

Eikelboom, D.H. 1975. Filamentous organisms observed in activated sludge. *Water Res.* 9: 365–388.

Emrich, S.J., Lowe, M. & Delcher A.L. 2003. PROBEmer: a web-based software tool for selecting optimal DNA oligos. *Nucleic Acids Res.* 31(13): 3746–3750.

Guschin, D.Y., Mobarry, B.K., Proudnikov, D., Stahl, D.A., Rittmann, B.E. & Mirzabekov, A.D. 1997. Oligonucleotide microchips as genosensors for determinative and environmental studies in microbiology. *Appl. Environ. Microbiol.* 63: 2397–2402.

Ihaka, R. & Gentleman, R. 1996. R: A language for data analysis and graphics. *J. Comput. Graph. Stat.* 5(3): 299–314.

Jenkins, D. 1992. Towards a comprehensive model of activated sludge bulking and foaming. *Water Sci. Technol.* 25: 215–230.

Jenkins, D., Richard, M.G. & Daigger, G.T. (eds.) 1993. *Manual on the Causes and Control of Activated Sludge Bulking and Foaming.* Edn. 2. Boca Raton, Fl: Lewis Publishers.

Liu, W. T., Mirzabekov, A.D. & Stahl D. A. 2001. Optimization of an oligonucleotide microchip for microbial identification studies: a non-equilibrium dissociation approach. *Environ. Microbiol.* 3: 619–629.

Small, J., Call, D.R., Brockman, F.J., Straub, T.M. & Chandler, D.P. 2001. Direct detection of 16S rRNA in soil extracts by using oligonucleotide microarrays. *Appl. Environ. Microbiol.* 67: 4708–4716.

Strunk, O., Gross, O., Reichel, B., May, M., Hermann, S., Stuckman, N., Nonhoff, B., Lenke, M., Ginhart, A., Vilbig, A., Ludwig, T., Bode, A., Schleifer, K.-H. & Ludwig, W., 1998. http://www.mikro.biologie.tumuenchen.de/pub/ARB).

van der Merwe, T., Wolfaardt, F. & Riedel, K.H. 2003. Analysis of the functional diversity of the microbial communities in a paper-mill water system. *Water SA* 29 (1): 31–34.

Wagner, M., Loy, A., Nogueira, R., Purkhold, U., Lee, N. & Daims, H. 2002. Microbial community composition and function in wastewater treatment plants. *Anton. Leeuw. Int. J. G.* 81(1–4): 665–680.

Wanner, J. 1994. *Activated sludge bulking and foaming control.* Lancaster, Pa.: Technomic Publishing Company, Inc.

European Symposium on Environmental Biotechnology, ESEB 2004 - Verstraete (ed)
© 2004 Taylor & Francis Group, London, ISBN 90 5809 653 X

Abundance of mobile genetic elements, antibiotic resistance traits, and catabolic genes in a BTEX contaminated aquifer measured by DNA microarray and culture-dependent analyses

J.R. de Lipthay, H. Jørgensen, & S.J. Sørensen
Department of General Microbiology, University of Copenhegen, Copenhagen K, Denmark

ABSTRACT: Several studies indicate that mobile genetic elements (e.g. plasmids and transposons) play a significant role in the adaptation of microbial communities to contaminant exposure. However, only few field studies have been done to demonstrate this influence. In the present study we measured the abundance of mobile genetic elements, antibiotic resistance traits and catabolic genes in BTEX contaminated and non-contaminated sediment samples from an aquifer at an industrial site in the Czech Republic by application of cultivation-dependent and cultivation-independent techniques. We included the analysis of antibiotic resistance as these traits are often found on catabolic plasmids. Generally we observed culturable streptomycin, ampicillin, and chloramphenicol resistant bacteria in all samples, kanamycin resistant bacteria in some samples, and tetracycline and mercury resistant bacteria in few samples. No correlation was found between the abundance of antibiotic resistant bacteria and the level of BTEX contamination, although high levels of kanamycin and chloramphenicol resistant bacteria were only found in the highly contaminated aquifer sediments. Experiments showed that the abundance of plasmid DNA was not significantly different in bacterial isolates retrieved from aquifer sediments with high or low levels of BTEX compounds. Also, there was no significant difference in the abundance of plasmid DNA isolated from kanamycin resistant bacteria compared to bacteria isolated from non-selective agar plates. In order to embrace the total indigenous bacterial population, microarray analyses are employed to investigate the possible correlation between the level of BTEX contamination and the incidence of plasmid, transposon, antibiotic resistance, and catabolic genes.

1 INTRODUCTION

The importance of horizontal gene transfer in the adaptation of natural microbial communities to contaminant exposure has long been suggested (van der Meer et al. 1992). From bacterial pure culture isolates it is known that a number of genes encoding catabolic degradation pathways and resistance to antibiotics and heavy metals are located on plasmids (Frantz & Chakrabarty 1986, Jacoby 1986). However, the involvement of mobile genetic elements (e.g. plasmids and transposons) in the microbial adaptation *in situ* has only been sparsely investigated. Horizontal gene transfer may be a mean for the bacterial community to maintain the gene pool in changing environments. If conditions suddenly change dramatically, due to pollution, human intervention etc., the bacterial community may permanently lose important genetic traits. This loss may be prevented if genes providing "resistance" to the perturbation can be horizontally distributed within the bacterial community.

In this study we want to correlate the abundance of mobile genetic elements, antibiotic resistance traits and catabolic genes to the level of BTEX (Benzene Toluene Ethylbenzene Xylene) contamination in a subsurface aquifer by application of cultivation-dependent and cultivation independent techniques.

2 MATERIALS AND METHODS

2.1 *Sediment samples*

BTEX contaminated and non-contaminated aquifer sediment was obtained from an industrial site in the Czech Republic. Samples were collected at locations A0, A1, A2, A3 and A4. Sediment from the unsaturated (X-samples), the occasionally saturated (Y-samples), and from the constantly water saturated (Z-samples) zones were retrieved. In Table 1 the number of sediment samples and their BTEX contamination level is described.

2.2 *Culturable antibiotic resistant bacteria*

The abundance of culturable antibiotic and heavy metal resistant bacteria was estimated by plating sediment suspensions on 10% PTYG (Peptone Tryptone Yeast extract Glucose) agar media containing either kanamycin (50 μg ml^{-1}), tetracycline (20 μg ml^{-1}), streptomycin (100 μg ml^{-1}), ampicillin (100 μg ml^{-1}), chloramphenicol (50 μg ml^{-1}) or mercury chloride (10 μg ml^{-1}). Plating on 10% PTYG agar without amendments served as control. All agar media were supplemented with nystatin (50 μg ml^{-1}) to prevent fungal growth. Five grams of sediment were diluted in 45 ml sterile NaCl (0.9% wt/vol). Samples were shaken for 15 min at 250 rpm, and tenfold dilutions were prepared. Each sample was plated in triplicate and agar plates were incubated at 24°C for 4 days before enumeration of colony forming units.

2.3 *Plasmid DNA preparation*

Plasmid DNA was purified from bacterial colonies isolated from the highly contaminated sediment samples A1-X, A1-Y and A1-Z (July 2001) and from the less contaminated sediments A3-X, A3-Y and A3-Z (October 2001). One hundred randomly chosen isolates were picked from each sediment sample and grown in 7 ml of 10% PTYG broth. Furthermore, 50 randomly chosen kanamycin resistant isolates were obtained from samples A1-X, A1-Y and A1-Z (July 2001), and from A4-Z (October 2001) and grown in kanamycin amended 10% PTYG broth.

Plasmid DNA was isolated from the bacterial isolates by alkaline cell lysis and phenol:chloroform:isoamyl alcohol extraction as described in protocols by Birnboim & Doly (1979) and Sambrook et al. (1989). Visualization of plasmid DNA was performed by agarose (0.7%) gel electrophoresis and staining in 1 μg ml^{-1} ethidium bromide. By this procedure plasmid DNA up to approximately 100 kb can be visualized.

3 RESULTS AND DISCUSSION

3.1 *Abundance of culturable antibiotic resistant bacteria*

The percentage abundance of culturable antibiotic and mercury resistant bacteria compared to the number of bacteria retrieved on non-selective 10% PTYG media is indicated in Table 1. Generally we observed culturable streptomycin, ampicillin, and chloramphenicol resistant bacteria in all samples, kanamycin resistant bacteria in some samples, and tetracycline and mercury resistant bacteria in few samples. No correlation was found between the abundance of

Table 1. The percentage abundance of culturable antibiotic and mercury resistant bacteria in a BTEX contaminated subsurface aquifer.*

Sediment sample	Level of BTEX contamination	Km (%)	Tc (%)	Sm (%)	Ap (%)	Cm (%)	Hg (%)
A0-X (July 2002)	+	0.060	0.000	0.174	3.816	0.067	0.000
A0-Y (July 2002)	+	0.190	0.000	0.471	5.142	0.233	0.000
A0-Z (July 2002)	+	0.050	0.000	0.079	1.487	0.103	0.008
A1-X (July 2001)	+ + + + +	0.233	0.002	1.065	6.969	0.311	0.000
A1-Y (July 2001)	+ + + + +	4.421	0.002	9.285	81.00	29.10	0.000
A1-Z (July 2001)	+ + + + +	0.809	0.000	1.268	90.61	25.65	0.000
A1-X (July 2002)	+ + + + +	2.236	0.001	0.063	33.32	4.571	0.321
A1-Y (July 2002)	+ + + + +	0.000	0.000	0.000	26.22	18.72	0.000
A1-Z (July 2002)	+ + + + +	0.000	0.000	0.009	55.15	36.23	0.000
A2-X (July 2001)	+ + + + +	1.511	0.000	0.934	49.77	4.529	0.000
A2-Y (July 2001)	+ + + + +	0.154	0.000	1.719	8.848	0.986	0.000
A2-Z (July 2001)	+ + + + +	1.569	0.002	2.043	8.477	22.39	0.000
A3-X (October 2001)	+ +	0.000	0.000	0.335	71.69	0.024	0.000
A3-Y (October 2001)	+ +	0.000	0.003	0.048	15.49	0.028	0.000
A3-Z (October 2001)	+ +	0.000	0.000	0.000	39.66	1.348	0.000
A3-X (July 2002)	+ + + +	0.102	0.037	0.083	3.918	0.028	0.003
A3-Y (July 2002)	+ + + +	0.003	0.000	0.019	1.457	0.030	0.001
A3-Z (July 2002)	+ + + +	0.005	0.000	0.009	0.931	0.054	0.029
A4-Z (October 2001)	+ + + +	0.741	0.000	0.478	55.74	10.28	0.000

* Level of BTEX contamination: the higher number of + indicates a higher concentration of BTEX compounds.
Km, kanamycin; Tc, tetracycline; Sm, streptomycin; Ap, ampicillin; Cm, chloramphenicol; Hg, mercury chloride.

antibiotic resistant bacteria and the level of BTEX contamination, although high levels of kanamycin (>0.7%) and chloramphenicol (>4.5%) resistant bacteria were only found in the most highly contaminated aquifer sediments.

3.2 *Abundance of plasmid DNA in culturable bacteria*

Experiments showed that the abundance of plasmid containing bacterial isolates from A1 sediment samples grown at non-selective conditions was 14–32%. For the A3 samples, 7–41% of the isolates contained plasmid DNA. Thus, the higher concentration of BTEX compounds in the A1 sediments did not have a significant effect on the frequency of plasmid containing bacteria. In contrast, Burton et al. (1982) and Campbell et al. (1995) found a higher abundance of plasmids in bacteria isolated from contaminated compared to non-contaminated environments.

When looking at the abundance of plasmid DNA among the kanamycin resistant bacterial isolates, no significant difference was found compared to the plasmid frequency among bacteria isolated at non-selective conditions. Thus, with the A1 samples, 12–36% of the bacterial isolates contained plasmid DNA, whereas 34% of the isolates from A4 did.

3.3 *Microarray analyses*

The data presented here only give information on the culturable fraction of the bacterial community. In order to embrace the total indigenous bacterial population of the BTEX contaminated aquifer we are performing microarray analyses in order to investigate the presence of plasmid genes (e.g. IncM, IncN, IncP, IncQ, IncT, IncW), transposon genes (e.g. *tnp*, IS elements), antibiotic resistance genes (e.g. *cat, nptII, str, tet*), mercury resistance genes (*mer*), and catabolic genes (e.g. *nah, phn, xyl, cat, bph*).

REFERENCES

Birnboim, H.C. & Doly, J. 1979. Arapid alkaline extraction procedure for screening recombinant plasmid DNA. *Nucleic Acids Research* 7: 1513–1523.

Frantz, B. & Chakrabarty, A.M. 1986. Degradative plasmids in *Pseudomonas*. In J.R. Sokatch & L.N. Ornston (ed.), *The Bacteria*, vol. 10, p. 295–323, Academic Press, USA.

Jacoby, G.A. 1986. Resistance plasmids of Pseudomonas. In J.R. Sokatch & L.N. Ornston (ed.), *The Bacteria*, vol. 10, p. 265–295, Academic Press, USA.

van der Meer, J.R., de Vos, W.M., Harayama, S. & A.J.B. Zehnder. 1992. Molecular mechanisms of genetic adaptation to xenobiotic compounds. *Microbiol. Rev.* 56: 677–694.

Sambrook, J., Fritsch, E.F. & Maniatis, T. 1989. *Molecular cloning: a laboratory manual.* Second edition. Cold Spring Harbor Laboratory Press, USA.

European Symposium on Environmental Biotechnology, ESEB 2004 - Verstraete (ed)
© 2004 Taylor & Francis Group, London, ISBN 90 5809 653 X

Plasmid isolation and curing in gram-positive *Enterococcus faecalis*

J. Keyhani
Laboratory for Life Sciences, Tehran, Iran

E. Keyhani
Laboratory for Life Sciences, Tehran, Iran, and Institute of Biochemistry and Biophysics, University of Tehran, Tehran, Iran

ABSTRACT: This research reports a novel and efficient method for rapid plasmid isolation from a clinical isolate of *Enterococcus faecalis* exhibiting resistance to kanamycin (MIC 2 mg/ml) and tetracycline (MIC 50 μg/ml). Results showed that the method which involves the use of sodium N-lauroylsarcosinate allowed for the reproducible isolation of two plasmids of molecular weight ~7 kb and 5.7 kb, respectively, as determined by gel electrophoresis in 1% agarose. The plasmid DNA thus prepared would transform *E. coli* and render it resistant to kanamycin (MIC 2 mg/ml) and tetracycline (MIC 50 μg/ml). Comparatively, the routine alkaline lysis procedure produced only one plasmid, barely detectable by agarose gel electrophoresis in one out of three preparations; the boiling procedure, although more efficient than the alkaline lysis procedure, also led to the detection of one plasmid only. Thus the method gave results superior to those from the conventional procedures used for gram-negative bacteria.

1 INTRODUCTION

As small circles of cytoplasmic DNA capable of autonomous replication and responsible for sexual differentiation and gene exchange in bacteria, plasmids were of limited interest. But when it was found that they would confer to their hosts a battery of functions extending the host ability to survive in new and aggressive environments, plasmids became the center of an ever-growing area of research and technology. Eventually, they became the choice tool for the manipulation of genetic information. Some of the properties conferred by plasmids, such as heavy metals resistance, have led to a number of beneficial environmental and industrial applications. On the other hand plasmids are also carrier of antibiotic resistance and their propagation and transmission from one organism to another may lead to major and unforeseen problems such as the spread of multiple antibiotic resistance.

An example is the growing concern over the spread of nocosomial infections essentially caused by *Enterococcus faecalis*, a gram-positive bacterium that has developed wide antibiotic resistance over the past 20 years (Chung et al. 1995, Huycke et al. 1998, Johnson et al. 2003, Ruggero et al. 2003, Scott et al. 2003, Waar et al. 2003).

E. faecalis is a normal inhabitant of the human gastrointestinal tract; it is also found in water, soil, vegetation, food and the gastrointestinal tract of a variety of other organisms besides humans. It has been used in probiotic cultures such as cheese and other milk product starter (Eaton & Gasson 2003). *E. faecalis* is a hardy bacterium that can grow at temperatures ranging from 10 to 45°C and over a broad pH range (McLean & Smith 1991). It is also able to grow in 6.5% NaCl (Hancock & Gilmore 1999) and has been reported to adapt to lethal levels of bile salts and sodium dodecyl sulfate (SDS) after brief exposure to sub-lethal levels of either of these detergents (Flahaut et al. 1996). Besides its ability to survive harsh conditions, it has acquired plasmid-mediated resistance to a number of antibiotics (Chung et al. 1995), an alarming fact given its omnipresence in our environment, not to mention that some *E. faecalis* strains also carry genes coding for antibiotic resistance in their genome (Huycke et al. 1998, Hancock & Gilmore 1999). It is important to be able to differentiate between plasmid-mediated and genome-mediated antibiotic resistance.

Unfortunately, although the increasing occurrence of antibiotic resistance has prompted research for methods to cure plasmid-bearing bacteria, the majority of cloning and genetic manipulations have been carried out in *Escherichia coli*, a gram-negative bacterium. Hence, the methods for plasmid curing, plasmid isolation and bacterial transformation that have been extensively developed are most efficient with *E. coli*. They become less efficient when applied to other gram-negative bacteria and give very poor results when used with gram-positive bacteria.

We have worked out a novel and efficient method for rapid plasmid isolation from *E. faecalis*. The method which involves the use of sodium N-lauroyl-sarcosinate (sarkosyl) allowed for the reproducible isolation of two plasmids while routine methods such as the alkaline lysis procedure or the boiling procedure were either unsuccessful or led to the detection of one plasmid only. The use of sarkosyl was also useful for curing *E. faecalis*.

2 MATERIALS AND METHODS

2.1 *Bacterial strains*

E. coli strains HB101, HB101(pBR322) and HB101 (pBR325) were provided to us by Dr. N.O. Keyhani, Department of Microbiology and Cell Science, University of Florida at Gainesville, U.S.A. The *E. faecalis* strain used in this study was a clinical isolate provided by Dr. S. Lauwers from the Microbiology Department of the VUB University Hospital, Brussels, Belgium.

pBR322 is a 4.3 kb plasmid carrying Amp^R and Tet^R (Bolivar et al. 1977); pBR325 is a 5.7 kb plasmid carrying Amp^R, Tet^R and Cm^R (Bolivar 1978).

2.2 *Culture media*

Luria broth (LB) and LB-agar plates were used throughout this study. Peptone, yeast extract and agar were from Merck. When supplied with antibiotics, the autoclaved media were allowed to cool to 48°C before adding the required amount of a stock antibiotic solution that had been sterilized by filtration; antibiotics were all from Sigma. All media and antibiotics preparations were according to Sambrook et al. (1989).

2.3 *Chemicals*

Sodium dodecyl sulfate (SDS) was from Sigma; acridine orange was from Searle; sodium N-lauroyl-sarcosinate (sarkosyl) was from Fluka. All other chemicals were from Merck.

2.4 *Plasmid DNA extraction and isolation*

Plasmid DNA was extracted from plasmid carrying-strains either according to the alkaline lysis method (Birnboim & Doly 1979) or the boiling method (Holmes & Quigley 1981). In addition, for *E. faecalis*, a novel method was used similar to the boiling method but replacing the lysozyme (10 mg/ml) by sarkosyl (10%).

2.5 *Transformation experiments*

E. coli strain HB101 was transformed with plasmid DNA isolated from the *E. faecalis* strain used in this study according to the calcium chloride procedure described in Sambrook (1989).

2.6 *Replica plating*

To test for curing or transformation, isolated colonies were picked with sterile toothpicks and transferred on a distinct position on grid plates containing specific antibiotics. Plates were incubated at 37°C. After 24 h, growth was compared with that on a control plate without antibiotic.

2.7 *Curing experiments*

Curing of plasmid-carrying *E. coli* and *E. faecalis* strains was attempted by growth in either acridine orange (at final concentrations ranging from 25 to 200 μg/ml), SDS (at final concentrations ranging from 0.01% to 5% for *E. coli* strains and from 0.001% to 0.2% for *E. faecalis* strain) or sarkosyl (at final concentrations ranging from 0.5% to 5% for *E. coli* strains and from 0.005% to 0.5% for *E. faecalis* strain).

2.8 *Determination of antibiotic resistance*

The minimum inhibitory concentration (MIC) of ampicillin, tetracycline, chloramphenicol and kanamycin against the strains used in this study was determined by following the growth of each strain at 37°C, with aeration, in LB medium supplied with increasing concentrations of antibiotic. Growth was monitored by measuring optical density at 550 nm. MIC was the minimum concentration required to observe no growth after 24 h culture.

2.9 *Agarose gel electrophoresis*

Plasmid DNA preparations were electrophoresed in 1% agarose gel at 7.0 V per cm for 1 to 3 h, using Tris-Borate-EDTA (TBE) buffer. The procedure was as described in Sambrook et al. (1989), except that ethidium bromide was added to the electrophoresis buffer only, at a final concentration of 0.5 μg/ml. The agarose was type-II, low-endo-osmotic agarose from Merck. Gels were 8 × 10 cm.

3 RESULTS AND DISCUSSION

3.1 *Antibiotic resistance*

Table 1 shows the MICs of various antibiotics against *E. coli* strains HB101, HB101(pBR322), HB101 (pBR325), and the *E. faecalis* strain used in this study.

E.coli strain HB101 was sensitive to all four antibiotics tested. As expected, plasmid pBR322 conferred resistance to tetracycline and ampicillin while plasmid pBR325 conferred resistance to tetracycline, ampicillin and chloramphenicol to strain HB101. On the other hand, *E. faecalis* exhibited resistance to kanamycin and tetracycline but not to ampicillin and chloramphenicol. *E. faecalis* is reputed sensitive to ampicillin making the latter a medication that was often used against the bacterium in a number of infections (Huycke et al. 1998). Tetracycline resistance in *E. faecalis* has been reported to be encoded within its genome (Huycke et al. 1998, Eaton & Gasson 2001) but also on a 58 kb conjugative plasmid (Chung et al. 1995). Resistance to aminoglycosides (e.g. kanamycin) has been attributed to the ability of enterococci to block the uptake of the drug at the cell wall (Hewitt et al. 1966). However, resistance to high levels of aminoglycosides has been reported to be plasmid-borne (Horodniceanu et al. 1979). Thus we investigated the presence of plasmid in the *E. faecalis* strain under study using rapid plasmid isolation techniques.

Table 1. Antibiotic resistance of the *E. coli* and *E. faecalis* strains used in this study.

	MIC			
Strain	Kan* (μg/ml)	Tet* (μg/ml)	Amp* (μg/ml)	Cm* (μg/ml)
HB101	50	2	10	10
HB101(pBR322)	50	100	3000	10
HB101(pBR325)	50	100	1000	100
E. faecalis	2000	50	5	5

* Kan: Kanamycin; Tet: Tetracycline; Amp: Ampicillin: Cm: Chloramphenicol.

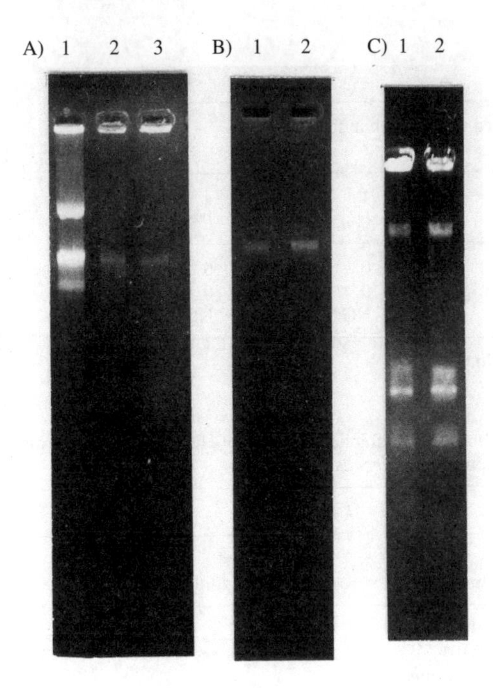

Figure 1. Agarose gel electrophoresis of plasmid DNA preparations. (A) Alkaline lysis method – lane 1: pBR322; lanes 2,3: *E. faecalis* plasmid. (B) Boiling method – lanes 1,2: *E. faecalis* plasmid. (C) Novel method using sarcosyl – lanes 1,2: *E. faecalis* plasmids (~7 kb and 5.7 kb,respectively).

3.2 *Rapid plasmid isolation*

Use of the alkaline lysis method or the boiling method as described in Sambrook et al. (1989) gave consistently good results for plasmid isolation from HB101(pBR322) and HB101(pBR325) as tested by electrophoresis in 1% agarose gels. When the alkaline lysis method was used for plasmid isolation from *E. faecalis*, the resulting preparation produced only a faint band in agarose gel in one out of three preparations (Fig.1A). When the boiling method was used, one band was almost always detectable in agarose gels (Fig. 1B).

However, when we modified the boiling method by replacing the lysozyme with sarkosyl, the plasmid DNA preparation obtained showed consistently two sets of bands, revealing the presence of two plasmids in the *E. faecalis* strain studied (Fig. 1C).

3.3 *Curing*

Under our experimental conditions, growth in various concentrations of acridine orange led to the cure of plasmid-carrying *E. coli* strains HB101(pBR322) and HB101(pBR325) but not of *E. faecalis* as shown in Table 2.

Growth in SDS led only to the cure of HB101 (pBR322); not one cured colony was found out of 500 HB101(pBR325) colonies or *E. faecalis* colonies that were tested (Table 3).

However, growth in sarkosyl produced a small number of cured *E. faecalis* colonies; no cured *E. coli* colony was found out of 500 tested for each strain (Table 4). Thus sarkosyl was much more efficient than SDS for curing *E. faecalis*.

Table 2. Curing effect of acridine orange.

Strain	% Cured colonies* produced after growth in following concentrations of acridine orange					
	25 (μg/ml)	50 (μg/ml)	75 (μg/ml)	100 (μg/ml)	150 (μg/ml)	200 (μg/ml)
HB101(pBR322)	0	26	26	35	0	0
HB101(pBR325)	0	0	15	15	0	0
E. faecalis	0	0	0	0	NG**	NG**

* 500 colonies tested for each strain; ** NG = "No growth".

Table 3. Curing effect of SDS.

Strain	% Cured colonies* produced after growth in following concentrations of SDS							
	0.001 (%)	0.005 (%)	0.01 (%)	0.05 (%)	0.25 (%)	0.5 (%)	1 (%)	2 (%)
HB101(pBR322)	–	–	0	0	27	27	35	0
HB101(pBR325)	–	–	0	0	0	0	0	0
E. faecalis	0	0	0	NG**	NG**		NG**	

* 500 colonies tested for each strain; ** NG = "No growth".

Table 4. Curing effect of sarkosyl.

Strain	% Cured colonies* produced after growth in following concentrations of sarkosyl.						
	0.005 (%)	0.02 (%)	0.1 (%)	0.5 (%)	1 (%)	2 (%)	4 (%)
HB101(pBR322)	0	0	0	0	0	0	0
HB101(pBR325)	0	0	0	0	0	0	0
E. faecalis	0	3	NG**	NG**	NG**		NG**

* 500 colonies tested for each strain; ** NG = "No growth".

3.4 *Transformation of E. coli strain HB101 with E. faecalis plasmid*

E. coli strain HB101 was transformed with the plasmid DNA preparation obtained from *E. faecalis* using the novel method described. The transformed colonies were resistant to kanamycin and tetracycline as shown in Table 5. The table also reports the MICs measured for *E. faecalis* cured by sarkosyl. Residual resistance to tetracycline was still observed.

Thus, using a novel rapid plasmid isolation method, two small plasmids (~7 kb and 5.7 kb) conferring kanamycin and tetracycline resistance were found in this strain of *E. faecalis* where conventional rapid plasmid isolation methods were unreliable.

Table 5. Antibiotic resistance of HB101 after transformation with *E. faecalis* plasmid.

Strain	MIC			
	Kan* (μg/ml)	Tet* (μg/ml)	Amp* (μg/ml)	Cm* (μg/ml)
HB101	50	2	10	10
HB101 transformed	2000	50	10	10
E. faecalis	2000	50	5	5
E. faecalis cured	50	25	5	5

* Kan: Kanamycin; Tet: Tetracycline; Amp: Ampicillin; Cm: Chloramphenicol.

REFERENCES

Birnboim, H.C. & Doly, J. 1979. A rapid alkaline extraction procedure for screening recombinant plasmid DNA. *Nucleic Acids Res.* 7: 1513–1523.

Bolivar, F., Rodriguez, R., Greene, P.J., Betlach, M., Heyneker, H.L., Boyer, H.W., Crosa, J. & Falkow, S., 1977. II. Construction and characterization of new cloning vehicles. A multipurpose cloning system. *Gene* 2: 95–113.

Bolivar, F. 1978. Construction and characterization of new cloning vehicles, III. Derivatives of plasmid pBR322 carrying unique EcoRI sites for selection of EcoRI generated recombinant molecules. *Gene* 4: 121–136.

Chung, J.W., Bensing, B.A. & Dunny, G.M. 1995. Genetic analysis of a region of the Enterococcus faecalis plasmid pCF10 involved in positive regulation of conjugative transfer functions. *J. Bacteriol.* 177: 2107–2117.

Eaton, T.J. & Gasson, M.J. 2001. Molecular screening of Enterococcus virulence determinants and potential for genetic exchange between food and medical isolates. *Appl. Environ. Microbiol.* 67: 1628–1635.

Flahaut, S., Frere, J., Boutibonnes, P. & Auffray, Y. 1996. Comparison of the bile salts and sodium dodecyl sulfate stress response in Enterococcus faecalis. *Appl. Environ. Microbiol.* 62: 2416–2420.

Hancock, L.E. & Gilmore, M.S. 1999. Pathogenicity of Enterococci. In V. Fischetti, R. Novich, J. Ferretti, D. Portnoy & J. Rood (eds), *Gram-Positive Pathogens*. AMS Publications.

Hewitt, W.L., Seligman, S.J. & Deigh, R.A. 1966. Kinetics of the synergism of penicillin-streptomycin and penicillin-kanamycin for enterococci and its relationship to L-phase variants. *J. Lab. Clin.* 67: 792–807.

Holmes, D.S. & Quigley, M. 1981. A rapid boiling method for the preparation of bacterial plasmids. *Anal. Biochem.* 114: 193–197.

Horodniceanu, T., Bougueleret, L., El-Solh, N., Bieth, G. & Delbos, F. 1979. High-level, plasmid-borne resistance to gentamicin in Streptococcus faecalis subsp. zymogenes. *Antimicrob. Agents Chemother.* 16: 686–689.

Huycke, M.M., Sahm, D.F. & Gilmore, M.S. 1998. Multiple-drug resistant Enterococci: the nature of the problem and an agenda for the future. *Emerg. Infect. Dis.* 4: 239–249.

Johnson, A.P., Henwood, C., Mushtaq, S., James, D., Warner, M. & Livermore, D.M. 2003. Susceptibility of gram-positive bacteria from ICU patients in UK hospitals to antimicrobial agents. *J. Hosp. Infect.* 54: 179–187.

McLean, D.M. & Smith, J.A. 1991. *Medical Microbiology Synopsis*. Philadelphia: Lea & Febiger.

Ruggero, K.A., Schroeder, L.K., Schreckenberger, P.C., Makin, A.S. & Quinn, J.P. 2003. Nosocomial superinfections due to linezolid-resistant Enterococcus faecalis: evidence for a gene dosage effect on linezolid MICs. *Diagn. Microbiol. Infect. Dis.* 47: 511–513.

Sambrook, J., Fritsch, E.F. & Maniatis, T. 1989. *Molecular Cloning. A Laboratory Manual, 2nd ed.* Cold Spring Harbour Laboratory Press.

Scott, I.U., Loo, R.H., Flynn, H.W. Jr. & Miller, D. 2003. Endophthalmitis caused by Entorococcus faecalis: antibiotics selection and treatment outcome. *Ophthalmology* 110: 1573–1577.

Waar, K., Willems, R.J., Slooff, M.J., Harmsen, H.J. & Degener, J.E. 2003. Molecular epidemiology of Enterococcus faecalis in liver transplant patients at University Hospital Groningen. *J. Hosp. Infect.* 55: 53–60.

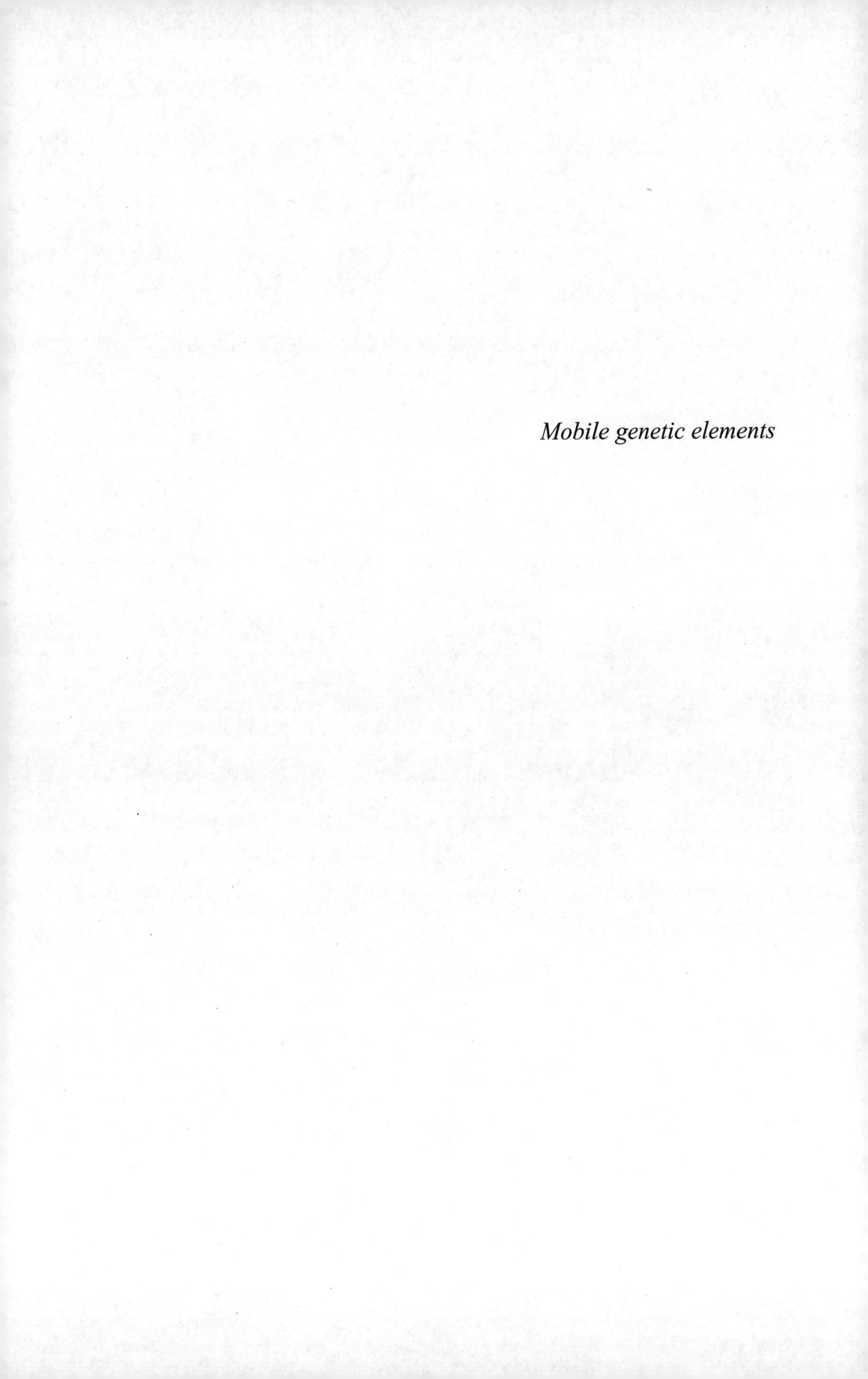

Mobile genetic elements

European Symposium on Environmental Biotechnology, ESEB 2004 - Verstraete (ed)
© 2004 Taylor & Francis Group, London, ISBN 90 5809 653 X

Bacterial antibiotic resistance levels in agricultural soils as a result of treatment with pig slurry and detection of sulphonamide resistance genes

K.G. Byrne, W. Gaze & E.M.H. Wellington
Department of Biological Sciences, University of Warwick, Gibbet Hill, Coventry, West Midlands, U.K.

P. Kay, P. Blackwell A. & Boxall
Cranfield Centre for EcoChemistry, Cranfield University, Shardlow Hall, Shardlow, Derby, U.K.

ABSTRACT: An investigation was carried out to assess the environmental impact of veterinary medicines released into the environment through the spreading of pig manure slurry. Concern has been growing about the use of antibiotics in livestock husbandry and the possible selection for resistance genes in bacteria. Slurry containing antibiotics and bacteria is released into the environment and may result in the dissemination of resistance genes through horizontal gene transfer.

Three antibiotic groups were used in this study, a macrolide, sulphonamide and a tetracycline. The fate of these antibiotics when added to soil via slurry and direct application was measured over a two-year period. Samples were taken over this period and are currently being analyzed for resistant bacteria. Resistance to the above antibiotics was analyzed by targeting *SulI*, *SulII* and *IntI* genes. Populations of resistant bacteria were isolated and the distribution of *SulI, SulII* and *IntI* genes investigated.

1 INTRODUCTION

The intensive use and misuse of antibiotics in both human medicine and agriculture has resulted in wide spread bacterial resistance. This has lead to concerns about a number of agricultural practices, one being the spread of pig manure slurry onto arable and livestock fields. Antibiotic resistance in pig manure slurry has been associated with antibiotic use (Schwarz & Chaslus-Dancla, 2001). There is thought to be a possibility that when manure slurry is spread on fields, antibiotics, resistant bacteria and antibiotic resistance are transferred to the environment (Boxall et al., 2003). The antibiotics and residues of antibiotics may be active in farmland or be leached into surface and subsequently ground water. These may impact on human and environmental health as well as conferring a selective advantage to resistant bacteria. Furthermore, the transfer of resistant bacteria and resistance genes into the environment may create the possibility for horizontal transfer of those genes to the indigenous soil bacterial population. Horizontal gene transfer has been shown to occur in terrestrial environments where antibiotic resistance is plasmid borne (Rosser & Young, 1999). This transfer has been shown to increase in the presence of pig manure, and pig manure has been shown to contain mobile plasmids conferring antibiotic resistance (Halling-Sorensen et al., 2001).

The aim of this study was to evaluate whether the use of tylosin fed pig manure slurry, amended with a sulphonamide and a tetracycline, used as fertilizer has created reservoirs of resistance in an agricultural clay soil.

2 METHODS

A 2 year study was carried out in which pig slurry from tylosin feed pigs was amended with sulphachloropyridazine (SCP) and oxytetracycline (OTC) before being spread onto a clay field. Soil cores were collected from the clay field to a depth of 30 cm, at 6 different sites, before spreading and at 10 time intervals each year thereafter.

CFU counts and bacterial isolations were carried out on ISO-Sensitest agar (Oxoid) containing 100 μg/ml cycloheximide (Sigma) and differing concentrations of tylosin tartrate (Sigma), OTC (Sigma) and SCP (Sigma). Plates were incubated at 28°C overnight and colonies counted. Resistance quotient values (RQ) were calculated to normalize the data. DNA, from bacterial isolates, was isolated from 510 bacterial soil isolates using a Qiagen DNA easy kit following the manufacturers instructions. Bacterial DNA was screened by PCR for IntI, SulI, II and 3 using the following primers; Int1F(5′-ATCATCGTCGTAGA GACGTCGG), Int1R(5′-GTCAAGGTTCTYG GACCAGTTGC) at an annealing temperature of 67°C; SulIF(5′-CTTCGATGAGAGCCGGCGGC) SulIR(5′-GCAAGGCGGAAACCCGCGCC) at an annealing temperature of 63°C (Sundstrom et al. 1988); SulIIF(5′-TCGTCAACATAACCTCGGACAG) SulIIR(5′-GTTGCGTTTGATACCGGCAC) at an annealing temperature of 60°C (Enne, personal communication); Sul3F(5′-GAGCAAGATTTTTG GAATCG) Sul3R(5′-CATCTGCAGCCTAACC TAGGGCTTTGGA) at an annealing temperature of 51°C (Perreten & Boerlin, 2003).

3 RESULTS AND DISCUSSION

The data showed high RQ values for bacteria isolated from pig manure slurry for SCP, OTC and tylosin. Bacteria resistant to SCP were seen throughout year 1 but not cultured after day 90 in year 2. This decrease in bacteria resistant to SCP may have been due to a decrease in selective pressure because sulphonamides do not sorb to soils and are leached into ground waters (Boxall et al., 2002). OTC resistant bacteria were not isolated from the soil cores after day 1 in both years, indicating a lack of faecal bacterial survival. Tylosin bacterial resistance is high throughout the 2-year study. Pig manure slurry from tylosin fed pigs has been continuously applied to the field for a number of years, suggesting a reservoir of resistance has accumulated due to a continual antibiotic pressure within the field and pig intestinal tract. The suggestion that usage patterns of antimicrobial agents had an effect on the distribution pattern of antimicrobial resistance was also proposed by Lanz (Lanz et al., 2003)

Our high frequency data of environmental isolates carrying *SulII*, with a small percentage carrying both *SulI* and *SulII* genes conflicts with a number of studies. Radstrom (Radstrom et al., 1991) carried out a study to determine the frequency of sulphonamide resistance genes in clinical and environmental isolates including pig slurry. Radstrom found that in clinical isolates *SulI* and *SulII* were found at approximately equal frequencies (Radstrom et al., 1991), however, *SulII* has been demonstrated to be less frequent in environmental isolates (Lanz et al., 2003)

Int1 screening was carried out to investigate the possible transfer of resistance genes within the environment. *SulI* was shown to be found on class I integrons (Sundstrom et al., 1988), a mobile genetic element implicated in horizontal gene transfer. *Int1* is thought to be encoded within the 3′ conserved region of class 1 integrons (Sundstrom et al., 1988). We found similar frequencies of class 1 integron occurrence to those found in an environmental study of Gram negative bacterial isolates, from an estuary (Rosser & Young, 1999), giving a figure of 3.6%. Although, the actual figure is thought to be significantly higher as it is thought that only approximately 1% of the bacterial population in soil is culturable. Studies of clinical bacterial populations have shown the frequency of isolates containing class 1 integrons to be between 59 and 75%

Our study also found the frequency of *Int1* linked with *SulI* to be zero. This conflicts with other environmental studies which have found frequencies between 42% (Rosser & Young, 1999) and 100% (Schmidt et al., 2001).

Further screening for the *Sul3* gene, along with a screening strategy for tetracycline and macrolide resistance genes will be carried out to further evaluate whether there is a reservoir of resistance which can be attributed to the soil application of pig manure slurry containing significant concentrations of antibiotics (Boxall et al., 2002). Exogenous isolations will also be carried out (Smalla et al., 2000), to evaluate frequencies of transfer for a number of antibiotic resistance genes.

It is clear that further studies need to be done with these soils to determine whether the resistance genes are mobilized or whether resistance is due to existing mechanisms. We also wish to confirm the impacts observed for antibiotic applications to soils.

ACKNOWLEDGEMENT

This work was funded by the BBSRC and Wyeth Pharmaceutical Company

REFERENCES

Boxall, A.B., P. Blackwell, R. Cavallo, P. Kay, and J. Tolls. 2002. The sorption and transport of a sulphonamide antibiotic in soil systems. *Toxicol Lett* 131:19–28.

Boxall, A.B., L.A. Fogg, P. Kay, P.A. Blackwel, E.J. Pemberton, and A. Croxford. 2003. Prioritisation of veterinary medicines in the UK environment. *Toxicol Lett* 142: 207–18.

Halling-Sorensen, B., J. Jensen, J. Tjornelund, and M. Monforts. 2001. Worst-case estimations of predicted environmental soil concentrations (PEC) pf selected veterinary antibiotics and residues in Danish agriculture., p. 171–82, *In* K. Kummerer, ed. *Pharmaceuticals in the environment.*, Vol. Chapter 13. Springer Verlag, Germany.

Lanz, R., P. Kuhnert, and P. Boerlin. 2003. Antimicrobial resistance and resistance gene determinants in clinical *Escherichia coli* from different animal species in Switzerland. *Vet Microbiol* 91: 73–84.

Perreten, V., and P. Boerlin. 2003. A new sulfonamide resistance gene (sul3) in *Escherichia coli* is widespread in the pig population of Switzerland. *Antimicrob Agents Chemother* 47: 1169–72.

Radstrom, P., G. Swedberg, and O. Skold. 1991. Genetic analyses of sulfonamide resistance and its dissemination in gram-negative bacteria illustrate new aspects of R plasmid evolution. *Antimicrob Agents Chemother* 35: 1840–8.

Rosser, S.J., and H.K. Young. 1999. Identification and characterization of class 1 integrons in bacteria from an aquatic environment. *J Antimicrob Chemother* 44: 11–8.

Schmidt, A.S., M.S. Bruun, J.L. Larsen, and I. Dalsgaard. 2001. Characterization of class 1 integrons associated with R-plasmids in clinical Aeromonas salmonicida isolates from various geographical areas. *J Antimicrob Chemother* 47: 735–43.

Schwarz, S., and E. Chaslus-Dancla. 2001. Use of antimicrobials in veterinary medicine and mechanisms of resistance. *Vet Res* 32: 201–25.

Smalla, K., H. Heuer, A. Gotz, D. Niemeyer, E. Krogerrecklenfort, and E. Tietze. 2000. Exogenous isolation of antibiotic resistance plasmids from piggery manure slurries reveals a high prevalence and diversity of IncQ-like plasmids. *Appl Environ Microbiol* 66: 4854–62.

Sundstrom, L., P. Radstrom, G. Swedberg, and O. Skold. 1988. Site-specific recombination promotes linkage between trimethoprim- and sulfonamide resistance genes. Sequence characterization of dhfrV and sulI and a recombination active locus of Tn21. *Mol Gen* 213: 191–201.

European Symposium on Environmental Biotechnology, ESEB 2004 - Verstraete (ed)

Investigation into microbial diversity and gene dissemination in the air of the International Space Station

P. De Boever & M. Mergeay
Laboratory for Microbiology, Belgian Nuclear Research Centre, SCK•CEN, Mol, Belgium

A. Toussaint
Service de Conformation de Macromolécules Biologiques et de Bioinformatique, ULB, Brussels, Belgium

J. Mahillon
Food and Environmental Microbiology, UCL, Louvain-la Neuve, Belgium

C. Lasseur
Life Support & Thermal Control, ESTEC/ESA, Noordwijk, The Netherlands

ABSTRACT: This research aims at a better understanding of the diversity of airborne bacterial species aboard the former Russian MIR space station and the current International Space Station (ISS) using molecular biology tools. Considerable attention is also paid to the presence and type of plasmids in the airborne bacteria using a culture-dependent and a culture-independent approach. It is envisioned that the results will contribute to a better understanding of possible microbiological risks (i.e. pathogenicity, biodeterioration, etc.) during long haul space flights. The latter risks may have an important impact on crew health and hardware integrity and may jeopardize the space flight.

1 INTRODUCTION

The monitoring of the biological air quality is of paramount importance in many different environments such as hospitals, pharmaceutical and food industry, offices and homes. The bacteria and fungi present in the air and air conditioning systems can often be held responsible for hospital-acquired infections, foreign body infections (associated with implant of a prothesis), lower product lifetime and the 'sick building' syndrome (Pareta, et al., 1997, Tegnell et al., 2002). The impact of airborne microorganisms is even more important in confined environments such as space vehicles and space stations. There, a limited number of people may be isolated for several years and in case of an emerging microbiological problem the treatment possibilities may be limited. Any infection and cross infection may compromise the health of the crew members seriously. Deposition of microorganisms on surfaces and electrical wiring may lead to biofilm formation, which may be at the origin of biodegradation and malfunctioning of the equipment. Hence, microorganisms may have a great impact on the health of the crew as well as on the space hardware and may thus jeopardize the success of long haul space flights. Up to now, few steps have been taken to characterize the biodiversity of airborne microorganisms in a confined environment such as a space station. The largest source of information is the Russian research performed aboard the former space station MIR. Microbial contaminants were identified virtually everywhere using the selective culture plate technique. High microbial virulence was observed and a number of allergic reactions were reported by the crew. Many pieces of hardware were contaminated with microorganisms and in a number of cases biodegradation occurred (Kawamura et al., 2001).

The recent revolution in molecular techniques for the analysis of the DNA composition of complex microbial communities has evidenced that the plate count technique gives a biased and largely incomplete view of the microbial community. The correct culture conditions of many organisms are often not met and many environmental organisms have a 'viable but not culturable' status. Some authors estimate that 80 to 99% of the microorganisms can remain unidentified in a complex environmental sample (Amann et al., 1994). It is therefore believed that the microbial presence in the MIR space station and the associated risk is being underestimated. The possible hazards associated with microbial presence in a confined environment may be aggravated when genetic information from one strain (e.g. pathogenicity, virulence, catabolic genes, etc.) is transferred to another species. In this respect, the type and transfer frequency of plasmids may be important (Top et al., 1994).

Plasmids are self-replicating circular DNA molecules that often carry genes conferring an advantage to their host *e.g.* resistance to antibiotics, ability to degrade a variety of carbon sources and especially recalcitrant organics and man-made chemicals. Genetic determinants for pathogenesis are also often plasmid-encoded (van Elsas et al., 2002). To our knowledge, little work has been done in relation to gene transfer in air and especially the air in a confined space station. Because gene transfer can be linked with pathogenicity (and hence biosafety) and biodegradation potential, it should clearly be taken into account when performing a risk assessment of the microbial impact during a long haul space flight.

2 RESEARCH PROJECT

The first objective of the research project is the use of molecular techniques for an in-depth analysis of the microbial species present in preserved air samples from MIR and the International Space Station (ISS). Furthermore, during the Taxiflights of the year 2003 to the ISS, air samples of the ISS have been collected on Petri dishes. Samples were retrieved within 48 h of collection and processed immediately in the laboratory.

All samples were incubated for a 7-day period and micro-organisms have been counted on regular intervals. Biomass from the Petri dishes has been recovered in a physiological solution. Aliquots of this suspension have been used to perform plate counting and a DNA extraction. DNA amplification was followed by restriction analysis and terminal Restriction Fragment Length Polymorphism (tRFLP), a recently developed technique for analyzing complex microbial communities (Marsh, 1999). In parallel, a cloning strategy has been applied and the obtained bacterial clones have been sequenced.

The second objective is the quantification of plasmid mobilization in MIR and ISS samples. The research aiming at the estimation of possible gene dissemination started with mastering the triparental exogenous isolation method as described earlier (Top et al., 1994). This culture-based approach will be complemented by a PCR-based detection of plasmids after total DNA extraction. The latter technique permits to study the prevalence of plasmids in culturable as well as non-culturable species bacterial species (Götz et al., 1996). Next, the triparental mating system and PCR-based approach will be applied to air samples collected on Petri dishes aboard MIR and ISS.

3 CONCLUSION

This research aims at a better understanding of the diversity of bacterial species aboard the former MIR station and the ISS using novel molecular techniques. The data will be matched with more classical approaches such as plate counting. Furthermore, considerable attention is paid to the presence and type of self-transferable DNA in the airborne bacteria using a culture-dependent and a culture-independent approach. Biodiversity and gene dissemination of airborne bacteria is not only of concern when estimating the microbial impact in a confined space vehicle, but also hospitals and industry deal with this question. Hence, it is believed that our study is also relevant for those environments.

ACKNOWLEDGEMENT

This project is financed by the European Space Agency (ESA). Contract N° 16370/02/NL/CK.

REFERENCES

Amann, R.I., Ludwig, W. & Schleifer, K.-H. 1994. Identification of uncultured bacteria: a challenging task for molecular taxonomists. *ASM News 60*: 360–365.

Götz, A., Pukall, R., Smit, E., Tietze, E., Tschäpe, H., van Elsas, J.D. & Smalla, K. 1996. Detection and characterization of broad-host-range plasmids in environmental bacteria by PCR, *Applied and Environmental Microbiology 62*: 2621–2628.

Kawamura, Y., Li, Y., Liu, H., Huang, X., Li, Z. & Ezaki, T. 2001. Bacterial population in Russian space station 'Mir', *Microbiology and Immunology 45*: 819–828.

Marsh, T.L. 1999 Terminal restriction fragment length polymorphism (T-RFLP): an emerging method for characterizing diversity among homologous populations of amplification products. *Current Opinion in Microbiology. 2*: 323–327.

Parata, S., Perdrixa, A., Fricker-Hidalgob, H., Saudec, I., Grillotb, R. & Baconnierd, P. 1997. Multivariate analysis comparing microbial air content of an air-conditioned building and a naturally ventilated building over one year, *Atmospheric Environment 31*: 441–449.

Tegnell, A., Saeediy, B., Isakssony, H., Granfeldtz & Öhman, L. 2002. A clone of coagulase-negative staphylococci among patients with post-cardiac surgery infections. *Journal of Hospital Infection 57*: 37–42.

Top, E.M., De Smet, I., Verstraete, W. & Mergeay, M. 1994. Exogenous isolation of mobilizing plasmids from polluted soils and sludges. *Applied and Environmental Microbiology 60*: 831–839.

van Elsas, J.D. & Bailey, M. J. 2002. The ecology of transfer of mobile genetic elements. *FEMS Microbiology Ecology 42*: 187–197.

European Symposium on Environmental Biotechnology, ESEB 2004 - Verstraete (ed)
© 2004 Taylor & Francis Group, London, ISBN 90 5809 653 X

Fate of endocrine disruptors in planted fixed bed reactors (PFR)

J. Müller, U. Kappelmeyer, P. Kuschk, H.H. Richnow & M. Kästner
Department of Remediation Research, UFZ-Leipzig-Halle GmbH, Leipzig, Germany

M. Möder
Department of Analytical Chemistry, UFZ-Leipzig-Halle GmbH, Leipzig, Germany

ABSTRACT: In this study the fate of the endocrine disrupting compounds like bisphenol A (BPA) and nonylphenols (NP) were investigated in a lab scale test system for constructed wetlands. Mass balances, removal rates and adsorption factors were determined in BPA and NP pretreated systems using ^{13}C-labelled BPA and NP. The results show that the elimination of BPA by microbial activity could be assessed with this approach. However, complete mass balances were not possible due to the respiratory activity of the plants. A half-live of ~6.6 d for BPA was measured in constructed wetlands and thus the compound can be degraded in constructed wetlands. A biodegradation of technical NP could not be proven. However, due to the high sorption capacity of the rhizosphere/gravel bed system the compounds were retained to a higher extent. The straight chain NP (*n*-NP) as model compound was rapidly degraded to more than 90% showing the inappropriateness of such compounds as a surrogate for technical mixtures. The test reactor combined with the stable isotope approach is a highly effective tool for analysing the fate of such compounds in constructed wetlands and in model reactors of complex environmental systems.

1 INTRODUCTION

Nonylphenols are persistent metabolites from nonylphenol ethoxylates in waste water, that are used in the formulation and the production of plastics, paints, pesticides and detergents. Bisphenol A is a recalcitrant anthropogenic organic pollutant in wastewater particularly from plastic industry and paper-recycling factories. Due to toxicity and endocrine disrupting activity, both compounds affect many organisms, e.g. plants, invertebrates, fishes, and mammalian and are hardly degradable under environmental conditions. To date, only a few aerobic cultures are known to degrade NP with branched and linear side chains.

To investigate the fate of endocrine disrupting compounds and other organic pollutants in sewage treatment with constructed wetlands, a new labscale system was designed and described by Kappelmeyer et al. (2002). This unit – called planted fixed bed reactor (PFR) – was designed as a finite element of a horizontal subsurface flow system. The PFR combined with the stable isotope ^{13}C-tracer technique (Richnow et al. 1999) is an experimental set up to balance the general metabolism of trace organic pollutants in such systems in detail.

The goal of the present work was to analyse the microbial degradation of BPA and NP's (*n*-NP and technical mixtures of branched NP) in a lab scale test system of constructed wetlands for wastewater treatment.

2 METHODS

2,2-Bis(4-hydroxyphenyl)propane (BPA, purity grade >97%, CAS 80-05-7) was purchased from Fluka (Zwijndrecht, Netherlands), technical nonylphenol (Pestanal, purity grade >94%, CAS 104-40-5) was obtained from Riedel de Haen (Seelze, Germany) and the linear single isomer 4-*n*-nonylphenol (*n*-NP) was obtained from Dr. Ehrenstorfer (Augsburg, Germany). Deuterium labelled [$^{2}H_{14}$] 2,2-bis(4-hydroxyphenyl) propane (^{2}H-BPA) was synthesised at the University of Leipzig (Germany) and used as internal standard. The uniform phenol-ring ^{13}C-labelled BPA (2,2-bis(4-hydroxy[U-^{13}C]phenyl)propane, ^{13}C-BPA) and *n*-NP (4-*n*-nonyl[U-^{13}C]phenol, ^{13}C-*n*-NP) were synthesised at the University of Hamburg.

The fate of BPA and NP was investigated in two parallel operated PFR planted with *Juncus effusus*. Artificial waste water containing 120 mg/l TOC (urea, peptone and meat extract) was used for the inflow. PFR comprise 10 kg gravel (size 4–8 mm) and 10 l liquid reaction volume. The hydraulic load was 2 l/d (~retention time 5 d). The gravel bed was initially inoculated by sewage sludge from a domestic waste water treatment plant. The microorganisms in the gravel bed were adapted for a period of 6 months with a continuous supply of 100 μg/l BPA and NP.

After the adaptation, the turnover was determined in pulse-chase experiments using ^{13}C-tracer compounds. The ^{13}C-BPA pulse was set for one week and the ^{13}C-*n*-NP pulse over a period of three weeks. ^{13}C-labelled NP with branched side chain was not available. Therefore ^{13}C-labelled NP with a straight alkyl chain was used as model compound.

After the pulse, a period of no supply followed, for four weeks in case of ^{13}C-BPA and 8 weeks in case of by ^{13}C-*n*-NP. During the experimental period, samples were taken from the outflow to monitor the concentration by GC/MS. In addition extracts from liquid liquid extraction (LLE) were derivatised with chlorotrimethylsilane and analysed for isotope composition by GC/isotope ratio – MS (GC/IR-MS). For quantification of BPA and NP ^{2}H-BPA and *n*-NP were used as internal standards. Outflow rates in μg/d were calculated using the concentrations and the discharged volumes. The concentration of ^{13}C bound to the biomass or in polar metabolites were analysed as a sum parameter. Therefore the water samples were taken from the outflow and were lyophilised and subjected to isotope ratio elementary analysis with mass spectrometry (IR-EA-MS) for determination. The concentrations and isotope composition of inorganic carbon (IC) were used to assess the mineralisation of ^{13}C-labelled compounds.

After the experiment, samples from the compartments of gravel, roots, and shoots were lyophilised and the dry weights were determined. For recovery of the parent compounds, the samples were extracted by soxhlet and accelerated solvent extraction (ASE). Tubes and pipes of the PFR were rinsed with methanol, acetone, and chloroform and the extracts were analysed with GC/MS. To describe the discharge of the parent compound, the flow rates were analysed mathematically by curve fitting. The first equation describes a reaction of first order within an ideal stirred tank reactor with completely remixing of the inflowing compound during the pulse application period (Fritzer & Fitz 1975). The second equation describes the same reaction without remixing and inflow term for the period of no supply after the pulse.

$$c_{A,o} = c_{A,i}\frac{1}{1+k\tau}\left(1 - e^{-(1/\tau+k)t}\right) \tag{1}$$

$$c_{A,o} = c_{A,end}\, e^{-(1/\tau+k\tau)(t-t_{pulse})} \tag{2}$$

$c_{A,o}$ concentration of A in outflow
$c_{A,i}$ concentration of A in inflow
$c_{A,end}$ concentration of A at the end of the pulse period
k removal constant of A
τ retention time of A
t experimental time
t_{pulse} used time for pulse of the ^{13}C-labelled compound

The parameters retention time of the compound and the removal rate were fitted and are presented in Tables 1, 2 and 3. The same fit was used for balancing of IC, suspended particles and lyophilised water residues.

Additional experiments were carried out to prove the microbial degradation of the endocrine disrupting compounds. In batch experiments, the biological oxygen demand (BOD) was determined from diluted PFR outflow samples with added BPA or NP's as sole C-sources. In similar experiments with ^{13}C-BPA or ^{13}C-*n*-NP (100 μg/l) the $^{13}C/^{12}C$-ratio of the resulting CO_2 was analysed by headspace-GC/IR-MS after acidifying the samples with concentrated HCl.

3 RESULTS

The data presented below are mean values of the two PFR operated in parallel.

3.1 *Results BPA*

At the end of the adaptation period, more than 50% of BPA was removed in PFRs. About 40% of ^{13}C-BPA was recovered in the outflow during the pulse operation period of 35 days. The label found in the mineralisation product CO_2 was only about 5% of the applied compound. That might be due to the fact that the rhizosphere was not sealed against the atmosphere. In the LLE extract known metabolites (2,2′-(4,4′-dihydroxydiphenyl)-1-propanol, 4-hydroxyacetophenone, 4-hydroxybenzoic acid (Lobos et al. 1992), 2-(4-hydroxyphenyl)-3-hydroxy-1-propanoic acid and lactone of 2-(4-hydroxyphenyl)-3-hydroxy-1-propanoic acid (Ben-Jonathan & Steinmetz 1998)) were found. The observed concentrations for these metabolites were approximately <1% of the applied ^{13}C-BPA. One percent of ^{13}C-BPA was found in gravel and only 0.007 percent were found in the plants.

The best fit curve to the concentration data revealed parameters as listed in Table 1.

Batch experiments with ^{13}C-BPA showed an increase of the $^{13}C/^{12}C$-ratio of the formed CO_2 from background values of $\delta C \approx -21\%$ to 50% indicating a mineralisation of this compound. Single strains able to degrade BPA were isolated from mixed cultures and were identified as *Sphingomonas* sp. and *Variovorax paradoxus* (*Alcaligenes paradoxus*).

Table 1 Results for the ^{13}C-BPA pulse experiment.

Parameter	Reactor A	Reactor B
Hydraulic retention time $t_{hydr. ret.}$ [d]	5.6	5.6
Retention time BPA $t_{retention BPA}$ [d]	11.6 ± 0.5	11.1 ± 0.5
removal constant $k_{removal BPA}$ [d^{-1}]	0.114 ± 0.007	0.097 ± 0.008
elimination [%]	57	52
elimination rate $v_{removal BPA}$ [μg/l/d]	12	11
Half-life $\tau_{1/2}$, normalised with τ_{hydr}/τ_{BPA} [d]	6.1, 2.9	7.1, 3.6
K_{Ads}, lg K_{BM}, lg K_{OC}	1.04, 2.1, 2.4	0.99, 2.1, 2.4

Table 2 Results for the ^{13}C-*n*-NP pulse experiment.

Parameter	Reactor A	Reactor B
Hydraulic retention time $t_{hydr. ret.}$ [d]	5.2	5.2
Retention time *n*-NP $t_{retention\ n\text{-}NP}$ [d]	1700 ± 130	700 ± 50
removal constant $k_{removal\ n\text{-}NP}$ [d^{-1}]	0.0047 ± 0.0036	0.022 ± 0.0038
elimination [%]	89	94
elimination rate $v_{removal\ n\text{-}NP}$ [μg/l/d]	19	20
Half-life $\tau_{1/2}$, normalised with $\tau_{hydr}/\tau_{n\text{-}NP}$ [d]	147, 0.5	32, 0.2
K_{Ads}, lg K_{BM}, lg K_{OC}	300 ± 31, 4.7, 4.4	120 ± 13, 4.4 4.1

3.2 *Results NP*

At the end of the adaptation time, only 30% of the applied branched NP was recovered in the outflow. The observed removal of the branched NP amounted to <70% in this experiment. In contrast only, 4% of linear ^{13}C-*n*-NP was recovered in the outflow. About 9% of the ^{13}C-*n*-NP was mineralised to CO_2. The ^{13}C-label found in the lyophilised water residues account to 60–80% of which 10–15% was bound to suspended particles. Except CO_2, no other metabolites could be found. No significant change in the ^{13}C/^{12}C-ratio was observed in the shoots. 10–20% of ^{13}C-*n*-NP were found in the gravel, 5%–10% in the microbial biomass and 0.3–0.7% in the plants. Tubes and sealings of the PFR contained 1% of the ^{13}C-*n*-NP.

The best fit curve to the concentration data revealed parameters as listed in Table 2. for ^{13}C-*n*-NP and in Table 3. for branched NP.

Table 3 Results for branched NP in the experiments.

Parameter	Reactor A	Reactor B
Hydraulic retention time $t_{hydr. ret.}$ [d]	5.2	5.2
retention time NP $t_{retention\ branched\ NP}$ [d]	⩽84 d	⩽140 d
removal constant $k_{removal}$ [d^{-1}]	⩾0.0049	⩾0.0017
elimination [%]	⩽30	⩽20
elimination rate $v_{removal}$ [μg/l/d]	6	4
Half-life $\tau_{1/2}$, normalised with $\tau_{hydr}/\tau_{branched\ NP}$ [d]	142, 8.8	412, 15.3
K_{Ads}, lg K_{BM}, lg K_{OC}	16, 3.4, 3.7	27, 3.7, 4.0

BOD experiments showed that, 70–100% of the theoretical oxygen demand was consumed in case of *n*-NP and in case of branched NP only 0–15% was consumed. However, the ^{13}C/^{12}C-ratio of the CO_2 formed during a ^{13}C-*n*-NP batch experiment was shifted from $\delta C \approx -21.2\%$ only to $+0.4\%$, which indicates only a slight mineralisation of the linear NP.

4 DISCUSSION AND CONCLUSIONS

The half-life of BPA was observed to be between 3 d and 7 d which is similar to the half-life in natural waters (Dorn et al. 1987). The adsorption factors (K_{BM} and K_{OC}) determined from the concentration curve are similar to the earlier reports from Dorn et al. (1987). Therefore, the fate of BPA in constructed wetlands can be described by sorption coefficient (K_{OC}) and the half-life ($\tau_{1/2}$), whereby the overall reaction is depending on water volume, hydraulic retention time, biomass and the ability of the microbial consortia to degrade BPA. About 60% of the degraded BPA will be transferred to CO_2, 20% will be used for cell growth and the rest has to be transformed to polar metabolites as described earlier by Lobos et al. (1992). This suggests that BPA can be mineralised easily under the conditions applied and formation of residues is not expected. As indicated by the half-life of BPA, the removal in constructed wetlands is dependent by the hydraulic retention time which is normally much higher in comparison to activated sludge waste water treatment systems.

Linear NP was rapidly degraded up to 90% in 5.2 d and adsorption interactions were even stronger than those of branched NP. The degradation pathways of *n*-NP may be different from those of the branched NPs because the alkyl chain in the linear NP was obviously degraded much faster than the branched side chain. Even with a high recovery of ^{13}C in the outflow, no primary metabolites could be found by LLE-GC/MS. The high amounts of the ^{13}C label in the water residues suggest the formation of hydrophilic, highly metabolised compounds. These metabolites may become easily converted to biogenic matter.

The degradation of branched NP was very low and could not be proven in the present study. High sorption effects in the gravel bed/rhizosphere were observed causing the removal from the water body. However, NP will continuously be released from the sorbed state into the water phase in trace amounts without any degradation. The results show that linear NP is not a representative model compound for the assessment of the environmental fate of technical (branched) NP. In future, emphasis should be laid on microbial degradation of branched NP, since degradation of technical NP mixtures have been demonstrated with cultures and in activated sludge experiments by Tanghe et al. (1998) and Fujii et al. (2001). In general, constructed wetlands are considered to have a higher elimination potential related to hardly degradable compounds of the waste water caused by the higher retention time and the higher sorption capacity of the gravel bed. Due to the significant affinity to sorb on hydrophobic surfactants complete mass balances are essential to prove biodegradation of highly sorptive compounds in constructed wetlands. Therefore, sorption processes may have pretend the elimination in the present experiment.

ACKNOWLEDGEMENTS

The work was funded by the German Ministry of Education and Research, grant # 02 WA 9982/1.

REFERENCE

Ben-Jonathan, N. & Steinmetz, R.. 1998. Xenoestrogens: The Emerging Story of Bisphenol A. *Trends in Endocrinology and Metabolism*, 9(3):124.

Dorn, P. B., Chou, C.-S. & Gentempo, J. J. 1987. Degradation of bisphenol a in natural waters. *Chemosphere*, 16(7): 1501–1507.

Fitzer, E. & Fritz, W. 1975. Technische Chemie – Eine Einführung in die Chemische Reaktionstechnik. Berlin-Heidelberg-New York: Springer-Verlag.

Fujii, K., Urano, N., Ushio, H., Satomi, M. & S. Kimura. 2001. *Sphingomonas cloacae* sp. nov., a nonylphenol-degrading bacterium isolated from wastewater of a sewage-treatment plant in Tokyo. *International Journal of Systematic and Evolutionary Microbiology*, 51:603–610.

Kappelmeyer, U., Wießner, A., Kuschk, P. & Kästner, M. 2002. Operation of a Universal Test Unit for planted Soil Filters – Planted Fixed Bed Reactor (PFR). *Engineering in Life,* 2:311–315.

Lobos, J.H., Leib, T.K. & Su, T.-M. 1992. Biodegradation of bisphenol A and other bisphenols by a gram-negative aerobic bacterium. *Applied and Environmental Microbiology,* 58(6):1823–1831.

Richnow, H.H., Eschenbach, A., Mahro, B., Kästner, M., Annweiler, E., Seifert, R. & Michaelis, W. 1999. Formation of nonextractable soil residues: A stable isotope approach. *Environmental Science & Technology*, 33:3761–3767.

Tanghe, T., Devriese, G. & W. Verstraete. 1998. Nonylphenol degradation in lab scale activated sludge units is temperature dependent. *Water Research*, 32(10):2889–2896.

European Symposium on Environmental Biotechnology, ESEB 2004 - Verstraete (ed)

Modelling the biodegradation of treated and untreated waste and risk assessment of landfill gas emissions – SIMGASRISK

Telemachus C. Koliopoulos
University of Strathclyde, Centre for Environmental Management Research, Scotland, UK

George Fleming
University of Strathclyde, Envirocentre Ltd., Centre for Environmental Management Research, Scotland, UK

ABSTRACT: Sanitary landfill remain an attractive disposal route for household, commercial and industrial waste, as it is more economical than alternative solutions. The landfill biodegradation processes are complex, including many factors that control the progression of the waste mass to final stage. The trends in biogas production and associated risk are presented as a result of change in waste composition and management. The main aim of this paper is to investigate the effects of different disposed waste compositions and variable waste management conditions in order to enhance-accelerate landfill stabilisation and control the associated risks. Projections of SIMGASRISK model are made between several experimental and big scale landfill cases giving useful conclusions.

1 INTRODUCTION

This paper is the state of the art of landfill gas production numerical modelling and associated risk assessment. Projections and are made of the numerical model SIMGASRISK (SIMulation of GAS RISK), which is modelling the waste biodegradation stages based on field data of landfill case studies. The validation and robustness of SIMGASRISK were based on the measurements of experimental landfill Mid Auchencarroch (MACH), which is a UK Environment Agency and industry funded research facility.

The Mid Auchencarroch experimental landfill is a field scale facility, constructed in order to assess a number of techniques that promote sustainable landfill. It has been capped since 1995. The experimental variables are waste pretreatment, leachate recirculation and co-disposal with inert material. The project consists of four cells each of nomimal volume of 4,200 m^3. This project attempts to develop and assess techniques to enhance the degradation, and pollutant removal processes for Municipal Solid Waste (MSW) landfill. The wet-flushing bioreactor landfill model is seen as the method of achieving the goal of sustainability (Koliopoulos, T. 1999; Koliopoulos *et al.* 2001). Also projections are made for particular LFG characteristics-emissions from big scale sites like Tokyo Metropolitan landfill and Greek landfills. Risk assessment of landfill gas production is presented for different landfill composition and waste management techniques.

2 MODEL DESCRIPTION

SIMGASRISK develops a primary risk assessment for lateral LFG migration based on the produced LFG pressure taking into account the mid-depth waste mass temperature and biogas generation. Also SIMGASRISK develops a secondary risk assessment for lateral LFG migration from landfill boundaries. It is focused on the calculation of biogas migration advection velocity taking into account the particular source and pathway risk factors. SIMGASRISK can be applied easily to particular landfill cases, giving satisfactory results as a diagnostic tool for future effective designs. There were selected the most important parameters for LFG generation and heat generation based on the waste input characteristics (biodegradation, moisture content, thermal properties of each waste material). The model input parameters were chosen to be: the weight of the particular disposed waste materials; the total disposed waste quantity; selection of wet or dry climate of the site. A flow chart of SIMGASRISK is presented in figure 1.

SIMGASRISK is a dynamic numerical model as it calculates the following: i) LFG peak production for

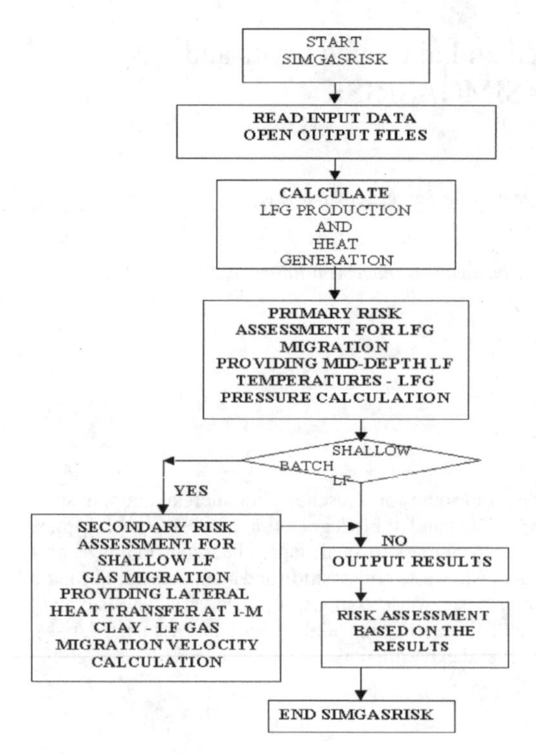

Figure 1. Flow chart of SIMGASRISK.

wet/dry site conditions; ii) Peak Temperatures for different waste inputs; iii) Risk assessment of lateral landfill gas migration based on the source and pathway risk factors of the examining site; iv) Heat transfer vertical & lateral distribution at shallow landfill bioreactor during its life cycle; v) Calculation of LFG advection velocity; vi) Threshold distances vs methane explosive levels in time; vii) Risk assessment of different waste inputs-management vs heat emissions-LFG emissions. SIMGASRISK uses several analytical and numerical solutions so as to solve the particular differential equations of heat transfer and lateral LFG migration from landfill boundaries.

SIMGASRISK's LFG generation module is based on Andreottola and Cossu LFG yield model, applying the waste characteristics, like moisture content, biodegradation and organic material of the particular waste materials (Koliopoulos *et al.* 2001). Moreover, SIMGASRISK's heat generation equations have been developed based on MACH field data. They are used for the module of the heat generation over landfill life. Yoshida *et al.* (Yoshida *et al.* 1999) found that the maximum temperatures in the waste mass exist at the mid-depth of landfills. SIMGASRISK assumes that the maximum temperature in the waste mass exists at the mid-depth of a landfill. The heat generation source term α of the waste mass at mid-depth of a landfill, could be described by the following formula (Koliopoulos, T. 2000).

$$\alpha = D_{LFG}\, D_{waste}\, Gt\, \Omega$$

where

α	heat generation source term at landfill mid-depth (Kcal/m^3 day)
D_{LFG}	average LFG density (kg LFG/m^3 LFG)
D_{waste}	waste density (kg/m^3)
Gt	LFG production in time (m^3 LFG/1,000 kg waste day)
$\Omega = \mu e^{-lt}$	μ heat generation (Kcal/kg LFG), l biodegradation rate (day^{-1}), t (day)

Long-term exponential heat generation curves, have been developed by SIMGASRISK based on MACH data taking into account different waste input characteristics and treatments. However, two functions describe landfill gas production (Koliopoulos *et al.* 2001):

i) increasing gas production

$$G_t = G_{tmax}\, e^{-k1(t1 - t)} \tag{1}$$

ii) decreasing gas production (after t1)

$$G_t = G_{tmax}\, e^{-k(t - t1)} \tag{2}$$

where

t	time since dumping (year)
t1	time period peak LFG production (year)
G_{max}	peak LFG production at time t1 (lt)
G_t	LFG production at time t (lt)
k	$= -\ln(0.5)/t_{0.5}$ (year^{-1})
k1	$= (\ln Gtmax - \ln 0.01)/t1$ (year^{-1})
$t_{0.5}$	half time, which approaches the time when half of the ultimate gas production is reached (year)

It has been reported in the literature that k biodegradation variable, takes values 0.288 for high biodegradable sites and 0.099 for stable big scale dry landfills (Christensen *et al.* 1996). The LFG peak production takes place at MACH site, in 105 days. For all the MACH cells, it was taken k = 0.278, after calibration based on the filed data. MACH's k parameter takes this value as MACH is shallow and the substrate can be degraded quickly based on the measurements and the field data. Furthermore, based on equation (1) and MACH's field data, the heat generation source terms have been developed in SIMGASRISK, for the particular three different waste types, which are presented below (Koliopoulos, T. 2000).

$$\alpha_{wt1} = 1.18\, D_{waste}\, Gt\, 0.41\, e^{-lt} \tag{3}$$

$$\alpha_{wt2} = 1.18\, D_{waste}\, Gt\, 0.68\, e^{-lt} \tag{4}$$

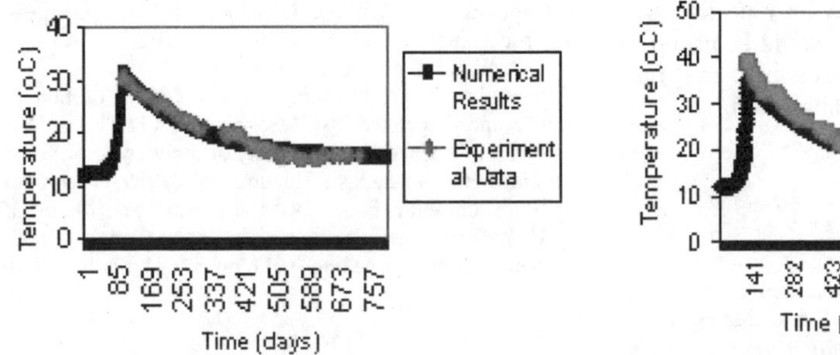

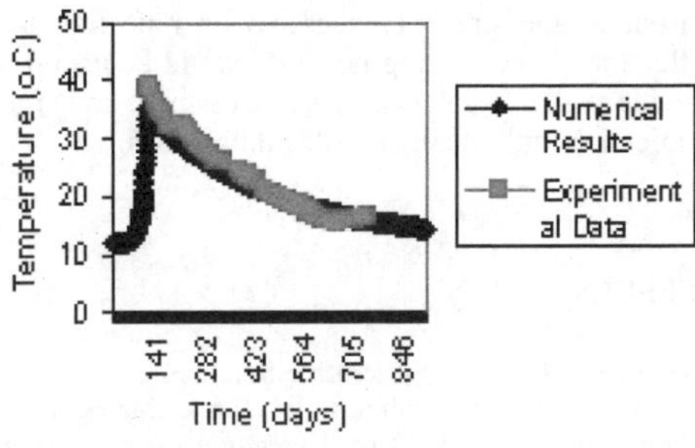

Figure 2,3. Calculated vs measured temperature emissions at MACH cell 1,3.

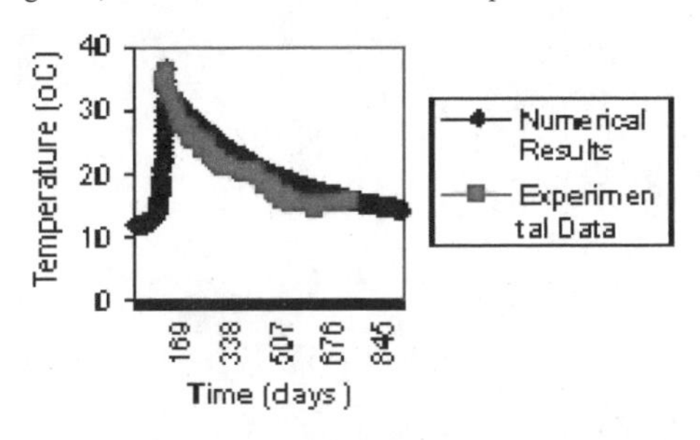

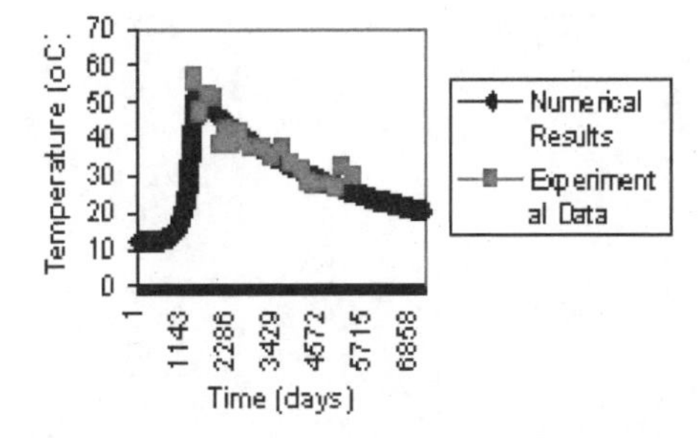

Figure 4,5. Calculated vs measured temperature emissions at MACH cell 4 (left) and Tokyo landfill site (right).

$$\alpha_{wt3} = 1.18\, D_{waste}\, Gt\, 0.89\, e^{-lt} \quad (5)$$

where

- α_{wt1} heat generation source term for co-disposal of pulverised waste with inert material (Kcal/m^3 day)
- α_{wt2} heat generation source term for waste type of pulverised waste (Kcal/m^3 day)
- α_{wt3} heat generation source term for waste type of untreated waste (Kcal/m^3 day)
- D_{waste} waste density (kg/m^3)
- Gt LFG production in time (m^3 LFG/1,000 kg waste day)
- l biodegradation rate (day^{-1})
- t (day)

The validation of the above equations is presented below in figures 2,3,4,5 i) MACH cells 1,3 (case studies of treated waste), ii) MACH cells 4 and Tokyo Metropolitan landfill (case studies of untreated waste). However, conversion of LFG yields, production to methane ones can be made by multiplying the LFG yield by the average methane percentage by volume. It has been estimated by local authoritie's reports that the LFG production in four years after dumbing for Athens and Thessaloniki landfills equals to 39 lt/year*kg MSW and 40 lt/year*kg MSW respectively or for the methane one 25.3 and 24 lt/year*kg MSW respectively with average methane percentage by volume 65% and 60% respectively (Koliopoulos, T. 2000). SIMGASRISK was applied for Athens and Thessaloniki sites and it calculated a methane production in four years after dumping 25.26 lt/year*kg MSW, 24.3 lt/year*kg MSW respectively and a LFG production 38.86 lt/year*kg MSW, 40.5 lt/year*kg MSW respectively. Greek case studies present a high value of LFG production due to the fact that high putrescible and fermentable waste fractions have been disposed into them. Hence, effective design and monitoring have to take place so as to minimise and control the associated risks of landfill emissions in time.

3 CONCLUSIONS

Risk assessment estimations of several environmental pollution subjects, must be site specific, yet no single preferred method is available. SIMGASRISK can be used as a risk assessment diagnostic tool for site specific case studies. Both primary and secondary risk assessments in SIMGASRISK are focused mainly on the quantification of the source and pathway risk factors as these are the most hazardous so as to take additional measures, like timing of risk assessment and communication of the results. However, the timing of risk assessment is promoted by recommending that it should be undertaken in parallel with frequent in situ measurements in time, monitoring the landfill emissions and prevent excessive or repeated data collection. A monitoring network would be useful to provide available data for a risk assessment and communication with the responsible authorities.

REFERENCES

Koliopoulos, T. (1999) Sustainable Solutions for the Most Pressing Problem within Solid Waste Management, *International Solid Waste Association Times Journal*, Copenhagen, Denmark, **3**, pp. 21–24.

Koliopoulos, T., Fleming, G. (2002) I.S.W.A Congress, Proceedings vol.2, pp. 1019–1028, (Ed. G. Kocasoy, T. Atabarut, I. Nuhoglu), Constantinople, Turkey.

Koliopoulos, T. (2000) Numerical Modelling of Landfill Gas and Associated Risk Assessment, Ph.D. Thesis, Dept. of Civil Engineering, University of Strathclyde, U.K.

Yoshida, H., Tanaka, N., Hozumi, H. (1999) Theoretical Study on Temperature Distribution in Landfills by 3D Heat Transport, Sardinia 7th International Landfill Symposium, Proceedings vol.I, pp. 85–94, Sardinia, Italy.

European Symposium on Environmental Biotechnology, ESEB 2004 - Verstraete (ed)
© 2004 Taylor & Francis Group, London, ISBN 90 5809 653 X

Modelling the biodegradation of treated and untreated waste and risk assessment of landfill leachate emissions

Telemachus Koliopoulos
University of Strathclyde, Dept. of Civil Engineering, Centre for Environmental Management Research, Glasgow, Scotland, UK

Vassilios Kollias
Dept. of History and Philosophy of Sciences, University of Athens, Greece

Panagiotis Kollias
Sanitary Civil Engineer, Naxou, Athens, Greece

Georgia Koliopoulou
University of Ioannina, Department of Medicine, Greece

Spyridon Kollias
[5]*Mathematician, University of Athens, Greece*

ABSTRACT: This paper examines several biodegradation factors in solid waste production, as a result of change in waste composition and management. This paper investigates and estimates the design and treatment of different solid waste pollution loads in leachates. Projections of the leachate emissions are made from the experimental Mid Auchencarroch landfill in the UK and case studies of landfill sites in Greece.
2 diagrammata kyria biologikoy me Dexamenh Aerismoy.

1 INTRODUCTION

Sanitary landfill remain an attractive disposal route for household, commercial and industrial waste, as it is more economical than alternative solutions. In this paper comparisons are made between several characteristic cases of biodegradation factors for different leachates compositions, and waste-input cases. Projections are made for UK and Greek landfill sites. Also a risk assessment is presented for the landfill emissions taking into account the experimental landfill Mid Auchencarroch (MACH) (UK) and Thessaloniki, Patras Greek case studies.

The first case study is the experimental landfill Mid Auchencarroch, which is a UK Environment Agency and industry funded research facility. The Mid Auchencarroch experimental landfill is a field scale facility, constructed in order to assess a number of techniques that promote sustainable landfill. It has been capped since 1995. The experimental variables are waste pretreatment, leachate recirculation and co-disposal with inert material. The project consists of four cells each of nominal volume of 4,200 m^3. This project attempts to develop and assess techniques to enhance the degradation, and pollutant removal processes for Municipal Solid Waste (MSW) landfill. The wet-flushing bioreactor landfill model is seen as the method of achieving the goal of sustainability. Also two more characteristic case studies are examined in this paper from Thessaloniki and Patras landfill sites in Greece.

2 LEACHATE EMISSIONS' CHARACTERISTICS – TREATMENT

The landfill biodegradation processes are complex, including many factors that control the progression of the waste mass to final stage (Kollias, P. 1993; Koliopoulos, T. 1999; Koliopoulos, T. *et al.* 1999, 2003). The estimations of the main leachate concentration parameters change with landfill age for the

Table 1. Landfill leachate characteristics in time.

Parameter	0–5 yr	5–10 yr	10–20 yr	<20 yr
BOD_5 (mg/l)	4,000–30,000	1,000–4,000	50–1,000	<50
COD (mg/l)	10,000–60,000	10,000–20,000	1,000–5,000	<100
Ammonia (mg/l)	100–1,500	300–500	50–200	<30
pH	3–6	6–7	7–7.5	6.5–7.5
Chloride (mg/l)	500–3,000	500–2,000	100–50	<100
Sulphate (mg/l)	50–2,000	200–1,000	50–200	<50

particular sites in time and they can be defined as presented below in table 1.

For the MACH the biodegradation rate has been evaluated according to the most indicative characteristic biodegradation parameters of the produced leachate emissions. COD could be characterized as the most hazardous leachate characteristic in relation to groundwater and site contamination.

Moreover, samples from leachate measurements there are from Thessaloniki and Patras landfill sites in Greece. High magnitudes of leachate emissions' characteristics there are from fresh samples at Tagarades site in Thessaloniki, i.e. for chloride concentrations average[Cl] = 3,255 mg/l, for C.O.D average = 70,857 mg/l; max = 115,000 mg/l; min = 44,000 mg/l, for B.O.D average = 29,720 mg/l; max = 80,795 mg/l; min = 9,500 mg/l. For Patras landfill C.O.D and B.O.D presented a maximum value 47400 and 41300 mg/l respectively and a minimum value 3260 and 2258 mg/l respectively (M.O.E., 1999). The COD/BOD ratio is different for fresh samples, middle-aged and old samples of leachates. The latter fact determines the method, which will be selected for leachates' treatment.

However, below the biodegradation for MACH cells 1,2,3 and 4 is presented based on the calculated least-square equations of the COD, TOC change in 22-month time period (Koliopoulos *et al.* 1999).

Cell 1: $COD1 = 0.804 + 0.246 * TOC1$, $SD = 3.97$, $CV = 0.689$, $R^2 = 0.8\ \forall\ TOC1 \in [2,100]$;

Cell 2: $COD2 = 0.668 + 0.264 * TOC2$, $SD = 1.94$, $CV = 0.43$, $R^2 = 0.84\ \forall\ TOC2 \in [1.7,62]$;

Cell 3: $COD3 = 0.62 + 0.194 * TOC3$, $SD = 0.6$, $CV = 0.273$, $R^2 = 0.96\ \forall\ TOC3 \in [1.85,68.5]$;

Cell 4: $COD4 = 3.96 + 0.0712 * TOC4$, $SD = 6.89$, $CV = 1.12$, $R^2 = 0.94\ \forall\ TOC4 \in [0.99,76.5]$.

where COD1,COD2,COD3,COD4 are the respective COD values per cell divided by 10^3 in mg/l; TOC1,TOC2, TOC3,TOC4 are the respective TOC values per cell divided by 10^2 in mg/l; SD = Standard Deviation; CV = Coefficient Variability.

Evaluating the above results it is clear that there was the greatest depletion of carbon and COD pollutants at cell 1. Moreover, cell 4 presents higher max COD concentrations due to the fact that there has been disposed higher waste fraction of biodegradable carbon content in it than at cell 3 and 2.

Cell 2 presents temporarily high risk between the 15th and 21st month. The latter can be explained due to the fact that leachate recirculation began in November 1996. After that period chloride was rising sharply, indicating flushing out of soluble salts, which had already occurred in the pulverized cells and they exhibited a greater electrical conductivity effecting further chemical reactions. However, it has been found that there is a linear relation of the chloride vs electrical conductivity concentrations for cell 2 in time (Koliopoulos *et al.* 1999). All the TOC and COD concentrations present great reduction after 1996. The latter fact certifies the quick and efficient Mid Achencarroch's site stabilization.

2 RISK ASSESSMENT OF LEACHATE EMISSIONS

The COD parameter was selected as a representative risk characteristic of the MACH leachate emissions, due to its high grade of hazard for groundwater-environmental contamination. The evaluation of the short-term involved risk was selected for the examining cells as it is the most crucial period for high leachate pollutant concentrations at any stable landfill. During this period there is a peak of leachate pollutant concentrations, which are hazardous for possible groundwater-environmental contamination. The short-term risk, which is involved within the MACH leachate emissions, is presented in figure 1, based on the produced COD concentration rates of the four cells. Furthermore, in figure 2, is presented the concentration X (mg/l) of microorganisms vs time (days) in a waste-water unit with aerobic treatment. In figure 3 is presented the Monod kinetic variable (μ) versus the organic material (F) (mg/l). For the examining

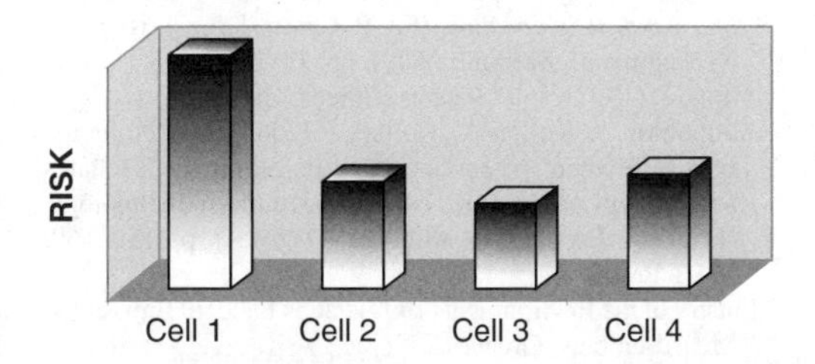

Figure 1. Involved risk of the COD leachate emissions at MACH cells.

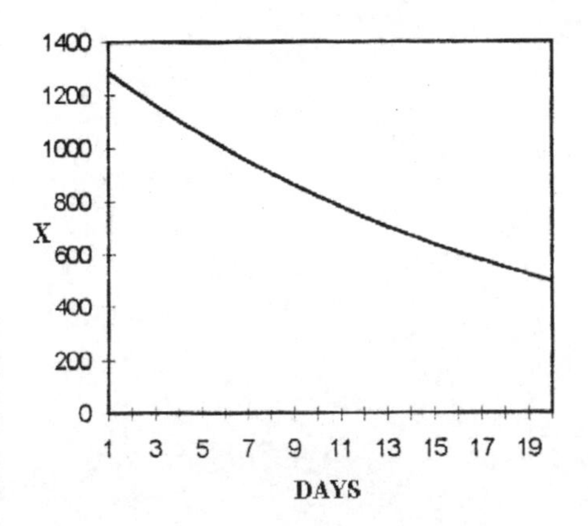

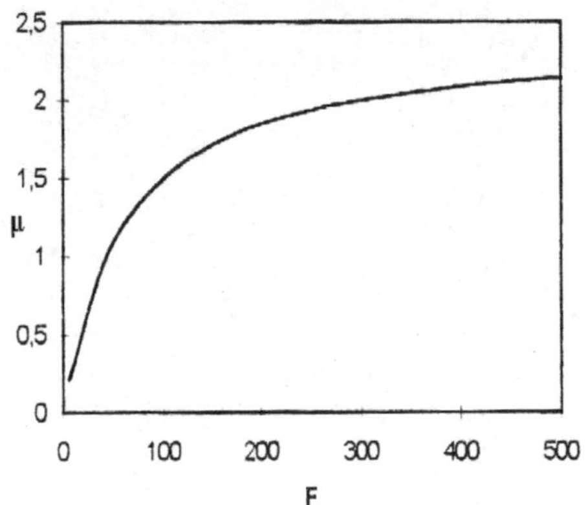

Figures 2, 3. Minimising the risk of microorganisms concentration X vs time (days) with efficient waste-water aerobic treatment – Monod kinetic variable (μ) vs the organic material (F).

waste-water aerobic treatment case study were taken Xo = 1000 mg/l, Fo = 500 mg/l, Y = 0,7 gr/gr, b = 0.05 day^{-1} and for the Monod kinetics μ_{max} = 2,4 $days^{-1}$, Ks = 60 mg/l. An efficient waste-water aerobic treatment unit can be used for fresh leachate quantities. Old leachates could be firstly treated in inverted-osmosis units and after could be recirculated in the treated waste mass. In this way can be minimised efficiently the particular risks of leachates' pollution loads in short time.

Evaluating and analysing MACH leachate emissions, is clear that higher short-term risk of environmental contamination by leachates present cell 1 and 4 than 2, 3 ones. Cell 1 presents the highest short-term risk, as greater carbon and COD depletion rate exists in it than at the rest of the cells. However, cell 4 presents higher short-term risk than at the rest of the cells from the point of view that it presents high constant COD values without any decrease in short-term. The latter exists due to the fact that not only there isn't leachate recirculation at cell 4 for quick carbon depletion but also there is the high disposed putrescible waste fraction into it.

However, for the Greek case studies there are high magnitudes of the BOD, COD and TOC concentrations in the leachate emissions. This can be explained due to the fact that high putrescible waste fractions have been disposed into the Greek landfills. The batch landfill bioreactor design and composting units could be applied efficiently for the Greek sites. This solution will biodegradate efficiently the organic fractions and leachate emissions in short time, minimising the risks of environmental pollution.

3 CONCLUSIONS

At Mid Auchencarroch it was clear that the co-disposal with inert material is sustainable as well as the pretreatment by wet pulverisation since the recirculation of leachate expedite the biodegradation and methanogenesis. The high CH_4 concentrations and the reduced CO_2 emissions after 1996, show that the methanogenesis and quick site stabilization was achieved. According to the TOC-COD equations the best waste biodegradation took place in cell 3, as well good organic depletion there was in cell 1, minimizing both their emissions in short time. However, leachates' environmental contamination control has to be improved in future, taking into account the particular different landfill conditions which exist and monitoring landfill emissions in time.

Long-term liability can be minimized when waste is quickly treated to a point where no further degradation will occur, protecting the environment from long-term biogas and leachate emissions. The waste treatment and the increase of the recycling rates will play a catalytic role in the reduction of the landfill emissions-contamination and the acceleration of waste biodegradation in time.

REFERENCES

Koliopoulos, T. (1999) Sustainable Solutions for the Most Pressing Problem within Solid Waste Management, *International Solid Waste Association Times Journal*, Copenhagen, Denmark, **3**, pp. 21–24.

Koliopoulos, T. (2000) Management and Risk Assessment of Mid Auchencarroch landfill, Scotland, *In: Young Researchers' Conference*, International Water Association, Environment Agency, Headquarters, Nottingham, Trentside, England, U.K.

Koliopoulos, T., Fleming, G., Skordilis, A. (1999) Evaluation of the Long Term Behaviour of Three Different Landfills in the UK and in Greece, *In: Proceedings of the 7th International Waste Management and Landfill Symposium*, (Ed. R. Cossu, T.H. Christensen, R. Stegmann), Sardinia, Vol. I, pp. 19–26, Italy.

Kollias, P. (1993) Solid Wastes, Athens, Greece.

Koliopoulos, T., Kollias, V., Kollias, P. (2003). Modelling the risk assessment of groundwater pollution by leachates and landfill gases, U.K. Wessex Institute of Technology, (Ed. C.A. Brebbia, D. Almorza, D. Sales), pp.159–169, W.I.T Press.

Ministry of the Environment (1999) Patras Landfill Emissions' Measurements, Greece.

Biosensing

European Symposium on Environmental Biotechnology, ESEB 2004 - Verstraete (ed)

Oral exposure to PAH: bioactivation processes in the human gut

T.R. Van de Wiele, L. Vanhaecke, C. Boeckaert & W. Verstraete
Laboratory Microbial Ecology and Technology (LabMET), Ghent University, Ghent, Belgium

S. Siciliano
Department of Soil Science, University of Saskatchewan, Saskatoon, Saskatchewan, Canada

ABSTRACT: We used the Simulator of the Human Intestinal Microbial Ecosystem (SHIME) to perform *in vitro* stomach, small intestine and colon digestion experiments on polycyclic aromatic hydrocarbons (PAH) as pure compounds or from contaminated environmental samples entering the gastrointestinal tract. To screen for possible toxicological aspects, two yeast bioassays were used on the digests. Binding of PAHs to the human aryl hydrocarbon receptor (hAR) and binding of probable PAH biotransformation products to the human estrogen receptor (hER) was evaluated in the aryl yeast and estrogen yeast assay, respectively. Upon incubation with colon microbiota, positive signals in the estrogen receptor test were observed. PAHs as such are not estrogenic, leading to our hypothesis that biotransformation reactions had occurred during incubation with colon microbiota in the SHIME reactor. The production of intermediates that have already been described to have estrogenic or genotoxic properties, was confirmed by the detection of 1-hydroxypyrene and 7-hydroxybenzo(a)pyrene using LC-ESI-MS.

1 INTRODUCTION

The human body can be exposed to PAHs through a number of pathways from which ingestion of contaminated soils appears to be a predominant exposure route with reported ingestion rates of 1–500 mg soil/d (Calabrese *et al.*, 1997; Staneck *et al.*, 1997; Vanwijnen *et al.*, 1990). Upon ingestion, PAHs are partly released from their soil matrix. When freely dissolved or present in small enough complexes, PAHs can be absorbed through the small intestinal wall and subsequently transported to enterocytes and hepatocytes, where they are biotransformed (Zhang *et al.*, 1997; Münzel *et al.*, 1999).

Phase I enzymes will form more polar metabolites by placing functional groups on the PAH molecule. In the phase II biotransformation, these metabolites are further conjugated by glucuronosyltransferase, sulfotransferase or glutathione, rendering the PAH derivatives more hydrophilic. These phase II metabolites are more prone to removal from the body by urinary or biliary excretion.

However, this detoxification mechanism is sometimes counter-productive as PAHs may be bioactivated during phase I biotransformation leading to the production of genotoxic, carcinogenic or even estrogenic compounds (Goldman *et al.*, 2001; Hirose *et al.*, 2001). Furthermore, excreted PAH conjugates from phase II biotransformation can again be deconjugated by bacterial enzymes (e.g. glucuronidase) in the large intestine, regenerating the more toxic intermediates. This could delay the excretion of many exogenous compounds (Weisburger 1971). Additionally, the high diversity of microorganisms in the large intestine provides a wide variety of enzymes that may be able to transform xenobiotics compounds (Ilett *et al.,* 1990). Some of these biotransformation products, for example PAH hydroxylated compounds are of high environmental concern as they can exhibit toxic, mutagenic and carcinogenic effects or may even possess (anti)-estrogenic activities.

In this research, we investigated the potency of a large intestinal microbiota to transform PAHs to hydroxylated derivatives. Two bioassays, the aryl hydrocarbon yeast test and the estrogen yeast test, were combined with an analytical tool, LC-ESI-MS (liquid chromatography – electrospray ionization – mass spectrometry). In a first phase using the bioassays, we monitored whether bioactivation of PAHs by a large intestinal micobiota was occurring upon incubation. In a second part of the research, we used LC-MS to screen for PAH metabolites, particularly hydroxylated compounds, since these compounds are known to bind the human estrogen receptor.

2 MATERIALS AND METHODS

2.1 *PAH incubations and HPLC-ESI-MS analysis*

The Simulator of the Human Intestinal Microbial Ecology consists of 5 vessels simulating the stomach, the small intestine, the colon ascendens, the colon transversum and the colon descencdens, respectively. The colon vessels contain a mixed microbial suspension that is representative for the microbial ecology of the human large intestine both in concentration as in composition (Molly *et al.*, 1994).

Colon suspension aliquots of 50 mL were taken from the SHIME reactor. Different PAH compounds were dosed to the colon suspension at a concentration of 0.2 mM and subsequently incubated for 24 hours at 37°C. After this colon incubation, samples were centrifuged at 3000 g for a duration of 10 minutes to remove the biomass and subsequently stored at −20°C prior to analysis for PAH hydroxylates and parent compounds.

PAH compounds that were incubated in colon suspension as well as control samples were analyzed for hydroxylated PAHs and parent PAH compounds with LC-ESI-MS. More details on the sample preparation and the analytical methodology are provided in Van de Wiele *et al.* (2003).

2.2 *Bioassays: human estrogen receptor and aryl receptor yeast test*

PAH parent components and PAH metabolites were extracted from the digests by performing a liquid/liquid extraction in which the digest and ethylacetate were mixed in a 1:1 ratio. The ethylacetate fraction was subsequently put in a rotavapor to remove most of the solvent. The remainder of the solvent was removed under a light stream of N_2 and finally replaced by DMSO (dimethylsulfoxide) which is a proper solvent to use in bioassay tests.

In the estrogen and aryl bioassays, two genetically engineered *Saccharomyces cerevisiae* strains were used that had received the human estrogen or aryl receptor. The yeast strains contained expression plasmids with estrogen or aryl responsive elements and the lacZ reporter gene, encoding the enzyme β-galactosidase. The β-galactosidase activity is quantified by the conversion of a chromogenic substrate chlorophenol-red-galactopyranoside (CPRG). We used a modified protocol from De Boever *et al.* (2001) that was based on the protocol developed by Routledge and Sumpter (1996) and Miller (1999) for the estrogen and aryl test, respectively. The response is expressed as absorbance at 540 nm divided by the optical density at 630 nm $(A540/A630)_{net}$.

3 RESULTS

PAH standards as such generated a positive response in the aryl hydrocarbon test in the order benzo (a)pyrene > pyrene > phenanthrene > naphthalene (Figure 1). No positive signals from the PAH standard were observed in the estrogen test (data not shown).

However, when these PAH standards had been incubated in a colon suspension from the SHIME reactor, an estrogenic response was detected, suggesting that the PAH parent compounds were partly bioactivated to compounds that bind the human estrogen receptor. Figure 2 plots the estrogenic signal for 4 PAH containing samples upon incubation in a colon microbial suspension. It is important to note that corrections for matrix background signals have been made and negative controls showed no significant estrogenic signals.

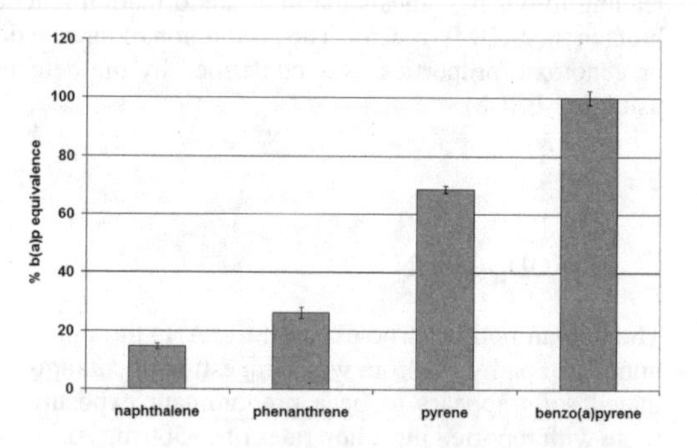

Figure 1. Response of 4 PAHs of different degrees of aromaticity in the aryl hydrocarbon yeast assay.

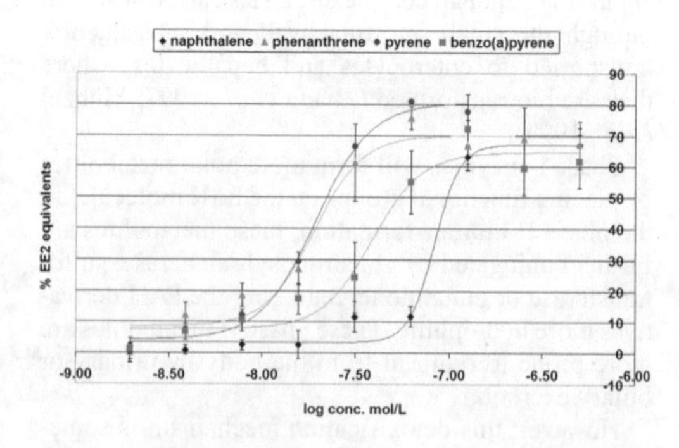

Figure 2. Dose-response curves of colon incubated PAH digests in the estrogen yeast assay, expressed as % EE2 equivalence of 6.96 nM EE2.

Aliquots of the PAH incubation mixtures were also analyzed by LC-ESI-MS.

Two hydroxyl metabolites of PAHs were identified and measured in the colon suspension incubated PAHs.

After 24 h of incubation, 1 of the 8 target PAH hydroxylates was detected namely 1-hydroxypyrene at a concentration of 2.5 μg/L. After incubation of a 1 mL aliquot of the sample in glucuronidase and aryl sulfatase – enzymes that deconjugate phase II like metabolites – a concentration of 4.4 μg/L was obtained for 1-hydroxypyrene. 7-hydroxybenzo-(a)pyrene was found at a 1.9 ppb concentration.

To see whether similar biotransformation were to be expected when oral exposure to PAH contaminated environmental samples was simulated, we subjected a former playground soil, contaminated with 50 mg PAH/kg soil DW, to stomach, small intestine and large intestine respectively. When entering the gastrointestinal tract, PAHs may release from the soil matrix and thus become available for intestinal absorption. We observed the highest PAH release, 18.1 μg/L, in the stomach compartment of the SHIME reactor. The small intestine digest and the colon digest of the PAH contaminated soil resulted in a lower PAH release with 3.2 and 1.9 μg PAH/L respectively. This decrease in released PAH fraction was explained by complexation effects with dissolved organic matter and microbial biomass as explained more in detail in Van de Wiele *et al.* (2003).

To investigate the effect of the gastrointestinal digestion processes on the soil with respect to estrogenic properties, the yeast estrogen bioassay was performed on the digests. No estrogenic signals were obtained for the stomach and duodenum digests of the PAH contaminated soil. However, we observed an induction in estrogen response for the soil sample that had gone through a digestion with active colon microbiota (Figure 3). The positive response was not as high as when pure PAH compounds had been incubated

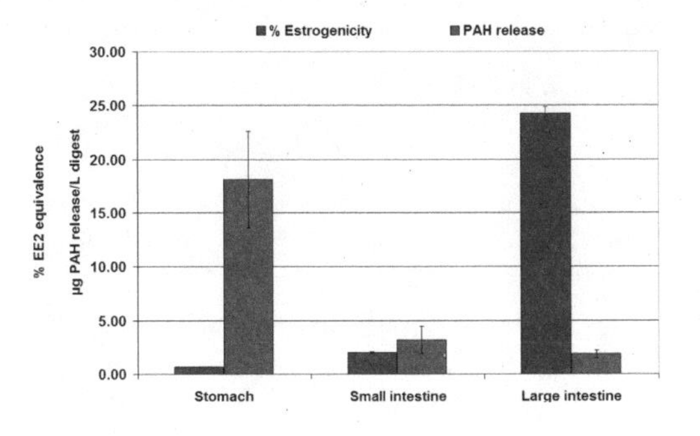

Figure 3. % estrogenicity and concentration of PAH in the stomach, small intestine and colon digest on PAH contaminated soil samples. Data show that with lower released PAH concentrations in the digests, estrogenicity increases.

with colon microbiota. Yet, an average EE2 equivalence of 24.3 ± 0.63% was seen.

4 DISCUSSION

In the bioassay tests, the potency of the released PAHs or their biotransformation products were evaluated for their potency to bind the human estrogen or aryl receptor. PAHs as such do not possess estrogenic activities, but some hydroxymetabolites do (Charles *et al.*, 2000). Positive signals in the estrogen test of PAH compounds that had been incubated in an *in vitro* colon suspension indicate that the large intestinal microbiota is indeed capable of performing a bioactivation reaction. Corrections for matrix background interference have been made when calculating the degree of estrogenic response.

Analysis of the colon digests in which PAHs had been incubated, confirmed the hypothesis that metabolites had been formed. Both 1-hydroxypyrene as 7-benzo(a)pyrene have been detected at ppb level. The fact that so few biotransformation compounds were detected, even at a low concentration, can be explained by the very short incubation time (24 hours) of the PAHs in the SHIME suspension, compared to other studies where PAH biodegradation was monitored for days and months in soils, sludges or sediments (Colombo *et al.*, 1996, Canet *et al.*, 1996). A significant PAH biotransformation yielding hydroxylated PAH metabolites may not be detected within this short incubation period. It is also possible that biotransformation products, other than the compounds that we envisaged, were present. However, the present study focused on PAH hydroxylates due to their general toxicity and putative carcinogenic and estrogenic properties. The fact that higher concentrations of 1-hydroxypyrene were found after enzymatic incubation with β-glucuronidase and aryl sulfatase may indicate that phase II type biotransformation products were present. Fungi and yeasts have been described to possess the P450 cytochrome complexes to perform conjugation reactions (Leipelt *et al.,* 2000) leading to the presumed phase II metabolites in this study. If the right conditions are prevailing in the gut, these or other organisms could perform such reactions.

The appearance of estrogenic properties was not exclusively observed with pure PAH compounds at elevated concentrations. The PAH contaminated former playground soil showed that comparable effects were obtained upon incubation with colon microbiota. This constitutes an additional risk besides the probable formation of hazardous PAH intermediates in the liver.

Current risk assessment only takes into account the hepatic biotransformation of intestinally absorbed compounds. We have shown that contaminants, for example mobilized in the gut from an ingested soil or food matrix do not necessarily have to be absorbed for biotransformation to occur. Further investigation on this aspect of the large intestinal microbiota is required, since a possible health risk can exist when colonocytes are subjected to estrogenic and possibly other toxic intermediates that are produced *in vivo.*

REFERENCES

Calabrese, E., Stanek, E., James, R., Roberts, S. 1997. Environ Health Perspect 105, 1354.

Canet, R., Birnstingl, J., Malcolm, D., Lopez-Real, J., Beck, A. 2001. Bioresource Technol. 76, 113.

Charles, G., Bartels, S., Zacharewski, T., Freshour, N., Carney, W. 2000. Tox. Sciences 55, 320.

Colombo, J., Cabello, M., Arambarri, A.M. 1996. Environ. Poll. 94, 355.

De Boever, P., Demare, W., Cooreman, K., Bossier, P., Verstraete, W. 2001. Env. Health Persp. 109, 691.

Goldman, R., Enewold, L., Pellizzari, E., Bowman, Z., Krishnan, S., Shields, P. 2001. Cancer Res. 61, 6367.

Hirose, T., Morito, H., Kizu, R., Toriba, A., Hayakawa, K., Ogawa, S., Inoue, S., Muramatsu, M., Masamune, Y. 2001. Health Sci. 47, 552.

Ilett, K., Tee, L., Reeves, P., Minchin, R. 1990. Pharmacol. Therapeut. 46, 67.

Miller, C.A. 1999. Tox and Appl. Pharmac. 160, 297.

Routledge, E., Sumpter, J. 1996. Env. Tox. and Chem 15, 241.

Stanek, E., Calabrese, E., Barnes, R., Pekow, P. 1997. Ecotox Environ Safe 36, 249.

Vanwijnen, J., Clausing, P., Brunekreef, B. 1990. Environ Res 51, 147.

Zhang, Q., He, W., Dunbar, D., Kaminsky, L. 1997. Biochem. Biophys. Res. Commun. 233, 623.

Weisburger, J. 1971. Cancer 28, 60.

Münzel, P., Schmohl, S., Heel, H., Kalberer, K., Bock, K. 1999. Drug Metab. Dispos 27, 569.

Molly, K., Vandewoestyne, M., Desmet, I., Verstraete, W. 1994. Microb. Ecol. Health D. 7, 191.

Van de Wiele, T., Peru, K, Siciliano, S., Verstraete, W., Headley, J. 2003. Analysis of PAH hydroxylates, formed in a simulator of the human gastrointestinal tract. Submitted.

European Symposium on Environmental Biotechnology, ESEB 2004 - Verstraete (ed)
© 2004 Taylor & Francis Group, London, ISBN 90 5809 653 X

Titration biosensors for estimating the biochemical nitrate demand of municipal and industrial wastes

A. Onnis, A. Carucci & G. Cappai
DIGITA, University of Cagliari, Italy

ABSTRACT: An anoxic titrimetric test was investigated for measuring denitrification potential of different wastewaters, both municipal and industrial, and to quantify the denitrifying activity in an activated sludge system. The method measures the amount of acid that is required to maintain the pH set-point value in a batch denitrification experiment, and it was performed using the DENICON (Denitrification CONtroller) biosensor. Good correlation between titration data and analyses was found.

1 INTRODUCTION

Denitrification is the biological process in which nitrate is reduced, acting as the terminal electron acceptor during the oxidation of organic substrates in the absence of oxygen. Temperature and pH are among the factors affecting this process, but the denitrification rate is also influenced by concentration and nature of the carbon source, where the highest rates are obtained with the most easily degradable forms (Isaacs & Henze, 1995; Lee & Welander, 1996).

To obtain a high nitrogen removal efficiency, in BNR systems not only the COD/N ratio needs to lie in the range of 5–10 gCOD/gN (Henze, 1991) but also a high denitrification rate must be guaranteed. The utilization rate of the internal carbon source of the wastewater used by the denitrifying biomass to reduce nitrate is often quite low. In these cases the controlled addition of external organics can improve process stability and flexibility.

In order to limit operating costs the availability and compatibility of organic residues from other processes (industrial wastewater) should be verified. This may allow to enhance denitrification and at the same time to reduce the amount of industrial waste to treat (Rozzi et al. 1997). Monteith et al. (1980) tested several industrial wastes as organic carbon sources; they found that some organic wastes such as formaldehyde and dextrose wastes were less efficiently degraded than distillery oils or brewery wort. Tsonis (1997) investigated the possibility of using an olive oil mill wastewater as a non-nitrogenous external carbon source in the second anoxic stage modified Bardenpho system for nutrients removal; he found that the addition of this waste is acceptable only up to a certain amount due to additional color problems in the treated effluents.

To identify possible organic wastes to use as external carbon sources to enhance the denitrification process an accurate characterization study of those wastes needs to be undertaken.

In this experimental study an anoxic titrimetric biosensor was used for the characterization of industrial wastewaters with respect to their denitrification potential expressed as BND (Biochemical Nitrate Demand) as well as the estimation of the denitrifying activity of activated sludges under different conditions.

This device is in fact able to measure the nitrate used to oxidize the biodegradable COD present in the wastewater from the amount of acid added to neutralize alkalinity produced in the denitrification reaction (approximately one mole of OH^- for every mole of nitrate removed); nitrate concentration is in this case kept in excess compared to the needs. At the same time the denitrifying activity can be derived from the acid consumption rate.

The BNDs of the wastewaters tested obtained with the titrimetric method, were compared with those obtained through analyzing the nitrate reduced during the tests.

2 MATERIALS AND METHODS

2.1 *Wastewaters and sludge: origin and preparation*

The titrimetric measurements for the assessment of the denitrification potential were conducted both on

Table 1. Conventional characterization of selected domestic and industrial wastewaters.

Waste	pH (–)	COD (mg)/l	SS (mg/l)	TN (mg)/l	TP (mg)/l
Domestic	7.5	160	110	26.3	3.9
Ice cream	4.4	2870	508	27.5	5.5
Beet-sugar	1.1	8000	–	230	6.2
Brewery	8.2	1250	–	15	–
Tuna cannery	6.5	1400	150	170	–

domestic sewage and industrial effluents including dairy (ice cream), beet-sugar, brewery, and tuna cannery wastewaters, exhibiting a wide spectrum of organic matter content and structure, reflected by COD concentrations varying from 1250 mg/l for brewery waste to 8000 mg/l for beet-sugar wastes.

The wastewater samples drawn at the different facilities were analyzed for pH, COD, TN, SS and TP according to the Standard Methods (APHA, 1998), and stored at 4°C until use. Table 1 outlines the results of the conventional characterization.

It should be noted that the samples were prepared to represent the expected quality of biological treatment influent. Therefore primary effluents were collected for domestic and dairy samples, chemical settling and the lagoon neutralized effluents were collected respectively for tuna cannery and beet sugar samples, because they are routinely subjected to these types of pretreatments before biological processes. The brewery sample was collected in the equalization tank at the factory. In the experiments the samples were also adjusted to a pH of 7–8, a range suitable for biological activity.

The experimental tests were carried out (batch tests) utilizing denitrifying biomass samples drawn at the WWTP of Cagliari located in the south of Sardinia, Italy, (later indicated with the acronym AS-1) that is a conventional activated sludge system, and from a laboratory scale SBR started up for the research, which was fed with a mixture of domestic and beet-sugar wastewater, later indicated as AS-2.

The samples of sludge drawn at the full scale plant were analysed for TSS and VSS and stored at 4°C for a period not longer than 15 days. Reagent grade chemicals were also used.

2.2 *Experimental procedure*

Tests were carried out utilizing DENICON (DENItrification CONtroller, AUSTEP, Milan), an automated titration device developed by Massone et al. (1996), which allows to determine alkalinity production following nitrate reduction. The anoxic titrimetric method records the cumulative amount of 0.05 M HCl that is added during the batch denitrification test in a 1 l thermo-stated and stirred vessel in order to maintain a constant pH (pH_{eq}).

To determine the pH_{eq}, samples of 0.5 l to 1 l of activated sludge were transferred into the pH-stat reactor, where they were stirred, thermostated (temperature between 20 and 25°C), and bubbled with a mixture of N_2 and CO_2 (in the ratio of 100:1). The pH was measured until its value did not change more than 0.02 units over a period of 10–20 min.

The pH_{eq} is set as the working pH of the pH-stat titration unit; then KNO_3 is added to the denitrifying sludge to give an initial concentration of 15 to 40 mg NO_3-N/l. Once a stable endogenous respiration rate was obtained, indicated by a constant acid addition, the organic waste (industrial wastewater) to be tested, was added to give a range of F/M ratios from 0.03 to 0.07 g COD/g VSS. The acid addition was monitored until the organic substrate had been consumed, and a steady endogenous rate had been obtained again.

During the pH_{eq} determination and the denitrification test, the sparging at a constant rate of the mixture of N_2 and CO_2 within the mixed liquor allows to keep the CO_2 concentration constant, so avoiding possible interferences caused by additional CO_2 production (e.g. by facultative heterotrophic microorganisms) which could shift the carbonates equilibrium and so modify the system pH.

The characterization tests of the beet-sugar waste were carried out using both the sludge AS-1 and AS-2, the latter is a biomass acclimated to the wastewater so those tests give information on the impact of the acclimatization on the denitrification rate and denitrification potential; whereas all the other tests were conducted only with the AS-1.

Samples of mixed liquor were taken during the batch experiments, filtered through 0.45 μm membranes and analyzed in order to determine nitrate/nitrite nitrogen concentrations.

2.3 *Calculation of BND of the wastes and denitrification activity of the biomass*

An example of the titration curve obtained with the dairy wastewater is shown in Figure 1, where significant parameters are also represented for the data evaluation.

The shape of the curve reflects the complex composition of the wastewaters: in fact when a pure compound is added as a carbon source (e.g. acetate) in an anoxic titrimetic test the response is acid dosage at a constant rate till all the substrate is used by the biomass to reduce nitrate; whereas, when a real wastewater is added into the pH-controlled reactor, the utilization rate of the organic carbon of the waste by the denitrifying biomass and thus the alkalinity production and neutralization rate strongly depends on the amount and the kind of COD (readily and slowly biodegradable COD) available. It is then possible to determine the

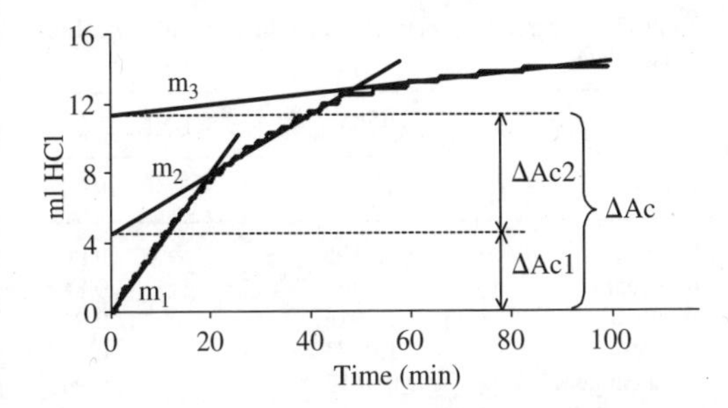

Figure 1. Example of titration curve obtained with the wastewater.

Table 2. Results of the anoxic tests.

Waste	k_{Dmax} (mgN/gSSV/h)		BND (mgN/l_{WW})	
	Mean	Conf.*	Mean	Conf.*
Domestic	4.05	±0.98	9.05	±1.52
Ice cream	6.02	±0.75	272	±30.2
Beet-sugar	5.84	±2.3	1020	±430
Brewery	7.23	±2.0	117	±18.1
Tuna cannery	3.53	±0.8	61	±40.0

* 95% Confidence.

different denitrification rates on the different COD fractions.

The maximum denitrification rate in mgN/gVSS/h was determined from the first slope of the titration curve (m_1 in Fig. 1) after the addition of the wastewater sample using the equation below:

$$k_{Dmax} = \frac{m_1 \cdot MW_N \cdot C_{Ac} \cdot 60}{X_V \cdot V} \quad (1)$$

where MW_N is the molecular weight of nitrogen, C_{Ac} is the acid concentration in Mol, X_V is the mixed liquor suspended solids concentration in g/l and V is the total volume of the batch. Using the equation (1) and substituting m_1 with the difference between m_1 and m_3 or between m_2 and m_3 (also in Fig. 1), it is possible to calculate the denitrification rates on the readily and slowly biodegradable fraction of the COD, respectively.

The denitrification potential of the wastewater expressed as the amount of nitrates, in mg NO_3-N/l, which can be denitrified on the COD of the wastewater was determined from the titration curve using the expression:

$$BND = \frac{\Delta Ac \cdot MW_N \cdot C_{Ac}}{V_{WW}} \quad (2)$$

where ΔAc is defined in Figure 1 and V_{WW} is the volume in liters of wastewater added to the reactor vessel. Even in this case it is possible to distinguish the contribution to the BND of the different fractions.

3 RESULTS AND DISCUSSION

The results of the anoxic tests for the assessment of the denitrification potential in selected industrial effluents and with the domestic sewage are outlined in Table 2.

The results obtained with the domestic and with the ice-cream wastewaters show very little variations.

The denitrification potential of the sewage is very low and this is probably due to the amount of readily biodegradable compounds in the wastewater available for denitrification; in a previous research using respirometric technique readily biodegradable fraction of this sewage was found to be 13% of the total COD; the ice-cream waste instead, has a quite high denitrification potential and could be considered as an alternative carbon source in denitrification. In this case a problem that can arise is the variation in quality and quantity due to the production cycle, which determines peak of flow rate and organic loading in definite periods of the year; so before selecting this kind of wastewater, a study of the annual variation should be carried out or pretreatment of the waste to equalize the loading should be considered.

The wide range obtained for beet-sugar waste is due to the different results obtained with the non acclimated and acclimated sludges. In the latter case the denitrification rates were considerably higher, and a second slope with a still high rate (i.e. 2.38 mgN/gSSV/h) was observed.

Furthermore it was noticed that using the not acclimated sludge an accumulation of nitrite took place, leading to a lesser alkalinity production and so slower acid addition. In fact, during denitrification it is the second step, i.e. the reduction of nitrite to nitrogen gas, that influences the pH in the reactor.

The different denitrification rates observed can be then explained with the presence in the acclimated biomass of microorganisms and enzymes able to reduce at higher rate the nitrites and to use more easily the different organic substances of the wastewater.

The range obtained for the maximum denitrification rate with the brewery waste could be explained considering the operating conditions of the tests; in particular different COD/N ratios were applied.

Figure 2 shows the relation that was found between the COD/N ratio applied and the maximum denitrification rates observed in the tests for this waste.

The tests with the tuna cannery waste were performed over a period of one week. It was observed that even if the maximum denitrification rate did not change significantly, the denitrification potential of the waste decreased from 120 mgN/l_{WW} obtained the

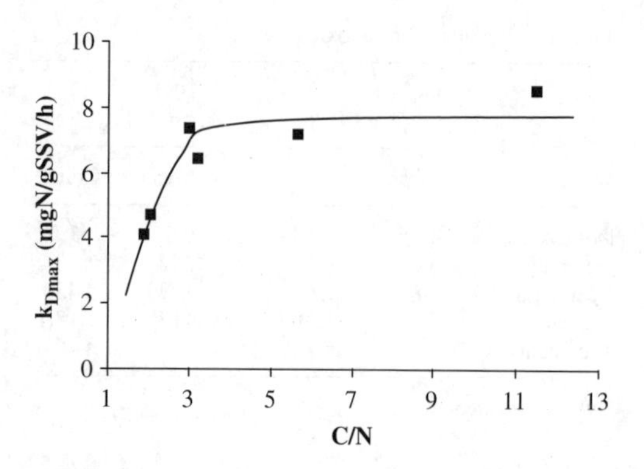

Figure 2. Denitrification rates of brewery waste at different C/N ratio.

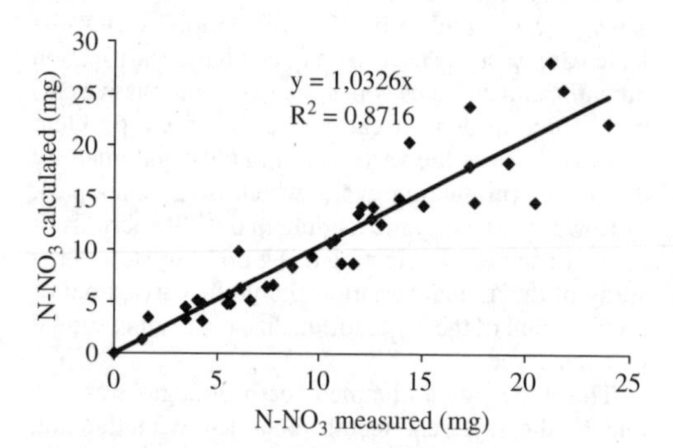

Figure 3. Correlation between nitrate measured and calculated from the acid added during the tests.

first day to 20 mgN/l_{WW} in the test performed after one week. This might means that the storage of the wastewater influenced its composition; in this case it is necessary to perform the test as soon as possible after sampling to obtain reliable results.

The nitrates reduced during the tests were deducted from the total amount of acid added by the titration unit, and correlated to those measured with the analytical methods; Figure 3 outlines the relation found.

The graph in Fig. 3 shows a very good correlation, indicating the titrimetric method as a useful tool to determine nitrate concentration in the wastewater or in the biological reactor (in this case with an excess of carbon) which can be integrated, in its on-line version, in an automatic control system, especially in alternated or intermittent aeration processes, in order to verify the end of nitrification and denitrification phases and so switch aeration off or on.

Similar relations were found for the single wastewaters, both for the nitrate reduced and the denitrification potential. Table 3 illustrates the correlation obtained for all the different wastes in regard to the BND.

Table 3. Correlation between analytical results and biosensor data.

	$BND_{analysis} = a \times BND_{biosensor}$	
Waste	a	R^2
Domestic	1.0185	0.909
Ice cream	0.9442	0.887
Beet-sugar	1.0916	0.914
Brewery	1.0873	0.973
Tuna cannery	0.9653	0.943

The data in the table show that the values of the denitrification potential were determined with a relative error smaller than 10%.

4 CONCLUSIONS

The main advantage of using the biosensor for denitrification monitoring during anoxic tests consists in the possibility of a continuous and instantaneous observation of the process; therefore it allows to identify the different slopes of the titration curve and so better evaluate the kinetics; moreover a cost saving derives from the reagents foe the analyses.

REFERENCES

APHA 1998. *Standard Methods for the Examination of water and wastewater*. 20th edn, American Public Health Association/American Water Works Association/Water Environment Federation, Washington DC, USA.

Henze, M. 1991. Capabilities of biological nitrogen removal processes from wastewater. *Water Science and Technology* 23: 669–679.

Isaacs, S.H. & Henze, M. 1995. Controlled carbon source addition to an alternating nitrification-denitrification wastewater treatment process including biological P removal. *Water Research* 29(1): 77–89.

Lee, M. & Welander, T. 1996. The effect of different carbon sources on respiratory denitrification in biological wastewater treatment. *Journal of Fermentation and Bioengineering* 82(3): 277–285.

Massone, A., Antonelli, M. & Rozzi, A. 1996. The DENICON: a novel biosensor to control denitrification in biological wastewater treatment plants. Mededelingen Faculteit Landbouwkundige, University of Gent, 1709–1714.

Monteith, H.D., Bridle, T.R. & Suthon, P.M. 1980. Industrial waste carbon sources for biological denitrification. *Progress Water Technology* 12: 127–141.

Rozzi, A., Massone, A. & Alessandrini, A. 1997. Measurement of rbCOD as biological nitrate demand using a biosensor: preliminary results. In Proceeding of 3rd International Symposium Environmental Biotechnology, Oostende, Belgium, April 21–23.

Tsonis 1997. Olive oil mill wastewater as carbon source in post anoxic denitrification. *Water Science and Technology* 36(2–3): 53–60.

Novel nitrogen removal

European Symposium on Environmental Biotechnology, ESEB 2004 - Verstraete (ed)
© 2004 Taylor & Francis Group, London, ISBN 90 5809 653 X

Activity of the granulated biomass of the mixed microbial culture for highly efficient carbon and nitrogen removal in the process wastewaters

T. Landeka Dragičević, M. Zanoški, M. Glancer-Šoljan
Faculty of Food Technology and Biotechnology, University of Zagreb, Zagreb

V. Šoljan, V. Matić
Eco-engineering, Kufci, Poreč, Croatia

J. Krajina
Ministry of Interior, Police Department, Forensic Center, Zagreb, Croatia

ABSTRACT: Present work describes the use of the selected granulated mixed microbial culture in simultaneous biodegradation, nitrification and denitrification of the simply and complexly structured ingredients with carbon and nitrogen from chemical, pharmaceutical and fermentation industries. The mixed microbial culture in the granulated form comprised the selected strains of autotrophic and heterotrophic microcolony-forming bacteria. The comparison is given between nitrification with pre-denitrification, and the sequencing batch reactor (SB-reactor). In addition to presenting chemical quality parameters of the untreated and treated wastewater, the work also shows the stability of the granules (microbiological quality) in the changed composition of wastewaters from various sources. Environmental parameters are shown that enable highly efficient wastewaters treatment with respect to their ingredients with carbon and nitrogen.

1 INTRODUCTION

Removal of the substances with carbon from municipal wastewater and wastewater of some industries (e.g. food industry) by biodegradation, or the so called activated sludge technology, has been a well-known practice for many years now (Grady et al. 1999).

However, given the required quality of the treated wastewater, legislation of the EU governs both the levels of substances with carbon and substances with nitrogen in the effluent released into environment (Pöpel 1995).

Bio-removal of nitrogen from wastewater has the advantages over physical–chemical processes (ion exchange and membranous processes) particularly in dealing with process wastewater that frequently varies in the volume and quality, due to the switches in the production program (Nielsen et al. 1992, Grunditz et al. 1998).

Biodegradation of organic substances is performed with many microorganisms. Most of these processes are taking the advantage of their known microbiological properties and enzymatic potential for degradation of wastewater ingredients (Pretorius 1987).

Nitrification is performed by chemolithoautotrophic ammonium-oxidizing and nitrite-oxidizing bacteria sensitive to many environmental factors (pH, temperature and the level of dissolved oxygen) (Surmacz-Gorska et al. 1995) and many wastewater ingredients acting as their growth-inhibitors (Tanaka et al. 2001). Removal of nitrogen from wastewater also requires denitrification in which biodegradable organic substances serve as electron donors (sources of carbon) or, in the absence of carbon sources, the addition of various organic substances, usually methanol, as the external source of carbon (Tam et al. 1992).

Wastewaters from chemical (Pascik 2001), pharmaceutical (Kabdashi et al. 1999) and fermentative industries (La Para et al. 2002), except for the readily biodegradable substances (carbohydrates, proteins, alcohols and short-chain fatty acids) also contain sparingly degradable organic substances of the complex chemical structure often with nitrogen in the molecule, e.g. aminonaphthalenesulfonic acid, di-methyl formamide and betaine (Fitz-Gibson et al. 1995). There is no literature data showing that these complexly structured substances that can be classified as xenobiotics (Gregier et al. 2002) can in wastewater biodegradation

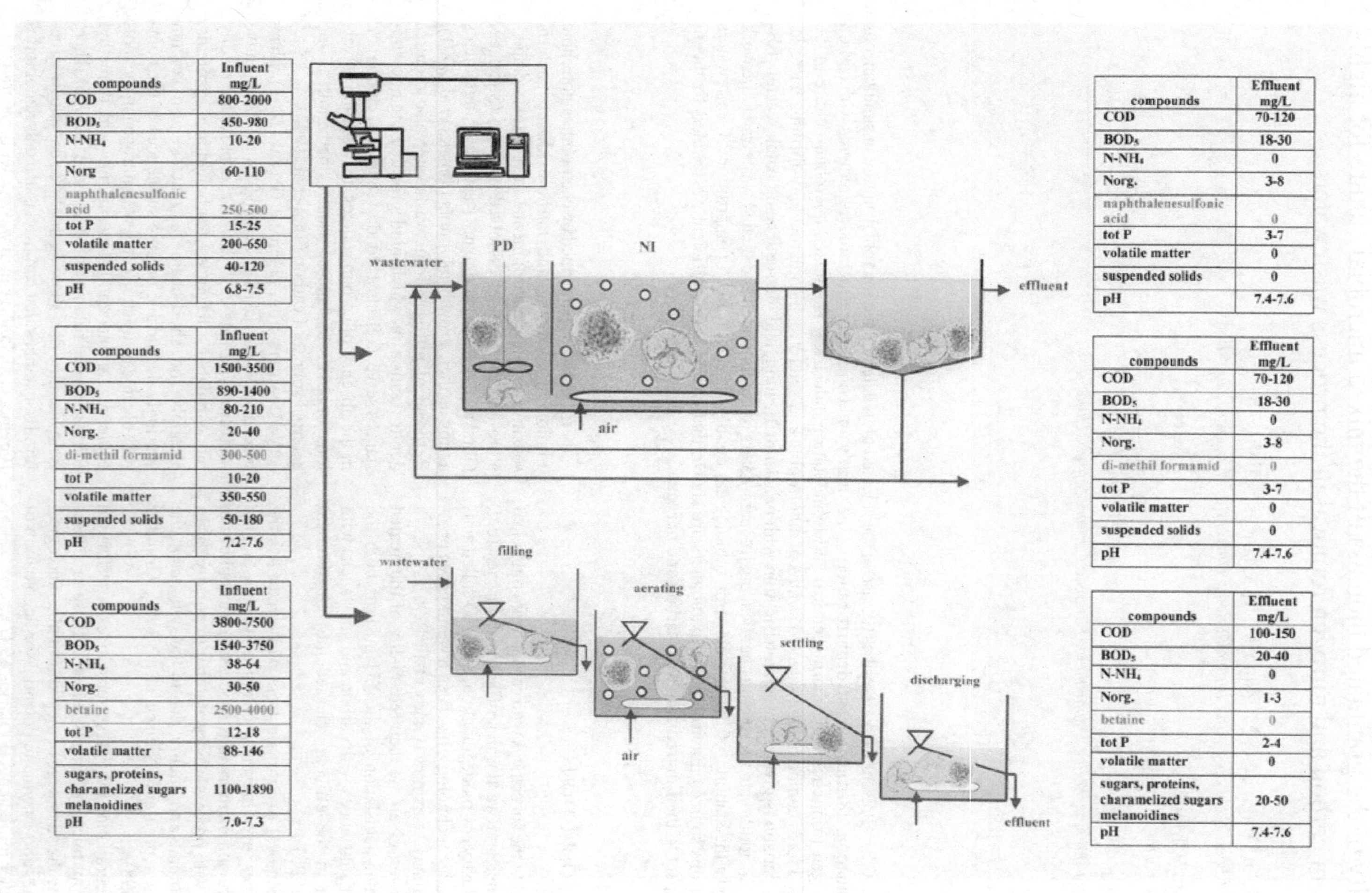

compounds	Influent mg/L
COD	800-2000
BOD_5	450-980
$N\text{-}NH_4$	10-20
Norg	60-110
naphthalenesulfonic acid	250-500
tot P	15-25
volatile matter	200-650
suspended solids	40-120
pH	6.8-7.5

compounds	Influent mg/L
COD	1500-3500
BOD_5	890-1400
$N\text{-}NH_4$	80-210
Norg.	20-40
di-methil formamid	300-500
tot P	10-20
volatile matter	350-550
suspended solids	50-180
pH	7.2-7.6

compounds	Influent mg/L
COD	3800-7500
BOD_5	1540-3750
$N\text{-}NH_4$	38-64
Norg.	30-50
betaine	2500-4000
tot P	12-18
volatile matter	88-146
sugars, proteins, charamelized sugars melanoidines	1100-1890
pH	7.0-7.3

compounds	Effluent mg/L
COD	70-120
BOD_5	18-30
$N\text{-}NH_4$	0
Norg.	3-8
naphthalenesulfonic acid	0
tot P	3-7
volatile matter	0
suspended solids	0
pH	7.4-7.6

compounds	Effluent mg/L
COD	70-120
BOD_5	18-30
$N\text{-}NH_4$	0
Norg.	3-8
di-methil formamid	0
tot P	3-7
volatile matter	0
suspended solids	0
pH	7.4-7.6

compounds	Effluent mg/L
COD	100-150
BOD_5	20-40
$N\text{-}NH_4$	0
Norg.	1-3
betaine	0
tot P	2-4
volatile matter	0
sugars, proteins, charamelized sugars melanoidines	20-50
pH	7.4-7.6

Figure 1.

serve as the source of carbon for denitrification. In addition to being sparingly biodegradable, they can inhibit autotrophic nitrifying bacteria, for which they are easily washed out from the wastewater treatment system.

Due to instability of nitrification, especially as regards process wastewaters, recent research has been channeled towards defining microbiological quality of the nitrifying autotrophic active sludge. The aim is to detect autotrophic nitrifying microorganisms in sludge and to classify them with respect to formation of distinguishable and specific microcolonies. New monitoring methods have been developed, particularly for autotrophic bacteria, while the FISH technique (fluorescent *in situ* hybridization) with the gene techniques and the state-of-the-art microscopes enables their recognition and detection in sludge i.e. their classification (Wilderer et al. 2002).

Such monitoring of microorganisms in the active sludge with the available "on-line" image analyses for the control of the wastewater treatment system enables evaluation of nitrification at any time point (Heine et al. 2002).

In parallel with this, the research has been made in granulation of the sludge biomass, so as to preserve in it major part of the nitrifying microorganisms. Also, the attempts are being made to increase by sludge bioaugmentation the portion of nitrifying microorganisms (Plaza et al. 2001), either flocculated or granulated. Recent discoveries and first results with the use of the granulated selective mixed microbial cultures show the efficiency of the process wastewater nitrification (Glancer et al. 2001, Glancer et al. 2003).

2 MATERIAL AND METHODS

Mixed microbial cultures comprising autotrophic nitrifying and heterotrophic denitrifying bacteria (*Nitrosomonas europaea*, *Nitrospira marina*, *Nitrosolobus multiformis*, *Pseudomonas putida*, *Pseudomonas cepacia* and *Flavobacterium marinum*) were used to remove specific substances from the wastewaters of chemical, pharmaceutical and fermentative industries.

Continuous experiments were performed with laboratory models comprising the reactor for nitrification with pre-denitrification and the SB-reactor.

The preset parameters were temperature 18–32°C, pH 7.4–7.6, dissolved oxygen 0.4–4 mg O_2/L, HRT 6–24 hrs and biomass level of 3.5–4.5 g/L.

For analytical control of influent and effluent APHA-methods (APHA), ION-chromatography, GC and HPLC were used.

3 RESULTS AND DISCUSSION

The selected mixed bacterial cultures exhibited high activity in biodegradation, nitrification and denitrification of wastewater containing naphthalenesulfonic acid, di-methyl formamide and betaine (Tables showing influent and effluent, and scheme showing the process (Figure 1)).

4 CONCLUSIONS

By biodegradation of the complex organic substances and by nitrification of the generated ammonium from the wastewaters of chemical, pharmaceutical and fermentative industries the selected mixed microbial cultures ensure appropriate source of carbon for denitrification and maintain their granulated readily precipitating form.

REFERENCES

APHA – Sandard methods for the examination of wastewater and wastewater treatment. American Public Health Association American Water Works Association and Water Pollution Control Federation, Washington, D.C.

Fitz-Gibson, F.J., Nigam, P., Singh, D. & Marchant, R. 1995. Biological treatment of destillery waste for pollution – remediation. *J. Basic Microbiol.* 35: 293–301.

Glancer, M., Ban, S., Landeka Dragičević, T., Šoljan, V. & Matić, V. 2001. Granulated mixed microbial culture suggesting successful employment of bioaugmentation in the treatment of process wastewater. *Chem. Biochem. Eng. Q.* 15: 87–94.

Glancer, M., Šoljan, V. & Matić, V. 2003. Granular gains, *Water 21* October: 31–32.

Grady, C.P.L., Daigger, G.T. & Lim, H.C. 1999. *Biological wastewater treatment.* 2nd ed., New York: Marcel Deker.

Gregier, P.H., Meier, H.M., Gerle, M., Vogt, U., Grath, T. & Knakmuss, H.J. 2002. Xenobiotic in the environment: present and future strategies to obviate the problem of biological persistence. *J. Biotechnol.* 94: 101–123.

Grunditz, C., Gumaelius, L. & Dalmammar, G. 1998. Comparison of inhibition assays using nitrogen removing bacteria: Application to industrial wastewater. *Wat. Res.* 32: 2995–3000.

Heine, W., Sekulov, I., Burkhardt, H., Bergen, L. & Behrendt, J. 2002. Early warming – system for operation failures in biological stages of WWTPs by on-line image analyses. *Wat. Sci. Tech.* 46: 117–124.

Kabdashi, J., Gürel, M. & Türedy, A. 1999. Pollution prevention and waste treatment in chemical synthesis processes for pharmaceutical industry. *Wat. Sci. Tech.* 39: 265–271.

La Para, M., Nakatsu, C.H., Pantea, L.M. & Alleman, J.E. 2002. Stability of the bacterial communities supported by a seven-stage biological process treating pharmaceutical wastewater as revealed by PCR-DGGE. *Wat. Res.* 36: 638–646.

Nielsen, P.H., Raunkjaren, K., Norsker, N.H., Jensen, N.A. & Hvitved-Jakobsen, T. 1992. Transformation of wastewater in sewer systems – a review. *Wat. Sci. Tech.* 25: 17–31.

Pascik, I. 2001. Umveltbiotechnologie: Immobilisierung von Mikroorganismen auf adsorbierenden PUR – Trägern. Wasser-/Abwassertechnik, *Luft und Boden* 1–2: 11–14.

Plaza, E., Trela, J. & Hitman, B. 2001. Impact of seeding with nitrifying bacteria on nitrification process efficiency. *Wat. Sci. Tech.* 43: 155–164.

Pöpel, H.J. 1995. Advanced wastewater treatment in Germany-effluent requirements, design standards, practical experience and problems. *EWPC* 5: 5–16.

Pretorius, W.A. 1987. A conceptual basis for microbial selection in biological wastewater treatment. *Wat. Res.* 21: 891–894.

Surmacz-Gorska, J. 1995. Nitrification process control in activated sludge oxygen uptake rate measurements. *Environ. Technol.* 16: 569–577.

Tam, N.F.Y., Wong, Y.S. & Leung, G. 1992. Effect of exogenous carbon sources on removal of inorganic nutrient by the nitrification–denitrification process. *Wat. Res.* 26: 1229–1236.

Tanaka, Y., Taguchi, K. & Utsumi, H. 2001. Toxicity assessment of 255 chemicals to pure culture nitrifying bacteria using biosensor. *Wat. Sci. Tech.* 46: 331–335.

Wilderer, P.A., Bungartz, J.H., Lemmer, H., Wagner, M., Keller, J. & Wueitz, S. 2002. Modern scientific methods and their potential in wastewater science and technology. *Wat. Res.* 36: 370–393.

European Symposium on Environmental Biotechnology, ESEB 2004 - Verstraete (ed)
© 2004 Taylor & Francis Group, London, ISBN 90 5809 653 X

Development and application of an Anammox activity test based on gas production

A. Dapena-Mora, J.L. Campos, A. Mosquera-Corral & R. Méndez
Department of Chemical Engineering. School of Engineering. University of Santiago de Compostela, Rúa Lope Gómez de Marzoa s/n, Santiago de Compostela, Spain

ABSTRACT: In order to assess the applicability of the Anammox process for the treatment of industrial wastewaters it is necessary to study the possible toxic effects of compounds commonly present in industrial effluents on the Anammox specific activity. This study is focused on the application of batch tests based on the monitoring of the gas production to determine the maximum specific Anammox activity (SAAm). The SAAm measured was around 0.28 g $N\text{-}NH_4^+$/g VSS·d and it did not depend on the biomass concentration or on the S_0/X_0 ratio in a wide range of values. Substrate and product inhibition effects were also evaluated, showing the Anammox process suitable for the treatment of wastewater with high ammonia concentrations. Nitrite concentration must be controlled under a limit value of 15 mM to maintain 100% of the SAAm.

1 INTRODUCTION

Recently a new autotrophic process called Anammox and involved in the nitrogen cycle was discovered. This process consists of the anaerobic oxidation of ammonia using nitrite as electron acceptor (equation 1). When the Anammox process is applied for the nitrogen removal the amounts of oxygen and organic matter required are reduced compared to the requirements of conventional nitrification/denitrification processes.

$$NH_4^+ + 1.31\ NO_2^- + 0.066\ HCO_3^- + 0.13\ H^+ \rightarrow N_2 + 0.26\ NO_3^- + 0.066\ CH_2O_{0.5}N_{0.15} + 2\ H_2O \quad (1)$$

Application of the Anammox process to the treatment of real wastewaters requires knowing the potential toxic effect on the process caused by the compounds that are usually present.

Batch activity tests are frequently used to assess biomass activities, as well as to measure kinetic parameters and the influence of toxic compounds.

The methods commonly used can be divided in two groups:

1.1 *Methods based in measuring the liquid phase*

(a) Some of these methods are based in the measurement of the substrate (NH_4^+, NO_2^-, NO_3^-) consumption rate monitoring the concentration of these compounds in the liquid phase along time. In case of aerobic processes, dissolved oxygen concentration can be followed with an oxygen electrode.

For the application of this method to the Anammox process some preliminary studies have been made to develop biosensors for the determination of NO_x^- and NO_3^- (Larsen et al. 1997, Revsbech et al. 2000). The determination of the evaluated compound is realized employing a biological culture that oxidises or reduces the measured compound, and measuring the redox potential of the reaction. When totally developed, the application of these biosensors to the Anammox process would permit its continuous following and control.

In the meanwhile, concentration of nitrogenous compounds in the liquid phase must be followed by spectrofotometric or cromatographic methods, requiring a high number of analysis, meaning time and reagents. Until now, this has been the most widely used method to measure Anammox activity (Strous et al. 2000, Kuai et al. 1998, Sliekers et al. 2002, Helmer et al. 2001, Toh et al. 2002…).

(b) Other methods are based on the alkalinity production or consumption. ANITA is an example of this kind of methods. For a nitrifying culture, the ammonium oxidising rate is measured by the flow rate of alkaline solution needed to neutralize the acidity produced by the metabolic activity of the biomass (Ficara et al. 2000). The dificulty to apply this method for measuring Anammox activity is its low production of protons (only 0.13 mol H^+ for each mole of ammonium consumed), so a large amount of sample would be needed for measuring an appreciable activity.

1.2 *Methods based in measuring the gas phase*

(c) Anaerobic batch tests are often performed by adding a large amount of substrate to the sludge and monitoring the resulting gas production with a Mariotte flask system (Soto et al. 1992). This system is not sensitive enough to measure small amounts of gas produced accurately and therefore demands the addition of large amounts of substrate. This is not possible in the case of Anammox process, as it is inhibited by substrate (nitrite).

(d) Another method to follow the gas production is to measure the pressure increase during the experiment, in a sealed vial, as well as the gas composition (Buys et al. 2000). By using a pressure sensor, small pressure increases may be measured accurately, so that the experiments can be performed at relatively low substrate concentrations.

On the other hand, the biomass concentration and the initial substrate concentration to initial biomass concentration ratio (S_0/X_0) can have a significant effect on the specific biomass activity measured (Moreno et al. 1999). Therefore, it would be interesting to establish standard conditions.

The present study is focused on the application of batch tests based on the monitoring of nitrogen gas production for the determination of the maximum Specific Anammox Activity (SAAm). The validity of the method and the influence of the S_0/X_0 ratio are tested. The developed method is afterwards applied to the study of the inhibition caused on the Anammox process by the substrates (ammonium and nitrite) and product (nitrate).

2 MATERIALS AND METHODS

2.1 *Batch experiments procedure*

Batch experiments to determine the SAA were performed according to the methodology described by Buys et al. (2000), based on the measurement along time of the overpressure generated in closed vials by the nitrogen gas produced.

The assays were performed in hermetically closed vials with a total volume of 38 mL and a volume of liquid of 25 mL. The vials were inoculated with sludge collected from a 5 L lab-scale Anammox reactor, washed with phosphate buffer. The gaseous and liquid phases were gasified with argon to remove the dissolved oxygen. The initial pH value was fixed at 7.8. The vials were placed in a thermostatic shaker, at 150 rpm and 30°C until stable conditions were reached. Then the substrates were added ($(NH_4)_2SO_4$ and $NaNO_2$) and pressure was equalized to the atmospheric one. The production of N_2 gas was tracked by measuring the overpressure in the headspace with a time frequency depending on the biomass activity rate in each test.

2.2 *Calculations*

The SAAm was calculated from the maximum slope of the curve describing the nitrogen gas production along the time according to Equation 2:

$$SAA_m \text{ (g } NH_4^+\text{-N/g VSS·d)} = \frac{dNH_4^+/dt}{X \cdot V_L} \quad (2)$$

where (dNH_4^+/dt) is the maximum NH_4^+ consumption rate, X the biomass concentration (g VSS/L) and V_L the volume of the liquid phase.

The maximum ammonium consumption rate is calculated from the slope of the line representing the overpressure increase with time, α (atm/min), using the ideal gas law and stoichiometric considerations:

$$dNH_4^+/dt = \alpha \cdot \frac{V_G}{R \cdot T} \cdot 28 \quad (3)$$

being V_G the volume of the gaseous phase(L), R the ideal gas coefficient (atm·L/mol·K) and T the temperature (K).

The percentage of activity maintained when the process was inhibited was calculated as:

$$\% \text{ act} = (SAA/SAAm)\ 100 \quad (4)$$

where SAAm is the specific activity on the blank assay and SAA the specific activity of the tests with inhibitory compounds.

2.3 *Analytical methods*

The overpressure in the headspace was measured using a differential pressure transducer 0-5 PSI, linearity 0.5% of full scale, Centerpoint Electronics.

Ammonium was analysed by the phenol-hypochloride method (Wheatherburn 1967). Nitrite and nitrate were analysed by spectrophotometry (APHA 1985). Biomass concentration was measured as g VSS/L, according to standard methods (APHA 1985). pH was measured with an electrode Ingold model U-455 connected to a pH/mv measurer Crison 506.

3 RESULTS AND DISCUSSION

3.1 *Effect of biomass concentration and substrate concentration to biomass concentration ratio*

A first set of assays was carried out using the previously detailed standard conditions (70 mg NH_4^+-N/L and 70 mg NO_2^--N/L and 1 g VSS/L) to check the accuracy of the method. The amounts of N_2 gas produced and NH_4^+-N and NO_2^--N consumed were measured in each vial. The mass balances, calculated for the gas and the liquid phases, showed an averaged difference of $6 \pm 3\%$ between the activity calculated using the liquid phase results or the pressure variations.

Batch tests may induce changes in the biomass concentration depending on the initial substrate to biomass ratio (S_0/X_0) applied.

Several S_0/X_0 between 0.018 and 0.14 g NO_2^--N/g VSS were tested at constant biomass concentration of 1 g VSS/L (Figure 1). The lower value of this interval of S_0/X_0 ratios was the one that produced the minimum quantity of nitrogen gas detectable by the pressure transducer. The maximum value corresponds to the higher nitrite concentration tolerated by the biomass, not causing substrate inhibition. The results showed a maximum difference between the SAAm of 15% for the range of S_0/X_0 ratios assayed.

On the other hand, the effect of initial biomass concentration was also tested. Assays with initial biomass concentrations of 0.25, 0.5, 1.0 and 2.0 g VSS/L were carried out at a constant S_0/X_0 ratio of 0.07 g NO_2^--N/g VSS. The SAAm was practically constant and around 0.28 g NH_4^+-N/g VSS·d for these conditions (Figure 2).

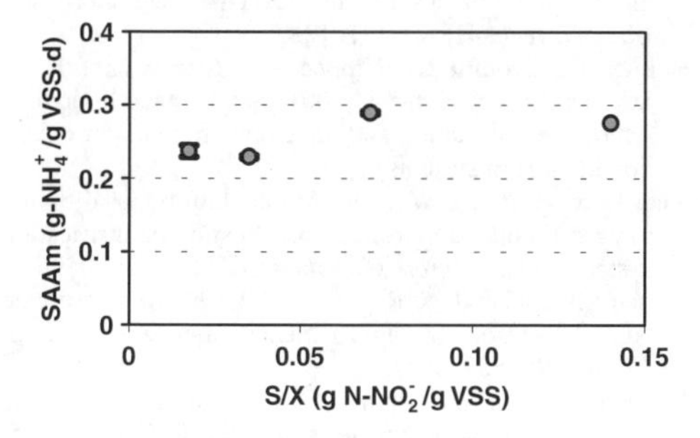

Figure 1. Maximum Anammox specific activity for tested S_0/X_0 ratios.

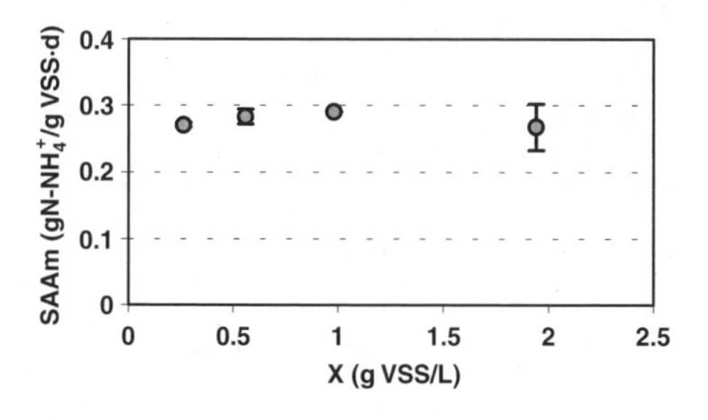

Figure 2. Maximum Anammox specific activity for different biomass concentrations.

Some authors report that the initial S_0/X_0 ratio used in batch experiments influences the obtained values of specific denitrifying activities (Chudoba et al. 1991; Chudoba et al. 1992). Moreno et al. (1999) and Buys et al. (2000) observed in batch experiments that the higher the S_0/X_0 ratio the higher the specific activity of the biomass. Chudoba et al. (1991) explained, in the case of aerobic activated sludge exposed to periodical anaerobic conditions, that if the initial S_0/X_0 ratio is low ($S_0/X_0 < 2$–4 g COD/g VSS) no cell multiplication occurs while substrate is removed. When the S_0/X_0 ratio is high biomass growth is produced and the substrate removal increases. These authors also observed that, when high initial S_0/X_0 ratios are used, an apparent lag phase on the curves of biomass production is observed due to the time necessary for multiplication of an original small amount of biomass, which causes substrate consumption. When the S_0/X_0 ratio is low no cell multiplication takes place during the exogenous substrate removal. Under these conditions, a biomass increase is mostly due to the synthesis of storage polymers.

In the case of Anammox biomass, it is observed that the S_0/X_0 ratio does not have a notable effect on the activity. As the growing rate of Anammox biomass is very low (Strous et al. (2000) reported a doubling time of 11 d), there is not a significant biomass production during the time needed for the batch test (five or six hours). Therefore, no lag phase and no specific activity increase are expected when the S_0/X_0 ratio is increased.

Initial S_0/X_0 ratios usually employed by authors measuring SAAm range from 0.0084 g NO_2^--N/g VSS (Strous et al. 2000) to 0.03 g N-NO_2^-/g VSS (Sliekers et al. 2002). SAA calculated for these authors were, 0.66 and 0.25 gNH_4^+-N/g VSS·d respectively.

The variations in SAA obtained by the different authors are most probably due to the different degree of enrichment of the biomass employed for the study.

3.2 *Inhibition by substrate and product*

In order to apply the Anammox process for the treatment of industrial wastewaters with high ammonia concentration it is necessary to know the effect of substrates and product on the activity. Experiments with different concentrations of ammonia, nitrite and nitrate were individually performed at concentrations between 5 and 100 mM (Figures 3–5).

Nitrite exerted the highest inhibitory effect on the Anammox specific activity. An inhibition of 50% (IC50) was measured at 25 mM for this compound, while the IC50 values for ammonia and nitrate corresponded to concentrations of 50 and 150 mM, respectively. These results show the availability of the Anammox process for the treatment of industrial wastewaters with high ammonia concentration.

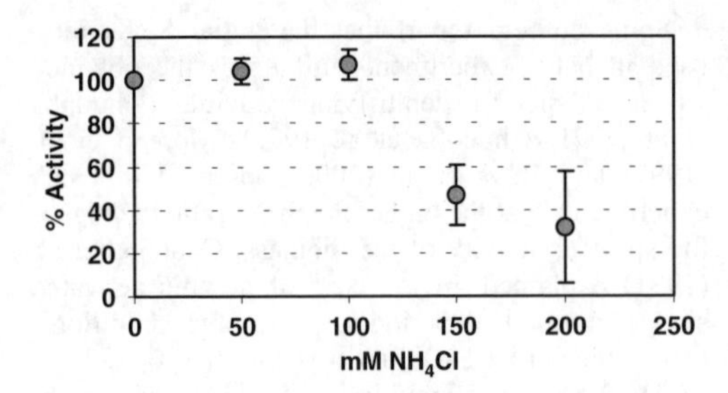

Figure 3. Inhibition caused by ammonium.

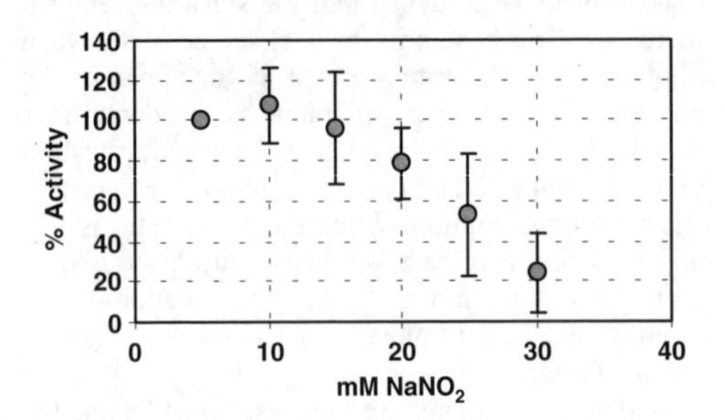

Figure 4. Inhibition caused by nitrite.

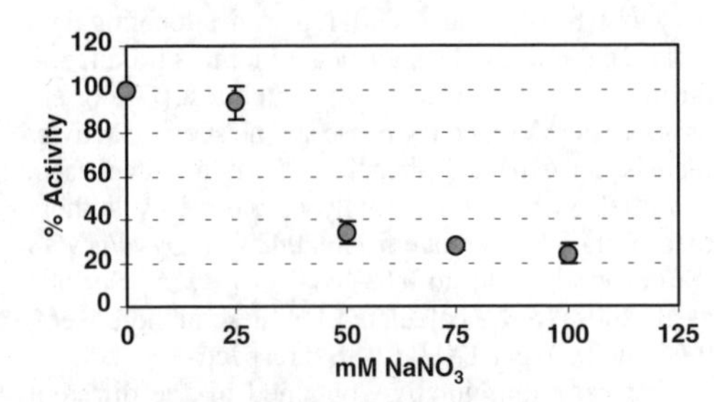

Figure 5. Inhibition caused by nitrate.

Strous (2000) exposed the Anammox biomass to concentrations up to 70 mM of ammonium and nitrate in a SBR during one week observing no negative effect on the activity. Nevertheless, at concentrations of nitrite higher than 7 mM the biomass completely lost its activity.

4 CONCLUSIONS

Batch tests based on the measurement of overpressure increase seem to be a reliable method for the determination of SAAm. The obtained results show an independence of the measured SAAm of the S_0/X_0 and biomass concentration in a usual range of values.

The IC50 results show the Anammox process suitable for the treatment of industrial wastewaters with high ammonia concentration. Nitrite concentration must be controlled under a limit value of 15 mM to maintain 100% of the SAAm.

ACKNOWLEDGEMENTS

This work was funded by the European Commission through the ICON project (Ref: EVK1-CT-2000-00054ICON), Xunta de Galicia (PGIDT10XI1209-04PM) and the Spanish CICYT through the Oxanammon project (PPQ-2002-00771).

REFERENCES

APHA-AWWA-WPCF. 1985. Standard Methods for examination of water and wastewater. 16th Ed. Washington.

Buys B.R., Mosquera-Corral A., Sánchez M. & Méndez R. 2000. Development and application of a denitrification test based on gas production. *Wat. Sci. Tech.* 41(12): 113–120.

Chudoba P., Chevalier J.J., Chang J. & Capdeville B. 1991. Effect of anaerobic stabilization of activated sludge on its production under batch conditions at various S_0/X_0 ratios. *Wat. Sci. Tech.*, 23(4–6): 917–926.

Chudoba P., Chevalier J.J., Chang J. & Capdevilee B. 1992. Explanation of biological meaning of the S_0/X_0 ratio on batch cultivation. *Wat. Sci. Tech* 23(4–6): 917–926.

Ficara E., Rocco A. & Rozzi A. 2000. Determination of nitrification kinetics by the ANITA-Dostat biosensor. *Wat. Sci. Tech.* 41(12): 121–128.

Helmer C., Tromm C., Hippen A., Rosenwinkel K.H., Seyfried C.F. & Kunst S. 2001. Single stage biological removal by nitritation and anaerobic ammonium oxidation in biofilm systems.

Kuai L. & Verstraete W. 1998. Ammonium removal by the oxygen-limited autotrophic nitrification-denitrification system. *Appl. Environ. Microbiol.* 64(11): 4500–4506.

Larsen L.H., Kjær T. & Revsbech N.P. 1997. A microscale NO_3^- biosensor for environmental applications. *Anal. Chem.*, 69: 3527–3531.

Moreno G., Cruz A. & Buitrón G. 1999. Influence of S_0/X_0 ratio on anaerobic activity tests. *Wat. Sci. Tech* 40(8): 9–15.

Revsbech N.P., Kjær T., Damgaard L. & Larsen L.H. 2000. Biosensors for analysis of water, sludge, and sediments with emphasis on microscale biosensors. In J. Buffle & G. Horvai (eds.) *In situ monitoring of aquatic systems: Chemical analysis and speciation.* Wiley.

Sliekers A.O., Derwort N., Campos J.L., Strous M., Kuenen J.G. & Jetten M.S.M. 2002. Completely autotrophic nitrogen removal over nitrite in one single reactor. *Wat. Res.* 36: 2475–2482.

Soto M., Méndez R. & Lema J. 1992. Methanogenic and non-methanogenic activity test. Theoretical basis and experimental set up. *Wat. Res.* 27: 1361–1376.

Strous M. 2000. Microbiology of anaerobic ammonium oxidation. Doctoral Thesis. Technical University Delft, Holanda.

Toh S.K. & Ashbolt N.J. 2002. Adaptation of anaerobic ammonium-oxidising consortium to synthetic coke-ovens wastewater. *Appl. Microbiol. Biotechnol.* 59: 344–352.

Wheatherburn M.W. 1967. Phenol-hypochlorite reaction for determination of ammonia. *Anal. Chem.* 28: 971–974.

European Symposium on Environmental Biotechnology, ESEB 2004 - Verstraete (ed)
© 2004 Taylor & Francis Group, London, ISBN 90 5809 653 X

An anammox-like process in the waste water treatment of the coke-chemical plant

A.Ye. Kouznetsov
D. Mendeleyev University of Chemical Technology of Russia, Moscow, Russia

V.A. Stepanenkov
JSC "Severstal", Cherepovets, Russia

ABSTRACT: With the purpose to improve the removal of nitrogen from the waste water of the coke-chemical plant a process of biological purification was studied in laboratory conditions and an opportunity of the nitrification–denitrification opposite to the classical one was shown. The process takes place in aerobic condition and without organic substrates. On the synthetic medium with ammonium ions and at the absence of organic substrates it is accompanied by decrease in ammonium concentration without accumulation of nitrite and nitrate in the medium and stimulated by addition of nitrate into the system.

After removal of phenolics, thiocyanates and cyanides from the waste water of the coke-chemical plant inorganic nitrogen has to be removed at the stage of biological nitrification and denitrification technological scheme of which is shown on Figure1.

For this stage of nitrification–denitrification it was found that along with the classical processes of nitrification–denitrification according to the scheme:

$$NH_4^+ + O_2 \rightarrow NO_2^- \rightarrow NO_3^-$$

$$C_6H_5OH + NO_3^- \rightarrow CO_2 + N_2 + H_2O$$

There is a presence of the anammox-like process (Mulder A. et al., 1995):

$$NH_4^+ + NO_3^- \rightarrow N_2(N_2O) + H_2O \quad \text{or}$$

$$NH_4^+ + O_2 \rightarrow N_2(N_2O).$$

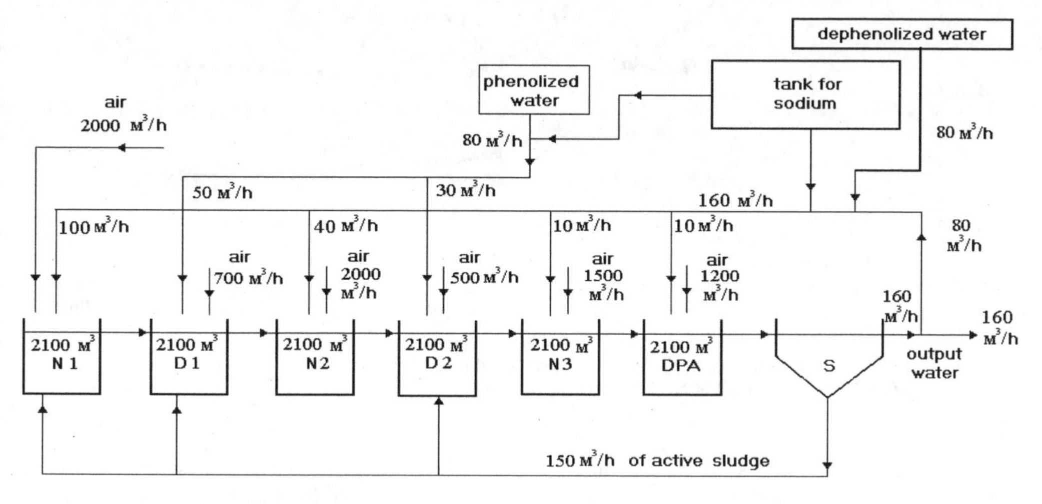

Figure 1. Nitrification–denitrification stage of the waste water treatment of the coke-chemical plant.
N – nitrificator, D – denitrificator, DPA – denitrificator–postaerator, S – settler.

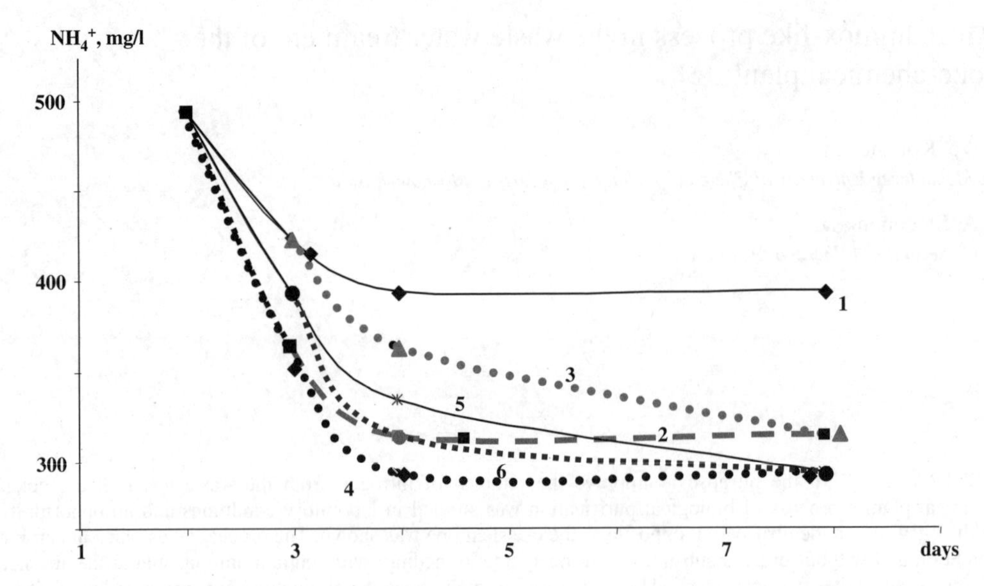

Figure 2. Influence of NO_3^--ions on denitrification in aerobic conditions.
1 – 0; 2 – 20; 3 – 40; 4 – 60; 5 – 80; 6 – 100 mg/l of NO_3^-.

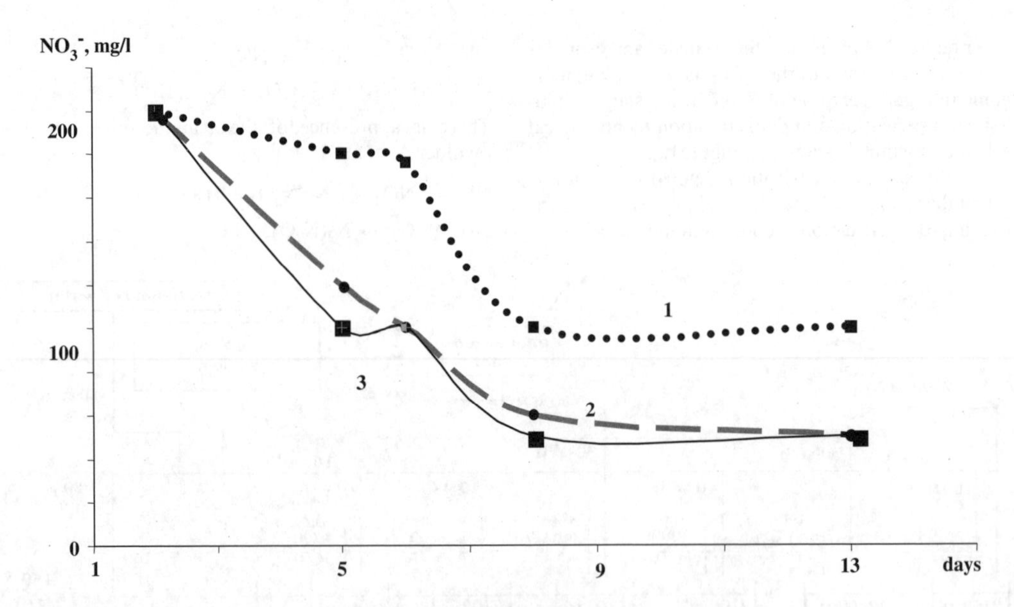

Figure 3. Influence of NH_4^+-ions on denitrification in anoxic conditions.
1. +500 mg/l of phenol; 2. +100 mg/l of NH_4^+; 3. +60 mg/l of NH_4^+.

Though we met this process occasionally in 1997 at study of waste water treatment using active sludge with nitrifiers and denitrifiers obtained from the coke-chemical plant of the Cherepovets metallurgical complex, today it seems no surprising taking into consideration (i) numerous works of various authors on the anammox process published since 1995 (review see Jetten M.S.M. et al., 2002) and

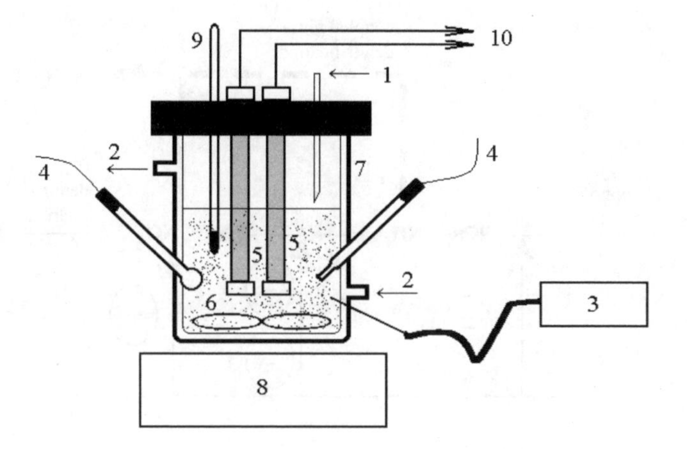

Figure 4. Bioreactor with dynamic membrane.
1. Inlet of medium; 2. Water for thermostating; 3. Air pump; 4. Measuring electrodes; 5. Macroporous filter; 6. Suspension of lignine; 7. Reactor shell; 8. Magnetic stirrer; 9. Thermometer; 10. Outlet of filtrate.

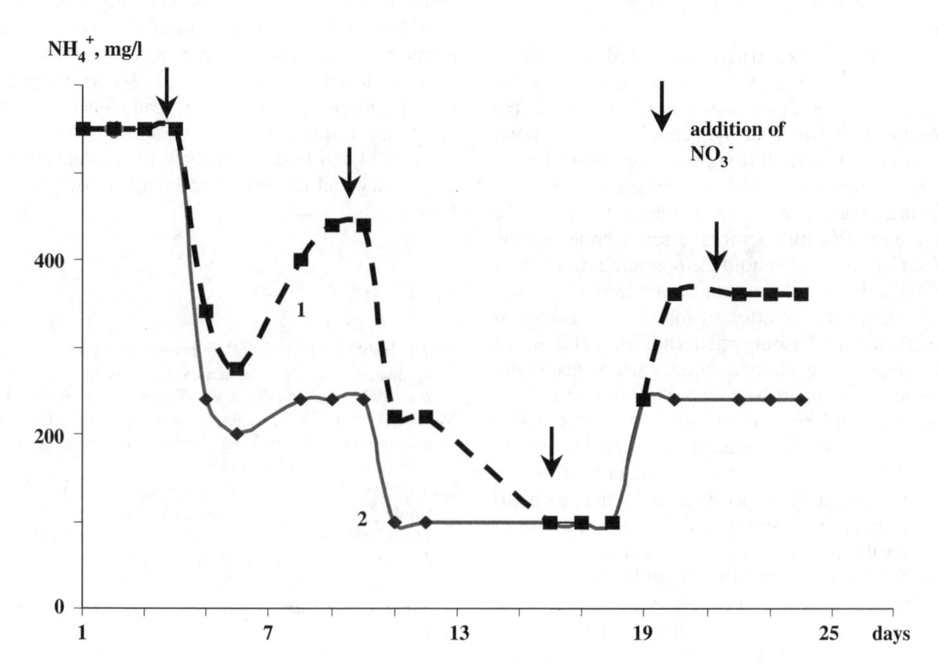

Figure 5. Influence of nitrate on oxidation of ammonium in continuous culture.
1 – cultivation in chemostat; 2 – cultivation in the bioreactor with dynamic membrane; hours 0–19 – dilution rate $0.2\,d^{-1}$; hours 19–25 – dilution rate $0.4\,d^{-1}$. Nitrites were absent in the culture liquid (below 1 mg/l).

(ii) the nitrification– denitrification ecosystem of the coke-chemical plant waste water treatment has suitable ecological conditions to develop such a process.

In our case the process takes place in aerobic condition at pH 6.5–8.5 and without organic substrates. On the synthetic medium with ammonium ions at the range of 100–500 mg/l and at the absence of organic substrate it is accompanied by decrease in ammonium concentration without accumulation of nitrite and nitrate in the medium and stimulated by addition of 20–100 mg/l nitrate ions into the system (Figures 2, 5).

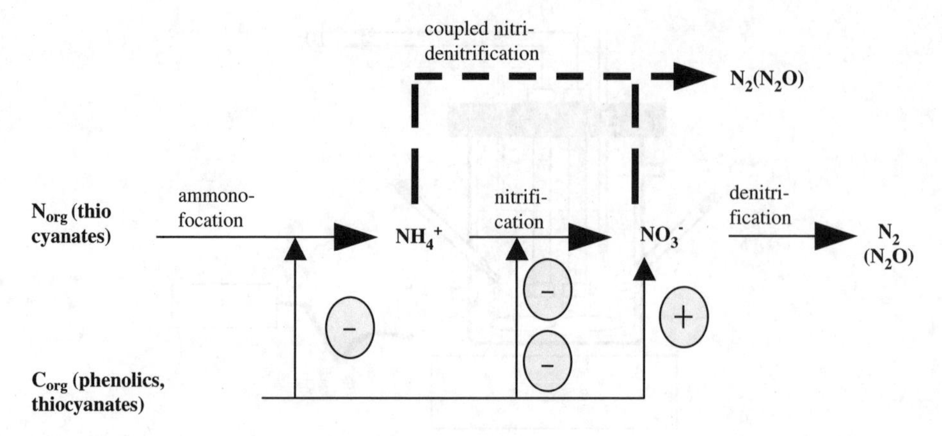

Scheme 1

Addition of NH_4^+ in the medium with NO_3^- without organic substrates stimulates denitrification as well (Figure 3).

The anammox-like process was studied in batch and chemostat conditions and in the bioreactor with filtration (membrane bioreactor, Figure 4). In the bioreactor with filtration the retention of microbial cells and an increase of its capacity were achieved using dynamic membrane formed on the macroporous filter by lignine added into reactor medium. Usage of the principle of dynamic membrane for retention of the biomass has allowed to hold the operating capacity of the filter and bioreactor during 3 months of the experiment without regeneration of the filter and renewal of the membrane forming agent (lignine). The rate of NH_4^+ oxidation in continuous conditions for anammox-like process was up to 10 mg/l per h (Figure 5).

Influence of the various streams of the coke-chemical plant as well as phenolized and dephenolized waters of the waste water treatment plant on nitrification and denitrification was studied. All the processes and the most probable substrate influences on the nitrogen removal can be described by scheme 1.

Nitrification is the most susceptible phase to inhibition by phenolics in the process of nitrogen removal from waste water of the coke-chemical plant.

Owing to the concurrent anammox-like process (coupled nitri–denitrification) removal of nitrogen from this waste water requires less amounts of organic substrate initially present or added into waste water. Unbalanced processes of removal of organic contaminants and inorganic nitrogen at the purification as well as lower in contrast with biodestruction of organic compounds rate of nitrification–denitrification can result in unsatisfactory operation of waste treatment plant and leads to increase of residual amounts of primary and secondary pollutants in output water stream.

REFERENCES

Jetten, M.S.M. et al. 2002. Improved nitrogen removal by application of new nitrogen-cycle bacteria. *Re/Views in Environmental Science & Bio/Technology* (1): 51–63.

Mulder, A. et al. 1995. Anaerobic ammonium oxidation discovered in a denitrifying fluidized bed reactor. *FEMS Microbiol. Ecol.* (16): 177–183.

Kouznetsov, A.Ye. et al. 1998. Normalization of the functioning of the stage of nitrogen removal from the waste water of the JSC "Severstal" coke-chemical plant. *Environment for us and future generations: ecology, business and ecological education; Proc. Conference, Samara-Astrakhan-Samara*, 6–13 September 1998 (in Russian).

European Symposium on Environmental Biotechnology, ESEB 2004 - Verstraete (ed)
© 2004 Taylor & Francis Group, London, ISBN 90 5809 653 X

Determination of the nitrification kinetic parameters of sludges using a new titration bioassay system: MARTINA

P. Artiga., J.M. Garrido & R. Méndez
Department of Chemical Engineering. School of Engineering. University of Santiago de Compostela
Rua Lope Gómez de Marzoa s/n, Santiago de Compostela, Spain

F. Rimoldi., E. Ficara & A. Rozzi
DIIAR, Department of Hydraulic, Environmental, Transportation and Surveying Engineering–Environmental Section
Politecnico di Milano, Piazza L. Da Vinci, Milan, Italy

ABSTRACT: A new titration system named MARTINA (Multiple Analyte Reprogrammable TItratioN Analyser) developed by Politecnico di Milano, which combines a pH-stat titration unit with a DO-stat titration unit has been used to estimate kinetic parameters of sludge samples taken from two Italian wastewater treatment plants (municipal and industrial). The nitrification kinetic parameters were estimated on nitrogen profiles obtained experimentally by means of the minimum square error criteria. By comparing ammonium oxidisers kinetic parameters determined by the two set-point titrations, differences lower than 12% were obtained, indicating a good correspondence between the two methods. Kinetic parameters of nitrite oxidisers could be obtained only by DO-stat titration. Results showed a good repeatability as indicated by a coefficient of variation (CV) lower than 30% for maximum reaction rate estimation. MARTINA makes it possible to estimate kinetic parameters of ammonium and nitrite oxidisers bacteria, demonstrating that by combining the two set point titrations a complete characterisation of the nitrification biomass is easily achieved.

1 INTRODUCTION

MARTINA is an improvement of the titration biosensor ANITA, that is used to determine ammonium oxidisers activity (Rozzi et al., 2000).

The concept of DO-stat respirometry with addition of hydrogen peroxide as oxygen source has been developed by Ficara et al. (2000). The advantage of this new titration system relies on the possibility of making independent measurements of pH-affecting reactions and dissolved oxygen (DO) affecting reactions by means of set-point titrations.

Several biological reactions, in aerobic or anaerobic conditions, may be monitored, e.g. nitrification and denitrification. Using MARTINA is possible the separate determination of the nitrification activity related to the two groups of nitrifying bacteria (ammonium and nitrite oxidisers) from the consumption of O_2 and products of H^+ (equations 1 and 2).

$$NH_4^+ + \mathbf{1.5\,O_2} \rightarrow NO_2^- + H_2O + \mathbf{2\,H^+} \quad (1)$$

$$NO_2^- + \mathbf{0.5\,O_2} \rightarrow NO_3^- \quad (2)$$

The objective of this work is to estimate and to compare the values of kinetic parameters (maximum oxidation rate, Vmax, and half saturation constant, Ks) for both ammonium oxidisers (AO) and nitrite oxidisers (NO) bacteria by pH-DO-stat titration.

2 MATERIALS AND METHODS

2.1 *Experimental set-up*

The titration system used in this work is named MARTINA and it is a pH-stat and DO-stat titrator prototype developed by the Politecnico di Milano in cooperation with SPES (Fabriano, AN, Italy) (Figure 1).

Depending on tests operating conditions, either aerobic or anoxic, the DO probe may be replaced by an Oxidation Reduction Potential (ORP) probe, which is also connected to the MARTINA unit.

During nitrification tests, two titration solutions (diluted solutions of NaOH and H_2O_2) were used to maintain constant both pH and dissolved oxygen concentration. The hydrogen peroxide solution (H_2O_2

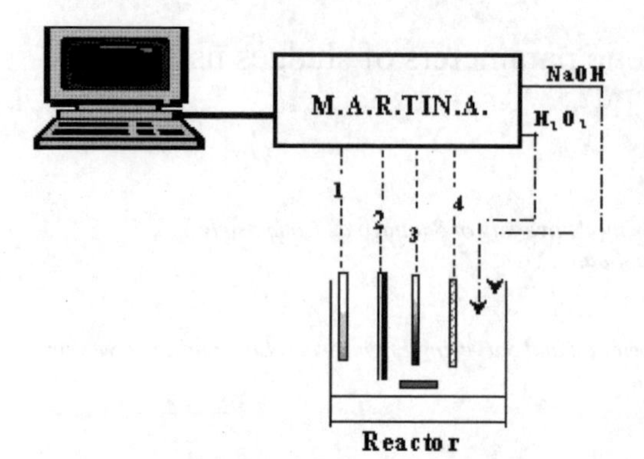

Figure 1. Scheme of titration unit: (1) ORP probe, (2) temperature probe, (3) pHmeter, (4) oxygen probe.

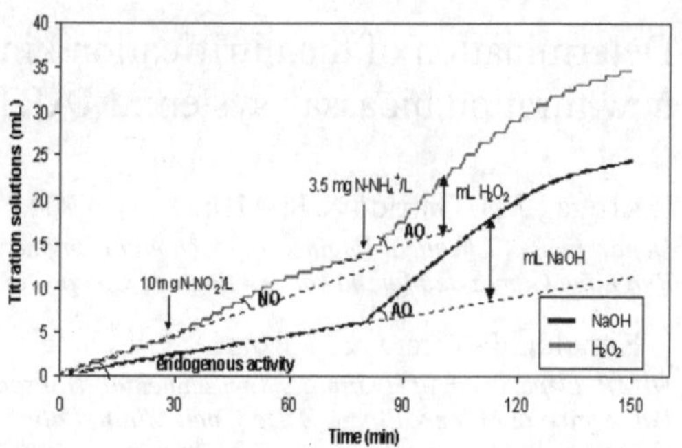

Figure 2. Typical profile of titration curves (NaOH and H_2O_2 solutions) to determine ammonium oxidisers activity (AO) and nitrite oxidisers activity (NO).

0.08 N) was added when the dissolved oxygen concentration decreased below a threshold value due to oxygen consuming bioreactions (ammonium and nitrite oxidation and heterotrophic respiration). The alkaline solution (NaOH 0.05 N) was added when the pH decreased due to acidity produced by ammonium oxidation or CO_2 production by heterotrophic respiration.

2.2 *Characteristics of the sludge*

Nitrifying sludge samples were drawn from two Italian wastewater treatment plant (WWTP): San Giuliano (Milano), fed with industrial and WWTP Peschiera di Garda (Verona) fed on domestic wastewaters sewage. After sampling, activated sludge was stored under anoxic conditions at a temperature of 4°C for less one month. Sludge preparation before testing was performed according to procedure of Ficara & Rozzi (2001).

2.3 *Nitrifying activity determination*

Tests were performed at 25°C, biomass concentration was kept within 2 and 3 g TSS/L, maintaining a set-point of 8.3 for pH and 8.3 mgO_2/L for DO concentration.

Figure 2, shows a typical titration curve obtained during a pH-DO-stat experiment for AO and NO kinetic parameters determination. Nitrites and ammonium are added as substrates for NO and AO bacteria.

In the first 30 min, endogenous titration rates are assessed, which have to be subtracted to the exogenous values obtained afterwards. Then, 10 mg N-NO_2^-/L were added as substrate to activate nitrite oxidisers bacteria. When nitrites are depleted, the slope of the H_2O_2 titration curve with diminished to the same endogenous value observed before nitrite addition. Afterwards, ammonium chloride was dosed. At this stage, both titration curves showed a major slope, due to the

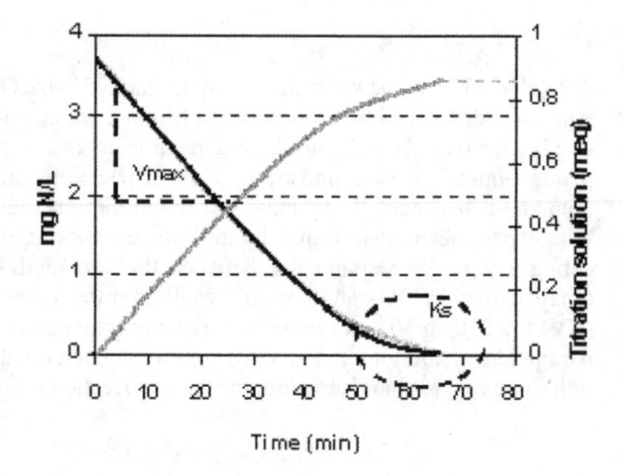

Figure 3. Titration curve and calculated experimental nitrogen profile fitted by a Monod model.

increased acidity production and oxygen consumption for the oxidation of ammonium to nitrite.

Activity of AO and NO bacteria are calculated from the slope of the titration curves, taking into account their reaction stoichiometry. Nitrite oxidisers activity was calculated from the maximum slope of the H_2O_2 titration curve obtained after the addition of nitrite. Ammonium oxidisers activity was calculated from the maximum slope of both NaOH and H_2O_2 titration curves assessed after ammonium addition.

2.4 *Kinetics parameters*

The volume the titration solutions consumed versus time was used to calculate the curves of the substrate oxidised versus time, using the Monod model (equation 3).

$$V = Vmax * \left(\frac{S}{S + Ks} \right) \tag{3}$$

where V = specific oxidation rate; Vmax = maximum specific oxidation rate; S = substrate concentration; and Ks = half saturation constant.

Kinetic parameters were estimated according to the minimum square error criteria, i.e. when the theoretical nitrogen profile closely reproduces the experimental (Figure 3).

With this method it is not necessary to determine specific activities at various substrate concentrations (Corman & Pave, 1983).

3 RESULTS AND DISCUSSION

3.1 *Determination of ammonium oxidisers activity*

Table 1 reports estimated values for the maximum specific ammonium oxidation rate and the half saturation constant obtained for both types of sludge, municipal and industrial from both titration curves. Comparing values for the maximum specific ammonium oxidation rate, an average difference of 12% was observed between estimates obtained from pH-stat and DO-stat titration, indicating a good correspondence between the two methods. As far as the half saturation constant is concerned, an average difference of 30% was obtained between the two methods which is a satisfactory value taking into account the well known variability of Ks estimation due to the poor observability of this parameter.

Activities values obtained are in the typical range of activated sludges: 1.5 to 5 $mgN\text{-}NH_4^+$/(g VSS h) (Andeattola et al., 1999).

Ficara & Rozzi (2001), using the titrimetric system ANITA, have obtained values for the half saturation constant of nitrifying sludge samples drawn from WWTPs within 0.11 and 0.57 mg $N\text{-}NH_4^+$/L, similar to the values obtained in this experimentation. Drtil et al., (1993), using a respirometric method have obtained values of Ks between 0.33 and 0.56 mg/L for the first nitrification stage.

3.2 *Determination of nitrite oxidisers activity*

Kinetics parameters for nitrite oxidisers were estimated from DO-stat titration data. In Table 2, the values obtained for Vmax and Ks for the two sludges are summarised. Results indicate a good repeatability as indicated by the low coefficient of variation, which was found to be close to 20% for both kinetic parameters, similar to the results obtained by Ficara et al., (2000) using H_2O_2 data. Values of Ks obtained in this assays are similar to 0.35 mg/L obtained by Hanaki et al., (1990).

Ks values obtained for two sludges were less than 0.7 mg N/L, indicating a good affinity for substrate and the possibility to obtain an effluent with low residual concentration of substrate.

Table 1. Comparison of kinetics parameters obtained for ammonium oxidisers bacteria from titration solutions.

	Vmax (mg N/gSST h)		Ks (mg N/L)	
Statistics	NaOH	H_2O_2	NaOH	H_2O_2
Industrial sludge				
Mean	2.4	2.5	0.34	0.36
Standard deviation	0.11	0.18	0.04	0.07
C.V. (%)	4.4	7.5	12.9	19.4
Number of determinations: 5				
Municipal sludge				
Mean	1.7	1.5	0.68	0.48
Standard deviation	0.35	0.30	0.15	0.14
C.V. (%)	20.5	26.6	22.2	29.7
Number of determinations: 5				

Table 2. Kinetics parameters of nitrite oxidisers bacteria estimated from H_2O_2 solution consumption.

Statistics	Vmax (mg N/gSST h)	Ks (mg N/L)
Industrial sludge		
Mean	2.8	0.4
Standard deviation	0.60	0.08
C.V. (%)	21.1	19.9
Number of determinations: 5		
Municipal sludge		
Mean	1.0	0.23
Standard deviation	0.17	0.02
C.V. (%)	17.3	6.71
Number of determinations: 4		

In the industrial sludge, due to the ammonium oxidizers activity lower than the nitrite oxidizers activity, the accumulation of nitrite is negligible and the nitrification may be characterized by means of determination only ammonium oxidizers activity. However, in the municipal sludge, the nitrite oxidizers activity is higher than ammonium oxidizers activity being necessary to determine the activity of both stages, due to possibility of accumulation of nitrite that could effect to the overall efficiency of the nitrogen removal.

4 CONCLUSIONS

The feasibility to estimate kinetic parameters for ammonium and nitrite oxidisers by combining pH-stat and DO-stat titration, performed with the MARTINA sensor, was demonstrating on two sludge samples.

Results of ammonium oxidisers activity obtained using both titration solutions showed a very good correspondence, coefficients of variation for the Monod model parameters (maximum specific reaction rate and half saturation constant) showing an acceptable repeatability of the measurements.

Comparing the methodology employed in this experimentation with other conventional methods, our set-point titration technique is faster, the analytical load is minimal and the estimation of kinetic parameters is obtained by using a single substrate concentration instead of determining specific activities at various substrate concentrations. Furthermore, a single assay lasting 2–3 hours allows for the determination of the kinetic parameters of both ammonium and nitrite oxidisers achieving a complete characterisation of nitrification biomass.

ACKNOWLEDGEMENTS

This work is dedicated to the memory of Professor Alberto Rozzi (Politecnico di Milano, ✝2003) for his contributions to the environmental sciences and by his human qualities. We also thanks the Spanish and Italian Ministeries for financing travel expenses through an Integrated Action Ref. (HI2002-0138) and the European Comunity for financing the MARTINA development (EOLI project ICA4-CT-2002-10012).

REFERENCES

Andreottola, G., Canziani, R. & Cossu, R. (1990) Rimozione Bilogica dei nutrient dalle acque di scarico. *Published by Istituto per l'Ambiente*, Milano *Published by Istituto per l'Ambiente*, Milano.

Corman, A. & Pave, A. (1983) On parameter estimation of Monod's bacterial growth model from batch culture data. *J. Gen. Appl. Microbiol.* 29, 91–101.

Hanaki, K., Wantawin, C. & Ohgaki, S. (1990) Effects of the activity of heterotrophs on nitrification in a suspended-growth reactor.

Ficara, E., Rocco, A. & Rozzi, A. (2000) Determination of nitrification kinetics by the ANITA-DOstat biosensor. *Water. Sci. Technol.* 41(12): 121–128.

Ficara, E. & Rozzi, A. (2001) pHstat titration to assess nitrification inhibition. *J. Environ. Eng.*, 127(8): 698–704.

Drtil, M., Németh, P. & Bodik, I. (1993) Kinetic constants of nitrification. *Wat. Res.* 27(1): 35–39.

Rozzi, A., Ficara, E., Massone, A. & Verstraete, W. (2000) Titration biosensors for treatment process control. *Water* 21(4): 50–55.

European Symposium on Environmental Biotechnology, ESEB 2004 - Verstraete (ed)
© 2004 Taylor & Francis Group, London, ISBN 90 5809 653 X

Effects of C/N value and DO on nitrification and denitrification in a novel hollow-fiber membrane biofilm reactor

Jeong-Hoon Shin, Byoung-In Sang & Yun-Chul Chung
Environment & Process Technology Division, Korea Institute of Science and Technology, Seoul, Korea

A novel hollow-fiber membrane biofilm reactor (HfMBR), which consisted of two polysulfone-membrane modules with supplying air/oxygen and hydrogen, respectively, for aerobic and anaerobic treatment, was used to investigate the nitrogen removal efficiency of wastewater under the different condition of C/N and DO value. In the aerobic HfMBR in which a mixture gas of air and O_2 was supplied, COD removal and nitrification occurred. Denitrification occurred in the anaerobic HfMBR using H_2 as the electron donor. Denitrification rate depended on the pressure of H_2 and the hydrogendissolution throughout the hollow-fiber made H_2 a viable alterative electron donor. The microbial community in the each reactor was analyzed by denaturing gradient gel electrophoresis (DGGE) of PCR-amplified 16S rDNA partial sequences (PCR-DGGE). Since dissolved oxygen must diffuse through the heterotroph layer before it reaches the nitrifiers, thick biofilm increases nitrifiers' susceptibility to oxygen limitation in the conventional biofilm reactor. However, in the aerobic HfMBR, because oxygen supplied from the membrane surface where biofilm formed, nitrification occurred even in low bulk-liquid DO concentration (0.03mg/l) and nitrifiers grew well in the biofilm against washout even in the low DO concentration. Using H_2 as an electron donor allowed nearly complete NO_3^- removal in the anaerobic HfMBR and the diversity-shift of denitrifiers by changing from the organic electron donor to inorganic electron donor was also observed.

Managerial aspects and policies

European Symposium on Environmental Biotechnology, ESEB 2004 - Verstraete (ed)
© 2004 Taylor & Francis Group, London, ISBN 90 5809 653 X

New strategies for the development of environmental biosensors

Stephen S. Koenigsberg
Regenesis, Calle Sombra, San Clemente, CA, USA

Jeremy G.A. Birnstingl
Regenesis, Suite Greenwood, Princes Way, London

Lance Laing
Regenesis, Belmont, MA, USA

David Weinkle
Eran Associates, Cypress, CA, USA

The environmental industry is now being impacted by a paradigm shift in analytical measurement that has been traditionally called the "bench to in-line shift", but in reality is a "bench to biosensor shift". The latter is tied to the recent biotechnology revolution and offers a promise for a host of "better, faster and cheaper" means of obtaining data in the field.

Biosensors at their most fundamental level are small-scale binding reactions between a sensor molecule and the target analyte. The term biosensor is invoked here because the sensor molecule in our system is a DNA-protein complex that can react with the target analyte. These reaction chemistries are then coupled to special detection and signaling platforms. Regenesis has completed proof-of-concept work that shows it is possible to detect inorganic species, such as arsenic, at very low levels with minimal interference and with an output measured in a few minutes. This technology can be extended to other inorganic species, as well as organic molecules. Ultimately these devices will be conveniently field portable and hand-held.

A discussion of basic binding, specific detection, and signaling interactions will be discussed in contrast to current options. Eventually, the goal is to multiplex the system to give a suite of results at one time in the field as an alternative to expensive and time-delayed results that require off-site laboratory services. Currently, the value of individual contaminant tests is great as witnessed by the recent need for arsenic monitoring under the lowered U.S. drinking water standard (10 μg/L).

European Symposium on Environmental Biotechnology, ESEB 2004 – Verstraete (ed.)
© 2004 Taylor & Francis Group, London, ISBN 90 5809 653 X

New strategies for the development of environmental biosensors

[illegible]

[illegible]

Solid waste and composting

European Symposium on Environmental Biotechnology, ESEB 2004 - Verstraete (ed)
© 2004 Taylor & Francis Group, London, ISBN 90 5809 653 X

Enrichment and characterization of iron and sulphur oxidizers from Indonesian and South African copper/gold mines

P.H.-M. Kinnunen & J.A. Puhakka
Institute of Environmental Engineering and Biotechnology, Tampere University of Technology, Tampere, Finland

ABSTRACT: Iron and chalcopyrite oxidizing enrichments at 50°C were obtained from Indonesian and South African copper/gold mines. Copper yield of 100% was obtained in 3 months with 2% solids concentration. Iron oxidation rates by Indonesian and South African cultures were 83 and 46 mg Fe^{2+} $L^{-1}h^{-1}$, respectively. Excision and sequencing of fragments from DGGE of the amplified partial 16S rDNA from enrichment cultures showed that the populations were related to *Sulfobacillus yellowstonensis* and *Sulfobacillus acidophilus*.

1 INTRODUCTION

Simplicity, no emissions to air, low cost and applicability to low-value ores are the main benefits of biohydrometallurgy compared to conventional pyrometallurgy (Bosecker 1997, Hsu & Harrison 1995, Krebs et al. 1997, Rossi 1990). Sulphidic mineral, such as economically valuable chalcopyrite, can be biologically leached by non-contact, contact or co-operative leaching mechanism (for a review, see Rawlings 2002) by Fe^{3+} and proton attack (Schippers & Sand 1999). This process is widely used in dump and heap leaching applications. Bacterial leaching is based on the iron and sulphur oxidation ability by a wide range of mesophilic, moderately thermophilic and thermophilic acidophiles (for a review, see Hallberg & Johnson 2001). Moderately and extremely thermophilic iron and sulphur oxidizers are of great interest, as the reaction rates increase and the costs associated with the provision of cooling for the exothermic process are reduced at elevated temperatures (Rossi 1990). The isolation of novel acidophilic bacterial strains from the environment is one possibility to improve the bioleaching process (Johnson 2001) and research is focusing on microorganisms from extreme environments (Horikoshi 1995). The heterogeneous conditions at mine sites with temperature and acidity gradients support a wide diversity of acidophiles (Norris et al. 2000). In the present work, iron and chalcopyrite oxidizers from copper/gold mines in Indonesia and South Africa were enriched and their microbial community structure was delineated with DGGE followed by partial 16S rDNA sequencing.

2 MATERIALS AND METHODS

The samples for enrichments were obtained from Freeport Grasberg copper and gold mine in Indonesia and Target Avgold gold mine in South Africa. Microbes were enriched in $CuFeS_2$ and Fe^{2+} media at 50 and 70°C.

The most numerous microbes in the cultures were obtained using decimal dilution series repeated 3 times. The most diluted microbial solution, in which growth was still observed, was selected for further dilutions. For microbial DNA extraction of original and diluted cultures, a 10 mL sample was filtered on a 0.2 μm pore size polycarbonate (hydrophilic) filter (CycloporeTM Track Etched Membrane, Whatman, UK). The remaining iron was washed from the filter with 0.9% (w/v) NaCl solution at pH 1.5. The filter was stored at −20°C until nucleic acid extraction according to the method by Jurgens et al. (2000). The crude DNA sample was used as a template for the PCR. Archaeal 16S rDNA was amplified using Ar3f (Giovannoni et al. 1988) and Ar958r (Jurgens et al. 2000) primers. Bacterial fragments corresponding to nucleotide positions 341–926 of the *E. coli* 16S rDNA sequence were amplified with the forward primer 341fGC (5′-CCT ACG GGA GGC AGC AG-3′) to which at 5′ end a GC clamp (5′-CGC CCG CCG CGC GCG GCG GGC

GGG GCG GGG GCA CGG GGG G-3′) was added to stabilize the melting behaviour of the DNA fragments in the DGGE, and the reverse primer 907r (5′-CCG TCA ATT CMT TTG AGT TT-3′) (Muyzer et al. 1996). PCR mixtures contained 5 μL of 10 × PCR buffer IV (200 mM $(NH_4)_2SO_4$, 750 mM Tris-HCl, 0.1% (v/v) Tween®, pH 8.8), 1.75 mM $MgCl_2$, 0.5 μM each primer, 100 μM each deoxynucleoside triphosphate, 1.25 U of Red Hot DNA polymerase (AB-gene, Advanced Biotechnologies Ltd), 400 ng μL^{-1} bovine serum albumine (Kreader 1996), and sterile water to a final volume of 50 μL, to which 1 μL of template was added. The following program was used for PCR amplification: 95°C for 5 min; 30 cycles of denaturation at 94°C for 30 s, annealing at 50°C for 1 min, and extension at 72°C for 2 min; and a single final extension at 72°C for 10 min.

PCR samples were loaded onto 8% (w/v) polyacrylamide gel (bisacrylamide gel stock solution, 37.5:1; Bio-Rad Laboratories, Inc.) in 1 X TAE (40 mM Tris, 20 mM acetic acid, 1 mM EDTA, pH 8.3) with denaturing gradients ranging from 20 to 70% (100% denaturant contains 7 M urea and 40% (v/v) formamide). The electrophoresis was run at 60°C with 100 V for 14 h. DNA fragments were frozen at −20°C and cut into small pieces. 20 μL of sterile water was added and the DNA was allowed to diffuse into the water for 30 min. Eluate was used as template in a PCR with the primers 341f (without GC clamp) and 907r, and PCR program described above. The sequencing of the purified products was performed at DNA Sequencing Facility, Institute of Biotechnology, Helsinki University.

3 RESULTS

At 50°C, moderate thermophilic iron and chalcopyrite oxidizers were obtained (Figure 1). Copper yield of 100% was obtained in 3 months with 2% solids concentration. Iron oxidation rates by Indonesian and South African cultures were 83 and 46 mg Fe^{2+} $L^{-1}h^{-1}$, respectively. The enrichments at 70°C were not successful. When archaeal primers were used for DNA amplification, no product was obtained. Bacterial DGGE profiles showed a fairly simple community structure in all enrichment cultures (Figure 2). Indonesian $CuFeS_2$ culture produced only one DGGE fragment showing 98% similarity with *Sulfobacillus yellowstonensis*. Although Indonesian Fe^{2+} culture produced several fragments, they were all related to *S. yellowstonensis* (98–99% similarity), but differed slightly from each other and from $CuFeS_2$ enrichment. The dominant population in Indonesian Fe^{2+} culture based on dilution series (fragment 7) differed from the dominant population based on the intensity of the DGGE fragment. The fragments obtained from South African Fe^{2+} cultures were related to *S. acidophilus* (99% similarity). As

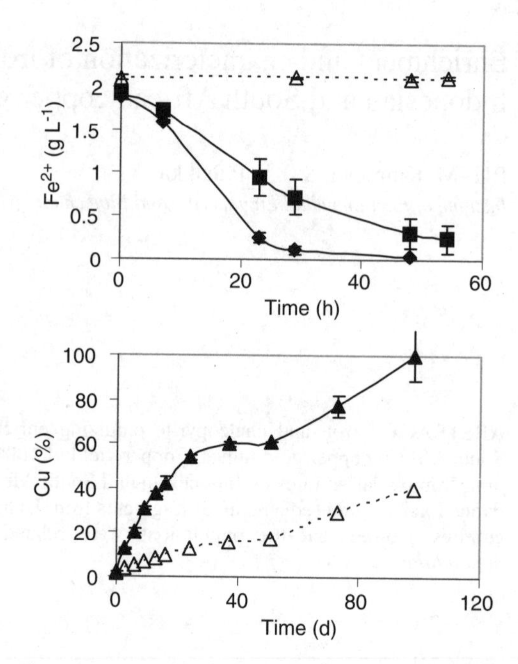

Figure 1. Iron oxidation by Indonesian (circles) and South African (squares) cultures and chalcopyrite leaching by Indonesian culture (triangles) with 2% solids concentration. White triangles represent controls.

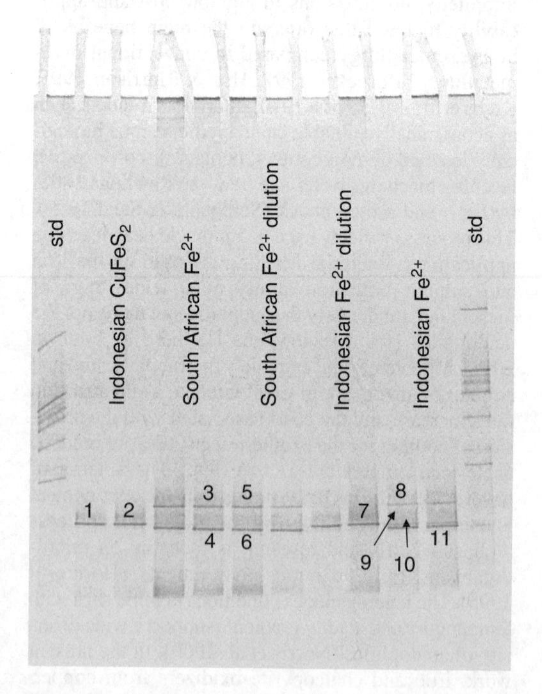

Figure 2. Duplicate DGGE profiles of Indonesian and South African $CuFeS_2$ and Fe^{2+} enrichments and dilutions.

both fragments present in the original mixed culture were present also in the diluted culture, their numbers are likely similar with each other.

4 DISCUSSION

DGGE followed by partial 16S rDNA sequencing demonstrated that the iron and chalcopyrite oxidizing cultures at 50°C enriched from copper/gold mines in Indonesia and South Africa were related to *S. yellowstonensis* and *S. acidophilus*, respectively. No archaea was enriched. Members of *Sulfobacillus* genera are found in many naturally acidic environments (Goebel et al. 2000) and in this study similar *Sulfobacillus* cultures were obtained from two geographically distinct areas. As DGGE has weaknesses in analysing populations because of potential differences in lysing or amplification efficiency among different populations, dilution series were done from mixed cultures showing several bands in DGGE to determine the predominant strain. The numbers of two South African iron oxidizing strains are likely similar with each other, whereas Indonesian iron oxidizers were predominated by *S. yellowstonensis* corresponding to fragment 7.

5 CONCLUSIONS

This study confirms the wide global distribution of iron and chalcopyrite oxidizing *Sulfobacillus* sp., which is a potential catalyst for the bioleaching of sulphidic minerals.

REFERENCES

Bosecker, K. 1997. Bioleaching: metal solubilization by microorganisms. *FEMS Microbiology Reviews* 20: 591–604.

Giovannoni, S.J., DeLong, G.J., Olsen, G.J. & Pace, N.R. 1988. Phylogenetic group-specific oligonucleotide probes for identification of single microbial cells. *Journal of Bacteriology* 170: 720–726.

Goebel, B.M., Norris, P.R. & Burton, N.P. 2000. Acidophiles in biomining. In F.G. Priest & M. Goodfellow (eds.), *Applied Microbial Systematics*: 293–214. Amsterdam: Kluwer Academic Publishers.

Hallberg, K.B. & Johnson, D.B. 2001. Biodiversity of acidophilic prokaryotes. *Advances in Applied Microbiology* 49: 37–84.

Horikoshi, K. 1995. Discovering novel bacteria, with an eye to biotechnological applications. *Current Opinion in Biotechnology* 6: 292–297.

Hsu, C.-H. & Harrison, R.G. 1995. Bacterial leaching of zinc and copper from mining wastes. *Hydrometallurgy* 37: 169–179.

Johnson, D.B. 2001. Importance of microbial ecology in the development of new mineral technologies. *Hydrometallurgy* 59: 147–157.

Jurgens, G., Glöckner, F.-O., Amann, R., Saano, A., Montonen, L., Likolammi, M. & Münster, U. 2000. Identification of novel archaea in bacterioplankton of a boreal forest lake by phylogenetic analysis and fluorescent *in situ* hybridization. *FEMS Microbiology Ecology* 34: 45–56.

Kreader, C.A. 1996. Relief of amplification inhibition in PCR with bovine serum albumin of T4 gene 32 protein. *Applied and Environmental Microbiology* 62(3): 1102–1106.

Krebs, W., Brombacher, C., Bosshard, P.P., Bachofen, R. & Brandl, H. 1997. Microbial recovery of metals from solids. *FEMS Microbiology Reviews* 20: 605–617.

Muyzer, G., Hottenträger, S., Teske, A. & Wawer, C. 1996. Denaturing gradient gel electrophoresis of PCR-amplified 16S rDNA – A new molecular approach to analyse the genetic diversity of mixed microbial communities. In A.D.L. Akkermans, J.D. Van Elsas & F. De Bruijn (eds), *Molecular Microbial Ecology Manual* 3.4.4: 1–23. Netherlands: Kluwer Academic Publishers.

Norris, P.R., Burton, N.P. & Foulis, N.A.M. 2000. Acidophiles in bioreactor mineral processing. *Extremophiles* 4: 71–76.

Rawlings, D.E. 2002. Heavy metal mining using microbes. *Annual Reviews in Microbiology* 56: 65–91.

Rossi, G. 1990. *Biohydrometallurgy*. Hamburg: McGraw-Hill.

Schippers, A. & Sand, W. 1999. Bacterial leaching of metal sulfides proceeds by two indirect mechanisms via thiosulfate or via polysulfides and sulfur. *Applied and Environmental Microbiology* 65(1): 319–321.

European Symposium on Environmental Biotechnology, ESEB 2004 - Verstraete (ed)

Influence of solid wastes in the mycelium growth of *Lentinula edodes* (berk.) Pegler

M. Regina & A.F. Eira
Faculdade de Ciências Agronômicas – UNESP, Botucatu, SP

N.B. Colauto
Instituto de Pesquisas Estudos e Ambiência Científica – UNIPAR – Umuarama – Paraná

J.R.S. Passos, F. Broetto & J.A. Marchese
Instituto de Biociências da FMVZ/Unesp, Botucatu, SP

ABSTRACT: In Brazil there was little research related to Shiitake axenic culture.The aim of this research was to understand the substratum effects in the kinetics of the Shiitake mycelium growth. It was used two Shiitake strains and two different base substrate (eucalyptus sawdust and sugar cane bagasse) varying in three proportions of the supplements. The supplements, a blend of rice and wheat brans, were added in the proportion of 0, 10 and 20% of the base substrate. The experiment was composed of six treatments. The mycelium growth kinetics in volume had no effect relation to the strains and substrate and it followed a mathematical model represented by logarithmic equation. Beta, gamma and delta parameters didn't show any correlation with the growth velocity in volume. The strain L55 was better adapted than L17.

1 INTRODUCTION

The *L. edodes* has been cultivated on a number of other materials either alone or in combination with sawdust. These include straw, corn-cobs and other agricultural wastes such as the sugar cane bagasse, citrus-peel wastes, and grain chaff. (Przybylowicz & Donoghue, 1988).

The use of another substrates, besides sawdust, as the agriculture-industrial wastes, in the axenic cultivation, is perfectly possible once the shiitake mycelium is capable to grow in a great variety of lignicelullosics residues (Lacaz et al., 1970).

Rossi (1999) verified that the sugar cane bagasse needs to be supplemented and that, when done with rice bran, in the shiitake axenic cultivation, it provided better speed of mycelium growth. According to Sturion (1994) the sugar cane bagasse with supplements can produce mushrooms.

This work had for objective to study the growth kinetics of *L. edodes* strains in substrate, in laboratory conditions, verifying the influence of the eucalyptus sawdust and sugar cane bagasse, with several brans percentages.

2 MATERIALS AND METHODS

The experiment was guided in Mushrooms Module of the Vegetable Production Department, situated in Experimental Farm Lageado, Campus of Botucatu of the São Paulo State University – UNESP.

2.1 *Determination of volume mycelium growth*

In this work the *L. edodes* strains were BMA-LE 96/17 and BMA-LE 99/55, identified here as L17 and L55. The material base used for the composition of the substratum was eucalyptus sawdust (S) and sugar cane bagasse (B) mixed with three rice brans percentages (0, 10 and 20%), a total of six substrates, and put on the tubes, after the inoculation, incubated in BOD for 25°C.

The measures were made until the complete colonization of the substrate. The first measure went 3 days after to inoculation and, the others, for 5, 6, 7, 8, 10, 11, 13, 14, 16, 18, 19 and 21 days after.

2.2 *Statistical analyses*

For the study of the growth in mycelium volume, according to the treatments already defined, the model of non lineal regression was used, adjusted for each one of the repetitions, of the treatments:

$$y_{ij} = \alpha + \beta \ln\left(\frac{x+\gamma}{-\gamma+\delta+x}\right) + e_{ij} \quad (1)$$

where, y_{ij} = growth volume in the repetition i and treatment j; x = days; e_{ij} = aleatory component or noise in the sample i and in the treatment j; α = location parameter; β, γ and δ are parameters in way of the model, related to the instantaneous velocity in a day.

The instantaneous velocity, it was obtained being analyzed the expression of the first derived in relation to x (time) $f'(x) = df/dx$.

$$f'(x) = -\frac{(2\gamma+\delta)\beta}{(x+\gamma)(-\gamma-\delta+x)} \quad (2)$$

This expression defines, therefore, the instantaneous growth velocity ($cm^3 dia^{-1}$) for each repetition in each treatment. It is verified that the instantaneous growth velocity is function beta, gama and delta. For comparison models parameters non lineal regression (beta, gama and delta), referring to each one of the treatments, it was made non parametric Variance Analysis with application of the Kruskal-Wallis test. In the cases in that there was significant difference for that test, were applied the following tests: for each lineage, Student-Newmam-Kills (SNK) at the level of 5%, for each substratum, Mann-Whitney at the level of 5%.

3 RESULTS

The deterministic component used for obtaining of the *L. edodes* growth volume in the treatments, is function beta, gama and delta as can observe for the equation (1). It was not found any direct relationship of the estimated parameters beta, gama and delta of L17 (Tables 1, 2 and 3), thus as the L55 (Tables 4, 5 and 6) with the instantaneous mycelium growth velocity in volume although gama has presented, in some cases, relationship with initial growth velocity and, delta with the final velocity, this was not observed so that if it could conclude a relationship among them.

Table 1. Betas comparison for the Student-Newmam-Kills (SNK) test, according to L17 and amount of brans, in sawdust and sugar cane bagasse.

	Substratos	
L17	S0% 1,21^{A*}	B0% 1,06^{B}
	S10% 0,97^{A}	B10% 0,98^{A}
	S20% 0,84^{A}	B20% 1,05^{B}

For line: medium followed by same letter, don't differ significantly at the level of 5% for the Student-Newmam-Kills test.

Table 2. Gamas comparison for the Student-Newmam-Kills (SNK) test, according to L17 and amount of brans, in sawdust and sugar cane bagasse.

	Substratos	
L17	S0% 1,21^{A*}	B0% 1,06^{B}
	S10% 0,97^{A}	B10% 0,98^{A}
	S20% 0,84^{A}	B20% 1,05^{B}

For line: medium followed by same letter, don't differ significantly at the level of 5% for the Student-Newmam-Kills test.

Table 3. Deltas comparison for the Student-Newmam-Kills (SNK) test, according to L17 and amount of brans, in sawdust and sugar cane bagasse.

	Substratos	
L17	S0% 27,9^{A*}	B0% 18,1^{B}
	S10% 20,6^{A}	B10% 20,0^{A}
	S20% 22,5^{A}	B20% 18,1^{B}

For line: medium followed by same letter, don't differ significantly at the level of 5% for the Student-Newmam-Kills test.

Table 4. Betas comparison for the Student-Newmam-Kills (SNK) test, according to L55 and amount of brans, in sawdust and sugar cane bagasse.

	Substratos	
L55	S0% 0,94^{A*}	B0% 1,05^{B}
	S10% 1,40^{A}	B10% 1,06^{B}
	S20% 1,11^{A}	B20% 0,96^{A}

For line: medium followed by same letter, don't differ significantly at the level of 5% for the Student-Newmam-Kills test.

Table 5. Gamas comparison for the Student-Newmam-Kills (SNK) test, according to L55 and amount of brans, in sawdust and sugar cane bagasse.

	Substratos	
L55	S0%	B0%
	0,94^{A}*	1,05^{B}
	S10%	B10%
	1,40^{A}	1,06^{B}
	S20%	B20%
	1,11^{A}	0,96^{A}

For line: medium followed by same letter, don't differ significantly at the level of 5% for the Student-Newmam-Kills test.

Table 6. Deltas comparison for the Student-Newmam-Kills (SNK) test, according to L55 and amount of brans, in sawdust and sugar cane bagasse.

	Substratos	
L55	S0%	B0%
	25,4^{A}*	22,6^{B}
	S10%	B10%
	24,5^{A}	23,7^{B}
	S20%	B20%
	23,3^{A}	22,3^{A}

For line: medium followed by same letter, don't differ significantly at the level of 5% for the Student-Newmam-Kills test.

The influence of the used substrate base, in the L17 instantaneous growth velocity, inside of the several brans percentages (Figure 1), showed that in 0% the sugar cane bagasse, promoted better velocitys, in 10% the velocity was independent of the used material base, and in 20%, the sawdust resulted as material best in terms of instantaneous growth velocity.

In L55, the instantaneous velocity, in the several brans percentages, showed that the sawdust promoted a larger instantaneous velocity of volume growth, except in the amount of 20% in that the sugar cane bagasse didn't differ of the sawdust (Figure 2). The instantaneous velocity curve in 20% had a reaction, to the low oxigenation index, softer than in the other percentages.

4 DISCUSSION

Second Klein (1996), the used material base has influence once it can determine a peculiarmycelium growth depending on its structure and chemical substances.

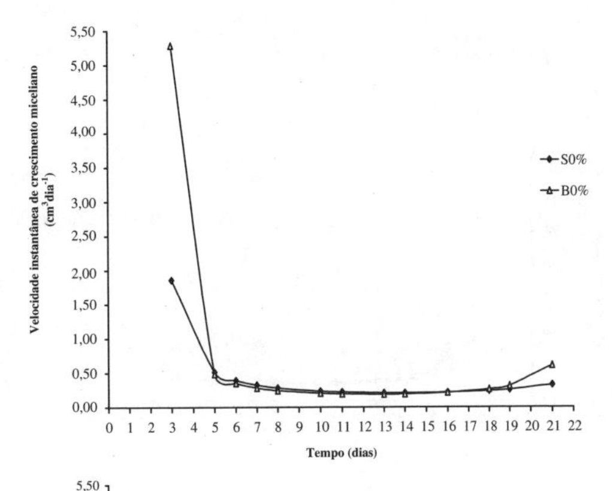

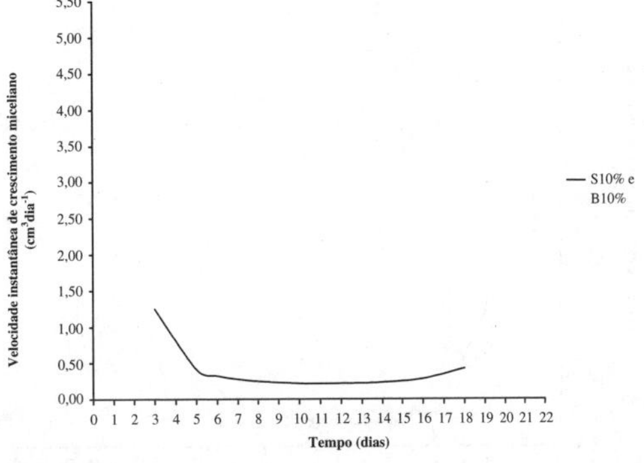

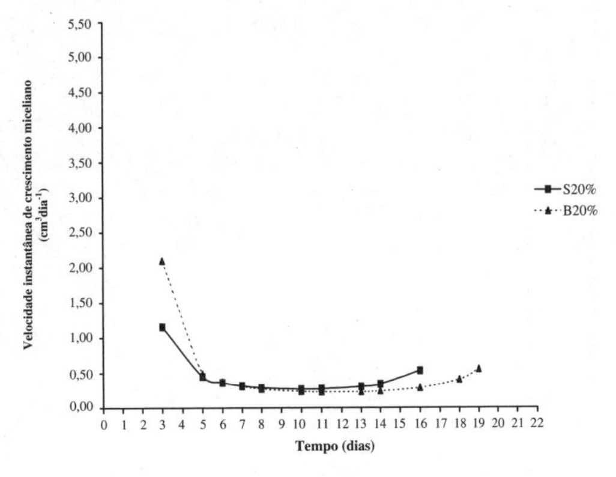

Figure 1. Instantaneous velocity curves of L17 mycelium growth, in the different brans percentages, sawdust and sugar cane bagasse.

As the sugarcane bagasse is composed struturally for laminate fibers and its density is inferior to sawdust it was necessary to the application of a light pressure so that all the tubes had the same substrate height. This procedure can have been creating different density zones, what can have been harming the mycelium growth velocity in some of the treatments with this material.

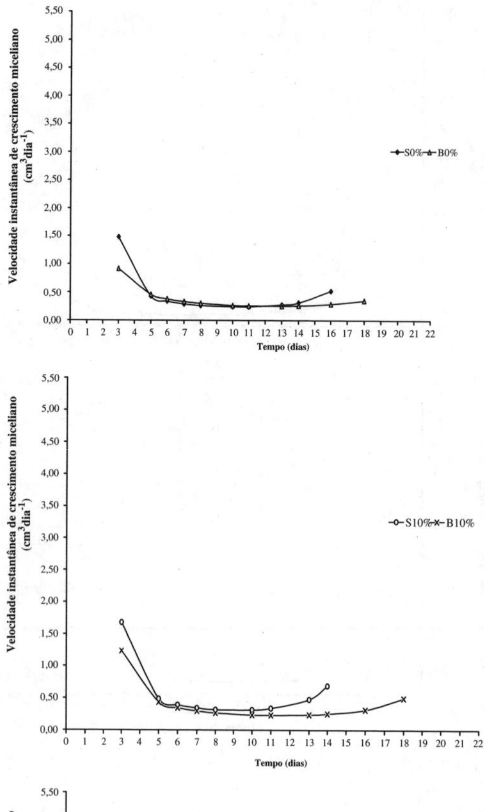

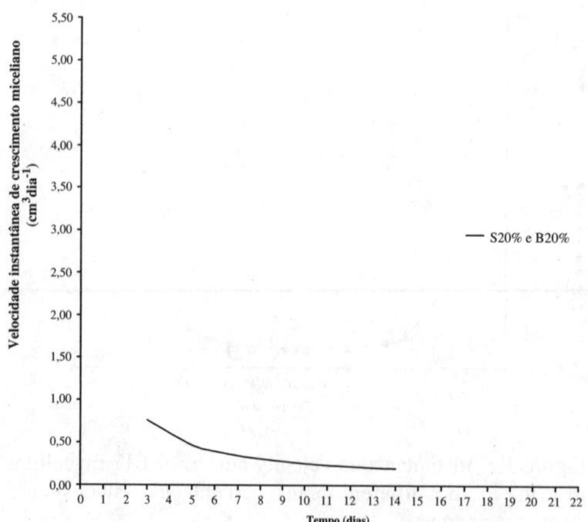

Figure 2. Instantaneous velocity curves of L55 mycelium growth, in the different brans percentages, sawdust and sugar cane bagasse.

The growth kinetics in rehearsal tube could verify the effect of the oxigenation lack, happened in the middle of the tube, that resembles to the that happens in the bags production. Donoghue & Denison (1995), they verified that the *L. edodes* mycelium growth is harmed in high concentrations of CO_2, altering its enzymatic system and decreasing the growth velocity. The authors could check that the mushroom inoculated in substrate, inside of bags with larger openings for gaseous change, had a faster mycelium growth mycelium.

The use of plastic sacks for the producing of seed, is quite similar to the condition of the cylinder (rehearsal tube). Usually, some days after the inoculation, the bags is agitated to distribute the inoculum, with that it can be obtained a growth velocity of to the cube of the incubation time, just as having emphasized for Stanier et al. (1969). The enzymes aids in the penetration of the solid substratum, but the penetration depth is dependent of the appropriate aeration (Dix & Webster, 1995).

Under those circumstances, it was observed in the present work that the mycelium adaptation (previous growth in flasks) and limitation of the tube diameter, where the mycelium was "addressed" to grow just down, can have been contributing, in the first days, for a great growth instantaneous velocity in volume, for all the treatments. However, soon after, it happened an accentuated fall of the instantaneous velocity, and then, it stayed practically constant in some treatments (middle of the tube). Starting from this point, the instantaneous velocity increased again, in direction at the end of the substrate where for the adopted methodology there was probably a oxygen concentration larger in the tube bottoms and larger possibility of gaseous changes.

It is believed that if a three-dimensional simulation was accomplished, relatively to the mycelium growth of the used cylinder, the values of the speeds (cm^{3}dia^{-1}) they could be larger than found four times them. The decrease of the mycelium growth instantaneous velocity, in the middle of the tube, it have been cause, probably, for the low oxigenation this point, as they verified Leatham & Stahmann (1987), that the inadequate gaseous change inhibited L.edodes mycelium growth or still for other factors as the accumulation of toxicant final products and production of secondary metabolits (Stanier et al., 1969; Prosser, 1994).

In the present work, the low oxigenation didn't cause the complete stop of growth, even so a strong decrease of the process. It is known that through the accumulated reservations, the mushrooms are capable to stay to find favorable conditions for the normal retaking of growth (Dickinson & Bottomley, 1980). this can be verified, with the increase of the instantaneous velocity of the middle for the end of the tube (increase of the oxigenation). The decline phases, in the growth of a culture can be caused, among other things, for the finished of some nutritious or for the residual products accumulated (Griffin, 1994). The factors that altered the growth miceliano was the low oxigenation in the middle of the tube and the substrate without supplemented.

ACKNOWLEDGEMENT

This work was supported by CAPES.

REFERENCES

Dickinson CH & Bottomley D (1980) Germination and growth of *Alternaria* and *Cladosporium* in relation to their activity in the phylloplane. Transactions of the Britsh Mycological Society 74:309–319.

Dix JN & Webster J (1995) The mycelium and substrates for growth. Pp. 12–26. *In*:_ *Fungal Ecology*. Chapman & Hall.

Donoghue JD & Denison WC (1995) Shiitake cultivation: gas phase during incubation influences productivity. Mycologia 87:239–44.

Griffin DH (1994) Growth. *In*:_. *Fungal Physyology*. 2ª ed. Wiley-Liss.

Han YH, Yeng WT, Chen LC & Chang S (1981) Physiology and ecology of *Lentinus edodes* (Berk). Mushroom Science 11:623–658.

Klein KK (1996) Pattern formation and development of the fungal mycelium. Pp.70–86. *In*: Chiu S, Moore D *Patterns ins fungal development*. Cambridge University Press.

Lacaz CS, Minami PS & Purchio A (1970) *O grande mundo dos fungos*. USP/ Polígono.

Leatham GF & Stahmann MA (1987) Effect of ligth and aeration on fruiting of *Lentinula edodes*. Transactions British Mycological Society 88:9–20.

Przybylowicz P & Donoghue J (1988) *Shiitake growers handbook: the art and science of mushroom cultivation*. kendall/Hunt.

Rossi IR (1999) *Suplementação de bagaço de cana-de-açúcar para cultivo axênico do cogumelo shiitake [Lentinula edodes (Berk.) Pegler]*. Jaboticabal, 129p. Dissertação (Mestrado em Agronomia/Microbiologia) – Faculdade de Ciências Agrárias e Veterinárias, Universidade Estadual Paulista.

Stanier RY, Doudoroff M & Adelberg EA (1969) *O mundo dos micróbios*. Edgard Blücher.

Sturion GL (1994) *Utilização da folha de bananeira como substrato para o cultivo de cogumelos comestíveis* (*Pleurotus* spp). Piracicaba, 147p. Dissertação (Mestrado em Agronomia/Microbiologia), Escola Superior de Agricultura Luiz de Queiroz, Universidade de São Paulo.

European Symposium on Environmental Biotechnology, ESEB 2004 - Verstraete (ed)

The role of structure in composting of sewage sludge and diatomite

K. Malinska[1], G. Malina[1], D. Krajewski[1,2], T. Recko[1], A. Veeken[2] & B. Hamelers[2]
[1]*Czestochowa University of Technology, Poland*
[2]*Wageningen University and Research Centre, The Netherlands*

ABSTRACT: The paper aims at determining the influence of green waste as a bulking agent on the structure of composted sewage sludge and, in consequence on the biodegradation rate and the decomposition degree. Mixes of sewage sludge with addition of 5%, 10%, 20%, 33% and 40% (w/w) of green waste were composted in a 5-L reactor. The most favourable physical parameters of the composting process were observed for the mixes with 33% and 40% (w/w) of green waste. The highest biodegradation rates, as well as the highest decomposition degree were observed during composting of mixes with 33% of green waste. Based on these results and the experimental set-up, the experiment was carried out to determine the decomposition degree and the biodegradation rate during composting of mixes consisted of sewage sludge and diatomite with grass as a bulking agent. Addition of diatomite to sewage sludge resulted in ca. 5 times higher biodegradation rates and twice higher decomposition rates, as compared to composting of sewage sludge alone.

1 INTRODUCTION

Significant increase in number of wastewater treatment plants, as well as upgrading the wastewater treatment technology, results in increasing the quantities of sewage sludge production (Bien at al. 1999). Depending on the properties of sewage sludge, composting is considered a beneficial way of sewage sludge utilization (Urbaniak et al. 2000). The process may result in production of compost useful for agricultural or horticultural purposes, and for land reclamation. What is more, composting is also regarded as a cheap and robust process resulting in reduction of the volume of waste materials (Smars et al. 2001). Therefore, it can be applied in different waste management schemes, such as production of high quality compost or reducing organic contents of waste prior to landfilling (Hamelers 2001). However, for optimal composting/co-composting of diverse organic substances, the knowledge of physical properties of composting materials plays an important role (Agnew & Leonard 2002).

The paper presents the research aimed at determination of the role of structure in composting of sewage sludge with green waste as a bulking agent. The influence of a bulking agent on composting parameters, such as: bulk density, permeability, porosity and compaction of material was investigated to optimise composting of sewage sludge. The special attention was paid to establish the most optimal ratio of green waste as a bulking agent and sewage sludge, when the highest biodegradation rate of organic matter occurs. The research was carried out at Wageningen University and Research Centre, the Netherlands. With reference to the project results, the similar research design was applied to conduct the preliminary experiment on composting of mixes of sewage sludge and diatomite with addition of 33% of grass as a bulking agent. The experiment carried out in the Institute of Environmental Engineering at Czestochowa University of Technology, Poland, was aimed at (1) testing and verifying an experimental set-up, (2) determination of the biodegradation rate and the decomposition degree during composting.

2 MATERIALS AND METHODS

2.1 *Materials*

2.1.1 *Experiment 1*

Digested, dewatered sewage sludge and green waste were used for preparation of composting mixes. Sewage sludge was collected from the wastewater treatment plant, Veendendal, the Netherlands. The samples were stored in the laboratory refrigerator prior to analysing. Green waste was selected as a bulking agent due to easy access and low costs, and was provided by Van Versel Composting & Recycling, Biezenmortel, the Netherlands.

2.1.2 *Experiment 2*

Digested and dewatered sewage sludge from the wastewater treatment plant in Blachownia (near Czestochowa), Poland, was used. Diatomite provided by the brewery from Czestochowa was selected as a composting substrate because: (1) it is used as a filtrating agent in beverage industry and deposited at landfills afterwards, (2) the content of organic matter is high, (3) microorganisms present may promote the process of composting. Grass was selected as a bulking agent due to easy access and low costs. The composting substrates were stored in the laboratory refrigerator prior to analysing.

2.2 *Analytical procedures*

2.2.1 *Experiment 1*

The composting mixes of sewage sludge with addition of: 5%, 10%, 33%, 40% (w/w) of green waste, were tested. Separate composting of sewage sludge and green waste alone was used as a control. Total Solids (TS), Volatile Solids (VS) and bulk density were determined in the composting mixes. Pycnometer was used to determine free air space and porosity. Permeability was determined with reference to the pressure drop measured during airflow through the composting material. Fresh sewage sludge and fresh green waste were subject to the OXITOP test in order to calculate the Oxygen Uptake Rate (OUR) (molO_2/kgVS/d). Prior to composting, in order to simulate composting conditions typical for an industrial reactor, the mixes were pressed with heavy ballast (weight 18.04 kg) that resulted in the stress on the bed of 1021 kg/m^3.

2.2.2 *Experiment 2*

Three composting mixes (w/w) were prepared as follows: sewage sludge with 33% of grass, diatomite with 33% of grass, and sewage sludge and diatomite (1:1 w/w) with 33% of grass. TS, VS and pH were determined in the composting substrates, and in the material after composting.

2.3 *Experimental set-up*

The experimental set-up consisted of a 5-L reactor (Figure 1), with a perforated plate, an air inlet at the bottom, an air outlet at the top and a temperature sensor. The air collected from the outlet was cleaned in a H_2SO_4 solution, and the oxygen content was measured by an O_2 sensor. The reactor was placed in a water bath container (Figure 2), equipped with a water pump maintaining temperature of ca. 55°C that is considered beneficial for the process rate (Hamelers, 2001). Composting of sewage sludge and green waste mixes was carried out simultaneously in two trials for one week. The aeration rate varied from 5 to 20 dm^3/h. The process was monitored by

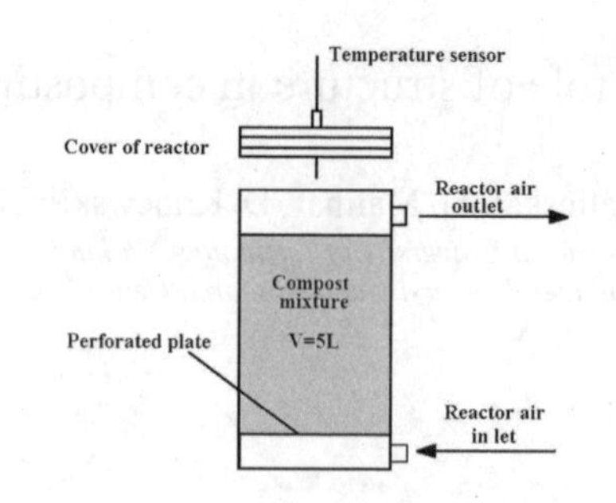

Figure 1. The composting reactor.

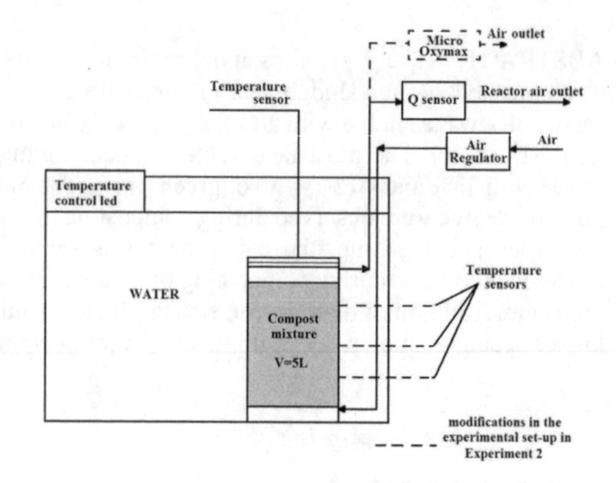

Figure 2. The experimental set-up for composting.

every day measurements of temperature, aeration and oxygen content in the outlet air. In the case of experiment 2, the set-up was equipped with the 10-channel respirometer Micro-Oxymax® (Columbus Ohio), which enables online measurements of the O_2/CO_2 contents in the outlet air. Temperature in a water bath was maintained at 32°C. Composting was run at the constant aeration rate of 50 dm^3/h. The experiment was terminated when no significant changes in the O_2 uptake and the CO_2 production were observed, i.e. after 13 days (diatomite + grass), 11 days (sewage sludge + grass and 16 days (sewage sludge + diatomite + grass).

3 RESULTS AND DISCUSSION

3.1 *Experiment 1*

3.1.1 *Total solids (TS) and volatile solids (VS)*

TS for sewage sludge and green waste composted separately amounted at ca. 21% and 86%, respectively (Figure 3), whereas VS was determined at ca. 64.5% and 95%, respectively (Figure 4). The values of TS and VS are high for green waste and low in the case of sewage sludge. Addition of green waste to sewage sludge increased both TS and VS in composting mixes. Addition of 20%, 33% and 40% of green

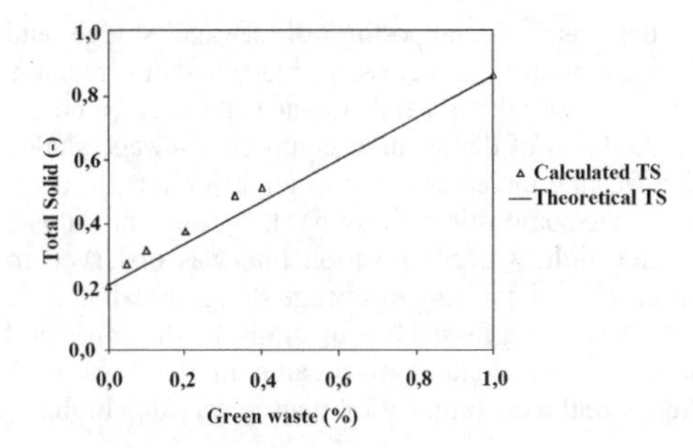

Figure 3. Total solids as a function of green waste.

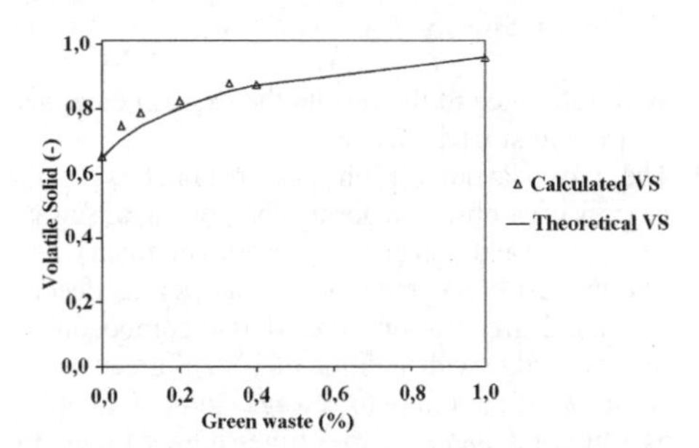

Figure 4. Volatile solids as a function of green waste.

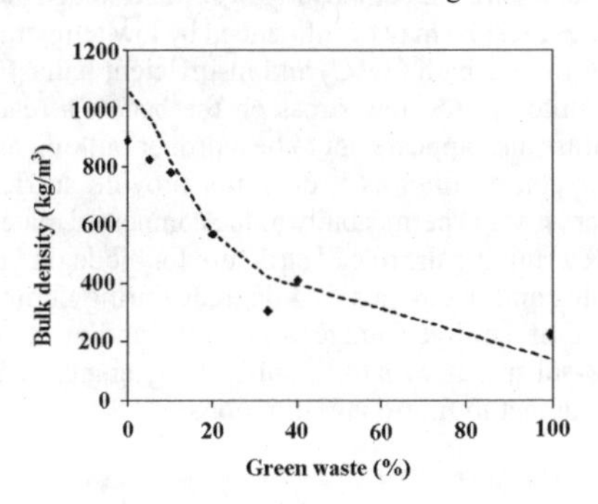

Figure 5. Bulk density as a function of green waste.

waste significantly improved bulk density of composting mixes (Figure 5).

3.1.2 *Porosity, permeability and bulk density*
Porosity and air permeability are crucial for airflow through the composting material. Green waste addition improved the structure of composting mixes, which resulted in a heterogeneous material with loose structure (Figure 6). The values of air permeability and porosity are considered the most desired for green waste (Figure 7).

3.1.3 *Cumulative oxygen uptake*
The ultimate oxygen demand for investigated composting mixes was predicted from cumulative oxygen

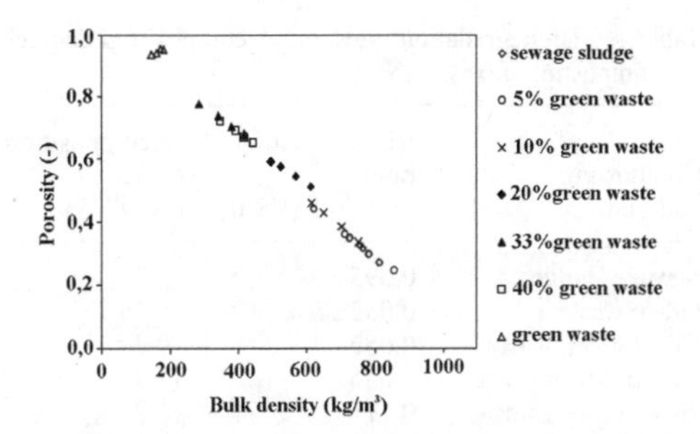

Figure 6. Porosity vs. bulk density.

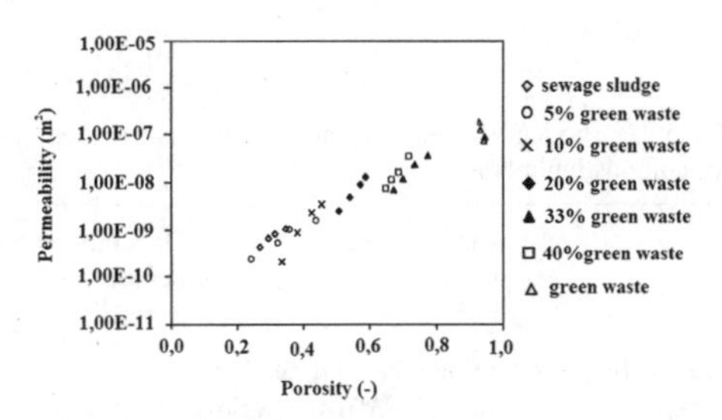

Figure 7. Permeability vs. porosity.

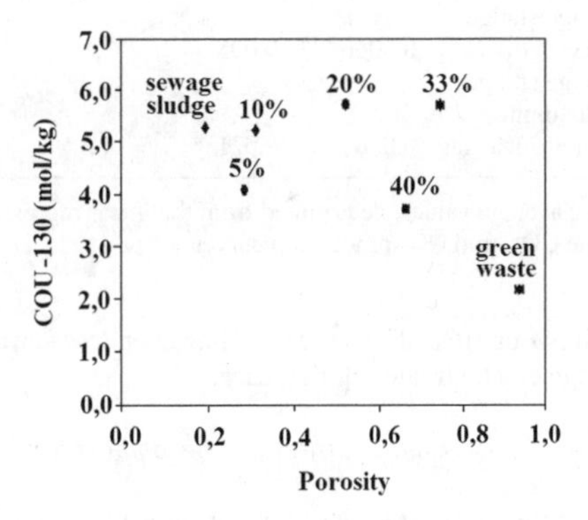

Figure 8. The COU values vs. porosity for the initial composting.

uptake (COU) curves vs. time. For the initial period of 130 h, the lowest values of COU were observed for green waste (2.5 $molO_2/kgVS$), whereas the highest were found for sewage sludge (6.2 $molO_2/kgVS$) (Figure 8). The highest values were observed for composting mixes with 20% and 33% of green waste that also showed the highest porosity (from 0.60 to 0.80). For the mixes with 5% and 10% of green waste with low porosity (0.3) the COU value was low, as well. An increase in porosity can result in higher values of COU. Therefore, composting mixes with 33% of green waste seem to be more biodegradable than the mixes

Table 1. Biodegradation rates and decomposition degrees for composting mixes.

Composting material	Biodegradation rate ($molO_2/kgVS/h$)	Decomposition degree* (% of VS)
Sewage sludge	0.095	13.2
Green waste	0.032	5.35
5% of green waste	0.059	9.7
10% of green waste	0.086	11.7
20% of green waste	0.111	12.7
33% of green waste	0.121	13.6
40% of green waste	0.074	9.65

* The average values.

Table 2. Biodegradation rates and decomposition degrees for composting mixes.

	Biodegradation rate*		Decomposition degree
Composting material	O_2 uptake (mg/gVS/d)	CO_2 production (mg/gVS/d)	(% of VS)
Diatomite + 33% grass	0.31	0.17	7.8
Sewage sludge + 33% grass	0.0066	0.005	7.5
Sewage sludge + diatomite (1:1) + 33% grass	0.036	0.021	14.5

* The average values, determined from the linear regression of the COU and CO_2 production curves.

with 5% or 10% of green waste, however, less sewage sludge can be treated in that case.

3.1.4 *Biodegradation rates and decomposition degrees*

The optimal biodegradation rates, as well as decomposition degree, were observed for the composting mix with 33% of green waste (Table 1).

3.2 *Experiment 2*

3.2.1 *Biodegradation rates and decomposition degrees*

Biodegradation rates were determined from the COU curves and compared with the values obtained from cumulative CO_2 production curves. The highest biodegradation rate was observed during composting of diatomite with 33% of grass (Table 2), due to a significant quantity of VS (83%). Composting of sewage sludge and 33% of grass showed the lowest biodegradation rate as porosity of the composting mix decreased. In the case of composting of sewage sludge and diatomite with 33% of grass, the biodegradation rate was influenced by addition of diatomite that increased porosity. Addition of diatomite to composted sewage sludge significantly increases the biodegradation rate, in comparison to composting of sewage sludge and grass alone.

The highest decomposition rate was observed in the case of composting of sewage sludge and diatomite with 33% of grass. Due to addition of grass and diatomite the content of organic matter increased, therefore the decomposition degree was also higher.

4 CONCLUSIONS

1. With reference to the results the experimental setup worked sufficiently.
2. The most favourable physical parameters of the process were observed during composting of sewage sludge with addition of 33% and 40% of green waste.
3. The highest biodegradation rate as well as decomposition degree was observed during composting of sewage sludge with addition of 33% of green waste.
4. Addition of diatomite to sewage sludge with grass as a bulking agent led to 5 times higher biodegradation rate and 2 times higher decomposition rate, as compared to composting of sewage sludge alone. These results may be influenced by low temperature of a water bath (32°C) and insufficient ballast that resulted in too low stress on the bed in a reactor. Grass also appears not to be a proper bulking agent for composting, as it does not provide sufficient porosity and permeability of the composted material.
5. Determining the role of structure for biodegradation rates and decomposition degrees during composting of sewage sludge, diatomite and/or diverse co-substrates with different bulking agents, will be a subject to future investigations.

REFERENCES

Agnew J.M. & Leonard J.J. 2003. Literature review: The physical properties of compost. *Compost Science & Utilization* 11(3): 238–264.

Bien J.B. et al. 1999. *Gospodarka odpadami w oczyszczalniach sciekow*. Czestochowa: Wydawnictwo Politechniki Czestochowskiej.

Hamelers H.V.M. 2001. A mathematical model for composting kinetics. PhD Thesis, Wageningen University, the Netherlands.

Smars S. et al. 2001. An advanced experimental composting reactor for systematic simulation studies. *Journal of Agricultural Engineering Research* 78(4): 415–422.

Urbaniak M. & Mokrzycka-Wieteska B. 2000. Kompostowanie osadów wstepnych, surowych i poddanych fermentacji metanowej. *Inżynieria i Ochrona Środowiska* 3 (1–2): 247–254.

European Symposium on Environmental Biotechnology, ESEB 2004 - Verstraete (ed)
© 2004 Taylor & Francis Group, London, ISBN 90 5809 653 X

Potential ammonia oxidation in a compost reactor treating source sorted organic household waste

Å. Jarvis, I. Nyberg, K. Steger & M. Pell
Department of Microbiology, SLU, Uppsala, Sweden

ABSTRACT: A method to measure potential ammonia oxidation (PAO) in soil was adapted to suit compost samples from a lab scale compost reactor treating source sorted organic household waste. PAO increased almost fourfold during the first eight days of composting, indicating the establishment of an abundant and potentially active population of ammonia oxidizing bacteria despite the thermophilic (55°C) conditions.

1 INTRODUCTION

Nitrification is an important process responsible for nitrogen cycling in many natural environments such as soil, sediments and organic waste treatment processes. In the first step of nitrification, ammonia (NH_3) is used as a substrate and oxidized to NO_2^- by ammonia oxidizing bacteria (AOB). The next step is the further oxidation of NO_2^- to NO_3^- by another group of bacteria, the nitrite oxidizers. During the first step of nitrification the strictly aerobic AOB may emit nitrous oxide (N_2O) as a by-product, especially at low oxygen pressures (Blackmer et al. 1980). This is of concern, since N_2O is a potential greenhouse gas and a contributor to the breakdown of the ozone layer in the stratosphere. Since these bacteria use NH_3 as a substrate, they are also involved in the regulation of NH_3 emissions from different environments. This emission represents not only a loss of nitrogen, but also an environmental hazard due to eutrophication and acidification.

Composting is an aerobic self-heated biological process for treatment and stabilization of different kinds of organic wastes. Many substrates are rich in nitrogen, and nitrification probably plays a central role for the transformation of nitrogen during composting. However, although some authors have showed the presence of AOB during composting (Kowalchuk et al. 1999, Tiquia et al. 2002), the process of nitrification and its impact on nitrogen turnover and nitrous gas emissions during composting has not yet been well described. The work presented here aims at gaining further insight into the process of nitrification, in particular the first step where AOB are active, at the different stages and temperatures during composting. A method that measures potential ammonia oxidation (PAO), initially developed for soil (Torstensson 1993, ISO 2001), was tested for this purpose.

2 MATERIALS AND METHODS

The short-incubation PAO method was adapted to suit compost samples from a lab scale (200 L) compost reactor fed with source sorted organic household waste mixed with straw. The temperature of the reactor was regulated at a maximum of 55°C, followed by a cooling phase down to 30°C (Smårs et al. 2001). The compost batch, which was well aerated at 16% oxygen pressure, was turned every day and run for 58 days.

PAO was determined in samples collected in connection to the daily turning of the reactor at different times during the process by measuring the production of NO_2^- under standardized conditions. Chlorate was added to inhibit further oxidation to NO_3^-. Triplicates of 5 g of compost were incubated in flasks containing 100 ml phosphate buffer amended with ammonium (NH_4^+) and chlorate on a rotary shaker, and 2 ml of the slurry was sampled on five occasions during the 180 min incubation period for further spectrophotometrically analysis with flow injection analysis (FIA). The linear production of NO_2^- was then used to calculate PAO. This activity test reflects the total amount of AOB present and potentially active at a certain moment in the compost. Since this method had not been applied to compost before, some initial experiments were carried out. Thus, the absorbance of the compost extract was measured to make sure that the absorbance peak would not interfere with the FIA measurements. In addition, we searched for the optimal concentration of phosphate buffer needed to neutralize pH during PAO measurements in the compost. The concentration of NH_4^+ needed to saturate the enzymes of AOB was also determined. All PAO measurements were carried out at 25°C.

3 RESULTS

3.1 *PAO optimization*

The absorbance peak of the compost extract appeared at 214 nm, whereas the interval for measuring NO_2^- with FIA was 540–720 nm. Thus, the absorbance of the extract would not interfere with the measurements. The concentration of NH_4^+ needed to saturate the enzymes was in the interval 100–500 mg/L (not shown), and a concentration of 200 mg/L was chosen for the experiments. PAO increased with increasing buffer concentration, at least significantly in the day 0 samples (Figure 1).

Table 1 also clearly shows the effect of increasing buffer concentration on pH during the FIA experiments. Thus, as a result of this test, we used a buffer concentration of 50 mM in the following PAO assays.

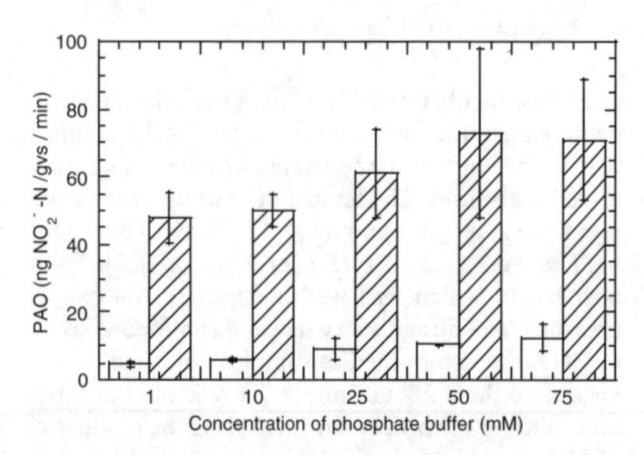

Figure 1. PAO (mean and standard deviation, n = 3) in samples taken from a compost reactor at day 0 (empty bars) and day 58 (filled bars) of composting.

Table 1. pH in slurry samples with different concentrations of phosphate buffer at the start (0 min) and end (180 min) of the PAO measurement.

	Phosphate buffer				
Day	1 mM	10 mM	25 mM	50 mM	75 mM
0 start	5.9	7.0	7.1	7.2	7.2
0 end	5.8	6.9	6.9	7.1	7.2
58 start	8.2	7.6	7.4	7.3	7.3
58 end	8.4	7.9	7.4	7.4	7.5

3.2 *Reactor experiment*

PAO increased almost fourfold during the first eight days of composting. After this, PAO showed a tendency to increase further, although this was not statistically significant (Figure 2).

4 DISCUSSION

The results from this study show that the PAO method originally designed for soil also seems suitable for determination of nitrification potential at different times and stages of the composting process. However, a stronger buffer (50 mM) seems to be needed to neutralize pH during FIA measurements of compost compared to the buffer (1 mM) recommended for soil by Torstensson (1993). This is not surprising, since compost probably has a strong buffering capacity in itself, due to its large content of organic matter (Brady 1984).

The significant increase in PAO during the initial thermophilic phase of composting indicated a population increase of AOB and/or a stimulated enzyme activity of the existing population. Thus, despite the high temperature, these bacteria, which are normally defined as mesophilic, slow-growing and sensitive to disturbances, seemed to thrive in the initial stage of composting. Indeed, DNA from nitrifiers has been found during high temperature (75°C) composting (Kowalchuk et al. 1999). It may be that AOB colonies are well protected, and thus potentially active, in the largely aggregated compost material. The pH increase during the initial phase of composting (pH 5.6 day 0 to pH 8.5 day 3; not shown) probably also enhanced the enzyme activity of AOB. The stimulating effect on AOB seemed to remain, since they were potentially active throughout the 58 days operation period. PAO

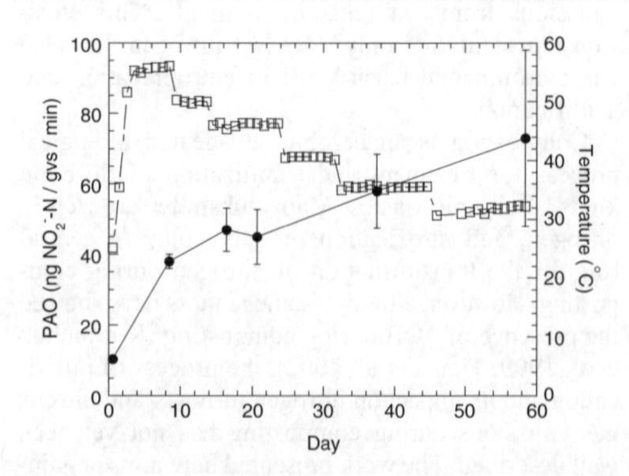

Figure 2. PAO (•) and regulated temperature (□) in a compost reactor run for 58 days with source sorted organic household waste as a substrate.

during the cooling phase, expressed as ng NO_2^- – N/gvs/min, was within the range of the average PAO measured in a number of Swedish soils (Stenberg et al. 1998), indicating the establishment of an abundant population of AOB in the compost.

5 CONCLUSION

The PAO method can be successfully used to determine the potential activity of ammonia oxidizers in compost.

AOB seem not only to survive but also to increase in number and/or specific activity during thermophilic composting conditions.

ACKNOWLEDGEMENT

This work was financed by the "Oscar and Lili Lamm Foundation", Uppsala, Sweden.

REFERENCES

Blackmer, A., Bremner, J. & Schmidt, E. 1980. Production of nitrous oxide by ammonia-oxidizing chemoautotrophic microorganisms in soil. *Appl. Environ. Microbiol.* 40(6): 1060–1066.

Brady, N.C. 1984. *The nature and properties of soils.* New Yourk: Macmillan Publishing Company.

ISO/DIS 15685. 2001. Soil quality – Determination of potential nitrification – Rapid test by ammonium oxidation.

Kowalchuk, G.A., Naoumenko, Z.S., Derikx, P.J.T., Felske, A., Stephen, J.R. & Arkhipchenko, I.A. 1999. Molecular analysis of ammonia-oxidizing bacteria of the ß subdivision of the class *Proteobacteria* in compost and composted materials. *Appl. Environ. Microbiol.* 65(2): 396–403.

Smårs, S., Beck-Friis, B., Jönsson, H. & Kirchmann, H. 2001. An advanced experimental composting reactor for systematic simulation studies. *J. Agric. Engng. Res.* 78: 415–422.

Stenberg, B., Pell, M. & Torstensson, L. 1998. Integrated evaluation of variation in biological, chemical and physical soil properties. *Ambio* 27: 9–15.

Tiquia, S., Wan, J. & Tam, N. 2002. Microbial population dynamics and enzyme activities during composting. *Compost Sci. Utilization* 10: 150–161.

Torstensson, L. 1993. Guidelines – Soil biological variables in environmental hazard assessment. *Swedish Environmental Protection Agency*. Report no 4262, 168 pp.

European Symposium on Environmental Biotechnology, ESEB 2004 - Verstraete (ed)

COD and colour removal of pulp and paper mill effluent by adsorption on formaldehyde treated Keekar bark (*Acacia nilotica*)

A. Kumar & M. Bajaj
Indian Institute of Technology, Delhi, New-Delhi, India

A. Gupta & T. Mangilal
Dept. of Environmental Sciences and Engineering GJ, University Hisar, Haryana, India

ABSTRACT: In the present investigation chemical oxygen demand and color removal of pulp and paper mill effluent, by adsorption on formaldehyde treated activated keekar bark was studied at varying the pollutants concentrations, adsorbents dose, pH, and agitation time in a batch reactor. The activated keekar bark was prepared thermally in size 0.3 mm. The COD reduction was found maximum at alkaline pH (10), where as color removal was observed at a low pH (2) with a adsorbent dose of 20 mg/L. The amount of pollutants adsorbed increased with increase in dose of both adsorbents and their agitation time, higher reduction of COD and color removal was possible provided the low initial concentration. Thus it is proposed that AKB can be potential adsorbents for pollutants removal from the pulp and paper mill effluent and easily available in country side at low cost.

1 INTRODUCTION

The major pollutants from pulp and paper mill are suspended solids, chlorinated organics, BOD, COD, color and other chemicals those impart color to the effluent. In India more than 55% of the pulp and paper industries do not have any type of effluent treatment plant and around 20% have partial treatment facilities. All of these mill drain off black liquor with out recovery/treatment of effluent due to economic reason (Srivastava *et al.*, 1994). This effluent is high in lignin content that is biodegradable even after long exposure to biological process. The color in effluent is due to lignin and its derivatives (Sharma, 1991). The color in the objectionable from aesthetic point of view, several treatment technologies are available for the treatment of pulp and paper mill effluent but suffer from one or other shortcomings, adsorption has been found to be a promising technique as it provides high efficiency in retaining organics contaminants from effluent which are main culprit for high pollution load and high color in pulp and paper mill effluent and creates less pollution etc. although activated carbon is very efficient in reduction of COD and Color, but very expensive and some times low availability, their use is not feasible it should be. So there is an urgent need of hrs to develop an easily available low cost adsorbent. The activated Keekar bark can be used as low cost adsorbent as an alternative to presently available adsorbents, since it is widely available in roadside in India.

2 METHODS

2.1 *Preparation of formaldehyde treated keekar bark*

The bark of keekar (A*cacia nilotica*) was obtained from the keekar plants subjected to thermal activation at a temp of 600°C. In muffle furnace for 30 minutes and grounded to fine powder to immobilize the color and water-soluble substances the ground powder was treated with 1% formaldehyde in ratio of 1:5 (KB: formaldehyde, w/v) at 50°C for 4 hrs. The powder was filtered out, washed with double distilled water to remove free formaldehyde and activated at 105.5°C for 24 hrs. The resulting material was ground followed by sieving in the size of 0.3 mm ASTM. The material was placed in air tight container for further use.

2.2 *Sample collection and characterization*

Sample was collected from Balarpur Industries limited Yamuna nagar, Haryana, India. All the necessary steps were taken during and after the collection of sample, and characterized for physico-chemical parameters, pH, EC, TDS, TSS, Total alkalinity, Total hardness,

Table 1. Characteristics of pulp and paper mill effluent.

Parameters	Characteristics
Color	1210
TDS (mg/L)	2310
TSS (mg/L)	1210
pH	5.89
Total alkalinity (mg/L)	205
Total hardness (mg/L)	412
BOD (mg/L)	2800
COD (mg/L)	9428
EC (mmho/cm)	0.47

BOD, COD and Color (Table 1). Analysis of the relevant parameters was done according to APHA, 1989.

2.3 *Experimental method for adsorption*

In each adsorption experiment known concentration of COD, Color, adsorbent was taken in a 150 ml round bottomed reactor at room temperature (26.1°C) and the mixture was stirred on a rotary orbital shaker at 160 rpm. The sample was withdrawn from the shaker at the pre determined time intervals, and adsorbent was removed from the solution by centrifugation at 450 rpm for 5 minutes. The residual concentration of COD and Color in supernatant was estimated. The experiments were done by varying the amount of adsorbent (10–60 mg/L), contact time (15–90 minutes), pH (2–12).

3 RESULTS AND DISCUSSION

3.1 *Effect of pH on COD reduction*

Maximum reduction of COD (36.1%) was found at pH 10, which decreased with decrease in pH. It also observed at an adsorbent dose of 20 gm/lit and at initial COD Concentration of 1440 mg/L (fig 1). Similar pattern of COD reduction with the increase in pH was observed by Singh and Srivastava (1999) and Sharma *et al.* (1999) when they used activated bagasse carbon and cited that increase in COD reduction at higher pH is due to greater Association of cation with the negatively charged adsorbents. Similar results were also obtained in the treatment of paper mill effluent by Activated Bagasse, Activated Jute Carbon and Fly ash (Kumar *et al.*, 2000). The sharp decrease in the reduction efficiency at a lower pH may be attributed due to inhibition of dissociation of the phenolic compound derived from the lignin to phenoxylate ions that binds themselves to the substrate surface (Das and Patnaik, 2001)

3.2 *Effect of pH on colour removal*

Activated Keekar Bark showed a maximum of colour reduction of 95.36% with an adsorbent dose of

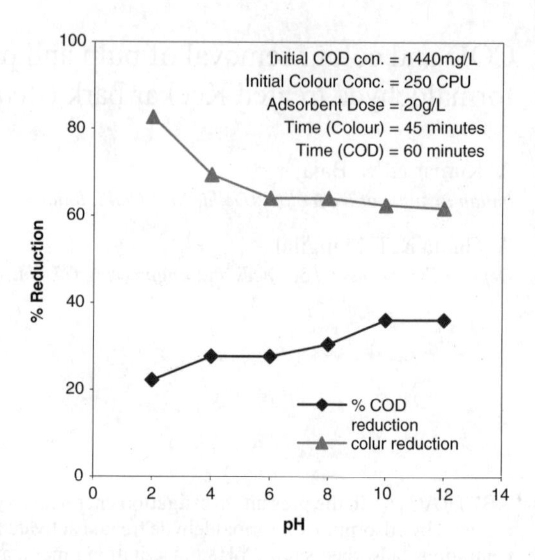

Figure 1. Effect of pH on COD and color removal.

20 mg/L at a pH of 2 and at initial colour concentration of 313.3 CPU (fig 1). The decrease in colour reduction at a higher pH may be due to solubility of colour causing species and the abundance OH^- ions, thereby, decreasing hindrance to diffusion of materials responsible for imparting colour (Boehm, 1996). Singh (2000) used different adsorbents viz, Activated Alumina, Activated saw dust and activated coconut Jute carbon for the reduction of colour from pulp and paper mill effluent and observed a maximum adsorption efficiency at low pH which decreased with increase in pH for all the adsorbents studied.

3.3 *Effect of AKB dose on COD and colour removal*

Adsorbent dose had a very profound effect on the reduction on COD and colour both. An increase in percent COD reduction from 22.2% to 64.4% was observed with increase in dose from 10 mg/L to 60 mg/L of AKB (fig 2). Since a significant increase in percent reduction of COD beyond 30 mg/L is not observed and beyond it may result in the production of more sludge, adsorbent dose of 30 mg/L was selected as optimum dose for further studies. Percent reduction of colour was also observed to increase from 49.6% to 85% with increasing dose from 4 gm/L to 20 gm/L of AKBC (fig 3) but it was observed that increase in percent reduction beyond 12 gm/L, so it was found as optimum dose.

The phenomenon of increases in the percent CODS and Colour reduction with increase in Adsorbent dose up to certain level and beyond that more or less constant reduction may be explained as, with the increase

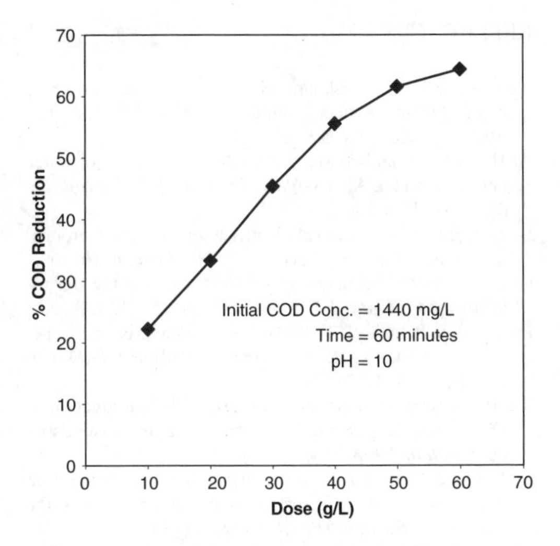

Figure 2. Effect of dose on COD reduction by AKB.

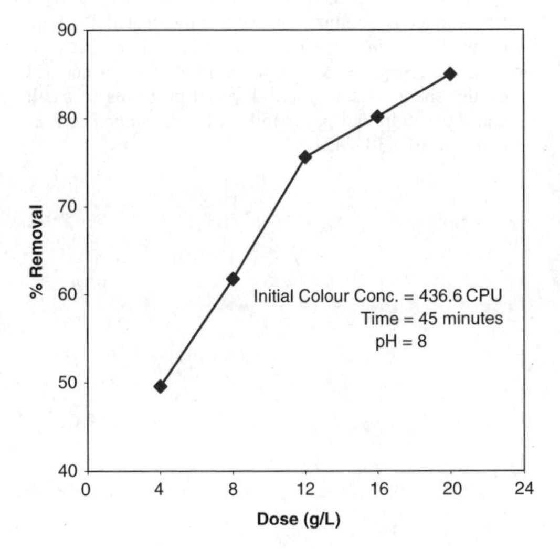

Figure 3. Effect of dose on color removal.

in adsorbent dose, more and more adsorbent surface becomes available for the solutes to adsorb and this increase in reduction beyond an optimum dose may be attributed to the attainment of equilibrium between adsorbent and adsorbate at the existing operating conditions (Kumar, 2000). Mall and Prasad (1998) also observed the increase in the COD reduction with an increase unto 15 mg/L on treating pulp and paper mill effluent with pyrolysed Bagasse Char. Singh (2000) also observed increase in reduction capacity of both COD and Colour with increase in dose of the adsorbent (activated Alumina, Activated Saw Dust, Activated Coconut jute carbon).

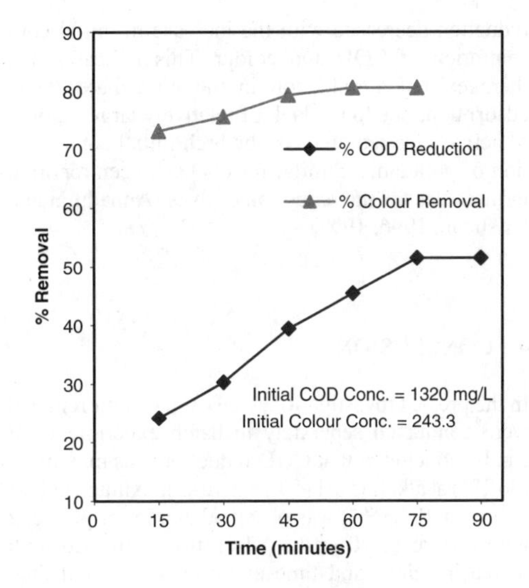

Figure 4. Effect of time on COD and color removal.

3.4 *Effect of time on % COD and colour removal*

A maximum of 51.5% of COD reduction with AKB was observed at pH 10 and adsorbent dose of 30 mg/L from the effluent having an initial COD concentration of 1320 mg/L in 75 minutes of contact time (fig 4). In case of reduction percent of colour was also increased with increase in contact time and found to be maximum in 60 minutes, 80.5% with initial color concentration of 273.3 CPU at pH 8 and adsorbent dose of 12 mg/L. The least color reduction under the same condition was observed to be 73.1% in 15 minutes. Because initially a large number of vacant surface sites may be available for the adsorption and after sometimes, the remaining vacant surface sites may be difficult to occupy due to repulsive forces between the solute molecules of the solids and bulk phase (Chand, 1999; Viswanathan *et al.*, 2000).

3.5 *Effect of initial COD and colour concentration on % removal*

Percent COD reduction was found to decrease with the increase in initial COD concentration by the AKB adsorbent was 75% on initial COD concentration of 640 mg/L with adsorbent dose of 30 mg/L and contact time of 60 minutes. Under identical operating condition it was decreased to 24.2%. Similar results were observed for the reduction of colour by the AKB 99.4% at an initial carbon concentration of 730 CPU, but at a slight increase in reduction percent from 72.3 to 73.3 was observed when the initial colour concentration was increased from 730 CPU to 863.3 CPU. The percent reduction decreased with the increase in initial concentration of COD and colour. This indicates that there existed a reduction in the immediate solute adsorption, due to the lack of relatively large number of active sites required for the high initial concentration of pollutants. Similar results have been reported in literature for the reduction of dyes (Annadurai and Krishnan, 1996, 1997).

4 CONCLUSION

In the present investigation, COD and Colour removal were conducted separately in batch experiments. It can be concluded that COD reduction was maximum (36.1%) at alkaline pH (10), where as maximum colour reduction (95.36%) was observed at a lower pH i.e. 2 with a dose of 20 gm/L. Adsorption increased with increasing dose and time at initial stages and then it becomes somewhat constant due to attainment of equilibrium. Higher removal of COD and Colour was possible provided the initial concentration. AKB is easily available in countryside, so can be used by Industries having low concentration of COD and colour in the wastewater using batch or stirred tank reactors after treating it with formaldehyde.

REFERENCES

Annadurai, G. & Krishnan, M.R.V. 1997. Adsorption of acidic dye from aqueous solution by Chitin-A Equilibrium Studies. *Indian. J. Chem. Tech*. 4: 217–222.

APHA, 1989. Standard Method for the Examination of water and wastewater, APHA-AWWTA, *Wat. Poll. Control. Fe. NewYork*, 17th Ed.

Boehm, H.P. 1996. Chemical Identification of Surface groups. *Advances in Catalysis. Acedemic Press., London*, 16: 169.

Chand, S. 1999. Reduction of Cr (VI) from wastewater by adsorption. *Indian. J. Env. Hlth*. 38 (3): 151–158.

Das, C.P. & Patnaik, L.N. 2001. Use of industrial waste for the reduction of COD from paper mill effluent. *Indian. J. Env. Hlth*. 43 (1): 21–27.

Kumar, V., Sharma, S. & Maheswari, R.C. 2000. Reduction of COD from the paper mill effluent using low cost adsorbent. *Indian J. Env. Prot*. 20 (2): 91–95.

Mall, I.D. & Prasad, J. 1998. Pyrolysed bagasse chars a low cost adsorbent for effective effluent treatment system for pulp and paper mill. *IPPTA*. 10 (2): 11–19.

Sharma, N. 1991. Color reduction in pulp mill waste. *Indian J. Env. Port*. 11 (9): 675–679.

Singh, D. & Srivastava, K. 1999. Reduction of basic dyes from aqueous solution by chemically treated Psidium Guava leaves. *Indian J. Env. helth*. 35 (3): 169–177.

Srivastava, S., Singh, K. & Sharma, A. 1994. Physicochemical on the characterization and disposal problems of small and large pulp and paper mill effluents. *Indian J. Env. Prot*. 10 (6): 438–442.

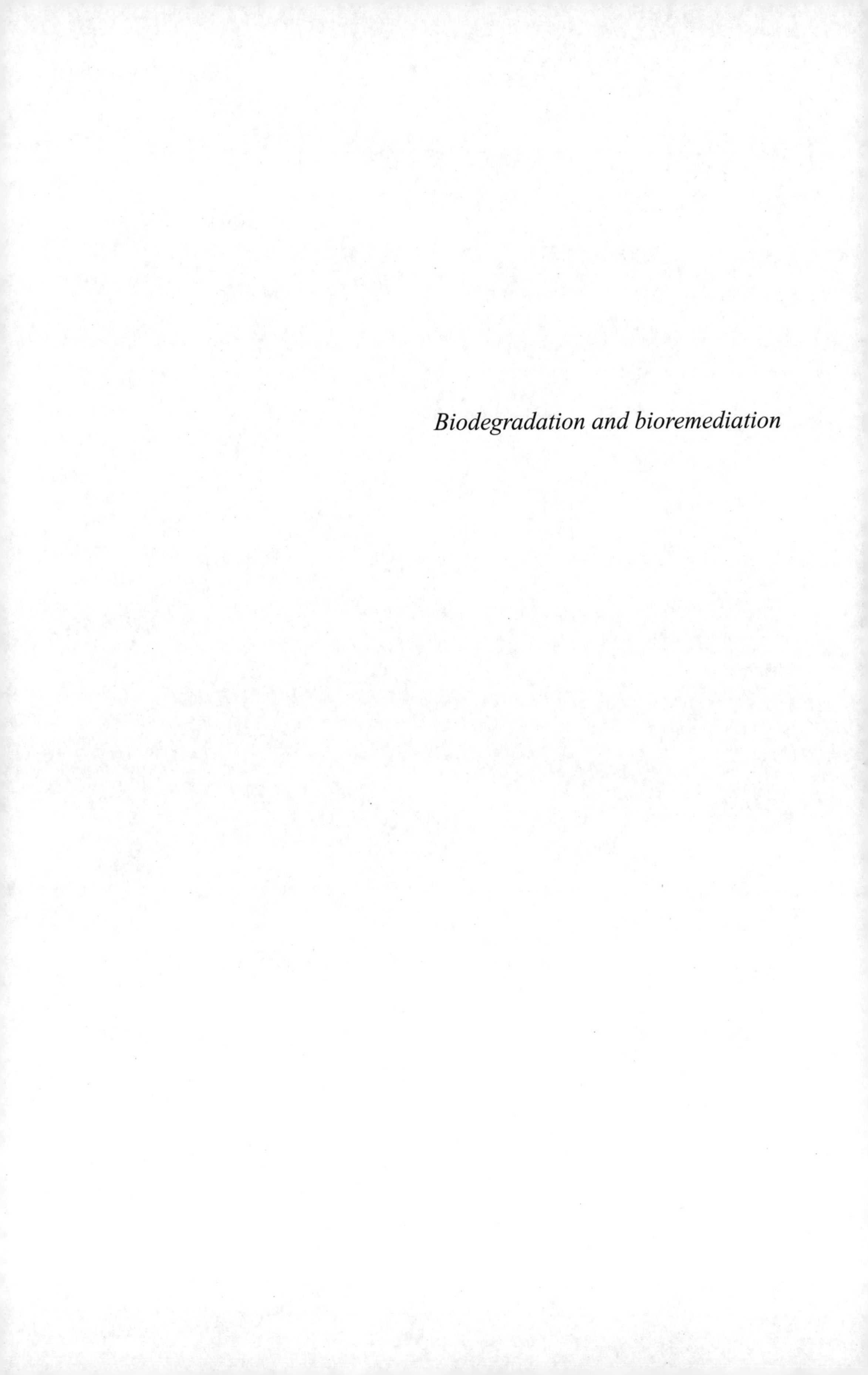

Biodegradation and bioremediation

European Symposium on Environmental Biotechnology, ESEB 2004 - Verstraete (ed)
© 2004 Taylor & Francis Group, London, ISBN 90 5809 653 X

Anaerobic 1,3-propanediol dissimilation by *Desulfovibrio alcoholivorans* and other *Desulfovibrio* species

A.I. Qatibi & R. Bennisse
Microbiological Engineering group, Sciences & Techniques Faculty, Cady Ayyad University, Marakesch, Morocco

C. Rodrigues & M.A.M. Reis
Department of Chemistru-CQFB, Faculdade de Ciencias e Tecnologia, Universidade Nova de Lisboa, Quinta da Torre, Caparica, Portugal

M. Labat
Post Harvest Microbial Biotechnology, Microbiology Laboratory, Universités de Provence et de la Méditerranée, Marseille, France

ABSTRACT: The specific growth rate of *Desulfovibrio alcoholivorans*, *D. carbinolicus* and *D. fructosivorans* on 1,3-propanediol in the presence of sulfate was affected by the soluble sulfides produced from sulfate reduction and the pH, which however did not change the products of oxidation, attesting presumably of diversity in the metabolism of these species. A comparative biochemical study with cell-free extracts of the three species showed that NAD-dependent 1,3-propanediol dehydrogenase played a key role in the catabolism of 1,3-propanediol. In contrast to *D. alcoholivorans*, no 3-hydroxypropionate dehydrogenase activity was detected in cell-free extracts of *D. carbinolicus* or *D. fructosivorans*, suggesting that 3-hydroxypropionate is an intermediate in the degradation of 1,3-propanediol into acetate only in *D. alcoholivorans*.

1 INTRODUCTION

The utilization of 1,3-propanediol by sulfate-reducing bacteria (SRB) in the presence of sulfate was conclusively demonstrated during the anaerobic degradation of wastewater rich in glycerol and sulfate produced by bioethanol-producing plants (Qatibi et al. 2001). Several strains of *Desulfovibrio* able to grow on 1,3-propanediol were isolated and included *Desulfovibrio alcoholivorans*, *Desulfovibrio* sp. DFG and *Desulfovibrio* sp. dQ. Both oxidize propanediol to form acetate, a property shared by other strains such as *Desulfovibrio* sp. OttPd1 (Oppenberg & Schink 1990). Some strains are also able to oxidize propanediol to 3-hydroxypropionate, e.g. *D. carbinolicus* (Nanninga & Gottschal 1987), *D. fructosivorans* (Ollivier et al. 1988) and *Desulfovibrio* sp. IsBd1 (Tanaka 1992). Other strains produce a mixture of acetate and 3-hydroxypropionate from propadiol, e.g. *Desulfovibrio vulgaris* strain Marburg (Tanaka 1990) and *Desulfovibrio burkinensis* (Ouattara et al. 1999). The results reported suggest that 1,3-propanediol degradation is species specific and/or depends on the experimental conditions. Recent experiments suggest that soluble sulfides and pH affect the growth of SRB (Reis et al. 1991a, 1991b, 1992).

The objective of the study reported here was to evaluate the effect of certain operational parameters such as the soluble sulfides produced from sulfate reduction and the pH on the anaerobic degradation of 1,3-propanediol by three species of genus *Desulfovibrio* (*D. alcoholivorans*, *D. carbinolicus*, and *D. fructosivorans*), and to compare the enzymes and probable pathways involved. The information obtained would help clarify how 1,3-propanediol is consumed by SRB and thus contribute to a better understanding of the anaerobic digestion of wastewater from bioethanol-producing plants rich in sulfate.

2 MATERIALS AND METHODS

2.1 *Media*

Use was made of the bicarbonate-buffered mineral medium described by Qatibi et al. (1991) except that the dithionite was omitted and 0.01% (w/v) yeast extract was added. After autoclaving at 120°C for 20 min, the medium was cooled under a stream of O_2-free Ar. $Na_2S \cdot 9H_2O$ (1.5 mM) and $NaHCO_3$ (30 mM) from separately sterilized anoxic solutions, and 1 ml l–1 of filter-sterilized vitamin solution (Pfennig 1978) were added. Sulfate and 1,3-propanediol were added from 1 M sterile anaerobic solutions.

2.2 *Reactor and operating procedure*

Batch experiments were conducted in a 0.5-liter reactor with magnetic stirring, automatic pH control, and temperature maintained at 35°C. Reactors were inoculated (10%, v/v) from stock cultures *of D. alcoholivorans* (DSM 5433), *D. carbinolicus* (DSM 3852), and *D. fructosivorans* (DSM 3604). Stock cultures were grown on 1,3-propanediol (20 mM) and sulfate (20 mM) medium at 35°C in 100-ml serum bottles containing 60 ml of medium under an atmosphere of pure Ar. Three series of experiments were conducted in batch conditions with sulfate to (1) compare the anaerobic 1,3-propanediol degradation by the three *Desulfovibrio* species at a constant pH (7.5), (2) check the effect of soluble sulfides ($pKa_{H_2S} = 6.80$) on specific growth rate and profile of end products at a constant pH (7.5), and (3) evaluate the effect of pH (6.2 and 7.5) on specific growth rate and end products from anaerobic 1,3-propanediol degradation.

2.3 *Analytical methods*

Samples to measure specific growth rates (μ_{max}) were collected anaerobically from the fermentors through a port closed with a rubber stopper, and, using sterilized syringes flushed with pure Ar. Culture growth was monitored turbidimetrically at 600 nm. The sulfide content (growth marker) was measured spectrophotometrically as colloidal CuS at 480 nm as described by Cord-Ruwisch (1985). Samples were then centrifuged (10 min at 5,000 × g and 4°C) and filtered prior to sulfate, acetate, 3-hydroxypropionate and 1,3-propanediol analysis. Sulfate concentrations was determined by segmented flow analysis as described by Reis et al. (1991b). Acetate, 3-hydroxypropionate, and 1,3-propanediol were analyzed with a high-performance liquid chromatograph as described by Reis et al. (1992).

2.4 *Preparation of cell-free extracts and enzyme assays*

The cells were harvested by centrifugation (20 min at 8,000 × g and 4°C in an N_2 atmosphere), washed twice with potassium phosphate buffer (50 mM, pH 7.5), suspended in the same buffer containing 2 mM dithiothreitol (DTT), and stored under N_2 at –20°C. Cell extracts and enzyme assays were prepared and conducted as descibed by Hensgens et al. (1995). The supernatants were referred to as cell-free extracts and used for enzyme assays. NAD-dependent 1,3-propanediol dehydrogenase and NAD-independent 1,3-propanediol dehydrogenase (MTT:3-(4′,5′-dimethylthiazol-2-yl)-2,4-diphenyltetrazolium bromide)-linked) were assayed as described by Qatibi et al. (1998) except that 1,3-propanediol (10 mM) was used as substrate. NAD-independent 1,3-propanediol dehydrogenase was also assayed in 50 mM Tris-HCl (pH 7.5) with 1 mM 2,6-dichlorophenolindophenol (DCPIP) as electron acceptor and 1,3-propanediol (10 mM). The reaction was started by adding cell-free extract; the 1,3-propanediol-dependent reduction of DCPIP was recorded at 600 nm ($E600 = 19\,mM^{-1}cm^{-1}$). 3-hydroxypropionate dehydrogenase was assayed in 50 mM Tris-HCl (pH 7.5) containing 1 mM DCPIP and 10 mM 3-hydroxypropionate. The reaction was started by adding cell-free extract; the 3-hydroxypropionate-dependent reduction of DCPIP was recorded at 600 nm. Dye-linked acetaldehyde dehydrogenase was assayed in the presence of CoA (2 mM) in 50 mM Tris-HCl (pH 7.5) containing 5 mM benzylviologen (BV^{2+}) and 5 mM acetaldehyde. The reaction was started by adding cell-free extract; the acetaldehyde-dependent reduction of BV^{2+} to BV^{+} was recorded at 500 nm ($E500 = 7.4\,mM^{-1}cm^{-1}$). The Following enzyme activities were measured according to their respective authors: lactate dehydrogenase (Stams & Hansen 1982), except that we have used Tris-HCl (50 mM, pH 7.5) as buffer and DCPIP (1 mM) as electron acceptor, pyruvate dehydrogenase (Odom & Peck 1981), phosphate acetate transferase and acetate kinase (Oberlies et al. 1980) and NADH-dehydrogenase and NADPH-dehydrogenase (Kremer & Hansen 1987). Enzyme units were defined as μmoles of product formed or substrate consumed per minute ($\mu mol\,min^{-1}$). Protein content was determined by the method of Bradford (1976) with bovine serum albumin as a standard. Experiments were conducted at least in duplicate.

3 RESULTS AND DISCUSSION

The final product formed from the degradation of 1,3-propanediol by *D. alcoholivorans* was mainly acetate, while *D. carbinolicus* and *D. fructosivorans* produced 3-hydroxypropionate. The profile of product formation in these species was unchanged when the pH (Table 1) was decreased from 7.5 to 6.2 and was unaffected by the presence or absence of H_2S (Table 2).

However, it was noted that growth rates were affected by changes in both pH and H_2S. In fact, specific growth rates almost doubled when H_2S was stripped from the medium, confirming that H_2S inhibits growth (Reis et al. 1991b, 1992).

Table 1. Anaerobic 1,3-propanediol degradation by *Desulfovibrio* species at different controlled pH* values.

Species	pH	μ_{max} (h^{-1})	End products (mM)		
			3-OHC3	Acetate	Sulfides
DA	6.2	0.026	0	16	16.4
	7.5	0.044	0	16.6	16.2
DC	6.2	0.005	19	0	9.2
	7.5	0.032	18.2	0	10.2
DF	6.2	0.052	18.2	0	9
	7.5	0.031	19	0	8.6

3-OHC3, 3-hydroxypropionate; DA, *D. alcoholivorans*; DC, *D. carbinolicus*; DF, *D. fructosivorans*; *Batch experiments were performed without stripping in the presence of 20 mM 1,3-propanediol and 20 mM sulfate at the pH values indicated in the table and at 35°C.

Table 2. Anaerobic 1,3-propanediol degradation by *Desulfovibrio* species, in the absence and presence of the soluble sulfides* produced.

Organism	Without stripping*		With stripping	
	μ_{max} (h^{-1})	Products* (mM)	μ_{max} (h^{-1})	Products* (mM)
DA	0.042	Acetate (17.6)	0.067	Acetate (16.4)
		3-OHC3 (0)		3-OHC3 (0)
		Sulfides (18)		Sulfides (<1)
DC	0.032	Acetate (0)	0.071	Acetate (0)
		3-OHC3 (0.87)		3-HC3 (18.2)
		Sulfides (11.4)		Sulfides (<1)
DF	0.030	Acetate (0)	0.075	Acetate (0)
		3-OHC3 (17.8)		3-OHC3 (17.4)
		Sulfides (10)		Sulfides (<1)

*This was done by continuously purging the medium with pure Ar to strip the sulfides produced (Reis et al. 1991a); DA, *Desulfovibrio alcoholivorans*; DC, *Desulfovibrio carbinolicus*; DF, *Desulfovibrio fructosivorans*; 3-OHC3, 3-hydroxypropionate; *Batch experiments were performed in the presence of 20 mM 1,3-propanediol and 20 mM sulfate at constant pH (7.5) and at 35°C.

Table 3. Enzyme activities (μmol min^{-1} mg^{-1} protein) in cell-free extracts of *D. alcoholivorans*, *D. carbinolicus* and *D. fructosivorans* grown on 1,3-OH in the presence of sulfate.

Enzymes	DA	DC	DF
NAD-dependent 1,3-OH DH	0.022	0.015	0.009
DCPIP-dependent 1,3-OH DH	0	0	0
MTT-dependent 1,3-OH DH	0	0	0
DCPIP-dependent 3-HOC3 DH	0.003	0	0
DCPIP-linked lactate DH	0	0	0
BV-linked pyruvate DH	0.038	0.023	0.045
Phosphate acetyl-transferase	0.006	0.004	0.005
Acetate kinase	0.082	0.047	0.052
BV-linked acetaldehyde DH	0.153	nd	nd
MTT-linked NADH DH	0.48	0.32	0.42
MTT-linked NADPH DH	0.28	0.23	0.12

nd, not determined; 1,3-OH, 1,3-propanediol; 3-OHC3, 3-hydroxypropionate; DH, Dehydrogenase; DA, *D. alcoholivorans*; DC, *D. carbinolicus*; DF, *D. fructosivorans.*

The ratio of sulfate consumption to 1,3-propanediol degradation by *D. alcoholivorans* (0.985) was twice that noted for *D. carbinolicus* (0.56) and *D. fructosivorans* (0.41).This stoichiometry was maintained in all the experiments, confirming that the metabolism was unaffected by the pH and the amount of H_2S. The nature of the product formed and reaction as well as the stoichiometry seem to be related to the enzymatic composition of the two bacterial groups. In order to confirm this hypothesis, the activities of the key enzymes involved in the degradation of 1,3-propanediol were measured (Table 3) and this showed that a key role was played by a NAD-dependent 1,3-propanediol dehydrogenase in the growth of *D. alcoholivorans*, *D. carbinolicus,* and *D. fructosivorans* on 1,3-propanediol.

The weak activities observed could explain the relatively low specific growth rates noted for these species. No 1,3-propanediol dehydrogenase activity was detected in cell-free extracts of *D. alcoholivorans* grown on lactate with sulfate. This indicates that the NAD-dependent 1,3-propanediol dehydrogenase measured with *D. alcoholivorans* was probably an inducible enzyme. In contrast to *D. carbinolicus* and *D. fructosivorans*, which produced 3-hydroxypropionate as an end product, *D. alcoholivorans* metabolized the product to acetate (Tables 1 & 2). This result is consistent with the presence of a 3-hydroxypropionate dehydrogenase in *D. alcoholivorans*. But the nature of the enzyme involved in 3-hydroxypropionate dehydrogenation is still unclear in *D. alcoholivorans* and this preliminary evidence for a clear role played by a DCPIP-linked 3-hydroxypropionate requires a more sensitive assay. The relatively high activity of the dye-linked aldehyde dehydrogenase during the weak growth of *D. alcoholivorans* on 1,3-propanediol suggested that it is involved in 1,3-propanediol degradation. Since indications were found that this activity is CoA-dependent, it appears likely that oxidation of acetaldehyde to acetate is coupled with substrate-level phosphorylation in *D. alcoholivorans* (Hansen 1994). A CoA-dependent, benzylviologen-reducing aldehyde dehydrogenase has also been reported for the 1,3-propanediol metabolizing *Desulfovibrio* sp. OttPd1 (Oppenberg & Schink 1990). A hypothetical scheme suggesting 3-hydroxypropionaldehyde, 3-hydroxypropionate, malonylsemialdehyde and acetaldehyde as intermediates in the oxidation of 1,3-propanediol to acetate by *D. alcoholivorans* can be proposed.

The generation of mainly 3-hydroxypropionate as the end product with 3-hydroxypropionaldehyde as the obvious intermediate in 1,3-propanediol degradation by *D. carbinolicus* and *D. fructosivorans* and the absence of a 3-hydroxypropionate dehydrogenase in cell-free extracts of these species, strongly indicate that 1,3-propanediol is first dehydrogenated to 3-hydroxypropionaldehyde which is then oxidized to 3-hydroxypropionate.

ACKNOWLEDGMENTS

The authors are indebted to C.M.H. Hensgens, Department of Microbiology, University of Gröningen, The Netherlands and P.C. Lemos, Department of Chemistry-CQFB, Faculdade de Ciencias e Tecnologia, Universidade Nova de Lisboa, Portugal for helpful discussions and technical assistance. Many thanks to A. El Asli, Al Akhawayn, Ifrane, Morocco for improving the manuscript. A.I. Qatibi's work was supported by International Foundation of Science (IFS), Sweden grants, and by International Foundation of Science and Technology (ICCTI), Portugal and CNCPRST, Morocco program.

REFERENCES

Bradford, M.M. 1976. A rapid and sensitive method for quantification of microgram quantities of protein utilizing the principle of protein-dye binding. *Anal Biochem* 72: 248–254.

Cord-Ruwisch, R. 1985. A quick method for the determination of dissolved and precipitated sulfides in cultures of sulfate-reducing bacteria. *J Microbiol Methods* 4: 33–36.

Hansen, T.A. 1994. Metabolism of sulphate-reducing prokaryotes. *Antonie Leeuwenhoek* 66: 165–185.

Hensgens, C.M.H., Jansen, M., Nienhuis-Kuiper, M.E., Boekema, E.J., Van Breemen, J.F.L. & Hansen, T.A. 1995. Purification and characterization of an alcohol dehydrogenase from 1,2-propanediol-grown *Desulfovibrio* strain HDv. *Arch Microbiol* 164: 265–270.

Kremer, D.R. & Hansen, T.A. 1987. Glycerol and dihydroxyacetone dissimilation in *Desulfovibrio* strains. *Arch Microbiol* 147: 249–256.

Nanninga, H.J. & Gottschal, J.C. 1997. Properties of *Desulfovibrio carbinolicus* sp. nov. and other sulfate-reducing bacteria isolated from an anerobic-purification plant. *Appl Environ Microbiol* 51: 572–579.

Oberlies, G., Fuchs, G. & Thauer, R.K. 1980. Acetate thiokinase and the assimilation of acetate in *Methanobacterium thermoautotrophicum*. *Arch Microbiol* 128: 248–252.

Odom, J.M. & Peck, H.D. 1981. Localization of dehydrogenase, reductases and electron transfer components in the sulfate-reducing bacterium *Desulfovibrio gigas*. *J Bacteriol* 147: 161–169.

Ollivier, B., Cord-Ruwisch, R., Hatchikian, E.C. & Garcia, J.L. 1988. Characterization of *Desulfovibrio fructosovorans* sp. nov. *Arch Microbiol* 149: 447–450.

Oppenberg, B. & Schink, B. 1990. Anaerobic degradation of 1,3-propanediol by sulphate-reducing and by fermenting bacteria. *Antonie Leeuwenhoek* 57: 205–213.

Ouattara, A.S., Patel, B.K.C., Cayol, J.L., Cuzin, N., Traore, A.S. & Garcia, J.L. 1999. Isolation and characterization of *Desulfovibrio burkinensis* sp. nov. from an african ricefield, and phylogeny of *Desulfovibrio alcoholivorans*. *Int J.Syst Bacteriol* 49: 639–643.

Pfennig, N. 1978. *Rhodocyclus purpureus* gen. nov. and sp. nov., a ring-shaped, vitamin B_{12}-requiring member of the family *Rhodospirillaceae*. *Int J Syst. Bacteriol* 23: 283–288.

Qatibi, A.I., Niviere, V. & Garcia, J.L. 1991. *Desulfovibrio alcoholovorans* sp. nov., a sulphate-reducing bacterium able to grow on glycerol, 1,2- and 1,3-propanediol. *Arch Microbiol* 155: 143–148.

Qatibi, A.I., Bennisse, R., Jana, H. & Garcia, J.L. 1998. Anaerobic degradation of glycerol by *Desulfovibrio fructosovorans* and *D. carbinolicus* and evidence for glycerol-dependent utilisation of 1,2-propanediol. *Curr Microbiol* 36: 283–290.

Qatibi, A.I., Bennisse, R. & Jana, M. 2001. Role of Sulfate-Reducing bacteria on anaerobic degradation of industrial waste water from bioethanol production plants. *L'Ea, l'Industrie les Nuisances* 243: 56–60.

Reis, M.A.M., Almeida, J.S., Lemos, P.C. & Carrondo, M.J.T. 1991a. Influence of sulfates and operational parameters on volatile fatty acids concentration profile in acidogenic phase. *Bioprocess Eng* 6: 145–151.

Reis, M.A.M., Almeida, J.S., Lemos, P.C. & Carrondo, M.J.T. 1991b. Evidence for the intrinsic toxicity of H_2S to sulphate-reducing bacteria. *Appl Microbiol Biotechnol* 36: 145–147.

Reis, M.A.M., Almeida, J.S., Lemos, P.C. & Carrondo, M.J.T. 1992. Effect of hydrogen sulfide on growth of sulfate reducing-bacteria. *Biotechnol Bioeng* 40: 593–600.

Stams, A.J.M. & Hansen, T.A. 1982. Oxygen labile L(+) lactate dehydrogenase in *Desulfovibrio desulfuricans*. *FEMS Microbiol Lett* 13: 384–394.

Tanaka, K. 1990. Several new substrates for *Desulfovibrio vulgaris* strain Marburg and a spontaneous mutant from it. *Arch Microbiol* 155: 18–21.

Tanaka, K. 1992. Anaerobic oxidation of 1,5-pentanediol, 2-butanol, and 2-propanol by a newly isolated sulphate-reducer. *J Ferment Bioeng* 73: 362–365.

European Symposium on Environmental Biotechnology, ESEB 2004 - Verstraete (ed)
© 2004 Taylor & Francis Group, London, ISBN 90 5809 653 X

Effect of MTBE on growth and metabolism of degrading microorganisms

J. Pazlarova, M. Vosahlikova & K. Demnerova
Department of Biochemistry and Microbiology, Institute of Chemical Technology in Prague, Prague, Czech Republic

ABSTRACT: Methyl *tert*-butyl ether (MTBE) is a synthetic compound that was developed as a technological solution to a technology-derived problem created by air pollution from vehicle emissions. Because of its undesirable effects on drinking water and ecologically harmful effects, MTBE removal has become a public health and environmental concern. Potential method used for MTBE removal from soils and aquifers is microbial decomposition. Genotoxicity of MTBE was monitored by Ames test with *Salmonella typhimurium* his^- (TA98, TA 100, YG1041, YG1042) as indicator strains. Concentrations of MTBE/plate tested were 3.0 mg and 1.5 mg. Obtained results did not prove the MTBE genotoxicity. Among the MTBE aerobic degradation products of indigenous bacterial strains *tert*-butyl alcohol (TBA) was found using GC. The genotoxicity of TBA was tested using the identical system. All so far measured data indicate that MTBE will be cometabolically oxidized by monooxygenase (MO) enzyme activity that is induced by oxidation of n-alkanes under aerobic conditions. The MO enzyme requires molecular oxygen to oxidize the target chemical; this process is called β-oxidation and is common mechanism for n-alkane biodegradation. Its role in MTBE biodegradation is evaluated.

1 INTRODUCTION

Methyl *tertiary*-butyl ether (MTBE) is a synthetic compound that was developed as technological solution to a technology-derived problem created by air pollution from vehicle emissions. Usual concentration of MTBE in gasoline is about 5–10% vol. The risk has not been well defined owing to a lack of conclusive studies regarding the human health risk of MTBE.

In USA the magnitude and remediation cost of methyl *tertiary*-butyl ether MTBE contamination in drinking water has rapidly become a national concern, Kane et al. 2001. The U.S. Environmental protection Agency has listed MTBE as a possible human carcinogen, whereas TBA is a known animal carcinogen, Cirvello et al. 1995.

In Europe since the 1990 Clean Air Act, MTBE has been used commonly as a gasoline oxygenate to increase fuel efficiency and lower vehicle emissions. It has become one of the most common pollutants of ground water and surface water.

Biodegradation of MTBE can offer an efficient and low-cost method of contaminated water treatment. As a group, alkyl ethers including MTBE are chemically stable and little information on their biodegradation is available. Salanitro et al. 1994 study was one of the papers dealing this task. They described a mixed microbial culture apparently growing on MTBE as its sole carbon source in continuous culture. Data from Hyman 1998 indicated that MTBE would be cometabolically oxidized by monooxygenase (MO) enzyme activity which is induced by oxidation of n-alkanes under aerobic conditions. The MO enzyme requires molecular oxygen to oxidize the target chemical; this process is called β-oxidation and is common mechanism for n-alkane biodegradation. Despite success with n-alkanes, Hyman 1998 noted that the presence of toluene inhibited biodegradation. There was a speculation that this is a result of toluene dioxygenase production for BTEX (benzene, toluene, ethylbenzene, and xylene) degradation, which is not ultimately used in the biodegradation mechanism for MTBE. However Hyman et al. 2001 observed cometabolic MTBE degradation with microbes grown on *ortho*-xylene and benzene. While biodegradation of these aromatics could be the result of MO enzyme, these results also suggest that a dioxygenase enzyme could be involved.

In summary, while the successful aerobic degradation of MTBE appears to be the result of a MO enzyme, there are likely other enzymes that significantly contribute. According to Deeb et al. 2000 several cultures from diverse environment have been shown to efficiently degrade MTBE in shake flask experiments under controlled laboratory conditions. This paper describes the MTBE genotoxicity and a part played by different types of cosubstrates in the process of its biodegradation by indigenous microorganisms.

2 MATERIALS AND METHODS

2.1 *Bacterial strains*

Pseudomonas C12B (NCIMB 11753) described by Payne & Feisal 1963 and *Pseudomonas* P2, Pazlarova et al. 1997 were used for growth experiments. Our own bacterial isolates from contaminated soil designated as 3A, 4A, and 6A were used for measurement of the specific growth rate (μ). Two cocultures named "Landa" and "Polepy" were used throughout.

2.2 *Ames test*

The Ames test employs several strains of *Salmonella typhimurium* which have been selected based on their sensitivity to mutation of: (1) an increased cell wall permeability, due to mutation which causes partial loss of the lipopolysaccharide barrier that coats the bacterial surface and results in increased permeability to large molecules (rfa mutation); (2) a mutation in the bacterial cell system to excise and repair defects in the DNA, resulting in the inability to repair damaged or mutated sections (uvrB mutation); and (3) R-factor plasmids (some strains) and a multicopy plasmid (some strains) which contain error-prone DNA repair systems. When the *Salmonella* tester strains are grown on a minimal media agar plate containing a trace of histidine, only those bacteria that revert to histidine (*his*$^-$) are able to form colonies. The tester strains employed were YG1041, YG1042, TA98 and TA100. YG strains were selected from TA strains and have enlarged enzymatic equipment enabling them to metabolize tested substances in a different way than TA strains and can be more sensitive. The test was performed both with and without liver homogenate S-9 activation. The S-9 activation system is designed to simulate mammalian liver enzyme systems and is used to detect substances which undergo metabolic activation from non-mutagenic forms, Mortelsmans & Zeiger 2000.

2.3 *Measurement of the growth by Bioscreen®*

Bioscreen® was used to measure the growth rate of tested bacterial strains on different types of substrates. Optical density (OD_{400nm} 22°C) is monitored in regular intervals in each well of plate. From these data the respective growth rates were calculated. In all experiments a simple mineral medium ABC was used (10 ml of a solution **A**, 0.1 ml of a solution **B**, and 0.1 ml solution **C** is added to 100 ml distilled H_2O). **A**: $K_2HPO_4.7\ H_2O$ 7 g, KH_2PO_4 3 g, NH_4Cl 1 g, NaCl 1 g in 800 ml. **B**: Na_2SO_4 4.2 mg in 30 ml. **C**: $MgCl_2.6H_2O$ in 30 ml H_2O. Beside MTBE following compounds were examined as carbon source: n-alkanes mixture, C10–C14 (Al), toluene (T), acetone (Ac), cyclohexane (C), and lactate (L).

3 RESULTS AND DISCUSSION

3.1 *Measurement of genotoxicity*

To measure the genotoxicity of MTBE two variants of Ames test were employed, classic plate test and test with preincubation. In these tests two solvents were used: H_2O and DMSO. Tables summarize measured results.

When DMSO (dimethyl sulfoxide) was used as 3 mg MTBE per plat solvent the resulting mixture

Table 1. MTBE solved in DMSO mutagenity – preincubation test.

		Concentration of MTBE		
Indicator strains	Mutation type	3 mg/pl.	1.5 mg/pl.	1.5 mg/pl. + S9 mix
TA98	shift-1	tox	no	no
TA100	AT/GC tr.	tox	no	no
YG1041	shift-1	no	no	/
YG1042	AT/GC tr.	tox	no	/

pl. – plate, tr. – transition, / – not done.

Table 2. MTBE solved in H_2O mutagenity – preincubation test source.

		Concentration of MTBE	
Indicator strains	Mutation type	3 mg/pl.	1.5 mg/pl. + S9 mix
TA98	shift-1	no	no
TA100	AT/GC tr.	no	no
YG1041	shift-1	no	/
YG1042	AT/GC tr.	no	/

pl. – plate, tr. – transition, / – not done.

Table 3. MTBE solved in DMSO mutagenity – by classic plate test.

		Concentration of MTBE	
Indicator strains	Mutation type	3 mg/pl.	1.5 mg/pl. + S9 mix
TA98	shift-1	no	no
TA100	AT/GC tr.	no	no
YG1041	shift-1	no	/
YG1042	AT/GC tr.	no	/

pl. – plate, tr. – transition, / – not done.

Table 4. MTBE solved in H_2O mutagenity – by classic plate test.

Indicator strains	Mutation type	Concentration of MTBE	
		3 mg/pl.	1.5 mg/pl. + S9 mix
TA98	shift-1	no	no
TA100	AT/GC tr.	no	no
YG1041	shift-1	no	/
YG1042	AT/GC tr.	no	/

pl. – plate, tr. – transition, / – not done.

was so toxic, that killed all tested bacteria. From this reason was also tested lower concentration (1.5 mg per plate).

Thorough analysis of MTBE genotoxicity revealed that it is harmless. This finding is rather comforting, but the presence of MTBE even in very low doses spoiles the quality of drinking water. The treshhold of tolerance is individual, and is oscillating about 20 ppm. These results will be compared with identically measurement of TBA genotoxicity.

3.2 *Effect of MTBE on growth*

Bioscreen® was used to measure the growth rate of bacterial strains that previously proved the ability to utilize n-alkanes. Both *Pseudomonas* C12B and Pseudomonas P2 possesed these qualities. Coculture "Landa" was used for clean up of oil contamination. Table 5 summarizes the specific growth rates on different types of carbon sources in the presence and absence of MTBE.

Even when these microorganisms were not selected in the MTBE presence, they proved the ability to utilize this carbon source alone, and when MTBE was added to media, the growth was not inhibited. The conditions of the growth in Bioscreen® plates are far to be optimal; the reason is a poor intensity of aeration in the wells. Therefore we measured the growth rate with chosen substrate mixtures on standard shaker in flasks. Enzymes attacking above tested substrates belong at least between two groups: monooxygenases and dioxygenases. The exact sequence of reactions leading to entire mineralization of MTBE has not been assigned.

From contaminated sites in the Czech Republic was isolated a collection of individual strains described "A" strains and coculture "Polepy". The growth rates of these cultures are shown in Table 6.

Table 6 presents data measured by BIOSCREEN® using as carbon source also lactate. In the presence of lactate, the all tested cultures (3A, 4A, 6A) showed diauxic growth, therefore two values of μ are presented. Lactate was chosen with regard to its nontoxic qualities and as an environmentally friendly compound.

Table 5. Growth rates of tested strains by BIOSCREEN®.

Strain	Carbon and energy source	$10^{-1}.\mu.h^{-1}$
P2	1% Ac	0.19
P2	1% Ac + 0.05% MTBE	0.33
P2	1% Ac + 0.5% MTBE	0.25
P2	0.5% MTBE	0.28
P2	1% Al	0.15
P2	1% Al + 0.05% MTBE	0.17
P2	1% Al + 0.5% MTBE	0.18
C12B	1% T	0.30
C12B	1% T + 0.05% MTBE	0.32
C12B	1% T + 0.5% MTBE	0.32
C12B	1% C	0.32
C12B	1% C + 0.05% MTBE	0.37
C12B	1% C + 0.5% MTBE	0.30
C12B	0.5% MTBE	0.48
Landa	0.5% Al	0.09
Landa	0.5% Al + 0.%% MTBE	0.08
Landa	0.5% MTBE	0.06

P2 – *Pseudomonas* P2, C12B – *Pseudomonas* C12B

Table 6. Growth rates of isolated strains by BIOSCREEN®.

Strain	Carbon and energy source	$10^{-1}.\mu.h^{-1}$
3A	1% Al	0.13
3A	1% Al + 0.05% MTBE	0.18
3A	1% Al + 0.5% MTBE	0.17
3A	0.5% MTBE	0.11
3A	1% L	0.32
3A	1% L + 0.05 MTBE	0.32/0.20
3A	1% L + 0. 5 MTBE	0.32/0.20
4A	1% Al	0.25
4A	1% Al + 0.05% MTBE	0.21
4A	1% Al + 0.5% MTBE	0.22
4A	0.5% MTBE	0.17
4A	1% L	0.11
4A	1% L + 0.05 MTBE	0.67/0.13
4A	1% L + 0.5 MTBE	0.67/0.16
6A	1% Al	0.07
6A	1% Al + 0.05% MTBE	0.10
6A	1% Al + 0.5% MTBE	0.12
6A	0.5% MTBE	0.08
6A	1% L	0.28
6A	1% L + 0.05 MTBE	0.20/0.08
6A	1% Al + 0.5% MTBE	0.28/0.10
Polepy	1% Al	0.11
Polepy	1% Al + 0.5% MTBE	0.11
Polepy	0.5% MTBE	0.10

Data about missing genotoxicity are very important, because we have previously confirmed its high toxicity, Pazlarova et al. 2003. Using Microtox (*Vibrio fisheri*) system the MTBE was found to be very toxic, because the concentration 33 mg/L was established as EC 50. The introduction of MTBE into the water supply represents significant threats to the water quality because of MTBE's persistence in the environment and because of the potential of MTBE to accumulate in the aquifers. According to our first analytical findings TBA was found in media after bioremediation by so far tested bacteria. Measurement of TBA toxicity and genotoxicity are on process and will be presented. Next task is the analytical confirmation of other MTBE degradation products. We expect to prove the presence of hydroxy isobutyric acid (HIBA), formaldehyd, 2-propanol and acetone. The environmental and toxic impact of these intermediates will be measured using different tests. The ultimate goal is the optimization of conditions for MTBE degradation by the most efficieny MTBE degradative strain.

4 CONCLUSIONS

This study exluded the genetoxicity of methyl *tertiary*-butyl ether (MTBE) using standard Ames test.

Bacterial strains capable of MTBE biodegradation were found and the corresponding specific growth rates were monitired.

ACKNOWLEDGEMENT

This work was supported by grant of Ministry of Education of Czech Republic No. LN 00 B030.

REFERENCES

Cirvello, J.D., Radovsky, A., Heath, J.E., Farnell, D.R. & Landamond, C. 1995 Toxicity and carcinogenity of t-butyl alcohol in rats and mice following chronic exposure in drinking water. *Toxicol. Ind. Health* 11:151–166

Deeb, R. & Scow, K.M. 2000 Aerobic MTBE biodegradation: an examination of past studies, current challenges and future research directions. *Biodegradation* 11:171–186

Hyman, M. 1998 Cometabolism of MTBE by alkane-utilizing microorganisms. *First International Conferrence on remediationof Chlorinated and Recalcitrant Compounds,* Monterey, California, May 18–21, 1998

Hyman, M., Smith, C., & O'Reilly, K.T. 2001 Cometabolism of MTBE by an aromatic hydrocarbon-oxidizing bacterium. In Bioremediation of MTBE, Alcohols, and Ethers, Eds.: Magar, V.S., Gibbs, J.C., O'Reilly, K.T., Hyman, M., Leeson, A. Battelle Press Online Bookstore 2001

Kane, S.R., Beller, H.R., Legler, T.C., Koester, C.J., Pinkart, H.C., Halden, R.U. & Happel, A.M. 2001 Aerobic biodegradation of methyl *tert*-butyl ether by aquifer bacteria from leaking underground storage tank sites. *Appl. Environ. Microbiol.* 67:5824–5829

Mortelsmans, K. & Zeiger, E. 2000 The Ames *Salmonella*/microsome mutagenicity assay. *MUtat. Res.* 455:29–60

Payne, W.J. & Feisal, V.E. 1963 Bacterial utilization of dodecyl sulfate and dodecyl benzene sulfonate. *Appl. Microbiol.* 11:339–344

Pazlarova, J., Demnerova, K., Mackova, M. & Burkhard, J. 1997 Analysis of PCB-degrading bacteria: physiological aspects. *Lett. Appl. Microbiol.* 24:334–336

Pazlarova, J., Vosahlikova, M. & Demnerova, K. 2003 Effect of MTBE on the growth and metabolism of *Pseudomonas* P2 and *Pseudomonas* C12B. First European Conferrence on MTBE, Eds: Bilitewski, B., Werner, P., Conferrence proceedings, The series of the Institute of Waste Management and Contaminated Site Treatment, Dresden University of Technology, Band 31, pp.165–172, ISBN 3-934253-24-5

Salanitro, J.P., Diaz, L.A., Williams, M.P. & Wisniewski, H.L. 1994 Isolation of a bacterial culture that degrades methyl t-butyl ether. *Appl. Environ. Microbiol.* 60: 2593–2596

European Symposium on Environmental Biotechnology, ESEB 2004 - Verstraete (ed)

Biodegradation of diesel oil by various environmental and specialized microflorae

S. Penet, R. Marchal & F. Monot
Institut Français du Pétrole, Rueil-Malmaison, France

R. Chouari, D. Le Paslier & A. Sghir
Genoscope, Evry, France

ABSTRACT: Biodegradation of diesel oil by various environmental microflorae was investigated in closed batch systems by hydrocarbon consumption and CO_2 production. Intrinsic biodegradation capacities of microflorae from polluted and unpolluted soils were determined and compared to those of an activated sludge from an urban wastewater treatment plant. In order to define the specific contributions of bacterial groups to diesel oil degradation, microbial culture enrichments were produced using model substrates representative of each hydrocarbon class of diesel oil as sole carbon sources. Degradation capacities of the enriched microflorae were measured and microbial compositions were determined by 16S rDNA sequence analysis. Results show that: (i) microflorae from polluted sites displayed higher degradation capacities (80% of the initial diesel oil amount) than unpolluted soil (67%) or sludge (64%), (ii) *n*-alkanes were totally degraded whatever the microflora used, (iii) biodegradation capacities and bacterial composition of enriched cultures were dissimilar and dependent on the substrate used.

The large use of petroleum products makes them a significant source of pollutants in ground water and soils. Biodegradation tests are essential either to evaluate the possibilities of natural attenuation or to define a bioremediation strategy in cases of accidental pollution. Information on diesel oil biodegradation is rather incomplete (Richard & Vogel 1999) because of the complexity of both substrate and environmental microflorae (Hueseman 1997). Actually, most data have been obtained from direct measurements on polluted sites where it was not possible to determine complete carbon balance (Salanitro 2001). In the present work, the biodegradation capacities of polluted and unpolluted soils towards commercial diesel oil were determined in liquid culture under favourable conditions. The degradation yields were compared to those of an activated sludge from an urban wastewater treatment plant, considered as a reference. In order to understand the degradation properties of bacterial groups, specialized microflorae were produced using a polluted soil as an inoculum and selected individual products as model substrates representative of the various hydrocarbon classes of diesel oil. Bacterial groups from each microflora were identified and compared to the microflora of the native soil.

1 MATERIALS AND METHODS

1.1 *Culture media and microflorae*

The vitamin-supplemented mineral salt medium described by Bouchez *et al.* (1995) was used as a nutrient solution. Diesel oil was added at 400 mg L^{-1} to the nutrient solution as the sole carbon and energy source. Microbial suspensions for biodegradation tests were prepared by dispersing directly either 5 g L^{-1} of soil sample or 100 mg dry weight L^{-1} of activated sludge into the nutrient solution. The hydrocarbon content of soils was determined by gas chromatography with flame ionisation detection (GC-FID) after cyclohexane:acetone (85/15 v/v) extraction. Soil granulometry was determined as described by Musy & Soutter (1991).

1.2 *Biodegradation tests*

Biodegradation tests were performed in 120 mL-flasks closed with Teflon-coated stoppers and sealed with aluminium caps. 5 μL of diesel oil were added to 10 mL of inoculated culture medium. After an incubation period of 28 days at 30°C under alternative shaking, 10 mL of CH_2Cl_2 were introduced into the flasks which were stored for one night at −20°C before extraction. The CH_2Cl_2 phase of each flask was analysed by GC-FID after evaporation to 1.5 mL. Dotriacontane (nC_{32}) at 50 mg L^{-1} in CH_2Cl_2, was used as an internal standard. Experiments were performed in triplicate and abiotic controls supplemented with $HgCl_2$ were run under similar conditions. The degradation yields were calculated as the ratio of the amount of substrate degraded in test flasks to the amount of recovered substrate in abiotic controls. Kinetics of CO_2 production during the biodegradation were determined at 30°C. Control tests without diesel oil were performed to monitor CO_2 production resulting from endogenous respiration.

Diesel oil was analysed and CO_2 was measured as previously described by Solano-Serena *et al.* (2000b).

1.3 *DNA extraction, amplification, cloning and sequencing of the 16S rDNA*

DNA was extracted from the reference soil (sample 14) and enriched cultures using a bead-beating protocol (Godon *et al.* 1997). 16S rDNA genes were amplified, cloned et sequenced as described by Chouari *et al.* (2003).

2 RESULTS AND DISCUSSION

2.1 *Methodology for assessment of diesel oil biodegradation*

The procedure used was based on the determination of hydrocarbon consumption in liquid cultures carried out in non-limiting conditions, in particular for oxygen supply. The end point of test periods (usually 28 days) was determined by monitoring CO_2 production in test-flask headspaces (Fig. 1).

CO_2 production was also measured in substrate-free flasks in order to quantify endogenous-like respiration resulting from biodegradation of the residual organic matter contained in the soil suspensions. At the end of test period, degradation capacity of microflorae was obtained by quantification of residual hydrocarbons in test and abiotic flasks using GC-FID (Fig. 2). Experiments were carried out at least in triplicate. Standard deviation of biodegradation yields (Table 1) showed that the methodology provided quite suitable assessment of biodegradation extents.

The mean-value of biodegradation capacities of unpolluted soils (67%) was close to that of activated sludge. However, significant disparities were observed within the data which varied from 50 ± 8% to 96 ± 1. Actually, degradation performances did not seem to be affected by soil granulometry but were most probably related to indigenous microflora composition

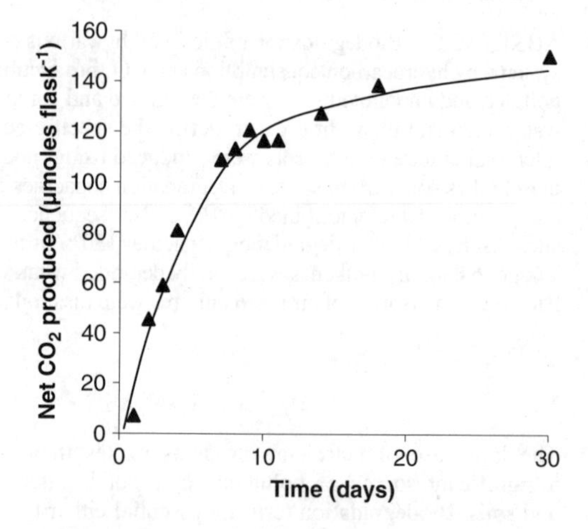

Figure 1. Kinetics of CO_2 production during diesel oil biodegradation by soil sample 14. Values from endogenous respiration (substrate-free flask) are deduced. Tests were performed in 120 mL closed flasks, with 10 mL of medium, 400 mg L^{-1} of commercial diesel oil and 5 g L^{-1} of polluted soil.

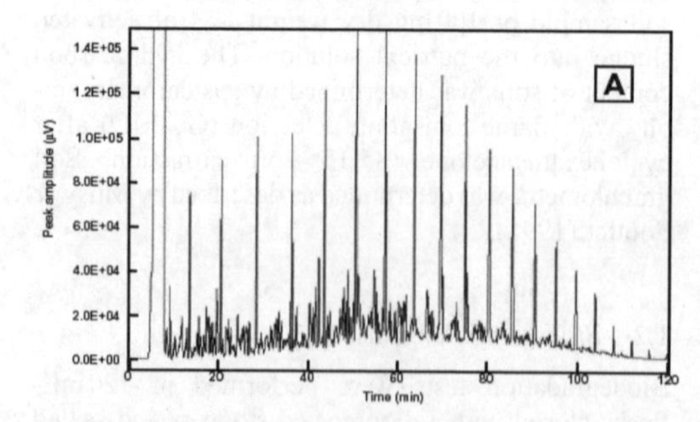

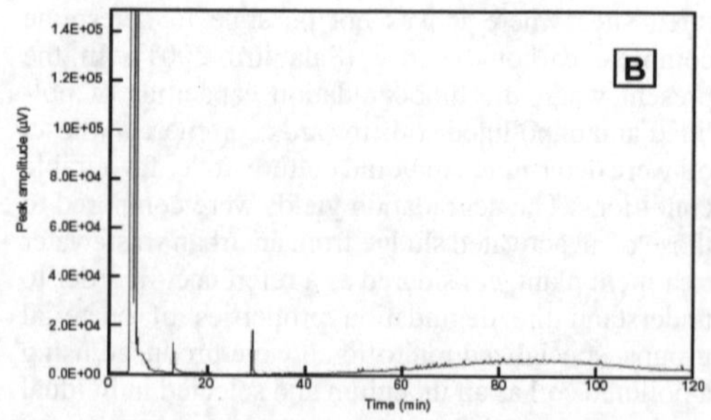

Figure 2. Chromatographic profiles of commercial diesel oil by GC-FID before biodegradation (**A**) and after biodegradation (**B**) by soil sample 14. Biodegradation tests have been performed in 120 mL-closed flasks for 28 days with 10 mL of medium, 400 mg L^{-1} of diesel oil and 5 g L^{-1} of polluted soil.

Table 1. Biodegradation capacities of microflorae from soil samples.

Sample	Origin	Pollution type*	Granulometry**	Degradation yield*** %
1	Garden	None	Sand:clay (79:19)	58 ± 5
2	Unknown	None	Sand:clay (64:21)	77 ± 1
3	Garden	None	Sand:clay (56:43)	58 ± 8
4	Forest	None	Sand:clay (68:27)	87****
5	Garden	None	Sand:clay (67:27)	31 ± 5
6	Garden	None	Sand:clay (53:43)	59 ± 5
7	Vineyard	None	Sand:clay (19:78)	87 ± 1
8	Forest	None	Sand:clay (14:82)	69 ± 3
9	Garden	None	Clay:gravel (43:23)	96 ± 1
10	Garden	None	Sand:clay (39:55)	50 ± 8 Mean Value: 87%
11	Polluted site	Kerosene (2.02)	Sand:clay (18:81)	82 ± 1
12	Polluted site	Kerosene (2.44)	Sand:clay (65:23)	81 ± 2
13	Polluted site	Kerosene (2.26)	Sand:clay (38:61)	70 ± 5
14	Polluted site	Diesel oil (10)	Gravel:clay (42:36)	93 ± 2
15	Polluted site	Diesel oil	Sand:clay (29:71)	81 ± 3
16	Polluted site	Diesel oil	Sand:clay (53:34)	83 ± 4
17	Polluted site	Crude oil (9)	Sand (100)	69 ± 2
18	Polluted site	Crude oil (10)	Sand (100)	65 ± 7
19	Polluted site	Diesel oil (3)	Sand:clay (28:48)	98 ± 0 Mean Value: 67%
20	Activated sludge	None		61****
21	Activated sludge	None		73****
22	Activated sludge	None		59 ± 5 Mean Value: 80%

* amounts of polluting hydrocarbons in mg per g of soil dry weight are indicated in parentheses.
** relative composition is indicated in parentheses.
*** with respect to the final amount of hydrocarbons in the abiotic flasks. Tests were performed in triplicate, standard deviation is indicated.
**** only one test was performed.

influenced by the organic matter content. Considering soil samples 4 and 9, conifer trees were present near the sampling site. They have probably released terpenic compounds which might have possibly acted as a selective substrate on soil microflora compositions.

The mean value of diesel oil degradation by microflorae from polluted soils (80%) was significantly higher than that of the unpolluted ones (67%). Large variations in biodegradation capacities were also noticed as for instance for polluted-marsh samples 17 and 18 (69% and 65%, respectively) which exhibited biodegradation capacities similar to those of unpolluted soil samples. Diesel oil-polluted soils (samples 14, 15, 16, 19) were most efficient for substrate biodegradation. Soil sample 19 that had been polluted for about 20 years by domestic-fuel leaks displayed a remarkable high biodegradation capacity (98%). Thus, hydrocarbon pollution has most likely resulted in a selective adaptation of the native soil microflora.

2.2 *Bacterial composition of diesel oil-degrading microflorae*

In order to assess contribution of bacterial groups to diesel oil degradation, specialized microflorae were produced from soil sample 14. Adaptation of the native soil microflora was performed with successive subcultures monitored by headspace CO_2 (Solano-Serena *et al.* 2000).

In the culture series, distinct model substrates were used as sole carbon sources. They were chosen as being representative of the main hydrocarbon classes of diesel oil *i.e.* hexadecane for *n*-alkanes, pristane for isoalkanes, decaline for cycloalkanes, and phenanthrene for aromatics. The compositions of the respective enriched microflorae obtained (noted MF1 to MF4, respectively) were then determined by 16S rDNA sequence analysis and compared to that of the native soil (Table 2).

Predominant groups were *Actinobacteria* in MF1, *Bacteroidetes* in MF2, *γ-proteobacteria* in MF3, and *β-proteobacteria* in MF4, indicating that overall microorganism composition closely depended on the growth substrates used for selective pressure. In addition, degradation efficiencies of enriched microflorae towards diesel oil were found dissimilar (data not shown). Some hydrocarbons such as hexadecane, pristane and phenanthene were totally biodegraded by each microflora. It was also found that the native soil microflora degraded diesel oil more extensively than the enriched microflorae.

Table 2. Effect of selective pressure on bacterial phylogenetics groups.

Bacterial group	Bacterial group presence in				
	Soil	MF1	MF2	MF3	MF4
Firmicutes	+				
Cyanobacteria	+				
Chloroflexi	+				
Bacteriodetes	+	+	++		
Actinobacteria		++			+
Acidobacteria	+				
α-proteobacteria	++	+	+	+	
β-proteobacteria	++	+	+	+	++
δ-proteobacteria	+				
γ-proteobacteria	++			++	
Uncultured	++	+	+		
Number of sequences analysed	328	87	50	90	85

Microflora MF1 to MF4 were adapted on hexadecane, pristane, decaline and phenanthrene respectively + presence, ++ predominant groups.

3 CONCLUSION AND PERSPECTIVES

The adaptation of microflorae to hydrocarbons has been mentioned in gasoline-polluted soils (Horowitz & Atlas 1977, Solano-Serena *et al.* 2000a). In the case of diesel oil, comparison between polluted and unpolluted soil samples shows that a similar adaptation of the native microflorae after pollution has occurred.

Detection of hydrocarbon-degradation genes (Widada *et al.* 2002) in specialized microflorae is needed to further describe microflora adaptation. It might help to elucidate the contribution of the different bacterial groups to diesel oil degradation.

We are developing a set of novel 16S RNA targeted hybridization probes. They will help us to quantify the dot blot hybridization and FISH activity of the predominant representative groups detected in the different libraries. Simultaneously, we will analyse expression of genes by RT-PCR especially those involved in the degradation of the substrates.

REFERENCES

Bouchez, M., Blanchet, D., Vandecasteele, J.P., 1995. Degradation of polycyclic aromatic hydrocarbons by pure strains and by defined strain associations: inhibition phenomena and cometabolism. Appl Microbiol Biotechnol 43, 156–164.

Chouari, R., Le Paslier, D., Daegelen, P., Ginestet, P., Weissenbach, J., Sghir, A., 2003. Molecular Evidence for Novel Planctomycete Diversity in a Municipal Wastewater Treatment Plant. Appl Envir Microbiol, in press.

Godon, J.J., Zumstein, E., Dabert, P., Habouzit, F., Moletta, R., 1997. Molecular microbial diversity of an anaerobic digestor as determined by small-subunit rDNA sequence analysis. Appl Environ Microbiol 63, 2802–2813.

Horowitz, A., Atlas, R.M., 1977. Response of microorganisms to an accidental gasoline spillage in an artic freshwater ecosystem. Appl Environ Microbiol, 1252–1258.

Hueseman, M.H., 1997. Incomplete hydrocarbon biodegradation in contaminated soils: limitations in bioavailability or inherent recalcitrance? Biorem J 1, 27–39.

Musy, A., Soutter, M., 1991. Physique du sol. Collection Gérer l'Environnement, Presse polytechniques et universitaires romandes.

Richard, J.Y., Vogel, T.M., 1999. Characterization of soil bacterial consortium capable of degrading diesel fuel. Int Biodeter Biodeg 44, 93–100.

Salanitro, J.P., 2001. Bioremediation of petroleum hydrocarbons in soil. Adv in Agron 72, 53–105.

Solano-Serena, F., Marchal, R., Lebeault, J.M., Vandecasteele, J.P., 2000a. Distribution in the environment of degradative capacities for gasoline attenuation. Biodegradation 11, 29–35.

Solano-Serena, F., Marchal, R., Lebeault, J.M., Vandecasteele, J.P., 2000b. Selection of microbial populations degrading recalcitrant hydrocarbons of gasoline by monitoring of culture-headspace composition. Lett Appl Microbiol 30, 19–22.

Widada, J., Nojiri, H., Kasuga, K., Yoshida, T., Habe, H., Omori, T., 2002. Molecular detection and diversity of polycyclic aromatic hydrocarbon-degrading bacteria isolated from geographically diverse sites. Appl Microbiol Biotechnol 58, 205–209.

European Symposium on Environmental Biotechnology, ESEB 2004 - Verstraete (ed)

Liquid and solid state NMR study of benzothiazole degradation by *Rhodococcus* isolates

N. Haroune, B. Combourieu, P. Besse, M. Sancelme & A.M. Delort
Laboratoire Synthèse Et Etudes de Systèmes à Intérêt Biologique, UMR CNRS-Université Blaise Pascal, Aubière cedex, France

ABSTRACT: In this paper, recent data on biodegradative pathways of benzothiazoles by various *Rhodococcus* isolates (*R. erythropolis*, *R. rhodochrous*, *R. pyridinovorans* PA) are presented. Benzothiazole derivatives belong to a large family of xenobiotics produced world-wide in chemical industry for various applications. These pollutants are quite recalcitrant and toxic, they are found in various environmental compartments, especially in water. We used *in situ* 1D and 2D liquid state NMR to investigate the biodegradative pathway of benzothiazole (BT), 2-hydroxybenzothiazole (OBT) and 2-aminobenzothiazole (ABT). A common step of the degradative pathway was found corresponding to the hydroxylation of benzothiazole derivatives on the aromatic ring in position 6. In the case of *R. pyridinovorans* PA a catechol 1,2-dioxygenase activity was shown using a specific inhibitor. In a more recent study, we used ^{1}H solid state NMR to study the degradation of benzothiazole-2-sulfonate ($BTSO_3$) by *Rhodococcus erythropolis* in the presence of anionic clays.

1 INTRODUCTION

Nuclear Magnetic Resonance (NMR) spectroscopy offers an interesting alternative to the most commonly used analytical techniques such mass spectrometry, HPLC and GPC to study microbial metabolism, and more specifically the biodegradation of organic pollutants (for review see Grivet et al. 2003). In the past, only few studies have been reported on microbial degradation of xenobiotics using NMR as they were limited to hetero-nuclei NMR studies (for review see Delort & Combourieu 2000, Combourieu et al. 2003). The recent development of ^{1}H NMR is quite promising in the field of environment, as molecules can be studied at natural abundance and at low concentration. Besides HPLC-NMR techniques (Wilson 2000) which are largely used in biofluids but rarely for microbial degradation, *in situ* 1D and 2D ^{1}H liquid state NMR can be used to monitor the degradation of organic pollutants by micro-organisms in aqueous media. More recently ^{1}H High-Resolution (HR) Magic Angle Spinning (MAS) NMR appears as a very innovative technique to study the degradation of organic pollutants in the presence of soil components such as clays (Grivet et al. 2003).

In this paper we shall focus on recent data we obtained on biodegradative pathways of benzothiazoles by various *Rhodococcus* isolates. Benzothiazole derivatives belong to a large family of xenobiotics containing a benzene ring fused with a thiazole ring, they are produced world-wide in chemical industry for various applications, the main one concerns the production of rubber (for review see De Wever et al. 2001). These pollutants are quite recalcitrant and toxic, therefore they are of concern for the environment when they are released in various comparments, especially in water (De Wever & Verachtert 1997). By using liquid state NMR, we could establish the biodegradative pathway of benzothiazole (BT) and 2-hydroxybenzothiazole (OBT) by *Rhodococcus erythropolis*, *Rhodococcus rhodochrous*, *Rhodococcus pyridinovorans* PA (Besse et al. 2001, Haroune et al. 2002), and 2-aminobenzothiazole (ABT) by *Rhodococcus erythropolis*, *Rhodococcus rhodochrous* (Haroune et al. 2001) (Figure 1). We found a common step of the degradative pathway corresponding to the hydroxylation of benzothiazole derivatives on the aromatic ring in position 6. In a more recent study we used ^{1}H HR-MAS NMR to study the degradation of benzothiazole-2-sulfonate ($BTSO_3$) by *Rhodococcus erythropolis* in the presence of anionic clays (Haroune 2003).

2 MONITORING DEGRADATION KINETICS BY ^{1}H NMR IN LIQUID MEDIA

To monitor kinetics of biodegradation, *in situ* ^{1}H NMR, directly performed on incubation media, is a very convenient and powerful tool. Samples (1ml) are

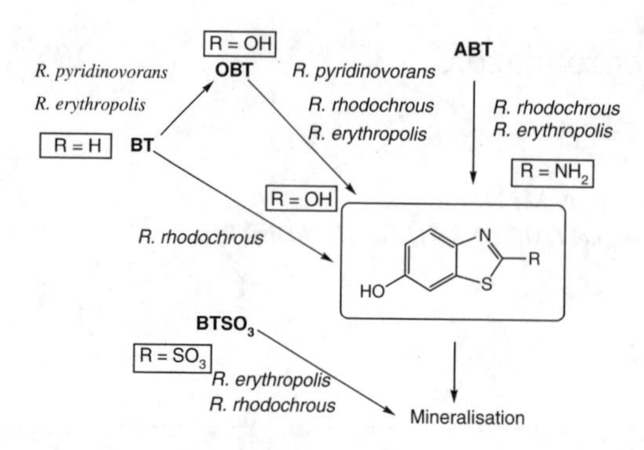

Figure 1. General degradative pathway of benzothiazoles by *Rhodococcus* isolates. BT: benzothiazole, OBT: 2-hydroxybenzothiazole, ABT: 2-aminobenzothiazole, $BTSO_3$: benzothiazole-2-sulfonate.

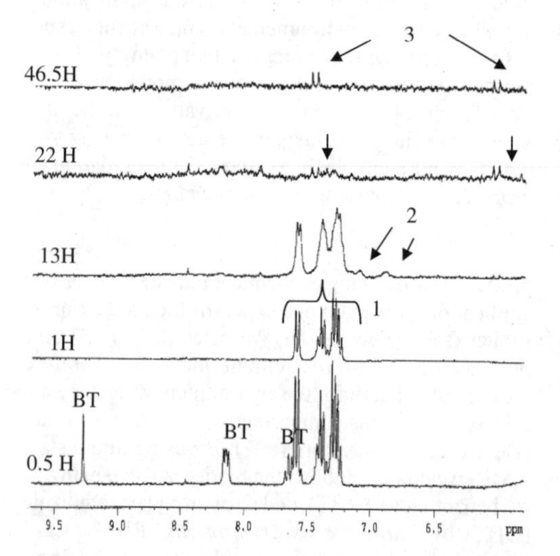

Figure 2. *In situ* ^{1}H NMR spectra of samples taken during the degradation of BT (benzothiazole) by *Rhodococcus pyridinovorans* PA. Metabolites: (1) OBT (2-hydroxybenzothiazole), (2) diOBT (2,6-dihydroxybenzothiazole), (3) diacid derivative.

taken at regular intervals, quickly centrifuged, then the supernatant is pH adjusted to avoid chemical shifts variations and supplemented by TSP*d4* which constitutes an internal reference for chemical shift (0 ppm) and quantification of metabolites (as previously described, Combourieu et al. 1998). The whole process takes less than 15 minutes and the lowest metabolite concentration detected under these conditions is around 50 μM. An example of *in situ* ^{1}H NMR spectra collected during the biodegradation of BT (3 mM) by *Rhodococcus pyridinovorans* PA resting cells is presented in Figure 2.

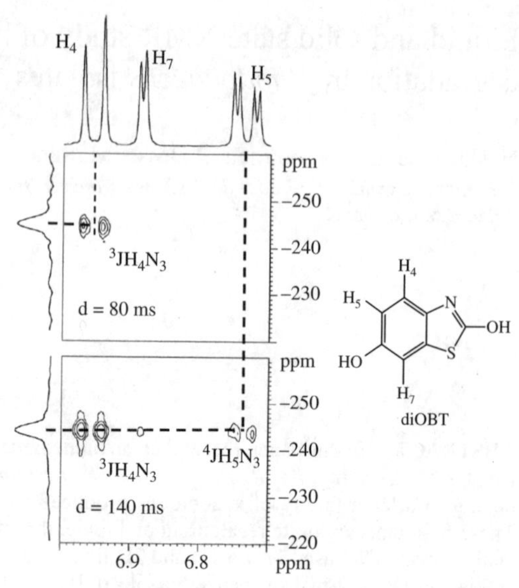

Figure 3. ^{1}H-^{15}N HMBC spectra of diOBT (2,6-dihydroxybenzothiazole) purified from the incubation medium of *Rhodococcus rhodochrous* incubated with BT (benzothiazole). d = delay used to select ^{3}J or ^{4}J coupling constants in the HMBC sequence.

^{1}H NMR signals corresponding to 4 compounds have been detected: the initial substrate BT and 3 new metabolites corresponding to 2-hydroxybenzothiazole, OBT (metabolite 1), 2,6-di-hydroxybenzothiazole, diOBT (metabolite 2) and a diacid compound (metabolite 3). In addition, integration of ^{1}H signal areas compared to TSP*d4* area gives quantitative data about the time course of benzothiazole degradation and metabolite synthesis.

3 METABOLITE STRUCTURE ELUCIDATION BY 2D NMR

The main difficulty when working with xenobiotics is that metabolic pathways are unknown and thus the structure of intermediates must be established using more sophisticated 2D NMR experiments. For instance ^{1}H-^{13}C HSQC and HMBC experiments were carried out at natural abundance to get information about $^1J_{13C\text{-}1H}$ or long range $^nJ_{13C\text{-}1H}$ couplings of the various benzothiazole metabolites (Haroune et al. 2001). ^{1}H-^{15}N HMBC experiments were used to characterize the structure of 6-OH derivatives of benzothiazoles, an example of spectra is given in Figure 3. More concentrated samples must be used in that case because of the low sensitivity of ^{13}C and ^{15}N nuclei: samples can be gathered and freeze dried or alternatively more metabolites can be produced using large amounts of

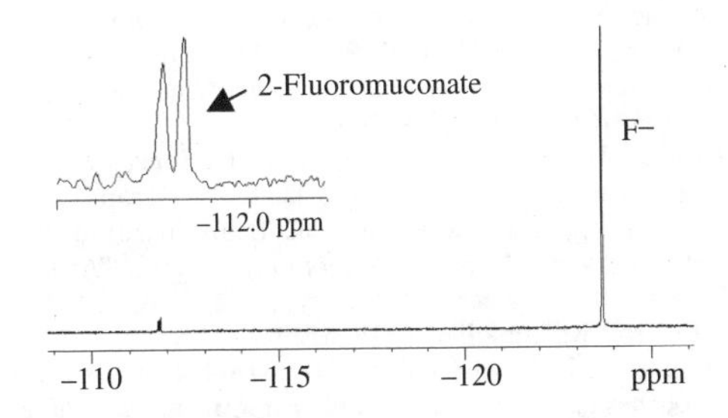

Figure 4. *In situ* ^{19}F NMR spectrum collected after 48 hours of incubation of 3-FC (3-fluorocatechol) with *Rhodococcus pyrinidinovorans* PA.

biomass, the metabolites of interest being then purified on silica-gel columns.

4 *IN SITU* NMR STUDY OF CATECHOL 1,2-DIOXYGENASE ACTIVITY

When metabolites are identified, hypotheses can be made about the nature of the enzymes involved in the different steps of the biodegradative pathway. In addition to enzymatic assays or genetic studies, *in situ* NMR can be a convenient tool to investigate the effect of a specific enzyme inhibitor. For instance, in the case of *Rhodococcus pyridinovorans* strain PA, the activity of a catechol 1,2-dioxygenase leading to the dicarboxylic intermediate issued from BT degradation was confirmed by using 3-fluorocatechol (3-FC) a specific inhibitor (Haroune et al. 2002). The time courses of BT degradation in the absence or presence of 3-FC were monitored by ^{1}H *in situ* NMR, and a clear decrease of the transformation rate of OBT into diOBT was observed. By using ^{19}F NMR it was also shown that 3-FC was biotransformed into 2-fluoromuconate (Figure 4), showing that 3-FC is a competitive substrate of the catechol 1,2-dioxygenase.

5 HR-MAS NMR STUDY OF DEGRADATION IN THE PRESENCE OF CLAYS

In addition to liquid state studies, NMR can be a powerful tool to investigate the fate of organic pollutants in soils (for review see Delort et al. 2003). Solid-state CP-MAS (Cross Polarization Magic Angle spinning) NMR spectroscopy allows to characterize bound residues to organic matter. However, these experiments, performed on dry samples, do not give the real fingerprint of an environmental material, which is usually highly hydrated with mobile pollutant at its surface, in particular they give no indication about the "bioavailability" of the pollutant. By using ^{1}H High-Resolution (HR) MAS NMR which allows the characterization of inhomogeneous compounds with liquid-like dynamics, we could distinguish adsorbed and intercalated MCPA (4-chloro-2-methylphenoxyacetic acid) pesticide at natural abundance on hydrated anionic clays (Combourieu et al. 2001). We recently applied this technique to investigate the degradation of $BTSO_3$ by *Rhodococcus erythropolis* in hydrated anionic clays (Haroune 2003). Anionic clays, namely LDHs (Layered Double Hydroxides) were used as very well characterized model of soils.

6 CONCLUSIONS

Liquid and solid state NMR allowed us to describe in details the metabolism of benzothiazoles by *Rhodococcus* isolates. Our future work will extend these studies to other benzothiazole derivatives, in particular to 2-mercaptobenzothiazole (MBT) used in rubber industry. Also complementary photo- and biodegradation approaches will be combined in order to reach complete mineralization of recalcitrant benzothiazoles. We recently showed it was very efficient to degrade the herbicide methabenzthiazuron (Malouki et al. 2003).

The present paper outlines the accuracy of *in situ* liquid state NMR for the assessment of microbial metabolism of xenobiotics: it allows studies at natural abundance, simultaneous detection of various metabolites without purification. Also this technique does not require any *a priori* hypothesis. The main limitation of this technique is its low sensitivity limits but recent developments of NMR technology will partially overcome this difficulty (magnetic fields of higher intensity; sensitivity-improving cold-metal NMR probes). Although limited to laboratory experiments, HR-MAS NMR is a very promising approach to investigate fundamental processes occurring during the biodegradation of pollutants in soil, particularly to assess pollutant mobility at the interface liquid/solid of hydrated matrices. It should give indications about the bioavailability of these pollutants.

REFERENCES

Besse, P., Combourieu, B., Boyse, G., Sancelme, M., De Wever, H. & Delort, A.M. 2001. Long-range ^{1}H-^{15}N heteronuclear shift correlation at natural abundance: a tool to study benzothiazole biodegradation by two *Rhodococcus* strains. Appl. Environ. Microbiol. 67:1412–1417.

Combourieu, B., Besse, P., Sancelme, M., Veschambre, H., Delort, A.M., Poupin, P. & Truffaut, N. 1998. Morpholine degradation pathway of *Mycobacterium aurum* MO1: direct evidence of intermediates by *in situ* ^{1}H nuclear magnetic resonance. Appl. Environ. Microbiol. 64: 153–158.

Combourieu, B., Haroune, N., Besse, P. Sancelme, M. & Delort, A.M. 2003. ^{1}H nuclear magnetic resonance: a tool to study biodegradative pathways of organic pollutants in *Mycobacterium* and *Rhodococcus* isolates. In R.M. Mohan (ed) *Research Advances in Microbiology*: 1–22. Global Research Network.

Combourieu, B., Inacio, J., Delort, A.M. & Forano, C. 2001. Differenciation of mobile and immobile pesticides on anionic clays by ^{1}H HR-MAS NMR spectroscopy. Chem. Commun. 2214–2215.

Delort, A.M. & Combourieu, B. 2000. Microbial degradation of xenobiotics. In J.N. Barbotin & J.C. Portais (eds) *NMR in microbiology:theory and applications*: 411–430. Horizon Scientific, UK.

Delort, A.M., Combourieu, B. & Haroune, N. 2003. NMR: a tool to study the interactions between organic pollutants and soil components. A mini-review. Environ. Chem. Lett. In press.

De Wever, H. & Verachtert, H. 1997. Biodegradation and toxicity of benzothiazoles. Water Res. 31:2673–2684.

De Wever, H., Besse, P. & Verachtert, H. 2001. Microbial transformations of 2-substituted benzothiazoles. Appl Microbiol. Biotechnol. 57:620–625.

Grivet, J.P., Delort, A.M. & Portais, J.C. 2003. NMR and microbiology: from physiology to metabolomics. Biochimie. In press.

Haroune, N., Combourieu, B., Besse, P., Sancelme, M. & Delort, A.M. 2001. ^{1}H NMR: a tool to study the fate of pollutants in the environment. CR Acad Sci Paris, Chimie/Chemistry 4:759–763.

Haroune, N., Combourieu, B., Besse, P., Sancelme, M., Reemtsma, .T., Kloepfer, A., Diab, A., Knapp, J.S., Baumberg, S. & Delort A.M. 2002. Benzothiazole degradation by *Rhodococcus pyridinovorans* strain PA: evidence of a catechol 1,2-dioxygenase activity. Appl. Environ. Microbiol. 68:6114–6120.

Haroune, N. 2003. Métabolisme de benzothiazoles par des souches de *Rhodococcus*: Etude par *RMN in situ*. Thèse, Université Blaise Pascal.

Malouki, M., Giry, G., Besse, P., Combourieu, B., Sancelme, M., Bonnemoy, F., Richard, C. & Delort, A.M. 2003. Sequential bio- and photo-transformation of the herbicide methabenzthiazuron in water. Environ. Toxicol. Chem. 22: 2013–2019.

Wilson, I.D. 2000. Multiple hyphenation of liquid chromatography with nuclear magnetic resonance spectroscopy, mass spectrometry and beyond. J. Chromatogr. A 892:315–27.

European Symposium on Environmental Biotechnology, ESEB 2004 - Verstraete (ed)
© 2004 Taylor & Francis Group, London, ISBN 90 5809 653 X

Inoculated bioreactors for treatment of water contaminated with MTBE

L. Bastiaens, Q. Simons, D. Moreels, H. De Wever & J. Gemoets
Vito (Flemish Institute for Technological Research), Mol, Belgium

ABSTRACT: A lab scale study has been set-up to evaluate the possibilities of an inoculated bioreactor for the on-site treatment of MTBE-contaminated groundwater. Activated carbon was inoculated with two different MTBE-degrading cultures, an axenic strain (*Rubrivivax* sp. PM-1) and an MTBE-degrading enrichment culture. Based on batch experiments the cultures were found to degrade MTBE in the presence of the activated carbon. Column-experiments have been started to evaluate MTBE-degradation in a continuous system. Oxygen uptake was detected in the inoculated column as well as in the non-inoculated column. Based on the available preliminary data it is difficult to make a distinction between sorption and degradation of MTBE. Formation of TBA indicates that at least a part of the MTBE is removed by biodegradadation.

1 INTRODUCTION

Methyl Tert-Butyl Ether (MTBE) is a widely used human-made substitute for lead in gasoline. Primarily because of its low odour and taste threshold MTBE-polluted groundwater is of concern. However, its high water solubility and low sorption onto organics make MTBE a difficult pollutant to treat and classical methods such as air stripping and sorption onto GAC are not efficient. MTBE is rather recalcitrant to biodegradation, although some (mostly aerobic) MTBE-degrading bacterial consortia and isolates have been reported (Salanitro et al., 1994; Mo et al., 1997; Hanson et al., 1999; Deep et al., 2000; Stocking et al., 2000; Foyolle et al., 2001). Biodegradation may be considered as an effective in-situ and on-site (pump & treat) remediation technology for MTBE-polluted groundwater. As an endogenous MTBE-degrading bacterial potential is often absent at contaminated sites (Moreels et al., 2002, Kane et al., 2001), inoculation and bio-augmentation with bacteria specially selected for their MTBE-degradation capacities may be promising.

The aim of the present study is to determine whether bioreactors inoculated with MTBE-degrading bacteria are efficient for on-site treatment of MTBE-contaminated groundwater. Activated carbon was chosen as filling material for the bioreactor, based on its (I) high specific surface area (good for biofilm formation) and (II) sorption capacities (concentration of MTBE).

2 MATERIALS AND METHODS

2.1 *Batch experiments*

The MTBE-degradation capacities of the axenic bacterial strain *Rubrivax* sp. PM1 (Church et al., 2000) and mixed cultures enriched at Vito (Moreels et al., 2002) were checked by growing the cultures in 250 ml flasks with minimal mineral medium supplied with MTBE. The evolution of MTBE and TBA concentration in time was followed by GC-MS analyses of the liquid phase as described by Moreels et al. (2003). MTBE and oxygen were spiked to the bottles when necessary. Once MTBE-degradation was observed, 25 g of Organosorb 10 (DESOTEC) was added to the cultures. A similar set-up without inoculation was included in the test as a negative control to make a distinction between sorption and biodegradation.

2.2 *Column experiment*

A lab-scale column test was set-up and a mixture of PM-1 and the MTBE-degrading enrichments decribed in section 1.1 was used to inoculate granular activated carbon, which was the carrier material in the bioreactor. A non-inoculated reference column was included as a control in order to be able to evaluate the benefit of the inoculum on MTBE removal, i.e. to make a distinction between sorption and biodegradation of MTBE (Figure 1). The MTBE removal efficiency was monitored in time by determination of MTBE-concentration

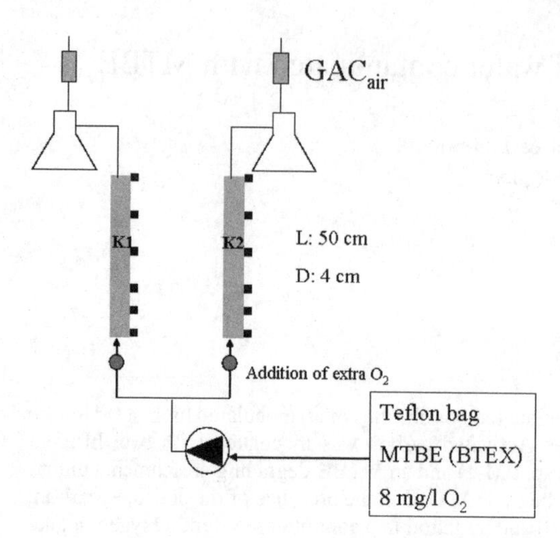

Figure 1. Schematic overview of a lab scale inoculated bioreactor (K2) and a non-inoculated control (K1).

and oxygen concentration profiles along the column. The hydraulic retention time was 1 day/column.

3 RESULTS

3.1 *MTBE-degradation in the presence of activated carbon*

In all inoculated flasks decreases in MTBE-concentration were observed before and after addition of activated carbon. Figure 2 gives as an example the results for the mixed culture. After addition of a high amount of MTBE (about 300 mg/l) the formation of TBA was observed. In the non-inoculated control the MTBE-concentration decreased also, but no TBA could be detected. This indicates that at least a part of the MTBE is biodegraded by the cultures.

Table 1 summarises the MTBE-concentrations that were measured 7 days (T1) and 50 days (T4) after the addition of activated carbon to the flasks, as also the total amount of MTBE that was added. The MTBE-concentrations in the inoculated flasks are generally lower than the concentrations measured in the non-inoculated control, although a comparable amount of MTBE was added. This indicates again that biodegradation of MTBE occurs by the tested cultures in the presence of activated carbon.

3.2 *Removal of MTBE in an inoculated bioreactor*

A column experiment was set-up in the lab to simulate bioreactors. In the inoculated column as well as in the control, oxygen was consumed quite rapidly. The oxygen concentration decreased from more than 8 mg/l in

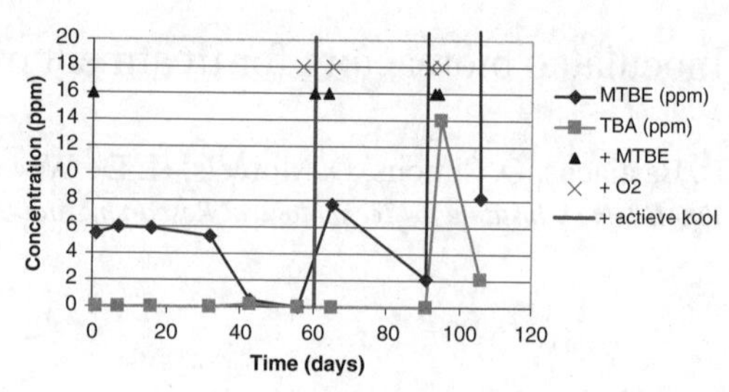

Figure 2. MTBE-degradation by a mixed culture before and after addition of activated carbon.

Table 1. Evolution of the MTBE-concentration (μg/l) 7 days (T1) and 50 days (T4) after addition of activated carbon to the MTBE-degrading cultures and the non-inoculated controls (C1, C2).

Culture	T1	T4	Total supplied MTBE (μl)
C1	24883	27185	390
C2	7916	22458	270
Vito 1	29234	10744	260
Vito 2	9219	7766	310
Vito 3	5384	12331	340
Vito 4	7644	8205	320
Vito 5	5648	7459	340
Vito 6	5675	1891	340
Vito 7	5932	1130	340
Vito 8	5944	1450	340
PM1	10690	14771	450
PM2	24651	26411	450

the influent to about 2 mg/l 3 cm after the input of the column. Addition of extra oxygen resulted in input concentration of 13 to 20 mg/l, but again the oxygen in the effluent of both columns was 1 to 2.5 mg/L. So no evidence for biodegradation of MTBE was obtained based on oxygen measurements (results not shown). The pH in the columns remained near neutral.

Based on the available chemical data it was found that the MTBE concentrations in the effluent of both columns were nearly the same. The MTBE-concentrations decreased along both columns from 15 mg/l to 2–4 g/l. This suggests that sorption was the main MTBE removal process. Only in the inoculated column TBA was detected, which points at bacterial MTBE-degradation in the column.

4 CONCLUSIONS

The axenic strain *Rubrivivax* sp. PM-1 and an MTBE-degrading enrichment culture were found to degrade MTBE in the presence of the activated carbon. Column-experiments have been started to evaluate the contribution of biodegradation in an inoculated continuous GAC-system in respect with MTBE-removal from groundwater. Oxygen consumption is being detected in the inoculated column as well as in the non-inoculated column. In prelimary batch-experiments decreases in the dissolved oxygen concentration in the presence of non-inoculated GAC has also been observed. Oxygen may be sorbed on the GAC. Based on the available preliminary data it is difficult to make a distinction between sorption and degradation of MTBE. In the first operating phase sorption seems to be the major MTBE-removal process in the columns. However, the detection of TBA, a degradation product of MTBE, in the inoculated column indicates that at least a part of the MTBE is removed from the water phase by microbial activity. Possibly the effect of the inoculation will be more clear on the longer term.

ACKNOWLEDGEMENT

This project was partially financed by OVAM.

REFERENCES

Moreels, D., L. Bastiaens, L. Diels, F. ollevier, R. Merckx, D. Springael. 2002. Third International Conference on Remediation of chlorinated and recalcitrant compounds, May 20–23, Monterey, Califormnia. Paper 2B-65.

Moreels, D., L. Bastiaens, D. Springael, F. Ollevier, R. Merckx, L. Diels. 2003. Seventh International Symposium on In Situ and On-Site Bioremediation, June 2–5, Orlando.

Bradley, P.M., J.E. Landmeyer, F.H. Chapelle. 2001. Widespread potential for microbial MTBE degradation in surface-water sediments. Environ. Sci. Technol. 35:658–662.

Church, C.D., P. Tratnyek, K. Scow. 2000. Pathways for the degradation of MTBE and other fuel oxygenates by isolate PM1. Am. Chem. Soc. 40:261–263.

Kane, S.R., H.R. Beller, T.C. Legler, C.J. Koester, H.C. Pinkart, R.U. Halden, A.M. Happel. 2001. Aerobic biodegradation of methyl tert-butyl Ether by aquifer bacteria from leaking underground storage tank sites. Appl. Environ. Microbiol. 67:5824–5829.

Fayolle, J. P. Vandecasteele, F. Monot. 2001. Microbial degradation and fate in the environment of methyl tert-butyl ether and related fuel oxygenates. Appl. Microbiol. Biotechnol. 56:339–349.

Stocking, A.J., R.A. Deeb, A.E. Flores, W. Stringfellow, J. Talley, R. Brownell, M.C. Kavanaugh. 2000. Bioremediation of MTBE: a review from a practical perspective.

Deeb, R.A., K.M. Scow, L. Alvarez-Cohen. 2000. Aerobic MTBE biodegradation: an examination of past studies, current challeges and future research directions. Biodegradation 11:171–186.

Salanitro, J.P., L.A. Diaz, M.P. Williams, H.L. Wisniewski. 1994. Isolation of a bacterial culture that degrades methyl-t-butyl ether. Appl. Environ. Microbiol. 60:2593–2596.

Mo, K., C.O. Lora, A.E.Wanken, M. Javanmardian, X. Yang, C.F. Kupla. 1997. Biodegradation of methyl t-butyl ether by pure bacterial cultures. Appl. Microbiol. Biotechnol. 47:69–72.

Hanson, J.R., C.E. Ackerman, K.M. Scow. 1999. Biodegradation of methyl tert-butyl ether by a bacterial pure culture. Appl. Environ. Microbiol. 65:4788–4792.

European Symposium on Environmental Biotechnology, ESEB 2004 - Verstraete (ed)
© 2004 Taylor & Francis Group, London, ISBN 90 5809 653 X

Selection of liquid medium for biodegradation of RBB-R by white-rot fungi

A.K. Omori, A.Z. Santos, C.R.G. Tavares & S.M. Gomes-da -Costa
State University of Maringá, Department of Chemical Engineering, Av. Colombo, Maringá-PR, Brazil

ABSTRACT: The aim of this work was to verify the best culture medium for biodegradation of the antraquinone dye remazol brilliant blue-R (RBB-R) by three white-rot fungi. The strains *Pleurotus pulmonarius* CCB019, *Phanerochaete chrysosporium* CCB478 and *Lentinus edodes* CCB047 were evaluated using three media: modified yeast extract, potato dextrose and minimum medium with cellulose, which were supplemented with 0,5 g/L RBB-R. The assays were accomplished under batch system, for fifteen days. The following parameters were analyzed: the color reduction, residual glucose, enzymatic activity and chemical oxygen demand (COD). All the fungi utilized degraded the dye; however, the best results of the decolorization were obtained in the modified yeast extract medium with 70.7% by *Pleurotus pulmonarius*, 93.3% by *Phanerochaete chrysosporium* and 39.4% by *Lentinus edodes*.

1 INTRODUCTION

Textile, paper and printing industries extensively use synthetic dyes. The wastewater of these industries is highly colored. Because direct discharge of these effluents may cause formation of toxic aromatic amines under anaerobic conditions, and color in wastewater is highly visible and affects esthetics, water transparency and gas solubility in water bodies, dye wastewaters have to be treated (Kapdan et al. 2000, Fu & Viraraghavan 2001).

Color can be removed from wastewater by chemical and physical methods, including adsorption, coagulation – flocculation and oxidation, and electrochemical methods (Lin & Peng 1994). These methods are quite expensive and have operational problems.

The use of white-rot fungi for decomposition of recalcitrant compounds and dyes presents several advantages. Lignolytic enzymes are non-specific to a substrate; therefore, they can degrade a wide variety of recalcitrant compounds and even complex mixtures of pollutants. In addition, the extracellular enzyme system enables white-rot fungi to tolerate high pollutant concentrations (Kapdan 2000).

Three extra-cellular enzymes involved in biodegradation of dyestuffs are lignin peroxidase (Lip), Mn peroxidase (MnP) and H_2O_2 dependent peroxidases. Laccase is another extracellular enzyme that takes place in the biodegradation of lignin and dyestuffs (Field et al. 1993, Reddy 1995). Some white-rot fungi produce all these enzymes while others produce only one or two of them (Zilly et al. 2002).

RBB-R is an industrially important dye that is frequently used as starting material in the production of polymeric dyes. It represents an important class of often toxic and recalcitrant organopollutants. Its structure resembles certain polymeric aromatic compounds, which are substrates of ligninolytic peroxidases. Therefore, we evaluate its decolorization by the isolates *Pleurotus pulmonarius* CCB019, *Phanerochaete chrysosporium* CCB478 and *Lentinus edodes* in three different media to select the best medium to be used for dye biodegradation.

2 MATERIALS AND METHODS

2.1 *Microorganisms and culture methods*

The microorganisms used in this study, *Pleurotus pulmonarius* CCB019, *Phanerochaete chrysosporium* CCB478 and *Lentinus edodes* CCB047, were obtained from the Culture Collection of the Botanical Institute of São Paulo. The stock cultures were maintained on malt extract agar (malt extract 20 g, peptone 1 g, dextrose 20 g and agar 15 g per liter) slants at 5°C. Transfers were made on malt extract agar plates and the strains were cultivated at 28°C, for 7–10 days. Agar discs (6 mm) were punched from these cultures and inoculated into 250 mL erlenmeyers flasks containing 50 mL of the liquid media described below.

After cultivation period of 7–10 days at 28°C, 10 mL of these cultures were used as inoculum.

2.2 *Media composition*

Modified yeast extract (MYE): yeast extract 1 g, peptone 1 g and glucose 10 g per liter of distilled water. Potato dextrose (PD): potato 140 g and dextrose 10 g per liter of distilled water. Minimum medium with cellulose (MMC): $NaNO_3$ 0,0243 g, KH_2PO_4 6 g, KCl 0,5 g, $MgSO_4{\cdot}7H_2O$ 1,025 g, $FeSO_4{\cdot}7H_2O$ 0,0367 g, $ZnSO_4{\cdot}7H_2O$ 0,01 g, cellulose 0,2 g per liter of distilled water.

2.3 *Dye*

The dye used it was Remazol Brilliant Blue-R (RBB-R). Spectrophotometric scanning of a solution with 0.1 g dye/L was performed and the maximum wavelength identified was 590 nm.

2.4 *Decolorization assays*

The inoculum was added to 125 mL erlenmeyers flasks containing 50 mL of the liquid media added by RBB-R 0,5 g/L. Experiments were carried out at 28°C, for fifteen days. The biotic controls were prepared using dead inoculum. The decolorization percent, at 590 nm, was determined through Eq. 1.

$$Decolorization(\%) = \frac{Abs_c - Abs_t}{Abs_c} \quad (1)$$

where: Abs_c = absorbance of the control experiment; Abs_t = absorbance of the test experiment.

2.5 *Enzymes assays*

Manganese peroxidase activity (MnP) was determined by oxidation of the manganese sulphate 4.5 mM in malonate sodium buffer 50 mM (pH 4.5) in the presence of 10 mM H_2O_2. The increase in absorbance at 270 nm was determined at ambient temperature for five minutes. Lignin peroxidase activity (LiP) was determined with veratryl alcohol 4 mM as substrate in the presence of 10 mM H_2O_2. The assay was performed in 0.5 M sodium tartrate buffer, pH 3.0. The change in absorbance was measured at 310 nm at 37°C for five minute. Laccase activity (Lcc) was determined spectrophotometrically as the absorbance increase at 525 nm at 50°C for five minutes. The reaction medium contained 0.5 mM syringaldazine in 0.1 mM phosphate buffer, pH 6.5 (Souza et al. 2002).

2.6 *Additional analysis*

After filtration, the mycelium was washed with distilled water and dried overnight at 105°C, for determination of biomass production. Residual glucose in the culture filtrate was determined using the glucose oxidase kit from Laborclin (Brazil). The chemical oxygen demand (COD) was determined according to Standard Methods (1995).

3 RESULTS AND DISCUSSION

Three different media were evaluated in order to determine the best medium for the biodegradation of remazol brilliant blue-R dye by three white-rot fungi.

The biomass growth (fig. 1a) in the MMC medium was less than in the other two tested media and COD (fig. 3a) remained practically constant. Despite the fungi have produced enzymes, which were detected by Lcc (fig. 4a), MnP (fig. 5a) and LiP (fig. 6a), there was not good decolorization percents. The isolate

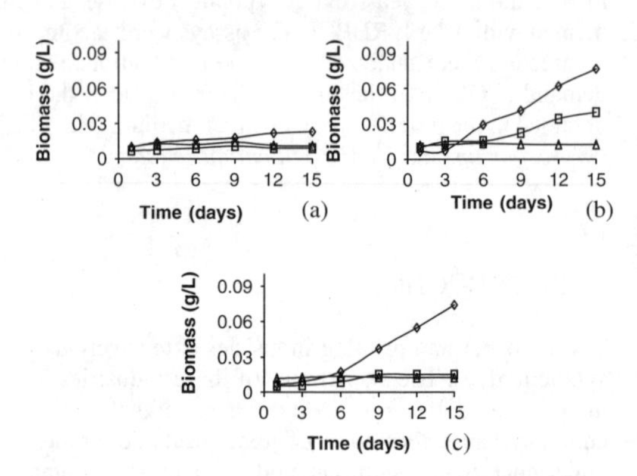

Figure 1. Results of biomass growth on (a) MMC, (b) PD and (c) MYE for the isolates (◇) *P. pulmonarius* CCB019, (□) *P. chrysosporium* CCB478 and (△) *L. edodes* CCB047.

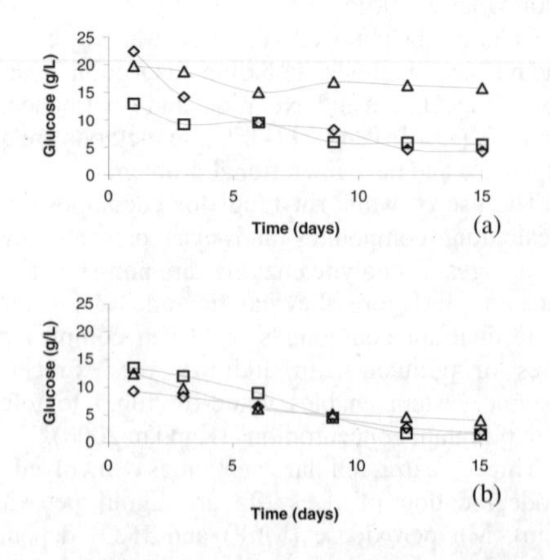

Figure 2. Residual glucose on (a) PD and (b) MYE. (◇) *P. pulmonarius* CCB019, (□) *P. chrysosporium* CCB478, (△) *L. edodes* CCB047.

P. pulmonarius CCB019 presented the best result, 34.1% (Table 1).

In the PD medium, the isolates *P. pulmonarius* CCB019 and *P. chrysosporium* CCB478 presented the best results of biomass growth (fig. 1b), glucose consumption (fig. 2a) and the COD reduction (fig. 3b). The three isolates produced MnP (fig. 5b) and LiP (fig. 6b), however, only the isolate *P. pulmonarius* produced laccase (fig. 4b). This isolate was responsible for the best dye decolorization result using PD medium, 81.1% (Table 1).

When the MYE medium was used, the fungus *Pleurotus pulmonarius* produced the greater biomass quantity (fig. 1c). The three fungi consumed glucose (fig. 2b). COD reduction (fig. 3c) was observed only for the isolates *P. pulmonarius* CCB019 and *P. chrysosporium* CCB478. In this medium, the enzymatic activity results (fig. 4c; fig. 5c; fig. 6c) were better than the ones of the others media, resulting in better decolorization percents, which were greater than 40% for the three fungi (Table 1).

According to Kapdan et al. (2000) the fungi consume and grow on readily available carbon sources at

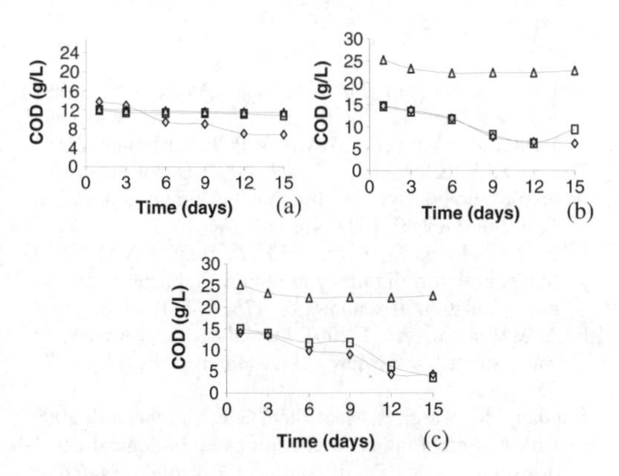

Figure 3. Chemical oxygen demand (COD) on (a) MMC, (b) PD and (c) MYE. (◇) *P. pulmonarius* CCB019, (□) *P. chrysosporium* CCB478, (△) *L. edodes* CCB047.

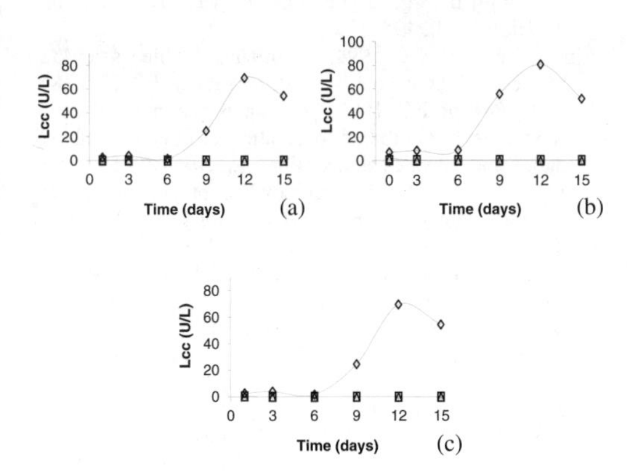

Figure 4. Laccase activity on (a) MMC, (b) PD and (c) MYE. (◇) *P. pulmonarius* CCB019, (□) *P. chrysosporium* CCB478, (△) *L. edodes* CCB047.

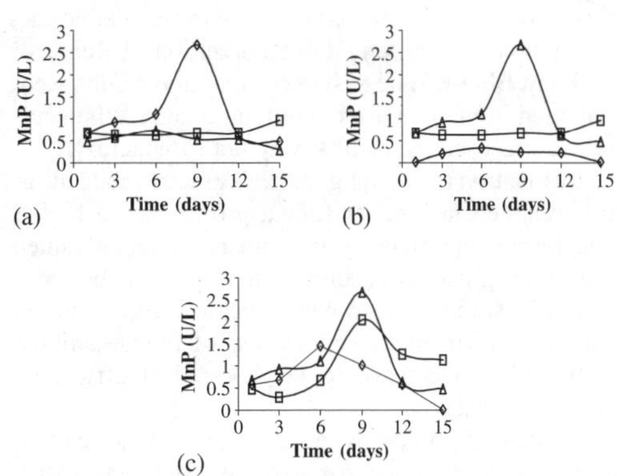

Figure 5. Manganese peroxidase activity on (a) MMC, (b) PD and (c) MYE. (◇) *P. pulmonarius* CCB019, (□) *P. chrysosporium* CCB478, (△) *L. edodes* CCB047.

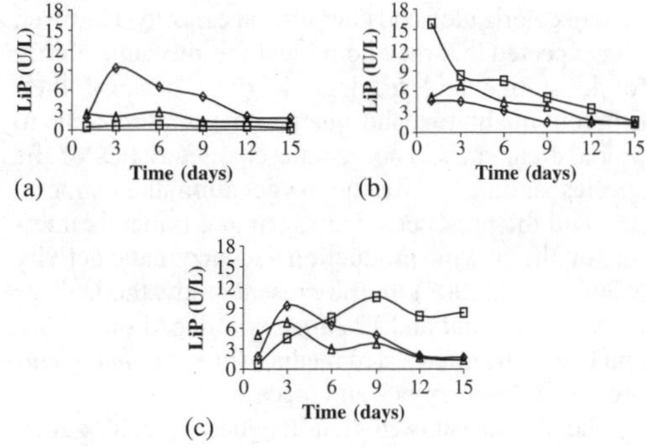

Figure 6. Lignin peroxidase activity on (a) MMC, (b) PD and (c) MYE. (◇) *P. pulmonarius* CCB019, (□) *P. chrysosporium* CCB478, (△) *L. edodes* CCB047.

Table 1. RBB-R decolorization percent.

Medium	Dye removal efficiency (%)	Fungi
Potato dextrose	81.1	*P. pulmonarius*
	15.9	*P. chrysosporium*
	7.8	*L. edodes*
Medium minimum with cellulose	34.6	*P. pulmonarius*
	11.7	*L. edodes*
	15.9	*P. chrysosporium*
Modified yeast extract	70.7	*P. pulmonarius*
	39.4	*L. edodes*
	93.3	*P. chrysosporium*

the initial stages of growth and then produce secondary metabolites and extracellular enzymes for biodegradation of dyestuffs at low concentrations of carbon or nitrogen. The most readily useable carbon source by white-rot fungi is glucose. This was also observed in this work. Glucose seemed to be the most suitable carbon source for the fungal decolorization of Remazol Brilliant Blue-R. The best decolorization results were obtained in the glucose containing media. When cellulose was used the results were not satisfactory.

Cultivation of the fungi at high glucose concentration resulted in a better fungal growth and a higher decolorization efficiency as compared to that obtained with a low glucose concentration. Lower glucose concentrations, either at growth or decolorization phase, caused insufficient growth or loss of fungal activity, leading to a decrease in color removal efficiency (Kapdan 2002).

Laccase, lignin peroxidase, manganese peroxidase and H_2O_2 dependent peroxidases are functional extracellular enzymes produced by fungi in biodegradation of lignin and dyes. Some white rot fungi produce all these enzymes while others produce only one or two of them (Balan 2001). Ligninolytic cultures of several white-rot fungi have been documented by their decolorization and degradation capacity. However, it is expected that they do not exhibit the same ability to decolorize and to degrade dyes because their different qualitative and quantitative requirements to produce enzymes. The genetic characteristics of the species/strains, the pH, the oxygenation, the temperature and the presence of nitrogen and minerals interfere on the enzyme production and enzymatic activity (Santos et al. 2002). In the present study, the isolates *P. chrysosporium* and *L. edodes* produced only MnP and LiP, in the three tested media, while *P. pulmonarius* produced this enzymes and laccase.

The results showed that the best decolorization (Table 1) percents, for the three fungi, were obtained using the modified yeast extract medium. The best COD reduction (fig. 3) was 75.5% that was obtained using the fungus *Phanerochaete chrysosporium*. It was not observed the dye adsorption by the inoculum and the enzymes Lcc, MnP and LiP was detected in the medium. This results indicated that the enzymatic system of the fungi were responsible by the dye degradation. The maximum enzymatic activities (fig. 4, 5 and 6) were 69.4 U/L of laccase for the isolate *Pleurotus pulmonarius* CCB019, 2.7 U/L of MnP for *Lentinus edodes* CCB047 and 10.6 U/L of LiP for *Phanerochaete chrysosporium* CCB478.

4 CONCLUSIONS

In this work, the selected medium, among the three tested media, was the modified yeast extract medium. Using it the best results of RBB-R decolorization was 70.7%, 93.3% and 39.4% using the fungi *Pleurotus pulmonarius, Phanerochaete chrysosporium* and *Lentinus edodes,* respectively. The results of this research indicated that *Pleurotus pulmonarius*, *Phanerochaete chrysosporium* and *Lentinus edodes* were able to decolorize and to degrade the remazol brilliant blue-R dye. This indicates their potential to be used in the treatment of effluents containing this dye.

ACKNOWLEDGEMENT

The authors acknowledge the financial support of CAPES and the Chemical Engineering Department – State University of Maringá.

REFERENCES

APHA – American Public Health Association. 1995. Standard Methods for the Examination for Water and Wastewater. 19th ed., AWWA, WPCF, Washington, D.C.

Balan, D.S.L. & Monteiro, R.T.R. 2001. Decolorization of textile indigo dye by ligninolytic fungi. *Journal of Biotechnology* 89: 141–145.

Field, J.A., Jong, E., Costa, G.F. & Bont, J.A.M. 1993. Screening for ligninolytic fungi applicable to the biodegradation of xenobiotics. *TIBTEC* 11: 44–8.

Fu, Y. & Viraraghavan, T. 2001. Fungal decolorization of dye wastewaters: a review. *Bioresource Technology* 79: 251–262.

Kapdan, I.K., Kargi, F., McMullan, G. & Marchant, R. 2000. Effect of environmental conditions on biological decolorization on textile dyestuff by *Coriolus versicolor*. *Enzyme and Microbial Technology* 26: 381–387.

Kapdan, I.K. 2002. Biological deodorization of textile dyestuff containing wastewater by *Coriolus versicolor* in a rotating biological contactor. *Enzyme and Microbial Technology* 30: 195–199.

Lin, S.H. & Peng, F.C. 1994. Treatment of textile wastewater by electrochemical methods. *Water Research* 2: 277–282.

Lin S.H. & Peng F.C. 1996. Continuous treatment of textile wastewater by combined coagulation, electrochemical oxidation and activated sludge. *Water Research* 30: 587–592.

Reddy, C.A. 1995. The potential for white-rot fungi in the treatment of pollutants. *Cur Opin Biotechnology* 6: 320–328

Santos, A. Z., Tavares, C.R.G., Costa, S.M.G. & Neto, J.M.C. 2002. Decolorization of a commercial reactive dye by white-rot basidiomicetes in liquid medium. *SHEB.*

Souza, C.G.M., Zilly, A. & Peralta, R.M. 2002. Production of laccase as the sole phenoloxidase by a Brazilian strain of *Pleurotus pulmonarius* in solid-state fermentation. *J. Basic Microbiol.* 42: 83–90.

Zilly, A., Souza, C.G.M., Barbosa-Tessmann, I.P. & Peralta, R.M. 2002. Decolorization of industrial dyes by a Brazilian isolate of *Pleurotus pulmonarius* producing laccase as the sole phenol-oxidizing enzyme. *Folia Microbiol.* 47: 273–277.

European Symposium on Environmental Biotechnology, ESEB 2004 - Verstraete (ed)
© 2004 Taylor & Francis Group, London, ISBN 90 5809 653 X

Degradation of MTBE by soil isolates

M. Vosahlikova, J. Pazlarova & K. Demnerova
Department of Biochemistry and Microbiology, Institute of Chemical Technology, Technicka, Prague, Czech Republic

Z. Krejcik
Institute of Molecular Genetics, Academy of Sciences of the Czech Republic, Flemingovo Prague, Czech Republic

T. Cajthaml
Department of Ecology, Institute of Microbiology, Academy of Sciences of the Czech Republic, Videnska, Prague, Czech Republic

ABSTRACT: Methyl tert-butyl ether (MTBE) has been used to improve engine performance and enhance air quality in the U.S. and Europe for over 20 years, and it is currently used around the world. Nowadays MTBE is frequently detected in surface water and groundwater. The aim of this study was to find a prospective MTBE-degrading bacterial strain or a consortium. The potential for aerobic MTBE degradation was investigated with isolates from contaminated soils (4A and 6A) and a consortium of microorganisms from contaminated water (Polepy) under laboratory conditions. Growth of studied bacterial strains was supported by n-alkane as a cosubstrate. Growth rate of tested bacterial strains was monitored and MTBE concentration after biodegradation was measured by gas chromatography (GC).

1 INTRODUCTION

Methyl tert-butyl ether (MTBE) has been used as a gasoline additive since the late 1970s in an effort to increase combustion efficiency and reduce air pollution. A major source of MTBE in environment comes leaking underground storage tanks at gas station. MTBE has emerged as an important water pollutant because of its persistence, toxicity, mobility and widespread use. A report by the U.S. Geological Survey identified MTBE as the second most common volatile organic contaminant of urban aquifers in the United States. The major aromatic components of gasoline, benzene, toluene, ethylbenzene, and xylene (BTEX) are degraded relatively rapidly under aerobic and anaerobic conditions. When MTBE containing gasoline is released, BTEX plumes may still stabilized on average within 100 m from the source of contamination but MTBE might continue to migrate. While much is known about the biodegradation of many gasoline components under both aerobic and anaerobic conditions, the process is not fully understood in the case of MTBE. However, recent studies have reported the ability of several bacterial and fungal cultures from various environmental sources to degrade MTBE under aerobic or anaerobic conditions either as the sole source of carbon and energy or cometabolically (Stocking, 2000).

2 MATERIALS AND METHODS

2.1 *Used bacterial strains*

There were used pure bacterial isolates (4A and 6A) from soil contaminated with gasoline, consortium Polepy (bacterial mixture from a contaminated groundwater) and consortium Landa (bacterial mixture from contaminated soil).

2.2 *16S rDNA analysis*

16S rDNA amplification was performed by PCR using the forward-primer 5′-AGA GTT TGA TCM TGG CTC AG-3′ and the reverse-primer 5′-TAC GGY TAC CTT GTT ACG ACT T-3′. The products of amplification were purified with PCR purifying kit (Sigma), and both strands were sequenced (Beckman Coulter). The percentage of identity was determined using the EMBL/GenBank database with the Blast alignment tool.

2.3 *Incubation medium and culture conditions*

All tested bacterial strains were grown in ABC medium (Kostal, 1998), and the carbon source was added before inoculation. The cultures were inoculated in 0.5 l flasks with 100 ml ABC and they were incubated on a rotary shaker at 28°C. The growth was aerobic. The volume of headspace was sufficient to prevent any limitation by O_2 during growth.

Different source of carbon and energy were tested: n-alkanes (Al) (max 1% v/v), MTBE (max 1% v/v), tert-butyl alcohol (TBA) (0.2 v/v), toluene (T) (0.5% v/v), cyclohexane (C) (0.5% v/v), casamino acids (CA) (100 mg/l), amino acids (AA) (100 mg/l), lactic acid (LA) (100 mg/l) and yeast extract (YE) (100 mg/l).

The growth was monitored with three different methods. The first was a monitoring of the optical density at 650 nm (OD_{650}) with Heλios β spectrophotometer. The flasks were inoculated to obtain an initial OD_{650} of 0.1–0.2 (Francois, 2002).

The second method was measuring of biochemical oxygen demand (BOD). The BOD was monitored by using system OxiTop®. OxiTop® system is based on pressure detection. It is realised by pressure difference measurement via piezoresistive electronic pressure sensor. The system does not use any mercury (Hg) and was developed by Merck Company (Instruction manual). The cultivation was carried out in special brown sample bottles with stirrings in a thermostat (20°C). The data were automatically daily stored in 5 days.

The third method was realised by monitoring of bacterial growth using system Bioscreen®. Cultivation in the Bioscreen® system took place on a special plastic plate with 100 cultivation wells (maximal filling volume was 330 μl). Bioscreen® provided measuring of optical density in each well in regular intervals for several days.

2.4 *Measurement of MTBE degradation*

The cultures were inoculated either with 0.5% Al (v/v) and 0.5% MTBE (v/v) or 5% MTBE (v/v). They were incubated on a rotary shaker at 28°C for 14 days. The samples were not modified after the degradation time.

2.5 *Analytical assay*

MTBE was quantified by gas chromatography (Agilent 6890N). The instrument was equipped with flame ionisation detector and a capillary column HP-5 (30 m × 0.32 mm × 0.25 μm). The analyses were performed at constant temperature 40°C. The time of a run was 3 min. Headspace of the samples was injected by autosampler Agilent 7683.

TBA was quantified on a gas chromatography (Varian 3400) with mass detector ITS 40 (Finnigan) and capillary column SE-624 (60 m × 0.25 mm × 0.25 μm). An isothermal program was applied at 40°C (11 min). Headspace of the samples was injected as well.

3 RESULTS AND DISCUSSION

3.1 *Characterization of the isolate 4A*

The strains 4A and 6A were isolated from contaminated soil cultivated on ABC medium with Al. 4A was a gram-positive, pink, strictly aerobic, oval rod. Comparison of the 16S rRNA gene sequence with sequences of the EMBL/GenBank database showed 99% identity with *Rhodococus pyridinivorans*.

3.2 *Growth of bacterial strains on MTBE*

All the tested bacterial strains or the mixture were isolated from industrial contaminated soils. First of all the capability of growth in the presence of MTBE under aerobic conditions was tested. The results are summarized in the Table 1. These experiments were performed in the system Bioscreen®. The effect of MTBE on the growth of all tested strains was considered to be positive. In the case when MTBE was only source of carbon and energy a specific growth rate (μ) was the lowest.

Table 1. The growth of the bacterial strains (4A, 6A, Polepy and Landa) in the presence of different concentrations of MTBE and Al.

Tester bacterial cultures	Source of carbon and energy	$\mu\ (h^{-1})10^{-1}$
4A	1% Al	0.25
4A	1% Al + 0.5% MTBE	0.22
4A	0.5% MTBE	0.17
6A	1% Al	0.07
6A	1% Al + 0.5% MTBE	0.12
6A	0.5% MTBE	0.08
Polepy	1% Al	0.11
Polepy	1% Al + 0.5% MTBE	0.11
Polepy	0.5% MTBE	0.10
Landa	1% Al	0.09
Landa	1% Al + 0.5% MTBE	0.08
Landa	0.5% MTBE	0.06

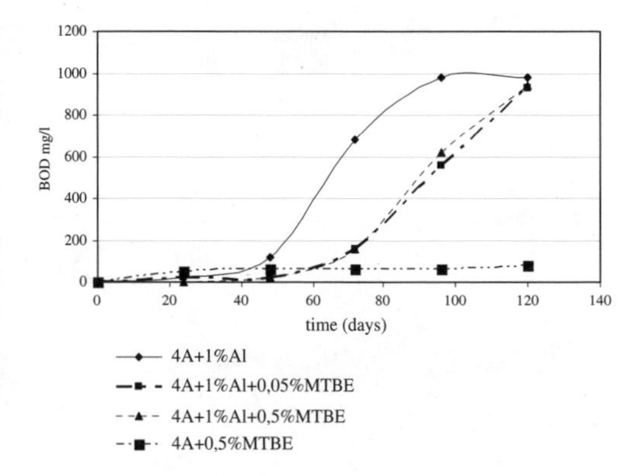

Figure 1. Influence of MTBE on BOD by soil isolate 4A.

3.3 *Monitoring of biochemical oxygen demand (BOD)*

The bacterial strain 4A was chosen for the next experiments. This strain was tested on biochemical oxygen demand. The method was optimized for this strain and the best results were obtained with combination of inoculation OD_{650} 0.3 and 100 ml volume of the medium in a bottle. The results are shown in Figure 1.

3.4 *Growth of bacterial strains on different sources of carbon*

The strain 4A was tested on its ability to growth with another source of carbon. The substrates TBA and lactic acids were proposed as potential metabolites, T and C as potential cometabolites, and CA, AA and YE as compounds supporting growth 4A in the presence of MTBE (Francois, 2002). These measurements were carried out with the Heλios β spectrophotometer. The results are summarized in Table 2.

3.5 *MTBE degradation by 4A*

The possible MTBE-degradation was tested by the strain 4A. GC analysis showed that decrease of

Table 2. Growth of bacterial strains 4A in the presence of different primary substrates.

Primary substrate	Growth
Toluene	no
Cyclohexane	no
TBA	no
Lactic acids	yes
Casamino acids	no
Amino acids	no
Yeast extract	no

MTBE is very low, although TBA was detected as the metabolite (Deeb, 2000). The results will be presented on poster.

ACKNOWLEDGEMENT

This work was supported by grant MSMT no. LN00B030.

REFERENCES

Deeb, R.A., Scow, K.M. & Alvarez-Cohen, L. 2000. Aerobic MTBE biodegradation: an examination of past studies, current challenges and future research directions. *Biodegradation* 11: 171–186.

Francois, A., Mathis, H., Godefroy, D., Piveteau, P., Fayolle, F. & Monot, F. 2002. Biodegradation of methyl *tert*-butyl ether and other fuel oxygenates by a *new strain, Mycobacterium austroafricanum* IFP 2012. *Applied and Environmental Microbiology* 68: 2754–2762.

Instruction manual, Manometric BOD Measuring Devices, BA31107/07.00/AS/OxiTop IS-6_IS-12-1.

Kostal, J. 1998. Degradation of n-alkanes by *Pseudomonas* C12B, *Dissertation thesis.* ICT Prague, Czech Republic.

Stocking, A.J., Deeb, R.A. & Kavanaugh, C. 2000. Biodegradation of MTBE: review from a practical perspective. *Biodegradation* 11: 187–201.

European Symposium on Environmental Biotechnology, ESEB 2004 - Verstraete (ed)
© 2004 Taylor & Francis Group, London, ISBN 90 5809 653 X

The role of plant peroxidases in metabolism of polychlorinated biphenyls

M. Mackova[1,3], P. Lovecka[1], E. Ryslava[1], L. Kochankova[2] & K. Demnerova[1,3]
[1]*Dept. of Biochemistry and Microbiology, Faculty of Food and Biochemical Technology*
[2]*Dept. of Environmental Chemistry, Institute of Chem. Technol. Prague, Technicka, Prague, Czech Republic*

J. Rezek[1,3] & T. Macek[1,3]
[3]*Department of Natural Products, Institute of Organic Chemistry and Biochemistry, Czech Academy of Sciences, Flemingovo, Prague, Czech Republic*

ABSTRACT: Plant cell cultures of three plant species (alfalfa, black nightshade, tobacco) were tested for extracellular and intracellular synthesis of peroxidases and ability of isolated enzymes transform polychlorinated biphenyls. Peroxidase from alfalfa exhibited the same activity with and without presence of PCBs and did not metabolized PCBs. Black nightshade and tobacco synthesized higher levels of enzyme after addition of PCBs. The transformation of PCBs by extracellular enzyme was less efficient than by intracellular one but the enzyme exhibited higher stability. The intracellular enzyme was concentrated and partially purified. After 22 hours it transformed PCB 2 by 35%, PCB 3 by 100% and PCB 4 and PCB 9 by 60%.

1 INTRODUCTION

Contamination of soils and waters with PCBs is often associated with the manufacturing, use, and disposal of these chemicals. Although the chemical stability had been a benefit from the standpoint of commercial use, it has created an environmental problem because it translated into extreme persistence when PCBs were eventually released into the environment. Despite the fact that industrial use of PCBs was severely reduced, their persistence in the environment and ability to bioconcentrate in the food chain mean human health concerns are still warranted. In fact PCBs are among the most widespread environmental pollutants. Exposure to low levels of PCBs is thought to cause various acute and chronic health effects. Currently available engineering-based remedial technologies for PCB contaminated soils are disruptive and expensive. Biological decontamination is a promising strategy for clean up of PCB contaminated areas, since it is considered as safer nondestructive and less expensive method to remove harmful pollutants compared to physical or chemical processes. PCBs can be degraded by microbes (bacteria, fungi) (Abramowicz, 1990), recently also the ability of plants to transform and metabolise PCBs was described (Schnoor et al., 1995, Mackova et al., 1997, Koeller et al., 2000, Bock et al., 2002). Several steps are necessary before actual field implementation of phytoremediation operation technique. Plants in contaminated soils always include transformation of pollutants, decreasing vertical and lateral migration of pollutants to ground water by extracting water and reversing the hydraulic gradient, improvement of aeration of the soil and stabilization against wind and erosion.

Enzymatic degradation of xenobiotics in plants usually involves several enzymatic steps in which detoxified product of some stability is formed. The first step is called activation and reactions might include ring hydroxylation of aromatics, oxidation, reduction, or hydrolysis of chemical bonds. Activation is in most cases catalysed by cytochrome P450 or peroxidases (POX) (Koeller et al., 2000, Chromá et al., 2002). These enzymes are localized in membrane fractions, in the apoplast, and in the cytosol (POX), sometimes they can be released into soils and waters. The degree of enzyme release into soils and sediments remains poorly understood but the measured half-life of these enzymes suggests, that they may actively degrade soil contaminants for days following release from plant tissues (Schnoor, 1995). *In vitro* POXs catalyse the oxidation of a variety of compounds in the presence of hydrogen peroxide.

Understanding the basic physiology and biochemistry that underlie various phytoremediation processes is very important to improve the applicability of this plant based method. Although organic pollutants are metabolised in plants, some xenobiotics can be toxic, limiting the applicability of phytoremediation. Thus more research is required to better understand and exploit the reactions in plants.

In our studies we used three plant species (alfalfa, tobacco, black nightshade), previously successfully used in laboratory phytoremediation studies investigating the abilities of plant cells to degrade PCBs. In described study the process of PCBs transformation was studied from the point of intracellular and extracellular peroxidase synthesis, products formed and ability of isolated POX to remove PCBs from the environment.

2 MATERIALS AND METHODS

2.1 *Plant cell cultures*

In vitro cultures of three species (alfalfa, black nightshade, tobacco) and of different morphology (non-differentiated amorphous callus strains crown-gall cultures and hairy root clones transformed by Ti or Ri plasmid of *Agrobacterium tumefaciens* and *A. rhizogenes*, respectively) (Mackova et al., 1997) were from the collection of the Dept. of Natural Products, IOCB, CAS, Prague.

2.2 *Cultivation conditions*

Plant cell cultures were incubated aseptically in Murashige and Skoog's nutrient medium for 14 days on rotary shaker (100 rpm) in the dark at 24°C. Usually 5 g of cells were used as inoculum, and the cultivation was carried out in 100 ml of liquid media in 250 ml Erlenmeyer flasks.

2.3 *PCBs*

A standard commercial mixture (previously produced in Czechoslovakia) of PCBs, Delor 103 (D), was used as methanol solution. This mixture contains about 60 individual congeners of PCBs substituted by 1–5 chlorine atoms per biphenyl molecule. The initial concentration of Delor 103 was 25 mg/l.

2.4 *Analysis of residual content of PCBs*

After the cultivations with PCBs, the cells were killed by boiling for 20 min., sonicated, homogenised with UltraTurax. The whole content of flasks was extracted with hexane for 2 hours. Samples were analyzed using the Hewlett-Packard 5890 gas chromatograph with an electron capture detector (ECD) and autosampler (Kucerova et al., 1999, 2000). Degradation was expressed as percentage of PCBs removed comparing to control flasks. For the calculation of the residual amount of PCBs, 22 main chromatographic peaks of Delor 103 were used. Controls comprising of heat-killed cells were used to establish that observed changes in the content of congeners of PCBs were dependent exclusively on the activity of living cells.

2.5 *Extraction and purification of plant peroxidases*

To estimate the changes in peroxidase activity, additional flasks with the same content of inoculum were incubated parallely with the samples for estimation of PCBs transformation. The flasks contained 2.5 or 5 mg (conc. 25 mg/l) of the mixture Delor 103 (+D), individual PCB congeners in concentration of 0.3 mg/100 ml in each flask, or no Delor 103 (−D). Cell extracts were prepared by homogenization of 1 g frozen fresh cells by pestle and mortar with 1 ml of 0.1 M phosphate buffer (pH 6.5), and centrifugation for 15 minutes at 5000 rpm.

2.6 *Estimation of POX activity*

The total peroxidase activity in the cell extract and in the medium was measured by the spectrophotometric method by using guaiacol and hydrogen peroxide at 470 nm as described by Chromá et al. (2002) or Strycharz and Shetty (2002). The isoenzyme pattern of peroxidases in the cell extract and in the medium was analysed by native electrophoresis in polyacrylamide gel after visualization of peroxidase isoenzymes with guaiacol and hydrogen peroxide (Kucerova et al., 1999).

2.7 *Analysis of proteins*

Protein content was measured by the method described by Bradford (1976).

2.8 *Concentration and partial purification of intracellular peroxidases*

Proteins from alfalfa, black nightshade and tobacco suspended in 0.1 M phosphate extraction buffer were precipitated and partially purified by salting out (40–80% saturation of ammonium sulphate). Following precipitation, protein suspension was centrifuged (10 min., 10 000 rpm) and desalted by gel chromatography on commercial columns PD-10 (Pharmacia) containing Sephadex G-25 as a resin.

2.9 *Reactions of isolated peroxidases with individual congeners of PCBs*

Concentrated peroxidase preparations were tested for their transformation potential towards individual PCB congeners. Four different individual congeners of monochlorobiphenyls (IUPAC No.: PCB 2, 3) and dichlorobiphenyls (PCB 4, 9) in concentrations of

Table 1. Degradation of PCB by intracellular plant peroxidases isolated and partially purified from the cultures of black nightshade and tobacco after 20 hours incubations.

IUPAC no.	Toxicity (%)*	Degradation of PCB by POX (%)	
		From nightshade	From tobacco
Delor	–	45 ± 5	20 ± 3
PCB 2	15.0	35 ± 4	–
PCB 3	8.0	100 ± 8	79 ± 10
PCB 4	24.4	56 ± 7	60 ± 8
PCB 9	20.1	66 ± 7	–

* Inhibition coefficient calculated by the difference between the growth of the control roots without toxicant and the growth of the roots with toxicant divided by the value of the growth without presence of the toxicant.

Table 2. Degradation of PCBs by extracellular plant peroxidases isolated from the culture medium of black nightshade SNC 90.

Incub. days	PCB degradation by extr. POX Delor 103 (%)	PCB 2 (%)	PCB 3 (%)	PCB 9 (%)	POX activity nkat/mg
0	–	–	0	0	5.2
3	34 ± 2.8	–	23 ± 2	15 ± 1.3	4.3
6	30 ± 1.9	–	27 ± 2.5	20 ± 2.4	3.7
9	38 ± 4.1	30 ± 2	34 ± 3	17 ± 1.8	1.6

5–10 μmol/l (see Table 1) were used as substrates for POX from black nightshade and tobacco cells (see Table 2). Reactions were started by the addition of hydrogen peroxide. After 22 hours of the reaction at 25°C the residual active POX was inactivated by adding 1 ml of sulphuric acid (10%) and the remaining PCBs were extracted by hexane (Mackova et al., 1997) and analysed by GC-ECD.

3 RESULTS AND DISCUSSION

In our studies, synthesis of intracellular and extracellular peroxidases by three plant species – alfalfa (*Medicago sativa*), black nightshade (*Solanum nigrum*) and tobacco (*Nicotiana tabacum*) and the ability of these enzymes to transform polychlorinated biphenyls were studied. These plants were previously used for the experiments on PCB transformation (Mackova et al., 1997) and their removal. In laboratory we used as models plant tissue cultures cultivated *in vitro* exhibiting faster growth independent of climate and weather conditions.

Comparing abilities of three plant species, black nightshade proved the best transformation of PCBs at laboratory conditions (Kucerova et al., 1999, Chroma et al., 2002). The cells of alfalfa metabolised very low quantities of PCBs and finally the lowest PCB removal, comparing to black nightshade and tobacco cells, was measured. Hydroxychlorobiphenyls were identified as the products of the initial phase of PCB metabolism in plant cells (Kucerova et al., 2000). Very low amounts of chlorobenzoic acids, benzoic acids were also detected (Kucerova et al., 2000, Macek et al., 2000). Hydroxychlorobiphenyls were probably products of the reactions catalysed by cytochrome P450. From the literature it is known that also involvement of plant peroxidases can not be excluded (Koller et al., 2000, Chroma et al., 2002). For this reason we paid our attention to analysis of plant peroxidases of above mentioned plant species. We studied extracellular and intracellular production, changes in isoenzyme pattern and ability of isolated peroxidases *in vitro* to transform PCBs.

All three cultures synthesized both intracellular and extracellular POX showing the same or similar isoenzyme patterns when incubated without PCBs. POX of tobacco and black nightshade changed in the presence of PCBs their total activities and some changes in isoenzyme patterns were also visible after native electrophoresis. Total POX activity of alfalfa exhibited the highest values without PCB presence, but its activity and isoenzyme pattern was not influenced by the presence of PCBs and this plant was generally less active in PCB transformation (data not shown). Thus in further experiments we tested only the extracellular and intracellular peroxidases from tobacco and black nightshade cells. Extracellular POX from both sources was not purified and the activity against PCBs was evaluated in liquid medium after the removal of plant cells. Concentrated crude extracts of intracellular enzymes were partly purified and concentrated by ammonium sulphate precipitation (40–80% saturation) prior to reactions with PCBs. Both concentrated intracellular POX activities per ml of preparations exhibited similar values – 0.89 nkat/ml for black nightshade and 0.87 nkat/ml for tobacco. Mixture of PCBs – Delor 103 and four different individual congeners were used as substrates for POX from black nightshade and tobacco cells (Table 1 and 2). Parallely reaction mixtures with the same PCB congeners treated the same way as described above were extracted with toluene and after silylation (Koller et al., 2000) of the reaction products analysed by GC-MS.

From Table 1 and 2 it can be seen that the efficiency of degradation of chosen congeners was similar to intracellular peroxidases from both sources. Activities of POXs decreased during whole reaction time (22 hours) with all chosen congeners to

0.009–0.013 nkat/ml, i.e. 9–10 times. Efficiency of degradation of structurally different PCB is probably connected with their chemical structure, physico-chemical properties and also toxicity (see Table 1). Comparing three monochlorobiphenyls the most efficient transformation of PCB was estimated in reaction with PCB 3. This congener exhibited also the lowest toxicity measured by the growth of plant hairy roots (Lovecka et al., 2002).

Small quantities of numerous products formed during reactions of POX, isolated from black nightshade, with PCBs were detected by GC-MS. Unfortunately reactions with tobacco POX did not give any detectable products. The concentration of products was probably under detection limit. In reaction mixtures containing POX from black nightshade, dechlorination was observed as the initial step. As metabolites less chlorinated biphenyls than the original ones or non-substituted biphenyl were detected. Koeller et al. (2000) documented presence of hydroxy-chloroderivatives. In our study surprisingly no hydroxylated chlorobiphenyls were found. This phenomenon can be explained by further fast reactions following the first dechlorination step without any accumulation of intermediates (Koller et al., 2000). Regarding oxidative degradation, the cleavage of the ring and subsequent reactions gave benzoic acids and hydroxy-benzoic acids as the products. As a result of POX radical mechanism higher chlorinated isomers were formed. Traces of phenylacylchlorides were also found in reaction mixtures.

To confirm the metabolic potential of extracellular POX against PCBs, enzymes present in incubation medium, after cultivation and separation of plant cells of black nightshade were tested for the ability to transform chosen congeners PCB 2 (3-chlorobiphenyl), PCB 3 (4-chlorobiphenyl), PCB 4 (2,2′-dichlorobiphenyl, PCB 9 (2,5-dichlorobiphenyl) and mixture Delor 103 in concentration of 25 mg/l. Hairy root culture was chosen due to its morphological similarity to normal roots, bio-chemical resemblance to the roots of the plant, from which they have been derived, and high efficiency of intracellular POX to metabolise PCBs. PCBs were added to incubation medium immediately after aseptic removal of plant cells. Table 2 shows the results obtained within 9 days of the reactions.

The total activity of extracellular POX was lower than that of intracellular ones, on the other hand extracellular enzyme was used without any previous treatment (saturation with ammonium sulphate). As expected transformation potential of extracellular POX was lower than that of intracellular POX but extracellular enzyme exhibited much higher stability. Generally use of pure or partly purified enzymes is questionable due to their limited stability and thus use of the whole plant parts or plants extracellularly active in soil and water has been recommended (de Araujo, 2002). In our case the transformation by extracellular enzymes was less efficient but could be prolonged much longer than with intracellular enzymes. POX from other plant sources (tobacco, black nightshade) than horse-radish (the most studied model peroxidase) can be involved not only in defence mechanisms to toxic compounds and increase of the resistance of plants to toxic compounds, but they even can be involved in direct transformation and degradation of xenobiotics. POX from tobacco and black nightshade exhibited remarkable changes in the presence of PCBs, intracellular enzymes from both sources could transform individual congeners of PCBs. From above mentioned results can be concluded that plants transform PCBs using different oxidative enzymes, cytochrome P450 and also peroxidases.

ACKNOWLEDGMENT

The experiments described in this paper were supported by the grant 526/01/1292 of the Grant Agency of the Czech Rep. and ME 498 (KONTAKT CZE/026) of Ministry of Education of the Czech Republic and 5FW EU QLK 3-CT-2001-00101.

REFERENCES

Abramowicz, M., 1990. Aerobic and anaerobic PCB biodegradation in the environment. *Environmental Health Perspectives* 103, 97–99.

Bock, C., Kolb, M., Bokern, M., Harms, H., Mackova, M., Chroma, L., Macek, T., Hughes, J., Just, C., Schnoor, J., 2002. Advances in phytoremediation: Phytotransformation. *PCB – approaches to possible removal from the environment.* (Reible, D., and Demnmerova, K., eds.), NATO ASI Series, Kluwer Academic Publishers, Dordrecht.

Bradford, M.M., 1976. A rapid and sensitive method for the quantitation of microgram quantities of protein utilizing the principle of protein-dye binding. *Analytical. Biochemistry* 72, 248–253.

Chroma, L., Mackova, M., Kucerova, P., in der Wiesche, C., Burkhard, J., Macek, T., 2002. Enzymes in plant metabolism of PCBs nad PAHs. *Acta Biotechnologica* 22, 34–41.

de Araujo, B.S., Charlwood, B.V., Pletsch, M., 2002. Tolerance and metabolism of phenol and chloroderivatives by hairy root cultures of *Daucus carota* L. *Environmental Pollution* 117, 329–335.

Koeller, G., Moeder, M., Czihal, K., 2000. Peroxidative degradation of selected PCB: a mechanistic study. *Chemosphere* 41, 1827–1834.

Kucerova, P., Mackova, M., Polachova, L., Burkhard, J., Demnerova, K., Macek, T., 1999. Correlation of PCB transformation by plant tissue cultures with their morphology and peroxidase activity changes. *Coll. Czech. Chem. Commun.* 64, 1497–1509.

Kucerova, P., Mackova, M., Chroma, L., Burkhard, J., Triska, J., Demnerova, K., Macek, T., 2000. Metabolism of polychlorinated biphenyls by *Solanum nigrum* hairy root clone SNC-9O and analysis of transformation products. *Plant and Soil* 225, 109–115.

Lovecka, P., Melenova, I., Kucerova, P., Nováková, H., Macková, M., Ruml, T., Demnerová, K., 2002. Evaluation of ecotoxicity of PCB and their bacterial and plant degradation products using different biological systems. (Holoubek, Holoubkova eds.) In: *Book of Abstracts, The Second PCB Workshop, Brno, May, 2002,* Masaryk University.

Macek, T., Macková, M., Káš, J., 2000. Exploitation of plants for the removal of organics in environmental remediation. *Biotechnology Advances* 18, 23–35.

Mackova, M., Macek, T., Ocenaskova, J., Burkhard, J., Demnerova, K., Pazlarova, J., 1997. Biodegradation of polychlorinated biphenyls by plant cells. *International Biodeterioration and Biodegradation* 39, 317–325.

Ryslava, E., Krejčík, Z., Macek, T., Novakova, H., Mackova, M., 2003. Study of PCB biodegradation in real contaminated soil. *Fresenius Environ. Bulletin* 12, 296–301.

Schnoor, J.L., Licht, L.A., McCutcheon, S.C., Wolfe, N.L., Carreira, L.H., 1995. Phytoremediation of organic and nutrient contaminants. *Environmental Science and Technology* 29, 318–323.

Strycharz, S., Shetty, K., 2002. Peroxidase activity and phenolic content in elite clonal lines of *Mentha pyulegium* in response to polymeric dye R-478 and *Agrobacterium rhizogenes*. *Process Biochemistry* 37, 805–812.

European Symposium on Environmental Biotechnology, ESEB 2004 - Verstraete (ed)
© 2004 Taylor & Francis Group, London, ISBN 90 5809 653 X

Simultaneous hybrid processes as a way to improve characteristics of cultivation of micro-organisms and biodestruction of pollutants

A.Ye. Kouznetsov & S.V. Kalyonov
D.Mendeleyev University of Chemical Technology of Russia, Moscow, Russia

ABSTRACT: Some systems for investigation of the biosynthesis, biotransformation and biodestruction processes with a simultaneous chemical or photochemical reactions has been developed. It is expected that such hybrid processes will improve biotechnological performances and a variety of the biotechnological methods for pollutant's destruction.

As a hybrid process we define that one where chemical or photochemical reactions and biocatalysis or biodestruction in the same volume and simultaneously proceed. They are typical for natural ecosystems. For the purposes of practical application such processes could have an interest to improve bioavailability of resistant substrate, characteristics of the systems of biopurification and biodestruction, microbial growth and biosynthesis. For scientific research an evolution of these systems in conditions of autoselection has some interest as well.

Among various abiotic systems of transformation for hybrid processes of cultivation and biodestruction there are:

- Fenton reagent (oxidation of organic substances)

$$Fe^{2+} + H_2O_2 \rightarrow Fe^{3+} + OH^- + OH^{\cdot}$$

- Kostelfranco reagent (oxidation of organic substances)

$$\text{ascorbate} + Cu^{2+}$$

- photochemical oxidation

$$C_{org.} \xrightarrow{h\nu\ 330\text{–}380\,nm} CO_2 + H_2O$$

- advanced photocatalytic oxidation

$$TiO_2 + C_{org.} \xrightarrow{h\nu\ 330\text{–}380\,nm} TiO_2 + CO_2 + H_2O$$

- hydrolysis of organic substances, catalyzed by TiO_2, FeOOH, Al_2O_3
- dechlorination by smectites with Cu(II)
- redox processes with oxides Fe(II, III) and Mn(III, IV).

Using model systems with addition of hydrogen peroxide or Fenton reagent ($H_2O_2 + Fe^{2+}$) into nutrient medium before or during fermentation as well as ultraviolet irradiation (UVA and UVB subrange) of cell suspensions during cultivation several hybrid processes have been studied:

- biodestruction of ion exchange resins as potential waste from atomic power station,
- biodestruction of phenol by an isolate with bacteria and yeast,
- cultivation of yeast *Candida tropicalis* on sucrose,
- cultivation of *Azotobacter* strains,
- biodestruction of phenol by consortia of yeast and bacteria *Azotobacter* on nitrogen free medium.

The summary of the investigations are following:

- pre-treatment of ion exchange resins by Fenton reagent improves their bioavailability;
- at the cultivation in fed batch conditions and at the addition of H_2O_2 the amount of extracellular metabolites and residual concentration of substrate were lower than in the control (without addition of H_2O_2); there was not an accumulation of metabolites inhibiting growth of micro-organisms – that was shown for the process of phenol biodestruction and for the growth of *C. tropicalis* on sucrose (Figs 1 and 2); these peculiarities kept for a long time after stoppage of H_2O_2 addition;
- in case of high density cultivation with fractional addition of dry sucrose and periodical addition of H_2O_2 the yeast cells of *C. tropicalis* accumulated

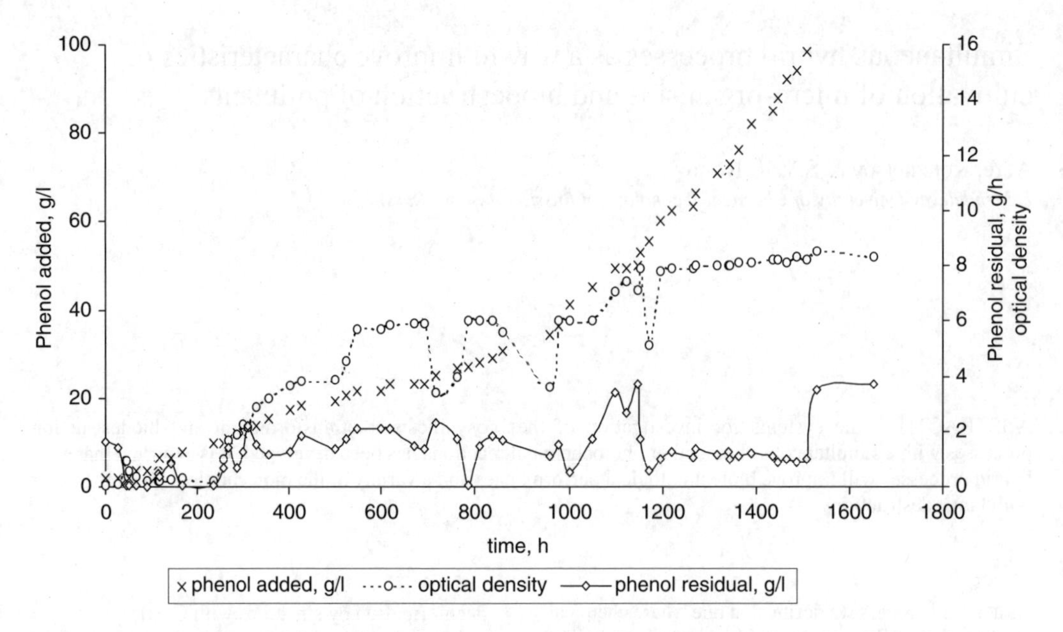

Figure 1. Biooxidation of phenol in fed batch conditions with fractional addition of phenol and small amounts of H_2O_2 (since 1 to 1000 h after start of the experiment) into the biooxidation zone and using yeast biocenosis preadapted to hydrogen peroxide.

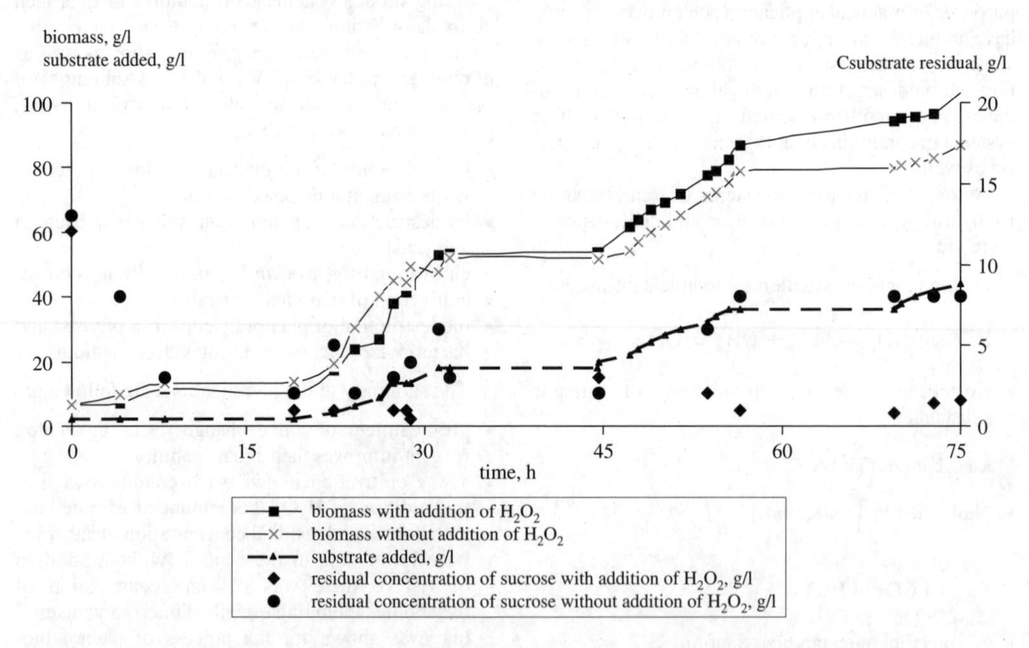

Figure 2. Growth of yeast *C. tropicalis* on sucrose in fed batch conditions with and without addition of H_2O_2.

up to the maximal level of 150–170 g of dry weight per one liter of culture liquid;

- the rate of oxidation of substrate in fed batch conditions was limited only by oxygen mass-transfer (for biodestruction of phenol and for the growth of *C. tropicalis*);
- diauxy in the consumption of sucrose and extracellular intermediates (acetic acid, ethanol)

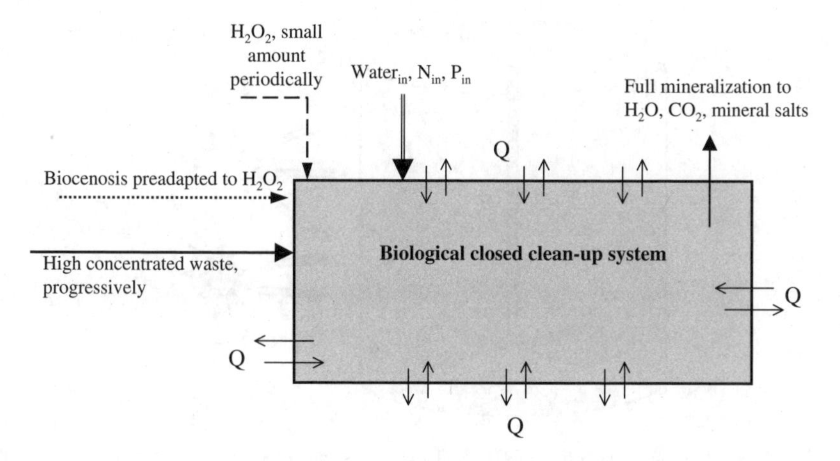

Figure 3. Closed intensive microbiological system for the high concentrated waste treatment.

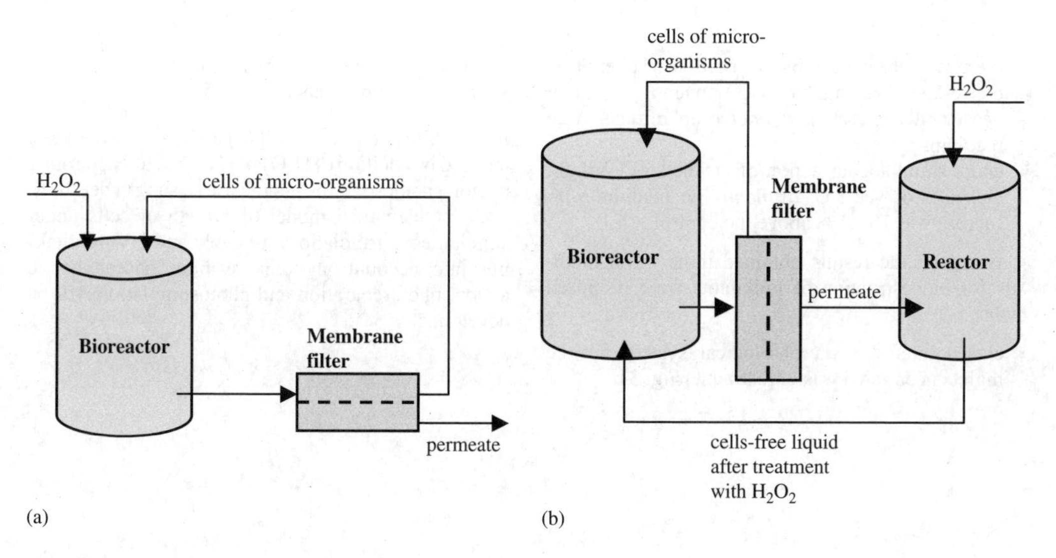

Figure 4. Artificial peroxisome.

temporarily accumulated in the culture liquid disappeared (for the growth of *C. tropicalis* on sucrose);

- there was not a dying phisiological activity of cells in fed batch conditions (for biodestruction of phenol and for the growth of *C. tropicalis*);
- cells are more resistant to H_2O_2 at the active stage of the growth (phenol biodestruction, growth of *C. tropicalis*);
- cells pre-adaptated to H_2O_2 were more resistant to various stresses:
  - their activity recovered better after substrate starvation;
  - they could grow at higher running concentration of substrate – up to 5 g of phenol per liter of medium – for phenol biodestruction; up to 700 g of sucrose per liter of medium – for *C. tropicalis* growth;
  - they were high resistant to hydrogen peroxide: population of phenol biodestructors survived after addition of 2% of H_2O_2; population of yeast *C. tropicalis* survived after one-time addition of 16.5% H_2O_2 (perhydrol diluted 2-fold and burning skin); the last one could be charged by addition up to 15 g of 100% H_2O_2 per 1 L of medium per 1 hour in continuous condition and could use H_2O_2 as a source of oxygen;
- addition of H_2O_2 to the populations with *Azotobacter* strains stimulated in some cases their growth on nitrogen free medium with sucrose;

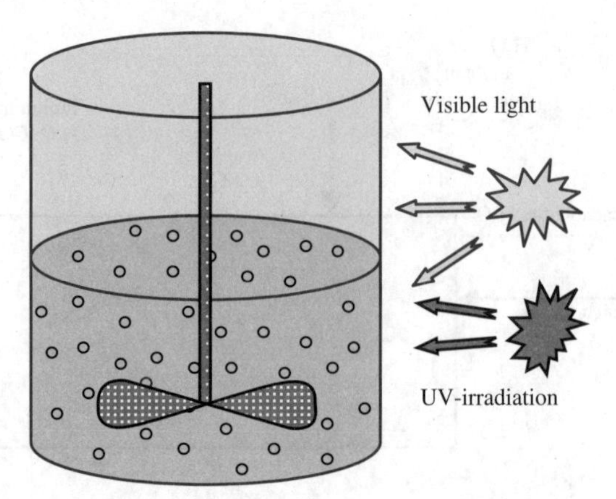

Figure 5. "Solar" photobioreactor. Possible effects: oxidative stress; photoreparation; sinchronisation of cell division; enhancing substrate bioavailability; increase of rate of oxidative processes; high-density culture.

- there was observed a biodestruction of phenol by consortia of resistant to H_2O_2 phenol-degrading yeast and bacteria *Azotobacter* on nitrogen free medium.
- under simultaneous action of visible and UV-light the yield of yeast *C. tropicalis* on medium with sucrose was 10–15% above the control.

Based on the results obtained three novel methods for biodestruction of pollutants were proposed by us:

- closed intensive microbiological system for the high concentrated waste treatment (Fig. 3);
- artificial peroxisome (Figs 4a, b);
- "Solar" photobioreactor (Fig. 5).

A possibility to protect cells against oxidative stress, induced by addition of H_2O_2 or UV-irradiation, through photoreparation way by photolyase is being under study.

A mathematical model of growth of cells under simultaneous irradiation of visible and UV-light taking into account photoinactivation, photodynamic action, photoreparation and photostimulation will be developed.

European Symposium on Environmental Biotechnology, ESEB 2004 - Verstraete (ed)
© 2004 Taylor & Francis Group, London, ISBN 90 5809 653 X

Preliminary studies on application of immobilized microorganisms to petroleum hydrocarbons biodegradation

K. Piekarska
Wrocław Universityl of Technology, Institute of Environmental Protection Engineering, Wrocław, Poland

ABSTRACT: Five strains of microorganisms are able to degrade diesel oil hydrocarbons isolated from water contaminated with petroleum compounds. These strains belonged to the following genera: *Pseudomonas sp.* (M1), *Bacillus sp.* (D1), *Rhodotorula sp.* (P16), *Bacillus sp.* (P17) and *Acinetobacter sp.* (A2). Selected strains of bacteria degraded the hydrocarbons with different efficiency from 2.2% to 72.3%. The highest efficiency of biodegradation, about 94%, was achieved in case of mixed culture containing two strains: *Bacillus sp.* (D1) and *Pseudomonas sp.* (M1). The mixed culture of these bacteria were immobilized in different polysaccharides carriers: agarose, carrageenan and alginic acid sodium. The process of biodegradation of diesel oil compounds by the free and immobilized cells was monitored with microbiological, biochemical and chemical methods. As a result of these tests the most appropriate carrier to immobilization and biodegradation of diesel oil hydrocarbons was chosen. 2% alginic acid sodium of high viscosity 14000 cps turned out to be this carrier. It was found that the immobilization of microorganism cells in 2% alginic acid sodium of high viscosity 14000 cps had a favourable effect on the biodegradation of diesel oil components, increasing the removal efficiency by 18.2% after 24 hours and by 24.3% after 72 hours of incubation to compare the removal efficiency by free cells.

1 INTRODUCTION

Pollution of the environment with petroleum products is particularly dangerous because hydrocarbons are not only toxic but they can also be mutagenic and carcinogenic. They inhibit natural decomposition processes taking place in surface and ground waters, which results in the deterioration of water quality and loss of its value for drinking, household and recreation purposes. Therefore, this problem has become a subject of extensive research. Among the processes of degradation of petroleum products, biological methods are worth great attention. Although petroleum and its products are used by microorganisms as food substrates, biodegradation process is very slow in the natural conditions. In case of heavy pollution it may take many years, therefore it should be accelerated by man (Zakrzewski 1995, Vargas et al. 1993, Vandermeulen 1991).

Immobilized cell technology has been widely applied in variety of research and industrial applications. Immobilization of cells offers a number of advantages over free-cells. On the basis of numerous tests it was observed that deposited microorganisms can demonstrate a far greater metabolic activity and thus higher efficiency of the biodegradation processes than the aggregate of free-cells suspended in the environment. Furthermore, the immobilization of microorganisms allows to reduce the size of equipment for biological purification of sewage as well as increase the concentration of microorganisms in the bioreactor. A variety of matrices have been used for cell immobilization, such as natural polymeric gels (agar, carrageenan, calcium alginate) and synthetic polymers (polyacrylamide, polyurethane, polyvinyl). Entrapment in natural polymeric gels has become a preferred technique for cell immobilization due to the toxicity problems associated with synthetic polymeric materials (Hartmeir 1988, Nicolov & Karamanev 1990, Cassidy et al. 1996) .

The subject of this research was the determination of biodegradation efficiency of petroleum hydrocarbons by immobilized active strains of microorganisms in polysaccharides carriers.

2 RESEARCH METHODS

Diesel oil used in the experiment was purchased at a filling station in Wrocław. Chromatographic analyses showed that it was a mixture of 85 different hydrocarbons containing 10–24 carbon atoms in the molecules. The prevailing group was the hydrocarbons with 16 carbon atoms. HPLC analysis revealed the presence of approximately 20% of aromatic compounds with signal spectra typical of monoaromatic, alkyl aromatic, naphtene alkyl aromatic, and di- and tricyclic aromatic compounds. The density of the oil was 0.804 g/cm^3.

The microorganisms degrading diesel oil hydrocarbons were isolated from water contaminated with petroleum compounds from the vicinity of filling stations and Kerosene Products Centre on the territory of the city of Wrocław.

In order to achieve highly active strains with high ability to utilize diesel oil the cultures were isolated after earlier growth of enrichment cultures. The method of growing enrichment cultures consists in inoculating with a suspension of microorganisms from the previous culture of fresh mineral medium with diesel oil hydrocarbons as the only source of carbon and energy. The strains of microorganisms isolated this way were subjected to further morphological, physiological and biochemical research in order to determine their taxonomic designation with the use of Bergey's key.

The ability to degrade diesel oil by microorganisms was tested during a 31-day-long growth of strains on the liquid Siskinej-Trocenko mineral medium with diesel oil hydrocarbons in the amount of 1%, 5%, 10%, 25% and 50% vol. of the medium i.e. 8.04; 4.02; 8.04; 20.1 and 40.2 g of oil/dm^3. 2 cm^3 of individual strains of microorganisms suspended in the mineral medium of absorbance of 0.2 measured with the 650 nm long wave were put in the bulbs. The optical density of the tested cultures was measured on SIMADZU UV-VIS 1202 spectrophotometer. The process of chemooxidation was controlled in the bulbs containing the same components but without microorganisms. All tests were performed at room temperature in a JW.ELECTRONIC WL-972 shaker machine. After 31 days the loss of diesel oil components was marked in all cultures with the hexane extract method (Grabińska- Łoniewska et al. 1996).

The immobilization of the selected microorganisms was conducted in the following polysaccharide carriers purchased in Sigma: agarose (type VI), carrageenan (type I), alginate of viscosity of 3500 cps and alginate of viscosity of 14000 cps. The carriers were prepared as described in (Maliszewska et al. 1999). The cells used in the immobilization process were prepared in the following way: the cultures of the selected microorganisms were grown on the liquid Siskinej-Trocenko medium in the amount of 100 cm^3 with 10% vol. diesel oil as the only source of carbon and energy. After a 6-day-long incubation the cultures were spun (15000 r/min, 15 min, 0°C). The suspensions of cells were suspended in 1 cm^3 of a mineral medium and used in immobilization.

The cultures of the immobilized bacteria were grown in sterile Erlenmajer bulbs on JW.ELECTRONIC WL-972 shaker machine. The total capacity of the culture was 500 cm^3. Biosorbents with the mounted biomass and 10% vol. diesel oil were put into the bulbs. At the same time cultures of free cells were grown in the same conditions and in the same concentration. After 24 and 72 hours the loss of diesel oil in the cultures was measured with the hexane extract method as well as emulsification activity was measured (Galas et al. 1997).

Next in the same way the cultures of the immobilized bacteria strains were prepared in 2% alginic acid sodium of viscosity of 14000 cps with 1%, 5%, 10%, 25% and 50% vol. diesel oil. Every day for 7 days the loss of diesel oil was measured in those cultures with the use of the hexane extract method, as well as emulsification activity and intensity of respiration processes with the use of WTW Oxi Top mercury-free measurement system.

3 RESEARCH FINDINGS

Five strains of microorganisms well adapted for degradation of hydrocarbon contaminants were isolated from the contaminated water. The strains include the following genera: *Pseudomonas sp.* (M1), *Bacillus sp.* (D1), *Rhodotorula sp.* (P16), *Bacillus sp.* (P17) and *Acinetobacter sp.* (A2). The tested strains were isolated with the use of enrichment cultures method. After two months it was found that in the medium of that culture there were only two of the tested strains, namely *Bacillus sp.* D1 and *Pseudomonas sp.* M1. The use of the hydrocarbon substrate by those strains was good (Table 1). A significant decrease of the concentration of diesel oil dependent on the concentration of hydrocarbon substances added to them after 31 days of the experiment was observed in all cultures. In case when the concentration of the substrate in the culture was 1% vol., the loss of hydrocarbons was 15.7%–72.3%, when the concentration of diesel oil was 5% vol. it was 21%–67.3%, and when the concentration of diesel oil in the culture was 10% vol. it was 2.2%–40.1%. The lowest ability to degrade hydrocarbon substances included in diesel oil was indicated by strain *Bacillus sp.* P17 where the percentage loss

Table 1. % loss of diesel oil after 31 days in the culture of microorganisms on liquid mineral medium.

	Concentration of diesel oil in the culture		
Strain	1%	5%	10%
Rhodotorula sp. P16	43.2	43.7	19.2
Bacillus sp. P17	15.7	21.0	2.2
Acinetobacter sp. A2	63.0	67.3	25.3
Pseudomonas sp. M1	51.2	60.0	23.1
Bacillus sp. D1	72.3	66.0	40.1

of oil observed in this strain was the smallest and it was, depending on the initial concentration of the substrate respectively: 15.7%, 21% and 2.2%. On the other hand, the biggest loss of hydrocarbons was observed in the cultures of strains *Bacillus sp.* D1 and *Acinetobacter sp.* A2. It was respectively 72.3%, 66%, 40.1% and 63%, 67.3% and 25.3%. At the same time the biggest percentage loss of diesel oil was observed in the cultures of strains with 1% vol. concentration of the medium; the smallest one with the concentration of 10% vol. of substrate. The best results were achieved, however, in the culture with two strains *Bacillus sp.* D1 and *Pseudomonas sp.* M1 (Table 2). After 31 days in the mixed culture the loss of hydrocarbons was between 22.3% (for 50% vol. of initial concentration of oil in the culture) and 93.8% (for 1% vol. of initial concentration of oil in the culture). This mixed culture of bacteria was used also in the next stage of the research in which the ability to degrade hydrocarbons of diesel oil in free bacteria cells was compared with such ability of the immobilized bacteria in various polysaccharide carriers.

Table 2. % loss of diesel oil after 31 days in the mixed culture of microorganisms on liquid mineral medium.

	Concentration of diesel oil in the culture				
Strains	1%	5%	10%	25%	50%
Pseudomonas sp. M1 and *Bacillus sp.* D1	93.8	720.1	78.0	47.5	22.3

In order to select the best carrier the immobilization was conducted in the following polysaccharide carriers: agarose, carrageenan, alginate of viscosity of 3500 cps, alginate of viscosity of 14000 cps, a mixture of alginate as well as carrageenan and alginate with polyvinyl alcohol. The 2% alginate of viscosity of 14000 cps (Table 3) proved to be the best carrier. The culture containing 10% of diesel oil and the immobilized cells in this carrier indicated 42.4% loss of oil after 24 hours and 58.4% after 72 hours. After the same period the emulsification activity of the immobilized cells was 0.987 and 1.758 (A_{540}). For comparison of those values for free cells were respectively: loss of oil – 24.2% and 34.1%, emulsification activity – 0.265 and 0.498 (A_{540}). With the use of this carrier and the mixed culture of strains *Bacillus sp.* D1 and *Pseudomonas sp.* M1 immobilized cultures were prepared in which every day for 7 days the loss of diesel oil was measured with the use of the hexane extract method (Table 4), as well as emulsification activity and intensiveness of respiration processes. Both the level of production of surfactants and the respiration activity of the immobilized

Table 3. Loss of diesel oil and emulsification activity in the mixed culture of free and immobilized cells.

		Loss of diesel oil [%]		Emulsification activity [A_{540}]	
Type of carrier	Time [hours]	Free	Immobilized	Free	Immobilized
Agarose 6%	24	28.3	32.4	0.372	0.410
	72	38.5	*39.1	0.619	*0.567
Alginate 1%	24	31.2	34.8	0.398	0.498
3500 cps	72	39.4	*42.7	0.521	*0.956
Alginate 2%	24	27.1	38.2	0.302	0.621
3500 cps	72	39.2	47.5	0.578	0.990
Alginate 3%	24	25.4	19.4	0.298	0.120
3500 cps	72	37.8	21.3	0.502	0.213
Alginate 4%	24	27.4	5.2	0.345	0.087
3500 cps	72	35.2	6.1	0.478	0.090
Alginate 1%	24	29.6	31.5	0.384	0.398
14000 cps	72	39.8	*43.7	0.593	*0.989
Alginate 2%	24	24.2	42.4	0.265	0.987
14000 cps	72	34.1	58.4	0.498	1.758
Alginate 3%	24	25.1	34.1	0.276	0.456
14000 cps	72	37.6	41.2	0.567	0.998
Alginate 4%	24	28.7	6.1	0.384	0.098
14000 cps	72	39.8	7.8	0.592	0.109
Carrageenan	24	21.3	29.4	0.198	0.234
1%	72	32.1	*39.2	0.304	*0.578
Carrageenan	24	28.9	37.2	0.398	0.504
2%	72	39.6	48.4	0.556	1.134

(*Continued*)

Table 3. (*Continued*)

Type of carrier	Time [hours]	Loss of diesel oil [%]		Emulsification activity [A_{540}]	
		Free	Immobilized	Free	Immobilized
Carrageenan	24	31.2	34.1	0.287	0.345
3%	72	39.2	42.7	0.567	1.087
Carrageenan	24	26.1	7.8	0.289	0.102
4%	72	34.5	9.2	0.432	0.123
alginate (1.5 g)	24	28.5	39.1	0.354	0.987
3500 cps and carrageenan (0.5 g)	72	38.6	*42.6	0.543	*0.987
alginate (0,75 g)	24	23.1	37.4	0.201	0.965
14000 cps and carrageenan (0.25 g)	72	34.5	48.2	0.387	1.234
alginate (1,5 g)	24	28.7	35.6	0.321	0.567
3500 cps and polyvinyl alcohol (0.5 g)	72	39.8	48.1	0.587	1.123
alginate (2 g)	24	28.9	**31.4	0.376	**0.398
3500 cps and polyvinyl alcohol (5 g)	72	38.2	**42.3	0.598	**0.976
alginate (1 g)	24	28.2	39.4	0.345	0.401
14000 cps and polyvinyl alcohol (5 g)	72	39.0	45.5	0.599	1.123
alginate (1 g)	24	23.1	36.1	0.287	0.456
14000 cps and polyvinyl alcohol (10 g)	72	38.5	49.2	0.602	1.328

* Carrier was destroyed. ** Carrier in the culture medium caused its foaming. Efficiency of extraction process 85.4%. Non-biological loss of diesel oil due to chemooxidation 4.5%.

Table 4. Loss of diesel oil in the mixed culture of free and immobilized cells in 2% alginic acid sodium of viscosity 14 000 cps.

Type of culture	Concentration of diesel oil [%]	Loss of diesel oil [%] Day of culture 1	2	3	4	5	6	7
Free	1	29.8	48.5	63.1	71.5	79.5	80.1	85.1
Immobilized	1	34.6	52.8	72.1	85.4	92.3	94.5	96.1
Free	5	22.6	36.1	42.7	48.5	61.5	68.4	72.1
Immobilized	5	36.8	40.2	69.4	72.4	88.5	90.2	91.4
Free	10	21.7	26.7	33.8	47.8	55.3	58.6	60.2
Immobilized	10	45.2	49.1	57.3	64.2	70.9	72.3	77.4
Free	25	12.4	20.7	24.7	30.4	35.5	36.1	37.2
Immobilized	25	19.4	29.8	31.2	39.8	44.5	47.0	54.2
Free	50	7.5	8.0	9.5	12.5	15.0	17.0	18.5
Immobilized	50	15.5	17.0	18.0	20.8	25.0	28.4	32.1

cells was greater than that of free cells in all tested concentrations of diesel oil. In time the value of the tested parameters also increased and reached the highest values at the end of research.

4 SUMMARY

The strains of microorganisms were isolated from the water contaminated with petroleum compounds enzymatically prepared for the use of various kinds of petroleum hydrocarbons as a source of carbon and energy. Those strains belonged to the following genera: *Pseudomonas* (M1), *Bacillus* (D1), *Rhodotorula* (P16), *Bacillus* (P17) and *Acinetobacter* (A2). All tested microorganisms demonstrated good biodegradation qualities in a wide range of concentrations of oil substrate in the cultures. Their efficiency of degradation of hydrocarbons was from 2.2% (10% vol. oil in the culture) to 72.3% (1% vol. oil in the culture), depending on the kind of strain. The best efficiency of biodegradation of hydrocarbons, 22.3% (50% vol. oil in the culture) and 93.8% (1% vol. oil in the culture), was achieved in case when a mixture of culture of strains *Bacillus sp.* D1 and *Pseudomonas sp.* M1 was used, and not individual active strains. This is caused probably by the complicated structure of diesel oil as it is an exceptionally difficult contaminant with a lot of components of various chemical composition and susceptibility to biodegradation. Each of them can be present in a different concentration and their reaction to microorganisms varies.

The tests performed on the cultures containing a mixed culture of bacteria *Bacillus sp.* D1 and *Pseudomonas sp.* M1 immobilized in various polysaccharide carriers, as well as diesel oil as the only source of carbon and energy, enabled to select the best gel for immobilization and biodegradation of hydrocarbons. The carrier was the 2% alginate of viscosity of 14000 cps. Immobilization of microorganism cells in the 2% alginate of viscosity of 14000 cps improved the ability to remove hydrocarbons by 18.2% after 24 hours and by 24.3% after 72 hours with the concentration of 10% vol. diesel oil. The immobilized cells in the selected gel improved also the efficiency of degradation of hydrocarbons in case of tests in a broad range of concentrations of diesel oil added to the culture (1%, 5%, 10%, 25% and 50%). In all cases the efficiency of degradation of diesel oil by the immobilized cells was higher, irrespective of the added concentration of hydrocarbons. In general during the whole experiment that took 7 days the biodegradation efficiency increased due to the immobilized cells, respectively for individual concentrations of diesel oil in the culture by 12.09%; 19.57%; 18.91%; 9.85% and by 9.83%. The tests of respiration activity and the tests regarding the ability to produce surfactants by free and immobilized cells of microorganisms confirmed a higher biodegradation activity for diesel oil hydrocarbons of the immobilized cells.

REFERENCES

Bergeys Manual of Determinative Bacteriology, 1974. Eight Edition Buchanan R.E. & Gibbons N.E. (eds). The Wilkins Company.

Cassidy, M.B. et al. 1996. Environmental applications of immobilized microbial cells: a review. *J. Ind. Microbiol.* 16: 79–101.

Galas, E. et al. 1997. Impact of the source of nitrogen on the speed of biodegradation of petroleum hydrocarbons and production of biosurfactants by the strains of *Pseudomonas sp.* BP and *Micrococcus sp.* R51 (polish). *Papers from the 5th National Scientific and Technical Symposium: Biotechnology of the Environment, Ustroń-Jaszowiec, Dec. 10–12, 1997.*

Grabińska-Łoniewska, et al. 1996. *Laboratory Exercises in General Microbiology* (polish). Oficyna Wydawnicza PWN (eds), Warszawa.

Hartmeir, W. 1988. *Immobilized Biocatalysts. An Introduction.* Springer Verlag (eds). Berlin.

Instructions for WTW Oxi Top mercury-free measurement system. POLEKO APARATURA, Wodzislaw Śląski.

Maliszewska, I. et al. 1999. Biodegradation of Chloroalkane Acids by the Immobilized Strains of Bacteria (polish). *Papers form the 6th National Scientific and Technical Symposium: Biotechnology of the Environment at the 1st National Congress of Biotechnology, Wrocław, Sept. 23–24, 1999.*

Nicolov, L. & Karamanev, D. 1990. Change of microbial activity after immobilisation of microorganisms. In J.A.M. DeBont (eds), *Physiology of Immobilised Cells.* Elsevier.

Vandermeulen, J.H. 1991.Toxicity and sublethal effects of petroleum hydrocarbons in freshwater biota. Oil in Freshwater. *Chemistry, Biology, Countermeasure Technology* 19: 267–303.

Vargas, V.M.F. et al. 1993. Mutagenic activity detected by the Ames test in river water under the influence of petrochemical industries. *Mutation Research* 319: 31–45.

Zakrzewski, S.F. 1995. *Basis of environmental toxicology* (polish). PWN (eds), Warszawa.

European Symposium on Environmental Biotechnology, ESEB 2004 - Verstraete (ed)
© 2004 Taylor & Francis Group, London, ISBN 90 5809 653 X

Growth physiology and biochemical storage characteristics of *Microlunatus phosphovorus*, a model organism for activated sludge systems

A. Akar, E. Ucisik Akkaya, S.K. Yesiladali, G. Çelikyilmaz, C. Tamerler & Z.P. Çakar
Department of Molecular Biology and Genetics, Faculty of Science and Letters, Istanbul Technical University, Maslak, Istanbul, Turkey

E. Ubay Çokgor & D. Orhon
Department of Environmental Engineering, Faculty of Civil Engineering, Istanbul Technical University, Maslak, Istanbul, Turkey

ABSTRACT: *Microlunatus phosphovorus* is a microorganism found in activated sludge systems. It belongs to the group of polyphosphate-accumulating organisms. Under anaerobic conditions, it utilizes the energy released by the hydrolysis of polyphosphate, for substrate uptake and conversion to internal carbon sources. Under aerobic conditions, however, the previously stored polyhydroxyalcanoate (PHA) is utilized for growth. Because of these properties, *M. phosphovorus* could be used in wastewater treatment plants for biological phosphorus removal.

In this study, the growth physiology and biochemical storage metabolism of *M. phosphovorus* was investigated. Pure culture studies in batch growth systems were conducted using chemically defined media. A batch growth system with anaerobic-aerobic cycles was also employed. PHB and polyphosphate staining was applied to microscopy samples to detect the presence of these metabolites. Other key metabolites like total protein, acetate were determined during different phases of growth in addition to cell dry weight and optical density measurements.

This study will help us understand the physiology and metabolism of *M. phosphovorus*, a model organism of activated sludge systems.

1 INTRODUCTION

Microlunatus phosphovorus is an activated sludge bacterium and it has been shown that it shows high levels of phosphorus-accumulating activity and phosphate uptake and release activities similar to the activated sludge conditions (Nakamura et al., 1995) Therefore, it has been used as a model organism to study biological phosphorus removal.

Under anaerobic conditions, the polyphosphate accumulating organisms can use the energy derived from polyphosphate hydrolysis for substrate uptake and conversion to internal carbon sources. Polyphosphate accumulating organisms take up short-chain fatty acids and store them as PHA and release phosphorus. Under aerobic conditions, the stored PHA are used for growth (Santos et al., 1999).

In this study, the growth physiology and biochemical storage metabolism of *M. phosphovorus* was investigated. Pure culture studies in batch growth systems were conducted using chemically defined media with glucose as the sole carbon source. A batch growth system with anaerobic-aerobic cycles was also employed. PHB and polyphosphate staining was applied to microscopy samples to detect if there is any accumulation of these metabolites.

2 MATERIALS AND METHODS

2.1 *Microorganisms and growth media*

The bacterial strain used in this study was *Microlunatus phosphovorus* (DSM no: 10555). It was purchased from Deutsche Sammlung von Mikroorganismen und Zellkulturen GmbH (DSMZ), Braunschweig, Germany. For all cultivations, M9 minimal medium with 4 g/l glucose as the sole carbon source was used. The other ingredients per 1 liter of this medium were $Na_2HPO_4 \cdot 7H_2O$ 12.8 g, K_2HPO_4 3.0 g, NaCl 0.5 g, NH_4Cl 1.0 g, 1 ml of 1 M $MgSO_4$ stock solution, and 1 ml of 0.1 M $CaCl_2$ stock solution. The pH of the medium was adjusted to 7.4.

2.2 *Culture conditions*

Batch cultivations in three parallel shaken flask sets were performed using 1 liter baffled Erlenmeyer flasks with 250 ml culture volume. The inoculum size was 4%. The cultivations were performed at 30°C and 150 rpm in an orbital shaker.

Aerobic batch cultivations in a bioreactor (Braun BIOSTAT B, Germany) were performed at 30°C and 200 rpm with 1.5 liters working volume and 20% inoculum size. The optical density (OD_{600}) of the preculture was 1.09.

Aerobic-anaerobic cycles in a batch growth system were performed according to Nikata et al., 2001 with some modifications. Briefly, 45 ml of culture was grown in 250 ml Erlenmeyer flasks at 250 rpm for 9 h during aerobic cultivation. Subsequently, the anaerobic cultivation was performed for 15 h in 50 ml Falcon flasks sparged with N_2 gas prior to cultiva tion. The aerobic-anaerobic cycles were repeated 8 times.

2.3 *Analytical methods*

Cell growth was monitored spectroscopically by OD_{600} measurements and by dry weight determination using 0.22 μ filters. Briefly, 5 ml culture samples were filtered on preweighed filters using a vacuum filtration system. The filters were heated at 105°C for one hour and cooled down in a desiccator for 30 min prior to weighing.

Residual glucose concentrations in filtered culture supernatants were determined using an HPLC system with refractive index measurements. The mobile phase was 5 mM H_2SO_4 and the column temperature was 65°C.

Acetate concentrations in filtered culture supernatants were determined using a gas chromatography system (Agilent Technologies, HP, USA) equipped with a DB-FFAP column.

Total protein determinations from whole cells were made by centrifuging 2 ml culture samples in a benchtop centrifuge and resuspending the pellet in 550 μl of a solution containing 0.9% (w/v) NaCl and 10 mM $MgSO_4$. After addition of 400 μl of 4 M NaOH, the suspension was incubated in sealed glass tubes in a boiling water bath for 10 min. Upon cooling to room temperature, 1.6 ml of solution A was added (Per liter, solution A consists of 2.5 g $CuSO_4.5H_2O$, 16.75 g Na-K tartrate.$4H_2O$, 10.0 g NaOH and 6.25 g KI). The resulting mixture was incubated at 37°C for 30 min which was then centrifuged in a benchtop centrifuge at 4000 rpm for 20 min. The absorbance of the supernatants was measured at 546 nm and the protein concentrations were calculated using a standard curve with known BSA concentrations in the range of 0–10 g BSA per liter.

2.4 *Microscopic staining methods*

PHB staining was done by using Sudan Black B (0.3% w/v in 60% ethanol) and Safranin O (0.5% w/v aqueous solution) solutions (Murray, R.G.E., 1981).

Poly-p (Neisser) staining was done as described previously (Gerhardt et al., 1994).

3 RESULTS AND DISCUSSION

3.1 *Batch cultivations in shaken flasks*

Batch cultivations in three parallel shaken flask sets were performed in 1 liter baffled Erlenmeyer flasks at 30°C and 150 rpm in an orbital shaker. Their growth behavior is represented in Figure 1.

Briefly, the cells have reached a final OD_{600} value of about 5.4 in M9 medium. The maximum specific growth rate was calculated as $0.22\,h^{-1}$.

3.2 *Batch cultivations in a bioreactor*

Aerobic batch cultivations in a bioreactor (Braun BIOSTAT B, Germany) were performed in a 2 liter-bioreactor with 1.5 liters working volume. The final dry weight concentration of the culture at the end of the cultivation (10 h) was 1.44 g/l. The growth curve of *M. phosphovorus* in the bioreactor is shown in Figure 2. Compared to the shaken flask system, the cells in the bioreactor have shown a rapid growth. The maximum specific growth rate was calculated as $0.29\,h^{-1}$. There was some acetate production during batch growth and after the sixth hour of growth (during late exponential phase), acetate concentrations began to decrease which indicates the possible utilization of acetate when glucose is depleted (Figure 3).

The total protein concentrations of the cells during their growth have shown a significant increase starting from the exponential phase of growth (Figure 4).

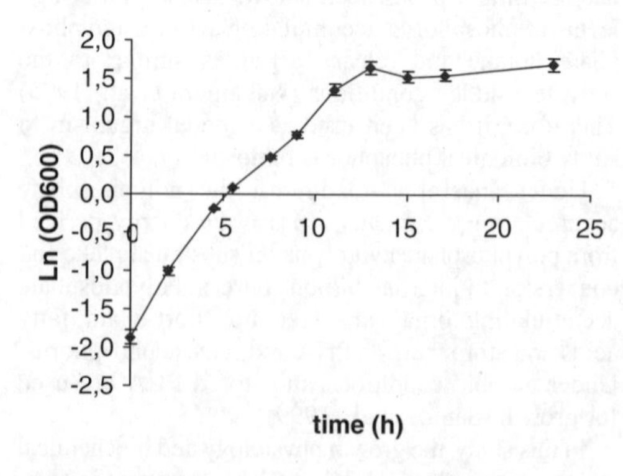

Figure 1. Growth curve of *M. phosphovorus* grown in shaken flasks. The curve represents the average values of three experiments.

Cells were also tested for the presence of poly-P and PHB during different phases of batch growth in the bioreactor. However, no accumulation of poly-P and PHB was observed under aerobic, batch conditions (data not shown).

3.3 *Aerobic-anaerobic cycles in a batch growth system*

Aerobic-anaerobic cycles in a batch growth system were performed as described before (Nikata et al., 2001) with some modifications. The aerobic-anaerobic cycles were repeated 8 times. At the end of each aerobic phase, cells were transferred to fresh medium for the anaerobic phase. The aerobic and anaerobic phases lasted for 9 and 15 h, respectively. Thus, the duration of each cycle was 24 h. The OD600 values were determined at the end of each aerobic/anaerobic step. The results are shown in Figure 5.

Figure 5 indicates that the aerobic phases of growth result in higher cell densities. The absence of oxygen during anaerobic phase decreases growth significantly. This finding is also supported by the residual glucose concentrations in the medium. There is much less glucose left in the medium at the end of the aerobic phases compared to the anaerobic phases of growth (data not shown).

The most interesting findings, however, result from the poly-P and PHB staining of the cells taken from different phases of the aerobic/anaerobic cycles. In contrast with the literature, it was found that there

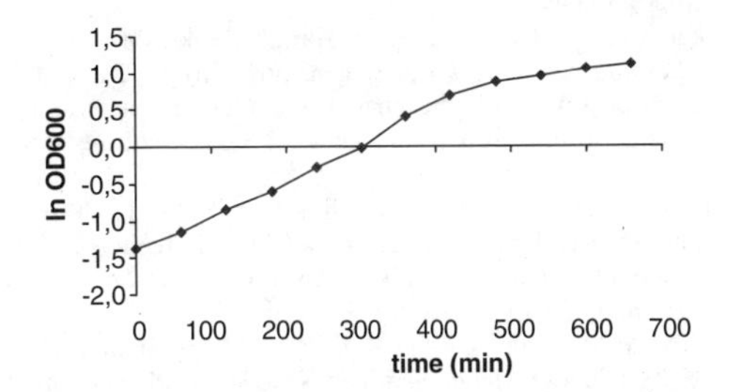

Figure 2. Growth curve of *M. phosphovorus* aerobically grown in a bioreactor. The curve represents the average values of two experiments.

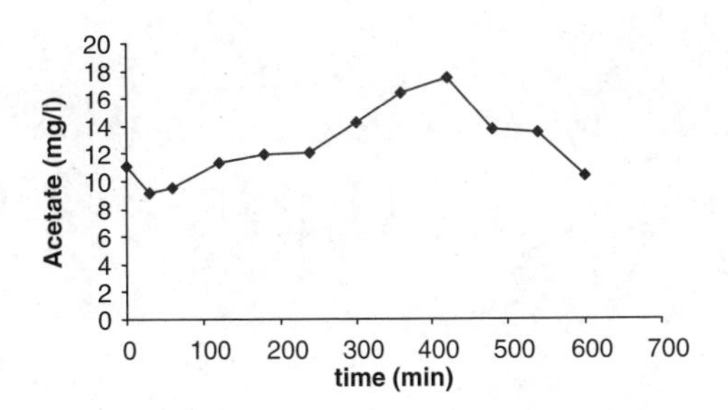

Figure 3. Acetate concentration profile of *M. phosphovorus* aerobically grown in a bioreactor. The curve represents the average values of two experiments.

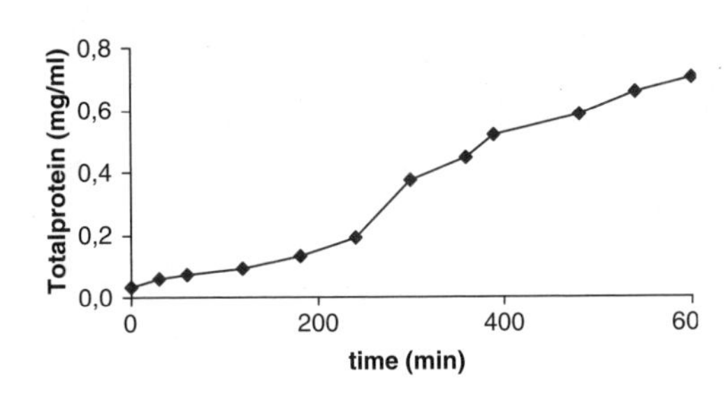

Figure 4. Total protein concentration profile of *M. phosphovorus* aerobically grown in a bioreactor. The curve represents the average values of two experiments.

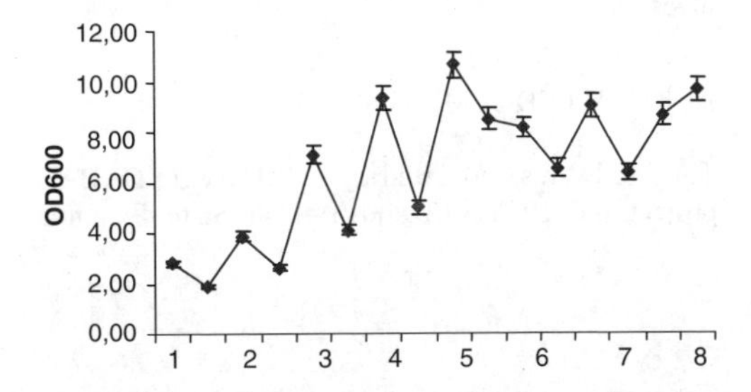

Figure 5. OD600 values of *M. phosphovorus* grown in eight repeating aerobic/anaerobic cycles. The values represent the average values of three experiments. The numbers on the X-axis represent cycle number. The first, third, etc. data points belong to the end of aerobic phases, and the second, fourth, etc. data points belong to the end of anaerobic phases.

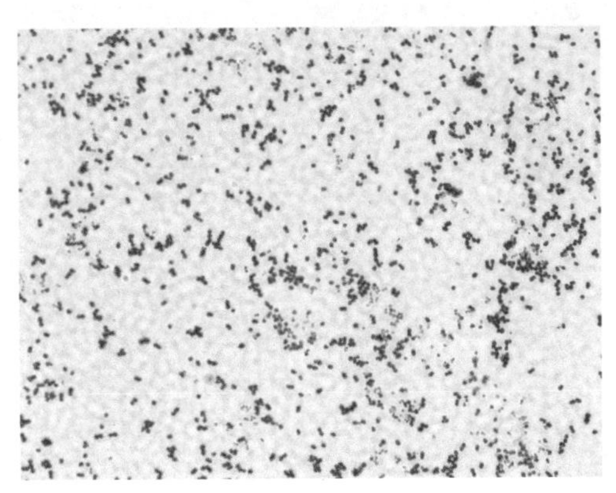

Figure 6. PHB accumulation of *M. phosphovorus* grown in aerobic/anaerobic cycles. PHB granules appear as intracellular, blue-black granules.

was no poly-P accumulation in the cells during any of the conditions investigated in this study (data not shown). However, PHB staining results have shown that the cells have accumulated PHB when there was repeating aerobic/anaerobic cycles (Figure 6).

4 CONCLUSIONS

In this study, aerobic growth behavior of pure cultures of *M. phosphovorus* on glucose as the sole carbon source was investigated including repeating aerobic/anaerobic growth cycles. Our results revealed that, under the conditions investigated, there was no polyphosphate accumulation. However, during aerobic/anaerobic cycles, there was significant PHA accumulation.

These results will help us to understand the physiology and metabolism of *M. phosphovorus*, a model organism of activated sludge systems. The physiology with different carbon sources remains to be investigated.

ACKNOWLEDGEMENTS

This study was supported by TUBITAK (YDABAG project no: 102Y060) and Turkish State Planning organization (DPT). We thank Tarik Ozturk for technical assistance.

REFERENCES

Gerhardt, P., Murray, R.G.E., Wood, W.A. & Krieg, N.R. (eds.) 1994. *Methods for General and Molecular Bacteriology.* Washington DC: American Society for Microbiology.

Murray, R.G.E. (ed.) 1981. *Manual of Methods for General Microbiology.* Washington DC: American Society for Microbiology.

Nakamura, K., Ishikawa, S. & Kawaharasaki, M. 1995. *Microlunatus phosphovorus* gen. nov., sp. nov., a new gram-positive polyphosphate-accumulating bacterium isolated from activated sludge. *Int. J. Syst. Bacteriol.* 45: 17–22.

Nikata, T., Natsui, M., Sato, K., Niki, E. & Kakii, K. 2001. Photometric Estimation of Intracellular Polyphosphate Content by Staining with Basic Dye. *Analytical Sciences* 17: Suppl. i1675–i1678.

Santos, M.M., Lemos, P.C., Reis, M.A.M. & Santos, H. 1999. Glucose Metabolism and Kinetics of Phosphorus removal by the fermentative bacterium Microlunatus phosphovorus. *Appl.Environ. Microbiol.* 65: 3920–3928.

European Symposium on Environmental Biotechnology, ESEB 2004 - Verstraete (ed)
© 2004 Taylor & Francis Group, London, ISBN 90 5809 653 X

An innovative bioprocess that rapidly and completely dechlorinates chlorinated ethenes

C.-S. Hwu
Department of Environmental Engineering, Hungkuang University, Taichung, Taiwan

ABSTRACT: Dechlorination of chlorinated ethenes, common groundwater contaminants, was investigated using anaerobic granular sludge exposed to O_2. The exposure established a synchronously anaerobic/aerobic bioconversion process, i.e. reductive dechlorination in combination with aerobic co-oxidation. Experimental results showed that the highest dechlorination rate of tetrachloroethene (PCE), trichloroethene (TCE), *cis*-dichloroethene (*c*DCE) and vinyl chloride (VC) was 6.44, 2.98, 1.70 and 0.97 nmol/g VSd, respectively, at an initial O_2 concentration of 10, 100, 5 and 0%. Strictly anaerobic conditions favored VC dechlorination and absolutely aerobic conditions was preferred for TCE dechlorination. Considering that both rapid and complete dechlorination are required in overall biodegradation of the chlorinated ethenes, microaerophilic conditions, i.e. 5–10% O_2, are suggested as for a potential process.

1 INTRODUCTION

Throughout the world, chlorinated ethenes are among the most common contaminants observed in the groundwater environment. Due to their widespread use as degreasing and dry-cleaning solvents for many industrial applications, chlorinated ethenes have been threats to human health and the environment. Of the chlorinated ethenes, tetrachloroethene (PCE), trichloroethene (TCE) appear in groundwater with the greatest frequency and highest concentration. In biological treatment processes, TCE, *cis*-dichloroethene (*c*DCE) and vinyl chloride (VC) are vulnerable to aerobic degradation. In contrast, PCE is resistant to aerobic degradation and usually is degraded via reductive dechlorination to the less chlorinated ethenes like TCE, *c*DCE or vinyl chloride (VC). Unfortunately, in the majority of anaerobic treatment processes reductive dechlorination apparently stops at DCE or VC as the result of incomplete reduction. In the groundwater environment, chlorinated ethenes often exist as a mixture, a treatment process that is capable of both rapid and complete dechlorination is warranted.

Previous studies demonstrated that synchronous anaerobic and aerobic biotransformation of chlorinated compounds was achievable by using gel beads with immobilized cells (Beunink & Rehm 1988, 1990). In the present study we used anaerobic granular sludge exposed to oxygen to establish a synchronous anaerobic/aerobic process. It was reasonable to expect that dechlorination of chloroethenes would become more rapid and complete with the combination of reductive dechlorination and the aerobic co-oxidation in a same microbial niche.

2 MATERIALS AND METHODS

Batch experiments were performed by using anaerobic granular sludge taken from a 1,000-m^3 upflow anaerobic sludge blanket reactor at a fructose syrup factory (Tainan, Taiwan). This sludge was not pre-exposed to PCE or other chlorinated compounds. Before use, the sludge was elutriated to remove floating matter and fine particulates. The diameters of the elutriated sludge granules ranged 1–3 mm. The volatile solid (VS) content of the granules was determined as 7.14% (w/w). In a 120-ml amber serum bottle was added 2 g VS/l of the granular sludge and 500 mg acetate/l.

The amount of PCE, TCE, *c*DCE and VC introduced individually to each set of bottles was 1,930, 580, 182 and 50 nmol, respectively. The final liquid volume including a basal medium (Hwu et al. 1996) was 25 ml, leaving ca. 95-ml headspace. The bottles were sealed with Viton-rubber septa and aluminum caps. The headspace was flushed with pure nitrogen for 3 min. All bottles of each set of chlorinated ethene were then supplied with designated volumes of pure oxygen, after replacing with the same volumes of nitrogen gas. The initial O_2 concentrations (%O_2, v/v) in the headspace were 0, 5.2, 10.4, 26.3 and 100%. Subsequently, the bottles were incubated in a reciprocating shaker water-bath with temperature controlled at 30°C.

Headspace gas was sampled at time intervals using 25-μl or 1,000-μl air-tight precision microsyringes. Compositions and concentrations of the gas samples were analyzed by gas chromatograph (GC) equipped with electron capture detector (ECD) for PCE and TCE and by GC equipped with flame ionization detector (FID) for *c*DCE, VC and methane. Acetate concentration in liquid phase was determined by HPLC. Operating conditions of the GC-ECD/FID and HPLC have been described elsewhere (Liu 2001). All experiments were conducted in triplicate and the mean values are used in this paper.

3 RESULTS AND DISCUSSION

The concentrations of chlorinated ethenes applied in the present study were far below their 50% inhibition concentrations on methanogenic activity of granular sludge (Sanz et al. 1997, van Eekert 1999). Methanogens are generally regarded as highly sensitive to toxicants (Hwu et al. 1996). Therefore, the possible inhibition effects on the methanogenic granular sludge used in this study could be negligible. The variety of degradation of the chlorinated ethenes described below would mainly be attributed to the exposure of microbes to various O_2 concentrations.

Figure 1 illustrates the degradation of the chlorinated ethenes over time under different O_2 concentrations. The three lower O_2 concentrations achieved better PCE dechlorination. This indicated that strictly anaerobic conditions might not be obligate for PCE degradation in granular sludge (Hwu et al. 2002). In contrast, the best TCE dechlorination was attained at the highest O_2 content. Partially oxidized conditions favored *c*DCE dechlorination. VC had better dechlorination at lower O_2 concentrations, i.e. completely dechlorinated to ethene and/or CO_2. Note that, in any cases, methane production was observed even at 100% O_2 concentration. This evidenced that the core region in a granule remained anaerobic. Consequently, a good variety of dechlorination of each chlorinated ethene can be regarded as overall results from anaerobic/aerobic microbial metabolic interactions.

Table 1 summarizes the dechlorination rates of each chlorinated ethene at the five O_2 concentrations tested. Although the higher PCE conversion rates were attained at the three O_2 concentrations lower than 10.4%, however, significant accumulation of TCE was also observed. After PCE degradation, TCE accumulated up to 20 nmol/bottle within 3, 20, and 30 days at 0, 5.2, and 10.4% O_2 concentration, respectively, and to 86, 50, and 20 nmol/bottle upon termination of the experiment.

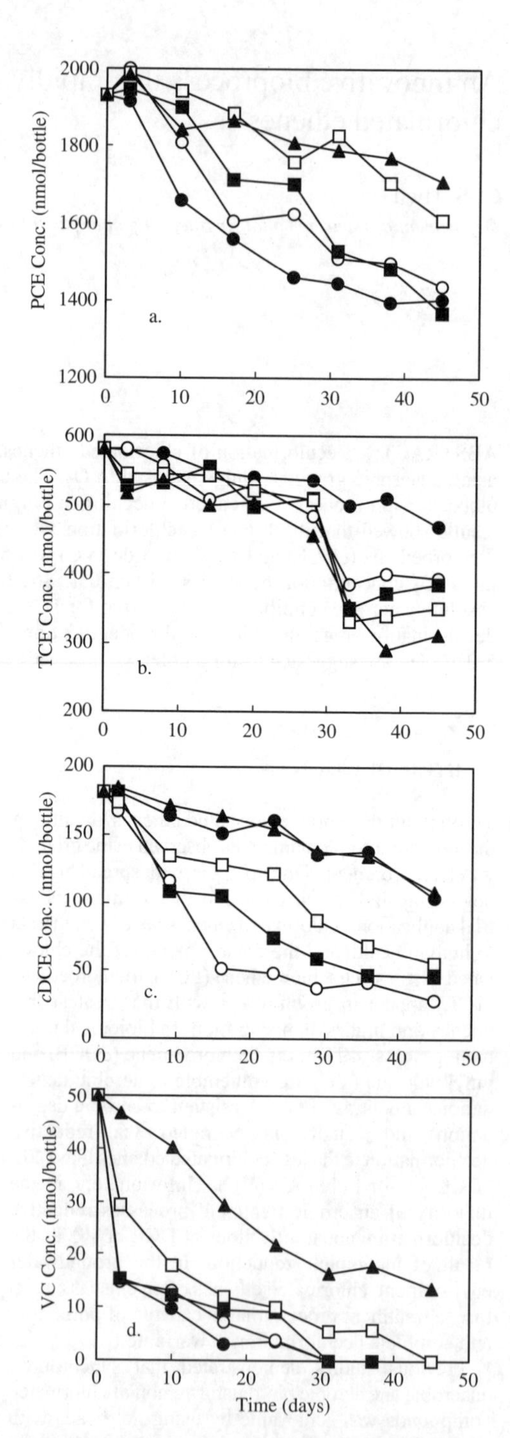

Figure 1. Dechlorination of (a) PCE, (b) TCE, (c) *c*DCE, and (d) VC with methanogenic granular sludge exposed to different oxygen contents (symbols: ● = 0%; ○ = 5.2%; ■ = 10.4%; □ = 26.3%; ▲ = 100%).

Table 1. Comparison of dechlorination rates (nmol/g VSd) of the chlorinated ethenes at different oxygen concentrations.

	Initial O_2 concentration in the headspace (%)				
Chloroethene	0	5.2	10.4	26.3	100
PCE	5.85	5.49	6.44	3.54	2.44
TCE	1.24	2.08	2.18	2.55	2.98
*c*DCE	0.86	1.70	1.51	1.39	0.81
VC	0.97	0.96	0.85	0.83	0.59

No TCE accumulation could be found at the two higher O_2 concentrations. Similarly to the dechlorination of PCE, the contradictory (rapid dechlorination vs. metabolite accumulation) was also observed for that of *c*DCE. The comparatively lower dechlorination rates of *c*DCE and VC reflect their higher inhibition to the granular sludge.

A study using anaerobic granular sludge, fed with PCE and ethanol for 1 year, to degrade PCE observed accumulation of *c*-DCE (Hörber et al. 1999). In the present study, DCEs were not detectable (GC/FID detection limit 6 nmol/bottle) in all headspace gas samples. The different observations in intermediate accumulation were possibly due to different microbial compositions in granular sludges, use of primary substrates, or both. It can also be seen that pre-exposure to PCE would not benefit the degradation, judging from accumulation of metabolic intermediates.

For practical application it is essential that all chlorinated ethenes present as a mixture in groundwater can be mineralized at high rates to low residual concentrations. Considering that both rapid and complete dechlorination are required, partially oxidative conditions, i.e. 5.2–10.4% O_2, are suggested by the results in this study. The synchronous anaerobic/aerobic dechlorination that fulfills the demands can be achieved by using granular sludge. This may imply that upflow anaerobic sludge blanket reactors, in which granular sludge is commonly formed (Lettinga et al. 1980), could serve as adequate facilities for on-site bioremediation of groundwater contaminated with chlorinated ethenes (Hwu & Lu in press). Further investigation on the optimum redox conditions in continuous-reactor treatment is recommended.

4 CONCLUSIONS

Synchronously anaerobic/anaerobic dechlorination of tetrachloroethene (PCE), trichloroethene (TCE), *cis*-dichloroethene (*c*DCE) and vinyl chloride (VC) was respectively achieved by anaerobic granular sludge exposed to oxygen at various concentrations. Considering that both rapid and complete dechlorination are required in overall degradation of chlorinated ethenes, partially oxidative conditions, i.e. 5–10% O_2, are suggested for groundwater bioremediation.

ACKNOWLEDGEMENTS

Funding for this study was provided by the National Science Council, Taiwan, under Project No. NSC-89-2211-E-241-006. Part of the batch work described in this study was financially supported by the grant HKHSC-91-02.

REFERENCES

Beunink, J. & Rehm, H.J. 1988. Synchronous anaerobic and aerobic degradation of DDT by an immobilized mixed culture system. *Appl. Microb. Biotechnol.* 29: 72–80.

Beunink, J. & Rehm, H.J. 1990. Coupled reductive and oxidative degradation of 4-chloro-2-nitrophenol by a co-immobilized mixed culture system. *Appl. Microb. Biotechnol.* 34: 108–115.

Hörber, C., Christensen, N., Arvin, E. & Ahring, B.K. 1999. Tetrachloroethene dechlorination kinetics by *Dehalospirillum multivorans* immobilized in upflow anaerobic sludge blanket reactors. *Appl. Microb. Biotechnol.* 51: 694–699.

Hwu, C.-S., Donlon, B. & Lettinga, G. 1996. Comparative toxicity of long-chain fatty acid to anaerobic sludge from various origins. *Water Sci. Technol.* 34: 351–358.

Hwu, C.-S., Lu, C.-J. & Liu, H.-P. 2002. Simultaneous anaerobic and aerobic dechlorination of PCE in granular sludge. Paper 2H-36, in: A.R. Gavaskar and A.S.C. Chen (Eds.), *Proc. 3rd Intl. Conf. on Remediation of Chlorinated and Recalcitrant Compounds (CD-ROM)*, May 20–23, Monterey, CA.

Hwu, C.-S. & Lu, C.J. (in press). Continuous dechlorination of tetrachloroethene to ethene in an ypflow anaerobic sludge blanket reactor. IWA Conf. Anaerobic Digestion 2004.

Lettinga, G., van Velsen, A.F.M., Hobma, S.W., de Zeeuw, W. & Klapwijk, A. 1980. Use of upflow anaerobic (USB) reactor concept for biological wastewater treatment, especially for anaerobic treatment. *Biotechnol. Bioeng.* 22: 677–734.

Liu, H.P. 2001. *Treatment of tetrachloroethylene by anaerobic granular sludge.* MSc thesis, National Chunghsing University, Taichung, Taiwan.

Sanz, J.L., Rodríguez, N. & Amils, R. 1997. Effect of chlorinated aliphatic hydrocarbons on the acetoclastic methanogenic activity of granular sludge. *Appl. Microb. Biotechnol.* 47: 324–328.

Van Eekert, M.H.A. 1999. *Transformation of chlorinated compounds by methanogenic granular sludge.* PhD Thesis. Wageningen University, Wageningen, the Netherlands.

European Symposium on Environmental Biotechnology, ESEB 2004 - Verstraete (ed)
© 2004 Taylor & Francis Group, London, ISBN 90 5809 653 X

Regulation of tetralin biodegradation genes

E. Moreno-Ruiz, O. Martínez-Pérez & E. Santero
C.A.B.D., Departamento de Ciencias Ambientales, Universidad Pablo de Olavide, Sevilla, Spain

ABSTRACT: *Sphingomonas macrogolitabida* strain TFA is able to grow using tetralin as sole carbon and energy source. Two divergent operons, containing the *thn* genes, have been sequenced. Their transcription is induced by tetralin and repressed by alternative carbon sources, like β-hydroxybutirate (catabolic repression). The effects of azide (respiratory chain inhibitor) and 2,4-dinitrophenol (respiratory chain uncoupler) in the presence of tetralin and β-hydroxybutirate were studied. ThnR and ThnY, encoded by two adjacent genes are regulatory proteins. Null *thnR* or *thnY* mutants were unable to grow on tetralin as the sole carbon and energy source and did not induce the *thn* genes. ThnR is a transcriptional activator similar to LysR-type regulators, while ThnY is homologous to ring hydroxylases ferredoxin reductases. Mini-Tn*5 Km* insertion mutants, inducible by tetralin in β-hydroxybutirate were isolated and initially characterized.

1 INTRODUCTION

The organic solvent tetralin (1,2,3,4-tetrahydronaphthalene) is a bicyclic molecule composed of an aromatic and an alicyclic moiety which share two carbon atoms. Tetralin is widely used as a degreasing agent and solvent for fats, resins and waxes, as a substitute for turpentine in paints, lacquers and shoe polishes, and also in the petrochemical industry in connection with coal liquefaction (Gaydos 1981).

The metabolism of tetralin in *Sphingomonas macrogolitabida* strain TFA has been widely characterized. Biodegradation of tetralin by the strain TFA involves initial oxidation of the aromatic ring to yield 1,2-dihydroxytetralin, through reactions catalyzed by a ring-hydroxylating dioxygenase and a dehydrogenase (Moreno-Ruiz et al. 2003). The catecholic intermediate is then successively metabolized by an extradiol dioxygenase, a hydrolase, a hydratase and an aldolase (Andújar et al. 2000, Hernáez et al. 2000, Hernáez et al. 2002). This set of enzymes, typically involved in metabolism of one aromatic ring, is able to cleave both the aromatic and the alicyclic rings of tetralin yielding pyruvate and pimelic semialdehyde (Hernáez et al. 2002). The genes coding for these enzymes have also been identified, and shown to cluster together in two closely linked operons, which are divergently transcribed (Fig. 1) (Hernáez et al. 1999, Moreno-Ruiz et al. 2003).

The expression of most catabolic operons is regulated by specific inducible systems of control in order to assure that the enzymes are only produced under appropriate environmental conditions. Additionally, expression of catabolic operons is very frequently subject to overimposed global regulatory controls, which prevent transcription of catabolic genes under conditions of nutritional excess, thus optimizing gene expression by connecting it to the metabolic and/or energetic status of the cell (Cases & de Lorenzo 1998, Díaz & Prieto 2000). Most of them fit within the category of carbon catabolite repression, which prevents expression of catabolic operons in the presence of preferential carbon and energy sources. Although carbon catabolite repression describes a similar phenomenon, the molecular mechanisms that exert the control may be completely different in distantly related bacteria (Saier 1996, Saier 1998). In fact, more than one global control may regulate expression of biodegradation genes within the same bacteria (Cases & de Lorenzo 2000, Dinamarca et al. 2003).

This paper reports on the regulated expression of the tetralin biodegradation operons of *Sphingomonas*

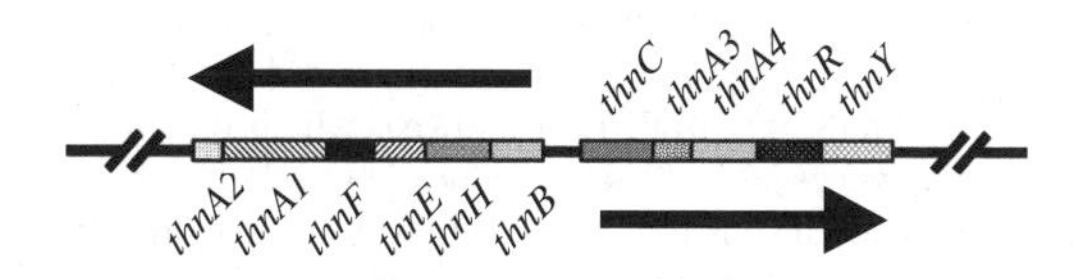

Figure 1. Representation of the genomic region of strain TFA involved in tetralin biodegradation.

macrogolitabida strain TFA, showing that it is induced in the presence of the pathway substrate and subject to carbon catabolite repression. Characterization of two regulatory genes whose products are essential for *thn* gene expression is also described.

2 RESULTS

2.1 *Inducible expression of* thn-lacZ *gene fusions*

To easily test expression of the tetralin catabolic operons, translational *lacZ* gene fusions were constructed to *thnC* and *thnB*, the first genes of each operon (Fig. 1), using an non-replicative plasmid in TFA. Therefore, transconjugants resulting from a single recombination event leading to integration of the plasmid into the TFA genome were isolated. This approach allows testing expression of the gene fusions in the same copy number and in the same genomic context as the original genes.

The obtained strains were grown in mineral medium containing β-hydroxybutirate (βHB) as the only carbon and energy source to exponential phase. Growing cells were then washed and resuspended in mineral medium with tetralin in the gas phase, and samples were taken at time intervals for testing β-galactosidase activity. Cells growing on βHB did not express the gene fusion at any growth phase. However, expression of the gene fusion was evident shortly after the cells were transferred to growing conditions on tetralin, thus showing that expression of both tetralin biodegradation operons is not constitutive but induced by the presence of the pathway substrat (not shown).

2.2 *Carbon catabolic repression of* thn *operons*

In order to test the effect of availability of alternative carbon sources on the induction of *thn* operons, similar induction kinetics by tetralin were carried out using mineral medium containing different concentrations of βHB, which allows a higher growth rate than tetralin. As shown for the *thnC-lacZ* translational fusion (TFA::1002 strain) in Figure 2, increasing the concentration of carbon in the mineral medium resulted in a proportional delay and reduced level of induction, thus showing that availability of βHB prevented expression of *thn* operons. The undefined rich medium MML also prevented induction although significant expression was observed at the end of the induction (Fig. 2).

Similar cultures to those used for the induction kinetics but lacking tetralin in the gas phase did not induce *thnC* expression at all (not shown), which confirms that *thn* genes are not simply induced by carbon-limited growth conditions but their expression is also strictly dependent on the presence of the pathway inducer. In addition, induction kinetics by tetralin in conditions where nitrogen availability was the growth-limiting

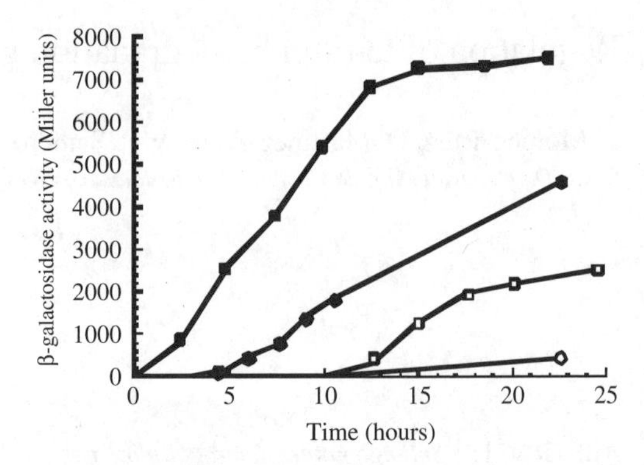

Figure 2. Induction kinetics in TFA::1002 in tetralin. ■: tetralin; ●: tetralin and 8 mM βHB.; □: tetralin in MML; ○: tetralin and 40 mM βHB.

factor clearly indicate that carbon limitation but not growth-limiting conditions *per se* allows induction of *thnC* by tetralin (not shown).

2.3 *Effects of altering electron transport chain function or ATP production on carbon catabolite repression*

Several reports suggested that the redox or the energetic status of the cells might be signals for the overimposed global control of catabolic operons in Pseudomonads (Collier et al. 1996, Dinamarca et al. 2002, Petruschka et al. 2001).

Tetralin induction of *thnC* under repressing conditions was tested in the presence of sublethal concentrations of sodium azide, which preferentially inhibits cytochrome o ubiquinol oxidase, or 2,4-dinitrophenol (2,4-DNP), a proton motive force uncoupler that does not affect electron transport but reduces the ATP content of the cell.

The presence of 1 mM sodium azide in the repressing medium very slightly increased *thnC* expression levels after 30 hours of induction with tetralin, when exponential growth rate slowed down (Fig. 3). In addition, when cultures showing significant induction of *thnC* at the end of the exponential phase were diluted in fresh repressing medium containing sodium azide and tetralin, the level of β-galactosidase activity was progressively reduced, thus suggesting that sodium azide did not significantly prevent catabolic repression.

In the presence of 1 mM 2,4-DNP, expression of *thnC* was 6-fold higher than that observed in its absence (Fig. 3). However, induction was only evident after 28 hours of induction, when exponential growth rate slowed down. Similarly, dilution of the induced cultures in fresh repressing medium containing 2,4-DNP and tetralin, resulted in progressive loss of β-galactosidase activity until the culture reached again the late

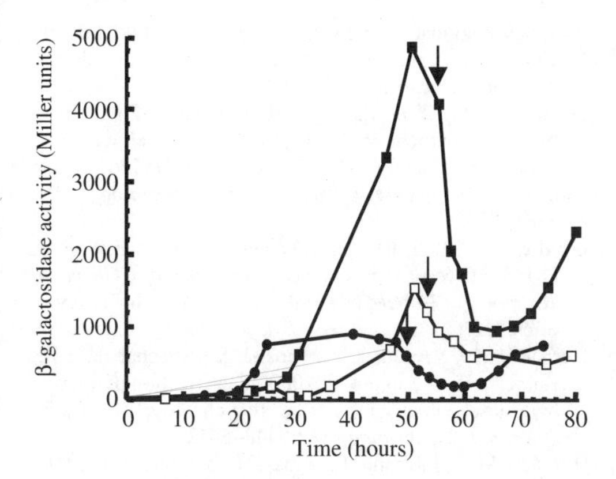

Figure 3. Azide and 2,4-DNP kinetics in 40 mM βHB with tetralin in TFA::1002. □ 1 mM Azide; ■ 1 mM 2,4-DNP; ● nothing added. Arrows mark the first point after dilution in new medium.

exponential phase, when activity started to increase, thus indicating that 2,4-DNP does not relieve catabolic repression during exponential phase.

2.4 *Identification of genes required for* thn *gene expression*

The product encoded by *thnR* showed high similarity to known LysR-type activators of operons involved in biodegradation of different aromatic pollutants. It showed highest similarity to DntR from *Burkholderia* sp. strain DNT and to NagR from *Ralstonia* sp. strain U2 (45% identity along the molecules) (Zhou et al. 2001). A possible evolutionary relationship of ThnR and activators of naphthalene biodegradation operons (NagR/NahR) could be suggested from a dendrogram resulting from the comparison of aminoacid sequences of similar LysR-type activators.

Seven nucleotides downstream of the stop codon of *thnR* is the start codon of *thnY*. BLAST comparison of the putative product to those in the databases showed significant similarity to ferredoxin reductases, which are components of electron transfer systems to dioxygenases or monooxygenases of different aromatic pollutants. It showed highest identity (36%) to the ferredoxin reductase component of naphthalene dioxygenases from different strains, including that from *Ralstonia* sp. strain U2 (Zhou et al. 2001). This type of ferredoxin reductases contains three domains. From their C-terminus, an NAD binding-1 domain, an FAD binding-6 domain and a fer2 domain, which binds a chloroplast-type Cys_4[2Fe-2S] iron sulfur center, are recognizable by sequence analysis. In a sequence analysis of ThnY the NAD binding-1 domain was not detected. Multialignment of C-terminal regions of ferredoxin reductases and the consensus NAD binding-1 domain showed two blocks of highly conserved residues. Interestingly, two residues of each block were substituted in the sequence of ThnY. These data clearly suggest that ThnY was originally a ferredoxin reductase whose NAD binding domain has degenerated and, therefore, it is not expected that ThnY could bind NADH.

Table 1. Expression of *thnC*::*lacZ* fusion in *thnR* or *thnY* mutants and complementation of the regulatory phenotype.

	β-galactosidase activity (Miller units)	
Strain	−tetralin	+tetralin
TFA::1002	78	4570
T656::1002	68	53
T669::1002	50	55
T601::1002	69	64
T656::1002/pIZ1017	157	3810*
T669::1002/pIZ698	263	4979*
T601::1002/pIZ698	84	4217*

* 1 mM IPTG added.

2.5 *Functional characterization of ThnR and ThnY*

Mutant T656 contains a non-polar KIXX insertion in *thnR* (Hernáez et al. 1999). Mutant T669 bears a non-polar KIXX insertion in *thnY* (Hernáez et al. 1999), while mutant T601 bears a polar kanamycin resistance cassette insertion, flanked by transcription terminators, in *thnY*. None of these mutants were able to grow using tetralin as the only carbon and energy source, indicating that both ThnR and ThnY are required for tetralin utilization.

The translational *thnC*::*lacZ* fusion was integrated into the genome of those mutants, yielding T656::1002, T669::1002 and T601::1002 strains, respectively. None of these mutants were able to induce *thnC* in response to tetralin (Table 1). *thnR* and *thnY* were cloned separately in pIZ1016 (Moreno-Ruiz et al. 2003) yielding pIZ1017 and pIZ698, respectively, where transcription proceeded from the IPTG-inducible *tac* promoter. Mutant T656 transformed with pIZ1017 was able to grow on tetralin and maximal levels of *thnC* induction were achieved in the presence of IPTG (Table 1). Similar positive complementation was observed in the mutant T601 transformed with pIZ698 (Table 1). Transformation of T656 with pIZ698 or T601 with pIZ1017 did not result in a change of the mutant phenotype (not shown).

2.6 *Characterization of new mutants*

Mini-Tn*5 Km* insertion mutants were constructed in the *thnC-lacZ* strain by triparental mating. Blue mutants were obtained in MML X-gal plates in the presence of tetralin. Induction kinetics in MML with tetralin, in 40 mM βHB with tetralin and in tetralin were then successively done in order to characterize the metabolism of tetralin of the obtained mutants.

Four mutants were finally selected. They were able to grow in tetralin as the sole carbon and energy source, so they were not affected in the *thn* genes. They were also able to induce the catabolic operons in the presence of tetralin and an alternative carbon source, avoiding the carbon catabolic repression. They could induce either in 40 mM βHB and in MML suggesting that repression in both nutrient sufficient media may operate trough the same overimposed control. The mini-Tn*5 Km* flanking regions of those mutants are being cloned and sequenced.

REFERENCES

Andújar, E., Hernáez, M. J., Kaschabek, S. R., Reineke, W. & Santero, E. 2000. Identification of an extradiol dioxygenase involved in tetralin biodegradation: gene sequence analysis, purification and characterization of the gene product. *Journal of Bacteriology* 182:789–795.

Cases, I. & De Lorenzo, V. 1998. Expression systems and physiological control of promoter activity in bacteria. *Current Opinion in Microbiology* 1:303–310.

Cases, I. & De Lorenzo, V. 2000. Genetic evidence of distinct physiological regulation mechanisms in the σ^{54} *Pu* promoter of *Pseudomonas putida*. *Journal of Bacteriology* 182:956–960.

Collier, D. N., Hager, P. W. & Phibbs, P. V. Jr. 1996. Catabolite repression control in the Pseudomonads. *14th Forum in microbiology; Research in Microbiology* 147:551–561.

Díaz, E. & Prieto, M. A. 2000. Bacterial promoters triggering biodegradation of aromatic pollutants. *Current Opinion in. Biotechnology* 11:467–475.

Dinamarca, A., Aranda-Olmedo, I., Puyet, A. & Rojo, F. 2003. Expression of the *Pseudomonas putida* OCT plasmid alkane degradation pathway is modulated by two different global control signals: evidence from continuous cultures. *Journal of Bacteriology* 185:4772–4778.

Dinamarca, A., Ruiz-Manzano, A. & Rojo, F. 2002. Inactivation of cytochrome *o* ubiquinol oxidase relieves catabolic repression of the *Pseudomonas putida* GPo1 alkane degradation pathway. *Journal of Bacteriology* 184: 3785–3793.

Gaydos, R. M. 1981. Naphthalene. In M. Grayson & D. Eckroth (ed.), *Kirk-Othmer Encyclopedia of Chemical Technology. 3rd ed.*: 698–719. New York: John Wiley & Sons, Inc.

Hernáez, M. J., Andújar, E., Ríos, J. L., Kaschabek, S.R., Reineke, W. & Santero, E. 2000. Identification of a serine hydrolase, which cleaves the alicyclic ring of tetralin. *Journal of Bacteriology* 182:5448–5453.

Hernáez, M. J., Floriano, B., Ríos, J. L. & Santero, E. 2002. Identification of a Hydratase and a Class II Aldolase Involved in Biodegradation of the Organic Solvent Tetralin. *Applied and Environmental Microbiology* 68: 4841–4846.

Hernáez, M. J., Reineke, W. & Santero, E. 1999. Genetic analysis of biodegradation of tetralin by a *Sphingomonas* strain. *Applied and Environmental Microbiology* 65:1806–1810.

Moreno-Ruiz, E., Hernáez, M. J., Martínez-Pérez, O. & Santero, E. 2003. Identification and functional characterization of *Sphingomonas macrogolitabida* strain TFA genes involved in the first two steps of the tetralin catabolic pathway. *Journal of Bacteriology* 185:2026–2030.

Petruschka, L., Burchhardt, G., Müller, C., Weihe, C. & Hermann, H. 2001. The *cyo* operon of *Pseudomonas putida* is involved in carbon catabolite repression of phenol degradation. *Mollecular and General Genomics* 266:199–206.

Saier, M. H. Jr. 1996. Catabolite repression. *14th Forum in microbiology; Research in Microbiology* 147:439–588.

Saier, M. H. Jr. 1998. Multiple mechanisms controlling carbon metabolism in bacteria. *Biotechnology and Bioengeneering* 58:170–174.

Zhou, N. Y., Fuenmayor, S. L., & Williams, P. A. 2001. *nag* genes of *Ralstonia* (formerly *Pseudomonas*) sp. strain U2 encoding enzymes for gentisate catabolism. *Journal of Bacteriology* 183:700–708.

European Symposium on Environmental Biotechnology, ESEB 2004 - Verstraete (ed)
© 2004 Taylor & Francis Group, London, ISBN 90 5809 653 X

Cloning of bacterial PCB-degrading gene into the plants

M. Surá[1,3], M. Mackova[1,3], R. Borovka[1], K. Franečová[1,3] M. Szekeres[2] & T. Macek[1,3]
[1] *Dept. of Biochemistry and Microbiology, Faculty of Food and Biochemical Technology, Institute of Chem. Technol. Prague, Prague, Czech Republic*
[2] *Institute of Plant Biology, BRC of Hungarian Academy of Sciences*
[3] *Department of Natural Products, Institute of Organic Chemistry and Biochemistry, Czech Academy of Sciences, Prague, Czech Republic*

ABSTRACT: The target of this work was cloning of bacterial degrading gene *bphC* to increase biodegradation potencial of polychlorinated biphenyls (PCB) by the plants. For this purpose the gene *bphC* encoding the enzyme 2,3-dihydroxybiphenyl-1,2-dioxygenase from bacteria *P. testosteroni B-356* was chosen to be cloned with the detection marker gene GFP. Because of the difficulties with the detection of expression of GFP in plant tissue, also other constructs were prepared. These contain luciferase gene next to the *bphC* gene, GUS gene and/or six His. The presence of *bphC*/GFP DNA and RNA was proved by PCR. Immunochemical analysis confirmed the presence of BphC/GFP fusion in several transformants. The constructs with *bphC*/LUC, *bphC*/GUS and *bphC*/His were transformed into the *N. tabacum* via agrobacterial infection.

1 INTRODUCTION

Polychlorinated biphenyls (PCBs) are lipophilic substances, which were widely used till the beginning of the eighties. PCBs have very good physical and chemical properties, but these properties negatively affected their persistence in the nature, which can further negatively act on fauna, flora and human health. There are several ways how to decrease the amount of PCBs from the environment. One possibility of the removal of contaminants from soil represents the physicochemical methods. These techniques are unfortunately highly economically demanding and often can further destroy the environment (Macek et al. 2003). Therefore the research has been oriented towards the use of biological remediation methods based on the fact that various organisms can degrade various xenobiotics. One of the biological method is the use of green plants for transfer, accumulation and removal of pollutants from the environment, or at least reduction of their spreading (Cunnigham et al. 1995, Macek et al. 2000) – phytoremediation. It was shown that plants have limited abilities to mineralise PCBs (Kucerova et al., 2000, Wilken et al., 1995). Unlike bacteria, plants generally transform PCBs to hydroxychlorobiphenyls without cleavage of the biphenyl ring. This limitation can be overcome by preparation of transgenic plants with known bacterial genes cleaving and destroying biphenyl structure.

The purpose of this investigation was to engineer genetically modified plants bearing the bacterial gene *bphC* coding enzyme 2,3-dihydroxybiphenyl-1,2-dioxygenase. This enzyme is responsible for opening of biphenyl ring and degradation of the molecular structure. It catalyses the conversion of 2,3-dihydroxychlorobiphenyl into 2-hydroxy-6-oxo-6-phenyl-hexa-2,4-dienoic acid.

Understanding the basic physiology and biochemistry that underlie various phytoremediation processes is very important to improve the applicability of this plant based method. Although organic pollutants are metabolised in plants, some xenobiotics can be toxic, limiting the applicability of phytoremediation. Thus more research is required to better understand and exploit the reactions in plants.

In our studies we used three plant species (alfalfa, tobacco, black nightshade), previously successfully used in laboratory phytoremediation studies investigating the abilities of plant cells to degrade PCBs. In described study the process of PCBs transformation was studied from the point of intracellular and extracellular peroxidase synthesis, products formed and ability of isolated POX to remove PCBs from the environment.

2 MATERIALS AND METHODS

2.1 *Bacterial and plant strains*

In our study following bacterial strains were used – *Comamonas testosteroni* B-356, *Escherichia coli* M15 harboring pREP4 plus hybrid plasmid pQE31 (carrying B356 *bphC*), *Escherichia coli* XL1-Blue, *E. coli* S17-I and *Agrobacterium* GV3101 (pPM90RK). Aseptic cultures of *Nicotiana tabacum* var. Wisconsin 38 and *Arabidopsis thaliana* var. Wassilewskiya were used for transformation by agrobacterial infection.

2.2 *Cloning of* bphC *gene to plants*

Gene *bphC* (882 bp) was originally isolated from the operon of *Comamonas testosteroni* B-356 (Hein et al.). Other genes were for fusion with gene *bphC*. These were gene GFP (gene for green fluorescent protein), GUS (gene for beta-glucuronidase), LUC (gene for luciferase), and also six histidine motives were fused with gene *bphC*. Plasmids, which were used in this study, are following: pBluescript, pQE31, pPCV812i (Konz et al., 1994) pPCV/LUC + -NOS [7]. Constitutive promoter CaMV 35S was used in each construct. In this study the construct with *bphC*/GFP was already transferred into the plant cells of *Nicotiana tabacum*. Other prepared constructs were first check for their correct primary structure and then transformed to *E. coli* S17-1 later conjugated with *Agrabacterium* GV3101 (pPM90RK). Agrobacteria bearing constructs with the gene *bphC* in cassete with GUS, LUC or His tail were used for transormation of plant cells by agrobacterial infection. Transformed regenerants were selected on media with selective antibiotic. Transgenes were recognized by root formation.

2.3 *Confirmation of the presence of* bphC *in plants*

Presence of *bphC*/GFP in plants was confirmed after isolation of plant DNA and PCR amplification of the gene using combination of specific primers bphC1/F, bphC2/R, bphC3/F, GFP/F and GFP/R (fig. 3). The following experiments included the isolation of plant RNA, RNA cleaning from DNA and RT-PCR. Also protein studies were done by Western blot analysis with GFP and BphC antibodies.

Presence of *bphC*/LUC in plants *Arabidopsis thaliana* was confirmed on the DNA level by PCR with specific primers bphC1/F, bphC2/R, LUC/F and LUC/R (fig. 4). Also histochemical studies were performed to prove the expression of the protein BphC/LUC.

2.4 *Analysis of residual content of PCBs*

After the cultivations with PCBs, the cells were killed by boiling for 20 min., sonicated, homogenised with UltraTurax. The whole content of flasks was extracted with hexane for 2 hours. Samples were analyzed using the Hewlett-Packard 5890 gas chromatograph with an electron capture detector (ECD) and auto-sampler (Kucerova et al., 2000). Degradation was expressed as percentage of PCBs removed comparing to control flasks. For the calculation of the residual amount of PCBs, 22 main chromatographic peaks of Delor 103 were used. Controls comprising of heat-killed cells were used to establish that observed changes in the content of congeners of PCBs were dependent exclusively on the activity of living cells.

3 RESULTS AND DISCUSSIONS

Several bacteria can degrade PCB to less toxic chlorobenzoic acid (Abramowicz 1990, Seeger et al., 1995). This degradation pathway of PCB contains four step, from which the third one (catalysed by enzyme BphC), responsible for the cleavage of the biphenyl ring, is crucial. However plant can also transform PCB and the monohydroxylated and dihydrohydroxylated chloroderivatives are occurring as a major product [5]. These experiments show the inability of plants to destroy biphenyl structure. Therefore the aim of this study is to engineer plant harboring gene *bphC* and thus create plant enzyme system that can destroy the biphenyl structure.

Gene *bphC* (882 bp) encodes the enzyme 2,3-dihydroxybiphenyl-1,2-dioxygenase and is originally isolated from *C. testosteroni* B-356.

Four designs to clone gene *bphC* into plants were proposed: To clone it in fusion with the gene for green fluorescent protein (GFP), beta-glucuronidase (GUS),

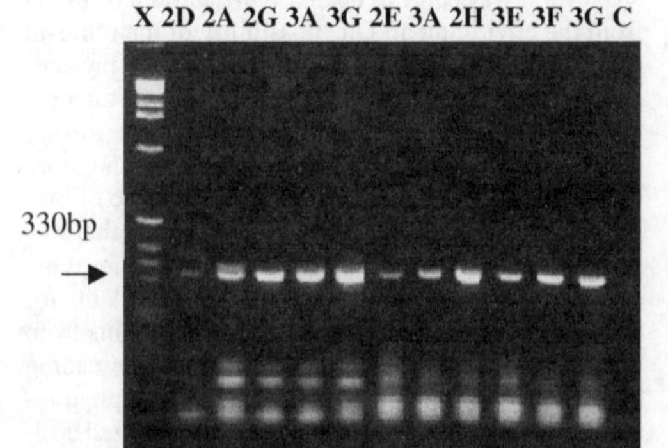

Figure 1. Detection of RNA in plant clones containing BphC/GFP.

Table 1. Germination of transgenic seeds on contaminated agar medium containing 150 ppm of PCBs.

Tobacco seeds of the culture	Number of germinated seeds on 150 ppm PCBs after	
	8 days	10 days
Nontransgenic	9	12
R2G	18	19
R3A	14	14

luciferase (LUC) and with histidine tail (His). All these constructs harbor the constitutive promoter CaMV 35S. The plants of *Nicotiana tabacum* transformed by cassette *bphC*/GFP were already prepared. The detection of presented gene was proved after isolation of plant DNA and PCR amplification of the selected sequences with specific primers. The plant RNA was isolated, treated by DNase I and RT-PCR was performed. The presence of appropriate RNA (330 bp) was confirmed. The expression of gene *bphC*/GFP was studied on the protein level by Western blot analysis with the BphC antibody and GFP antibody. Experiments were successful only in one case of using GFP antibody. Band about 56 kDa was visible, which size corresponds with the fusion protein BphC/GFP.

Seeds obtained from plant cultures exhibited higher resistance to PCB concentration (see Table 1). Detection of the fusion protein fluorescent microscopy of protein GFP did not give satisfactory results, mainly because of the significant autofluorescence of the tobacco tissue.

Because of the difficulties with detection of protein BphC/GFP, additional constructs bearing gene *bphC* were designed. Gene *bphC* was fused with gene for beta-glucuronidase, luciferase and also with histidine tail. These constructs were prepared in plasmids pPCV812i (containing GUS gene) and pPCV/LUC + -NOS (containing LUC gene). The promoter CaMV 35S had to be inserted into these plasmids. These constructs were cloned in bacteria *E. coli* XL-1Blue, primary structure was proved after isolation of plasmid DNA. Further experiments were directed to transfer of transgenes to plant cells. First the constructs were cloned to *Escherichia coli* S17-I (helper strain in transfer of the prepared plasmids into the cells of agrobacteria, which is described in the article Konz et al. 1994)) and then to bacterial cells of the plant pathogen – *A. tumefaciens*.

Agrobacteria GV3103 (pPM90RK) bearing three different constructs were used for transformation of *N. tabacum* (with constructs CaMV35S/*bphC*/GUS, CaMV35S/*bphC*/LUC, CaMV35S/His/*bphC*) and *A. thaliana* (with construct CaMV35S/*bphC*/LUC) by agrobacterial infection.

In the case of transformation of *Nicotiana tabacum* the regenerants nowadays grow on selective media, where the transgenes are recognized by root formation. In the case of transformation of *Arabidopsis thaliana* by construct CaMV35S/*bphC*/LUC the transgene plants were already selected. To detect the presence of appropriate gene histochemical detection was followed. Plants were exposed to the effect of luciferine and the occurred luminescence was measured by luminometer. Expression of BphC/LUC was confirmed by detection of the strong luminescence.

3 CONCLUSIONS

- Different constructs containing gene *bphC* in fusion with gene for green fluorescent protein, luciferase, beta-glucuronidase and histidine tail were prepared.
- The correct primary structure of the fusion genes was proved.
- Fusion genes were transformed into plant cells by agrobacterial infection.
- In transformed plants the presence of DNA (*bphC*/GFP, *bphC*/LUC) and RNA (*bphC*/GFP) was detected.
- Expressed protein BphC/GFP was detected by Western blot analysis.
- Positive expression of BphC/LUC was detected by histochemical assays of luciferase.

ACKNOWLEDGEMENTS

The work described in above was sponsored by the grants of FRVS No. 1581/G4 and 5FW QLK 3-CT-2001-00101.

REFERENCES

Ambramowicz M., 1990. Aerobic and anaerobic biodegradation of PCBs: a review, Crit. Rev. Biotechnol. 10 241–245.

Cunningham S., W. Berti, J. Huang, 1995. Phytoremediation of contaminated soils, *TIBTECH* 13 393–397.

Demnerova K., T. Macek, 2000. Metabolism of polychlorinated biphenyls by *Solanum nigrum* hairy root clone SNC-90 and analysis of transformation products, *Plant and Soil* 225 109–115.

Koncz C., N. Martini, L. Szabados, M. Hrouda, A. Bachmair, J. Schell, 1994. Specialized vectors for gene tagging and expression studies, in: *Plant Molecular Biology Manual B2*, Kluwer Academic Publishers, Printed in Belgium, pp. 1–22.

Wilken A., C. Bock, M. Bokern, H. Harms, 1995. Metabolism of different PCB congeners in plant cell cultures, *Environ. Chem. Toxicol.* 14 2017–2022.

Macek T., J. Kas, M. Mackova, 2000. Exploitation of plants for the removal of organics in environmental remediation, *Biotechnol. Advances* 18 23–34.

Macek T., K. Francova, M. Sura, M. Mackova, 2003. Genetically modified plants with improved properties for phytoremediation purposes, in: J.Morel, (Ed.), *Phytoremediation of metals, NATO Science Series*, Kluwer Academic Publishers, Dordrecht, , in press.

Hein P., J. Powlowski, D. Barriault, Y. Hurtubise, D. Ahmad, M. Sylvestre, 1998. Biphenyl-associated meta-cleavage dioxygenases from *Comamonas testosteroni* B-356, *Can. J. Microbiol*. 44 42–49.

Seeger M., K.N. Timmis, B. Hofer, 1995. Conversion of chlorobiphenyls into phenylhexadienoates and benzoates by the enzymes of the upper pathway for polychlorobiphenyl degradation encoded by the *bph* locus of *Pseudomonas* sp strain LB400. *Appl. Environ. Microbiol*. 61 2654–2658.

European Symposium on Environmental Biotechnology, ESEB 2004 - Verstraete (ed)
© 2004 Taylor & Francis Group, London, ISBN 90 5809 653 X

Kinetics of phenol biodegradation by selected strains of bacteria

Anna Trusek-Holownia
Wroclaw University of Technology, Institute of Chemical Engineering, Norwida Wroclaw, Poland

1 INTRODUCTION

An alternative to chemical methods for removing phenol compounds (sorption, ozonation) is degradation in the presence of properly selected microorganisms.

A source of a large amount of wastewater containing aromatic compounds is mainly petrochemical industry that processes petroleum to fuels and oils, and produces semi-finished products for chemical-organic syntheses. High phenol concentration (0.26–0.3 g/l) occurs also in coke plant wastewater where phenol is produced mainly in the process of coke production and tar processing. Phenols along with other aromatic compounds occur also in the wastewater produced in gas works, timber processing plants, plants producing plastics, dyes, pesticides and pharmaceuticals.

The main problem of microbiological utilisation of wastewater is high sensitivity of degrading strains to the wastewater parameters (pH, T, total concentration of aromatic substances, presence of inhibiting substances, etc.). So, numerous research projects concerning the characteristics of particular microbial strains or mixed cultures in view of their future applications are carried out.

The aim of research presented in this paper was to recognise the range of concentration of phenol degraded by selected microorganism strains, determination of its effect on the duration of lag-phase, determination of growth kinetics and the influence of other cyclic compounds present in the wastewater on cell growth. Investigations were carried out using *Pseudomonas fluorescens*, *Rhodococcus erythropolis* and *Serratia marcescens* strains in batch cultures.

2 METHODS

Tests were made in 500 ml flasks containing 100 ml culture broth with phenol at the concentration 0.02–0.70 g/l and mineral salts suitable for a given strain. The process was monitored by spectrophotometric measurement of cell concentration at $\lambda = 550$ nm and by chromatographic (HPLC, Waters) measurement of the concentration of aromatic compounds (RP18 column, mobile phase acetonitrile-water 60:40 v/v, 0.8 ml/min, detection UV (254 nm)). The culture was grown at the temperature 30°C (estimated as an optimum for all tested strains), at slight flask shaking.

The composition of culture media selected for particular strains – per 1 litre:

Pseudomonas fluorescens	
KH_2PO_4	1.0 g
K_2HPO_4	1.0 g
KNO_3	1.0 g
$MgSO_4$	0.2 g
NaCl	1.0 g
$CaCl_2$	0.02 g
$FeCl_3$	0.001 g
Rhodococcus erythropolis	
KH_2PO_4	2.0 g
Na_2HPO_4	3.0 g
NH_4NO_3	2.0 g
NaCl	3.0 g
$MgSO_4$	0.2 g
Na_2CO_3	0.2 g
$CaCl_2$	0.01 g
$MnSO_4$	0.02 g
$FeSO_4$	0.01 g
Serratia marcescens	
KH_2PO_4	0.27 g
K_2HPO_4	0.35 g
$(NH_4)_2SO_4$	0.53 g
$CaCl_2$	0.001
$MgCl_2$	0.1 g
$FeCl_2$	0.02 g
$MnCl_2$	0.005 g
H_3BO_3, $ZnCl_2$, $CoCl_2$, $NiSO_4$	0.0005 g
$CuCl_2$, $NaWO_4$	0.0003 g
NaMoO	0.0001 g

3 RESULTS AND THEIR DISCUSSION

3.1 *Strain origin and adaptation*

The tested strains came from laboratory cultures in the case of *Pseudomonas fluorescens* and *Rhodococcus erythropolis* (pure cultures, grown on full media with sugars as a carbon source), while *Serratia marcescens* strain was isolated from surface water taken from industrial environment which contained cyclic compounds including phenol. The laboratory strains were adopted to the medium containing phenol by gradual increase of phenol concentration in batch cultures initiated on two carbon sources: in the case of *Pseudomonas fluorescens* on glucose and phenol, and in the case of *Rhodococcus erythropolis* on yeast extract and phenol. Phenol was used by microorganisms after exhausting the more available carbon source. After growing several cultures on mixed source the adopted strains acquired how to use phenol as the only carbon source.

3.2 *Duration of the lag-phase versus phenol concentration*

It was observed that with an increase of the initial phenol concentration the lag-phase of microorganism growth was prolonged. This time for low concentrations, i.e. up to 0.15 g/l was 5–10 hours depending on strain, and from the concentration equal to 0.3 g/l it increased rapidly – Figure 1. For all tested strains no growth was observed when phenol concentration in the flask exceeded 0.70 g/l.

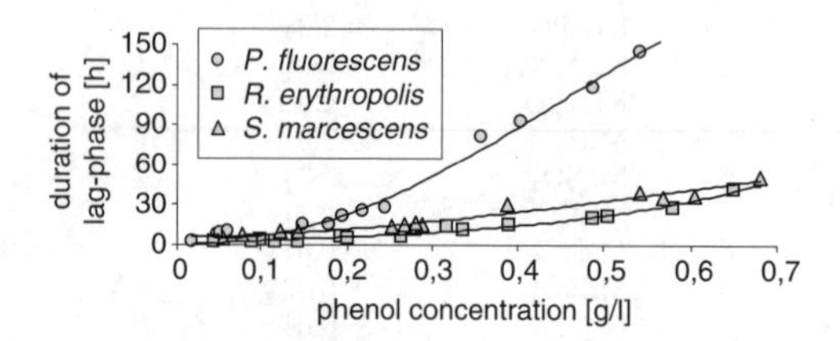

Figure. 1. Duration of lag- phase in dependency on the initial phenol concentration.

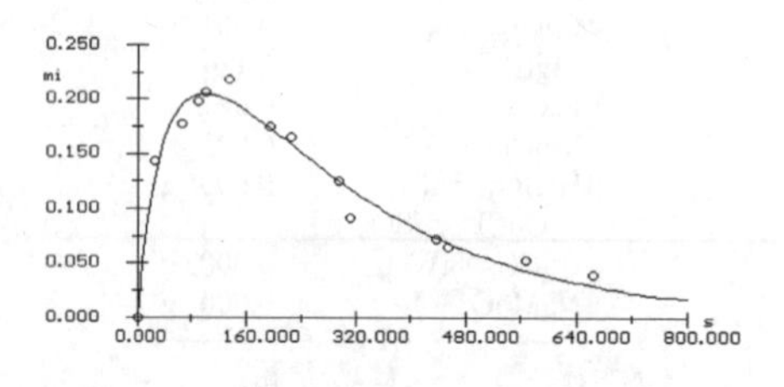

Figure. 2. Dependence of growth rate of *Rhodococcus erythropolis* on phenol concentration.

3.3 *Growth kinetics*

Attempts were made to describe the growth kinetics of the tested microorganisms by equations available in literature. As for high phenol concentrations a decrease of conversion rate with an increase of its concentration was observed, the equations were chosen in which substrate inhibition was taken into account. The following equations were considered:

Haldan's $\mu = \mu_{max}(S/(K_S + S + S^2/K_I))$

Aiba's $\mu = \mu_{max}(S/(K_S + S)\exp(-S/K_I)$

Yamane's $\mu = \mu_{max}(S/(K_S + S + (S/K_I)^n))$

The best solution was found by using the program Polymath on the basis of discrepancy sum square of experimental and model values. Figure 2 shows an example of the dependence of growth rate on microorganism concentration described by a model curve.

To describe the growth rate of *Pseudomonas fluorescens* and *Serratia marcescens* strains Haldan's equation was applied, while for *Rhodococcus erythropolis* Aiba's equation was used. The obtained kinetic constants are summarised in Table 1.

On the basis of the obtained dependencies and determined kinetic constants the best phenol utilisation properties from the kinetics point of view were shown by *Rhodococcus erythropolis*.

3.4 *The effect of other cyclic compounds on phenol degradation*

In naturally occurring wastewater, beside phenol there are also other cyclic compounds whose presence is not indifferent to the growth of microorganisms and phenol utilisation rate. The effect of such substances as cresol, benzene, toluene, resorcin and naphthol was considered. Due to the type of investigation and low solubility of these reagents in water, benzene, toluene and naphthol were used at the concentration corresponding to the concentration of their saturation in water at process temperature, i.e. 30°C, while cresol and resorcin at the concentration 0.08 g/l.

It was found that cresol could also be used as a carbon source; in the case of *Pseudomonas fluorescens* and *Serratia marcescens* simultaneously with phenol and in the case of *Rhodococcus erythropolis* even prior to phenol degradation. This provides an evidence that in the two first cases at least the microorganisms probably do not distinguish between phenol and cresol

Table 1. The values of kinetics constant.

	Pseudomonas fluorescens (Haldane's Eq.)	*Rhodococcus erythropolis (Aiba's Eq.)*	*Serratia marcescens (Haldane's Eq.)*
μ_{max} [1/h]	0.468	0.550	0.405
K_S [g/l]	0.008	0.075	0.078
K_I [g/l]	0.071	0.236	0.092

and they utilise these compounds at similar or equal rates. Other studies showed that all tested strains grew on cresol, as the only source of carbon and energy.

After long-lasting trials, only the cells of *Serratia marcescens* managed to adapt to resorcin utilisation but the cells of *Pseudomonas fluorescens* and *Rhodococcus erythropolis* could grow under these conditions. The resorcin chemical structure is richer only in one hydroxide group in position 2 as compared to phenol.

It is worth noting that benzene and toluene preclude the growth of *Pseudomonas flurescens* strain when naphthol which often occurs beside phenol in the wastewater, has a strong inhibiting effect on all tested strains.

4 CONCLUSION

A possibility of phenol wastewater degradation by the three selected microbial strains: *Pseudomonas fluorescens*, *Rhodococcus erythropolis*, *Serratia marcescens* was reported. A similar level of phenol concentration tolerance was observed at which strains effectively degrade phenol. At high phenol concentrations, i.e. above 0.3 g/l the lag-phase time increases remarkably and the growth rate assumes small values. Hence the process kinetics has been described by the equations with substrate inhibition, i.e. Haldan's and Aiba's equations available in literature. For kinetic reasons, the best parameters are shown by *Rhodococcus erythropolis*.

Analysis of the effect of other cyclic compounds present in the wastewater showed that some of them, e.g. cresol, may be degraded along with phenol or after its removal, resorcin could be utilised by *Serratia marcescens* strain and the others, like for instance naphthol, have a strongly inhibiting effect and then these compounds should be separated before the bioreaction zone or utilised by mixed cultures.

European Symposium on Environmental Biotechnology, ESEB 2004 - Verstraete (ed)
© 2004 Taylor & Francis Group, London, ISBN 90 5809 653 X

The effect of lead (Pb) on the growth and lignin degradation activity of selected litter-decomposing fungi

M. Tuomela, K.T. Steffen & A. Hatakka
Department of Applied Chemistry and Microbiology, University of Helsinki, Finland

M. Hofrichter
International Graduate School Zittau, Zittau, Germany

ABSTRACT: Shooting ranges are often heavily contaminated with lead (Pb) originating from shots and bullets, and the pollution inhibits the growth and metabolism of microorganisms, although lead is mostly in non-bioavailable form. We studied the influence of lead pollution to selected litter-decomposing fungi, which are typical inhabitants of forest soils. Polluted soil was collected from a shooting range, and the upper part of the soil comprising mainly litter, was used in the experiments. The growth of eleven fungal species was monitored in polluted and in non-polluted soil. Fungal metabolism of three selected species was also monitored by measuring their lignin biodegradation capability. The growth of *Stropharia coronilla* was substantial in both polluted and non-polluted litter, but lignin mineralization in polluted litter was only 52% of the mineralization in non-polluted litter. However, the lignin degradation in polluted soil by *Lepista nebularis* ceased almost totally, and degradation by *Collybia dryophila* was also notably diminished.

1 INTRODUCTION

The soil in shooting ranges contains considerable amount of shotgun pellets or shots and bullets, which are mostly comprised of lead (Pb). The behavior of lead varies according to the environmental conditions on site, but there are some common characteristics for all sites. The lead concentration of contaminated soil is usually very high, occurring mostly in metallic form (even up to 90%) (Astrup et al. 1999, Lin et al. 1995). Lead remains predominantly in the upper layer of soil, which is rich in humic substances, and it is generally bound to humic substances and other organic matter (Charlatchka & Cambier 2000, Manninen & Tanskanen 1993). Lead in soil and the surface layer of the shots has transformed mostly to lead carbonates and lead sulphates which may be stable in the environment (Lin et al. 1995). However, due to the remarkably high total concentration of lead, the bioavailable fraction is probably high enough to interact with the organisms at a polluted site, although lead is mostly non-soluble, and thus in non-bioavailable form. Weathering and solubilization of the ammunition material will nevertheless release lead constantly to the environment for a long time, and if pH decreases the solubility and mobility of lead in environment increases (Astrup et al. 1999, Charlatchka & Cambier 2000, Lin et al. 1995, Turpeinen et al. 2000).

Heavy metals inhibit growth and metabolism of all microorganisms, and their biomass decreases in polluted areas (Chander et al. 2001, Kuperman & Carreiro 1997). However, fungi in general tolerate higher concentrations of heavy metals than bacteria, and the bacterial population will cease more due to heavy metal contamination than the fungal population (Chander et al. 2001). Thus, the fungal to bacterial ratio in the population increases, sometimes dramatically (Chander et al. 2001, Shi et al. 2002).

Some heavy metals are essential for fungi in trace amounts (e.g. Fe, Cu, Mn), but some have only toxic effects (e.g. Hg, Pb, Cd), and mercury and cadmium are generally the most toxic heavy metals to fungi (Baldrian et al. 2000, Baldrian 2003). Heavy metals often have an influence on growth and morphology of fungi, the production of ergosterol is enhanced as fungi are affected to heavy metals, and usually heavy metals inhibit the enzymatic activity of the fungi (Baldrian 2003, Chander et al. 2001, Fomina et al. 2003). However, the activity of laccase – one of the ligninolytic enzymes – may increase in the presence of copper or cadmium (Baldrian & Gabriel 2002).

Fungi often accumulate heavy metals by biosorption, but the biosorptive capability of fungi as well as the toxic effect of each individual heavy metal varies greatly between fungal species, even between strains of a species (Baldrian 2003, Dey et al. 1995). The accumulation of heavy metals to fungi is probably connected to fungal melanins as they are bio-sorptive (Fogarty & Tobin 1996). Lead has been observed to accumulate especially to *Boletus* sp., *Lycoperdon perlatum*, *Macrolepiota mastoidea*, *Macrolepiota procera* and *Russula foetus* in two screening studies in lead polluted areas (Demirbaş 2001, Pokorny & Ribarič-Lasnik 2002). Of these fungi *Boletus* sp. and *R. foetus* are mycorrhizal fungi, and *L. perlatum*, *M. mastoidea* and *M. procera* litter-decomposing fungi. It is not known if the biosorption capability of heavy metals has an impact on metal tolerance (Dey et al. 1995).

In this work we studied the influence of lead pollution to selected litter-decomposing fungi, which are typical inhabitants of forest soils. These fungi are known to produce lignin degrading enzymes, and therefore fungal metabolism was monitored by measuring lignin biodegradation capability in polluted and in non-polluted soil.

2 MATERIALS AND METHODS

2.1 *Soil*

Polluted soil was collected from a shooting range in Hälvälä, Hollola, Finland (see Turpeinen et al. 2000 for details of the sampling site). The upper part of the soil, the litter layer, containing partially degraded twigs, bark etc., and considerable amount of shots was used in studies. Polluted soil contained 32 000 mg Pb/kg soil after shots were removed. Non-polluted soil was collected from a similar forest nearby the shooting range, and it contained 120 mg Pb/kg soil. The lead concentration of non-polluted soil was also elevated, and considered slightly polluted, as the average concentration in Finland is 17 mg Pb/kg soil (Dahlbo & Regelin 1995). However, this soil was chosen as non-polluted soil because the soil type and environmental conditions on site were similar to polluted site, and the lead content was nevertheless drastically lower than in polluted soil.

2.2 *Fungi*

Eleven basidiomycetous litter-decomposing fungi were selected for a preliminary experiment, in which fungal growth was visually monitored in lead contaminated litter, and compared to growth in similar non-polluted litter collected near the shooting range. The fungal species were *Agrocybe praecox* TM70.84, *Clitocybe gibba* K32, *Collybia dryophila* K209, *Collybia peronata* K220, *Lepista nebularis* K103, *Marasmius scorodonius* Ho1, *Mycena epipterygia* K72, *Stropharia aeruginosa* K218, *Stropharia coronilla* TM47-1, *Stropharia rugosoannulata* B DSM11372, and an unidentified fungus K23 isolated from a compost environment.

2.3 *Lignin*

Side chain labelled synthetic lignin was used as substrate in a lignin degradation experiment. The β-^{14}C-labelled dehydrogenation polymer (^{14}C-DHP) was synthesized according to Brunow et al. (1998) with some modifications. Solid ^{14}C-DHP was dissolved to *N, N*-dimethylformamide (DMF), and the obtained ^{14}C-DHP solution was added to a vigorously stirred DMF-aqueous suspension (1:20) (Kirk et al. 1978; Vares et al., 1994).

2.4 *Lignin degradation experiment*

C. dryophila, L. nebularis and *S. coronilla* were selected for a lignin biodegradation experiment, which was performed in 120 ml flasks with 7 g of soil litter. The amount of shots was adjusted to a constant (4 g/flask) in flasks with polluted soil. The flasks were autoclaved (20 min, 121°C), and ^{14}C-DHP was added from a stock suspension (approximately 4.88 kBq/flask). Soil litter was inoculated with three agar plugs of active mycelium, and uninoculated flasks were used as controls. Moisture content of the medium was adjusted to 60% by adding sterile deionized water. Culture conditions and experimental design were analogous to those explained by Steffen et al. (2000). Cultures and control flasks with three replicates were incubated for 91 days, and evolved gases were trapped by bubbling through two sequential flasks in order to separate radioactive volatile organic compounds and $^{14}CO_2$. Radio-activity was measured with a liquid scintillation counter (Wallac 1411).

3 RESULTS AND DISCUSSION

The capability to grow in soil contaminated with lead varied notably between the selected fungal species (Table 1). Three species showed no significant difference between growth in polluted soil and non-polluted soil. On the other hand four species did not grow at all in soil polluted with lead. In earlier studies with white-rot fungi the variation in capability to tolerate heavy metals among species has been observed, and variation may exist even between fungal strains (Baldrian 2003). It is no surprise that basidiomycetous litter-decomposing fungi that are physiologically rather similar to white-rot fungi show similar characteristics in metal tolerance. Both sensitive and tolerant organisms for heavy metals can be found in

Table 1. The growth of basidiomycetous litter-decomposing fungal species in Pb polluted and in non-polluted soil.

Fungus	Growth on non-polluted litter	Growth on Pb polluted litter	Difference
Agrocybe praecox	+++[a]	+++[e]	0[f]
Clitocybe gibba	+	+[c]	–
Collybia dryophila	++	+	–
Collybia peronata	+++	+[c]	–
Lepista nebularis	+++[a]	+	–
Marasmius scorodonius	+++[a]	+[c]	–
Mycena epipterygia	+++[a]	+	–
Stropharia aeruginosa	++	++	0
Stropharia coronilla	+++	+++[b]	0
Stropharia rugosoannulata	+++[a]	++	–
Unidentified fungus[g]	+++[a]	no[d]	–

[a] Very good growth (thick white mycelium throughout the litter).
[b] Growth only on the top layer.
[c] Growth only on the agar plug.
[d] No growth.
[e] Growth through the entire soil-litter layer.
[f] Thinner mycelium on the top compared to uncontaminated soil.
[g] Basidiomycetous fungus isolated from compost.
0 = no difference.
– = Pb had a negative impact on mycelial growth.

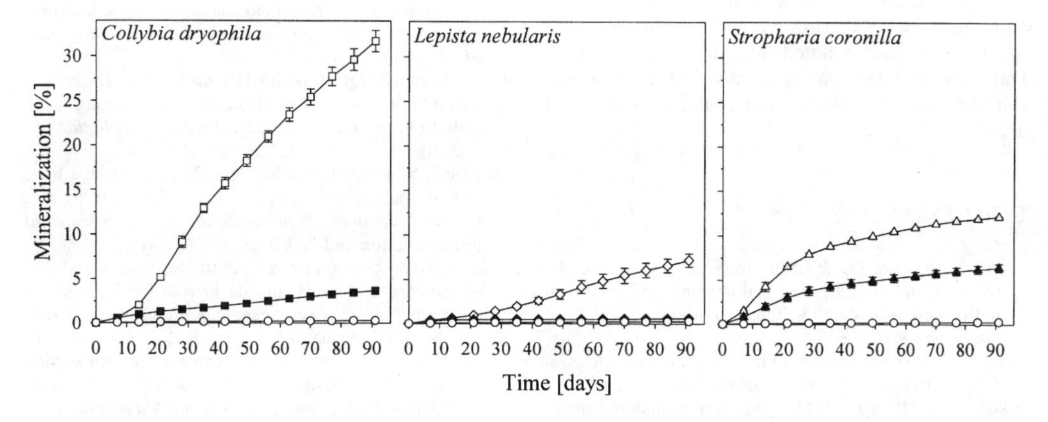

Figure 1. Mineralization of β-^{14}C-labelled synthetic lignin (DHP) by three litter-decomposing fungi, open circles represent uninoculated controls, other open symbols represent fungus in non-polluted soil, closed symbols represent fungus in soil polluted with lead (Pb) originating from shotgun pellets.

polluted environments, probably because the pollution is not evenly distributed (Baldrian 2003, Shi et al. 2002). Also, the correlation between toxic effect and concentration of heavy metals is not linear (Chander et al. 2001).

For a lignin biodegradation experiment we selected three fungi, namely *Collybia dryophila* (growth was slightly reduced in polluted soil), *Lepista nebularis* (growth was strongly reduced), and *Stropharia coronilla* (growth was substantial in both polluted and non-polluted soil) (Table 1).

Most litter-decomposing fungi studied so far produce manganese peroxidase (MnP) and laccase, the same ligninolytic enzymes produced also by white-rot fungi (Steffen et al. 2000). Thus, litter-decomposing fungi are capable to degrade lignin almost as efficiently as white-rot fungi, and the decomposing activity of these fungi in soil probably correlates to their lignin degradation capability (Steffen et al. 2000). In lead containing environment the mineralization of ^{14}C-DHP by *C. dryophila* and *L. nebularis* was strongly reduced (Fig. 1). *S. coronilla* mineralized 12.2% of added ^{14}C-DHP in non-polluted and 6.4% in polluted soil (Fig. 1). Lignin mineralization in polluted soil by *S. coronilla* was accordingly reduced by 48%, although the growth was not significantly reduced. Thus, the activity of fungus in polluted soil is not governed only by growth. Chander et al. (2001) observed high respiration rate in sites contaminated with heavy metals, even in heavily polluted areas, with a great diversity of microorganisms. However, populations in contaminated sites are significantly different from populations in non-contaminated sites (Shi et al. 2002). It can be expected that elevated concentration of lead in the environment will change the incidence of various fungal species, also among litter-decomposing fungi, which comprise a notable fraction of the microbial population of boreal forests. The degradation capability of litter-decomposing fungi probably decreases in such environment, hence the ecological impact of lead pollution would be the retardation of carbon cycle in the forest.

4 CONCLUSIONS

The effect of lead contamination to litter-decomposing fungi varies considerably according to the fungal species. The growth and lignin mineralization of *Collybia dryophila* and *Lepista nebularis* were significantly suppressed by lead. However, *Stropharia coronilla* grew well in lead polluted litter, but lignin mineralization in polluted litter was only half of the mineralization in non-polluted litter, indicating reduced activity.

REFERENCES

Astrup T., Boddum J.K. & Christensen T.H. 1999. Lead distribution and mobility in a soil embankment used as a bullet stop at a shooting range. *Journal of Soil Contamination* 8: 653–665.

Baldrian P. 2003. Interactions of heavy metals with white-rot fungi. *Enzyme and Microbial Technology* 32: 78–91.

Baldrian P. & Gabriel J. 2002. Copper and cadmium increase laccase activity in *Pleurotus ostreatus*. *FEMS Microbiology Letters* 206: 69–74.

Baldrian P., in der Wiesche C., Gabriel J., Nerud F. & Zadražil F. 2000. Influence of cadmium and mercury on activities of ligninolytic enzymes and degradation of polycyclic aromatic hydrocarbons by *Pleurotus ostreatus* in soil. *Applied and Environmental Microbiology* 66: 2471–2478.

Brunow G., Raiskila S. & Sipilä J. 1998. The incorporation of 3,4-dichloroaniline, a pesticide metabolite, into dehydrogenation polymers of coniferyl alcohol (DHPs). *Acta chemica Scandinavica* 52: 1338–1342.

Chander K., Dyckmans J., Joergensen R.G., Meyer B. & Raubuch M. 2001. Different sources of heavy metals and their long-term effects on soil microbial properties. *Biology and Fertility of Soils* 34: 241–247.

Charlatchka R. & Cambier P. 2000. Influence of reducing conditions on solubility of trace metals in contaminated soils. *Water, Air, and Soil Pollution* 118: 143–167.

Dahlbo H. & Regelin E. 1995. Ampumarata-alueen kartoitus kannettavalla metallianalysaattorilla. *Ympäristö ja terveys* 26(6): 12–19 (in Finnish).

Demirbaş A. 2001. Concentrations of 21 metals in 18 species of mushrooms growing in the East Black Sea region. *Food Chemistry* 75: 453–457.

Dey S., Rao P.R.N., Bhattacharyya B.C. & Bandyopadhyay M. 1995. Sorption of heavy metals by four basidiomycetous fungi. *Bioprocess Engineering* 12: 273–277.

Fogarty R.V. & Tobin J.M. 1996. Fungal melanins and their interactions with metals. *Enzyme and Microbial Technology* 19: 311–317.

Fomina M., Ritz K. & Gadd G.M. 2003. Nutritional influence on the ability of fungal mycelia to penetrate toxic metal-containing domains. *Mycological Research* 107: 861–871.

Kirk T.K., Schultz E., Connors W.J., Lorenz L.F. & Zeikus J.G. 1978. Influence of culture parameters on lignin metabolism by *Phanerochaete chrysosporium*. *Archives of Microbiology* 117: 277–285.

Kuperman R.G. & Carreiro M.M. 1997. Soil heavy metal concentrations, microbial biomass and enzyme activities in a contaminated grassland ecosystem. *Soil Biology and Biochemistry* 29: 179–190.

Lin Z., Comet B., Qvarfort U. & Herbert R. 1995. The chemical and mineralogical behaviour of Pb in shooting range soils from central Sweden. *Environmental Pollution* 89: 303–309.

Manninen S. & Tanskanen N. 1993. Transfer of lead from shotgun pellets to humus and three plant species in a Finnish shooting range. *Archives of Environmental Contamination and Toxicology* 24: 410–414.

Pokorny B. & Ribarič-Lasnik C. 2002. Seasonal variability of mercury and heavy metals in roe deer (*Capreolus capreolus*) kidney. *Environmental Pollution* 117: 35–46.

Shi W., Becker J., Bischoff M., Turco R.F. & Konopka A.E. 2002. Association of microbial community composition and activity with lead, chromium, and hydrocarbon contamination. *Applied and Environmental Microbiology* 68: 3859–3866.

Steffen K.T., Hofrichter M. & Hatakka A. 2000. Mineralisation of ^{14}C-labelled synthetic lignin and ligninolytic enzyme activities of litter-decomposing basidiomycetous fungi. *Applied Microbiology and Biotechnology* 54: 819–825.

Turpeinen R., Salminen J. & Kairesalo T. 2000. Mobility and bioavailability of lead in contaminated boreal forest soil. *Environmental Science and Technology* 34: 5152–5156.

Vares T., Niemenmaa O. & Hatakka A. 1994. Secretion of ligninolytic enzymes and mineralization of ^{14}C-ring-labelled synthetic lignin by three *Phlebia tremellosa* strains. *Applied and Environmental Microbiology* 60: 569–575.

European Symposium on Environmental Biotechnology, ESEB 2004 - Verstraete (ed)
© 2004 Taylor & Francis Group, London, ISBN 90 5809 653 X

Glutathione reductase of *Xanthomonas campestris*: a unique enzyme and its physiological role

S. Loprasert, W. Whangsuk, R. Sallabhan & S. Mongkolsuk
Laboratory of Biotechnology, Chulabhorn Research Institute, Bangkok, Thailand

S. Mongkolsuk
Department of Biotechnology, Faculty of Science, Mahidol University, Bangkok, Thailand

ABSTRACT: The thiol-containing tripeptide glutathione is one of the most prevalent reducing thiols in living cells, where it participates in many biological processes such as the detoxification of xenobiotics and free radicals. The glutathione reductase gene (*gor*) of the plant pathogen *Xanthomonas campestris* pv. *phaseoli* (*X. p.*) has been cloned and found to encode a protein of 457 amino acid residues with a calculated molecular mass of 50 kDal. Generally, Gor proteins isolated from varying sources contain a highly conserved NADPH binding motif and most Gor proteins prefer NADPH as their reductant. Interestingly, the Gor protein of *X. p.* has a unique NADPH binding site in which two, normally highly conserved, arginine residues are replaced by glutamine and glutamic acid. The *X. p.* Gor protein was purified to homogeneity and found to be an atypical Gor enzyme in that it is able to use NADH.

Glutathione is the major free thiol-containing compound that is very widespread and is present in animals, plants, fungi, and a large number of prokaryotic species. Glutathione can react with a variety of compounds containing electrophilic centers. Apart from its function as an antioxidant, it is responsible for the maintenance of the intracellular thiol redox status and thus contributes to the function of many biological processes within the cell (Herouart et al. 1993, Wingsle & Karpinski 1996). For most of its functions glutathione must be in the reduced form. Glutathione reductase (Gor) is the enzyme that reduces the oxidized form of glutathione, glutathione disulfide (GSSG), to reduced glutathione (GSH). It is involved in redox cycles which are important in maintaining the anti-oxidative capacity of cells engaged in a wide variety of functions in which reactive oxygen species may be produced. Gor is considered to be a key enzyme in conserving the redox status of the cell during oxidative stress. It is a member of the flavoprotein disulfide oxidoreductases whose member enzymes share a considerable degree of structural similarity implying that they have arisen by divergent evolution from a common ancestor (Perham et al. 1991). The reaction of GSSG to GSH, catalyzed by Gor, exhibits a considerable preference for NADPH over NADH as its reducing cofactor. *Xanthomonas* belongs to an important family of plant bacterial pathogens. The bacterial enzymes and genes involved in oxidative stress and redox status regulation are likely to play important roles in disease development.

In this report, we describe the isolation and characterization of the *gor* gene from the phytopathogen *Xanthomonas campestris* pv. *phaseoli*. Furthermore, the Gor enzyme was overproduced and characterized. The *gor* gene was interrupted and the knockout mutant's sensitivity to oxidative stress was determined.

1 MATERIALS AND METHODS

1.1 *Bacterial cultures and media*

All *Xanthomonas* strains were grown aerobically at 28°C in SB medium as previously described (Ou 1987, Mongkolsuk et al. 1998). All *Escherichia coli* strains were grown aerobically in Luria-Bertani (LB) broth at 37°C.

1.2 *Nucleic acid extraction and analysis, cloning and nucleotide sequencing*

Genomic DNA extraction from *Xanthomonas* and electroporation were done as previously described (Mongkolsuk et al. 1996). Multiple alignment of amino acid sequences was performed using ClustalW (Thompson et al. 1994).

1.3 *High-level production and purification of Gor*

Overproduction of Gor was achieved by using a His-tagged gene fusion expression vector system in *E. coli*. The oligonucleotide primers generated from the 5′ and 3′ coding regions of *gor* (5′ CGGCATG-CAT-GAGTGCGCGTTA 3′ and 5′ CGAAGCTTCG-CAACCAACCAT 3′) were used to amplifiy *gor* from pZL-G1. A PCR product of 1400 bp was then digested with *Sph*I and *Hin*dIII, and cloned into pQE30 (Qiagen Inc.) and named pQEG. A 200-ml mid-exponential phase culture of *E. coli* harboring pQEG was induced with 2 mM IPTG for 2 h. The cells were subsequently pelleted, and sonicated. Gor fusion protein was purified using nickle affinity columns according to the manufacturer's recommendations (Clontech).

1.4 *Gor enzyme assay*

Gor activity was determined by monitoring the reduction of 5,5′-dithiobis (2-nitrobenzoic acid) to thiobis (2-nitrobenzoic acid) by GSH, which is produced by Gor (Smith et al. 1988).

1.5 *Disruption of* gor *gene*

A *gor* mutant was created by single recombination between the suicide plasmid, pBX170 and the chromosomal *gor* gene. Specifically, a 1800-bp *Sph*I-*Hin*dIII fragment from pZl-G1 was cloned into similarly digested pUC18 resulting in pGR1800. A 800 bp fragment was deleted from pGR1800, by digestion with *Bst*EII, *Hin*dIII followed by end-filling with Klenow polymerase and ligation, to form pBX1000. A further deletion of pBX1000 by *Xba*I digestion followed by re-ligation generated plasmid pBX170. The plasmid pBX170 was electroporated into *X. p.* and an ampicillin resistant mutant was selected.

2 RESULTS AND DISCUSSIONS

2.1 *Cloning of a* gor *gene homologue*

Multiple amino acid sequence alignment analysis of many Gor proteins revealed two conserved regions, EVSDDFF and GVRLHFG (Jiang et al. 1995). Degenerate oligonucleotide primers I (5′ GAXGTXTCX-GACGAXTTXTT 3′) and II (5′ CCGAAGTGXAGXCGXACXCC 3′) (where X is mixed bases G and C), corresponding to the conserved regions were synthesized. PCR of the *X. p.* chromosome with primers I and II yielded a 220-bp DNA fragment that was cloned and sequenced. The DNA sequence was identified to have high homology to part of the *gor* gene. Thus, the 150-bp *gor* fragment was used as a probe to screen a *X. campestris pv. phaseoli* genomic library constructed in a ZipLox vector (BRL Life technology). One positive clone G1 harboring pZL-G1 with a 2.6-kb inserted fragment was characterized and the entire nucleotide sequence determined.

2.2 *Amino acid sequence analysis*

Analysis of the nucleotide sequence revealed an open reading frame coding for a polypeptide of 457 amino acid residues (49 kDa) with high homology to Gor. Amino acid sequences of Gor from various sources including bacteria, plant, animal and human were compared and analyzed (the alignment is not shown). *Xanthomonas* Gor showed moderate homology to Gor from *E. coli* (44%), *Pseudomonas aeruginosa* (38%), *Haemophilus influenzae* (43%), *Burkholderia cepacia* (40%), *Streptococcus thermophilus* (45%), soybean (31%), rat (35%), and human (41%). All of the amino acid residues involved in FAD binding, GSSG binding, and the active sites were conserved among Gor from various sources. Generally, Gor proteins isolated from varying sources contain a highly conserved NADPH binding motif (GxGYIAx_{18}**R**x_5**R**)(Danielson et al. 1999). Consequently most Gor proteins prefer NADPH as their reducing coenzyme with only one exception, Gor from *Chromatium vinosum*, that has been shown to use NADH (Chung & Hurlbert 1975). Interestingly, the Gor protein of *X. p.* has a unique NADPH binding site in which two, normally highly conserved among all Gor, arginine residues are replaced by glutamine (Gln 200) and glutamic acid (Glu 206) (GxGYIAx_{18}**Q**x_5**E**) suggesting that *Xanthomonas* Gor may have a different coenzyme specificity. In a previous study of human glutathione reductase, it was found that NADH also binds to Gor but with less affinity, i.e. 60 times higher K_m, than that for NADPH due to its lack of the 2′-phosphate group (Scrutton et al. 1990). The prominent positively charged residues Arg 218 and Arg 224 make several contacts with this phosphate group carrying two negative charges (Scrutton et al. 1990). Arg 218 and Arg 224 of human glutathione reductase are both conserved in virtually all the NADPH-requiring enzymes of the flavoprotein oxidoreductase family. Using site-directed mutagenesis on the *E. coli gor* gene, Arg 218 and 224 were converted to Met and Leu, respectively. The effects of the substitution were a substantially decreased affinity for NADPH and a catalytically less favourable configuration for bound NADPH (Scrutton et al. 1990). In the NADH-dependent enzymes, like dihydrolipoamide dehydrogenase, conserved Glu residues replace Arg in equivalent positions of the coenzyme binding site is conserved and these were suggested to be involved in binding the 2′-OH group of the ribose of NADH (Scrutton et al. 1990).

Therefore, Glu 206 in *Xanthomonas* Gor may facilitate NADH ribose group binding thus allowing the enzyme to use NADH as a cofactor. This unusual and interesting coenzyme binding motif prompted us to investigate the enzyme kinetics of *Xanthomonas* Gor.

2.3 *Enzyme kinetic study of Gor*

Highly purified Gor enzyme, as judged by SDS-PAGE (data not shown), was assayed for enzyme activity using GSSG as a substrate and either NADPH or NADH as the coenzyme. Surprisingly, preliminary result indicated that, while NADPH is the preferred reductant, *Xanthomonas* Gor is also able to utilize NADH as a reductant. The specific activity of Gor, when assayed at a GSSG concentration of 1 mM and a coenzyme concentration of 100 μM, was 30 unit/mg with NADPH and 14 unit/mg with NADH. Certainly more kinetic parameters such as K_m, k_{cat} and V_{max} must be determined in order to elucidate the coenzyme binding affinities between NADPH, NADH and Gor. A determination of *X. p.* Gor enzyme kinetics is in progress. Our finding that Gor from *Xanthomonas* has a unique coenzyme binding site, which differs from Gor of other sources, and also that it has the ability to use both NADPH and NADH as a coenzyme is interesting both in terms of enzymology and physiology. It raises questions as to how *Xanthomonas* Gor, is able to display a coenzyme specificity that is unlike that of other organisms and what is its precise physiological role? To answer these questions we inactivated the *gor* gene in the *X. p.* chromosome.

2.4 *Gor mutant construction and characterization*

The *gor* knockout mutant G26 was shown by southern analysis to have the desired disruption of the *gor* gene. A Southern blot of genomic DNA of both the mutant G26 and the wild type strain, digested with *Cla*I and *Bst*EII, was probed with a 494-bp *Bst*XI fragment spanning a portion of the *gor* coding region (Fig. 1A). The results of the hybridization revealed that mutant G26 contained a *Cla*I-digested DNA fragment that was 2.8 kb larger than that obtained from similarly digested DNA from the wild type strain (Fig. 1B, lane 1 and 2). The Southern hybridization also revealed a 3.7-kb *Bst*EII fragment that was present in both strains. An additional 600 bp *Bst*EII fragment was present in the wild type strain while a 3.1-kb *Bst*EII fragment, indicative of the insertion of pBX170 into *gor*, was observed in G26 (Fig. 1B, lane 3 and 4). We would like to study the defensive role of Gor during oxidative stress. Therefore, the levels of resistance to various killing concentrations of oxidative stress compounds expressed by mutant G26 and wild type strain were determined. The mutant did not exhibit any sensitivity alteration to H_2O_2, superoxide-generating compound menadione, and organic hydroperoxide tert-butylhydroperoxide compared to the wild type strain under our experimental conditions. This is consistent with the previous observation that a Gor-deficient mutant strain of *E. coli* showed no reduction in growth rate or increased sensitivity to methyl viologen (Kunert et al. 1990). We are now investigating the sensitivity of the *gor* mutant to additional compounds that induce oxidative and nitrosative stress.

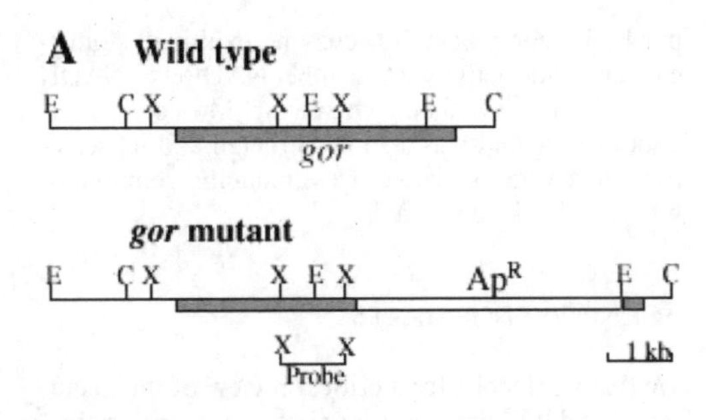

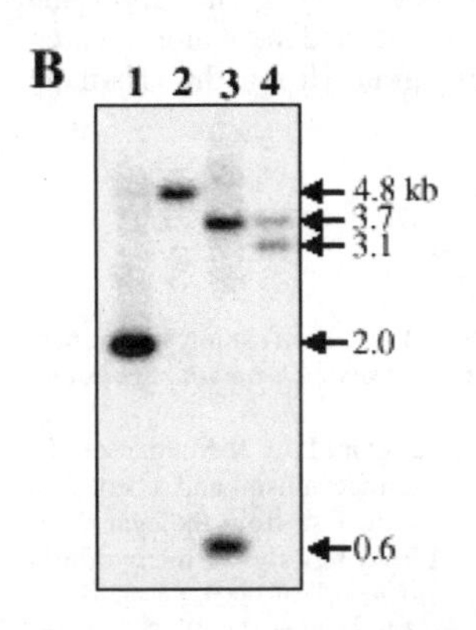

Figure 1. Genetic characterization of the *X. campestris gor* mutant. (A) Diagrams of the genetic organization of *gor* in wild type and mutant strains. E, *Bst*EII; C, *Cla*I; X, *Bst*XI. (B) Genomic Southern analysis of *gor*. Genomic DNA of wild type strain was digested with *Cla*I (lane 1) and *Bst*EII (lane 3), and *gor* mutant genomic DNA was digested with *Cla*I (lane 2) and *Bst*EII (lane 4), separated by electrophoresis, transferred to a membrane and probed with a 494-bp *Bst*XI *gor* gene internal fragment. Arrows indicate the size of the positively hybridizing bands.

3 CONCLUSIONS

A homologue of the glutathione reductase gene *gor* was identified in *X. campestris*. Analysis of the deduced amino acid sequence revealed that Gln and Glu residues had replaced two normally highly conserved Arg residues in the NADPH-binding site. The purified recombinant Gor enzyme exhibited significant enymatic activity when either NADPH or NADH was used as the reducing coenzyme. Moreover, a *gor* knockout mutant was also constructed and its sensitivity to several oxidative stress inducing compounds was found to be unaltered.

ACKNOWLEDGEMENTS

We thank J. Dubbs for a critical review of the manuscript, and P. Munpiyamit for photograph preparation. This research was supported by grants from the Chulabhorn Research Institute and the senior research scholar RTA 4580010 grant from the Thailand Research Fund to SM.

REFERENCES

Chung, Y.C. & Hurlbert, R.E. 1975. Purification and properties of the glutathione reductase of *Chromatium vinosum. J Bacteriol* 123: 203–211.

Danielson, U.H., Jiang, F., Hansson, L.O. & Mannervik, B. 1999. Probing the kinetic mechanism and coenzyme specificity of glutathione reductase from the cyanobacterium *Anabaena* PCC 7120 by redesign of the pyridine-nucleotide-binding site. *Biochemistry* 38: 9254–9263.

Herouart, D., Van Montagu, M. & Inze, D. 1993. Redox-activated expression of the cytosolic copper/zinc superoxide dismutase gene in *Nicotiana*. *Proc Natl Acad Sci USA* 90: 3108–3112.

Jiang, F., Hellman, U., Sroga, G.E., Bergman, B. & Mannervik, B. 1995. Cloning, sequencing, and regulation of the glutathione reductase gene from the cyanobacterium *Anabaena* PCC 7120. *J Biol Chem* 270: 22882–22889.

Kunert, K.J., Cresswell, C.F., Schmidt, A., Mullineaux, P.M. & Foyer, C.H. 1990. Variations in the activity of glutathione reductase and the cellular glutathione content in relation to sensitivity to methylviologen in *Escherichia coli*. *Arch Biochem Biophys* 282: 233–238.

Mongkolsuk, S., Loprasert, S., Vattanaviboon, P., Chanvanichayachai, C., Chamnongpol, S. & Supsamran, N. 1996. Heterologous growth phase- and temperature-dependent expression and H_2O_2 toxicity protection of a superoxide-inducible monofunctional catalase gene from *Xanthomonas oryzae* pv. *oryzae*. *J Bacteriol* 178: 3578–3584.

Mongkolsuk, S., Praituan, W., Loprasert, S., Fuangthong, M. & Chamnongpol, S. 1998. Identification and characterization of a new organic hydroperoxide resistance (*ohr*) gene with a novel pattern of oxidative stress regulation from *Xanthomonas campestris* pv. *phaseoli*. *J Bacteriol* 180: 2636–2643.

Ou, S.H. 1987. *Bacterial disease*. Tucson, Arizona: CAB International.

Perham, R.N., Scrutton, N.S. & Berry, A. 1991. New enzymes for old: redesigning the coenzyme and substrate specificities of glutathione reductase. *Bioessays* 13: 515–525.

Scrutton, N.S., Berry, A. & Perham, R.N. 1990. Redesign of the coenzyme specificity of a dehydrogenase by protein engineering. *Nature* 343: 38–43.

Smith, I.K., Vierheller, T.L. & Thorne, C.A. 1988. Assay of glutathione reductase in crude tissue homogenates using 5,5′-dithiobis(2-nitrobenzoic acid). *Anal Biochem* 175: 408–413.

Thompson, J.D., Higgins, D.G. & Gibson, T.J. 1994. CLUSTAL W: improving the sensitivity of progressive multiple sequence alignment through sequence weighting, position-specific gap penalties and weight matrix choice. *Nucleic Acids Res* 22: 4673–4680.

Wingsle, G. & Karpinski, S. 1996. Differential redox regulation by glutathione of glutathione reductase and CuZn-superoxide dismutase gene expression in *Pinus sylvestris* L. needles. *Planta* 198: 151–157.

European Symposium on Environmental Biotechnology, ESEB 2004 - Verstraete (ed)
© 2004 Taylor & Francis Group, London, ISBN 90 5809 653 X

Effect of bioaugmentation and supplementary carbon sources on degradation of PAHs by a soil derived culture: a chemostat study

R. van Herwijnen[1], B. Joffe[2], A. Ryngaerts[3], N.M. Leys[3], S. Wuertz[2,5], A. Schnell[2], M. Hausner[2], D. Springael[3,4], H.A.J. Govers[1] & J.R. Parsons[1]

[1] *University of Amsterdam, Department of Environmental and Toxicological Chemistry, Nieuwe Achtergracht, Amsterdam, the Netherlands*

[2] *Technical University of Munich, Institute of Water Quality Control and Waste Management, Am Coulombwall, Garching, Germany*

[3] *Vlaamse Instelling voor Technologisch Onderzoek (Vito), Boeretang Mol, Belgium*

[4] *Catholic University of Leuven, Laboratory of Soil Fertility and Soil Biology, Kasteelpark, Arenberg Heverlee, Belgium*

[5] *University of California, Department of Civil & Environmental Engineering, Davis, CA, USA*

ABSTRACT: We examined the performance of PAH-degrading bacterial mixtures under different carbon-limited conditions, including mixtures of 3 bacterial strains that are able to utilize one or more PAHs as sole carbon and energy substrate, an undefined culture obtained from a PAH-polluted soil solely and the latter mixture bioaugmented with the 3 strains. The results revealed that the composition of the supplied carbon sources had an important effect on the steady state bacterial composition of the mixed cultures.

1 INTRODUCTION

Bioaugmentation and biostimulation are methods that are applied in bioremediation of contaminated soils. These methods require understanding of the effects of environmental conditions on the behaviour of contaminant-degrading bacteria. We examined the performance of PAH-degrading bacterial mixtures under different carbon-limited conditions, including mixtures of 3 bacterial strains that are able to utilize one or more PAHs as sole carbon and energy substrate, an undefined culture obtained from a PAH-polluted soil solely and the latter mixture bioaugmented with the 3 strains.

2 MATERIALS AND METHODS

The bacteria used for augmentation were *Sphingomonas* sp. strain LB126, *Sphingomonas* sp. strain. LH128 (Bastiaens et al., 2000) and *Mycobacterium gilvum* strain VM552 (Springael, unpublished results). These bacteria were isolated from PAH-contaminated soil by selective enrichment using fluorene, phenathrene and pyrene, respectively. Bacteria were grown in chemostat cultures as described previously (Van Herwijnen et al., 2003). The effects of augmentation and biostimulation with easily utilisable carbon sources (EUC) on the degradation of PAHs and on the composition of the mixed culture were examined in three different experiments in chemostat cultures. The exact compositions of the EUC and PAH mixture supplied during the experiments are given in Table 1. All combinations of the total carbon sources were supplied at a total atomic carbon concentration of 6 mM. Culture samples were analysed for PAHs as described (van Herwijnen et al., 2003). The methods used for microbial analyses using plate counting, FISH and DGGE will be described elsewhere (van Herwijnen et al., in preparation).

3 RESULTS AND DISCUSSION

Chemostat cultures were used in this study of the response of bacteria to mixed substrates as has been proposed previously (van Herwijnen et al., 2003).

Table 1. Composition of carbon source mixtures supplied to the chemostats in the 3 experiments.

	Experiment 1			Experiment 2		Experiment 3	
	EUC + PAHs	EUC	PAHs	EUC + PAHs	EUC	PAHs + EUC + salicylic acid	PAHs + EUC + nonadecane
Compound	Concentration in medium (mg l^{-1})						
Sodium acetate	33.0	54.9		33.0	54.9	39.6	34.6
Glucose	25.2	42.0		25.2	42.0	30.2	26.5
Benzoic acid	14.7	24.4		14.7	24.4	17.6	15.4
Sodium citrate	18.1	30.2		18.1	30.2	21.7	19.0
Salicylic acid	4.1	6.9		4.1	6.9	5.0	–
Nonadecane*	9.9	11.3		9.9	11.3	–	9.9
Phenanthrene*	12.7		36.9	21.2		21.2	21.2
Fluorene*	12.8		37.0	4.2		4.2	4.2
Pyrene*	1.9		2.8	1.9		1.9	1.9
Chrysene*	–		–	–		1.3	1.3

EUC: Easy Utilizable Carbon
* Nominal concentrations

Mixed cultures of the three strains LB126, LH128 and VM552 showed similarly high removal of PAHs, irrespective of the carbon sources (EUC only, PAHs only or a mixture of PAHs and EUC). Nevertheless, a clear stimulating effect was observed by biostimulation with EUC on the concentration of VM552.

The soil culture degraded the PAHs efficiently at steady state. Bioaugmentation did not significantly improve the PAH degrading performance of the soil culture and the augmenting strains fell rapidly in numbers during the experiment to 10% or less of the total. However, the augmenting strains remained present in the soil culture and the cell concentrations of LB126 and LH128 were higher in the presence of PAHs. The presence of PAHs influenced the steady state bacterial composition in both the bioaugmented and non-bioaugmented soil cultures, resulting in increased cell concentration of sphingomonad strains. In contrast, EUC appeared to stimulate the cell numbers of mycobacteria.

Initial effects of biostimulation with EUC on the degradation of the examined PAHs largely disappeared after a steady state was established. Of the additional carbon sources, only the presence of salicylate had a slightly stimulating effect on the degradation of phenanthrene.

The results revealed that the composition of the supplied carbon sources had a significant effect on the steady state bacterial composition of the mixed cultures. Therefore, the right choice of additional carbon sources may stimulate the biodegradation of PAHs in contaminated soil by supporting PAH-degrading bacteria. More research on the stimulation of specific bacterial groups and specific bacterial capacities by additional carbon sources is therefore recommended.

REFERENCES

Bastiaens, L., Springael, D., Wattiau, P., Harms, H., de Wachter, R., Verachtert, H., and Diels, L. (2000) Isolation of adherent polycyclic aromatic hydrocarbon (PAH) degrading bacteria using PAH sorbing carriers. Appl.Env.Microbiol. 66: 1834–1843.

van Herwijnen, R., van de Sande, B.F., van der Wielen, F.W.M., Govers, H.A.J., and Parsons, J.R. (2003) Influence of phenanthrene and fluoranthene on the degradation of fluorene and glucose by Sphingomonas sp. strain LB126 in chemostat cultures. FEMS Microbiol.Ecol. 46: 105–111.

van Herwijnen, R., Joffe, B., Ryngaerts, A., Leys, N.M., Wuertz, S., Schnell, A., Hausner, M., Springael, D., Govers, H.A.J., and Parsons, J.R., in preparation.

European Symposium on Environmental Biotechnology, ESEB 2004 - Verstraete (ed)
 ISBN 90 5809 653 X

Fungal biofilms: upgrading the performance of biodegraders

V. Jirků, J. Masák & A.Čejková
Department of Fermentation Chemistry and Bioengineering, Institute of Chemical Technology, Prague, Czech Republic

ABSTRACT: Fundamental limitations of any bioremediation are the effect of environmental (stress) conditions, cytotoxicity of pollutants and slow mass transfer of hydrophobic compounds to the degrading microorganism. Therefore, efforts are focused on optimizing technological applications of microbial degraders by modulating their susceptibility as well as by enhancing their accessibility to poorly soluble compounds. A new approach that does cope with these requirements is the biofilm based reactors and an application of additives increasing the pollutant bioavailability. Using different techniques, we have compared the phenotypes of suspended and attached fungal biodegraders, investigating a wide range of technologically significant markers. The results obtained revealed that the cell attachment of fungal degrader brings significant increase of cell tolerance to environmental stressors and pollutants damaging cell structures. Moreover, the contribution deals with the capacity of humic acids to enhance the capability of yeast populations to degrade polycyclic aromatic hydrocarbons.

Technological application of attached biodegraders and a detailed knowledge of cell attachment effect appear to offer a large scale of opportunities not only to upgrade bioreactor (permeable bio-barrier) design, but also to reduce markedly the amount of treatment equipment as well as to enhance the spectrum of biodegradation activity through flexible co-attachment of complementary modulated cell populations, among others. Unlike microbial cells freely dispersed in an aqueous phase, attached (biofilm) cells associated with surfaces develop spatial relationship to each other that permit interactions approaching those of multicellular organisms (O'Toole et al. 2000). In mixed species biofilm communities thus provides intra-population variations as well as inter-population interactions offering technologically significant phenotype modulations. This cell attachment effect, together with now recognized capacity of microbial cells to sense and respond to physical contacts (Junter et al. 2002), is a potential tool to optimize the biodegradative function and behavior of bio-treatments microflora through changes in metabolic activity, survival and fitness of microbial population. In this context, the knowledge obtained is shown not only in view of the application of fungal biodegraders, but also in view of one of the bottlenecks of water bio-treatment systems, i.e. an insufficient capability of their microflora to take up hydrophobic pollutants (Ressler et al. 1999).

To tentatively clarify the nature and scope of (fungal) cell attachment effect, a wide range of markers has been used to compare suspended and attached (viable) yeast cells. Using a monolayer of covalently attached yeasts, the phenotypes of suspended and attached cells were compared. The data obtained proved no significant differences regarding tested markers of primary metabolic activity. Attached yeast cells show, however, increased concentrations of mannan and β-1,3-, β-1,6-alkali-insoluble glucans as well as a higher intensity of biosynthesis of all main lipids, a higher proteosynthetic capacity, and a decreased level of glycogen and trehalose. A comparison of total content of cell wall proteins, lipids, and amino sugars in suspended and attached cells shows that the cell walls of attached unicellular fungi contain increased amounts of each component. A different composition of mannoproteins in the walls of attached cells was detected in parallel. A comparison of the sterol/phospholipid ratio, between suspended and attached cells, exhibited no notable difference, however, cell attachment stimulates an increase in the total amounts of both of these lipids. Moreover, cell attachment stimulates a significant shift up in the excretion patterns of acid phosphatase and β-glucanases; an enhancement of the extracellular production of alkaline protease, specifically induced by four different starvation conditions, was found as well.

The aforementioned results encouraged the idea that the cell attachment, stimulating compositional changes in the cell surface structures, could allow us to manipulate, among others, the susceptibility of fungal biodegraders both to the pollutants and extracellular conditions damaging these structures. In this connection, the uniform pattern of attached cells resistance to respective effect of acetone, benzene, dimethyl sulfoxide, ethanol, ethylene glycol, phenol, mercaptoethanol, dithiothreitol and 5-bromo-6-azauracil proves that the yeast cell attachment induces a pleiotropic tolerance to cytotoxic agents. Similarly, the effect of reductions in water activity (a_w) on the growth activity of the suspended and attached yeasts was found different and indicating that attached yeast cells can tolerate a growth inhibiting (osmotic) effects of the environment. Moreover, attached (more resistant) cell populations show an uniform increase in the proportion of saturated and a decrease in the proportion of unsaturated fatty acids, as well as an increase in the content of three phospholipids and ergosterol. In addition, the attached (more osmo-resistant) yeast populations show an uniform decrease in the trehalose level. The results obtained support the rising hypothesis that a multipoint contact sensing of fungal cell surface brings significant modulation of cellular structures, functions and properties. The long-term stability of compositional alterations and the resistance observed suggest that their development is not stimulated by transient stimuli in the cell surface microenvironment. The paucity of information about the existence of a transductive system participating in a cell response to cell (physical) contacts, and a range of diverse effects potentially acting in an attached cell microenvironment, make any clear-cut explanation of the above phenomenon impossible. Nevertheless, the detected compositional changes must primarily affect (enhance) the rigid membrane/wall framework as well as the biochemical functions that are connected with these structures. The technological potential of the protective effect of fungal cell attachment illustrates a column reactor simulation of phenol and acetone removal performed, respectively, by *Candida. maltosa* and *Fusarium. proliferatum* natural biofilm. In comparison with suspended biodegraders this model shows that the removal efficiency is significantly enhanced if natural biofilms are used. Moreover, the inhibitory effect of the pollutant concentration, which almost or significantly inactivate the reactor degradative function, is reversible, if the level of pollutant in the recirculating medium is considerably lowered within next 12 h., and the physiological adaptation of both biofilms to each higher concentration of model pollutant (or a higher flow rate of the media) is accompanied with a transient suppression of its uptake.

In addition, quantitative data obtained for yeast biofilm formation under different flow rates of growth medium suggest that the intensity of fungal biofilm formation can be controlled by medium flow rate as well. Limitations in the mass-transfer rate of a phase, in which biodegradation can occur, is widely accepted as the cause of some pollutants persistence. Therefore to enhance their bioavailability and accordingly biodegradation, the search for additives affecting the pollutant/biodegrader contact is wanted. The similarities between the effect of micelle – forming surfactants and humic acid on the surface tension (Ressler et al. 1999), has prompted the suggestion that humic acid is a potentially useful agent enhancing the accessibility of some pollutant to be degraded. In this connection, the use of humic acids was investigated with a prerequisite that fungal degraders can tolerate humic compounds and take up above pollutants to degrade them more efficiently and less costly to meet clean up standards.

In view of the fact that natural biofilm formation is always determined by complex regulation of surface attachment and consequent biofilm maturation, another field of research of the above project is focused on the effect of the cell surface hydrophobicity on the (fungal) cell adhesion correlating this effect with the influence of environmental stressors and carrier hydrophobicity. In practical terms, the biofilm formation detected was found to be affected in a complex manner by the carrier modification, cell type as well as by physiologically acting factors of extracellular environment. It was shown that, e.g. for materials and cells that are mutually repellent, the effect of some stressors can enhance the adhesion capacity of microbial cell. Moreover, the presence of humic acid evokes a considerable enhancement of yeast capacity to utilize, respectively, naphthalene and phenanthrene as sole source of carbon; (yeast biodegraders were found to be not capable of utilizing humic acids as sole source of carbon).

The parallel results, indicating that some effects of humic substances on yeast cell could consist in their ability to aggregate on the cell surface, as well as different chemical nature of individual humic compounds, suggest that any upgrading of biodegradation efficiency through a humic compound effect will be determined, among others, by the binding capacity of this compound in the aqueous phase. In this context, the "binding coefficient" and "critical concentration" values have been proved to be useful characteristics to make a proper choice of a humic additive.

Wastewater treatment systems R&D, implementing above knowledge on a module basis (below), has been successfully applied in the Czech Republic (Aquatest SG, contact: e-mail: sima@aquatest.cz), using a cascade of reactor vessels loaded, respectively,

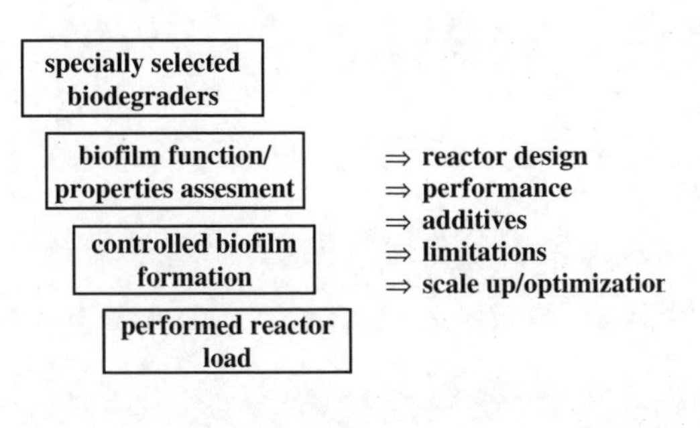

with organic and inorganic carriers colonized by specifically acting fungal and bacterial degraders.

REFERENCES

Junter, G.A., Coquet, L., Vilain, S. & Jouenne, T. 2002. Immobilized – cell physiology: current data and the potentialities of proteomics. *Enzyme Microb. Technol.* 31: 201–212.

O'Toole, G., Kaplan, H.B. & Kolter, R. 2000. Biofilm formation as microbial development. *Ann. Rev. Microbiol* 54: 49–79.

Ressler, B.P., Kneifel, H. & Winter, J. 1999. Bioavailability of polycyclic aromatic hydrocarbons and formation of humic acid-like residues during bacterial PAH degradation. *Appl. Microbiol. Biotechnol.* 53: 85–91.

European Symposium on Environmental Biotechnology, ESEB 2004 - Verstraete (ed)

Bioseparation of suspended solid and oil from palm oil mill effluent and secondary treatment by photosynthetic bacteria

Poonsuk Prasertsan, Aran H-Kittikun & Preecha Muneesri

Department of Industrial Biotechnology, Faculty of Agro-Industry, Prince of Songkla University, Hatyai, Thailand

ABSTRACT: Palm oil mill effluent (POME) is generated in large quantity with the average of 0.87 t/t fresh fruit bunches. The characteristics of POME are its brownish color, low pH (3.5–4.5) and high content of suspended solid (SS) and oil that are difficult to separate due to its emulsion property. Bioseparation was proposed and investigated. The process involved screening of thirteen strains of thermotolerant lipase-producing microorganism in which the fungal isolates ST4 and ST29 exhibited higher lipase activities than the others. In comparison to the treatment of the decanter effluent by *Aspergillus niger* ATCC 6275, *A. oryzae, Candida tropicalis* F-129 and *C. palmeoliophila* Y-128, the two isolates gave better results as 98.9% and 88.9% oil removal and 61.9% and 43.6% COD reduction were achieved respectively at 45°C after 4 days cultivation with the biomass of 44.6 g/l. The high treatment efficiency was the consequence of the entrapment of most suspended solids and residual oil of POME in the polymer excreted from these strains after 2 days cultivation only. It is very interesting to note that this phenomena occurred only at high temperature (45°C), but not at room temperature (~30°C). Secondary treatment of the two culture filtrates using *Rhodocyclus gelatinosus* R7 was conducted under anaerobic-light (3000 lux) and aerobic-dark conditions. The bacterium could grow only in the culture filtrate of the isolate ST29. Optimization studies for this treatment was aerobic-dark condition at the initial pH of 7.0 and the optimum COD:N ratio of 100:2.5. This resulted in the 71% COD removal with the generation of 1.67 g/l biomass.

INTRODUCTION

High quantity of both solid and liquid wastes were generated from palm oil milling process. The average quantity of palm oil mill effluent (POME) in Thailand was found to be 0.87 m^3 per ton of fresh fruit bunch (FFB) (H-Kittitun *et al.*, 1994). POME, the mixed effluent discharged mainly from sterilizer and decanter or separator during the extraction stage, contained very high organic matter (COD), suspended solids (SS) and oil with the values of 32–112, 11–20 and 5–25 g/l, respectively (Prasertsan *et al.*, 2001). The unique characteristics of POME is the emulsion property which make the solid and oil difficult to settle or separate despite the use of physical method (heating, etc.), chemical method ($FeCl_3$, $Al_2(SO_4)_3$, $Ca(OH)_2$) (H-Kittikul *et al.*, 1994). Biological treatment of POME using enzymes from *Aspergillus niger* ATCC 6275 was previously tested with successful results (Prasertsan *et al.*, 2001). However, due to the economic reason, using microorganisms would be more appropriate and it is the aim of this investigation to propose an alternative approach for treatment of POME.

MATERIALS AND METHODS

Screening of high lipase-producing thermotolerant microorganisms and comparison on oil and suspended solids removal from POME with other strains

Thirteen strains of thermotolerant microorganism, comprised of 11 fungal isolates and 2 bacterial isolates, were screened for the highest lipase producer by cultivating in synthetic media for bacteria and fungi (Suwansontichai *et al.*, 1991) and yeast (Lee *et al.*, 1993) using palm oil as carbon source at 45°C on a shaker (200 rpm) for 3 days. After centrifugation, each culture filtrate (enzyme) was dropped into a hole (0.5 cm dia.) on agar plate and incubated at 45°C for 3 days. The appearance of clear zone around the hole was observed every 24 h. Two isolates giving high clear zone were selected for further studies.

The ability of the selected strains to remove oil from the decanter effluent of a palm oil mill was compared to those of two fungi and two yeast strains. Starter culture (10% v/v) of each stain was inoculated into 500 ml Erlenmeyer flasks containing 100 ml each of the decanter effluent with the addition of 0.06% NH_4NO_3 (Prasertsan *et al.*, 1997). They were cultivated on a shaker (200 rpm) at room temperature (~30°C), and at 45°C as well for the two thermotolerant strains. Samples were taken every 24 h for 4 days to determine for pH, oil & grease, SS and COD (APHA, AWWA and WEF, 1998). Growth (biomass) was measured by modification method of Rossi and Clement (1985).

Secondary treatment of POME by photosynthetic bacteria

Since the result from preliminary studies indicated that *Rhodocyclus gelatinosus* R7 could grow in the culture filtrate of the isolate ST29 but not in that of the isolate ST4, therefore only the culture filtrate of the isolate ST29 (pH 5.4) was used for further studies. Effects of initial pH (5.4 and 7.0), cultivation condition (aerobic-dark and anaerobic-light), and COD:N ratio were investigated.

RESULTS AND DISCUSSION

Screening of high lipase-producing thermotolerant microorganisms and comparison on oil and suspended solids removal from POME with other strains

Among thirteen strains of lipase-producing thermotolerant microorganisms tested, only six of them exhibited lipase production, giving the clear zone on the agar plates (data not shown). The isolates ST4 and ST29 were selected due to their higher lipase activities as the clear zone appeared within 24 h incubation.

Cultivation of these two selected isolates in the decanter effluent was compared with those of *Candida tropicalis* F-129, *C. palmeoliophila* Y-128, *Aspergillus niger* ATCC 6275 and *A. oryzae*. Decanter effluent had the pH of 4.7 and contained (g/l): COD 35.5, oil & grease 24.9, nitrogen 0.9, total solids 53.0, SS 33.1, P 0.25, K 4.14, Mg 0.63, and Ca 0.39. At room temperature (~30°C), the maximum oil removal values were found to be 97.5%, 53.4%, 81%, 78.5%, 90.5%, 67.0%, respectively after 4 days cultivation. The ability of *C. tropicalis* F-129 to assimilate oil (81%) in the decanter effluent was similar to that in the synthetic medium (2% oil) at 38°C (Lee *et al.*, 1993). At 45°C, the isolates ST4 and ST29 not only removed oil at high percentages (98.9% and 88.9%, respectively) but also could harvest suspended solids from the POME by entrapping in the polymer excreted within 2 days cultivation. It should be noted that this phenomena occurred only at high temperature (45°C), but not at room temperature (~30°C). This highest oil removal (98.9%) was similar to the value of oil removal (99.8%) from the separator effluent treated for 10 days under anaerobic condition at the same temperature (Edewor, 1986). This is the first investigation of biopretreatment using polymer-producing fungi to remove suspended solids and oil from the wastewater.

The thermotolerant fungi, fungi and yeast could remove organic matter (COD) by 72, 48, and 24%, respectively. Hence, the thermotolerant strains were 2 and 3 folds more efficient than the fungi and yeast strains, respectively. Treatment efficiency by these two strains was equivalent to the treatment by thermophilic microorganism under anaerobic condition (Bojia and Banks, 1993) at the same duration. The biomass obtained from the yeast (16–22 g/l), fungi (32–35 g/l) and thermotolerant fungi at 30°C (30–32 g/l) were lower than that achieved from thermotolerant fungi at 45°C (43–44 g/l).

Secondary treatment of POME by photosynthetic bacteria

Since the culture filtrate from cultivation of the isolate ST29 in POME had the pH of 5.4 which may not be suitable for growth of the photosynthetic bacteria, the effect of pH adjustment was studied. Results indicated that the neutral pH (7.0) gave a better growth of *R. gelatinosus* R7 under anaerobic-light and aerobic-dark conditions with 1.5 and 3 folds increase in biomass and 32–33% increase in COD removal. Although the biomass concentration was higher under anaerobic-light condition which agreed with the previous investigation (Prasertsan *et al.*, 1993), the growth rate was higher under aerobic condition.

Effect of COD:N, at the ratio of 100:0, 100:0.5, 100:1.5 and 100:2.5, on treatment of biopretreated POME (7.44 g/l COD) under anaerobic-light and aerobic-dark condition was investigated. Under anaerobic-light condition, the COD:N ratio of 100:1.5 was optimal for growth of *R. gelatinosus* R7 (4.3 g/l) and correlated to the highest reduction of nitrogen. COD removal decreased with the increase of nitrogen supplementation. This may be due to the ability to fix nitrogen or use nitrogen from the photosynthesis process. For treatment under aerobic-dark condition, the optimal COD:N ratio was 100:1.5 for growth (1.67 g/l) and 100:2.5 for COD removal (71% or 89.8% for total COD removal). The optimal COD:N ratio of 100:1.5 for growth under both cultivation in this study was higher than the optimal value of 100:0.5 for photosynthetic bacteria grown under aerobic-dark condition (Morikawa *et al.*, 1971).

CONCLUSION

Treatment of palm oil mill effluent using two-stage process consisting of separation of suspended solid and oil by the polymer-producing fungi, the isolate ST29,

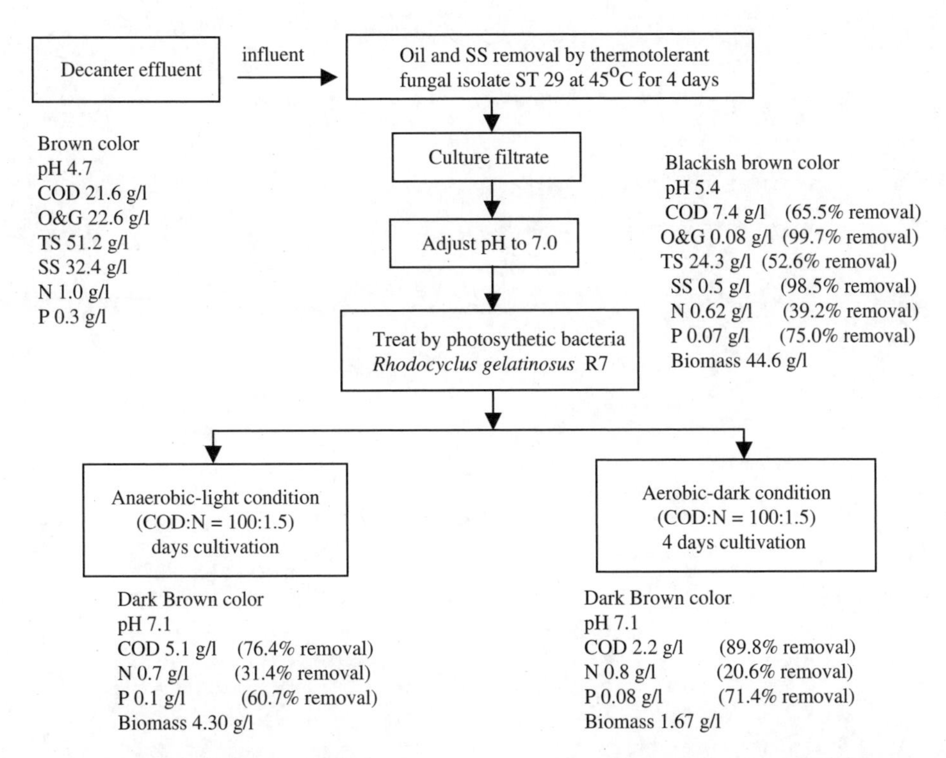

Figure 1. Diagram of two-stage treatment of decanter effluent from palm oil mill by thermotolerant fungal isolate ST 29 and photosynthetic bacteria *Rhodocyclus gelatinosus* R7.

at 45°C under aerobic condition for 4 days followed by treatment with the photosynthetic bacteria, *R. gelatinosus* R7, at room temperature under anaerobic-light condition and aerobic-dark condition was proposed in this investigation and results were summarized in Figure 1.

REFERENCES

APHA, AWWA and WPCF. 1985. 16th ed. American Public Health Association, Washington, D.C.

Borja, P.R. and Banks, C.L. 1993. Biotechnol. Lett. 15(7): 761–766.

Edewor, J.O. 1986. J. Chem. Technol. Biotechnol. 36: 213–218.

H-Kittikul, A., Prasertsan, P., Srisuwan, K., Jitbunjerdkul, S. and Tonglim, V. 1994. Seminar on Reduction of Oil Loss in Palm Oil Industry. 7 April, 1994. Suratthanee Province, Thailand.

Lee, C., Yamakawa, T. and Kodama, T. 1993. World J. Microbiol. Biotechnol. 9: 187–190.

Pokorny, D., Friedrich, J. and Cimerman, A. 1994. Biotechnol. Lett. 16(4): 363–366.

Prasertsan, P., H-Kittikul, A. and Chantapaso, S. 2001. J. Songklanakarin. 23(Suppl): 797–806.

Prasertsan, P., H-Kittikul, A., Kunghae, A., Maneesri, J. and Oi, S. 1997. World J. Microbiol. Biotechnol. 15(5): 555–559.

Prasertsan, P., Choorit, W. and Suwanno, S. 1993. World J. Microbiol. Biotechnol. 9(5): 593–596.

Rossi, J. and Clement, A. 1985. J. Food Technol. 20: 319–330.

Suwansontichai, K., Kitprechavanich, V., Kraidet, L. and Jitdon, L. 1991. J. Kasetsart (Sci.) 25: 162–168.

European Symposium on Environmental Biotechnology, ESEB 2004 - Verstraete (ed)
© 2004 Taylor & Francis Group, London, ISBN 90 5809 653 X

Biodegradation of petroleum hydrocarbons by indigenous thermophilic bacteria carrying an alkane hydroxylase gene

C.K. Meintanis, K.I. Chalkou & A.D. Karagouni
University of Athens, Department of Biology, Section of Botany, Microbiology Lab., Athens, Greece

ABSTRACT: The use of thermophiles for biodegradation of long chain alkanes is of interest as bioavailability of hydrocarbons at high temperatures is enhanced, due to solubility increase. One hundred fifty different thermophilic bacteria isolated from volcanic sediments were screened for detection of an alkane hydroxylase gene using degenerated primers developed to amplify genes related to the Pseudomonas putida and Pseudomonas oleovorans alkane hydroxylases. Isolates found to be carrying the *alk*J gene were further examined for their ability to degrade long chain crude oil alkanes and characterized by 16s rDNA gene sequencing. Results indicate that indigenous thermophilic hydrocarbon degraders are of special significance as they could be efficiently used for bioremediation of oil-polluted desert soil and composting processes.

Bioremediation has been evaluated in several studies as an option to treat the oil pollution resulting from the spillage or leakage of crude oil and fuels in the environment. Temperature plays an important role in controlling the nature and efficiency of microbial hydrocarbon degradation, which is of major significance for *in situ* bioremediation. A temperature increase leads to decrease of viscosity of petroleum hydrocarbons, reducing the degree of their distribution and therefore the risk of contamination. Degradation of long chain alkanes by mesophiles has been extensively studied in contrast with biodegradation pathways in thermophiles that are not well characterized. Yet, it is recognized that enzymes from thermophiles are more resistant to physical and chemical denaturation, while another advantage of using thermophiles in bioremediation processes would be faster growth rates. Although thermophiles growing on medium and long chain alkanes constitute a great biotechnology perspective, little is known about the main alkane hydroxylase systems of the thermophilic degradation pathways. Homologues of the *alk* genes of *Pseudomonas* that allow alkanes degradation, found in thermophilic bacteria can be correlated with high biodegradation rates exhibited at high temperatures.

In this study we report the isolation of indigenous thermophilic bacteria from an extreme volcanic environment by virtue of their ability to utilize petroleum hydrocarbons as a sole carbon source. An *alk*J gene probe has been used to detect the presence of the wide-range distribution gene of *Pseudomonas* hydrocarbon catabolism in the thermophilic isolates. Thermophiles found to be carrying the *alk*J gene were grown in liquid cultures with crude oil as a sole carbon source in order to estimate their biodegradation potential. The isolates were further characterized by 16s rDNA gene sequencing.

MATERIALS AND METHODS

Thermophilic bacteria isolation

Sediment samples were taken in May 2000, October 2000 and February 2001, from the bay of Agios Nikolaos at Palaia Kameni island (Santorini, Greece, 25° 25′ N, 36° 25′ E).

Nutrient Agar (NA), Thermophilic Bacillus Medium (TBM) and Mineral (L – Salts) for Thermophiles (ATCC 1554) were the growth media used for the isolation. Pure strains were isolated at 30°C and 60°C. In addition ten grams of sediment mixed with 90 ml sterilized and deionized water and homogenized for 30 min. The suspension was filtered through a 0.22 μm filter. The filter was placed on Nutrient Agar plates at 60°C.

Growth and oil degradation conditions

For the biodegradation studies a mineral salts medium was used, supplied after autoclaving with 2% (w/v) crude oil. Flasks were incubated at 55°C on an orbital shaker set to 180 rpm. All assays which included uninoculated controls were performed in triplicate.

Biological degradation of selected crude oil alkanes was calculated from the differences between concentrations in inoculated and uninoculated samples analysed on the same day.

Table 1. Thermophilic bacteria isolated from the volcano of Santorini (Greece) found to be carrying the *alk*J gene.

Isolate	Strain	% Identity
O114	*Bacillus eolicus*	95.08
O080	*Geobacillus thermodenitrificans*	97.04
O089	*Geobacillus anatolicus*	96.40
N017	*Geobacillus stearothermophilus*	98.31
N065	*Geobacillus bogazici*	98.03
N005	*Geobacillus stearothermophilus*	97.04
A083	*Bacillus pallidus*	97.61
H128	*Bacillus thermoleovorans*	96.48
A015	*Geobacillus stearothermophilus*	97.53
N057	*Bacillus thermoleovorans*	96.85
R123	*Bacillus pallidus*	99.42
N022	*Geobacillus stearothermophilus*	98.28
H093	*Bacillus* sp	95.73

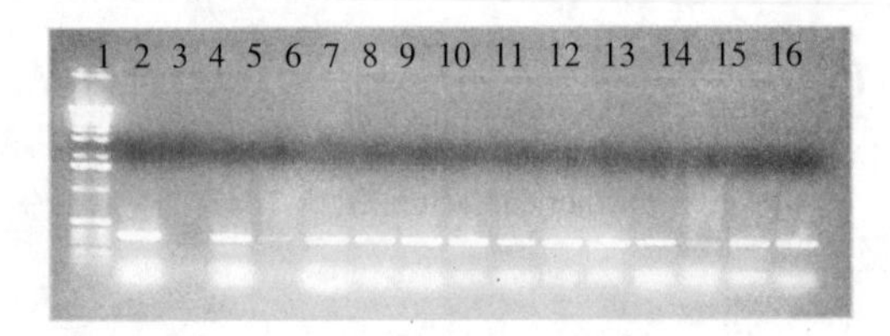

Figure 1. Detection of *alk*J gene using primers *alk*JF 5′ – tcggcc(ct)aatttgcagtttc – 3′ and the reverse *alk*JR 5′ – tttacccat(ct)ctacaagtacc – 3′. (1) 1 kb DNA ladder, (2) *E.coli* DSM8830-positive control, (3–16) isolates carrying the *alk*J gene.

Oil extraction and biodegradation measurements by GC-MS

10 days cultures were extracted using equal volume of n-hexane. The n-hexane soluble fraction was analyzed by gas chromatography – mass spectrometry (GC-MS hp6890), using a 30 m × 0.25 mm i.d. fused silica column (DB5) and a flame ionization detector. Initial column temperature was 60°C; temperature was increased to 290°C at a rate of 15°C per minute.

*alk*J *gene probe preparation and hybridization*

Primers to amplify a 352 bp region of *alk*J gene of *P. putida* were designed on the basis of published sequences. Genomic DNA of *E. coli* strain ATCC8830 carrying *Pseudomonas putida* OCT plasmid was used as template in PCR reaction with the *alk*J primers described above. The resulting PCR product was digested with *Hpa*II and *Msp*I to give a 290 bp region within the amplified area of the coding sequence of the *alk*J gene. The digested PCR product was used as a probe. Probe was labeled with ^{32}P – dCTP by random priming, according to the vendor's instructions (GibcoBRL). The Southern blot membranes were hybridized overnight at 62°C.

Sequencing

Genomic DNA (40 ng) of each isolate was used for amplification of fragments (968–1401 *E. coli* numbering) of the rRNA encoding gene. PCR products were separated from free PCR primers using the NucleoSpin Kit (MN) and were sequenced directly with the SequiTherm Excel Kit (Epicentre, Madison, Wis., USA) using an automated infrared laser fluorescence sequencer (Li Cor Model 4000 DNA sequencer;

Table 2. % remaining of high molecular weight alkanes from crude oil after 10 days culture, compared to the uninoculated control.

	Number of carbon atoms in aliphatic compound								
Isolate	C18	C20	C22	C24	C26	C28	C30	C32	C34
O114	12,34	36,55	24,06	35,58	28,85	29,14	29,28	38,95	39,50
O080	21,86	30,41	24,50	23,80	24,56	28,75	27,62	33,47	50,69
O089	29,61	33,69	27,87	32,45	29,69	30,83	29,15	32,80	53,36
N017	32,48	36,56	29,68	35,24	33,42	34,73	38,41	43,11	40,25
N065	14,61	34,34	17,13	37,78	26,00	29,15	32,83	38,10	50,36
N005	27,66	35,23	32,35	32,52	33,65	33,29	35,66	46,19	44,41
A083	11,46	29,91	15,94	31,16	23,40	26,73	28,48	32,57	39,70
H128	32,82	43,67	40,11	40,14	42,11	39,75	41,50	52,57	42,63
A015	28,16	30,70	28,17	21,81	26,28	27,01	27,21	35,46	47,19
N057	28,84	34,68	26,15	33,10	31,18	29,96	30,84	53,84	45,14
R123	35,54	45,33	39,03	32,67	37,27	38,38	38,48	52,17	41,44
N022	15,55	36,57	17,25	39,20	28,23	28,91	30,47	37,77	48,58
H093	17,88	40,19	17,14	40,56	28,52	33,06	30,18	37,43	41,05

Li Cor., Neb., USA). Sequences were initially compared to the available databases by using the BLAST network service.

RESULTS AND DISCUSSION

- The selected isolates are *Bacillus* and *Geobacillus* strains that have been reported to occur with high frequency in oilfields. Additionally, the ability to degrade octane and longer chain alkanes has been described as a taxonomic characteristic of *Geobacillus* genus.
- Homologues of the *alk* genes of *Pseudomonas* that allow alkanes degradation, found in the studied thermophilic bacteria can be correlated with high biodegradation rates exhibited at high temperatures.
- The results of the biodegradation studies indicate that thermophilic hydrocarbon degraders are of special significance as they could be efficiently used for bioremediation of oil-polluted desert soil, sediments in semi-arid climates with long hot summers and in composting processes.

REFERENCES

Zhuang W.Q., Tay J.H. and Tay S.T. (2002) Bacillus napthovorans sp. From oil-contaminated tropical marine sediments and its role in napthalene biodegradation. Applied Microbiology Biotechnology **58**:533–547.

Shimura M., Kimbara K., Nagato H. and Hatta T. (1999) Isolation and characterization of a thermophilic Bacillus sp. JF8 capable of degrading polychlorinated biphenyls and napthalene. FEMS Microbiology Letters **178**:87–93.

Schinner F. and Margesin R. (2001) Biodegradation and bioremediation of hydrocarbons in extreme environments. Applied Microbiology and Biotechnology **56**:650–663.

Nazina T.N., Tourova T.P., Poltaraus A.B., Novikova E.V, Grigoryan A.A., Ivanova A.E., Lysenko A.M., Petrunyaka V.V., Osipov G.A., Belyaev S.S. and Ivanov M.V. (2001) Taxonomic study of aerobic thermophilic bacilli: descriptions of Geobacillus subterraneus gen. nov., sp. nov. and Geobacillus uzenensis sp. nov. from petroleum reservoirs and transfer of B.stearothermophilus, B.thermocatenulatus, B.thermoleovorans, B.kaustophilus, B.thermodenitrificans to Geobacillus as the new combinations G. stearothermophilus, Int J Syst Evol Microbiol. Mar;51(Pt 2):433–46.

European Symposium on Environmental Biotechnology, ESEB 2004 - Verstraete (ed)
© 2004 Taylor & Francis Group, London, ISBN 90 5809 653 X

Transformation of dichlobenil to 2,6-dichlorobenzamide (BAM) by soil bacteria harboring nitrile hydratases and amidases or nitrilases

M.S. Holtze, R.K. Juhler & J. Aamand
Geological Survey of Denmark and Greenland, Copenhagen, Denmark

H.C.B. Hansen
The Royal Veterinary and Agricultural University, Copenhagen, Denmark

ABSTRACT: We tested the degradation of dichlobenil by ten bacteria, which harbor enzymes relevant for the degradation of nitriles. Hypothetically dichlobenil can be degraded either by (i) nitrile hydratases to BAM and further to 2,6-dichlorobenzoic acid by amidases or (ii) directly to 2,6-dichlorobenzoic acid by nitrilases. In addition we tested the degradation of benzonitrile, a non-halogenated reference compound. The degradation was tested adding 5-mg l^{-1} of dichlobenil or benzonitrile to the cultures. The nitriles were quantified using HPLC. We showed that after 3–16 days dichlobenil was fully degraded to BAM by seven out of ten soil bacteria. Hence, the transformations presumably were catalyzed by nitrile hydratases even though some of the bacteria are known to degrade other nitriles using nitrilases. In contrast to benzamide BAM was persistent in the bacterial cultures, which is in accordance with BAM being considered persistent to degradation in soil and ground water.

1 INTRODUCTION

Dichlobenil (2,6-dichlorobenzonitrile) is the active ingredient of the herbicides Prefix G and Casoron G that have been prohibited in Denmark since 1997 (Anonymous 2002); but still used in other countries. In soil dichlobenil is mainly degraded to the persistent metabolite 2,6-dichlorobenzamide (BAM) which now has been detected in 22% of the 4.771 investigated ground water supply wells in Denmark; in 10% of these wells the concentration exceeds the EU limit of 0.1 $\mu g\ l^{-1}$. Denmark has as the only country included measurements of this metabolite in the national ground water monitoring on a large scale. In Germany the concentration of the metabolite exceeds the EU limit in 15% of 359 ground water samples and in Sweden BAM has been detected below the EU limit in 10 out of 24 ground water samples (Wolter et al. 2001, Kreuger et al. 2003).

Dichlobenil is an aromatic nitrile. Two different enzyme systems for the hydrolysis of nitriles are known; (i) a direct hydrolysis of the nitrile to a carboxylic acid by nitrilases or (ii) a two-step reaction via an amide (e.g. BAM) by means of nitrile hydratases and further to the carboxylic acid by amidases (McBride et al. 1986, Banerjee et al. 2002). Originally, it was thought that nitrilases predominantly hydrolyzed aromatic nitriles, and that nitrile hydratases mainly converted aliphatic nitriles (Wyatt & Linton 1988, Nagasawa & Yamada 1990, Nawaz et al. 1992, Stevenson et al. 1992). However, it has been demonstrated that this is not always the case since for instance benzonitrile is transformed by both enzyme systems (Hjort et al. 1990). It (Layh et al. 1997, Hoyle et al. 1998) has also been claimed that nitrilases do not hydrolyze halogenated nitriles (Mahadevan & Thimann 1964); however, nitrilases have been shown to transform 3,5-dibromo-4-hydroxybenzonitrile (bromoxynil), a halogenated aromatic nitrile, to 3,5-dibromo-4-hydroxybenzoic acid (McBride et al. 1986).

BAM is the main metabolite of dichlobenil in soil and in three bacterial cultures tested so far (Briggs & Dawson 1970, Verloop & Nimmo 1970, Montgomery et al. 1972, Miyazaki et al. 1975, Blakey et al. 1995, Vosáhlová et al. 1997, Meth-Cohn & Wang 1997, Hoyle et al. 1998). Hence, investigations of the degradation of dichlobenil indicate that nitrile hydratase is involved. Since other aromatic nitriles, halogenated or not, can be transformed by nitrilases this enzyme system might be involved in dichlobenil degradation as well.

To investigate whether the activities of the two types of enzyme systems degrade dichlobenil we tested several pure bacterial cultures harboring the two enzyme systems. We hypothesized that a degradation of dichlobenil could be catalyzed either by (i) nitrile hydratase via BAM and further by amidase to 2,6-dichlorobenzoic acid or (ii) nitrilase producing 2,6-dichlorobenzoic acid without producing BAM. Furthermore, we included the degradation of benzonitrile as a reference compound since dichlobenil is the halogenated analogue to benzonitrile with two chlorine substituents in the ortho-positions. We expected the non-chlorinated analogue to be more easily degraded by the bacteria than dichlobenil.

2 RESULTS AND DISCUSSION

2.1 *Degradation of the nitriles*

All of the added dichlobenil was degraded in three to 16 days by seven out of ten bacteria namely: *R. radiobacter* 9674, *R. erythropolis* 9675, *R. erythropolis* 9685, *R. erythropolis* 11397, *P. putida* 11388, *Rhizobium* sp. 11401 and *Variovorax* sp. 11402 (Fig. 1, the numbers following the species name refer to the culture collection DSMZ). The dichlobenil-degrading bacteria were all isolated from a German soil using aliphatic nitriles for the enrichment indicating a large potential for dichlobenil degradation in soil.

In addition to dichlobenil six of the seven dichlobenil-degrading bacteria also degraded benzonitrile, namely: *R. radiobacter* 9674, *R. erythropolis* 9675, *R. erythropolis* 9685, *R. erythropolis* 11397, *Rhizobium* sp. 11401 and *Variovorax* sp. 11402 (Fig. 1). These strains all degraded benzonitrile much faster than dichlobenil. *S. chlorophenolicum* 6824 was not able to degrade dichlobenil but degraded benzonitrile (Fig. 1). These facts indicate that the two chlorine substituents in the *ortho*-positions hinder or retard the degradation as seen previously for *R. rhodochrous* AJ270 (Meth-Cohn & Wang 1997). In contrast, *P. putida* 11388 degraded dichlobenil but did not degrade benzonitrile. This could be due to some sort of enzyme specificity as shown with *Klebsiella pneumoniae* ssp. *ozaenae*, which transformed bromoxynil without being able to transform benzonitrile (McBride et al. 1986).

2.2 *Production of the amides*

All of the seven dichlobenil-degrading bacteria produced BAM (Fig. 1) and six of the seven benzonitrile-degrading bacteria produced benzamide when benzonitrile was added (Fig. 1). The seventh benzonitrile-degrader, *R. erythropolis* 11397, degraded 5 mg-l^{-1} benzamide in 1 h (data not shown). We therefore assume that this metabolite may be an intermediate metabolite in the benzonitrile metabolism as suggested by others (Nawaz et al. 1992). Due to the production of amides the following bacteria were assumed to harbor nitrile hydratases: *S. chlorophenolicum* 6824, *R. radiobacter* 9674, *R. erythropolis* 9675, *R. erythropolis* 9685, *R. erythropolis* 11397, *P. putida* 11388, *Rhizobium* sp. 11401 and *Variovorax* sp. 11402. A degradation of the nitriles via the amides was not expected for *P. putida* 11388, *Rhizobium* sp. 11401 and *Variovorax* sp. 11402 since these bacteria have been reported to produce nitrilases when degrading one of the aliphatic nitriles 2-phenylpropionitrile and ketoprofen nitrile (Layh et al. 1997). Either the nitrilases were not induced by dichlobenil or they somehow failed to degrade dichlobenil to 2,6-diclorobenzoic acid. It has previously been shown that 2-chlorobenzonitrile was unaffected by nitrilases which converted non-halogenated nitriles to their corresponding acids (Mahadevan & Thimann 1964). If this were true no degradation of dichlobenil would have been seen. In some cases nitrilases can transform small amounts of nitrile to amide instead of acid because the tetrahedral intermediate formed can break down anomalously (Stevenson et al. 1992). Since approximately 100% of the dichlobenil were transformed to BAM this explanation might not be valid (see below). We therefore assume that the bacteria shown to harbor nitrilases by Layh et al. (1997) were able to produce both nitrile hydratases and nitrilases as observed for different species of *Rhodococcus* before (Nagasawa et al. 1988, Nagasawa et al. 1991, Kato et al. 2000). Neither did we expect a production of benzamide in *S. chlorophenolicum* 6824, which is known not to produce aromatic metabolites (Topp et al. 1992).

Our findings are in contrast with the former proposal that the degradation of aromatic nitriles is mainly catalyzed by nitrilases (Wyatt & Linton 1988, Nagasawa & Yamada 1990, Nawaz et al. 1992). This is also shown by e.g. Layh et al. (1997) who enriched bacteria harboring both types of enzyme systems on aliphatic nitriles. Instead, Layh et al. (1997) propose that Gram-negative bacteria predominantly hydrolyze nitriles via nitrilases, while in Gram-positive bacteria the nitrile hydratase/amidase system dominates. E.g. the Gram-positive genus *Rhodococcus* is well known for its commercial synthesis of acrylamide from acrylonitrile (Warhurst & Fewson 1994, Bunch 1998). In contrast, we showed that the Gram-negative bacteria *S. chlorophenolicum* 6824, *R. radiobacter* 9674, *P. putida* 11388, *Agrobacterium* sp. 11401 and *Variovorax* sp. 11402 all degraded the tested nitriles via the amide and hence nitrile hydratases appears to be active. In concert with Bunch (1998) we speculate that a combination of

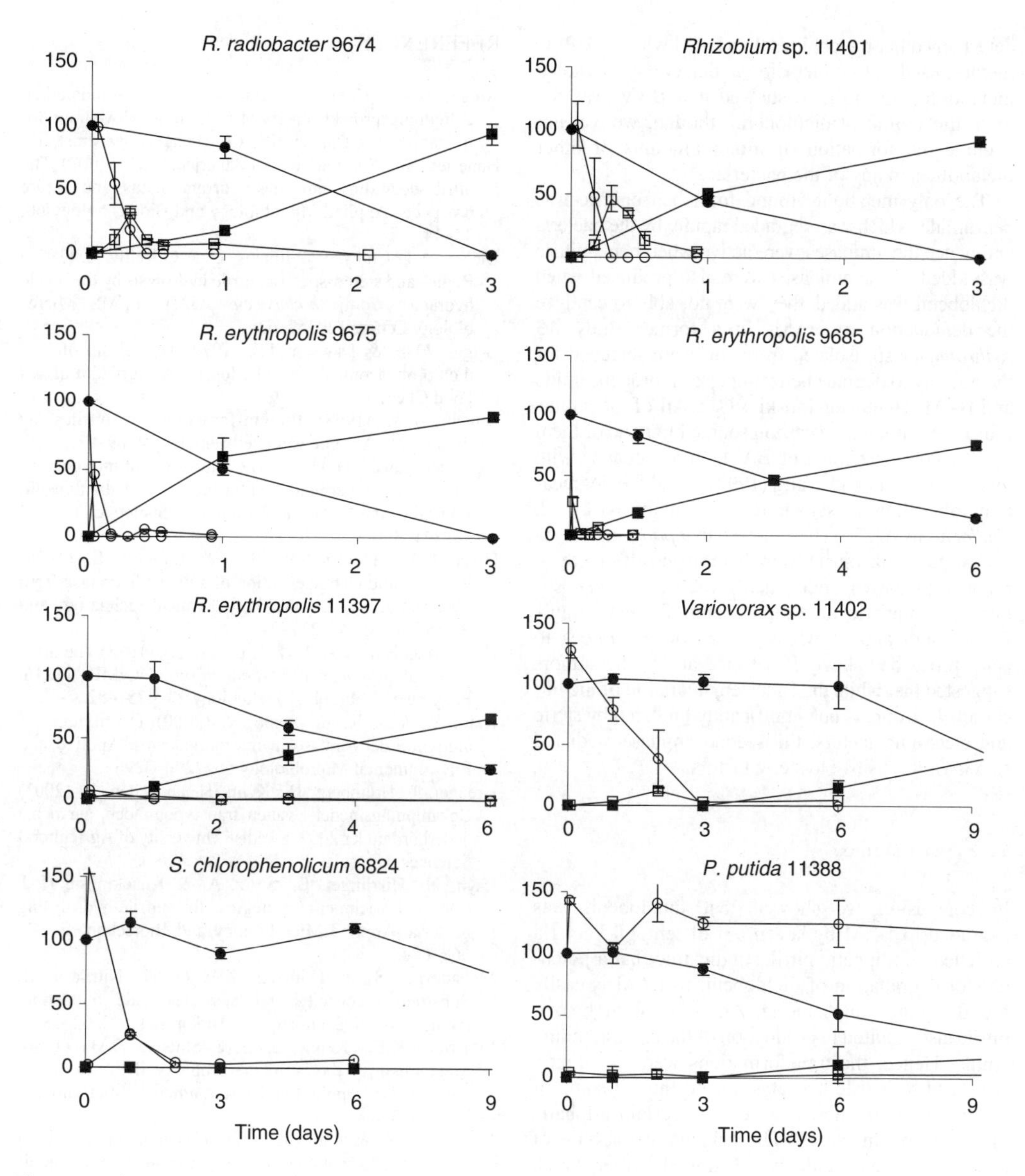

Figure 1. The percentages of the two nitriles (dichlobenil • and benzonitrile ○) compared to the sterile control are shown versus time in days. The percentages of the two amides (BAM ■ and benzamide □) as the percentages of added nitrile. Data represent mean ± SE of triplicate samples.

the nitrile being degraded and the type of bacterial degrader determine the type of enzyme system used rather than the type of nitrile or bacteria *per se*.

2.3 *Degradation of the amides*

We obtained values of r^2 close to 1 for most of the dichlobenil-degrading bacteria when correlating the percentages of degraded dichlobenil to the percentages of BAM produced (data not shown). This indicates that most of the bacteria did not degrade any of the BAM produced from dichlobenil. The only exception is *P. putida* 11388 with an r^2 of 0.85. Therefore this bacterium might have produced other metabolites directly from dichlobenil or alternatively it degraded some of the BAM. Unfortunately, we were not able to detect any other metabolites than BAM with the HPLC method used but neither did we detect any unknown metabolites. Overall we showed that BAM was the main metabolite of dichlobenil, though, we can not exclude the formation of minor amounts of other metabolites in any of the bacteria.

The only metabolite found from benzonitrile was benzamide, which was degraded rapidly by the bacteria and hence the amidases were active when benzonitrile was added. If the amidases were also produced when dichlobenil was added, they were not able to catalyze the degradation of BAM. In a former study, 13 *Arthrobacter* sp. isolated from soil were screened for their ability to degrade benzamide, 2-chlorobenzamide and BAM (Heinonen-Tanski 1981). All of the strains utilize benzamide as a carbon source but none of them use 2-chlorobenzamide or BAM. In accordance with this, Meth-Cohn and Wang (1997) tested the degradation of different substituted benzonitriles by *R. rhodochrous* AJ270. They showed that *para*- and *meta*-substituted benzonitriles were hydrolyzed efficiently to produce the corresponding acids whereas *ortho*-substituted benzonitriles were rapidly and efficiently transformed into amides while conversion of amides to acids proceeded slowly if detected at all. The authors suggested that while the first step, hydration of the linear nitrile group, is not significantly hindered by steric and electronic factors, the second amidase-mediated step is very sensitive to steric factors.

3 CONCLUSION

In conclusion we showed that dichlobenil was degraded to BAM by seven out of ten soil bacteria enriched on aliphatic nitriles indicating that a potential for degradation of dichlobenil to BAM is easily found among soil bacteria. An addition of benzonitrile also resulted in production of the corresponding amide. Hence, the transformations were apparently catalyzed by nitrile hydratases even though some of the bacteria are known to degrade certain aliphatic nitriles using nitrilases. The primary metabolite of benzonitrile, benzamide, was degraded further, which was in contrast with BAM being persistent. Further investigations are needed on soil bacteria enriched directly on dichlobenil or BAM, since this might reveal a potential for a further transformation of BAM.

ACKNOWLEDGEMENTS

The Immunonalysis Project (Grant no. 9901188) and the Danish Research Council (RECETO) supported this work.

REFERENCES

Anonymous (2002) Pesticider og vandværker. Udredningsprojekt om BAM-forurening. Hovedrapport. In Danish. The Danish EPA, Copenhagen, www.mst.dk.

Banerjee, A., Sharma, R. & Banerjee, U.C. (2002) The nitrile-degrading enzymes: current status and future prospects. Applied Microbiology and Biotechnology 60, 33–44.

Blakey, A.J., Colby, J., Williams, E. & O'Reilly, C. (1995) Regio- and stereo-specific nitrile hydrolysis by the nitrile hydratase from *Rhodococcus* AJ270. FEMS Microbiology Letters 129, 57–62.

Briggs, G.G. & Dawson, J.E. (1970) Hydrolysis of 2,6-dichlorobenzonitrile in soils. Journal of Agricultural and Food Chemistry 18, 97–99.

Bunch, A.W. (1998) Biotransformation of nitriles by rhodococci. Antonie van Leeuwenhoek 74, 89–97.

Heinonen-Tanski, H. (1981) The interaction of microorganisms and the herbicides chlorthiamid and dichlobenil. Journal of the Scientific Agricultural Society of Finland 53, 341–390.

Hjort, C.M., Godtfredsen, S.E. & Emborg, C. (1990) Isolation and characterisation of a nitrile hydratase from a *Rhodococcus* sp. Journal of Chemical Technology and Biotechnology 48, 217–226.

Hoyle, A.J., Bunch, A.W. & Knowles, C.J. (1998) The nitrilases of *Rhodococcus rhodochrous* NCIMB 11216. Enzyme and Microbial Technology 23, 475–482.

Kato, Y., Ooi, R. & Asano, Y. (2000) Distribution of aldoxime dehydratase in microorganisms. Applied and Environmental Microbiology 66, 2290–2296.

Kreuger, J., Holmberg, H., Kylin, H. and Ulén, B. (2003) Bekämpningsmedel i vatten från typområden, åar och i nederbörd under 2002. Swedish University of Agricultural Sciences, Uppsala, Sweden, *in Swedish*.

Layh, N., Hirrlinger, B., Stolz, A. & Knackmuss, H.-J. (1997) Enrichment strategies for nitrile-hydrolysing bacteria. Applied Microbiology and Biotechnology 47, 668–674.

Mahadevan, S. & Thimann, K.V. (1964) Nitrilase. II. Substrate specificity and possible mode of action. Archives of Biochemistry and Biophysics 107, 62–68.

McBride, K.E., Kenny, J.W. & Stalker, D.M. (1986) Metabolism of the herbicide bromoxynil by *Klebsiella pneumoniae*. Applied and Environmental Microbiology 52, 325–330.

Meth-Cohn, O. & Wang, M.-X. (1997) An in-depth study of the biotransformation of nitriles into amides and/or acids using *Rhodococcus rhodochrous* AJ270. Journal of the Chemical Society Perkin Transactions I 8, 1099–1104.

Miyazaki, S., Sikka, H.C. & Lynch, R.S. (1975) Metabolism of dichlobenil by microorganisms in the aquatic environment. Journal of Agricultural and Food Chemistry 23, 365–368.

Montgomery, M., Yu, T.C. & Freed, V.H. (1972) Kinetics of dichlobenil degradation in soil. Weed Research 12, 31–36.

Nagasawa, T., Kobayashi, M. & Yamada, H. (1988) Optimum culture conditions for the production of benzonitrilase by *Rhodococcus rhodochrous* J1. Archives of Microbiology 150, 89–94.

Nagasawa, T., Takeuchi, K., Nardi-Dei, V., Mihara, Y. & Yamada, H. (1991) Optimum culture conditions for the production of cobalt-containing nitrile hydratase by *Rhodococcus rhodochrous* J1. Applied Microbiology and Biotechnology 34, 783–788.

Nagasawa, T. & Yamada, H. (1990) Application of nitrile converting enzymes for the production of useful compounds. Pure and Applied Chemistry 62, 1441–1444.

Nawaz, M.S., Heinze, T.M. & Cerniglia, C.E. (1992) Metabolism of benzonitrile and butyronitrile by *Klebsiella pneumoniae*. Applied and Environmental Microbiology 58, 27–31.

Stevenson, D.E., Feng, R., Dumas, F., Groleau, D., Mihoc, A. & Storer, A.C. (1992) Mechanistic and structural studies on *Rhodococcus* ATCC 39484 nitrilase. Biotechnology and Applied Biochemistry 15, 283–302.

Topp, E., Xun, L.Y. & Orser, C.S. (1992) Biodegradation of the herbicide bromoxynil (3,5-dibromo-4-hydroxybenzonitrile) by purified pentachlorophenol hydroxylase and whole cells of *Flavobacterium* sp. strain ATCC 39723 is accompanied by cyanogenesis. Applied and Environmental Microbiology 58, 502–506.

Verloop, A. & Nimmo, W.B. (1970) Metabolism of dichlobenil in sandy soil. Weed Research 10, 65–70.

Vosáhlová, J., Pavlù, L., Vosáhlo, J. & Brenner, V. (1997) Degradation of bromoxynil, ioxynil, dichlobenil and their mixtures by *Agrobacterium radiobacter* 8/4. Pesticide Science 49, 303–306.

Warhurst, A.M. & Fewson, C.A. (1994) Biotransformations catalysed by the genus *Rhodococcus*. Critical Reviews in Biotechnology 14, 29–73.

Wolter, R., Rosenbaum, S. & Hannappel, S. (2001) The German groundwater monitoring network. Proceedings Monitoring Taylor Made III, 277–282.

Wyatt, J.M. & Linton, E.A. (1988) The industrial potential of microbial nitrile biochemistry. Ciba Foundation Symposium 140, 32–48.

European Symposium on Environmental Biotechnology, ESEB 2004 - Verstraete (ed)
 ISBN 90 5809 653 X

Degradation of selected pesticides in aquatic ecosystem model

J. Kalka, E. Grabińska-Sota & K. Miksch
Environmental Biotechnology Department Silesian University of Technology, Gliwice, Poland

ABSTRACT: The objective of this study was to evaluate the distribution of β-cyflutrin, λ-cyhalotrin, esfenvalerate and permethrin in an aquatic ecosystem model. The investigations were carried out in three parallel series at a temperature 22 ± 2°C. Concentrations of pyrethroids were close to those which were found in surface waters after spraying. It was shown, that half life of investigated pyrethroids changed within the range 2.87 to 13.86 d. It was proved that degradation rate of pyrethroids was dependent on the preparation concentration and form they were introduced to water.

1 INTRODUCTION

Due to massive use of pyrethroids in agriculture production, animal husbandry, public health and welfare of mankind, they contribute to environmental pollution. The substances are neuropoisons which act on the axons in the peripheral and central nervous systems by interacting with sodium channels in mammals and insects (Kamrin 1997, Angerer & Ritter 1997). They are considered to be safe for mammals (Różański 1992) Nevertheles, the long-term toxic effect in humans is a subject of controversy in the literature (Müller-Mohnssen 1999, Pogoda & Preston-Martin 1997, Straube et al. 1999). The lack of dose-response data in internal exposure from biological monitoring make it difficult to show a clear correlation between the symptoms and current internal exposure (Angerer & Ritter 1997). Pyrethroids were found in surface waters in concentratins from 0.05 to 2.5 $\mu g/dm^3$. The maximal amounts were recorded immediately after spraying and ranged from 0.0 to 15.7 $\mu g/dm^3$ (Lutnicka et al. 1999). Monitoring of pyrethroid residue levels in vegetables and drinking water is of particular concern for human health. The aim of this study was investigation differences in kinetics and dynamics of degradation of four pyrethroids in the dependence on formulation and concentration introduced to water.

2 MATERIALS AND METHODS

2.1 *Preparations*

Four pyrethroid formulations and active substances were investigated during this study. Table 1 presents names of pyrethroids and used technical formulations.

Table 1. Preparations used during the study.

Active substance	Formulation
Permethrin	Ambusz 250 EC
λ-cyhalotrin	Karate 025 EC
β-cyflutrin	Bulldock 025 EC
Esfenvalerate	Sumi-Alpha 050 EC

Active substances were obtained from commercial supplier in a best available purity (Promochem, Poland) Formulations were obtained from common agricultural market.

2.2 *Biodegradation tests*

Biodegradation tests were conducted under conditions simulating natural aquatic environment. Glass tanks were prepared, filled up with 30 dm^3 of standardized water and inoculated with activated sludge microorganisms at a concentration of 1 cm^3/dm^3. Characteristic of standardized water prepared at the base of distilled water was as follows: total concentration of ions of Ca and Mg 2.5 $mmol/dm^3$; ratio Ca:Mg = 4:1, Na:K = 10:1; acidic capacity 0.8 $mmol/dm^3$. Examined substances were added in such quantity to obtain concentrations of active substances:

- I series – pure active substances – 0.05 mg/dm^3
- II series – active substances in formulations – 0.05 mg/dm^3
- III series – active substances in formulations – 0.5 mg/dm^3.

In series I active substances were introduced to water in small volume (0.5 cm^3) of acetone. After that aeration was started. Water solutions in tanks were protected from evaporation by coverage tanks with film of foil.

Analyses
Biochemical oxidation was evaluated on the basis of measurement:

- concentration of active substances with gas chromatography GC-ECD
- pH and concentration of dissolved oxygen in water.

Analysis were conducted in 0, 2, 4, 7, 9, 14, 21, 29, 37 and 40 day of the process. Water samples were extracted two times with hexane (10:1 and 20:1). The extracts were combined and dried with anhydrous sodium sulfate, evaporated to 10 ml and concentrated at ambient temperature to 1 cm^3. Pyrethroids concentration were determinated with gas chromatography GC-ECD according to method described by Chen & Wang (1996).

3 RESULTS

Results of biodegradation tests are presented on Figures 1–4 as function

$$c = f(t) \tag{1}$$

where: c = concentration of active substance; t = time of research.

3.1 *Determination of half-lives*

Statistical analysis of obtained results has proved that biological degradation of pytethroids in river-water test was in accordance with first order reaction. Concentration of preparations decreased exponentially according to equation:

$$\frac{dc}{dt} = -K \cdot c \tag{2}$$

where: t = time of research; c = concentration of preparations; K = reaction constant.

The confidence level calculated with Student's test for variables c and t showed statistically important correlation between those variables with probability above 95%. Reaction constants (K) and half-lives (t_{50}) were calculated according equations 2 and 3.

$$K = 2.3 \cdot \bar{k} \tag{3}$$

$$t_{50} = \frac{\ln 2}{K} \tag{4}$$

where: $\bar{k}$ = slope of curve $\log \frac{c}{c_0} = f(t)$.

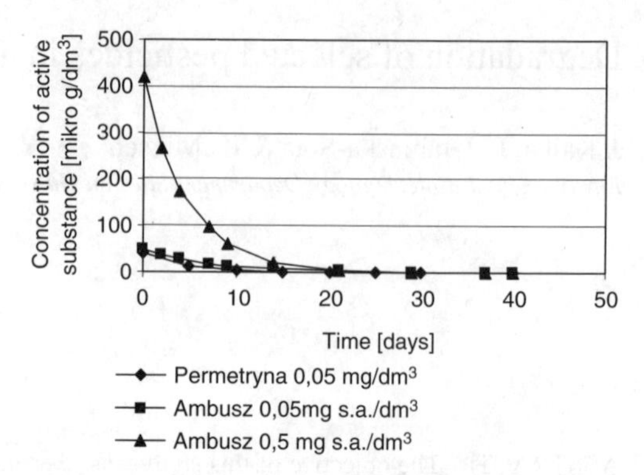

Figure 1. Disappearance of permethrin in water.

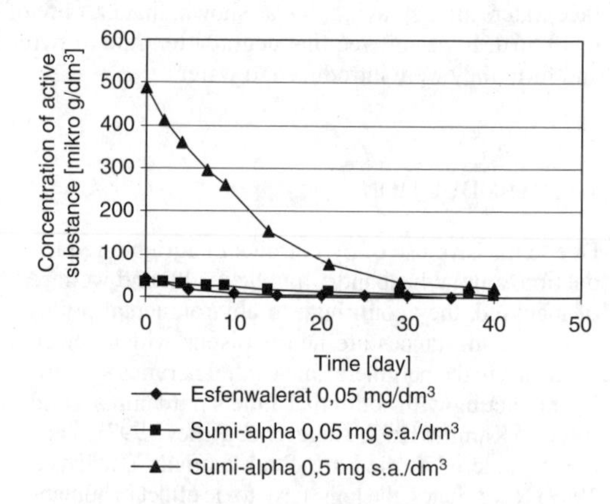

Figure 2. Disappearance of esfenvalerate in water.

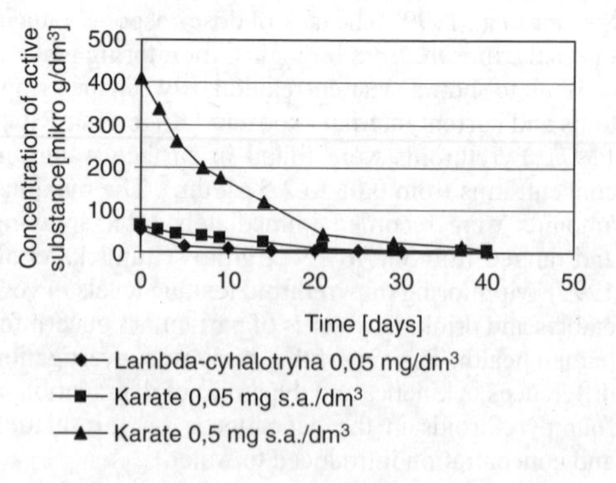

Figure 3. Disappearance of λ-cyhalotrine in water.

Calculated half-lives were within the range 2.87 to 13.86 and depended on the form of preparation and its concentration (Table 2). The fastest degradation was observed for pure active substances. Introduction of pyrethroids formulations in II and III series caused

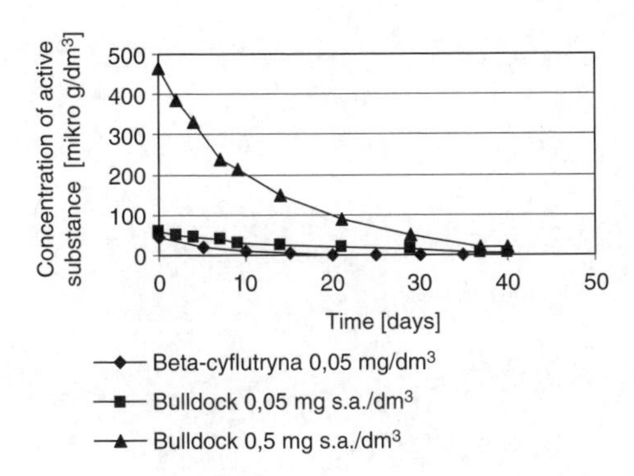

Figure 4. Disappearance of β-cyflutrine in water.

Table 2. Calculated values of reactions kinetics.

Preparation	Concentration (mg/dm^3)	n	k	SD	t_{50} (d)
Permethrin	0.05[1]	7	0.105	0.0153	2.87
	0.05[2]	9	0.064	0.0029	4.71
	0.5[2]	9	0.095	0.0031	3.16
λ-cyhalotrin	0.05[1]	6	0.066	0.0264	4.57
	0.05[2]	9	0.033	0.0123	9.12
	0.5[2]	9	0.044	0.0035	6.86
β-cyflutrin	0.05[1]	7	0.063	0.0117	4.78
	0.05[2]	9	0.024	0.0040	12.38
	0.5[2]	9	0.036	0.0026	8.25
Esfenvalerate	0.05[1]	7	0.054	0.0140	5.54
	0.05[2]	9	0.022	0.0020	13.86
	0.5[2]	9	0.035	0.0029	8.60

elongation of t_{50}. Similar phenomenon was observed by Lutnicka (1999). Cypermethrin in the commercial product Nurelle D 550 EC was more stable in water than introduced as active substance. It should also been noted that for higher concentration of pyrethroid formulation shorter half-lives were obtained. It is supposed that higher concentrations of preparations (which were the only source of carbon and energy for microorganisms) could increase metabolic activity of microorganisms and as a result shorter half-lives were obtained. Commercial formulations except active substances contain also other supporting substances – adjuvants. It is supposed, that during degradation of formulations cometabolism phenomenon was observed. According to Alexander (1980) cometabolism is one of the main processes during detoxication of pesticide contaminated soil. Cometabolism of pesticides had also been observed in surface waters. Yei-Shung (1984) had pointed that biological decomposition of 2,4-D was accelerated by addition of yeast extract. Results obtained during this study showed, that some pesticides were degradable only by cometabolic pathway.

4 CONCLUSIONS

From the obtained results it could be pointed, that there were differences between degradation of active substances and technical formulations of pyrethroids. The fastest degradation was observed for pure active substances. Cometabolic processes were observed during degradation of commercial formulations.

REFERENCES

Alexander M. 1980. Biodegradation of chemicals of environmental concern. *Science* 211: 132–138.

Angerer J., Ritter A. 1997. Determination of metabolites of pyrethroids in human urine using solid-phase extraction and gas chromatography – mass spectrometry. *Journal of Chromatography* B 695: 217–226.

Chen Z-M., Wang Y-H. 1996. Chromatographic methods for the determination of pyrethrin and pyrethroid pesticide residues in crops, foods and environmental samples. *Journal of Chromatography* A 754: 367–395.

Kamrin M. (ed) 1997. *Pesticide profiles*. New York: Lewis Publishers.

Lutnicka H., Bogacka T., Wolska L. 1999. Degradation of pyrethroids in an aquatic ecosystem model. *Water Research* 33 (16): 3441–3446.

Müller-Mohnssen H. 1999. Chronic sequel and irreversible injuries following acute pyrethroid intoxication. *Toxicology Letters* 107 (1–3): 161–176.

Pogoda J.M., Preston-Martin S. 1997. Household pesticides and risk of pediatric brain tumors. *Environmental Health Perspectives* 105 (11): 1214–1220.

Różański L. (ed) 1992. *Przemiany pestycydów w organizmach żywych i środowisku*. Warszawa: PWRiL.

Straube E., Straube W., Krüger E., Bradatsch M., Jakob Meisel M., Rose H.J. 1999. Disruption of male sex hormone with regard to pesticides. *Toxicology Letters* 107 (1–3): 225–231.

Yei-Shung W., Madsden E.L., Alexander M. 1985. Microbial degradation by mineralization or cometabolism determined by chemical concentration and environment. *Appl. Environ. Microbiol.* 33 (3): 495–499.

European Symposium on Environmental Biotechnology, ESEB 2004 - Verstraete (ed)
© 2004 Taylor & Francis Group, London, ISBN 90 5809 653 X

Determination of some oil oxidizing microorganisms from the Caspian water

S. Kaffarova
Institute of Botany, Azerbaijan National Academy of Sciences, Azerbaijan

ABSTRACT: We isolated bacteria from more polluted sites where concentration of oil hydrocarbons might be higher. Thus, bacteria were isolated that possess high and constant oil-oxidizing activity. The influence of source of nitrogen, temperature and pH was determined on destruction of oil hydrocarbons by microorganisms. The character of petroleum destruction under cultivation of selected cultures was also studying.

Preliminary identification of active cultures and strains to gender and , partially, to has been done.

Most of naphthalene oxidizing strains lost the activity very quickly. It was necessary to select conditions for successful growth.

1 INTRODUCTION

Many oil production sites have become polluted hydrocarbons. Frequently, these sites are the subject to remediation programs. One such remediation strategy employs the use of microorganisms *in situ* to oxidize these hydrocarbons to carbon dioxide. The successful implementation of a bioremediation process requires a detailed characterization of the *in situ* biodegradative activity. The majority of bioactivity assessment methods employed to date based on the *ex situ* measurement of microbial metabolism.

We have started the investigating the possibility to apply bioremediative techniques to the Caspian Sea region, which is affected by very serious environmental problems due to oil-wells, refineries and pipelines which are concentrated in the Baku area.

Initial work gave availability of a large number of microorganisms which have been isolated from soil, sediments and water samples and which were selected for their capacity to use oil components as carbon source.

2 MATERIALS AND METHODS

The object of investigation: microorganisms isolated from Caspian water, and oil-oxidizing microorganisms the sediment. Isolates were cultured in Hutner's minimal medium prepared.

The freshly isolated cultures have been checked up concerning the spectrum of oil hydrocarbon assimilability. We used Hutners minimal medium with the desired substrate. The analysis of residual hydrocarbon was carried out by means of adsorptive chromatography on silica gel. The solvents hexanes, benzene, chloroform, ethanol in different relationship were used for elution. Bacterial growth in liquid culture was monitored spectrophotometrically by measuring the optical density (OD660)of the culture.

Optimal pattern of nitrogen supply was established by comparison of the oil destruction on the media with different nitrogen sources NH_4Cl, NH_4NO_3, KNO_3, $(NH_4)_2SO_4$; all salts were taken in the equivalent amounts. The results of experiments have been estimated in accordance with quantity of residual hydrocarbons.

The influence of cultivation temperature on the growth and oxidation of oil were studying at 17, 26 and 36°C with shaking. The culture was growth up to stationary phase on each temperature regime. At the end of each experiment the analysis had been carried out for total quantity of residual hydrocarbons and their grouped hydrocarbons composition. The biomass was determined by use of nephelometric method.

The influence of medium acidity on the oil destruction was studying under the periodical growing on shakers at 230 rpm, 28°C, and pH 5,0–9,0.

The petroleum destruction under long term cultivation of selected cultures was studying by the use of periodical growth in shaker (220 rpm) at 28°C. The petroleum destruction content in the samples was 5 g/l. The sampling have been carried out after 1, 3, 7, 14, 28 and 42 days of growth.

The biomass was determined by use of nephelometric method. The total quantity of residual hydrocarbons

Table 1. Growth of bacterial isolates on Hutner's minimal medium with different carbon source.

Strain	Carbon source: Benzene	Hexane	Nonane	Isooctane	Undecan	Hexadecane	Diesel	Oil
Casks1	– –	– –	+ +	+ +	+ + +	+ + + +	+ + + +	+ + +
Casks2	– –	±	+ +	+ + +	+ + + +	+ + + +	+ + + +	+ + + +
Casks3	– –	– –	+ +	+	+ + + +	+ + + +	+ + + +	+ + +
Casks4	– –	–	+ +	+ +	+ + + +	+ + + +	+ + + +	+ + + +
Caskw1	– –	– –	+	+ +	+ + + +	+ + + +	+ + + +	+ + + +
Caskw2	– –	– –	+	+	+ +	+ + +	+ + +	+ +
Caskw3	– –	±	±	– –	+ +	+ + +	+ +	+
Caskw4	– –	– –	– –	+	+	+ +	+ + +	+

– – *no growth*, + *low activity*, + + *moderate*, + + + *high activity*, + + + + *very high activity.*

Table 2. Oxidation of petroleum products during the growth of isolated strains on mediums with different nitrogen sources.

No strain	Of source nitrogen	Of residual hydrocarbons (g/l)	Oxidized hydrocarbons (%)	Oxidized hydrocarbons on a medium with NH_4NO_3
Casks4	NH_4CL	0.8776	56.1	109.6
	NH_4NO_3	0.9756	51.22	100
	KNO_3	0.9156	54.22	105.9
	$(NH_4)_2SO_4$	1.0236	48.82	95.3
Casks1	NH_4CL	1.0106	49.47	113.3
	NH_4NO_3	1.127	43.65	100
	KNO_3	1.3036	39.82	91.2
	$(NH_4)_2SO_4$	0.836	58.2	133.3
Caskw3	NH_4CL	1.0576	47.12	119.3
	NH_4NO_3	1.2106	39.47	100
	KNO_3	1.289	35.55	90.1
	$(NH_4)_2SO_4$	0.9658	51.57	130.7
Caskw1	NH_4CL	0.7918	60.41	103.7
	NH_4NO_3	0.8446	58.27	100
	KNO_3	1.1315	43.43	74.5
	$(NH_4)_2SO_4$	0.5520	72.41	124.3

was determined in the samples and their grouped hydrocarbon composition as well.

3 RESULTS AND DISCUSSION

After sowing on mineral media with petroleum products and appropriate growing, 8 strains of microorganisms which have actively utilized the oil hydrocarbons were isolated. As related to these strains it was checked up their growth on the individual hydrocarbons (Table 1).

As it was shown during the studying of isolated strains on the media with different nitrogen sources, ammonium or mixed sources of nitrogen supply were the preferable ones for all isolated groups of micro organisms. Nitrate salts were used in the worse way.

It would be interesting to study the hydrocarbons consumption under the temperatures which are usual for given water body (17–26°C), and to establish the influence of temperature oscillation on the bacteria growth and their ability for oxidation of petroleum products.

The oxidation on Hutners medium with NH_4NO_3 at 26°C was taken for 100%. Under these conditions consumption patterns for studied cultures of strain Caskw3 and casks1 were the similar ones: lag-phase was continuing for 4–6 hours, petroleum consumption was insignificant one and made up 6–10% of applied quantity. The active petroleum consumption and biomass increase were beginning after 7 hours of

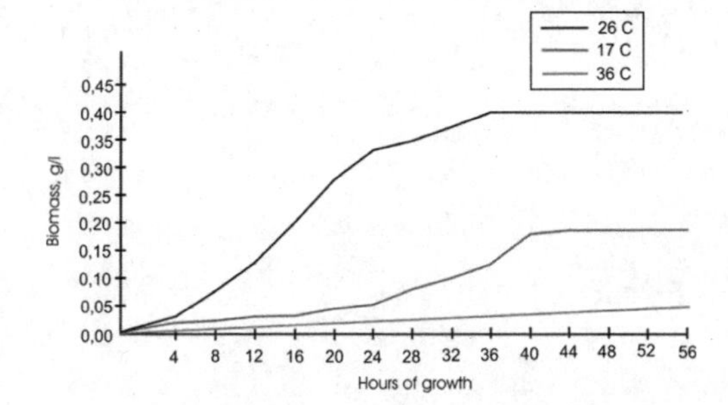

Figure 1. Growth of strain Caskw3 on oil hydrocarbons under different temperatures.

growth. Caskw3 was passing to stationary phase by 36th hour of grows, with simultaneous stoppage of biomass increase and petroleum products consumption (fig.1).

The growth pattern of cultures was not changing under the temperature decrease.

In fact, there is no date in literature on the influence of medium acidity on hydrocarbon degradation. As it is known, pH is not the unspecific inhibition factor which acts, chiefly, on the cell enzymes. The entering of high-mineralized sewage in the water body can shift strongly the pH value to one or another side which may affect the water auto-purification.

	Strain Caskw3		Strain Casks1	
pH	Biomass, g/L	% of oxidized petroleum products	Biomass, g/L	% of oxidized petroleum products
5.0	0.15	27	0.34	39
6.0	0.37	33	0.68	53
7.0	0.68	47	1.1	64
8.0	0.54	41	1.01	58
9.0	0.50	39	0.42	47

We have established that the best condition for growth of microorganisms and destruction of petroleum products by them takes place at neutral pH values. The change of medium pH value to the acid or basic region has influenced, actually, on biomass growth, meanwhile the petroleum products oxidation is changing by negligible way. Low pH values inhibit the biomass growth and hydrocarbon oxidation in a higher extent than basic pH values.

European Symposium on Environmental Biotechnology, ESEB 2004 - Verstraete (ed)

Bacteria and the bioremediation of stone: the potential for saving cultural heritage

A.M. Webster, D. Vicente & E. May
University of Portsmouth, Portsmouth, UK

ABSTRACT: Crusts on stone result in loss and disfiguration of cultural heritage. Acids formed from fossil fuel gases react with carbonates in stone leading to the formation of sulphates and nitrates which dissolve in water and can be lost during rainfall. In addition black crusts form from gypsum crystals mixed with atmospheric dust particles. Certain bacteria are able to mineralise sulphates, nitrates and organic pollutants and thus destroy crusts, while others are able to deposit calcite and consolidate stone surfaces. BIOBRUSH, an EU-funded project, aims to combine these various bacterial attributes so that mineralisation of crusts can be followed by consolidation of the stone surface. The University of Portsmouth has isolated some forty isolates from an environmental source, and compared them for their biocalcifying capabilities. A selected number have been tested for temperature sensitivity. Suitable isolates will be used in *in situ* trials of the BIOBRUSH project.

1 INTRODUCTION

The problem of a crumbling cultural heritage at historical sites is a very real problem in European cities and urbanised developments. Alternations are caused by inorganic atmospheric pollutants such as nitrogen oxides and sulphur dioxide produced by the combustion of petroleum and its derivatives; these compounds are oxidized in the air into nitric and sulphuric acid, respectively, that, after deposition on the surface of stone carbonates, are converted into sulphates and highly soluble nitrates, which are easily washed away by rain. The binding material of limestone is first dissolved and this leads to the loss of the large grains and ultimately the integrity of the surface (Rao et al., 1996).

Attempts to limit or retard the deterioration processes on historic buildings have been made in many ways by application of conservation treatments using both inorganic and organic products. Conventional treatments however, can result in changes in the penetration and functional adherence of stone (Koestler, 1999), loss of original surfaces, colour changes in the rocks or excessively remove of the original rock material. Several treatments can act also on the unaltered substrates inducing an acceleration of degradation phenomena (Heselmeyer et al., 1991) while the use of inappropriate treatment methodologies can often induce more final damage compared with the original.

Until now, the remediation of surfaces altered by pathologies such as black crusts, the presence of sulphate and nitrate and uncombusted hydrocarbons, has been carried out using chemical or physical methods, often associated with detrimental effects e.g. surface scraping. The substitution of potentially toxic chemicals and harsh physical methods, possibly based on chemicals, by biological techniques is in line with EU policies. These direct particular attention to the protection of our cultural heritage, representing the history and the culture of European countries, but also to the utilisation of methods that reduce the risks for the artworks, the operators and the environment. In addition, chemicals are products that usually require a great quantity of energy for their synthesis, often based on petroleum. In contrast, microorganisms can be cultivated and produced in large amounts by opportune use of waste by-products from food and agricultural processes which otherwise can contaminate the environment.

There is growing evidence that microorganisms can be used to reverse the deterioration processes through bioremediation. In particular, the ability and potential of microorganisms to utilize sulphate, nitrate or organic residues has been impressively demonstrated (Atlas, 1988; Lal Gauri, 1989; Orial, 1992, Ranalli et al., 1999). Bacteria are also capable of inducing or mediating calcium carbonate (calcite) production, a widely reported activity of soil and marine bacteria (Boquet et al., 1973; Morita, 1980). Several groups have already explored the use of biocalcifying bacteria in stone conservation (Le Métayer-Levrel et al., 1999; Rodriguez-Navarro et al., 2003) but it is the novel aim of the BIOBRUSH project to sequentially link mineralisation processes of desulfurication, denitrification and organic removal to the consolidation phenomenon of biocalcification. A collection of biocalcifiers has been made using bacteria from established culture collections and from an environmental source.

2 METHODS

Calcified material from a freshwater source in Somerset, England was ground in a pestle and mortar to produce a paste, which was then inoculated onto a selection of liquid and solid media. Over 40 isolates were isolated from the original material using a variety of culture media; these were then screened for their biocalcification ability in modified B4 liquid media (Boquet et al. 1973) containing 5 g calcium acetate, 1 g yeast extract, 1 g glucose dissolved in 1 l distilled water adjusted to pH 8. Isolates were ranked according to their ability, over a three week period, to remove calcium ions from solution, maintain a pH close to neutrality and increase the weight of limestone discs placed in the media. Crystal formation on solid media over time was recorded by light microscopy combined with digital photography. Scanning Electron Microscopy (SEM) was used to observe biocalcification of limestone discs and sintered glass discs incubated in liquid media with candidate isolates. SEM in combination with EDAX was also used to determine the constituents of the crystals produced. The top ten ranking bacteria were then assessed for their ability to produce calcium carbonate crystals over a three week period on solid and in liquid modified B4 media at 10°, 20°, 30° and 40°C. Crystals were recovered by washing.

Initial identification of isolates was made by use of standard characterisation tests and the API 20NE system (BioMerieux). Sequencing of candidate isolates is ongoing. The resistance or susceptibility of isolates to antibiotics is also been tested.

3 RESULTS

SEM images of limestone discs after 21 days incubation with biocalcifying bacteria demonstrate the production of presumptive calcified bacterial cells (Figure 1) similar to those seen by other workers (Rivadeneyra et al., 1996; Rodriguez-Navarro et al., 2003). Calcified cells were also observed on sintered glass discs incubated with biocalcifying bacteria. EDAX measurements from these calcified cells and from individual washed crystals recovered from liquid media detected the presence of calcium only. The use of sintered glass

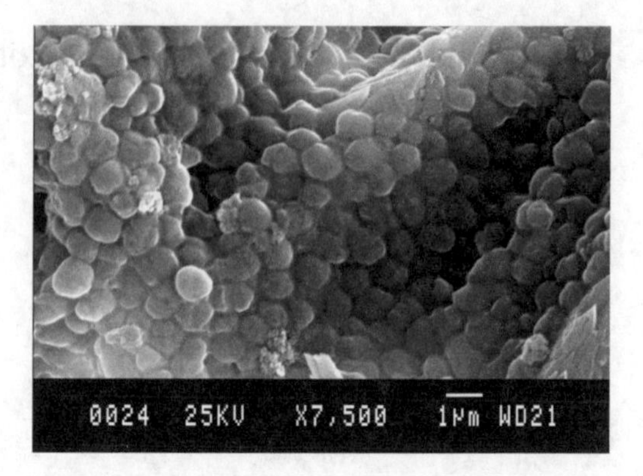

Figure 1. Calcified bacterial cells on limestone disc.

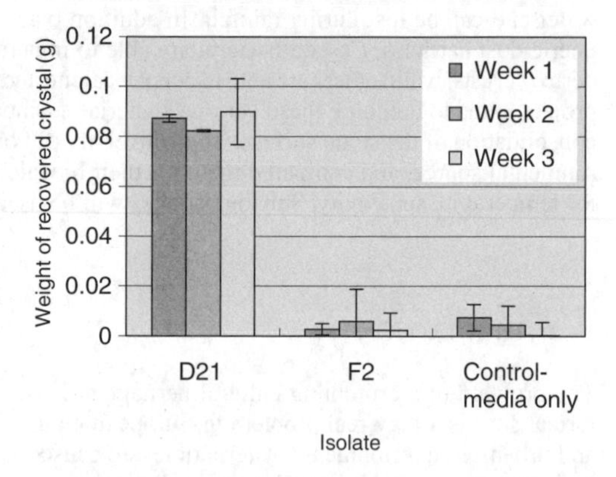

Figure 2. Weight of recovered calcite from isolates grown on solid media over 3 weeks at 40°C.

discs reduced the possibility of misinterpretation of EDAX readings, which might occur when taking readings of calcified cells on a limestone surface.

Candidate isolates demonstrated differences in their ability to produce calcite over a range of temperatures. All of the ten isolates produced crystals at 20° and 30°C however significant differences in weights were observed at 10° and 40°C (Figure 2).

Isolates streaked onto modified B4 media showed evidence of crystal production within 16 hours of streaking when viewed by the light microscope at a magnification of ×40.

Using the API system the following isolates were among those identified from the environmental source – *Aeromonas hydrophila, Aeromonas salmonicida salmonicid, Acinetobacter juni, Brevibacillus laterosporus, Chryseomonas luteola, Pseudomonas fluorescens, Pseudomonas putida and Sphingomonas paucimobilis.* Some of these bacteria are classified Hazard Group 1 “unlikely to cause human disease”, others fall into Hazard Group 2 “can cause human disease" (Advisory Commission on Dangerous Pathogens).

4 DISCUSSION

The ability of the isolates recovered from the environmental source to carry out biocalcification support the view that this is a common phenomenon amongst bacteria (Boquet et al., 1973).

Experiments conducted at high temperature indicate that some of the isolates are better able to function at higher temperatures than other, despite being isolated from the same source. Other experiments have also been undertaken to determine the capacity of the selected isolates to carry out biocalcification in colder environments.

EDAX results confirm the calcified nature of the bacterial cells on the limestone and sintered disc surfaces. No magnesium was detected during EDAX analysis, this contrasts with work by other authors (Rivadeneyra et al., 1996; Rivadeneyra et al., 2000) where varying levels of magnesium have been measured.

Microscopy results also suggest that crystal production is rapid (visible by light microscopy in 16 hours) and that calcified bacterial cells will form a layer on suitable substrates. The layer of cells follows the contours of the stone substrate and leaves the porous structures intact thus allowing water vapour transfer within the stone

The variety of bacteria found belong to Hazard Groups 1 and 2, this illustrates the need to definitively identify unknown species in order to eliminate those that are potential human pathogens. Sequencing of those isolates considered suitable for use in the BIOBRUSH project is already underway; any species considered potentially harmful will not be considered for further use.

5 SUMMARY

The goal of BIOBRUSH is to provide an effective, environmental-friendly biotechnological tool for restoration and conservation of artistic stoneworks central to the European heritage. The emphasis is on practical treatment options based on bioremediation, which are safe for humans, heritage and the environment and yet flexible enough to be adapted to the prevailing material, exposure and climatic conditions.

A primary task of the BIOBRUSH consortium is the establishment of culture collection of bacteria capable of mineralisation and biocalcification that will provide a database of information for conservators working to arrest or control the different forms of damage suffered by cultural heritage sites in European cities. This culture collection has drawn from established culture collections in Europe and the USA but also from environmental sources, the latter being an ongoing process.

Analysis of biocalcifying bacteria so far isolated has determined that many have the capacity to produce a calcified layer on a stone substrate. Experiments also suggest that certain of these produce calcite in a matter of hours and that some can continue to do this even at 40°C. The potential therefore exists for these bacteria to be used in bioremediation as part of the BIOBRUSH project. Determination of the species isolated by molecular means is vital in excluding potential harmful microorganism from the use in the proposed technology. Analysis is also required to ensure that the use of these microorganisms does not further compromise already damaged stone, and such work is already in progress.

The biological methods proposed by BIOBRUSH may be less harsh than the chemical and physical ones, which could be considered to be destructive methods. The removal of the altered material from the surface mediated by microorganisms takes place in a more natural way given that microorganisms have an active role in the environment and contribute to the closure of the biogeochemical cycles and to the stabilisation of dynamic equilibria. The consolidation of the stone surface by biocalcifiers is also a natural process albeit one that normally occurs in other locations. Together mineralisation and consolidation constitute a biotechnological approach utilising bacterial properties in a controlled manner.

REFERENCES

Atlas, RM., Chowdhury, Ahad N. and Lal Gauri, K. (1988). Microbial calcification of gypsum-rock and sulfated marble. *Studies in Conservation*. 33: 149–153.

Boquet, E., Boronat, A., Ramos-Cormenza, A. (1973). Production od calcite (calcium carbonate) crystals by soil bacteria is a general phenomenon. *Nature*. 246: 527–529.

Castanier, S., Le Metayer-Levrel, G., Orial, G., Loubiere, J.F., Perthuisot, J.P. (1999). Bacterial carbonatogenesis and applications to preservation and restoration of historic property. *Proceedings of International Conference on Microbiology and Conservation (ICMC 99)*. Florence, pp 246–252.

Heselmeyer, K., Fischer, U., Krumbein, W.E. and Warscheid, Th. (1991). Application of *Desulfovibrio vulgaris* for the bioconservation of rock gypsum crusts into calcite. *Bioforum* 1/2, 89.

Koestler, R.J. (1999). Polymers and resins as food for microbes. *Proceedings of International Conference on Microbiology and Conservation (ICMC 99)*. Florence, pp 193–196.

Lal Gauri, K., Chowdhury, Ahad N., Kulshreshtha, Niraj P., Adinarayana, R., Punuru. (1989). The sulfatation of marble and treatment of gypsum crusts. *Studies in Conservation*. 34: 201–206

Le Métayer-Levrel, Castanier, S., Orial, G., Loubière, J.-F., Perthuisot, J.-P. (1999). Applications of bacterial carbonatogenesis to the protection and regeneration of limestones in buildings and historical patrimony. *Sedimentary Geology*. 126: 25–34.

Morita, R. Y. (1980). Calcite precipitation by marine bacteria. *Geomicrobiology Journal.* 2(1): 63-82.

Orial, G., Castanier, S., Le Metayer, G., Loubiere, J.F. (1992). The biomineralization: a new process to protect calcareous stone: applied to historic monuments. *Proceedings of II nd International Conference on Biodeterioration of Cultural Heritage property.* (Eds.: H. Arai, T. Kenjo, K. Yamano), Japan., pp 98–116.

Ranalli, G., Matteini, M., Pizzigoni, G:, Zanardini, E. and Sorlini, C. (1999). Bioremediation on cultural heritage: removal of sulphates, nitrates and organic substances. *Proceedings of International Conference on Microbiology and Conservation (ICMC 99).* Florence, 257–260.

Rao, S.M.,Brinker, C.J., Ross, T.J. (1996). Enviromental microscopy in stone conservation. *Scanning.* 18: 508–514.

Rivadeneyra, M.A., Ramos-Cormenzana, A., Delgado, G., Delgado, R. (1996). Process of carbonate precipitation by *Deleya halophila*. *Current Microbiology*. 32: 308–313.

Rivadeneyra, M.A., Delgado, G., Ramos-Cormenzana, A., Delgado, R. (2000).Pecipitation of carbonates by *Nesterenkonia halobia* in liquid media. *Chemosphere*. 41: 617–624.

Rodriguez-Navarro, C., Rodriguez-Gallego, M., Ben Chekroun, K., Gonzalez-Mũnoz, M.T. (2003). Conservation of ornamental stone by *Myxococcus xanthus*-induced carbonate precipitation. *Applied and Environmental Microbiology*. April:2182–2193.

European Symposium on Environmental Biotechnology, ESEB 2004 - Verstraete (ed)
© 2004 Taylor & Francis Group, London, ISBN 90 5809 653 X

Engineered endophytic bacteria improve phytoremediation of water-soluble volatile organic pollutants

B. Borremans & A. Provoost
Flemish Institute for Technological Research (Vito), Mol, Belgium

T. Barak, L. Oeyen & J. Vangronsveld
Limburgs Universitair Centrum (LUC), Department of Environmental Biology, Universitaire Campus building, Diepenbeek, Belgium

D. van der Lelie & S. Taghavi
Brookhaven National Laboratory (BNL), Biology Department, Upton, New York, USA

ABSTRACT: Phytoremediation of highly water-soluble and volatile organic xenobiotics often is limited because plants, and their rhizospheres, do not degrade them sufficiently. This can result in phytotoxicity, and/or volatilization of the compounds through the leaves, causing new environmental problems. We demonstrate that endophytic bacteria equipped with the appropriate degradation pathway improve the *in planta* degradation of the model compound toluene. We introduced, by conjugation, the pTOM toluene-degradation plasmid of *Burkholderia cepacia* G4 into *B. cepacia* L.S.2.4, a natural endophyte of yellow lupine. After successfully inoculating surface-sterilized lupine seeds with the recombinant strain, the engineered endophytic bacteria strongly degraded toluene, resulting in a marked decrease in its phytotoxicity, and a 3–4 times reduction of its evapotranspiration through the leaves. This strategy promises to greatly improve the efficiency of phytoremediating volatile organic contaminants, and to promote the acceptance of phytoremediation by regulatory agencies and the public.

1 INTRODUCTION

Phytoremediation, the use of vegetation for the *in situ* treatment of contaminated soils and sediments, is an emerging technology that promises effective and inexpensive clean-up of certain contaminated soil and groundwater. Phytoremediation of organic xenobiotics is based on collaboration between plants and their associated microorganisms. Degradation of organic contaminants can occur in the plant rhizosphere and *in planta*. The problem that occurs during the phytoremediation process is an insufficient degradation of the organic pollutants by the plants, leading to poisoning of the plants or volatilization of the compounds through the leaves, causing a new environmental problem. Endophytic bacteria are defined as bacteria that reside within the living plant tissue without doing substantive harm to the plant. The plant/endophyte association expresses very close interaction where plants provide nutrients and residency for bacteria, which in exchange can improve plant growth and health.

This study aims to demonstrate the contribution of endophytic bacteria in the degradation of organic pollutants during their transport in the plant vascular system. It seems reasonable to hypothesize that endophytic bacteria, possessing the genetic information required for the efficient degradation of a xenobiotic, can promote degradation as the pollutant moves through the plant vascular system. We chose toluene as a model of a moderately hydrophobic (logKo/w 2.69 at 20°C) and volatile compound. In order to demonstrate the use of genetically modified endophytic bacteria to improve the *in planta* degradation of toluene *Burkholderia cepacia* L.S.2.4, which has yellow lupine (*Lupinus luteus L.*) as a host, was selected as endophytic strain. The toluene degradation pathway on the pTOM plasmid of *B. cepacia* G4 (Shields *et al.* 1995) was introduced into strain L.S.2.4. The resulting strain,

B. cepacia VM1330, was used to inoculate the *Lupinus luteus L.* This model system was subsequently used to evaluate phytotoxicity and toluene release through the leaves.

2 MATERIALS AND METHODS

2.1 *Construction of a toluene degrading endophytic B. cepacia strain*

The nickel-kanamycin marked derivative of *Burkholderia cepacia* L.S.2.4, named strain BU0072 (Taghavi et al., 2001) was used to introduce the degradation pathway encoded on the pTOM conjugative plasmid of *Burkholderia cepacia* G4 by natural gene transfer. A representative transconjugant, *B. cepacia* strain VM1330, which had the correct genetic background, i.e. Ni^R and Km^R, and was able to grow under the appropriate selective conditions with toluene as sole carbon source, was selected for further studies.

2.2 *Inoculation of yellow lupine with B. cepacia*

B. cepacia VM1330 was grown in 284 medium (250 ml culture), with 0.2% gluconate added as carbon source, at 22°C on rotary shaker for approximately 7 days until a density of 10^{+9} cfu/ml was reached (OD_{660} of 1). The cells were collected by centrifugation, washed twice in 10 mM $MgSO_4$ and suspended in 1/10 of the original volume 10 mM $MgSO_4$ to obtain an inoculum with a cell density of 10^{+10} cfu/ml.

Seeds of *Lupinus luteus L.* were surface-sterilized for 30 minutes at room temperature in a solution containing 1% active chloride (added as a NaOCl solution) and 1 droplet Tween 80 per 100 ml solution. The seeds then were rinsed 3 times for 1 minute in sterile water and dried on sterile filter paper. To test the efficiency of sterilization, the seeds were incubated on 869 medium for 3 days at 30°C. Seeds were considered as sterile when no bacterial growth was observed. Five surface sterile seeds of *Lupinus luteus L.* were planted in a sterile plastic jar (800 ml), completely filled with sterilized perlite and saturated with 400 ml of a half-strength sterile Hoagland's nutrient solution. Subsequently, the bacterial inoculum was added to each jar at a final concentration of 10^{+8} cfu/ml Hoagland's solution. The jars were covered with sterile tinfoil to facilitate bacterial colonization and prevent contamination and dispersion of the inoculated bacteria through the air. After the seeds had germinated, perforations were made in the tinfoil and plants were allowed to grow through them over 21 days in a growth chamber (constant temperature of 22°C, relative humidity 65%, and 14/10 hour light and dark cycle, PAR (photosynthetic active radiation) 165 μmol/m^2s). The same procedure was used to inoculate *Lupinus luteus L.* with the *B. cepacia* strains BU0072 and G4.

2.3 *Recovery of endophytic bacteria*

Plants were harvested after 21 days. Roots and shoots were treated separately. Fresh root and shoot material was vigorously washed in distilled water for 5 minutes end surface-sterilized. After sterilization, the roots and shoots were macerated in 10 ml 10 mM $MgSO_4$ using a Polytron PT1200 mixer. 100 μl samples were plated on different selective and non-selective media to test for the presence of the endophytes and their characteristics.

2.4 *Effect of toluene on the growth of inoculated and non-inoculated plants*

Three weeks old *Lupinus luteus* L. plants were grown hydroponicaly, settled in two compartment glass cuvets, so that shoots in the upper compartment and roots in the lower compartment were completely separated with no gas exchange between them, except through the stem (figure 3 and 4). The lower compartment was filled with sterile, half-strength Hoagland's solution. Toluene was added to a final concentration of 0, 100, 500 or 1000 mg/ml to the lower compartment of the cuvet. The cuvets with plants were placed in a growth chamber with constant temperature 22°C and 14/10 hours light/dark cycle. Each compartment was connected with a synthetic air source with an inflow of 1 liter per hour. After 96 hours the plants were harvested and the phytotoxic effects of toluene under the different conditions were examined by determining the increase in plant biomass. The growth indexes were calculated as the difference in plant fresh weight between the onset of experiment and after 96 hours exposure to different concentrations of toluene. Figure 2 shows the obtained results. The shoots and roots of the plants were separately sterilized, rinsed and macerated and samples of 100 μl were plated on different media.

2.5 *Toluene degradation and evapotranspiration*

In the experiment were toluene was added at a concentration of 100 mg/l, the amount of toluene evapotranspirated through the aerial parts of the plant and toluene removal from the Hoagland's solution were measured. In order to capture any transpired or volatilized toluene, two-serial linked Tenax traps were inserted in the out flow of each compartment. Between the cuvets and the Tenax traps, a column filled with $CaCl_2$ was installed as a water trap in order to prevent condensation of water in the Tenax traps. The whole experiment was running for 96 hours, and toluene concentration in the traps was determined by

GC-MS. All experiments were performed in triplicate to allow statistical analysis of the data using ANOVA.

3 RESULTS AND DISCUSSION

3.1 *Inoculation of yellow lupine with B.cepacia*

To test if *B. cepacia* VM1330, BU0072 and G4 were able to colonize yellow lupine as their host plant, inoculation experiments were performed. After 21 days, plants were harvested, roots and shoots were separated, surface sterilized, rinsed and macerated. The total number of specific bacteria in the mixture as well as their specific growth characteristics were determined on different selective media. The results are summarized in table 1.

All three *B. cepacia* strains could be isolated from yellow lupine. Bacteria isolated from the shoots and roots of yellow lupine inoculated with VM1330 were, as expected, able to grow on all three selective media. When the plants were inoculated with BU0072 (Tol^-) no toluene degrading bacteria were found. From plants inoculated with strain G4, bacteria could grow on the medium with toluene, but only in the absence of Ni and Km. For the control plants without inoculum no bacteria were found on the media used, except on non-selective medium (284 + gluc). This shows that despite the surface sterilization endogenous endophytes remain present in the plants.

Using REP-PCR (results not shown) we demonstrated that the bacteria isolated on the selective media had the same genetic fingerprints as *B. cepacia* G4, BU0072, and VM1330, respectively. In addition, the presence of *nre* (strain BU0072), pTOM (strain G4), and both *nre* and pTOM (strain VM1330) was determined by PCR, confirming that the three strains had colonized the plants. Bacteria isolated from the control plants showed a different genetic fingerprint from strains BU0072, G4, and VM1330. These bacteria, which were not further characterized, also were found after inoculation with *B. cepacia* BU0072 and G4.

3.2 *Effect of toluene on the growth of inoculated and non-inoculated plants*

We examined the effect of toluene on the growth of plants inoculated with *B. cepacia* VM1330, BU0072, or G4 and compared it with that of non-inoculated controls. The growth indexes were calculated as the difference in plant's fresh weight between the onset of the experiment and after 96 hours exposure to different concentrations of toluene (figure 1).

A clear effect of the toluene concentration on the colonization efficiency by the different endophytic strains could not be observed (results not shown).

In the absence of toluene, plants inoculated with *B. cepacia* G4 show a significant decrease in biomass production as compared to the control plants and plants inoculated with *B. cepacia* BU0072 and VM1330 (figure1). This indicates that high numbers of the environmental *B. cepacia* G4 strain, which is not a natural endophyte of yellow lupine, has a negative effect on plant development. When the plants and bacteria are incubated in the presence of toluene, we

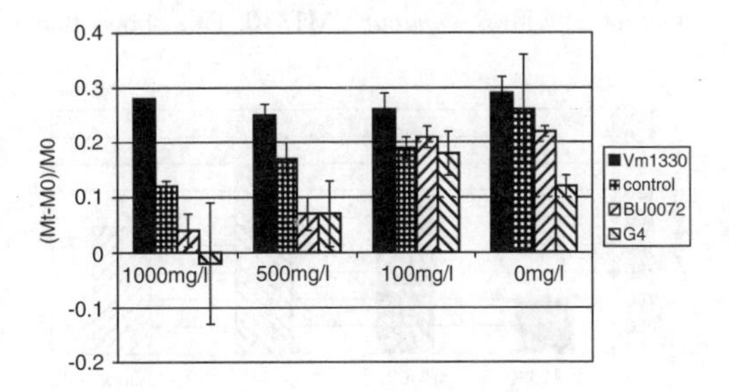

Figure 1. Difference in biomass between reinoculated and control plants before and after addition of toluene.

Table 1. Number of bacterial colonies isolated from roots and shoots from Lupinus luteus plants inoculated with *B. cepacia* strains VM1330, BU0072 and G4. As control plants without inoculum were analyzed. The number of bacteria is expressed per gram fresh weight. Numbers between brackets are the numbers of different morphological types of bacteria as observed visually. Data are the average of 3 experiments.

Inoculum	Plant part	284 + gluc	284 + Ni + Km + gluc	284 + Ni + Km + tol	284 + tol
No	Shoot	$2.3\ 10^2$ (2)	0	0	0
	Root	$1.7\ 10^3$ (3)	0	0	0
VM1330	Shoot	$6.9\ 10^3$ (1)	$3.8\ 10^2$ (1)	$5.8\ 10^2$ (1)	$4.3\ 10^2$ (1)
	Root	$9.5\ 10^3$ (1)	$2.2\ 10^2$ (1)	$1.7\ 10^2$ (1)	$1.9\ 10^2$ (1)
BU0072	Shoot	$1.3\ 10^4$ (2)	$2.2\ 10^2$ (1)	0	0
	Root	$1.5\ 10^3$ (3)	$1.5\ 10^3$ (1)	0	0
G4	Shoot	$5.7\ 10^4$ (2)	0	0	$1.0\ 10^4$ (1)
	Root	$7.8\ 10^4$ (2)	0	0	$1.5\ 10^2$ (1)

observe that increasing levels of toluene result in increasing phytotoxicity, as measured by the growth indexes. For plants inoculated with the endophytic *B. cepacia* strain VM1330, which is able to efficiently metabolize toluene, no decrease in growth is observed as compared to the control situation without toluene. This suggests that strain VM1330 is able to assist its host plant in overcoming the phytotoxicity of toluene. Plants inoculated with *B. cepacia* BU0072 and G4 suffer strongly from toluene toxicity, indicating that the combination of natural endophytic behavior plus toluene degradation is required to protect the plant against toluene phytotoxicity.

3.3 *Toluene degradation and evapotranspiration*

After adding toluene at a sub-phytotoxic concentration of 100 mg/l, we measured the amount of it evapotranspirated through the aerial parts of the plant (upper compartment) and also its removal from Hoagland's nutrient solution (lower compartment) using GC-MS (figures 2 and 3).

Compared to the control plants as well as plants inoculated with *B. cepacia* BU0072 or G4, the amount of toluene that was released in the upper compartment was 3 to 4 times lower for those plants that were inoculated with *B. cepacia* VM1330. This shows that

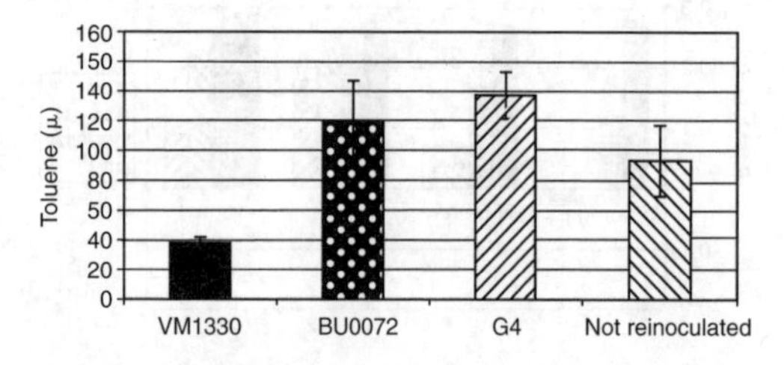

Figure 2. Amount of toluene in μg detected in Tenax traps connected with the upper compartment (containing the aerial part of the plant) determined by GC-MS.

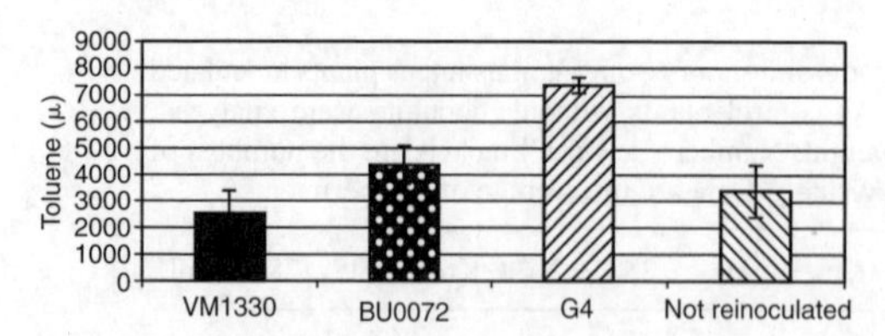

Figure 3. Amount of toluene in μg detected in Tenax traps connected with the lower compartment (containing the roots of the plant) determined by GC-MS.

this toluene degrading endophytic strain not only protects its host plant against toluene phytotoxicity, but also is accompanied by a significant decrease in toluene evapotranspiration. No significant differences in the concentrations of evapotranspirated toluene were observed between plants inoculated with BU0072 or G4, and the non-inoculated control plants.

The amount of toluene that evaporated from the Hoagland's solution in the gas-phase of the lower compartment turned out to be significantly higher for those plants inoculated with *B. cepacia* G4. The lowest amount of residual toluene was found for plants inoculated with *B. cepacia* VM1330. Therefore we conclude that the combination of the endophytic strain *B. cepacia* VM1330 and its host plant, yellow lupine, results in improved degradation of toluene and a decrease of toluene phytotoxicity and toluene release by evapotranspiration.

4 CONCLUSIONS

Our results show that an engineered endophytic bacterium can improve the phytoremediation of an organic xenobiotic. We demonstrated that an endophytic bacterium equipped with the appropriate degradation pathway not only protects its host plant against the phytotoxic effect of an environmental pollutant, but also improves its overall degradation, resulting in a decreased evapotranspiration of the contaminant to the environment.

ACKNOWLEDGEMENTS

This work was supported by the European Commission, Fifth Framework Program, Quality of Life, grant n° QLK3CT200000164 entitled "Endegrade" and by Ford Motor Co, in Genk, Belgium.

REFERENCES

Shields, M.S., Reagin, M.J., Gerger, R.R., Campbell, R. and Somerville, C. *TOM, a New Aromatic Degradative Plasmid from Burkolderia (Pseudomonas) cepacia G4*. Applied Environmental Mikrobiology 61, 1352–1356 (1995).

Taghavi, S., Delanghe, H., Lodewyckx, C., Mergeay, M. and van der Lelie, D. *Nickel-Resistance-Based Minitransposons: New Tools for Genetic Manipulation of Environmental Bacteria.* Applied and Environmental Microbiology 1015–1019 (2001).

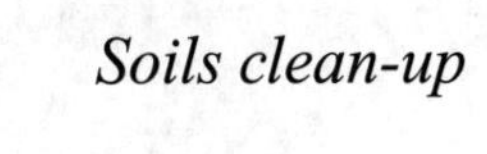
Soils clean-up

European Symposium on Environmental Biotechnology, ESEB 2004 - Verstraete (ed)

Benzene degradation coupled with chlorate reduction

N.C.G. Tan, W. van Doesburg & A.J.M. Stams
Laboratory of Microbiology, Wageningen University and Research center, Hesselink van Suchtelenweg, Wageningen, The Netherlands

A.A.M. Langenhoff & J. Gerritse
TNO Environment Energy and Process Innovation, Apeldoorn, The Netherlands

ABSTRACT: A chlorate-reducing enrichment culture was able to degrade benzene. The initial amount of 150-μM benzene was degraded in 100 days. The rate of benzene degradation increased after re-addition of benzene, and an active enrichment culture was obtained. This culture did not degrade benzene if chlorate was omitted and did not use nitrate or perchlorate as electron acceptor for the benzene degradation. Molecular analyses of a part of the 16S rDNA with DGGE revealed at least 10 amplicons from various bacterial strains. Cloning and sequencing of the dominant DGGE band revealed the presence of 16S rDNA of two strains. One clone had 99% homology with the 16S rDNA with three different aerotolerant nitrate reducers, i.e. *Acidovorax* and *Acidophilus* spp. The other clone had 92% homology with the 16S rDNA of *Dechloromonas*, a chlorate-reducing organism.

1 INTRODUCTION

Mobile aromatic hydrocarbons like benzene, toluene, ethylbenzene and the xylenes are often found in nature and cause pollution of soil and groundwater. The contamination of the environment with aromatic hydrocarbons is observed at many sites, especially those related with petrochemical activities like refinery and gasoline stations. Due to the relative high solubility of these aromatic hydrocarbons environmental contamination occurs mainly in the anoxic zones of the environment. Therefore, anaerobic bioremediation is an attractive remediation technique for such polluted soil sites. The bottleneck in the application of anaerobic techniques is the supposed poor anaerobic biodegradability of benzene. However, evidence for anaerobic benzene degradation is growing (Anderson et al. 1998; Burland & Edwards 1999; Kazumi et al. 1997; Lovley 2000; Phelps et al. 1996; Ulrich & Edwards 2003; Weiner & Lovley 1998).

Under chlorate-reducing conditions, chlorate is converted into chlorite if a suitable electron donor is available. Chlorite is dismutated into chloride and molecular oxygen and therefore chlorate reduction yields valuable oxygen under "anoxic" conditions (Logan 1998). Microorganisms can use the formed oxygen to activate the aromatic ring of benzene and produce an easier degradable substrate for anaerobic microorganisms. The first isolated anaerobic benzene-degrading organism is a nitrate- and chlorate-reducing microorganism (Coates et al. 2001).

2 AIM

The aim of this research is to get insight into the occurrence of anaerobic benzene oxidation at polluted sites, isolate (or enrich) microorganisms involved in benzene biodegradation, and study their physiological and phylogenetic properties. For this purpose many different (polluted) sources were tested under several different terminal electron-accepting conditions, like methanogenic, sulfate, nitrate and chlorate reducing conditions. Here, we describe anaerobic benzene degradation with chlorate as electron acceptor.

3 MATERIALS AND METHODS

Batch experiments were performed in 117-ml bottles filled with 40-ml mineral medium especially designed for chlorate reducing organisms (Wolterink et al. 2002). The medium did not contain any reducing agents like sodium sulfide, and instead sodium sulfate was added in equal amount based on molar concentration. The gas phase of the batches consisted of N_2/CO_2 (80/20).

Benzene was added from an anaerobic sterile stock solution (20 mM), and final concentrations ranged from 0.1 to 1.0 mM. The initial batches were inoculated with an enrichment culture (2.5% v/v) obtained by J. Gerritse at TNO. Heat sterilized, non-inoculated controls, and controls without chlorate were included. Furthermore, benzene degradation was also tested with different electron acceptors like nitrate or perchlorate to see if these electron acceptors were used by the enrichment culture to degrade benzene.

Benzene was measured using a gaschromatograph (Chrompack 436, Chrompack Packard BV, the Netherlands) with a capillary SIL 5CB (10 m, 0.53 mm, 2-μm bead size) column and a FID detector. The column temperature was 50°C. Chlorate and chloride were analyzed by HPLC as described before (Scholten & Stams 1995).

4 RESULTS AND DISCUSSION

This paper presents the results of an anaerobic culture, which degraded benzene coupled with chlorate. This culture originated from a TNO enrichment culture. The added benzene was degraded within hundred days (Fig. 1). As expected, re-addition of benzene led to faster degradation of benzene, clearly indicating the development of an enrichment culture.

According to equation (1), 5 mole of chlorate would be consumed and 5 mole of chloride produced per mole of benzene degraded. At the end of the experiment (day 161) this ratio was respectively 6.3 and 3.4 (Fig. 1). The chlorate consumption and chloride production is shown in Figure 2.

$$C_6H_6 + 5\ ClO_3^- + 3\ H_2O \rightarrow 6\ HCO_3^- + 5\ Cl^- + 6\ H^+ \quad (1)$$

No degradation of benzene was observed in sterile controls (data not shown), without chlorate, with nitrate or perchlorate as electron acceptor (Fig. 3). These experiments were preformed with a fourth generation of the enrichment culture. Active batches with a similar amount of inoculum showed also benzene degradation within the first four days, due to the presence of some chlorate originating from the inoculated material. After this chlorate was consumed no further degradation of benzene was observed. The results from this experiment clearly show that chlorate is required for the degradation of benzene and cannot be replaced by nitrate or perchlorate which are closely related terminal electron acceptors.

The benzene degradation rate in an actively degrading first generation enrichment was also determined (Fig. 4). The slope of the active batch was calculated and a degradation rate of 1.65-mM benzene per day was determined. This value is 22–1650 times higher than anaerobic degradation rates reported in the literature (Ulrich & Edwards 2003; Coates et al. 2003) and comparable with aerobic degradation rates (Reardon et al. 2000).

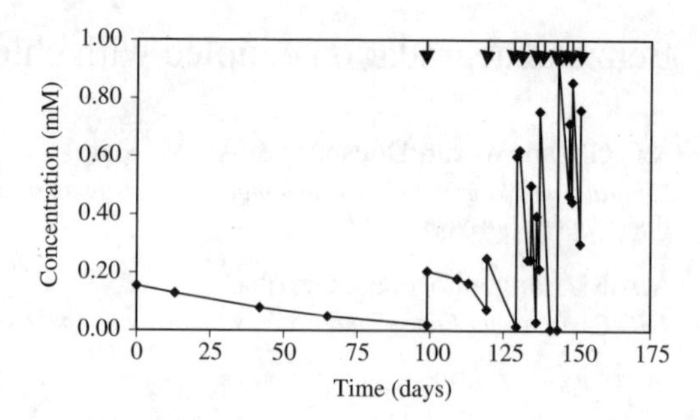

Figure 1. Benzene degradation coupled with chlorate reduction (arrows indicate re-addition of benzene; averages of duplicates).

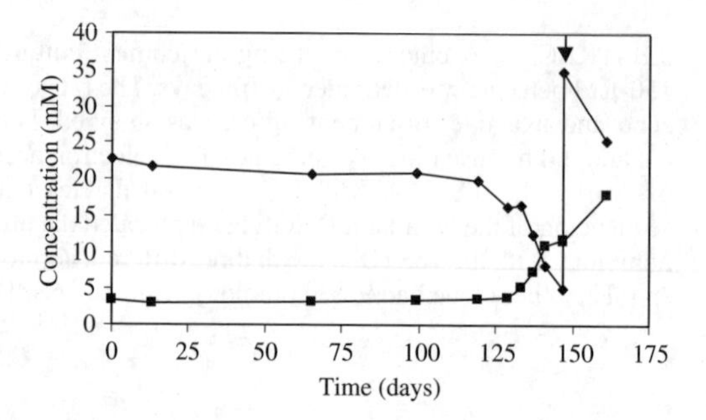

Figure 2. Chlorate (◆) reduction and chloride (■) production coupled to benzene degradation in time (arrow indicates re-addition of chlorate; averages of duplicates).

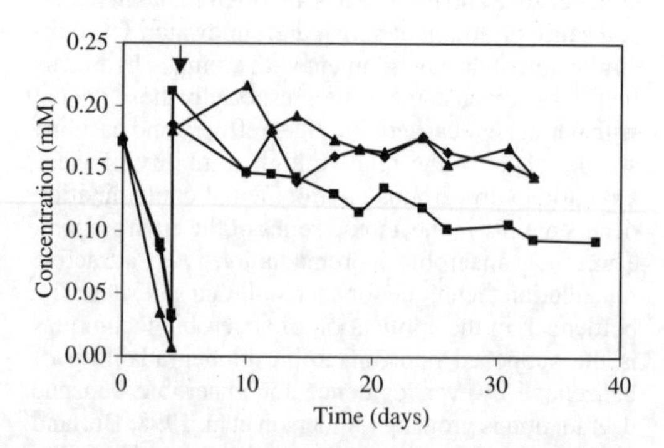

Figure 3. Benzene degradation in the controls with omission of chlorate (◆); with nitrate (■) and perchlorate (▲) as electron acceptor (arrow indicates re-addition of benzene).

Molecular analyses of the 16S rDNA showed the presence of at least 10 amplicons from the bacteria in the enrichment culture, with one dominant band on

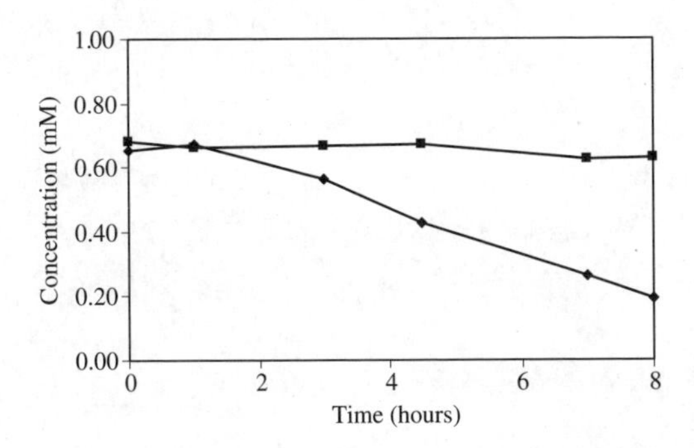

Figure 4. Benzene degradation followed during a day (average of six active batches (◆); omission of chlorate (■)).

DGGE. Analyses of this band revealed that it consisted of 16S rDNA fragments of two organisms. Both fragments were cloned and sequenced. One of them had 99% homology with three organisms, *Acidovorax avenae* (an oxygen tolerant denitrifying phenol degrading bacteria), *Acidovorax aerodenitrificans* (an aerobic nitrate reducing bacteria) and *Alicycliphilus denitrificans* (a cyclohexanol-degrading, nitrate-reducing betaproteobacterium (Mechichi et al. 2003)). The other clone showed 92% homology with a *Dechloromonas* sp. LT-1.

ACKNOWLEDGEMENT

The project (835.80.009) was financed by NWO-ALW TRIAS.

REFERENCES

Anderson, R.T., Rooney-Varga, J.N., Gaw, C.V. & Lovley, D.R. 1998 Anaerobic benzene oxidation in the Fe(III) reduction zone of petroleum-contaminated aquifers *Environmental Science and Technology* 32, 1222–1229

Burland, S.M. & Edwards, E.A. 1999 Anaerobic benzene biodegradation linked to nitrate reduction *Applied and Environmental Microbiology* 65, 529–533

Coates, J.D., Chakraborty, R., Lack, J.G., O'Connor, S.M., Cole, K.A., Bender, K.S. & Achenbach, L.A. 2001 Anaerobic benzene oxidation coupled to nitrate reduction in pure culture by two strains of Dechloromonas *Nature* 411, 1039–1043

Coates, J.D., Chakraborty, R. & McInerney, M.J. (2002) Anaerobic benzene biodegradation – a new era. *Research in Microbiology* 153, 621–628

Kazumi, J., Caldwell, M.E., Suflita, J.M., Lovley, D.R. & Young, L.Y. 1997 Anaerobic degradation of benzene in diverse anoxic environments *Environmental Science and Technology* 31, 813–818

Logan, B.E. 1998 A review of chlorate- and perchlorate-respiring microorganisms *Bioremediation Journal* 2, 69–79

Lovley, D.R. 2000 Anaerobic benzene degradation *Biodegradation* 11, 107–116

Mechichi, T., Stackebrandt, E. & Fuchs, G. 2003 *Alicycliphilus denitrificans* gen. nov., sp nov., a cyclohexanol-degrading, nitrate-reducing beta-proteobacterium *International Journal of Systematic and Evolutionary Microbiology* 53, 147–152

Phelps, C.D., Kazumi, J. & Young, L.Y. 1996 Anaerobic degradation of benzene in BTX mixtures dependent on sulfate reduction *FEMS Microbiology Letters* 145, 433–437

Reardon, K.F., Mosteller, D.C. & Rogers, J.D.B. 2000 Biodegradation kinetics of benzene, toluene, and phenol as single and mixed substrates for *Pseudomonas putida* F1. *Biotechnology and Bioengineering* 69, 385–400

Scholten, J.C.M. & Stams, A.J.M. 1995 The effect of sulfate and nitrate on methane formation in a freshwater sediment *Antonie van Leeuwenhoek* 68, 309–315

Ulrich, A.C. & Edwards, E.A. 2003 Physiological and molecular characterization of benzene-degrading mixed culture *Environmental Microbiology* 5, 92–102

Weiner, J.M. & Lovley, D.R. 1998 Rapid benzene degradation in methanogenic sediments from a petroleum-contaminated aquifer *Applied and Environmental Microbiology* 64, 1937–1939

Wolterink, A.F.W.M., Jonker, A.B., Kengen, S.W.M. & Stams, A.J.M. 2002 Pseudomonas chloritidismutans sp. nov., a non-denitrifying, chlorate-reducing bacterium *International Journal of Systematic and Evolutionary Microbiology* 52, 2183–2190

European Symposium on Environmental Biotechnology, ESEB 2004 - Verstraete (ed)
© 2004 Taylor & Francis Group, London, ISBN 90 5809 653 X

Biodegradation of xenobiotics

M. Magony, K. Perei, I. Kákonyi & G. Rákhely
Department of Biotechnology, University of Szeged, Szeged, Hungary

K.L. Kovács
Institute of Biophysics, Biological Research Centre, Hungarian Academy of Sciences, Szeged, Hungary

ABSTRACT: A bacterium capable to grow on sulfanilic acid as sole carbon, nitrogen and sulfur source has been isolated. A unique feature of this strain is that it contains the full set of enzymes necessary for the biodegradation of sulfanilic acid. Taxonomical analysis determined our isolate as *Sphingomonas subarctica sp.* The biodegradation pathway of sulfanilic acid was investigated at the molecular level. Screening the substrate specificity of the strain disclosed its capability to completely degrade six analogue aromatic compounds and oil contamination. *Sphingomonas sp.* seemed to use distinct enzyme cascades to utilize these molecules, since alternative enzymes were induced in cells grown on various substrates. However, the protein patterns appearing upon induction by sulfanilic acid and sulfocatechol were very similar to each other indicating common pathways for the degradation of these substrates. From the soluble fraction the enzyme capable to oxidize 4-sulfocatechuate could be partially purified and characterized.

1 INTRODUCTION

The hazardous wastes released into the environment are usually a mixture of pollutants and their partially degraded derivatives. The appearance of these chemicals in the natural world generally causes public health concern. Many of these compounds are incompatible with the life and their natural degradation is very slow or it does not take place at all. Efficient bioremediation technologies should therefore require a mixture of microorganisms forming synergistic consortia.

The biodegradation of toxic aromatic compounds may occur either anaerobically (Heider & Fuchs, 1997) or aerobically (Furukawa, 2000). The central compound of the former metabolism is the benzoyl-CoA, while the core intermediers in the aerobic degradations are usually dihydroxy derivatives of an aromatic ring. Only few studies are available about the degradation of sulfonated aromatic compounds. The known bacterial strains can aerobically degrade and utilize certain sulfonated azo dyes or couple of sulfonated benzene derivatives as sole carbon source (Blumel et al., 1998; Dangmann et al., 1996).

1.1 *Role of dioxygenases, hydroxylases in the biodegradation process*

Both aerobic and anaerobic microorganisms capable to decompose toxic aromatic hydrocarbons have been isolated, but much more is known about the aerobic pathways. In general, degradation proceeds in several successive steps ending in the TCA cycle (Harayama & Kok, 1992). First, para substituted, hydroxylated aromatic compounds are prepared, which are further metabolized by ring cleavage reactions. Similar reaction is hypothesized in the case of sulfanilic acid by an unknown hydroxylase enzyme. The second step is a ring fission reaction catalyzed by various dioxygenases, which may take place either between the hydroxyl groups (intradiol cleavage) or outside of these groups (extradiol splitting) (Harwood et al., 1996). The oxidative degradation of carboxylated aromatic compounds (for example: protocatechuate) may go through various pathways, like ortho (intradiol) or meta cleavage (extradiol) pathways. Usually, dioxygenases use mononuclear metal (frequently iron) cofactor. The iron requirement of effective sulfanilic acid decomposition in *Sphingomonas sp.*, was previously evidenced (Perei et al., 2001).

1.2 *Sulfanilic acid*

Sulfanilic acid or p-amino-benzenesulfonate (henceforth SA) is a typical representative of aromatic sulfonated amines widely used and manufactured as an important intermediate in the production of azo dyes, plant protective and pharmaceuticals. Its natural degradation is slow and incomplete, because of the sulfonate group, which is a strongly charged anion; hence its penetration into the intact bacteria is restricted. Sulfanilic acid is the intermedier for the synthesis of various sulfonamide drugs, noted for their strong bactericide effect. Their physiological influence is based on their ability to inhibit nucleotide biosynthesis. So far, description of efficient decomposition of sulfanilic acid has been reported for a bacterial consortium, only (Dangmann et al., 1996). The bacterial co-culture consisted of *Hydrogenophaga pallaronii* and *Agrobacterium radiobacter* strain. Protocatechuate dioxygenase type II, being able to convert sulfocatechuate to sulfomuconate was purified from both strains (Hammer et al., 1996; Contzen et al., 2001), but the other components involved in the sulfanilic acid degradation are still unknown. Recently, a single strain capable to use sulfanilic acid as sole carbon, nitrogen and sulfur source has been isolated (Perei et al., 2001).

2 MATERIALS AND METHODS

2.1 *Bacterial media and identification*

The strain was routinely grown either in LB or in minimal medium containing various organic compounds as described (Perei et al., 2001).

2.2 *DNA manipulations*

DNA manipulations were performed according to the standard methods.

2.3 *Chemicals*

Dr. Andreas Stolz kindly provided synthetic 4-sulfocatechute (Institute für Mikrobiologie der Universtät Stuttgart, Germany). The other chemicals were purchased from the standard chemical manufacturers.

2.4 *Preparation of cell free extracts*

Cells grown in media containing 10 mM SA as sole carbon source were harvested by centrifugation at 13.000 rpm for 30 min at 4°C, and washed in equal volume 50 mM Tris-HCl (pH = 8.0). Cell suspensions in 50 mM Tris-HCl puffer (pH = 8.0) were disrupted using sonicator. Cells and cell debris were removed by centrifugation as described above. Membrane and soluble fractions were separated by ultracentrifugation.

2.5 *Enzyme assays*

Conversion of SA was followed by measuring the absorbance at 248 nm in 50 mM Tris-HCl buffer (pH = 8.0). Decomposition of sulfocatechuate was followed by recording the spectra between 200–350 nm, and two peaks (at 244 nm and 288 nm) were assigned to sulfocatechol. Dioxygenase activity assay buffer contained 50 mM Tris-HCl (pH = 8.0) and 100 μM 4-sulfocatechuate.

2.6 *Protein purification*

Soluble proteins were prepared by ultracentrifugation and proteins precipitated between 25–30 m/m% $(NH_4)_2SO_4$, were collected, redissolved and dialyzed. The proteins obtained were further purified at room temperature by FPLC system (Bio-Rad Duo-Flow equipment) using Q-Sepharose fast-flow (1 ml volume) column. Purification protocol: column was washed by 6 ml buffer A, then 1 ml sample was loaded, proteins were eluted with linear gradient of 12 ml 100–80% buffer A (50 mM Tris-HCl, pH = 8.0), and 0–20% buffer B (50 mM Tris-HCl, pH = 8.0, 1 M NaCl).

3 RESULTS AND DISCUSSIONS

3.1 *Strain identification*

The previously isolated strain identified as *Pseudomonas paucimobilis* (Perei et al., 2001) was taxonomically reinvestigated. The sequenced region (between the 27. and 1492. nt) of the 16S rDNA gene of our isolate was identical to the corresponding region of *Sphingomonas subarctica* NKF1 strain. That strain was described to be able to grow on trichloro-, and pentachloro-phenols, but our bacteria could not utilize these substrates. On other hand, *Sphingomonas subarctica* NKF1 strain could not degrade sulfanilic acid, so it was concluded, that the two bacteria differ and – in addition to the similarities – they have distinct metabolic pathways.

3.2 *Substrate specificity*

AROMATIC MOLECULES. Several substituted aromatic compounds were tested as potential carbon sources. The results indicated, that the bacterium was able to mineralize SA, protocatechuate (PC), p-aminobenzoic acid (PABA), 4-sulfocatechuate (4SC), 4-hydroxybenzoate, 3,5-dihydroxybenzoate, and phthalic acid. Analysis of the soluble protein patterns obtained from cells grown in various aromatics and standard reach medium disclosed alternative enzymes induced by the various substrates (Fig. 1). Sulfanilic acid and 4-sulfocatechuate strongly induced proteins with the same mobility indicating common enzymes involved in their metabolisms. Samples obtained with p-aminobenzoic acid and 4-hydroxy benzoic acid also contained identical bands, their metabolisms likely

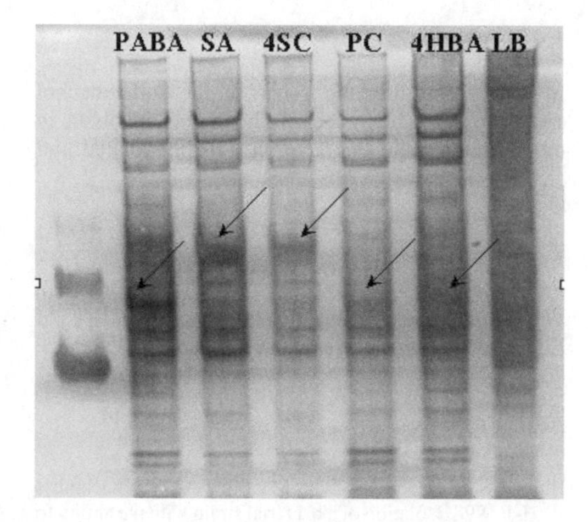

Figure 1. Soluble protein patterns of *Spingomonas sp.* cells grown on various aromatic compounds and complex medium. The protein extracts were examined on 10% native polyacrylamide gel. PABA: p-aminobenzoic acid, SA, sulfanilic acid, SC: sulfocatechuate, PC: protocatechuate, 4HBA: 4-hydroxy-benzoic acid.

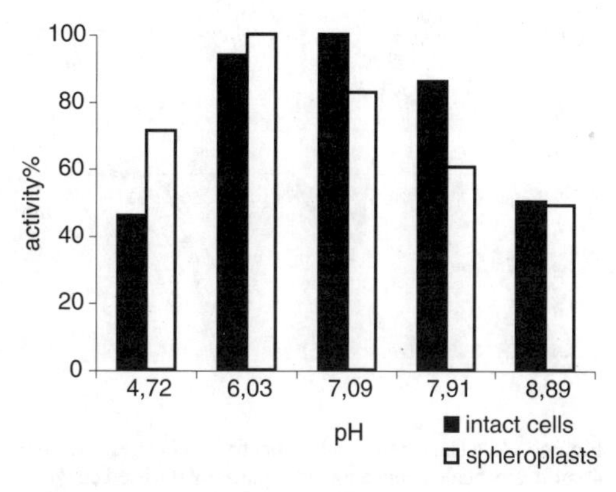

Figure 2. Effect of the pH on the sulfanilic acid conversion by intact cells and spheroplasts. Relative activities are compared, where the highest activity was considered as 100%.

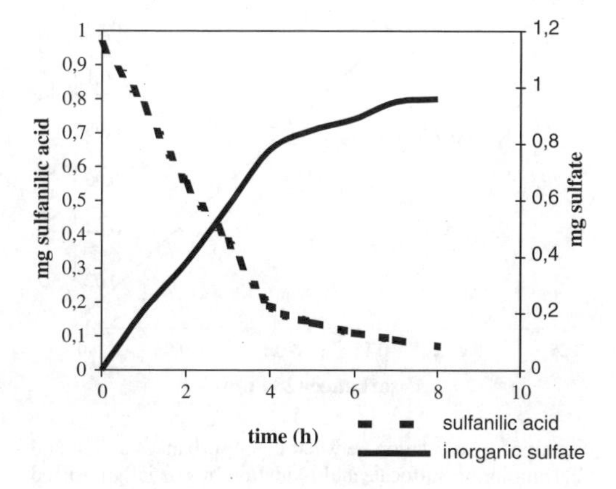

Figure 3. Sulfanilic acid decomposition is accomplished by inorganic sulfate release into the medium.

passed along a protocatechuate pathway. From these results, it was concluded, that alternative metabolic pathways were present in the cells for the degradation of the various substrates.

OTHER CARBON SOURCES. In addition to aromatic molecules, bioremediation of oil-contaminated soil was investigated by our *Sphingomonas subarctica sp.* and other soil microorganisms (*Pseudomonas, Rhodococcus sp.*) in symbiosis. Good results were obtained for couple of strain combinations, when the oil decontamination was performed under aerobic conditions. The exact biodegradation process is still under investigation. It is supposed, that the bacterial consortia produced surface activated molecules, which solubilized the apolar oil components into hydrophilic media, which process is indispensable for decontamination. Without activity, the oil aggregatum stick to flask wall even after 1 month, while in the presence of active bacteria it moved into the a stable emulsion (data not shown).

3.3 *Enzyme activity measurement*

3.3.1 *Sulfanilic acid conversion*

SA could only be decomposed with either intact cells or spheroplasts, but unfortunately, disrupted cells was incapable to convert SA. In Fig. 2. the pH dependence of the SA decomposition activity of the intact cells and spheroplasts was shown to be parallel. It is known, that NADH is and important cofactor of dioxygenases and hydroxylase enzymes. Indeed, addition of NADH and Mg^{2+} significantly increased the conversion rate (data not shown). As Fig. 3 shows, during the degradation sulfate is released into the medium, which means, that the cells utilize only small part of the sulfur taken up, the excess is released into the environment in the form of sulfate.

3.3.2 *Reaction with 4-sulfocatechuate*

The ring of 4-sulfocatechuate (the first hypothetic intermedier) is supposed to be cleaved by a second – dioxygenase type – enzyme of the SA degradation pathway (Knackmuss, 1996). The enzyme was partially purified from the soluble fraction, and its activity was measured by spectrophotometer (Fig. 4). Activity was high in the first 2 minutes, and then product seemed to inhibit the enzymatic reaction. During substrate

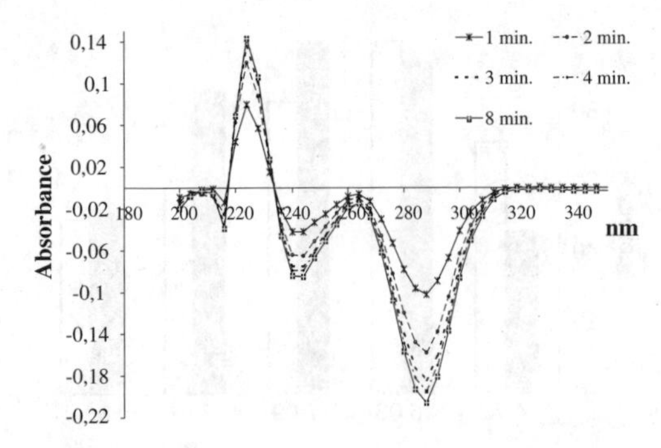

Figure 4. The temporal spectral changes during 4-sulfocatechuate conversion with partially purified enzyme.

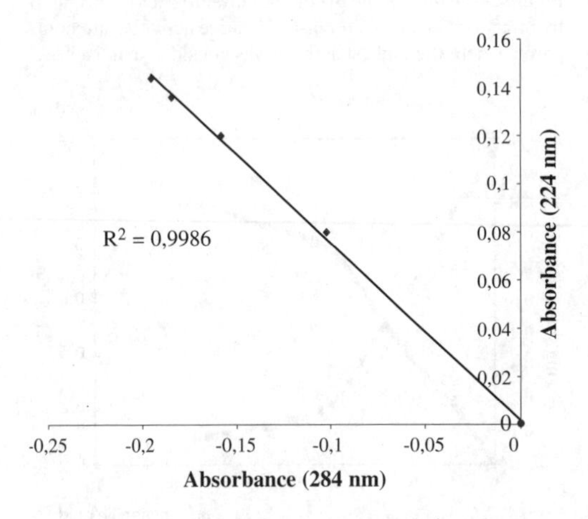

Figure 5. Correlation between the absorbances at 284 and 224 nm during sulfocatechol oxidation by partially purified enzyme.

consumption (followed at 244 and 284 nm) a product formation could be observed (at 224 nm) and the area of the peaks showed good correlation (Fig. 5) indicating a substrate – product linkage between the two peaks.

4 PLANS

In the near future, the genes of the sulfocatechol dioxygenase, other genes and proteins involved in the sulfanilic acid degradation will be isolated and characterized.

ACKNOWLEDGEMENTS

We would like thank to Hungarian Ministry of Education for financial support (OMFB-00299/2002).

REFERENCES

Blumel S., Contzen M., Lutz M., Stolz A., & Knackmuss H-J. 1998. Isolation of a bacterial strain with the ability to utilize the sulfonated azo compound 4-carboxy-4′-sulfoazobenzene as the sole source of carbon and energy. *Appl Environ Microbiol.,* 64:2315–2317.

Contzen M., Burger S., & Stolz A. 2001. Cloning of the genes for a 4-sulphocatechol-oxidizing protocatechuate 3,4-dioxygenase from Hydrogenophaga intermedia S1 and identification of the amino acid residues responsible for the ability to convert 4-sulphocatechol. *Mol Microbiol.,* 41:199–205.

Dangmann E., Stolz A., Kuhm A.E., Hammer A., Feigel B., Noisommit-Rizzi N., Rizzi M., Reuss M., & Knackmuss H-J. 1996. Degradation of 4-aminobenzenesulfonate by a two-species bacterial coculture. *Biodegradation,* 7:223–229.

Furukawa K. 2000. Engineering dioxygenase for efficient degradation of environmental pollutants. *Curr. Opin. Biotechnol.* 11:244–249.

Hammer A., Stolz A., & Knackmuss H-J. 1996. Purification and characterization of novel type of protocatechuate 3,4-dioxygenase with the ability to oxydase 4-sulfocatechol. *Arch. Microbiol.,* 166:92–100

Harayama S., Kok M., & Neidle E.L. 1992. Functional and evolutionary relationships among diverse oxygenases. *Annu. Rev. Microbiol.,* 46:565–601.

Heider J., & Fuchs G. 1997. Anaerobic metabolism of aromatic compounds. *Eur. J. Biochem.* 243:577–596.

Perei K., Rákhely G., Kiss I., Polyák B., & Kovács K.L. (2001). Biodegradation of sulfanilic acid by *Pseudomonas paucimobilis. Appl. Microbiol. Biotechnol.,* 55:101–107

European Symposium on Environmental Biotechnology, ESEB 2004 - Verstraete (ed)
© 2004 Taylor & Francis Group, London, ISBN 90 5809 653 X

Degradation of polycyclic aromatic hydrocarbons by actinomycetes

L. Pizzul, J. Stenström & M.d.P. Castillo
Department of Microbiology, SLU, Uppsala, Sweden

ABSTRACT: Fifteen strains belonging to the group actinomycetes were screened and selected for their ability to degrade the PAH phenanthrene (metabolically or cometabolically) and to produce biosurfactants. Five selected strains were then inoculated into spiked soil contaminated with anthracene, phenanthrene, pyrene and benzo(a)pyrene. Rapeseed oil was used as cosubstrate. After 15 days *Rhodococcus* sp. DSM 44126 degraded 92% of the phenanthrene and 30% of the anthracene without cosubstrate addition. The content of pyrene and benzo(a)pyrene did not change. When rapeseed oil was present, the contents of anthracene and benzo(a)pyrene decreased more than 45% for all the treatments. The amount of phenanthrene and pyrene did not vary in presence of the oil.

1 INTRODUCTION

Polycyclic aromatic hydrocarbons (PAH) are hazardous compounds often found in high concentrations at sites associated with petroleum, coal tar, gas and wood-preserving industries. Some PAHs have carcinogenic and mutagenic properties and they are therefore considered environmental priority pollutants by the United States Environmental Protection Agency and the European Community (Wattiau 2002).

Microbial degradation and transformation are the principal processes in the removal of PAHs from the environment and bioremediation has proved to be a useful method to clean up contaminated sites (Wilson & Jones 1993).

PAHs can be degraded both metabolically and cometabolically. Low molecular weight compounds are used as the sole source of carbon and energy by numerous microorganisms (Wattiau 2002, Weisenfels et al 1990). Mineralization of PAHs containing more than three fused aromatic rings is less common (Hwang & Cutright 2002, Kanaly & Harayama 2000, Walter et al 1991) and in some cases only possible in the presence of a cosubstrate (Hwang 2002, Kanaly & Harayama 2000).

PAHs have low water solubility, bind tightly to soils and are trapped within soil micropores. This results in a reduction of their bioavailability, often the rate-limiting factor for biodegradation. One of the means to stimulate desorption and dissolution of these compounds in the soil is the use of surfactants, which increase their apparent solubility by reducing surface tension and forming emulsions. Surfactants can be synthetic or produced by microorganisms (biosurfactants).

Actinomycetes have an important role as agents of biodegradation in the environment (Mc Carthy & Williams 1992). The ability of members of this group to degrade PAHs has been studied mostly in the genus *Mycobacterium* (Wattiau 2002, Wick 2002, Cerniglia & Heitkamp 1990), but other genera might be of interest in bioremediation. Besides, some genera are able to produce biosurfactants, especially when growing on water-immiscible substrates (Cristofi & Ivshina 2002). These two characteristics of the group could be combined and optimised for the successful cleaning up of contaminated soils.

The present study consisted of two parts:

a) an initial screening and selection of different actinomycetes for their ability to degrade PAHs (metabolically and/or cometabolically) and to produce biosurfactants.
b) the use of the selected strains for degradation of a mixture of anthracene, phenanthrene, pyrene and benzo(a)pyrene in spiked soil, with and without the presence of rapeseed oil.

2 MATERIALS AND METHODS

2.1 *Microorganisms*

Gordonia rubripertincta DSM 46038, *Mycobacterium vanbaalenii* DSM 7251, *Rhodococcus* sp. DSM 44126, *R. erythropolis* DSM 1069, DSM 312, DSM 43066,

DSM 43135 and DSM 44306, *R. opacus* DSM 43251, *R. rhodochrous* DSM 11097 and DSM 46045, and *Streptomyces setonii* DSM 41780 were obtained from the German Collection of Microorganisms and Cell Cultures (DSMZ). *R. erythropolis* TA57 and *Arthrobacter globiformis* were isolated from soil in Denmark (Andersen et al. 2001) and England, respectively. *Gordonia* sp. APB was isolated from soil contaminated with chlorophenols.

2.2 *Screening for degradation of phenanthrene*

Fifteen grams of glass beads (5 mm diameter) were added to 50-ml tubes, and autoclaved for 20 min at 121°C. Phenanthrene in acetone was added on the glass beads to reach a final concentration in the medium of 100 ppm. Acetone was allowed to evaporate.

Bacteria were first grown in GYM Streptomyces medium at 30°C and 150 rpm for 3 days. A portion of the culture was transferred to the minimal salt medium (MSM) to a final dilution of 1:1000 (resulting in a bacterial concentration of 10^5 CFU/ml). MSM was adapted from Mandelbaum et al. (1993), using 1 g of KNO_3 instead of atrazine. Glucose (0.2%) or hexadecane (0.1%) was used as the additional carbon source for testing cometabolic degradation. Ten ml of the medium was added to the tubes containing phenanthrene. The tubes were placed on a shaking table at 40 ppm, in horizontal position leaning slightly to avoid contact between the liquid and the cotton plug and incubated during 14 days at 30°C. Phenanthrene was extracted by adding 10 ml of toluene and shaking the tubes for one hour. A 1-ml sample was analyzed by gas chromatography.

2.3 *Screening for biosurfactants production*

Microorganisms were grown as mentioned above, but the tubes were replaced by 100 ml-Erlenmeyers, containing 200 g of glass beads and 50 ml of MSM. Rapeseed oil (0.1%) was used as the only carbon source. The biosurfactant production was estimated by measuring the decrease in surface tension and the production of emulsions of the whole culture and the supernatant after 24 and 168 h.

Surface tension was measured using a Krüss Educational Tensiometer K6. The emulsification activity was tested by mixing 2 ml of sample with 2 ml of hexane, shaking during 2 min and registering the formation of an emulsion at the interface.

2.4 *Degradation of PAHs in soil by selected strains*

Agricultural soil was contaminated with 50 ppm of each of the following PAHs: anthracene (ANT), phenanthrene (PHE), pyrene (PYR) and benzo(a)pyrene (BaP). Rapeseed oil (0.1 g oil in 10 g of soil) was used as cosubstrate.

Microorganisms grew during 2 days in GYM *Streptomyces* medium at 150 ppm at 30°C. The culture was diluted once in sterile tap water and added to the soil (2 ml in 10 g).

The soil was incubated during 14 days at 30°C and the water content was kept at 60% of the water holding capacity.

PAHs were extracted adding 10 ml toluene and 10 ml 0.05 M sodium pyrophosphate and shaking during 16 h. The extracts were centrifuged during 20 min at 2000 rpm and cleaned in an alumina column. A 1-ml sample was analysed by gas chromatography.

3 RESULTS AND DISCUSSION

3.1 *Screening and selection of microorganisms*

Nine of the strains degraded PHE to some extent (Table 1) and degradation was usually enhanced by the presence of a cosubstrate. Only *Rhodococcus* sp. DSM 44126 and *R. erythropolis* DSM 44306 used PHE as the sole source of carbon. A specificity for the cosubstrate was observed: strains that were able to break down the PAH in the presence of glucose, did not degrade it with hexadecane and vice versa. *Rhodococcus* sp. DSM 44126 metabolized completely the PHE after 14 days when growing on glucose.

The strains differed in their characteristics associated to biosurfactant production (Table 2). The decrease in surface tension for *R. erythropolis* TA 57 was more than 20 mN/m, a value often considered as the threshold for selection of biosurfactant producing bacteria. A large emulsification activity was observed for *G. rubripertincta* DSM 46038, which could indicate the production of bioemulsifiers. *Gordonia* sp. APB produced a thick layer of foam at the interface when the sample included the whole culture but no emulsification was observed in the supernatant, indicating activity related to the cell surface.

3.2 *Degradation of PAHs in soil*

Figure 1 shows the initial concentration of PHE at time 0 and after 14 days for the treatments without rapeseed oil. *Rhodococcus* sp. DSM 44126 showed a reduction of approximately 92% of the initial concentration.

The other strains did not differ from the control without inoculum. This was not surprising because those strains had been selected for improving bioavailability and not for degradation.

The lower concentration in these treatments compared to the original level could be attributed to the activity of the indigenous microflora. The same

Table 1. Degradation of PHE (%) after 14 days incubations with and without the addition of cosubstrate.

Strain	Cosubstrate *		
	None	Glu	Hex
Arthrobacter globiformis	–	–	–
Gordonia sp. APB	–	–	21,1
G. rubripertincta DSM 46038	–	–	–
Mycobacterium vanbaalenii DSM 7251	–	–	12,5
Rhodococcus sp. DSM 44126	79,4	100	80,6
R. erythropolis DSM 1069	–	15,0	–
R. erythropolis DSM 312	–	7,6	–
R. erythropolis DSM 43066	–	–	–
R. erythropolis DSM 43135	–	–	–
R. erythropolis DSM 44306	13,9	–	9,1
R. erythropolis TA 57	–	15,5	–
R. opacus DSM 43251	–	15,9	–
R. rhodochrous DSM 11097	–	25,8	–
R. rhodochrous DSM 46045	–	–	–
Streptomyces setonii DSM 41780	–	–	–

* no significant degradation was observed.

Table 2. Production of biosurfactants measured as emulsification activity and decrease of surface tension in the supernatant by actinomycetes growing on rapeseed oil.

Strain	Emulsif. activity*	Decrease in ST (mN/m)
Arthrobacter globiformis	–	–
G. rubripertincta DSM 46038	+++	–
Gordonia sp. APB	–	–
Mycobacterium vanbaalenii DSM 7251	–	–
R. erythropolis TA 57	+	21
R. erythropolis DSM 1069	–	15,5
R. erythropolis DSM 312	–	14
R. erythropolis DSM 43066	–	–
R. erythropolis DSM 43135	–	–
R. erythropolis DSM 44306	+	–
R. opacus DSM 43251	–	11,5
R. rhodochrous DSM 11097	++	5
R. rhodochrous DSM 46045	–	–
Rhodococcus sp. DSM 44126	–	–
Streptomyces setonii DSM 41780	–	–

* no emulsification, + to +++ : increase in emulsification.

tendency was found for ANT and no changes were observed for PYR and BaP.

When rapeseed oil was added to the soil, the different treatments did not differ from the control, but about 45 to 70% of the ANT (fig 2) and BaP disappeared from the soil. This reduction coincided with the appearance of a new peak in the chromatogram that was identified as anthraquinone and it was not produced when the oil was absent. The concentration

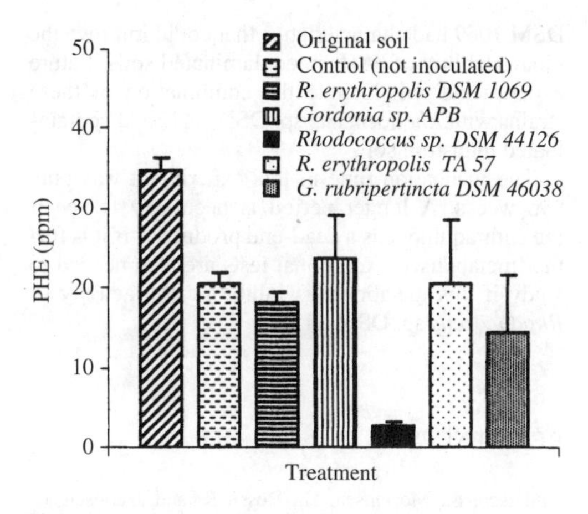

Figure 1. Concentration of PHE in soil after 14 days of incubation inoculated with the selected strains without addition of cosubstrate. The first column shows the original content of the PAH.

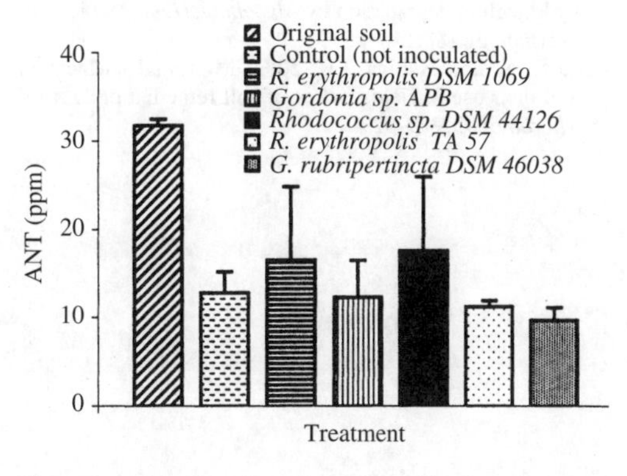

Figure 2. Concentration of ANT in soil after 14 days of incubation inoculated with the selected strains with addition of rapeseed oil. The first column shows the original content of the PAH.

of PHE and PYR remained as at the beginning of the experiment. Thus, *Rhodococcus* sp. DSM 44126 did not degrade PHE when rapeseed oil was used as a cosubstrate.

4 CONCLUSIONS

Rhodococcus sp. DSM 44126 was selected because of its ability to degrade PAHs in liquid medium. The following inoculation of a contaminated soil with this strain was successful for the degradation of phenanthrene and anthracene.

Gordonia sp. APB and *G. rubripertincta* DSM 46038, *R. erythropolis* TA 57 and *R. erythropolis*

DSM 1069 had characteristics that could improve the bioavailability of PAHs in contaminated soils. Future experiments will include the combination of these strains with *Rhodococcus* sp. DSM 44126 in contaminated industrial soil.

The incubation time in this experiment was only two weeks. A longer period is necessary to see if the anthraquinone is a dead-end product or if it is further metabolised. Additional tests are also needed to study if this metabolite is inhibiting the activity of *Rhodococcus* sp. DSM 44126.

REFERENCES

Andersen, S., Mortensen, H., Bossi, R. and Jacobsen, C. 2001. Isolation and characterisation of *Rhodococcus erythropolis* TA57 able to degrade the triazine amine product from hydrolysis of sulfonylurea pesticides in soils. *System Appl Microbiol* 24:262–266.

Cerniglia, C.E. and Heitkamp, M. 1990. Polycyclic aromatic hydrocarbon degradation by *Mycobacterium*. *Methods in enzymology* 188:148–153.

Cristofi, N. and Ivshina, I.B. 2002. Microbial surfactants and their use in field studies of soil remediation. *J Appl Microbiol* 93:915–929.

Hwang, S. and Cutright, T.J. 2002. Biodegradability of aged pyrene and phenanthrene in a natural soil. *Chemosphere* 4:891–899.

Kanaly, R. and Harayama, S. 2000. Rapid mineralization of benzo(a)pyrene by a microbial consortium growing on diesel fuel. *J Bacteriol* 182(8):2059–2067.

Mandelbaum, R.T., Wackett, L.P. and Allan, D.L. 1993. Mineralization of the s-triazine ring of atrazine by stable bacterial mixed cultures. *App Environ Microbiol* 59(6): 1695–1701.

Mc Carthy, A.J. and Williams, S.T. 1992. Actinomycetes as agents of biodegradation in the environment – a review. *Gene* 115:189–192.

Walter, U., Beyer, M., Klein, J. and Rehm, H.J. 1991. Degradation of pyrene by *Rhodococcus* sp. UW1. *Appl Microbiol Biotechnol* 34:671–676.

Wattiau, P. 2002. Microbial aspects in bioremediation of soils polluted by polyaromatic hydrocarbons. *Focus on Biotechnology* 3A:2–22.

Weisenfels, W., Beyer, M. and Klein, J. 1990. Degradation of phenanthrene, fluorene and fluoranthene by pure bacterial cultures. *Appl Microbiol Biotechnol* 32:479–484.

Wick, L.Y., de Munain, A.R., Springael, D. and Harms, H. 2002. Responses ok *Mycobacterium* sp. LB501T to the low bioavailability of solid anthracene. *Appl Microbiol Biotechnol* 58:378–385.

Wilson, S. and Jones, K. 1993. Bioremediation of soil contaminated with polynuclear aromatic hydrocarbons (PAHs): a review. *Environmental Pollution* 81:229–249.

European Symposium on Environmental Biotechnology, ESEB 2004 - Verstraete (ed)
© 2004 Taylor & Francis Group, London, ISBN 90 5809 653 X

Monitoring of dehalogenating communities in contaminated soils and in combined anaerobic/aerobic soil biotreatment reactors

F. de Ferra, C. Marsilli & G. Carpani
EniTecnologie, San Donato Milanese, Italy

M. Marzorati, S. Borin & D. Daffonchio
DISTAM, Dipartimento di Scienze e Tecnologie alimentari e Microbiologiche, Milano, Italy

ABSTRACT: We are evaluating the use of sequence markers specific for dehalogenating species to assess the dehalogenating potential of contaminated soils and to optimize soil bioremediation process parameters. Two systems contaminated with toxic and persistent chlorinated hydrocarbons have been investigated: a soil contaminated by dichloroethane (DCA) and a laboratory-scale, sequential anaerobic-aerobic process for the treatment of a soil contaminated with DDTs, chlorinated benzenes (CBs) and heavy metals, (mainly Hg and As). The microbial population profile of samples from these soils was analyzed by ARISA (Amplified Ribosomal Intergenic Spacer Analysis) and DGGE (Denaturing gradient gel electrophoresis) of 16S rRNA genes. DGGE band sequencing detected the presence of bacteria previously associated with soils polluted by halogenated hydrocarbons or present in consortia active in the removal of these compounds. A PCR specific assay using primers for *Dehalococcoides* sp. showed the presence of this genus in the anaerobic phase of the soil treatment process.

1 INTRODUCTION

Halogenated compounds are important industrial chemicals that tend to persist in the environment since they are relatively resistent both to biotic and abiotic degradation (Sims et al., 1991). These contaminants have a great toxicity and tendency to accumulate in food chains. DCA is used as an intermediate of industrial polyvinyl chloride production. This compound is the most abundant chlorinated C_2 groundwater pollutant on earth; however a reductive in situ detoxification technology for this compound does not exist (De Wildeman et al., 2003). Despite the recalcitrance of chlororganics to biodegradation, specialized dehalogenating natural strains have been retrieved from contaminated soils and are being considered for bioremediation of soils and water with low levels of contamination.

Sequence markers specific for dehalogenating species can potentially offer a tool to evaluate the dehalogenating potential of contaminated soils and to monitor and optimize soil bioremediation processes. In this work preliminary microbial analysis by culture-independent techniques was utilized on DCA contaminated soil samples and on samples from a laboratory-scale, sequential anaerobic-aerobic reactors for the treatment of soil contaminated with a mixture of chlorinated aromatics and heavy metals (mainly DDT (1,1,1-trichloro-2-2-bis(p-chlorophenyl)ethane), chlorinated benzenes (CBs) and heavy metals, (Hg and As)). In this system the enrichment of indigenous anaerobic soil microflora with discontinuous addition of the appropriate electron donor resulted in good dechlorination levels of DDT and CBs, heavy metal sequestration as sulfides and silicates and in a strong reduction of the whole soil toxicity (solid phase Microtox™ toxicity assay) in several months time (Camilli et al., 2002). Further degradation of halogenated intermediates occurs in the aerobic phase where the activity of aerobic degraders is stimulated by air sparging into the soil slurry. The microbial species involved in the different phases of the process are completely uncharacterized.

Molecular methods like ARISA (Fisher and Triplett, 1999; Ranjard et al., 2000), DGGE (Muyzer et al., 1998; Iwamoto et al., 2000) and DGGE bands sequencing have been used to investigate the microbial community shifts during the contaminated soil treatment and for the detection of strains potentially involved in the degradation of the pollutants.

2 RESULT AND DISCUSSION

Molecular analysis of microbial populations was performed on soil samples collected from two sites contaminated by organochlorinated compounds. In the first case soil in contact with DCA contaminated subsurface water was sampled from a well at two diverse depths below the surface (samples A, B). In the second case soil samples were collected from anaerobic reactors for the biological treatment of soil contaminated by a mixture of organochlorinated compounds mostly represented by DDT isomers and chlorosubstituted benzenes (hexa, penta and tetrachlorobenzenes). Analysis of the contaminants during the treatment indicated the progressive dehalogenation of the mixture of pollutants in the anaerobic phase driven by the continuous addition of a source of electron donor (Camilli et al., 2002). D, E and F samples pertain to three reactors in which treatment conditions were varied in terms of water to soil ratio in the slurry, and level of contamination. H, I and L samples were taken at a different time points from the same reactor.

Total DNA was extracted from these samples and utilized as a template for direct amplification with specific primers designed on 16S rDNA sequences specific for dehalogenating strains, and for the bacterial domain (by ARISA and DGGE analysis).

In samples derived from both sites sequences clustering with *Desulfitobacteria* (using primers DsF3 – 5′ TTAA/GTAGATGGATCCGCGTCTG 3′ and DsR5 – 5′ TTTCCGATGCAGTCCCAGG 3′) showing highest homology with a strain present in a TCE degrading soil sample, *Dehalococcoides*-like bacteria (primers DtF1 – 5′ CGCTAGCGGCGTG CCTTATGC 3′ and DtR1 – 5′ CACCCTC GGCGACTGCCTCCTT 3′) and *Clostridium* sp. (using primers ClF2 – 5′ AGCCGCGGTAATACGTAGGGAAGC 3′, ClR2 – 5′ACAAGGCCCGAGA ACGTATTCACC 3′ and CldF – 5′ TAGGTATAGGGAGTATCGAC 3′) have been identified by direct PCR and sequencing of the amplification products. All of these species have been described in association with natural processes of degradation of halogenated hydrocarbons (among others De Wildeman et al., 2003; Freeborn et al., unpubl.; Reiss and Guerra, unpubl.). Amplification with sets of primers designed on short conserved sequences of dehalogenases specific for the forementioned species did not result in the detection of any amplification product; this result is not surprising since these enzymes are known to be highly variable in aminoacidic and nucleotidic sequences, while sharing a common function, as reported by Suyama et al. (2002).

The ARISA patterns from diverse sets of samples indicated this technique as a good tool for the detection of changes in the complexity of the populations and in the identification of detectable and quantifiable trends in complex populations.

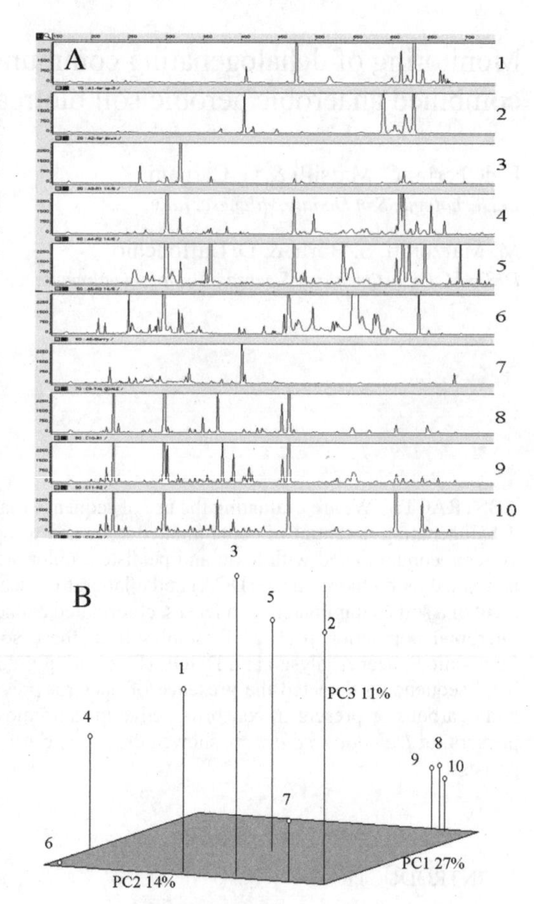

Figure 1. (A) Electropherograms derived by ARISA analysis on the following samples: (1) Sample A, (2) Sample B (DCA contaminated soil), (3) Sample D, (4) Sample E, (5) Sample F (reactors – June 2002), (6) Sample G (reference Slurry), (7) Sample H, (8) Sample I, (9) Sample L (reactors – May 2002), (10) Time zero; X axis: ITS lenghts in bp, Y axis: fluorescence intensity. (B) Result of the PCA statistical analysis: the axes represent the main three components.

Figure 1 shows the results of the ARISA analysis that conveys a picture of the abundance of species in the microbial communities.

The laboratory scale simulation apparatus of treatment of organochlorinated compounds had a richer population than the DCA contaminated sample and Principal Component Analysis (PCA) underscored the diversity of samples from the two sites.

DGGE analysis of eubacterial 16S rDNA gene sequences and further band sequencing from DGGE gels detected relevant changes in the microbial community composition depending on time and conditions for treatment. Band sequencing identified the presence of some microorganisms potentially correlated with

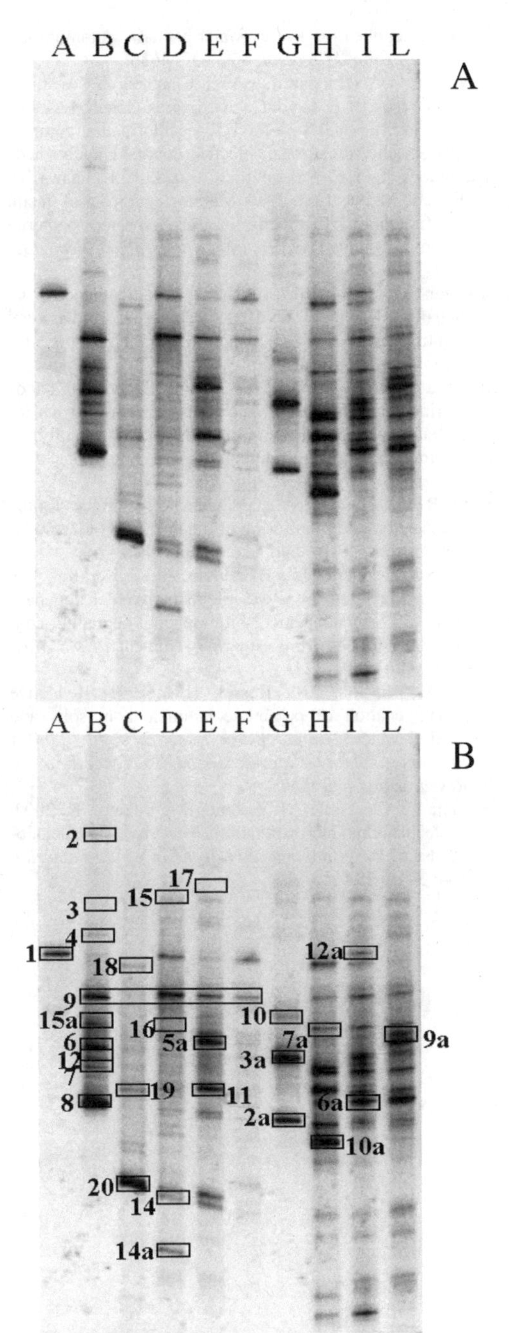

Figure 2. (A) DGGE patterns on a 7% polyacrylamide denaturing gel (30–60% urea and formamide); samples: A, B (DCA contaminated site) C. T_0, D. Reactor 1 6/02, E. Reactor 2 6/02, F. Reactor 3 6/02, G. reference Slurry, H. Reactor 1 5/02; I. Reactor 2 5/02, L. Reactor 3 5/02. (B) Same DGGE pattern in which bands cut for sequencing have been boxed.

Table 1. Summary of the sequencing results with band number, most correlated species and reference code in NCBI.

Band	Closest relative	%	Reference
1	*Janthinobacterium* sp.	92	AY247410
2	Mollicutes	98	AY133091
3	Uncultured Comamonadaceae	94	AY214207
4	Uncultured β Proteobacteria	90	AY145619
7	β Proteobacteria	97	AY146668
8	Uncultured β Proteobacteria	97	AY214206
9	Uncultured *Clostridium* sp.	95	AY122598
10	Uncultured δ Proteobacteria	94	AF050536
11	Agricultural soil clone	92	AJ252638
12	Comamonadaceae	96	AJ505858
14	Uncultured *Holophaga* sp.	93	AY289393
15	Uncultured CFB group	91	AF314430
16	Uncultured SRB	95	AF050536
17	Uncultured Bacteroides	96	AJ295642
18	*Clostridium termitidis*	94	X71854
19	Uncultured eubacterium	95	AJ292598
20	*Mycobacterium* sp.	84	AY184225
2a	Uncultured Clone AF1	90	AF143840
6a	Uncultured *Holophaga* sp.	94	AJ519376
9a	Uncultured Clone DSS20	92	AY328719
10a	Uncultured WB1	96	AF317758
12a	Uncultured β Proteobacteria	95	AF423284
14a	Uncultured Clone C0184	90	AF507400

the natural degradation of halogenated hydrocarbons. The microbial population profiling showed peculiar population structures in the DCA-contaminated soil and different microbial population composition in the last two phases of the DDT polluted soil treatment process (see lanes D, E, F and H, I, L).

Bands 2 and 9 from the DCA contaminated soil sample B were found to have the highest (>95%) sequence similarity to microorganisms identified in sites polluted by TCE and dihaloethanes, respectively. In the anaerobic reactors bands 9, 10, 16, 17, 19 and 6a were found to have a high homology with microorganisms present in soils contaminated with chlorinated solvents and polychlorinated biphenyls. Table 1, summarizes the sequencing results.

3 CONCLUSIONS

Preliminary work on microbial population composition and changes in DCA- and DDT-contaminated soils at different stages in the decontamination process has led to the detection of the presence of strains potentially involved in the dehalogenation step of the processes. The species identified by PCR-based techniques are: *Desulfitobacteria* found in chlororganics polluted soils, *Dehalococcoides*-like bacteria that can metabolically degrade PCE, TCE and TCA (Rhee et al., 2003) and *Clostridium* sp. (most similar to a strain belonging to community extracted from a TCE contaminated soil).

ARISA and DGGE analysis detected different bacterial populations in the two sets of samples, some components of which were previously identified in soils contaminated with halogenated compounds and shown to be directly involved in the decontamination process. We are currently investigating the role of these microorganisms during the decontamination processes in dependence of varying parameters known to affect the process of dehalogenation. Real time PCR quantitative assays have been designed to follow the species actively involved in the decontamination process to aid, optimize and monitor the process itself.

REFERENCES

Camilli, M. Sisto, R. D'Addario, E. Bernardi, A. Franzosi, G. 2002. Anaerobic/aerobic bioremediation of chlorinated organic- and mercuri polluted sites. In *Remediation of chlorinated and recalcitrant compounds AR Gavaskar and ESC Chen eds, 2002 Battelle Press*, Columbus, OH.

De Wildeman, S. Diekert, G. Van Langenhove, H. Verstraete, W. 2003. Stereoselective microbial dehalorespiration with vicinal dichlorinated alkanes. *Appl Environ Microbiol.* 69(9): 5643–7.

De Wildeman, S. Linthout, G. Van Langenhove, H. Verstraete, W. 2003. Lab-scale detoxification of groundwater containing 1,2-dichloroethane. *Appl Microbiol Biotechnol.* In press.

Fisher, M.M. & Triplett, E.W. 1999. Automated approach for ribosomal intergenic spacer analysis of microbial diversity and its application to freshwater bacterial communities. *Appl. Environ. Microbiol.* 65: 4630–4636.

Freeborn, R.A. Bhupathiraju,V.K. Chauhan, S. West, K. Richardson, R.E. Goulet, T.A. Alvarez-Cohen, L. 2003. Phylogenetic analysis of a TCE-dechlorinating community enriched on alternate electron donors. Unpublished.

Iwamoto, T. Tani, K. Nakamura, K. Suzuki, Y. Kitagawa, M. Eguchi, M. Nasu, M. 2000. Monitoring impact of in situ biostimulation treatment on groundwater bacterial community by DGGE. *FEMS Microbiol Ecol.* 32: 129–141.

Muyzer, G. Brinkhoff, T. Nubel, U. Santagoeds, C. Schafer, H. Wawer, C. 1998. Denaturing gradient gel electrophoresis (DGGE) in microbial ecology. *Molecular Microbial Ecology Manual.* 3.4.4: 1–27.

Ranjard L, Brothier E, Nazaret S. 2000 "Sequencing bands of ribosomal intergenic spacer analysis fingerprints for characterization and microscale distribution of soil bacterium populations responding to mercury spiking." *Appl Environ Microbiol.* 66: 5334–9.

Reiss, R.A. & Guerra, P.A. 2002. Genetic Techniques for the verification and monitoring of dihaloethane biodegradation in New Mexico aquifers. Unpublished.

Rhee, S.K. Fennell, D.E. Häggblom, M.M. Kerkhof, L.J. 2003. Detection by PCR of reductive dehalogenase motifs in a sulfidogenic 2-bromophenol-degrading consortium enriched from estuarine sediment *FEMS Micr. Ecol.* 43(3): 317–324.

Sims, J.L. Suflita, J.M. Russell H.H. 1991. Reductive dehalogenation of organic contaminants in soils and ground water. *Ground Water Issue EPA/540/4-90/054 Environmental Protection Agency and Emergency Response Research and Development.*

Suyama, A. Yamashita, M. Yoshino, S. Furukawa, K. 2002. Molecular characterization of the PceA reductive dehalogenase of *Desulfitobacterium* sp. strain Y51. *J Bacteriol.* 184: 3419–25.

European Symposium on Environmental Biotechnology, ESEB 2004 - Verstraete (ed)
© 2004 Taylor & Francis Group, London, ISBN 90 5809 653 X

Elaboration of methods of bioremediation of contaminated soils on former military locations and proving grounds in Georgia

G. Khatisashvili, G. Adamia, N. Gagelidze, L. Sulamanidze, D. Ugrekhelidze & M. Ghoghoberidze
The Durmishidze Institute of Biochemistry and Biotechnology, Georgian Academy of Sciences, David Agmashenebeli Ave., Tbilisi, Georgia

ABSTRACT: The type and level of chemical pollution of military proving grounds of Georgia has been investigated. The pollution of local sites of soil of proving grounds by lead, 2,4,6-trinitrotoluene (TNT) and waste mineral oil has been revealed. The plants, capable to accumulate of lead and compounds, promoting this process have been identified. The plants actively absorbing TNT have been revealed. The reductive and oxidative enzymes, participating in TNT degradation have been studied. The correlation between the plant nitroreductase activity and plant ability to uptake TNT from water solutions has been revealed. The strains of microorganisms, destructors of TNT and mineral oil are selected and degree of degradation of these toxicants has been studied. The cultivation condition for strains of microorganisms with the high detoxification ability has been selected.

1 INTRODUCTION

The remediation of soils, polluted by military activity, is a one of the basic problems of environmental safety. Bioremediation of the environment is an effective contemporary approach based on living organisms abilities to assimilate and transform a wide spectrum of toxicants (Salt et al., 1998, Sadowsky, 1999). Unlike the chemical methods of purification, bioremediation presents a cost-effective technology, secures maximal detoxification and long-term protection of the environment maintaining the ecological balance (Korte et al., 2000; Kvesitadze et al., 2001). From this point plants and microorganisms are particularly distinguished.

Recent study of the molecular mechanisms of xenobiotics detoxification by plant and microorganisms implies the strategy of perfection of the methods of bioremediation.

The goal of the presented research is to work out the methods of bioremediation for detoxification of contaminated by military activities territories from organic toxicants and heavy metals.

2 MATERIALS AND METHODS

2.1 *Plants*

Investigations were carried out on the annual agricultural and decorative plants: soybean (*Glycine max*), barley (*Hordeum sativum*), alfalfa (*Medicago sativa*), chickpea (*Cicer arietinum),* pea (*Pisum sativum*), ryegrass *(Lolium multiflorum)*, sunflower (*Helianthus annuus*) maize (*Zea mays*), capsicum and Bulgarian peppers (*Capsicum annuum*). Plants were cultivated under daytime illumination at 20–25°C, on running water. The 5 day old seedlings were placed in a solution of toxicant (TNT or Pb^{2+}) during 5 days.

2.2 *Microorganisms*

Microorganisms isolated from soils of military proving grounds of Georgia, contaminated with organic toxicants, were served as objects of investigation. During the search typical strains of yeasts and genera *Rhodococcus* and *Mycobacterium* from collection of microorganisms of the Laboratory of Yeasts' Biochemistry and Microbial Detoxification of the Institute of Biochemistry and Biotechnology, Georgian Academy of Sciences also have been used.

2.3 *Electron-microscopic investigation*

For studying the influence of TNT on cell ultrastructure root and leaf tips of experimental plants were excised and fixed in 2.5% glutaraldehyde, and then in 1% OsO_4. After dehydratation in ethanol of increasing concentrations the samples were embedded in Epon-Araldite resin (1.5:1.0) and poured into gelatin capsules. Thin serial sections were made using LKB III ultra microtome, stained with uranil acetate and examined in a Tesla BS 500 electron microscope.

Intracellular distribution of (1-^{14}C)-TNT was studied by a modified method of autoradiography (Zaalishvili et al., 2000). Ultra thin sections were incubated in a 0.1 mM $NiCl_2$ for 30 s, twice. Development was carried out in 2% hydroquinone for 40 sec., twice.

2.4 *Determination of enzymes activity*

The nitroreductase activity was determined according to rate of TNT reduction. For this purpose was measured untransformed TNT by spectrophotometric method at 447 nm in high alkaline area (pH > 12.2) (Oh et al., 2000).

The peroxidase activity was determined spectrophotometrically at 470 nm, according to the rate of H_2O_2-dependent oxidation of guaiacol (Gregory, 1966).

The phenoloxidase activity was determined spectrophotometrically at 420 nm according to the rate of catechol oxidation (Lanzarini et al., 1972).

2.5 *Study of TNT oxidation*

For study of TNT oxidation by plant oxidizing enzymes and chemical oxidizing agents the (C^3H_3)-TNT was used. As oxidizers applied microsomal fraction, horseradish peroxidase, tea leaves phenoloxidase, Fenton's reagent and $KMnO_4$. As a result of oxidation of methyl group, tritium atoms incorporate into the water molecules and the radioactivity of the organic phase decreases. After termination of oxidation process untransformed (C^3H_3)-TNT and their oxidation products were extracted by benzene. The benzene extract was concentrated and radioactivity measured on a scintillation counter.

2.6 *Estimation of plants ability to absorb TNT*

From TNT-containing medium on which the plants grew, every 24 hours the concentration of TNT was determined by spectrophotometric method at 447 nm in high alkaline area.

2.7 *Investigation of distribution of assimilated TNT*

Distribution of assimilated TNT and its metabolites in the plant organs and in the fractions of low-molecular and high-molecular compounds has been studied. The 5 day old seedlings of soybean and maize were placed in the nutrient area containing 0.5 mM of (1-^{14}C)-TNT. After 5 days exposition roots and upper parts were separately fixed in the boiling 80% ethanol and low-molecular weight compounds were extracted from the biomass. Insoluble residue containing high-molecular weight compounds was rinsed in alcohol, dried, and incinerated. Formed $^{14}CO_2$ was absorbed by a 30% KOH solution. Radioactivity of alkaline solution and alcoholic extract was estimated on a scintillation counter.

2.8 *Synthesis of TNT radioactive preparations*

(1-^{14}C)-TNT and (C^3H_3)-TNT were synthesized from commercial preparations of (1-^{14}C)-toluene and (C^3H_3)-toluene by a standard method. The products were purified by double recrystallization in water. Specific radioactivity of (1-^{14}C)-TNT 500 Bq/mg, and of (C^3H_3)-TNT 33 Bq/mg.

2.9 *Estimation of lead accumulation in plants*

Contents of lead was measured in air-dried upper parts and roots of plants grown on medium with Pb^{2+}, and coefficient of bioaccumulation (ratio between concentration of lead in the plant tissue and the initial lead-containing nutrient solution) was determined

2.10 *Determination of detoxification capability of microorganisms*

For the revealing of capability of the microorganism cultures to assimilate and degrade organic toxicants strains were grown on liquid or solid (agar or soil) media on circular rotator (180 r/min) at 28–30°C. As toxicants TNT (100, 200 and 300 mg/l or mg/kg) and waste motor oil (2–5% by volume) was used.

Vegetative culture grown up to the exponential phase of growth served as inoculants. The nutrient medium was inoculated with 10% of bacterial suspension.

Intensity of growth was estimated visually according to 4+ point system.

Oil degradation capability was determined according to residual contents of hydrocarbons by gravimetric method after extraction of organic fraction with light petroleum. In incubation medium TPH was determined by standard methods of gas-liquid chromatography and IR spectroscopy.

Ability of microorganisms to assimilate TNT was determined according to residual concentration of TNT by spectrophotometric method at 447 nm in high alkaline area.

2.11 *Investigation of (1-^{14}C)-TNT metabolites*

Products of transformation of TNT were identified after incubation of plants or microorganisms with (1-^{14}C)-TNT. Radioactive compounds in plant material (fractions soluble and insoluble in 80% ethanol) and biomass of microorganisms or cultivation area was analyzed by standard methods of paper chromatography and autoradiography.

2.12 *Determination of toxicants in polluted soils*

The soil samples were collected on territories of tank directrix lines, artillery shooting grounds and autodromes of military units (Adamia et al., 2003). Contents of heavy metals (Pb, Hg, Cu) in soil samples were measured by standard method of atomic absorption spectroscopy. From soil TNT was extracted by methanol and determined by standard method of HPLC. From soil waste mineral oil was extracted by chloroform and determined by gravimetric method.

3 RESULTS AND DISCUSSION

3.1 *Contamination of proving grounds*

In the first stage of investigation works were carried out for to state the type and rate of chemical contamination of the territories of locations and proving grounds of the former Soviet Army in Georgia. The results of the analyses showed that lead, copper and 2,4,6-trinitrotoluene (TNT) content in local sites significantly exceeds the limited concentrations. Besides these toxicants, the investigated territories are contaminated by mercury and waste mineral oil (Adamia et al., 2003).

3.2 *Absorption and enzymatic transformation of TNT in plants*

The ability of above mentioned plants to assimilate TNT has been investigated. According to the obtained data, all plants greatly decrease the toxicant concentration in the nutrient area (water solution of 0.1 mM TNT) during 3–5 days. Such ability was more strongly expressed by soybean which was characterized by rapid assimilation of higher concentrations of this explosive (0.6 mM).

The penetration and localization of (1-^{14}C)-TNT in cells of plants roots and leaves via electron-microscopic autoradiography has been studied. It was shown that TNT is mainly localized on membrane structures participating in transportation of reductive equivalents (membrane of endoplasmic reticulum, mitochondria, plastids), also in nuclei, nucleotides and vacuoles.

Distribution of assimilated (1-^{14}C)-TNT and its metabolites through the plant organs and between the fractions of low-molecular and high-molecular weight compounds has been studied. The radioactive label of TNT in roots was mainly incorporated in low-molecular weight metabolites, but in stems and leaves its basic part was incorporated in biopolymers. About 60% of assimilated TNT by soybean seedlings is bound with biopolymers of upper part of a plant. By the methods of paper chromatography and radioautography have been identified main metabolites formed in the process of toxicant biodegradation. It appeared that two types of metabolites conjugated with biopolymers. About 80% of metabolites contain amino groups and are the reduction products of TNT nitro groups (2-amino-4,6-dinitrotoluene, 4-amino-2,6-dinitrotoluene, 2,4-diamino-6-nitrotoluene, and 2,6-diamino-4-nitrotoluene. Other type of individual compounds is formed via TNT methyl group oxidation and basically contains carboxyl group (for example 2,4,6-trinitro-benzoic acid). Proceeding of the obtained results, in plants main part of TNT metabolites maintain benzene ring, and as they are chemically bound with biopolymers, can remain in plant in a stable and unchanged form. By distribution in the environment or food chain these metabolites can be released via hydrolysis and thus be hazardous for living organisms. Therefore upper parts of plants, used for cleaning contaminated with TNT soils, must be strictly gathered and incinerated.

The enzymatic transformation of TNT in roots was studied. It was stated that in degradation mainly takes part the nitroreductase which catalyzes reduction of TNT nitro groups. The process is intensified at presence of electron donors – NADH and NADPH. Activity of nonspecific NAD(P)H-dependent nitroreductase is basically revealed in soluble phase of cell (cytosol) and has strongly expressed inductive character at plant cultivation on TNT-containing medium.

Oxidation of (C^3H_3)-TNT by peroxidase and phenoloxidase has been studied. Comparison of activities and inductive characters of oxidative enzymes with nitroreductase, also the ratio between quantities of reductive and oxidative metabolites shows that the main pathway of TNT transformation in plant cell is a reduction of nitro groups.

The correlation between the plant nitroreductase activity and plant ability to uptake TNT from water solutions has been revealed. It was observed that the higher is the nitroreductase activity the faster is the assimilation of TNT by plant. The obtained results allow supposing that plant nitroreductase activity may serve as the biochemical criterion to select plants for phytoremediation of soils contaminated with TNT.

Thus, when realizing the results of investigations on selection of plants with the maximum potential for TNT detoxification, should be mentioned some characteristics of soybean, that allow application of this plant in phytoremediation technology: high TNT uptake potential; tolerance to high TNT concentrations; ability to accumulate most metabolites (65–70%) in upper parts of a plant; high activity of enzymes, participating in TNT detoxification.

3.3 *Lead accumulation in plants*

Due to results of experimental works it was selected plants, having high coefficient of bioaccumulation of lead (pepper – 133, ryegrass – 95, alfalfa – 95). It has been revealed that EDTA effectively promotes accumulation of lead in upper part of ryegrass: in roots contents of lead decreases for 90%, and in leaves – increased 10 times. It has been shown, that bioactive preparations – Fosnutren and Humiforte significantly increase (approximately 2.5 times) of lead accumulation in roots, also enhance of plant ability of large biomass formation and resistance to toxic effect of heavy metals.

Therefore, EDTA can be recommended for the application in technology of phytoextraction of lead, and Fosnutren and Humiforte – in technology of rhyzofiltration.

3.4 *Assimilation of organic toxicants by microorganisms*

Autochthonic microorganisms adapted to contaminated soils of military proving grounds were isolated. According to the results of analyses the number of culture colonies in 1 g dry soil was between $2.4 \cdot 10^4$–$3.2 \cdot 10^7$, which was 10-fold higher than the number of microorganisms in control samples. From freshly isolated microorganisms are received pure cultures of about 200 different strains among which cultures of rhodococcus, mycobacteria and microscopic fungi (strains of *Aspergillus, Mucor, Trichoderma, Trichothecium*) prevailed. From these cultures, also from collection of microorganisms it was revealed strains, having ability of grown on soil, contaminated by TNT and waste motor oil. We have selected more then 20 strains (bacteria, similar to *nocardia*, microscopic fungi, yeast) and their associations with the high detoxification ability. Part of these cultures can degrade 80–90% TNT in medium with 200 mg/ml toxicant; other part are destructors of mineral oil – they assimilate more of 80% waste motor oil in liquid medium and more of 50% in soil. It has been chosen cultivation conditions (temperature, nutrients, pH-optimum etc.) for selected microorganisms.

It has been carried out identification of individual low-molecular weight compounds, formed as a result of biodegradation of (1-^{14}C)-TNT by above mentioned microorganisms, having high detoxification ability. The obtained results show, that the main radioactive products of TNT transformation are organic acids and amino acids. Among amino acids prevail aromatic amino acids, as for organic acids, in most cases radioactive label of TNT is detected in fumaric and succinic acids. As known, fumaric acid is one of the products of biodegradation of benzene ring, and it is easily metabolized into succinic acid.

Proceeding from the analysis of the current data, it can be concluded that carbon skeleton of TNT molecules adsorbed by test cultures undergoes deep degradation. The initial stage of this process must be reduction of nitro groups, after which the aromatic ring of TNT molecule is used in biosynthesis of aromatic amino acids. After reduction of main part of assimilated toxicant molecules it follows their oxidation which leads to removal of amino groups and cleavage of aromatic ring and as a result organic acids are formed, they are standard cell metabolites. Thus, as a result of successive reduction and oxidation reactions complete detoxification of TNT occurs, and the atoms of this toxicant are involved in the vital processes of an organism.

ACKNOWLEDGMENT

ISTC Grant G-369 funded this work.

REFERENCES

Adamia, G., Khatisashvili, G., Varazashvili, T., Ananiashvili, T., Gvakharia, V., Adamia, T., Gordeziani, M. 2003. Determination of the Type and Rate of Soil Contamination with Heavy Metals and Organic Toxicants on the Territories of Military Proving Grounds in Georgia. *Bulletin of the Georgian Academy of Sciences* 167: 155–158.

Gregory, R.P.F. 1966. A rapid assay for peroxidase activity. *Biochem J* 101: 582–583.

Lanzarini, G., Pifferi, P.G., Zamorani, A. 1972. Specificity of an *o*-diphenol oxidase from *Prunus avium* fruits. *Phytochemistry* 11: 89–94.

Korte, F., Kvesitadze G., Ugrekhelidze D., Gordeziani M., Khatisashvili, G., Buadze, O., Zaalishvili, G., Coulston, F. 2000. Review: organic toxicants and plants. *Ecotoxicol Environ Saf* 47: 1–26.

Kvesitadze, G., Gordeziani, M., Khatisashvili, G., Sadunishvili, T., Ramsden, J.J. 2001. Some aspects of the enzymatic basis of phytoremediation. *J Biol Physics and Chemistry* 1: 49–57.

Oh, B., Sarath, G., Drijber, R.A., Comfort, S.D. 2000. Rapid spectrophotometric determination of 2,4,6-trinitrotoluene in a Pseudomonas enzyme assay. *Microbiol Methods* 42(2): 149–158.

Sadowsky, M.J. (1999). Phytoremediation: past promises and future practices. *In: Microbial Biosystems: New Frontiers. proceedings of the 8th International Symposium on Microbial Ecology*. Bell, C.R., Brylincky, M., Johnson-Green, P. (eds). Atlantic Canada Society for Microbial Ecology, Halifax, Canada.

Salt, D.E., Smoth, R.D., Raskin, I. 1998. Phytoremediation. *A Rev. Plant Physiol Mol Biol* 49: 643–668.

Zaalishvili, G. Lomidze, E., Buadze, O., Sadunishvili, T., Tkhelidze, P., Kvesitadze, G. 2000. Electron microskopic investigation of benzidine effect on maize root tip cells ultrastructure, DNA synthesis and calcium homeostasis. *Int Biodete. Biodegrad* 46: 133–140.

European Symposium on Environmental Biotechnology, ESEB 2004 - Verstraete (ed)

The determination of filamentous fungi in polluted soils as a first step in bioremediation process

M. Kacprzak, A. Szewczyk & G. Malina
Częstochowa University of Technology, Institute of Environmental Engineering, ul. Brzeźnicka Częstochowa, Poland

ABSTRACT: Bioremediation is the one of the *in situ* technologies, depending on the use of living organisms to reduce or to eliminate toxic chemicals and other hazardous wastes. The process is based on the addition of bioagents to enhance a specific biological activity or stimulation of indigenous micoflora. The paper presents the investigations focused on the occurrence of filamentous fungi in highly heavy metal contaminated soil. The soil samples were taken from different areas (vicinity of chemical plants, fast road, steel mill, wastewater treatment plant). The obtained results indicate few fungal strains presence in soils with high contents of heavy metals. *Mortierella exigua*, *Trichoderma atroviride*, *T. harzianum*, *Gliocladium roseum* and yeast-like fungi were most frequently present in Zn, Ba and Fe highly contaminated soil in vicinity of chemical plant. The *Mucor hiemalis, Penicillium chrysogenum*, *Cladosporium cladosporioides* were dominant in soil near fast road (high concentration of Fe, Pb, Ni and Mn). *P. chrysogenum*, *Chrysosporium pannorum*, *Verticillium chlamydoporium* and *T. atroviride* were strong present in soil in vicinity of steel mill with high concentration of Fe, Zn, Cr. *P. waksmani*, *P. chrysogenum*, *C. cladosporioides*, *G. roseum*, *Rhodotorula* sp., *P. canescens* and *M. hiemalis* in soil near wastewater treatment plant with high concentration of Fe, Pb, Cd and Mn.

1 INTRODUCTION

Contaminated and polluted terrains – "brownfields" present usually real environmental problem, but from the other point of view are very attractive for potential investors. Brownfield redevelopment is a complex process based on environment remediation, revitalization and finally reintegration into life cycle. Applied procedures are sometimes much compiled, time-consuming and expensive.

Bioremediation is the one of the *in situ* technologies, depending on the use of living organisms to reduce or to eliminate toxic chemicals and other hazardous wastes. In this approach very important role play native microbial populations, which are characterized by wide adaptation to stress conditions caused by pollutant factors. These are the reasons, why the use of biomass, especially bacteria, fungi, or algae, for biodegradation, stable immobilization and/or recovery of toxic contaminants, is now being considered more seriously. For bioremediation can involve indigenous microbial populations with or without nutrient supplementation (i.e. biostimulation) or selected organisms inoculated into the soil (i.e. bioaugmentation), (Vogel 1996). Fungi are known, to be able to accumulate of heavy metals, such as Pb, Cd, Cu, Zn, Ni or U, and also to degrade of PAHs or even pesticides. The fungal walls are composed of polysaccharides, proteins and lipids, which contain functional groups with potential complexion capacities. Moreover, extracellular nature of the degradative enzymes enables fungi to tolerate higher concentrations of toxic chemicals than would be possible if these compounds had to be brought into the cell.

The filamentous fungi belonging to *Mucorales* (Remacle 1990), *Rhizpous* (Volesky 1994), the species such as: *Aspergillus niger* (Castro et al. 2000, Price et al. 2001), *Fusarium flocciferum* (Delgado et al. 1998), *Trichoderma atroviride* (Errasquin and Vázquez 2003), *T. harzianum*, *Penicillium spinulosum* and *Mortierella isabelina* (Krantz-Rülcker et al. 1996), were used as a biosorbent of contaminants. The second group – white rot fungi has been also intensively studied (Baldrian 2003, Bending et al. 2002).

The paper presents the investigations focused on the occurrence of filamentous fungi in highly contaminated soil. Local strains of fungal species with high pollutant-tolerance abilities provide an opportunity for application of them as bioagents in bioremediation process.

2 MATERIALS AND METHODS

The soil samples were taken at the depth from 20 to 120 cm from the contaminated areas around damping sites at Silesia Region, Poland:

- chemical plant (Tarnowskie Góry),
- waste-water treatment plant (Częstochowa),
- steel mill (Częstochowa),
- fast road (Częstochowa).

Visible pieces of plant materials, stones and visible soil fauna were picked out and then soil was sieved (<2 mm). Metal contents in samples were measured by the atomic absorption spectrophotometer (AAS ZEEnit 60) using the standard method.

Fungi were isolated using the dilution plate method: 1 ml of 10^3 dilution was plated on Sabouraud dextrose agar containing gentamicin-chloramphenicol (Scharlau). Enumeration (Colony Forming Units – CFU) and identification of species was achieved according to the method described by Gams et al. (1980).

3 RESULTS

3.1 *Contents of metals*

The mean contents of analysed metals in the soil samples from the contaminated area are shown in Table 1.

The Zn, Ba and Fe high concentration was achieved in vicinity of chemical plant. The soil near fast road was characterized by high concentration of Fe, Pb, Ni and Mn, near steel mill – Fe, Zn, Cr, and soil in vicinity of wastewater treatment plant were characterized by high concentration of Fe, Pb, Cd and Mn.

Table 1. The mean contents of analysed metals in the soil.

	The place of sampling			
	The chemical plant	The fast road	The steel mill	The wastewater treatment plant
Metal	The metal concentration mg/g dry matter			
Cu	no data	0,0259	0,0065	0,0253
Fe	2.29	8,85	9,65	10,35
Pb	no data	0,0725	0,02605	0,0735
Cd	no data	0,0008	0,0007	0,00115
Ni	no data	0,172	0,00485	0,191
Zn	0.165	0,0081	0,081	0,0192
Mn	0.045	0,1985	0,00485	0,4305
Cr	no data	0,01655	0,204	0,0665
Ba	0.907	0,037	0,012	0,0147

3.2 *The fungal community*

The fungi recorded in tested soils are shown in Tables 2–5. In studied soils, mean number of propagules ranged from 386, 7 (CFU/g) in vicinity of chemical

Table 2. The soil fungal communities present in soil samples in vicinity of the chemical plant.

Species of fungi	CFU/g
Mortierella exigua Linnem	176,7
Trichoderma atroviride Bissett	39,3
T. harzianum Rifai	26,7
Gliocladium roseum Bain	26,7
Yeast and yeast like fungi	24,0
Alternaria alternata Keissler	10,0
Mucor hiemalis Wehmer	10,0
Fusarium sp. 1	6,7
Aspergillus flavus van Tieghem	6,7
Penicillium commune Thom	6,7
P. expansum Link ex Gray	6,7
Oidiodendron griseum Robak	6,7
Verticillium sp. 1	6,7
Chrysosporium merdiarum Link ex Grev.	3,3
Sporothrix schenckii Hektoen & Parkins	3,3
Mycelium radicis atrovirens Melin	3,3
P. canescens Soop	3,3
P. vulpinum Cook & Massee	3,3
P. montanense Christensen & Beckus	3,3
Paeciliomyces marquandii Hughes	3,3
Penicillium sp. 1	3,3
Truncatella truncata Lev. (Stayaert)	3,3
V. nigrescens Pethybr.	3,3
Total	386,7

Table 3. The soil fungal communities present in soil samples in vicinity of the wastewater treatment plant.

Species of fungi	CFU/g
P. waksmanii Zaleski	700
P. chrysogenum Bain	563
Cladosporium cladosporioides (Fres.) de Vries	413
G. roseum Bain	250
Rhodotorula sp.1	225
P. canescens Sopp	200
M. hiemalis Wehmer	138
H. grisea Traaen	125
F. solani (Mart.) Sacc.	113
P. expansum Link ex Gray	88
F. oxysporum Schlecht.	88
T. atroviride Bissett	38
A. alternata Keissler	38
G. virens Miller	25
F. culmorum Sacc.	25
Zygorhynchus moelleri Vuill.	25
T. viride Pers. ex Gray	13
M. sylvaticus (Hagem) Schipper	13
Geotrichum candidum Link.	13
Total	3093

plant to 3093 (CFU/g) in terrain of wastewater treatment plant. Together 44 fungal species were identified, from 16 to 23 species in each studied community. The species belonging to genera *Mortierella exigua, Trichoderma atroviride*, *T. harzianum*, *Gliocladium roseum* and yeast and yeast-like fungi were most frequently present in Zn, Ba and Fe highly contaminated soil in vicinity of chemical plant (Tab. 2). The *Penicillium waksmanii*, *P. chrysogenum*, *Cladosporium cladosporioides*, *Gliocladium roseum*, *Mucor hiemalis*, *T. atroviride*, *Chrysosporium pannorum*, *Verticillium chlamydoporium*, *Rhodotorula* sp., *Humicola grise* and *P. canescens* are dominant in soil near wastewater treatment plant (Tab. 3), fast road (Tab. 4) and steel mill (Tab. 5) with high concentration of Pb, Zn and Cd.

Table 4. The soil fungal communities present in soil samples in vicinity of the fast road.

Species of fungi	CFU/g
M. hiemalis Wehmer	256
P. chrysogenum Bain	222
C. cladosporioides (Fres.) de Vries	189
P. expansum Link ex Gray	78
T. atroviride Bissett	67
T. viride Pers. ex Gray	44
P. marquandii Hughes	44
Ch. pannorum (Link) Hughes	44
P. funiculosum Thom	33
Mortierella sp. 1	22
Aspergillus flavus van Tieghem	22
P. waksmanii Zaleski	22
V. chlamydosporium Goddard	11
Saccharomyces sp. 1	11
H. grisea Traaen	11
M. sylvaticus (Hagem) Schipper	11
G. candidum Link.	11
Total	1098

Table 5. The soil fungal communities present in soil samples in vicinity of the steel mill.

Species of fungi	CFU/g
P. chrysogenum Bain	950
V. chlamydospoirum Goddard	237.5
Ch. pannorum (Link) Hughes	212.5
T. atroviride Bissett	125
P. jensenii Zaleski	100
Verticillium sp. 1	100
P. waksmanii Zaleski	87.5
H. grisea Traaen	62.5
T. harzianum Rifai	50
T. viride Pers. ex Gray	37.5
M. hiemalis Wehmer	37.5
C. cladosporioides (Fres.) de Vries	25
G. roseum Bain	25
M. sylvaticus (Hagem) Schipper	12.5
F. solani (Mart.) Sacc.	12.5
F. oxysporum Schlecht.	12.5
A. glauca Hagem	12.5
Total	2100

4 DISCUSSION

It is common knowledge that bioremediation is a safe clean-up technology which does not cause negative ecological effects. The use of microorganisms: bacteria, yeast, fungi or algae to decontaminate non-organic and organic pollutants in soils is an emerging technique that offers many benefits of being efficient, economic and ecologically accepted alternative to traditional technologies. Identification of microorganisms indigenous to high contaminated environments ought to be first step of bioremediation process. The different species of bacteria have been used to different technologies, such as immobilization, bioaugmentation, biosorption, biostabilization. At present there is growing interest in application of fungi for bioremediation. The fungi have a strong potential to use them in non-sterile open environment:

- the mycelial growth gives a competitive advantage over single cells such as bacteria and yeasts, especially with respect to the colonization of insoluble substrates;
- fungi can rapidly ramify through substrates, literally digesting their way along by secreting a battery of extracellular degradative enzymes;
- the high surface-to-cell ratio characteristics of filaments maximizes both mechanical and enzymatic contact with the environment;
- the extracellular nature of the degradative enzymes enables fungi to tolerate higher concentrations of toxic chemicals than would be possible if these compounds had to be brought into the cell;
- insoluble compounds that cannot cross a cell membrane are also susceptible to attack;
- induction of relevant enzymes by nutritional signs and resulting in non-specific activity on diverse substrates and independent of their concentrations (Bennetta et al. 1999).

Bioremediation is still developing technology. One of the main difficulty is that bioremediation is carried out in the natural environment, which contains many different microorganisms. Hence, most-promising bioagents, isolated and characterized in the laboratory must bring these properties to natural environment.

ACKNOWLEDGEMENTS

This work was financially supported by the 5 EU Project **W**ater, **E**nvironment, **L**andscape **M**anagement at **C**ontaminated **M**egasite – WELCOME (EVK1-CT-2001-00103), and the Polish Committee for Scientific Research (Project no.: 154/E-358/SPB/ 5PRUE/ DZ165/2002-2004).

We wish to thank Lucyna Gadek for technical assistance.

REFERENCES

Baldrian P. 2003. Interaction of heavy metals with white-root fungi. *Enzyme and Microbial Technology* (32): 78–91.

Bending G.D., Friloux M., Walker A. 2002. Degradation of contrasting pesticides by white rot fungi and its relationship with lignolytic potential. FEMS Microbiology Letters 212: 59–63.

Castro I.M., Fietto J.L., Vieira R.X.,Tropia M.J.M., Campos L.M.M., Paniago E.B., Brandao R.L. 2000. Bioleaching of zinc and nickel from silicates using *Aspergillus niger* cultures. *Hydrometallurgy* 57: 39–49.

Errasquin E.L., Vázquez C. 2003. Tolerance and uptake of heavy metals by *Trichoderma atroviride* isolated from sludge. *Chemosphere* 50: 137–143.

Gams W., Anderson T.H., Domsch W. 1980. *Compendium of soil fungi*. London: Academic Press Ltd.

Krantz-Rülcker C., Allard B., Schnürer J. 1996. Adsorption of IIB-metals by three common soil fungi – comparison and assessment of importance for metal distribution in natural soil systems. *Soil Biol. Biochem*. 28 (7): 967–975.

Price M.S., Classen J.J., Payne G.A. 2001. *Aspergillus niger* absorbs copper and zinc from swine wastewater. *Biores. Technolog*. 77: 41–49.

Remacle J. 1990. The cell wall and metal binding. In: Volesky B (ed.) *Biosorption of heavy metals*. CRC Press, Boca Raton, FL, pp. 83–92.

Vogel T.M. 1996. Bioaugmentation as a soil bioremediation approach. *Current Option in Biotechnology*. 7: 311–316.

Volesky B. 1994. Advances In biosorption of metals: selection of biomass types. *FEMS Microbiol. Rev*. 14: 291–302.

European Symposium on Environmental Biotechnology, ESEB 2004 - Verstraete (ed)
© 2004 Taylor & Francis Group, London, ISBN 90 5809 653 X

Assessment of chemical and microbiological signatures during natural attenuation of gasoline-contaminated groundwater

Y. Takahata
Taisei Corporation, Yokohama, Japan
Y. Kasai & K. Watanabe
Marine Biotechnology Institute, Kamaishi, Japan

ABSTRACT: A gasoline (mainly benzene, toluene and xylene [BTX])-contaminated groundwater plume had been treated by the pumping and aeration system for more than 10 years, and the treatment has recently been changed to natural attenuation according to the cost effectiveness consideration. In order to gain information for assessing the potential of natural attenuation, chemical and microbiological signatures of groundwater have been observed in several wells along the contaminant gradient in the plume. We found that the BTX contamination has caused changes in chemical and microbiological signatures of groundwater. In recent contaminated groundwater, the depletion of DO, nitrate and sulfate was apparent concomitant with the increase of iron and manganese concentrations. Between the contaminated and non-contaminated zones, there was an interfacial zone, where BTX were not detected but total cell counts were high. Laboratory incubation of groundwater supplemented with BTX showed that BTX disappeared most rapidly in the groundwater obtained from the interfacial zones, while the BTX-degradation capacity in groundwater obtained from the center of the contaminated zone was marginal. Supplementation of contaminated groundwater with oxygen, nitrate, or sulfate stimulated BTX biodegradation. Bacterial populations occurring in groundwater and those after biodegradation of BTX in the laboratory incubation were analyzed by denaturing gradient gel electrophoresis (DGGE) of PCR-amplified 16S rDNA fragments. Dominant species on DGGE profiles after incubation of groundwater were possibly important for evaluating natural attenuation of BTX. Based on the results, we have identified some bacterial populations that are possibly important for natural attenuation of BTX in the groundwater.

1 INTRODUCTION

Contamination of groundwater with gasoline is a serious environmental problem, as it may affect drinking water resources and have an impact on oligotrophic environments. BTX are major components of gasoline and of great concern, because they are toxic and relatively soluble in water. BTX are known to be easily biodegradable in aerobic environments, while their persistence in the anaerobic environments has also been reported.

MNA (monitored natural attenuation) has been considered as one of treatment strategies for contaminated groundwater, in which careful monitoring of the behavior of contaminants is conducted for increasing the reliability of natural attenuation (NA) processes (Richmond et al. 2001). MNA has become an attractive treatment strategy, because it is generally less costly and therefore more practical than engineered cleanup solutions in many cases. The monitoring information can be used for predicting the potential of NA and for evaluating the on-going degradation processes (Bhupathiraju et al. 2002). Analyses of microbial populations occurring in the MNA zone may also be useful to know how BTX are degraded. It has been known that BTX can be degraded by many different types of microorganisms, including aerobes, nitrate reducers, sulfate reducers, Fe(III) reducers, and methanogens (Kao et al. 2001).

In this study, we have monitored groundwater in several wells located at a gasoline-polluted zone close to the city of Kumamoto, Japan before and after terminating the active remediation. Here we report relationships between hydrochemistry and microbial ecology in the BTX contaminated plume during natural attenuation.

2 MATERIALS AND METHODS

2.1 *Site description and sampling method*

The studied area is located at the Higashino region in Kumamoto, Japan (Fig. 1). Groundwater provided from the shallow dug wells had been used for drinking water of residence in this area. In January 1991, groundwater in these wells had first been detected to be contaminated with fuel hydrocarbons leaked from an underground storage tank of a gas station. Soil type of contaminated aquifer is volcanic ash soil which has high water permeability. Downslope of groundwater level is about 0.5% from north-northeast to south-southwest. The velocity of groundwater flow is high, approximately 50 to 100 meters per year as determined by water permeability test (data not shown). The seasonal fluctuation of water table is approximately 1 meter.

Since February 1992, gasoline-contaminated groundwater had been treated by the pumping and aeration system located on 14 well. The time course of the BTX concentrations in the groundwater located on the center (29 well) and the edge (60 well) of contaminated plume are shown in Figure 2. In April 2002, monitored natural attenuation was introduced as an alternative method after terminating the active remediation. Groundwater samples from the 12 monitor wells located inside and outside contaminated plume were obtained by sampling tubes. They were collected into the sterile glass bottles and analyzed within 24 hours.

2.2 *Chemical and microbial analyses*

Dissolved oxygen (DO), redox potential (Eh), pH of groundwater samples were measured in the field using multi-monitoring system (U-22, Horiba, Japan). BTX analyses were performed using purge-and-trap with GC-FID. Concentrations of HCO3-, NO3- and SO42- were analyzed with an HPLC equipped with a conductivity detector. Dissolved total ferric and manganese ions in the samples were measured with an atomic absorption spectrophotometer. Total cell count of microorganisms was determined by the acridine orange direct count method (Hobbie et al. 1977).

2.3 *Biodegradable potential of BTX in groundwater*

In order to evaluate the potential for aerobic and anaerobic biodegradation of BTX in gasoline-contaminated groundwater, batch culture tests were performed using the samples obtained from wells 29 and 60 in March 2003. Soon after sampling, sterile 125-ml vials was completely filled with groundwater. They were sealed with sterile teflon-coated butyl rubber stoppers and aluminum caps. Benzene, toluene, m-, p-xylene, and o-xylene were then inoculated into each of the vials at an initial concentration of 4.0, 1.0, 0.43, and 0.29 mg/l, respectively. The vials were incubated at 20°C for 6 weeks in a chamber. After appropriate periods, aliquot samples were taken and the BTX concentrations were determined.

2.4 *Molecular analyses by DGGE profiling*

Microorganisms in cultures were collected on a GV membrane by filtration, and DNA was extracted. The variable V3 region of bacterial 16S rDNA (corresponding to positions 341 to 534 in the E. coli sequence) was analyzed by denaturing gradient gel electrophoresis (DGGE) after PCR amplification with primers P2 and P3 (Muyzer et al., 1996). DGGE was performed with a D-CodeTM instrument (Bio-Rad) and DNA

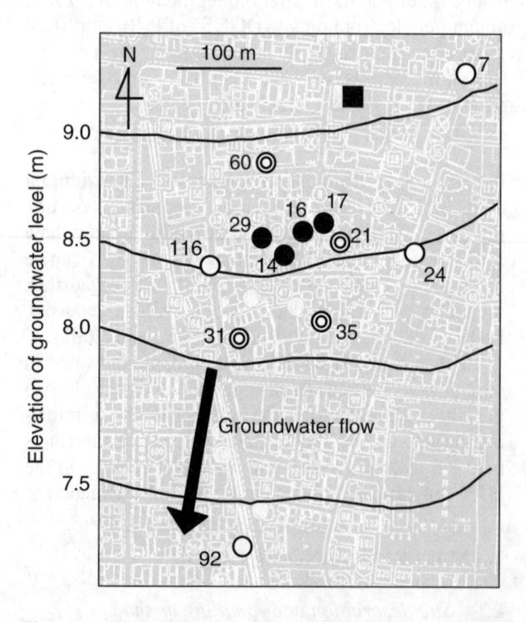

Figure 1. Location and map of the study area. ■, Contaminated source (gas station); ●, Wells in the contaminated zone; ◎, Wells in the interfacial zone; ○, Wells in the background (non-contaminated zone).

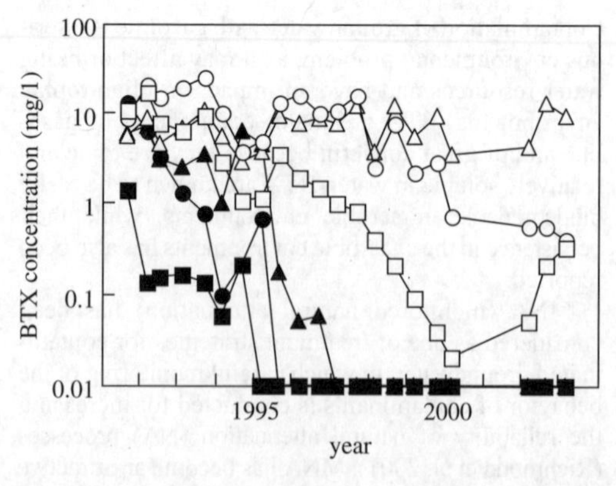

Figure 2. Changes in BTX concentrations in groundwater. □, Benzene (29 well); ○, Toluene (29 well); △, Xylene (29 well); ■, Benzene (60 well); ●, Toluene (60 well); ▲, Xylene (60 well). Xylene = m-, p-xylene + o-xylene.

sequences of DGGE bands were determined by using previously described method (Kasai et al., 2001).

3 RESULTS AND DISCUSSION

3.1 *Chemical and microbial signatures*

BTX in the groundwater have been measured at one or two month intervals after starting MNA. BTX was always detected in the contaminated zone (wells 14, 16, 17, and 29) with seasonal fluctuations, and significant decay of contaminants was not observed during the first year of MNA. On the other hand, none of BTX has been detected in the wells located at the interfacial zone during the investigation period, suggesting that the contaminated plume did not expand to the downstream region.

The chemical and microbiological signatures in the background (well 7), contaminated zone (well 29), and interfacial zone (well 60) before and after starting MNA were shown in Table 1. The depleted oxygen, nitrate, sulfate, and elevated iron, manganese, carbonate in groundwater from the contaminated zone compared to those in the background suggest that significant biodegradation has occurred at this site (Bhupathiraju et al. 2002). While most of the chemical signatures in the interfacial zone were similar to those in the background, lower concentrations of dissolved oxygen and higher total cell counts suggested that microbial activities were high.

3.2 *Biodegradable potential of BTX in groundwater*

After incubation of the groundwater obtained from well 60 without any additives, BTX completely degraded within 2 weeks (Fig. 3, A), indicating that

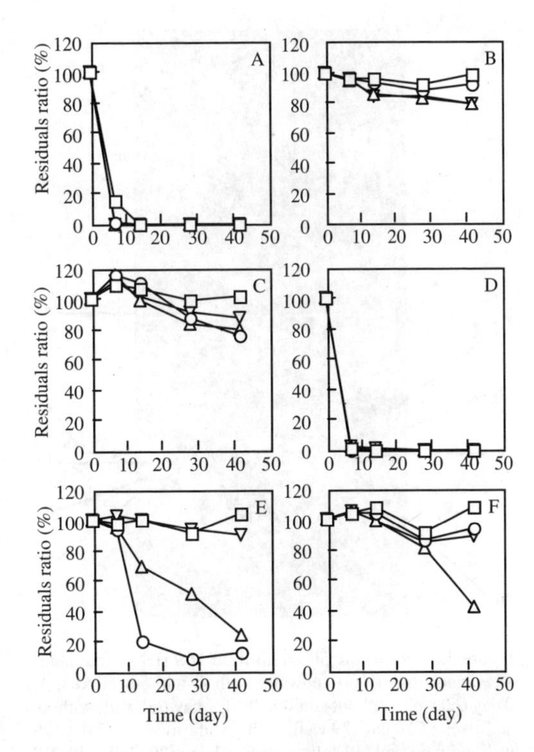

Figure 3. Laboratory incubation for biodegradable potential of BTX in the groundwater obtained from 60 and 29 wells. A, groundwater without additives (60 well); B, groundwater without additives (29 well); C, groundwater with NP solution (1.0 mg/l of NH_4^+ and 0.2 mg/l of PO_4^-) (29 well); D, groundwater with air (liquid:air = 1:3) and NP solution (29 well); E, groundwater with 0.12 mg/l of NO_3^- and NP solution (29 well); F, groundwater with 19.2 mg/l of SO_4^{2-} and NP solution (29 well). □, Benzene; ○, Toluene; △, m-, p-xylene; ▽, o-xylene.

Table 1. Chemical and microbial signatures for monitor wells.

	7 well		29 well		60 well	
	Mar. 02	Mar. 03	Mar. 02	Mar. 03	Mar. 02	Mar. 03
Benzene (mg/l)	<0.01	<0.01	0.05	0.33	<0.01	<0.01
Toluene (mg/l)	<0.01	<0.01	0.59	0.42	<0.01	<0.01
m-, p-xylene (mg/l)	<0.01	<0.01	3.00	7.40	<0.01	<0.01
o-xylene (mg/l)	<0.01	<0.01	1.20	2.30	<0.01	<0.01
DO (mg/l)	7.1	6.8	0.1	0.5	4.7	5.3
Eh (mV)	345	151	−36	−125	307	162
pH	6.0	5.8	6.5	6.5	6.1	5.9
NO_3^- (mg/l)	2.6	2.7	<0.1	<0.1	5.7	5.2
SO_4^{2-} (mg/l)	16.0	15.0	0.3	<0.1	14.0	13.0
T-Fe (mg/l)	<0.1	<0.1	12.0	17.0	<0.1	<0.1
T-Mn (mg/l)	<0.1	<0.1	22.0	23.0	<0.1	0.7
HCO_3^- (mg/l)	46	46	217	276	51	46
Total cell density ($\times 10^5$ cells/ml)	1.6	1.1	44.0	9.6	6.5	4.6

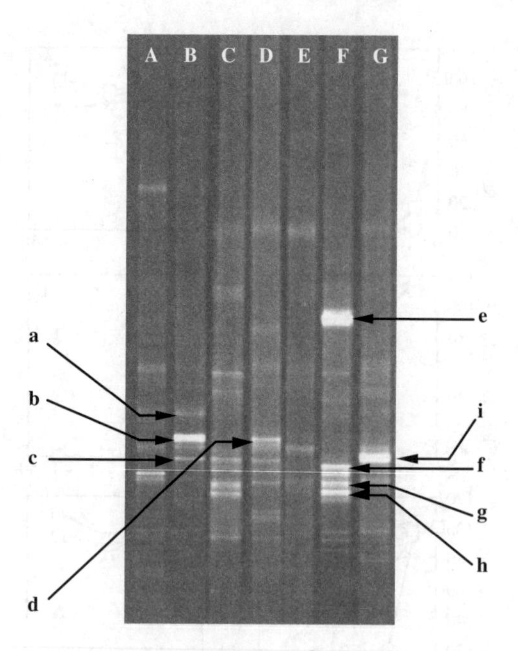

Figure 4. DGGE profiles amplified 16S rDNA fragments after incubation of groundwater with or without TEAPs. A, 0 day (60 well, without additives); B, 7 day (60 well, without additives); C, 0 day (29 well, without additives); D, 7 day (29 well, with air and NP solution); E, 42 day (29 well, with NP solution); F, 42 day (29 well, with NO_3^- and NP solution); G, 42 day (29 well, with SO_4^{2-} and NP solution).

Table 2. Identities of DGGE bands in Figure 4 as determined by partial 16S rDNA sequencing.

DNA band	Phylogenetic group	Closest 16S rDNA sequence	Identity (%)
a	β-proteobacteria	*Rubrivivax gelatinosus*	98
b	α-proteobacteria	*Sphingomonas* sp.	99
c	β-proteobacteria	*Leptothrix mobilis*	97
d	α-proteobacteria	*Sphingomonas*	100
e	ε-proteobacteria	Uncultured sp.	89
f	β-proteobacteria	*Azoarcus evansii*	100
g	Bacteria	Uncultured sp.	97
h	β-proteobacteria	*Nitrosospira* sp.	97
i	Firmicutes	*Desulfotomaculum* sp.	99

The identity of the closest relative in the GenBank database.

the groundwater at the interfacial zone had the high potential to degrade BTX. These results seem to be ascribable to the existence of terminal electron acceptors (TEAPs) and the appropriate microbial consortium that adapted to BTX degradation. On the other hand, BTX in the groundwater obtained from well 29 decreased slowly in the presence of oxygen (Fig. 3, F), suggesting that aerobic biodegradation of BTX at the contaminated zone was limited due to the lack both of dissolved oxygen and trace nutrients such as phosphorus.

The supply of nitrate and sulfate stimulated the degradation of toluene and m-, p-xylene, suggesting that denitrification and sulfate reduction occurred when these electron acceptors are supplied to the contaminated site. Significant methanogenesis or iron reduction were not observed within 6 weeks (data not shown). These results indicated that benzene and o-xylene were more persistent than toluene and m-, p-xylene at this site.

3.3 *Molecular analyses by DGGE profiling*

Identification of pollutant-degrading bacteria in contaminated site may help develop tools for predicting natural attenuation (Röing et al. 2001). The DGGE profiles of the bacterial communities in groundwater obtained from monitor wells were however complex, and it was difficult to identify specific species contributing to BTX degradation (data not shown). On the other hand, DGGE profiles after the bottle incubation of groundwater with TEAPs were relatively simple, which allowed to identify bacterial populations possibly involved in the BTX biodegradation in groundwater (Fig. 4). Sequential analyses (Table 2) showed that some major bands are closely related to BTX degrading species reported previously (Röing et al. 2001).

4 CONCLUSIONS

1) The increases of carbonate, iron, manganese concentrations and decreases of DO, nitrate, sulfate concentrations in the contaminated plume compared from those in the background indicates that microbial mineralization of BTX has occurred.
2) Total cell counts in the interfacial zone were approximately 5-fold higher than those in the background, suggesting they presumably grew by mineralizing BTX.
3) Biodegradation potential for BTX in the interfacial zone was higher than that at the center of the contaminated zone, because terminal electron acceptors and trace minerals such as phosphorus were present.
4) DGGE bands appearing after the bottle incubation were related to known aromatics-degrading bacteria that may have contributed to BTX biodegradation in groundwater.
5) The remediation strategy of transition from the active remediation to MNA is considered to be acceptable at the study site, because the contaminated zone is not spread resulting from the presence of the barrier of the high activity zone (interfacial zone) around the contaminated zone.

ACKNOWLEDGEMENTS

We thank the Kumamoto city hall for help in the sampling and Hiromi Awabuchi for the DGGE analyses. This work was supported by the New Energy and Industrial Technology Development Organization (NEDO).

REFERENCES

Bhupathiraju, V. K., Krauter, P., Holman, H. N., Conrad, M. E., Daley, P. F., Termpleton, A. S., Hunt, J. R., Hernandez, M. & Alvarez-Cohen, L. 2002. Assessment of in-situ bioremediation at a refinery waste-contaminated site and aviation gasoline contaminated site. *Bioremediation* 13:79–90.

Hobbie, J. E., Daley, R. J. & Jasper, S. 1977. Use of nuclepore filters for counting bacteria by fluorescence microscopy. *Appl. Environ. Microbiol.* 33:1225–1228.

Kasai, Y., Kishira, H. Syutsubo, K. & Harayama, S. 2001. Molecular detection of marine bacterial populations on beaches contaminated by Nakhodka tanker oil-spill accident. *Environ. Microbiol.* 3:1–10.

Kao, C. M. & Prosser, J. 2001. Evaluation of natural attenuation rate at a gasoline spill site. *J. Hazardous Materials.* B82:275–289.

Muyzer, G., Hottentrâer, S. Teske, A. & Wawer, C. 1996. Denaturing gradient gel electrophoresis of PCR-amplified 16S rDNA – A new molecular approach to analyse the genetic diversity of mixed microbial communities, pp. 1–23. In: A. D. L. Akkermans, J. D. van Elsas, and F. J. de Bruijn (eds.), Molecular Microbial Ecology Manual 3.4.4. Kluwer Academic Publishers, Dordrecht.

Richmond, S. A., Lindstrom, J. E. & Braddock, J. F. 2001. Assessment of natural attenuation of chlorinated aliphatics and BTEX in subarctic groundwater. *Environ. Sci. Technol.* 35:4038–4045.

Röling, W. F. M., Breukelen, B. M., Braster, M., Lin, B. & Verseveld, H. W. 2001. Relationships between microbial community structure and hydrochemistry in a landfill leachate-polluted aquifer. *Appl. Environ. Microbiol.* 67:4619–4629.

European Symposium on Environmental Biotechnology, ESEB 2004 - Verstraete (ed)
© 2004 Taylor & Francis Group, London, ISBN 90 5809 653 X

The effect of vegetation on decrease of PAH and PCB content in long-term contaminated soil

J. Rezek[1,2], T. Macek[1,2] & M. Macková[1,2]
[1] *Dept. of Natural Products, Institute of Organic Chemistry and Biochemistry, Czech Academy of Sciences, Prague, Czech Republic*
[2] *Dept. of Biochemistry and Microbiology, Faculty of Food and Biochem. Technology, ICT Prague, Prague, Czech Republic*

C. in der Wiesche & F. Zadrazil
Inst. of Plant Nutrition and Soil Sciences, FAL, Braunschweig, Germany

ABSTRACT: The described research aimed to investigate the ability of biodegradation of PAHs and PCBs in microecosystems containing long-term contaminated soil and different plant species. A combination of plants growing in region where the phytoremediation process would be done has been tested. Birch, mulberry, ryegrass, alfalfa, tobacco and egg-plant were used and phytoremediation process was compared among them and with non vegetated contaminated soil. Soil extracts were analysed by HPLC and GC. The plant species grown in the contaminated soil showed a significant effect of the presence of vegetation on PCB and PAH removal, although differences among individual plant species were not clearly visible especially in the case of trees.

1 INTRODUCTION

The task of this research was to investigate some new facts about cooperation between plants and soil microflora to improve the biodegradation of selected persistent toxic compounds. We addressed the polycyclic aromatic hydrocarbons (PAHs) and polychlorinated biphenyls (PCBs) as toxic persistent compounds cumulating in soil (Fewson 1988, Jones et al. 1989), and tried to test the improvement of degradation of these compounds by some plant species in conditions very similar to local environment, nearly natural using indigenous plants species. This process called phytoremediation (Harvey et al. 2002, Macek et al. 2000) already had some success and looks efficient regarding its social acceptability and cost comparison with traditional techniques, which are usually more expensive and not so environmental-friendly. Anyway this process shows high complexity and it is necessary to think about system plant – degrading microorganisms supporting each other in the remediation process (Kas et al. 1997). Metabolic cooperation between plants and microorganisms in PCB degradation was described by Francova et al. (2004).

In the case of PAH experiments we exploited plant species which live in region where the phytoremediation process would be done. Soil, contaminated since World War II (when a chemical company in Germany has been destroyed by bombing and the area polluted by different chemicals, including PAHs) was used for our experiment. It was already shown that some plant cell cultures are able to degrade PAHs (Kolb & Harms 2000, Kucerová et al. 2001), and that different plants can enhance degradation of PAHs in their root system (Liste & Alexander 2000, Myia & Firestone, 2000). As the most promising plants we used birch (*Betula pendula*), the tree known to grow in that area, mulberry (*Morus rubrum*) which is one of the typical trees of the country and was successful in other PAH degrading experiments (Olson & Fletcher 1999), and ryegrass (*Lolium perene*), also a typical plant of the area and successful in some phytoremediation experiments (Binet et al. 2000). As well as with PAHs long-term contaminated soil, also experiments with PCB long-term contaminated soil have been performed. This soil from a dumpsite in the Czech Republic was contaminated during 30 years by commercial mixture of PCBs DELOR 103, which contains 59 congeners of PCBs. The selection of plant species for PCB experiments was based on studies with tissue cultures successful in biodegradation of PCBs (Kucerová et al. 1999, 2000, 2001, Macková et al. 1997). Also the aspect of survival in stress conditions, fast biomass production or dense root system was important. Because of this, alfalfa (*Medicago sativa*), tobacco (*Nicotiana tabacum*) and egg-plant (*Solanum nigrum*) were our candidates. Also in these experiments we tried to operate at the conditions as close as possible to natural ones.

2 MATERIAL AND METHODS

2.1 *Monitored PAHs and construction of microecosystems*

In the experiment with long-term PAH contaminated soil were monitored these PAHs: naphthalene (NAP), acenaphthene (ACE), fluorene (FLU), phenanthrene (PHE), anthracene (ANT), fluoranthene (FLT), pyrene (PYR), benzo(a)anthracene (B(a)A), Chrysene (CH), benzo(b)fluoranthene (B(b)F), benzo(k)fluoranthene (B(k)F), benzo(a)pyrene (B(a)P), dibenzo (a,h) anthracene (D(ah)A), benzo-(g,h,i)perylene (B(ghi)P), iIndeno(1,2,3-c,d)pyrene (I(cd)P). For experiments we used microecosystems (MES) containing different combinations of long-term PAH contaminated soil with natural microorganisms, plants birch (*Betula pendula*), mullberry (*Morus rubrum*) and ryegrass (*Lolium perene*). Some microecosystems were fertilised with usual N-P-K fertiliser and all were cultivated in greenhouse at natural photoperiod for 12 and 18 months. Average temperature was 27/12°C (day/night) during summer time and 10/5°C during wintertime with cold shock for 10 days at -10°C.

2.2 *Sample preparation and PAHs HPLC analysis*

After harvest the soil from each pot was divided into 3 horizontal layers for mutual comparison and each layer was divided into four equal samples, which were extracted with methanol by Soxhlet extraction at 90°C for 20 hours. Diluted extracts were analysed using HPLC with PAH reverse phase column, acetonitrile/water gradient and fluorescent detector with changing wavelengths. Concentrations of all 15 PAHs present were estimated in samples. Obtained data were statistically analysed using Duncan test.

2.3 *Construction of experiment with PCB contaminated soil*

In the experiment with PCB contaminated soil two parallel experiments were done. One experiment was done straight at the dumpsite with alfalfa (*Medicago sativa*), tobacco (*Nicotiana tabacum*) and egg-plan (*Solanum nigrum*). The plants were pricked out at the beginning of vegetation period and 5 samples of soil vegetated by each tested plant and the non-vegetated soil were taken out after 5 months cultivation. Similar experiment was performed with the same plants and PCB contaminated soil at cultivation vessels for the same time of cultivation. Plants in pots were cultivated for 5 months at natural conditions with regular watering.

2.4 *Sample preparation and GC analysis of PCBs*

All of the soil from pot experiment was used up for sample preparation. Tested soil was air-dried overnight and sieved through a mesh with 1 mm pore size. Then 1 g of the soil was Soxhlet extracted with hexane for 4 hours. The extract was concentrated to approximately 1 ml in volume by nitrogen flow, purified on a Florisil column, diluted with hexane to exactly 10 ml volume, than diluted 1/100 and analysed using GC with capillary column, temperature gradient and electron capture detector. Because DELOR 103 is a mixture of 59 PCB congeners the sum of content was used for calculations. Results were calculated using US EPA (Environmental Protection Agency) methods 8089/8081 for estimating total PCB concentration as a sum of recommended indicator congeners.

3 RESULTS

3.1 *PAH contaminated soil*

In the experiment with PAH contaminated soil 8 microecosystems containing different combinations of birch, ryegrass and fertiliser were analysed after 12 months cultivation. Other 8 microecosystems of

Table 1. Values of PAH concentrations in microecosystem constructed with morus, ryegrass and fertiliser cultivated 18 months.

	Residual content of PAHs in layers (ppm)			
PAHs	Top layer	Middle layer	Bottom layer	Original soil
NAP	2,7	3,0	2,6	3,3
ACE	1,9	2,1	1,9	4,3
FLU	1,4	1,6	1,4	2,6
PHE	10,3	11,4	12,9	18,9
ANT	2,4	2,6	2,3	6,0
FLT	26,2	29,1	28,0	103,5
PYR	17,8	20,1	19,3	83,3
B(a)A	10,5	11,9	11,2	21,8
CH	8,6	9,6	8,8	18,6
B(b)F	30,9	34,3	31,2	40,8
B(k)F	6,4	6,9	6,3	10,3
B(a)P	14,2	14,4	13,1	19,1
D(ah)A	2,0	2,2	2,1	3,1
B(ghi)P	8,9	9,9	8,9	9,7
I(cd)P	9,8	11,1	9,9	10,6
SUM	154,2	170,1	159,6	355,7

the same composition were analysed after 18 months cultivation and next 4 microecosystems with morus, ryegrass and fertiliser were analysed only after 18 months cultivation. After 12 months cultivation was visible decrease of concentration of all PAHs except high molecular PAHs benzo(g,h,I)perylene and indeno(1,2,3-c,d)pyrene compared with concentrations in original soil, where microbial activity was stopped by freezing at −20°C. All microecosystems had good capability for degradation even microecosystems without plants. After one year cultivation the content of the followed compounds declined to 50%. Mostly there have been no significant differences between the microecosystems. Best degraded were fluoranthen and pyrene, present in original soil in highest concentrations. Also other compounds were successfully degraded, even benzo(a)pyrene and dibenzo(a,h) anthracene. Benzo(g,h,i)perylene and indeno(1,2,3-c,d)pyrene were the only PAHs, which have almost not been degraded. One example of total concentrations of PAHs and residual content of PAHs (%) in soil samples from microecosystem with morus, ryegrass and fertiliser is shown in Table 1.18 and 12 moths cultivated microecosystems of the same composition were compared statistically using Duncan's test. Significantly lower concentrations of some PAHs in 18 months cultivated microecosystems in regard of 12 months cultivated microecosystems are represented by the X letters in Table 2.

Table 2. Significantly lower concentrations of some PAHs in microecosystems after 18 months cultivation in regard to the same constructed microecosystems after 12 months cultivation are represented by X letters. In other cases the decrease of PAH concentrations (empty fields) did not show significant difference. S (soil), F (fertiliser), R (ryegrass), B (birch).

MES containing	NAP	ACE	FLU	PHE
S			X	X
SF			X	X
SR		X		X
SRF			X	X
SB	X	X	X	X
SBF	X	X	X	X
SBR	X	X	X	X
SBRF				

MES containing	ANT	FLT	PYR	B(a)A
S				
SF				
SR				X
SRF			X	X
SB			X	X
SBF			X	X
SBR			X	X
SBRF				

MES containing	CH	B(a)F	B(k)F	B(a)P
S	X			X
SF	X			
SR	X	X	X	X
SRF	X	X	X	X
SB	X	X	X	X
SBF	X	X	X	X
SBR	X			X
SBRF				X

MES containing	D(ah)A	B(ghi)P	I(cd)P	SUM
S				
SF	X			
SR				
SRF	X			X
SB	X			X
SBF	X			X
SBR				
SBRF				

There is not visible any unambiguous tendency, but microecosystems containing plant have higher number of PAHs with significantly lower concentration, the sum of PAHs concentrations is significantly lower mostly in microecosystems containing ryegrass and birch. Differences between the soil layers did not show any statistically significant trend, although significantly lower concentrations of PAHs were estimated mostly in the bottom layer samples.

3.2 *PCB contaminated soil*

In the experiment with PCB contaminated soil a visible decrease of PCB content appeared in soil vegetated by all used plant species. The best results were obtained with tobacco and egg-plant, where the PCB content decreased below 80%. With alfalfa the decrease was smaller but also appreciable. Similar results were obtained both in field experiments (see Table 3) and in pot experiments (see Table 4).

4 CONCLUSIONS

After one year cultivation the content of nearly all PAHs declined to 25–80%. After one and half

Table 3. Residual content of PCBs in soil vegetated by plants for 5 months in field experiment.

Plant	Rediual content of PCBs (%)	Concentration of PCB (mg/kg dry soil)
Alfalfa	90	425
Tobacco	76	358
Egg-plant	78	365
Non-vegetated soil	100	468

Table 4. Residual content of PCBs in soil vegetated by plants for 5 months in pot experiment.

Plant	Redial content of PCBs (%)	Concentration of PCB (mg/kg dry soil)
Alfalfa	82	88
Tobacco	77	83
Egg-plant	87	93
Non-vegetated soil	100	107

year cultivation changes in the content of these compounds were significant only in some microecosystems containing ryegrass and birch. It could be due to winter period, when plant and microbial activity is lower than during summer period and this winter period represented the 6-month difference in cultivation between the same composed microecosystems. Other possible reason is the difference in homogeneity of long-term contaminated soil, where the binding of PAHs to soil particles can cause partial non-homogeneity of soil (Pignatello & Xing, 1996) and it can cover small decrease of PAH concentrations during such "short" time from decontamination point of view. In all cases best degraded were fluoranthene and pyrene. It can be easily due to their presence in original soil in highest concentrations and their easier bioavailability. This could be another possible reason for slow degradation, that other PAHs are present in soil in very low concentrations and perhaps bound to soil particles and not available for plant root system and present microorganisms. Dibenz(a,h) anthracene and indeno(1,2,3-c,d) pyrene were the only PAHs which have not been degraded. This is not surprising because these high molecular weight PAHs are more recalcitrant than the low ones. Finding of degradation of benzo(a)pyrene was a success, because this is one of prime interest PAHs due to its carcinogenity.

In the experiment with PCB contaminated soil all used plant species showed good capability for biodegradation of PCBs mixture in natural conditions. The best results were obtained with tobacco, which is also very useful for its fast production of biomass. Anyway all tested plants accelerated biodegradation of PCBs in the soil and enable possible use in remediation technologies.

ACKNOWLEDGEMENTS

The authors thank for support of the grant GACR 526/01/1292, grant ME 498 of the Czech Republic, grant MSMT CZE 01/026, EU 5FW grant no. QLK3-CT-2001-00101.

REFERENCES

Binet P., Portal J.M. & Leyval C. 2000. Dissipation of 3-6-ring polycyclic aromatic hydrocarbons in the rhizosphere of ryegrass. *Soil Biology and Biochemistry* 32: 2011–2017.

Fewson C.A. 1988. Biodegradation of xenobiotic and other persistent compounds: the causes of recalcitrance. *Trends in Biotechnology* 6: 148–153.

Francova K., Macková M., Macek T. & Sylvestre M. 2004. Ability of bacterial biphenyl dioxygenases from *Burkholderia sp.* LB400 and *Comamonas testosteroni* B-356 to catalyse oxygenation of ortho-hydroxybiphenyls formed from PCBs by plants. *Environmental Pollution*: 127(1): 41–48.

Harvey P.J., Campanella B.F., Castro P.M.L., Harms H., Lichtfouse E., Schäffner A.R., Smrcek S. & Werck-Reichhart D. 2002. Phytoremediation of polyaromatic hydrocarbons, anilines and phenols. *Environmental Science and Pollution Research* 9: 29–47.

Jones K.C., Stratford J.A., Tidridge P., Waterhouse K.S. & Johnston A.E. 1989. Polynuclear aromatic hydrocarbons in an agricultural soil: long-term changes in profile distribution. *Environmental Pollution* 56: 337–351.

Kas J., Burkhard J., Demnerová K., Kostal J., Macek T., Macková M. & Pazlarova J. 1997. Perspectives in biodegradation of alkanes and PCBs. *Pure and Applied Chemistry* 69: 2357–2369.

Kolb M. & Harms H. 2000. Metabolism of fluoranthene in different plant cell cultures and intact plants. *Environmental Toxicology and Chemistry* 19: 1304–1310.

Kucerová P., Macková M., Polachova L., Burkhard J., Demnerová K., Pazlarova J. & Macek T. 1999. Correlation of PCB transformation by plant tissue cultures with their morphology and peroxidase activity changes. *Coll. Czech Chemical Commun.* 64: 1497–1509.

Kucerová P., & Macková M., Chromá L., Burkhard J., Triska J., Demnerová K. & Macek T. 2000. Metabolism of polychlorinated biphenyls by *Solanum nigrum* hairy root clone SNC-9O and analysis of transformation products. *Plant and Soil* 225: 109–115.

Kucerová P., in der Wiesche C., Wolter M., Macek T., Zadrazil F. & Macková M. 2001. The ability of different plant species to remove polycyclic aromatic hydrocarbons and polychlorinated biphenyls from incubation media. *Biotechnology Letters* 23: 1355–1359.

Liste H.H. & Alexander M. 2000a. Plant promoted pyrene degradation in soil. *Chemosphere* 40: 7–10.

Liste H-H. & Alexander M. 2000b. Accumulation of phenathrene and pyrene in rhizosphere soil. *Chemosphere* 40: 11–14.

Macek T., Macková M. & Kas J. 2000. Exploitation of plants for the removal of organics in environmental remediation. *Biotechnology Advances* 18: 23–34.

Macek T., Macková M., Kucerová P., Chromá L., Burkhard J. & Demnerová K. 2002. Phytoremediation. In: *Focus on Biotechnology*, (Hofman M. and Anne J., series eds.), Vol. 3, Kluwer Academic Publishers, Dordrecht, pp. 115–137.

Macková M., Macek T., Ocenaskova J., Burkhard J., Demnerová K. & Pazlarova J. 1997. Biodegradation of polychlorinated biphenyls by plant cells. *International Biodeterionation and Biodegradation* 39: 317–325.

Myia R.K. & Firestone M.K. 2000. Phenathrene-degrader community dynamics in rhizosphere soil from a common annual grass. *Journal of Environmental Quallity* 29: 584–592.

Olson P.E. & Fletcher J.S. 1999. Field evaluation of mullberry root structure with regard to phytoremediation. *Bioremediation Journal* 3(1): 27–33.

Pignatello J.J. & Xing B. 1996. Mechanisms of slow sorption of organic chemicals to natural particles. *Environmental Science and Technology* 30: 1–11.

in der Wiesche C., Rezek J., Wolter M., Macková M., Macek T. & Zadrazil F. 2001. The effect of mullberry and birch on the degradation of polycyclic aromatic hydrocarbons in long term contaminated soil. *Proceedings of the 3rd Symposium on Natural Attenuation*, Frankfurt, 4–5. December, 2001, Frankfurt: DECHEMA. pp. 142–144.

European Symposium on Environmental Biotechnology, ESEB 2004 - Verstraete (ed)
© 2004 Taylor & Francis Group, London, ISBN 90 5809 653 X

Detection of polychlorinated biphenyl-degrading bacteria in soil

E. Ryslava, T. Macek & M. Macková
Department of Biochemistry and Microbiology, Institute of Chemical Technology, Prague, Czech Republic

Z. Krejcik
Institute of Molecular Genetics, Academy of Sciences of the Czech Republic, Prague, Czech Republic

ABSTRACT: This work is based on research which focuses on the study of the possible mutual relationship of microorganisms and plants in PCB metabolism. Three different plant species – *Nicotiana tabacum* (tobacco), *Solanum nigrum* (black nightshade) and *Medicago sativa* (alfalfa) – were cultivated in real soil from an industrial site. After six months the decrease of PCB content in soil and the PCB accumulation in plant tissue was analysed. The highest decrease of PCB concentration was measured from the soil vegetated with tobacco. From this polluted soil, bacteria which potentially participate in the biodegradation of PCBs were isolated and were screened for the presence of *bphA1* and *bphC* genes. These genes are a part of the *bph* operon, which is very important for PCB degradation. The PCR primers were designed for conserved regions of *bphA1* and *bphC* genes from an alignment of previously published sequences. The degradative capabilities of the isolates were confirmed by growing in the presence of biphenyl as a cosubstrate. For the extraction of DNA directly from soil a commercial DNA purification kit and a laboratory-devised method based on mechanical lysis were used.

1 INTRODUCTION

Chlorinated organic chemicals constitute half of the environmental organic pollutant problems in the world. This broad contamination is a consequence of the huge amounts of chlorinated organic chemicals synthetized by industries during the last decades. Polychlorinated biphenyls (PCBs) are one of the most widely distributed classes of chlorinated chemicals in the environment. During last 30 years several research groups have successfully isolated different strains of PCB-degrading bacteria from environmental samples and characterized their properties. PCB-degrading bacteria do exist in the environment but persistence of PCBs indicates that under conditions prevailing at PCB-sites the indigenous bacteria are ineffective in degrading PCBs, even though some of them are genetically capable of metabolizing PCBs. Thus the primary challenge for successful bioremediation is to devise ways to encourage the growth and PCB metabolism of a select group of microbes. At laboratory conditions addition of biphenyl helps to enrich bacterial cultures degrading PCBs and even PCB-degradation when provided as cosubstrate. Provision of biphenyl in a similar manner to soil organisms at terrestrial sites poses some problems.

One solution which has been shown to efficiently stimulate microbial growth and even microbial degradation abilities is vegetating of proper plants which produce organic compounds, released to the soil and selectively foster the growth of PCB-degrading bacteria[4,5]. Generally there are three imporatnt plant driven mechanisms that may operate within rhizosphere to degrade recalcitrant compounds – direct metabolism by either endogenous or exogenous plant enzymes, indirect stimulation of the rhizospere microflora to degrade pollutants and release of plant compounds that solubilize contaminants whereby they are more available to both plant and microbial degradative enzymes[6].

In our study we cultivated three different plant species in PCB contaminated soil. After period of 6 months we analysed content of PCBs in vegetated and nonvegetated soil, isolated soil microorganisms growing in rhizosphere and identified them.

2 MATERIALS AND METHODS

2.1 *Pot experiments*

For pot experiments 20 l buckets were filled with contaminated soil containing about 300 μg/g soil of PCBs..

Three different plant species – *Nicotiana tabacum* (tobacco), *Solanum nigrum* (black nightshade) and *Medicago sativa* (alfalfa) were cultivated in contaminated soil and bulk soil was used as the control. Plants grew in summer time for 5 months in open-air conditions. At the end of cultivation several samples from each pot were collected and the residual amount of PCB was analysed. Also, the total number of microorganisms and bacteria, growing on the minimal medium, were estimated in the rhizosphere and rhizoplane areas of cultivated plants.

2.2 *PCB content analysis*

Collected soil samples were dried overnight at room temperature and sieved through a mesh with 1 mm pore size. Then 1 g of soil were extracted with hexane for 4 hours. The extract was concentrated to 1 ml in volume by nitrogen flow, purified on a florisil column, diluted with hexane to the same volume as was used for extraction and analysed by GC. Samples were analysed using a Hewlett-Packard 5890 gas chromatograph with an electron capture detector and a fused silica capillary column (30 m, 0.2 mm inner diameter) coated with 0.25 μm immobilised phase SE-54 with nitrogen as the carrier gas (flow rate 1 ml/min). Results were calculated from the residual amounts of congener peaks present in the sample, compared to the value recommended by US EPA for expressing of the total content of PCBs as a sum of recommended indicator congeners.

2.3 *Isolation of soil bacteria*

10 g samples of bulk soil, rhizospheric soil and root surface were extracted with 90 ml of medium with 1% peptone. Flasks with samples were shaken for 2 hours with glass beads, then aliquots of the extracts were diluted with saline solution and spread on Petri dishes containing Plate count agar (Oxoid) for estimation of the total number of bacteria and minimal medium with biphenyl as the sole carbon source were used for cultivation of bacteria and evaluation of their number. Chosen individual strains were tested for their ability to degrade PCBs.

2.4 *PCR amplification of* bph-*genes*

Bacterial colonies were scraped off agar plates and dissolved in 50 μl of TE buffer. The samples were heated (95°C for 10 min) and centrifuged. *BphA1* and *bphC* genes were amplified by PCR using forward and reverse primers forming fragments of 500 bp, 340 bp, 209 bp, 640 bp and180 bp. PCR reaction mixture consisted of 5 μl of template DNA, 1 μl of each primer (10 mM), 1 μl of deoxynucleoside triphosphate mixture (10 mM), 5 μl of 10 × PCR buffer, 3 μl of $MgCl_2$, 0.2 μl of *Taq*-polymerase and sterile distilled water up to 50 μl of each reaction volume. PCR amplifications were performed in an automated thermal cycler Techne (Progene) with an initial denaturing (94°C for 5 min), followed by 30 cycles of denaturation (94°C, 30 s), annealing (55–60°C, 30 s), and extension (72°C, 1 min) and concluded by a single final extension (72°C, 10 min). The nucleotide sequences of the primers were as follows:

A 5′-TTCACCTGCASCTAYCACGGC-3′
B 5′-GGTACATGTCRCTGCAGAAYTGC-3′
C 5′-CGCGTSGMVACCTACAAR G -3′
D 5′-CARTTCTGCAGYGACATGTACCACG-3′
E 5′-ACCCAGTTYTCDCCRTCGTCCTGCC-3′
F 5′-ATCGCCGTTCAGCAGGGCGA-3′
G 5′-CTCCAGCCATACTCGACCTC -3′
H 5′-CTGCACTGCAACGAACGCCAC -3′
I 5′-GACACCATGTGGTGGTTGGT-3′.

For the PCR amplification of *bphA1* gene three set of primers, A and B, C and B, D and E, were used. For the PCR amplification of *bphC* gene another two pairs of primers, F and G, H and I, were used. The presence of amplified parts of *bphA1* (500, 340, 209 bp) and *bphC* (640, 180 bp) genes was confirmed by gel electrophoresis on a 2% agarose gel with TAE buffer.

2.5 *Analysis of PCB degradation*

Bacterial strains or consortia were cultivated in 250 ml Erlenmeyer flasks in mineral medium with biphenyl as cosubstrate and 50 ppm of PCB mixture Delor 103 for 14 days at 28°C on a rotary shaker. In laboratory experiments the standard commercial PCB mixture, Delor 103, containing 59 individual congeners substituted with 3-5 chlorines per biphenyl molecule was used. After 14 days cultivation the flasks were heated (95°C for 10 min), sonicated and the contents of the flasks were extracted by 10 ml of hexane on a rotary shaker for 2 hours. Following phase separation, the hexane layer was subjected to GC analysis. 22 of Delor 103 congeners were assigned to peaks with areas larger than 0.5% of the total area of all 59 individual chromatographic peaks. For the calculation of the residual amount of PCBs the above mentioned 22 chromatographic peaks were used, the residual amounts of each congener peak of the sample were compared to the respective peaks of the controls.

3 RESULTS AND DISCUSSION

3.1 *PCB assay*

Soil from pot experiments, planted with three different species – *Nicotiana tabacum* (tobacco), *Solanum*

Table 1. Comparison of PCB content in non-vegetated soil and soil cultivated with tobacco, alfalfa and black nightshade.

	PCB content (μg PCB/g soil)		
	Start	End	**Decrease of PCBs**
Bulk soil	299,8	275,4	8%
Tobacco	330,6	217,2	**33,70%**
Black nightshade	345,5	254,5	23,20%
Alfalfa	341,3	297,3	18,50%

Table 2. Microbial analysis of the contaminated soil, vegetated with plants.

Plant		Total amount of MO (cfu/g soil)	MM with biphenyl (cfu/g soil)
Nicotiana	rhizosphere	$5{,}1 \times 10^4$	$9{,}9 \times 10^3$
tabacum	rhizoplane	$12{,}7 \times 10^5$	$14{,}2 \times 10^4$
Solanum	rhizosphere	$11{,}4 \times 10^4$	10×10^3
nigrum	rhizoplane	3×10^5	$2{,}5 \times 10^4$
Medicago	rhizosphere	$7{,}8 \times 10^4$	$1{,}9 \times 10^3$
sativa	rhizoplane	$4{,}1 \times 10^5$	$4{,}4 \times 10^4$
Bulk soil		$7{,}6 \times 10^4$	6×10^3

nigrum (black nightshade) and *Medicago sativa* (alfalfa), was analysed for PCB content after 5 months of plant' growth. The data in Table 1 show significant decline of PCB content. In pots cultivated with tobacco the final PCB content decreased to 66% in comparison with the start amount, 74% in soil cultivated black nightshade and 81% in soil cultivated with alfalfa.

3.2 *Microbial analysis*

Plants alone are able to accumulate and transform PCBs but they also play an important role in supporting of nutrients and increasing of bioavailability of pollutants for rhizosphere microorganisms living on or near the roots of plants. Number of microorganisms and their properties can differ with the plant species and they can significantly affect PCB removal from the soil. Total counts of bacteria present in samples from pot experiments are shown in Table 2. The higher numbers of bacteria were detected in the rhizoplane of all plants in comparison with the bulk soil, i.e. the presence of plants and root exudates had a beneficial effect on microbial growth.

3.3 *Detection of* bph-*genes*

Bacteria isolated from contaminated soil were tested for the ability to grow on minimal medium with biphenyl as cosubstrate and to degrade PCBs. This

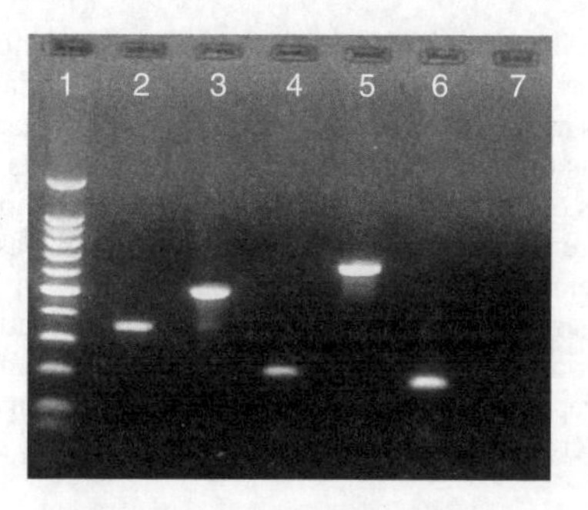

Figure 1. Detection of *bphA1* and *bphC* genes in strain JAB 1. 1 – marker 100 bp DNA ladder, 2 – PCR product of *bphA1* gene – 500 bp, 3 – PCR product of *bphA1* gene – 350 bp, 4 – PCR product of *bphA1* gene – 209 bp, 5 – PCR product of *bphC* gene – 640 bp, 6 – PCR product of *bphC* gene – 180 bp, 7 – negative control.

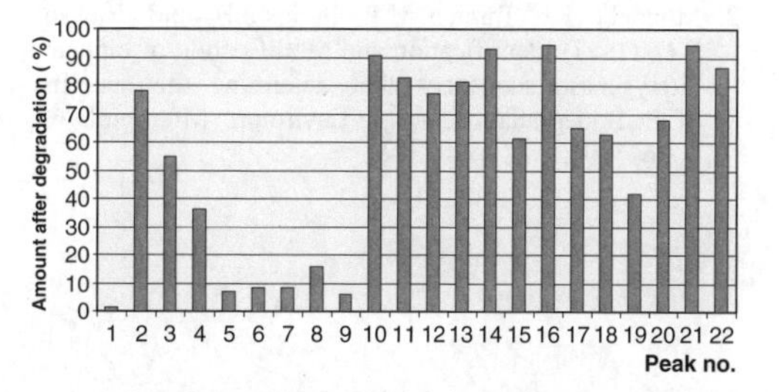

Figure 2. PCB degradation by the strain JAB 1 in the presence of biphenyl.

property can be quickly proved in presence of dibenzofuran which is cleaved for visible color product. Presence of biphenyl operon in bacteria giving positive reaction with dibenzofuran was detected using specific primers for the part of *bphA1* gene (500, 350 and 209 bp), encoding dioxygenase enzyme that converts biphenyl to dihydrodiol and the other dioxygenase coding by the *bphC* gene (640,180 bp), which cleaves 2,3-dihydroxybiphenyl to yield a colored *meta* cleavage product. Presence of amplified fragments of *bph* genes and their proper size indicated successful isolation of PCB degraders (data not shown). The best results (see Figure 1) and degradation abilities were obtained with the strain JAB 1, originally isolated from highly polluted site in Nothern Bohemia.

The degradative capabilities were confirmed in presence of biphenyl as cosubstrate under the laboratory conditions. Strain JAB 1 proved 56% PCB degradation after 14 days cultivation, as shown in Figure 2.

4 CONCLUSIONS

Our experiments with real contaminated soil proved that plants help to remediate soil contaminated with

PCBs. The action may result from the stimulation and support of indigenous PCB-degrading bacterial consortia in the rhizosphere and rhizoplane of plants, as well as from the transformation ability of plants themselves. Our study showed that phytoremediation and rhizoremediation are interrelated processes, based on cooperation of different organisms.

Also fast screening method for identification of bacterial strains containing genes of biphenyl operon was verified. It can be used in following investigations of microbial communities in polluted soils.

REFERENCES

1. Bedard D.L. (1990) In: Biotechnology and Biodegradation, Kamely D.M., Chakrabarty A., Omenn G.S. (eds), Advances in Applied Biotechnology Series, V.4, Portfolio Pub. Co., The Woodlands, TX.
2. Mondello F.J., Turcich M.P., Lobos J.H. and Erickson B.D. (1997) Identification and Modification of biphenyl dioxygenase sequences that determine the specifity of PCB degradation. Appl. Environm. Microbiol. 63, 3096–3103.
3. Bedard D.L., Wagner R.E., Brennan M.J., Haberl M. and Brown J.F. (1987) Extensive degradation of aroclors and environmentally transformed polychlorinated biophenyls by *Alcaligenes eutrophus* H850. Appl. Environm. Microbiology 53, 1094–1102.
4. Fletcher J.S., Donnelly P.K. and Hegde R.S. (1995) Biostimulation of PCB-degrading bacteria by compounds released from plant roots. In Bioremediation of Recalcitrant Organics. Battelle Press, Columbus Ohio. pp.131–136.
5. Hegde R.S. and Fletcher J.S. (1996) Influence of plant growth stage and season on the release of root phenolics by mulberry as related to development of phytoremediation technology. Chemosphere 32, 2471–2479.
6. Macek T., Macková M. and Káš, J. (2000) Exploitation of plants for the removal of organics in environmental remediation. Biotechnol. Advances 18(1), 23–35.
7. Novakova H. (2002) Dissertation Thesis, ICT Prague.
8. Kučerová P., Macková M., Chromá L., Burkhard J., Tříska J., Demnerová K. and Macek T. (2000) Metabolism of polychlorinated biphenyls by *Solanum nigrum* hairy root clone SNC-9O and analysis of transformation products. Plant and Soil 225, 109–115.

European Symposium on Environmental Biotechnology, ESEB 2004 - Verstraete (ed)
© 2004 Taylor & Francis Group, London, ISBN 90 5809 653 X

Biodegradation of diesel fuel in soil at low temperature by *Rhodococcus erythropolis*

E.M.B. Michel, I. Sokolovská, S.N. Agathos
Catholic University of Louvain, Faculty of Bioengineering, Agronomy and Environment, Unit of Bioengineering, Louvain-la-Neuve, Belgium

ABSTRACT: Diesel fuel degradation by a psychrotrophic *Rhodococcus erythropolis* in liquid minimal medium and in sandy soil at 12°C was tested. In liquid culture, 80% of diesel fuel was eliminated in 10 days. In soil, the degradation was monitored by a respirometric system and it was found that degradation yield depended on the inoculum size. The optimal dose for inoculation to ensure predominance of the desired bacterial population and high degradation yields was about 1011 cells kg^{-1} soil dw. The soil used was an uncontaminated sandy soil. Our results showed that the degradation yield depended on biostimulation with N and P sources. *Rhodococcus erythropolis* appears to have significant potential – 85 $mg.kg^{-1}$ soil $dw.d^{-1}$ – to degrade diesel fuel at 12°C in sandy soil and it is a promising strain for large-scale bioremediation processes.

1 INTRODUCTION

Organic pollutants appear in all climates. With increasing attention towards environmental preservation, the clean-up of contaminated sites is gaining increasing interest. Most investigations on biodegradation of organic pollutants concern petroleum hydrocarbons because oil spills represent a widespread problem. Bioremediation of hydrocarbon contaminated soils, aiming to cleanup and detoxify these hazardous pollutants, has been established as an efficient, economic, versatile and environmentally sound treatment. Field temperature plays a significant role in controlling the nature and extent of hydrocarbons metabolism. Temperature affects the rate of biodegradation (limitation of microbial growth and enzymatic activity), as well as the physical nature and the chemical composition of hydrocarbons.

The goal of our research was to select and characterise psychrotrophic bacteria isolated from cold environments that possess the ability to degrade efficiently diesel fuel, derived from distillation fraction of crude oil. Because the range of growth temperature of cold-adapted bacteria is wide (4–30°C) the application of such cold-adapted bacteria in the form of starters is a promising approach in the decontamination of polluted soils and wastewater.

2 MATERIALS AND METHODS

2.1 *Bacterial strain*

The bacteria were maintained on phosphate buffer mineral minimal agar plates containing 0.1% glucose and 0.1% sodium acetate. The phosphate buffer mineral minimal medium composition (PB-MM 284 Mergeay et al. 1985) was as follows per litre of distilled water: 100 ml phosphate buffer (pH 7.2, 0.5 $mol.l^{-1}$), 4.68 g NaCl, 1.49 g KCl, 1.07 g NH_4Cl, 0.43 g Na_2SO_4, 0.2 g $MgCl_2.6H_2O$, 40 mg Na_2HPO_4 $2H_2O$, 30 mg $CaCl.2H_2O$, 10 ml of ferric ammonium citrate solution (480 $mg.l^{-1}$) and 1 ml of trace elements solution (in 1 litre of distilled water: 1.3 ml HCl (25% v/v), 144 mg $ZnSO_4.7H_2O$, 100 mg $MnCl_2.4H_2O$, 62 mg H_3BO_3, 190 mg $CoCl_2.6H_2O$, 17 mg $CuCl_2.2H_2O$, 24 mg $NiCl_2.6H_2O$, 36 mg $Na_2MoO_4.2H_2O$).

2.2 *Inoculum preparation*

For liquid culture studies, *R. erythropolis* grown at 20°C on solid rich medium 869 (per liter of distilled water: 10 g of caseine peptone, 5 g of yeast extract, 5 g of NaCl, 1 g of dextrose, 0.345 g of $CaCl_2.2H_2O$, pH 7 and for solid medium 12 g of agar) was harvested and resuspended in PB-MM 284 medium without carbon source. The cell concentration was determined in terms of Colony Forming Units (CFU) as $3.4.10^8$ cells.ml^{-1}. This cell suspension was used to seed the cultivation medium at an inoculum concentration of 1% (v/v). For soil studies, the *R. erythropolis* cultivation was done at 20°C and 180 rpm in PB-MM 284 medium, with sodium acetate (2 g.l^{-1}) or pristane (1 g.l^{-1}) as sole source of carbon. When the culture reached the late exponential phase, cells were harvested, washed and resuspended in PB-MM 284 medium to obtain an optical density of 2.25 ± 0.15 (5 ± 2.10^{10} CFU).

2.3 *Soil*

The soil selected for this study was an uncontaminated subsoil (C-horizon) from Meix in the Belgium Ardennes. The soil pH was 4.96 (measured in H_2O) and 4.56 (measured in KCl 1M). It contained 0.04% total N, <0.5% total C, 0.033% CaO, 0.030% P_2O_5, 0.276% K_2O. The particle size distribution was 92.4% sand, 4.5% silt and 3.1% clay. The density was 1.4 and the maximum water holding capacity was 261 ml·kg^{-1} soil dry weight (dw). The native soil hydrocarbon content was negligible and did not interfere with the biodegradation studies.

2.4 *Biodegradation studies*

2.4.1 *Biodegradation studies in liquid culture*

Kinetics of diesel fuel degradation was studied in 5 ml of liquid PB-MM 284 medium (pH = 7). The cultivation was performed in sterile glass tubes with screw caps (20 ml). The sterile diesel fuel was supplied to obtain an initial concentration of 1600 ± 53 ppm. All cultivation tubes were shaken in the dark at 12°C at 180 rpm. At each sampling point, two tubes were sacrificed. Monitoring of residual concentration was performed by gas chromatography and bacterial growth via optical density measurement at 600 nm and CFU determination.

2.4.2 *Biodegradation studies in soil*

Kinetics of diesel fuel degradation in soil was studied using respirometric measurement system which consists in hermetic bottles of 1000 ml equipped with OxiTop®-C (WTW) measuring heads. 50 g dw of non-sterile soil was added into the previously sterilized bottles and the water content was adjusted to 70% of the soil maximum water holding capacity with sterile saline solution (NaCl 0.9% w/v) complemented with 10% of Tris-HCl buffer (0.5 M, pH 8). Subsequently, diesel fuel previously filtered (PALL, 0.2 μm) was added to obtain an initial concentration of 10000 mg·kg^{-1} soil dw. In terms of chemical oxygen demand (COD) this value represents 5943 mg O_2.kg^{-1} soil dw and it was determined using COD kit (Merck). A part of samples were supplemented with mineral nutrients (NH_4Cl, K_2HPO_4) so that a C/N/P ratio of 100/10/1 was obtained (Braddock et al. 1997). Finally, the soil was inoculated with *R. erythropolis* at an initial concentration of 5 . 10^{11} cells.kg^{-1} soil dw. A positive control with acetate as sole carbon source (10 g.l^{-1}) was inoculated by the strain pregrown on acetate. In order to determine abiotic hydrocarbon loss, previously inoculated soil was poisoned using NaN_3 at a final concentration of 0.04% (w/w). Based on our preliminary study, the content of bioavailable carbon source present in the soil with and without nitrogen and phosphorus supplementations did not interfere with the repirometric measurements. Furthermore, indigenous microorganisms were unable to significantly degrade diesel fuel with or without nutrient supplementation. All experiments were performed at 12°C in the dark. To avoid anaerobic conditions, the bottle contents were mixed thoroughly every second day.

The OxiTop® Control System was applied for respirometric measurements during bacterial diesel fuel degradation. The principle of this method consists in developing negative pressure during oxygen consumption in a closed vessel at a constant temperature. The OxiTop®-C measuring head quantifies and records a pressure developed during the respirometric test. The pressure values from the measuring heads are collected and processed by the OxiTop® OC110 controller. The BOD values are calculated using the values from the OxiTop®-C measuring head by the following formula (System OxiTop® Control. Operating Manual. WTW GmbH, D-82362 Weilheim, 1998):

$$BOD = \frac{M(O_2)}{R \cdot Tm} \cdot \left(\frac{Vt - Vl}{Vl} + \alpha \frac{Tm}{To} \right) \cdot \Delta p(O_2) \quad (1)$$

where BOD = biological oxygen demand (mg O_2.l^{-1}); M (O_2) = molecular weight of oxygen (32 mg.mmol^{-1}); R = gas constant (83.144 mbar.l.mol^{-1} K^{-1}); T_0 = reference temperature (273.15 K); T_m = temperature of measurement (K); V_t = bottle volume (nominal volume in ml); V_l = sample volume (ml); α = Bunsen absorption coefficient (0.03103); and Δp (O_2) = difference of the oxygen partial pressure (mbar).

3 RESULTS AND DISSCUSION

3.1 *Biodegradation studies in liquid culture*

In liquid medium, 80% of diesel fuel initially introduced into the culture broth was consumed in 10 days (Fig. 1). The maximal specific degradation rate was 0.51 . 10^{-2} mg·10^6 cells^{-1}.d^{-1} and the specific growth rate 0.02 h^{-1}. In Fig. 2, GC-chromatograms of cyclohexane extracts of medium on day 10 of cultivation

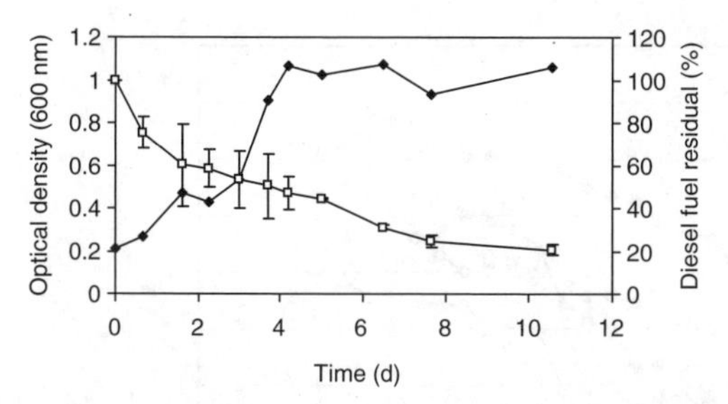

Figure 1. Kinetics of diesel fuel degradation by *R. erythropolis* in liquid PB-MM284 medium at 12°C, 180 rpm. Initial diesel fuel concentration: 1600 ppm. (□) Diesel fuel. (◆) Bacterial growth.

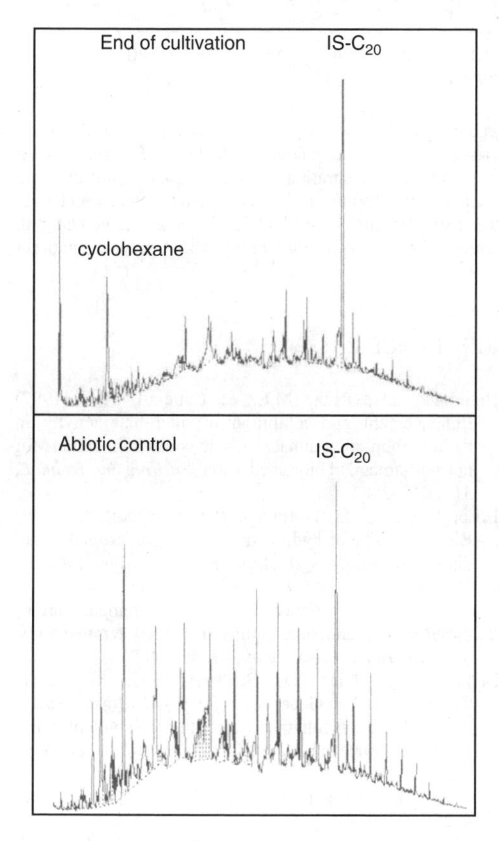

Figure 2. GC-chromatograms of culture broth extract on day 10 of culture and abiotic control during diesel fuel degradation by *R. erythropolis* in liquid PB-MM284 medium at 12°C, 180 rpm. Initial diesel fuel concentration: 1600 ppm. Eicosan (C20) was used as internal standard (IS).

and an abiotic control are shown. We can see that aliphatic hydrocarbons of about 15 carbons eluting in 15 minutes as well as hydrocarbons with retention time higher than 20 minutes were almost completely consumed at the end of culture. As our bacterial strain lacks in capacity to degrade aromatic hydrocarbons, the residual peaks may represent this fraction and the most recalcitrant aliphatic hydrocarbons. At 25°C, diesel fuel degradation reached 90%.

3.2 *Biodegradation studies in soil*

It was considered that oxygen depletion is correlated with diesel fuel assimilation. In our study, the yields of hydrocarbon degradation are expressed in terms of the ration $BOD/COD_{initial}$. $COD_{initial}$ is COD value of diesel fuel added into the soil. As shown in Table 1and Fig. 3, 9.4% of diesel fuel was degraded in 18 days without any mineral medium addition. N and P supplementation enhanced the biodegradation from 9.4 to 14.1% with an acetate preculture, and from 10.4 to 15.3% with a pristane preculture. Generally N and P supplemented cultures performed better than the corresponding unsupplemented cultures. It was possible that mineral nutrients may have become limiting in our soil and demonstrated the importance of C/N/P adjustment. A 10-fold concentrated N/P supplementation was tested and showed that such a nutrient level reduced microbial activity and became toxic for the microorganisms. Due to a low water holding capacity of our soil, soluble fertilizer salts partition into a limited supply of pore water, thus provoking a relatively high salt concentration in the soil solution. It has been suggested that due to osmotic potential changes following the addition of fertilizer salts, the calculated fertilizer necessary for complete degradation of petroleum contaminants may prove to be inhibitory when the fertilizer is added in a single application (Walworth & Reynolds 1995).

Within 18 days, the 1530 mg diesel fuel.kg^{-1} soil dw was degraded by a preculture on pristane when used with a N/P supplementation. As it can be seen in Fig. 3, the respirometric activity correlating with biodegradation did not reach a stationary phase. The inoculum precultured on pristane apparently enhanced

Table 1. Effects of preculture and nutrient supplementations on diesel fuel bioremediation efficiency.

	Biodegradation rate ($mg.kg^{-1}$ soil $dw.d^{-1}$)	ΔBOD/COD %
Control DF*	6	>1
Control DF + NP	6	>1
RE (acetate) + DF	52	9.4
RE (acetate) + DF + NP	78	14.1
RE (pristane) + DF	58	10.4
RE (pristane) + DF + NP	85	15.3

* Abbreviations as Figure 3.

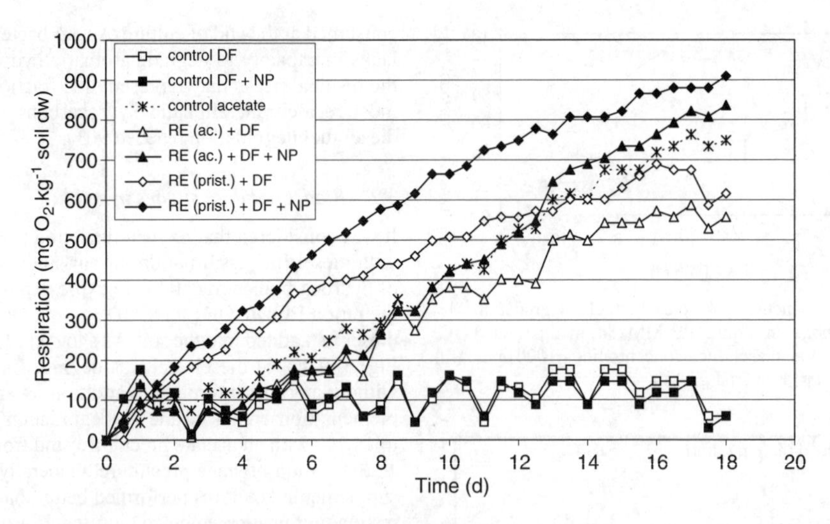

Figure 3. Effects of preculture and nutrient-supplementations on the respiratory activity of *R. erythropolis* at 12°C with an initial diesel fuel concentration of 10000 mg.kg^{-1} soil dw. (□) Diesel fuel (DF) abiotic control. (■) Diesel fuel abiotic control with nutrient supplementations (NP). (✳) Acetate control at an initial concentration of 10000 mg.kg^{-1} soil dw with *R. erythropolis* (RE) pregrown on acetate. (△) Diesel fuel bioaugmented with *R. erythropolis* pregrown on acetate. (▲) Diesel fuel bioaugmented with *R. erythropolis* pregrown on acetate, nutrient supplementations. (◇) Diesel fuel bioaugmented with *R. erythropolis* pregrown on pristane. (◆) Diesel fuel bioaugmented with *R. erythropolis* pregrown on pristane, nutrient supplementations.

the degradation rate by an enzymatic machinery stimulation and/or an adaptation of the bacteria to the hydrophobic compounds.

The conditions of hydrocarbon degradation in liquid and in soil medium are quite different. The shaking of liquid culture is favorable for oxygen transfer, but on the other hand, this parameter can be limiting in the soil.

Also, substrate bioavailability in the soil can be a crucial factor for soil bioremediation because of the hydrocarbon adsorption into the soil matrix (Löser et al. 1999). For microorganisms introduced into the soil to be effective in contaminant degradation, they must be transported to the zone of contamination, attach to the surface matrix, survive, grow and maintain their degradative abilities (Margesin & Schinner 1997).

4 CONCLUSION

Our results demonstrate that the potential of *R. erythropolis* to degrade diesel fuel at 12°C in liquid culture is very efficient and in sandy soil is significant, so that it is a promising strain to be used in degradation trials at large scales.

REFERENCES

Braddock, J.F., Ruth, M.L. & Catterall, P.H. (1997) Enhancement and inhibition of microbial activity in hydrocarbon-contaminated Arctic soils: implications for nutrient-amended bioremediation. *Environ. Sci. Technol.*, 31: 2078–2084.

Löser, C., Seidel, H., Hoffmann, P. & Zehnsdorf, A. (1999) Bioavailability of hydrocarbons during microbial remediation of sandy soil. *Appl. Environ. Microbiol.*, 51: 105–111.

Margesin, R. & Schinner, F. (1997) Bioremediation of diesel-oil-contaminated alpine soils at low temperatures. *Appl. Microbiol. Biotechnol.*, 47: 462–468.

Mergeay, M., Nies, D., Schlegel, H.G., Gerits, J., Charles, P., & Van Gijsegem, F. (1985) *Alcaligenes eutrophus* CH34 is a facultative chemolithotroph with plasmid-bound resistance to heavy metals. *Journal of Bacteriology*, 162: 328–334.

Walworth, J.L. & Reynolds, C.M. (1995) *J. Soil Contam.*, 4, 299.

System OxiTop® Control. Operating Manual. WTW GmbH, D-82362 Weilheim, (1998).

European Symposium on Environmental Biotechnology, ESEB 2004 - Verstraete (ed)
© 2004 Taylor & Francis Group, London, ISBN 90 5809 653 X

Combined chelate induced phytoextraction and *in situ* soil washing of Cu

B. Kos & D. Leštan
Agronomy Department, Biotechnical Faculty, University of Ljubljana, Ljubljana, Slovenija

ABSTRACT: In this study we examined the feasibility of new combined method for remediation of Cu contaminated soil. In a soil columns experiment, we investigated the effect of 5 mmol kg^{-1} soil addition of citric acid, ethylenediamine tetraacetate (EDTA), diethylenetriamine-pentaacetate (DTPA) and EDDS on phytoextraction of Cu from a vineyard soil with 162.6 mg kg^{-1} Cu, into the test plant *Brassicca rapa* var. *pekinensis*. We also examined the use of a horizontal permeable barrier for reduction of chelate induced Cu leaching and *in situ* soil washing. The addition of all chelates, except citric acid, enhanced Cu mobility and caused leaching. However, Cu plant uptake did not increase accordingly; the most effective was the EDDS treatment, in which the test plants accumulate only 0.85% of the total chelate mobilized Cu. Plant Cu concentration reached 37.8 ± 1.3 mg kg^{-1} Cu and increased by 3.3-times over the control treatment. The addition of none of the chelates in the concentration range from 5 to 15 mmol kg^{-1} exerted any toxic effect on respiratory soil microorganisms. When EDDS was applied into the columns with horizontal permeable barriers, only 0.53 ± 0.32% of the initial total Cu was leached. 36.7% of Cu was washed from the 18 cm soil layer above the barrier and accumulated in the barrier.

1 INTRODUCTION

The pollution of soils with heavy metals in the EU and associated European states is widespread. Heavy metals are one of the most prevalent agents causing public health problems. Some heavy metals: V, Cr, Mn, Fe, Cu, Zn, Mo and Ni are considered to be essential micronutrients for at least some forms of life; others have no known biological function. All heavy metals at high concentrations have strong toxic effects and are an environmental threat. Cu enters the soil by deposition from local foundries and smelters, through manuring with contaminated sludges, and from application of fungicides. Cu is an essential element: it forms organic complexes and metalloproteins, especially haemoglobin. With its known antifungal and algicidal properties, elevated levels of Cu in soil adversely affect microbially mediated soil processes (Wright and Welbourn, 2002). In EU countries the warning and critical limits of Cu in soil are set at 50 and 140 mg kg^{-1}, respectively (Council Directive 86/278/EEC, 1986).

A variety of approaches have been suggested for remediating of heavy metal contaminated soils. Solidification/stabilization contains the contaminants in an area by mixing or injecting agents such are cement and lime. Electrokinetic processes involve passing a low intensity electric current between a cathode and an anode imbedded in the contaminated soil. Soil washing involves the addition of water with chelates such as ethylenediaminetetraacetic acid (EDTA), or acids in combination with chelates. Phytoextraction of metals is biological treatment that uses metal-accumulating plants to clean up contaminated soils. The availability of some heavy metals for plants is restricted by the complexation of metals mainly within solid soil fractions. Chelates have been therefore used to artificially enhance heavy metals solubility in soil solution from the soil solid phase and thus to increase heavy metals phytoavailability. One of the main drawbacks of chelate induced phytoextraction is that most synthetic chelates form chemically and microbiologically stable complexes with heavy metals that pose a threat of groundwater contamination (Grčman et al., 2001).

We recently proposed a new method of combined induced phytoextraction and *in situ* soil washing technique by using biodegradable [S,S]-stereoisomer of ethylenediamine-disuccinate ([S,S]-EDDS) as a chelate, and horizontal permeable reactive barrier (Kos and Leštan, 2003). Barrier is placed below the layer of contaminated soil. Water soluble, biodegradable heavy metal-chelate complexes are than phytoextracted and accumulated into the plant biomass and washed from the layer of contaminated soil into the barrier by irrigation. In the barrier heavy metal-chelate complexes are microbialy degraded. Released ions of heavy metals chemically react with the sorbents in the barrier to form insoluble products and consequently accumulate in the barrier.

2 MATERIALS AND METHODS

2.1 *Soil properties*

Soil samples were collected from the 0–20 cm surface layer of a vineyard with a four-decade history of soil contamination with Cu containing pesticides. The following soil properties were determined: pH ($CaCl_2$) 7.4, organic matter 4.3%, total N 0.25%, sand 18.9%, coarse silt 16.0%, fine silt 36.4%, clay 28.7%, P (as P_2O_5) 49.8 mg 100 g^{-1}, K (as K_2O) 43.1 mg 100 g^{-1}, Cu 162.6 mg kg^{-1}. The soil texture was silty loam. After being air-dried, the soil was passed through a 4-mm sieve.

2.2 *Experimental set up*

The influence of 5 mmol kg^{-1} EDDS, EDTA, DTPA and citrate on Cu plant uptake and leaching were tested in a soil column experiment with four replicates for each treatment. Four thousand g of air-dried soil was placed into 18 cm high 15 cm diameter columns. Three-week old seedlings of *Brassica rapa* L. var. *pekinensis* (Nagaoka F1) were transplanted into columns and were grown for 5 weeks. EDDS, EDTA, DTPA and citrate were applied in 200 ml of deionised water in a single dose of 5 mmol kg^{-1} soil on the 35th day of cultivation. The aboveground tissues of test plants were harvested on the 40th day of cultivation, by cutting the stem 1 cm above soil surface. We continued with irrigation and leachate collection for 6 weeks after harvesting.

Horizontal permeable barriers were installed into four 28 cm high columns. Each barrier was positioned 18 cm below the soil surface and was composed of a 3 cm wide layer consisting of sawdust (74.25 g), soya meal (74.25 g) and vermiculite (16.5 g), followed by a 2 cm wide layer of soil, and 3 cm wide layer composed of apatite ($Ca_5(PO_4)_3OH$) (14.5 g), vermiculite (46.5 g) and soil (270 g). The apatite layer was followed by a 2 cm soil layer in the bottom of the column. Columns were treated with 5 mmol kg^{-1} EDDS.

All columns were equipped with trapping devices for leachate collection. Plastic meshes (D = 1.5 cm) were placed between layers to separate them, and at the bottom of the columns (D = 0.2 mm) to retain the soil. Leachates were sampled on the 14th, 28th and 42nd day after applying chelates.

2.3 *Cu determination*

For the analysis of Cu content, the soil samples were ground in an agate mill for 10 min and then passed through a 150 μm sieve. After the digestion of soils in *aqua regia*, AAS was used for the determination of Cu concentrations.

Shoot tissues were collected and thoroughly washed with deionized water. They were dried to a constant weight and ground in a titanium centrifugal mill. Cu concentrations in plant tissue samples (290–310 mg dry weight) were determined using an acid (65% HNO3) dissolution technique with microwave heating and analyzed by Flame-AAS. The Cu concentration in leachates was determined by Flame-AAS.

2.4 *Glucose-induced respiration of soil microorganisms*

A modified procedure described by Fredrickson and Balkwill (1998) was used to measure glucose-induced respiration in soil, to assess the potential toxicity of EDDS, EDTA, DTPA and citrate soil additions on soil microorganisms. One hundred g (dry weight) of 5 mm sieved soil was placed into 1L glass jars. The soil was treated with 0 (control), 5, 10 and 15 mmol kg^{-1} of EDDS, EDTA, DTPA and citrate. Sterile glucose solution was added to achieve a glucose concentration of 10 μmol g^{-1}, and a final water content of 90% of the maximum water holding capacity of the soil. A plastic beaker, filled with 10 mL of 25% NaOH was placed into the jars, crimp-sealed with rubber lids equipped with manometric measuring heads (OxiTop, WTW, Weilheim). Glucose-induced respiration was expressed in μmols of oxygen consumed by microorganisms in 1 g of soil after 3 days of incubation at 22°C. Five replicates for each treatment were made.

3 RESULTS AND DISCUSSION

3.1 *Cu plant uptake*

EDDS, which was the most effective chelate for induced phytoextraction of Cu into the aboveground parts of the test plant *Brassicca rapa* var. *pekinensis*, increased plant Cu concentrations by 3.3-times over the control treatment (Table 1). Additions of EDTA and DTPA were almost equally effective, increasing Cu plant uptake by 1.9 and 2.1 times, respectively. The addition of citrate had no effect on the Cu plant concentration. The absence of Cu in leachates in citrate treatments could indicate citrate degradation (Figure 1). The argument against degradability as a factor of inefficiency is that EDDS, the most effective chelate in our study, is also easily degradable. The reported complete mineralisation of EDDS in sludge amended soil was 28 days, with a calculated half life of only 2.5 days (Jaworska et al., 1999).

Cu in plants did not increase in accordance with the chelate mediated increase of phytoavailable Cu in soil. Less than 1% of total Cu available in the soil solution was removed into the plant biomass (Table 1). We

Table 1. Biomass dry weight and concentration of Cu in the leaves of the test plant *Brassica rapa* var. *pekinensis* and the portion of initial total Cu phytoextracted by the test plant in treatments with a 5 mmol kg^{-1} addition of different chelates.

Treatment	Dry plant biomass (g)	Cu uptake (mg kg^{-1})	Cu accumulated (%)
Control	5.41 ± 0.69	11.28 ± 2.17[a]	0.054 ± 0.027
Citrate	5.88 ± 1.09	11.23 ± 1.14[a]	0.066 ± 0.008
EDTA	5.21 ± 0.99	21.71 ± 1.55[b]	0.114 ± 0.025
DTPA	5.33 ± 0.62	24.28 ± 0.83[b]	0.129 ± 0.013
EDDS	4.87 ± 0.40	37.81 ± 1.34[c]	0.185 ± 0.021

[a, b, c] Statistically different treatments according to Tukey test (P = 0.05).

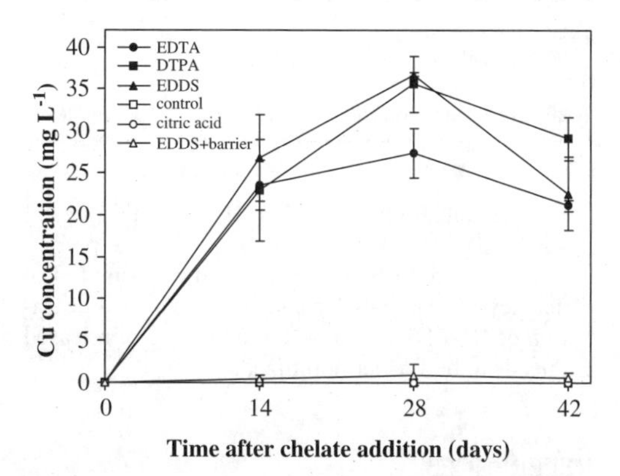

Figure 1. The effect of 5 mmol kg^{-1} addition of citric acid, EDTA, DTPA and EDDS on Cu leaching from soil columns during phytoextraction, and leaching of Cu from columns.

observed necrotic lesions on the leaves of *Brassicca rapa* var. *pekinensis* in EDTA treatments, but no reduction in plant biomass (Table 1).

The mass balance of Cu showed that even in EDDS treatments, the percentage of Cu phytoextracted in one phytoextraction cycle was only 0.185 ± 0.021% of the total initial Cu in the soil (Table 1). For efficient soil remediation within a reasonable time span, plant Cu concentrations exceeding 2000 mg kg^{-1} of dry plant biomass would be required to reduce soil Cu concentrations by approximately 110 mg kg^{-1}, below the warning limit of 50 mg kg^{-1} (Council Directive 86/278/EEC, 1986), over 10 years using plants with a high biomass yield (20 tons ha^{-1} of dry matter). This concentration is more than 50-times higher than concentrations in our low biomass test plant.

3.2 *Toxicity of chelate addition to soil microorganisms*

Soil microorganisms are critically important for normal functioning of the soil and are common indicators

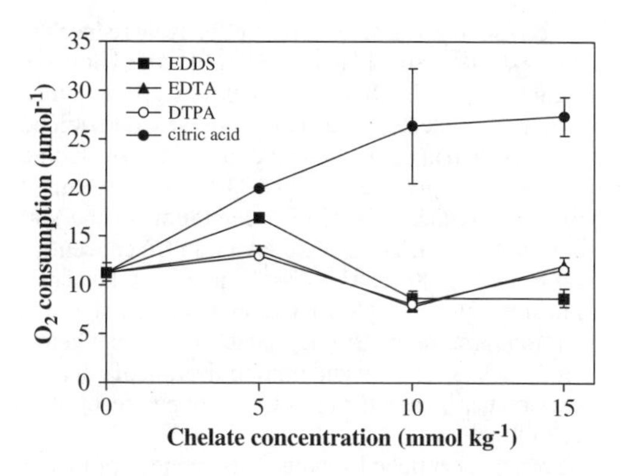

Figure 2. Glucose-induced microbial respiration, in µmol of consumed O_2 per g of dry soil, in soils treated with 5, 10 and 15 mmol kg^{-1} of citric acid, EDTA, DTPA and EDDS, after 24 hours of incubation at 21°C. The means of five replicates are presented; error bars represent standard deviation.

of soil quality. Soil microorganisms depend directly on the soil solution for uptake of food and water, and chelate elevated metal concentrations could lead to toxic effects for these organisms. The possible toxicity of citrate, EDTA, DTPA and EDDS soil additions on soil microorganisms was examined with a glucose-induced respiration method. The addition of different concentrations of citrate increased soil respiration, presumably due to the microbial use of citrate as an additional carbon and energy source (Figure 2). Low (5 mmol kg^{-1}) concentrations of other chelates (especially EDDS) also slightly increase respiration, perhaps because they increase the bioavaliability of trace metals essential for microbial growth. At higher concentrations, the respiration was not significantly lower than in soil with no chelate addition, indicating that chelates and metal-chelate complexes exert no toxic effect on respiratory soil microorganisms.

3.3 *Permeable horizontal barrier*

Horizontal permeable barriers were used in the EDDS treatment to prevent Cu leaching by trapping Cu, released from Cu-EDDS complex, into the barrier. The barrier was composed of a layer of soya meal enriched sawdust and vermiculite for enhanced microbial degradation of the Cu-EDDS complex, and a layer of apatite, vermiculite and soil mixture for immobilization of released Cu. Vermiculite was used to increase the water holding capacity of the sawdust and apatite layer to 3.10 and 1.20 g g^{-1} substrate, respectively. The water holding capacity of the soil was 0.68 g g^{-1}. A high water holding capacity of the barrier was important to retain the soil solution with the Cu-EDDS complex in the barrier and thus to prolong the time available for microbial degradation of Cu-EDDS and for Cu binding.

Horizontal permeable barriers effectively reduce the concentration of Cu in leachates in EDDS treatments (Figure 1). The distribution of Cu through the columns is shown in Table 2. Cu concentrations were determined at seven different points along the soil profile, including in the sawdust and apatite layer. The EDDS treatment uniformly reduced the Cu concentration in the soil above the barrier. The steep increase in Cu concentration from 0 to 289 ± 41 mg kg^{-1} in the sawdust layer indicates intensive Cu-EDDS degradation and Cu sorption processes. Bio-degradability of the heavy metal-chelate complex and immobilization of released heavy metals is essential for this type of permeable barrier to function.

Soya meal enriched sawdust was intended primarily as an energy source for microbes and as a surface for development of microbial films. Data from Table 2 indicate that sawdust (and probably vermiculite), served also for adsorption of Cu.

To catch the remaining released Cu, a second, apatite layer was set in the barrier. The sorption mechanisms of apatite for heavy metals other than Pb are less clear. Cu phosphates have low solubility, e.g. the solubility constant (log K_{sp}) for $Cu_5(PO_4)_3OH$ is −51.62 (Eighmy et al., 1997), which indicates that apatite could be used for Cu immobilization. A decreased concentration of Cu in the apatite layer after EDDS treatment (Table 2) was therefore unexpected. A possible explanation is that essentially all Cu released from the Cu-EDDS was complexed into the sawdust layer. Since the phosphate matrix strongly interferes with metal determination by AAS, it is also possible that we measured lower than actual Cu concentrations in the apatite layer. This is also a possible explanation for unaccounted Cu in the total balance of Cu in columns.

After EDDS addition and irrigation, 36.7% of total initial Cu was washed from 18 cm of soil above the permeable barrier and accumulated there (Table 2). Assuming a constant efficiency of repeating EDDS treatments (soil washing cycles), these data indicate that 3 cycles would be required to clean the top 18 cm soil from an initial 160 to 50 mg kg^{-1} Cu, the warning limit set by the European Council Directive on protection of the environment (Council Directive 86/278/EEC, 1986).

Results of our study indicate that rather than for the prevention of Cu leaching during chelate induced phytoextraction, horizontal permeable barriers, together with the use of biodegradable chelate EDDS, are potentially important for *in situ* soil washing of Cu, as proposed earlier for Pb (Kos and Leštan, 2003). Instead of EDDS, other commercially available biodegradable chelates could be used, e.g. nitrilotriacetic acid (NTA). To implement *in situ* soil washing of Cu, further research is needed to assess the possible risks associated with a permanently installed permeable barrier saturated with Cu. Alternatively, after the soil remediation process has been completed, the permeable barrier could be excavated and deposited. The efficiency of materials for enhanced microbial degradation of Cu-EDDS complex and binding of released Cu needs to be further optimized.

Table 2. Cu concentration through the profile in soil columns with permeable barriers after single cycle of EDDS induced soil washing, and percentages of total initial Cu removed from the contaminated soil and leached from the soil columns with and without permeable barrier. Means of four replicates ± SD are presented.

Soil Layer	Cu (mg kg^{-1})
0–6 cm	103.05 ± 6.73
6–12 cm	96.02 ± 13.33
12–18 cm	114.39 ± 14.27
Barrier	289.05 ± 41.35
21–23 cm	114.49 ± 10.34
Barrier	87.80 ± 11.75
26–28 cm	118.20 ± 12.30
Total Cu removed	35.36 ± 7.08%
Total Cu leached (barrier) 0.528 ± 0.323%	
Total Cu leached (control)	21.64 ± 2.69%

REFERENCES

Council Directive 86/278/EEC. 1986. On the Protection of the Environment, and In Particular of the Soil, When Sewage Sludge is Used in Agriculture. EC Official Journal, L181.

Eighmy, T.T., Crannell, B.S., Butler, L.G., Cartledge, F.K., Emery, E.F., Oblas, D., Krzanowski, J.E., Eusden, J.D., Shaw, E.L. & Francis, C.A. 1997. Heavy metal stabilization in municipal solid waste combustion dry scrubber residue using soluble phosphate. Environ. Sci. Technol., 31: 3330–3338.

Fredrickson, J.K. & Balkwill, D.L. 1998. Sampling and Enumeration Techniques. In Burlage, S.B., Atlas, R., Stahl, D., Geesey, G., Sayler, G. (Eds.), *Techniques in Microbial Ecology.* Oxford University Press Inc., New York.

Grčman, H., Velikonja Bolta, Š., Vodnik, D., Kos, B., & Leštan, D. 2001. EDTA enhanced heavy metal phytoextraction: metal accumulation, leaching and toxicity. Plant and soil, 235: 105–114.

Jaworska, J.S., Schowanek, D. & Feijtel, T.C.J. 1999. Environmental risk assessment for trisodium [S,S]-ethylene diamine disuccinate, a biodegradable chelator used in detergent application. Chemosphere, 38: 3597–3625.

Kos, B. & Leštan, D. 2003. Induced phytoextraction/soil washing of lead using biodegradable chelate and permeable barriers. Environ. Sci.Technol., 37: 624–629.

Wright, D.A. & Welbourn, P. 2002. *Environmental toxicology*. Cambridge University Press, Cambridge.

European Symposium on Environmental Biotechnology, ESEB 2004 - Verstraete (ed)
© 2004 Taylor & Francis Group, London, ISBN 90 5809 653 X

Combination of respirometry and indirect impedancemetry to evaluate the microbial activity of soils

T. Ribeiro
Department of Agro-Industrial Sciences and Techniques, Institut Supérieur Agricole de Beauvais, rue Pierre Waguet, Beauvais Cedex, France

R. Lehtihet, E. Lux, G. Couteau, O. Schoefs, N. Cochet* & A. Pauss
*Department of Chemical Engineering, *Department of Biological Engineering, UMR CNRS, Université de Technologie de Compiègne, Compiègne Cedex, France*

ABSTRACT: A combination of existing methods used for agrofood industry (impedancemetry) and wastewater treatment (respirometry) was evaluated to assess the microbial activity of soils. On a model soil, we showed it is possible to measure easily simultaneously the oxygen consumption and the carbon dioxide production. A C/O balance can be calculated rapidly in the first hours of experimentation, allowing therefore a rapid evaluation of microbial activity or even a detection of inhibitors.

1 INTRODUCTION

How much organics (pollutants or from natural origin) that a soil or compost contain, can be digested by microbial activity? This question can be answered by means of this modern measurement set-up. Important application is to determine remediation (natural or man-initiated) activity of micro-organisms of polluted soils eg. hydrocarbons. Practically all organics in soil or compost will be digested by micro-organisms in a certain period. Micro-organisms may be present in the soil naturally or initiated or stimulated by man. They may be available in the aerobe as well as in the anaerobe zone of the soil. They can be present as wide broad communities, exhibiting several different activities. The micro-organism numeration in soil is somewhat difficult, very often partial and not being reflected their activity in situ. Only aspecific methods allow the evaluation of global activity, to realize this activity measurement without altering the soil structure. Some techniques allow the activity estimation of a microbial population. For example the degradative capacity or ability of a pollutant can be evaluated by measuring the isotopic amount incorporated from a radiolabelled precursor. Other indirect techniques, like respirometry, allow the quantification of a weak microbial activity.

The respirometry is a technique based on quantification of gas produced by microbial metabolism or activity. A specific case of respiration measurement is the BOD assay, where the carbon dioxide produced reacts with a KOH solution. The depression formed by the KOH reaction was detected by a pressure sensor, expressed finally in mg of oxygen consumed per Liter.

This impedancemetry technique is dedicated to the measurement of growth and/or microbial activities, by quantification of carbon dioxide produced by micro-organism metabolism. The CO_2 from microbial growth is absorbed by a KOH solution, whose impedance is measured; the impedance variation is so quantitatively and qualitatively related with CO_2 production.

We developped a combination of both to evaluate microbial activity in soil in our laboratory and/or to determine microbial respiration according to DIN-standard 19737, in order to realize and correlate carbon and oxygen balances. This method is being optimized for a model soil.

This method, based on gas measurement, is not specific and does not allow the biodiversity knowledge of the soil. But the comparison between different soils and the estimation of several factors (weather, pollution, *etc.*) can be realized, because of the global aspect.

2 MATERIAL AND METHODS

2.1 *Soil*

The soil was sampled from a commercial product: a compost for gerania COMPO SANA® (sphaignes, barks of conifer stamped and fertilizers of cattle), and was chosen as a soil model.

The soil was stored in a semi-hermetic trash at room temperature (1–3 months). The soil has the following characteristics: Dry Matter 33%, Organic Matter 24%, pH 6.5, water retention ability 250 g per 100 g dry matter, resistivity 1500 Ohm cm^{-1}.

2.2 *BOD-meter*

The BOD measurements were carried out by the OxiTop® with IR-pressure sensor (WTW, Germany). The flasks were glass bottles with a volume of 625 mL, closed with rubber. The acquisition of data was carried out by a controller. The carbon dioxide produced reacts with a KOH solution. The depression formed by the KOH reaction was detected by a pressure sensor, expressed finally in mg of oxygen consumed per Liter.

2.3 *Impedancemeter*

The device consists of a software system enabling the data acquisition via a connected bus which is wired to the measuring cells. Each measuring cell is an individual impedancemeter, able to measure the conductance change in the potassium hydroxide solution for a given sample. These cells are composed of a tube containing a KOH solution and two electrodes. The conductance values are recorded every 15 minutes or less, although their acquisition is permanent. Each cell can be activated independently of the others, and gives a result expressed finally in mg of carbon dioxide produced per Liter. This device was developped for determination of microbiological contaminations in agrofood industries (Ribeiro et al. 2003) and is adapted for the measurement of carbon dioxide in soils and heterogeneous media.

2.4 *Respirometric assays*

The soil sample was located in the BOD bottle equipped with an infrared-pressure sensor on the top of the sample flask and an impedance cell measurement. The bottle contains 116.9 g of soil sample, corresponding in a volume of 165 mL. The series of experiments were performed at room temperature, approximatively 295 K. The concentration of KOH solution was 15 gL^{-1}.

3 RESULTS AND DISCUSSION

Figure 1 shows the conductance response of the impedancemeter, after the reaction between the CO_2 produced due to the soil respiration and the KOH solution, and the Biological Oxygen Demand of Oxygen measured with the pressure sensor. Between 0 and 2.5 hours, the conductance curve gives quickly an estimation of the soil activity, reflecting by the carbon dioxide production, while the BOD curve starts to increase slowly. After 2.5 h, no variations of

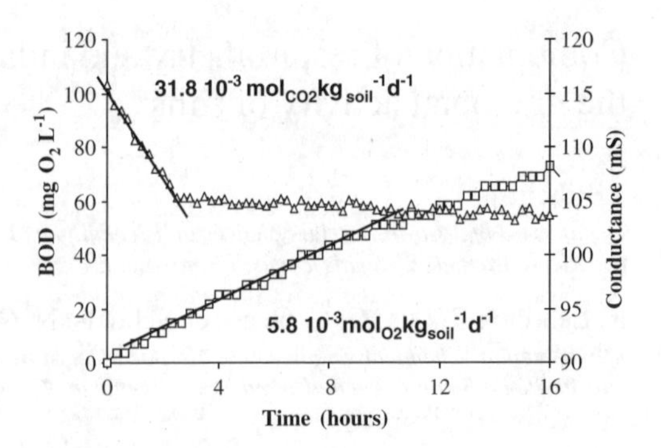

Figure 1. Relationship between BOD (squares) and indirect impedancemetry (triangles).

conductance were observed due to the KOH saturation by the CO_2, but the BOD increases in a constant way to reach a plate after 16 hours. The O_2 consumption and the CO_2 production rates were calculated for the first hours. The O_2 consumption rate is equal to 0.18 g_{O2} $kg_{soil}^{-1} d^{-1}$, while the CO_2 production rate is equal to 1.4 g_{CO2} $kg_{soil}^{-1} d^{-1}$, providing an important Q_R ratio of 7.8. The CO_2 production rate appears important as compared to the few available data in the literature – (Wang et al. 2003) determined a carbon dioxide production rate equal to 0.88 g_{CO2} $kg_{soil}^{-1} d^{-1}$ for pre-incubated soil – but our soil is rich of organic matter. During the 16 hours of experimentation, the oxygen was not limited in the headspace.

4 CONCLUSIONS

The combination of indirect impedancemetry and respirometry for soil evaluation is feasible. The indirect impedancemetry method is currently more sensitive than respirometry and a C/O balance can only be calculated during the first hours. This system is quite interesting for the rapid determination of the metabolism or activity of microorganisms communities in soil. This method is being optimized for this model soil and currently tested with several different soils.

REFERENCES

Deutsches Institut für Normung e.V. 2001. Soil quality – Laboratory methods for determination of microbial soil respiration. Norme DIN 19737.

Ribeiro, T. et al. 2003. Development, validation, and applications of a new laboratory-scale indirect impedancemeter for rapid microbial control. Applied Microbiology & Biotechnology 63(1): 35–41.

Wang, W.J. et al. 2003. Relationships of soil respiration to microbial biomass, substrate availability and clay content. Soil Biol. Biochem. 35: 273–284.

Gas treatment

European Symposium on Environmental Biotechnology, ESEB 2004 - Verstraete (ed)
© 2004 Taylor & Francis Group, London, ISBN 90 5809 653 X

The two-phase reactor water/silicon-oil: prospects in the off-gas treatment

J.M. Aldric, J. Destain & P. Thonart
Faculté Universitaire des Sciences Agronomiques de Gembloux, Unit of Bioindustries,
Centre Wallon de Biologie Industrielle

ABSTRACT: A research was carried out to develop a biphasic biological reactor able to cleanse the gas effluents polluted by volatile organic compounds. Initially, *Rhodococcus erythropolis* T 902.1 had been selected on the basis of its good capacity to degrade a compound selected as model (isopropyl-benzene). The "Doehlert" statistical model was selected to estimate the effect of gas flow and the IPB concentration on the biodegradation of IPB. The two factors chosen in this first step are the gas flow and the concentration in IPB of this one. The results shows that the use of silicon-oil makes it possible to absorb large quantities of IPB within the medium of biological abatement. On the other hand, the rate of biodegradation is directly related to the inlet flow of IPB. Thus, the reactor is able to degrade more than 7 mg/min.L and presents interesting opportunities in the biological treatment of gas effluents.

1 INTRODUCTION

A research is carried out within the framework of gaseous treatment. It aims to develop a biphasic reactor "water/silicon- oil". Silicon-oil will be used to allow a better biological abatement of the aromatic organic compounds by improving their solubility within the biphasic reactor. Initially a bacterial strain (*Rhodococcus erythropolis T 902.1*) was selected on the basis of its good capacity to degrade isopropyl-benzene (IPB), compound selected as model such as Benzene, Toluene and Xylene. Various research was carried out in order to improve degradation of VOC in gas effluents. In particular by improving the gas transfer of VOC within the reactor. Thus, Budwill and Coleman (1997), developed a bio-filter with silicone oil addition. This addition makes it possible to reach a clear improvement hexane biodegradation.

In addition, Yeom and Daugulis (2001) recommend the use of a biphasic biological reactor whose organic phase (hexadecane) constitutes 1/3 of the reactional medium. Hexadecane presents the property to be slightly toxic for the micro-organisms. A similar process showed its effectiveness on the biodegradation of a mixture of organic pollutants (BTEX) by a strain of *Pseudomonas sp.* (Collins and Daugulis, 1999). Lastly, Aldric (2001) showed the interest of the use of the silicone oil at a rate of 10% in a biphasic bioreactor. The silicone oil allows an important improvement of the gas retention and gas transfer.

2 MATERIALS AND METHODS

2.1 *Strain and chemicals*

The *Rhodococcus erythropolis* strain was obtained from the collection of the Walloon Center of Industrial Biology (C.W.B.I.).

All the substrates and other chemicals were purchased at VWR internationnal (Leuven, Belgium) or Aldrich (Bornem, Belgium).

2.2 *Bioreactor and assembly*

The bioreactor used (Biolafitte BL06.1) and the composition of the two-phase medium were described by Aldric (2001). The assembly is schematized in Figure 1.

2.3 *Statistical method*

The "Doehlert" statistical model was selected to estimate the effect of gas flow and the concentration of IPB on biodegradadation of IPB. The ten experiments carried out are included and characterized in Table 1. Blackened experiments represent the repetitions of central experiment.

2.4 *Sampling and analytical methods*

Gas samples are taken regularly from each bubble of sampling like in the liquid reactional medium. Estimation of Isopropylbenzene was performed with a Perkin Elmer headspace sampler HS 40 XL (for liquid samples) and a gaz chromatograph Hewlett Packard 5890 equiped with a Altech INC. Deerfield EC-WAX column and flame ionisation detector, the température of injector, column and detector were 153, 150 and 250°C respectively.

2.5 *Implementation of the biomass*

The inoculum of the biological reactor is obtained by centifugation of 2,25 L of a culture of *Rhodococcus erythropolis* in medium 868 (glucose 20 g/l.; Casein peptone 20 g/l; yeast extract 10 g/l).

The pellet obtained is washed twice and diluted in 200 ml of water (6 g/l NaCl). Inoculum is then introduced into the bioreactor. The medium for biodegradation is composed of silicon oil (10%) and M284 (90%) as described by Aldric (2001).

2.6 *Determination of concentrations and flows*

Only the data corresponding to a stabilization of the IPB concentration within the liquid medium are retained, the following relation is indeed correct:

$$K_L a = \frac{Q_{in} - Q_{out}}{\left(C_{L^0} - C_L\right)}$$

where $K_l a$ is the total coefficient of mass transfer for IPB (min^{-1}); Q_{in} and Q_{out} = the inlet flow and outlet flow of IPB respectively (mg/min.L of reactional medium); C_L^0 = saturating concentration of IPB in the biphasic liquid medium (ppm); C_L = concentration of IPB in the biphasic liquid medium (ppm).

Inlet flow and outlet flow of IPB are given as follows:

$$Q_{IN} = \frac{Conc.IN_{moy} \times flow}{vol}$$

$$Q_{OUT} = \frac{Conc.OUT_{moy} \times flow}{vol}$$

with

$$Conc_{moy} = \sum_{i=1 \, \grave{a} n} \frac{Conc_i \times (t_i - t_{i-1})}{t_{tot}}$$

where $Conc._{moy}$ = weighted average concentration during an equilibrium phase (ppm). $Conc_i$ = specific concentration measured by gas injection at time t = i. flow = flow of gas effluent charged in IPB t = time (h).

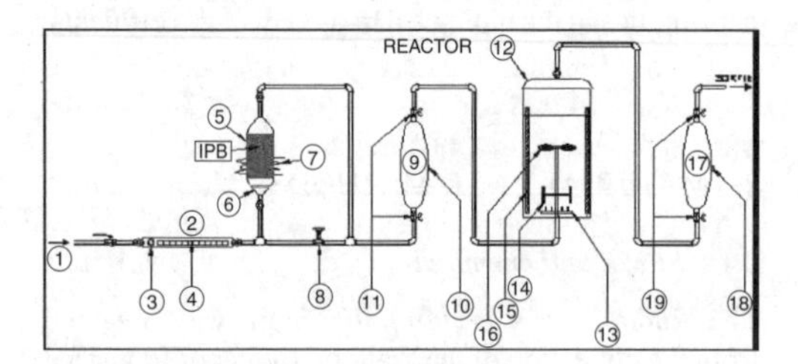

Figure 1. Assembly used.

1) Compressed air
2) Rothametre
3) Adjustment
4) Graduation
5) Stripping of IPB
6) Sintered glass
7) Temperature control
8) Regulation of polluted air flow
9) 10) 17) and 18) bubble for sampling
11) Isolating valve
12) Biological reactor
13) Gas diffuser
14) and 16) Stirring
15) Baffle.

Table 1. Experiments carried out according to Dhoelert statistical model.

Experiment	Concentration (mg/m³)	Gas flow (1/min)
1	3050	4,5
2	1575	1,47
3	3050	4,5
4	1575	7,53
5	4525	7,53
6	4525	1,47
7	3050	4,5
8	100	4,5
9	3050	4,5
10	6000	4,5

3 RESULTS AND DISCUSSION

3.1 *Follow-up of experimentation*

As example, evolution of IPB concentrations in inlet gas, outlet gas and within the liquid medium is given in Figures 2 and 3 for expériments 2 and 5.

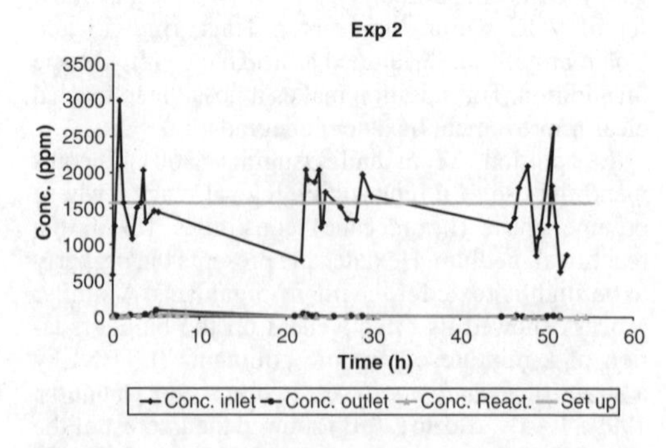

Figure 2. Evolution of IPB concentrations in inlet gas, outlet gas and within the liquid medium for experiment 2. Gas flow = 1.47 l/min; Concentration of inlet gas = 1575 ppm.

Figure 2 shows the evolution of the IPB concentration as described above. Experiment 2 corresponds to a IPB flow of 0.60 mg/min.L of reactional medium. As well concentration within the liquid as the concentration in outlet gas of bioreactor are very weak, which represents a very high rate of biodegradation (96% on average). Experiment 5 corresponds to the highest flow of IPB (9.47 mg/min.L). Figure 3 shows the evolution of the IPB concentrations for experiment 5.

It is clear that the concentration in outlet gas is high (1727 ppm on average), but comparatively to concentration of inlet gas, and the gas flow, the rate of biodegradation remains very acceptable (65% on average).

3.2 *Rate of biodegradation*

The rate of biodegradation for each experiment is shown in Figure 4.

The rates of biodegradation are high for the whole of the experiment, except for experiment 5 and 7 (respectively 65% and 64%). The relatively low rate of biodegradation observed in experiment 5 can be explained by a too significant flow of IPB. For experiment 7, this can be explained by a less cellular concentration (Fig. 8). For a low concentration of IPB in inlet gas (100 ppm; exp 8) the rate of biodegradation is also less. This can be explained by a weaker biodisponibility of IPB for the micro-organisms. Indeed, IPB is strongly solubilized in the silicon oil., what weakens the biodisponibility.

The rate of biodegradation is very high for experiments 2 and 6 (96%), corresponding to low flow of IPB (0.60 and 1.77 mg/min.l respectively). The high rates observed do not make it possible to highlight the independent influence of the flow or the concentration on the biodegradation.

However, a correlation can be established between the rate biodegradation and the inlet flow of IPB (Fig. 5). The correlation is low but it translates the good response of bioreactor with respect to an increase in the flow of pollutant.

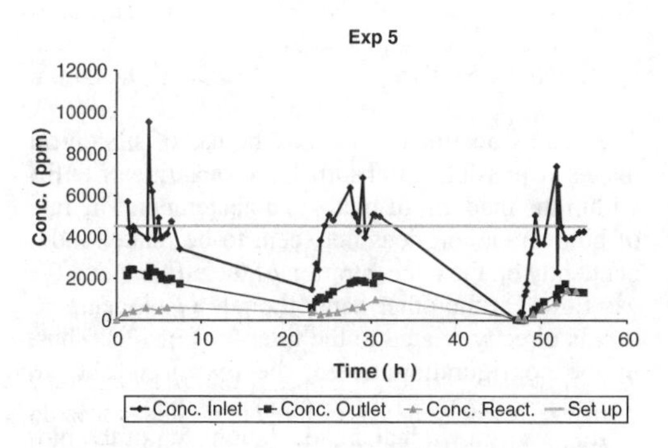

Figure 3. Evolution of IPB concentrations in inlet gas, outlet gas and within the liquid medium for experiment 5. gas flow = 7.53 l/min; concentration of inlet gaz = 4525 ppm.

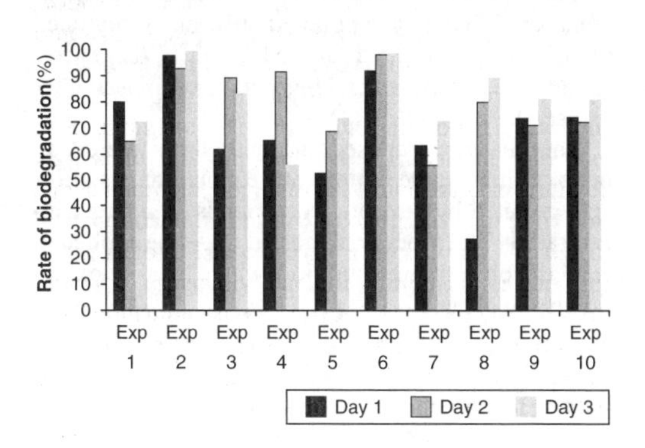

Figure 4. Average rate of biodegradation observed during experiments.

3.3 *Flow of biodegradation for the experiments*

In order to visualize the absolute quantities of degraded IPB, Figure 6 presents the flows of biodegradation.

$Q_0 = Q_{in} - Q_{out}$

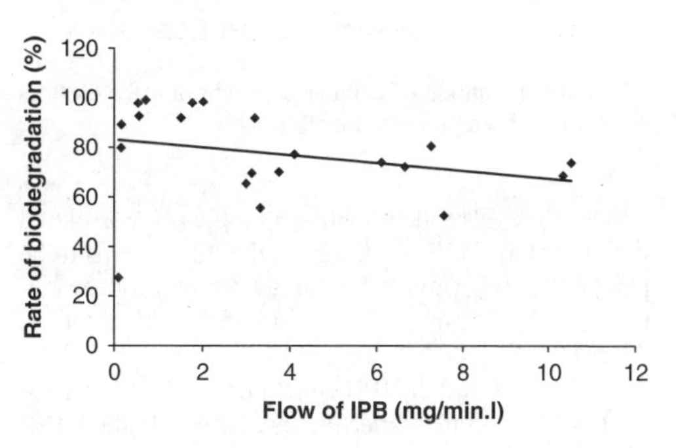

Figure 5. Correlation between rate of biodegradation and inlet flow of IPB.

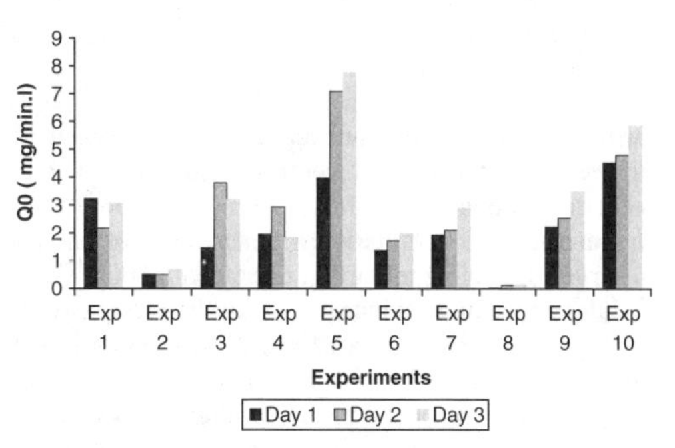

Figure 6. Flow of biodegradation.

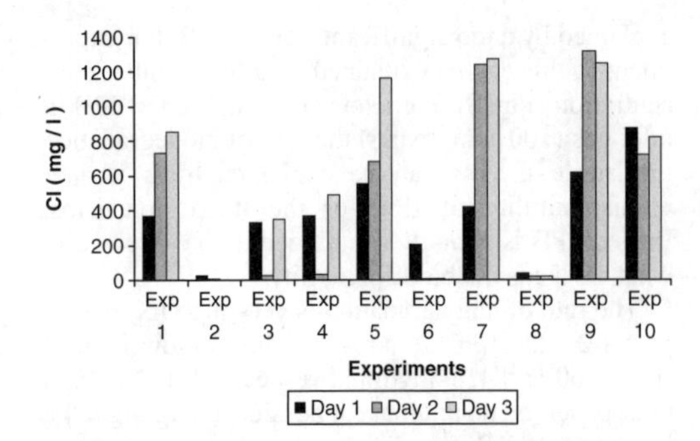

Figure 7. IPB concentration in the reactional medium.

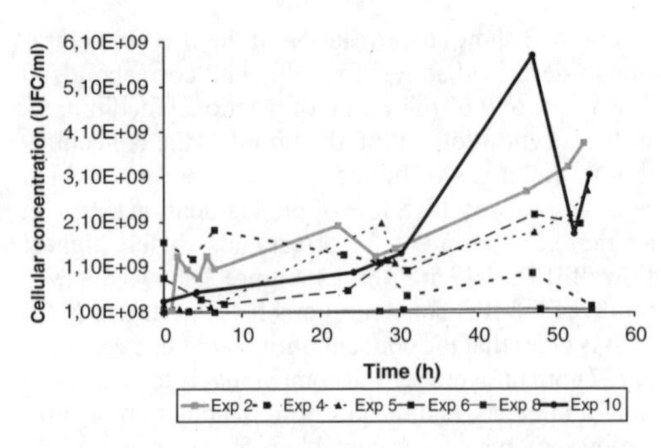

Figure 9. Evolution of cellular concentration for experiments.

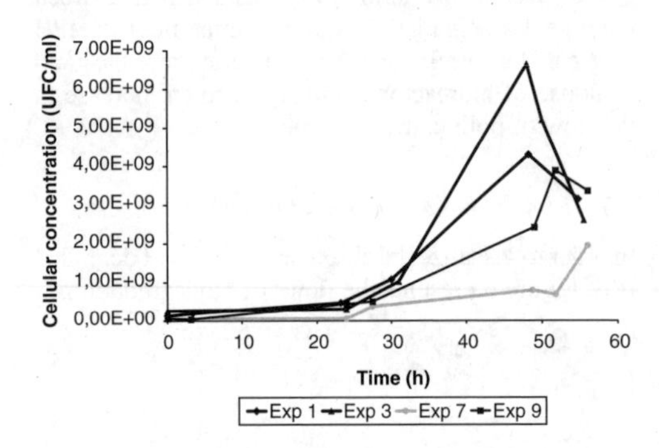

Figure 8. Evolution of cellular concentration for the four repetitions of central experiment.

The flow of biodegradation is logically correlated with the flows of IPB. However, the limit seems to be reached in experiment 5 with a flow of biodegradation of 7.5 mg/min.L. This limit can be explained by Figure 7.

Figure 7 shows the IPB concentration in the biphasic liquid medium. Experiments 5, 7 and 9 show that the concentration in the liquid medium is maximum with approximately 1200 ppm. The gas transfer is thus limiting for the biodegradation.

3.4 *Evolution of cellular concentration*

Evolution of cellular concentration is presented in Figures 8 (repetitions of central experiment) and 9 (other experiments).

Generally, a adaptation phase of biomass is observed at the beginning of experiment (24 h). Then, cellular concentration increases slightly during experiments. This growth is maximum for flows of IPB ranging between 3.5 and 6.5 mg/min.L. (Exp 1, 3, 9 and 10). However, a fall of the cellular concentration is observed at the end of the experiment, which probably represents an exhaustion of the medium. For a very low flow of IPB (exp 8; $Q_{in} = 0.15$ mg/min.l) the cell multiplication is strongly slowed down. What is explained logically by a small feed of carbonaceous substrate. Lastly, it is important to note that the IPB supply, including with strong concentration, does not induce a cellular mortality.

4 CONCLUSIONS

The results obtained show that the use of silicon-oil makes it possible to absorb large quantities of IPB within the medium of biological abatement. The rate of biodegradation does not seem to be related independently on the concentration of the effluent and the gas flow. On the other hand, the rate of biodegradation is directly related to the inlet flow of IPB. Thus, in the configuration tested, the reactor is able to degrade 7.5 mg/min.L of reactional media.

It is also shown that an adaptation step of the biomass is necessary to reach rates of substancial abatement. However, the biomass is maintained and presents growth during the experimentation.

The bioscrubber usually used in the off-gas treatment are able to treat a polluted effluent concentrated with 1000 ppm at a rate of 1.5 L/min.L. What corresponds to a flow of 1.5 mg/min.L of bioscrubber (Solstys).

The reactor proposed presents thus interesting opportunities in the biological treatment of gas effluents polluted by aromatic compounds in strong concentration. The process suggested might be applied in field of concentration and flow where thermal oxidation is too expensive.

REFERENCES

Aldric J.M. (2001). Contribution à la mise au point d'un réacteur biphasique destiné à la dégradation des composés organiques volatiles par voie biologique. Mémoire de fin d'études présenté en vue de l'obtention du diplôme d'études approfondies en sciences agronomiques et ingénierie biologique. Faculté Universitaire des Sciences Agronomiques de Gembloux. 75 p.

Budwill K., Coleman R.N. (1997). Effect of silicone oil on biofiltration of n-hexane vapours. Med. Fac. Univ. Gent. 62/4b, 1521–1528.

Solstys N. Technique de l'ingénieur, traité Génie des procédés J 3 928

Yeom S.H., Daugulis A.J. (2001). Benzene degradation in a two-phase partioning bioreactor by alcaligenes xylosoxidans Y234. Process biochemistry. 36, 765–772.

Yeom S.H., Daugulis A.J. (2001). Development of a novel bioreactor system for treatment of gazeous benzene. Biotechnology and bioengineering. 72, 156–165.

European Symposium on Environmental Biotechnology, ESEB 2004 - Verstraete (ed)
© 2004 Taylor & Francis Group, London, ISBN 90 5809 653 X

Systematical selection of a support material for hydrogen sulphide biofiltration

I. Cano, A. Barona, A. Elías & R. Arias
Department of Chemical and Environmental Engineering, Faculty of Engineering
University of the Basque Country, Alda Urkijo Bilbao, Spain

ABSTRACT: The success of biofilters relies on the biodegradability of the contaminant to be degraded and on the optimal selection of the support material (also called carrier, packing or filter bed material). Although the selection of a suitable material involves the determination of many physical-chemical parameters, we found that the pH, moisture content and particle size distribution were useful for determining the initial performance of a biofilter for hydrogen sulphide treatment.

1 INTRODUCTION

Odorous gas treatment is crucial to reduce the impact of industry on the environment and on the society as odorous compound emissions are frequent targets of public complaint. One of those odorous compounds is hydrogen sulphide, which has a very disgusting smell of rotten eggs and can be smelled at concentrations as low as 0.5 ppb_v. At 100 ppm_v, it can no longer be smelled and at 500 ppm_v, it is very toxic (Busca & Pistarino, 2003).

Biofiltration has been used as an Air Pollution Control (APC) technology to effectively degrade waste air with relatively low concentrations of biodegradable compounds at competitive operational and maintenance costs.

Biofilters are packed bed biochemical reactors whose bed or carrier material can be considered the "heart" of the system. In addition to providing a physical support for the microorganisms, certain features of the material as well as its own activity, will guarantee a long operation.

The selection of new support materials for biofiltration purposes must be carried out taking into account the organic or inorganic origin of the material. The organic materials may provide the active microorganisms and the nutrients for the biodegradation to take place, may control small pH changes during operation and are usually made of disposal wastes with low cost (Kim et al., 1998; Ramírez-López et al., 2003). However, physical disgregation after long operation and consequent high pressure drop and variable composition of the material are disadvantages for proper biofilter performance. The inorganic materials may present some advantages compared to the organic ones, such as a homogenous and well defined composition and the possibility to design the initial biofilter pore structure, which minimises pressure drop and channelling problems (Schwarz et al., 2001). However, due to the absence of microorganisms and nutrients, the operational and maintenance costs should be taken into account.

It has been suggested that a good filter should have the following characteristics (Bohn, 1996; Elías et al., 2002): (1) high moisture retention capacity, (2) high buffer capacity, (3) high available nutrient content, (4) a diverse and adaptable microbial population, (5) large specific surface area, (6) low bulk density (7) high void fraction (8) low cost, (9) and long life (mechanically resistant and chemically inert and stable).

In this work, four different organic materials were selected and investigated as potential carrier materials in biofilters. A quick characterization of the materials was carried out in order to foresee their use in biofiltration.

2 MATERIALS AND METHODS

2.1 *Support materials*

The four materials used in this work are mainly made of pig manure, horse manure, wastewater sludge compost and a mixture of soil and algae respectively. The pig manure material was mixed with sawdust in order to increase the porosity and the moisture retention capacity of the final product. The four materials have been selected because of two main reasons. On the one hand, they provide their own biomass to degrade the contaminants and on the other hand, their possible use in bioreactors will reduce their disposal costs.

2.2 *Filter bed characterization*

The following general parameters were measured in the organic materials as a quick characterization of the bed material for biofiltration purposes. Thus, total C, H and N contents were measured in a CHN-600 LECO analyser and the total S content was determined in a LECO SC 132 analyser. The moisture content was determined by thermogravimetry in a LECO TGA 500 thermobalance.

In order to determine the metal content in the bed materials, they were fully digested with a $HF/HNO_3/HClO_4$ mixture (Medved et al., 1996). The metal content in these acid digestions were measured in a Perkin Elmer 1100B atomic absorption spectrophotometer. Standard solutions were prepared with the same matrix as the samples. The metals measured were Fe, Ca, Mg, K, Na, Cu, Pb, Ni, and Zn.

The pH value was measured in water at a 1:9 dry material weight:water volume ratio (CRISON micro-pH 2002).

Table 1. Characterization of the four filter or support materials.

Parameter	Pig manure	Horse manure	Wastewater sludge	Soil and algae
pH	6.5	4.3	7.2	6.9
C (%)	20	53	28	21
H (%)	2	5	4.8	2.7
N (%)	0.6	0.5	4.3	3.1
Moisture cont. (%)	23.2	57	8	42.9
Total sulphur (%)	3.3	<1	<1	<1
Particle size distribution (%)				
>8 mm	0.35	86.3	0	8.94
8–1 mm	99.2	13.5	71.13	59.8
1–0.2 mm	0	0	27.9	29.4
<0.2 mm	0.45	0.20	0.33	1.83
Metal content (μg/g)				
Cu	160	70	230	36
Fe	30600	1410	15200	7650
Ni	45	11	130	31
Pb	38	7	170	19
Zn	430	30	1600	29
Mg	6190	390	7000	4840
Ca	61860	4900	42000	41030
K	54500	720	2800	7330
Na	9690	360	110	1920

2.3 *Operation in the biofilter*

The gas flow containing hydrogen sulphide was fed into four PVC columns, which were filled up with each material in order to study their response to increasing the contaminant loading rate (ranging from 40 to 300 ppm_v). The contaminated gas flow to be treated was obtained in the laboratory by mixing H_2S (99.7% purity) and moisture saturated air. The H_2S flow from the gas cylinder was controlled by a 5850S Brooks Mass Flow Controller. The air stream from the compressor was regulated by a Platon Rotameter and, subsequently, this stream was bubbled into two columns filled with deionized water in order to saturate the air stream. No nutritive solution to feed microorganisms was provided.

All the experimentation was carried out by triplicate and removal efficiency higher than 90% was established as good performance criteria. The preliminary response of the materials at each concentration was studied for 4 days of operation.

3 RESULTS AND DISCUSSION

The characterization results of the four organic carriers are shown in Table 1. The pH value for the horse manure material is acid, which is an important drawback above all if the degradation process renders acidic products. Besides, its particle size distribution showed that about 86.3% of this material is constituted by extremely big particles, which would generate preferential channelling and heterogenous distribution of the gas into the bed. Only based on those two parameters, it can be concluded that the horse manure material would not render good results as filtering bed.

The wastewater sludge material showed an initial moisture content extremely low and considering that this parameter is somehow related to the water holding capacity, a proper development of the liquid phase in the biofilter is not ensured. The material made of soil and algae and the pig manure material are similar as far as pH and elemental analysis are concerned. However, the metal content in both carriers is very different. Certain metals as K and Fe are nutrients but others metals such as Pb may be toxic for the biomass. In general, metal content in the pig manure carrier is high, which is an advantage as far as nutrients are concerned but it is also a drawback as far as possible lixiviates and toxic effects are concerned.

In order to verify the preliminary conclusions, the four materials were tested in a laboratory scale biofilter for the treatment of H_2S. Discrimination among the materials was carried out by comparing their removal efficiency in the biodegradation of the contaminant. The removal efficiency was calculated.

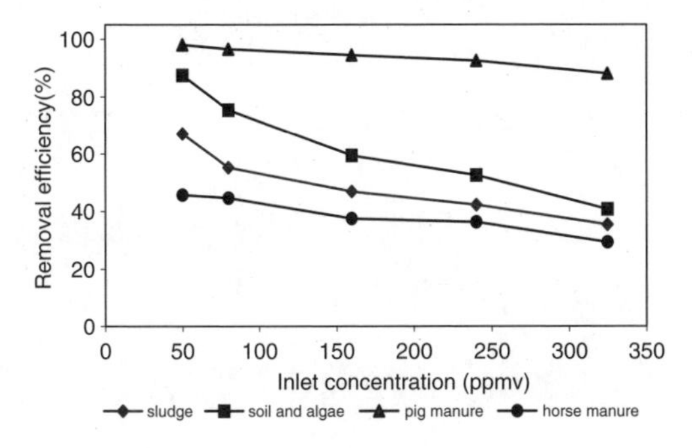

Figure 1. H_2S removal efficiency of the four organic materials.

according to the following expression:

$$\text{Removal efficiency} = \frac{S_{in} - S_{out}}{S_{in}} 100 \qquad (1)$$

where S_{in} = inlet contaminant concentration (ppm_v); S_{out} = outlet contaminant concentration (ppm_v).

The experimental results are shown in Figure 1. As predicted by the brief chemical characterization, the horse manure material presented the lowest removal efficiency whereas the pig manure showed a removal efficiency higher than 90% for the concentration interval studied. Concerning the other two materials (sludge and soil + algae), degradation rate quickly decreased up to 30% as H_2S concentration was increased. Consequently it can be concluded that the use of both materials for biofiltration purposes is not advisable.

Although the proper selection of a suitable carrier material involves the determination of al least 8 different physical-chemical parameters, we found that the pH, moisture content and particle size distribution were useful for rapidly determine the initial performance of a biofilter for H_2S treatment.

ACKNOWLEDGEMENT

The authors acknowledge the financial support of the Spanish Ministry of Science and Technology (MCYT PPQ2002-01088 with FEDER funding) and of the University of the Basque Country (UPV 00112.345-T-13929/2001).

REFERENCES

Bohn, H.L. 1996. Biofilter media. In *Air & Waste Management Association 89th Annual Meeting & Exhibition, Nashville, Tenessee* 98-Wp87A-01.

Busca, G. & Pistarino, C. 2003. Technologies for the abatement of sulphide compounds from gaseous streams: a comparative overview. *Journal of Loss Prevention in the Process Industries* 16: 363–371.

Elías, A., Barona, A., Arreguy, A., Ríos, J., Aranguiz, I. & Peñas, J. 2002. Evaluation of a packing material for the biodegradation of H_2S and product analysis. *Process Biochemistry* 37: 813–820.

Kim, N.J., Hirai, M. & Shoda, M. 1998. Comparison of organic and inorganic carriers in removal of hydrogen sulfide in biofilters. *Environmental Technology* 19: 1233–1241.

Medved, J., Stresko, V., Kubová, J. & Polakovicová, J. 1998. Efficiency of decomposition procedures for the determination of some elements in soils by atomic spectroscopic methods. *Fresenius Journal of Anaytical Chemisty* 360: 219–224.

Ramirez-López, E., Corona-Hernandez, J., Dendooven, L., Rangel, P. & Thalasso, F. 2003. Characterization of five agricultural by-products as potential biofilter carriers. *Bioresource Technology* 88: 259–263.

Schwarz, B.C.E., Devinny, J.S. & Tsotsis, T.T. 2001. A biofilter network model – importance of the pore structure and other large-scale heterogeneities. *Chemical Engineering Science* 56: 475–483.

European Symposium on Environmental Biotechnology, ESEB 2004 - Verstraete (ed)
© 2004 Taylor & Francis Group, London, ISBN 90 5809 653 X

The monolith reactor – a novel reactor concept for biological gas purification

S. Ebrahimi, R. Kleerebezem, J.J. Heijnen & M.C.M. van Loosdrecht
Department of Biotechnology, Delft University of Technology, Julianalaan BC Delft, The Netherlands

M.T. Kreutzer
Department of Chemical Engineering, Reactor and Catalysis Engineering section, Delft University of Technology, Julianalaan BL Delft, The Netherlands

ABSTRACT: Biological gas treatment has been shown to be an effective and inexpensive technique for the removal of low concentrations of contaminants from waste gas streams. A number of multiphase contacting devices have been developed to improve the contacting efficiency in gas liquid systems. The most widely used technologies are: biofilters, biotrickling filters, and bioscrubbers. This work suggests a novel reactor technology for biological gas treatment; the monolith reactor. The monolith offers some clear advantages over conventional bioreactors such as high mass transfer and a low pressure drop, making it a competitive alternative for conventional biological gas treatment bioreactors. The main potential problem of monoliths is clogging due to biofilm formation. This was experimentally investigated, in a pilot-scale monolith reactor.

1 INTRODUCTION

Biological gas treatment is based on the absorption of volatile contaminants in an aqueous phase or biofilm followed by biological oxidation by bacteria and fungi. Air treatment bioreactors are similar to liquid purification bioreactors in many ways. Like in water treatment bioreactors, microorganisms utilize the pollutants as a source of carbon or energy for growth and maintenance purposes. The main difference is that for air phase bioreactors the organic contaminants must first be absorbed into the liquid phase prior to biodegradation (Edwards and Nirmalakhandhan, 1996).

Biological gas treatment is relatively inexpensive (low capital and operating costs) compared with conventional physico-chemical techniques. Biological treatment is achieved at normal operating conditions of temperature and pressure, does not generate secondary pollutants and is positively perceived by the general public. In general biological treatment can be an effective and relatively inexpensive technique for the removal of relatively low concentrations ($<1\ g\ m^{-3}$) of contaminates from large waste gas streams (Weber and Hartmans, 1996).

Obviously, transfer of poorly soluble pollutant compounds from large gas volumes at dilute concentration to small liquid volumes as well as transfer of oxygen required for aerobic degradation is the most important duty of the bio-contactors.

General problems of existing techniques are too high pressure drop creating high energy consumption and large area requirement. This implies the use of non-conventional multiphase contacting devices. We are studying a new multiphase contactor, the monolith reactor. The incentives for its application in biotechnological gas treatment are the low-pressure drop, high mass transfer rate, and small footprint. Herewith high capacities can be achieved at elevated gas-loading rates under energetic and size conditions compatible with industrial constraints.

2 MONOLITH REACTOR

A monolithic support consists of ceramic structure with a large number of narrow, parallel channels separated by thin walls. Liquid is distributed on the top of the monolith structure and the reactors can either be operated co- or counter current with respect to the gas. The monolith reactor is widely used as catalyst support for gas treatment applications such as cleaning of automotive exhaust gases and industrial off gases. In these applications, in which large volumetric gas flows must be handled, monoliths offer certain advantages, such as a low pressure drop and high mechanical strength. Recently, the use of monoliths has been extended to include applications for performing multiphase reactions. There is growing interest in the chemical industries to find new applications for monoliths as catalyst support in three phase catalytic reactions (Irandoust and Andersson, 1988). However, the potential application in multiphase biological reaction systems has hardly been explored.

The monolith reactor offers several potential advantages in multiphase reactions in which, due to the presence of different phases, the interfacial transport is of major importance for the reactor designs. It can be an alternative option for biotechnological process where microorganisms are used for the conversion of material in large gas streams. The main features of monolith support for application in gas-liquid systems are (Irandoust et al., 1998):

- Very low pressure drop
- Large specific surface area
- Good mass transfer properties
- Ease for reactor scale up
- Good liquid distribution within the matrix at low liquid flow rates.

3 MONOLITH AS A COMPETITIVE ALTERNATIVE FOR CONVENTIONAL CONTACTORS

A comparison between conventional bioreactors (biofilters, biotrickling filters and bioscrubbers) shows that each of the treatment systems is considered to be most effective at specific operating conditions. The application range of each technique is determined by the solubility of the compounds to be removed, the inlet concentrations and the technical hindrances and limitations.

The monolith reactor can be applied in several mode of operation to be an alternative for each type of conventional biological gas purification techniques; such as biofilters and biotrickling filters or those using suspended microorganisms such as bioscrubbers. In doing so, the application range for each type of the conventional technique can be improved and a more compact installation would be possible.

3.1 *Monolith as a biofilter*

In essence, biofilters have a distinct applicability and region of operation for poorly soluble compounds; high mass transfer can be achieved because of the absence of a separate liquid phase. By application of the ceramic monolith as biofilter packing indeed the geometry of packing will be changed which improves feasibility and range of operation and much lower pressure drop can be obtained.

In monolith reactors, due to the high suitable surface area per unit volume for the microorganisms (at least one order of magnitude higher than the biofilters one) high biodegradation capacity can be achieved as a result, the reactor volume can be smaller in size than a biofilter. In monolith reactors air velocities in the region of 1 m/s can be applied, which is approximately 30 times the air velocity of biofilters (Hansen and Rindel, 2000). The pressure drop in co-currently operated monolith reactors is very low and monolith reactors can be built much taller than biofilters. As a result, the footprint of biofilters can be reduced by two orders of magnitude. In the monolith reactor higher mass transfer rates can be achieved due to the larger allowable superficial gas velocities.

Potential limitations of biofilters include excessive growth of biomass and subsequent preferential channelling of the gas. Furthermore, nutrient levels (e.g., nitrogen) and the pH are difficult to control and the humidification of moisture content of the biofilter can also be problematic (Weber and Hartmans, 1996). Contrary to biofilters monolithic bed can be easily and regularly washed to overcome these problems, in particular to avoid acidification therefore, it can be superior in eliminating high contaminate concentration and acid-producing pollutants.

Monolithic bed can be further improved to attain high adsorption capacity, good biomass adhesion and easy start up, by coating with activated carbon.

3.2 *Monolith as a biotrickling filters*

Biofilters usually are not designed for a continuous flow of water. Yet some processes need a continuous liquid flow either to supply fresh nutrients or to wash out non-volatile products that would otherwise accumulate to inhibitory levels. In such cases biotrickling filters can be a suitable option.

Although trickling filters have certain advantages compared to biofilters, a major disadvantage can be a reduction of reactor performance due to the formation of excessive amount of biomass. Clogging problems can occur due to poor biofilm formation and biofilm detachment from the packing material. Weber and Hartmans noted that the most of the biomass was entrapped between the packing materials (Weber and Hartmans, 1996). The monolith reactor can be applied in the same mode of operation as biotrickling filters, thin biofilm on the wall with small liquid recycling flow along the wall and gas flow in the central core (annular flow). The straight channels feature of the monolith reactor limits the clogging problem because entrapment of the detachment biomass between packing materials will not occur. Monolithic bed combines a large geometrical surface with low-pressure drop and a high degree of uniformity, which makes it suitable for a process characterized by high gas flow rates.

3.3 *Monolith as a competitive alternative for bioscrubbers*

Many of the problems and limitations associated with biofilters and biotrickling filters could be overcome with suspended growth bioreactors (i.e. bioscrubbers). A bioscrubber consists of an absorber and a separate bioreactor, normally operated as activated sludge reactor. The most used absorption contactor is packed column. The packing materials mostly have specific surface area between 100–300 m^2/m^3. The size of packing elements ranges from 2 to 8 cm (Kennes and Veiga, 2001). It is hypothesized that the suspended growth microorganisms in a monolith reactor could be a feasible configuration for biological gas purification.

The desired flow pattern for this purpose is cocurrent downflow operation in a Taylor flow regime. The monolith reactor with the proposed mode of operation offers several advantages over the conventional contactors such as: lower pressure drop (up to two orders of magnitude), and higher volumetric mass transfer rates. Therefore, high capacity for gas loading rate in a small footprints and low energy consumption. For a monolith reactor operating in downflow mode, it is possible to balance the frictional pressure drop with the hydrostatic pressure of the liquid inside the channels. The essentially zero net pressure drop provides as an opportunity to operate the monolith reactor without the need of a compressor or blower.

In suspended growth application of monolith reactor, formation of biofilm is unfavorable. Therefore, in this study some experiments were conducted to study formation of undesired biofilm.

4 RESEARCH QUESTIONS

Numerous physical-chemical characteristics of monolith reactors have been determined, but with regard to the behaviour of biomass no information is available. Formation of biofilms in monolith reactors can either be regarded as an unwanted side effect (option 3.3) or an operational requirement (options 3.1 and 3.2), depending on the treatment objectives to be met. Operation as biofilm reactor is desired if uncoupling of the solid and liquid retention time is required.

In case of sufficiently long liquid retention times in gas treatment systems the formation of biofilms can be regarded as an unwanted side effect that may increase the pressure drop and decrease the mass transfer rates. Biofilm formation was proposed to be controlled by the liquid residence time as compared to the bacterial doubling time (Tijhuis, 1994).

The main potential problem of monoliths is clogging due to biofilm formation. This was experimentally investigated in a pilot-scale monolith reactor. The principle objective of the experimental work was to determine if biofilm formation and maintenance could be controlled in monolith type reactors. The operational variables investigated were the hydraulic residence time, substrate surface loading rate and the biomass concentration in the system. In order to keep the experimental procedure as simple as possible it was decided to work with aerobic heterotrophic mixed culture grown on glucose. The cocurrent operation was chosen because of the low pressure-drop and high mass transfer rate.

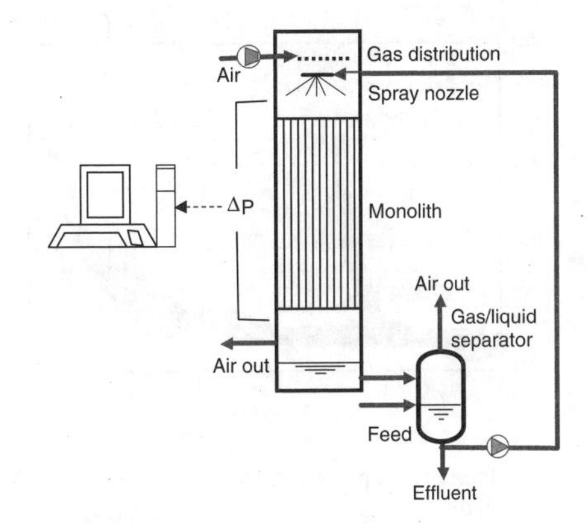

Figure 1. Experimental setup used for investigation of biofilm formation in monolith columns.

5 EXPERIMENTS

Experiments have been conducted in short monolith columns (length 30–40 cm, diameter 10 cm) placed in a Plexiglas column. A schematic representation of the experimental setup is shown in Figure 1. The reactor was operated continuously for gas and liquid in a cocurrent mode of operation. Influent is distributed over the top of the monolith by means of a spray nozzle. A vessel was used to separate gas-liquid at the bottom of the reactor. Liquid was circulated with a centrifugal pump.

Gas and liquid recirculation flow rates were approximately 50 l/min. Influent liquid flow rates were chosen to obtain a high (30 h) and short (0.5 h) liquid residence time.

6 RESULTS AND DISCUSSIONS

6.1 *Experiments at long liquid residence time*

By applying long hydraulic residence times of 2 to 3 days, conditions favorable for suspended growing biomass are maintained. Liquid was recirculated over the monolith column at a flow rate of 50 l/min, and air was supplied in volumetrically equal amounts by means

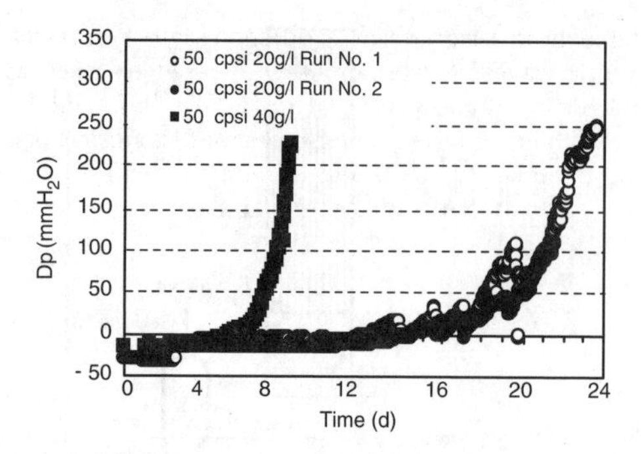

Figure 2. Pressure drop as a function of time in the 50 cpsi monolith column.

of a mass flow controller. This approach enabled operation of the total system under aerobic conditions.

The experiments were conducted with 50 cpsi monolith with an influent glucose concentration of 10 to 80 g/l. The pressure drop as a function of time is shown in Figure 2. In the beginning there is a small negative pressure drop (suction) whereas after approximately 20 days of operation at a low pressure drop ($<20\,cmH_2O$), the pressure drop sharply increased in time.

Every week the reactor was dismantled and pictures were made from the biofilms on the monolith. The pictures clearly illustrated that biofilm formation is started from the centre of the monolith channels. After initial formation the biofilm thickness increases and finally the channel gets clogged. Distribution of the channels that are clogged appeared to be in clusters that were randomly distributed over the monolith cross section.

Biofilms were furthermore found all over the length of the monolith. The observation that biofilm formation clearly occurs starting from the middle in the monolith channel is counterintuitive, because shear forces are lowest in the corners of the channels. Future work on the formation of biofilms as a function of the flow pattern in the monolith and mass transfer considerations may help to clarify this point.

The biomass concentration in the reactor amounted approximately 10 and 20 g/l when concentration of glucose in the influent is 20 and 40 g/l respectively, corresponding to a biomass yield of 0.5 g biomass-COD/g glucose-COD, which is a normal value at the operational temperature of approximately 30°C.

6.2 *Experiments at short liquid residence time*

At short liquid residence time growth is favored of biomass that is capable of biofilm formation. By applying high hydraulic dilution rates corresponding to a HRT of 30 min, conditions favorable for biofilm formation are maintained. The experiments were conducted with 50 cpsi monolith with an influent glucose concentration of 20 to 40 g/l. Liquid and gas flow rates were maintained equal to previous experiments.

In this case in a few hours pressure drop sharply increases due to biofilm development. The $1/HRT > \mu^{max}$ hypothesis for biofilm formation was verified and the pressure drop is dramatically increased due to the biofilm formation.

7 CONCLUSIONS

First insight into the application of monoliths for biological gas purifications was given. Compared to the conventional bioreactors, the monolith reactor offers some clear advantages such as a high mass transfer rate and a low pressure drop. Therefore high treatment capacities can potentially be achieved at high gas loading rates and a small footprint. Herewith the monolith can be considered to become a competitive alternative to conventional biological gas treatment. Preliminary experimental results have demonstrated the wide operational window for monoliths in biological processes.

Biological processes are achieved at ambient operational conditions of temperature and pressure. Therefore, other inexpensive material and design can be utilized to increase the potential use of monoliths. For instance, development of new monoliths from plastic materials, with other geometries at lower prices can be desirable.

REFERENCES

Edwards, F.G. and Nirmalakhandan, N., 1996. Biological treatment of airstreams contaminated with VOCs: an overview. Water Science and Technology, 34(3–4): 565–571.

Hansen, N.G. and Rindel, K., 2000. Bioscrubbing, an effective and economic solution to odour control at wastewater treatment plants. Water Science and Technology, 41: 155–164.

Irandoust, S. and Andersson, B., 1988. Mass transfer and liquid-phase reactions in a segmented two-phase flow monolithic catalyst reactor. Chemical Engineering Science, 43(8): 1983–1988.

Irandoust, S., Cybulski, A. and Moulijn, T.A., 1998. The use of monolith catalysts for three phase reaction. Structured catalysts and reactors. Marcel Dekker, New York.

Kennes, C. and Veiga, M.C., 2001. Bioreactor for Waste gas treatment.

Tijhuis, L., 1994. The biofilm airlift suspension reactor; biofilm formation, detachment and heterogeneity, TUDelft.

Weber, F.J. and Hartmans, S., 1996. Prevention of clogging in a biological trickle-bed reactor removing toluene from contaminated air. Biotechnology and Bioengineering, 50: 91–97.

European Symposium on Environmental Biotechnology, ESEB 2004 - Verstraete (ed)
© 2004 Taylor & Francis Group, London, ISBN 90 5809 653 X

Formulation of biofiltration packing materials: essential nutriment release for the biological treatment of odorous emissions

F. Gaudin, Y. Andres & P. Le Cloirec
Ecole des Mines de Nantes, GEPEA UMR CNRS 6144, Nantes Cedex 3, France

ABSTRACT: Biofiltration is a current biological gas treatment technology extensively used for the treatment of polluted air. The advantages of this cleaning technique are high – superficial area best-suited for poor water solubility compounds treatment, ease of operation and low operating costs. It involves a filter bed serving both as carrier for the microorganisms and as nutrient supplier. In this study, two types of material were formulated with an organic binder to compare two different nitrogen sources: ammonium phosphate or urea phosphate. The obtained supports were tested in terms of cohesion capacities in water, bulk density, moisture retention capacity, dissolving rate of the mineral elements (carbonate, phosphorus, calcium, nitrogen) and for the pH regulation. The results show that the binder gives important cohesion capacities to the material even in drastic conditions (material submerged in water) and allows a low diffused release for different elements. Therefore, formulated packing materials present interesting properties for biofiltration processes.

1 INTRODUCTION

Biofiltration is increasingly used as a method to decontaminate gas streams containing low concentrations of biodegradable volatile organic and inorganic compounds. This process has gained worldwide acceptance as an economical air pollution control technology for low concentration gases treatment. In biofiltration the gas stream containing the pollution is injected through a packed bed that contains an unsaturated solid medium that supports a biofilm. The pollutant substances transfer from the air flow to the biofilm, where they are degraded by microorganisms mostly into innocuous products such as carbon dioxide, water and biomass. In many cases the gas stream and the water flow used to humidify the biofilm brings all the nutriments requested for the development of the microorganisms. But, the contaminant concentrations in most waste gas streams could vary with time due to the inherent nature of processes that generate them, and during holidays could totally be stopped. Moreover, some air flows could contain only some of the elemental compounds needed for the microorganisms growth and make the biofiltration process non-efficient.

To ensure a high removal efficiency of the biofilter, the support must have a high specific surface, between 300 and 1000 m^2/m^3 (Herrygers et al., 2000), high buffer capacities, a low bulk density to avoid the bed packing down (Devinny et al., 1999), a bed porosity between 0.5 and 0.9 (Herrygers et al., 2000) and a bed humidity rate between 20 and 60% (Williams et al., 1992). It must also provides the elemental compounds not contained in the gas stream to the biomass. Moreover the porosity of the material must ensure the fixation of the microorganisms (Cohen, 2000).

Organic materials are the most currently supports used for biofiltration because of their low cost and their high nutrients content. The most common are peat, soil, compost but also sugar cane bagasse (Sene et al., 2000) or peanuts shell (Ramirez-Lopez et al., 2003). However, these materials lead to bed packing down troubles and cause pressure drops decreasing the biofilter efficiency. Thanks to their good hydrodynamic properties and a higher strength, mineral materials can avoid this problem (Gemeiner et al., 1994). The most currently used are metal oxides as porous ceramic or calcinated cristobalite (Hirai et al., 2000). However, they are high-cost materials and does not provide any nutriments to the biomass.

The aim of the present work is to develop original solids used as biofiltration support which provides a nutriments release to complete the ionic composition of the treated gas stream. In a first step, eight materials were formulated and produced by extrusion using calcium carbonate, urea phosphate or di-ammonium phosphate and an organic binder. In a second step, they have been characterized in terms of nutriments release and pH.

2 MATERIALS AND METHODS

2.1 *Supports formulation*

Eight materials containing various compounds and an organic binder have been extruded in a cylinder shape following the same procedure. First, the salts were mixed, then the organic binder was added to water and finally the mixture of compounds was added. Extrusion was realized with a Zyliss meat mincer and the granules were dried at 50°C during 20 hours. The Table 1 shows the main characteristics of each formulated support. .

These 8 mixtures were chosen in order to study two different nitrogen sources and the influence of the binder rate on the nutriments release. The organic binder is used in the building industry and the Table 2 gives some properties of this compound.

2.2 *Physical properties*

Cohesion capacities in water, bulk density and moisture retention capacity were tested on each support. Cohesion capacities were evaluated with the time the granules started to break up. Bulk density was determined as the mean value of a 30 granules sample for each support. Moisture retention capacity was determined as the mass of water per mass of dry granules after a 24 hours immersion.

2.3 *Release kinetics*

Release kinetics were achieved to evaluate the release capacities of the 8 supports. A definite mass of granules was introduced in 500 mL distilled water glasses and a floculator (Bioblock scientific floculator 11199) was carried out to keep a constant homogeneity of the solutions. The release kinetics have been studied on a 24 hours period. After the first kinetics, the materials were washed with distilled water and dried 24 hours at 45°C. 4 kinetics were successively performed on each material.

2.4 *Samples analysis*

Calcium was measured by atomic absorption (Analyst 200 Perkin Elmer). Samples were first diluted in a Lanthanum solution to avoid interferences with phosphate ions. Carbonate and urea were measured out using a TOCmeter (Shimadzu Total Organic Carbon 5000 A). Phosphate was measured by spectrophotometry (Shimadzu UV-1601) according to the European norm EN 1189:1996. Ammonium was measured by spectrophotometry using the Spectroquant kit 1.14752 (Berthelot's reaction).

pH was measured using a Bioblock 90431 electrode connected to a C-835 Bioblock multi-parameters analyzer.

Table 1. Supports formulation.

Name	Compounds	C/N/P molar ratio	Binder mass (%)
UP 10	Urea phosphate, calcium carbonate	100/10/5	10
UP 15	Urea phosphate, calcium carbonate	100/10/5	15
UP 20	Urea phosphate, calcium carbonate	100/10/5	20
UPt 20	Urea phosphate, calcium carbonate, phosphate buffer	100/10/5	20
Am	Di-ammonium phosphate, calcium carbonate	100/10/5	0
Am 1	Di-ammonium phosphate, calcium carbonate	100/10/5	1
Am 5	Di-ammonium phosphate, calcium carbonate	100/10/5	5
Am 10	Di-ammonium phosphate, calcium carbonate	100/10/5	10

Table 2. Properties of the organic binder.

Appearance	Fluid white powder
Composition	Ethylene, vinyl acetate
pH after dispersion	7.0–8.0
Film formation minimum temperature	+3°C

3 RESULTS AND DISCUSSION

3.1 *Physical properties*

3.1.1 *Cohesion capacities*

Table 3 presents the influence of the organic binder mass rate on the cohesion of the different formulations when they are submerged in water. For both formulation type (Am and UP) the cohesion capacities were significantly increased using a higher binder mass rate. Moreover, for similar cohesion capacities Am formulations require a low binder mass rate (<10%) than UP formulations.

3.1.2 *Bulk density and moisture retention capacity*

The Table 4 compares the bulk density and moisture retention capacity values for the 8 supports with classical values obtained with common biofilters material such as porous ceramics, cristobalite (Hirai et al, 2000) and peat (Wu et al, 1999).

The eight new materials have a slightly higher bulk density than the common ones. Their moisture retention capacity show values between 47 and 69% which are comparable with porous ceramics and peat values.

Table 3. Influence of the binder on the cohesion capacities of the supports submerged in water.

Am formulations	UP formulations	Binder mass (%)
Slight break up after 24 hrs (Am)	–	0
Very slight break up after 24 hrs (Am 1)	–	1
–	Complete break up after 5 min.	2
No break up after 24 hrs (Am 5)	Complete break up after 5 min.	5
No break up after 24 hrs (Am 10)	Slight break up after 24 hrs (UP 10)	10
–	Very slight break up after 24 hrs (UP15)	15
–	No break up after 24 hrs (UP 20)	20

Table 4. Bulk density and moisture retention capacity.

Material	Bulk density (g/cm^3)	Moisture retention capacity (%)
Am	1.10	64
Am 1	1.18	68
Am 5	0.97	69
Am 10	1.05	65
UP 10	1.24	67
UP 15	1.35	59
UP 20	1.32	47
UPt 20	1.35	48
Porous ceramics	0.47	62
Cristobalite	0.91	35
Peat	0.72	64

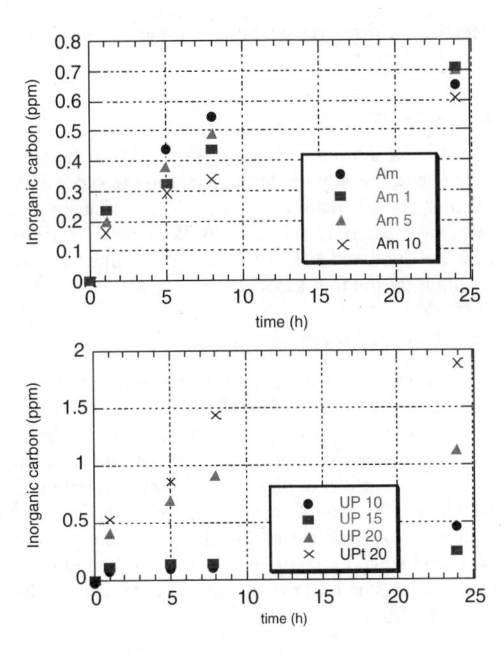

Figure 1. First carbonate release kinetics.

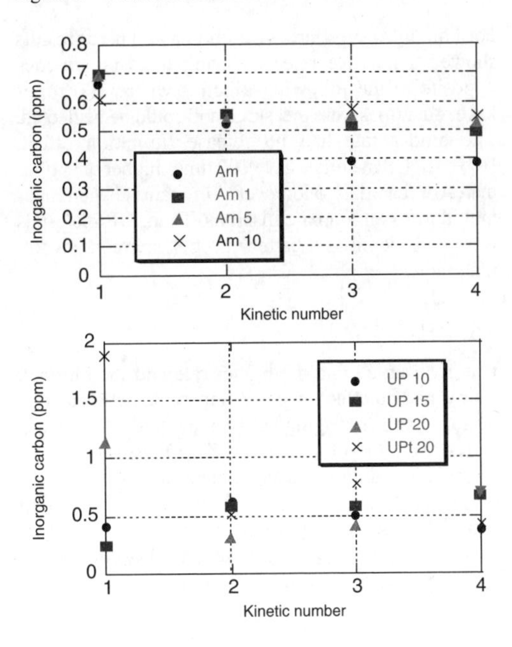

Figure 2. Amount of dissolved carbonate obtained after four successive kinetics.

Moreover, Am materials present a lower bulk density and a higher moisture retention capacity than UP materials. It seems that an increase of the binder mass rate in the formulations induces an increase of the bulk density and a decrease of the moisture retention capacity.

3.2 *Release kinetics*

3.2.1 *Carbonate release*

Figure 1 presents carbonate release first kinetics on a 24 hours period for Am and UP formulations. Figure 2 presents the amount of dissolved carbonate successively obtained after four 24 hours kinetics. All values are given per gramme of material.

For Am materials, first kinetics gave similar values and the final amount of dissolved carbonate ranged from 0.6 to 0.7 ppm per gramme of material. UP materials kinetics showed a larger range of values that can be explained by a higher variation of the binder mass rate in their different formulations.

For both materials type, a maximum amount of dissolved carbonate was observed during the first kinetics. Then, the successive kinetics seemed to reach a constant value after 24 hours that can be called 'the minimum dissolution level'. Am material, which did not contain any binder, showed the straightest decrease of dissolved carbonate along the four successive kinetics.

3.3 *Calcium release*

The results showed a maximum release for the first kinetic and a straighter decrease along the successive kinetics for UP materials with a lower binder mass rate. Am materials showed similar kinetic profiles and the dissolved calcium amount remained nearly constant for the four kinetics.

3.4 *Phosphorus release*

It was found that the maximum release occured during the first kinetics and the highest quantity of dissolved phosphorus was observed with the lower binder mass rate materials. Opposite to carbonate and calcium release, the amount of dissolved phosphorus obtained for kinetics 2, 3 and 4 showed similar values with all materials. The mean value of this minimum dissolution level equal to 0.020 μg/L/g of material and is 10 to 25 times smaller than the values obtained for the first kinetics.

3.5 *Nitrogen release*

For both nitrogen sources (ammonia and urea) results showed a massive release during the first kinetics opposite to the following kinetics where 200 times lower quantities were measured and could be neglected. The binder rate had no release limitation effect. However, UP materials gave 100 times higher quantities than Am materials. Moreover, Am materials generates malodorous ammonia emissions. Then, UP materials were considered as a better nitrogen source than Am materials.

3.6 *pH*

It was observed that the binder rate and the nitrogen source had no effect on pH. All materials generated at basic pH values ranging between 8.0 and 9.0. The use of K_2HPO_4/KH_2PO_4 buffer (3% in mass) in UPt 20 formulation didn't induce a neutral pH as it was expected. Although the basic pH of the different materials was not an optimum condition, it could be an advantage to neutralize acidity generated by compounds like H_2S, mercaptans and fatty acids.

4 CONCLUSIONS

The physical characterization of the 8 chosen supports showed acceptable values concerning, bulk density and moisture retention capacity when they are compared to data obtained with common materials used for biofiltration.

Cohesion capacity experiments demonstrated the positive influence of the organic binder on the mechanical viability of the different supports.

Release kinetics did not simulate the real conditions of a biofilter but they were easy experiments to set up and to follow the dissolution of the main nutriments in drastic conditions (supports submerged in water). It was observed that a higher binder mass rate induced a more homogenous carbonate and calcium release and limited a massive dissolution during the first kinetic. For phosphorus, the minimum dissolution level was nearly constant with all the materials and was reached during kinetic 2. Nitrogen release showed the less interesting results because of the massive dissolution of this element during the first kinetic with all materials. As Am materials generated malodorous ammonia emissions, UP materials were considered as a better nitrogen source.

Every materials generated a basic pH whereas a neutral pH had been an optimum condition for bacterial growth.

The authors would like to thank the French Environment and Energy Agency (ADEME).

REFERENCES

Cohen, Y. 1996. Biofiltration – the treatment of fluids by microorganisms immobilized into the filter bedding material: a review, *Bioresource Technology*, 77: 257–274.

Devinny, J.S., Deshusses, M., Webster, T.S. 1999. Biofiltration for air pollution control, *CRC-Lewis Publishers*, Boca Raton, FL, U.S.A., 299 p.

Gemeiner, P., Rexova, L., Svec, F., Norrlow, O. 1994. Natural and synthetic carriers suitable for immobilization of viable cells, active organelles and molecules, In: Veliky, I.A., McLean, R.J.C. (Eds.), *Immobilized Biosystems*. Chapman & Hall, London.

Herrygers, V., Van Langenhove, H., Smet, E. 2000. Biological treatment of gases polluted by volatile sulfur compounds, in Environmental Technologies to treat Sulfur Compounds, *Ed Piet Lens et Look Hulshoff Pol*, 12, 281–304.

Hirai, M., Kamamoto, M., Yani, M., Shoda, M. 2000. Comparison of the Biological H_2S Removal Characteristics among Four Inorganic Packing Materials, *Journal of Bioscience and Bioengineering*, Vol. 91, No. 4, 396–402.

Ramirez-Lopez, E., Corona-Hernandez, Dendooven L., Range P., Thalasso F. 2003. Characterization of five agricultural by-products as potential biofilter carriers, *Bioressource Technology*, 88, 259–263.

Sene, L., Converti, A., Felipe, M.G.A., Zilli, M. 2002. Sugarcane bagasse as alternative packing material for biofiltration of benzene polluted gaseous streams: a preliminary study, *Bioresource Technology*, 83, 153–157.

Williams, T.O., Miller, F.C. 1992. Biofilters and facility operations, *Biocycle*, 75–79.

Wu, G., Conti, B., Leroux, A., Brzezinski, R., Viel, G., Heitz, M. 1999. A high performance Biofilter for VOC emission control, *Journal of the Air & Waste Management Association*, Vol. 49, 185–192.

European Symposium on Environmental Biotechnology, ESEB 2004 - Verstraete (ed)
© 2004 Taylor & Francis Group, London, ISBN 90 5809 653 X

NO removal from flue gases: the kinetics of the oxidation of Fe^{2+} (EDTA) under typical BiodeNOx conditions

F. Gambardella, S. Fabianti, L.M. Galan, K.J. Ganzeveld, H.J. Heeres
University of Groningen, Chemical Engineering Department, Nijenborgh, Groningen

ABSTRACT: BiodeNOx is a chemical–biological method used to remove nitrogen monoxide from flue gases. It is based on the reaction of NO gas with an Fe^{2+} chelete solution. However, Fe^{2+} (EDTA) can be easily reduced in presence of oxygen, decreasing the efficiency of the removal process. The kinetics of the oxidation of Fe^{2+} (EDTA) has been studied in typical BiodeNOx conditions (T = 329 K, [Fe^{2+} (EDTA)] = 50 mol/m^3) and the influence of the pH on the oxidation rate in the range 5–8 is tested. Moreover, the influence of the presence of typical BiodeNOx components (NaCl and biomass) on the kinetics is measured.

1 INTRODUCTION

The BiodeNOx process is an alternative method for the removal of NO from waste flue gas. It consists of two integrated processes: a wet absorption, taking place in a scrubber and a biological regeneration in a bioreactor. The nitrogen monoxide is not easily soluble in water and for this reason, a solution of Fe^{2+} (EDTA) is present in the scrubber: the NO reacts with the iron chelate to give the nitrosyl complex (Fe^{2+} (EDTA)-NO). In flue gas, other typical components are present which can interfere with the process and cause unwanted side reactions with the iron chelate. The main problem is oxygen: as a matter of fact, the oxidation of the iron chelate produces Fe^{3+} (EDTA), which is not able of binding NO. Of course, the presence of oxygen cannot be avoided and the oxidation rate has to be limited as much as possible by an optimization of the scrubber working conditions.

The aim of this study is to investigate the kinetics of the reaction between Fe^{2+} (EDTA) and oxygen in typical BiodeNOx conditions. The kinetics of the oxidation of Fe(EDTA) was studied in the past by several authors, as shown in Table 1. A general mechanism for the reaction was proposed by Wubs (1993). The reaction is usually represented in the following way:

$$4Fe^{II}(EDTA)^{2-} + O_2 + 2H_2O \rightarrow \rightarrow 4Fe^{III}(EDTA)^{2-} + 4OH^- \quad (1)$$

In Wubs' research, the reaction was studied in presence of high concentrations of oxygen and at a pH of 7.5. Instead, in BiodeNOx, the oxygen is present in lower concentration, between 2 and 20% of the flue gas. Zang and Van Eldik (1990) reported kinetic results for [Fe^{2+} (EDTA)] < 20 mol/m^3 and for T = 298 K. No kinetic data are available for higher temperatures. In this study, the oxidation reaction is studied in case of Fe^{2+} (EDTA) concentrations of 50 mol/m^3 with a concentration of oxygen of 20% in the gas phase (worst BiodeNOx case) and at high temperature

Table 1. Literature overview on the kinetic studies of the oxidation of Fe^{2+}(EDTA).

Authors	TK	pH	System	[Fe(EDTA)] mol/m^3	$[02]^i$ mol/m^3
Travin, 1981	295	6	Stop-flow unit	0.1	0.25
Sada, 1987	293–333	6–8	Bubble column	<20	0.018–0.064
Brown, 1987	298	4–9	Bubble column	2.5	0.125–1.5
Zang, 1990	298	2–7	Stop-flow unit	2.5–20	0.125
Wubs, 1993	293–333	7.5	Stirred cell	100	0.1–1
Seibig, 1997	298	1–6	Stop-flow unit	0.03–2	0.125–0.625

(T = 328 K). In particular, the possible influence of a pH between 5 and 8 is taken into consideration. Moreover, the influence of typical BiodeNOx components (NaCl and biomass), normally present in the industrial process, on the kinetics is tested in the present work.

2 EXPERIMENTAL

Every kinetic experiment is run in a thermostated glass stirred reactor with a stainless steel cap. The stirrer has six blades for the gas phase and six smaller ones for the liquid phase. The stirrer speed is regulated to maintain a flat surface of the liquid phase. Inside the reactor, four-baffles are present to avoid any disturbance of the liquid surface. The pressure and the temperature inside the reactor are measured and recorded continuously during the experiments. Every experiment is composed of two parts: first a constant pressure and then a decreasing pressure measurement. The first measurement is performed with oxygen. Before the start, the gas cap of the reactor is evacuated to remove the air present, until the vapor pressure of the water is established. At this point, the oxygen gas is supplied to the reactor until the pressure reaches the desired pre-set maximum pressure. Subsequently, the inlet valve is closed, the stirrer is activated and the absorption starts. The experiment goes on till the pressure reaches a pre-set minimum experimental pressure and after that, some extra gas is supplied until the original pre-set maximum pressure is reached again. By repeating this procedure several times, the absorption profile appears as a sawtooth curve. When the measurement is concluded, the vapor pressure is established again keeping the reactor under vacuum. Subsequently, a physical absorption experiment characterized by a decreasing pressure profile, is performed with nitrous oxide (NO_2 does not react with the iron chelate solution). The N_2O gas is supplied to the reactor until

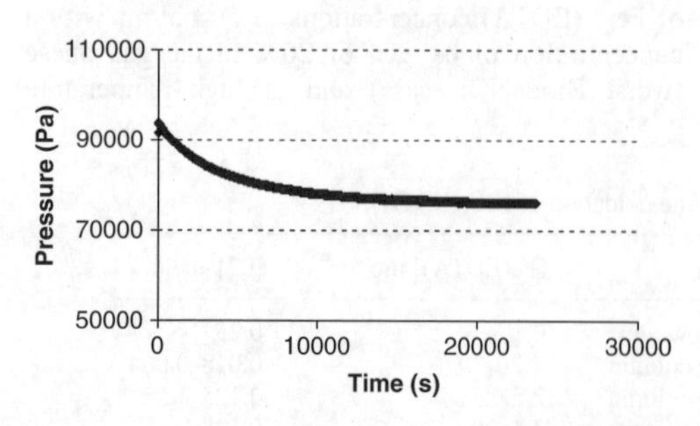

Figure 1. Absorption profile of N_2O in Fe(EDTA) solution. T = 328 K, [Fe(EDTA)] = 60 mol/m^3, $p^0_{N_2O}$ = 77390 Pa, p_{H_2O} = 14180 Pa.

the pressure reaches a pre-set value. At this point, the inlet valve is closed, the stirrer is activated and the absorption starts. The absorption goes on until the pressure inside the reactor is constant. A typical profile of absorption is shown in Figure 1. The solution of Fe^{2+} (EDTA) is prepared using the procedure described by Wubs (1993). The content of Fe^2 in the solution is measured before and after every experiment by titration with $CeSO_4$, Vogel (1978).

3 RESULTS AND DISCUSSION

3.1 *Kinetic considerations*

Knowledge of the regime in which the absorption takes place is necessary to determine the kinetics of the reaction. The oxidation of Fe(EDTA) is reported in literature as first order in oxygen and second order in iron chelate in the experimental conditions (Wubs, (1994); Zang and Van Eldik, (1990); Brown and Mazzarella, (1987)). In this case, the Hatta number can be calculated as:

$$Ha = \frac{C_{Fe(EDTA)}}{k_L}\sqrt{D_{O_2}k_{1,2}} \tag{2}$$

where k_L is the mass transfer coefficient in the liquid phase (m/s), D_{O_2} is the diffusion coefficient of oxygen in Fe^{2+} (EDTA solution (m^2/s) and k_{12} is the reaction rate constant (m^6/mol^2 s). Using the kinetic data reported by Wubs (1993), it is possible to show that:

$$2 < Ha \le E_{A\infty} \tag{3}$$

with

$$E_{A\infty} = 1 + \frac{D_{Fe(EDTA)}C_{Fe(EDTA)}}{\nu_B D_{O2} C^i_{O2}} \tag{4}$$

where ν_B is the stoichiometric coefficient of the reaction and $C^i_{O_2}$ is the concentration of oxygen at the gas–liquid interface (mol/m^3). The calculated value of the Hatta number indicates that the reaction takes place in an intermediate regime between fast and instantaneous.

3.2 *pH influence*

The influence of the pH on the oxidation of Fe(EDTA) was studied by Brown and Mazzarella (1987) and Seibig and van Eldik (1997). They observed that the oxidation rate is not pH dependent in a range between 5 and 7 and T = 298 K. In the present study, the influence of the pH on the physical properties of the solution, like the solubility (He) and the mass transfer coefficient (k_L), is examined with physical absorption experiments using N_2O. The experiments are run at T = 328 K and different concentration of iron chelate are used (between 15 and 60 mol/m^3).

The results in Table 2 show that there is no significant influence of the pH on the physical properties of the solution.

The enhancement factor for the absorption of oxygen in iron chelate solutions can be calculated as:

$$E_A = \frac{J_{reaction}}{J_{physical}} = \frac{J_{O_2}}{k_L C_{O_2}^{\ i}} = \frac{1}{k_L a}\cdot\frac{He_{O_2}}{p_{O_2}}\cdot\frac{V_G}{RTV_L}\cdot\left(-\frac{dp_{O_2}}{dt}\right) \qquad (5)$$

where J = molar flux(mol/m^{2}s), a = interfacial area (m^2/m^3), p = partial pressure of the gas (Pa). Different expressions are available in the literature for the calculation of the kinetic constant for an intermediate regime. The implicit approximated relation for the enhancement factor of De Coursey (1989), valid for an irreversible reaction of finite rate, can be used for our purpose:

$$\frac{1}{Ha^2} + \frac{1}{q\left(E_A\sqrt{r_B}+1\right)} = \frac{1}{E_A^{\ 2}-1} \qquad (6)$$

with:

$$r_B = \frac{D_{FeEDTA}}{D_{O_2}},\ q = \frac{C_{FeEDTA}}{v_B\, C_{O_2}}, \qquad (7)$$

The values of the enhancement factor obtained experimentally are compared with those calculated using equation 6 for two different values of the pH (Figures 2 and 3).

Table 2. Solubility and mass transfer coefficient values for oxygen in Fe(EDTA) solution, calculated through physical absorption experiments using N_2O.

Exp.	pH	Fe(EDTA) (mol/m^3)	He (Pa *m^3/mole)	k_L (m/s)
1	8	60	1.23E + 05	2.33E − 05
2	5	60	1.17E + 05	2.85E − 05
3	8	30	1.17E + 05	2.22E − 05
4	5	30	1.18E + 05	2.31E − 05
5	8	15	1.17E + 05	2.51E − 05
6	5	15	1.37E + 05	2.13E − 05

The kinetic constants calculated for pH 5 and 8 are equal and the value is k_{12} = 0.02 m^6/mol^{2}s ± 0.01 m^6/mol^{2}s. Wubs (1993) calculated $k_{1,2}$ = 0.020 m^6/mol^{2}s for T = 331 K and [Fe(EDTA)] = 100 mol/m^3. It can be concluded that the pH does not have a significant effect on the kinetics of the oxidation reaction.

3.3 *Salt addition*

Some kinetic experiments were performed in the presence of NaCl, in a concentration ranging between 5 and 15 kg/m^3. The physical properties of the solution are calculated and reported in the Table 3. It can be concluded that the salt presence does not have a remarkable effect on the properties of the solution.

The experimental values of the enhancement factor are calculated and compared with the values calculated for a Fe(EDTA) solution, according to equation 6. It is possible to observe a variation of the slope proportional to the concentration of salt present. The reaction rate constants are calculated and reported in Table 4. It can be concluded that the salt presence can have an effect on the kinetics of the

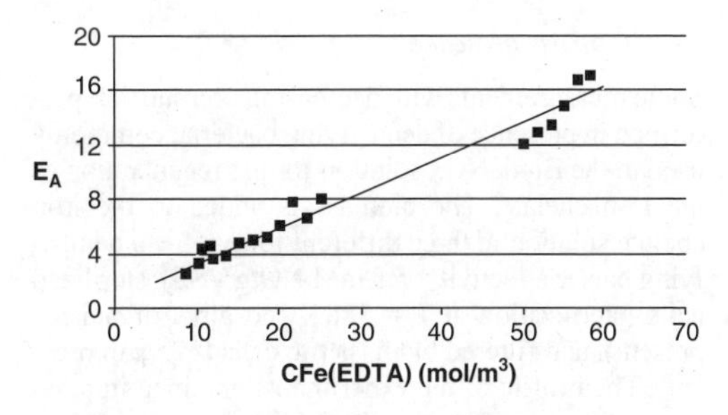

Figure 2. Comparison of the experimental values of the enhancement factor with the theoretical values (line) according to equation 6. T = 328 K, p_{O_2} = 0.18 * 10^5 Pa, pH = 5.

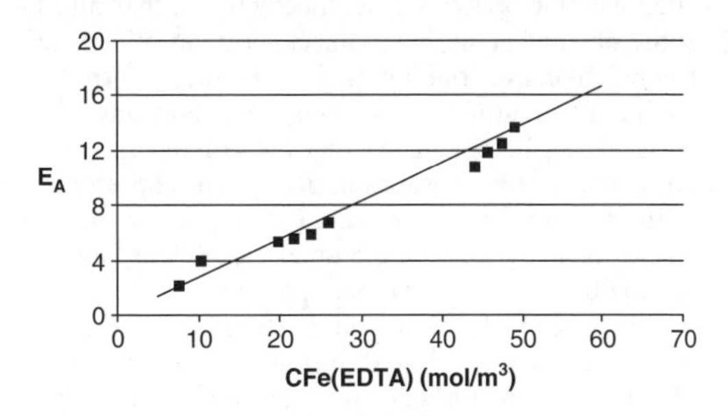

Figure 3. Comparison of the experimental values of the enhancement factor with the theoretical values (line) according to equation 6. T = 328 K, p_{O_2} = 0.18 * 10^5 Pa, pH = 8.

Table 3. Salt effect on the physical properties of the solution.

[Nacl] (kg/m^3)	k_L (m/s)	He ($Pa\,m^3/mol$)
0	1.98E − 05	1.01E + 05
5	2.29E − 05	1.01E + 05
10	2.31E − 05	1.04E + 05
15	2.11E − 05	1.05E + 05

Table 4. Kinetic constant values of the Fe(EDTA) oxidation calculated inpresence of NaCl.

[NaCl] (kg/m^3)	Fe(EDTA) mol/m^3	k_{12} ($mol^6/m^3\,s$)
0	50	0.02
5	50	0.02
10	50	0.02
15	50	0.03

reaction, but only when it is present in high concentrations ([Nacl] > $10\,kg/m^3$).

3.4 *Biomass presence*

Some measurements with decreasing pressure are performed in presence of denitrifying bacteria, commonly used in the BiodeNOx solution for the regeneration of the iron chelate. The biomass is added to the iron chelate solution in three different forms: fresh denitrifying bacteria (activity: 5.5 mM NO/g VSS), sterilized cells (sterilization at T = 388 K, no alive organisms present), and filtered biomass (no cells >25 μm present). The profile of the experiments obtained in presence of oxygen for the three different cases were compared and the results are shown in Figure 5.

In absence of biomass, the iron chelate solution absorbs an amount of oxygen which is easily calculated with the pressure difference between the values at the start and at the end of the experiment. When the filtered biomass solution is present, the absorption profile is similar to the one obtained without any biomass. It can be concluded that the filtered solution does not affect the absorption of oxygen. In presence of fresh denitrifying bacteria, the final pressure in the reactor is higher than in the other cases. This can be due to different reasons. First, a reduction of the interfacial area can take place due to the particle presence on the surface of the liquid. In this situation, the flux of the oxygen through the gas–liquid interface decreases. Moreover, in presence of cells, the viscosity of the solution increases, causing an increase of the mass transfer resistance in the liquid phase. Furthermore, it cannot be excluded the presence of

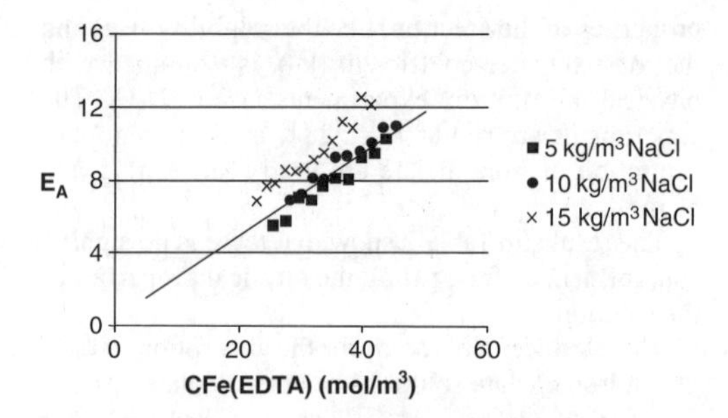

Figure 4. Comparison of the theoretical enhancement factor of the reaction of Fe(EDTA) solution with oxygen (line) with the values obtained experimentally in the presence of NaCl. T = 328 K, $p_{O_2} = 0.18 * 10^5$ Pa, pH = 5.

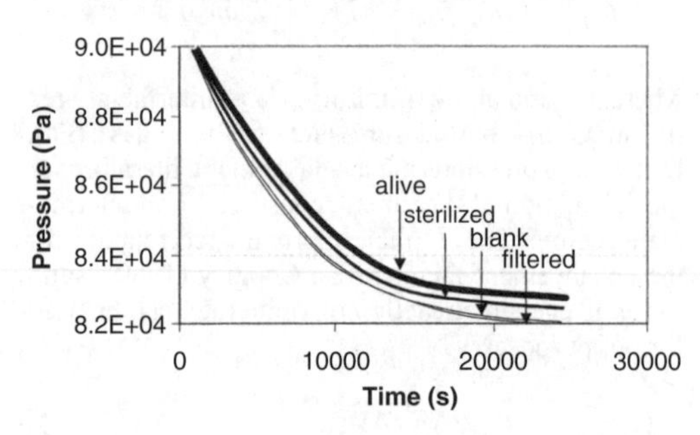

Figure 5. Oxidation profiles of different kinds of Fe(EDTA) solutions.

some extra gas formation as a result of the biomass respiration: another possible factor responsible for the increase of the final pressure. In case of sterilized biomass, the gas absorbed increases compared to the case of alive microorganisms: this effect may be due to the foam formation and to the presence of decomposition products, responsible for a change of the physical properties.

4 CONCLUSIONS

The oxidation of Fe^{2+} (EDTA) solution is not influenced by the pH of the solution in the range 6–8 in the experimental conditions. Moreover, the addition of salt seems to have a small effect on the kinetics of the reaction. The presence of bacteria changes the absorption profile of oxygen, influencing the physical characteristics of the solution.

REFERENCES

Brown, E.R., Mazzarella, J.D., (1987): Mechanism of oxidation of ferrous polydentate complexes by dioxygen, *J. Electroanal. Chem.*, 222: 173–192.

Dagaonkar, M., (2001): Effect of a Microphase on gas-liquid mass transfer, *PhD Thesis*, Rijksuniversitait Groningen.

De Coursey, W.J.,Thring, R.W., (1989): Effect of unequal diffusivities on enhancement factors for reversible and irreversible reaction, *Chem. Eng. Sc.* 44: 1715–1721.

Sada, E., Kumazawa, H., (1987): Oxidation kinetics of Fe^{II}EDTA and Fe^{II}NTA chelates by dissolved oxygen, *Ind. Eng. Chem. Res.* 26: 1468–1472.

Seibig, S., van Eldik, R., (1987): Kinetics of [Fe^2(EDTA)] oxidation by molecular oxygen revisited. New evidence for a multistep mechanism, *In. Chem.* 36: 4115–4120.

Travin, S.O., Skurlatov, Yu.I., (1981): Kinetics and mechanism of the reaction of iron (II)ethylenediaminotetra-acetate (Fe^{2+} EDTA) with molecular oxygen in the presence of ligands, *Rus. J. of Phys. Chem.* 55: 815–818.

Vogel, A.I., (1978): *Vogel's textbook of quantitative chemical analysis*, Longman.

Wubs, H.J., Beenackers, A.A.C.M., (1993): Kinetics of the oxidation of ferrous chelates of EDTA and HEDTA in aqueous solution, *Ind. Eng. Chem.Res.* 32(11): 2580–2594.

Zang, V., Van Eldik, R., (1990): Kinetics and mechanism of the autoxidation of iron (II) induced trough chelation by ethylenediaminetetraacetate and related ligands, *Inorg. Chem.* 29: 1705–1711.

European Symposium on Environmental Biotechnology, ESEB 2004 - Verstraete (ed)
© 2004 Taylor & Francis Group, London, ISBN 90 5809 653 X

Control of biomass growth by nematode-grazing bacteria in a biofilter treating chlorobenzenes

C. Seignez, J. Calvelo, N. Adler & C. Holliger
Swiss Federal Institute of Technology Lausanne, Laboratory for Environmental Biotechnology Lausanne, Switzerland

ABSTRACT: The effect of nematodes on biomass accumulation in a biotrickling filter treating a waste gas containing monochlorobenzene (CB) and 1,2-dichlorobenzene (*o*-DCB) was studied. The diversity of microfauna in the biofilm included bacteria, fungi, protozoa, and nematodes. At mass loading rates higher than $4550\,g\,m^{-3}\,d^{-1}$, the protozoa disappeared and at $8400\,g\,m^{-3}\,d^{-1}$ also the nematodes died off. Between day 68 and 92, the mass loading rate was maintained at $3200\,g\,m^{-3}\,d^{-1}$ which resulted in a high removal efficiency of 85–90% and the biomass increased at a rate of $69\,g\,d^{-1}$. After 2 days of nematode inhibition with ivermectin, when 69–76% of the nematodes were inactive, the biomass increased at a rate of $125\,g\,d^{-1}$. Hence, nematode-grazing apparently reduced the biomass growth rate by 45% showing that nematodes can be an efficient mean to reduce biomass increase and to replace backwashing.

1 INTRODUCTION

Biofilters are used successfully for the purification of industrial waste gases containing specific pollutants such as chlorobenzenes. The waste gas is forced through a packed column that is continuously wetted by recycled water phase. After inoculation with the mixed microbial consortium, aeration, chlorobenzenes addition, the microbial degradation occurs and the biofilm forms on the packing surface. The excessive biomass development on supports, especially at higher inlet concentrations, can cause pressure drop and clogging.

Clogging is a complex phenomenon determined by diverse factors such as the characteristic of the pollutants, their microbial degradation rates and Henry coefficients, the morphology of the formed biofilm and the characteristics of the packing material. To avoid clogging, several solutions have been proposed. The inert packing material can be cleaned by regular backwashing with water or with various chemicals, by physical removal of the accumulated biomass and cleaning packing elements (Okkerse et al., 1999).

Recently, the processes that reduce the amount of immobilized biomass on the packing increase in importance. It was observed that the addition of protozoa or nematodes to the biotrickling filter resulted in a decrease of the biomass accumulation rate and an increase of pollutant mineralization (Selivanovskaya et al., 1997, Cox and Deshusses, 1999). Nematodes are among the primary grazers of bacteria. Besides, the development and activity of nematodes inside the biofilm, results in a higher diffusion of oxygen and nutrients and generally, increase microbial activity.

In the present study, we investigate the performance of biotrickling filter and the nematode impact on biomass growth and biotrickling elimination capacity degrading the mixture of chlorobenzene (CB) and dichlorobenzene (DCB). To assess the efficiency of nematodes, the biomass growth on the support was followed with and without an inhibitor of nematodes, ivermectine, which has an anthelmintic activity.

2 EXPERIMENTAL

Experiments were performed in a laboratory scale downflow co-current biotrickling filter (BTF). (Seignez et al., 2002) The BTF column of 40 l total volume, 1.27 m height and 0.2 m internal diameter was made of glass. The column was filled with three cylindrical elements with a height of 0.3 m and a diameter of 0.19 m each, and made of structured PVC packing material (Biodek® FB 10.12, Munters Euroform, Aachen, Germany) that has a specific packing area of $240\,m^2\,m^{-3}$ and a porosity of 96%.

At the bottom of the reactor, the trickling liquid was gathered and automatically neutralized before recirculation or waste out. The mineral salt medium

(Seignez et al., 2002) used to replace the trickling liquid contained nitrate as nitrogen source. The ivermectin ($200 \, mg \, l^{-1}$) in the experiment of nematode inhibition was added to the nutrient medium.

The empty bed residence time (EBRT) was maintained at 1.9 min, liquid recirculation flow rate was $20 \, l \, h^{-1}$ and the rate of nutrient medium addition was $0.47 \, l \, h^{-1}$ at $1800 \, g \, m^{-3} \, d^{-1}$. The waste gas was produced by injection of a mixture of pure CB and *o*-DCB (mass ratio CB: *o*-DCB = 4.2:1) in the air flow up to a maximal concentration of $8.9 \, g$ CB m^{-3} and $2.8 \, g$ *o*-DCB m^{-3}. The biotrickling filter was inoculated with an adapted culture (Seignez et al., 2001) of bacteria cultivated in a batch reactor with CB and *o*-DCB as sole source of carbon and energy.

The chlorobenzene concentration of the gaseous outlet was measured on-line with a flame ionization detector (Kull Instruments, Oftrigen, Switzerland). Biomass growth was evaluated by continuous weighing of the biotrickling filter. The microfauna and the nematode vitality were observed by an optical microscope. The identification of nematode was performed on the basis of scanning electronic microscope pictures.

3 RESULTS AND DISCUSSION

3.1 *Biotrickling filter performance*

The evolution of total load, corresponding removal efficiency and residual load during the biofiltration experiment of 90 days is shown in Fig. 1. In the course of colonization of 14 days, the constant load was maintained of about $1134 \, g \, m^{-3} \, d^{-1}$ with the goal to reach the maximal removal efficiency at a given load. From day 15 to 27, we increased loading, up to about $2040 \, g \, m^{-3} \, d^{-1}$ which followed by removal efficiency increase up to about 95%. Maintaining the constant load for certain time was the decisive factor in the biofilter performance enhancement. From 27 to 44 days, the load was decreased to $1134 \, g \, m^{-3} \, d^{-1}$ in order to enhance the biomass degradation activity i.e. to permit further biomass adaptation. In the course of experimentation, up to 87th day the load was stepwise increased to $7260 \, g \, m^{-3} \, d^{-1}$. During this period the removal efficiency remained constantly high. By this strategy, we were able to increase the load without the removal efficiency decrease, which stayed stable during the whole working period and even reached a maximum of about 99%.

Further load increase to $9525 \, g \, m^{-3} \, d^{-1}$ during last three days of the experimentation resulted in the immediate fall of removal efficiency under 80%.

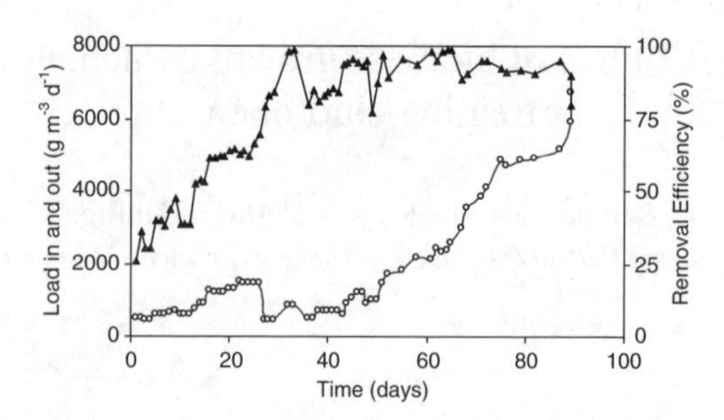

Figure 1. Mass loading (○) and removal efficiency (▲) during biotrickling operation.

3.2 *Mass transfer*

In order to control the purification process, oxygen-liquid and chlorobenzene-liquid were studied. The overall liquid mass transfer coefficient (K_La) is convenient as it represents the interfacial area per unit volume of the filter (a) and the liquid phase and gas phase mass transfer coefficients (k_L and k_G).

The oxygen-liquid mass transfer is important to get optimal performance of pollutant degradation in biofilm. If oxygen became a limiting factor, the micro-organisms are not able to degrade the chlorobenzenes properly. The mass transfer coefficient for oxygen was measured in the absence of biomass. With the gas flow rate of $15 \, L \, min^{-1}$ after 15 min the oxygen concentration in the liquid phase attain 95% and average oxygen K_La obtained was $11.1 \, h^{-1}$. The maximal concentration measurements of CB and DCB in the liquid phase were 82% for a load of $1300 \, g \, m^{-3} \, d^{-1}$ using $18.4 \, L \, h^{-1}$ flow rate and not taking into account the repartition between the two phases according to Henry's law.

Knowing the diffusivity factor of chlorobenzene ($D_{COV} = 0.6 \, m^2 s^{-1}$) and oxygen ($D_{O2} = 1.6 \, m^2 s^{-1}$) and $k_La_{(O2)}$ we can calculate the $k_La_{(COV)}$ of chlorobenzene:

$$k_La_{(COV)} = k_La_{(O2)} \frac{\sqrt{D\,cov}}{\sqrt{Do2}}$$

We obtained chlorobenzene mass transfer coefficient value of $7 \, h^{-1}$.

The values between 6 and $10 \, h^{-1}$ have been established previously in our laboratory for the same biotrickling filter, but filled with another support (Seignez et al., 2002), while Pederson and Arvin (1977) found chlorobenzene mass transfer coefficient being $2–10 \, h^{-1}$ depending on the specific system.

3.3 *Microfauna*

Both *Procaryotes*: bacteria and *Eucaryotes*: mycetes, protozoan and metazoan were detected in the biofilm. The bacterial consortium consists of different types of bacilli and cocci (Fig. 2).

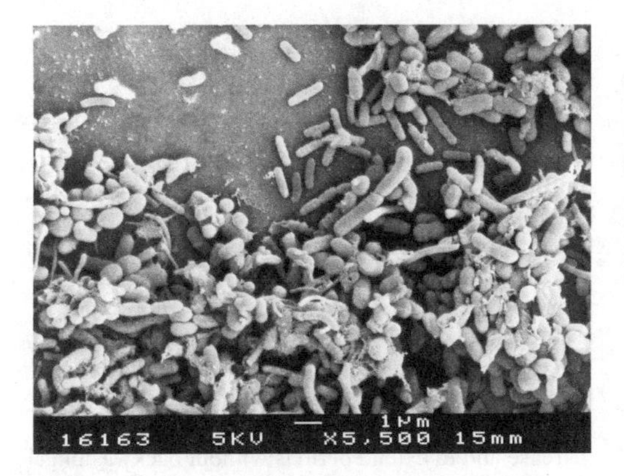

Figure 2. SEM picture of bacteria in biofilm.

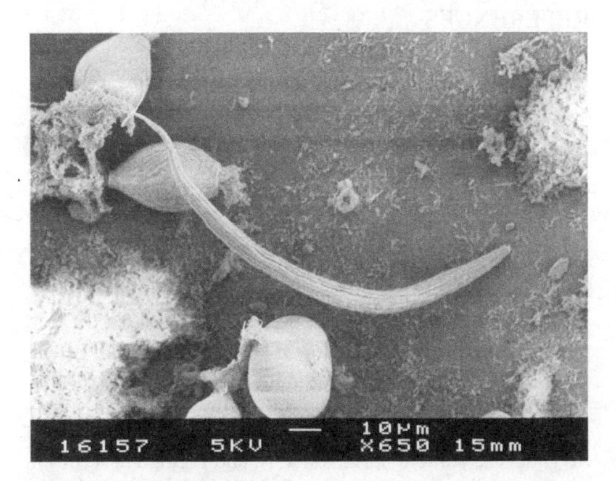

Figure 3. SEM picture of nematode and protozoa in biofilm.

In our study special attention has been paid to a particular specie of metazoan: nematodes. They were found in the biofilm together with protozoan and bacteria using optical and scanning electronic microscope (Fig. 3).

The identification of nematodes, based on SEM pictures, has been made by P. Arpin from Musée National d'Histoire Naturelle, Paris (Fig. 4). These nematodes belonged to the diplogasterideae, i.e. more precisely to *Diplogaster nudicapitatus*. Up to a mass loading rate of 4500 $g\,m^{-3}\,d^{-1}$, the diversity of microfauna in the biofilm was large and included bacteria, fungi, protozoa, and nematodes. At higher mass loading rates, the protozoa disappeared and at 8400 $g\,m^{-3}\,d^{-1}$ also the nematodes died off.

3.4 *Nematode impact on biotrickling performance*

In the course of continuous biofiltration, due to rapid biomass accumulation in the packed bed, interfacial

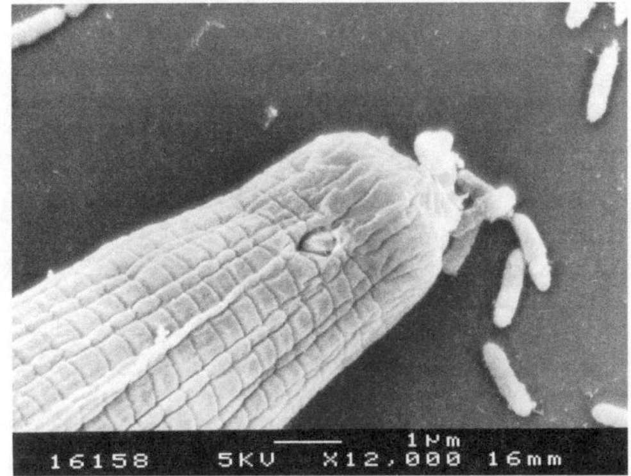

Figure 4. SEM picture of nematode-grazing bacteria.

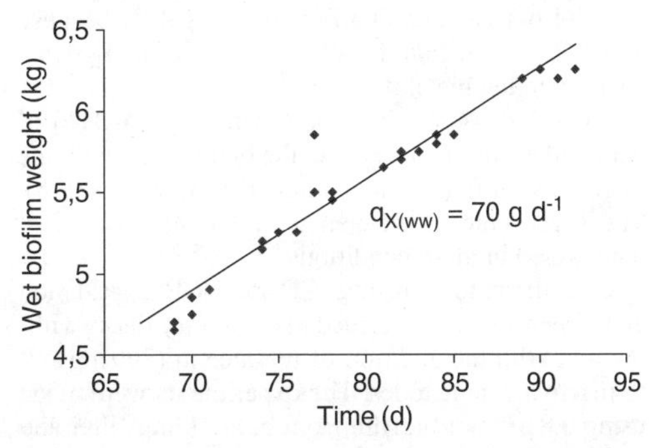

Figure 5. The biofilm weight in biotrickling filter without nematode inhibition.

area for mass transfer decrease, the pressure drop and the pollutant removal decline and the biofilter clog. The prolonged continuous biofiltration results in formation of biomass not actively involved in pollutant degradation (Zuber, 1995; Huub et al., 1999).

To investigate the influence of the nematodes on biomass accumulation, the experiments of biofiltration were made, during a period of 90 days, in the presence and in the absence of nematodes at the mass loading rate of 3200 $g\,m^{-3}\,d^{-1}$. In that purpose, the nematode inhibitor, ivermectin, solubilized in the medium was added to the biotrickling filter at a concentration of 200 $mg\,l^{-1}$.

Without the nematode inhibitor, between day 68 and 92, the biomass increased at a constant rate of 70 $g\,d^{-1}$ and a high removal efficiency of 85–90% was maintained (Fig. 5). In the presence of inhibitor the biomass growth and nematode mortality was measured during six days. The activity of ivermectin was not instantaneous, only 20–33% of the nematodes were dead two days after the inhibitor addition, and 69–76% of the nematodes were inactive during the following

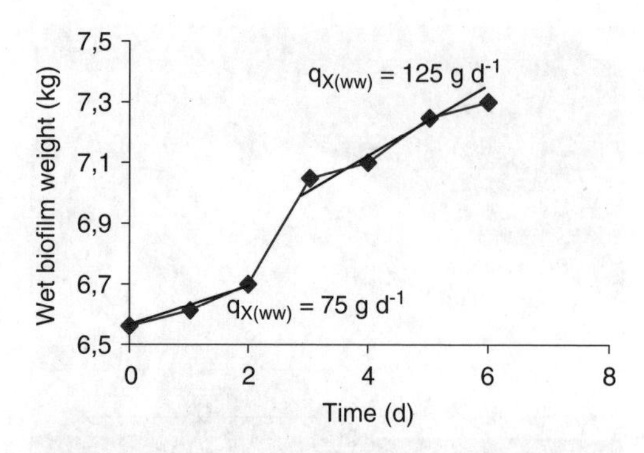

Figure 6. The biofilm weight in biotrickling filter with ivermectin nematode inhibitor.

four days. During the first period, the biomass growth rate did not change significantly ($75\,g\,d^{-1}$) because 70 to 80% of nematodes were still active and grazed bacteria in the biofilm.

After 3 days of inhibition when only 24–31% of nematodes present are active, the biomass growth rate increase significantly and attained $125\,g\,d^{-1}$ (Fig. 6). The high removal efficiency of 85–90% was also maintained in these conditions.

To confirm these results, CB and DCB degradation experiments were performed in laboratory flasks without and with the addition of ivermectin ($200\,mg\,l^{-1}$) to inactivate nematodes. The experiments were made using $0.8\,g\,l^{-1}$ of biofilm from biotrickling filter and a CB and DCB concentration of 8 and $1.9\,\mu g\,l^{-1}$, respectively.

It was found that after about six hours, the chlorobenzenes were completely degraded in both samples and there was no difference between CB and DCB degradation rate in experiments with (control) and without active nematodes (ivermectin addition). So, we concluded that there was no influence of nematodes on the chlorobenzenes degradation.

4 CONCLUSIONS

The influence of nematode presence on biomass accumulation in the biofilm was presented. It was shown that the nematodes-grazing activity can help in biofilter clogging problems. The nematode activity on biomass accumulation was evaluated in the biofilm by biofiltration experiments in the presence and in the absence of nematodes at the identical loading rate. To suppress the nematode activity, ivermectin was added as inhibitor. The biomass growth rate was almost twice higher when nematodes were inhibited, showing that application of organic solvent resistant nematodes can be an efficient mean to control biomass increase and to extend periods without backwashing.

REFERENCES

Cox, H.H.J. and Deshusses, M.A. (1999). Chemical removal of biomass from waste air biotrickling filters: screening of chemicals of potential interest. Water Research 33: 2383–2391.

Okkerse, W.J.H., Ottengraf, S.P.P., Diks, R.M.M., Osinga-Kuipers, B. and Jacobs, P. (1999). Long term performance of biotrickling filters removing a mixture of volatile organic compounds from an artificial waste gas:dichloromethane and methylmethacrylate. Bioprocess Engineering 20: 49–57

Seignez, C., Atti, A., Adler, N.and Péringer, P. (2002). Effect of Biotrickling Filter Operating Parameters on Chlorobenzenes Degradation. Journal of Environmental Engineering 128: 360–366.

Seignez, C., Vuillemin, A., Adler, N. and Peringer, P. (2001). A procedure for production of adapted bacteria to degrade chlorinated aromatics. Journal of Hazardous Materials 84: 265–277

Selivanovskaya, S.Y., Petrov, A.M., Egorova, K.V. and Naumova, R.P. (1997). Protozoan and metazoan communities treating a simulated petrochemical industry wastewater in a rotating disc biological reactor. World Journal of Microbiology & Biotechnology 13: 511–517.

European Symposium on Environmental Biotechnology, ESEB 2004 - Verstraete (ed)
© 2004 Taylor & Francis Group, London, ISBN 90 5809 653 X

A novel three-phase circulating-bed biofilm reactor for removing toluene from gas stream

B. Sang
Water Environment & Remediation Research Center, Korea Institute of Science and Technology, Seoul, Korea

B.E. Rittmann
Department of Civil and Environmental Engineering, Northwestern University, Evanston, IL, U.S.A.

ABSTRACT: For removing toluene from gas streams, a series of steady-state and short-term experiments on a three-phase circulating-bed biofilm reactor (CBBR) was conducted. The goal was to investigate the effect of macroporous-carrier size (1 mm cubes versus 4 mm cubes) on CBBR performance over a wide range of toluene and oxygen loading. We hypothesized that the smaller biomass accumulation with 1 mm carriers would minimize dissolved oxygen limitation and improve toluene removal, particularly when the dissolved-oxygen loading is constrained. The CBBR with 1 mm carriers overcame the performance limitation observed with the CBBR with 4 mm carriers. The 1 mm carriers consistently gave superior removal of toluene and COD, and the advantage was greatest for the lowest oxygen loading and the greatest toluene loading. The 1 mm carriers achieved superior performance because they minimized the negative effects of oxygen depletion, while continuing to provide protection from excess biomass detachment and inhibition from toluene. The enhanced toluene-degradation performance of the CBBR with 1 mm carrier was described by the flux analysis of oxygen, toluene, and intermediates in the bulk phase and inside the biofilm. Finally, 1 mm CBBR achieved volumetric removal capacities up to 300 times greater than demonstrated by other biofilters treating toluene and related volatile hydrocarbons.

1 INTRODUCTION

Among the many water and air pollutants that can be biodegraded aerobically are those whose biodegradation is inhibited by the pollutant itself. Excellent examples are the aromatic hydrocarbons that comprise a major fraction of gasoline: benzene, toluene, and xylenes (BTX). These aromatic hydrocarbons are among the most common contaminants in groundwaters, and they also commonly contaminate gases, such as at refineries, during fuel-storage, and with soil-vapor extraction of contaminated groundwaters. Although BTX can be completely biodegraded in aerobic systems, their removal and the growth of the degrading bacteria are seriously slowed by inhibition from BTX themselves. If inhibition is too serious, conventional biological treatment can fail or require large, expensive systems.

Microorganisms that grow very slowly, perhaps due to inhibition, must be retained in the treatment system with high efficiency. In general, good retention is accentuated when the bacteria are retained as biofilms, which are layer-like aggregates attached to solid surfaces. Biofilm accumulation provides an added benefit when the mass-transport resistance in the biofilm lowers the concentration of the inhibitor inside the biofilm.

The three-phase circulating-bed biofilm rector (CBBR) was developed by Yu and co-workers to treat inhibitory pollutants, and it was extensively tested with BTX removal from contaminated gas streams. Despite the CBBR's success using the standard 4 mm carriers, Yu and co-workers (Yu et al. 2001) identified a significant limitation: Oxygen depletion inside the biofilm slowed biodegradation rates, particularly the first step of BTX dioxygenation. When the BTX loading was high and/or the dissolved oxygen concentration was well below saturation, the overall removal rate for BTX reached a plateau. This was due to oxygen depletion inside the biofilm, and the effect was to force the initial dioxygenation reactions more and more to the small concentration of bacteria suspended in the liquid phase. Oxygen limitation was accentuated when the carriers contained more biomass, which resulted from using a higher pollutant-loading rate. In this case, it was possible to have "too much biomass," because additional biomass consumed more oxygen for its endogenous respiration.

The goal of the work presented here was to preclude creating the serious oxygen limitation found with the standard 4 mm AQUACEL carriers because of having too much biomass, while still maintaining the protection role of the porous carriers in biomass retention. To achieve this two-pronged goal, we employed 1 mm AQUACEL carriers, their smaller size should reduce the buildup of excess biomass, but the internal porosity should maintain the benefits of protection. This study documents that a prototype CBBR using 1 mm porous carriers met the goal for the model compound toluene. The CBBR with 1 mm carriers showed outstanding removal efficiency, stability, and, to our knowledge, by far the highest loading rates obtained by any biological process treating gas-phase BTX.

2 MATERIALS AND METHODS

2.1 *The circulating-bed biofilm reactor*

Figure 1 illustrates the CBBR employed for this study, as well as for Yu and coworkers (Yu et al. 2001). The working volume was 2.78 L. The pH in the reactor was 6.8 throughout the experiments. The temperature was maintained at a constant 22°C. To improve the removal kinetics for the VOCs without harming biofilm accumulation, we selected a smaller AQUACEL carrier size, 1 mm, than the standard 4 mm used before. To achieve a comparable amount of biofilm carrier, we matched the external surface area of the 1 mm carriers to that with 4 mm carriers used before. The 1 mm (dry) cubes had a side dimension of 1–1.2 mm, an average porosity of 0.93, and a wet density of 1.04 g/cm^3. The average dry weight and wet volume of one of these carriers were 0.13 mg and 1.6 mm^3, respectively.

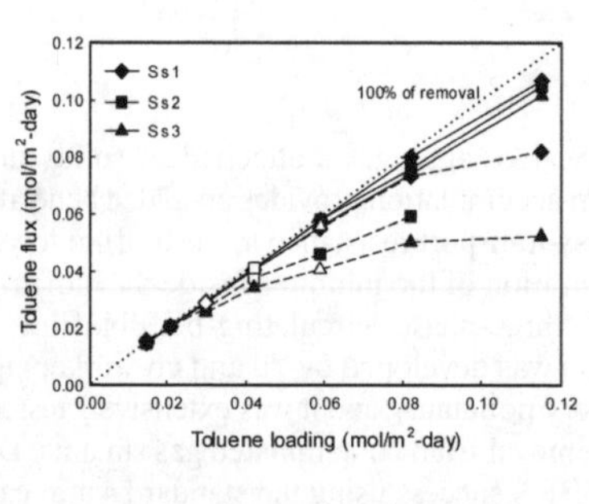

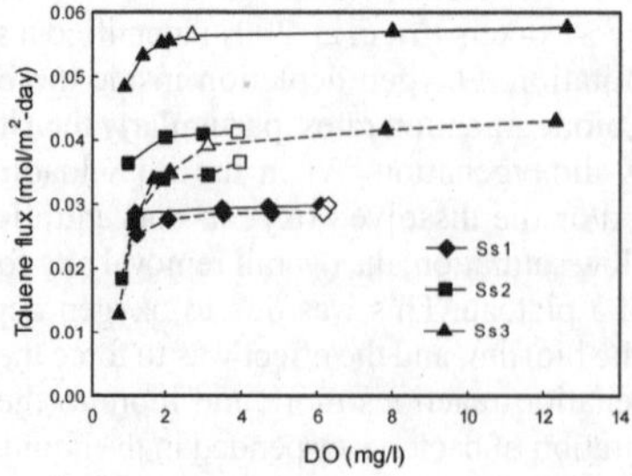

Figure 1. Toluene flux with toluene loading and dissolved oxygen.

2.2 *Analytical methods*

The amount of biomass in the suspended phase was measured by its optical density, which was calibrated to dry weight per liquid volume (mg/L). For the porous carrier, biomass was immobilized inside the pores and difficult to remove. Therefore, the biomass in the carriers was measured as the weight difference between the carriers taken from the reactor and clean carriers. Carriers taken from the reactor were first dried at 104°C for 2 h and then weighed immediately. An HP 5890 gas chromatograph (GC) equipped with a flame-ionization detector (FID) and a glass capillary column (Model DB-5, J&W Scientific, Inc., Folsom, CA) was used to analyze the concentrations of toluene in the gas or liquid phase. The GC/FID output signals were analyzed by an HP integrator (Model 3396). The DO (dissolved oxygen) concentration was measured with a high-resolution digital DO meter and probe (Martek Instruments, Inc., Raleigh, NC; Model Mark XVIII). Because the electrolyte in the probe was consumed continuously during measurements, the probe needed to be calibrated before each use. We used the same reactor, microbial inoculum, steady-state loading conditions, and experimental protocols as did Yu and co-workers when they documented the performance of the CBBR with 4 mm carriers (Yu et al. 2001). The only difference from the prior work is that we used 1 mm porous carriers, instead of 4 mm carriers. Thus, our experimental work directly tested the hypothesis that the smaller biofilm carrier would improve the removal of toluene without sacrificing good biofilm retention.

3 KINETIC ANALYSIS

3.1 *Mass balances in the gas phase*

Because toluene was fed into the reactor through the gas stream, its mass balance in the gas phase is expressed by the following equation:

$$M_{GL,T} = r_{pump,T}\rho_T - S_{1,T,G}Q_G \quad (1)$$

where $M_{GL,T}$ is the rate of toluene mass transferred from the gas into the liquid phase ($M_{tol}T^{-1}$), $S_{1,T,G}$ is the toluene concentration in the gas effluent ($M_{tol}L^{-3}$), and Q_G is the volumetric gas flow rate (L^3T^{-1}). Since all the variables on the right side of Eq. (1) were experimentally measured or are known for a given temperature, $M_{GL,T}$ can be calculated directly. The mass balance for oxygen is:

$$M_{GL,O} = (k_l a)_O (O^*_{L,i} - O_L)V \quad (2)$$

where $M_{GL,O}$ is the rate of oxygen mass transferred from the gas into the liquid phase ($M_{oxy}T^{-1}$), $(k_l a)_O$ is the oxygen gas-liquid mass-transfer coefficient (T^{-1}), V is the reactor volume (L^3), $O^*_{L,i}$ is the liquid-phase concentration that would be in equilibrium with the bulk liquid phase ($M_{oxy}L^{-3}$), and O_L is the oxygen concentration in the bulk liquid phase ($M_{oxy}L^{-3}$).

3.2 *Removals in the bulk liquid*

When toluene was fed into the reactor as the sole substrate, the rates of removals for toluene, $R_{1,T,L}$ ($M_{tol}T^{-1}$) and its intermediate $R_{2,T,L}$ ($M_{mcatl}T^{-1}$) by suspended biomass are expressed as:

$$R_{1,T,L} = \frac{q_{1,\max,T}S_{1,T,L}}{K_{1,T} + S_{1,T,L}} X_a \frac{O_L}{K_O + O_L} V_L \quad (3)$$

$$R_{2,T,L} = \frac{q_{2,\max,T}}{1 + \frac{S_{1,T,L}}{K_{I,12,T}}} \frac{S_{2,T,L}}{K_{2,T} + S_{2,T,L}} X_a \frac{O_L}{K_O + O_L} V_L \quad (4)$$

in which $S_{1,T,L}$ is the concentration of toluene in the liquid, X_a is the concentration of active biomass, O_L is the concentration of dissolved oxygen, V_L is the volume of the liquid, and K_O is the half-maximum-rate concentration for oxygen in oxygenation. A dual-limitation term for oxygen was used to incorporate the effects of oxygen. The oxygen consumption in the bulk liquid includes utilization associated with the degradation of toluene and its intermediate and the decay of suspended biomass:

$$R_{O,L} = \alpha_{1O,T}R_{1,T,L} + \alpha_{2O,T}R_{2,T,L} + \alpha_{O,X}bX_a V_L \frac{O_L}{K_{O,r} + O_L} \quad (5)$$

where $K_{O,r}$ is the half-maximum-rate concentration for oxygen during respiration. In Eq. (5), $\alpha_{O,X}$ is the oxygen stoichiometric coefficient for complete oxidation of biomass and equals 1.42 g-O_2/g-X_a based on the molecular formula of biomass $C_5H_7O_2N$. $\alpha_{1O,T}$ is the stoichiometric coefficient of oxygen in the first-step of toluene degradation. Based on the toluene degradation pathway, one mole of oxygen reacts with one mole of toluene to form one mole of 3-methylcatechol in the first-step reaction. Thus, $\alpha_{1O,T}$ equals to 0.348 g-O_2/g-toluene. In the second-step reaction, one mole of oxygen is used as a cosubstrate for every mole of 3-methylcatechol utilized. In addition, oxygen is used during respiration as the electron acceptor. Thus, $\alpha_{2O,T}$ in Eq. (5) is the amount of oxygen consumed per unit of 3-methlycatechol oxidized, which can be calculated based on the biomass true yield of 3-methylcatechol ($S_{2,T}$) (0.38 mg-X_a/mgCOD). $\alpha_{2O,T}$ was 0.46 g-O_2/g-COD. In Eq. (3), the value of K_O equals 2.0 mg/l. The value of $K_{O,r}$ for oxygen used as the terminal electron acceptor was reported as equals 0.02 mg/l.

3.3 *Removals by the biofilm*

At steady state, the mass flux of toluene into the biofilm, $J_{1,T}$ ($M_{tol}L^{-2}T^{-1}$), can be obtained from the mass balance of toluene in the reactor:

$$J_{1,T} = \frac{M_{GL,T} - R_{1,T,L} - Q_L S_{1,T,L}}{A} \quad (6)$$

where $M_{GL,T}$ and $R_{1,T,L}$ are defined in Eqs. (1) and (3), respectively, $S_{1,T,L}$ is the toluene concentration in the liquid effluent ($M_{tol}L^{-3}$), and Q_L is the volumetric liquid flow rate of the effluent (L^3T^{-1}). The concentration of toluene intermediate in the bulk liquid, $S_{2,T,L}$, is generated either from the reaction of toluene in the liquid phase or outward diffusion from the biofilm, where much of the toluene is degraded. The mass flux of toluene intermediate into biofilm, $J_{2,T}$ ($M_{mcat}L^{-2}T^{-1}$), is then:

$$J_{2,T} = \frac{\alpha_{12,T}R_{1,T,L} - R_{2,T,L} - Q_L S_{2,T,L}}{A} \quad (7)$$

The sign of $J_{2,T}$ could be negative in Eq.(7), which indicates the flux of toluene intermediate is from the biofilm to the liquid, or it could be positive, which means that the intermediate's flux is into the biofilm. The oxygen transferred from the gas phase into the liquid phase was utilized in the first- and second-step reactions, utilized in biomass decay, diffused into the biofilm, or advected out in the effluent. The mass flux of oxygen into the biofilm, J_O ($M_{oxy}L^{-2}T^{-1}$), is:

$$J_O = \frac{M_{GL,O} - R_{O,L} - Q_L O_L}{A} \quad (8)$$

where $M_{GL,O}$ and $R_{O,L}$ are defined in Eqs. 4 and 7, respectively.

4 RESULTS AND DISCUSSION

Figures 1 to 3 show the dependencies of mass fluxes of toluene, its intermediate, and oxygen on the

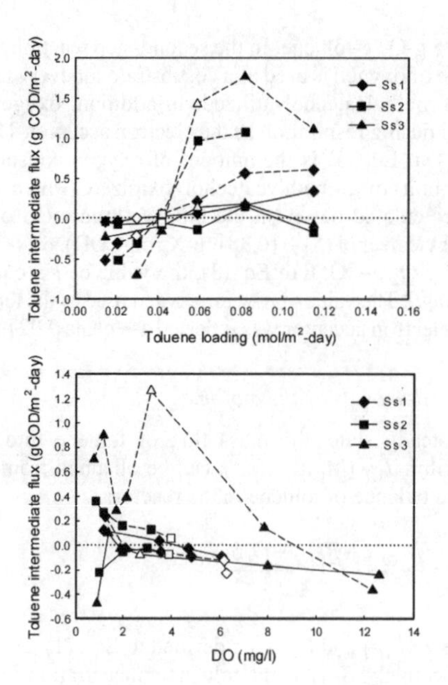

Figure 2. Flux of toluene intermediates with toluene loading and dissolved oxygen.

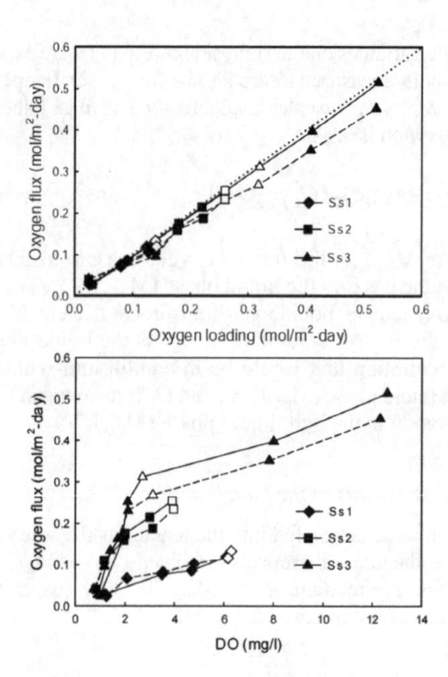

Figure 3. Oxygen flux with oxygen loading and dissolved oxygen.

concentrations of oxygen in the liquid. Figure 1 shows that the toluene flux (i.e., biofilm surface removal rate) follows a relationship similar to the Monod-type with oxygen concentrations. At low concentrations of oxygen in the liquid, the flux increases linearly (first order) with the concentrations of oxygen. At higher concentrations, the reaction order decreases to zero, indicating that the other substrate is rate limiting. At the same DO concentration, a higher toluene flux was observed in 1 mm CBBR, particularly, for the high biomass accumulation, Ss3. This supports our hypothesis of strong alleviation of oxygen depletion in the 1 mm CBBR, particularly at the highest biomass accumulation. This conclusion is further supported by the small change in intermediate flux (Figure 2) and the high oxygen fluxed (Figure 3). The dependency of oxygen flux on the oxygen loading is shown in Figure 3. When toluene loading was kept the same, oxygen flux continued to increase with DO loading, which shows that the "capacity effect" was minimized by the smaller carriers (Sang et al. 2003).

During Ss1, the biofilm concentration in the reactor was 1.8 kg-X_a/m^3-reactor volume, or 1.8 g-X_a/m^2-biofilm surface (Table 1), where g-X_a stands for gram of attached dry weight. With the amount of carriers loaded in this study, the reactor had a specific biofilm surface area of 1,010 m^2/m^3, while that for Yu and coworkers (Yu et al. 2003) was 263 m^2/m^3. The average biofilm density inside the carriers can be calculated from the biofilm dry weight and wet volume of each carrier based on the assumption that the pores were evenly filled with biomass. For example, the average biofilm density was 9.3 mg-X_a/cm^3-carrier during Ss1, and it increased to 16.0 mg-X_a/cm^3-carrier by Ss3. The corresponding densities for the previous work with the 4 mm carriers were 12.0, 16.4, and 23.6 mg-X_a/cm^3-carrier during Ss1, Ss2, and Ss3, respectively. Although

Table 1. Comparison of biomass with 1-mm carriers and 4-mm carriers.

Parameter	Steady state	1-mm carriers	4-mm carriers
Biomass attached to the porous carriers (g-X_a/L)	Ss1	1.8	2.2
	Ss2	2.4	3.0
	Ss3	3.1	4.2
Biofilm-biomass density (mg-X_a/cm^3-carrier)	Ss1	9.3	12.0
	Ss2	12.4	16.4
	Ss3	16.0	23.6
Biomass in biofilm per m^2-biofilm surface (g-X_a/m^2-biofilm surface)*	Ss1	1.8	8.2
	Ss2	2.4	11.2
	Ss3	3.1	16.1
Suspended bacteria concentration (mg-X_a/L)**	Ss1	17 (11–25)	11 (7–15)
	Ss2	23 (17–27)	28 (22–33)
	Ss3	27 (24–32)	36 (25–44)
Fraction of suspended bacteria to the total bacteria concentration (%)	Ss1	0.94	0.50
	Ss2	0.95	0.92
	Ss3	0.86	0.85

the total biofilm mass in the reactor volume was only slightly larger for the 4 mm carriers, higher biomass per m^2-carrier surface accumulated with the 4 mm carrier. If the biomass accumulated to the same average density in both carriers, the ratio of biofilm depth in 4 mm carrier to 1 mm carrier can be calculated as:

$$\frac{L_{f,4-mm\ carrier}}{L_{f,1-mm\ carrier}} = 6.8\ (Ss1);\ 4.7\ (Ss2);\ 5.2\ (Ss3) \quad (9)$$

where $L_{f,4\text{-}mm\ carrier}$ and $L_{f,1\text{-}mm\ carrier}$ are the biofilm depths in 4 mm and 1 mm carriers, respectively. These results demonstrate that the depth of biofilm in a 1 mm carrier was much thinner than in a 4 mm carrier. This result supports the hypothesis underlying our research with the 1 mm carriers: namely, that diffusion resistance in the biofilm should be reduced by using the smaller carriers. In this case, the average biofilm depth was 4.7 to 6.8 times smaller with the 1 mm carriers.

REFERENCES

Yu, H., Kim, B.J. & Rittmann, B.E. 2001. Contributions of biofilm versus suspended bacteria in an aerobic, circulating-bed biofilm reactor. Water Sci Technol 43: 303–310.

Sang, B.-I. Yoo, E.S., Kim, B.J. & Rittmann, B.E. 2003. The trade-offs and effect of carrier size and oxygen-loading on gaseous toluene removal performance of a three-phase circulating-bed biofilm reactor. Appl. Microbiol. Biotechnol 61: 214–219.

TBT (Tributyltin)

European Symposium on Environmental Biotechnology, ESEB 2004 - Verstraete (ed)
© 2004 Taylor & Francis Group, London, ISBN 90 5809 653 X

Alternative remediation techniques to cope with TBT

O. Eulaerts
DEC nv, Zwijndrecht, Belgium

Within the framework of the European LIFE program, DEC explored techniques for remediation of TBT-containing sediments. In collaboration with the port of Antwerp, initiator of the project, and other partners, we focused our study on three promising remedia-tion concepts for TBT contaminated sediments: bio-remediation and phyto-remediation, to cope with TBT concentration up to few ppm, and electro oxidation to treat the heavy polluted sediments (up to 100 ppm TBT).

Bio-remediation was studied in laboratory using *Trametes Versicolor*. The influence of oxygen, nutrients, UV, and surfactants on the rate of degradation was investigated. Results from this study in laboratory were used to start large-scale remediation trials (few hundred cubic meters). In parallel to these tests, large-scale lagunation experiments were also conducted to assess the role of natural bacterial activity, UV light, aeration and drying off of the sediments in the TBT degradation process.

Prior to large-scale phyto-remediation trials, a laboratory study was conducted to select salt tolerant plants specifically for the sediments from Antwerp. This was done in collaboration with the Technical University of Denmark and the Limburg University Centre. Several species were tested and sorted in function of their ability to growth on the contaminated sediments. The most promising amongst them were then used for large-scale trials.

Interesting results were obtained regarding oxidation techniques within the framework of another Life project held in Germany. The two phases of the electro oxidation tests were therefore conducted in collaboration with the Technical University Hamburg and its German spin-off company Bluewater. The first phase consisted in a feasibility study: TBT containing sediments were treated for several hours at relatively high current density. This technique appeared so promising (>99% reduction) that we started real-scale trials to optimise the treatment. The results of this cost-effectiveness study will be detailed.

European Symposium on Environmental Biotechnology, ESEB 2004 - Verstraete (ed)
© 2004 Taylor & Francis Group, London, ISBN 90 5809 653 X

Remediation of TBT contaminated sediments – sludge dewatering as a pre-treatment before the thermal treatment

L. Goethals
Envisan nv, Hofstade-Aalst, Belgium

Due to the extensive use of Tributyltin (TBT) as anti-fouling agent in shippaints during the last decades, this chemical has been accumulating up to significant concentrations in sediments, especially in harbours and marinas with high shipping activity.

Tributyltin (TBT) is an aggressive biocide that is slowly released from the ship hulls. Because of the potential biocidal properties and adverse affects on non-target organisms, a world-wide ban has been put on the use of this anti-fouling agent since January 1, 2003. Also the dumping of dredging sludges from harbours with high industrial activities in sea is prohibited since that date.

Next to this stringent legislation regarding the primary release via TBT-containing anti-fouling paints, the development of an effective remediation method of (historically) TBT – contaminated sediment is equally important. In the framework of the TBT – CLEAN research project, funded by the European Commission within the LIFE – environment program, different sediment remediation techniques are currently being investigated.

Because of the chemical properties of Tributyltin (TBT), i.e. the high volatility and the relatively low boiling point of 170°C, thermal treatment was considered as one of the possible remediation techniques. Labscale tests, at different temperatures, have shown that thermal treatment is indeed a promising technology for the treatment of TBT contaminated sediments. Therefore full scale tests will be executed with a thermal desorption unit by ENVISAN n.v. During the full scale tests, the process parameters such as temperature, residence time, etc. will be optimised with emphasis on the treatment of TBT.

The feasibility of thermal treatment of TBT – contaminated sediments however is strongly influenced by the pre-conditioning of the sediments. The limit-ing (and cost-determining) factor for this treatment appears to be the heat necessary for water vaporisation. Therefore a dewatering step in advance will probably be needed. The dewatering step will first be optimised on a pilot scale and will further be tested on full scale.

In relation with these dewatering tests, the treatment of the filtrate water will be investigated in view of the TBT – release from the solid phase to the liquid phase. It is in fact known that TBT desorption is largely influenced by a high pH.

Our LIFE – TBT-team consists of 6 partners: Port of Antwerp authorities, Antwerp Port Consultancy (APEC), an analytical and research laboratory (ERC), two dredging and sediment remediation companies (ENVISAN & DEC) and an external environmental consultant (ERM). Through close cooperation between the project partners, an integrated approach will be provided for the removal of organotins from waterways and harbours, which can also be applicable for remediation of other major ports in Europe. More information on the project can be found on www.portofantwerp.be/tbtclean.

Late posters

European Symposium on Environmental Biotechnology, ESEB 2004 - Verstraete (ed)
© 2004 Taylor & Francis Group, London, ISBN 90 5809 653 X

Remediation of oil shale chemical industry solid wastes using phytoremediation and bioaugmentation

J. Truu, E. Heinaru, E. Vedler, M. Viirmäe, J. Juhanson, E. Talpsep & A. Heinaru
Institute of Molecular and Cell Biology, University of Tartu, Estonia

ABSTRACT: Oil shale thermal processing has resulted in solid waste dump sites containing up to 100 million tons of solid waste. The processed oil shale contains several organic and inorganic compounds and is highly toxic. Laboratory and field experiments were carried out in order to test the effect of phytoremediation and bioaugmentation for remediation of pollutants in semi-coke. Microbial community of aged (ca 10 years) semi-coke is characterized by few dominant populations and possesses low diversity. The phytoremediation increased the number of bacteria and diversity of microbial community in semi-coke. Within a two and half year period starting from establishment of test plots, the concentration of phenolic compounds decreased up to 100% and oil products up to three times at plots with vegetation compared to control. Bioaugmentation increased biodegradation intensity of oil products up to 50% compared to untreated planted controls and enhanced plant growth.

1 INTRODUCTION

More than 70 years of oil shale thermal processing has resulted in huge dump sites of the coke ash (semi-coke) from semi-coking of oil shale in the areas surrounding oil shale chemical industry plants in northeastern part of Estonia. The semi-coke mounds cover an area about 200 ha and contain up to 100 million tons of solid waste. Currently, about 600 000 tons of processed semi-coke is disposed annually. Semi-coke solid wastes contain several organic and inorganic compounds (oil products, bitumens, phenols, PAHs, sulfuric compounds), while liquid wastes (leachate) from depository area are characterized by high concentration of oil products, phenol, cresols, dimethylphenols and resorcinols. The liquid pollution from semi-coke dump area deteriorates surface water as well as the underlying aquifers (Truu et al. 2002).

The aim of current study is to assess the suitability of phytotechnological approach, which includes phytoremediation in combination with bioaugmentation for remediation of semi-coke dump area.

2 MATERIALS AND METHODS

2.1 *Phytoremediation experiment*

Four test plots (each 50 m^2) were established at semi-coke depository in July 2001. Plant treatment was based on a grass mixture of four species. In addition to plants, four different treatments were utilized. The following treatments were applied: (1) plot – no treatment (grass seeds in semi-coke), (2) plot – seeds in semi-coke were covered by sand layer (1–2 cm), (3) plot – seeds in semi-coke were covered by peat layer (1–2 cm), (4) plot – semi-coke was covered with the layer of pre-grown grass (sod rolls). In October 2001, 2002 and 2003 soil sampling was performed on plots and control area. We analyzed semi-coke samples collected from the test plots at the depository area for chemical and microbiological parameters.

2.2 *Bioaugmentation experiment*

For the bioaugmentation experiment the set of bacteria consisting of three strains isolated from nearby area was selected. These three bacterial strains *Peudomonas mendocina* PC1, *P. fluorescens* PC24 and *P. fluorescens* PC18 degrade phenols via catechol meta, catechol or protocatechuate ortho or via the combination of catechol meta and protocatechuate ortho pathways, respectively (Heinaru et al. 2000). In bioaugmentation experiments the biomass of these bacteria was supplied to the part experimental plots (each 10 m^2) in July 2002. Each treatment received 20 L of bacterial suspension with concentration 108 CFU ml^{-1}. The ratio of bacterial strains PC1, PC18 and PC24 was in suspension 3:1:1.

2.3 *Microbiological methods*

The microbial communities were removed from semi-coke and plant root surface by vortexing in sterile tap water. Heterotrophic plate count was enumerated by the spread plate method in triplicate on R2A agar (Difco). The number of phenol-degrading bacteria was determined in triplicate sets on M9-salts agar plates supplemented with trace elements and phenol (2.5 mM). The heterotrophic activity and diversity of microbial community was measured using Biolog EcoPlates (Biolog, Inc.). Results of Biolog profiles are presented as total activity (average well color development (AWCD) of all 31 wells) and by Shannon diversity index. Color development data of Biolog EcoPlates was also subject to kinetic data analysis according to method of Lindstrom et al. 1998. Microbial DNA was extracted from soil samples with UltraClean Mega Soil DNA kit (Mo Bio Laboratories, Inc.). Bacterial community structure was assessed with two primer pairs, 318f-GC/535r and 968f-GC/1401r. A denaturating gradient gel electrophoresis (DGGE) system DCode (Bio Rad, Inc.) was used to separate the amplified gene fragments.

3 RESULTS

Chemical properties of the semi-coke from experimental area are shown in the Table 1. After retorting at 500°C, processed oil shale is highly saline, alkaline, biologically sterile, nutrient deficient material with no structure. Semi-coke is characterized by high organic carbon content, nearly half of which is bitumen. According to ecotoxicological tests the fresh semi-coke is classified as of "high acute toxic hazard", whereas aged semi-coke is classified as of "acute toxic hazard" (Põllumaa et al. 2001).

Table 1. Chemical properties of semi-coke from the control plot.

Parameter	Measured value
pH	8.0
Total nitrogen (%)	0.08
P-PO_4^{3-} mg/kg	12.27
K^+ mg/kg	799.1
Ca^{2+} mg/kg	18673
Mg^{2+} mg/kg	826
Total organic carbon (%)	15.0–18.0
Oil products mg/kg	340
Volatile phenols mg/kg	0.30–0.34

The chemical analysis of soil samples showed impact of the plant treatment on degradation rate of pollutants. Within a two and half year period starting from the establishment of test plots in July 2001, the concentration of volatile phenols was reduced up to 100%, the concentration of oil products 3 times (from 340 mg kg^{-1} to 100 mg kg^{-1}), and the total content of organic carbon decreased by 10 to 30 g per kg (from 18% to 15%). The best results were obtained on the plots with peat amendment and pre-grown grass with the highest root density in semi-coke. In upper layer samples (0–10 cm) the reduction of oil products and phenols was even bigger being in the range from 83% to 98%.

Bacterial biomass consisting of three bacterial strains was applied to three experimental plots in June 2002. Within a three months period the concentration of residual shale oil in semi-coke decreased by 13.6% to 53.6% at plots treated with bacterial biomass compared to untreated parts of experimental plots. In our experiments we observed the increase in root biomass and length on the plots amended with bacterial biomass, which in turn may led to enhanced rhizodegradation.

The number of phenol-degrading bacteria increased by order of magnitude, while the number of heterotrophic aerobic bacteria remained on the same level compared to the untreated plot. Samples from the second year (2002) of experiment showed lower values of aerobic heterotrophic bacteria, which could be due to extremely dry vegetation period. The general trend was the increase of proportion of biodegradable bacterial numbers within microbial community due to the treatment. Highest values for all measured microbiological parameters were found in rhizosphere samples. While bacterial total numbers increased by order of magnitude compared to control, the number of phenol-degrading bacteria was more than 100 times higher in the rhizospheric soil. Addition of bacterial biomass to semi-coke resulted in increase both in absolute number (up to 7.8×10^6 CFU g^{-1} dw) and relative abundance (up to 30%) of phenol-degrading bacteria in the studied samples. The highest values for microbial activity and diversity measured with Biolog EcoPlates were recorded in rhizosphere samples.

Partial fragments of gene for the largest subunit of multicomponent phenol hydroxylases (LmPHs) were amplified from total DNA, separated by DGGE and sequenced. Obtained sequences were compared with published sequences. In the plots with plants dominated two different multicomponent phenol hydroxylases (LmPHs) belonging to low- and moderate-K_s kinetics groups, indicating more efficient degradation of aromatics at these plots. There was only one dominant LmPH at untreated plot, belonging to high-K_s group (Truu et al. 2003).

We compared microbial communities from treatments using kinetic model based on average well color development of Biolog EcoPlates (Fig. 1). Model parameters estimated from these fitted curves were statistically different (ANOVA), indicating changes in microbial community structure due to phytoremediation

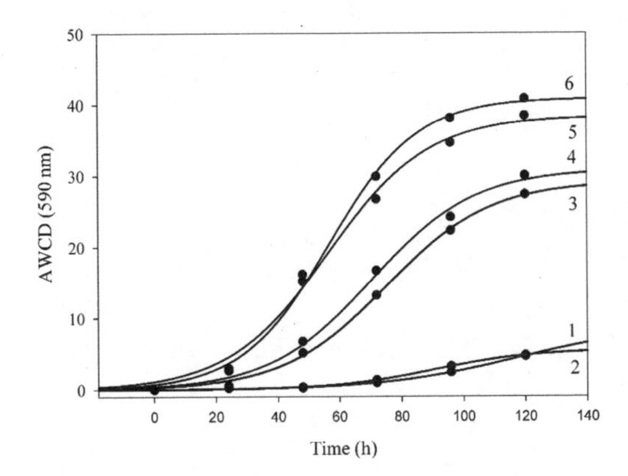

Figure 1. Kinetics of average well color development of Biolog plates for different treatments. 1 – untreated semi-coke, 2 – semi-coke with added bacterial biomass, 3 – planted semi-coke with peat layer, 4 – semi-coke with pre-grown grass, 5 – semi-coke with pre-grown grass with added bacterial biomass, 6 – planted semi-coke with peat layer with added bacterial biomass.

and bioaugmentation. Analysis of RAPD and DGGE patterns also indicated changes in bacterial community due to phytoremediation and bioaugmentation.

4 CONCLUSIONS

Our results indicate that phytoremediation and bioaugmentation could be considered as an alternative management option for remediation of oil shale solid waste. Additional beneficial side effects of the establishment of vegetation on the semi-coke deposit are reduction of leachate amount and toxicity, and surface erosion. Also, plant cover would diminish the dispersion of pollutants into adjacent area including Kohtla-Järve city by air. When semi-coke toxicity is decreased, it becomes more suitable substrate for trees, as *mycorrhizal* fungi of tree roots are more sensitive to pollutants compared to bacteria. The better growth of trees in turn favors the remediation of deeper layers of semi-coke as well stabilizes the soil structure. Our findings support the prospect to use integrated environmental biotechnology approach for remediation of semi-coke dump area. The goals of integrated approach could be achieved through sequential application of phytotechnology, bioaugmentation, composting of semi-coke for creating soil amendment on site and constructed wetlands. Constructed wetlands decrease the amount of recalcitrant pollutants in leachate and reduce the load of wastewater to local water treatment plant.

ACKNOWLEDGEMENTS

The study was funded by the Maj and Tor Nessling Foundation (Finland) and partly by the Estonian Science Foundation grant No. 4344.

REFERENCES

Heinaru, E., Truu, J., Stottmeister, U., Heinaru, A. 2000. Three types of phenol and p-cresol catabolism in phenol- and p-cresol-degrading bacteria isolated from river water continuously polluted with phenolic compounds. FEMS Microbiology Ecology 31: 195–205.

Lindstrom, J.E., Barry, R.P., Braddock, J.F. 1998. Microbial community analysis: a kinetic approach to constructing potential C source utilization patterns. Soil Biol. Biochem 30: 231–239.

Põllumaa, L., Maloveryan, A., Trapido, M., Sillak, H., Kahru, A. 2001. A study of the environmental hazard caused by the oil shale industry solid waste ATLA 29: 259–267.

Truu, J., Heinaru, E., Talpsep, E., Heinaru, A. 2002. Analysis of river pollution data from low-flow period by means of multivariate techniques: A case study from the oil-shale industry region, northeastern Estonia. Environmental Science and Pollution Research 1: 8–14.

Truu, J., Kärme, L., Talpsep, E., Heinaru, E., Vedler, E., Heinaru, A. 2003. Application of phytoremediation for enhanced bioremediation of oil shale chemical industry solid wastes. Acta Biotechnologica 23 (2–3): 301–307.

European Symposium on Environmental Biotechnology, ESEB 2004 - Verstraete (ed)

Fate of the genetic diversity of a population of *V. cholerae* non-O1/ non-O139 through two wastewater treatment systems

S. Baron, S. Chevalier, N. Boisgontier & J. Lesne
Ecole Nationale de la Santé Publique – Laboratoire d'Etude et de Recherche en Environnement et Santé CS *Rennes cedex, France*

N. Grosset & M. Gauthier
Laboratoire microbiologie – UNR Pôle agronomique de Rennes – France

Vibrio cholerae is a bacterium of sanitary interest, the particularity of which is to accommodate to two quite distinctive lifestyles: long term residence in aquatic ecosystems and colonization of the human intestine.

Among the about 200 O-serogroups described, only O1 and the newly emerged O139 are the agents of cholera. The other serogroups so called non-O1/non-O139 are considered as either occasional pathogens or saprophytes. Now that recent work has shown that the main virulence factors (cholera toxin and pilus tcpA) of the agents of cholera are brought by bacteriophages, the interest for non-O1/non-O139 *V. cholerae* is increasing. Indeed, environmental strains of non-O1/non-O139 *V. cholerae* may constitute a reservoir for the emergence of new toxinogenic or other pathogenic strains.

The presence of non-O1/non-O139 *V. cholerae* in wastewaters, which constitute ecosystems combining some characteristics from the two habitats of the bacteria, has been described. However very little is known about their behaviour during a wastewater treatment process. Absence of decrease of the abundances of *V. cholerae* between the inlet and the outlet of wastewater treatment plants has been reported in few studies. Nevertheless, the question about a possible difference between the behaviour of O1/O139 serogroups and the others is left open, although of considerable significance. This question can be approached indirectly by the study of the fate of the genetic diversity of populations of *Vibrio cholerae* non-O1/non-O139 through these plants.

The aim of this study is to compare the genetic diversity of two populations of *V. cholerae* non-O1/ non-O139 at the inlet and the outlet of a biological treatment plant and of stabilization ponds. Four representative collections of isolates were typed for this purpose by ERIC-PCR and/or PFGE.

European Symposium on Environmental Biotechnology, ESEB 2004 - Verstraete (ed)

Anaerobic biodegradability of mercaptans and effect of salt concentration

R.C. van Leerdam, F.A.M. de Bok & P.N.L. Lens
Wageningen University, Sub-department of Environmental Technology, Wageningen, The Netherlands

A.J.H. Janssen
Wageningen University, Sub-department of Environmental Technology, Wageningen, The Netherlands & Shell Global Solutions International BV, Amsterdam, The Netherlands

ABSTRACT: Because of their potential negative effects on the environment, organosulfur compounds present in petroleum and fossil fuels (more than 200 compounds) are receiving considerable attention. Strong alkaline solutions are used to scrub hydrogen sulfide and acidic organic sulfur compounds, such as thiols (R-SH), from hydrocarbon streams. Once hydrogen sulfide is absorbed in sodium hydroxide, the solution becomes known as a spent sulfidic caustic. The objective of this study was to investigate if methanethiol containing spent sulfidic caustics can be pretreated biologically. Batch tests for assessing anaerobic biodegradability of methanethiol (MT), ethanethiol (ET) and propanethiol (PT) were performed for seven (anaerobic) sludge samples. A continuous Upflow Anaerobic Sludge Bed (UASB) reactor (1.6 L) was fed with MT (2–6 mM) and operated at 30°C and pH 7.2–7.5. Volatile organic sulfur compounds (VOSC) and sulfide were monitored. Additional batch tests were performed with the sludge used for the lab scale reactor and with adapted sludge from the lab scale reactor to investigate the adaptation of the sludge to MT and to examine the effect of the salt concentration on the conversion rate. Several sludges from wastewater treatment plants were capable of degrading MT. None of them was able to convert ET or PT. In the reactor the removal efficiency of VOSC was >90% for almost the entire experiment, while 70–85% of the organic sulfur was recovered as sulfide in the effluent. Adaptation of the sludge to MT in the reactor caused an increase of the sulfide production rate by 3–7 times in batch tests. The IC_{50} of NaCl and $NaHCO_3$ for both adapted and unadapted sludge was close to the maximum concentration tested for NaCl (25 g/L) and $NaHCO_3$ (29 g/L).

European Symposium on Environmental Biotechnology, ESEB 2004 - Verstraete (ed)
© 2004 Taylor & Francis Group, London, ISBN 90 5809 653 X

Author Index

Science Dossier
Dioxins and furans in the environment
January 2003
Euro Chlor

Euro Chlor
representing the chlor-alkali industry

Better drugs for a better world
We aim for the highest quality in the development, production and marketing of drugs. We want to achieve this with the greatest possible care to the well-being of our employees, our fellow citizens and our environment.
The best possible health for as many people as possible is our objective. And our responsibilities do not stop at the company gate. We are fully aware that a company like ours is helping to shape the future of the world. That is the challenge for everyone in our company – a challenge that we gladly take upon ourselves day after day.
Turnhoutseweg 30, B-2340 Beerse, Belgium

JANSSEN
PHARMACEUTICA

Johnson & Johnson
PHARMACEUTICAL RESEARCH
& DEVELOPMENT

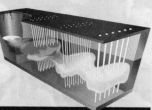